To Lyn

... who witnessed m
and gave me hospitality after
the Hay festival. I look forward
to visiting you again in this
and many other universes!

With love from

Bernard

UNIVERSE OR MULTIVERSE?

Recent developments in cosmology and particle physics, such as the string landscape picture, have led to the remarkable realization that our universe – rather than being unique – could be just one of many universes. The multiverse proposal helps to explain the origin of our universe and some of its observational features. Since the physical constants can be different in other universes, the fine-tunings that appear necessary for the emergence of life may also be explained. Nevertheless, many physicists remain uncomfortable with the multiverse proposal, since it is highly speculative and perhaps untestable.

In this volume, a number of active and eminent researchers in the field – mainly cosmologists and particle physicists but also some philosophers – address these issues and describe recent developments. The articles represent the full spectrum of views, from enthusiastic support of the multiverse to outright scepticism, providing for the first time a valuable overview of the subject. Contributions are written at varying academic levels, providing an engaging read for everyone. To preserve accessibility, mathematical equations are used in only a few chapters.

BERNARD CARR is Professor of Mathematics and Astronomy at Queen Mary, University of London (QMUL). His research interests include general relativity, the early universe, primordial black holes, dark matter and the anthropic principle. He regularly appears on television, for example on *The Sky at Night* and *Horizon*, and has published several dozen popular science articles. He is the author of nearly 200 technical papers and has worked in the USA, UK, Japan and Canada. He is a member of several learned societies, most notably the Royal Astronomical Society and the Institute of Physics. In 1984 he was awarded the Adams Prize by the University of Cambridge, one of the UK's most prestigious mathematical awards.

UNIVERSE OR MULTIVERSE?

Edited by

BERNARD CARR
Queen Mary, University of London

CAMBRIDGE UNIVERSITY PRESS
Cambridge, New York, Melbourne, Madrid, Cape Town, Singapore, São Paulo

Cambridge University Press
The Edinburgh Building, Cambridge CB2 8RU, UK

Published in the United States of America by Cambridge University Press, New York

www.cambridge.org
Information on this title: www.cambridge.org/9780521848411

© Cambridge University Press 2007

This publication is in copyright. Subject to statutory exception
and to the provisions of relevant collective licensing agreements,
no reproduction of any part may take place without
the written permission of Cambridge University Press.

First published 2007

Printed in the United Kingdom at the University Press, Cambridge

A catalogue record for this publication is available from the British Library

ISBN 978-0-521-84841-1 hardback

Cambridge University Press has no responsibility for the persistence
or accuracy of URLs for external or third-party internet websites referred to
in this publication, and does not guarantee that any content on such
websites is, or will remain, accurate or appropriate.

Contents

List of contributors		*page* viii
Preface		xi
Acknowledgements		xiv
Editorial note		xv
Part I	**Overviews**	1
1	Introduction and overview Bernard Carr	3
2	Living in the multiverse Steven Weinberg	29
3	Enlightenment, knowledge, ignorance, temptation Frank Wilczek	43
Part II	**Cosmology and astrophysics**	55
4	Cosmology and the multiverse Martin J. Rees	57
5	The anthropic principle revisited Bernard Carr	77
6	Cosmology from the top down Stephen Hawking	91
7	The multiverse hierarchy Max Tegmark	99
8	The inflationary multiverse Andrei Linde	127
9	A model of anthropic reasoning: the dark to ordinary matter ratio Frank Wilczek	151

10	Anthropic predictions: the case of the cosmological constant *Alexander Vilenkin*	163
11	The definition and classification of universes *James D. Bjorken*	181
12	M/string theory and anthropic reasoning *Renata Kallosh*	191
13	The anthropic principle, dark energy and the LHC *Savas Dimopoulos and Scott Thomas*	211

Part III   Particle physics and quantum theory — 219

14	Quarks, electrons and atoms in closely related universes *Craig J. Hogan*	221
15	The fine-tuning problems of particle physics and anthropic mechanisms *John F. Donoghue*	231
16	The anthropic landscape of string theory *Leonard Susskind*	247
17	Cosmology and the many worlds interpretation of quantum mechanics *Viatcheslav Mukhanov*	267
18	Anthropic reasoning and quantum cosmology *James B. Hartle*	275
19	Micro-anthropic principle for quantum theory *Brandon Carter*	285

Part IV   More general philosophical issues — 321

20	Scientific alternatives to the anthropic principle *Lee Smolin*	323
21	Making predictions in a multiverse: conundrums, dangers, coincidences *Anthony Aguirre*	367
22	Multiverses: description, uniqueness and testing *George Ellis*	387
23	Predictions and tests of multiverse theories *Don N. Page*	411
24	Observation selection theory and cosmological fine-tuning *Nick Bostrom*	431

25	Are anthropic arguments, involving multiverses and beyond, legitimate?	445
	William R. Stoeger, S. J.	
26	The multiverse hypothesis: a theistic perspective	459
	Robin Collins	
27	Living in a simulated universe	481
	John D. Barrow	
28	Universes galore: where will it all end?	487
	Paul Davies	
Index		507

List of Contributors

Anthony Aguirre
Department of Physics, University of California, Santa Cruz, California 95064, USA

John D. Barrow
DAMTP, Centre for Mathematical Sciences, Cambridge University, Wilberforce Road, Cambridge CB3 0WA, UK

Nick Bostrom
Philosophy Faculty, Oxford University, 10 Merton Street, Oxford OX1 4JJ, UK

James D. Bjorken
Stanford Linear Accelerator Center, 2575 Sand Hill Road, Menlo Park, CA 94025, USA

Bernard Carr
Astronomy Unit, Queen Mary, University of London, Mile End Road, London E1 4NS, UK

Brandon Carter
Département d'Astrophysique Relativiste et Cosmologie, Observatoire de Paris, 5 Place J. Janssen, F-92195 Meudon Cedex, France

Robin Collins
Department of Philosophy, Messiah College, P.O. Box 245, Grantham, PA 17027, USA

Paul Davies
Beyond: Center for Fundamental Concepts in Science, Arizona State University, Temple, AZ 85281, USA

Savas Dimopoulos
Varian Physics Building, Stanford University, Stanford, CA 94305-4060, USA

John F. Donoghue
Department of Physics, University of Massachusetts, Amherst, MA 01003, USA

George Ellis
Department of Mathematics and Applied Mathematics, University of Cape Town, 7700 Rondebosch, South Africa

James B. Hartle
Physics Department, University of California, Santa Barbara, CA 93106, USA

Stephen Hawking
DAMTP, Centre for Mathematical Sciences, Cambridge University, Wilberforce Road, Cambridge CB3 0WA, UK

Craig J. Hogan
Astronomy and Physics Departments, University of Washington, Seattle, WA 98195-1580, USA

Renata Kallosh
Varian Physics Building, Stanford University, Stanford, CA 94305-4060, USA

Andrei Linde
Varian Physics Building, Stanford University, Stanford, CA 94305-4060, USA

Viatcheslav Mukhanov
Sektion Physik, Ludwig-Maximilians-Universtät, Theresienstr. 37, D-80333 Munich, Germany

Don N. Page
Department of Physics, University of Alberta, Edmonton, Alberta T6G 2J1, Canada

Martin J. Rees
Institute of Astronomy, Madingley Road, Cambridge CB3 0HA, UK

Lee Smolin
Perimeter Institute for Theoretical Physics, 35 King Street North, Waterloo, Ontario N2J 2W9, Canada

William R. Stoeger
Vatican Observatory Research Group, Steward Observatory, University of Arizona, Tucson, AZ 85719, USA

Leonard Susskind
Varian Physics Building, Stanford University, Stanford, CA 94305-4060, USA

Max Tegmark
Department of Physics, Massachusetts Institute of Technology, Cambridge, MA 02139, USA

Scott Thomas
Department of Physics and Astronomy, University of Rutgers, Piscataway, NJ 08854-8019, USA

Alexander Vilenkin
Department of Physics and Astronomy, Tufts University, Medford, MA 02155, USA

Steven Weinberg
Physics Department, University of Texas at Austin, Austin, TX 78712, USA

Frank Wilczek
Center for Theoretical Physics, MIT 6-305, 77 Massachusetts Avenue, Cambridge, MA 02139, USA

Preface

This book grew out of a conference entitled 'Universe or Multiverse?' which was held at Stanford University in March 2003 and initiated by Charles Harper of the John Templeton Foundation, which sponsored the event. Paul Davies and Andrei Linde were in charge of the scientific programme, while Mary Ann Meyers of the Templeton Foundation played the major administrative role. The meeting came at a critical point in the development of the subject and included contributions from some of the key players in the field, so I was very pleased to be invited to edit the resulting proceedings. All of the talks given at the Stanford meeting are represented in this volume and they comprise about half of the contents. These are the chapters by James Bjorken, Nick Bostrum, Robin Collins, Paul Davies, Savas Dimopoulos and Scott Thomas, Renata Kallosh, Andrei Linde, Viatschelav Mukhanov, Martin Rees, Leonard Susskind, Max Tegmark, Alex Vilenkin, and my own second contribution.

Several years earlier, in August 2001, a meeting on a related theme – entitled 'Anthropic Arguments in Fundamental Physics and Cosmology' – had been held in Cambridge (UK) at the home of Martin Rees. This was also associated with the Templeton Foundation, since it was partly funded out of a grant awarded to myself, Robert Crittenden, Martin Rees and Neil Turok for a project entitled 'Fundamental Physics and the Problem of Our Existence'. This was one of a number of awards made by the Templeton Foundation in 2000 as part of their 'Cosmology & Fine-Tuning' research programme. In our case, we decided to use the funds to host a series of workshops, and the 2001 meeting was the first of these.

The theme of the Cambridge meeting was somewhat broader than that of the Stanford one – it focused on the anthropic principle rather than the multiverse proposal (which might be regarded as a particular interpretation of the anthropic principle). Nevertheless, about half the talks were on the multiverse theme, so I was keen to have these represented in the current volume. Although I had published a review of the Cambridge meeting in *Physics World* in October 2001, there had been no formal publication of the talks. In 2003 I therefore invited some of the Cambridge participants to write up their talks, albeit in updated form. I was delighted when almost everybody accepted this invitation, and their contributions represent most of the rest of the volume. These are the chapters by John Barrow, Brandon Carter, John Donoghue, George Ellis, James Hartle, Craig Hogan, Don Page, Lee Smolin, William Stoeger and Frank Wilczek.

We organized two further meetings with the aforementioned Templeton support. The second one – entitled 'Fine-Tuning in Living Systems' – was held at St George's House, Windsor Castle, in August 2002. The emphasis of this was more on biology than physics, and we were much helped by having John Barrow on the Programme Committee. Although this meeting was of great interest in its own right – representing the rapidly burgeoning area of astrobiology – there was little overlap with the multiverse theme, so it is not represented in this volume. Also, the proceedings of the Windsor meeting have already been published as a special issue of the *International Journal of Astrobiology*, which appeared in April 2003.

The third meeting was held at Cambridge in September 2005. It was again hosted by Martin Rees, but this time at Trinity College, Martin having recently been appointed Master of Trinity. The title of the meeting was 'Expectations of a Final Theory', and on this occasion David Tong joined the Programme Committee. Most of the focus was on the exciting developments in particle physics – in particular M-theory and the string landscape scenario, which perhaps provide a plausible theoretical basis for the multiverse paradigm. Many of the talks were highly specialized and – since this volume was already about to go to press – it was anyway too late to include them. Nevertheless, the introductory talk by Steven Weinberg and the summary talk by Franck Wilczek were very general and nicely complemented the articles already written. I was therefore delighted when they both agreed – at very short notice – to produce write-ups for this volume. The article by Stephen Hawking also derives from his presentation at the Trinity meeting, although he had previously spoken at the 2001 meeting as well. It is therefore gratifying that both Cambridge meetings – and thus all three Templeton-supported meetings – are represented in this volume.

Although I have described the history behind this volume, I should emphasize that the articles are organized by topic rather than chronology. After the overview articles in Part I, I have divided them into three categories. Part II focuses on the cosmological and astrophysical aspects of the multiverse proposal; Part III is more relevant to particle physics and quantum cosmology; and Part IV addresses more general philosophical aspects. Of course, such a clean division is not strictly possible, since some of the articles cover more than one of these areas. Indeed, it is precisely the amalgamation of the cosmological and particle physical approaches which has most powered the growing interest in the topic. Nevertheless, by and large it has been possible to divide articles according to their degree of emphasis.

Although this book evolved out of a collection of conference papers, the articles are intended to be at semi-popular level (for example at the level of *Science* or *Scientific American*) and most of the contributions have been written by the authors with that in mind. However, there is still some variation in the length and level of the articles, and some more closely resemble in technicality the original conference presentations. Where papers are more technical, I have elaborated at greater length in my introductory remarks in order to make them more accessible. In my view, the inclusion of some technical articles is desirable, because it emphasizes that the subject is a proper branch of science and not just philosophy. Also it will hopefully broaden the book's appeal to include both experts and non-experts.

As mentioned in my Introduction, the reaction of scientists to the multiverse proposal varies considerably, and some dispute that it constitutes proper science at all. It should therefore be stressed that this is not a proselitizing work, and this is signified by the question mark in the title. I did briefly consider the shorter title 'Multiverse?' or even 'Multiverse' (without the question mark), but I eventually discarded these as being too unequivocal. In fact, the authors in this volume display a broad range of attitudes to the multiverse proposal – from strong support through open-minded agnosticism to strong opposition. The proponents probably predominate numerically and they are certainly more represented in Parts II and III. However, the balance is restored in Part IV, where many of the contributors are sceptical. Therefore readers who persevere to the end of this book are unlikely to be sufficiently enlightened to answer the question raised by its title definitively. Nevertheless, it is hoped that they will be stimulated by the diversity of views expressed. Finally, it should be stressed that perhaps the most remarkable aspect of this book is that it testifies to the large number of eminent physicists who now find the subject interesting enough to be worth writing about. It is unlikely that such a volume could have been produced even a decade ago!

Bernard Carr

Acknowledgements

This volume only exists because of indispensable contributions from various people involved in the three conferences on which it is based. First and foremost, I must acknowledge the support of the John Templeton Foundation, which hosted the Stanford meeting in 2003 and helped to fund the two Cambridge meetings in 2001 and 2005. I am especially indebted to Charles Harper, the project's initiator, and his colleague Mary Ann Meyers, director of the 'Humble Approach Initiative' programme, who played the major administrative role in the Stanford meeting and subsequently helped to oversee the progress of this volume. Special credit is also due to Paul Davies and Andrei Linde, who were in charge of the scientific programme for the Stanford meeting and conceived the title, which this book has inherited. The Templeton Foundation indirectly supported the Cambridge meetings, since these were partly funded from a Templeton grant awarded to myself, Robert Crittenden, Martin Rees and Neil Turok. I would like to thank my fellow grant-holders for a most stimulating collaboration. They undertook most of the organizational work for the Cambridge meetings, along with David Tong, who joined the Programme Committee for the 2005 meeting. I am especially indebted to Martin Rees, not only for hosting the two Cambridge meetings, but also for triggering my own interest in the subject nearly thirty years ago and for encouraging me to complete this volume. I am very grateful to various people at Cambridge University Press for helping to bring this volume to fruition: the editor Simon Capelin, who first commissioned the book; the editor John Fowler, who made some of the later editorial decisions and showed great diplomacy in dealing with my various requests; the production editors Jacqui Burton and Bethan Jones; and especially the copy-editor Irene Pizzie, who went though the text so meticulously, suggested so many improvements and dealt with my continual stream of changes so patiently. Most indispensable of all were the contributors themselves, and I would like to thank them for agreeing to write up their talks and for dealing with all my editorial enquiries so patiently. Finally, I would like to thank my dear wife, Mari, for her love and support and for patiently putting up with my spending long hours in the office in order to finish this volume.

Editorial note

Although the term 'universe' is usually taken to mean the totality of creation, the theme of this book is the possibility that there could be other universes (either connected or disconnected from ours) in which the constants of physics (and perhaps even the laws of nature) are different. The ensemble of universes is then sometimes referred to as the 'multiverse', although not everybody likes that term and several alternatives are used in this volume (for example, megaverse, holocosm, and parallel worlds).

This lack of consensus on what term to use is hardly surprising, since the concept of a multiverse has arisen in many different contexts. Therefore, in my role as editor, I have not attempted to impose any particular terminology and have left authors to use whatever terms they wish. However, in so much as most authors use the word 'universe', albeit in different contexts, I have tried to impose uniformity in whether the first letter is upper or lower case. Although this might be regarded as a minor and rather pedantic issue, I feel that a book entitled *Universe or Multiverse?* should at least address the problem, and this distinction in notation can avoid ambiguities.

I have adopted the convention of using 'Universe' (with a big U) when the author is (at least implicitly) assuming that ours is the only one. When the author is (again implicity) referring to a general member of an ensemble (or just an abstract mathematical model), the term 'universe' (with a small u) is generally used. The particular one we inhabit is then described as 'our universe', although the phrase 'the Universe' (with a big U) is also sometimes used. This mirrors the way in which astronomers refer to 'our galaxy' as 'the Galaxy', and allows a useful distinction to be drawn (for example) between 'the visible Universe' (i.e. the visible part of our universe) and 'the visible universe' (i.e. the universe of which a part is visible to us). The word 'multiverse' is always spelt with a small m, since the idea arises in different ways, so there could be more than one of them.

Some authors prefer to reserve the appellation 'Universe' for the ensemble itself, perhaps preserving the term 'multiverse' for some higher level ensemble. In this case a capital U is used. In the inflationary scenario, for example, the term 'Universe' would then be used to describe the whole collections of bubbles rather than any particular one. This issue also arises in the context of quantum cosmology, which implicitly assumes the 'many worlds' interpretation of quantum mechanics. The literature in this field commonly refers to the 'wave-function of the Universe', although one might argue that wave-function is really being taken over a multiverse. The title of this book can therefore be understood to refer not only to the ontological issue of whether other universes exist, but also to the etymological issue of what to call the ensemble!

Cover picture

The picture on the cover is a tri-dimensional representation of the quadri-dimensional Calabi-Yau manifold. This describes the geometry of the extra 'internal' dimensions of M-theory and relates to one particular (string-inspired) multiverse scenario. I am grateful to Dr Jean-Francois Colonna of CMAP/Ecole Polytechnique, FT R&D (whose website can be found at http://www.lactamme.polytechnique.fr) for allowing me to use this picture. The orange background represents the 'fire' in the equations and is a modification of a design originally conceived by Cindy King of King Design Group. A similar image was first used in the poster for the second meeting on which this book is based (at Stanford in 2003).

Part I

Overviews

1
Introduction and overview

Bernard Carr
Astronomy Unit, Queen Mary, University of London

1.1 Introducing the multiverse

Nearly thirty years ago I wrote an article in the journal *Nature* with Martin Rees [1], bringing together all of the known constraints on the physical characteristics of the Universe – including the fine-tunings of the physical constants – which seemed to be necessary for the emergence of life. Such constraints had been dubbed 'anthropic' by Brandon Carter [2] – after the Greek word for 'man' – although it is now appreciated that this is a misnomer, since there is no reason to associate the fine-tunings with mankind in particular. We considered both the 'weak' anthropic principle – which accepts the laws of nature and physical constants as given and claims that the existence of observers then imposes a selection effect on where and when we observe the Universe – and the 'strong' anthropic principle – which (in the sense we used the term) suggests that the existence of observers imposes constraints on the physical constants themselves.

Anthropic claims – at least in their strong form – were regarded with a certain amount of disdain by physicists at the time, and in some quarters they still are. Although we took the view that any sort of explanation for the observed fine-tunings was better than none, many regarded anthropic arguments as going beyond legitimate science. The fact that some people of a theological disposition interpreted the claims as evidence for a Creator – attributing teleological significance to the strong anthropic principle – perhaps enhanced that reaction. However, attitudes have changed considerably since then. This is not so much because the status of the anthropic arguments themselves have changed – as we will see in a later chapter, some of them have become firmer and others weaker. Rather, it is because there has been a fundamental shift in the epistemological status of the anthropic principle. This arises because cosmologists have come to realize that there are many

Universe or Multiverse?, ed. Bernard Carr. Published by Cambridge University Press.
© Cambridge University Press 2007.

contexts in which our universe could be just one of a (possibly infinite) ensemble of 'parallel' universes in which the physical constants vary. This ensemble is sometimes described as a 'multiverse', and this term is used pervasively in this volume (including the title). However, it must be stressed that many other terms are used – sometimes even in the same context.

These multiverse proposals have not generally been motivated by an attempt to explain the anthropic fine-tunings; most of them have arisen independently out of developments in cosmology and particle physics. Nevertheless, it now seems clear that the two concepts are inherently interlinked. For if there *are* many universes, this begs the question of why we inhabit this particular one, and – at the very least – one would have to concede that our own existence is a relevant selection effect. Indeed, since we necessarily reside in one of the life-conducive universes, the multiverse picture reduces the strong anthropic principle to an aspect of the weak one. For this reason, many physicists would regard the multiverse proposal as providing the most natural explanation of the anthropic fine-tunings.

One reason that the multiverse proposal is now popular is that it seems to be necessary in order to understand the origin of the Universe. Admittedly, cosmologists have widely differing views on how the different worlds might arise. Some invoke models in which a single universe undergoes cycles of expansion and recollapse, with the constants being changed at each bounce [3]. In this case, the different universes are strung out in *time*. Others invoke the 'inflationary' scenario [4], in which our observable domain is part of a single 'bubble' which underwent an extra-fast expansion phase at some early time. There are many other bubbles, each with different laws of low-energy physics, so in this case the different universes are spread out in *space*. As a variant of this idea, Andrei Linde [5] and Alex Vilenkin [6] have invoked 'eternal' inflation, in which each universe is continually self-reproducing, since this predicts that there may be an infinite number of domains – all with different coupling constants. The different universes then extend in *both* space and time.

On the other hand, Stephen Hawking prefers a quantum cosmological explanation for the Universe and has objected to eternal inflation on the grounds that it extends to the infinite past and is thus incompatible with the Hartle–Hawking 'no boundary' proposal for the origin of the Universe [7]. This requires that the Universe started at a finite time but the initial singularity of the classical model is regularized by requiring time to become imaginary there. If one uses the path integral approach to calculate the probability of a particular history, this appears to favour very few expansion *e*-folds, so the Universe would recollapse too quickly for life to arise.

However, anthropic selection can salvage this, since one only considers histories containing observers [8].

This sort of approach to quantum cosmology only makes sense within the context of the 'many worlds' interpretation of quantum mechanics. This interpretation was suggested by Hugh Everett [9] in the 1950s in order to avoid having to invoke collapse of the quantum mechanical wave-function, an essential feature of the standard Copenhagen interpretation. Instead, our universe is supposed to split every time an observation is made, so one rapidly generates a huge number of parallel worlds [10]. This could be regarded as the earliest multiverse theory. Although one might want to distinguish between classical and quantum multiverses, Max Tegmark [11] has emphasized that there is no fundamental distinction between them.

Quantum theory, of course, originated out of attempts to explain the behaviour of matter on small scales. Recent developments in particle physics have led to the popularity of yet another type of multiverse. The holy grail of particle physics is to find a 'Theory of Everything' (TOE) which unifies all the known forces of physics. Models which unify the weak, strong and electomagnetic interactions are commonly described as 'Grand Unified Theories' (GUTs) and – although still unverified experimentally – have been around for nearly 30 years. Incorporating gravity into this unification has proved more difficult, but recently there have been exciting strides, with superstring theory being the currently favoured model.[1] There are various versions of superstring theory but they are amalgamated in what is termed 'M-theory'.

Unlike the 'Standard Model', which excludes gravity and contains several dozen free parameters, M-theory might conceivably predict all the fundamental constants uniquely [12]. That at least has been the hope. However, recent developments suggest that this may not be the case and that the number of theories (i.e. vacuum states) could be enormous (for example 10^{500} [13]). This is sometimes described as the 'string landscape' scenario [14]. In this case, the dream that all the constants are uniquely determined would be dashed. There would be a huge number of possible universes (corresponding to different minima of the vacuum energy) and the values of the physical constants would be *contingent* (i.e. dependent on which universe we happen to occupy). Trying to predict the values of the constants would then be

---

1 String theory posits that the fundamental constituents of matter are string-like rather than point-like, with the various types of elementary particle corresponding to different excitation states of these strings. This was originally proposed as a model of strong interactions but in the 1980s it was realized that it could be extended to a version called 'superstring' theory, which also includes gravity.

as forlorn as Kepler's attempts to predict the spacing of the planets in our solar system based on the properties of Platonic solids.

A crucial feature of the string landscape proposal is that the vacuum energy would be manifested as what is termed a 'cosmological constant'. This is a term in the field equations of General Relativity (denoted by Λ) originally introduced by Einstein to allow a static cosmological model but then rejected after the Universe was found to be expanding. For many subsequent decades cosmologists assumed Λ was zero, without understanding why, but a remarkable recent development has been the discovery that the expansion of the Universe is accelerating under the influence of (what at least masquerades as) a cosmological constant. One possibility is that Λ arises through quantum vacuum effects. We do not know how to calculate these, but the most natural value would be the Planck density (which is 120 orders of magnitude larger than the observed value). Indeed in the string landscape proposal, one might expect the value of Λ across the different universes to have a uniform distribution, ranging from minus to plus the Planck value. The observed value therefore seems implausibly small.

There is also another fine-tuning problem, in that the observed vacuum density is currently very similar to the matter density, a coincidence which would only apply at a particular cosmological epoch. However, as first pointed out by Steven Weinberg [15, 16], the value of Λ is constrained anthropically because galaxies could not form if it were much larger than observed. This is not the only possible explanation for the smallness of Λ, but there is a reluctant acceptance that it may be the most plausible one, which is why both string landscape and anthropic ideas are rather popular at present. The crucial issue of whether the number of vacuum states is sufficiently large and their spacing sufficiently small to satisfy the anthropic constraints is still unresolved.

It should be noted that M-theory requires there to be extra dimensions beyond the four familiar ones of space and time. Some of these may be compactified, but others may be extended, in which case, the Universe would correspond to a 4-dimensional 'brane' in a higher-dimensional 'bulk' [17, 18]. In the first versions of this theory, the cosmological constant was negative, which was incompatible with the observed acceleration of the Universe. A few years ago, however, it was realized that M-theory solutions with a positive cosmological constant are also possible [19], and this has revitalized the collaboration between cosmologists and string theorists. The notion that our universe is a brane in a higher-dimensional bulk also suggests another multiverse scenario, since there might be many other branes in the bulk. Collisions between these branes might even generate big bangs of the kind which initiated the expansion of our own universe [20]. Indeed, some people have envisaged successive collisions producing cyclic models, and it has been claimed that this could provide another (non-anthropic) explanation for why Λ naturally tends to a value comparable to the matter density [21].

1.2 Historical perspective

We have seen how a confluence of developments in cosmology and particle physics has led to a dramatic improvement in the credibility of the multiverse proposal. In this section, we will put these developments into a historical perspective, by showing how the notion of the multiverse is just the culmination of attempts to understand the physics of the largest and smallest scales. For what we regard as the 'Universe' has constantly changed as scientific progress has extended observations outwards to ever larger scales and inwards to ever smaller ones. In the process, it has constantly revealed new levels of structure in the world, as well as interesting connections between the laws operating at these different levels. This section will also provide an opportunity to review some of the basic ideas of modern cosmology and particle physics, which may be useful for non-specialists.

1.2.1 The outward journey

Geocentric view

Early humans assumed that the Earth was the centre of the Universe. Astronomical events were interpreted as being much closer than they actually are, because the heavens were assumed to be the domain of the divine and therefore perfect and unchanging. The Greeks, for example, believed the Earth was at the centre of a series of 'crystal spheres', these becoming progressively more perfect as one moves outwards. The last one was associated with the immovable stars, so transient phenomena (like meteors and comets) were assumed to be of terrestrial origin. Even the laws of nature (such as the regularity of the seasons) seemed to be human-centred, in the sense that they could be exploited for our own purposes, so it was natural to regard them as a direct testimony to our central role in the world.

Heliocentric view

In 1542 Nicolaus Copernicus argued in *De Revolutionis Orbis* that the heliocentric picture provides a simpler explanation of planetary motions than the geocentric one, thereby removing the Earth from the centre of the Universe. The heliocentric picture had earlier been suggested by Artistarchus, although this was regarded as blasphemous by most of his fellow Greeks, and Nicholas de Cusa, who in 1444 argued that the Universe had no centre and looks the same everywhere. Today this notion is called the Copernican or Cosmological Principle. Then in 1572 Tycho Brahe spotted a supernova in the constellation of Cassiopeia; it brightened suddenly and then dimmed over the course of a year, but the fact that its apparent position did not change as the Earth moved around the Sun implied that it was well beyond the Moon. Because this destroyed the Aristotelian view that the heavens never change, the claim was at first received sceptically. Frustrated by those who had eyes but would not see, Brahe wrote in the preface of *De Nova Stella*: 'O crassa ingenia. O coecos coeli spectators.' (Oh thick wits. Oh blind watchers of the sky.)

Galactocentric view

The next step occurred when Galileo Galilei used the newly invented telescope to show that not even the Sun is special. His observations of sunspots showed that it changes, and in 1610 he speculated in *The Sidereal Message* that the Milky Way – then known as a band of light in the sky but now known to be the Galaxy – consists of stars like the Sun but at such a great distance that they cannot be resolved. This not only cast doubt on the heliocentric view, but also vastly increased the size of the Universe. An equally profound shift in our view of the Universe came a few decades later with Isaac Newton's discovery of universal gravity. By linking astronomical phenomena to those on Earth, Newton removed the special status of the heavens, and the publication of his *Principia* in 1687 led to the 'mechanistic' view in which the Universe is regarded as a giant machine. In the following century, the development of more powerful telescopes – coupled with Newton's laws – enabled astronomers to understand the structure of the Milky Way. In 1750 Thomas Wright proposed that this is a disc of stars, and in 1755 Immanuel Kant speculated that some nebulae are 'island universes' similar to the Milky Way, raising the possibility that even the Galaxy is not so special. However, the galactocentric view persisted for several more centuries, with most astronomers still assuming that the Milky Way comprised the whole Universe. Indeed this was Einstein's belief when he published his theory of General Relativity in 1915 and started to study its cosmological implications.

Cosmocentric view

Then in the 1920s the idea anticipated by Kant – that some of the nebulae are outside the Milky Way – began to take hold. For a while this was a matter of intense controversy. In 1920 Heber Curtis vigorously defended the island universe theory in a famous debate with Harlow Shapley. The controversy was finally resolved in 1924 when Edwin Hubble announced that he had measured the distance to M31 using Cepheid stars. An even more dramatic revelation came in 1929, when Hubble obtained radial velocities and distance estimates for several dozen nearby galaxies, thereby discovering that all galaxies are moving away from us with a speed proportional to their distance. This is now called 'Hubble's law' and it has been shown to apply out to a distance of 10 billion light-years, a region containing 100 billion galaxies. The most natural interpretation of Hubble's law is that space itself is expanding, as indeed had been predicted by Alexander Friedmann in 1920 on the basis of general relativity. Friedmann's model suggested that the Universe began in a state of great compression at a time in the past of order the inverse of the Hubble constant, now known to be about 14 billion years. This is the 'Big Bang' picture, and it received decisive support in 1965 with the discovery that the Universe is bathed in a sea of background radiation. This radiation is found to have the same temperature in every direction and to have a black-body spectrum, implying that the Universe must once have been sufficiently compressed for the radiation to have interacted with the matter. Subsequent studies by the COBE satellite confirmed that it has a perfect black-body spectrum, which firmly established the Big Bang theory as a branch of mainstream physics.

Multiverse view

Further studies of the background radiation – most notably by the WMAP satellite – have revealed the tiny temperature fluctuations associated with the density ripples which eventually led to the formation of galaxies and clusters of galaxies. The angular dependence of these ripples is exactly as predicted by the inflationary scenario, which suggests that our observable domain is just a tiny patch of a much larger universe. This was the first evidence for what Tegmark [11] describes as the 'Level I' multiverse. A still more dramatic revelation has been the discovery – from observations of distant supernovae – that the expansion of the Universe is accelerating. We don't know for sure what is causing this, but it is probably related to the vacuum energy density. As described in Section 1.1, the low value of this density may indicate that there exist many other universes with different vacuum states, so this may be evidence for Tegmark's 'Level II' multiverse.

This brief historical review of developments on the outer front illustrates that the longer we have studied the Universe, the larger it has become. Indeed, the multiverse might be regarded as just one more step in the sequence of expanding vistas opened up by cosmological progress (from geocentric to heliocentric to galactocentric to cosmocentric). More conservative cosmologists might prefer to maintain the cosmocentric view that ours is the only Universe, but perhaps the tide of history is against them.

1.2.2 The inward journey

Equally dramatic changes of perspective have come from revelations on the inward front, with the advent of atomic theory in the eighteenth century, the discovery of subatomic particles at the start of the twentieth century and the advent of quantum theory shortly thereafter. The crucial achievement of the inward journey is that it has revealed that everything in the Universe is made up of a few fundamental particles and that these interact through four forces: gravity, electromagnetism, the weak force and the strong force. These interactions have different strengths and characteristics, and it used to be thought that they operated independently. However, it is now thought that some (and possibly all) of them can be unified as part of a single interaction.

Figure 1.1 illustrates that the history of physics might be regarded as the history of this unification. Electricity and magnetism were combined by Maxwell's theory of electromagnetism in the nineteenth century. The electromagnetic force was then combined with the weak force in the (now experimentally confirmed) electroweak theory in the 1970s. Theorists have subsequently merged the electroweak force with the strong force as part of the Grand Unified Theory (GUT), although this has still not been verified experimentally. As discussed in Section 1.1, the final (and as yet incomplete) step is the unification with gravity, as attempted by string theory or M-theory.

A remarkable feature of these theories is that the Universe may have more than the three dimensions of space that we actually observe, with the extra dimensions being compactified on the Planck scale (the distance of 10^{-33} cm at which quantum gravity effects become important), so that we do not notice them. In M-theory itself, the total number of dimensions (including time) is eleven, with 4-dimensional physics emerging from the way in which the extra dimensions are compactified (described by what is called a Calabi–Yau manifold). The discovery of dark dimensions through particle physics shakes our view of the nature of reality just as profoundly as the discovery of dark energy through cosmology. Indeed, we saw in Section 1.1 that there may be an intimate link between these ideas.

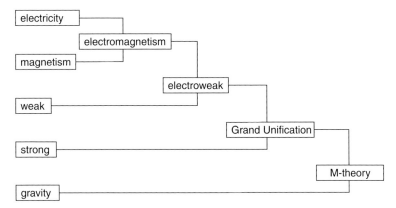

Fig. 1.1. This shows the successive steps by which physics has attempted to unify the four known forces of nature. Time runs to the right.

1.2.3 The cosmic uroborus

Taken together, scientific progress on both the outer and inner fronts can certainly be regarded as a triumph. In particular, physics has revealed a unity about the Universe which makes it clear that everything is connected in a way which would have seemed inconceivable a few decades ago. This unity is succinctly encapsulated in the image of the uroborus (i.e. the snake eating its own tail). This is shown in Fig. 1.2 (adapted from a picture originally presented by Sheldon Glashow) and demonstrates the intimate link between the macroscopic domain (on the left) and the microscopic domain (on the right).

The pictures drawn around the snake represent the different types of structure which exist in the Universe. Near the bottom are human beings. As we move to the left, we encounter successively larger objects: a mountain, a planet, a star, the solar system, a galaxy, a cluster of galaxies and finally the entire observable Universe. As we move to the right, we encounter successively smaller objects: a cell, a DNA molecule, an atom, a nucleus, a quark, the GUT scale and finally the Planck length. The numbers at the edge indicate the scale of these structures in centimetres. As one moves clockwise from the tail to the head, the scale increases through 60 decades: from the smallest meaningful scale allowed by quantum gravity (10^{-33} cm) to the scale of the visible Universe (10^{27} cm). If one expresses these scales in units of the Planck length, they go from 0 to 60, so the uroborus provides a sort of 'clock' in which each 'minute' corresponds to a factor of 10 in scale.

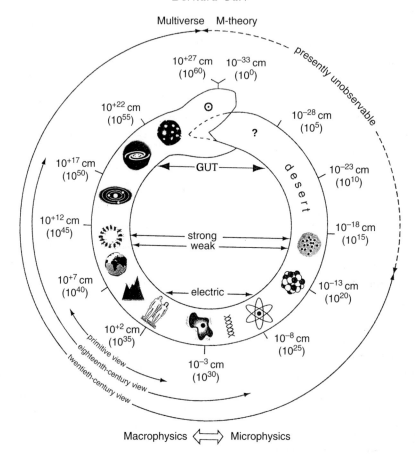

Fig. 1.2. The image of the uroborus summarizes the different levels of structure in the physical world, the intimate link between the microphysical and macrophysical domains and the evolution of our understanding of this structure.

A further aspect of the uroborus is indicated by the horizontal lines. These correspond to the four interactions and illustrate the subtle connection between microphysics and macrophysics. For example, the 'electric' line connects an atom to a planet because the structure of a solid object is determined by atomic and intermolecular forces, both of which are electrical in origin. The 'strong' and 'weak' lines connect a nucleus to a star because the strong force, which holds nuclei together, also provides the energy released in the nuclear reactions which power a star, and the weak force, which causes nuclei to decay, also prevents stars from burning out too soon. The 'GUT' line connects the grand unification scale with galaxies and clusters because the density fluctuations which led to these objects originated when the temperature of the Universe was high enough for GUT interactions to be important. Indeed the Big Bang theory suggests that these features arose when the current observable Universe had the size of a grapefruit!

The significance of the head meeting the tail is that the entire Universe was once compressed to a point of infinite density (or, more strictly, the Planck density). Since light travels at a finite speed, we can never see further than the distance light has travelled since the Big Bang, about 10^{10} light-years; more powerful telescopes merely probe to earlier times. Cosmologists now have a fairly complete picture of the history of the Universe: as one goes back in time, galaxy formation occurred at a billion years after the Big Bang, the background radiation last interacted with matter at a million years, the Universe's energy was dominated by its radiation content before about 10 000 years, light elements were generated through cosmological nucleosynthesis at around 3 minutes, antimatter was abundant before about a microsecond (before which there was just a tiny excess of matter over antimatter), electroweak unification occurred at a billionth of a second (the highest energy which can be probed experimentally), grand unification and inflation occurred at 10^{-35} s and the quantum gravity era (the smallest meaningful time) was at 10^{-43} s.

Perhaps the most striking aspect of the top of the uroborus is its link with higher dimensions. On the microscopic side, this arises because the various versions of superstring theory all suppose that the Universe has more than the three dimensions of space which we actually observe but with the extra dimensions being compactified. On the macroscopic side, the higher-dimensional link arises because we have seen that some versions of M-theory suggest that the Universe could be a 4-dimensional 'brane' in a higher-dimensional 'bulk' [17, 18]. This suggests that there might be many other branes in the bulk, although we have seen there are multiverse proposals which do not involve extra dimensions.

Figure 1.2 also has an historical aspect, since it shows how humans have systematically expanded the outermost and innermost limits of his awareness. Thus primitive humans were aware of scales from about 10^{-2} cm (mites) to 10^7 cm (mountains); eighteenth century humans were aware of scales from about 10^{-5} cm (bacteria) to 10^{17} cm (the solar system); and twentieth-century humans were aware of scales from about 10^{-13} cm (atomic nuclei) to 10^{27} cm (the most distant galaxies). Indeed it is striking that science has already expanded the macroscopic frontier as far as possible, although experimentally we may never get much below the electroweak scale in the microscopic direction. We might therefore regard the uroborus as representing the blossoming of human consciousness.

1.3 But is the multiverse science?

Despite the growing popularity of the multiverse proposal, it must be admitted that many physicists remain deeply uncomfortable with it. The reason is clear: the idea is highly speculative and, from both a cosmological and a particle physics perspective, the reality of a multiverse is currently untestable. Indeed, it may always remain so, in the sense that astronomers may never be able to observe the other universes with telescopes a and particle physicists may never be able to observe the extra dimensions with their accelerators. The only way out would be if the effects of extra dimensions became 'visible' at the TeV scale, in which case they might be detected when the Large Hadron Collider becomes operational in 2007. This would only be possible if the extra dimensions were as large as a millimetre. However, it would be very fortunate (almost anthropically so) if the scale of quantum gravity just happened to coincide with the largest currently accessible energy scale.

For these reasons, some physicists do not regard these ideas as coming under the purvey of science at all. Since our confidence in them is based on faith and aesthetic considerations (for example mathematical beauty) rather than experimental data, they regard them as having more in common with religion than science. This view has been expressed forcefully by commentators such as Sheldon Glashow [22], Martin Gardner [23] and George Ellis [24], with widely differing metaphysical outlooks. Indeed, Paul Davies [25] regards the concept of a multiverse as just as metaphysical as that of a Creator who fine-tuned a single universe for our existence. At the very least the notion of the multiverse requires us to extend our idea of what constitutes legitimate science.

In some people's eyes, of course, cosmology has always bordered on metaphysics. It has constantly had to battle to prove its scientific respectability, fighting not only the religious, but also the scientific orthodoxy. For example, the prevalent view until well into the nineteenth century (long after the demise of the heliocentric picture) was that speculations about things beyond the Solar System was not proper science. This was reflected by Auguste Comte's comments on the study of stars in 1859 [26]:

Never, by any means, will we be able to study their chemical compositions. The field of positive philosophy lies entirely within the Solar System, the study of the Universe being inaccessible in any possible science.

However, Comte had not foreseen the advent of spectroscopy, triggered by Gustav Kirchhoff's realization in the same year that the dark lines in the solar spectrum were absorption features associated with chemical elements.

For the first time this allowed astronomers to probe the composition of distant stars.

Cosmology attained the status of a proper science in 1915, when the advent of general relativity gave the subject a secure mathematical basis. The discovery of the cosmological expansion in the 1920s then gave it a firm empirical foundation. Nevertheless, it was many decades before it gained full scientific recognition. For example, when Ralph Alpher and Robert Herman were working on cosmological nucleosynthesis in the 1940s, they recall [27]: 'Cosmology was then a sceptically regarded discipline, not worked in by sensible scientists.' Only with the detection of the microwave background radiation in 1965 was the hot Big Bang theory established as a branch of mainstream physics, and only with the recent results from the WMAP satellite (postdating the Stanford meeting which led to this book) has it become a *quantitative* science with real predictive power.

Nevertheless, cosmology is still different from most other branches of science; one cannot experiment with the Universe, and speculations about processes at very early and very late times depend upon theories of physics which may never be directly testable. Because of this, more conservative physicists still tend to regard cosmological speculations as going beyond the domain of science. The introduction of anthropic reasoning doubtless enhanced this view. On the other hand, other physicists have always held a more positive opinion, so there has developed a polarization of attitudes towards the anthropic principle. This is illustrated by the following quotes. The first is from the protagonist Freeman Dyson [28]:

I do not feel like an alien in this Universe. The more I examine the Universe and examine the details of its architecture, the more evidence I find that the Universe in some sense must have known we were coming.

This might be contrasted with the view of the antagonist Heinz Pagels [29]:

The influence of the anthropic principle on contemporary cosmological models has been sterile. It has explained nothing and it has even had a negative influence. I would opt for rejecting the anthropic principle as needless clutter in the conceptual repertoire of science.

An intermediate stance is taken by Brandon Carter [2], who might be regarded as one of the fathers of the anthropic principle:

The anthropic principle is a middle ground between the primitive anthropocentrism of the pre-Copernican age and the equally unjustifiable antithesis that no place or time in the Universe can be privileged in any way.

The growing popularity of the multiverse picture has encouraged a drift towards Carter's view, because it suggests that the anthropic fine-tunings can at least have a 'quasi-physical' explanation. To the hard-line physicist, the multiverse may not be entirely respectable, but it is at least preferable to invoking a Creator. Indeed anthropically inclined physicists like Susskind and Weinberg are attracted to the multiverse precisely because it seems to dispense with God as the explanation of cosmic design.[2]

In fact, the dichotomy in attributing anthropic fine-tunings to God or the multiverse is too simplistic. While the fine-tunings certainly do not provide unequivocal evidence for God, nor would the existence of a multiverse preclude God since – as emphasized by Robin Collins [30] – there is no reason why a Creator should not act through the multiverse. Neverethless, the multiverse proposal certainly poses a serious challenge to the theological view, so it is not surprising that it has commended itself to atheists. Indeed, Neil Manson has described the multiverse as 'the last resort for the desperate atheist' [31].

By emphasizing the scientific legitimacy of anthropic and multiverse reasoning, I do not intend to deny the relevance of these issues to the science–religion debate [32]. The existence of a multiverse would have obvious religious implications [33], so contributions from theologians are important. More generally, cosmology addresses fundamental questions about the origin of matter and mind, which are clearly relevant to religion, so theologians need to be aware of the answers it provides. Of course, the remit of religion goes well beyond the materialistic issues which are the focus of cosmology. Nevertheless, in so much as religious and cosmological truths overlap, they must be compatible. This has been stressed by Ellis [34], who distinguishes between Cosmology (with a big C) – which takes into account 'the magnificent gestures of humanity' – and cosmology (with a small c), which just focuses on physical aspects of the Universe. In his view, morality is embedded in the cosmos in some fundamental way. Similar ideas have been expounded by John Leslie [35].

On the other hand, science itself cannot deal with such issues, and it seems unlikely that – even in the extended form required to accommodate the multiverse – science will ever prove or disprove the existence of God. Some people may see in the physical world some hint of the divine, but this can only provide what John Polkinghorne describes as 'nudge' factors [36].

---

2 It should be cautioned that the concept of 'cosmic design' being described here has nothing to do with the 'Intelligent Design' movement in the USA. Nevertheless, atheists might hope that the multiverse theory will have the same impact in the context of cosmic design as the theory of evolution did in the context of biological design.

Convictions about God's existence must surely come from 'inside' rather than 'outside' and even those eminent physicists who are mystically inclined do not usually base their faith on scientific revelations [37]. For this reason, theology receives rather short shrift in this volume. The contributors are nearly all physicists, and even those of a theological disposition have generally restricted their remarks to scientific considerations.

1.4 Overview of book

Part I contains articles deriving from two talks at the symposium *Expectations of a Final Theory*, which was held in Cambridge in September 2005. These provide appropriate opening chapters for this volume because of their historical perspective and because they illustrate the way in which the subject has been propelled by a combination of developments in cosmology and particle physics. Starting with contributions from two Nobel laureates also serves to emphasize the degree of respectability that the topic has now attained!

In the first contribution, 'Living in the multiverse', based on his opening talk at the Cambridge meeting, Steven Weinberg argues that the idea of the multiverse represents an important change in the nature of science, a radical shift in what we regard as legitimate physics. This shift is prompted by a combination of developments on the theoretical and the observational fronts. In particular, he highlights the anthropic constraint on the value of the vacuum energy or cosmological constant, a constraint which he himself first pointed out in 1987 and might be regarded as one of the few successful anthropic predictions. He also highlights the string landscape scenario, which is perhaps the most plausible theoretical basis for the multiverse proposal and is the focus of several later chapters.

Frank Wilczek's contribution, aptly entitled 'Enlightenment, knowledge, ignorance, temptation', is based on his summary talk at the Cambridge meeting. In this, he discusses the historical and conceptual roots of reasoning about the parameters of fundamental physics and cosmology based on selection effects. He describes the developments which have improved the status of such reasoning, emphasizing that these go back well before string theory. He is well aware of the downside of this development, but accepts it as part of the price that has to be paid. Such reasoning can and should be combined with arguments based on symmetry and dynamics; it supplements them, but does not replace them. This view is cogently encapsulated in Wilczek's eponymous classification of physical parameters.

1.4.1 Cosmology and astrophysics

Part II contains chapters whose emphasis is primarily on cosmology and astrophysics. The opening chapter, 'Cosmology and the multiverse', is by Martin Rees, one of the foremost champions of the multiverse concept and the host of the two Cambridge meetings represented in this volume. He points out that the parts of space and time that are directly observable (even in principle) may be an infinitesimal part of physical reality. Rejecting the unobservable part as a suitable subject for scientific discourse at the outset is unjustified because there is a blurred transition – what he describes as a 'slippery slope' – between what is observable and unobservable. After briefly addressing some conceptual issues, he discusses what the Universe would be like if some of the key cosmological numbers were different, and how one can in principle test specific hypotheses about the physics underlying the multiverse.

Although the focus of this volume is the multiverse rather than the anthropic principle, it is important to recall the fine-tunings which the multiverse proposal is purporting to explain. Indeed, in the absence of *direct* evidence for other universes, these might be regarded as providing the only *indirect* evidence. This motivates the inclusion of my own chapter, 'The anthropic principle revisited', in which I reconsider the status of some of the arguments presented in my 1979 *nature* paper with Rees [1]. Although I also veer into more philosophical issues, I have included my chapter here because most of the anthropic relationships are associated with cosmology and astrophysics. I emphasize that the key feature of the anthropic fine-tunings is that they seem necessary for the emergence of *complexity* during the evolution of the Universe from the Big Bang. The existence of conscious observers is just one particular manifestation of this and may not be fundamental.

In 'Cosmology from the top down', Stephen Hawking contrasts different approaches to the central questions of cosmology: why is the Universe spatially flat and expanding; why is it 4-dimensional; why did it start off with small density fluctuations; why does the Standard Model of particle physics apply? Some physicists would prefer to believe that string theory, or M-theory, will answer these questions and uniquely predict the features of the Universe. Others adopt the view that the initial state of the Universe is prescribed by an outside agency, code-named God, or that there are many universes, with ours being picked out by the anthropic principle. Hawking argues that string theory is unlikely to predict the distinctive features of the Universe. But neither is he is an advocate of God. He therefore opts for the last approach, favouring the type of multiverse which arises naturally within the context of his own work in quantum cosmology.

Several other contributors regard quantum cosmology as providing the most plausible conceptual framework for the multiverse, so the book returns to this theme later. However, the multiverse hypothesis comes in many different guises, and these are comprehensively summarized in Max Tegmark's chapter, 'The multiverse hierarchy'. Indeed, Tegmark argues that the key question is not *whether* parallel universes exist but on *how many* levels they exist. He shows that physical theories involving parallel universes form a four-level hierarchy, allowing progressively greater diversity. Level I is associated with inflation and contains Hubble volumes realizing all possible initial conditions. This is relatively uncontroversial, since it is a natural consequence of the cosmological 'concordance' model. Level II assumes that different regions of space can exhibit different effective laws of physics (i.e. different physical constants, different dimensionality and different particle content). For example, inflation models in the string landscape scenario subdivide into four increasingly diverse sublevels: IIa involves the same effective laws but different post-inflationary bubbles; IIb involves different minima in the effective supergravity potential; IIc involves different fluxes (of particular fields) for a given compactification; and IId involves different compactifications. Level III corresponds to the 'many worlds' of quantum theory. Tegmark argues that the other branches of the wave-function add nothing qualitatively new, even though historically this level has been the most controversial. Finally, Level IV invokes other mathematical structures, associated with different fundamental equations of physics. He then raises the question of how multiverse models can be falsified and argues that there is a severe 'measure problem' that must be solved to make testable predictions at levels II–IV. This point is addressed in more detail by later contributors.

Tegmark's classification emphasizes the central role of inflation, which postulates an era in the very early Universe when the expansion was accelerating. Inflation is invoked to explain two of the most striking features of the Universe – its smoothness and flatness – and to many physicists the theory still provides the most natural basis for the multiverse scenario. One of the prime advocates of the anthropic aspects of inflation is Andrei Linde, so it is most appropriate that he contributes the next chapter, 'The inflationary multiverse'. He first places the anthropic principle in an historical context: although anthropic considerations can help us understand many properties of our world, for a long time many scientists were ashamed to use the principle in their research because it seemed too metaphysical. However, the 'chaotic' inflationary scenario – which Linde pioneered and describes here – provides a simple justification for it. He especially favours 'eternal' inflation and links this to developments in string theory. He then discusses the implications of this idea for dark energy, relic axions and electroweak symmetry-breaking. These implications are explored in more detail in several later chapters, but Linde's article serves as an excellent introduction to these ideas and brings them all together.

One of the issues raised by Linde is the prevalence of dark matter, and this is the focus of the second contribution by Frank Wilczek, 'A model of anthropic reasoning: the dark to ordinary matter ratio'. He focuses on a dark matter candidate called the axion, which is a particle associated with the breaking of Peccei–Quinn (i.e. strong CP) symmetry in the early Universe. Large values of the symmetry-breaking energy scale (associated with large values of the Peccei–Quinn 'misalignment' angle) are forbidden in conventional axion cosmology. However, if inflation occurs after the breaking of Peccei–Quinn symmetry, large values are permitted providing we inhabit a region of the multiverse where the initial misalignment is small. Although such regions may occupy only a small volume of the multiverse, they contain a large fraction of potential observers. This scenario therefore yields a possible anthropic explanation of the approximate equality of the dark matter and baryon densities.

We have seen that another striking feature of the Universe is that its expansion appears to be accelerating under the influence of some form of 'dark energy'. The source of this energy is uncertain, but it may be associated with a cosmological constant. Indeed, we have seen that one of the most impressive successes of anthropic reasoning is that it may be able to explain the present value of the cosmological constant. Several contributions touch on this, but the most comprehensive treatment is provided by Alex Vilenkin, whose chapter, 'Anthropic predictions: the case of the cosmological constant', reviews the history and nature of this prediction. He also discusses the inclusion of other variable parameters (such as the neutrino mass) and the implications for particle physics. In anticipation of a theme which emerges later in the book, he emphasizes that anthropic models give testable predictions, which can be confirmed or falsified at a specified confidence level. However, anthropic predictions always have an intrinsic variance, which cannot be reduced indefinitely as theory and observations progress.

The cosmological constant also plays a central role in James Bjorken's chapter, 'The definition and classification of universes'. If the concept of a multiverse makes sense, one needs a specific, standardized definition for member universes which are similar to our own. Crucial to this description is the definition of the 'size' of the universe and, for the de Sitter model, Bjorken takes this to be the asymptotic value of the inverse Hubble constant. This is directly related to the value of the cosmological constant, so this parameter plays a natural role in his classification. He further proposes that the vacuum parameters and coupling constants of the Standard Model in any universe are dependent upon this size. Anthropic considerations then limit the size of habitable universes (as we understand that concept) to be within a factor of 2 of our own. Implications of this picture for understanding the 'hierarchy problem' in the Standard Model are discussed, as are general issues of falsifiability and verifiability.

Bjorken does not attempt to provide a physical basis for models with different cosmological constants, but a possible motivation comes from string theory, or M-theory. This point is discussed by several contributors, but the most thorough discussion of the cosmological applications of the idea is provided in Renata Kallosh's chapter, 'M/string theory and anthropic reasoning'. Here she outlines some recent cosmological studies of M/string theory and gives a couple of examples where anthropic reasoning – combined with our current incomplete understanding of string theory and supergravity – helps to shed light on the mysterious properties of dark energy. This is a rather technical article, but it is very important because it describes the results of her famous paper with A. Linde, S. Kachru and S. Trivedi, which shows that M/string theory allows models with a positive cosmological constant. This was a crucial development because string theorists used to assume that the constant would have to be *negative*, so this is an example of how cosmology has led to important insights into particle physics.

Closely related to Kallosh's theme is the final chapter in Part II by Savas Dimopoulos and Scott Thomas, 'The anthropic principle, dark energy and the LHC'. Here they argue that – in a broad class of theories – anthropic reasoning leads to a time-dependent vacuum energy with distinctive and potentially observable characteristics. The most exciting aspect of this proposal is that it leads to predictions that might be testable with the Large Hadron Collider, due to start operating in 2007. This illustrates the intimate link between cosmology and particle physics, so this naturally leads into the next part of the volume, which focuses on particle physics aspects of the multiverse hypothesis.

1.4.2 Particle physics and quantum theory

Part III starts with two articles on the values of the constants of particle physics, then moves onto the link with string theory, and concludes with articles concerned with quantum theory. There is a two-fold connection with quantum theory, since the 'many worlds' interpretation of quantum mechanics provided one of the earliest multiverse scenarios (i.e. Tegmark's Level III) and quantum cosmology provides one of the latest.

That the multiverse wave-function can explore a multitude of vacua with different symmetries and parameters is the starting point of Craig Hogan's chapter, 'Quarks, electrons and atoms in closely related universes'. In the context of such models, he points out that properties of universes closely related to ours can be understood by examining the consequences of small departures of physical parameters from their observed values. The masses of the light fermions that make up the stable matter of which we comprise – the up and down quarks and the electron – have values in a narrow window that allows the existence of a variety of nuclei other than protons and also atoms with stable shells of electrons that are not devoured by their nuclei. Since a living world with molecules needs stable nuclei other than protons and neutrons, these fundamental parameters of the Standard Model are good candidates for quantities whose values are determined through selection effects within a multiverse. Hogan also emphasizes another possible link with observation. If the fermion masses are fixed by brane condensation or compactification of extra dimensions, there may be an observable fossil of this 'branching event' in the form of a gravitational-wave background.

In the second chapter, 'The fine-tuning problems of particle physics and anthropic mechanisms', John Donoghue emphasizes that many of the classic problems of particle physics appear in a very different light when viewed from the perspective of the multiverse. Parameters in particle physics are regarded as fine-tuned if the size of the quantum corrections to their values in perturbation theory is large compared with their 'bare' values. Three parameters in the Standard Model are particularly puzzling because they are unnaturally small. Two of these – the Higgs vacuum expectation value and the cosmological constant – constitute the two great fine-tuning problems that motivate the field. The third is the strong CP violating factor, already highlighted in Wilczek's second contribution. All of these fine-tunings are alleviated when one accounts for the anthropic constraints which exist in a multiverse. However, the challenge is to construct a realistic physical theory of the multiverse and to test it. Donoghue describes some phenomenology of the quark and lepton masses that may provide a window on the multiverse theory.

The main reason that particle physicists have become interested in the multiverse proposal is the development in string theory. In particular, the possibility that M-theory may lead to a huge number of vacuum states – each associated with a different universe – is a crucial feature of Leonard Susskind's string landscape proposal. In 'The anthropic landscape of string theory', he makes some educated guesses about the landscape of string theory vacua and – based on the recent work of a number of authors – argues that the landscape could be unimaginably large and diverse. Whether we like it or not, this is the kind of behaviour that gives credence to the anthropic principle. He discusses the theoretical and conceptual issues that arise in a cosmology based on the diversity of environments implicit in string theory. Some of the later stages of his exposition are fairly technical, but these ideas are of fundamental importance to this volume. Indeed Susskind's chapter has already been on the archives for several years and is one of the most cited papers in the field.

As already stressed, the 'many worlds' interpretation of quantum theory provided one of the earliest versions of the multiverse scenario, and this is particularly relevant to quantum cosmology, which is most naturally interpreted in terms of this proposal. This view is advocated very cogently in 'Cosmology and the many worlds interpretation of quantum mechanics' by Viatschelav Mukhanov. Indeed, he argues that the wave-function of the Universe and the cosmological perturbations generated by inflation can *only* be understood within Everett's interpretation of quantum mechanics. The main reason it has not been taken seriously by some physicists is that it predicts we each have many copies, which may seem unpalatable. However, Mukhanov argues that these copies are not 'dangerous' because we cannot communicate with them.

The link with quantum cosmology is probed further by James Hartle in 'Anthropic reasoning and quantum cosmology'. He stresses that anthropic reasoning requires a theory of the dynamics and quantum initial condition of the Universe. Any prediction in quantum cosmology requires both of these. But conditioned on this information alone, we expect only a few general features of the Universe to be predicted with probabilities near unity. Most useful predictions are of conditional probabilities that assume additional information beyond the dynamics and quantum state. Anthropic reasoning utilizes probabilities conditioned on our existence. Hartle discusses the utility, limitations and theoretical uncertainty involved in using such probabilities, as well as the predictions resulting from various levels of ignorance of the quantum state.

The link between Everett's picture and the multiverse proposal is explored in depth by Brandon Carter. His chapter, 'Micro-anthropic principle for quantum theory', is somewhat technical but very valuable since it provides an excellent historical perspective and leads to an interpretation of the many worlds picture which goes beyond the original Everett version. Probabilistic models, developed by workers such as Boltzmann on foundations due to pioneers such as Bayes, were commonly regarded as approximations to a deterministic reality before the roles were reversed by the quantum revolution under the leadership of Heisenberg and Dirac. Thereafter, it was the deterministic description that was reduced to the status of an approximation, with the role of the observer becoming particularly prominent. In Carter's view, the lack of objectivity in the original Copenhagen interpretation has not been satisfactorily resolved in newer approaches of the kind pioneered by Everett. The deficiency of such interpretations is attributable to their failure to allow for the anthropic aspect of the problem, in the sense that there is *a priori* uncertainty about the identity of the observer. Carter reconciles subjectivity with objectivity by distinguishing the concept of an *observer* from that of a *perceptor*, whose chances of identification with a particular observer need to be prescribed by a suitable anthropic principle. It is proposed that this should be done by an entropy ansatz, according to which the relevant micro-anthropic weighting is taken to be proportional to the logarithm of the relevant number of Everett-type branches.

1.4.3 More general or philosophical aspects

The final part of the book addresses more philosophical and epistemological aspects of the multiverse proposal – especially the issue of its scientific legitimacy. The chapters in this part are also written from a different standpoint from those in the earlier parts. Whereas the contributors in Parts I–III are mainly positive about the idea of the multiverse (otherwise they would presumably not be exploring it), some of the contributors in Part IV are rather critical – either preferring more theological interpretations of the anthropic coincidences or regarding multiverse speculations as going beyond science altogether.

The most sceptical of the critics is Lee Smolin. His chapter, 'Scientific alternatives to the anthropic principle', is the longest contribution in the volume and plays a crucial role in bringing all the criticisms of the multiverse proposal together. He first argues that the anthropic principle cannot be considered a part of science because it does not yield any falsifiable predictions. Claimed successful predictions are either uncontroversial applications of selection principles in one universe or they depend only on observed facts which are logically independent of any assumption about life or intelligence. The Principle of Mediocrity (first formulated by Vilenkin) is also examined and claimed to be unreliable, as arguments for true conclusions can easily be modified to lead to false conclusions by reasonable changes in the specification of the ensemble in which we are assumed to be typical. However, Smolin shows that it is still possible to make falsifiable predictions from multiverse theories if the ensemble predicted has certain specified properties and he emphasizes his own favoured multiverse proposal – Cosmological Natural Selection – which involves the generation of descendant universes through black hole formation. This proposal remains unfalsified, but it is very vulnerable to falsification, which shows that it is a proper scientific theory. The consequences for recent applications of the anthropic principle in the context of string theory (as described in Part III) are also discussed.

Several other contributions in this part address the question of whether the multiverse proposal is scientifically respectable, although they do not all share Smolin's negative conclusion. In 'Making predictions in a multiverse: conundrums, dangers, coincidences', Anthony Aguirre accepts that the notion of many universes with different properties is one answer to the question of why the Universe is so hospitable to life. He also acknowledges that this notion naturally follows from current ideas in eternal inflation and M/string theory. But how do we test a multiverse theory and which of the many universes do we compare to our own? His chapter enumerates what would seem to be essential ingredients for making testable predictions, outlines different strategies one might take within this framework, and then discusses some of the difficulties and dangers inherent in these approaches. Finally, he addresses the issue of whether the predictions of multiverse theories share any general, qualitative features.

The issue of testing also features in the contribution of George Ellis, 'Multiverses: description, uniqueness and testing', who concludes that the multiverse proposal is not really proper science. He emphasizes that a multiverse is determined by specifying first a possibility space of potentially existing universes and then a distribution function on this space for actually existing universes. Ellis is sceptical because there is a lack of uniqueness at both these stages and we are unable either to determine observationally the specific nature of any multiverse that is claimed to exist or to validate experimentally any claimed causal mechanism that will create one. Multiverses may be useful in explanatory terms, but arguments for their existence are ultimately of a philosophical nature. Ellis is not against metaphysics – indeed he has written extensively on philosophical and theological issues – but he feels it should not be confused with science.

The importance of testing is also explored by Don Page in 'Predictions and tests of multiverse theories'. Page is also of a religious persuasion, but he comes to a somewhat different conclusion from Ellis. A multiverse usually includes parts unobservable to us, but if the theory for it includes suitable measures for observations, what is observable can be explained by the theory even if it contains unobservable elements. Thus good multiverse theories *can* be tested. For Bayesian comparisons of different theories that predict more than one observation, Page introduces the concept of 'typicality' as the likelihood given by a theory that a random result of an observation would be at least as extreme as the result of one's actual observation. He also links this to the interpretations of the quantum theory. Some multiverse theories can be regarded as pertaining to a *single* quantum state. This obeys certain equations, which raises the question of why those equations apply. Other theories can be regarded as pertaining to more than one quantum state, and these raise another question: why is the measure for the set of different universes such as to make our observations not too atypical?

The importance of a good probabilistic basis for assessing multiverse scenarios is also highlighted by Nick Bostrom's chapter, 'Observation selection theory and cosmological fine-tuning'. His title refers to a methodological tool for dealing with observation selection effects. Such a tool is necessary if observational consequences are to be derived from cosmological theory. It also has applications in other domains, such as evolution theory, game theory and the foundations of quantum mechanics. Bostrum shows that observation selection theory needs a probabilistic anthropic principle, which can be formalized in what he terms the 'Observation Equation'. Some implications of this for the problem of cosmological fine-tuning are discussed.

The next two contributions tackle the religious issue explicitly. 'Are anthropic arguments, involving multiverses and beyond, legitimate?' is particularly welcome because it comes from William Stoeger, who is both a working scientist and a Jesuit priest. After reviewing the history of the anthropic principle, he discusses the two main versions of the strong form – a divine creator or a multiverse. The latter strives to confine anthropic arguments within the realms of science and invokes an actually existing ensemble of universes or universe domains. He critically examines the scientific status of this proposal, briefly indicating what is needed for the definition and testability of a multiverse, and then describes some purely scientific applications of anthropic arguments. After discussing the key philosophical presumption on which the strong anthropic principle rests – that the Universe could have been different – and its relationship to a possible final theory, he summarizes his main conclusions concerning the two 'transcendent' explanations of the strong anthropic principle. Even if a multiverse is proved to exist, Stoeger would not regard this as providing an *ultimate* explanation and it would certainly not exclude the existence of God. However, he cautions that such considerations go beyond science itself.

As suggested by its title, the chapter by Robin Collins, 'The multiverse hypothesis: a theistic perspective', also takes an explicitly theological stance. Many people have promoted the multiverse hypothesis as the atheistic alternative to a theistic explanation of the fine-tuning of the cosmos for the existence of life. However, Collins argues that the multiverse hypothesis is also compatible with theism – indeed he claims that the generation of many universes by some physical process fits in well with the traditional belief that God is infinitely creative. Since such a process would have to be structured in just the right way to produce even one life-sustaining universe, this version of the multiverse hypothesis does not completely avoid the suggestion of design. Finally, he consider other pointers to a theistic explanation of the Universe, such as the beauty and elegance of the laws of nature, and argues that Tegmark's multiverse hypothesis – that all possible laws of nature are actualized in some universe or another – does not adequately account for this aspect of the laws of nature.

There are, of course, alternative interpetations of the multiverse hypothesis which are neither anthropic nor theistic. One example of this is Smolin's Cosmological Natural Selection proposal. Another (more exotic) version – which has been explored by Bostrom (though not in this volume) – is that the Universe is a computer simulation. This is the theme of John Barrow's chapter, 'Living in a simulated universe'. He explains why, if we live in a simulated reality, we might expect to see occasional glitches and small drifts in the supposed constants and laws of nature over time. There may even be evidence for this from astronomical observations, although the interpretation of these remains controversial.

Another possible interpetation of the anthropic tunings is provided in the final chapter, 'Universes galore: where will it all end?', by Paul Davies, who is also somewhat sceptical of the multiverse proposal. He argues that, although 'a little bit of multiverse is good for you', invoking multiverse explanations willy-nilly is a seductive slippery slope. Followed to its logical extreme, it leads to conclusions that are at best bizarre, at worst absurd. After reviewing several shortcomings of indiscriminate multiverse explanations, including the simulated multiverse discussed by Barrow, he challenges the false dichotomy that fine-tuning requires the existence of either a multiverse or some sort of traditional cosmic architect. Instead, he explores the possibility of a 'third way', involving a radical reappraisal of the notion of physical law, and presents a toy illustration from the theory of cellular automata.

References

[1] B. J. Carr and M. J. Rees. *Nature*, **278** (1979), 605.
[2] B. Carter. In *Confrontation of Cosmological Theory with Observational Data*, ed. M. S. Longair (Dordrecht: Reidel, 1974), pp. 291–298.
[3] R. C. Tolman. *Relativity, Thermodynamics and Cosmology* (Oxford: Clarendon Press, 1934).
[4] A. H. Guth. *Phys. Rev.* **D 23** (1981), 347.
[5] A. D. Linde. *Phys. Lett.* **B 175** (1986), 395.
[6] A. Vilenkin. *Phys. Rev.* **D 27** (1983), 2848.
[7] J. B. Hartle and S. W. Hawking. *Phys. Rev.* **D 28** (1983), 2960.
[8] S. W. Hawking and T. Hertog. *Phys. Rev.* **D 73** (2006), 123527.
[9] H. Everett. *Rev. Mod. Phys.* **29** (1957), 454.
[10] D. Deutsch. *The Fabric of Reality* (London: Penguin, 1998).
[11] M. Tegmark, In *Science and Ultimate Reality: From Quantum to Cosmos*, eds. J. D. Barrow, P. C. W. Davies & C. L. Harper (Cambridge: Cambridge University Press, 2003), p. 459.
[12] G. L. Kane, M. J. Perry and A. N. Zytkow. *New Astron.* **7** (2002), 45.
[13] R. Bousso and J. Polchinski. *JHEP*, **06** (2000), 006.
[14] L. Susskind. This volume (2007).
[15] S. Weinberg. *Phys. Rev. Lett.* **59** (1987), 2607.
[16] S. Weinberg. *Rev. Mod. Phys.* **61** (1989), 1.
[17] L. Randall and R. Sundrum. *Phys. Rev. Lett.* **83** (1999), 4690.
[18] L. Randall. *Warped Passages: Unravelling the Universe's Hidden Dimensions* (London: Penguin, 2005).
[19] S. Kachru, R. Kallosh, A. Linde and S. R. Trivedi. *Phys. Rev.* **D 68** (2003), 046005.
[20] J. Khoury, B. A. Ovrut, P. J. Steinhardt and N. Turok. *Phys. Rev.* **D 64** (2001), 123522.
[21] A. R. Liddle and L. A. Urena-Lopez. *Phys. Rev. Lett.* **97** (2006), 161301.
[22] G. Brumfiel. *Nature*, **439** (2006), 10.
[23] M. Gardner. *Are Universes Thicker than Blackberries?* (New York: W. W. Norton & Co. Ltd, 2003).
[24] G. F. R. Ellis. This volume (2007).
[25] P. C. W. Davies. This volume (2007).
[26] A. Comte. *Cours de Philosophie Positive* (Paris: Bachelier, 1842).
[27] R. A. Alpher and R. Herman. *Phys. Today*, **41** (1988), 24.
[28] F. Dyson. *Rev. Mod. Phys.* **51** (1979), 447.
[29] H. R. Pagels. *Perfect Symmetry* (New York: Simon & Schuster), p. 359.
[30] R. Collins. This volume (2007).
[31] N. A. Manson. *God and Design* (London: Routledge, 2003).
[32] B. J. Carr. In *The Oxford Handbook of Religion and Science*, eds. P. Clayton and Z. Simpson (Oxford: Oxford University Press, 2006), p. 139.
[33] R. Holder. *God, the Multiverse and Everything: Modern Cosmology and the Argument from Design* (Aldershot: Ashgate, 2004).
[34] G. F. R. Ellis. *Before the Beginning: Cosmology Explained* (London: Boyars/Bowerdean, 1993).
[35] J. Leslie. *Universes* (London: Routledge, 1989).
[36] J. Polkinghorne. *The Faith of a Physicist* (Princeton, NJ: Princeton University Press, 1994).
[37] K. Wilbur. *Quantum Questions: Mystical Writings of the Worlds Greatest Physicists* (Boston: Shambhala, 2001).

2
Living in the multiverse

Steven Weinberg

Physics Department, University of Texas at Austin

Opening talk at the symposium *Expectations of a Final Theory* at Trinity College, Cambridge, 2 September 2005

2.1 Introduction

We usually mark advances in the history of science by what we learn about nature, but at certain critical moments the most important thing is what we discover about science itself. These discoveries lead to changes in how we score our work, in what we consider to be an acceptable theory.

For an example, look back to a discovery made just one hundred years ago. Before 1905 there had been numerous unsuccessful efforts to detect changes in the speed of light, due to the motion of the Earth through the ether. Attempts were made by Fitzgerald, Lorentz and others to construct a mathematical model of the electron (which was then conceived to be the chief constituent of all matter) that would explain how rulers contract when moving through the ether in just the right way to keep the apparent speed of light unchanged. Einstein instead offered a symmetry principle, which stated that not just the speed of light, but all the laws of nature are unaffected by a transformation to a frame of reference in uniform motion. Lorentz grumbled that Einstein was simply assuming what he and others had been trying to prove. But history was on Einstein's side. The 1905 Special Theory of Relativity was the beginning of a general acceptance of symmetry principles as a valid basis for physical theories.

This was how Special Relativity made a change in science itself. From one point of view, Special Relativity was no big thing – it just amounted to the replacement of one 10-parameter spacetime symmetry group, the Galileo group, with another 10-parameter group, the Lorentz group. But never

Universe or Multiverse?, ed. Bernard Carr. Published by Cambridge University Press.
© Cambridge University Press 2007.

before had a symmetry principle been taken as a legitimate hypothesis on which to base a physical theory.

As usually happens with this sort of revolution, Einstein's advance came with a retreat in another direction: the effort to construct a classical model of the electron was permanently abandoned. Instead, symmetry principles increasingly became the dominant foundation for physical theories. This tendency was accelerated after the advent of quantum mechanics in the 1920s, because the survival of symmetry principles in quantum theories imposes highly restrictive consistency conditions (existence of antiparticles, connection between spin and statistics, cancellation of infinities and anomalies) on physically acceptable theories. Our present Standard Model of elementary particle interactions can be regarded as simply the consequence of certain gauge symmetries and the associated quantum mechanical consistency conditions.

The development of the Standard Model did not involve any changes in our conception of what was acceptable as a basis for physical theories. Indeed, the Standard Model can be regarded as just quantum electrodynamics writ large. Similarly, when the effort to extend the Standard Model to include gravity led to widespread interest in string theory, we expected to score the success or failure of this theory in the same way as for the Standard Model: string theory would be a success if its symmetry principles and consistency conditions led to a successful prediction of the free parameters of the Standard Model.

Now we may be at a new turning point, a radical change in what we accept as a legitimate foundation for a physical theory. The current excitement is, of course, a consequence of the discovery of a vast number of solutions of string theory, beginning in 2000 with the work of Bousso and Polchinski [1].[1] The compactified six dimensions in Type II string theories typically have a large number (tens or hundreds) of topological fixtures (3-cycles), each of which can be threaded by a variety of fluxes. The logarithm of the number of allowed sets of values of these fluxes is proportional to the number of topological fixtures. Further, for each set of fluxes one obtains a different effective field theory for the modular parameters that describe the compactified 6-manifold, and for each effective field theory the number of local minima of the potential for these parameters is again proportional to

---

[1] Smolin [2] had noted earlier that string theory has a large number of vacuum solutions, and explored an imaginative possible consequence of this multiplicity. Even earlier, in the 1980s, Duff, Nilsson and Pope had noted that $D = 11$ supergravity has an infinite number of possible compactifications, but of course it was not then known that this theory is a version of string theory. For a summary, see ref. [3].

the number of topological fixtures. Each local minimum corresponds to the vacuum of a possible stable or metastable universe.

Subsequent work by Giddings, Kachru, Kallosh, Linde, Maloney, Polchinski, Silverstein, Strominger and Trivedi (in various combinations) [4–6] established the existence of a large number of vacua with positive energy densities. Ashok and Douglas [7] estimated the number of these vacua to be of order 10^{100} to 10^{500}. String theorists have picked up the term 'string landscape' for this multiplicity of solutions from Susskind [8], who took the term from biochemistry, where the possible choices of orientation of each chemical bond in large molecules lead to a vast number of possible configurations. Unless one can find a reason to reject all but a few of the string theory vacua, we may have to accept that much of what we had hoped to calculate are environmental parameters, like the distance of the Earth from the Sun, whose values we will never be able to deduce from first principles.

We lose some and win some. The larger the number of possible values of physical parameters provided by the string landscape, the more string theory legitimates anthropic reasoning as a new basis for physical theories. Any scientists who study nature must live in a part of the landscape where physical parameters take values suitable for the appearance of life and its evolution into scientists.

An apparently successful example of anthropic reasoning was already at hand by the time the string landscape was discovered. For decades there seemed to be something peculiar about the value of the vacuum energy density ρ_V. Quantum fluctuations in known fields at well understood energies (say, less than 100 GeV) give a value of ρ_V larger than observationally allowed by a factor 10^{56}. This contribution to the vacuum energy might be cancelled by quantum fluctuations of higher energy, or by simply including a suitable cosmological constant term in the Einstein field equations, but the cancellation would have to be exact to fifty-six decimal places. No symmetry argument or adjustment mechanism could be found that would explain such a cancellation. Even if such an explanation could be found, there would be no reason to suppose that the remaining net vacuum energy would be comparable to the *present* value of the matter density, and since it is certainly not very much larger, it was natural to suppose that it is very much less, too small to be detected.

On the other hand, if ρ_V takes a broad range of values in the multiverse, then it is natural for scientists to find themselves in a subuniverse in which ρ_V takes a value suitable for the appearance of scientists. I pointed out in 1987 that this value for ρ_V cannot be too large and positive, because then galaxies and stars would not form [9]. Roughly, this limit is that ρ_V should be less than the mass density of the universe at the time when galaxies first condense. Since this was in the past, when the mass density was larger than at present, the anthropic upper limit on the vacuum energy density is larger than the present mass density, but not many orders of magnitude greater.

But anthropic arguments provide not just a bound on ρ_V, they give us some idea of the value to be expected: ρ_V should be not very different from the mean of the values suitable for life. This is what Vilenkin [10] calls the 'Principle of Mediocrity'. This mean is positive, because if ρ_V were negative it would have to be less in absolute value than the mass density of the universe during the whole time that life evolves (otherwise the universe would collapse before any astronomers come on the scene [11]), while if ρ_V were positive, it would only have to be less than the mass density of the universe at the time when most galaxies form, giving a much broader range of possible positive than negative values. In 1997–98 Martel, Shapiro and I [12] carried out a detailed calculation of the probability distribution of values of ρ_V seen by astronomers throughout the multiverse, under the assumption that the *a priori* probability distribution is flat in the relatively very narrow range that is anthropically allowed (for earlier calculations, see refs. [13] and [14]). At that time, the value of the primordial root-mean-square (rms) fractional density fluctuation σ was not well known, since the value inferred from observations of the cosmic microwave background depended on what one assumed for ρ_V. It was therefore not possible to calculate a mean expected value of ρ_V, but for any assumed value of ρ_V we could estimate σ and use the result to calculate the fraction of astronomers that would observe a value of ρ_V as small as the assumed value. In this way, we concluded that if Ω_Λ (the dimensionless density parameter associated with ρ_V) turned out to be much less than 0.6, anthropic reasoning could not explain why it was so small. The editor of the *Astrophysical Journal* objected to publishing papers about anthropic calculations, and we had to sell our article by pointing out that we had provided a strong argument for abandoning an anthropic explanation of a small value of ρ_V if it turned out to be too small.

Of course, it turned out that ρ_V is not too small. Soon after this work, observations of type Ia supernovae revealed that the cosmic expansion is accelerating [15, 16] and gave the result that $\Omega_\Lambda \simeq 0.7$. In other words, the ratio of the vacuum energy density to the present mass density ρ_{M0} in *our* subuniverse (which I use just as a convenient measure of density) is about 2.3, a conclusion subsequently confirmed by observations of the microwave background [17].

This is still a bit low. Martel, Shapiro and I had found that the probability of a vacuum energy density this small was 12%. I have now recalculated the probability distribution, using WMAP data and a better transfer function, with the result that the probability of a random astronomer seeing a value as small as $2.3\rho_{M0}$ is increased to 15.6%.[2] Now that we know σ, we can also calculate that the median vacuum energy density is $13.3\rho_{M0}$.

I should mention a complication in these calculations. The average of the product of density fluctuations at different points becomes infinite as these points approach each other, so the rms fractional density fluctuation σ is actually infinite. Fortunately, it is not σ itself that is really needed in these calculations, but the rms fractional density fluctuation averaged over a sphere of comoving radius R taken large enough so that the density fluctuation is able to hold on efficiently to the heavy elements produced in the first generation of stars. The results mentioned above were calculated for R (projected to the present) equal to 2 Mpc. These results are rather sensitive to the value of R; for $R = 1$ Mpc, the probability of finding a vacuum energy as small as $2.3\rho_{M0}$ is only 7.2%. The estimate of the required value of R involves complicated astrophysics, and needs to be better understood.

2.2 Problems

Now I want to take up four problems we have to face in working out the anthropic implications of the string landscape.

What is the shape of the string landscape?

Douglas [18] and Dine and co-workers [19, 20] have taken the first steps in finding the statistical rules governing different string vacua. I cannot comment usefully on this, except to say that it would not hurt in this work if we knew what string theory is.

What constants scan?

Anthropic reasoning makes sense for a given constant if the range over which the constant varies in the landscape is large compared with the anthropically allowed range of values of the constant; for then it is reasonable to assume that the *a priori* probability distribution is flat in the anthropically allowed range. We need to know what constants actually 'scan' in this sense. Physicists would like to be able to calculate as much as possible, so we hope that not too many constants scan.

---

2 This situation has improved since the release of the second and third year WMAP results. Assuming flat space, the ratio of the vacuum energy density to the matter density is now found to be about 3.2 rather than 2.3.

The most optimistic hypothesis is that the only constants that scan are the few whose dimensionality is a positive power of mass: the vacuum energy and whatever scalar mass or masses set the scale of electroweak symmetry-breaking. With all other parameters of the Standard Model fixed, the scale of electroweak symmetry-breaking is bounded above by about 1.4 to 2.7 times its value in our subuniverse, by the condition that the pion mass should be small enough to make the nuclear force strong enough to keep the deuteron stable against fission [21]. (The condition that the deuteron be stable against beta decay, which yields a tighter bound, does not seem to me to be necessary. Even a beta-unstable deuteron would live long enough to allow cosmological helium synthesis; helium would be burned to heavy elements in the first generation of very massive stars; and then subsequent generations could have long lifetimes burning hydrogen through the carbon cycle.) But the mere fact that the electroweak symmetry-breaking scale is only a few orders of magnitude larger than the QCD scale should not in itself lead us to conclude that it must be anthropically fixed. There is always the possibility that the electroweak symmetry-breaking scale is determined by the energy at which some gauge coupling constant becomes strong, and if that coupling happens to grow with decreasing energy a little faster than the QCD coupling, then the electroweak breaking scale will naturally be a few orders of magnitude larger than the QCD scale.

If the electroweak symmetry-breaking scale is anthropically fixed, then we can give up the decades long search for a natural solution of the hierarchy problem. This is a very attractive prospect, because none of the 'natural' solutions that have been proposed, such as technicolor or low-energy supersymmetry, were ever free of difficulties. In particular, giving up low-energy supersymmetry can restore some of the most attractive features of the non-supersymmetric Standard Model: automatic conservation of baryon and lepton number in interactions up to dimension 5 and 4, respectively; natural conservation of flavors in neutral currents; and a small neutron electric dipole moment. Arkani-Hamed and Dimopoulos [22] and others [23–25] have even shown how it is possible to keep the good features of supersymmetry, such as a more accurate convergence of the $SU(3) \times SU(2) \times U(1)$ couplings to a single value, and the presence of candidates for dark matter, WIMPs. The idea of this 'split supersymmetry' is that, although supersymmetry is broken at some very high energy, the gauginos and higgsinos are kept light by a chiral symmetry. (An additional discrete symmetry is needed to prevent lepton-number violation in higgsino–lepton mixing, and to keep the lightest supersymmetric particle stable.) One of the nice things about split supersymmetry is that, unlike many of the things we talk about these days, it makes predictions that can be checked when the LHC starts operation. One expects a single neutral Higgs with a mass in the range 120 to 165 GeV, possible winos and binos, but no squarks or sleptons, and a long-lived gluino. (Incidentally, a Stanford group [26] has recently used considerations of Big Bang nucleosynthesis to argue that a 1 TeV gluino must have a lifetime less than 100 seconds, indicating a supersymmetry breaking scale less than 10^{10} GeV. But I wonder whether, even if the gluino has a longer lifetime and decays after nucleosynthesis, the universe might not thereby be reheated above the temperature of helium dissociation, giving Big Bang nucleosynthesis a second chance to produce the observed helium abundance.)

What about the dimensionless Yukawa couplings of the Standard Model? If these couplings are very tightly constrained anthropically, then we might reasonably suspect that they take a wide range of values in the multiverse, so that anthropic considerations can have a chance to affect the values we observe. Hogan [27, 28] has analyzed the anthropic constraints on these couplings, with the electroweak symmetry-breaking scale and the sum of the u and d Yukawa couplings held fixed, to avoid complications due to the dependence of nuclear forces on the pion mass. He imposes the following conditions: (1) $m_d - m_u - m_e > 1.2$ MeV, so that the early universe does not become all neutrons; (2) $m_d - m_u + m_e < 3.4$ MeV, so that the pp reaction is exothermic; and (3) $m_e > 0$. With three conditions on the two parameters $m_u - m_d$ and m_e, he naturally finds these parameters are limited to a finite region, which turns out to be quite small. At first sight, this gives the impression that the quark and lepton Yukawa couplings are subject to stringent anthropic constraints, in which case we might infer that the Yukawa couplings probably scan.

I have two reservations about this conclusion. The first is that the pp reaction is not necessary for life. For one thing, the pep reaction $p + p + e^- \to d + \nu$ can keep stars burning hydrogen for a long time. For this, we do not need $m_d - m_u + m_e < 3.4$ MeV, but only the weaker condition $m_d - m_u - m_e < 3.4$ MeV. The three conditions then do not constrain $m_d - m_u$ and m_e separately to any finite region, but only constrain the single parameter $m_d - m_u - m_e$ to lie between 1.2 MeV and 3.4 MeV, not a very tight anthropic constraint. (In fact, He^4 will be stable as long as $m_d - m_u - m_e$ is less than about 13 MeV, so stellar nucleosynthesis can begin with helium burning in the heavy stars of Population III, followed by hydrogen burning in later generations of stars.) My second reservation is that the anthropic constraints on the Yukawa couplings are alleviated if we suppose (as discussed above) that the electroweak symmetry-breaking scale is not fixed, but free to take whatever value is anthropically necessary. For instance, according to the results of ref. [21], the deuteron binding energy could be made as large as about 3.5 MeV by taking the electroweak breaking scale much less than it is in our universe, in which case even the condition that the pp reaction be exothermic becomes much looser.

Incidentally, I do not set much store by the famous 'coincidence', emphasized by Hoyle, that there is an excited state of C^{12} with just the right energy to allow carbon production via α–Be^8 reactions in stars. We know that even–even nuclei have states that are well described as composites of α-particles. One such state is the ground state of Be^8, which is unstable against fission into two α-particles. The same α–α potential that produces that sort of unstable state in Be^8 could naturally be expected to produce an unstable state in C^{12} that is essentially a composite of three α-particles, and that therefore appears as a low-energy resonance in α–Be^8 reactions. So the existence of this state does not seem to me to provide any evidence of fine tuning.

What else scans? Tegmark and Rees [29] have raised the question of whether the rms density fluctuation σ may itself scan. If it does, then the anthropic constraint on the vacuum energy becomes weaker, resuscitating to some extent the problem of why ρ_V is so small. But Garriga and Vilenkin [30] have pointed out that it is really ρ_V/σ^3 that is constrained anthropically, so that, even if σ does scan, the anthropic prediction of this ratio remains robust.

Arkani-Hamed, Dimopoulos and Kachru [31], referred to below as ADK, have offered a possible reason for supposing that most constants do not scan. If there are a large number N of decoupled modular fields, each taking a few possible values, then the probability distribution of quantities that depend on all these fields will be sharply peaked, with a width proportional to $1/\sqrt{N}$. According to Distler and Varadarajan [32], it is not really necessary here to make arbitrary assumptions about the decoupling of the various scalar fields; it is enough to adopt the most general polynomial superpotential that is stable, in the sense that radiative corrections do not change the effective couplings for large N by amounts larger than the couplings themselves. Distler and Varadarajan emphasize cubic superpotentials, because polynomial superpotentials of order higher than cubic presumably make no physical sense. But it is not clear that even cubic superpotentials can be plausible approximations, or that peaks will occur at reasonable values in the distribution of dimensionless couplings rather than of some combinations of these couplings.[3] It also is not clear that the multiplicity of vacua in this kind of effective scalar field theory can properly represent the multiplicity of flux values in string theories [33], but even if it cannot, it presumably can represent the variety of minima of the potential for a given set of flux vacua.

3 M. Douglas, private communication.

If most constants do not effectively scan, then why should anthropic arguments work for the vacuum energy and the electroweak breaking scale? ADK point out that, even if some constant has a relatively narrow distribution, anthropic arguments will still apply if the anthropically allowed range is even narrower and near a point around which the distribution is symmetric. (ADK suppose that this point would be at zero, but this is not necessary.) This is the case, for instance, for the vacuum energy if the superpotential W is the sum of the superpotentials W_n for a large number of decoupled scalar fields, for each of which there is a separate broken R symmetry, so that the possible values of each W_n are equal and opposite. The probability distribution of the total superpotential $W = \sum_{n=1}^{N} W_n$ will then be a Gaussian peaked at $W=0$ with a width proportional to $1/\sqrt{N}$, and the probability distribution of the supersymmetric vacuum energy $-8\pi G|W|^2$ will extend over a correspondingly narrow range of negative values, with a maximum at zero. When supersymmetry-breaking is taken into account, the probability distribution widens to include positive values of the vacuum energy, extending out to a positive value depending on the scale of supersymmetry-breaking. For any reasonable supersymmetry-breaking scale, this probability distribution, though narrow compared with the Planck scale, will be very wide compared with the very narrow anthropically allowed range around $\rho_V = 0$, so within this range the probability distribution can be expected to be flat, and anthropic arguments should work. Similar remarks apply to the μ-term of the supersymmetric Standard Model, which sets the scale of electroweak symmetry-breaking.

How should we calculate anthropically conditioned probabilities?

We would expect the anthropically conditioned probability distribution for a given value of any constant that scans to be proportional to the number of scientific civilizations that observe that value. In the calculations described above, Martel, Shapiro and I took this number to be proportional to the *fraction* of baryons that find themselves in galaxies, but what if the total number of baryons itself scans? What if it is infinite?

How is the landscape populated?

There are at least four ways in which we might imagine the different 'universes' described by the string landscape actually to exist.

(i) The various subuniverses may be simply different regions of space. This is most simply realized in the chaotic inflation theory [34–38]. The scalar fields in different inflating patches may take different values, giving rise to different values for various effective coupling constants. Indeed, Linde speculated about the application of the anthropic principle to cosmology soon after the proposal of chaotic inflation [39, 40].

(ii) The subuniverses may be different eras of time in a single Big Bang. For instance, what appear to be constants of nature might actually depend on scalar fields that change very slowly as the universe expands [41].

(iii) The subuniverses may be different regions of spacetime. This can happen if, instead of changing smoothly with time, various scalar fields on which the 'constants' of nature depend change in a sequence of first-order phase transitions [42–44]. In these transitions, metastable bubbles form within a region of higher vacuum energy; then within each bubble there form further bubbles of even lower vacuum energy; and so on. In recent years this idea has been revived in the context of the string landscape [45, 46]. In particular, it has been suggested [47] that in this scenario the curvature of our universe is small for anthropic reasons, and hence possibly large enough to be detected.

(iv) The subuniverses could be different parts of quantum mechanical Hilbert space. In a reinterpretation of Hawking's earlier work on the wave-function of the universe [48, 49],[4] Coleman [51] showed that certain topological fixtures known as wormholes in the path integral for the Euclidean wave-function of the universe would lead to a superposition of wave-functions in which any coupling constant not constrained by symmetry principles would take any possible value.[5] Ooguri, Vafa and Verlinde [57] have argued for a particular wave-function of the universe, but it escapes me how anyone can tell whether this or any other proposed wave-function is *the* wave-function of the universe.

These alternatives are by no means mutually exclusive. In particular, it seems to me that, whatever one concludes about the first three alternatives, we will still have the possibility that the wave-function of the universe is a superposition of different terms representing different ways of populating the landscape in space and/or time.

---

4 Some of this work is based on an initial condition for the origin of the universe proposed by Hartle and Hawking [50].
5 It has been argued by Hawking and others that the wave-function of the universe is sharply peaked at values of the constants that yield a zero vacuum energy at late times [52–55]. This view has been challenged in ref. [56]. I am assuming here that there are no such peaks.

2.3 Conclusion

In closing, I would like to comment on the impact of anthropic reasoning within and beyond the physics community. Some physicists have expressed a strong distaste for anthropic arguments. (I have heard David Gross say 'I hate it'.) This is understandable. Theories based on anthropic calculations certainly represent a retreat from what we had hoped for: the calculation of all fundamental parameters from first principles. It is too soon to give up on this hope, but without loving it we may just have to resign ourselves to a retreat, just as Newton had to give up Kepler's hope of a calculation of the relative sizes of planetary orbits from first principles.

There is also a less creditable reason for hostility to the idea of a multiverse, based on the fact that we will never be able to observe any subuniverses except our own. Livio and Rees [58] and Tegmark [59] have given thorough discussions of various other ingredients of accepted theories that we will never be able to observe, without our being led to reject these theories. The test of a physical theory is not that everything in it should be observable and every prediction it makes should be testable, but rather that enough is observable and enough predictions are testable to give us confidence that the theory is right.

Finally, I have heard the objection that, in trying to explain why the laws of nature are so well suited for the appearance and evolution of life, anthropic arguments take on some of the flavour of religion. I think that just the opposite is the case. Just as Darwin and Wallace explained how the wonderful adaptations of living forms could arise without supernatural intervention, so the string landscape may explain how the constants of nature that we observe can take values suitable for life without being fine-tuned by a benevolent creator. I found this parallel well understood in a surprising place, a *New York Times* article by Christoph Schönborn, Cardinal Archbishop of Vienna [60]. His article concludes as follows.

Now, at the beginning of the 21st century, faced with scientific claims like neo-Darwinism and the multiverse hypothesis in cosmology invented to avoid the overwhelming evidence for purpose and design found in modern science, the Catholic Church will again defend human nature by proclaiming that the immanent design evident in nature is real. Scientific theories that try to explain away the appearance of design as the result of 'chance and necessity' are not scientific at all, but, as John Paul put it, an abdication of human intelligence.

It is nice to see work in cosmology get some of the attention given these days to evolution, but of course it is not religious preconceptions like these that can decide any issues in science.

It must be acknowledged that there is a big difference in the degree of confidence we can have in neo-Darwinism and in the multiverse. It is settled, as well as anything in science is ever settled, that the adaptations of living things on Earth have come into being through natural selection acting on random undirected inheritable variations. About the multiverse, it is appropriate to keep an open mind, and opinions among scientists differ widely. In the Austin airport on the way to this meeting I noticed for sale the October issue of a magazine called *Astronomy*, having on the cover the headline 'Why You Live in Multiple Universes'. Inside I found a report of a discussion at a conference at Stanford, at which Martin Rees said that he was sufficiently confident about the multiverse to bet his dog's life on it, while Andrei Linde said he would bet his own life. As for me, I have just enough confidence about the multiverse to bet the lives of both Andrei Linde *and* Martin Rees's dog.

Acknowledgements

This material is based upon work supported by the National Science Foundation under Grants nos. PHY-0071512 and PHY-0455649 and with support from The Robert A. Welch Foundation, Grant no. F-0014, and also grant support from the US Navy, Office of Naval Research, Grant nos. N00014-03-1-0639 and N00014-04-1-0336, Quantum Optics Initiative.

References

[1] R. Bousso and J. Polchinski. *JHEP*, **0006** (2000), 006.
[2] L. Smolin. *Life of the Cosmos* (New York: Oxford University Press, 1997).
[3] M. J. Duff, B. E. W. Nilsson and C. N. Pope. *Phys. Rep.* **130** (1986), 1.
[4] S. B. Giddings, S. Kachru and J. Polchinski. *Phys. Rev.* **D 66** (2002), 106006.
[5] A. Strominger, A. Maloney and E. Silverstein. In *The Future of Theoretical Physics and Cosmology*, eds. G. W. Gibbons, E. P. S. Shellard and S. J. Ranken (Cambridge: Cambridge University Press, 2003), pp. 570–91.
[6] S. Kachru, R. Kallosh, A. D. Linde and S. P. Trivedi. *Phys. Rev.* **D 68** (2003), 046005.
[7] S. K. Ashok and M. Douglas. *JHEP*, **0401** (2004), 060.
[8] L. Susskind. This volume (2007) [hep-th/0302219].
[9] S. Weinberg. *Phys. Rev. Lett.* **59** (1987), 2607.
[10] A. Vilenkin. *Phys. Rev. Lett.* **74** (1995), 846.
[11] J. D. Barrow and F. J. Tipler. *The Anthropic Cosmological Principle* (Oxford: Clarendon, 1986).
[12] H. Martel, P. Shapiro and S. Weinberg. *Astrophys. J.* **492** (1998), 29.
[13] G. Efstathiou. *Mon. Not. Roy. Astron. Soc.* **274** (1995), L73.

[14] S. Weinberg. In *Critical Dialogues in Cosmology*, ed. N. Turok (Singapore: World Scientific, 1997).
[15] A. G. Riess, A. V. Filippenko, P. Challis *et al. Astron. J.* **116** (1998), 1009.
[16] S. Perlmutter, G. Aldering, G. Goldhaber *et al. Astrophys. J.* **517** (1999), 565.
[17] D. N. Spergel, L. Verde, H. V. Peiris *et al. Astrophys. J. Supp.* **148**, (2003), 175.
[18] M. R. Douglas. *Compt. Rend. Phys.* **5** (2004), 965 [hep-ph/0401004].
[19] M. Dine, D. O'Neil and Z. Sun. *JHEP*, **0507** (2005), 014.
[20] M. Dine and Z. Sun. *JHEP* **0601** (2006), 129 [hep-th/0506246].
[21] V. Agrawal, S. M. Barr, J. F. Donoghue and D. Seckel. *Phys. Rev.* **D 57** (1998), 5480.
[22] N. Arkani-Hamed and S. Dimopoulos. *JHEP*, **0506** (2005), 073.
[23] G. F. Giudice and A. Romanino. *Nucl. Phys.* **B 699** (2004), 65.
[24] N. Arkani-Hamed, S. Dimopoulos, G. F. Giudice and A. Romanino. *Nucl. Phys.* **B 709** (2005), 3.
[25] A. Delgado and G. F. Giudice. *Phys. Lett.* **B 627** (2005), 155 [hep-ph/0506217].
[26] A. Arvanitaki, C. Davis, P. W. Graham, A. Pierce and J. G. Wacker. *Phys. Rev.* **D 72** (2005), 075011 [hep-ph/0504210].
[27] C. Hogan. *Rev. Mod. Phys.* **72**, 1149 (2000).
[28] C. Hogan. This volume (2007) [astro-ph/0407086].
[29] M. Tegmark and M. J. Rees. *Astrophys. J.* **499** (1998), 526.
[30] J. Garriga and A. Vilenkin. *Prog. Theor. Phys. Suppl.* **163** (2006), 245 [hep-th/0508005].
[31] N. Arkani-Hamed, S. Dimopoulos and S. Kachru [hep-th/0501082].
[32] J. Distler and U. Varadarajan (2005) [hep-th/0507090].
[33] T. Banks (2000) [hep-th/0011255].
[34] A. D. Linde. *Phys. Lett.* **129 B** (1983), 177.
[35] A. Vilenkin. *Phys. Rev.* **D 27** (1983), 2848.
[36] A. D. Linde. *Phys. Lett.* **B 175** (1986), 305.
[37] A. D. Linde. *Phys. Script.* **T 15** (1987), 100.
[38] A. D. Linde. *Phys. Lett.* **B 202** (1988), 194.
[39] A. D. Linde. In *The Very Early Universe*, eds. G. W. Gibbons, S. W. Hawking and S. Siklos (Cambridge: Cambridge University Press, 1983).
[40] A. D. Linde. *Rep. Prog. Phys.* **47** (1984), 925.
[41] T. Banks. *Nucl. Phys.* **B 249** (1985), 332.
[42] L. Abbott. *Phys. Lett.* **B 150** (1985), 427.
[43] J. D. Brown and C. Teitelboim. *Phys. Lett.* **B 195** (1987), 177.
[44] J. D. Brown and C. Teitelboim. *Nucl. Phys.* **B 297** (1987), 787.
[45] J. L. Feng, J. March-Russel, S. Sethi and F. Wilczek. *Nucl. Phys.* **B 602** (2001), 307.
[46] H. Firouzjahi, S. Sarangji and S.-H. Tye. *JHEP*, **0409** (2004), 060.
[47] B. Freivogel, M. Kleban, M. R. Martinez and L. Susskind. *JHEP* **3** (2006), 39 [hep-th/0505232].
[48] S. W. Hawking. *Nucl. Phys.* **B 239** (1984), 257.
[49] S. W. Hawking. *Relativity, Groups, and Topology II*, NATO Advanced Study Institute Session XL, Les Houches, 1983, eds. B. S. DeWitt and R. Stora (Amsterdam: Elsevier, 1984), p. 336.

[50] J. Hartle and S. W. Hawking. *Phys. Rev.* **D 28** (1983), 2960.
[51] S. Coleman. *Nucl. Phys.* **B 307** (1988), 867.
[52] S. W. Hawking. In *Shelter Island II – Proceedings of the 1983 Shelter Island Conference on Quantum Field Theory and the Fundamental Problems of Physics*, eds. R. Jackiw *et al.* (Cambridge: MIT Press, 1985).
[53] S. W. Hawking. *Phys. Lett.* **B 134** (1984), 403.
[54] E. Baum. *Phys. Lett.* **B 133** (1984), 185.
[55] S. Coleman. *Nucl. Phys.* **B 310** (1985), 643.
[56] W. Fischler, I. Klebanov, J. Polchinski and L. Susskind. *Nucl. Phys.* **B 237** (1989), 157.
[57] H. Ooguri, C. Vafa and E. Verlinde. *Lett. Math. Phys.* **74** (2005), 311 [hep-th/0502211].
[58] M. Livio and M. J. Rees. *Science*, **309** (2003), 1022.
[59] M. Tegmark. *Ann. Phys.* **270** (1998), 1.
[60] C. Schönborn. *N.Y. Times* (7 July 2005), p. A23.

3
Enlightenment, knowledge, ignorance, temptation

Frank Wilczek
Center for Theoretical Physics, Massachusetts Institute of Technology

Modified version of summary talk at the symposium *Expectations of a Final Theory* at Trinity College, Cambridge, 4 September 2005

3.1 A new zeitgeist

Our previous 'Rees-fest' *Anthropic Arguments in Fundamental Physics and Cosmology* at Cambridge in 2001 had much in common with this one, in terms of the problems discussed and the approach to them. Then, as now, the central concerns were apparent conspiracies among fundamental parameters of physics and cosmology that appear necessary to ensure the emergence of life. Then, as now, the main approach was to consider the possibility that significant observational selection effects are at work, even for the determination of superficially fundamental, universal parameters.

That approach is loosely referred to as anthropic reasoning, which in turn is often loosely phrased as the anthropic principle: the parameters of physics and cosmology have the values they do in order that intelligent life capable of observing those values can emerge. That formulation upsets many scientists, and rightly so, since it smacks of irrational mysticism.

On the other hand, it is simply a fact that intelligent observers are located only in a miniscule fraction of space, and in places with special properties. As a trivial consequence, probabilities conditioned on the presence of observers will differ grossly from probabilities per unit volume. Much finer distinctions are possible and useful; but I trust that this word to the wise is enough to it make clear that we should not turn away from straightforward logic just because it can be made to sound, when stated sloppily, like irrational mysticism.

For all their commonality of content, the spirit pervading the two gatherings seemed quite different, at least to me. One sign of the change is the

Universe or Multiverse?, ed. Bernard Carr. Published by Cambridge University Press.
© Cambridge University Press 2007.

different name attached to the present gathering. This time it is *Expectations of a Final Theory*. The previous gathering had a defensive air. It prominently featured a number of physicists who subsisted on the fringes, voices in the wilderness who had for many years promoted strange arguments about conspiracies among fundamental constants and alternative universes. Their concerns and approaches seemed totally alien to the consensus vanguard of theoretical physics, which was busy successfully constructing a unique and mathematically perfect universe.

Now the vanguard has marched off to join the prophets in the wilderness. According to the new zeitgeist, the real world of phenomena must be consulted after all, if only to position ourselves within a perfect, but inaccessible, multiverse. Estimating selection effects, in practice, requires considerations of quite a different character than what we have become accustomed to in the recent practice of theoretical (i.e. hep-th) physics: looser and more phenomenological, less precise but more accurate.

3.2 Sources

What caused the change? In his opening talk, Steve Weinberg [1] ascribed the change in attitude to recent developments in string theory, but I think its deep roots mostly lie elsewhere and go much further back in time. Those of us who attended *Anthropic Arguments* lived through an empirical proof of that point. I would like to elaborate on this issue a little, not only as a matter of accurate intellectual history, but also to emphasize that the main arguments do not rely on narrow, delicate (I might venture to say fragile) technical developments; rather, they are broadly based and robust.

(1) *The standardization of models*

With the extraordinary success of the standard model of fundamental physics, brought to a new level of precision at LEP through the 1990s, and with the emergence of standard model(s) of cosmology, confirmed by precision measurements of microwave background anisotropies, it became clear that an excellent working description of the world as we find it is in place. This remarkable success is graphically illustrated in Figs. 3.1–3.3. In particular, the foundational laws of physics that are relevant to chemistry and biology seem pretty clearly to be in place.

The standard model(s) are founded upon broad principles of symmetry and dynamics, assuming the values of a handful of numerical parameters as inputs. Given this framework, we can consider in quite an orderly way

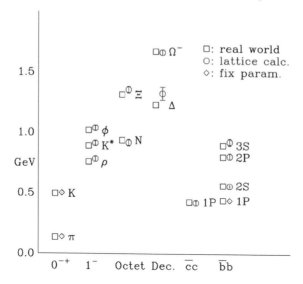

Fig. 3.1. The calculation of particle masses in QCD. N denotes the nucleon, K and K* denote pseudoscalar and vector K mesons, and the Greek letters are standard particle designations. The spectroscopic labels 1P, 2S, etc. refer to heavy quark–antiquark states; their energy relative to the ground state is displayed. These calculations, which employ the full power of modern computers, account for the bulk of the mass of ordinary matter on the basis of a conceptually based yet fully algorithmic theory. Figure courtesy of D. Toussaint, compiling results from the MILC collaboration and others.

the effect of a broad class of plausible changes in the structure of the world: namely, change the numerical values of those parameters! When we try this we find, in several different cases, that the emergence of complex structures capable of supporting intelligent observation appears quite fragile.

On the other hand, valiant attempts to derive the values of the relevant parameters, using symmetry principles and dynamics, have not enjoyed much success. Thus, life appears to depend upon delicate coincidences that we have not been able to explain. The broad outlines of that situation have been apparent for many decades. When less was known, it seemed reasonable to hope that better understanding of symmetry and dynamics would clear things up. Now that hope seems much less reasonable. The happy coincidences between life's requirements and nature's choices of parameter values might be just a series of flukes, but one could be forgiven for beginning to suspect that something deeper is at work. That suspicion is the first deep root of anthropic reasoning.

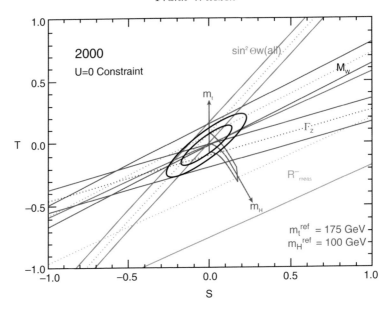

Fig. 3.2. Overdetermined, precision comparison of theory and experiment in electroweak theory, including radiative corrections. The calculations make ample use of the intricate rules for dealing with virtual particles in quantum field theory. Successful confrontations between theory with experiment, of the sort shown here and in Fig. 3.1, established our standard model of fundamental interactions. T and S are dimensionless parameters describing the deviations from this model. Figure courtesy of M. Swartz; for up-to-date information, consult http://lepewwg.lep.cern.ch/LEPEWWG.

(2) The exaltation of inflation

The most profound result of observational cosmology has been to establish the Cosmological Principle: that the same laws apply to all parts of the observed Universe, and moreover matter is – on average – uniformly distributed throughout. It seems only reasonable, then, to think that the observed laws are indeed universal, allowing no meaningful alternative, and to seek a unique explanation for each and every aspect of them. Within that framework, explanations invoking selection effects are moot. If there is no variation, then there cannot be selection.

Inflationary cosmology challenges that interpretation. It proposes a different explanation of the Cosmological Principle: that the observed universe originated from a small patch and had its inhomogeneities ironed out dynamically. In most theoretical embodiments of inflationary cosmology, the currently observed universe appears as a small part of a much larger multiverse. In this framework observed universal laws need not be multiversal, and it is valid – indeed necessary – to consider selection effects.

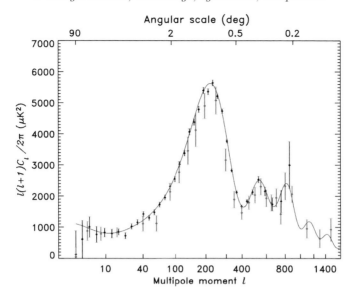

Fig. 3.3. Comparison of standard cosmological model, including dark matter, dark energy, and scale-invariant, adiabatic, Gaussian fluctuation spectrum, with observed microwave anisotropies. C_l gives the anistropy on the angular scale $180°/l$. The successful confrontation of theory and experiment in this case established our new standard model of cosmology. It traces the origin of all macroscopic structure to the growth of simply characterized, tiny seed fluctuations through gravitational instability. Figure courtesy of WMAP collaboration.

The success of inflationary cosmology is the second deep root of anthropic reasoning.

(3) The unbearable lightness of spacetime

Among the coincidences between life's requirements and nature's choices of parameter values, the smallness of the cosmological term, relative to its natural value, is especially clear and striking. Modern theories of fundamental physics posit an enormous amount of structure within what we perceive as empty space: quantum fluctuations, quark–antiquark condensates, Higgs fields, and more. At least within the framework of General Relativity, gravity responds to every sort of energy-momentum, and simple dimensional estimates of the contributions from these different sources suggest values of the vacuum energy, or cosmological term, many orders of magnitude larger than what is observed. Depending on your assumptions, the discrepancy might involve a factor of $10^{60}, 10^{120}$ or ∞. Again, attempts to derive an unexpectedly small value for this parameter, the vacuum energy, have not met with success. Indeed, most of those attempts aimed to derive the value zero, which now appears to be the wrong answer.

In 1987, Weinberg proposed to cut this Gordian knot by applying anthropic reasoning to the cosmological term. On this basis, he predicted that the cosmological term, rather than being zero, would be as large as it could be, while remaining consistent with the emergence of observers. The numerical accuracy of this prediction is not overwhelmingly impressive (the computed probability to observe a cosmological term as small as we do is roughly 10%), though this might be laid to the vagaries of sampling a statistical distribution just once. Also the original calculation was based on the hypothesis that one should consider variations in the vacuum energy alone, keeping all other parameters fixed, which might be too drastic a simplification. In any case, the apparent observation of vacuum energy that is ridiculously small from a microphysical perspective, but importantly large from a cosmological perspective, certainly encourages explanation based on selection.

(4) The superabundance of string theory

After a brief, heady period around 1984/5, during which it seemed that simple general requirements (for example $N=1$ supersymmetry and three light fermion generations) might pick out a unique Calabi–Yau compactification as the description of observed reality, serious phenomenological application of string theory was forestalled by the appearance of a plethora of candidate solutions. The solutions all exhibited unrealistic features (for example, unbroken supersymmetry, extraneous massless moduli fields), and it was anticipated that – when those problems were fixed – some degree of uniqueness might be restored. It was also hoped that string theory would provide a dynamical understanding for why the cosmological term is zero.

Recent constructions have provided a plethora of approximate solutions with broken supersymmetry and few or no moduli fields. They are not stable, but it is plausible that some of them are metastable, with very long lifetimes indeed. As yet none (among $\gtrsim 10^{\text{hundreds}}$) appears to be entirely realistic, but there is still plenty of scope for investigation in that direction, and even for additional constructions. In these new constructions, the cosmological term can take a wide range of values, positive or negative. So if cosmology provides a multiverse in which a significant sample of these metastable solutions are realized, then the stage might be set for selection effects to explain (roughly) the value we actually observe, as I just sketched.

3.3 Losses

Einstein expressed the traditional, maximally ambitious vision of mathematical physics with characteristic lucidity as follows [2].

I would like to state a theorem which at present can not be based upon anything more than upon a faith in the simplicity, i.e. intelligibility, of nature: there are no arbitrary constants... that is to say, nature is so constituted that it is possible logically to lay down such strongly determined laws that within these laws only rationally completely determined constants occur (not constants, therefore, whose numerical value could be changed without destroying the theory).

Over the course of the twentieth century, that programme has worked remarkably well. Rather than waste words to belabour the point, I will just present you with three icons.

What is most characteristic of these icons is their richness of detail and their quantitative precision. They confront profound theoretical ideas and complex calculations with concrete, precise observations. The fact that we physicists can worry over possible discrepancies at the level of parts per billion, in the case of the muon's magnetic moment, is our unique glory. Such examples epitomize what, traditionally, has distinguished fundamental physics from softer, 'environmental' disciplines such as history and biology.

With those words and images in mind, let me lament our prospective losses, if we adopt anthropic or statistical selection arguments too freely.

(1) Loss of precision

I do not see any realistic prospect that anthropic or statistical selection arguments – applied to a single sampling! – will ever lead to anything comparable in intellectual depth and numerical precision to what these icons represent. In that sense, intrusion of selection arguments into foundational physics and cosmology really does, to me, represent a genuine lowering of expectations.

(2) Loss of targets

Because the standard models of fundamental physics and cosmology describe the world so well, a major part of what ideas going beyond those standard models could aspire to achieve, for improving our understanding of the world, would be to fix the values of their remaining free parameters. If we compromise on that aspiration, there will be much less about the physical world for fundamental theory to target.

3.4 A classification

Of course, physicists have had to adjust their expectations before. In the development of Copernican–Newtonian celestial mechanics, attractive *a priori* ideas about the perfect shape of planetary orbits (Ptolemy) and their origin in pure geometry (Kepler) had to be sacrificed. In the development of quantum mechanics, ideas of strict determinism (Einstein) had to be sacrificed. In those cases, sacrifice of appealing philosophical ideas was compensated for by the emergence of powerful theories that described many specific features of the natural world and made surprising, impressive predictions. In the USA we have the saying 'No pain, no gain'.

There is a big difference, however, between those episodes and the present one. Resort to anthropic reasoning involves plenty of pain, as I have lamented, but so far the gain has been relatively meagre, to say the least. Even if we cannot be precise in our predictions of fundamental parameters, we can still aspire to clear thinking. Specifically, we can try to be clear concerning what it is we can or cannot be precise about. In this way we can limit our losses, or at least sharpen our discussion. In that spirit, I would like to suggest a chart, shown in Figs. 3.4 and 3.5, that draws some helpful boundaries.

The chart provides four boxes wherein to house parameters, or salient combinations of parameters, as in Fig. 3.4. On the horizontal axis, we have a binary distinction: is the parameter selected for, in the sense of anthropic reasoning, or not? In other words, is it relevant to the emergence of intelligent life or not? On the vertical axis, we have a different binary distinction: is the parameter one about whose values we have promising ideas based on symmetry and dynamics, or not? In that way we divide up the parameters into four classes. I have named the different classes in Fig. 3.5.

(1) Enlightenment

This class contains salient combinations of parameters that are both crucial to life and at least *significantly* understood. Its box is rather sparsely populated. I have entered the tiny ratio of the proton mass m_p to the Planck mass M_{Pl}. That small ratio is what allows the pull and tug of nuclear physics and chemistry, with attendant complexity, to dominate the relentless crunch of gravity. It can be understood as a consequence of the logarithmic running of the strong coupling, and the $SU(2) \times U(1)$ veto of quark and electron masses, modulo the weak-scale hierarchy problem (which opens a can of worms).

(2) Knowledge

This class contains parameters or regularities that do not appear to be crucial for life, but have been interpreted to have profound theoretical significance. Among these are the tiny θ parameter of QCD, the relationship among low-energy $SU(3)$, $SU(2)$ and $U(1)$ couplings that enables their unification at high energy, and the extremely long lifetime of the proton (τ_p).

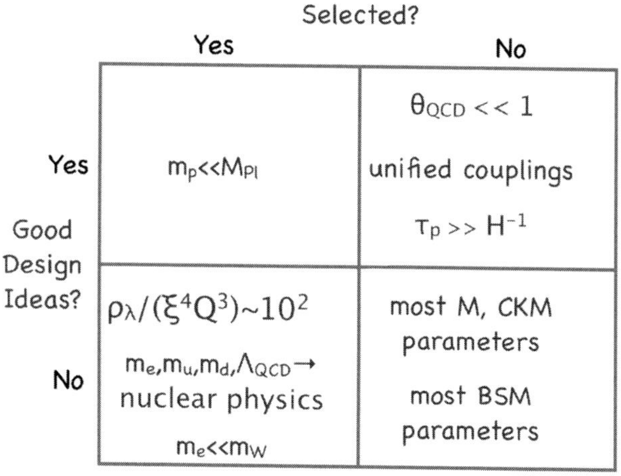

3 Enlightenment, knowledge, ignorance, temptation

	Selected? Yes	Selected? No
Good Design Ideas? Yes	$m_p \ll M_{Pl}$	$\theta_{QCD} \ll 1$ unified couplings $T_p \gg H^{-1}$
Good Design Ideas? No	$\rho_\lambda / (\xi^4 Q^3) \sim 10^2$ $m_e, m_u, m_d, \Lambda_{QCD} \to$ nuclear physics $m_e \ll m_W$	most M, CKM parameters most BSM parameters

Fig. 3.4. Classification of various parameters by criteria of selective pressure and theoretical insight. For definitions and discussion, see the text.

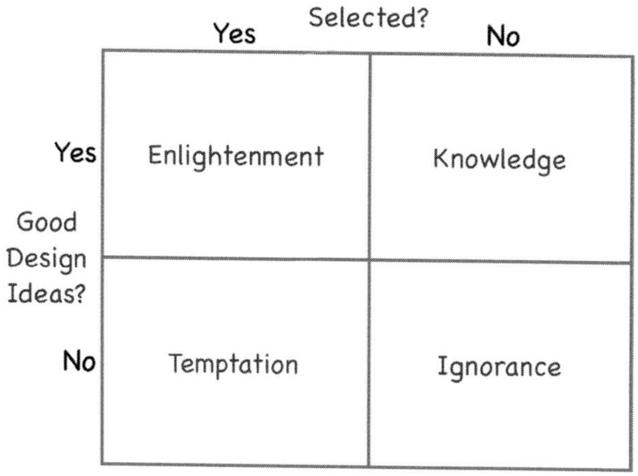

	Selected? Yes	Selected? No
Good Design Ideas? Yes	Enlightenment	Knowledge
Good Design Ideas? No	Temptation	Ignorance

Fig. 3.5. Appropriate names for the different classes of parameters. 'Temptation' labels the natural habitat for anthropic reasoning.

The θ parameter encodes the possibility that QCD might support violation of parity (P) and time reversal (T) in the strong interaction. It is a pure number, defined modulo 2π. Experimental constraints on this parameter require $|\theta| \lesssim 10^{-9}$, but it is difficult to imagine that life requires better than $|\theta| \lesssim 10^{-1}$, if that, since the practical consequences of nuclear P and T violation seem insignificant. On the other hand, there is a nice theoretical idea, Peccei–Quinn symmetry, that could explain the smallness of θ. Peccei–Quinn symmetry requires expansion of the Standard Model, and implies the existence of a remarkable new particle, the axion, of which more below.

The unification relationship among gauge couplings encourages us to think that the corresponding gauge symmetries are aspects of a single encompassing symmetry which is spontaneously broken, but would become manifest at asymptotically large energies or short distances. Accurate quantitative realization of this idea requires expanding the Standard Model even at low energies. Low-energy supersymmetry in any of its forms (including focus point or split supersymmetry) is very helpful in this regard. Low-energy supersymmetry, of course, requires a host of new particles, some of which should materialize at the LHC.

It is difficult to see how having the proton lifetime very much greater than the age of the Universe ($H^{-1} \sim 10^{18}$ s) could be important to life. Yet the observed lifetime is at least $\tau_\mathrm{p} \gtrsim 10^{40}$ s. Conventional anthropic reasoning is inadequate to explain that observation. (Perhaps including potential *future* observers in the weighting will help.) On the other hand, the unification of couplings calculation implies a very high energy scale for unification, which supplies a natural suppression mechanism. Detailed model implementations, however, suggest that if gauge unification ideas are on the right track, proton decay should occur at rates not far below existing limits.

(3) Ignorance

This class contains parameters that are neither important to life, nor close to being understood theoretically. It includes the masses M and weak mixing angles of the heavier quarks and leptons (encoded in the Cabibbo–Kobayashi–Maskawa, or CKM, matrix) and the masses and mixing angles of neutrinos. It also includes most of the prospective parameters of models beyond the Standard Model (BSM), such as low-energy supersymmetry, because only a few specific properties of those models (for example the rate of baryogenesis) are relevant to life. Of course, if the multiverse supports enough variation to allow selection to operate among a significant fraction of the parameters that are relevant to life, there is every reason to expect variation also among some parameters that are not relevant to life. In an abundant multiverse, wherein any particular location requires specification of many independent coordinates, we might expect this box to be densely populated, as evidently it is.

(4) Temptation

This class contains parameters whose values are important to life, and are therefore subject to selection effects, but which look finely tuned or otherwise odd from the point of view of symmetry and dynamics. It is in understanding these parameters that we are tempted to invoke anthropic reasoning. This class includes the smallness of the dark energy (ρ_λ), mentioned previously, and several other items indicated in Fig. 3.5.

Life in anything close to the form we know it requires both that there should be a complex spectrum of stable nuclei, and that the nuclei can be synthesized in stars. As emphasized by Hogan [3] and many others, those requirements imply constraints, some quite stringent, relating the QCD parameters $\Lambda_{\text{QCD}}, m_u, m_d$ and m_e and α. On the other hand, these parameters appear on very different footings within the Standard Model and in existing concrete ideas about extending the Standard Model. The required conspiracies *among* the masses m_u, m_d and m_e are all the more perplexing because each of the masses is far smaller than the 'natural' value, 250 GeV, set by the Higgs condensate. An objective measure of the degree of unnaturalness is that pure-number Yukawa couplings of order 10^{-6} underlie these masses.

More recent is the realization that the emergence of user-friendly macrostructures, that is stable planetary systems, requires rather special relationships among the parameters of the cosmological standard model. Here again, no conventional symmetry or dynamical mechanism has been proposed to explain those relationships; indeed, they connect parameters whose status within existing microscopic models is wildly different. Considerations of this sort have a rich literature, beginning with ref. [4]. Detailed discussion of these matters, which brings in some very interesting astrophysics, can be found in ref. [5]. (In this regard, this paper greatly improves on refs. [6] and [7] and on my summary talk as actually delivered.) A major result of the paper is a possible anthropic explanation of the observed abundance ξ of dark matter (normalized to photon number and thereby rendered time-independent), conditioned on the density of dark energy ρ_λ and the amplitude Q of primeval density fluctuations.

Dynamical versus anthropic reasoning is not an either/or proposition. It may be that some parameters are best understood dynamically and others anthropically (and others not at all). In my chart, no box is empty. Indeed, there is much potential for fertile interaction between these different modes of reasoning. For example, both axion physics and low-energy supersymmetry provide candidates for dark matter, and dark matter has extremely important anthropic implications [5]. Nor is the situation necessarily static. We can look forward to a flow of parameters along the paths from 'ignorance' to 'enlightenment' as physics progresses.

3.5 A new zeitgeist?

Actually, it is quite old. Earlier I discussed 'losses'. There could be a compensating moral gain, however, in well-earned humility. What we are 'losing', we never really had. Pure thought did not supersede creative engagement with phenomena as a way of understanding the world twenty years ago, it has not in the meantime, and will not anytime soon. I think it has been poetic to witness here at Trinity a re-emergence of some of the spirit of Newton. Perhaps not yet 'Hypothesis non fingo', but I hope the following applies [8]:

I know not how I seem to others, but to myself I am but a small child wandering upon the vast shores of knowledge, every now and then finding a small bright pebble to content myself with while the vast ocean of undiscovered truth lay before me.

Acknowledgement

The work of FW is supported in part by funds provided by the US Department of Energy under cooperative research agreement DE-FC02-94ER40818.

References

[1] S. Weinberg. This volume (2007) [hep-th/0511037].
[2] A. Einstein. In *Albert Einstein, Philosopher-Scientist*, ed. P. A. Schilpp (New York: Harper Torchbooks, 1959), p. 63.
[3] C. Hogan. *Rev. Mod. Phys.* **72** (2000), 4.
[4] B. Carr and M. Rees. *Nature*, **278** (1979), 605.
[5] M. Tegmark, A. Aguirre, M. Rees and F. Wilczek. *Phys. Rev.* **D 73** (2006), 23505 [astro-ph/0511774].
[6] F. Wilczek. *Phys. Rev.* **D 73** (2006), 023505 [hep-ph/0408167].
[7] F. Wilczek. This volume (2007).
[8] I. Newton. Letter to a friend, published by Bishop Hawsley, archivist; see http://www.newtonproject.ic.ac.uk.

Part II

Cosmology and astrophysics

4
Cosmology and the multiverse

Martin J. Rees

Institute of Astronomy, University of Cambridge

4.1 Do the 'special' values of the constants of physics and cosmology need an explanation?

In his book *Galaxies, Nuclei and Quasars* [1], Fred Hoyle wrote that 'one must at least have a modicum of curiosity about the strange dimensionless numbers that appear in physics'. Hoyle was among the first to conjecture that the so-called 'constants of nature' might not be truly universal. He outlined two possible attitudes to them. One is that 'the dimensionless numbers are all entirely necessary to the logical consistency of physics'; the second possibility is that the numbers are not in the broadest sense universal, but that 'in other places their values would be different'. Hoyle favoured this latter option because then

the curious placing of the levels in C^{12} and O^{16} need no longer have the appearance of astonishing accidents. It could simply be that, since creatures like ourselves depend on a balance between carbon and oxygen, we can exist only in the portions of the universe where these levels happen to be correctly placed.

Whatever one thinks of its motivation, Hoyle's conjecture is now even more attractive. The 'portions of the universe' between which the variation occurs must now, we realise, be interpreted as themselves vastly larger than the spacetime domain our telescopes can actually observe – perhaps even entire 'universes' within a multiverse.

If we ever established contact with intelligent aliens, how could we bridge the 'culture gap'? One common culture (in addition to mathematics) would be physics and astronomy. We and the aliens would all be made of atoms, and we would all trace our origins back to the big bang 13.7 billion years ago. We would all share the potentialities of a (perhaps infinite) future. But our existence (and that of the aliens, if there are any) depends on our universe being rather special. Any universe hospitable to life – what we

Universe or Multiverse?, ed. Bernard Carr. Published by Cambridge University Press.
© Cambridge University Press 2007.

might call a 'biophilic' universe – has to be 'adjusted' in a particular way. The prerequisites for any life of the kind we know about – long-lived stable stars, stable atoms such as carbon, oxygen and silicon, able to combine into complex molecules etc. – are sensitive to the physical laws and to the size, expansion rate and contents of the universe in which they exist. Indeed, even for the most open-minded science fiction writer, 'life' or 'intelligence' requires the emergence of some generic complex structures: it cannot exist in a homogeneous universe or in a micro-universe containing only a few dozen particles. Many recipes would lead to stillborn universes with no atoms, no chemistry and no planets; or to universes too short-lived or too empty to allow anything to evolve beyond sterile uniformity.

Consider, for example, the role of gravity. Stars and planets depend crucially on this force; but nothing remotely like us could exist if gravity were much stronger than it actually is. In an imaginary 'strong-gravity' universe, stars (gravitationally bound fusion reactors) would be small; gravity would crush anything larger than an insect. But what would preclude a complex ecosystem even more would be the limited time. The mini-Sun would burn faster and would have exhausted its energy before even the first steps in organic evolution had got under way. A large, long-lived and stable universe depends quite essentially on the gravitational force being exceedingly weak. Gravity also amplifies linear density contrasts in an expanding universe; it then provides a negative specific heat so that dissipative bound systems heat up further as they radiate. There's no thermodynamic paradox in evolving from an almost structureless fireball to the present cosmos, with huge temperature differences between the 3 K of the night sky and the blazing surfaces of stars. So gravity is crucial, but the weaker it is, the grander and more prolonged are its consequences. Newton's constant G need not be fine-tuned – one just needs gravity to be exceedingly weak on the atomic scale compared with the electrical force, so that the famous large number e^2/Gm_p^2 is indeed very large.

However, the natural world is much more sensitive to the balance between other basic forces. If nuclear forces were slightly stronger than they actually are relative to electric forces, two protons could stick together so readily that ordinary hydrogen would not exist, and stars would evolve quite differently. Some of the details – such as the carbon and oxygen abundances first noted by Hoyle – are still more sensitive, requiring some seeming 'tuning' in the nuclear forces.

Even a universe as large as ours could be very boring: it could contain no atoms at all – just black holes or inert dark matter. Even if it had the same ingredients as ours, it could be expanding so fast that no stars or galaxies had time to form; or it could be so turbulent that all the material formed vast black holes rather than stars or galaxies – an inclement environment for life. And our universe is also special in having three spatial dimensions. A four-dimensional world would be unstable, there are constraints on complex networks in two dimensions, and so forth.

The distinctive and special-seeming recipe characterizing our universe seems to me a fundamental mystery that should not be brushed aside merely as a brute fact. Rather than re-addressing the classic examples of fine-tuning in the fundamental forces, I shall in this chapter focus on the parameters of the big bang – the expansion rate, the curvature, the fluctuations and the material content. Some of these parameters (perhaps even all) may be explicable in terms of a unified theory, or somehow derivable from the microphysical constants. On the other hand, they may – in some still grander perspective – be mere 'environmental accidents'. But, irrespective of how that may turn out, it is interesting to explore the extent to which the properties of a universe – envisaged here as the aftermath of a single big bang – are sensitive to the cosmological parameters.

4.2 Is it scientific to enquire about other universes?

A semantic digression is necessary in order to pre-empt irrelevant criticism. The word 'universe' traditionally denotes 'everything there is'. Therefore if we envisage that physical reality could embrace far more than traditionally believed – for instance, other domains of spacetime originating in other big bangs, or domains embedded in extra spatial dimensions – we should really define the whole ensemble as 'the universe', and introduce a new word – 'meta-galaxy' for instance – to denote what observational cosmologists traditionally study. However, so long as this whole idea remains speculative, it is probably best to continue to denote what cosmologists observe as 'the universe' and to introduce a new term, 'multiverse', for the whole hypothetical ensemble.

If our existence – or, indeed, the existence of any 'interesting' universe – depends on a seemingly special cosmic recipe, how should we react? There seem three lines to take: we can dismiss it as happenstance, we can invoke 'providence', or we can conjecture that our universe is a specially favoured domain in a still vaster multiverse.

4.2.1 Happenstance or coincidence

Maybe a fundamental set of equations, which some day will be written on T-shirts, fixes all key properties of our universe uniquely. It would then be an unassailable fact that these equations permitted the immensely complex evolution that led to our emergence.

But I think there would still be something to wonder about. It is not guaranteed that simple equations permit complex consequences. To take an analogy from mathematics, consider the Mandelbrot set. This pattern is encoded by a short algorithm, but has infinitely deep structure; tiny parts of it reveal novel intricacies however much they are magnified. In contrast, you can readily write down other algorithms, superficially similar, that yield very dull patterns. Why should the fundamental equations encode something with such potential complexity, rather than the boring or sterile universe that many recipes would lead to?

One hard-headed response is that we could not exist if the laws had boring consequences. We manifestly are here, so there is nothing to be surprised about. I think we would need to know why the unique recipe for the physical world should permit consequences as interesting as those we see around us (and which, as a by-product, allowed us to exist).

4.2.2 Providence or design

Two centuries ago, William Paley introduced the famous metaphor of the watch and the watchmaker – adducing the eye, the opposable thumb and so forth as evidence of a benign Creator. These ideas fell from favour, even among most theologians, in the post-Darwinian era. However, the apparent fine-tuning in physics cannot be so readily dismissed as Paley's biological 'evidences': we now view any biological contrivance as the outcome of prolonged evolutionary selection and symbiosis with its surroundings; but so far as the biosphere is concerned, the physical laws are given and nothing can react back on them.

Paley's view of astronomy was that it was not the most fruitful science for yielding evidence of design, but 'that being proved, it shows, above all others, the scale of [the Creator's] operations'. Paley might have reacted differently if he had known about the providential-seeming physics that led to galaxies, stars, planets and the ninety-two natural elements of the periodic table. Our universe evolved from a simple beginning – a big bang – specified by quite a short recipe, but this recipe seems rather special. Different 'choices' for some basic numbers would have a drastic effect, precluding the hospitable cosmic habitat in which we emerged. A modern counterpart of Paley, the clergyman and ex-mathematical physicist John Polkinghorne, interprets our fine-tuned habitat as 'the creation of a Creator who wills that it should be so'.

4.2.3 A special universe drawn from an ensemble

But there is another perspective that, as the present book testifies, is gaining more attention: the possibility that there are many 'universes', of which ours is just one. In the others, some laws and physical constants would be different. But our universe would not be just a random one. It would belong to the unusual subset that offered a habitat conducive to the emergence of complexity and consciousness. If our universe is selected from a multiverse, its seemingly designed or fine-tuned features would not be surprising.

Some might regard other universes – regions of space and time that we cannot observe (perhaps even in principle and not just in practice) – as being in the province of metaphysics rather than physics. Science is an experimental or observational enterprise, and it is natural to be troubled by invocations of something unobservable. But I think other universes (in this sense) already lie within the proper purview of science. It is not absurd or meaningless to ask 'Do unobservable universes exist?', even though no quick answer is likely to be forthcoming. The question plainly cannot be settled by direct observation, but relevant evidence can be sought, which could lead to an answer.

There is actually a blurred transition between the readily observable and the absolutely unobservable, with a very broad grey area in between. To illustrate this, one can envisage a succession of horizons, each taking us further than the last from our direct experience, as illustrated in Fig. 4.1.

Limit of present-day telescopes

There is a limit to how far out into space our present-day instruments can probe. Obviously there is nothing fundamental about this limit; it is constrained by current technology. Many more galaxies will undoubtedly be revealed in the coming decades by bigger telescopes now being planned. We would obviously not demote such galaxies from the realm of proper scientific discourse simply because they have not been seen yet.

Limit in principle at present era

Even if there were absolutely no technical limits to the power of telescopes, our observations are still bounded by a horizon, set by the distance that any signal, moving at the speed of light, could have travelled since the big bang. This horizon demarcates the spherical shell around us at which the redshift would be infinite. There is nothing special about the galaxies on this shell, any more than there is anything special about the circle that defines your horizon when you are in the middle of an ocean. On the ocean, you can see farther by climbing up your ship's mast. But our cosmic horizon cannot be

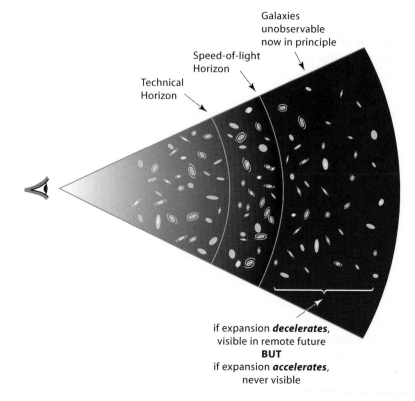

Fig. 4.1. Extending horizons beyond the directly observable, as discussed in the text.

extended unless our universe changes, so as to allow light to reach us from galaxies that are now beyond it. If our universe were decelerating, then the horizon of our remote descendants would encompass extra galaxies that are beyond our horizon today. It is, to be sure, a practical impediment if we have to await a cosmic change taking billions of years, rather than just a few decades (maybe) of technical advance before a prediction about a particular distant galaxy can be put to the test. But does that introduce a difference of principle? Surely the longer waiting time is merely a quantitative difference, not one that changes the epistemological status of these faraway galaxies?

Never-observable galaxies from 'our' big bang

But what about galaxies that we can never see, however long we wait? It is now believed that we inhabit an accelerating universe. As in a decelerating universe, there would be galaxies so far away that no signals from them have yet reached us; but if the cosmic expansion is accelerating, we are now receding from these remote galaxies at an ever-increasing rate, so if their light has not yet reached us, it never will. Such galaxies are not merely unobservable now – they will be beyond our horizon forever. But if a galaxy is now unobservable, it hardly seems to matter whether it remains unobservable for ever, or whether it would come into view if we waited a trillion years. (And as I have argued above, the latter category should certainly count as 'real'.)

Galaxies in disjoint universes

The never-observable galaxies discussed above would have emerged from the same big bang as we did. But suppose that, instead of causally disjoint regions emerging from a single big bang (via an episode of inflation), we imagine separate big bangs. Are spacetimes completely disjoint from ours any less real than regions that never come within our horizon in what we would traditionally call our own universe? Surely not – so these other universes should count as real parts of our cosmos too.

Whether other universes exist or not is a scientific question. Those who are prejudiced against the concept should regard the above step-by-step argument as an exercise in 'aversion therapy'. In this technique, someone terrified of spiders is first reconciled to a small one a long way away and then, stage by stage, to a tarantula crawling all over him. Likewise, from a reluctance to deny that galaxies with redshift 10 are proper objects of scientific enquiry, you are led towards taking seriously quite separate spacetimes, perhaps governed by quite different 'laws'.

Some theorists envisage an 'eternal' inflationary phase, where many universes sprout from separate big bangs into disjoint regions of spacetimes – each such region itself vastly larger than our observational horizon. Others have, from different viewpoints, suggested that a new universe could sprout inside a black hole, expanding into a new domain of space and time inaccessible to us. As a further alternative, other universes could exist, separated from us in an extra spatial dimension; these disjoint universes may interact gravitationally or they may have no effect whatsoever on each other. (Bugs crawling on a large sheet of paper – their two-dimensional universe – would be unaware of other bugs on a separate sheet of paper. Likewise, we would be unaware of our counterparts on another 'brane' separated in an extra dimension, even if that separation were only by a microscopic distance.) Other universes could be separate domains of space and time. We could not even meaningfully say whether they existed before, after or alongside our own, because such concepts make sense only insofar as we can impose a single measure of time, ticking away in all the universes.

None of these scenarios has been simply dreamed up out of the air; each has a serious, albeit speculative, theoretical motivation. However, one of them, at most, can be correct. Quite possibly none is; there are alternative theories that would lead just to one finite universe. Firming up any of these ideas will require a theory that consistently describes the extreme physics of ultra-high densities, how structures on extra dimensions are configured, etc. But consistency is not enough; there must be grounds for confidence that such a theory is not a mere mathematical construct, but applies to external reality. We would develop such confidence only if the theory accounted for things we can observe that are otherwise unexplained.

At the moment, the formulae of the 'Standard Model' involve numbers which cannot be derived from the theory but have to be inserted from experiment. Perhaps, in the twenty-first century, physicists will develop a theory that yields insight into (for instance) why there are three kinds of neutrinos and the nature of the nuclear and electric forces. Such a theory would thereby acquire credibility. If the same theory, applied to the very beginning of our universe, were to predict many big bangs, then we would have as much reason to believe in separate universes as we now have for believing inferences from particle physics about quarks inside atoms, or from relativity theory about the unobservable interior of black holes.

4.3 Universal laws or mere by-laws?

Are the laws of physics unique? This is a less poetic version of the famous question that Einstein once posed to his assistant, Ernst Strauss: 'Did God have any choice when he created the universe?' Offering an answer is a key scientific challenge for the new century. The answer determines how much variety the other universes – if they exist – might display. If there were something uniquely self-consistent about the actual recipe for our universe, then the aftermath of any big bang would be a re-run of our own universe. But a far more interesting possibility (which is certainly tenable in our present state of ignorance of the underlying laws) is that the underlying laws governing the entire multiverse may allow variety among the universes. Some of what we call 'laws of nature' may in this grander perspective be local by-laws, consistent with some overarching theory governing the ensemble, but not uniquely fixed by that theory. Many things in our cosmic environment – for instance, the exact layout of the planets and asteroids in the Solar System – are accidents of history. Likewise, the recipe for an entire universe may be arbitrary.

More specifically, some aspects may be arbitrary and others not. There could be a complementarity between chance and necessity, just as arises in biology, where our basic development – from embryo to adult – is encoded in our genes, but many aspects of our development are moulded by our environment and experiences. And there are far simpler examples of the same dichotomy. As an analogy (which I owe to Paul Davies), consider the form of snowflakes. Their ubiquitous six-fold symmetry is a direct consequence of the properties and shape of water molecules. But snowflakes display an immense variety of patterns because each is moulded by its microenvironments; how each flake grows is sensitive to the fortuitous temperature and humidity changes during its growth.

If physicists achieved a fundamental theory, it would tell us which aspects of nature were direct consequences of the bedrock theory (just as the symmetrical template of snowflakes is due to the basic structure of a water molecule) and which are (like the distinctive pattern of a particular snowflake) the outcome of accidents. Some of the accidental features could be imprinted during the cooling that follows the big bang, rather as a piece of red-hot iron becomes magnetised when it cools down, but with an alignment that may depend on chance factors. They could have other contingent causes, such as the influence of another nearby universe separated from ours in a fourth spatial dimension. Or they could simply depend on which particular oasis in the 'cosmic landscape' (to use Susskind's phrase) we happen to inhabit.

At the moment, as is evident from other chapters in this book, there is no consensus on the answer to Einstein's question: there could be a unique physics; there could, alternatively, be googles of alternative laws. Some theorists have strong preferences and prejudices favouring the former; they want as many features as possible of our universe to be 'explained' by neat formulae – indeed they yearn to discover these formulae themselves. But there is no reason why our universe has to accord with our aesthetic taste; the rational stance now is surely to be open-minded on this basic issue. The outcome determines which side of the 'fork' is taken in the decision tree of Fig. 4.2: in the one case, anthropic reasoning is irrelevant; in the other, it is unavoidable.

The cosmological numbers in our universe, and perhaps some of the so-called constants of laboratory physics as well, could be environmental accidents, rather than uniquely fixed throughout the multiverse by some final theory. Some seemingly fine-tuned features of our universe could then only be explained by anthropic arguments, as indicated by the right fork of Fig. 4.2. Although this style of explanation raises hackles among some

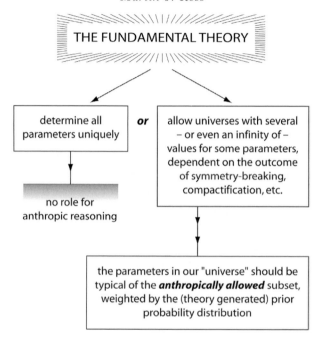

Fig. 4.2. Decision tree. Progress in twenty-first-century physics should allow us to decide whether anthropic explanations are irrelevant or the best we can ever hope for.

physicists, it is analogous to what any observer or experimenter does when they allow for selection effects in their measurements; if there are many universes, most of which are not habitable, we should not be surprised to find ourselves in one of the habitable ones!

The entire history of our universe could be just an episode of the infinite multiverse; what we call the laws of nature (or some of them) may be just parochial by-laws in our cosmic patch. Such speculations dramatically enlarge our concept of reality. Putting them on a firm footing must await a successful fundamental theory that tells us whether there could have been many big bangs rather than just one, and (if so) how much variety they might display. We will not know whether anthropic arguments are irrelevant or unavoidable until this fundamental issue is settled one way or the other.

4.4 Testing multiverse theories here and now: the value of Λ

While we are waiting for that theory – and it could be a long wait – we can check whether anthropic selection offers a tenable explanation for the apparent fine-tuning. Some hypotheses can even be refuted; this would happen if our universe turned out to be even more specially tuned than our presence requires. Let me give two quite separate examples of this style of reasoning.

(i) Boltzmann argued that our entire universe was an immensely rare 'fluctuation' within an infinite and eternal time-symmetric domain. Even when it was proposed, one could already have argued powerfully against it by noting that fluctuations in large volumes are far more improbable than in smaller volumes. If Boltzmann were right, we would be in the smallest fluctuation compatible with our awareness – indeed, the overwhelmingly most likely configuration would be a universe containing nothing but a single brain with external sensations fed into it. Whatever our assessment of the prior probability of Boltzmann's theory, its probability would plummet as we came to believe non-solipsistically in the extravagant scale of the cosmos.

(ii) Even if we knew nothing about how stars and planets formed, we would not be surprised to find that the Earth's orbit was moderately close to circular; had it been highly eccentric, water would boil when the Earth was at perihelion and freeze at aphelion – a harsh environment unconducive to our emergence. However, a modest orbital eccentricity, up to 0.1 or so, is plainly not incompatible with life. Had it turned out that the Earth moved in a near-perfect circle with eccentricity 0.000 001, then this would need some explanation: anthropic selection from orbits whose eccentricities had a Bayesian prior that was uniform in the range 0–1 could plausibly account for an eccentricity of 0.1, but not for one as tiny as this.

In Section 4.5, I will mention several applications of this line of reasoning. But first let us recall the one that has already been extensively discussed in the literature: the cosmological constant Λ. Interest in Λ has, of course, been hugely boosted recently through the convergence of several lines of evidence on a model where the universe is flat, but with about 4% in baryons, 25% in dark matter and the remaining (dominant) component in dark energy or quintessence.[1]

Most physicists would consider the 'natural' value of Λ to be large, because it is a consequence of a very complicated Planck-scale microstructure of

---

[1] The resurrection of Λ would be a great 'coup' for de Sitter. His model, dating from the 1920s, not only describes inflation, but would then also describe future aeons of our cosmos with increasing accuracy. Only for the fifty or so decades of logarithmic time between the end of inflation and the present would it need modification!

space. There might then be only a rare subset of universes where Λ was below the threshold that allows galaxies and stars to form before the cosmic repulsion takes over. In our universe, Λ obviously had to be below that threshold.

On the specific hypothesis that our universe is drawn from an ensemble in which Λ was equally likely to take any value, we would not expect it to be too far below the anthropic upper limit. Current evidence suggests that, if Λ constitutes the dark energy, its actual value is five to ten times below that threshold. That would put our universe between the 10th or 20th percentile of universes in which galaxies could form. In other words, our universe would not be significantly more special, with respect to Λ, than our emergence demanded. But suppose that (contrary to current indications) observations showed that Λ made no discernible contribution to the expansion rate, and was thousands of times below the threshold. This 'overkill precision' would raise doubts about the hypothesis that Λ was equally likely to have any value. It would suggest instead that it was zero for some fundamental reason or that the physics favoured values very close to zero or that Λ had a discrete set of possible values and all the others were well above the 'anthropic limit'.

In this example, one is essentially asking if our actual universe is 'typical' of the subset in which we could have emerged. The methodology requires us to decide what domain (in some multi-parameter space) is compatible with our emergence. But it requires something else as well: a specific theory that gives the relative Bayesian priors for any particular point within that domain (for example in the case of Λ, whether there is a uniform probability density, whether low values are favoured, whether there is a set of discrete values). When applied to the important numbers of physics and cosmology, this style of reasoning can test whether our universe is (under specific theoretical assumptions about the ensemble) typical of the subset that could harbour complex life. If our universe turns out to be a grossly atypical member, even of the anthropically allowed subset (not merely of the entire multiverse), then we would not necessarily need to abandon the hypothesis of anthropic selection, but we would certainly be forced to modify our model of the underlying physics and to seek an alternative theory that had a different distribution of priors. This involves subtle and still controversial issues that I will skate over here. In particular, what relative 'weight' does one give different volumes and how are infinities handled? (See ref. [2] for a brave attempt to confront these issues in the context of inflationary cosmology.)

4.5 Anthropic constraints on other cosmological numbers

Traditionally, cosmology was the quest for a few numbers. The first were the Hubble parameter H and the deceleration parameter q. Since the discovery of the microwave background in 1965, we have had another: the baryon/photon ratio of about 10^{-9}. This is believed to result from a small favouritism for matter over antimatter in the early universe – something that was addressed in the context of 'grand unified theories' in the 1970s. (Indeed, baryon non-conservation seems a prerequisite for any plausible inflationary model. Our entire observable universe, containing at least 10^{79} baryons, could not have inflated from something microscopic if the baryon number were strictly conserved.)

In the 1980s, non-baryonic dark matter became almost a natural expectation and $\rho_{\rm dm}/\rho_{\rm bar}$ is another fundamental number. We now seemingly have the revival of the cosmological constant (or some kind of dark energy, with negative pressure, which is generically equivalent to this). Another specially important dimensionless number tells us how smooth our universe is. It is measured by: (a) the amplitude of the gravitationally induced fluctuations in the microwave background; (b) the gravitational binding energy of clusters as a fraction of their rest mass; or (c) the square of the typical scale of mass-clustering as a fraction of the Hubble scale. It is, of course, somewhat oversimplified to represent this by a single number,[2] but insofar as one can, its value (let us call it Q) is pinned down to be around 10^{-5}.

We can make a list of what would be required for a big bang to yield an anthropically allowed universe – a universe where some kind of generic complexity could unfold, whether it were humanoid or more like Fred Hoyle's fictional 'Black Cloud'. The list would include the following.

Some inhomogeneities (i.e. a non-zero Q): clearly there is no potential for complexity if everything remains dispersed in a uniform ultra-dilute medium.

Some baryons: complexity would be precluded in a universe solely made of dark matter, with only gravitational interactions.

At least one star: nucleosynthesis is a precondition for complex chemistry, though perhaps superfluous for Black-Cloud-style complexity.

---

2 Detailed modelling of the fluctuations introduces further numbers: the ratio of scalar and tensor amplitudes and quantities such as the 'tilt', which measure the deviation from a pure scale-independent Harrison–Zeldovich spectrum.

Some second-generation stars: only later-generation stars would be able to have orbiting planets, unless heavy elements were primordial.

It is interesting to engage in 'counterfactual history' and ask what constraints these various requirements would impose not only on Λ (as discussed in the previous section) but on other key cosmological parameters, such as the fluctuation amplitude Q (about 10^{-5} in our universe); the baryon/photon ratio (about 10^{-9} in our universe); the baryon/dark matter density ratio (about 0.2 in our universe).

4.5.1 The fluctuation amplitude

What structures might emerge in a universe that was initially smoother (Q smaller) or rougher (Q larger) than ours? Were Q of order 10^{-6}, there would be no clusters of galaxies; moreover, the only galaxies would be small and anaemic. They would form much later than galaxies did in our actual universe. Because they would be loosely bound, processed material would be expelled from shallow potential wells; there may therefore be no second-generation stars and no planetary systems. If Q were even smaller than 10^{-6}, there would be no star formation at all; very small structures of dark matter would turn around late and their constituent gas would be too dilute to undergo the radiative cooling that is a prerequisite for star formation.[3]

Hypothetical astronomers in a universe with $Q = 10^{-4}$ might find their cosmic environment more varied and interesting than ours. Galaxies and clusters would span a wider range of masses. The biggest clusters would be 1000 times more massive than any in our actual universe. There could be individual galaxies – perhaps even disc galaxies – with masses up to that of the Coma cluster and internal velocity dispersions up to $2000 \,\mathrm{km\,s^{-1}}$. These would have condensed when the age of our universe was only 3×10^8 y and when Compton cooling on the microwave background was still effective.

However, a universe where Q were larger still – more than (say) 10^{-3} – would be a violent and inhospitable place. Huge gravitationally bound systems would collapse, trapping their radiation and being unable to fragment, soon after the epoch of recombination. (Collapse at, say, 10^7 y would lead to sufficient partial ionization via strong shocks to recouple the baryons and the primordial radiation.) Such structures, containing the bulk of the material,

---

[3] In a Λ-dominated universe, isolated clumps could survive for an infinite time without merging into a larger scale in the hierarchy. So eventually, for any $Q > 10^{-8}$, a 'star' could form – but by that time it might be the only bound object within the horizon.

would turn into vast black holes. It is unlikely that galaxies of any kind would exist; nor is it obvious that much baryonic material would ever go into stars. Even if it did so, they would be in very compact highly bound systems.[4]

According to most theories of the ultra-early universe, Q is imprinted by quantum effects: microscopic fluctuations, after exponential expansion, give rise to the large-scale irregularities observed in the microwave background sky, which are the seeds for galaxies and clusters. In a wide class of theories, Q depends on the detailed physics during an inflationary era. But, as yet, no independent evidence constrains such theories, so we cannot pin down Q.

4.5.2 The baryon/photon and baryon/dark matter density ratios

Baryons are anthropically essential. They need not be the dominant constituent (indeed they are far from dominant in our actual universe), but there must be enough of them to allow a gas cloud of at least a few solar masses to accumulate in some of the gravitationally bound halos of dark matter. However, a more restrictive lower limit may come from the requirement that this gas should be dense enough to cool. (The cooling timescale of a gas at a given temperature depends inversely on its density.) Lower ratios of baryons to dark matter would reduce the 'efficient cooling' domain shown in Fig. 4.3 [3].

If the photons outnumbered the baryons and the dark matter particles by a still larger factor than in our actual universe, then the universe would remain radiation-dominated for so long that the gravitational growth of fluctuations would be inhibited [4].[5] Suppose, on the other hand, that the baryon/photon ratio has its actual value, but the dark matter density is higher. Wilczek [5] has offered an interesting motivation for exploring this option; he suggests that, if axions constitute the dark matter, their density is lower than one would expect in a 'typical' universe. A higher value of ρ_{dm}/ρ_{bar} reduces the

---

4 Note that, irrespective of these anthropic constraints on its value, Q has to be substantially less than unity in order to make cosmology a tractable subject, separate from astrophysics. This is because the ratio of the length scales of the largest structures to the Hubble radius is of order $Q^{1/2}$. As an analogy, contrast a view from mid-ocean with a mountain landscape. On the ocean, we can define averages because even the biggest wave is small compared with the horizon distance; but we cannot do this in the mountain landscape. Quantities like ρ, ρ_{dm} and H are only well defined insofar as our universe possesses 'broad brush' homogeneity – so that our observational horizon encompasses many independent patches, each big enough to be a fair sample. This would not be so, and the simple Friedmann models would not be useful approximations, if Q were not much less than unity.

5 Note also that the mechanism that gives rise to baryon favouritism may be linked to the strong interactions, and therefore correlate with key numbers in nuclear physics.

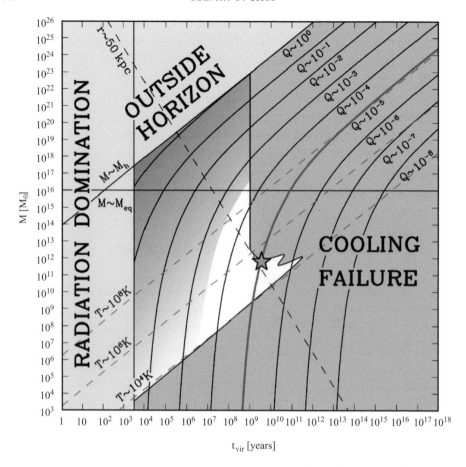

Fig. 4.3. The mass and time domains in which bound structures can form for different values of Q [3].

baryon fraction in dark halos. On the other hand, the enhanced density of dark matter compared to radiation reduces the time t_{eq} before which radiation mass–energy is dominant, thereby allowing gravitational clustering to start earlier. This reduces the minimum Q required for emergence of non-linear structures (see ref. [6]). Even if Q were 10^{-8}, dwarf galaxies could form in a universe where the dark matter density was 100 times higher (relative to the baryon density) than it is in our universe.

4.5.3 Delineating the anthropically allowed domain

In the above, I have envisaged changing just one parameter at a time, leaving the others with their actual values. But, of course, there may be correlations

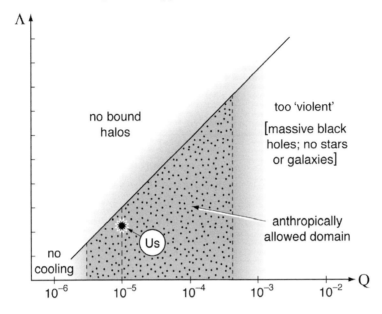

Fig. 4.4. This shows the two-dimensional parameter space associated with Λ and Q. The upper and lower limits to Q are discussed in ref. [3]. The upper limit to Λ stems from the requirement that galactic-mass bound systems should form. Our universe (obviously) lies in the anthropically allowed domain. But we cannot say whether it is at a typical location without a specific model for the probability distributions of Q and Λ in the ensemble.

between them. As a two-dimensional example, consider the joint constraints on Λ and Q, as illustrated in Fig. 4.4. There is an anthropically allowed area in the $\Lambda - Q$ plane. There are (rather vaguely defined) upper and lower limits to Q (as discussed in Section 4.5.1), but within that range we do not know the probability distribution of different values.

Suppose that there were big bangs with a whole range of Q-values. Structures form earlier (when the matter density is higher) in universes with larger Q, so obviously a higher Q is anthropically compatible with a higher Λ (indeed the limit to Λ scales as Q^3). We cannot decide whether our observable universe is typical without a theory that tells us what 'measure' to put on each part of the two-dimensional parameter space. If high-Q universes were more probable, and the probability density of Λ were uniform, then we should be surprised not to find ourselves in a universe with higher Λ and higher Q. We would be led to seek an alternative theory that led to a distribution of priors that made our universe less surprising and one in which the probabilities were steeply weighted in favour of low Q.

We can carry out the exercise in as many dimensions as we wish – including, for instance, the ratios of the photon, baryon and dark matter densities, as discussed above. Other parameters could be analyzed similarly – testing in a multi-parameter space whether our universe is a typical member within the anthropically allowed domain.

4.6 Conclusions

The examples in Section 4.5 show that some claims about other universes may, in principle, be refutable, as any good hypothesis in science should be. To delineate the anthropically allowed domains is procedurally uncontroversial, but what about the motivation? It obviously depends on believing that the laws of nature could have been otherwise and that there is some scientific validity in imagining 'counterfactual universes'. It obviously is predicated on the hope that theoretical ideas may sometime become sufficiently well developed to allow us to put some probability measure on the ensemble.

We cannot confidently assert that there were many big bangs – we just do not know enough about the ultra-early phases of our own universe. Nor do we know whether the underlying laws are 'permissive'. Settling this issue is a challenge to twenty-first-century physicists. But if they are, then so-called anthropic explanations would become legitimate – indeed they would be the only type of explanation we will ever have for some important features of our universe.

Moreover, the outcome of this issue (which path in the decision tree in Fig. 4.2 is the correct one) affects our attitude to our actual observable universe. Models with low Ω, non-zero Λ, two kinds of dark matter etc. may seem ugly. Some theorists are upset by these developments, because it frustrates their craving for maximal simplicity. There is perhaps an analogy with cosmological debates in the seventeenth century. Galileo and Kepler were upset that planets moved in elliptical orbits, not in perfect circles. Newton later showed, however, that all elliptical orbits could be understood by a single unified theory of gravity – something which would surely have elated Galileo. We have learnt that the Solar System is just one of a vast number (many millions even within our Galaxy). Likewise, what we have traditionally called our universe may be an infinitesimal part of physical reality – no more than one twig on one tree in a geometrical structure as complex as a biosphere. Our capacity to explain the cosmic parameters may then be limited, but to regard this outcome as ugly may be as myopic as Kepler's infatuation with circles.

References

[1] F. Hoyle. *Galaxies, Nuclei and Quasars* (New York: Harper and Row, 1965).
[2] M. Tegmark, A. Aguirre, M. J. Rees and F. Wilczek. *Phys. Rev.* **D 73** (2006), 23505.
[3] M. Tegmark and M. J. Rees. *Astrophys. J.* **499** (1998), 526.
[4] M. J. Rees. *Phys. Script.* **21** (1980), 614.
[5] F. Wilczek. This volume (2007).
[6] A. Aguirre. *Phys. Rev.* **D 64** (2001), 3508.

5
The anthropic principle revisited

Bernard Carr
Astronomy Unit, Queen Mary, University of London

5.1 Introduction

My brief historical overview in Chapter 1 alluded to the crucial influence of the Newtonian mechanistic picture on the development of our view of the Universe. According to this, the cosmos operates likes a giant machine, oblivious to whether life or any form of consciousness is present, i.e. the laws of physics and the characteristics of the Universe are independent of whether anybody actually observes them. In the last fifty years, however, the Anthropic Principle has developed [1], and this might be regarded as a reaction to the mechanistic view. This claims that, in some respects, the Universe has to be the way that it is because otherwise it could not produce life and we would not be here speculating about it. Although the term 'anthropic' derives from the Greek word for 'man', it should be stressed that most of the arguments pertain to life in general.

As a simple example of an anthropic argument, consider the following question: why is the Universe as big as it is? The mechanistic answer is that, at any particular time, the size of the observable Universe is the distance travelled by light since the Big Bang, which is about 10^{10} light-years. There is no compelling reason the Universe has the size it does; it just happens to be 10^{10} y old. There is, however, another answer to this question, which Robert Dicke [2] first gave in 1961. In order for life to exist, there must be carbon, and this is produced by cooking inside stars. The process takes about 10^{10} y, so only after this time can stars explode as supernovae, scattering the newly baked elements throughout space, where they may eventually become part of life-evolving planets. On the other hand, the Universe cannot be *much* older than 10^{10} y, else all the material would have been processed into stellar remnants. Since all the forms of life we can envisage require stars, this suggests that it can only exist when the

Universe or Multiverse?, ed. Bernard Carr. Published by Cambridge University Press.
© Cambridge University Press 2007.

Universe is aged about 10^{10} y. So the very hugeness of the Universe, which seems at first to point to our insignificance, is actually a prerequisite of our existence. This is not to say that the Universe itself could not exist with a different size, only that we could not be aware of it then.

Dicke's argument is an example of what is called the 'Weak Anthropic Principle' and is no more than a logical necessity [3]. This accepts the constants of nature as given and then shows that our existence imposes a selection effect on when (and where) we observe the Universe. Finding that we live at a particular *time* is no more surprising than finding that we live at a particular *place* (e.g. on a planet near a star). Much more controversial is the 'Strong Anthropic Principle', which – in the sense that I will use the term – says that there are relationships between the coupling constants (i.e. the dimensionless numbers which characterize the strengths of the four forces) and other physical quantities which are necessary in order for observers to arise. Some of these relationships are remarkably 'fine-tuned' and do not seem to be predicted by standard physics.

Chapter 1 also referred to the paper on the subject I wrote for *Nature* in 1979 with Martin Rees [4]. This turned out to be quite an influential article because it brought together all the anthropic arguments that were known at the time. In this chapter I will revisit some of these arguments to see how their status has changed. However, I will not give the full details since they can they found in our original paper and also in ref. [1]. I will then consider how one might interpret the anthropic relationships and discuss whether the multiverse proposal provides the best conceptual basis for understanding them. Naturally, other contributors will consider this point – since it is one of the main themes of the book – but without coming to any general consensus.

5.2 Status of the anthropic coincidences

My focus here will be entirely on the strong anthropic arguments, since I have always regarded the weak ones as relatively uncontroversial. The first set of fine-tunings involved the four dimensionless coupling constants. These were taken to be $\alpha \sim 10^{-2}$ for electromagnetism, $\alpha_G \sim 10^{-40}$ for gravity, $\alpha_W \sim 10^{-10}$ for the weak force and $\alpha_S \sim 10$ for the strong force. The second set of fine-tunings was associated with the formation of galaxies and their subsequent fragmentation into stars. These involved various cosmological parameters, such as the matter density in units of the critical density Ω, the amplitude of the density fluctutations Q on entering the cosmological particle horizon and the photon-to-baryon ratio S. At the time, it was not clear which of these parameters were determined by processes in the early Universe rather than being prescribed freely as part of the initial conditions.

- One of the most striking anthropic tunings was associated with the existence of stars with convective and radiative envelopes. Both types of star can exist only if α_G is roughly the 20th power of α [5]. This is because the critical mass which divides these types of stars is roughly $\alpha_G^{-2} \alpha^{10} m_p$, whereas the expected masses of stars span a few decades around $\alpha_G^{-3/2} m_p$ (m_p being the proton mass). The relationship $\alpha_G \sim \alpha^{20}$ is clearly satisfied numerically, but physics does not explain *why* this relationship should pertain. Its anthropic significance is that only radiative stars can end their lives as supernovae, which is required to disseminate heavy elements, whereas only convective stars may generate winds in their early phase, and this may be associated with the formation of rocky planets. To my mind, this is still the most striking coincidence because of the high power of α involved. Recently, Page has discussed the argument in more detail [6] and has shown that it constrains the electron charge ($e \propto \sqrt{\alpha}$) to 3%.

- We also found that α_G must be roughly the 4th power of α_W in order for neutrinos to eject the envelope of a star in a supernova explosion. If the weak force were weaker, the neutrinos would stream through the stellar surface unimpeded; if it were much stronger, they would be trapped inside the core and never reach the surface. At the time, it was not certain that neutrinos were responsible for supernovae, but this is now the standard view, although the full details are still not understood. We pointed out that the same coincidence explains why an interesting amount of helium (roughly 23% by mass) is produced by cosmological nucleosynthesis. This scenario is now undisputed and provides one of the main pillars of support for the Big Bang theory. However, the amount of helium produced is very sensitive to the temperature at which the weak interactions 'freeze out'. If $\alpha_G < \alpha_W^4$, this would occur later and the amount of helium would be drastically reduced. If $\alpha_G > \alpha_W^4$, it would occur earlier and almost all the nucleons would burn into helium, preventing the formation of hydrogen-burning stars. At least the latter condition might be anthropically excluded, since helium-burning stars may be too short-lived for life to evolve on surrounding planets.

- Perhaps the most famous anthropic tuning concerned the generation of carbon (a prerequisite for our form of life) in the helium-burning phase of red giant stars via the triple-alpha reaction: two alpha particles first combine to form beryllium and this then combines with a third alpha particle to form carbon. However, as Hoyle first pointed out [7], the beryllium would decay before interacting with another alpha particle were it not for the existence of a remarkably finely tuned resonance in this interaction. This is sometimes regarded as an anthropic 'prediction' because Hoyle's paper prompted nuclear physicists to look for the resonance and they did indeed find it. At the time, we were unable to quantify this coincidence, but recent work by Oberhummer and colleagues – calculating the variations in oxygen and carbon production in red giant stars as one varies the strength and range of the nucleon interactions – indicates that the nuclear interaction strength must be tuned to at least 0.5% if one is to account for this [8].

- We discussed several constraints involving α_S which are associated with chemistry, although these were subsequently examined in more detail by Barrow and Tipler [1]. For example, if α_S were increased by 2%, all the protons in the Universe would combine at Big Bang nucleosynthesis into diprotons. In this case, there could be no hydrogen-burning stars, so – as mentioned above – there might not be time for life to arise. If α_S were increased by 10%, the situation would be even worse because everything would go into nuclei of unlimited size and there would be no interesting chemistry. This would also apply if α_S were decreased by 5% because all deuterons would then be unbound and one could only have hydrogen. In addition, there are chemistry-related fine-tunings involving the electron and proton masses ($m_e/m_p \sim 10\alpha^2$) and the neutron–proton mass difference ($m_n - m_p \sim 2m_e$). Of course, from the modern perspective, α_S, m_p and m_e are no longer such fundamental quantities; the QCD interaction strength and quark masses would be regarded as more important [9]. Nevertheless, fine-tuning is still required at some level.

- We stressed the (already well known) anthropic reasons for why the total density parameter Ω must lie within an order of magnitude of unity. If it were much larger than unity, the Universe would recollapse on a timescale much less than the main-sequence time of a star. On the other hand, if it were much smaller than unity, density fluctuations would stop growing before galaxies could bind. This argument required that Ω be in the range 0.1 to 10. However, one year later early Universe studies were revolutionized by the introduction of the inflation scenario [10]. This *required* that Ω be very close to unity (so that the geometry of the Universe must be very nearly 'flat'), and observations of the microwave background radiation have subsequently confirmed this [11]. Therefore anthropic considerations may no longer seem relevant. On the other hand, it must be stressed that the inflationary explanation for flatness only works if the form of the vacuum potential $V(\phi)$ allows a sufficient number of expansion e-folds, and this form may itself itself be constrained anthropically [12]. Similar considerations apply in quantum cosmology, where the Universe is expected to collapse very quickly unless one imposes anthropic selection effects [13].

- We also obtained various anthropic constraints on the photon-to-baryon ratio $S \sim 10^9$. In the standard Big Bang model, the formation of galaxies cannot occur until the background radiation density falls below the matter density, but this occurs after the time invoked in the Dicke argument (i.e. the main-sequence lifetime of a star) for $S > \alpha_G^{-1/4} \sim 10^{10}$. On the other hand, if one requires that the Universe be radiation-dominated at cosmological nucleosynthesis, to avoid all the hydrogen going into helium, one requires $S > (m_p/m_e)^{4/3}(\alpha_W^4/\alpha_G)^{1/6} \sim 10^4$. Nowadays we believe the value of S results from of baryon-violating processes in the early Universe – possibly at the GUT epoch around 10^{-34} s after the Big Bang. However, in most GUT models, S is predicted to be of order α^{-n}, where n is an integer, so the anthropic constraint $S < \alpha_G^{-1/4}$ merely translates into the constraint $\alpha_G < \alpha^{4n}$. If $n = 5$, this just gives the convective star condition [14]. An interesting twist on these arguments has been provided by Aguirre [15], who describes anthropic constraints on 'cold' cosmological models (i.e. models with an initial photon-to-baryon ratio much smaller than currently observed). He points out that such models could provide life-supporting conditions with very different values of the cosmological parameters.

- We gave no anthropic constraints on the cosmological constant Λ since this was assumed to be zero, perhaps for fundamental physical reasons. Nowadays, observations indicate that the cosmic expansion is accelerating, and this may be attributed to a positive cosmological constant. Indeed, there is a remarkable coincidence in that the vacuum and matter densities are comparable at the present epoch, even though their ratio is strongly time-dependent. As first emphasized by Weinberg [16] and later studied by Efstathiou [17] and Vilenkin [18], this may provide the strongest anthropic fine-tuning of all, since *a priori* Λ could be 120 orders of magnitude larger than observed. This is because the growth of density perturbations is quenched once Λ dominates the cosmological density, so if bound systems have not formed by then, they never will. The precise form of the anthropic upper limit on Λ depends on the the amplitude of the density fluctuations, Q, and has been discussed by Tegmark and Rees [19]. These arguments have recently been refined [20, 21]. It should be stressed that a cosmological constant is not the only possible explanation for the cosmic acceleration. An alternative explanation is to invoke 'dark energy' in the form of a scalar field – termed 'quintessence' [22] – and this may better explain the near-equality of the vacuum and matter densities. However, some anthropic fine-tuning may be required even in this case [23].

- Various other anthropic constraints have been investigated since my 1979 paper with Rees, and some of these are discussed in this volume. For example, the existence of dark matter has been much more firmly established and this is now known to have about 25% of the total cosmological density. There are still many possible dark matter candidates, but most of them have been associated with such constraints. For example, if WIMPs provide the dark matter, then their density will be comparable to the baryon density provided the aforementioned relationship $\alpha_G \sim \alpha_W^4$ is satisfied [24]. If axions provide the dark matter, then anthropic arguments again explain why their density is comparable with the baryon density, providing the strong CP-violating factor θ (associated with Peccei–Quinn symmetry-breaking at a time of order 10^{-30} s) varies across the different inflationary domains [25].

The crucial role of these fine-tunings in the evolution of the Universe is summarized in Table 5.1. This summarizes the times of various key steps in the history of the Big Bang and indicates the various anthropic fine-tunings associated with each of them. The times are expressed logarithmically in seconds and the quantities appearing under the heading 'Condition' are defined in the text.

Table 5.1. *Cosmological anthropic constraints*

$\log(t/\text{s})$	Event	Condition	Anthropic significance
+17.5	present epoch	$\Omega < 10$	else premature recollapse
+17.0	planet formation	$\alpha_G > \alpha^{20}$	need convective stars
+16.5	star formation	$\alpha_G \sim \alpha_W^4$	need supernovae
		$\alpha_G < \alpha^{20}$	need radiative stars
		$|\Delta\alpha_S| < 0.005\alpha_S$	need triple-alpha resonance
+16	galaxy formation	$\Omega > 0.1, \Omega_\Lambda < 1$	overdense regions must bind
+11	end of radiation era	$S < \alpha_G^{-1/4}$	must precede galaxy formation
+2	Big Bang nucleosynthesis	$\alpha_G < \alpha_W^4$	else all hydrogen goes to helium
		$\Delta\alpha_S < 0.02\alpha_S$	else all hydrogen goes to diprotons
		$\Delta\alpha_S > 0.02\alpha_S$	else deuterons unbound
−30	axion production	$\theta \ll 1$	need enough baryons
−34	baryosynthesis	$\alpha_G > \alpha^{4n}$	need enough photons
−35	inflation	$V'' \ll V$	need enough inflation

5.3 Universe or multiverse?

Although the status of the anthropic constraints involving cosmological parameters was not completely clear in 1979, the anthropic relationships involving the parameters of fundamental physics were certainly not predicted by any theories at the time and this remains the case today. Even if such relationships do transpire to be predicted by some 'final theory', it would be remarkable that the theory should yield exactly the coincidences required for life. One must therefore turn to more radical interpretations of these coincidences.

The first possibility is that the coincidences reflect the existence of a 'beneficent being' who tailor-made the Universe for our benefit. Such an interpretation is logically possible and appeals to theologians [26]. Indeed, some people now use the term 'Strong Anthropic Principle' to imply that the Universe was created with the *purpose* of creating life. However, Rees and I certainly did not have this teleological interpretation in mind at the time of our paper. In any case, most physicists are uncomfortable with this interpretation.

Another possibility, proposed by Wheeler [27], is that the Universe does not properly *exist* until consciousness has arisen. This is based on the notion that the Universe is described by a quantum mechanical wave-function and that consciousness is required to collapse this wave-function. Once the Universe has evolved consciousness, one might think of it as reflecting back on its Big Bang origin, thereby forming a closed circuit which brings the world into existence. Even if consciousness really does collapse the wave-function (which is far from certain), this explanation is also somewhat metaphysical.

The third possibility (and the one that is the focus of this book) is that there is not just one universe but a large ensemble of them, all with different (possibly random) coupling constants. As mentioned in Chapter 1 and reviewed in more detail by Tegmark [28], there are many versions of the multiverse proposal. Not all of these necessarily entail a variation in the physical constants across the ensemble. Therefore, as stressed by Rees [29, 30], a key issue in assessing the multiverse proposal is whether some of the physical constants are contingent on accidental features of symmetry-breaking and the initial conditions of our universe, or whether the future 'Theory of Everything' will determine all of them uniquely.

In the first case, there would be room for the Anthropic Principle and one could envisage the Universe as occupying a point in some multi-dimensional space of coupling constants. In the second case, there would be no room for the Anthropic Principle and any fine-tunings would have to be regarded as coincidental. (The only way out, as emphasized by Tegmark [28], would be to consider worlds with different physical laws or different mathematical foundations and argue that only some of these can permit anthropic relationships.) There might in principle be other universes in the second case, but they would all have the same values for the constants, so there would be little point in invoking them. Therefore the two cases correspond essentially to the multiverse and single universe options, respectively.

If one grants the existence of a multiverse, the question then arises of whether our universe is typical or atypical within the ensemble. Anthropic advocates usually assume that life-forms similar to our own will be possible in only a tiny subset of universes. More general life-forms may be possible in a somewhat larger subset (e.g. if one envisages cold cosmological models of the kind discussed by Aguirre [15]), but life will not be possible everywhere. One may not have the *same* anthropic relation in every universe, but one will have *some* relation.

On the other hand, by invoking a Copernican perspective, Smolin has argued that *most* of the universes should have properties like our own, so that ours is typical. His own approach invokes a form of cosmological natural selection; the formation of black holes is supposed to generate new baby universes in which the constants are slightly mutated [31, 32]. In this way, after many generations, the parameter distribution will be peaked around those values for which black hole formation is maximized. This proposal involves very speculative physics, since we have no understanding of how the baby universes are born, but it has the virtue of being *testable* since one can calculate how many black holes would form if the parameters were different. Note that Smolin's proposal makes no reference to observers, so it would not be regarded as anthropic in the usual sense of the term.

A new twist arises if the (so-called) constants vary in time, even in our universe. This is expected in some higher-dimensional theories since the constants should be related to the size of the compact dimensions and this would be expected to change during at least part of our universe's history. Recently, some astronomers claim to have found positive evidence for a variation in α – of about seven parts in a million – by studying absorption lines in several hundred quasars [33]. If so, one might also expect the relationship $m_e/m_p \sim 10\alpha^2$ to imply that the electron–proton mass ratio varies, and there may indeed be evidence for this as well. Sandvik and colleagues attempt to model this effect and suggest that α should remain constant during both the early radiation-dominated phase of the universe and late curvature-dominated or Λ-dominated phases [34]. However, it could still vary over the intermediate matter-dominated phase, and this would make it difficult to satisfy the anthropic constraints on α for an extended period if the curvature or cosmological constant were very small.

At first sight, this suggests that a variation in the coupling constants would make the anthropic constraints harder to understand. However, it is interesting that the brane cosmology paradigm (discussed in Chapter 1) may provide a natural explanation for the sort of power-law relations between the coupling constants which arise in the anthropic arguments. This is because the variation in the gravitational coupling constant would be associated with the change in the bulk volume, whereas the variation in the other coupling constants would be associated with the change in the volume of the brane or some manifold of intermediate dimensionality. In this case, relationships like $\alpha_G \sim \alpha_W^4 \sim \alpha^{20}$ could just reflect the relative number of internal and external dimensions, so these could themselves be constrained anthropically.

Many contributors to this volume consider whether the multiverse proposal can be tested and ask how legitimate it is to invoke the existence of other universes for which there may never be any direct evidence? Lee Smolin stresses [35] that the multiverse proposal is legitimate only if one has a theory which independently predicts it and that such a theory, to be scientific, must be falsifiable. He argues very forcefully that the notion of a multiverse is neither falsifiable nor testable. However, not everybody concedes this point. For example, Rees points out [36] that one way of testing the multiverse proposal is to calculate the probability distribution for various parameters across the different universes. In particular, if the distribution for the amplitude of the density fluctuations fell off too slowly, we would be surprised to be in a universe with a value as small as that observed.

5.4 How do we interpret the anthropic coincidences?

Even if one accepts that a multiverse exists and – contrary to Smolin's picture – gives rise to anthropic selection effects, there is still considerable ambiguity in how one interprets this. What determines the selection, or, more precisely, what qualifies as an observer? Is it just human beings, or life in general? Is some minimum threshold of intelligence required, or does the mere existence of consciousness suffice? In addressing these questions, I will necessarily veer into more philosophical domains.

As mentioned earlier, although 'anthropos' is the Greek word for 'man', the arguments have nothing to do with humans in particular. Indeed, Brandon Carter (who coined the term) admits that its introduction was unfortunate. Therefore anthropic arguments do not necessarily enhance the status of human beings or support the religious view that we have a special role in the Universe. This interpretation may still be possible if humans turn out to be the only form of life in the Universe. In this context, it is interesting that Carter has argued that we may be the only site of life within our cosmological horizon [37]. He infers this from the remarkable coincidence that the time for life to arise on Earth seems to have been comparable to the cosmological time.

Even if this were true, most cosmologists would still be reluctant to attribute great significance to humans in particular. Therefore it is more traditional to associate the anthropic constraints with life in general. In fact, Davies explicitly associates them with a 'life principle' [38]. Until recently, science would have regarded the existence of life as an incidental rather than fundamental feature of the Universe. Indeed, in the nineteenth century, the second law of thermodynamics was taken to imply that the Universe must eventually undergo a 'heat death', with life and all other forms of order inevitably deteriorating. However, recent developments in cosmology have led to a reversal of this view. According to the Big Bang theory, the history of the Universe reveals an increasing rather than decreasing degree of organization, and modern physics – without any recourse to divine intervention and without any violation of the second law of thermodynamics – is able to explain this. Heat death is avoided because local pockets of order can be purchased at the expense of a global increase in entropy, and, if the Universe continues to expand forever, intelligent beings may be able to delay their disintegration indefinitely [39].

Some of the types of organization which exist in the Universe are summarized in the so-called 'Pyramid of Complexity', introduced by Reeves [40] and reproduced in Fig. 5.1. This shows the different levels of structure as one goes from quarks to nucleons to atoms to simple molecules to biomolecules to cells and finally to living organisms. This hierarchy of structure reflects the existence of the strong force at the lower levels and the electric force at the higher ones. As one ascends the pyramid, the structures become more complex – so that the number of different patterns becomes larger – but they also become more fragile. The pyramid becomes narrower as one rises, and this reflects the fact that the fraction of matter incorporated into the objects decreases as the degree of organization increases.

The Big Bang theory explains *when* these structures arise because the Pyramid of Complexity only emerges as the Universe expands and cools. At early times, the Universe is mainly in the form of quarks. Neutrons and protons appear at a few microseconds, light nuclei at several minutes, atoms at a million years, and – following the formation of stars and planets – molecules and cells at ten billion years. The Big Bang theory also explains

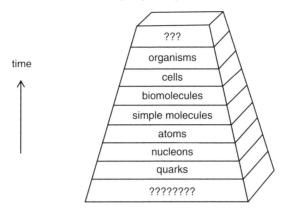

Fig. 5.1. Summary of the different levels of structure which exist in the Universe and how this has arisen during the Big Bang.

why the pyramid came about. The key point is that structures arise because processes cannot occur fast enough in an expanding universe to maintain equilibrium. If it had its way, each type of force would always form the objects which are most stable from its own perspective (e.g. the strong force would turn all nuclei into iron; the electric force would turn all atoms into noble gases; and gravity would turn all matter into black holes). However, all variety would be lost if this were the case, and it is only the disequilibrium entailed by the rapid expansion of the Universe which prevents this.

For example, the reason all nucleons do not go into iron as a result of cosmological nucleosynthesis is that the Universe is expanding too fast for most nuclei to interact with each other at this time. The reason gravity does not turn all stars into black holes is because the pressure associated with nuclear energy release and eventually quantum effects support them against gravity. The forces may eventually attain their goal, but only after an enormous length of time and, even then, only for a limited period. (For example, if the Universe exists long enough, everything may eventually end up in black holes, but on a still longer timescale these black holes will evaporate into radiation.) As emphasized in Table 5.1, it is only the anthropic fine-tuning of the coupling constants that allows the ascension of the lower levels of the pyramid. Therefore, the Pyramid of Complexity can only arise in a small subset of the ensemble of universes.

Note that there is an important difference between the structures which exist at the top and the bottom of the pyramid. Those at the bottom are stable and need large amounts of energy to destroy them, while those at the top must be constantly maintained by exchanging energy with the outside world. More precisely, they must extract information from the world, and the second law of thermodynamics requires that this process is inevitably accompanied by the release of entropy. A store of information arises whenever there exists a source of entropy which has not been released by previous processes. For example, nuclear information is contained in nuclei other than iron, and this can be extacted by nuclear burning inside the Sun, the ultimate source and sustainer of all life-forms on Earth. Similarly, living organisms can feed on plants, and humans can exploit fossil fuels, because these things contain complex molecules with consumable electromagnetic information. Thus there is an inevitable link between complexity and life, and the key to this link is *information*.

Another crucial ingredient at the top level of the pyramid is competitiveness. This is a vital factor in evolution because, as a population grows, the competition for food leads to predation and increasingly sophisticated survival strategies. The proliferation of life-forms due to mutation plays a crucial role in this process. Different modes of perception and motor activity are also required, and this leads to the development of organisms with a central nervous system. From this perspective, brains – certainly the most complex structures on Earth – are merely data integration systems, and the main purpose of intelligence is to increase survival efficiency. Minds, of course, might be regarded as the ultimate storers and extractors of information.

Figure 5.1 suggests that the anthropic fine-tunings are more related to the emergence of complexity than life or minds; they could equally well be regarded as prerequisites for inanimate objects such as motor cars or TV sets. However, here on Earth at least, the development of minds seems to have occurred relatively quickly once the first signs of life arose, so it is conceivable that this applies more generally. Provided there are no extra 'biological' fine-tunings required for the higher levels of the pyramid to arise, the evolution of complexity may inevitably (and fairly rapidly) lead to life and consciousness. In this case, the distinction between life and complexity is not so clear-cut. The former is just a particular realization of the latter and may naturally emerge from it. Therefore the question of what constitutes an observer may be rather incidental. Complexity appears to be the key, and that encompasses everything. From this perspective, the term 'Complexity Principle' would be preferable to 'Anthropic Principle'.

References

[1] J. D. Barrow and F. J. Tipler. *The Anthropic Cosmological Principle* (Oxford: Oxford University Press, 1986).
[2] R. H. Dicke. *Nature*, **192** (1961), 440.
[3] A. J. M. Garret and P. Coles. *Comments Astrophys.* **17** (1993), 23.

[4] B. J. Carr and M. J. Rees. *Nature*, **278** (1979), 605.
[5] B. Carter. In *Confrontation of Cosmological Models with Observations*, ed. M. S. Longair (Dordrecht: Reidel, 1974), pp. 291–298.
[6] D. N. Page (2002) [hep-th/0203051].
[7] F. Hoyle. *Astrophys. J.* **118** (1953), 513.
[8] H. Oberhummer, A. Csoto and H. Schlattl. *Science*, **289** (2000), 88.
[9] C. J. Hogan. *Rev. Mod. Phys.* **72** (2000), 1149.
[10] A. H. Guth. *Phys. Rev.* **D 23** (1981), 347.
[11] D. N. Spergel, L. Verde, H. V. Peiris *et al. Astrophys. J. Supp.* **148** (2003), 175.
[12] A. D. Linde. *Particle Physics and Inflationary Cosmology* (Chur, Switzerland: Harwood, 1990).
[13] S. W. Hawking. This volume (2007).
[14] B. J. Carr. *J. Brit. Interplan. Soc.* **44** (1991), 63.
[15] A. Aguirre. *Phys. Rev.* **D 64** (2001), 83508.
[16] S. Weinberg. *Rev. Mod. Phys.* **61** (1989), 1.
[17] G. Efstathiou. *Mon. Not. Roy. Astron. Soc.* **274** (1995), L73.
[18] A. Vilenkin. *Phys. Rev. Lett.* **74** (1995), 846.
[19] M. Tegmark and M. J. Rees. *Astrophys. J.* **499** (1998), 526.
[20] M. Tegmark. *Ann. Phys.* **270** (1998), 1.
[21] M. Tegmark. In *Science and Ultimate Reality*, eds. J. D. Barrow, P. C. W. Davies and C. L. Harper (Cambridge: Cambridge University Press, 2004).
[22] P. J. Steinhardt and N. Turok. *Science*, **296** (2002), 1436.
[23] R. Kallosh and A. Linde. *Phys. Rev.* **D 67** (2003), 023510.
[24] M. S. Turner and B. J. Carr. *Mod. Phys. Lett.* **A 2** (1987), 1.
[25] F. Wilczek. This volume (2007).
[26] J. Polkinghorne. *The Faith of a Physicist* (Princeton, NJ: Princeton University Press, 1994).
[27] J. Wheeler. In *Foundational Problems in the Special Sciences*, eds. R. Butts and J. Hintikka (Dordrecht: Reidel, 1977), p. 3.
[28] M. Tegmark. *Ann. Phys.* **270** (1998), 1.
[29] M. J. Rees. *Our Cosmic Habitat* (Princeton, NJ: Princeton University Press, 2001).
[30] M. J. Rees. *Just Six Numbers: The Deep Forces that Shape the Universe* (London: Weidenfeld and Nicholson, 1999).
[31] L. Smolin. *Class. Quant. Grav.* **9** (1992), 173.
[32] L. Smolin. *The Life of the Cosmos* (Oxford: Oxford University Press, 1997).
[33] J. K. Webb, M. T. Murphy, V. V. Flambaum *et al. Phys. Rev. Lett.* **87** (2001), 091301.
[34] H. B. Sandvik, J. D. Barrow and J. Magueijo. *Phys. Rev. Lett.* **88** (2002), 031302.
[35] L. Smolin. This volume (2007).
[36] M. J. Rees. This volume (2007).
[37] B. Carter. *Phil. Trans. Roy. Soc.* **A 310** (1983), 347.
[38] P. C. W. Davies. This volume (2007).
[39] F. Dyson. *Rev. Mod. Phys.* **51** (1979), 447.
[40] H. Reeves. *The Hour of Our Delight* (New York: W. H. Freeman, 1991).

6

Cosmology from the top down

Stephen Hawking
Centre for Mathematical Sciences, University of Cambridge

6.1 Problems with bottom-up approach

The usual approach in physics could be described as building from the bottom up. That is, one assumes some initial state for a system and then evolves it forward in time with the Hamiltonian and the Schrödinger equation. This approach is appropriate for laboratory experiments like particle scattering, where one can prepare the initial state and measure the final state. The bottom-up approach is more problematic in cosmology, however, because we do not know what the initial state of the Universe was, and we certainly cannot try out different initial states and see what kinds of universe they produce.

Different physicists react to this difficulty in different ways. Some – generally those brought up in the particle physics tradition – just ignore the problem. They feel the task of physics is to predict what happens in the laboratory, and they are convinced that string theory or M-theory can do this. All they think remains to be done is to identify a solution of M-theory, a Calabi–Yau or G2 manifold that will give the Standard Model as an effective theory in four dimensions. But they have no idea why the Universe should be 4-dimensional and have the Standard Model, with the values of the forty or so parameters that we observe. How can anyone believe that something so messy is the unique prediction of string theory? It amazes me that people can have such blinkered vision – that they can concentrate just on the final state of the Universe and not ask how and why it got there.

Those physicists that try to explain the Universe from the bottom up mostly belong to one of two schools, these being associated with either the inflationary or pre-big-bang scenarios. I now discuss these approaches in turn and show that neither of them is satisfactory.

Universe or Multiverse?, ed. Bernard Carr. Published by Cambridge University Press.
© Cambridge University Press 2007.

6.1.1 Inflationary scenarios

In the case of inflation, the idea is that the exponential expansion obliterates the dependence on the initial conditions [1], so we would not need to know exactly how the Universe began – just that it was inflating. To lose all memory of the initial state would require an infinite amount of exponential expansion, which leads to the notion of ever-lasting or eternal inflation [2, 3]. The original argument for eternal inflation went as follows. Consider a massive scalar field in a spatially infinite expanding universe. Suppose the field is nearly constant over several horizon regions on a spacelike surface. In an infinite universe, there will always be such regions. The scalar field will have quantum fluctuations. In half the regions, the fluctuations will increase the field; in the other half, they will decrease it. In the half where the field jumps up, the extra energy density will cause the universe to expand faster than in the half where the field jumps down. After a certain proper time, more than half the regions will have the higher value of the field, because the high-field regions will expand faster than the low-field regions.

Thus the volume-averaged value of the field will rise. There will always be regions of the universe in which the scalar field is high, so inflation is eternal. The regions in which the scalar field fluctuates downwards will branch off from the eternally inflating region and exit inflation. Because there will be an infinite number of exiting regions, advocates of eternal inflation get themselves tied in knots on what a typical observer would see. So even if eternal inflation worked, it would not explain *why* the Universe is the way it is.

In fact, the argument for eternal inflation that I have outlined has serious flaws. First, it is not gauge-invariant. If one takes the time surfaces to be surfaces of constant volume increase rather than surfaces of constant proper time, the volume-averaged scalar field does not increase. Second, it is not consistent. The equation relating the expansion rate to the energy density is an integral of motion. Neither side of the equation can fluctuate because energy is conserved. Third, it is not covariant. It is based on a $3+1$ split. From a 4-dimensional view, eternal inflation can only be de Sitter space with bubbles. The energy-momentum tensor of the fluctuations in a single scalar field is not large enough to support a de Sitter space, except possibly at the Planck scale, where everything breaks down. For these reasons – lack of gauge-invariance and covariance and inconsistency – I do not believe the usual argument for eternal inflation. However, as I shall explain later, I think the Universe may have had an initial de Sitter stage considerably longer than the Planck timescale.

6.1.2 Pre-big-bang scenarios

I now turn to pre-big-bang scenarios, which are the main alternative to inflation. I shall take them to include the ekpyrotic [4] and cyclic models [5], as well as the older pre-big-bang model [6]. The observations of the microwave background fluctuations show that there are correlations on scales larger than the horizon size at decoupling. These correlations could be explained if there had been inflation, because the exponential expansion would have meant that regions that are now widely separated were once in causal contact with each other. On the other hand, if there were no inflation, the correlations must have been present at the beginning of the expansion of the Universe. Presumably, they arose in a previous contracting phase and somehow survived the singularity or brane collision.

It is not clear if effects can be transmitted through a singularity, or if they will produce the right signature in the microwave background fluctuations. But even if the answer to both of these questions is yes, the pre-big-bang scenarios do not answer the central question of cosmology: why is the Universe the way it is? All the pre-big-bang scenarios can do is shift the problem of the initial state from 13.7 Gy ago to the infinite past. But a boundary condition is a boundary condition, even if the boundary is at infinity. The present state of the Universe would depend on the boundary condition in the infinite past. The trouble is, there is no natural boundary condition, such as the Universe being in its ground state. The Universe does not have a ground state. It is unstable and is either expanding or contracting. The lack of a preferred initial state in the infinite past means that pre-big-bang scenarios are no better at explaining the Universe than supposing that someone wound up the clock and set the Universe going at the big bang.

6.2 Sum over histories

The bottom-up approach to cosmology – supposing some initial state and evolving it forward in time – is basically classical, because it assumes that the universe began in a way that was well defined and unique. But one of the first acts of my research career was to show with Roger Penrose that any reasonable classical cosmological solution has a singularity in the past [7]. This implies that the origin of the Universe was a quantum event. This means that it should be described by the Feynman 'sum over histories'. The Universe does not have just a single history, but every possible history, whether or not they satisfy the field equations. Some people make a great mystery of the 'many worlds' interpretation of quantum theory, but to me these are just different expressions of the Feynman path integral.

One can use the path integral to calculate the quantum amplitudes for observables at the present time. The wave-function of the Universe, or amplitude for the metric h_{ij} on a surface S of co-dimension one, is given by a path integral over all metrics, g, that have S as a boundary. Normally, one thinks of path integrals as having two boundaries: an initial surface and a final one. This would be appropriate in a proper quantum treatment of a pre-big-bang scenario, like the ekpyrotic model. In this case, the initial surface would be in the infinite past.

But there are two big objections to the path integral for the Universe having an initial surface. The first is the G question (i.e. the issue of whether one needs God). What was the initial state of the Universe and why was it like that? As I said earlier, there does not seem to be a natural choice for the initial state. It cannot be flat space since that would remain flat space.

The second objection is equally fundamental. In most models, the quantum state on the final surface will be independent of the state on the initial one. This is because there will be metrics in which the initial surface is in one component and the final surface is in a separate, disconnected component of the 4-dimensional manifold. Such metrics will exist in the Euclidean regime. They correspond to the quantum annihilation of one universe and the quantum creation of another. This would not be possible if there were something that was conserved that propagated from the initial surface to the final surface. But the trend in cosmology in recent years has been to claim that a universe has no conserved quantities. Things like baryon number are supposed to have been created by grand unified or electroweak theories, together with CP violation. So there is no way one can rule out the final surface from belonging to a different universe than the initial surface. In fact, because there are so many different possible universes, they will dominate and the final state will be independent of the initial state. It will be given by a path integral over all metrics whose only boundary is the final surface. In other words, it is the so-called 'no boundary' quantum state [8].

6.3 Top-down approach

If one accepts that the no boundary proposal is the natural prescription for the quantum state of the Universe, one is led to a profoundly different view of cosmology and the relationship between cause and effect. One should not follow the history of the Universe from the bottom up, because that assumes there is a single history, with a well defined starting point and evolution. Instead, one should trace the histories from the top down – in other words, backwards from the measurement surface, S, at the present time. The histories that contribute to the path integral do not have an independent existence, but depend on the amplitude that is being measured. As an example of this, consider the apparent dimension of the Universe. The usual idea is that spacetime is a 4-dimensional nearly flat metric cross a small 6- or 7-dimensional internal manifold. But why are there not more large dimensions? Why are any dimensions compactified? There are good reasons to think that life is possible only in four dimensions, but most physicists are very reluctant to appeal to the anthropic principle. They would rather believe that there is some mechanism that causes all but four of the dimensions to compactify spontaneously. Alternatively, maybe all dimensions started small, but for some reason four dimensions expanded and the rest did not.

I am sorry to disappoint these hopes, but I do not think there is a dynamical reason for the Universe to appear 4-dimensional. Instead, the no boundary proposal predicts a quantum amplitude for every number of large spatial dimensions from 0 to 10. There will be an amplitude for the Universe to be 11-dimensional Minkowski space, i.e. with ten large spatial dimensions. However, the value of this amplitude is of no significance, because we do not live in eleven dimensions. We are not asking for the probabilities of various dimensions for the Universe. As long as the amplitude for three large spatial dimensions is not exactly zero, it does not matter how small it is compared with that for other numbers of dimensions. The Universe appears to be 4-dimensional, so we are interested only in amplitudes for surfaces with three large dimensions. This may sound like the anthropic principle argument that the reason we observe the Universe to be 4-dimensional is that life is possible only in four dimensions. But the argument here is different, because it does not depend on whether four dimensions is the only arena for life. Rather, it is that the probability distribution over dimensions is irrelevant, because we have already measured that we are in four dimensions.

The situation with the low energy effective theory of particle interactions is similar. Many physicists believe that string theory will uniquely predict the Standard Model and the values of its forty or so parameters. The bottom-up picture would be that the Universe begins with some grand unified symmetry, like E8 × E8. As the Universe expanded and cooled, the symmetry would break to the Standard Model, maybe through intermediate stages. The hope would be that string theory would predict the pattern of symmetry-breaking, the masses of particles, couplings and mixing angles. However, personally I find it difficult to believe that the Standard Model is the unique prediction of fundamental theory. It is so ugly and the mixing angles etc. seem accidental rather than part of a grand design.

In string/M-theory, low energy particle physics is determined by the internal space. It is well known that M-theory has solutions with many different internal spaces. If one builds the history of the Universe from the bottom up, there is no reason it should end up with the internal space for the Standard Model. However, if one asks for the amplitude for a spacelike surface with a given internal space, one is interested only in those histories which end with that internal space. One therefore has to trace the histories from the top down, backwards from the final surface.

One can calculate the amplitude for the internal space of the Standard Model on the basis of the 'no boundary' proposal. As with the dimensionality, it does not matter how small this amplitude is relative to other possibilities. It would be like asking for the amplitude that I am Chinese. I know I am British, even though there are more Chinese. Similarly, we know the low energy theory is the Standard Model, even though other theories may have a larger amplitude.

Although the relative amplitudes for radically different geometries do not matter, those for neighbouring geometries are important. For example, the fluctuations in the microwave background correspond to differences in the amplitudes for spacelike surfaces that are small perturbations of flat 3-space cross the internal space. It is a robust prediction of inflation that the fluctuations are Gaussian and nearly scale-independent. This has been confirmed by the recent observations by the WMAP satellite [9]. However, the predicted amplitude is model-dependent.

The parameters of the Standard Model will be determined by the moduli of the internal space. Because they are moduli at the classical level, their amplitudes will have a fairly flat distribution. This means that M-theory cannot predict the parameters of the Standard Model. Obviously, the values of the parameters we measure must be compatible with the development of life. I hesitate to say 'intelligent' life, but – within the anthropically allowed range – the parameters can have any values. So much for string theory predicting the fine-structure constant. However, although the theory cannot predict the value of the fine-structure constant, it *will* predict that it should have spatial variations, like the microwave background. This would be an observational test of our ideas of M-theory compactification.

How can one get a non-zero amplitude for the present state of the Universe if, as I claim, the metrics in the path integral have no boundary apart from the surface at the present time? I cannot claim to have the definitive answer, but one possibility would be if the 4-dimensional part of the metric went back to a de Sitter phase. Such a scenario is realized in trace-anomaly driven inflation, for example [10]. In the Lorentzian regime, the de Sitter phase would extend back into the infinite past. It would represent a universe that contracted to a minimum radius and then expanded again. But we know that Lorentzian de Sitter can be closed off in the past by half the 4-sphere. One can interpret this in the bottom-up picture as the spontaneous creation of an inflating universe from nothing. Some pre-big-bang or ekpyrotic scenarios, involving collapsing and expanding universes, can probably be formulated in no boundary terms with an orbifold point. However, this would remove the scale-free perturbations which, it is claimed, develop during the collapse and carry on into the expansion. So again it is a 'no no' for pre-big-bang and ekpyrotic universes.

6.4 Conclusions

In conclusion, the bottom-up approach to cosmology would be appropriate if one knew that the Universe was set going in a particular way, in either the finite or infinite past. However, in the absence of such knowledge, it is better to work from the top down, by tracing backwards from the final surface the histories that contribute to the path integral. This means that the histories of the Universe depend on what is being measured, contrary to the usual idea that the Universe has an objective, observer-independent, history. The Feynman path integral allows every possible history for the Universe, and the observations select out the sub-class of histories that have the property that is being observed. There are histories in which the Universe eternally inflates or is 11-dimensional, but they do not contribute to the amplitudes we measure. I would call this the 'selection principle' rather than the 'anthropic principle' because it does not depend on intelligent life. Life may, after all, be possible in eleven dimensions, but we know we live in four.

The results are disappointing for those who hoped that the ultimate theory would predict everyday physics. We cannot predict discrete features such as the number of large dimensions or the gauge symmetry of the low energy theory. Rather, we use them to select which histories contribute to the path integral. The situation is better with continuous quantities, such as the temperature of the cosmic microwave background or the parameters of the Standard Model. We cannot measure their probability distributions, because we have only one value for each quantity. We cannot tell whether the Universe was likely to have the values we observe, or whether it was just a lucky chance. However, it is noteworthy that the parameters we measure seem to lie in the interior of the anthropically allowed range rather than at the edge. This suggests that the probability distribution is fairly flat – not like the exponential dependence on the density parameter, Ω, in the open inflation model that Neil Turok and I proposed [11]. In that model, Ω would have had the minimum value required to form a single galaxy, which is all that is anthropically necessary. All the other galaxies which we see are superfluous.

Although the theory advocated here cannot predict the average values of these quantities, it will predict that there will be spatial variations – such as the fluctuations in the microwave background. However, the size of these variations will probably depend on moduli or parameters that we cannot predict. So even when we understand the ultimate theory, it will not tell us much about how the Universe began. It cannot predict the dimension of spacetime, the gauge group or other parameters of the low energy effective theory. On the other hand, the theory will predict that the total energy density will be exactly the critical one, though it will not determine how this energy is divided between conventional matter and a cosmological constant or quintessence. The theory will also predict a nearly scale-free spectrum of fluctuations, but it will not determine the amplitude.

So, to come back to the question with which I began this chapter: does string theory predict the state of the Universe? The answer is that it does not. It allows a vast landscape of possible universes in which we occupy an anthropically permitted location [12]. But I feel we could have selected a better neighbourhood.

References

[1] A. H. Guth. *Phys. Rev.* **D 23** (1981), 347.
[2] A. Vilenkin. *Phys. Lett.* **D 27** (1983), 2848.
[3] A. D. Linde. *Phys. Lett.* **B 175** (1986), 395.
[4] J. Khoury, B. A. Ovrut, P. J. Steinhardt and N. Turok. *Phys. Rev.* **D 61** (2001), 123522.
[5] P. J. Steinhardt and N. Turok. *Phys. Rev.* **D 65** (2002), 126003.
[6] M. Gasperini, M. Maggiore and G. Veneziano. *Nucl. Phys.* **B 494** (1997), 315.
[7] S. W. Hawking and R. Penrose. *Proc. Roy. Soc. Lond.* **A 314** (1970), 529.
[8] J. Hartle and S. W. Hawking. *Phys. Rev.* **D 28** (1983), 2960.
[9] D. N. Spergel, L. Verde, H. V. Peiris *et al. Astrophys. J. Suppl.* **148** (2003), 175.
[10] S. W. Hawking, T. Hertog and H. S. Reall. *Phys. Rev.* **D 63** (2001), 083504.
[11] S. W. Hawking and N. Turok. *Phys. Lett.* **B 425** (1998), 25.
[12] L. Susskind. This volume (2007) [hep-th/0302219].

7
The multiverse hierarchy

Max Tegmark

Department of Physics, Massachusetts Institute of Technology

7.1 Introduction

Parallel universes are now all the rage, cropping up in books, movies and even jokes: 'You passed your exam in many parallel universes – but not this one.' However, they are as controversial as they are popular, so it is important to ask whether they are within the purview of science or merely silly speculation. They are also a source of confusion, since many people fail to distinguish between the different types of parallel universes proposed.

In the big bang model, the farthest one can observe is the distance that light has travelled during the 14 billion years since the expansion began. The most distant visible objects are now about 4×10^{26} m away.[1] A sphere of this radius defines our observable universe or our *horizon volume*. We will sometimes loosely refer to this as 'our universe', although this may be part of a region which extends much further. In this article, I will survey theories of physics involving what are termed 'parallel universes' or 'multiverses'. These form a four-level hierarchy, allowing progressively greater diversity.

- **Level I** A generic prediction of cosmological inflation is an infinite 'ergodic' space, which contains Hubble volumes realizing all initial conditions – including one with an identical copy of you about $10^{10^{29}}$ m away.

- **Level II** Given the *fundamental* laws of physics that physicists one day hope to capture with equations on a T-shirt, different regions of space can exhibit different *effective* laws of physics (physical constants, dimensionality, particle content, etc.), corresponding to different local minima in a landscape of possibilities.

---

[1] After emitting the light that is now reaching us, the most distant things we can see have receded because of the cosmic expansion, and are now about about 40 billion light-years away.

Universe or Multiverse?, ed. Bernard Carr. Published by Cambridge University Press.
© Cambridge University Press 2007.

- **Level III** In unitary quantum mechanics, other branches of the wavefunction add nothing qualitatively new, which is ironic given that this level has historically been the most controversial.
- **Level IV** Other mathematical structures give different fundamental equations of physics for that T-shirt.

The key question is therefore not *whether* there is a multiverse (since Level I is the rather uncontroversial cosmological concordance model), but rather *how many* levels it has. The different levels of the multiverse are illustrated in Fig. 7.1.

This chapter will discuss at length whether there can be evidence for other universes and whether such speculations are science or philosophy. For now, the key point to remember is that *parallel universes are not a theory, but a prediction of certain theories.* For a theory to be falsifiable, we do not need to be able to observe and test *all* its predictions, merely at least one of them. By analogy, consider Einstein's theory of General Relativity. Because this has successfully predicted many things that we *can* observe, we also take seriously its predictions for things we *cannot* observe, for example that space continues inside a black hole event horizon and that (contrary to early misconceptions) nothing funny happens at the horizon itself. Likewise, successful predictions of the theories of cosmological inflation and unitary[2] quantum mechanics have made some scientists take more seriously other predictions of these theories, including various types of parallel universes. This is summarized in Table 7.1.

Let us make two cautionary remarks before delving into the details. Hubris and lack of imagination have repeatedly caused humans to underestimate the vastness of the physical world, and dismissing things merely because we cannot observe them from our vantage point is reminiscent of the ostrich with its head in the sand. Moreover, recent theoretical insights have indicated that nature may be tricking us. Einstein taught us that space is not merely a boring static void, but a dynamic entity that can stretch (the expanding universe), vibrate (gravitational waves) and curve (gravity). Searches for a unified theory also suggest that space can 'freeze', transitioning between different phases in a landscape of possibilities, just like water can be solid, liquid or gas. In different phases, the effective laws of physics (particles, symmetries, etc.) could vary. A fish never leaving the ocean might mistakenly conclude that the properties of water are universal, not realizing that there is also ice and steam. We may be smarter than a fish, but we

---

[2] As described below, the mathematically simplest version of quantum mechanics is 'unitary', lacking the controversial process known as wave-function collapse.

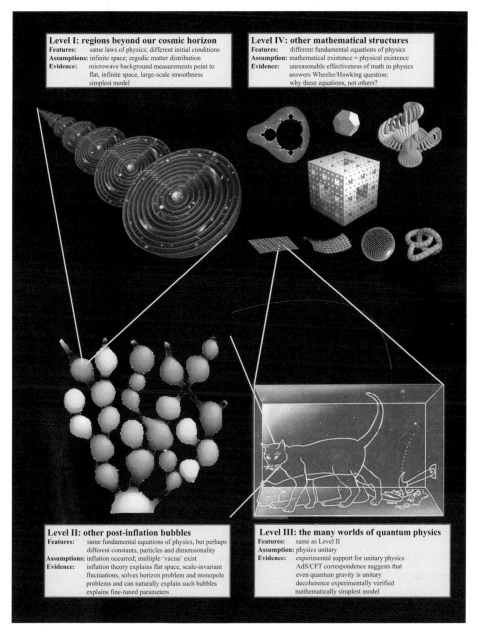

Fig. 7.1. Four different levels of multiverse.

could be similarly fooled; cosmological inflation has the deceptive property of stretching a small patch of space in a particular phase so that it fills our entire observable universe, potentially tricking us into misinterpreting our local conditions for the universal laws that should go on that T-shirt.

Table 7.1. *Theories with unobservable predictions*

Theory	Prediction
General relativity	Black hole interiors
Inflation	Level I parallel universes
Unitary quantum mechanics	Level III parallel universes

7.2 Level I: regions beyond our cosmic horizon

If space is infinite and the distribution of matter is sufficiently uniform on large scales, then even the most unlikely events must take place somewhere. In particular, there are infinitely many other inhabited planets, including not just one but infinitely many copies of you – with the same appearance, name and memories. Indeed, there are infinitely many other regions the size of our observable universe, where every possible cosmic history is played out. This is the Level I multiverse.

7.2.1 Evidence for Level I parallel universes

Although the implications may seem counter-intuitive, this spatially infinite cosmological model is currently the simplest and most popular. Yet the Level I multiverse idea has been always controversial. Indeed, its proposal was one of the heresies for which the Vatican had Giordano Bruno burned at the stake in 1600. In recent times his ideas have been elaborated by various people [1–3].

Let us first review the status of the assumption of infinite space. Observationally, the lower bound on the size of space has grown dramatically, as indicated in Fig. 7.2, with no indication of an upper bound. We all accept the existence of things that we cannot see but could see if we moved or waited, such as ships beyond the horizon. Objects beyond the cosmic horizon have similar status, since the observable universe grows by a light-year every year. If anything, the Level I multiverse sounds obvious. How could space not be infinite? If space comes to an end, what lies beyond it? In fact, Einstein's theory of gravity calls this intuition into question, since space could be finite if it has a convex curvature or an unusual topology. For example, a spherical, doughnut-shaped or pretzel-shaped universe would have a limited volume and no edges. The cosmic microwave background radiation allows sensitive tests of such scenarios, but so far the evidence is against them. Infinite models fit the data better and strong limits have been placed on the alternatives [4,5]. In addition, a spatially infinite universe is a generic prediction of the cosmological theory of inflation [2],

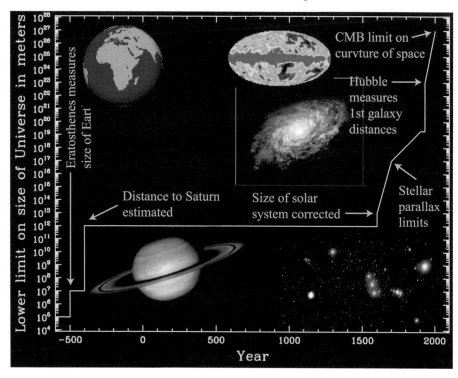

Fig. 7.2. Although an infinite universe has always been a possibility, the lower limit on the size of our universe has kept growing.

so the striking success of inflation lends further support to the idea that space is infinite.

Let us next review the status of the assumption that matter has a uniform distribution. It is possible that space is infinite, but with matter confined to a finite region around us, as in the historically popular 'island universe' model. In a variant on this model, matter thins out on large scales in a fractal pattern. In both cases, almost all universes in the Level I multiverse would be empty. However, recent observations of the 3-dimensional galaxy distribution and the microwave background have shown that the arrangement of matter gives way to dull uniformity on large scales, with no coherent structures larger than about 10^{24} m. Assuming that this pattern continues, space beyond our observable universe teems with galaxies, stars and planets.

7.2.2 What are Level I parallel universes like?

The physics description of the world is traditionally split into two parts: initial conditions and laws of physics specifying how these initial conditions evolve. Observers living in parallel universes at Level I observe the same laws of physics as we do, but with different initial conditions. The currently favoured theory is that the initial conditions were created by quantum fluctuations during inflation. This generates initial conditions that are essentially random, producing density fluctuations described by an *ergodic* random field. This means that if you imagine generating an ensemble of universes, each with its own random initial conditions, then the probability distribution of outcomes in a given volume is identical to the distribution that you get by sampling different volumes in a single universe. In other words, everything that could in principle have happened here did happen somewhere else.

Inflation, in fact, generates all possible initial conditions with non-zero probability, the most likely ones being almost uniform with fluctuations at the 10^{-5} level. These were then amplified by gravitational clustering to form galaxies, stars, planets and other structures. This means that all imaginable matter configurations should occur in some Hubble volume and that we should expect our own Hubble volume to be fairly typical – at least typical among those that contain observers. A crude estimate suggests that the closest identical copy of you is about $10^{10^{29}}$ m away. About $10^{10^{91}}$ m away, there should be a sphere of radius 100 light-years identical to the one centred here, so all perceptions that we have during the next century will be identical to those of our counterparts over there. About $10^{10^{115}}$ m away, there should be an entire Hubble volume identical to ours.[3]

This raises an interesting philosophical point that we will reconsider later: if there are indeed many copies of 'you' with identical past lives and memories, you would not be able to compute your own future even if you had complete knowledge of the entire cosmos! The reason is that there is no way for you to determine which of these copies is 'you'. Yet their lives will necessarily differ eventually, so the best you can do is predict probabilities for what you will experience from now on. This kills the traditional notion of determinism.

---

[3] This is an extremely conservative estimate, 10^{115} being roughly the number of protons that the Pauli exclusion principle would allow you to pack into a Hubble volume at a temperature of 10^8 K. Each of these slots can be either occupied or unoccupied, giving $2^{10^{115}} \sim 10^{10^{115}}$ possibilities, so the expected distance to the nearest identical Hubble volume is $10^{10^{115}}$ Hubble radii or $10^{10^{115}}$ m. Your nearest copy is likely to be much closer than $10^{10^{29}}$ m, since the planet formation and evolutionary processes that have tipped the odds in your favour are at work everywhere. There may be 10^{20} habitable planets in our own Hubble volume alone.

7.2.3 How a multiverse theory can be tested and falsified

Is a multiverse theory one of metaphysics rather than physics? As emphasized by Karl Popper, this depends on whether the theory is empirically testable and falsifiable. Containing unobservable entities does not itself render a theory non-testable. For instance, a theory stating that there are 666 parallel universes, all of which are devoid of oxygen, makes the testable prediction that we should observe no oxygen here, and is therefore ruled out by observation.

In fact, the Level I multiverse is routinely used to rule out theories in modern cosmology, although this is rarely spelled out explicitly. For instance, cosmic microwave background (CMB) observations have recently shown that space has almost no curvature. Hot and cold spots in CMB maps have a characteristic size that depends on the curvature of space, and the observed spots appear too large to be consistent with the previously popular 'open' model. However, the average spot size varies randomly from one Hubble volume to another, so it is important to be statistically rigorous. When cosmologists say that the open universe model is ruled out at 99.9% confidence, they really mean that – if the open universe model were true – then fewer than one out of every 1000 Hubble volumes would show CMB spots as large as those we observe. It is inferred that the entire model (with its infinitely many Hubble volumes) is ruled out, even though we have only mapped the CMB in our own particular Hubble volume.

Thus multiverse theories *can* be tested and falsified only if they predict what the ensemble of parallel universes is and specify a probability distribution and *measure* over it. As we will see later, the measure problem can be quite serious and is still unsolved for some multiverse theories.

7.3 Level II: other post-inflation bubbles

Imagine an infinite set of distinct Level I multiverses, each represented by a bubble in Fig. 7.1, some perhaps with different dimensionality and different physical constants. We will refer to this as the Level II multiverse, and it is predicted by most currently popular models of inflation. These other domains are so far away that you would never get to them even if you travelled at the speed of light forever. The reason is that the space between our Level I multiverse and its neighbours is still undergoing inflation, which creates space faster than you can travel through it. In contrast, you could travel to an arbitrarily distant Level I universe, providing the cosmic expansion decelerates. In fact, astronomical evidence suggests that the cosmic expansion is currently accelerating and, if this acceleration continues indefinitely, then even some Level I universes will remain forever separate. However, at least some models predict that our universe will eventually stop accelerating and perhaps even recollapse.

7.3.1 Evidence for Level II parallel universes

Inflation is an extension of the big bang theory, in which a rapid stretching of space at an early time explains why our universe is so big, so uniform and so flat [6, 7]. Such stretching is predicted by a wide class of theories of particle physics, and all available evidence bears it out. Much of space will continue to stretch forever, but some regions stop stretching and form distinct bubbles. Infinitely many bubbles may emerge, as shown in the lower left of Fig. 7.1, each being an infinite embryonic Level I multiverse filled with matter deposited by the energy field that drove inflation.[4] Recent cosmological measurements have confirmed two key predictions of inflation: that space has negligible curvature and that the clumpiness in the cosmic matter distribution was approximately scale-invariant.

7.3.2 What are Level II parallel universes like?

The prevailing view is that the physics we observe today is merely a low-energy limit of a much more general theory that manifests itself at extremely high temperatures. For example, this underlying fundamental theory may be 10-dimensional, supersymmetric and involve a grand unification of the four fundamental forces of nature. A common feature in such theories is that the potential energy of the field relevant to inflation has many different minima ('metastable vacuum states'), these corresponding to different effective laws of physics for our low-energy world. For instance, all but three spatial dimensions could be curled up ('compactified') on a tiny scale, resulting in a space like ours, or fewer could curl up, leaving a 5-dimensional space. Quantum fluctuations during inflation can therefore cause different post-inflation bubbles in the Level II multiverse to end up with different effective laws of physics, different dimensionality and different numbers of generations of quarks.

In addition to these discrete properties, our universe is characterized by a set of dimensionless *physical constants*, for example the electron/proton mass ratio, $m_p/m_e \approx 1836$, and the cosmological constant, which appears to be

---

[4] Surprisingly, it has been shown that inflation can produce an infinite Level I multiverse even in a bubble of finite volume. This is because the spatial directions of spacetime curve towards the (infinite) time direction [8].

about 10^{-123} in Planck units. There are also models where such non-integer parameters can vary from one post-inflationary bubble to another.[5] So the Level II multiverse is likely to be more diverse than the Level I multiverse, containing domains where not only the initial conditions differ, but also the physical constants.

This is currently a very active research area. In string theory 'landscape' [9, 10], the potential has perhaps 10^{500} different minima, so this may offer a specific realization of the Level II multiverse. This would have four sub-levels of increasing diversity.

- **IId** Different ways in which space can be compactified, allowing both different dimensionality and different symmetries and elementary particles.
- **IIc** Different 'fluxes' (generalized magnetic fields) that stabilize the extra dimensions, this being where the largest number of choices enter.
- **IIb** Once these two choices have been made, there may still be a handful of different minima in the effective supergravity potential.
- **IIa** The same minimum and effective laws of physics can be realized in many different post-inflationary bubbles, each constituting a Level I multiverse.

Let us briefly comment on a few closely related multiverse notions. An idea proposed by Tolman [11] and Wheeler [12], and recently elaborated by Steinhardt and Turok [13], is that the Level I multiverse is cyclic, going through an infinite series of big bangs. If it exists, the ensemble of such incarnations would also form a multiverse, arguably with a diversity similar to that of Level II. An idea proposed by Smolin [14] involves an ensemble similar in diversity to that of Level II, but mutating and sprouting new universes through black holes rather than inflation. This predicts a form of natural selection favouring universes with maximal black hole production. In braneworld scenarios, another 3-dimensional world could be literally parallel to ours but offset in a higher dimension. However, it is unclear whether such a world should be regarded as separate from our own, since we may be able to interact with it gravitationally.

---

5 Although the fundamental equations of physics are the same throughout the Level II multiverse, the approximate effective equations governing the low-energy world that we observe will differ. For instance, moving from a 3-dimensional to a 4-dimensional (non-compactified) space changes the observed gravitational force equation from an inverse square law to an inverse cube law. Likewise, breaking the underlying symmetries of particle physics differently will change the line-up of elementary particles and the effective equations that describe them. However, we will reserve the terms 'different equations' and 'different laws of physics' for the Level IV multiverse, where it is the fundamental rather than effective equations that change.

Note that, if one Level II multiverse can exist, eternally self-reproducing in a fractal pattern, then there may well be infinitely many other ones that are completely disconnected. However, this variant appears to be untestable, since it would neither add any qualitatively different worlds nor alter the probability distribution for their properties. All possible initial conditions and symmetry-breakings are already realized within each one.

7.3.3 Fine-tuning and selection effects

Physicists like to explain as much as possible, and some features of our universe seem to be explained by the presence of life. For example, consider the mass of the Sun. The mass of a star determines its luminosity, and using basic physics, one can compute that life as we know it on Earth is possible only if the Sun's mass falls into the narrow range between 1.6×10^{30} kg and 2.4×10^{30} kg. Otherwise Earth's climate would be colder than that of present-day Mars or hotter than that of present-day Venus. The actual mass of the Sun is 2.0×10^{30} kg, which at first glance appears to be a wild stroke of luck. Stellar masses run from 10^{29} kg to 10^{32} kg, so if the Sun's mass were chosen at random, it would have only a small chance of falling into the habitable range. But one can explain this coincidence by postulating an ensemble of planetary systems, together with the selection effect that we must find ourselves living on a habitable planet. Such observer-related selection effects are referred to as 'anthropic' [15]. Although the 'A-word' is notoriously controversial, physicists broadly agree that these selection effects cannot be neglected when testing fundamental theories. In this weak sense, it is obligatory.

What applies to planetary systems also applies to parallel universes. For example, as illustrated in Fig. 7.3, one can consider other universes in which the number of time and space dimensions is different from observed and show that this is inconsistent with life. Most, if not all, of the attributes set by symmetry-breaking appear to be fine-tuned. Changing their values by modest amounts would have resulted in a qualitatively different universe, in which we probably would not exist. If protons were 0.2% heavier, they would decay into neutrons, destabilizing atoms. If the electromagnetic force were 4% weaker, there would be no hydrogen and no normal stars. This is illustrated by Fig. 7.4, which shows constraints on the strong and electromagnetic coupling constants. If the weak interaction were much weaker, hydrogen would not exist; if it were much stronger, supernovae would fail to seed interstellar space with heavy elements. If the cosmological constant were much larger, our universe would have blown itself apart before

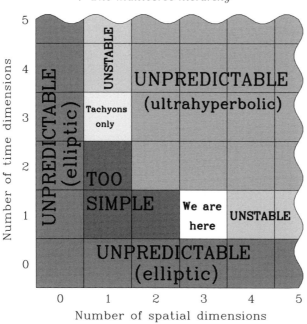

Fig. 7.3. Why we should not be surprised to find ourselves living in 3+1-dimensional spacetime. When the partial differential equations of nature are elliptic or ultrahyperbolic, physics has no predictive power for an observer. In the remaining (hyperbolic) cases, $n>3$ admits no stable atoms and $n<3$ may lack sufficient complexity for observers (no gravitational attraction, topological problems). From ref. [16].

galaxies could form. Indeed, most, if not all, of the parameters affecting low-energy physics appear fine-tuned, in the sense that changing them by modest amounts results in a qualitatively different universe.

Although the degree of fine-tuning is still debated [17–19], these examples suggest the existence of parallel universes with other values of some physical constants. If this is the case, physicists will never be able to determine the values of all physical constants from first principles. All they can do is compute probability distributions for what they should expect to find, taking selection effects into account. The result should be as generic as is consistent with our existence.

7.4 Level III: the 'many worlds' of quantum physics

If the fundamental equations of physics are *unitary*, as they so far appear to be, then the universe keeps dividing into parallel branches. Whenever a quantum event appears to have a random outcome, all outcomes should occur, one in each branch. This is illustrated by the bottom cartoon

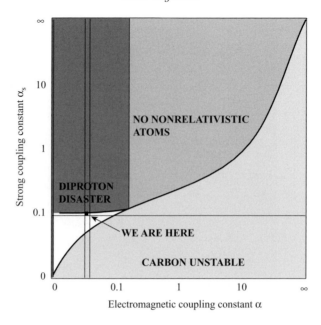

Fig. 7.4. Hints of fine-tuning for the parameters α and α_s, which determine the strengths of the electromagnetic force and the strong nuclear force, respectively. The observed values $(\alpha, \alpha_s) \approx (1/137, 0.1)$ are indicated by a filled square. Grand unified theories rule out everything except the narrow strip between the two vertical lines, and deuterium becomes unstable below the horizontal line. In the narrow shaded region to the very left, electromagnetism is weaker than gravity and therefore irrelevant. From ref. [16].

in Fig. 7.5 and corresponds to the Level III multiverse. Although more controversial than Level I and Level II, we will see that (surprisingly) this level adds no new types of universes.

7.4.1 The quantum conundrum

Despite the obvious successes of quantum theory, a heated debate rages about what it really means. The theory specifies the state of the universe not in classical terms, such as the positions and velocities of its particles, but in terms of a 'wave-function'. According to the Schrödinger equation, the state evolves over time in a fashion termed 'unitary', meaning that it rotates in an abstract, infinite-dimensional Hilbert space. Although quantum mechanics is often described as inherently random and uncertain, the wave-function evolves deterministically.

The difficulty is how to connect this wave-function with what we observe. Many legitimate wave-functions correspond to counter-intuitive situations,

Fig. 7.5. Difference between Level I and Level III. Level I parallel universes are far away in space, whereas those of Level III are right here, with quantum events causing classical reality to split into parallel storylines. Yet Level III adds no new storylines beyond Levels I or II.

such as a cat being dead and alive at the same time in a so-called 'superposition'. In the 1920s physicists explained away this weirdness by postulating that the wave-function 'collapses' into some definite classical state whenever someone makes an observation. This had the virtue of explaining observations, but it turned an elegant (unitary) theory into a messy (non-unitary) one, since there was no equation specifying when or how the collapse occurred. The intrinsic randomness commonly ascribed to quantum mechanics is the result of this postulate, triggering Einstein's objection that 'God does not play dice'.

Over the years, many physicists have abandoned this view in favour of one developed in 1957 by Hugh Everett [20]. He showed that the collapse postulate is unnecessary, so that unadulterated quantum theory need not pose any contradictions. Although it predicts that one classical reality gradually splits into superpositions of many such realities, observers subjectively experience this splitting merely as a slight randomness, with probabilities in exact agreement with those predicted by the old collapse postulate [21, 22].

This superposition of classical worlds, which is illustrated in Fig. 7.5, is the Level III multiverse.

7.4.2 What are Level III parallel universes like?

Everett's 'many worlds' interpretation has been puzzling physicists and philosophers for more than four decades. But the theory becomes easier to grasp when one distinguishes between two ways of viewing a physical theory: the outside view of a physicist studying its mathematical equations and the inside view of an observer living in the world described by the equations.[6]

From the outside perspective, the Level III multiverse is simple. There is only one wave-function, which evolves smoothly and deterministically without any splitting. The abstract quantum world described by this evolving wave-function contains a vast number of parallel classical storylines, as well as some quantum phenomena that lack a classical description. From the inside perspective, observers perceive only a tiny fraction of this full reality. They can view their own Level I universe, but a process called decoherence [23, 24] – which mimics wave-function collapse while preserving unitarity – prevents them from seeing Level III copies of themselves.

Whenever observers make a decision, quantum effects in their brains lead to a superposition of outcomes, such as 'Continue reading this article' and 'Put down this article'. From the outside perspective, the act of making a decision causes an observer to split into one person who keeps on reading and another one who does not. From their inside perspective, however, each of these *alter egos* is unaware of the other and notices the branching merely as a slight uncertainty in whether or not they continue to read.

Strangely, the same situation occurs in the Level I multiverse. You have evidently decided to keep on reading this article, but one of your *alter egos* in a distant galaxy put it down after the first paragraph. The only difference between Level I and Level III is where your doppelgängers reside.

---

6 The standard picture of the physical world corresponds to an intermediate viewpoint, that could be termed the *consensus view*. From your subjectively perceived internal perspective, the world turns upside down when you stand on your head and disappears when you close your eyes. Yet you subconsciously interpret your sensory inputs as though there is an external reality that is independent of your orientation, your location and your state of mind. Although this third view involves censorship (rejecting dreams), interpolation (between eye-blinks) and extrapolation (attributing existence to unseen cities), independent observers nonetheless appear to share this consensus view. Although the inside view looks black and white to a cat, iridescent to a bird seeing four primary colors, and even more different to a bee seeing polarized light or a bat using sonar, all agree on whether the door is open. The key challenge in physics is to derive this semiclassical consensus view from the fundamental equations specifying the internal perspective. Understanding the nature of human consciousness is an important challenge in its own right, but it may not be necessary for a fundamental theory of physics.

As illustrated in Fig. 7.5, they live elsewhere in 3-dimensional space in Level I, but on another quantum branch in infinite-dimensional Hilbert space in Level III.

7.4.3 Level III parallel universes: evidence and implications

The existence of the Level III multiverse depends on the crucial assumption that the time evolution of the wave-function is unitary. So far, experimenters have encountered no departures from unitarity. Indeed, in the past few decades they have confirmed it for ever larger systems, including carbon-60 'buckyball' molecules and kilometre-long optical fibres. On the theoretical side, the case for unitarity has been bolstered by the discovery of decoherence [25]. Some theorists who work on quantum gravity have questioned unitarity; one concern is that evaporating black holes might destroy information, which would be a non-unitary process. But a recent breakthrough in string theory – known as AdS/CFT correspondence – suggests that even quantum gravity is unitary. If so, black holes do not destroy information but merely transmit it elsewhere.

If physics is unitary, then the standard picture of how quantum fluctuations operated early in the big bang must change. Instead of generating initial conditions at random, these fluctuations generated a quantum superposition of all possible initial conditions, which coexisted simultaneously. Decoherence then caused these initial conditions to behave classically in separate quantum branches. The crucial point is that the distribution of outcomes on different quantum branches in a given Hubble volume (Level III) is identical to the distribution of outcomes in different Hubble volumes within a single quantum branch (Level I). This property of the quantum fluctuations is known in statistical mechanics as ergodicity.

The same reasoning applies to Level II. The process of symmetry-breaking did not produce a unique outcome but rather a superposition of all outcomes, which rapidly went their separate ways. So, if physical constants and spacetime dimensionality can vary among parallel quantum branches at Level III, then they will also vary among parallel universes at Level II. Thus the Level III multiverse adds nothing new beyond Levels I and II, just more indistinguishable copies of the same universes. The debate about Everett's theory therefore seems to be ending in a grand anticlimax, with the discovery of less controversial multiverses (Levels I and II) that are equally large.

Physicists are only beginning to explore the implications of this. For instance, consider the long-standing question of whether the number of universes exponentially increases over time. The surprising answer is no. From the outside perspective, there is of course only one quantum universe. From the inside perspective, what matters is the number of universes that are distinguishable at a given instant – that is, the number of noticeably different Hubble volumes. At the quantum level, there are $10^{10^{118}}$ universes with temperatures below 10^8 K, which is vast but finite. The evolution of the wave-function therefore corresponds to a never-ending sliding from one of the $10^{10^{118}}$ states to another. First you are in the universe in which you are reading this sentence. Next you are in the universe in which you are reading another sentence. The observer in the second universe is identical to the one in the first except for an extra instant of memories. All possible states exist at every instant, so the passage of time may be in the eye of the beholder – an idea explored by various authors [26–28]. The multiverse framework may thus prove essential to understanding the nature of time.

7.4.4 Two world views

Figure 7.6 illustrates that the debate over how classical mechanics emerges from quantum mechanics is just a small piece of a larger puzzle. Indeed, the debate over the interpretation of quantum mechanics – and the broader issue of parallel universes – is in a sense the tip of an iceberg. For there is a still deeper question that arguably goes as far back as Plato and Aristotle. This concerns the status of mathematics and how it relates to physical reality.

Aristotelian paradigm The internal perspective is physically real, while the external perspective and all its mathematical language is merely a useful approximation.

Platonic paradigm The external perspective (the mathematical structure) is physically real, while the internal perspective and all the human language we use to describe it is merely a useful approximation for describing our subjective perceptions.

What is more basic – the internal or external perspective, human language or mathematical language? Your answer will determine how you feel about parallel universes. Our feeling that the Level III multiverse is 'weird' merely reflects the extreme difference between the internal and external perspectives. We may break the symmetry by calling the latter weird, because we were all indoctrinated with the Aristotelian paradigm as children, long before we even heard of mathematics. If this is true, there can never be a 'Theory of Everything' (TOE), since one is ultimately just explaining certain verbal statements by other verbal statements. This is known as the infinite regress problem [29].

7 The multiverse hierarchy

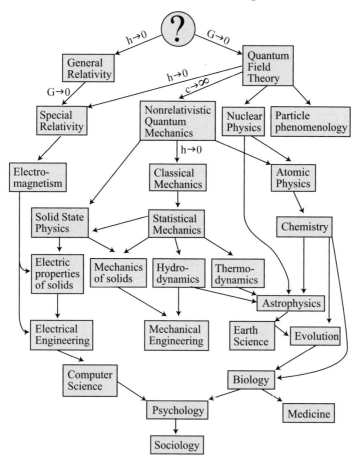

Fig. 7.6. Theories can be crudely organized into a family tree where each might, at least in principle, be derivable from more fundamental ones above it. For example, classical mechanics can be obtained from special relativity in the approximation that the speed of light c is infinite. Most of the arrows are less well understood. All these theories have two components: mathematical equations and words that explain how they are connected to what we observe. At each level in the hierarchy of theories, new words (e.g., protons, atoms, cells, organisms, cultures) are introduced because they are convenient, capturing the essence of what is going on without recourse to the more fundamental theory above it. It is important to remember, however, that it is humans who introduce these concepts and the words for them; in principle, everything could have been derived from the fundamental theory at the top of the tree, although such an extreme reductionist approach would be useless in practice. Crudely speaking, the ratio of equations to words decreases as we move down the tree, dropping to near zero for highly applied fields, such as medicine and sociology. In contrast, theories near the top are highly mathematical, and physicists are still struggling to understand the concepts, if any, in terms which we can understand. The Holy Grail of physics is to find a 'Theory of Everything' from which all else can be derived. If such a theory exists at all, it should replace the big question mark at the top of the theory tree. However, something is missing here, since we lack a consistent theory unifying gravity with quantum mechanics.

On the other hand, if you prefer the Platonic paradigm, you should find multiverses natural. In this case, all of physics is ultimately a mathematics problem, since an infinitely intelligent mathematician – given the fundamental equations of the cosmos – could in principle *compute* the internal perspective, i.e. what self-aware observers the universe would contain, what they would perceive, and what language they would invent to describe their perceptions to one another. In other words, there is a TOE at the top of the tree in Fig. 7.6, whose axioms are purely mathematical.

7.5 Level IV: other mathematical structures

If one accepts the Platonist paradigm and believes that there really is a TOE at the top of Fig. 7.6, even though we have not found the correct equations yet, then this question arises: why these particular equations and not others? The Level IV multiverse involves the idea of mathematical democracy, in which universes governed by other equations are equally real. This implies the notion that a mathematical structure and the physical world are in some sense identical. It also means that mathematical structures are 'out there', in the sense that mathematicians discover them rather than create them.

7.5.1 What is a mathematical structure?

Many of us think of mathematics as a bag of tricks that we learned in school for manipulating numbers. Yet most mathematicians have a very different view of their field. They study more abstract objects, such as functions, sets, spaces and operators and try to prove theorems about the relations between them. Indeed, some modern mathematics papers are so abstract that the only numbers you will find in them are the page numbers! Despite the plethora of mathematical structures with intimidating names like orbifolds and Killing fields, a striking underlying unity has emerged in the twentieth century: *all* mathematical structures are just special cases of one and the same thing, so-called formal systems. A formal system consists of abstract symbols and rules for manipulating them, specifying how new strings of symbols referred to as theorems can be derived from given ones referred to as axioms. This historical development represented a form of deconstructionism, since it stripped away all meaning and interpretation that had traditionally been given to mathematical structures and distilled out only the abstract relations capturing their very essence. As a result, computers can now prove theorems about geometry without having any physical intuition whatsoever about what space is like.

7.5.2 Is the physical world a mathematical structure?

Although traditionally taken for granted by many theoretical physicists, the notion that the physical world (specifically, the Level III multiverse) is a mathematical structure is deep and far-reaching. It means that mathematical equations describe not merely some limited aspects of the physical world, but *all* aspects of it, leaving no freedom for, say, miracles or free will in the traditional sense. Thus there is some mathematical structure that is *isomorphic* (and hence equivalent) to our physical world, with each physical entity having a unique counterpart in the mathematical structure and vice versa.

Let us consider some examples. A century ago, when classical physics still reigned supreme, many scientists believed that physical space was isomorphic to the three-dimensional Euclidean space R^3. Moreover, some thought that all forms of matter in our universe corresponded to various classical *fields*: the electric field, the magnetic field and perhaps a few undiscovered ones, mathematically corresponding to functions on R^3. In this view (later proven incorrect), dense clumps of matter such as atoms were simply regions in space where some fields were strong. These fields evolved deterministically according to some partial differential equations, and observers perceived this as things moving around and events taking place. However, fields in 3-dimensional space cannot be the mathematical structure corresponding to our universe, because a mathematical structure is an abstract, immutable entity existing outside of space and time. Our familiar perspective of a 3-dimensional space, where events unfold, is equivalent to a 4-dimensional spacetime, so the mathematical structure must be fields in 4-dimensional space. In other words, if history were a movie, the mathematical structure would not correspond to a single frame of it, but to the entire videotape.

Given a mathematical structure, we will say that it has *physical existence* if any self-aware substructure (SAS) within it subjectively perceives itself as living in a physically real world. In the above classical physics example, an SAS would be a tube through spacetime, a thick version of its worldline. Within the tube, the fields would exhibit certain complex behaviour, corresponding to storing and processing information about the field-values in the surroundings, and at each position along the tube these processes would give rise to the familiar but mysterious sensation of self-awareness. The SAS would perceive this 1-dimensional string of perceptions along the tube as passage of time.

Although this example illustrates how our physical world can *be* a mathematical structure, this particular structure (fields in 4-dimensional space) is now known to be the wrong one. After realizing that spacetime could be curved, Einstein searched for a unified field theory where the universe was a four-dimensional pseudo-Riemannian manifold with tensor fields. However, this failed to account for the observed behaviour of atoms. According to quantum field theory, the modern synthesis of special relativity theory and quantum theory, our universe (in this case the Level III multiverse) is a mathematical structure with an algebra of operator-valued fields. Here the question of what constitutes an SAS is more subtle [30]. However, this fails to describe black hole evaporation, the first instance of the big bang and other quantum gravity phenomena. So the true mathematical structure isomorphic to our universe, if it exists, has not yet been found.

7.5.3 Mathematical democracy

Suppose that our physical world really is a mathematical structure and that you are an SAS within it. This means that this particular structure enjoys physical as well as mathematical existence. What about all the other possible mathematical structures? Do they too enjoy physical existence? If not, there would be a fundamental, unexplained ontological asymmetry built into the very heart of reality, splitting mathematical structures into two classes: those with and without physical existence. As a way out of this philosophical conundrum, I have suggested [18] that complete mathematical democracy holds – that mathematical existence and physical existence are equivalent, so that *all* mathematical structures exist physically as well. This is the Level IV multiverse. It can be viewed as a form of radical Platonism, asserting that the mathematical structures in Plato's *realm of ideas* and Rucker's *mindscape* [31] exist 'out there' in a physical sense [32]. This casts the so-called modal realism theory of David Lewis [33] in mathematical terms, akin to what Barrow [34, 35] refers to as 'pi in the sky'. If this theory is correct, then – since it has no free parameters – all properties of all parallel universes (including the subjective perceptions of every SAS) could in principle be derived by an infinitely intelligent mathematician.

7.5.4 Evidence for a Level IV multiverse

Why should we believe in Level IV? Logically, it rests on the two following separate assumptions.

Assumption 1 The physical world (specifically our Level III multiverse) is a mathematical structure.

Assumption 2 All mathematical structures exist 'out there' in the same sense (mathematical democracy).

In a famous essay, Wigner [36] argued that 'the enormous usefulness of mathematics in the natural sciences is something bordering on the mysterious' and that 'there is no rational explanation for it'. This argument can be taken as support for Assumption 1; here the utility of mathematics for describing the physical world is a natural consequence of the fact that the latter *is* a mathematical structure, which we are uncovering bit by bit. The various approximations that constitute our current physics theories are successful because mathematical structures can provide good approximations to how an SAS will perceive more complex mathematical structures. In other words, our successful theories are not mathematics approximating physics but mathematics approximating mathematics. Wigner's observation is unlikely to be based on fluke coincidences, since far more mathematical regularity has been discovered in nature in the decades since he made it.

A second argument supporting Assumption 1 is that abstract mathematics is so general that *any* TOE that is definable in purely formal terms is also a mathematical structure. For instance, a TOE involving a set of different types of entities (words, say) and relations between them (additional words) is a set-theoretical model, and one can generally find a formal system of which it is a model. This argument also makes Assumption 2 more appealing, since it implies that *any* conceivable parallel universe theory can be described at Level IV. The Level IV multiverse, termed the 'ultimate ensemble theory' in ref. [16] since it subsumes all other ensembles, therefore brings closure to the hierarchy of multiverses, and there cannot be a Level V. Considering an ensemble of mathematical structures does not add anything new, since this is still just another mathematical structure.

What about the notion that our universe is a computer simulation? This idea occurs frequently in science fiction and has been substantially elaborated [37,38]. The information content (memory state) of a digital computer is a long string of bits, equivalent to some large but finite integer n written in binary form (e.g. $1001011100111001\ldots$). The information-processing of a computer is a deterministic rule for repeatedly changing each memory state into another one. Mathematically, it is a function f mapping the integers onto themselves that is iterated: $n \mapsto f(n) \mapsto f(f(n)) \mapsto \cdots$. In other words, even the most powerful computer simulation is just another special case of a mathematical structure, and this is already included in the Level IV multiverse.

A second argument for Assumption 2 is that if two entities are isomorphic, then there is no meaningful sense in which they are not the same [39]. This applies when the entities in question are a physical universe and the mathematical structure describing it. To avoid the conclusion that mathematical and physical existence are equivalent, one would need to argue that our universe is somehow made of stuff perfectly described by a mathematical structure, but which also has other properties that are not described by it. However, this violates Assumption 1 and implies that either it is isomorphic to a more complicated mathematical structure or it is not mathematical at all.

Having universes dance to the tune of all possible equations also resolves the fine-tuning problem, even at the fundamental equation level. Although many (if not most) mathematical structures are likely to be devoid of an SAS, failing to provide the complexity, stability and predictability that it requires, we know we must inhabit a mathematical structure capable of supporting life. Because of this selection effect, the answer to the question 'what is it that breathes fire into the equations and makes a universe for them to describe?' [40] would then be 'you, the SAS'.

7.5.5 What are Level IV parallel universes like?

We can test and potentially rule out any theory by computing probability distributions for our future perceptions – given our past perceptions – and comparing the predictions with the observed outcome. In a multiverse theory, there is typically more than one SAS that has experienced a past life identical to yours, so there is no way to determine which one is you. To make predictions, you therefore have to compute what fractions of them will perceive what in the future. This leads to the following possibilities.

Prediction 1 The mathematical structure describing our world is the most generic one that is consistent with our observations.

Prediction 2 Our future observations are the most generic ones that are consistent with our past observations.

Prediction 3 Our past observations are the most generic ones that are consistent with our existence.

We will return to the problem of what 'generic' means (i.e. the measure problem) later. However, one striking feature of mathematical structures, discussed in detail in ref. [16], is that the sort of symmetry and invariance properties that are responsible for the simplicity and orderliness of our universe tend to be more the rule than the exception – mathematical structures have them by default and complicated additional axioms must be added to make them go away. Because of this, as well as selection effects, we should not necessarily expect life in the Level IV multiverse to be disordered.

7.6 Discussion

We have seen that scientific theories of parallel universes form a four-level hierarchy, in which universes become progressively more different from our own. They might have different initial conditions (Level I), different effective physical laws, constants and particles (Level II), or different fundamental physical laws (Level IV). It is ironic that Level III is the one that has drawn most criticism, because it is the only one that adds no qualitatively new types of universe. Whereas the Level I universes join seemlessly, there are clear demarcations between those within Level II (caused by inflation) and Level III (caused by decoherence). The Level IV universes are completely disconnected and need to be considered together only for predicting your future, since 'you' may exist in more than one of them.

7.6.1 Future prospects

There are ample future prospects for testing and perhaps ruling out these multiverse theories. In the coming decade, dramatically improved cosmological measurements of the microwave background radiation and the large-scale matter distribution will test both Level I (by further constraining the curvature and topology of space) and Level II (by providing stringent tests of inflation). Progress in both astrophysics and high-energy physics should also clarify the extent to which various physical constants are fine-tuned, thereby weakening or strengthening the case for Level II. If the current effort to build quantum computers succeeds, it will provide further evidence for Level III, since they would essentially exploit the parallelism of the Level III multiverse for parallel computation [27]. Conversely, experimental evidence of unitarity violation would rule out Level III. Finally, success or failure in the grand challenge of modern physics, unifying general relativity and quantum field theory, will shed more light on Level IV. Either we will eventually find a mathematical structure which matches our universe, or the unreasonable effectiveness of mathematics will be found to be limited and we will have to abandon Level IV.

7.6.2 The measure problem

There are also interesting theoretical issues to resolve within the multiverse theories, in particular the *measure problem*. As multiverse theories gain credence, the sticky issue of how to compute probabilities in physics is growing from a minor nuisance into a major embarrassment. If there are indeed many identical copies of you, the traditional notion of determinism evaporates. You could not compute your own future even if you had complete knowledge of the entire state of the multiverse, because there is no way for you to determine which of these copies is you. All you can predict are probabilities for what you would observe. If an outcome has a probability of 50%, this means that half the observers observe that outcome.

Unfortunately, it is not an easy task to compute what fraction of the infinitely many observers perceive what. The answer depends on the order in which you count them. By analogy, the fraction of the integers that are even is 50% if you order them numerically (1, 2, 3, 4,...), but approaches 100% if you sort them digit by digit, the way your word processor would (1, 10, 100, 1000,...). When observers reside in disconnected universes, there is no obviously natural way in which to order them. Instead one must sample from the different universes with some statistical weights referred to as a 'measure'.

This problem crops up in a mild manner at Level I, becomes severe at Level II [41], has caused much debate at Level III [21, 22, 42] and is horrendous at Level IV. At Level II, for instance, several people have published predictions for the probability distributions of various cosmological parameters. They have argued that the different universes that have inflated by different amounts should be given statistical weights proportional to their volume [2]. On the other hand, $2 \times \infty = \infty$, so there is no objective sense in which an infinite universe that has expanded by a factor of two has become larger. Moreover, a finite universe with the topology of a torus is equivalent to a periodic universe with infinite volume, both mathematically and from the perspective of an observer within it. So why should its infinitely smaller volume give it zero statistical weight? After all, even in the Level I multiverse, Hubble volumes start repeating (albeit randomly rather than periodically) after about $10^{10^{118}}$ m. The problem of assigning statistical weights to different mathematical structures at Level IV is even more difficult. The fact that our universe seems relatively simple has led many people to suggest that the correct measure must somehow involve complexity.

7.6.3 The pros and cons of parallel universes

So should you believe in parallel universes? The principal arguments against them are that they are wasteful and that they are weird. The wastefulness argument is that multiverse theories are vulnerable to Ockham razor because they postulate the existence of other worlds that we can never observe. Yet this argument can be turned around. For what precisely would nature be wasting? Certainly not space, mass or atoms – the uncontroversial Level I multiverse already contains an infinite amount of all three.

The real issue here is the apparent reduction in simplicity. One might worry about all the information necessary to specify all those unseen worlds. But an entire ensemble is often much simpler than one of its members. This principle can be stated more formally using the notion of algorithmic information content. The algorithmic information content in a number is, roughly speaking, the length of the shortest computer program that will produce that number as output. For example, consider the set of all integers. Naïvely, you might think that a single number is simpler than the whole set of numbers, but the set can be generated by a trivial computer program, whereas a single number can be hugely long. Therefore, the whole set is actually simpler. Similarly, the set of all solutions to Einstein's field equations is simpler than a specific solution. The former is described by a few equations, whereas the latter requires the specification of vast amounts of initial data on some hypersurface.

The lesson is that complexity increases when we restrict our attention to one particular element in an ensemble, thereby losing the symmetry and simplicity that were inherent in the totality of all the elements taken together. In this sense, the higher-level multiverses are simpler. Going from our universe to the Level I multiverse eliminates the need to specify initial conditions, upgrading to Level II eliminates the need to specify physical constants, and the Level IV multiverse eliminates the need to specify anything at all. The opulence of complexity is all in the subjective perceptions of observers [43].

The weirdness objection is aesthetic rather than scientific and only makes sense in the Aristotelian worldview. Yet when we ask a profound question about the nature of reality, we surely expect an answer that sounds strange. Evolution provided us with intuition for the everyday physics that had survival value for our distant ancestors, so whenever we venture beyond the everyday world, we should expect it to seem bizarre. Thanks to clever inventions, we have glimpsed slightly beyond our normal subjective view and thereby encountered bizarre phenomena (e.g. at high speeds, small and large scales, low and high temperatures).

A common feature of all four multiverse levels is that the simplest and arguably most elegant theory involves parallel universes by default. To deny the existence of those universes, one needs to complicate the theory by adding experimentally unsupported processes and *ad hoc* postulates: finite space, wave-function collapse, ontological asymmetry, etc. Our judgement therefore comes down to which we find more wasteful and inelegant: many worlds or many words. Perhaps we will gradually become more used to the weird ways of our cosmos, and even find its strangeness to be part of its charm.

Acknowledgements

The author wishes to thank Anthony Aguirre, Aaron Classens, Marius Cohen, Angelica de Oliveira-Costa, Alan Guth, Shamit Kachru, Andrei Linde, George Musser, David Raub, Martin Rees, Harold Shapiro, Alex Vilenkin and Frank Wilczek for stimulating discussions. This work was supported by NSF grants AST-0071213 and AST-0134999, NASA grants NAG5-9194 and NAG5-11099, a fellowship from the David and Lucile Packard Foundation and a Cottrell Scholarship from Research Corporation.

References

[1] G. B. Brundrit. Life in the infinite universe. *Quart. J. Roy. Astron. Soc.* **20** (1979), 37.
[2] J. Garriga and A. Vilenkin. A prescription for probabilities in eternal inflation. *Phys. Rev.* **D 64** (2001), 023507.
[3] G. F. R. Ellis. *Are We Alone?* (New York: Perseus, 1995).
[4] A. de Oliveira-Costa, M. Tegmark, M. Zadamiaga and A. Hamilton. Significance of the largest scale CMB fluctuations in WMAP. *Phys. Rev.* **D 69** (2004), 063516.
[5] N. J. Cornish, D. N. Spergel, G. D. Starkman and E. Kometsu. Constraining the topology of the universe. *Phys. Rev. Lett.* **92** (2004), 201302.
[6] A. Linde. The self-reproducing inflationary universe. *Sci. Am.* **271** (1994), 32.
[7] A. H. Guth and D. I. Kaiser. Inflationary cosmology: Exploring the universe from the smallest to the largest scales. *Science*, **307** (2005), 884.
[8] M. A. Bucher and D. N. Spergel. Inflation in a low-density universe. *Sci. Am.* **280** (1999), 62.
[9] R. Bousso and J. Polchinski. Quantization of four-form fluxes and dynamical neutralization of the cosmological constant. *JHEP*, **0006** (2000), 006.
[10] L. Susskind. This volume (2007) [hep-th/0302219].
[11] R. C. Tolman. *Relativity, Thermodynamics, and Cosmology* (Oxford: Clarendon Press, 1934).
[12] J. A. Wheeler. Beyond the end of time. In *Black Holes, Gravitational Waves and Cosmology: An Introduction to Current Research*, eds. M. Rees, R. Ruffini and J. A. Wheeler (New York: Gordon and Breach, 1974).
[13] P. J. Steinhardt and N. Turok. A cyclic model of the universe. *Science*, **296** (2002), 1436.
[14] L. Smolin. *The Life of the Cosmos* (Oxford: Oxford University Press, 1997).
[15] B. Carter. Number coincidences and the anthropic principle in cosmology. In *Confrontation of Cosmological Theory with Observational Data*, ed. M. S. Longair (Dordrecht: Reidel, 1974), p. 291.
[16] M. Tegmark. On the dimensionality of spacetime. *Class. Quant. Grav.* **14** (1997), L69.

[17] J. D. Barrow and F. J. Tipler. *The Anthropic Cosmological Principle* (Oxford: Clarendon, 1986).
[18] M. Tegmark. Is 'the theory of everything' merely the ultimate ensemble theory? *Ann. Phys.* **270** (1998), 1.
[19] C. J. Hogan. Why the Universe is Just So. *Rev. Mod. Phys.* **72** (2000), 1149.
[20] H. Everett. Relative state formulations of quantum theory. *Rev. Mod. Phys.* **29** (1957), 454.
[21] H. Everett. In *The Many-Worlds Interpretation of Quantum Mechanics*, eds. B. S. DeWitt and N. Graham (Princeton, NJ: Princeton University Press, 1973).
[22] B. DeWitt. The Everett interpretation of quantum mechanics. In *Science and Ultimate Reality: From Quantum to Cosmos*, eds. J. D. Barrow, P. C. W. Davies and C. L. Harper (Cambridge: Cambridge University Press, 2003), pp. 167–98.
[23] H. D. Zeh. On the interpretation of measurement in quantum theory. *Found. Phys.* **1** (1970), 69.
[24] D. Giulini, E. Joos, C. Kiefer, J. Kupsch, I. O. Stamatescu and H. D. Zeh. *Decoherence and the Appearance of a Classical World in Quantum Theory* (Berlin: Springer, 1996).
[25] M. Tegmark and J. A. Wheeler. 100 years of quantum mysteries. *Sci. Am.* **284** (2001), 68.
[26] G. Egan. *Permutation City* (New York: Harper, 1995).
[27] D. Deutsch. *The Fabric of Reality* (New York: Allen Lane, 1997).
[28] J. B. Barbour. *The End of Time* (Oxford: Oxford University Press, 2001).
[29] R. Nozick. *Philosophical Explanations* (Cambridge, MA: Harvard University Press, 1981).
[30] M. Tegmark. The importance of quantum decoherence in brain processes. *Phys. Rev.* **E 61** (2000), 4194.
[31] R. Rucker. *Infinity and the Mind* (Boston: Birkhauser, 1982).
[32] P. C. W. Davies. *The Mind of God* (New York: Simon and Schuster, 1992).
[33] D. Lewis. *On the Plurality of Worlds* (Oxford: Blackwell, 1986).
[34] J. D. Barrow. *Theories of Everything* (New York: Ballantine, 1991).
[35] J. D. Barrow. *Pi in the Sky* (Oxford: Clarendon, 1992).
[36] E. P. Wigner. *Symmetries and Reflections* (Cambridge, MA: MIT Press, 1967).
[37] J. Schmidthuber. A computer scientist's view of life, the universe, and everything. In *Foundations of Computer Science: Potential Theory Cognition*, ed. C. Freksa. Lecture Notes in Computer Science (Berlin: Springer, 1997).
[38] S. Wolfram. *A New Kind of Science* (New York: Wolfram Media, 2002).
[39] M. Cohen. Master's thesis, Department of Philosophy, Ben Gurion University of the Negev (2003).
[40] S. Hawking. *A Brief History of Time* (New York: Touchstone, 1993).
[41] M. Tegmark. What does inflation really predict? *J. Cosmol. Astropart. Phys.* **0504** (2004), 001.
[42] V. F. Mukhanov. This volume (2007).
[43] M. Tegmark. Does the universe in fact contain almost no information? *Found. Phys. Lett.* **9** (1996), 25.

8
The inflationary multiverse

Andrei Linde
Department of Physics, Stanford University

8.1 Introduction

At the beginning of the 1980s, when the inflationary theory was first proposed, one of our main goals was to explain the amazing uniformity of the Universe. We were trying to find out why the Universe looks approximately the same in all directions. Of course, locally the Universe does not look uniform – there are such large deviations from uniformity as planets, stars and galaxies. But if one considers the density of matter on scales comparable to the size of the observable Universe, $l_{\text{obs}} \sim 10^{28}$ cm, one finds that this is uniform to an accuracy better than one part in 10 000. The most surprising thing about this is that, according to the standard big bang theory, the distant parts of the Universe which we can see with a powerful telescope were not in causal contact at the time of the big bang and could not have been in such contact until very late stages of cosmic evolution. So one could only wonder what made these distant parts of the Universe so similar to each other.

In the absence of any reasonable explanation, cosmologists invented the so-called 'cosmological principle', which claims that the Universe *must* be uniform. But the Universe is not perfectly uniform, since it contains inhomogeneities – such as stars and galaxies – which are crucial for life. Because of these small but important violations, the cosmological principle cannot be a true principle of nature, just like a person who takes only small bribes cannot be called a man of principle.

Even though the cosmological principle could not explain the observed properties of the Universe, it was taken for granted by almost all scientists. We believed that the Universe looks the same everywhere and that the physical laws in all of its parts are identical to those in the vicinity of the Solar System. We were looking for a unique and beautiful theory that

Universe or Multiverse?, ed. Bernard Carr. Published by Cambridge University Press.
© Cambridge University Press 2007.

would unambiguously predict the observed values for all parameters of all elementary particles, not leaving any room for pure chance.

However, most of the parameters describing elementary particles look more like a collection of random numbers than a unique manifestation of some hidden harmony of nature. Also it was pointed out long ago that a minor change (by a factor of two or three) in the mass of the electron, the fine-structure constant, the strong-interaction constant or the gravitational constant would lead to a world in which life as we know it could never have arisen. Adding or subtracting even a single spatial dimension of the same type as the usual three dimensions would make planetary systems impossible. Indeed, in spacetimes with dimensionality $d > 4$, the gravitational force between bodies falls off faster than r^{-2}, while in spacetimes with $d < 4$, general relativity tell us that such forces are absent altogether. This rules out the existence of stable planetary systems for $d \neq 4$. Furthermore, in order for life as we know it to exist, it is necessary that the Universe be sufficiently large, flat, homogeneous and isotropic. These facts, as well as a number of other observations, lie at the foundation of the so-called anthropic principle [1–3]. According to this principle, we observe the Universe to be as it is because only in one like ours could observers exist.

Many scientists are still ashamed of using the anthropic principle. Just as the friends of Harry Potter were afraid of saying the name 'Voldemort', the opponents of the anthropic principle often say that they do not want to use the 'A' word in their research. This critical attitude is quite understandable. Historically, the anthropic principle was often associated with the idea that the Universe was created many times until the final success. It was not clear who did it and why it was necessary to make the Universe suitable for our existence. Moreover, it would be much simpler to create proper conditions for our existence in a small vicinity of the Solar System rather than in the whole Universe. Why would one need to work so hard?

Fortunately, most of the problems associated with the anthropic principle were resolved [4–7] soon after the invention of inflationary cosmology [8–12]. Inflationary theory was able to explain the homogeneity of our part of the Universe, while simultaneously predicting that on a very large scale, much greater than $l_{\text{obs}} \sim 10^{28}$ cm, the Universe can be completely inhomogeneous, looking not like a sphere but like a huge growing fractal. The different parts of this fractal are enormous and may have dramatically different properties. They are connected to each other, but the distance between them is so large that for all practical purposes they look like separate universes.

Thus, although inflationary theory was able to explain the local homogeneity of the Universe, many of its versions predicted that on super-large scales one has a 'multiverse', consisting of many universes with different properties. In the context of this scenario, we were able for the first time to make sense of the basic premise of the anthropic principle: there is not just one copy of the Universe – we actually have a choice!

8.2 Chaotic inflation

In order to explain this picture in more detail, I will first describe the basic features of inflation. I will concentrate on the simplest version – the chaotic inflation scenario [11]. To explain the main idea of chaotic inflation, let us consider the simplest model of a scalar field ϕ, with a mass m and potential energy density $V(\phi) = \frac{1}{2}m^2\phi^2$, as shown in Fig. 8.1. Since this function has a minimum at $\phi = 0$, one may expect the scalar field to oscillate near this minimum. This is indeed the case if the Universe does not expand. However, one can show that – in a rapidly expanding Universe – the scalar field moves

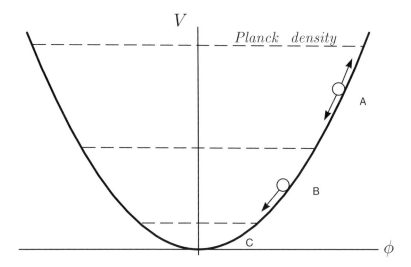

Fig. 8.1. Motion of the scalar field in the theory with $V(\phi) = \frac{1}{2}m^2\phi^2$. If the potential energy density of the field is greater than the Planck density, $\rho_p \sim M_p^4 \sim 10^{94}$ g cm^{-3}, quantum fluctuations of spacetime are so strong that one cannot describe it in the usual terms. Such a state is called spacetime foam. At a somewhat smaller energy density (region A: $mM_p^3 < V(\phi) < M_p^4$) quantum fluctuations of spacetime are small but quantum fluctuations of the scalar field ϕ may be large. Jumps of the scalar field due to quantum fluctuations lead to eternal self-reproduction of the inflationary universe. At even smaller values of $V(\phi)$ (region B: $m^2 M_p^2 < V(\phi) < mM_p^3$) fluctuations of the field ϕ are small; it moves down slowly like a ball in a viscous liquid. Inflation occurs in both regions A and B. Finally, near the minimum of $V(\phi)$ (region C) the scalar field rapidly oscillates, creates pairs of elementary particles, and the Universe becomes hot.

down very slowly, like a ball in a viscous liquid, with the viscosity being proportional to the speed of expansion.

There are two equations which describe evolution of a homogeneous scalar field in our model: the field equation,

$$\ddot{\phi} + 3H\dot{\phi} = -m^2\phi, \tag{8.1}$$

and the Einstein equation,

$$H^2 + \frac{k}{a^2} = \frac{8\pi}{3M_\text{p}^2}\left(\frac{1}{2}\dot{\phi}^2 + V(\phi)\right). \tag{8.2}$$

Here a dot denotes d/dt, $M_\text{p} = G^{-1/2}$ is the Planck mass (using units with $\hbar = c = 1$), $a(t)$ is the cosmic scale factor, $H = \dot{a}/a$ is the Hubble parameter and $k = -1, 0, 1$ for an open, flat or closed universe, respectively. The first equation is similar to the equation of motion for a harmonic oscillator, where instead of $x(t)$ we have $\phi(t)$, so the term $3H\dot{\phi}$ is like a friction effect.

If the scalar field ϕ is initially large, the Hubble parameter H is also large from Eq. (8.1). This means that the friction term is large, so the scalar field is moving very slowly. At this stage, the energy density of the scalar field remains almost constant and the expansion of the Universe continues much faster than in the old cosmological theory. Due to the rapid growth of the scale of the Universe and slow motion of the field, soon after the beginning of this regime one has $\ddot{\phi} \ll 3H\dot{\phi}$, $H^2 \gg k/a^2$ and $\dot{\phi}^2 \ll m^2\phi^2$, so the system of equations can be simplified to

$$3\frac{\dot{a}}{a}\dot{\phi} = -m^2\phi, \qquad H = \frac{\dot{a}}{a} = \frac{2m\phi}{M_p}\sqrt{\frac{\pi}{3}}. \tag{8.3}$$

The second equation shows that the scale factor in this regime grows approximately as

$$a \sim e^{Ht}, \qquad H = \frac{2m\phi}{M_\text{p}}\sqrt{\frac{\pi}{3}}. \tag{8.4}$$

This stage of exponentially rapid expansion of the universe is called inflation.

When the field ϕ becomes sufficiently small, H and the viscosity become small, inflation ends and the scalar field begins to oscillate near the minimum of $V(\phi)$. As any rapidly oscillating classical field, it loses its energy by creating pairs of elementary particles. These particles interact with each other and come to a state of thermal equilibrium with some temperature T. From this time on, the Universe can be described by the standard hot big bang theory.

8 The inflationary multiverse

The main difference between inflationary theory and the old cosmology becomes clear when one calculates the size of a typical domain at the end of inflation. Even if the initial size of the inflationary Universe was as small as the Planck scale, $l_p \sim 10^{-33}$ cm, one can show that after 10^{-30} s of inflation this acquires a huge size of $l \sim 10^{10^{12}}$ cm. This makes the Universe almost exactly flat and homogeneous on the large scale, because all inhomogeneities were stretched by a factor of $10^{10^{12}}$. This number is model-dependent, but in all realistic models the size of the Universe after inflation appears to be many orders of magnitude greater than the size of the part of the Universe which we can see now, $l_{\text{obs}} \sim 10^{28}$ cm. This solves most of the problems of the old cosmological theory [13].

If the Universe initially consisted of many domains, with a chaotically distributed scalar field ϕ, then the domains where the scalar field was too small never inflated, so they remain small. The main contribution to the total volume of the Universe will be given by the domains which originally contained a large scalar field. Inflation of such domains creates huge homogeneous islands out of the initial chaos, each one being much greater than the size of the observable part of the Universe. That is why I call this scenario 'chaotic inflation'.

In addition to the scalar field driving inflation, realistic models of elementary particles involve many other scalar fields ϕ_i. The final values acquired by these fields after the cosmological evolution are determined by the position of the minima of their potential energy density $V(\varphi_i)$. In the simplest models, the potential $V(\varphi_i)$ has only one minimum. However, in general, $V(\varphi_i)$ may have many different minima. For example, in the simplest supersymmetric theory unifying weak, strong and electromagnetic interactions, the effective potential has dozens of different minima of equal depth with respect to the two scalar fields, Φ and φ. If the scalar fields fall to different minima in different parts of the Universe (a process called spontaneous symmetry-breaking), the masses of elementary particles and the laws describing their interactions will be different in these parts. Each of the parts becomes exponentially large because of inflation. In some of them, there will be no difference between weak, strong and electromagnetic interactions, and life of our type will be impossible. Other parts will be similar to the one where we live [14].

This means that, even if we are able to find the final theory of everything, we will be unable to determine uniquely properties of elementary particles; the Universe may consist of different exponentially large domains where the properties of elementary particles are different. This is an important step towards the justification of the anthropic principle. A further step can be made if one takes into account quantum fluctuations produced during inflation.

8.3 Inflationary quantum fluctuations

According to quantum field theory, empty space is not entirely empty. It is filled with quantum fluctuations of all types of physical fields. The wavelengths of all quantum fluctuations of the scalar field ϕ grow exponentially during inflation. When the wavelength of any particular fluctuation becomes greater than H^{-1}, the fluctuation stops oscillating and its amplitude freezes at some non-zero value $\delta\phi(x)$ because of the large friction term $3H\dot\phi$ in the equation of motion of the field. The amplitude of this fluctuation then remains almost unchanged for a very long time, whereas its wavelength grows exponentially. Therefore, the appearance of such a frozen fluctuation is equivalent to the appearance of a classical field $\delta\phi(x)$ produced from quantum fluctuations.

Because the vacuum contains fluctuations of all wavelengths, inflation leads to the continuous creation of new perturbations of the classical field with wavelengths greater than H^{-1}. The average amplitude of perturbations generated during a time interval H^{-1} (in which the Universe expands by a factor e) is given by $|\delta\phi(x)| \approx H/(2\pi)$ [15, 16]. These quantum fluctuations are responsible for galaxy formation [17–21]. But if the Hubble constant during inflation is sufficiently large, quantum fluctuations of the scalar fields may lead not only to the formation of galaxies, but also to the division of the Universe into exponentially large domains with different properties.

As an example, consider again the simplest supersymmetric theory unifying weak, strong and electromagnetic interactions. Different minima of the effective potential in this model are separated from each other by a distance $\sim 10^{-3} M_\mathrm{p}$. The amplitude of quantum fluctuations in the fields ϕ, Φ and φ at the beginning of chaotic inflation can be as large as $10^{-1} M_\mathrm{p}$. This means that, at the early stages of inflation, the fields Φ and φ could easily jump from one minimum of the potential to another. Therefore, even if these fields initially occupied the same minimum everywhere, after the stage of chaotic inflation the Universe becomes divided into many exponentially large domains, corresponding to all possible minima of the effective potential [6, 14].

8.4 Eternal chaotic inflation and string theory landscape

The process of the division of the Universe into different parts becomes even easier if one takes into account the process of self-reproduction of inflationary domains. The basic mechanism can be understood as follows. If quantum fluctuations are sufficiently large, they may locally increase the value of the potential energy of the scalar field in some parts of the Universe. The probability of quantum jumps leading to a local increase of the energy density can be very small, but the regions where it happens start expanding much faster than their parent domains, and quantum fluctuations inside them lead to the production of new inflationary domains which expand even faster.

Self-reproduction of inflationary domains was first established in the context of the new inflation scenario, which is based on inflation near a local maximum of the potential [4, 22, 23]. The existence of this regime was used for justification of the anthropic principle in ref. [4]. However, nobody paid any attention to this possibility until the discovery of self-reproduction of the Universe in the chaotic inflation scenario [7].

In order to understand this effect, let us consider an inflationary domain of initial radius H^{-1} containing a sufficiently homogeneous field with initial value $\phi \gg M_p$. Equations (8.3) tell us that, during a typical time interval $\Delta t = H^{-1}$, the field inside this domain will be reduced by $\Delta \phi = M_p^2/(4\pi\phi)$. Comparing this expression with the amplitude of quantum fluctuations,

$$\delta\phi \sim \frac{H}{2\pi} = \frac{m\phi}{\sqrt{3\pi}M_p}, \tag{8.5}$$

one can easily see that for $\phi \gg \phi^* \sim M_p\sqrt{M_p/m}$, one has $|\delta\phi| \gg |\Delta\phi|$, i.e. the motion of the field ϕ due to its quantum fluctuations is much more rapid than its classical motion.

During the typical time H^{-1}, the size of the domain of initial size H^{-1} containing the field $\phi \gg \phi^*$ grows e times, its volume increases $e^3 \sim 20$ times, and in almost half of this new volume the field ϕ jumps up instead of down. Thus the total volume of inflationary domains with $\phi \gg \phi^*$ grows approximately ten times. During the next time interval H^{-1}, this process continues, so the Universe enters an eternal process of self-reproduction. I call this process 'eternal inflation'.

In this scenario, the scalar field may wander for an indefinitely long time as the density approaches the Planck density. This induces quantum fluctuations of all other scalar fields, which may jump from one minimum of the potential to another for an unlimited time. The amplitude of these quantum fluctuations can be extremely large, $\delta\varphi \sim \delta\Phi \sim 10^{-1} M_p$. As a result, quantum fluctuations generated during eternal chaotic inflation can penetrate through any barriers, even if they have Planckian height, and the Universe after inflation becomes divided into an indefinitely large number of exponentially large domains. These contain matter in all possible states, corresponding to all possible mechanisms of spontaneous symmetry-breaking, i.e. to all possible laws of low-energy physics [7, 24].

A rich spectrum of possibilities may appear during inflation in Kaluza–Klein and superstring theories, where an exponentially large variety of vacuum states and ways of compactification is available for the original 10- or 11-dimensional space. The type of compactification determines the coupling constants, the vacuum energy, the symmetry-breaking scale and, finally, the effective dimensionality of the space in which we live. As shown in ref. [25], chaotic inflation near the Planck density may lead to a local change in the number of compactified dimensions. This means that the Universe becomes divided into exponentially large parts with different dimensionality.

In some theories one may have a continuous spectrum of possibilities. For example, in the context of the Brans–Dicke theory, the effective gravitational constant is a function of the Brans–Dicke field, which also experienced fluctuations during inflation. As a result, the Universe after inflation becomes divided into exponentially large parts with all possible values of the gravitational constant G and the amplitude of density perturbations $\delta\rho/\rho$ [26, 27]. Inflation may divide the Universe into exponentially large domains with continuously varying baryon-to-photon ratio n_B/n_γ [28] and with galaxies having vastly different properties [29]. Inflation may also continuously change the effective value of the vacuum energy (the cosmological constant Λ), which is a prerequisite for many attempts to find an anthropic solution of the cosmological constant problem [6, 30–39]. Under these circumstances, the most diverse sets of parameters of particle physics (masses, coupling constants, vacuum energy, etc.) can appear after inflation. One can say that, in a certain sense, the Universe becomes a multiverse.

Recently, the multiverse scenario has attracted special attention because of the discovery that string theory admits many metastable de Sitter vacua with different properties, and different domains of the Universe may unceasingly jump between these vacua [40–42]. The lifetime of each of these states is typically much greater than the age of our part of the Universe. The total number of metastable vacuum states in string theory may be as large as 10^{1000} [43, 44].

Once this 'string landscape' became part of the string theory description of the world, it became very difficult to forget about it and return to the old idea that the theory must have only one vacuum state, with the goal of physics being to find it. One can either like this new picture or hate it, but it cannot be discarded purely on the basis of ideological considerations. If this scenario is correct, then physics alone cannot provide a complete explanation for all properties of our part of the Universe. The same physical theory may yield large parts of the Universe that have diverse properties. According to this scenario, we find ourselves inside a 4-dimensional domain with our kind of physical laws, not because domains with different dimensionality and with alternate properties are impossible or improbable, but simply because our kind of life cannot exist in other domains.

This scenario provides a simple justification of the anthropic principle and removes the standard objections against it. One does not need anymore to assume that some supernatural cause created the Universe with the properties specifically fine-tuned to make our existence possible. Inflation itself, without any external intervention, may produce exponentially large domains with all possible laws of low-energy physics. And we should not be surprised that the conditions necessary for our existence appear on a very large scale rather than only in a small vicinity of the solar system. If the proper conditions are established near the solar system, inflation ensures that similar conditions appear everywhere within the observable part of the Universe.

The new possibilities that appear due to the self-reproduction of the Universe may provide a basis for what I call the 'Darwinian' approach to cosmology [33, 45, 46]. Mutations of the laws of physics may lead to the formation of domains with the laws of physics that allow a greater speed of expansion of the Universe; these domains will acquire greater volume and may host a greater number of observers. On the other hand, the total volume of domains of each type grows indefinitely large. This process looks like a peaceful coexistence and competition, and sometimes even like a fruitful collaboration, with the fastest growing domains producing many slower growing brothers. In the simplest models of this type, a stationary regime is reached, and the speed of growth of the total volume of domains of each type becomes equally large for all of the domains [24].

8.5 Some problems addressed by the anthropic principle
8.5.1 The cosmological constant and dark energy

According to the most recent data, vacuum energy (be it a cosmological constant or some other form of 'dark energy') with density Λ (or ρ_Λ) constitutes 74% of the total energy density of the Universe ρ_0, dark matter with density ρ_{DM} constitutes 22% of ρ_0, and normal matter with density ρ_M contributes only 4% of ρ_0. One of the most challenging problems of theoretical physics is to explain why the vacuum energy is so small, $\Lambda \sim \rho_M \sim 10^{-120} M_p^4$, and, at the same time, why it is of the same order as the total energy density of the Universe.

The first attempt to solve the cosmological constant problem using the anthropic principle in the context of inflationary cosmology was made in ref. [6]. In this paper it was argued that, if one considers antisymmetric tensor fields F, they give a time-independent contribution to the vacuum energy density of the Universe, depending on the value of these fields. The total vacuum energy density is given by the sum $V(\phi) + V(F)$. According to quantum cosmology, which is based on the tunnelling wave-function of the Universe, the probability of quantum creation of the Universe is $O(1)$ for $V(\phi) + V(F) \sim 1$ in units of the Planck energy density.

Consider, for example, the theory $V(\phi) = m^2\phi^2/2 + V_0$. In this case, all models emerge with equal probability at the moment when $m^2\phi^2/2 + V_0 + V(F) \sim 1$, but then the vacuum energy density in each region relaxes to $\Lambda = V_0 +, V(F)$. The sum $V_0 + V(F)$ does not itself affect the probability of the quantum creation of the Universe, since all models are equally probable for $m^2\phi^2/2 + V_0 + V(F) \sim 1$. Thus one comes to the conclusion that models with all values of Λ are equally probable. It was argued [6] that life of our type can exist only in the models with $|\Lambda| \lesssim O(10)\rho_0 \sim 10^{-28}$ g cm$^{-3} \sim \rho_M$, where ρ_0 is the present density in our part of the Universe and $\rho_M \sim 0.3\rho_0$ is the density of matter (including dark matter). This, together with the flat probability distribution for creation of the universes with different Λ, may solve the cosmological constant problem.

There is another way to solve the cosmological constant problem [8]. One may consider inflation driven by the scalar field ϕ (the inflaton field) and mimic the cosmological constant by the very flat potential of a second scalar field Φ. The simplest potential of this type is linear:

$$V(\Phi) = \alpha M_{\mathrm{p}}^3 \Phi. \tag{8.6}$$

For a sufficiently small α, this potential can be so flat that the field Φ practically does not change during the last 10^{10} y, so at the present epoch its total potential energy $V(\Phi)$ acts exactly as a cosmological constant. This model was one of the first examples of what later became known as quintessence, or dark energy.

Even though the energy density of the field Φ hardly changes at the present time, it changed substantially during inflation. Since Φ is a massless field, it experienced quantum jumps with amplitude $H/(2\pi)$ during each timescale H^{-1}. These jumps moved the field Φ in all possible directions. In the context of the eternal inflation scenario, this implies that the field became randomized by quantum fluctuations. The Universe broke up into an infinite number of exponentially large parts, containing all possible values of the field Φ, i.e. into an infinite number of infinitely large 'regions' with different values of the effective cosmological constant $\Lambda = V(\Phi) + V(\phi_0)$, where $V(\phi_0)$ is the energy density of the inflaton field ϕ in the minimum of its effective potential. This quantity may vary from $-M_{\mathrm{p}}^4$ to $+M_{\mathrm{p}}^4$ in different parts of the Universe, but we can live only in universes with $|\Lambda| \lesssim O(10)\rho_0 \sim 10^{-28}$ g cm^{-3}, where ρ_0 is the present density in our part of the Universe [8].

This last statement requires an explanation. If Λ is large and negative, $\Lambda \lesssim -10^{-28}$ g cm^{-3}, the Universe collapses within a timescale much smaller than the present age of the Universe [1,6]. On the other hand, if $\Lambda \gg 10^{-28}$ g cm^{-3}, the present Universe would expand exponentially fast, the energy density of matter would become exponentially small and life as we know it would be impossible [6,8]. This means that we can live only in those parts of the Universe where the cosmological constant does not differ too much from its presently observed value, $|\Lambda| \sim \rho_0$.

The constraint $\Lambda \gtrsim -10^{-28}$ g cm^{-3} still remains the strongest one for a negative cosmological constant; for recent developments related to this constraint, see refs. [37] and [39]. The constraint for a positive cosmological constant, $\Lambda \leq 10^{-28}$ g cm^{-3}, was made much more precise and accurate in subsequent works. In particular, Weinberg pointed out that the process of galaxy formation occurs only up to the moment when the cosmological constant begins to dominate the density of the Universe, after which the Universe enters the late stages of inflation [31]. For example, galaxies which formed at $z \geq 4$, when the density of the Universe was two orders of magnitude greater than it is now, would not have done so for $\Lambda \gtrsim 10^2 \rho_0 \sim 10^{-27}$ g cm^{-3}.

The next important step was made in a series of works [32–38] which considered not only our own galaxy, but also all other ones that could harbour life of our type. This would include not only galaxies formed in the past but also those forming at the present epoch. Since the density at later stages of cosmic evolution always decreases, even a very small cosmological constant may disrupt late-time galaxy formation or prevent the growth of existing galaxies. This strengthens the constraint on the cosmological constant. According to ref. [34], the probability that an astronomer in any of the universes would find the presently observed ratio Λ/ρ_0 as small as 0.7 ranges from 0.05 to 0.12, depending on various assumptions. For some models based on extended supergravity, the anthropic constraints can be strengthened even further [39].

It would be most important to obtain a solution of the cosmological constant problem in string theory. Surprisingly, despite many attempts, for a long time we did not even know how to formulate this problem, because all existing string theories were unable to describe a model with a stable vacuum and a positive cosmological constant. This problem was solved only recently [2]; we have already mentioned the solution in our discussion of the string theory landscape. The solution involved investigation of stabilized flux vacua in string theory and suggested that there may be many different vacua with different values of Λ. An investigation of this issue demonstrated that the total number of different de Sitter vacua in string theory can be astonishingly large, of the order 10^{100} or perhaps even 10^{1000}, which created the notion of the vast string theory landscape [41–44]. Simple dimensional estimates suggest that the vacuum energy density in the stringy vacua may vary from $-O(1)$ to $+O(1)$ in Planck (or string) units. Therefore it is possible that there are many vacuum states with $|\Lambda| \sim 10^{-120}$ and that the total number of vacua has a relatively flat dependence on the energy density in this range This would provide an anthropic solution of the cosmological constant problem in string theory.

One should emphasize an important assumption made in all of these considerations. In order to solve the cosmological constant problem, it is necessary to assume that the prior probability to have non-vanishing cosmological constants is practically independent of Λ. Indeed, if larger values of the cosmological constant were much more probable, one would conclude that we must live when $\Lambda \gg 10^{-120}$. On the other hand, if Λ has to be zero because of some symmetry, but appears due to some non-perturbative effect, then one could expect $\Lambda \sim e^{-\alpha}$ where α is some field or random parameter. If this parameter, rather than Λ, has a flat probability distribution, then the probability that $\Lambda \sim 10^{-120}$ will be as large as the probability that $\Lambda \sim 10^{-121}$ or $\Lambda \sim 10^{-1000}$. This would make it very difficult to explain the observed value of the cosmological constant.

The situation appears even worse if one calculates the probability for a given point to be in a state with a particular value of Λ, corresponding to some set of vacua. This probability is given by the square of the Hartle–Hawking wave-function:

$$P \sim \exp\left(-\frac{24\pi^2}{\Lambda}\right) = e^{-S_\Lambda}, \tag{8.7}$$

where S_Λ is the entropy of de Sitter space [47–49]. If one uses this probability, one may conclude that we must live in the state with the smallest possible value of Λ, independently of any anthropic considerations. On the other hand, one may argue that this distribution should not be used, since the probability distribution $P \sim e^{-S_\Lambda}$ is established by the continuous tunnelling back and forth between different vacua. This takes a ridiculously large time, $t \sim e^{S_\Lambda} \sim 10^{10^{120}}$ y, which is much greater than the age of the Universe. Moreover, it is not obvious that it makes any sense to consider a typical situation at a given point. Rather, one may want to try to find a typical situation in a given volume at a given time [24].

However, here we face a new problem. An eternally self-reproducing Universe consists of an indefinitely large number of regions, where all kinds of processes may occur, even if their probability is very small. To compare the total volume of the parts of the Universe with different properties, one should compare infinities, which may lead to ambiguities. Different methods of calculations produce different results [24, 27, 33, 48, 50–53]. We believe that all of these different answers are in a certain sense correct; it is the choice of the questions that remains problematic.

To explain our point of view, let us study an example related to demographics. One may want to know the average age of a person living now on the Earth. In order to find it, one should take the sum of the ages of all people and divide it by their total number. Naïvely, one would expect the result of the calculation to be half the life expectancy. However, the actual result will be much smaller. Because of the exponential growth of the population, the main contribution to the average age will be given by very young people. Both answers (the average age of a person and half the life expectancy) are correct, despite the fact that they are different. Neither answer is any better; they are different because they address different questions. Economists may want to know the average age in order to make projections, but individuals – as well as the insurance industry – may be more interested in the life expectancy.

Similarly, the calculations performed in refs. [24], [27], [33] and [50–53] dissect all possible outcomes of the evolution of the Universe (or multiverse) in many different ways. Each of these ways is legitimate and leads to correct results, but some additional input is required in order to understand which of these results, if any, is most closely related to the anthropic principle.

This ambiguity may suggest that one should abandon the anthropic principle and replace it by something more predictive. For example, Smolin has suggested that universes could be formed inside black holes. By finding the parameters which maximize the production of black holes and, consequently, the creation of new universes, one could find the set of universes favoured by evolution [54, 55].

From my point of view, this suggestion is not an alternative to the anthropic principle, but a very speculative version of it, which does not offer any advantages and has a major drawback. It does not offer any advantages because the total number of black holes produced during inflation is exponentially sensitive to the duration of inflation and to the amplitude of density perturbations produced then. None of these issues have been considered in refs. [54] or [55]. In order to do so, one needs to resolve the problem of measure discussed above. The drawback of this suggestion is the absence of any reliable theoretical description of the creation of new universes with different properties inside black holes. On the other hand, the aforementioned theory for the creation of new parts of the inflationary Universe is based on processes which are rather well understood.

If the theory of all fundamental interactions indeed possesses a plethora of different vacuum states, we should learn how to live with this new scientific paradigm. We must find out which of the questions we are asking may have an unambiguous answer and which ones are meaningless. I have a strong suspicion that, in order to answer the question of why we live in our part of the Universe, we must first learn the answers to the questions 'What is life?' and 'What is consciousness?' [13, 46, 50, 56].

Until these problems are solved, one may take a pragmatic point of view and consider this investigation as a kind of 'theoretical experiment'. We may try to use probabilistic considerations in a trial-and-error approach. If we get unreasonable results, this may serve as an indication that we are using quantum cosmology incorrectly. However, if some particular proposal for the probability measure allows us to solve certain problems which could not be solved in any other way, then we will have a reason to believe that we are moving in the right direction.

We can also use the new cosmological paradigm in a rather modest way. For example, we may not exactly know the prior probability distribution for the cosmological constant Λ. However, if we do not feel that the assumption of a flat probability distribution near $\Lambda = 0$ is outrageous, then anthropic considerations will tell us that there is nothing outrageously unnatural in the possibility that the Universe has $\Lambda \sim 10^{-120}$ in Planck units. In other words, anthropic considerations allow us to find *possible* explanations of some facts which would otherwise look absolutely miraculous. In the next section, we will give another illustration of this way of thinking.

8.5.2 The anthropic principle and axions

Now we have a possible reason for why the vacuum density has the same order as the total matter density in the Universe, can we go further and understand why dark matter is five times more abundant than ordinary matter? Let us assume that dark matter is represented by the axion field θ, which was introduced in order to solve the problem of strong CP violation. The potential of the axion field has the following form:

$$V(\theta) \sim m_\pi^4 \left(1 - \cos \frac{\theta}{\sqrt{2} f}\right). \tag{8.8}$$

The field θ can take any value in the range $-\sqrt{2}\pi f$ to $\sqrt{2}\pi f$. A natural estimate for the initial value of the axion field would therefore be $\theta = O(f)$, and the initial value of V(θ) should be of order m_π^4. An investigation of the rate at which the energy of the axion field ρ_θ falls off as the Universe expands shows that, for $f \gtrsim 10^{12}$ GeV, most of the energy density would presently be contributed by axions, while the baryon energy density would be considerably lower than its presently observed value of $\rho_B \sim 0.05 \rho_0$. This information was used to derive the constraint $f \lesssim 10^{12}$ GeV [57–59].

This is a very strong constraint, especially since the astrophysical considerations lead to a constraint $f \gtrsim 10^{11}$ GeV. Note also that the standard scale for f in string theory is $f \sim M_p \sim 10^{18} - 10^{19}$ GeV. This means that one should construct theories with an unnaturally small value of f, and even this may not help unless the parameter f is in the very narrow 'axion window' $10^{11} \lesssim f \lesssim 10^{12}$ GeV.

Let us now take a somewhat closer look at whether one can actually obtain the constraint $f \lesssim 10^{12}$ GeV in the context of inflationary cosmology. Long-wave fluctuations of the axion field θ are generated during inflation if Peccei–Quinn symmetry-breaking, resulting in the potential given by Eq. (8.8), takes place before the end of inflation. By the end of inflation, therefore, a quasi-homogeneous distribution of the field θ will have appeared in the Universe, with the field taking on all values from $-\sqrt{2}\pi f$ to $\sqrt{2}\pi f$ at different points in space with a probability that is almost independent of θ. This means that one can always find exponentially large regions of space within which $\theta \ll f$. The energy of the axion field always remains relatively low in such regions and there is no conflict with the observational data.

This feature does not itself remove the constraint $f \lesssim 10^{12}$ GeV. Indeed, when $f \gg 10^{12}$ GeV, only within a very small fraction of the volume of the Universe is the axion field energy density small enough by comparison with the baryon density. It might therefore seem extremely improbable that we live in one of these particular regions.

Consider, for example, those regions initially containing a field $\theta_0 \ll f$, for which the present ratio of the energy density of the axion field to the baryon density is consistent with the observational data (i.e. where the density of dark matter is about five times greater than the baryon density). It can be shown that the total number of baryons in regions with $\theta \sim 10\theta_0$ should be ten times the number in regions with $\theta \sim \theta_0$. One might therefore expect the probability of randomly ending up in a region with $\theta \sim 10\theta_0$ (incompatible with the observational data) to be ten times that of ending up in a region with $\theta \sim \theta_0$.

However, closer examination of this problem indicates that the properties of galaxies formed in such a region should be very different from the properties of our galaxy. This makes it unclear whether life can exist in the regions with $\theta \geq 10\theta_0$ [29]. Let us compare the domains with $\theta = \theta_0$ and $\theta = N\theta_0$ at the same cosmological time t in an early universe dominated by hot matter. Since the Universe after inflation becomes flat, the total density during this post-inflationary stage is proportional to t^{-2}, practically independent of the relative fraction of matter in axions and in baryons. This has two interesting implications. The first is that at $t \sim 10^{10}$ y the total density in both domains will be the same but the baryon density will be N^2 times smaller in the domain with $\theta = N\theta_0$. In other words, in a domain with $\theta \sim 10\theta_0$, the observable region after 10^{10} y will contain one hundred times fewer baryons than a domain with $\theta \sim \theta_0$. As discussed below, this alone may reduce the probability of the emergence of life.

The second implication is related to the properties of galaxies in domains with $\theta \sim N\theta_0$. The ratio $n_B/n_\gamma \sim 10^{-10}$ is fixed by some processes in the early Universe, which are not expected to depend on the axion abundance. The main difference between the two domains discussed above is that the relative energy density of non-relativistic particles is N^2 times higher in the second domain. Also, at the same time t, the ratio of the energy density of photons and cold dark matter will be N^2 smaller, i.e. this domain is colder. The cold dark matter energy density decreases as $t^{-3/2}$, whereas the energy density of photons decreases as t^{-2}, i.e. $t^{-1/2}$ times faster. Therefore the period of cold dark matter dominance occurred N^4 times earlier in the second domain. The energy density of cold dark matter at that moment was $\sim N^8$ times higher than in the first domain.

Note that the beginning of cold dark matter dominance is the time when density perturbations $\delta\rho/\rho \sim 10^{-4}$ start growing. Since they start growing earlier, the moment when they reach $O(1)$ – i.e. the stage when overdense regions separate into galaxies – also occurs earlier. The density of matter inside galaxies in the future remains of the same order as the density of the Universe at the time of the galaxy formation. This means that the density of matter in the first (smallest) galaxies to be formed in the second domain will be N^8 higher than in the first domain, and the density of baryons there will be N^6 times higher.

The matter density in large galaxies, which formed later in the evolution of the Universe, should be less sensitive to θ. However, if most of the matter is packed into superdense dwarf galaxies formed in the very early Universe, the total amount of remaining matter – which would be distributed more smoothly like in our own galaxy – may be relatively small. Also, any galaxy of a given mass M will contain N^2 times fewer baryons than our galaxy.

Naïvely, it would seem ten times more probable to live in domains with $\theta = 10\,\theta_0$ because the total volume of such domains is ten times bigger. However, since the properties of galaxies in a universe with $\theta = 10\,\theta_0$ are very different from those of our galaxy, it well may happen that domains with $\theta \sim \theta_0$ provide much better conditions for the emergence of life than domains with $\theta = 10\,\theta_0$ [29]; see also ref. [60].

In order to study this situation quantitatively, one should perform a detailed investigation of galaxy formation in a model with $\rho_M \gg \rho_B$, similar to the investigation of galaxy formation in a model dominated by a cosmological constant ($\rho_\Lambda \gg \rho_M$), as discussed in Section 8.5.1. This investigation has been performed very recently [61]. The results obtained confirmed the expectations of ref. [29]: if dark matter is represented by axions with $f \gg 10^{12}$ GeV, then one is most likely to live in a domain where the density of dark matter is about one or two orders of magnitude greater than the density of ordinary matter. This is quite consistent with the observed value $\rho_{DM} \sim 5\rho_M$.

This result has two interesting implications. First, it will not be too surprising to find that the standard constraints $10^{11} \lesssim f \lesssim 10^{12}$ GeV are violated. Second, in the context of the axion cosmology with $f \gg 10^{12}$ GeV, it is not surprising that we live in a domain with $\rho_{DM} \sim 5\rho_M$. In this respect, such a theory has an important advantage with respect to many other dark matter theories, where one must fine-tune the parameters to obtain $\rho_{DM} \sim 5\rho_M$.

8.5.3 An anthropic explanation of the electroweak symmetry-breaking scale and the hierarchy problem

The situation with the Higgs boson mass and the amplitude of spontaneous symmetry-breaking in electroweak theory is even more interesting and impressive. One of the main problems of particle physics is the extremely small ratio of the Higgs field expectation value, $v \approx 246$ GeV, to the Planck mass, $M_p \approx 2.4 \times 10^{18}$ GeV. Assuming that the coupling constant λ of the Higgs boson is $O(1)$ or smaller, this leads [62] to an incredibly small ratio of the Higgs mass, $m_H \sim \sqrt{\lambda} v$, to the Planck mass: $m_H/M_p \sim v/M_p \sim 10^{-16}$.

A popular attempt to address this problem is based on supersymmetry. If supersymmetry is broken on a very small scale, then many particles, including the Higgs boson, acquire comparable masses $\sim 10^2$ GeV, and these masses do not acquire large radiative corrections. However, this would be a true solution of the problem only if we were able to understand the origin of the anomalously small scale of supersymmetry-breaking.

The anthropic principle allows one to look at this problem from a different point of view. Agrawal and colleagues [63, 64] have shown that all nuclei would be unstable for v five times larger than observed, whereas protons would be unstable and hydrogen would not exist for v less than half the observed value. This explains the origin of the ratio $m_H/M_p \sim 10^{-16}$.

The strongest anthropic constraints on the scale of spontaneous symmetry-breaking v can be obtained if one studies production of carbon and oxygen in the Universe [64]. These two elements are formed during helium-burning at late stages of stellar evolution. As noted by Hoyle and colleagues [65, 66], this process depends crucially on the existence of a certain resonance level in carbon nuclei. The existence and properties of this resonance was one of the first successful predictions based on the anthropic principle. However, these properties depend on the quark masses, which in turn depend on v. This leads to strong anthropic constraints on v.

This question has been studied by many authors, for example Livio et al. [67] and Hogan [68]. The most detailed investigation was carried out by Oberhummer et al. [69, 70], Jeltema and Sher [71] and Schlattl et al. [72]. They found that a change of v by 1% would lead to a strong suppression of the production of carbon (if v were smaller than 246 GeV) or oxygen (if v were greater than 246 GeV). Since both carbon and oxygen are necessary for our existence, this result strongly indicates that the otherwise unexplained value of v, as well as the small number m_H/M_p, can be determined by anthropic considerations. (The accuracy of this determination is much better than that with which the anthropic principle fixes the value of the cosmological constant Λ.)

If this is the case, supersymmetry is not required to explain the smallness of the Higgs boson mass. If this is small because of anthropic considerations, supersymmetry may become manifest at much higher energies, as suggested by Arkani-Hamed and colleagues [73, 74]. This may have important implications for attempts to find supersymmetry at the Large Hadron Collider (LHC). The main motivation for these attempts was the standard idea that there should be many supersymmetric particles with masses similar to the Higgs boson mass. This idea has guided theoretical investigations for the last twenty years, and the total cost of the experimental search for low-energy supersymmetry is billions of dollars. Thus, as in the case of the axion search, ignoring anthropic considerations can be expensive.

If no light supersymmetric particles are found at LHC, it will be an additional argument in favour of anthropic reasoning. On the other hand, anthropic arguments do not preclude low-energy SUSY-breaking. If light supersymmetric particles *were* discovered at LHC, then it would imply that SUSY is indeed broken at the low-energy scale associated with the Higgs mass. But since the Higgs mass can itself be explained by anthropic considerations, the discovery of light supersymmetric particles may imply that the low scale of SUSY-breaking also has an anthropic explanation.

8.6 Conclusions

For a long time physicists have believed that there is only one world and that a successful description of this world should eventually predict all of its parameters, such as the coupling constants and the masses of elementary particles. The fundamental theory was supposed to be beautiful and natural. This was a noble, but perhaps excessively optimistic, hope. One could call this period 'the age of innocence'.

I believe we are now entering 'the age of anthropic reasoning'. Inflationary cosmology – in combination with string theory – leads to a picture of a multiverse consisting of an infinite number of exponentially large domains ('universes') with an exponentially large number of different properties. In addition to a somewhat subjective notion of beauty and naturalness, we are adding the simple and obvious criterion that the part of the Universe where we live must be consistent with the possibility of our existence. This super-selection rule sometimes considerably improves our intuitive judgement about what is natural. For example, naïvely, the most natural scale for the Higgs boson mass is $O(M_\mathrm{p})$ and the most natural value of the vacuum energy density is $O(M_\mathrm{p}^4)$. However, it is unnatural and in fact impossible for us to live in a universe (or even part of a universe) with such parameter values. In a certain sense, one may consider anthropic reasoning as a way to improve the naïve use of the concept of naturalness.

The concept of beauty may also play an important role in the selection of the vacuum state. One possible idea is that symmetry implies that the properties of the world do not change under certain transformations. This may mean that, if one can live in a given vacuum state, one can live in a whole family of states related to each other by symmetry. The existence of many equivalent states may increase the probability of living in one of them, by effectively increasing the phase volume of the anthropically allowed vacua possessing the symmetry. In addition, states with large symmetry (i.e. beauty) are sometimes dynamically attractive, behaving as trapping points in the space of all possible vacua [75]. In this way, one may try to unify anthropic reasoning with the principles of naturalness and beauty, which have always guided our search for the fundamental theory describing our world.

References

[1] J. D. Barrow and F. J. Tipler. *The Anthropic Cosmological Principle* (New York: New York University Press, 1986).

[2] I. L. Rozental. *Big Bang, Big Bounce* (New York: Springer Verlag, 1988).

[3] M. Rees. *Just Six Numbers, The Deep Forces that Shape the Universe* (New York: Basic Books, Perseus Group, 2000).

[4] A. D. Linde. Non-singular regenerating inflationary universe. Cambridge University preprint-82-0554 (1982).

[5] A. D. Linde. The new inflationary universe scenario. In *The Very Early Universe*, eds. G. W. Gibbons, S. W. Hawking and S. Siklos (Cambridge: Cambridge University Press, 1983), pp. 205–49.

[6] A. D. Linde. The inflationary universe. *Rep. Prog. Phys.* **47** (1984), 925.

[7] A. D. Linde. Eternally existing self-reproducing chaotic inflationary universe. *Phys. Lett.* **B 175** (1986), 395.

[8] A. A. Starobinsky. A new type of isotropic cosmological model without singularity. *Phys. Lett.* **B 91** (1980), 99.

[9] A. H. Guth. (1981) The inflationary universe: a possible solution to the horizon and flatness problems. *Phys. Rev.* **D 23** (1981), 347.

[10] A. D. Linde. A new inflationary universe scenario: a possible solution of the horizon, flatness, homogeneity, isotropy and primordial monopole problems. *Phys. Lett.* **B 108** (1982), 389.

[11] A. D. Linde. Chaotic inflation. *Phys. Lett.* **B 129** (1983), 177.

[12] A. Albrecht and P. J. Steinhardt. Cosmology for grand unified theories with radiatively induced symmetry breaking. *Phys. Rev. Lett.* **48** (1982), 1220.

[13] A. D. Linde. *Particle Physics and Inflationary Cosmology* (Chur, Switzerland: Harwood Academic Publishers, 1990).

[14] A. D. Linde. Inflation can break symmetry in SUSY. *Phys. Lett.* **B 131** (1983), 330.

[15] A. Vilenkin and L. H. Ford. Gravitational effects upon cosmological phase transitions. *Phys. Rev.* **D 26** (1982), 1231.

[16] A. D. Linde. Scalar field fluctuations in expanding universe and the new inflationary universe scenario. *Phys. Lett.* **B 116** (1982), 335.

[17] V. F. Mukhanov and G. V. Chibisov. Quantum fluctuation and 'non-singular' universe. *JETP Lett.* **33** (1981), 532. (*Pisma Zh. Eksp. Teor. Fiz.* **33** (1981), 549.

[18] S. W. Hawking. The development of irregularities in a single bubble inflationary universe. *Phys. Lett.* **B 115** (1982), 295.

[19] A. A. Starobinsky. Dynamics of phase transition in the new inflationary universe scenario and generation of perturbations. *Phys. Lett.* **B 117** (1982), 175.

[20] A. H. Guth and S. Y. Pi. Fluctuations in the new inflationary universe. *Phys. Rev. Lett.* **49** (1982), 1110.

[21] J. M. Bardeen, P. J. Steinhardt and M. S. Turner. Spontaneous creation of almost scale-free density perturbations in an inflationary universe. *Phys. Rev.* **D 28** (1983), 679.

[22] P. Steinhardt. Natural inflation. In *The Very Early Universe*, eds. G. W. Gibbons, S. W. Hawking and S. Siklos (Cambridge: Cambridge University Press, 1983).

[23] A. Vilenkin. The birth of inflationary universes. *Phys. Rev* **D 27** (1983), 2848.

[24] A. D. Linde, D. A. Linde and A. Mezhlumian. From the big bang theory to the theory of a stationary universe. *Phys. Rev.* **D 49** (1994), 1783 [gr-qc/9306035].
[25] A. D. Linde and M. I. Zelnikov. Inflationary universe with fluctuating dimension. *Phys. Lett.* **B 215** (1988), 59.
[26] A. D. Linde. Extended chaotic inflation and spatial variations of the gravitational constant. *Phys. Lett.* **B 238** (1990), 160.
[27] J. Garcia-Bellido, A. D. Linde and D. A. Linde. Fluctuations of the gravitational constant in the inflationary Brans–Dicke cosmology. *Phys. Rev.* **50** (1994), 730 [astro-ph/9312039].
[28] A. D. Linde. The new mechanism of baryogenesis and the inflationary universe. *Phys. Lett.* **B 160** (1985), 243.
[29] A. D. Linde. Inflation and axion cosmology. *Phys. Lett.* **B 201** (1988), 437.
[30] A. D. Linde. Inflation and quantum cosmology. In *Three Hundred Years of Gravitation*, eds. S. W. Hawking and W. Israel (Cambridge: Cambridge University Press, 1987), pp. 604–30.
[31] S. Weinberg. Anthropic bound on the cosmological constant. *Phys. Rev. Lett.* **59** (1987), 2607.
[32] G. Efstathiou. An anthropic argument for a cosmological constant. *Mon. Not. Roy. Astron. Soc.* **274** (1995), L73.
[33] A. Vilenkin. Predictions from quantum cosmology. *Phys. Rev. Lett.* **74** (1995), 846 [gr-qc/9406010].
[34] H. Martel, P. R. Shapiro and S. Weinberg. Likely values of the cosmological constant. *Astrophys. J.* **492** (1998), 29 [astro-ph/9701099].
[35] J. Garriga, and A. Vilenkin. On likely values of the cosmological constant. *Phys. Rev.* **D 61** (2000), 083502 [astro-ph/9908115].
[36] J. Garriga and A. Vilenkin. Solutions to the cosmological constant problems. *Phys. Rev.* **D 64** (2001), 023517 [hep-th/0011262].
[37] J. Garriga and A. Vilenkin. Testable anthropic predictions for dark energy. *Phys. Rev.* **D 67** (2003), 043503 [astro-ph/0210358].
[38] S. A. Bludman and M. Roos. Quintessence cosmology and the cosmic coincidence. *Phys. Rev.* **65** (2002), 043503 [astro-ph/0109551].
[39] R. Kallosh and A. D. Linde. M-theory, cosmological constant and anthropic principle. *Phys. Rev.* **67** (2003), 023510 [hep-th/0208157].
[40] S. Kachru, R. Kallosh, A. Linde and S. P. Trivedi. De Sitter vacua in string theory. *Phys. Rev.* **D 68** (2003), 046005 [hep-th/0301240].
[41] L. Susskind. This volume (2007) [hep-th/0302219].
[42] L. Susskind. *The Cosmic Landscape, String Theory and the Illusion of Intelligent Design* (New York: Little, Brown and Company, 2006).
[43] R. Bousso and J. Polchinski. Quantization of four-form fluxes and dynamical neutralization of the cosmological constant. *JHEP*, **0006** (2000), 006 [hep-th/0004134].
[44] M. R. Douglas. The statistics of string/M theory vacua. *JHEP*, **0305** (2003), 046 [hep-th/0303194].
[45] A. D. Linde. Particle physics and inflationary cosmology. *Physics Today*, **40** (1987), 61.
[46] J. Garcia-Bellido and A. D. Linde. Stationarity of inflation and predictions of quantum cosmology. *Phys. Rev.* **D 51** (1995), 429 [hep-th/9408023].
[47] S. W. Hawking. The cosmological constant is probably zero. *Phys. Lett.* **B 134** (1984), 403.

[48] A. A. Starobinsky. Stochastic de Sitter (inflationary) stage in the early universe. In *Current Topics in Field Theory, Quantum Gravity and Strings*, eds. H. J. de Vega and N. Sanchez (Heidelberg: Springer, 1986), p. 107.

[49] A. D. Linde. Quantum creation of an open inflationary universe. *Phys. Rev.* **D 58** (1998), 083514 [gr-qc/9802038].

[50] A. D. Linde and A. Mezhlumian. On regularization scheme dependence of predictions in inflationary cosmology. *Phys. Rev.* **D 53** (1996), 4267 [gr-qc/9511058].

[51] V. Vanchurin, A. Vilenkin and S. Winitzki. Predictability crisis in inflationary cosmology and its resolution. *Phys. Rev.* **D 61** (2000), 083507 [gr-qc/9905097].

[52] J. Garriga and A. Vilenkin. A prescription for probabilities in eternal inflation. *Phys. Rev* **D 64** (2001), 023507 [gr-qc/0102090].

[53] M. Tegmark. What does inflation really predict? *J. Cosmol. Astropart. Phys.*, **0504** (2005), 001 [astro-ph/0410281].

[54] L. Smolin. *The Life of the Cosmos* (NewYork: Oxford University Press, 1997).

[55] L. Smolin. This volume (2007).

[56] A. D. Linde, D. A. Linde and A. Mezhlumian. Non-perturbative amplifications of inhomogeneities in a self-reproducing universe. *Phys. Rev.* **D 54** (1996), 2504 [gr-qc/9601005].

[57] J. Preskill, M. B. Wise and F. Wilczek. Cosmology of the invisible axion. *Phys. Lett.* **B 120** (1983), 127.

[58] L. F. Abbott and P. Sikivie. A cosmological bound on the invisible axion. *Phys. Lett.* **B 120** (1983), 133.

[59] M. Dine and W. Fischler. The not-so-harmless axion. *Phys. Lett.* **B 120** (1983), 137.

[60] F. Wilczek. This volume (2007) [hep-ph/0408167].

[61] M. Tegmark, A. Aguirre, M. Rees and F. Wilczek. Dimensionless constants, cosmology and other dark matters. *Phys. Rev.* **D 73** (2006), 023505 [astro-ph/0511774].

[62] J. F. Donoghue. This volume (2007).

[63] V. Agrawal, S. M. Barr, J. F. Donoghue and D. Seckel. The anthropic principle and the mass scale of the standard model. *Phys. Rev.* **D 57** (1998), 5480.

[64] V. Agrawal, S. M. Barr, J. F. Donoghue and D. Seckel. Anthropic considerations in multiple-domain theories and the scale of electroweak symmetry breaking. *Phys. Rev. Lett.* **80** (1998), 1822 [hep-ph/9801253].

[65] F. Hoyle, D. N. F. Dunbar, W. A. Wenzel and W. Whaling. A state in C^{12} predicted from astronomical evidence. *Phys. Rev.* **92** (1953), 1095.

[66] F. Hoyle. On nuclear reactions occurring in very hot stars: synthesis of elements from carbon to nickel. *Astrophys. J. Suppl.* **1** (1954), 121.

[67] M. Livio, D. Hollowell, A. Weiss and J. W. Truran. The anthropic significance of the existence of an excited state of ^{12}C. *Nature* **340** (1989), 281.

[68] C. J. Hogan. Why the universe is just so. *Rev. Mod. Phys.*, **72** (2000), 1149 [astro-ph/9909295].

[69] H. Oberhummer, A. Csoto, and H. Schlattl. Stellar production rates of carbon and its abundance in the universe. *Science*, **289** (2000), 88 [astro-ph/0007178].

[70] H. Oberhummer, A. Csoto and H. Schlattl. Bridging the mass gaps at A = 5 and A = 8 in nucleosynthesis. *Nucl. Phys.* **A 689** (2001), 269 [nucl-th/0009046].
[71] T. S. Jeltema and M. Sher. Triple-alpha process and the anthropically allowed values of the weak scale. *Phys. Rev.* **D 61** (2000), 017301.
[72] H. Schlattl, A. Heger, H. Oberhummer, T. Rauscher and A. Csoto. Sensitivity of the C and O production on the 3-alpha rate. *Astrophys. Space Sci.* **291** (2004), 27 [astro-ph/0307528].
[73] N. Arkani-Hamed and S. Dimopoulos. Supersymmetric unification without low energy supersymmetry and signatures for fine-tuning at the LHC. *JHEP*, **0506** (2005), 073 [hep-th/0405159].
[74] N. Arkani-Hamed, S. Dimopoulos, G. F. Giudice and A. Romanino. Aspects of split supersymmetry. *Nucl. Phys.* **B 709** (2005), 3 [hep-ph/0409232].
[75] L. Kofman, A. Linde, X. Liu, A. Maloney, L. McAllister and E. Silverstein. Beauty is attractive: moduli trapping at enhanced symmetry points. *JHEP*, **0405** (2004), 030 [hep-th/0403001].

9
A model of anthropic reasoning: the dark to ordinary matter ratio

Frank Wilczek
Department of Physics, Massachusetts Institute of Technology

9.1 Methodology and anthropic reasoning

There are good reasons to view attempts to deduce basic laws of matter from the existence of mind with scepticism.[1] Above all, it seems gratuitous. Physicists have done very well indeed at understanding matter on its own terms, without reference to mind. We have found that the governing principles take the form of abstract mathematical equations of universal validity, which refer only to entities – quantum fields – that clearly do not have minds of their own. Working chemists and biologists, for the most part, are committed to the programme of understanding how minds work under the assumption that it will turn out to involve complex orchestration of the building blocks that physics describes [1]; and while this programme is by no means complete, it has not encountered any show-stopper and it is supporting steady advances over a wide front. Computer scientists have made it plausible that the essence of mind is to be found in the operation of algorithms that in principle could be realized within radically different physical embodiments (cells, transistors, tinkertoys) and in no way rely on the detailed structure of physical law [2].

To put it shortly, the emergence of mind does not seem to be the sort of thing we would like to postulate and use as a basic explanatory principle. Rather, it is something we would like to understand and explain by building up from simpler phenomena. So there is a heavy burden to justify use of anthropic reasoning in basic physics. And yet there are, it seems to me, limited, specific circumstances under which such reasoning can be correct, unavoidable and clearly appropriate.

---

1 By 'basic' I mean irreducible and I am consciously avoiding the loaded term 'fundamental'. No doubt there are extremely profound insights about how complex systems behave and develop to be derived from the existence of mind, and in particular from its concrete emergence in history. Such insights will be fundamental by any reasonable standard, but not basic in the sense used here.

Universe or Multiverse?, ed. Bernard Carr. Published by Cambridge University Press.
© Cambridge University Press 2007.

Here is a simple, but I think instructive and far from trivial, example. Why is Earth at the distance it is from the Sun? At one time (for example, to Kepler) the size and shape of the Solar System – as yet not clearly distinguished from the cosmos as a whole – might have seemed like a major question for physics, that one might hope would have a unique answer closely related to basic principles. Now, of course, the question appears quite different. We know that the Universe contains many broadly similar systems with planets orbiting around stars. We know that such systems come in various sizes and shapes, and that their structure depends sensitively on details of the complicated conditions under which they formed. We can be confident of all these assertions because they emerge from a rich background involving astronomical observations, the success of Newtonian mechanics and modern developments in cosmology and chaos theory. Given the two key features of many independent realizations and effectively random variation over the realizations, we cannot address our question in the context of planets in general, or universal laws. If we are going to address it at all, we have to refer specifically to Earth; and what makes Earth special, in this context, is that it is where we, the question-askers, find ourselves. Once we accept this starting point, we can go on to have an edifying discourse about why life would be difficult if the distance from the Earth to the Sun were quite different. We can even imagine that normal, testable scientific predictions will emerge from this discourse about where we will find life in other planetary systems – or even elsewhere in our own.

A psychological weakness of this example is that we have come a long way since Kepler's time, and it is hard to put ourselves back in the frame of mind to regard the Earth–Sun distance as a serious question for physics; but however we regard the question, it seems clear that the final step in a serious answer must involve anthropic reasoning.

The main thing I want to do here is to demonstrate that there is a choice of assumptions that, while somewhat speculative, lie well within the mainstream of present-day ideas about basic physics and cosmology, which leads to a situation whose logical structure is quite similar to this simple example, but where the question that is addressed by anthropic reasoning is one that is open, topical and widely believed to be basic. I speak of the question of the ratio r of axion dark matter to baryon matter in cosmology. To be specific, I will demonstrate, given certain reasonably conventional *physical* assumptions, the following: that r varies over the ensemble of effective homogeneous universes within a spatially gigantic multiverse that is inhomogeneous on superhorizon scales; that its variation is random (with a well characterized, non-singular measure); and – an important refinement – that it varies essentially independently of other parameters. If we measure the probability by volume, we find that overly large values of r are most probable, but if we measure by number of potential observers, this conclusion is changed. It appears instead that the observed value of r is at least qualitatively, and perhaps semi-quantitatively, in accord with what these ideas suggest. That is intriguing, because – according to alternative, more conventional, ideas about r – the numerator and denominator arise from widely different physical causes and depend upon widely different parameters, so it has appeared as something of a mystery why the observed value is near unity. There are several additional implications of the assumptions that can be explored in future experiments.

The possibility of avoiding the bound that arises in conventional axion cosmology by having a small misalignment after inflation was mentioned in the earliest papers in axion cosmology in refs. [3–5]. It was exploited in the context of a specific inflationary cosmology in ref. [6], where the variation in the axion dark matter to baryon density ratio over different regions of the Universe was noted. Anthropic considerations were brought into the discussion in ref. [7]. Some constraints on the scenario, due to the axion field supplying an additional source of fluctuations, were derived in ref. [8].

9.2 Conventional and unconventional axion cosmology

I now very briefly review the relevant aspects of axion physics and cosmology.

9.2.1 Axion physics

Quantum chromodynamics (QCD) is well established as the basic (and fundamental) theory of the strong interaction [9]. When we combine QCD with the electroweak interactions, however, a subtle but I believe quite profound puzzle arises. (For reviews of axion physics, see refs. [10] and [11].) The general principles that define QCD – relativistic quantum field theory and gauge symmetry – specify its structure extremely tightly. The continuously adjustable parameters of the theory are a single overall coupling constant, a mass for each quark and one other much more obscure parameter, the so-called θ parameter. Mountains of data described by QCD precisely determine and vastly overdetermine the coupling and masses, and the description it affords is more than satisfactory.

Amidst this otherwise splendid party, the θ parameter appears as an empty chair, an invited guest whose absence is cause for concern. The θ parameter is a periodic variable whose possible values range from 0 to 2π.

It specifies the phase $e^{i\theta}$ which accompanies the occurrence of special topological features in the colour gluon field. One measure of its subtlety is that θ cannot be detected in perturbation theory. Under space inversion (P) or time reversal (T) it changes sign, so that for $\theta \neq 0$ or π these symmetries are violated. There are very stringent experimental constraints on P or (especially) T violation in the strong interaction, especially from the upper limit on the neutron's electric dipole moment. They indicate $|\theta| \leq 10^{-8}$. (The possibility that θ is near π requires separate consideration, but is excluded on other grounds.) If we were to regard QCD in isolation, we could simply impose P or T symmetry, thus naturally enforcing $\theta = 0$. But in a complete world-theory we must acknowledge that P and T are not exact symmetries of the world, and we cannot invoke them to justify $\theta = 0$. We must look for another way of explaining the smallness of θ.

Peccei and Quinn (PQ) introduced the idea that there is a special sort of approximate symmetry, valid asymptotically at short distances, that could be used to address this challenge. The PQ symmetry transformations allow translations of θ. If the symmetry were exact, all values of θ would be physically equivalent, and of course they would all preserve P and T for the strong interaction (some field redefinitions might be required to make the symmetries manifest). In reality, PQ symmetry must be spontaneously broken, since in its unbroken form it is inconsistent with non-zero quark masses. To capture this dynamics, we introduce a complex scalar order parameter field ϕ. The average value $\langle \phi \rangle$ will vanish when PQ symmetry is unbroken, but will take the form $\langle \phi \rangle = F e^{i\theta} \equiv F e^{ia/F}$ in the unbroken phase. Here F is a real scalar field, whose kinetic energy is inherited from that of ϕ and normalized in the canonical way. Furthermore, PQ symmetry is not exact, but only asymptotic, even before its spontaneous breakdown. The potential for ϕ is presumably of the general form $(|\phi|^2 - F^2)^2$ in the amplitude direction, but depends on the phase only through non-perturbative effects in QCD, in roughly the form $(1 - \cos\theta)\Lambda^4$, where $\Lambda \sim 200$ MeV is the QCD scale, here assumed to be much less than F. The PQ symmetry is responsible for this structure. There is a difference in energy densities of order Λ^4 as one varies over the range of θ. The minimum energy occurs very near $\theta = 0$. The scalar field a will tend to relax to zero, thus rendering $\theta = 0$ and solving our puzzle.

The field a introduced in this way is called the axion field, and of course its quanta are called axions. The phenomenology of axions is essentially controlled by the parameter F, which specifies the amplitude of the condensate; F has dimensions of mass. The mass $m_a = \Lambda^2/F$ of the axion and the strength of its basic couplings to matter are both proportional to $1/F$. Various laboratory and phenomenological constraints appear to require $F \geq 10^9$ GeV; the axion must be both extremely light and extremely weakly coupled.

9.2.2 Standard axion cosmology

Now let us consider the cosmological implications [3–5]. PQ symmetry is unbroken at temperatures $T \gg F$. When the symmetry breaks, the initial value of the phase, that is $e^{ia/F}$, is random beyond the then-current particle horizon scale. One can analyze the fate of these fluctuations by solving the equations for a scalar field in an expanding universe. The only unusual feature is that the effective mass of the axion field depends on temperature. The axion mass is very small for $T \gg \Lambda$, even relative to its zero-temperature value, because the non-perturbative QCD effects that generate it involve coherent gluon field fluctuations (instantons) which are suppressed at high temperature. It saturates, of course, for $T \ll \Lambda$. The full temperature dependence of the mass can be pretty reliably estimated, although the necessary calculations are technically demanding.

From standard treatments of scalar fields in an expanding universe, we learn that there is an effective cosmic viscosity, which keeps the field frozen so long as the expansion parameter is large compared to the mass, $H \equiv \dot{R}/R \gg m$. In the opposite limit, $H \ll m$, the field undergoes lightly damped oscillations, which result in an energy density that decays as $\rho \propto 1/R^3$. At intermediate times there is a period of quasi-adiabatic damping. This damping has a consequence that is very important for the present discussion, namely that the final mass density, normalized to the ambient T^3, varies roughly proportional to $F\theta^2$. The qualitative feature, that the final density decreases with decreasing F, may appear paradoxical, since the axions are getting heavier, but it is not hard to understand heuristically. For smaller values of F, corresponding to larger mass, the temperature at which the axion field begins to feel the effect of cosmic viscosity sets in earlier, and there are more damping cycles. However, the initial energy density depends only on the mismatch angle θ and is independent of F. The time-oscillating field can be interpreted as pressureless matter or dust (note that spatial inhomogeneities on small scales, which would provide pressure, begin to be damped as they enter the horizon). In simple words, we can say that the initial misalignment in the axion field, compared with what later turns out to be the favoured value, relaxes by emission of axions in a very cold coherent state, or Bose–Einstein condensate. It is *not* in thermal equilibrium with ordinary matter; the interactions are far too weak to enforce that equilibrium.

If we ignore the possibility of inflation, then – for the large values of F of interest – the horizon scale at the PQ transition at $T \approx F$ corresponds to a spatial region today that is negligibly small on cosmological scales. Thus, in calculating the axion density we are justified in performing an average over the initial mismatch angle. This allows us to calculate a unique prediction for the density, given the microscopic model. The result of the calculation is usually quoted in the following form:

$$\rho_{\text{axion}}/\rho_{\text{dark}} \approx F/(10^{12} \text{ GeV}), \tag{9.1}$$

where ρ_{dark} is the dark energy density. In this way, we would deduce that axions form a good dark matter candidate for $F \sim 10^{12}$ GeV and that larger values of F are forbidden. These conclusions are unchanged if we allow for the possibility that an epoch of inflation preceded the PQ transition.

9.2.3 Alternative axion cosmology

Things are very different, however, if inflation occurs after the PQ transition [12].[2] For then, the effective Universe accessible to present-day observation, instead of containing many horizon-volumes from the time of the PQ transition, is contained well within just one. It is therefore not appropriate to average over the initial mismatch angle. We have to restore it as a *contingent universal constant*. That is, it is a pure number that characterizes the observable Universe as a whole, but which clearly cannot be determined from any more basic quantities, even in principle – indeed it is a different number elsewhere in the multiverse!

In that case, it is appropriate to replace Eq. (9.1) by:

$$r \equiv \rho_{\text{axion}}/\rho_{\text{baryon}} \approx 12 \left(\frac{F}{10^{12} \text{ GeV}}\right) \sin^2(\theta/2). \tag{9.2}$$

This differs from the earlier form in that I have normalized the axion density relative to baryon density rather than dark matter density. This change is completely trivial at a numerical level, of course. (For concreteness, I have taken $\rho_{\text{dark}}/\rho_{\text{baryon}} = 6$.) It reflects, however, two important ideas. First, changes in the mismatch angle θ do not significantly affect baryogenesis, so that the baryon density is a fixed proportion of the photon density at high temperature, and provides an appropriate gauge for measuring the

---

2 The logical possibility of axion cosmology based on that large F and small initial mismatch in inflationary cosmology has been known since the publication of ref. [3]. The present discussion extends a portion of ref. [12].

aspects of the cosmic environment apart from F and θ. This is true in many, but perhaps not all, plausible models of baryogenesis. Second, I have reinstated the θ dependence. The exact formula for this dependence is more complicated, but Eq. (9.2) has the correct qualitative features.

In this alternative axion cosmology, values of $F \geq 10^{12}$ GeV are no longer necessarily inconsistent with existing observations. An 'over-large' value of F can be compensated for by a small value of the initial mismatch θ.

9.3 Application of anthropic reasoning

In the alternative axion cosmology, r – through its dependence on the initial mismatch θ – becomes a contingent universal constant. Furthermore, it varies in a statistically well categorized manner over the multiverse; its variation can be considered in isolation from possible changes in other universal constants; and it has significant impact upon the possible emergence of intelligent observers. Altogether it appears to be an ideally favourable case for the application of anthropic reasoning.

Since in practice we only get to sample one effective universe, there is no question of checking statements about the probability distribution of effective universes by normal sampling methods. The best we can do is to calculate the probability that the outcome fits what we observe, given some measure. In our problem, one possible measure that suggests itself is simply unit weight per unit volume within the multiverse, corresponding to the question: what does an average place look like? Another possible measure is unit weight per unit observer within the multiverse, corresponding to the question: what does an average observer observe? The first (measure V) is quite straightforward, in our immediate case, while the second (measure A, for anthropic) involves challenging issues, both practical and conceptual. Can we really tell which parameters support the emergence of observers, much less calculate how many? Do vastly more observers later count as much as relatively few today?[3] Should we really try to estimate the number of intelligent entities with distinct 'selfs' who actually form the notions of dark matter and baryons and measure r – or what?

Here I will briefly indicate a few key issues and tentative conclusions. A more definitive treatment can be found in ref. [13]. First, we consider the

---

3 This dynamic question, it seems to me, is especially relevant to anthropic reasoning about the dark energy. Universes with a smaller value of the effective cosmological term can support intelligent life for longer, and – plausibly – populations of intelligent life, once established, grow exponentially. It arises even for measure V: should we take spatial volume, spacetime volume, or something else?

situation with respect to measure V. If we define $F_R \sim 10^{12}$ GeV to be the value of F that leads to the observed dark matter density in the reference cosmology, then the probability of observing less than or equal to the density we do is, taking the $\sin^2(\theta/2)$ dependence literally, given by

$$L \equiv \frac{2}{\pi} \sin^{-1} \sqrt{\frac{F_R}{2F}}, \qquad (9.3)$$

while the probability of seeing more is one minus this. Note that any $F \geq F_R/2$ is allowed at some level, so that F could even be slightly smaller than F_R. We might claim victory, following measure V, if neither of these probabilities is terribly small. For $F/F_R = 10^2, 10^4, 10^6$ (the latter two roughly representing the unification and Planck scales, respectively), we find $L = 0.045$, $0.0045, 0.000\,45$. Viewed this way, really large values of F look unlikely.

Things appear quite different from the perspective of measure A. For a first pass, I will suppose that the number of observers is proportional to the number of baryons. In the relevant part of universal history, when the cosmological term is subdominant or nearly so, the baryon density at a fixed Hubble parameter – or, to an adequate approximation, fixed age of the Universe – depends on r as $\rho_b/(\rho_a + \rho_b) = 1/(1+r)$. Using Eq. (9.2), the probability that r is equal to or less than s, according to measure A, is then given by

$$L(s,u) = \frac{\int_0^w 1/(1 + 12u \sin^2 \phi) \mathrm{d}\phi}{\int_0^{\pi/2} 1/(1 + 12u \sin^2 \phi) \mathrm{d}\phi}, \qquad (9.4)$$

with $w \equiv \sin^{-1} \sqrt{s/(12u)}$ and $u \equiv F/F_R$. Half the probability is covered by $r \leq 1$, but there is plenty of weight around $r = 6$, even for very large values of F. The probability that r lies between 2 and 10 is very nearly 20%, whether u is 10 or 10^6!

Both very large and very small values of r may not be smart places to live [13]. At large r, it becomes difficult to make stars: in these baryon-poor universes the largest objects that cool and fragment, as opposed to relaxing into diffuse virial clouds, are too small to make stars efficiently. At small r, we have baryon-dominated universes, and we get Silk damping and slow growth of structure. These effects (and others) are hard to survey with confidence, at least for me; but I think they can only make a pretty good situation better. Indeed, we know everything works out nicely for r a little bigger than unity, so in that range we saturate the preceding estimate; these other complications will mainly suppress the competition.

9.4 Implications

The assumptions underlying the alternative axion cosmology I have pursued above have significant implications for axion physics, supersymmetry and cosmology. By pursuing these implications, we might be able either to enhance the credibility of their application to describe reality or to demolish that credibility.

Let us first consider axion physics. Laboratory searches for solar axions within the 'astrophysical window', or for cosmic background axions as dark matter, have been predicated on smaller values of F than assumed here. Large values of F imply weaker coupling to matter and render direct detection more difficult. So, unfortunately, the anthropically interesting scenario is incompatible with direct detection of axions in the foreseeable future. Of course, I would be quite happy to see it ruled out in this particular way! On the other hand, large values of F would appear to have some theoretical advantages. It might be possible to identify the PQ scale with the scale of gauge symmetry unification indicated by the successful calculation of the running of the coupling constants, for example. Independent of any particular model, the general idea that a single condensate might trigger breaking of several symmetries is quite attractive. There is also some advantage to having inflation occur after PQ symmetry-breaking, in that axion strings, which certainly complicate and might ruin the cosmology of axion dark matter, are diluted away.

Let us next consider supersymmetry. If axions dominate the dark matter density, then of course the dark matter candidate that arises in many models of low-energy supersymmetry does not. This candidate is often referred to interchangeably as the WIMP (weakly interacting massive particle) or the LSP (lightest supersymmetric particle), but it is convenient here to make a distinction. The framework in which the properties of LSP/WIMP particles are discussed is most often, either explicitly or in effect, the minimal supersymmetric extension of the Standard Model. In that framework, the lightest R-parity odd particle, the LSP, is stable on cosmological timescales, and for an otherwise plausible range of parameters – notoriously, several of the phenomenologically crucial parameters in models of low-energy supersymmetry are at present poorly constrained – one finds that it is indeed a weakly interacting particle whose density is predicted in big bang cosmology to be compatible with what astronomers find for dark matter. So the LSP can provide the cosmological WIMP. On the other hand, even within this framework there is an equally plausible range of parameters such that the LSP is produced with too small a density to provide the cosmological WIMP. The scenario discussed above therefore favours that range.

Along this line, if we accept the approximate equality of supersymmetric dark matter to baryonic matter as a *fait accompli* arising from a coincidence among disparate microscopic parameters, say, for concreteness, $\rho_{\text{LSP}}/\rho_{\text{baryon}} = 3$, then our anthropic scenario would at least make the additional coincidence $\rho_{\text{axion}}/\rho_{\text{baryon}} = 3$ appear less conspiratorial.

Another possibility, which I find especially intriguing and not at all implausible, is that the lightest supersymmetric particle is *not* the partner of a Standard Model particle. It could be the gravitino, the dilatino, the axino, a modulino or a combination of these. In these cases, the true LSP is generally a very feebly interacting particle, with coupling strength similar to a graviton, axion, etc. The pseudo-LSP that will be observed (at the LHC, presumably) as a Standard Model partner will decay into this true LSP. The decay will be rapid on cosmological timescales, so the pseudo-LSP sort of WIMP cannot supply the cosmological dark matter. Since the true LSP is very feebly interacting and relatively light, direct production of the true LSP during the big bang will not yield a cosmologically significant density of dark matter. It might be produced at a cosmologically significant level at relatively late times through decays of the pseudo-LSP, but it requires some special adjustments both to avoid wreaking cosmological havoc with these decays and to reproduce the observed dark matter abundance.

It is at least equally plausible to suppose that LSPs are not produced enough to make the observed dark matter, and that is an important independent motivation to consider axions as an alternative. An especially spectacular possibility is that the pseudo-LSP might be electrically charged. Cosmologically stable charged matter in the form of mass ~ 100 GeV particles produced with cosmological density comparable to the observed dark matter density is a phenomenological disaster, but I am emphasizing that the pseudo-LSP need not be stable. There are large, otherwise attractive regions of the parameter space for low-energy supersymmetry that have been excluded on these grounds, maybe prematurely. There is a wonderful signature for this possibility: the charged pseudo-LSP, produced at LHC, though unstable on cosmological timescales, could be stable on laboratory timescales.

Let us now consider cosmology. The most distinctive features of axions as dark matter, to wit that they are produced cold, in fact so cold that they fill out a very small region of phase space and form caustics, continue to hold in the alternative axion cosmology. If anything, their derivation is cleaner, since there are no axion strings, there is a clean specification of very simple initial conditions, and in the post-inflation period, since temperatures are well below F, axions have only very feeble non-gravitational interactions.

The initial misalignment angle, which eventually materializes as the dark matter density, can provide an independent source of cosmological density perturbations, apart from ambient temperature fluctuations. If we take the inflationary origin of fluctuations at face value, we find that this additional field provides a source of isocurvature fluctuations, whose amplitude depends on the scale of inflation. Recent observations put significant constraints on the amplitude of isocurvature fluctuations, so the scale of inflation cannot be too large; but perhaps the present scenario sharpens the motivation to search for them down to low levels.

Finally, it is tempting to connect the line of thought pursued here with the other context in which anthropic reasoning has been applied to cosmology recently, that is Weinberg's discussion of the cosmological term [14]. He framed his discussion rather abstractly, without specifying a microscopic model. It is quite simple to make a model along the lines discussed here. We can go back to the comforting – but of course totally unproved! – assumption that the asymptotic value of the cosmological term, in the distant future, is zero, and that what we are observing at present is residual energy frozen into a scalar axion-like field whose value is effectively uniform over the observable Universe. This requires a small value of Λ and a large value of F, relative to the axion that plays a role in the strong P and T problems, in order that the mass $m \sim \Lambda^2/F$ should be of order the inverse Hubble time, to ensure that the field is 'stuck': $\Lambda \sim 10^{-12}$ GeV, $F \sim 10^{19}$ GeV will do the job. These values also (barely!) assure, respectively, that the vacuum energy controlled by our field can supply enough for the observed cosmological term, and that it is associated with Planck-scale physics. The closeness of this call might be considered a small bit of encouragement for observational programmes to check whether the dark energy might have become unstuck and have started to evolve in recent cosmological times.

It could appear highly unnatural, upon first sight, that symmetry-breaking at such a large scale could be associated with so little energy. It generally would be, but in axion physics it is not so unreasonable. The point is that all effects of θ-like parameters are non-perturbative, and in weak coupling they contain explicit suppression factors such as $\mathrm{e}^{-8\pi^2/g^2}$. In QCD, that suppression is obscured, since g is not uniformly small, but it does not take much smallness in the g governing the relevant gauge theory to render this supression factor quite small.

Having made these alterations, we can repeat our cosmological story and – within this circle of ideas – justify Weinberg's hypothesis of effectively random variation of the cosmological term and its independence from other parameters. A minor difference is that negative values of the cosmological term do not appear. It would be logical, of course, and very interesting, to consider from an anthropic perspective the implications of allowing both the dark matter to baryon matter ratio r and the cosmological term Λ to vary independently but simultaneously. One must keep in mind that the inflated PQ horizon might be quite different from the corresponding horizon for the dark energy 'axion'. If that occurs, then we should vary one mismatch angle over the multiverse corresponding to the smaller horizon before varying the other over the 'multi-multiverse' associated with the larger horizon. The most probable value may be different, since the y that maximizes $f(x, y)$ is not the same as the y that maximizes the average $\langle f(x, y) \rangle_x$ taken over x, in general. A virtue of explicit dynamical models is that they bring subtleties such as this into the foreground.

References

[1] F. Crick. *The Astonishing Hypothesis: The Scientific Search for the Soul* (New York: Scribner, 2004).
[2] E. Baum. *What is Thought?* (Cambridge, MA: The MIT Press, 2004).
[3] J. Preskill, M. Wise and F. Wilczek. *Phys. Lett.* **B 120** (1983), 127.
[4] L. Abbott and P. Sikivie. *Phys. Lett.* **B 120** (1983), 133.
[5] M. Dine and W. Fischler. *Phys. Lett.* **B 120** (1983), 137.
[6] S-Y. Pi. *Phys. Rev. Lett.* **52** (1984), 1725.
[7] A. Linde. *Phys. Lett.* **B 201** (1988), 437.
[8] M. Turner and F. Wilczek. *Phys. Rev. Lett.* **66** (1991), 5.
[9] M. Shifman, ed. *Handbook of QCD* (Singapore: World Scientific, 2001).
[10] J. E. Kim. *Phys. Rep.* **150** (1987), 1.
[11] M. Turner. *Phys. Rep.* **197** (1990) 67.
[12] F. Wilczek. Four big questions with pretty good answers. In *Fundamental Physics – Heisenberg and Beyond*, eds. G. Buschhorn and J. Wess (Berlin: Springer 2004) [hep-ph/0201222, MIT-CTP-3236].
[13] M. Tegmark, A. Aguirre, M. J. Rees and F. Wilczek. *Phys. Rev.* **D 73** (2006), 023505 [astro-ph/0511774].
[14] S. Weinberg. *Phys. Rev. Lett.* **59** (1987), 2067.

10
Anthropic predictions: the case of the cosmological constant

Alexander Vilenkin
Department of Physics and Astronomy, Tufts University

10.1 Introduction

The parameters we call constants of nature may, in fact, be stochastic variables taking different values in different parts of the Universe. The observed values of these parameters are then determined by chance and by anthropic selection. It has been argued, at least for some of the constants, that only a narrow range of their values is consistent with the existence of life [1–5].

These arguments have not been taken very seriously and have often been ridiculed as handwaving and unpredictive. For one thing, the anthropic worldview assumes some sort of a 'multiverse' ensemble, consisting of multiple universes or distant regions of the same Universe, with the constants of nature varying from one member of this ensemble to another. Quantitative results cannot be obtained without a theory of the multiverse. Another criticism is that the anthropic approach does not make testable predictions; thus it is not falsifiable, and therefore not scientific.

While both of these criticisms had some force a couple of decades ago, much progress has been made since then, and the situation is now completely different. The first criticism no longer applies, because we now do have a theory of the multiverse. It is the theory of inflation. A remarkable feature of inflation is that, generically, it never ends completely. The end of inflation is a stochastic process; it occurs at different times in different parts of the Universe, and at any time there are regions which are still inflating [6,7]. If some 'constants' of nature are related to dynamical fields and are allowed to vary, they are necessarily randomized by quantum fluctuations during inflation and take different values in different parts of the Universe. Thus, inflationary cosmology gives a specific realization of the multiverse ensemble and makes it essentially inevitable. (For a review, see ref. [8].)

Universe or Multiverse?, ed. Bernard Carr. Published by Cambridge University Press.
© Cambridge University Press 2007.

In this chapter, I am going to address the second criticism: that anthropic arguments are unpredictive. I will try to dispel this notion and outline how anthropic models can be used to make quantitative predictions. These predictions are of a statistical nature, but they still allow models to be confirmed or falsified at a specified confidence level. I will focus on the case of the cosmological constant, whose non-zero value was predicted anthropically well before it was observed. This case is of great interest in its own right and is well suited to illustrate the issues associated with anthropic predictions.

10.2 Anthropic bounds versus anthropic predictions

For terminological clarity, it is important to distinguish between anthropic bounds and anthropic predictions. Suppose there is some parameter X, which varies from one place in the Universe to another. Suppose further that the value of X affects the chances for intelligent observers to evolve, and that the evolution of observers is possible only if X is within some interval:

$$X_{\min} < X < X_{\max}. \qquad (10.1)$$

Clearly, values of X outside this interval are not going to be observed, because such values are inconsistent with the existence of observers. This statement is often called the 'anthropic principle'.

Although anthropic bounds, like Eq. (10.1), can have considerable explanatory power, they can hardly be regarded as predictions: they are guaranteed to be right. And the 'anthropic principle', as stated above, hardly deserves to be called a principle; it is trivially true. This is not to say, however, that anthropic arguments cannot yield testable predictions.

Suppose we want to test a theory according to which the parameter X varies from one part of the Universe to another.[1] Then, instead of looking for the extreme values $X_{\min}$ and $X_{\max}$ that make observers impossible, we can try to predict what values of X will be measured by *typical* observers. In other words, we can make statistical predictions, assigning probabilities $P(X)$ to different values of X. ($P(X)$ is the probability that an observer randomly picked in the Universe will measure a given value of X.) If any principle needs to be invoked here, it is what I call 'the principle of mediocrity' [9] – the assumption that we are typical among the observers in the Universe. Quantitatively, this can be expressed as the expectation that we should find ourselves, say, within the 95% range of the distribution. This can be regarded as a prediction at a 95% confidence level. If instead we measure

---

[1] I assume, for simplicity, that X is variable only in space, not in time.

a value outside the expected range, this should be regarded as evidence against the theory.

10.3 The cosmological constant problem

The cosmological constant is (up to a numerical factor) the energy density of the vacuum, ρ_v. Below, I do not distinguish between the two and use the terms 'cosmological constant' and 'vacuum energy density' interchangeably. By Einstein's mass–energy relation, the energy density is simply related to the mass density, and I will often express ρ_v in units of g cm^{-3}.

The gravitational properties of the vacuum are rather unusual: for positive ρ_v, its gravitational force is repulsive. This can be traced to the fact that, according to Einstein's General Relativity, the force of gravity is determined not solely by the energy (mass) density ρ, but rather by the combination $(\rho+3P)$, where P is the pressure and I put $c=1$. In ordinary astrophysical objects, such as stars or galaxies, the pressure is much smaller than the energy density, $P \ll \rho$, and its contribution to gravity can be neglected. But, in the case of vacuum,[2] the pressure is equal and opposite to ρ_v,

$$P_v = -\rho_v, \tag{10.2}$$

so that $\rho_v + 3P_v = -2\rho_v$. Pressure not only contributes significantly to the gravitational force produced by the mass, it also changes its sign.

The cosmological constant was introduced by Einstein in his 1917 paper [10], in which he applied the newly developed theory of General Relativity to the Universe as a whole. Einstein believed that the Universe was static, but to his dismay he found that the theory had no static cosmological solutions. He concluded that the theory had to be modified and he introduced the cosmological term, which amounted to endowing the vacuum with a positive energy density. The magnitude of ρ_v was chosen so that its repulsive gravity exactly balanced the attractive gravity of matter, resulting in a static world. More than a decade later, after Hubble's discovery of the expansion of the Universe, Einstein abandoned the cosmological constant, calling it the greatest blunder of his life. But once the genie was out of the bottle, it was not so easy to put it back.

Even if we do not introduce the vacuum energy 'by hand', fluctuations of quantum fields, like the electromagnetic field, would still make this energy non-zero. Adding up the energies of quantum fluctuations with shorter and shorter wavelengths gives a formally infinite answer for ρ_v. The sum has

---

2 Since the vacuum energy is proportional to the volume V it occupies, $E = \rho_v V$, the pressure is $P_v = -dE/dV = -\rho_v$.

to be cut off at the Planck length, $l_P \sim 10^{-33}$ cm, where quantum gravity effects become important and the usual concepts of space and time no longer apply. This gives a finite, but absurdly large value: $\rho_v \sim 10^{94}$ g cm^{-3}. A cosmological constant of this magnitude would cause the Universe to expand with a stupendous acceleration. If indeed our vacuum has energy, it should be at least 120 orders of magnitude smaller in order to be consistent with observations. In supersymmetric theories, the contributions of different fields partially cancel, and the discrepancy can be reduced to 60 orders of magnitude. This discrepancy between the expected and observed values of ρ_v is called the cosmological constant problem. It is one of the most intriguing mysteries that we are now facing in theoretical physics.

10.4 The anthropic bound

A natural resolution to the cosmological constant problem is obtained in models where ρ_v is a random variable. The idea is to introduce a dynamical dark energy component X whose energy density ρ_X varies from place to place, due to stochastic processes that occurred in the early Universe. A possible model for ρ_X is a scalar field with a very flat potential [11, 12], such that the field is driven to its minimum on an extremely long timescale, much longer than the present age of the Universe. Another possibility is a discrete set of vacuum states. Transitions between different states can then occur through nucleation and expansion of bubbles bounded by domain walls [13, 14]. The effective cosmological constant is given by $\rho_v = \rho_\Lambda + \rho_X$, where ρ_Λ is the constant vacuum energy density, which may be as large as 10^{94} g cm^{-3} (with either sign). The cosmological constant problem now takes a different form: the puzzle is why we happen to live in a region where ρ_Λ is nearly cancelled by ρ_X.

The key observation, due to Weinberg [15] (see also refs. [3], [11] and [16]), is that the cosmological constant can have a dramatic effect on the formation of structure in the Universe. The observed structures – stars, galaxies and galaxy clusters – evolved from small initial inhomogeneities, which grew over eons of cosmic time by gravitationally attracting matter from surrounding regions. As the Universe expands, matter is diluted, so its density goes down as follows:

$$\rho_M = (1+z)^3 \rho_{M0}, \qquad (10.3)$$

where ρ_{M0} is the present matter density and z is the redshift.[3] At the same time, the density contrast $\sigma \equiv \delta\rho/\rho$ between overdense and underdense

---

3 The redshift z is defined so that $(1+z)$ is the expansion factor of the Universe between a given epoch and the present (earlier times correspond to larger redshifts).

regions keeps growing. Gravitationally bound objects form where $\sigma \sim 1$. In the standard model the first stars form in relatively small matter clumps of mass $\sim 10^6 M_\odot$. The clumps then merge into larger and larger objects, leading to the formation of giant galaxies like our own and galaxy clusters.

How is this picture modified in the presence of a cosmological constant? At early times, when the density of matter is high, $\rho_M \gg \rho_v$, the vacuum energy has very little effect on structure formation. But, as the Universe expands and the matter density decreases, the vacuum density ρ_v remains constant and eventually becomes greater than ρ_M. At this point, the character of cosmic expansion changes. Prior to vacuum-domination, the expansion is slowed down by gravity, but afterwards it begins to accelerate, due to the repulsive gravity of the vacuum. Weinberg showed that the growth of density inhomogeneities effectively stops at that epoch. If no structures were formed at earlier times, then none will *ever* be formed.

It seems reasonable to assume that the existence of stars is a necessary prerequisite for the evolution of observers. We also need to require that the stars belong to sufficiently large bound objects – galaxies – so that their gravity is strong enough to retain the heavy elements dispersed in supernova explosions. These elements are necessary for the formation of planets and observers. An anthropic bound on the vacuum energy can then be obtained by requiring that ρ_v does not dominate before the redshift z_{max} when the earliest galaxies are formed. With the aid of Eq. (10.3), this yields

$$\rho_v \leq (1 + z_{max})^3 \rho_{M0}. \tag{10.4}$$

The most distant galaxies observed at the time when Weinberg wrote his paper had redshifts $z \sim 4.5$. Assuming that $z_{max} \sim 4.5$, Eq. (10.4) yields the bound $\rho_v \leq 170 \rho_{M0}$. A more careful analysis by Weinberg showed that in order to prevent structure formation, ρ_v needs to be three times greater than suggested by Eq. (10.4), so a more accurate bound is given by

$$\rho_v \leq 500 \rho_{M0} \tag{10.5}$$

(see ref. [15]). Of course, the observation of galaxies at $z \sim 4.5$ means only that $z_{max} \geq 4.5$, and Weinberg referred to Eq. (10.5) as 'a lower bound on the anthropic upper bound on ρ_v'. At present, galaxies are observed at considerably higher redshifts, up to $z \sim 10$. The corresponding bound on ρ_v would be

$$\rho_v \leq 4000 \rho_{M0}. \tag{10.6}$$

For negative values of ρ_v, the vacuum gravity is attractive, and vacuum-domination leads to a rapid recollapse of the Universe. An anthropic lower bound on ρ_v can be obtained in this case by requiring that the Universe does not recollapse before life had a chance to develop [3, 17]. Assuming that the timescale for life evolution is comparable to the present cosmic time, one finds[4] $\rho_v \geq -\rho_{M0}$.

The anthropic bounds are narrower, by many orders of magnitude, than the particle physics estimates for ρ_v. Moreover, as Weinberg noted, there is a prediction implicit in these bounds. He wrote [18]:

...if it is the anthropic principle that accounts for the smallness of the cosmological constant, then we would expect a vacuum energy density $\rho_v \sim (10-100)\rho_{M0}$, because there is no anthropic reason for it to be any smaller.

One has to admit, however, that the anthropic bounds fall short of the observational bound, $(\rho_v)_{obs} \leq 4\rho_{M0}$, by a few orders of magnitude. If all the values in the anthropically allowed range were equally probable, an additional fine-tuning by a factor of 100–1000 would still be needed.

10.5 Anthropic predictions

The anthropic bound given by Eq. (10.4) specifies the value of ρ_v which makes galaxy formation barely possible. However, if ρ_v varies in space, then most of the galaxies will not be in regions characterized by these marginal values, but rather in regions where ρ_v dominates after a substantial fraction of matter had already clustered into galaxies.

To make this quantitative, we define the probability distribution $\mathcal{P}(\rho_v)d\rho_v$ as being proportional to the number of observers in the Universe who will measure ρ_v in the interval $d\rho_v$. This distribution can be represented as a product [9]:

$$\mathcal{P}(\rho_v)d\rho_v = n_{obs}(\rho_v)\mathcal{P}_{prior}(\rho_v)d\rho_v. \tag{10.7}$$

Here, $\mathcal{P}_{prior}(\rho_v)d\rho_v$ is the prior distribution, which is proportional to the volume of those parts of the Universe where ρ_v takes values in the interval $d\rho_v$, and $n_{obs}(\rho_v)$ is the number of observers that are going to evolve per unit volume. The distribution given by Eq. (10.7) gives the probability that a randomly selected observer is located in a region where the effective cosmological constant is in the interval $d\rho_v$.

Of course, we have no idea how to calculate n_{obs}, but what comes to the rescue is the fact that the value of ρ_v does not directly affect the physics and chemistry of life. As a rough approximation, we can then assume that

---

[4] An important distinction between positive and negative values of ρ_v is that, for $\rho_v > 0$, galaxies that formed prior to vacuum-domination can survive indefinitely in the vacuum-dominated Universe.

$n_{\text{obs}}(\rho_v)$ is simply proportional to the fraction of matter f clustered in giant galaxies like ours (with mass $M \geq M_G = 10^{12} M_\odot$),

$$n_{\text{obs}}(\rho_v) \propto f(M_G, \rho_v). \tag{10.8}$$

The idea is that there is a certain number of stars per unit mass in a galaxy and a certain number of observers per star. The choice of the galactic mass M_G is an important issue; I will comment on it in Section 10.6.

The calculation of the prior distribution $\mathcal{P}_{\text{prior}}(\rho_v)$ requires a particle physics model which allows ρ_v to vary and a multiverse model that would generate an ensemble of sub-universes with different values of ρ_v. An example of a suitable particle theory is superstring theory, which appears to admit an incredibly large number of vacua (possibly as large as 10^{1000} [19–21]), characterized by different values of particle masses, couplings and other parameters, including the cosmological constant. When this is combined with the cosmic inflation scenario, one finds that bubbles of different vacua copiously nucleate and expand during inflation, producing exponentially large regions with all possible values of ρ_v. Given a particle physics model and a model of inflation, one can in principle calculate $\mathcal{P}_{\text{prior}}(\rho_v)$. Examples of calculations for specific models have been given in refs. [12], [22] and [23].[5] Needless to say, the details of the fundamental theory and the inflationary dynamics are too uncertain for a definitive calculation of $\mathcal{P}_{\text{prior}}$. We shall instead rely on the following general argument [28, 29].

Suppose some parameter X varies in the range ΔX and is characterized by a prior distribution $\mathcal{P}_{\text{prior}}(X)$. Suppose further that X affects the number of observers in such a way that this number is non-negligible only in a very narrow range $\Delta X_{\text{obs}} \ll \Delta X$. Then one can expect that the function $\mathcal{P}_{\text{prior}}(X)$ with a large characteristic range of variation should be very nearly constant in the tiny interval ΔX_{obs}.[6] In the case of ρ_v, the range $\Delta \rho_v$ is set by the Planck scale or by the supersymmetry-breaking scale, and we have $(\Delta \rho_v)_{\text{obs}}/\Delta \rho_v \sim 10^{-60}$–$10^{-120}$. Hence, we expect

$$\mathcal{P}_{\text{prior}}(\rho_v) \approx \text{const.} \tag{10.9}$$

---

5 There are still some open issues regarding the definition of $\mathcal{P}_{\text{prior}}$ for models with a discrete spectrum of variable 'constants' [24–26]. For a recent discussion and a proposed resolution, see ref. [27].
6 A very different model for the prior distribution was considered by Rubakov and Shaposhnikov [30]. They assumed that $\mathcal{P}_{\text{prior}}(X)$ is a sharply peaked function with a peak outside the anthropic range $\mathcal{A}$ and argued that the observed value of X should then be very close to the boundary of $\mathcal{A}$. We note that, in this case, the peak of the full distribution is likely to be in a life-hostile environment, where both $\mathcal{P}_{\text{prior}}(X)$ and $n_{\text{obs}}(X)$ are very small. In the case of the cosmological constant, this would mean that the number density of galaxies is very low. This is not the case in our observable region, indicating that the model of ref. [30] does not apply.

I emphasize that the assumption here is that the value $\rho_{\rm v} = 0$ is not in any way special, as far as the fundamental theory is concerned, and is, therefore, not a singular point of $\mathcal{P}_{\rm prior}(\rho_{\rm v})$.

Combining Eqs. (10.7) to (10.9), we obtain

$$\mathcal{P}(\rho_{\rm v}) \propto f(M_{\rm G}, \rho_{\rm v}). \tag{10.10}$$

In ref. [9], where I first introduced the anthropic probability distributions of the form given by Eq. (10.7), I did not attempt a detailed calculation of the distribution for $\rho_{\rm v}$, resorting instead to a rough estimate. If we denote by $z_{\rm G}$ the redshift at the epoch of galaxy formation, then most of the galaxies should be in regions where the vacuum energy dominates only after some redshift $z_{\rm v} \leq z_{\rm G}$. Regions with $z_{\rm v} \gg z_{\rm G}$ will have very few galaxies, while regions with $z_{\rm v} \ll z_{\rm G}$ will be rare, simply because they correspond to a very narrow range of $\rho_{\rm v}$ near zero. Hence, we expect a typical galaxy to be located in a region where

$$z_{\rm v} \sim z_{\rm G}. \tag{10.11}$$

The expected value of $\rho_{\rm v}$ is then given by

$$\rho_{\rm v} \sim (1 + z_{\rm G})^3 \rho_{\rm M0}. \tag{10.12}$$

The choice of the galaxy formation epoch $z_{\rm G}$ is related to the choice of the galactic mass $M_{\rm G}$ in Eq. (10.8). I used $z_{\rm G} \sim 1$, obtaining $\rho_{\rm v} \sim 8\rho_{\rm M0}$.

A similar approach was later developed by Efstathiou [31]. The main difference is that he calculated the fraction of clustered matter f at the time corresponding to the observed value of the microwave background temperature, $T_0 = 2.73$ K, while my suggestion was to use the asymptotic value of f as $t \to \infty$. The two approaches correspond to different choices of the reference class of observers, among whom we expect to be typical. Efstathiou's choice includes (roughly) only observers that have evolved until the present, while my choice includes all observers – present, past and future. If we are truly typical, and live at the time when most observers live, the two methods should give similar results. Indeed, one finds that the probability distributions calculated by these methods are nearly identical.[7]

10.6 Comparison with observations

Despite a number of observational hints that the cosmological constant might be non-zero (see, for example, ref. [32]), its discovery still came as a

---

[7] L. Pogosian (private communication). The original calculation by Efstathiou gave a different result, but that calculation contained an error, which was later pointed out by Weinberg [29].

10 Anthropic predictions: the cosmological constant

great surprise to most physicists and astronomers. Observations of distant supernovae by two independent groups in 1997–98 provided strong evidence that the expansion of the Universe is accelerating [33]–[35]. The simplest interpretation of the data was in terms of a cosmological constant with $\rho_V \sim 2.3\rho_{M0}$. Further evidence came from the cosmic microwave background and galaxy clustering observations, and by now the case for the cosmological constant is very strong.

The discovery of the cosmological constant was particularly shocking to particle physicists, who almost universally believed that it should be zero. They assumed that something so small could only be zero and searched for a new symmetry principle or a dynamical adjustment mechanism that would force ρ_V to vanish. The observed value of ρ_V brought yet another puzzle. The matter density ρ_M and the vacuum energy density ρ_V scale very differently with the expansion of the Universe. In the early Universe the matter density dominates, while in the asymptotic future it becomes negligible. There is only one epoch in the history of the Universe when $\rho_M \sim \rho_V$. It is difficult to understand why we happen to live in this very special epoch. This is the so-called cosmic coincidence problem.

The coincidence is easily understood in the framework of the anthropic approach [36,37]. The galaxy formation epoch, $z_G \sim 1$–3, is close to the present cosmic time, and the anthropic model predicts that vacuum-domination should begin at $z \sim z_G$ from Eq. (10.11). This explains the coincidence.

The probability distribution for ρ_V based on Eq. (10.10) was extensively analyzed in ref. [38]. The distribution depends on the amplitude of galactic-scale density perturbations, σ, which can be specified at some suitably selected epoch (for example, the epoch of recombination). Until recently, significant uncertainties in this quantity complicated the comparison of anthropic predictions with the data [23,38]. These uncertainties appear now to have been mostly resolved [39]. In Fig. 10.1 we plot, following ref. [40], the resulting probability distribution per logarithmic interval of ρ_V. Only positive values of ρ_V are considered, so this can be regarded as a conditional distribution, given that $\rho_V > 0$. On the horizontal axis, ρ_V is plotted in units of the observed vacuum energy density, $\rho_V^* = 7 \times 10^{-30}$ g cm^{-3}. The ranges of the distribution excluded at the 68% and 95% confidence levels are indicated by light and dark shading, respectively.

We note that the confidence level ranges in Fig. 10.1 are rather broad. This corresponds to a genuine large variance in the cosmic distribution of ρ_V. The median value of the distribution is about twenty times greater than the observed value. But still, the observed value ρ_V^* falls well within the range of anthropic prediction at the 95% confidence level.

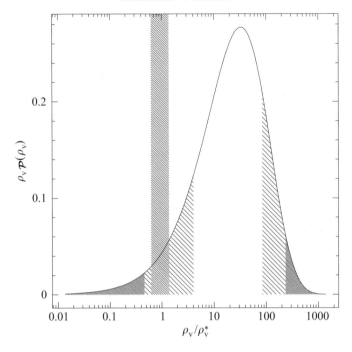

Fig. 10.1. The probability density per logarithmic interval of $\rho_{\rm v}$. The lightly and densely shaded areas are the regions excluded at 68% and 95% levels, respectively. The uncertainty in the observed value $\rho_{\rm v}^*$ is indicated by the vertical strip.

At this point, I would like to comment on some important assumptions that went into the successful prediction of the observed value of $\rho_{\rm v}$. First, we assumed the flat prior probability distribution given by Eq. (10.9). Analysis of specific models shows that this assumption is indeed valid in a wide class of cases, but it is not as automatic as one might expect [12, 22, 41, 42]. In particular, it is not clear that it is applicable to superstring-inspired models of the type discussed in refs. [19]–[21]. (We discuss this further in Section 10.8.)

Second, we used the value of $M_{\rm G} = 10^{12} M_\odot$ for the galactic mass in Eq. (10.10). This amounts to assuming that most observers live in giant galaxies like the Milky Way. We know from observations that some galaxies existed already at $z = 10$, and the theory predicts that some dwarf galaxies and dense central parts of giant galaxies could form as early as $z = 20$. If observers were as likely to evolve in galaxies that formed early as in those that formed late, the value of $\rho_{\rm v}$ indicated by Eq. (10.12) would be far greater than observed. Clearly, the agreement is much better if we assume that the conditions for civilizations to emerge arise mainly in galaxies which form at lower redshifts, $z_{\rm G} \sim 1$.

Following ref. [42], I will now point to some directions along which the choice of $z_G \sim 1$ may be justified. As already mentioned, one problem with dwarf galaxies is that their mass may be too small to retain the heavy elements dispersed in supernova explosions. Numerical simulations suggest that the fraction of heavy elements retained is $\sim 30\%$ for a $10^9 M_\odot$ galaxy and is negligible for much smaller galaxies [43]. Hence, we have to require that the structure formation hierarchy evolves up to mass scales $\sim 10^9 M_\odot$ or higher prior to the vacuum energy dominating. This gives the condition $z_G \leq 3$, but falls short of explaining $z_G \sim 1$.

Another point to note is that smaller galaxies, formed at earlier times, have a higher density of matter. This may increase the danger of nearby supernova explosions and the rate of near encounters with stars, large molecular clouds or dark matter clumps. Gravitational perturbations of planetary systems in such encounters could send a rain of comets from the Oort-type clouds towards the inner planets, causing mass extinctions.

Our own galaxy has definitely passed the test for the evolution of observers, and the principle of mediocrity suggests that most observers may live in galaxies of this type. The Milky Way is a giant spiral galaxy. The dense central parts of such galaxies were formed at a high redshift, $z \geq 5$, but their discs were assembled at $z \leq 1$ [44]. Our Sun is located in the disc, and if this situation is typical, then the relevant epoch to use in Eq. (10.12) is the redshift $z_G \sim 1$ associated with the formation of the discs of giant galaxies.[8]

It should be clear from this discussion that the confidence ranges in Fig. 10.1 are not to be taken too literally. The distribution in Fig. 10.1 is based on Eq. (10.8), which is only a very rough model for the density of observers. It assumes that all galaxies of mass $M > M_G$, regardless of when they are formed, will have the same average number of observers per unit mass. This probably overestimates the density of observers at high values of ρ_v, corresponding to denser galaxies which are more hazardous for life. More accurate models for $n_{obs}(\rho_v)$ will require a better understanding of galactic evolution and the conditions necessary to sustain habitable planetary systems.

10.7 Predictions for the equation of state

A generic prediction of anthropic models for the vacuum energy is that the vacuum equation of state given by Eq. (10.2) should hold with very high accuracy [42]. In models of discrete vacua, this equation of state is

---

8 These remarks may or may not be on the right track, but if the observed value of ρ_v is due to anthropic selection, then, for one reason or another, the evolution of intelligent life should require conditions which are found mainly in giant galaxies, which completed their formation at $z_G \sim 1$. This is a prediction of the anthropic approach.

guaranteed by the fact that in each vacuum the energy density is a constant and can only change by nucleation of bubbles. If ρ_X is a scalar field potential, it must satisfy the slow-roll condition, i.e. the field should change slowly on the timescale of the present age of the Universe. The slow-roll condition is likely to be satisfied by many orders of magnitude. Although it is possible to adjust the potential so that it is only marginally satisfied, it is satisfied by a very wide margin in generic models. This implies the equation of state given by Eq. (10.2).

There is also a related prediction, which is not likely to be tested anytime soon. In anthropic models, ρ_v can take both positive and negative values, so the observed positive dark energy will eventually start decreasing and turn negative, and our part of the Universe will recollapse to a big crunch. Since the evolution of ρ_v is expected to be very slow on the present Hubble scale, we do not expect this to happen sooner than a trillion years from now [42].

It should be noted that the situation may be different in more complicated models, involving more than one scalar field. It has been shown in ref. [23] that the equation of state in such models may significantly deviate from Eq. (10.2), and the recollapse may occur on a timescale comparable to the lifetime of the Sun. Observational tests distinguishing between the two types of models have been discussed in refs. [45–47]. Recent observations [39] yield $P_v/\rho_v = -1 \pm 0.1$, consistent with the simplest models.

10.8 Implications for particle physics

Anthropic models for the cosmological constant have non-trivial implications for particle physics. Scalar field models require the existence of fields with extremely flat potentials. Models with a discrete set of vacua require that the spectrum of values of ρ_v should be very dense, so that there are many such values in the small anthropically allowed range. This points to the existence of very small parameters that are absent in familiar particle physics models. Some ideas on how such small parameters could arise have been suggested in refs. [12], [41] and [48–51].

A different possibility, which has now attracted much attention, is inspired by superstring theory. This theory presumably has an enormous number of different vacua, scattered over a vast 'string theory landscape'. The spectrum of ρ_v (and of other particle physics constants) can then be very dense without any small parameters, due to the sheer number of vacua [19–21]. This picture, however, entails a potential problem. Vacua with close values of ρ_v are not expected to be close to one another in the 'landscape', and there seems to be no reason to expect that they will be chosen with equal probability by the inflationary dynamics. Hence, we can no longer argue that the prior probability distribution is flat. In fact, since inflation is characterized by an exponential expansion of the Universe, and the expansion rate is different in different parts of the landscape, the probabilities for well separated vacua are likely to differ by large exponential factors. If indeed the prior distribution is very different from flat, this may destroy the successful anthropic prediction for ρ_v. This issue requires further study, and I am sure we are going to hear more about it.

10.9 Including other variables

If the cosmological constant is variable, then it is natural to expect that some other 'constants' could vary as well, and it has been argued that including other variables may drastically modify the anthropic prediction for ρ_v [4, 52, 53]. The idea is that the adverse effect on the evolution of observers due to a change in one variable may be compensated by an appropriate change in another variable. As a result, the peak of the distribution may drift into a totally different area of the parameter space. While this is a legitimate concern, specific models with more than one variable that have been analyzed so far suggest that the anthropic prediction for ρ_v is rather robust.

Suppose, for example, that ρ_v and the primordial density contrast σ (specified at recombination) are both allowed to vary. Then we are interested in the joint distribution

$$\mathcal{P}(\rho_\text{v}, \sigma) \mathrm{d}\rho_\text{v} \mathrm{d}\sigma. \tag{10.13}$$

Using the same assumptions[9] as in Section 10.6 and introducing a new variable $y = \rho_\text{v}/\sigma^3$, one finds [42] that this distribution factorizes to the following form:[10]

$$\sigma^3 \mathcal{P}_\text{prior}(\sigma) \mathrm{d}\sigma \cdot f(y) \mathrm{d}y, \tag{10.14}$$

---

9 The assumption that the number of observers is simply proportional to the fraction of matter clustered into galaxies may not give a good approximation in regions where σ is very large. In such regions, galaxies form early and are very dense, so the chances for life to evolve may be reduced. A more accurate calculation should await better estimates for the density of habitable stellar systems.

10 Note that there is no reason to expect the prior distribution for σ to be flat. The amplitude of density perturbations is related to the dynamics of the inflaton field that drives inflation and is therefore strongly correlated with the amount of inflationary expansion. Hence, we expect $\mathcal{P}_\text{prior}$ to be a non-trivial function of σ. In fact, it follows from Eq. (10.14) that $\mathcal{P}_\text{prior}(\sigma)$ should decay at least as fast as σ^{-3} in order for the distribution to be integrable [36].

where $f(y)$ is the fraction of matter clustered in galaxies (which depends only on the combination ρ_v/σ^3).

After integration over σ, we obtain essentially the same distribution as before, but for a new variable y. The prediction now is not for a particular value of ρ_v, but for a relation between ρ_v and σ. Comparison of the predicted and observed values of y is given by the same graph as in Fig. 10.1, with a suitable rescaling of the horizontal axis. As before, the 95% confidence level prediction is in agreement with the data.

Another example is a model where the neutrino masses are assumed to be anthropic variables. Neutrinos are elusive light particles, which interact very weakly and whose masses are not precisely known. The current astrophysical upper bound on the neutrino mass is $m_\nu \leq 0.5$ eV [39] and the lower bound from the neutrino oscillation data is $m_\nu \geq 0.05$ eV [54]. (In what follows, m_ν denotes the sum of the three neutrino masses.) It has been suggested in ref. [55] that small values of the neutrino masses may be due to anthropic selection. A small increase of m_ν can have a large effect on galaxy formation. Neutrinos stream out of overdense regions, slowing the growth of density perturbations. The fraction of mass that neutrinos contribute to the total density of the Universe is proportional to m_ν. Thus, perturbations will grow slower, and there will be fewer galaxies, in regions with larger values of m_ν. A calculation along the same lines as in Section 10.5 yields a prediction 0.07 eV $< m_\nu < 5.7$ eV at the 95% confidence level.

In ref. [40] this model was extended, allowing both m_ν and ρ_v to be anthropic variables. The resulting probability distribution $\mathcal{P}(\rho_v, m_\nu)$ is concentrated in a localized region of the parameter space. Its peak is not far from the peaks of the individual distributions for ρ_v and m_ν. In fact, inclusion of m_ν somewhat improves the agreement of the prediction for ρ_v with the data.

The parameters ρ_v, σ and m_ν share the property that they do not directly affect life processes. Other parameters of this sort include the mass of dark matter particles and the number of baryons per photon. The effects of varying these parameters have been discussed in refs. [4] and [52]. In particular, Aguirre [52] argued that values of the baryon-to-photon ratio much higher than observed may be anthropically favoured. What he showed, in fact, is that this proposition cannot at present be excluded. This is an interesting issue and certainly deserves further study. Extensions to parameters such as the electron mass or charge, which do affect life processes, are on much shakier ground. Until these processes are much better understood, we will have to resort to qualitative arguments, as in refs. [1–3] and [5].

10.10 Concluding remarks

The case of the cosmological constant demonstrates that anthropic models can be subjected to observational tests and can be confirmed or ruled out at a specified confidence level. It also illustrates the limitations and difficulties of anthropic predictions.

The situation we are accustomed to in physics is that the agreement between theory and observations steadily improves, as the theoretical calculations are refined and the accuracy of measurements increases. This does not apply in anthropic models. Here, predictions are in the form of probability distributions, having an intrinsic variance which cannot be further reduced.

However, there is ample possibility for anthropic models to be falsified. This could have happened in the case of the cosmological constant if the observed value turned out to be much smaller than it actually is. And this may still happen in the future, with improved understanding of the prior and anthropic factors in the distribution given by Eq. (10.7). Also, there is always a possibility that a compelling non-anthropic explanation for the observed value of ρ_v will be discovered. As of today, no such explanation has been found, and the anthropic model for ρ_v can certainly be regarded a success. This may be the first evidence that we have for the existence of a vast multiverse beyond our horizon.

Acknowledgements

I am grateful to Jaume Garriga and Ken Olum for comments and to Levon Pogosian for his help with numerical calculations and Fig. 10.1. This work was supported in part by the National Science Foundation.

References

[1] B. Carter. In *Confrontation of Cosmological Theories with Observational Data*, ed. M. Longair (Dordrecht: Reidel, 1974), p. 291.
[2] B. J. Carr and M. J. Rees. *Nature* **278** (1979), 605.
[3] J. D. Barrow and F. J. Tipler. *The Anthropic Cosmological Principle* (Oxford: Clarendon, 1986).
[4] M. Tegmark and M. J. Rees. *Astrophys. J.* **499** (1998), 526.
[5] C. J. Hogan. *Rev. Mod. Phys.* **72** (2000), 1149.
[6] A. Vilenkin. *Phys. Rev.* **D 27** (1983), 2848.
[7] A. D. Linde. *Phys. Lett.* **B 175** (1986), 395.
[8] A. D. Linde. *Particle Physics and Inflationary Cosmology* (Chur, Switzerland: Harwood Academic, 1990).
[9] A. Vilenkin. *Phys. Rev. Lett.* **74** (1995), 846.
[10] A. Einstein. *Sitz. Ber. Preuss. Akad. Wiss.* **142** (1917).

[11] A. D. Linde. In *300 Years of Gravitation*, eds. S. W. Hawking and W. Israel (Cambridge: Cambridge University Press, 1987).
[12] J. Garriga and A. Vilenkin. *Phys. Rev.* **D 61** (2000), 083502.
[13] J. D. Brown and C. Teitelboim. *Nucl. Phys.* **B 279** (1987), 787.
[14] L. Abbott. *Phys. Lett.* **B 195** (1987), 177.
[15] S. Weinberg. *Phys. Rev. Lett.* **59** (1987), 2607.
[16] P. C. W. Davies and S. Unwin. *Proc. Roy. Soc.* **377** (1981), 147.
[17] R. Kallosh and A. D. Linde. *Phys. Rev.* **D 67** (2003), 023510.
[18] S. Weinberg. *Rev. Mod. Phys.* **61** (1989), 1.
[19] R. Bousso and J. Polchinski. *JHEP*, **0006** (2000), 006.
[20] M. R. Douglas. *JHEP* **0305**, (2003), 046.
[21] L. Susskind. This volume (2007) [hep-th/0302219].
[22] J. Garriga and A. Vilenkin. *Phys. Rev.* **D 64** (2001), 023517.
[23] J. Garriga, A. D. Linde and A. Vilenkin. *Phys. Rev.* **D 61** (2004), 063521.
[24] A. D. Linde, D. A. Linde and A. Mezhlumian. *Phys. Rev.* **D 49** (1994) 1783.
[25] J. Garcia-Bellido and A. D. Linde. *Phys. Rev.* **D 51** (1995), 429.
[26] J. Garriga and A. Vilenkin. *Phys. Rev.* **D 64** (2001), 023507.
[27] J. Garriga, D. Schwartz-Perlov, A. Vilenkin and S. Winitzki. *J. Cosmol. Astropart. Phys.*, **0601** (2006), 017.
[28] A. Vilenkin. In *Cosmological Constant and the Evolution of the Universe*, eds. K. Sato *et al.* (Tokyo: Universal Academy Press, 1996).
[29] S. Weinberg. In *Critical Dialogues in Cosmology*, ed. N. G. Turok (Singapore: World Scientific, 1997).
[30] V. A. Rubakov and M. E. Shaposhnikov. *Mod. Phys. Lett.* **A 4** (1989), 107.
[31] G. Efstathiou. *Mon. Not. Roy. Astron. Soc.* **274** (1995), L73.
[32] M. Fukugita. In *Critical Dialogues in Cosmology*, ed. N. G. Turok (Singapore: World Scientific, 1997)
[33] S. Perlmutter, S. Gabi, G. Goldhaber *et al. Astrophys. J.* **483** (1997), 565.
[34] B. P. Schmidt, N. B. Suntzeff, M. M. Phillips *et al. Astrophys. J.* **507** (1998), 46.
[35] A. G. Riess, A. V. Filippenko, P. Challis *et al. Astron. J.* **116** (1998), 1009.
[36] J. Garriga, M. Livio and A. Vilenkin. *Phys. Rev.* **D 61** (2000), 023503.
[37] S. Bludman. *Nucl. Phys.* **A 663** (2000), 865.
[38] H. Martel, P. R. Shapiro and S. Weinberg. *Astrophys. J.* **492** (1998), 29.
[39] M. Tegmark, M. A. Strauss, M. R. Blanton *et al. Phys. Rev.* **D 69** (2004), 103501.
[40] L. Pogosian, M. Tegmark and A. Vilenkin. *J. Cosmol. Astropart. Phys.* **0407** (2004), 005.
[41] S. Weinberg. *Phys. Rev.* **D 61** 103505 (2000) [astro-ph/0005265].
[42] J. Garriga and A. Vilenkin. *Phys. Rev.* **D 67** (2003), 043503.
[43] M. MacLow and A. Ferrara. *Astrophys. J.* **513** (1999), 142.
[44] R. G. Abraham and S. van der Bergh. *Science*, **293** (2001), 1273.
[45] S. Dimopoulos and S. Thomas. *Phys. Lett.* **B 573** (2003), 13.
[46] R. Kallosh, J. Kratochvil, A. Linde, E. Linder and M. Shmakova. *J. Cosmol. Astropart. Phys.*, **0310** (2003), 015.
[47] J. Garriga, L. Pogosian and T. Vachaspati. *Phys. Rev.* **D 69** (2004), 063511.
[48] J. L. Feng, J. March-Russell, S. Sethi and F. Wilczek. *Nucl. Phys.* **B 602** (2001), 307.
[49] T. Banks, M. Dine and L. Motl. *JHEP*, **0101** (2001), 031.

[50] J. Donoghue. *JHEP*, **0008** (2000), 022.
[51] G. Dvali and A. Vilenkin. *Phys. Rev.* **D 64** (2001), 063509.
[52] A. Aguirre. *Phys. Rev.* **D 64** (2001), 083508.
[53] T. Banks, M. Dine and E. Gorbatov. *JHEP*, **0408** (2004), 058.
[54] J. N. Bahcall. *JHEP*, **0311** (2003), 004.
[55] M. Tegmark, L. Pogosian and A. Vilenkin. *Phys. Rev.* **D 71** (2005), 103523.

11
The definition and classification of universes
James D. Bjorken
Stanford Linear Accelerator Center, Stanford University

11.1 Introduction

When discussing the concept of multiple universes, it is a major challenge to keep the discourse within the bounds of science. There is an acute need to define what is being talked about. The issues include the following questions. How does one in general define a universe? Should one entertain different laws of physics in different universes? What are the most important parameters and/or features that characterize a universe? Once a parametrization has been attained, what is the differential probability of finding a universe with specified parameters? Is the integral of this distribution function finite or infinite?

A useful and familiar analogy is to consider planet Earth as a universe. It is, after all, not so long ago that this was mankind's paradigm. Then one may take the ensemble of universes, or multiverse, to be the set of all compact massive objects within the solar system which orbit the Sun and/or each other. Alternatively, one may take as a toy multiverse the set of all planets in the Galaxy or our universe. In either case, it is clear that the characterization of individual members of the ensemble is a very difficult task and requires a sophisticated understanding of much of planetary science, especially the experimental side of the subject.

One simplification of the general problem of classification is to restrict the consideration to that subset of universes (or – in the analogy – planets) which are nearly the same as our own. And this restriction can be naturally expressed in anthropic terms, by asking that the subset in question be that which admits in principle the existence of life as we know it. Even so, in the case of planets, this restricted problem is still very difficult. It is not clear whether our universe contains a large set of such planets, or whether

Universe or Multiverse?, ed. Bernard Carr. Published by Cambridge University Press.
© Cambridge University Press 2007.

the set only consists of our own planet. The problem is nicely described in the book *Rare Earth* by Brownlee and Ward [1].

If the problem of characterizing habitable universes within the multiverse is as complex as that of characterizing habitable planets within our universe, it is not at all clear that scientific methods will ever lead to noticeable progress in understanding. We can only hope that the case of universes is simpler than that of planets. In what follows, we will assume that the characterization of universes can be made concise, and we will shape our further working assumptions with this in mind.

11.2 What is a universe?

The characterization of universes naturally begins with the characterization of our own. In what follows, we shall adopt uncritically the Standard Model of contemporary cosmology. We assume that the spacetime geometry within each universe, as defined below in more detail, is spatially flat and can be described (in the large) by the Friedmann–Robertson–Walker (FRW) metric throughout its history. We assume that there is a non-vanishing cosmological constant, with dark energy comprising at present 70% of the energy of the universe, with most of the remainder contributed by dark matter. We also assume that prior to the ignition of the (radiation-dominated) big bang, there was an inflationary epoch characterized by a quasi-de Sitter spacetime, during which the FRW scale factor inflated by at least thirty powers of ten.

Strictly speaking, the size of such a universe is infinite, because the FRW spatial volume is infinite. But even if the FRW metric can be extrapolated to arbitrarily large distances, it still remains the case that almost all of this spacetime region is causally disconnected from us and will remain so to our descendants – provided the cosmological constant remains non-vanishing in the future (something which we will assume, for better or worse). So we place a box with comoving walls, centred on us, into the FRW spacetime. The dimension of the box at the present time is taken to be of order ten times the nominal size of our universe, which is 30 000 Mpc or 10^{29} cm. For practical purposes, we assume periodic boundary conditions at the surface of the box. Sensitivity of physical phenomena to this artifice can be tested by varying the (comoving) dimensions of the box. We do not expect significant sensitivity, because the box surface will be causally disconnected from us and our descendants.

Just as we bound by hand the spatial extent of our universe, we shall also bound the temporal extent. We choose as the initial time a value for which the physical size of the comoving box is not too much larger than the Planck

11 The definition and classification of universes

scale. Clearly, if the initial time were to be chosen much smaller, the size of the box would become small in comparison to the Planck scale, and the uncertainties in the underlying physics would grow considerably. And again there is good reason to believe that the phenomenology accessible to us is not strongly dependent upon details of the assumed initial state – although this question is still under vigorous debate.

It is also tempting to put a bound on the future as well as the past. Assuming the cosmological constant to be truly constant, there will be in the future a landmark time, at about 10^{12} y, when the temperature of the primordial black-body radiation decreases to the Hawking temperature of our de Sitter spacetime and almost all of the matter entropy has disappeared behind the de Sitter horizon. This occurs after the universe has inflated by a factor of about 10^{30} more than at present. Beyond this time the universe within the de Sitter horizon (the part causally connected to us and our descendants) is a truly quantum system. Again the uncertainties in the theoretical description are sure to be much higher, so we draw the line at this point.

In summary, we define our universe as the region of spacetime within the spacetime box defined above. Theoretical physics restricted to 'inside the box' has a chance of being within the realm of physical science, i.e. verifiable or falsifiable by conceivable experiments. Outside the box it is much less likely that this is the case. I prefer to stay inside the box and, at most, compare the physics in our box with that in other hypothetical boxes, similarly constructed, located elsewhere in spacetime.

The most gross features of the universe 'inside the box' can be best seen in the limit of ignoring small details, such as the difference between the GUT and Planck scales, the difference between the 'reheat' temperature and the Planck scale, and the very existence of ordinary and dark matter. Each of these simplifications involves replacing factors of 10^5 or so by factors of order unity. While this is extreme, it is not a big deal in comparison with the remaining factor of 10^{30} in the description, which is ubiquitous. The history of this 'universe in a box for dummies' divides itself into three epochs: quasi-de Sitter inflation, radiation-dominated big bang and dark-energy-dominated de Sitter expansion. Each epoch is characterized by an increase in the FRW scale factor of about 10^{30}. After the first epoch, the entropy of the radiation quanta accessible to our observation, and that of our progeny (of order 10^{90}), is enclosed in a volume with dimensions of millimetres, a scale of order 10^{30} times larger than the Planck scale. And after the second epoch (roughly now), the temperature of the primordial radiation is 10^{30} larger than the Hawking temperature of the future de Sitter spacetime and a factor 10^{30} smaller than the Planck temperature. All these factors of 10^{30} can be traced to the ratio of the dark-energy scale to the Planck-energy scale, i.e. to the value of the cosmological constant.

From this viewpoint, the cosmological constant is the most robust parameter characterizing the spacetime architecture of our universe and the sub-ensemble of universes which have properties similar to ours. Indeed, we will define the size of a universe in these terms: it is the value of the inverse Hubble constant in the future, when dark energy overwhelms matter in the FRW evolution.

11.3 Standard model and cosmological parameters in the multiverse

When considering the multiverse at large, it is possible that different universes have different gross histories. Do other universes go through inflation and radiation-dominated big bangs? We shall assume that there is a subset which does so, in a manner similar to our own universe, and limit our attention to that subset. It is also possible that different universes have different laws of physics. For example, the pattern of symmetries in other universes may not be the same as what is expressed by the Standard Model group; even the degrees of freedom might differ. Again, we do not here entertain such possibilities, but restrict our attention to those universes which have the same Standard Model effective action as our own. However, we shall allow the parameters in that effective action to differ in different universes.

The candidate parameters which are necessary (but quite probably not sufficient) to specify a universe fall into different classes. One class of Standard Model parameters consists of those relevant to vacuum structure. This includes the scale of the dark energy ($\mu = 2$ meV), the QCD condensate ($\Lambda_{QCD} = 200$ MeV), the electroweak vacuum condensate ($v = 250$ GeV) and perhaps the grand unification (GUT) scales, ranging from 10^{13} GeV (neutrino condensates?) to 10^{16} GeV (coupling constant unification). Another parameter set consists of the magnitudes of the dimensionless coupling constants characterizing the strengths of the Standard Model forces. Another distinct set is the large number of parameters characterizing the masses and mixings of quarks and leptons (including neutrinos). Finally, there is the set of parameters emergent from cosmological considerations, which may or may not be intimately associated with particle physics questions lying 'beyond the Standard Model'. These include the parameters defining the properties of dark matter, the baryon-to-photon ratio and the magnitude

11 The definition and classification of universes

and spectrum of the temperature fluctuations of the primordial black-body photons.

If all of these parameters are independent, and determined at best by anthropic lines of argument, the description and characterization of universes will be very difficult. As promised in the preceding sections, we shall take a more optimistic point of view and assume that at least some of them are strongly correlated. Once one of the parameters in the correlated set is specified, the others are determined.

The material that follows is a brief summary of recent work based on this point of view [2, 3]. Our starting point is the consideration of the vacuum parameters defined above. One of those parameters is the coefficient of the dark-energy term in the effective action. By definition, it is different in universes of different sizes, with size defined in the previous section. As the size of a universe decreases, the dark-energy scale increases as the inverse square root of this. For Planck-sized universes, the dark energy reaches the Planck scale. For an idealized universe of infinite size, the dark energy would vanish.

We now make the basic assumption that the other vacuum parameters have a similar behaviour. For example, the QCD vacuum parameter Λ_{QCD} is assumed to vary as the inverse cube root of the size of the universe, and the electroweak condensate value v is assumed to vary as the inverse fourth root of the size. With this assumed behaviour, all these vacuum scales converge, or flow, toward a fixed point, which is of order the Planck/GUT scale in energy, when the size of the universe becomes of order the Planck/GUT size.

(A caveat: the assumed dependence of Standard Model parameters on size is with respect to the time-independent cosmological constant, *not* the FRW scale factor. We do *not* entertain here time-dependence of Standard Model parameters within our own universe, but only compare the parameters in different universes. However, a possible exception is mentioned in the final paragraph of this chapter.)

Given this assumption, more can be deduced. The dimensionless strong coupling constant α_s of QCD is determined by Λ_{QCD}. It is a 'running coupling', dependent upon the momentum scale probed. Since the size-dependence of this running coupling constant is determined, this is also true when the coupling constant is evaluated at the GUT scale, where it is presumably 'unified' with the electroweak couplings, which also 'run'. Therefore, the size-dependence of electroweak and electromagnetic coupling constants at low-energy scales is also determined. The result is that the inverse 'fine structure constants', including the famous 1/137 of quantum electrodynamics, depend linearly on the logarithm of the size of the universe, and vanish for small Planck/GUT-scale universes. This means that the Standard Model forces are very strong and non-perturbative for such small universes, and vanish in the limit of an infinite universe. What is strongly suggested is that an infinite universe, as we have defined it, is nearly trivial, containing no Standard Model interactions whatsoever other than gravity.

It is also reasonable, albeit more uncertain, to expect that the mass parameters for the quarks and leptons follow a similar pattern, i.e. they flow to values of order the Planck/GUT scale for small universes and to zero for very large universes. But the details become increasingly fuzzy as the masses become small. In particular, the largest uncertainties occur for the anthropically significant masses, such as the electron and up-quark and down-quark masses, important for the details of nuclear physics that condition our existence. It is the lack of theoretical understanding of the basic origin of these small masses which is the roadblock.

11.4 Anthropic considerations

For better or worse, the scaling rules enunciated above allow detailed study of the physical properties of universes with sizes different from our own. This is the main content of the aforementioned ref. [2]. What is found is that the existence of chemistry is robust; the size of the universe can be varied by thirty powers of ten without major effects. In broad terms the same is true for nuclear physics. However, as is well known to the anthropic community, there are details, essential for the existence of life as we know it, which are not robust. Examination of the anthropic constraints shows that, in the context of our assumed scaling rules, the strongest limitation on the size of universes which can support life as we know it comes from the famous triple-α process, responsible for the synthesis of carbon in stars. The overall strength of the nuclear force cannot vary by more than a fraction of one per cent without causing trouble. In the case of interest, this variation is effected only via chiral symmetry-breaking, i.e. by the non-vanishing masses of the up, down and perhaps strange quarks. Those mass parameters have a different dependence on the size of the universe than does Λ_{QCD}, and it is this disparity which destroys the delicate balance of parameters which allows the triple-α reaction to proceed.

When the dust settles, the bottom line is that the size variation allowed by the existence of life as we know it is of order of a factor 2. In a universe twice as large or twice as small as ours, the Standard Model parameters would arguably be different enough to block the production of carbon in stars and hence the evolution of life as we know it. The above estimate is quite uncertain – perhaps off by a factor of 3 or even 10. But it is accurate enough to draw a variety of tentative inferences. The most important inference has to do with the 'hierarchy problem', and it provides an *a posteriori* reason to take the scaling assumption seriously.

What is generally denoted by the hierarchy problem is the large disparity between the electroweak scale, characterized by the vacuum parameter v, and the Planck/GUT scale. However, it also includes the notorious 'cosmological constant problem': why the cosmological constant scale is thirty powers of ten smaller than the Planck/GUT scale. In addition, there is the 'problem of mass', which includes the issue of why the electron mass, say, is so much smaller than the top-quark mass. If the scaling behaviour is assumed, all of these questions are rendered moot. For small universes, there is no hierarchy problem; all these parameters arguably take values of order the Planck/GUT scale. And it may well be the case that the typical universe is, in fact, small. It is only because we live in such a large universe that we see these huge hierarchies of scale. The above anthropic considerations also require us to live in a large universe; the conditions for life as we know it only exist in this situation.

While this argument falls short of a full resolution of the hierarchy problem, it does provide a different way of viewing it. The problem of divergences and renormalization, which is part of the usual statement of the problem, is now expressed as the question of why the renormalized vacuum parameters should be dependent upon the size of the universe, and in the specified way. In other words, one must understand the scaling exponents such as 1/3 and 1/4 for the strong and electroweak sectors, respectively. This author has some ideas about the 1/3 [3]. Others, in particular Tom Banks, have already speculated about the 1/4 [4].

11.5 The distribution function for universes

The above considerations embolden us, perhaps foolishly, to speculate on the question of the size distribution of the universes in the ensemble, or sub-ensemble, that we have been looking at. It seems most reasonable to assume that the total number in the ensemble is finite. It also seems reasonable to assume, given the hierarchy arguments of Section 11.4, that the distribution peaks at small sizes, of order the Planck/GUT size. Given all that, the remaining question is the asymptotic behaviour at large sizes. Two natural classifications are power law and exponential. Anthropic considerations argue that the integral of the distribution over the habitable interval should give a number large compared with unity, in order to make the universes which are in principle habitable (as we understand the term) non-unique. For any reasonable fall-off with size, such an estimate gives essentially the same result as integrating over all sizes as large or larger than our own. And it is also clear that if the fall-off is exponential or faster, the total number of universes in the sample must be gigantic, for example of order $10^{10^{60}}$. Modesty therefore suggests the alternative choice of a power-law tail. If the distribution function, $R\,\mathrm{d}N/\mathrm{d}R$, falls off at large sizes as R^{-n}, the total number of universes in the ensemble will be bounded below by a number of order 10^{60n}. This is a big number to be sure, but not very far beyond other big numbers encountered in the study of our own universe. In addition, power-law behaviour is often associated with the notion of criticality and/or scale-invariant behaviour. The feature of spatial flatness of our universe may suggest criticality as an underlying feature of a future, better theory and/or of the subset of universes with features similar to our own.

It is unrealistic to expect the ensemble of universes to be characterized by only one parameter, the size. It is therefore of interest to look at the remaining candidate parameters and search for those most likely to be 'independently anthropic'. Amongst the Standard Model candidates, the light-quark (up and down) masses are strong ones, as forcefully advocated by Craig Hogan [5, 6]. Also, at least some subset of the three cosmological parameters mentioned in Section 11.3 seem to be strong candidates, in order that there is the right amount of large-scale structure in our universe. However, we do not have anything very new to add to this problem. Better understanding of the nature of dark matter would be of obvious help.

11.6 Concluding comments

It should be abundantly clear that the above discussion skirts dangerously close to the edge of legitimate science. Are any of these speculations falsifiable or, even better, verifiable? With regard to falsifiability, there is an answer: if the cosmological constant is eventually found to be zero, the scaling ideas die an unambiguous death. Likewise, if the cosmological constant is not constant, and exhibits a lot of quintessence, it could well be that the implied time-variation in other Standard Model parameters would exceed experimental limits. More interesting is the question of verifiability. Probably the best chance lies in finding a microscopic theory consistent with the scaling rules which has predictive power above and beyond what we now have. As mentioned in Section 11.4, there are some reasons for optimism in this regard.

It would also be advantageous if individual universes were in causal contact, which might admit experimental investigation, at least in principle. This is unlikely, and goes against the grain of almost all contemporary thinking. But it is perhaps not completely out of the realm of possibility. A few individuals, including this author, now and then entertain the notion of black hole interiors being non-singular static de Sitter space, as is the case to a good approximation for our own universe. This invites a model of the multiverse as nested black holes, with the remote possibility of two-way communication through the horizons.

Another possibility is that the 'universe in a box' described in Section 11.2 really consists of two universes. The first one is the inflationary universe present before ignition of the big bang. This is characterized by a huge cosmological constant (in the approximation of 'no-roll' instead of slow-roll). Perhaps the Standard Model parameters should take the values appropriate to the interpretation of that piece of spacetime as a 'small universe', with size parameter (Hubble scale) of order 10^{13} GeV (or 10^{-27} cm). It is interesting that, were this to be done, the dark-energy scale is of order the GUT scale, with the QCD and electroweak vacuum energy scales, naïvely estimated from the power-law rules, somewhat higher. It is easy to imagine that, in fact, these three scales become synthesized, and that the interpretation of the inflaton field could be in terms of QCD and/or electroweak condensates, which in turn might be more appropriately re-interpreted as a GUT condensate. The big bang would then be ignited by the 'decay' of the 'small' universe into our 'big' universe, accompanied somehow by a large amount of entropy production. This idea has not yet been pursued in detail. But the risk is not that this line of thinking has no phenomenological consequences, but rather that it has too many.

Acknowledgement

This work was supported by the Department of Energy contract DE–AC02–76SF00515.

References

[1] D. Brownlee and P. Ward. *Rare Earth* (New York: Copernicus, Springer-Verlag, 1999).
[2] J. Bjorken. *Phys. Rev.* **D 67** (2003), 043508 [hep-th/0210202].
[3] J. Bjorken. The classification of universes (2004) [astro-ph/0404233].
[4] T. Banks. Breaking SUSY on the horizon (2002) [hep-th/0206117].
[5] C. Hogan. *Rev. Mod. Phys* **72** (2000), 1149 [astro-ph/9909295].
[6] C. Hogan. This volume (2007).

12
M/string theory and anthropic reasoning

Renata Kallosh
Department of Physics, Stanford University

12.1 Introduction

After the development of inflationary cosmology, anthropic reasoning (AR) became one of the most important methods in theoretical cosmology. However, until recently it was not in the toolbox of many high-energy physicists studying 11- or 10-dimensional M/string theory and supergravity. The attitude of high-energy physicists changed dramatically in 1998, when the physics community was shocked by the new cosmological observations suggesting that we may live in a world with a tiny cosmological constant, $\Lambda \sim 10^{-120} M_\mathrm{P}^4$, with a weird combination of matter and dark energy.

The recent WMAP observations seem to confirm the earlier data and also support the existence of an inflationary stage in the very early Universe. In view of the accumulating observational evidence, the level of tolerance towards AR is currently increasing. More people are starting to take it into consideration when thinking about cosmology from the perspective of M/string theory and particle physics. I belong to this group, and I recently had two rather impressive encounters with AR that I would like to discuss in this chapter.

In the first encounter, Andrei Linde and I considered a model of maximal supergravity related to the 11-dimensional M-theory, which has a 4-dimensional de Sitter (dS) solution with spontaneously broken supersymmetry [1]. We found that this model offers an interesting playground for the successful application of AR. This model follows from maximal supersymmetry, and the potential $V(\phi)$ has the following important properties: (1) uniqueness; (2) $V''/V = -2$ (where a prime denotes differentiation with respect to ϕ); and (3) predictable future collapse. We found that this model suggests a possible anthropic explanation for both the present value of the cosmological constant and the observed ratio of the densities associated

---

Universe or Multiverse?, ed. Bernard Carr. Published by Cambridge University Press.
© Cambridge University Press 2007.

with dark energy and ordinary matter. Our conclusion was based on the calculation of the lifetime of the Universe in this model and the requirement that this should be no shorter than the present age of the Universe. Although the model is far from realistic, it may still be useful as an example of how one may think about applications of AR in combination with dynamical models. In particular, if supersymmetry is discovered by the LHC in 2007, we may have to apply supersymmetric models to cosmology or impose some other bias for a dynamical model in combination with AR.

In the second (more recent) encounter with AR, Shamit Kachru, Andrei Linde, Sandip Trivedi and I [2] have looked for dS solutions in string theory that would allow us to describe the present exponential expansion of the Universe in a more traditional context, with a cosmological constant instead of quintessence. For this purpose, it was necessary to drop certain conditions for the 'no-go' theorems, which predicted that 4-dimensional dS space is not allowed in perturbative compactification of 11- or 10-dimensional M/string theory. We were able to achieve this through a combination of various ingredients of perturbative and non-perturbative string theory. Each of the separate contributions to the potential responsible for the appearance of dS space entered into our construction with a parameter not strictly specified by string theory but expected to lie in some plausible range. Therefore, when all the ingredients are combined to produce a dS solution, the value of the cosmological constant and the number of possible dS vacua is not prescribed uniquely. The most we can say is that dS solutions are possible and that there are many dS vacua with many values of the cosmological constant.

In any case, our current partial knowledge may lead to attempts to apply AR to the dS solutions found in string theory. Serious attempts to explain the value of the cosmological constant in string theory, using the fact that there are multiple vacua, were initiated by Bousso and Polchinski [3]. These ideas were then developed by Susskind [4] and Douglas [5] and his collaborators, in a series of papers reviewed in ref. [6]. However, Douglas prefers the term 'vacuum selection problem', and this includes our recently discovered dS vacua.

12.2 Anthropic constraints on Λ in $N = 8$ supergravity

As a warm-up, let us first assume that the cosmological constant is large and negative, $\Lambda \ll -10^{-29}$ g cm^{-3}, as studied by Barrow and Tipler [7] and Linde [8] a long time ago. Such a model, even if flat, would collapse well before the current age of the Universe, $t_0 \sim 14$ Gy, which would make life

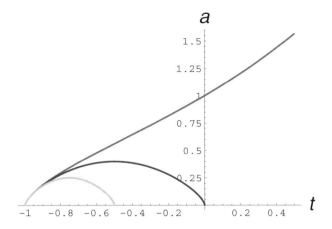

Fig. 12.1. Evolution of the cosmic scale factor a in a flat ΛCDM model for various values of Λ. Time is in units of the present age $t_0 \approx 14$ Gy. The present moment is $t=0$ and the big bang corresponds to $t=-1$. The upper line corresponds to a flat model with $\Omega_{\rm tot}=1$, $\Omega_\Lambda=0.7$ (i.e. $\Lambda=+0.7\rho_0$) and $\Omega_{\rm M}=0.3$. The middle line corresponds to a flat model with $\Lambda=-4.7\rho_0$, which collapses at the current epoch. The total lifetime of a model with $\Lambda=-18.8\rho_0$ (the lowest line) is only 7 Gy.

impossible. One may wonder whether intelligent life could emerge within 7 or 5 Gy, but we have no reason to believe this.

One may improve the earlier order-of-magnitude estimates and obtain a numerical constraint on negative Λ. The investigation is straightforward, so we simply show the results in Fig. 12.1. We find that the anthropic constraint on a negative cosmological constant is slightly less stringent than anticipated. If 7 Gy suffices for emergence of human life, then $\Lambda \gtrsim -18.8\rho_0 \sim -2 \times 10^{-28}$ g cm^{-3}. If we really need 14 Gy, the constraint is somewhat stronger: $\Lambda \gtrsim -4.7\rho_0 \sim -5 \times 10^{-29}$ g cm^{-3}.

However, the present observational data suggest that $\Lambda > 0$. In this case, the use of anthropic considerations becomes more involved, as discussed in several other contributions in this book. Here we will show that a constraint based on the total lifetime of a flat model can be derived in a class of theories based on $N=8$ supergravity that can describe the present stage of acceleration. This may allow us to avoid the fine-tuning that is usually required to explain the density parameter of the observed dark energy $\Omega_{\rm D}$.

12.2.1 Maximal supergravity as the dark energy hidden sector

No known compactifications of the fundamental M/string theory to four dimensions leads to potentials with dS solutions corresponding to $\Lambda > 0$.

Even the dS solution studied in ref. [2] does not come from the compactification of the supergravity solution alone; it requires the addition of an extended object, the anti-D3 (or $\overline{D3}$) brane, to provide a positive cosmological constant.

However, there are versions of maximally extended gauged $d=4$, $N=8$ supergravity which have dS solutions. They are also known to be solutions of $d=11$ supergravity with thirty-two supersymmetries, corresponding to M/string theory. Note that dS solutions of $d=4$, $N=8$ supergravity correspond to solutions of M/string theory with a non-compact internal 7- or 6-dimensional space. The relation between states of the higher-dimensional and four-dimensional theories in such backgrounds is complicated, since the standard Kaluza–Klein procedure is not valid in this context. It is nevertheless true that the class of $d=4$ supergravities with dS solutions that we will consider below as dark energy candidates has a direct link to M/string theory, unlike almost any other model of dark energy. Moreover, theories with maximal supersymmetry are perfectly consistent from the point of view of $d=4$ theory, all kinetic terms for scalars and vectors being positive definite.

All supersymmetries are spontaneously broken for the dS solutions of $N=8$ supergravity. These solutions are unstable; they correspond either to a maximum of the potential for the scalar fields or to a saddle point. In all known cases, one finds that there is a tachyon and the ratio between $V'' = m^2$ and $V = \Lambda$ at the extremum of $V(\phi)$ is -2. The simplest (and typical) representative of $d=4$, $N=8$ supergravity originating from M-theory with a dS maximum has the following action:

$$g^{-1/2}L = -\frac{1}{2}R - \frac{1}{2}(\partial\phi)^2 - \Lambda(2 - \cosh\sqrt{2}\phi). \tag{12.1}$$

Here we use units in which the Planck mass $M_p = 1$. At the critical point, we have $V' = 0$, $V_{cr} = \Lambda$ and $\phi_{cr} = 0$. This corresponds to $d=4$ supergravity with the gauged $SO(4,4)$ non-compact group. At the dS vacuum it breaks down to its compact subgroups, $SO(4) \times SO(4)$. The value of the cosmological constant is related to the current Hubble constant H_0 and the gauge coupling g by

$$\Lambda = 3H_0^2 = 2g^2. \tag{12.2}$$

The gauge coupling and cosmological constant in $d=4$ supergravity have the same origin in M-theory; they come from the flux of an antisymmetric tensor gauge field. The corresponding 4-form, $F_{\mu\nu\lambda\rho}$, in $d=11$ supergravity is proportional to the volume-form of the dS space:

$$F_{0123} \sim \sqrt{\Lambda}V_{0123}. \tag{12.3}$$

Here $F = dA$, where A is the 3-form potential of $d=11$ supergravity. According to this model, the small value of the cosmological constant is due to the

4-form flux, which has an inverse timescale of order the age of the Universe. Note that, in our model, there is no reason for flux quantization since the internal space is not compact. The 11-dimensional origin of the scalar field ϕ in the potential can be explained as follows: $d=4$, $N=8$ gauged supergravity has thirty-five scalars and thirty-five pseudo-scalars, together forming a coset space $E_{7(7)}/SU(8)$. The field ϕ is an $SO(4) \times SO(4)$ invariant combination of these scalars and it may also be viewed as part of the $d=11$ metric.

One's first reaction would be to discard this model altogether, because its potential is unbounded from below. However, the scalar potential in this theory remains positive for $|\phi| \lesssim 1$ and, for small Λ, the time for the development of the instability can be much greater than the present age of the Universe, which suffices for our purposes. In fact, we will see that this instability allows us to avoid the standard fine-tuning/coincidence problem which plagues most versions of quintessence theory. To use these theories to describe the present stage of acceleration (late inflation), one should take $\Lambda \sim 10^{-120} M_P^4$. This implies that the tachyonic mass is ultra-light, $|m^2| \sim -(10^{-33}\,\text{eV})^2$.

In the early Universe the ultra-light scalar fields may stay away from the extrema of their potentials; they 'sit and wait' and begin moving only when the Hubble constant decreases enough to become comparable with the scalar mass. This may result in noticeable changes of the effective cosmological constant during the last few billion years. Since the potential of $N=8$ supergravity with a dS solution is unbounded from below, the Universe will eventually collapse. However, if the initial position of the field is not far from the top of the potential, the time for collapse may be very long.

From the perspective of $d=11$ theory, it is natural to consider a large ensemble of possible values for the fields $F \sim \sqrt{\Lambda}$ and ϕ. One may also study such an ensemble in the context of $d=4$ theory. Consider a theory of a scalar field ϕ with the effective potential given by

$$V(\phi) = \Lambda(2 - \cosh\sqrt{2}\phi) \tag{12.4}$$

in $N=8$ theory. This is illustrated in Fig. 12.2. In order to understand the cosmological consequences of this theory, let us first consider this potential at $|\phi| \ll 1$. In this limit it has the following very simple form:

$$V(\phi) = \Lambda(1 - \phi^2) = 3H_0^2(1 - \phi^2). \tag{12.5}$$

The main property of this potential is that $m^2 = V''(0) = -2\Lambda = -6H_0^2$. One can show that a homogeneous field $\phi \ll 1$ with $m^2 = -6H_0^2$ in a model with

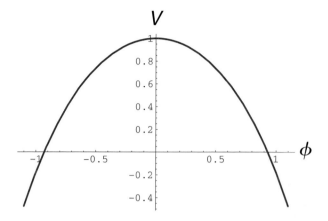

Fig. 12.2. The scalar potential $V(\phi) = \Lambda(2 - \cosh\sqrt{2}\phi)$ in $d=4$, $N=8$ supergravity. The value is in units of Λ and the field is in units of M_P.

Hubble constant H_0 grows as

$$\phi(t) = \phi_0 \exp(cH_0 t), \quad c = (\sqrt{33} - 3)/2 \approx 1.4. \qquad (12.6)$$

Consequently, if the energy density is dominated by $V(\phi)$, it takes a time $t \sim 0.7 H_0^{-1} \ln(\phi_0^{-1})$ for the scalar field to roll down from ϕ_0 to the region $\phi \gg 1$, where $V(\phi)$ becomes negative and the Universe collapses.

This means that one cannot take Λ too large without making the total lifetime of the Universe too short to support life, unless the scalar field ϕ_0 is exponentially small. But if the potential is always very flat, then the field just after inflation, ϕ_0, can take any value with equal probability, so there is no reason to expect it to be very small. This means that, for $\phi_0 \lesssim 1$, the typical lifetime of the Universe is $t_{\text{tot}} \sim H_0^{-1} \sim \Lambda^{-1/2}$. Therefore the universe can live longer than 14 Gy only if the cosmological constant is extremely small, $\Lambda \lesssim \rho_0$. On the other hand, for $\phi \gg 1$ the potential decreases, $V(\phi) \sim -\Lambda \exp(\sqrt{2}|\phi|)$, so the Universe collapses almost instantly, even if $\Lambda \lesssim \rho_0$. Figure 12.3 shows the expansion of the Universe for $\phi_0 = 0.25$ and for various values of Λ ranging from $0.7\rho_0$ to $700\rho_0$. The time is given in units of 14 Gy. One finds, as expected, that the total lifetime of the Universe for a given ϕ_0 is proportional to $\Lambda^{-1/2}$, which means that large values of Λ are anthropically forbidden.

Figure 12.4 shows the expansion of the Universe for $\Lambda = 0.7\rho_0$. The upper line corresponds to the fiducial model with $\phi_0 = 0$. In this case, the field does not move and all cosmological consequences are as in the standard theory with the cosmological constant $\Lambda = 0.7\rho_0$. The difference will appear only in the very distant future, at $t \sim 10^2 H_0^{-1} \sim 10^3$ Gy, when the unstable state

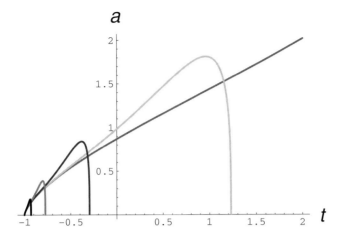

Fig. 12.3. Expansion of the Universe for $\phi_0 = 0.25$. Going from right to left, the lines correspond to $\Lambda = 0.07\rho_0$, $0.7\rho_0$, $7\rho_0$, $70\rho_0$ and $700\rho_0$.

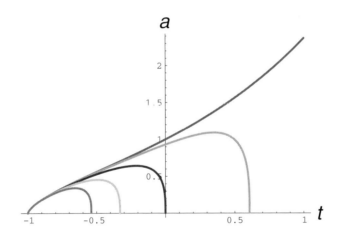

Fig. 12.4. Expansion of the Universe for $\Lambda = 0.7\rho_0$. The upper line corresponds to the fiducial model with $\phi_0 = 0$ (i.e. a cosmological constant and motionless field). The lines below this correspond successively to $\phi_0 = 0.5$, 1, 1.5 and 2.

$\phi_0 = 0$ will decay due to the destabilizing effect of quantum fluctuations. For $\phi_0 > 1$ the total lifetime of the Universe becomes unacceptably small, which means that large values of ϕ_0 are anthropically forbidden.

Further conclusions will depend on various assumptions about the probability of the parameters (Λ, ϕ_0). We will make the simplest assumption that all values of Λ and ϕ_0 are equally probable. We will discuss alternative assumptions and their consequences in Section 12.2.2.

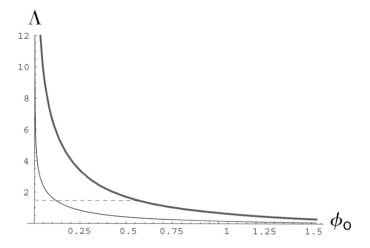

Fig. 12.5. The region below the top line corresponds to values of Λ and ϕ_0 such that the lifetime of the Universe exceeds 14 Gy. The dashed line $\Lambda \approx 1.5\rho_0$ separates this region into two equal areas. The region below the lower curve corresponds to universes with lifetime exceeding 28 Gy.

The values of Λ and ϕ_0 for which the total lifetime of the Universe exceeds 14 Gy are shown in Fig. 12.5 as the region under the thick (upper) line. If all values of Λ and ϕ_0 are equally probable, the measure of probability is given by the total area under this curve, $S_{\text{tot}} \approx 3.5$. One can estimate the probability of being in any region of the phase space (Λ, ϕ_0) by dividing the corresponding area by S_{tot}.

The dashed line $\Lambda \approx 1.5\rho_0$ separates the anthropically allowed region into two equal area parts. This implies that the average value of Λ in this theory is about $1.5\rho_0$. It is obvious that Λ can be somewhat larger or somewhat smaller than $1.5\rho_0$, but the main part of the anthropically allowed area corresponds to

$$\Lambda = O(\rho_0) \sim 10^{-120} M_{\text{P}}^4. \tag{12.7}$$

This is one of the main results of our investigation. It is a direct consequence of the relation $m^2 = -6H_0^2$, which is valid for all known versions of $d = 4$, $N = 8$ supergravity that allow dS solutions.

The region below the lower curve corresponds to all models with lifetimes greater than 28 Gy, i.e. to those that would live a further 14 Gy from now. The area below this curve is one-third of that between the lower and upper curves. This means that the 'life expectancy' of a typical anthropically allowed model (the time from the present moment until the global collapse) is smaller than the present age of the Universe. The prognosis becomes

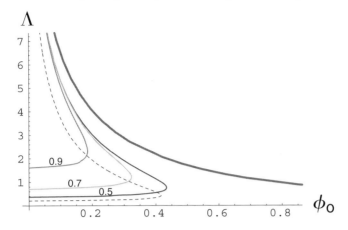

Fig. 12.6. The lifetime of the Universe exceeds 14 Gy below the top line. The other (progressively lower) curves correspond to $\Omega_D = 0.5$, 0.7 and 0.9. The region to the right of the dashed line corresponds to universes that are accelerating now.

more optimistic if one takes into account that the Universe has $\Omega_D = 0.7$: the probability that it will survive a further 14 Gy becomes more than 50%.

The average value of Λ does not immediately tell us the most probable value of Ω_D. In order to do this, we plot in Fig. 12.6 the curves corresponding to $\Omega_D = 0.5$, 0.7 and 0.9. The region to the right of the dashed line corresponds to models that are currently accelerating.

The total probability of living in an accelerating model 14 Gy after the big bang is determined by the area bounded by the thin dashed line in Fig. 12.6. Dividing this by $S_{tot} \approx 3.5$, one can find that this probability is about 35%. About half this area corresponds to $\Omega_D > 0.9$. The most interesting part of the accelerating region is bounded by the curves corresponding to $\Omega_D = 0.5$ and $\Omega_D = 0.9$. All points inside this region correspond to accelerating models with $0.5 < \Omega_D < 0.9$. One can see from Fig. 12.6 that the area of this region is about 0.4. Dividing this by the total area of the anthropically allowed region, $S_{tot} \approx 3.5$, one finds that the probability of living in such a model is about 10%. These results resolve the usual fine-tuning/coincidence problem.

12.2.2 More on dark energy

Most of the theories of dark energy face two problems. First, it is necessary to explain why the bare cosmological constant vanishes. Then one must find a dynamical mechanism imitating a small cosmological constant and explain why $\Omega_D \sim 0.7$ at the present cosmological epoch.

We have the studied cosmological consequences of the simplest toy model of dark energy based on $N=8$ supergravity and found that this can completely resolve the cosmological constant and coincidence problems plaguing most quintessence models. Indeed, one cannot simply add a cosmological constant to this theory. The only way to introduce something similar to the cosmological constant is to put the system close to the top of the effective potential. If the potential is very high, then it is also very curved since $V''(0) = -2V(0)$. We have found that the Universe can live long enough only if the field ϕ is initially within the Planck distance of the top, $|\phi| \lesssim M_P$, which is reasonable, and if $V(0)$ (which plays the role of Λ in this theory) does not much exceed the critical value $\rho_0 \sim 10^{-120} M_P^4$.

We made the simplest assumption that the values of Λ and ϕ_0 are uniformly distributed. However, in realistic models the situation may be different. For example, as already mentioned, $\Lambda^{1/2}$ is related to the 4-form flux in $d=11$ supergravity from Eq. (12.3). This suggests that the probability distribution should be uniform, not with respect to Λ and ϕ_0, but with respect to $\Lambda^{1/2}$ and ϕ_0. We studied this possibility and found that the numerical results change, but the qualitative features of the model remain the same.

The probability distribution for ϕ_0 may be non-uniform even if $V(\phi)$ is very flat at $\phi < 1$. First, the fields with $\phi \gg 1$ (i.e. $\phi \gg M_P$) may be forbidden or the effective potential at large ϕ may blow up. This is often the case in $N=1$ supergravity. Second, interactions with other fields in the early Universe may create a deep minimum, capturing the field at some time-dependent point $\phi < 1$. This also often happens in $N=1$ supergravity, which is one of the features of the cosmological moduli problem. If it does so in our model, one can ignore the region with $\phi_0 > 1$ (the right part of Figs. 12.5 and 12.6) in the calculation of probabilities. This will increase the probability of living in an accelerating model with $0.5 < \Omega_D < 0.9$.

Our estimates have assumed that the Universe must live as long as 14 Gy, so that human life can appear. One could argue that the first stars and planets were formed long ago, so we may not need much more than 5–7 Gy for the development of life. This would somewhat decrease our estimate for the probability of living in an accelerating model with $0.5 < \Omega_D < 0.9$, but it would not alter our results qualitatively. On the other hand, most of the planets were probably formed very late in the history of the Universe, so one may argue that the probability of the emergence of human life becomes much greater at $t > 14$ Gy, especially if one keeps in mind how many other coincidences have made life possible. If one assumes that human life is extremely improbable (after all, we do not have any indications of its existence elsewhere in the Universe), then one may argue that the probability of its emergence becomes significant only if the total lifetime of the Universe can be much greater than 14 Gy. This would increase our estimate for the probability of living in an accelerating model with $0.5 < \Omega_D < 0.9$.

So far, we have not used any considerations based on the theory of galaxy formation, as developed by Weinberg [9], Efstathiou [10], Vilenkin [11], Martel [12] and Garriga [13]. If we do so, the probability of the emergence of life for $\Lambda \gg \rho_0$ will be additionally suppressed, which will increase the probability of living in an accelerating model with $0.5 < \Omega_D < 0.9$.

To the best of my knowledge, only in models based on extended supergravity are the relation $|m^2| \sim H^2$ and the absence of freedom to add the bare cosmological constant properties of the theory rather than of a particular dynamical regime. That is why the increase of $V(\phi)$ in such models entails the increase in $|m^2|$. This, in turn, speeds up the development of the cosmological instability, which leads to anthropically unacceptable consequences.

The $N=8$ theory discussed here is just a toy model. In this case, we have been able to find a complete solution to the cosmological constant and coincidence problems (explaining why $\Lambda \sim \rho_0$ and why Ω_D noticeably differs from both zero and unity at the present stage of cosmological evolution). This model has important advantages over many other theories of dark energy, but – to make it fully realistic – one would need to construct a complete theory of all fundamental interactions, including the dark energy sector described above. This is a very complicated task, which goes beyond the scope of the present investigation. However, most of our results are not model-specific.

It would be interesting to apply our methods to models unrelated to extended supergravity. A particularly interesting model is axion quintessence. The original version had the potential given by

$$V(\phi) = \Lambda[\cos(\phi/f) + C], \tag{12.8}$$

where it was assumed that $C=1$. The positive definiteness of the potential, and the fact that it has a minimum at $V=0$, could then be motivated by global supersymmetry arguments. In supergravity and M/string theory, these arguments are no longer valid and the value of the parameter C is not specified.

In the axion model of quintessence based on M/string theory, the potential had the form $V = \Lambda \cos(\phi/f)$ without any constant. This has a maximum at $\phi = 0$, $V(0) = \Lambda$. The Universe collapses when the field ϕ rolls to the minimum of its potential, $V(f\pi) = -\Lambda$. The curvature of the effective potential at its maximum is given by

$$m^2 = -\Lambda/f^2 = -3H_0^2/f^2. \tag{12.9}$$

For $f = M_\mathrm{P} = 1$, one has $m^2 = -3H_0^2$ and, for $f = M_\mathrm{P}/\sqrt{2}$, one has $m^2 = -6H_0^2$, exactly as in $N = 8$ supergravity. Therefore the anthropic constraints on Λ based on the investigation of the collapse of the Universe in this model are similar to the constraints obtained in our $N = 8$ theory. However, in this model, unlike the ones based on extended supergravity, one can easily add or subtract any value of the cosmological constant. In order to obtain useful anthropic constraints on the cosmological constant, one should use a combination of our approach and the usual theory of galaxy formation.

In this sense, our main goal is not to replace the usual anthropic approach to the cosmological constant problem, but to enhance it. We find it very encouraging that our approach may strengthen the existing anthropic constraints on the cosmological constant in the context of the theories based on extended supergravity. One may find it hard to believe that, in order to explain the results of cosmological observations, one should consider theories with an unstable vacuum state. However, one should remember that exponential expansion of the Universe during inflation, as well as the process of galaxy formation, are themselves the result of the gravitational instability, so we should learn how to live with the idea that our world may be unstable.

12.3 de Sitter space in string theory

While the model discussed above is quite interesting, it is only partially related to a consistent $d = 10$ string theory. After many unsuccessful attempts to find a dS solution in string theory, we have recently come up with a class of such solutions [2]. We next outline the construction of metastable dS vacua of type IIB string theory and discuss their relation to AR.

Our starting point is the highly warped IIB compactifications with non-trivial NS and RR 3-form fluxes.[1] By incorporating known corrections to the superpotential from Euclidean D-brane instantons or gaugino condensations, one can make models with all moduli fixed, yielding a supersymmetric anti-de Sitter (AdS) vacuum. Inclusion of a small number of $\overline{D3}$ branes in the resulting warped geometry allows one to raise the AdS minimum and make it a metastable dS ground state. The lifetime of our metastable dS vacuum is much greater than the cosmological timescale of 10 Gy. We have also proven

---

[1] NS stands for (Neveu–Schwarz) bosonic closed string states whose left- and right-moving parts are bosonic. RR stands for (Ramond–Ramond) bosonic closed string states whose left- and right-moving parts are fermionic.

that, under certain conditions, the lifetime of dS space in string theory will always be shorter than the recurrence time.

Our basic strategy is to first freeze all the moduli present in the compactification, while preserving supersymmetry. We then add extra effects that break supersymmetry in a controlled way and lift the minimum of the potential to a positive value, yielding dS space. To illustrate the construction, we work in the specific context of IIB string theory compactified on a Calabi–Yau (CY) manifold in the presence of flux. Such constructions allow one to fix the complex structure moduli but not the Kähler moduli of the compactification. In particular, to leading order in α' and g_s, the Lagrangian possesses a no-scale structure which does not fix the overall volume.[2] (Henceforth we shall assume that this is the *only* Kähler modulus; it is plausible that one can construct explicit models which have this property.) In order to achieve the first step of fixing all moduli, we therefore need to consider corrections which violate the no-scale structure. Here we focus on quantum non-perturbative corrections to the superpotential, which are calculable, and show that these can lead to supersymmetry-preserving AdS vacua in which the volume modulus is fixed in a controlled manner.

Having frozen all moduli, we then introduce supersymmetry-breaking by adding a few $\overline{D3}$ branes in the compactification. The extent of supersymmetry-breaking, and the resulting cosmological constant of the dS minimum, can be varied in our construction – within certain limits – in two ways. One may vary the number of $\overline{D3}$ branes which are introduced or one may vary the warping in the compactification (by tuning the number of flux quanta through various cycles). It is important to note that this corresponds to freedom in tuning *discrete* parameters, so while fine-tuning is possible, one should not expect to be able to tune to arbitrarily high precision.

12.3.1 Flux compactifications of IIB string theory plus corrections

We now study a CY orientifold with flux. In such a model, one has the 'tadpole' consistency condition:

$$\frac{\chi(X)}{24} = N_{D3} + \frac{1}{2\kappa_{10}^2 T_3} \int_M H_3 \wedge F_3. \quad (12.10)$$

---

[2] α' is the coupling in the Nambu–Goto–Polyakov world-sheet action of the string and g_s is the string coupling itself. These two parameters control the two types of quantum corrections in string theory. The first is related to the world-sheet corrections; these correspond to higher derivative terms in the effective gravitational theory and are calculated via loop diagrams in the sigma model. The second is related to the vacuum expectation value (vev) of the dilaton which controls the corrections due to string loops; these are due to higher genus Riemann surfaces on which the string propagates.

Here T_3 is the tension of a D3 brane, N_{D3} is the net number of (D3 – $\overline{\text{D3}}$) branes one has inserted to fill the non-compact dimensions, and H_3 and F_3 are the 3-form fluxes which arise in the NS and RR sectors, respectively. We assume that we are working with a model with only one Kähler modulus, so that $h^{1,1}(M) = 1$. (In taking the F-theory limit, where one shrinks the elliptic fibre, one has $h^{1,1}(X) = 2$ and one modulus is frozen.) Such models can be explicitly constructed, for example by using the examples of CY 4-folds or by explicitly constructing orientifolds of known CY 3-folds with $h^{1,1} = 1$.

In the presence of the non-zero fluxes, one generates a superpotential for the CY moduli,

$$W = \int_M G_3 \wedge \Omega, \quad (12.11)$$

where $G_3 = F_3 - \tau H_3$ and τ is the IIB axiodilaton. Combining this with the tree-level Kähler potential,

$$K = -3\ln[-i(\rho - \bar{\rho})] - \ln[-i(\tau - \bar{\tau})], \quad (12.12)$$

where ρ is the single volume modulus given by

$$\rho = b/\sqrt{2} + i e^{4u - \phi}, \quad (12.13)$$

and using the standard $N = 1$ supergravity formula for the potential, one obtains

$$V = e^K \left(\sum_{a,b} g^{a\bar{b}} D_a W \overline{D_b W} - 3|W|^2 \right) \rightarrow e^K \left(\sum_{i,j} g^{i\bar{j}} D_i W \overline{D_j W} \right). \quad (12.14)$$

Here a and b run over all moduli fields, while i and j run over all moduli fields except ρ; we see that, because ρ does not appear in Eq. (12.11), it cancels out of the potential energy given by Eq. (12.14), leaving the positive semi-definite potential characteristic of no-scale models. These models are not satisfactory, as they lead to the cosmological decompactification of the internal space during the cosmological evolution.

One can use two known corrections to the no-scale models, both parametrizing possible corrections to the superpotential.

(i) In type IIB compactifications of this type, there can be corrections to the superpotential coming from Euclidean D3 branes:

$$W_{\text{inst}} = T(z_i) \exp(2\pi i \rho) \quad (12.15)$$

where $T(z_i)$ is a complex structure-dependent one-loop determinant and the leading exponential dependence comes from the action of a Euclidean D3 brane wrapping a 4-cycle in M.

(ii) In general models of this sort, one finds non-Abelian gauge groups arising from geometric singularities in X, or (in type IIB language) from stacks of D7 branes wrapping 4-cycles in M. This theory undergoes gluino condensation, which results in a non-perturbative superpotential. This leads to an exponential superpotential for ρ similar to the one above (but with a fractional multiple of ρ in the exponent, since the gaugino condensate looks like a fractional instant on effect in W):

$$W_{\text{gauge}} = T(z_i) \exp(2\pi i \rho/N). \tag{12.16}$$

So effects (i) and (ii) have rather similar consequences for our analysis; we will simply assume that there is an exponential superpotential for ρ at large volumes. There are some interesting possibilities for cosmology if there are multiple non-Abelian gauge factors. Using the 4-folds, it is easy to construct examples which could yield gauge groups of total rank up to ~ 30. However, much larger ranks should be possible.

The corrections to the superpotential discussed above can stabilize the volume modulus, leading to a supersymmetry-preserving AdS minimum. We analyze the vacuum structure, just keeping the tree-level Kähler potential,

$$K = -3\ln[-i(\rho - \bar{\rho})], \tag{12.17}$$

and a superpotential,

$$W = W_0 + A e^{ia\rho}, \tag{12.18}$$

where W_0 is a tree-level contribution which arises from the fluxes. The exponential term arises from either of the two sources above and the coefficient a can be determined accordingly. At a supersymmetric vacuum, we have $D_\rho W = 0$. We simplify the situation by setting the axion in the ρ modulus to zero and letting $\rho = i\sigma$. In addition, we take A, a and W_0 to be *real* and W_0 to be negative. The condition $DW = 0$ then implies that the minimum lies at

$$W_0 = -A e^{-a\sigma_{\text{cr}}} \left(1 + \frac{2}{3} a \sigma_{\text{cr}}\right). \tag{12.19}$$

The potential at the minimum is negative and equal to

$$V_{\text{AdS}} = (-3 e^K W^2)_{\text{AdS}} = -\frac{a^2 A^2 e^{-2a\sigma_{\text{cr}}}}{6 \sigma_{\text{cr}}}. \tag{12.20}$$

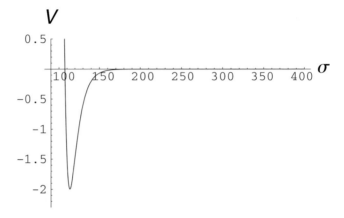

Fig. 12.7. The potential (multiplied by 10^{15}) for an exponential superpotential with $W_0 = -10^{-4}$, $A = 1$ and $a = 0.1$. In this case, there is an AdS minimum.

We see that we have stabilized the volume modulus, while preserving supersymmetry. It is important to note that the AdS minimum is quite generic. For example, if $W_0 = -10^{-4}$, $A = 1$ and $a = 0.1$, the minimum is at $\sigma_{\text{cr}} \sim 113$, as shown in Fig. 12.7.

Another possibility to obtain a minimum for large volumes is to consider a situation where the fluxes preserve supersymmetry and the superpotential involves multiple exponential terms, i.e. 'racetrack potentials' for the stabilization of ρ. Such a superpotential could arise from multiple stacks of 7-branes wrapping 4-cycles, which cannot be deformed into each other in a supersymmetry-preserving manner. In this case, by tuning the ranks of the gauge groups appropriately, one can obtain a parametrically large value of σ at the minimum.

Now we lift the supersymmetric AdS vacua to obtain the dS vacua of string theory. In the consistency condition given by Eq. (12.10), there are contributions from both localized D3 branes and fluxes. To find the AdS vacua with no moduli, of the kind discussed in Section 12.2, we assumed that the condition was saturated by turning on fluxes in the compact manifold.

Next we assume that, in fact, we turn on *too much* flux, so that Eq. (12.10) can only be satisfied by introducing one $\overline{D3}$ brane. The consistency equation is now satisfied due to the presence of the anti-D3 brane, but there is an extra bit of energy density from the 'extra' flux and $\overline{D3}$ brane. In general, we obtain a term in the potential which takes the following form:

$$\delta V = \frac{8D}{(\text{Im}\,\rho)^3}, \qquad (12.21)$$

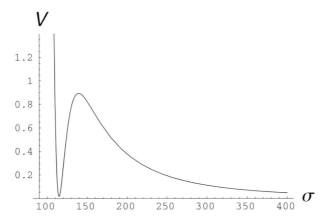

Fig. 12.8. The potential (multiplied by 10^{15}) for the case of an exponential superpotential, including a D/σ^3 correction with $D = 3 \times 10^{-9}$, which uplifts the AdS minimum to a dS minimum.

where the factor of 8 is added for later convenience. The coefficient D depends on the number of $\overline{D3}$ branes and on the warp factor at the end of the throat. These parameters can be altered by discretely changing the fluxes. This allows us to vary the coefficient D and the supersymmetry-breaking in the system, while still keeping them small. (Strictly speaking, since the flux can only be discretely tuned, D cannot be varied with arbitrary precision.) We will see that, by tuning the choice of D, one can perturb the AdS vacua to produce dS vacua with a tunable cosmological constant.

We now add to the potential a term of the form D/σ^3, as explained above. For suitable choices of D, the AdS minimum will become a dS minimum, but the rest of the potential does not change too much. However, there is one new important feature: a dS maximum separating the dS minimum from the vanishing potential at infinity. The potential is given by

$$V = \frac{aAe^{-a\sigma}}{2\sigma^2}\left(\frac{1}{3}\sigma a A e^{-a\sigma} + W_0 + A e^{-a\sigma}\right) + \frac{D}{\sigma^3}. \tag{12.22}$$

By fine-tuning D, it is easy to have the dS minimum very close to zero. For the model $W_0 = -10^{-4}$, $A = 1$, $a = 0.1$ and $D = 3 \times 10^{-9}$, the potential is as indicated in Fig. 12.8.

Note that, if one does not require the minimum to be so close to zero, D does not have to be so fine-tuned. A dS minimum is obtained as long as D lies within certain bounds, eventually disappearing for large enough D. If one does fine-tune to bring the minimum very close to zero, the resulting potentials are quite steep around the dS minimum. In this circumstance, the new term effectively uplifts the potential without changing the shape too much around the minimum, so the ρ field acquires a surprisingly large mass (relative to the final value of the cosmological constant).

If one wants to use this potential to describe the present stage of acceleration of the Universe, one needs to fine-tune the value of the potential in the dS minimum to be $V_0 \sim 10^{-120}$ in units of the Planck density. In principle, one could achieve this, for example, by fine-tuning D. However, the tuning achievable by varying the fluxes in microscopic string theory is limited, though it may be possible to tune well if there are enough 3-cycles in M.

12.4 Discussion of the anthropic landscape of string theory

It is difficult to construct realistic cosmologies in string theory if the moduli fields are not frozen. We have found that it is possible to stabilize all moduli in a controlled manner in the general setting of compactification with flux. This opens up a promising arena for the construction of realistic cosmological models based on string theory. More specifically, we have seen that it is possible to construct metastable dS vacua by including anti-branes and incorporating non-perturbative corrections to the superpotential from D3 instantons or low-energy gauge dynamics.

In the simplest possible case, our examples require knowledge of at least six parameters: two to specify the distinct electric and magnetic fluxes required to fix the dilaton; three to specify the non-perturbative corrections to the superpotential; and one to specify the anti-brane contribution. Moduli stabilization in more complicated models may depend on many more parameters, which means there are many ways to realize these vacua.

One may hope that the number of vacua in string theory is very large, at least of the order $N \geq 10^{120}$. In this case, it may be possible that some of these vacua have a positive cosmological constant of order $\Lambda \sim M_{\rm P}^4/N$, so the selection of a vacuum with $\Lambda \sim 10^{-120} M_{\rm P}^4$ could then be anthropic. The basic estimate for the number of flux vacua, satisfying the tadpole consistency condition of Eq. (12.10), is given by Douglas [5, 6] as

$$\mathcal{N}_{\rm vac} \sim \frac{(2\pi L)^{K/2}}{(K/2)^2}. \qquad (12.23)$$

Here K is the number of distinct fluxes and $L = \chi/24$ is the 'tadpole charge' on the left-hand side of Eq. (12.10). The estimates are $K \sim 100$–400 and $L \sim 500$–5000, which lead to $\mathcal{N}_{\rm vac} \sim 10^{500}$. This number is extremely large, even larger than the number 10^{120} required for the anthropic solution of the cosmological constant problem. Each of these vacua will have a different vacuum energy density and each part of the Universe with a particular positive cosmological constant will be exponentially large. Particles living in the different vacua will have dramatically different properties.

It is interesting that all of these conclusions have been reached after the recent discovery that the Universe is accelerating. Attempts to describe this acceleration in string theory forced us to invent a way to describe dS vacua. As a result, we have found that the solution of this problem is not unique and the same string theory there could have an incredibly large number of different vacua. This explains the sudden increased attention of cosmologists and string theorists towards the concept of the multiverse and anthropic reasoning.

Acknowledgements

It is a pleasure to thank T. Banks, M. Dine, N. Kaloper, A. Klypin, L. Kofman, L. Susskind, A. Vilenkin and S. Weinberg and all my collaborators for useful discussions. This work was supported by NSF grant PHY-9870115.

References

[1] R. Kallosh and A. Linde. M-theory, cosmological constant and anthropic principle. *Phys. Rev.* **D 67** (2003), 023510.
[2] S. Kachru, R. Kallosh, A. Linde and S. R. Trivedi. de Sitter vacua in string theory. *Phys. Rev.* **D 68** (2003), 046005 [hep-th/0301240].
[3] R. Bousso and J. Polchinski. Quantization of four-form fluxes and dynamical neutralization of the cosmological constant. *JHEP*, **0006** (2000), 006 [hep-th/0004134].
[4] L. Susskind. This volume (2007) [hep-th/0302219].
[5] M. R. Douglas. The statistics of string/M theory vacua. *JHEP*, **0305** (2003), 046 [hep-th/0303194].
[6] M. R. Douglas. Basic results in vacuum statistics. *Compt. Rend. Phys.* **5** (2004), 965 [hep-th/0409207].
[7] J. D. Barrow and F. J. Tipler. *The Anthropic Cosmological Principle* (New York: Oxford University Press, 1986).
[8] A. D. Linde. Inflation and quantum cosmology. In *300 Years of Gravitation*, eds. S. W. Hawking and W. Israel (Cambridge: Cambridge University Press, 1987), pp. 604–30.
[9] S. Weinberg. Anthropic bound on the cosmological constant. *Phys. Rev. Lett.* **59** (1987), 2607.
[10] G. Efstathiou. An anthropic argument for a cosmological constant. *Mon. Not. Roy. Astron. Soc.* **274** (1995), L73.

[11] A. Vilenkin. Predictions from quantum cosmology. *Phys. Rev. Lett.* **74** (1995), 846 [gr-qc/9406010].

[12] H. Martel, P. R. Shapiro and S. Weinberg. Likely values of the cosmological constant. *Astrophys. J.* **492** (1998), 29 [astro-ph/9701099].

[13] J. Garriga and A. Vilenkin. On likely values of the cosmological constant. *Phys. Rev.* **D 61** (2000), 083502 [astro-ph/9908115].

13

The anthropic principle, dark energy and the LHC

Savas Dimopoulos
Department of Physics, Stanford University
Scott Thomas
Department of Physics, Rutgers University

13.1 Naturalness versus the anthropic principle

The cosmological constant problem (CCP) is one of the most pressing problems in physics. It has eluded traditional approaches based on symmetries or dynamics. In contrast, the anthropic principle has scored a significant success in accounting for both the smallness of the cosmological constant (CC) and the proximity of the vacuum and matter energies in our universe [1].[1] Once we accept the anthropic principle as a legitimate approach for solving the CCP, it is natural to ask whether it might be applicable to other problems that can also be addressed with traditional methods. In this case, nature would have the interesting dilemma of choosing between an anthropic and a normal solution. An example is the gauge hierarchy problem (GHP). Like the CCP, it is a naturalness problem characterized by a small dimensionless number. Unlike the CCP, it can be solved with traditional symmetries, such as low-energy supersymmetry. As we will argue later, the GHP can also be addressed via anthropic arguments. So does nature choose the supersymmetric or the anthropic solution to the GHP? This question is far from academic, since the answer will be revealed experimentally by 2007 at the Large Hadron Collider (LHC).

The rise of naturalness as a principle for physics in the late 1970s led to the apparent need for a natural solution to the GHP and has convinced the majority of particle physicists that the LHC will discover either supersymmetry or another 'natural' theory that solves GHP. So if the LHC discovers nothing beyond the Standard Model, it will be a surprise. In our opinion, such a (non-)discovery would significantly strengthen the case for the anthropic principle and would cause a shift away from the usual

---

1 For earlier related work see refs. [2] and [3]. This constraint was sharpened in ref. [4], and good reviews may be found in refs. [5] and [6].

Universe or Multiverse?, ed. Bernard Carr. Published by Cambridge University Press.
© Cambridge University Press 2007.

naturalness-driven paradigm of attempting to understand the parameters of the Standard Model via symmetries or string theory. Instead, the nature of the dynamics that leads to the multiverse, and consequently provides a home for the anthropic principle, will become a primary question of physics. We now turn to this question.

13.2 Continuum and discretum

An essential hypothesis of the anthropic principle is the existence of an enormous number of universes, each with different physical laws. The collection of all these universes is called the multiverse. A major challenge is to build a compelling theory of the multiverse. There are several ideas on this and we will only mention a few. Perhaps the simplest is that there are several parallel universes, or 3-branes, all embedded inside the enormous 'bulk' of the space inside the large new dimensions that exist in theories with TeV-scale gravity. So the bulk is one possible home of the multiverse. It turns out that we can fit at most 10^{32} universes in the bulk, not enough to address the CCP.

Another possibility is that the fundamental theory has an enormous (but countable) number of almost stable 'vacua', with different values of the physical parameters. This has been coined the 'discretum' and may arise naturally in some versions of string theory [7–9]. A direct consequence of this discreteness is that the values of anthropically determined physical parameters, e.g. the CC, remain constant in this framework.

The third and most developed possibility, which we focus on here, is the 'continuum' scenario [10–13]. The idea is that a physical quantity of interest, such as the CC or the weak interaction scale, depends on a scalar field ϕ (called a 'modulus') which varies very slowly over length and time scales relevant to our observable universe. Nevertheless, because of the vast extent of the multiverse, the modulus (as well as the physical quantity of interest that depends on it) varies over a large continuous range of values. The canonical archetype of this sort of approach to the CCP is described by the following Lagrangian:

$$\mathcal{L} = (\partial_\mu \phi)^2 - \alpha\phi - \Lambda_0, \qquad (13.1)$$

where Λ_0 is the bare CC and α has to be very small to ensure that ϕ varies slowly over spacetime scales of order those for our observable universe [14]. As a result, for any Λ_0, there is always a hospitable domain where the effective CC,

$$\Lambda_{\text{eff}} = \alpha\phi + \Lambda_0, \qquad (13.2)$$

is adequately small for galaxies and life to form.

Possible concerns in this approach are that the Lagrangian given by Eq. (13.1) is *ad hoc* and that the extreme smallness of α is unexplained and potentially unnatural. So it is fair to wonder if this approach amounts to nothing but a complicated restatement of the CCP – even after accepting the assumption of the existence of the multiverse. A related issue is the degree of fine-tunings that we have to perform to keep the modulus ultra-light after quantum corrections are taken into account (so as to ensure that it remains overdamped and essentially motionless throughout the history of our universe). Still other concerns are the tunings necessary to ensure that the modulus remains extremely weakly coupled to matter, as dictated by the high-precision observational tests of the Equivalence Principle. To address all these issues at once, it is convenient to consider a more general class of theories, and we discuss these next.

13.3 Large-Z moduli

Consider a modulus with the following Lagrangian [15]:

$$\mathcal{L} = Z(\partial_\mu \phi)^2 - V(\phi) + \mathcal{L}_{\text{rest}}(\phi, \text{SM}), \tag{13.3}$$

where Z is the wave-function factor of ϕ, $V(\phi)$ is its potential and $\mathcal{L}_{\text{rest}}$ is the rest of the Lagrangian (which includes all other fields of the Standard Model, as well as their couplings to ϕ). Neither $V(\phi)$ nor $\mathcal{L}_{\text{rest}}$ contains mass scales exceeding the Planck mass or any abnormally large or small new scales or dimensionless numbers.

Note that Z acts as the friction for the field ϕ. In the limit of enormous Z, ϕ freezes and does not move even on a Hubble time scale; similarly, its spatial gradients are suppressed and consequently ϕ becomes homogeneous over cosmological scales. In addition, all its couplings become enormously suppressed, including those to the Standard Model particles. Therefore, rapid exchange of ϕ-particles does not lead to violations of the Equivalence Principle. One way to see this is to rescale,

$$\phi \to \frac{1}{\sqrt{Z}} \phi, \tag{13.4}$$

and to note that all of the couplings of ϕ are now suppressed by $Z^{-1/2}$. Consequently, in the limit of enormous Z, ϕ decouples from ordinary matter and freezes, as required in order for ϕ to be a viable anthropic modulus.

Next, we come to the fundamental question of whether it is technically natural to have a large-Z modulus. This is not just an aesthetic question, since the essence of the CCP is the absence of any symmetry that could explain the vanishing (or smallness) of the CC and consequently protect it from large radiative corrections. If the hugeness of Z were unstable against radiative corrections, then ϕ would not be a useful anthropic modulus for the CCP and this approach would just be a restatement of the CCP. However, the large value of Z *is* stable against radiative corrections since, in the limit of infinite Z, all but the first (kinetic) term of the Lagrangian can be neglected and the theory becomes symmetric under the global shift symmetry:

$$\phi \to \phi + C. \tag{13.5}$$

Therefore the anthropic approach allows the CC to be protected by a symmetry.

Since this is a global symmetry, it is, in principle, possible that non-perturbative quantum gravity effects break it significantly. Whether this happens or not is model-dependent and hinges on how gravity is embedded into a fundamental theory, as well as the mechanism that causes Z to be large. For example, if the string length is an order of magnitude (or more) larger than the Planck length, the violation of global symmetries is expected to be small. This is because the violation is caused by black-hole-related Planckian physics, and is screened and softened by string effects.

Finally, although this mechanism is already technically natural, it would be appealing to construct models where the large value of Z emerges from calculable dynamics. This dynamics, to be reliable, should involve physics below the Planck scale. This would ensure that the approximate global shift symmetry, $\phi \to \phi + C$, is not much affected by quantum gravity.

13.4 Large-Z moduli, the CCP and dark energy

Consider a modulus with the following Lagrangian:

$$\mathcal{L} = Z(\partial_\mu \phi)^2 - V(\phi) - \Lambda_0 - \mathcal{L}(\text{SM}), \tag{13.6}$$

where Λ_0 is the bare CC and $\mathcal{L}(\text{SM})$ is the Standard Model Lagrangian. For an anthropically viable universe, the present value of the effective CC, $\Lambda_{\text{eff}} = V(\phi) + \Lambda_0$, must not exceed the present energy density by more than an order of magnitude. Furthermore, it must not have changed much during the recent history of our universe, else the conditions in our recent past would not have been anthropically viable.

For Z much larger than this minimum value, ϕ (and the effective CC) is frozen, resulting in a dark energy equation of state $w = -1$. This coincides with the prediction of the discretum, where again Λ_{eff} is constant. However, when Z is close to the minimum anthropically allowed value Z_{min}, ϕ can

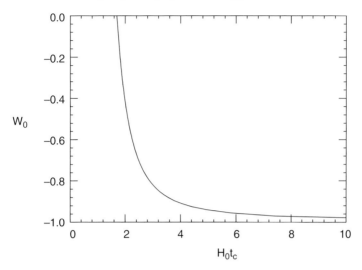

Fig. 13.1. Current value of the equation of state parameter $w_0 \equiv (p_\phi/\rho_\phi)_0$ as a function of crunch time in units of current Hubble time scale H_0^{-1} for $\Omega_m = 0.3$ and $\Omega_\phi = 0.7$.

evolve, leading to a time-dependence in Λ_{eff} and to $w \neq -1$. So, in this case, the anthropic modulus ϕ for the CCP would lead to a time-dependent equation of state for the dark energy which would have potentially testable predictions.[2] The premise that Z is near Z_{min} is valid for any theory in which the probability distribution for Z favours small Z.

It is easy now to see why this theory is very predictive: the hugeness of Z_{min} guarantees that ϕ does not move much during the recent history of the Universe. Therefore we can Taylor-expand the function Λ_{eff} and keep just the constant and linear parts. The constant is fixed by the magnitude of the observed dark energy. The linear part is determined by a single number (the slope), and this determines the complete time evolution of ϕ. As ϕ evolves, the vacuum energy eventually becomes negative, leading to a future big crunch. The crunch time is correlated with the current time-dependence of ϕ and therefore with the current equation of state, as illustrated in Fig. 13.1. The rate of change of the equation of state is also correlated with the current equation of state, as shown in Fig. 13.2. This correlation will be tested in future high-precision measurements of cosmic evolution, such as those from SNAP [16]. In this way, the continuum and discretum realizations of the anthropic principle can be distinguished and tested experimentally.

---

2 General models of quintessence can also lead to a time-dependent equation of state but are not *a priori* predictive.

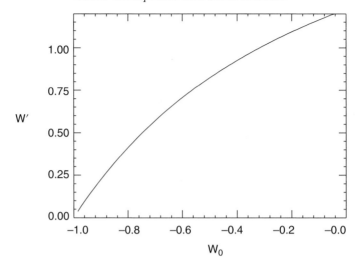

Fig. 13.2. Rate of change of the equation of state parameter $w' \equiv dw/d\ln(1+z)$, evaluated at redshift $z=1$, as a function of the current equation of state parameter $w_0 = (p_\phi/\rho_\phi)_0$ for $\Omega_m = 0.3$ and $\Omega_\phi = 0.7$.

13.5 Large-Z moduli and the hierarchy problem

Consider the Lagrangian

$$\mathcal{L} = Z(\partial_\mu \phi)^2 + \left[m_1^2 f(\phi) + m_2^2\right] H^* H, \qquad (13.7)$$

where H is the usual electroweak Higgs field, the masses m_1 and m_2 are of order of the Planck scale, and $f(\phi)$ is a non-constant function of ϕ. The quantity

$$v_{\text{eff}} = \sqrt{m_1^2 f(\phi) + m_2^2} \qquad (13.8)$$

is proportional to the Higgs vacuum expectation value, which in turn is proportional to the masses of all quarks, leptons and weak bosons. The smallness of the observed Higgs vacuum expectation value, $v_{\text{obs}} \sim 200$ GeV, compared with the Planck mass, $M_{\text{Pl}} \sim 10^{19}$ GeV, is the usual hierarchy problem.

Typical members of the multiverse have v_{eff} of order of the Planck mass and are consequently very different from our own. Those with v_{eff} a few times smaller than v_{obs} have protons heavier than neutrons and consequently do not have stable hydrogen atoms. Those with v_{eff} a few times larger than v_{obs} have neutrons much heavier than protons and consequently do not have any long lived 'heavy' nuclei beyond hydrogen. More precisely, even small deviations of v_{eff} from v_{obs} (at the few per cent level) would change the standard carbon production in red giants. So anthropically viable universes have v_{eff} very close to v_{obs}.

The large-Z moduli anthropic approach to the hierarchy problem eliminates the need for supersymmetry, low-scale gravity, technicolour or any other natural solution to the hierarchy problem. If the LHC discovers just the Higgs, it will be concrete evidence – over and above what we now have from the CCP – in favour of the anthropic approach. It will shift the paradigm of particle physics away from explaining the parameters of the Standard Model by short-distance symmetries or string theory and towards an understanding of the enormous multiverse.

References

[1] S. Weinberg. Anthropic bound on the cosmological constant. *Phys. Rev. Lett.* **59** (1987), 2607.
[2] T. Banks. TCP, quantum gravity, the cosmological constant and all that. *Nucl. Phys.* **B 249** (1985), 332.
[3] A. D. Linde. Inflation and quantum cosmology. In *300 Years of Gravitation*, eds. S. Hawking and W. Israel (Cambridge: Cambridge University Press, 1987), p. 604.
[4] A. Vilenkin. Predictions from quantum cosmology. *Phys. Rev. Lett.* **74** (1995), 846 [gr-qc/9406010].
[5] C. J. Hogan. Why the universe is just so. *Rev. Mod. Phys.* **72** (2000), 1149 [astro-ph/9909295].
[6] M. J. Rees. Numerical coincidences and 'tuning' in cosmology. In *Fred Hoyle's Universe*, eds. C. Wickramasingheb, G. Burbidge and J.Narlikar (Dordrecht: Kluwer, 2003), pp. 95–108 [astro-ph/0401424].
[7] R. Bousso and J. Polchinski. Quantization of four-form fluxes and dynamical neutralization of the cosmological constant. *JHEP*, **0006** (2000), 006 [hep-th/0004134].
[8] J. L. Feng, J. March-Russell, S. Sethi and F. Wilczek. Saltatory relaxation of the cosmological constant. *Nucl. Phys.* **B 602** (2001), 307 [hep-th/0005276].
[9] S. Kachru, R. Kallosh, A. Linde and S. P. Trivedi. de Sitter vacua in string theory. *Phys. Rev.* **D 68** (2003), 046005 [hep-th/0301240].
[10] A. Strominger, A. Maloney and E. Silverstein. de Sitter space in noncritical string theory. In *The Future of Theoretical Physics and Cosmology*, eds. G. W. Gibbons, E. P. S. Shellard and S. J. Rankin (Cambridge: Cambridge University Press, 2003), pp. 570–591 [hep-th/0205316].
[11] M. R. Douglas. The statistics of string/M theory vacua. *JHEP*, **0305** (2003), 046 [hep-ph/0303194].
[12] F. Denef and M. R. Douglas. Distributions of flux vacua. *JHEP*, **0405** (2004), 072 [hep-th/0404116].

[13] L. Susskind. This volume (2007) [hep-th/0302219].
[14] A. D. Linde. Life after inflation and the cosmological constant problem. *Phys. Lett.* **B 227** (1989), 352.
[15] S. Weinberg. The cosmological constant problems. In *Sources and Detection of Dark Matter and Dark Energy in the Universe*, ed. D. Cline (New York: Springer-Verlag, 2001) [astro-ph/0005265].
[16] R. Kallosh, J. Kratochvil, A. Linde, E. V. Linder and M. Shmakova. Observational bounds on cosmic doomsday. *J. Cosmol. Astropart. Phys.*, **0310** (2003), 015 [astro-ph/0307185].

Part III

Particle physics and quantum theory

14

Quarks, electrons and atoms in closely related universes

Craig J. Hogan
Astronomy and Physics Departments, University of Washington

14.1 Introduction

We know that nature is governed by mathematics and symmetries. Not very long ago, it was an article of faith among most physicists that everything about physics would eventually be explained in terms of fundamental symmetries – that nothing in the make-up of physical laws is accidental, that nature ultimately has no choices, and that all the properties of particles and fields are fixed purely by mathematics.

In the thirty years since modern anthropic reasoning was introduced into cosmology [1, 2], the competing idea that anthropic selection might have an indispensable role in fundamental physical theory has gradually become, if not universally accepted, at least mainstream. There are now concrete physical models for realizing anthropic selection in nature. Cosmology has provided not only a concrete mechanism (inflation) for manufacturing multiple universes, but also a new phenomenon (dark energy) whose value is most often explained by invoking anthropic explanations. String theory has uncovered a framework by which many different symmetries and parameters for fields can be realized in the low-energy, 4-dimensional universe; this depends on the topology and size of the manifold of the other seven (truly fundamental) dimensions and on the configurations of p-branes within it, especially the local environment of the 3-brane on which our own Standard Model fields live. The number of locally metastable configurations of manifold and branes, and therefore the number of options for low-energy physics, is estimated to be so large that, for all practical purposes, there is a continuum of choices for the fundamental parameters that we observe [3–5].

Of course, the details of how this works in the real world are still sketchy. Cosmology unfolds in a series of phase transitions and symmetry-breakings. For example, it is now part of standard inflation that the quantum

Universe or Multiverse?, ed. Bernard Carr. Published by Cambridge University Press.
© Cambridge University Press 2007.

wave-function of the Universe branches early into various options for the zero-point fluctuations of the inflaton field, different branches of which correspond to different distributions of galaxies. String theory opens up a scenario in which the multiverse wave-function may also branch very early into a variety of whole universes, each of which has different physics. If things happen this way, it is natural for us to find ourselves in a branch with physics remarkably well suited to make the stuff of which we are made.

It then makes sense to ask new questions about the world: how would things change if this or that aspect of physics were altered? If a small change in a certain parameter changes the world a great deal, in a way that matters to our presence here, then that is a clue that this particular parameter is fixed by selection rather than by symmetry. The following arguments along these lines are elaborated more fully in ref. [6].

Now we may be faced with a situation where some seemingly fundamental features of physics might not ever be derived from first principles. Even the particular gauge group in our Grand Unified Theory (that is, the one in our branch of the wave-function) might be only one group selected out of many options provided by the Theory of Everything. We may have to adjust our scientific style to this larger physical reality, which forces cosmology and fundamental physics into a new relationship. For example, although we cannot look inside the other universes of the multiverse ensemble and cannot predict the branching outcome from first principles, cosmological experiments now under development might reveal relict gravitational waves from the same symmetry-breaking that fixed the parameters.

14.2 Changing Standard Model parameters

Evaluating changes in the world in response to changes in the fundamental physics is actually a difficult programme to carry out. For the most fundamental theory we have, the Standard Model, the connection of many of its parameters with generally observable phenomena can only be roughly estimated. First-principle calculations of the behaviour of systems such as nuclei and molecules are possible only for the simplest examples.

The traditional minimal Standard Model has nineteen 'adjustable' parameters [7,8]: Yukawa coefficients fixing the masses of the six quark and three lepton flavors ($u, d, c, s, t, b, e, \mu, \tau$); the Higgs mass and vacuum expectation value v (which multiplies the Yukawa coefficients to determine the fermion masses); three angles and one phase of the CKM (Cabibbo–Kobayashi–Maskawa) matrix, which mixes quark weak and strong interaction eigenstates; a phase for the quantum chromodynamic (QCD) vacuum; and three coupling constants g_1, g_2, g_3 of the gauge group $U(1) \times SU(2) \times SU(3)$. We now know experimentally that the neutrinos are not massless, so there are at least seven more parameters to characterize their behaviour (three masses and another four CKM matrix elements). Thus, twenty-six parameters, plus Newton's constant G and the cosmological constant Λ of general relativity, are enough to describe the behaviour of all experimentally observed particles, except perhaps those related to dark matter. If, in addition, the Standard Model is extended by supersymmetry, the number of parameters exceeds 100.

Imagine that you are sitting at a control panel of the Universe. It has a few dozen knobs – one for each of the parameters. Suppose you start twiddling the knobs. For all but a few of the knobs, you find nothing changes very much; the mass of the top quark for example (that is, its Yukawa coupling coefficient in the Standard Model equations) has little direct effect on everyday stuff.

Which knobs matter for the stuff we care about most – atoms and molecules? Some knobs are clearly important, but their exact value does not seem too critical. The fine structure constant α, for example, controls the sizes of all the atoms and molecules, these scaling like the Bohr radius $(\alpha m_e)^{-1}$. If you twiddle this knob, natural phenomena dominated by this physics – which include all of familiar chemistry and biology – grow or shrink in size. On the other hand, they all grow or shrink by roughly the same fractional amount, so the structural effect of changes is hard to notice; the miraculous fit of base-pairs into the DNA double-helix would still work quite well, for example. There are, however, subtle changes in structural relationships and molecular reaction rates. Our complicated biochemistry would probably not survive a sudden big change in α, but if you turned the knob slowly enough, living things would probably adapt to the changing physics. Simulations of cellular reaction networks show that their behaviour is remarkably robust with respect to changes in reaction rates, and mostly depend on network topology [9].

It turns out that a few of the knobs have a particularly large qualitative effect, even with a very small amount of twiddling. Three knobs stand out for their particularly conspicuous effects: the Yukawa coefficients controlling the masses of the electron, the up-quark and the down-quark. These are the light fermions that dominate the composition and behaviour of atoms and molecules. Changing them by even a small fractional amount has a devastating effect on whether molecules can exist at all. The most dramatic sensitivity of the world on their values seems to be in the physics of atomic nuclei.

14.3 Effects of changing u,d,e masses on atoms and nuclei

The light fermion masses are all very small compared with the mass of a proton (less than one per cent). (Ironically, the mass of protons and neutrons, which comprise the bulk of the mass of ordinary matter, is dominated not by the 'real mass' of their constituent quarks, but almost entirely by the kinetic energies of the quarks and the massless gluons mediating the colour forces.) However, the light fermion masses are critical because they determine the energy thresholds for reactions that control the stability of nucleons.

In the 3-dimensional parameter space formed by these masses, the most reliable phenomemological statements can be made about changes within the 2-dimensional surface defined by holding the sum of the u and d masses constant. (That is because many complicated features of nuclear physics remain constant if the pion mass, which is proportional to $(m_u + m_d)^{1/2}$, is constant.) In this plane, some properties of worlds with different values of the masses are summarized in Figs. 14.1 and 14.2, the latter having been taken from ref. [2]. The figures also show a constraint for a particular $SO(10)$ grand unified scenario, to illustrate that likely unification schemes probably do not leave all these parameters independent – at least one relationship between them is likely fixed by unification symmetry.

In the lower part of Fig. 14.1, towards larger up-quark mass, there are 'neutron worlds'. As one turns the knobs in this direction, a threshold is soon crossed where it is energetically favourable for the electron in a hydrogen atom to join with its proton to make a neutron. If you turn it farther, even a free proton (without any nearby electron) spontaneously decays into a neutron. In the upper part of the figure, there are 'proton worlds'. Moving up from our world, a threshold is soon crossed where a deuteron in a plasma is no longer energetically favoured over a pair of protons. If you go farther, even an isolated deuteron spontaneously decays into a pair of protons.

In the neutron world, there are nuclei, but not atoms with electrons around them, so chemistry does not happen. In the proton world, there are hydrogen atoms, but they are the only kind of atoms because the other nuclei do not form or are not stable. Fortunately for us there is a world in between, where a few dozen stable nuclei are both possible and are actually produced in stars, and are endowed with electron orbitals leading to chemistry with arbitrarily large and complex molecules. This world would disappear with only a few per cent fractional change in the quark mass difference in either direction. It does not exist in some closely related branches of the multiverse wave-function.

14 *Quarks, electrons and atoms in closely related universes* 225

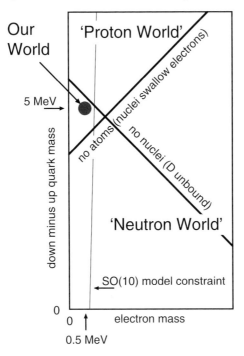

Fig. 14.1. An overview of the simple nuclear physics of the neutron, proton and deuteron, in other universes closely related to ours. Thresholds for various reactions are shown in terms of the mass difference between the down and up quarks and the electron mass, in the plane where the sum of the up and down masses does not change. Our world is shown by the dot. The $SO(10)$ constraint shown imposes the restriction that the ratio of electron to down quark mass is fixed by a symmetry to have the same value as it does in the real world; the region to the right of this is excluded for positive down-quark mass.

One can estimate roughly the effects of leaving this plane. In that case, nuclear physics is changed in new ways, since the mass of the pion changes. It appears that if the masses are increased by more than about 40%, the range of nuclear forces is reduced to the point where the deuteron is unstable; and if they are reduced by a similar amount, the nuclear forces are strengthened to the point where the diproton is stable. On the other hand, the latter change also reduces the range of nuclear forces, so that there are fewer stable elements overall. The sum of the quark masses in our world appears roughly optimized for the largest number of stable nuclei. Again, the situation would change qualitatively (for example, far fewer stable elements) with changes in summed quark masses at the 10% level.

Why is it even possible to find parameters balanced between the neutron world and the proton world? For example, if the $SO(10)$ model is

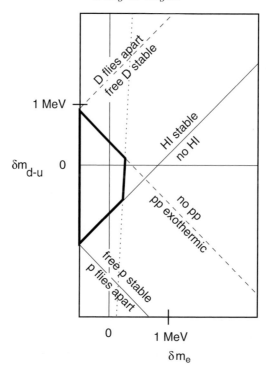

Fig. 14.2. A more detailed view, from ref. [2], of the changes in thresholds of nuclear reactions, as a function of the change in the u, d mass difference and the change in the electron mass. Our world is at the origin in these quantities.

the right one, it seems that we are lucky that its trajectory passes through the region that allows for molecules. The answer could be that even the gauge symmetries and particle content have an anthropic explanation. A great variety of compact 7-manifolds and 3-brane configurations solve the fundamental M-theory. Each one of them has dimensional scales, corresponding to parameter values such as particle masses, as well as topological and geometrical relationships corresponding to symmetries. Many of these configurations undergo inflation and produce macroscopic universes. In this situation, it is not surprising that we find ourselves in one where atoms and nuclei can exist.

14.4 Quantum mechanics and anthropic selection

Discussions of anthropic selection have sometimes differentiated between the kind that selects whole universes (with different values of the electron mass etc.) and the kind that selects a congenial environment (why we do not

14 Quarks, electrons and atoms in closely related universes

live on an asteroid or a quasar etc.) While these seem very different from a quantum-mechanical perspective, they do not differ in kind. Both involve selections of a congenial branch of the wave-function of the Universe.

In the original formulation of quantum mechanics, it was said that an observation collapsed a wave-function to one of the eigenstates of the observed quantity. The modern view is that the cosmic wave-function never collapses, but only appears to collapse from the point of view of observers who are part of the wave-function. When Schrödinger's cat lives or dies, the branch of the wave-function with the dead cat also contains observers who are dealing with a dead cat, and the branch with the live cat also contains observers who are petting a live one.

Although this is sometimes called the 'many worlds' interpretation of quantum mechanics, it is really about having just one world, with one wave-function obeying the Schrödinger equation: the wave-function evolves linearly from one time to the next, based on its previous state. Anthropic selection in this sense is built into physics at the most basic level of quantum mechanics. Selection of a wave-function branch is what drives us into circumstances in which we thrive. Viewed from a disinterested perspective outside the Universe, it is as though living beings swim like salmon up their favourite branches of the wave-function, chasing their favourite places.

The selection of a planet or a galaxy is a matter of chance. In quantum mechanics, this means that a branch of the wave-function has been selected. The binding energy of our galaxy was determined by an inflaton fluctuation during inflation; that was when the branching occurred that selected the large-scale gravitational potential that set the parameters for our local cosmic environment. We can achieve statistical understanding about this kind of selection because we can observe other parts of the ensemble, by observing galaxy clustering, the microwave background and so on. In this way, we understand the physics of the symmetry-breaking. We even know something about the formation of the different galaxy distributions in other universes that we will never see. These are regarded as being just different by chance.

If the quark and electron masses are also matters of chance, the branching of the wave-function occurred along with the symmetry-breaking that fixed their masses. There may be ways to observe aspects of the statistical ensemble for this event also, by studying the gravitational-wave background rather than the microwave background.

We do not know when all the choices of parameters and symmetries were made. Some of these branchings may leave traces of other choices observable

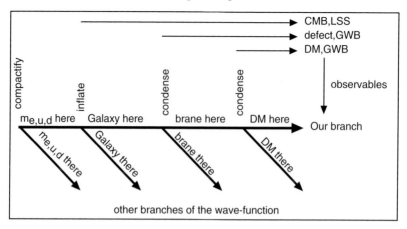

Fig. 14.3. A schematic sketch of the branching history of the wave-function to which we belong. At various points in cosmic history, symmetry-breaking (for example compactification, inflation, condensation) made random choices, which were frozen into features such as Standard Model parameters, the galaxy distribution, or the dark matter (DM) density. In some cases, these events left other observables which can be observed directly in other ways, such as cosmic microwave background (CMB) anisotropy, large-scale structure (LSS), gravitational-wave backgrounds (GWB) or cosmic defects. Thus, although the other branches of the wave-function cannot be observed directly, the physics of the branching events in some situations may be independently observable.

in our past light cone, as illustrated in Fig. 14.3. It could be that some parameters are spatially varying even today, in response to spatial variations in scalar or dark matter fields. For example, one model of dark energy predicts large variations in the masses of neutrinos, depending on the local density of the neutrino component of dark matter [10]. (Indeed, the basic idea that effective neutrino masses depend on the local physical environment is now part of the standard theory of solar neutrino oscillations.) Thus the properties of stars can be spatially modulated, depending on the dark matter density – a quantity determined, in many theories, by a branching event that occurred recently enough to have an observable effect. Such ideas provide a new motivation for observational programmes to quantify the extent to which the constants of nature are really constant in spacetime. (A thriving example of this can be found in studies of varying α.)

In some models, events connected with fixing the local quark and electron masses may have happened late enough to leave fossil traces. This could happen during the final compactification of some of the extra dimensions, or the condensation of our own Standard Model 3-dimensional brane within a larger-dimensional space. If compactification happens in a sufficiently catastrophic symmetry-breaking, it can lead to a background of gravitational waves. Because they are so penetrating, gravitational waves can carry information directly from almost the edge of our past light-cone, well beyond recombination, even beyond weak decoupling – indeed back to the edge of 3-dimensional space as we know it.

If the extra dimensions are smaller than the Hubble length at dimensional compactification or brane condensation, their collapse can appear as a first-order phase transition in our 3-dimensional space, leading to relativistic flows of mass-energy. If the extra dimensions are larger than or comparable to the Hubble scale, our 3-dimensional brane may initially condense with warps and wiggles that lead to a gravitational wave background. Either way, the mesoscopic, classical motion of branes settling down to their final equilibrium configuration could lead to a strong gravitational-radiation background in a frequency range detectable by detectors now under development [11–14]. Thus, instruments designed to observe the early boundary of spacetime may also explore the early boundary of physics as we know it, and directly test ideas concerning the separation of various branches of the multiverse having different fundamental parameters.

This blending of empirical cosmology and fundamental physics is reminiscent of our Darwinian understanding of the tree of life. The double-helix, the four-base codon alphabet and the triplet genetic code for amino acids, any particular gene for a protein in a particular organism – all these are frozen accidents of evolutionary history. It is futile to try to understand or explain these aspects of life, or indeed any relationships in biology, without referring to the way the history of life unfolded. In the same way that – in Dobzhansky's phrase [15] – 'nothing in biology makes sense except in the light of evolution', physics in these models only makes sense in the light of cosmology.

Acknowledgement

This work was supported by NSF grant AST-0098557 at the University of Washington.

References

[1] B. Carter. Large number coincidences and the anthropic principle in cosmology. In *Confrontation of Cosmological Theory with Observational Data*, ed. M. S. Longair (Dordrecht: Reidel, 1974), p. 291.
[2] B. J. Carr and M. J. Rees. The anthropic principle and the structure of the physical world. *Nature*, **278** (1979), 605.

[3] R. Bousso and J. Polchinski. Quantization of four-form fluxes and dynamical neutralization of the cosmological constant. *JHEP*, **0006** (2000), 006 [hep-th/0004134].
[4] S. Kachru, R. Kallosh, A. Linde and S. P. Trivedi. de Sitter vacua in string theory. *Phys. Rev.* **D 68** (2003), 046005 [hep-th/0301240].
[5] L. Susskind. This volume (2007) [hep-th/0302219].
[6] C. J. Hogan. Why the universe is just so. *Rev. Mod. Phys.* **72** (2000), 1149 [astro-ph/9909295].
[7] R. N. Cahn. The eighteen arbitrary parameters of the standard model in your everyday life. *Rev. Mod. Phys.* **68** (1996), 951.
[8] M. K. Gaillard, P. D. Grannis and F. J. Sciulli. The standard model of particle physics. *Rev. Mod. Phys.* **71** (1999), 596.
[9] E. M. Munro and G. Odell. Morphogenetic pattern formation during ascidian notochord formation is regulative and highly robust. *Development*, **129** (2002), 1.
[10] R. Fardon, A. E. Nelson and N. Weiner. Dark energy from mass varying neutrinos. *J. Cosmol. Astropart. Phys.*, **0410** (2004), 005 [astro-ph/0309800].
[11] C. J. Hogan. Scales of the extra dimensions and their gravitational wave backgrounds. *Phys. Rev.* **D 62** (2000), 121302 [astro-ph/0009136].
[12] C. J. Hogan. Gravitational waves from mesoscopic dynamics of the extra dimensions. *Phys. Rev. Lett.* **85** (2000), 2044 [astro-ph/0005044].
[13] C. J. Hogan and P. L. Bender. Estimating stochastic gravitational wave backgrounds with Sagnac calibration. *Phys. Rev.* **D 64** (2001), 062002 [astro-ph/0104266].
[14] K. Ichiki and K. Nakamura. Stochastic gravitational wave background in brane world cosmology. (2004) [astro-ph/0406606].
[15] T. Dobzhansky. Nothing in biology makes sense except in the light of evolution. *Am. Biol. Teacher*, **345** (1973), 125.

15

The fine-tuning problems of particle physics and anthropic mechanisms

John F. Donoghue
Department of Physics, University of Massachusetts

15.1 Open questions in particle physics

Each field has a set of questions which are universally viewed as important, and these questions motivate much of the work in the field. In particle physics, several of these questions are directly related to experimental problems. Examples include questions such as: Does the Higgs boson exist and, if so, what is its mass? What is the nature of the dark matter in the Universe? What is the mechanism that generated the net number of baryons in the Universe? For these topics, there is a well posed problem related to experimental findings or theoretical predictions. These are problems that must be solved if we are to achieve a complete understanding of the fundamental theory.

There also exists a different set of questions which have a more aesthetic character. In these cases, it is not as clear that a resolution is required, yet the problems motivate a search for certain classes of theories. Examples of these are the three 'naturalness' or 'fine-tuning' problems of the Standard Model; these are associated with the cosmological constant Λ, the energy scale of electroweak symmetry-breaking v and the strong CP-violating angle θ. As will be explained more fully below, these are free parameters in the Standard Model that seem to have values 10 to 120 orders of magnitude smaller than their natural values and smaller than the magnitude of their quantum corrections. Thus their 'bare' values plus their quantum corrections need to be highly fine-tuned in order to obtain the observed values. Because of the magnitude of this fine-tuning, one suspects that there is a dynamical mechanism at work that makes the fine-tuning natural. This motivates many of the theories of new physics beyond the Standard Model. A second set of aesthetic problems concern the parameters of the Standard Model, i.e. the coupling constants and masses of

Universe or Multiverse?, ed. Bernard Carr. Published by Cambridge University Press.
© Cambridge University Press 2007.

the theory. While the Standard Model is constructed simply using gauge symmetry, the parameters themselves seem not to be organized in any symmetric fashion. We would love to uncover the principle that organizes the quark and lepton masses (sometimes referred to as the 'flavour problem'), for example, but attempts to do so with symmetries or a dynamical mechanism have been unsuccessful.

These aesthetic questions are very powerful motivations for new physics. For example, the case for low energy supersymmetry, or other TeV scale dynamics to be uncovered at the Large Hadron Collider (LHC), is based almost entirely on the fine-tuning problem for the scale of electroweak symmetry-breaking. If there is new physics at the TeV scale, then there need not be any fine-tuning at all and the electroweak scale is natural. We are all greatly looking forward to the results of the LHC, which will tell us if there is in fact new physics at the TeV scale. However, the aesthetic questions are of a different character from direct experimental ones concerning the existence and mass of the Higgs boson. There does not *have* to be a resolution to the aesthetic questions – if there is no dynamical solution to the fine-tuning of the electroweak scale, it would puzzle us, but would not upset anything within the fundamental theory. We would just have to live with the existence of fine-tuning. However, if the Higgs boson is not found within a given mass range, it would falsify the Standard Model.

The idea of a multiverse will be seen to change drastically the way in which we perceive the aesthetic problems of fine-tuning and flavour. In a multiverse, the parameters of the theory vary from one domain to another. This naturally leads to the existence of anthropic constraints – only some of these domains will have parameters that reasonably allow the existence of life. We can only find ourselves in a domain which satisfies these anthropic constraints. Remarkably, the anthropic constraints provide plausible 'solutions' to two of the most severe fine-tuning problems: those of the cosmological constant and the electroweak scale. Multiverse theories also drastically reformulate some of the other problems – such as the flavour problem. However, at the same time, these theories raise a new set of issues for new physics. My purpose in this chapter is to discuss how the idea of the multiverse reformulates the problems of particle physics.

It should be noted up front that the Anthropic Principle [1–3] has had a largely negative reputation in the particle physics community. At some level this is surprising – a community devoted to uncovering the underlying fundamental theory might be expected to be interested in exploring a suggestion as fundamental as the Anthropic Principle. I believe that the problem really lies in the word 'Principle' more than in the word 'Anthropic'.

The connotation of 'Principle' is that of an underlying theory. This leads to debates over whether such a principle is scientific, i.e. whether it can be tested. However, 'anthropics' is not itself a theory, nor even a principle. Rather, the word applies to constraints that naturally occur within the full form of certain physical theories. However, it is the theory itself that needs to be tested, and to do this one needs to understand the full theory and pull out its predictions. For theories that lead to a multiverse, anthropic constraints are unavoidable. As we understand better what types of theory have this multiverse property, the word anthropic is finding more positive applications in the particle physics community. This article also tries to describe some of the ways that anthropic arguments can be used to positive effect in particle physics.

15.2 The golden Lagrangian and its parameters

The Lagrangian of the Standard Model (plus General Relativity) encodes our present understanding of all observed physics except for dark matter [4]. The only unobserved ingredient of the theory is the Higgs boson. The Standard Model is built on the principle of gauge symmetry – that the Lagrangian has an $SU(3) \otimes SU(2)_\mathrm{L} \otimes U(1)$ symmetry at each point of spacetime. This, plus renormalizability, is a very powerful constraint and uniquely defines the structure of the Standard Model up to a small number of choices, such as the number of generations of fermions. General Relativity is also defined by a gauge symmetry – local coordinate invariance. The resulting Lagrangian can be written in compact notation:

$$\mathcal{L} = -\frac{1}{4}F^2 + \bar{\psi}\mathrm{i}D\psi + \frac{1}{2}D_\mu\phi D^\mu\phi$$
$$+ \bar{\psi}\Gamma\psi\phi + \mu^2\phi^2 - \lambda\phi^4 - \frac{1}{16\pi G_\mathrm{N}}R - \Lambda. \tag{15.1}$$

Experts recognize the various terms here as indications of the equations governing the photon, gluons and W-bosons (the F^2 terms), quarks and leptons (the ψ terms), the Higgs field (ϕ) and gravity (R), along with a set of interactions constrained by the gauge symmetry. Of course, such a simple form belies a very complex theory, and tremendous work is required to understand the predictions of the Standard Model. But the greatest lesson of particle physics of the past generation is that nature organizes the Universe through a simple set of gauge symmetries.

However, the story is not complete. The simple looking Lagrangian given by Eq. (15.1), and the story of its symmetry-based origin, also hide a far less beautiful fact. To *really* specify the theory, we need not only the Lagrangian, but also a set of twenty-eight numbers which are the parameters of the theory. These are largely hidden underneath the compact notation of the Lagrangian. Examples include the masses of all the quarks and leptons (including neutrinos), the strengths of the three gauge interactions, the weak mixing angles describing the charge current interactions of quarks and leptons, the overall scale of the weak interaction, the cosmological constant and Newton's gravitational constant. None of these parameters is predicted by the theory. The values that have have been uncovered experimentally do not obey any known symmetry pattern, and the Standard Model provides no principle by which to organize them. After the beauty of the Standard Model Lagrangian, these seemingly random parameters reinforce the feeling that there is more to be understood.

15.3 Fine-tuning

Three of the twenty-eight parameters are especially puzzling, because their values appear to be unnaturally small. Naturalness and fine-tuning have very specific technical meanings in particle physics. These meanings are related to, but not identical to, the common usage in non-technical settings. The technical version is tied to the magnitude of quantum corrections. When one calculates the properties of any theory using perturbation theory, quantum mechanical effects give additive corrections to all its parameters. Perturbation theory describes the various quantities of a theory as a power series in the coupling constants. The calculation involves summing over the effects of all virtual states that are possible in the theory, including those at high energy. The quantum correction refers to the terms in the series that depend on the coupling constants. The 'bare' value is the term independent of the coupling constants. The physical measured value is the sum of the bare value and the quantum corrections.

The concept of naturalness is tied to the magnitude of the quantum corrections. If the quantum correction is of the same order as (or smaller than) the measured value, the result is said to be natural. If, on the contrary, the measured value is much smaller than the quantum correction, then the result is unnatural because the bare value and the quantum correction appear to have an unexpected cancellation to give a result that is much smaller than either component. This is an unnatural fine-tuning.

In fact, the quantum correction is often not precisely defined. The ambiguity can arise due to possible uncertainties of the theory at high energy. Since physics is an experimental science, and we are only gradually uncovering the details of the theory as we probe higher energies, we do not know the high energy limits of our present theory. We expect new particles and interactions to be uncovered as we study higher energies. Since the quantum correction includes effects from high energy, there is an uncertainty about their extent and validity. We understand the theory up to some energy – let us call this $E_{\max}$ – but beyond this new physics may enter. The quantum corrections will typically depend on the scale $E_{\max}$. We will see below that, in some cases, the theory may be said to be natural if one employs low values of $E_{\max}$ but becomes unnatural for high values.

The Higgs field in the Standard Model takes a constant value everywhere in spacetime. This is called its 'vacuum expectation value', abbreviated as vev, which has the magnitude $v = 246$ GeV. This is the only dimensionful constant in the electroweak interactions and hence sets the scale for all dimensionful parameters of the electroweak theory. For example, all of the quark and lepton masses are given by dimensionless numbers Γ_i (the Yukawa couplings) times the Higgs vev, $m_i = \Gamma_i v/\sqrt{2}$. However, the Higgs vev is one of the parameters which has a problem with naturalness. While it depends on many parameters, the problem is well illustrated by its dependence on the Higgs coupling to the top quark. In this case, the quantum correction grows quadratically with $E_{\max}$. One finds

$$v^2 = v_0^2 + \frac{3\Gamma_t^2}{4\pi^2 \lambda} E_{\max}^2, \tag{15.2}$$

where Γ_t is the Yukawa coupling for the top quark, v_0 is the bare value, λ is the self-coupling of the Higgs and the second term is the quantum correction. Since $v = 246$ GeV and $\Gamma_t \sim \lambda \sim 1$, this would be considered natural if $E_{\max} \sim 10^3$ GeV, but it would be unnatural by twenty-six orders of magnitude if $E_{\max} \sim 10^{16}$ GeV (characteristic of the Grand Unified Theories which unite the electroweak and strong interactions) or thirty-two orders of magnitude if $E_{\max} \sim 10^{19}$ GeV (characteristic of the Planck mass, which sets the scale for quantum gravity).

If we philosophically reject fine-tuning and require that the Standard Model be technically natural, this requires that $E_{\max}$ should be around 1 TeV. For this to be true, we need a new theory to enter at this scale that removes the quadratic dependence on $E_{\max}$ in Eq. (15.2). Such theories do exist – supersymmetry is a favourite example. Thus the argument against fine-tuning becomes a powerful motivator for new physics at the scale of 1 TeV. The LHC has been designed to find this new physics.

An even more extreme violation of naturalness involves the cosmological constant Λ. Experimentally, this dimensionful quantity is of order $\Lambda \sim (10^{-3}\ \text{eV})^4$. However, the quantum corrections to it grow as the fourth power of the scale E_{max}:

$$\Lambda = \Lambda_0 + cE_{\text{max}}^4, \tag{15.3}$$

with the constant c being of order unity. This quantity is unnatural for all particle physics scales by a factor of 10^{48} for $E_{\text{max}} \sim 10^3$ GeV to 10^{124} for $E_{\text{max}} \sim 10^{19}$ GeV.

It is unlikely that there is a technically natural resolution to the cosmological constant's fine-tuning problem – this would require new physics at 10^{-3} eV. A valiant attempt at such a theory is being made by Sundrum [5], but it is highly contrived to have new dynamics at this extremely low scale which modifies only gravity and not the other interactions.

Finally, there is a third classic naturalness problem in the Standard Model – that of the strong CP-violating parameter θ. It was realized that QCD can violate CP invariance, with a free parameter θ which can, in principle, range from zero up to 2π. An experimental manifestation of this CP-violating effect would be the existence of a non-zero electric dipole moment for the neutron. The experimental bound on this quantity requires $\theta \leq 10^{-10}$. The quantum corrections to θ are technically infinite in the Standard Model if we take the cut-off scale E_{max} to infinity. For this reason, we would expect that θ is a free parameter in the model of order unity, to be renormalized in the usual way. However, there is a notable difference from the two other problems above in that, if the scale E_{max} is taken to be very large, the quantum corrections are still quite small. This is because they arise only at a very high order in perturbation theory. So, in this case, the quantum corrections do not point to a particular scale at which we expect to find a dynamical solution to the problem.

15.4 Anthropic constraints

The standard response to the fine-tuning problems described above is to search for dynamical mechanisms that explain the existence of the fine-tuning. For example, many theories for physics beyond the Standard Model (such as supersymmetry, technicolour, large extra dimensions, etc.) are motivated by the desire to solve the fine-tuning of the Higgs vev. These are plausible, but as yet have no experimental verification. The fine-tuning problem for the cosmological constant has been approached less successfully; there are few good suggestions here. The strong CP problem has motivated the theory of axions, in which an extra symmetry removes the strong CP violation, but requires a very light pseudo-scalar boson – the axion – which has not yet been found.

However, theories of the multiverse provide a very different resolution of the two greatest fine-tuning problems, that of the Higgs vev and the cosmological constant. This is due to the existence of anthropic constraints on these parameters. Suppose for the moment that life can only arise for a small range of values of these parameters, as will be described below. In a multiverse, the different domains will have different values of these parameters. In some domains, these parameters will fall in the range that allows life. In others, they will fall outside this range. It is then an obvious constraint that we can only observe those values that fall within the viable range. For the cosmological constant and the Higgs vev, we can argue that the anthropic constraints only allow parameters in a very narrow window, all of which appears to be fine-tuned by the criteria of Section 15.3. Thus the observed fine-tuning can be thought to be required by anthropic constraints in multiverse theories.

The first application of anthropic constraints to explain the fine-tuning of the cosmological constant – even before this parameter was known to be non-zero – was due to Linde [6] and Weinberg [7]; see also refs. [8–10]. In particular, Weinberg gave a physical condition – noting that, if the cosmological constant was much different from what it is observed to be, galaxies could not have formed. The cosmological constant is one of the ingredients that governs the expansion of the Universe. If it had been of its natural scale of $(10^3 \text{ GeV})^4$, the Universe would have collapsed or been blown apart (depending on the sign) in a fraction of a second. For the Universe to expand slowly enough that galaxies can form, Λ must lie within roughly an order of magnitude of its observed value. Thus the 10^{124} orders of magnitude of fine-tuning is spurious; we would only find ourselves in one of the rare domains with a tiny value of the cosmological constant.

Other anthropic constraints can be used to explain the fine-tuning of the Higgs vev. In this case, the physical constraint has to do with the existence of atoms other than hydrogen. Life requires the complexity that comes from having many different atoms available to build viable organisms. It is remarkable that these atoms do not exist for most values of the Higgs vev, as has been shown by my collaborators and myself [11, 12]. Suppose for the moment that all the parameters of the Standard Model are held fixed, except for v which is allowed to vary. As v increases, all of the quark masses grow, and hence the neutron and proton masses also increase. Likewise, the neutron–proton mass-splitting increases in a calculable fashion. The most model-independent constraint on v then comes from the value when the neutron–proton mass-splitting becomes larger than the 10 MeV per nucleon that binds the nucleons into nuclei; this occurs when v is about five times the observed value. When this happens, all bound neutrons will decay to protons [11,12]. However, a nucleus of only protons is unstable and will fall apart into hydrogen. Thus complex nuclei will no longer exist.

A tighter constraint takes into account the calculation of the nuclear binding energy, which decreases as v increases. This is because the nuclear force, especially the central isoscalar force, is highly dependent on pion exchange and, as v increases, the pion mass also increases, making the force of shorter range and weaker. In this case, the criteria for the existence of heavy atoms require v to be less than a few times its observed value. Finally, a third constraint – of comparable strength – comes from the need to have deuterium stable, because deuterium was involved in the formation of the elements in primordial and stellar nucleosynthesis [11,12]. In general, even if the other parameters of the Standard Model are not held fixed, the condition is that the weak and strong interactions must overlap. The masses of quarks and leptons arise in the weak interactions. In order to have complex elements, some of these masses must be lighter than the scale of the strong interactions and some heavier. This is a strong and general constraint on the electroweak scale. All of these constraints tell us that the viable range for the Higgs vev is not the thirty or so orders of magnitude described above, but only the tiny range allowed by anthropic constraints.

15.5 Lack of anthropic constraints

While anthropic constraints have the potential to solve the two greatest fine-tuning problems of the Standard Model, similar ideas very clearly fail to explain the naturalness problem of the strong CP-violating parameter θ [4]. For any possible value of θ in the allowed range from 0 to 2π, there would be little influence on life. The electric dipole moments that would be generated could produce small shifts in atomic energy levels but would not destabilize any elements. Even if a mild restriction could be found, there would be no logical reason why θ should be as small as 10^{-10}. Therefore the idea of a multiverse does nothing to solve this fine-tuning problem.

The lack of an anthropic solution to this problem is a very strong constraint on multiverse theories. It means that, in a multiverse ground state that satisfies the other anthropic constraints, the strong CP problem must *generically* be solved by other means. Perhaps the axion option, which appears to us to be an optional addition to the Standard Model, is in fact required to be present for some reason – maybe in order to generate dark matter in the Universe. Or perhaps there is a symmetry that initially sets θ to zero, in which case the quantum corrections shift it only by a small amount. This can be called the 'small infinity' solution, because – while the quantum correction is formally infinite – it is small when any reasonable cut-off is used. Thus the main problem in this solution is to find a reason why the bare value of θ is zero rather than some number of order unity. In any case, in multiverse theories the strong CP problem appears more serious than the other fine-tuning problems and requires a dynamical solution.[1]

15.6 Physical mechanisms

The above discussion can be viewed as a motivation for multiverse theories. Such theories would provide an explanation of two of the greatest puzzles of particle physics. However, this shifts the focus to the actual construction of such physical theories. So far we have just presented a 'story' about a multiverse. It is a very different matter to construct a proper physical theory that realizes this story.

The reason that it is difficult to construct a multiverse theory is that most theories have a single ground state, or at most a small number of ground states. It is the ground state properties that determine the parameters of the theory. For example, the Standard Model has a unique ground state, and the value of the Higgs vev in that state determines the overall scale for the quark masses etc. Sometimes theories with symmetries will have a set of discretely different ground states, but generally just a few. The utility of the multiverse to solve the fine-tuning problems requires that there be *very* many possible ground states. For example, if the cosmological constant has a fine-tuning problem of a factor of 10^{50}, one would expect that one needs of order 10^{50} different ground states with different values of the cosmological constant in order to have the likelihood that at least one of these would fall in the anthropically allowed window.

In fact, such theories do exist, although they are not the norm. There are two possibilities: one where the parameters vary continuously and one where they vary in discrete steps. In the former case, the variation of the parameters in space and time must be described by a field. Normally such a field would settle into the lowest energy state possible, but there is a mechanism whereby the expansion of the Universe 'freezes' the value of the field and does not let it relax to its minimum [14–16]. However, since

---

[1] In Chapter 3 of this volume, Wilczek [13] suggests a possible anthropic explanation in the context of inflationary models for why θ should be very small.

the present expansion of the Universe is very small, the forces acting on this field must be exceptionally tiny. There is a variant of such a theory which has been applied to the fine-tuning of the cosmological constant. However, it has proven difficult to extend this theory to the variation of other parameters.

A more promising type of multiverse theory appears to be emerging from string theory. This originates as a 10- or 11-dimensional theory, although in the end all but four of the spacetime dimensions must be rendered unobservable to us, for example by being of very tiny finite size. Most commonly, the extra dimensions are 'compact', which means that they are of finite extent but without an endpoint, in the sense that a circle is compact. However, solutions to string theory seem to indicate that there are very many low energy solutions which have different parameters, depending on the size and shape of the many compact dimensions [17–21]. In fact, there are so many that one estimate puts the number of solutions that have the properties of our world – within the experimental error bars for all measured parameters – as of order 10^{100}. There would then be many more parameters outside the possible observed range. In this case, there are astonishingly many possible sets of parameters for solutions to string theory. This feature of having fantastically many solutions to string theory, in which the parameters vary as you move through the space of solutions, is colloquially called the 'landscape'.

There are two key properties of these solutions. The first is that they are discretely different and not continuous [22]. The different states are described by different field values in the compact dimensions. These field values are quantized, because they need to return to the same value as one goes around the compact dimension. With enough fields and enough dimensions, the number of solutions rapidly becomes extremely large.

The second key property is that transitions between the different solutions are known [23–25]. This can occur when some of the fields change their values. From our 4-dimensional point of view, what occurs is that a bubble nucleates, in which the interior is one solution and the exterior is another one. The rate for such nucleations can be calculated in terms of string theory parameters. In particular, it apparently always occurs during inflation or at finite temperature. Nucleation of bubbles commonly leads to large jumps in the parameters, such as the cosmological constant, and the steps do not always go in the same direction.

These two properties imply that a multiverse is formed in string theory if inflation occurs. There are multiple states with different parameters, and transitions between these occur during inflation. The outcome is a model in which the different regions – the interior of the bubble nucleation regions – have the full range of possible parameters.

String theorists long had the hope that there would be a unique ground state of the theory. It would indeed be wonderful if one could prove that there is only one true ground state and that this state leads to the Standard Model, with exactly the parameters seen in nature. It would be hard to imagine how a theory with such a high initial symmetry could lead only to a world with parameters with as little symmetry as seen in the Standard Model, such as $m_u = 4$ MeV, $m_d = 7$ MeV, etc. But if this were in fact shown, it would certainly prove the validity of string theory. Against this hope, the existence of a landscape and a multiverse seems perhaps disappointing. Without a unique ground state, we cannot use the prediction of the parameters as a proof of string theory.

However, there is another sense in which the string theory landscape is a positive development. Some of us who are working 'from the bottom up' have been led by the observed fine-tuning (in both senses of the word) to desire the existence of a multiverse with exactly the properties of the string theory landscape. From this perspective, the existence of the landscape is a strong motivation in favour of string theory, more immediate and pressing even than the desire to understand quantum gravity.

Inflation also seems to be a necessary ingredient for a multiverse [26–28]. This is because we need to push the boundaries between the domains far outside our observable horizon. Inflation neatly explains why we see a mostly uniform Universe, even if the greater multiverse has multiple different domains. The exponential growth of the scale factor during inflation makes it reasonable that we see a uniform domain. However, today inflation is the 'simple' ingredient that we expect really does occur, based on the evidence of the flatness of the Universe and the power spectrum of the cosmic microwave background temperature fluctuations. It is the other ingredient of the multiverse proposal – having very many ground states – that is much more difficult.

15.7 Testing through a full theory

Let us be philosophical for a moment. Anthropic arguments and invocations of the multiverse can sometimes border on being non-scientific. You cannot test for the existence of other domains in the Universe outside the one visible to us – nor can you find a direct test of the Anthropic Principle. This leads some physicists to reject anthropic and multiverse ideas as being outside of the body of scientific thought. This appears to me to be unfair. Anthropic consequences appear naturally in some physical theories. However, there are nevertheless non-trivial limitations on what can be said in a scientific manner in such theories.

The resolution comes from the realization that neither the anthropic nor the multiverse proposal constitutes a concrete theory. Instead there are real theories, such as string theory, which have a multiverse property and lead to our domain automatically satisfying anthropic constraints. These are not vague abstractions, but real physical consequences of real physical theories. In this case, the anthropic and multiverse proposals are not themselves a full theory but rather the output of such a theory. Our duty as scientists is not to give up because of this but to find other ways to test the original theory. Experiments are reasonably local and we need to find some reasonably local tests that probe the original full theory.

However, it has to be admitted that theories with a multiverse property, such as perhaps the string landscape – where apparently 'almost anything goes' – make it difficult to be confident of finding local tests. Perhaps there are some consequences which always emerge from string theory for all states in the landscape. For example, one might hope that the bare strong CP-violating θ angle is always zero in string theory and that it receives only a small finite renormalization. However, other consequences would certainly be of a statistical nature that we are not used to. An example is the present debate as to whether supersymmetry is broken at low energy or high energy in string theory. It is likely that both possibilities are present, but the number of states of one type is likely to be very different (by factors of perhaps 10^{100}) from the number of states of the other type – although it is not presently clear which is favoured. If this is solved, it will be a good statistical prediction of string theory. If we can put together a few such statistical predictions, we can provide an effective test of the theory.

15.8 A test using quark and lepton masses

Of the parameters of the Standard Model, none are as confusing as the masses of the quarks and leptons. From the history of the periodic table and atomic/nuclear spectroscopy, we would expect that the masses would show some pattern that reveals the underlying physics. However, no such pattern has ever been found. In this section, I will describe a statistical pattern, namely that the masses appear randomly distributed with respect to a scale-invariant weight, and I will discuss how this can be the probe of a multiverse theory.

Fig. 15.1. The quark and lepton masses on a log scale. The result appears to be qualitatively consistent with a random distribution in $\ln m$, and quantitative analysis bears this out.

In a multiverse or in the string theory landscape, one would not expect the quark and lepton masses to exhibit any pattern. Rather, they would be representative of one of the many possible states available to the theory. Consider the ensemble of ground states which have the other parameters of the Standard Model held fixed. In this ensemble, the quark and lepton masses are not necessarily uniformly distributed. Rather we could describe their distribution by some *weight* [29,30]. For example, perhaps this weight favours quarks and leptons with small masses, as is in fact seen experimentally. We would then expect that the quark masses seen in our domain are not particularly special but are typical of a random distribution with respect to this weight.

The quark masses appear mostly at low energy, yet extend to high energy. To pull out the range of weights that could lead to this distribution involves a detailed study of their statistical properties. Yet it is remarkably easy to see that they are consistent with being scale-invariant. A scale-invariant weight means that the probability of finding the masses in an interval dm at any mass m scales as dm/m. This in turn means that the masses should be randomly distributed when plotted as a function of $\ln m$. It is easy to see visually that this is the case; Fig. 15.1 shows the quark and lepton masses plotted on a logarithmic scale. One can readily see that this is consistent with being a random distribution. The case for a scale-invariant distribution can be quantified by studying the statistics of six or nine masses distributed with various weights [30]. When considering power-law weights of the form dm/m^δ, one can constrain the exponent δ to be greater than 0.8. The scale-invariant weight ($\delta = 1$) is an excellent fit. One may also discuss the effects of anthropic constraints on the weights [30].

What should we make of this statistical pattern? In a multiverse theory, this pattern is the visible remnant of the underlying ensemble of ground states of different masses. An example of how this distribution could appear from a more fundamental theory is given by the Intersecting Brane Worlds solutions of string theory [31, 32]. In these solutions, our 4-dimensional world appears as the intersection of solutions (branes) of higher dimension, much as a 1-dimensional line can be described as the intersection of two 2-dimensional surfaces. In these theories, the quark and lepton masses are determined by the area between three intersections of these surfaces. In particular, the distribution is proportional to the exponential of this area, $m \sim e^{-A}$. In a string landscape there might not be a unique area, but rather a distribution of areas. The mathematical connection is that, if these areas are distributed uniformly (i.e. with a constant weight), then the masses are distributed with a scale-invariant weight. In principle, the distribution of areas is a calculation that could be performed when we understand string theory better. Thus, we could relate solutions of string theory to the observed distribution of masses in the real world. This illustrates how we can test the predictions of a multiverse theory without a unique ground state.

15.9 Summary

The idea of a multiverse can make positive contributions to particle physics. In a multiverse, some of our main puzzles disappear, but they are replaced by new questions.

We have seen how the multiverse can provide a physical reason for some of the fine-tuning that seems to be found in nature. We have also stressed that two distinct meanings of the phrase 'fine-tuning' are used in different parts of the scientific literature. One meaning, often encountered in discussions of anthropic considerations, relates to the observation that the measured parameters seem to be highly tuned to the narrow window that allows life to exist. The other meaning is the particle physics usage described above, which concerns the relative size of the quantum corrections compared with the measured value. The latter usage has no *a priori* connection to the former. However, the idea of the multiverse unites the two uses – the requirement of life limits the possible range of the particle physics parameters and can explain why the measured values are necessarily so small compared with the quantum effects.

However, in other cases, the multiverse makes the problems harder. The strong CP problem is not explained by the multiverse. It is a clue that a dynamical solution to this problem has to be a generic feature of the underlying full theory.

The flavour problem of trying to understand the properties of the quarks and leptons also becomes reformulated. I have described how the masses appear to be distributed in a scale-invariant fashion. In a multiverse theory, it is possible that this is a reflection of the dynamics of the underlying theory and that this feature may someday be used as a test of the full theory.

We clearly have more to discover in particle physics. In answering the pressing experimental questions on the existence of the Higgs boson and the nature of dark matter etc., we will undoubtably learn more about the underlying theory. We also hope that the new physics that emerges will shed light on aesthetic questions concerning the Standard Model. The idea of the multiverse is a possible physical consequence of some theories of physics beyond the Standard Model. It has not been heavily explored in particle physics, yet presents further challenges and opportunities. We clearly have more work to do before we can assess how fruitful this idea will be for the theory of the fundamental interactions.

Acknowledgements

I am pleased to thank my collaborators on these topics, Steve Barr, Dave Seckel, Thibault Damour, Andreas Ross and Koushik Dutta, as well as my long-term collaborator on more sensible topics, Gene Golowich, for discussions that have helped shape my ideas on this topic. My work has been supported in part by the US National Science Foundation and by the John Templeton Foundation.

References

[1] J. Barrow and F. Tipler. *The Anthropic Cosmological Principle* (Oxford: Clarendon Press, 1986).
[2] C.J. Hogan. Why the universe is just so. *Rev. Mod. Phys.* **72** (2000), 1149 [astro-ph/9909295].
[3] R. N. Cahn. The eighteen arbitrary parameters of the standard model in your everyday life. *Rev. Mod. Phys.* **68** (1996), 951.
[4] J. F. Donoghue, E. Golowich and B. R. Holstein. *Dynamics of the Standard Model* (Cambridge: Cambridge University Press, 1992).
[5] R. Sundrum. Towards an effective particle-string resolution of the cosmological constant problem. *JHEP*, **9907** (1999), 001 [hep-ph/9708329].
[6] A. Linde. Inflation and quantum cosmology. In *Results and Perspectives in Particle Physics*, ed. M.Greco (Gif-sur-Yvette, France: Editions Frontières, 1989), p. 11.
[7] S. Weinberg. Theories of the cosmological constant. In *Critical Dialogues in Cosmology*, ed. N. Turok (Singapore: World Scientific, 1997).
[8] H. Martel, P. R. Shapiro and S. Weinberg. Likely values of the cosmological constant. *Astrophys. J.*, **492** (1998), 29 [astro-ph/9701099].
[9] T. Banks, M. Dine and L. Motl. On anthropic solutions of the cosmological constant problem. *JHEP*, **0101** (2001), 031 [hep-th/0007206].
[10] J. D. Bjorken. Standard model parameters and the cosmological constant. *Phys. Rev.* **D 64** (2001), 085008 [hep-ph/0103349].

[11] V. Agrawal, S. M. Barr, J. F. Donoghue and D. Seckel. Anthropic considerations in multiple-domain theories and the scale of electroweak symmetry breaking. *Phys. Rev. Lett.* **80** (1998), 1822 [hep-ph/9801253].
[12] V. Agrawal, S. M. Barr, J. F. Donoghue and D. Seckel. The anthropic principle and the mass scale of the standard model. *Phys. Rev.* **D57** (1998), 5480 [hep-ph/9707380].
[13] F. Wilczek. This volume (2007).
[14] J. Garriga and A. Vilenkin. On likely values of the cosmological constant. *Phys. Rev.* **D 61** (2000), 083502 [astro-ph/9908115].
[15] J. Garriga and A. Vilenkin. Solutions to the cosmological constant problems. *Phys. Rev.* **D 64** (2001), 023517 [hep-th/0011262].
[16] J. F. Donoghue. Random values of the cosmological constant. *JHEP* **0008** (2000), 022 [hep-ph/0006088].
[17] M. R. Douglas. The statistics of string M theory vacua. *JHEP*, **0305** (2003), 046 [hep-th/0303194].
[18] S. Ashok and M. R. Douglas. Counting flux vacua. *JHEP*, **0401** (2004), 060 [hep-th/0307049].
[19] L. Susskind. This volume (2007) [hep-th/0302219].
[20] T. Banks, M. Dine and E. Gorbatov. Is there a string theory landscape? *JHEP*, **0408** (2004), 058 [hep-th/0309170].
[21] R. Kallosh and A. Linde. M-theory, cosmological constant and anthropic principle. *Phys. Rev.* **D 67** (2003), 023510 [hep-th/0208157].
[22] R. Bousso and J. Polchinski. Quantization of four-form fluxes and dynamical neutralization of the cosmological constant. *JHEP*, **0006** (2000), 006 [hep-th/0004134].
[23] J. D. Brown and C. Teitelboim. Neutralization of the cosmological constant by membrane creation. *Nucl. Phys.* **B 297** (1988), 787.
[24] J. D. Brown and C. Teitelboim. Dynamical neutralization of the cosmological constant. *Phys. Lett.* **B 195** (1987), 177.
[25] J. F. Donoghue. Dynamics of M-theory vacua. *Phys. Rev.* **D 69** (2004), 106012; erratum 129901 [hep-th/0310203].
[26] A. D. Linde. Eternally existing self-reproducing chaotic inflationary universe. *Phys. Lett.* **B 175** (1986), 395.
[27] A. D. Linde. Eternal chaotic inflation. *Mod. Phys. Lett.* **A 1** (1986), 81.
[28] A. H. Guth. Inflation and eternal inflation. *Phys. Rep.* **333** (2000), 555 [astro-ph/0002156].
[29] J. F. Donoghue. The weight for random quark masses. *Phys. Rev.* **D 57** (1998), 5499 [hep-ph/9712333].
[30] J. F. Donoghue, K. Dutta and A. Ross. Quark and lepton masses in the landscape. *Phys. Rev.* **D 73** (2006), 113002.
[31] D. Cremades, L. E. Ibanez and F. Marchesano. Towards a theory of quark masses, mixings and CP-violation. (2002) [hep-ph/0212064].
[32] D. Cremades, L. E. Ibanez and F. Marchesano. Yukawa couplings in intersecting D-brane models. *JHEP*, **0307** (2003), 038 [hep-th/0302105].

16
The anthropic landscape of string theory

Leonard Susskind
Department of Physics, Stanford University

16.1 The landscape

The world-view shared by most physicists is that the laws of nature are uniquely described by some special action principle that completely determines the vacuum, the spectrum of elementary particles, the forces and the symmetries. Experience with quantum electrodynamics and quantum chromodynamics suggests a world with a small number of parameters and a unique ground state. For the most part, string theorists bought into this paradigm. At first, it was hoped that string theory would be unique and explain the various parameters that quantum field theory left unexplained. When this turned out to be false, the belief developed that there were exactly five string theories with names like 'type 2a' and 'heterotic'. This also turned out to be wrong. Instead, a continuum of theories were discovered that smoothly interpolated between the five and also included a theory called 'M-theory'. The language changed a little. One no longer spoke of different theories, but rather of different solutions of some master theory.

The space of these solutions is called the 'moduli space of supersymmetric vacua'. I will call it the 'supermoduli-space'. Moving around on this supermoduli-space is accomplished by varying certain dynamical 'moduli'. Examples of moduli are the size and shape parameters of the compact internal space that 4-dimensional string theory always needs. These moduli are not parameters in the theory, but are more like fields. As you move around in ordinary space, the moduli can vary and have their own equations of motion. In a low-energy approximation, the moduli appear as massless scalar fields. The beauty of the supermoduli-space point of view is that there is only one theory but many solutions, these being characterized by the values of the scalar field moduli. The mathematics of string theory is so precise that it is hard to believe that there is not a consistent mathematical framework underlying the supermoduli-space vacua.

Universe or Multiverse?, ed. Bernard Carr. Published by Cambridge University Press.
© Cambridge University Press 2007.

However, the continuum of solutions in the supermoduli-space are all supersymmetric, with exact super-particle degeneracy and vanishing cosmological constant. Furthermore, they all have massless scalar particles – the moduli themselves. Obviously, none of these vacua can possibly be our world. Therefore the string theorist must believe that there are other discrete islands lying off the coast of the supermoduli-space. The hope now is that a single non-supersymmetric island or at most a small number of islands exist and that non-supersymmetric physics will prove to be approximately unique. This view is not inconsistent with present knowledge (indeed it is possible that there are *no* such islands), but I find it completely implausible. It is much more likely that the number of discrete vacua is astronomical, measured not in millions or billions but in googles or googleplexes.[1]

This change in viewpoint is demanded by two facts, one observational and one theoretical. The first is that the expansion of the Universe is accelerating. The simplest explanation is a small but non-zero cosmological constant. Evidently we have to expand our thinking about vacua to include states with non-zero vacuum energy. The incredible smallness and apparent fine-tuning of the cosmological constant makes it absurdly improbable to find a vacuum in the observed range unless there are an enormous number of solutions with almost every possible value. It seems to me inevitable that, if we find one such vacuum, we will find a huge number of them. I will from now on call the space of all such string theory vacua the *landscape*.

The second fact is that some recent progress has been made in exploring the landscape [1,2]. Before explaining the new ideas, I need to define more completely what I mean by the landscape. The supermoduli-space is parameterized by the moduli, which we can think of as a collection of scalar fields Φ_n. Unlike the case of Goldstone bosons, points in the moduli space are not related by a symmetry of the theory. Generically, in a quantum field theory, changing the value of a non-Goldstone scalar involves a change of potential energy. In other words, there is a non-zero field potential $V(\Phi)$. Local minima of V are what we call vacua. If the local minimum is an absolute minimum, the vacuum is stable. Otherwise it is only meta-stable. The value of the potential energy at the minimum is the cosmological constant for that vacuum.

To the extent that the low-energy properties of string theory can be approximated by field theory, similar ideas apply. Bearing in mind that the low-energy approximation may break down in some regions of the landscape,

---

[1] A google is defined to be 10 to the power 100; that is $\mathcal{G} = 10^{100}$. A googleplex is $10^{\mathcal{G}}$.

I will assume the existence of a set of fields and a potential. The space of these fields is the landscape.

The supermoduli-space is a special part of the landscape where the vacua are supersymmetric and the potential $V(\Phi)$ is exactly zero. These vacua are marginally stable and can be excited by giving the moduli arbitrarily small time derivatives. On the supermoduli-space the cosmological constant is also exactly zero. Roughly speaking, the supermoduli-space is a perfectly flat plain at exactly zero altitude.[2] Once we move off the plain, supersymmetry is broken and a non-zero potential develops, usually through some non-perturbative mechanism. Thus, beyond the flat plain, we encounter hills and valleys. We are particularly interested in the valleys, where we find local minima of V. Each such minimum has its own vacuum energy. The typical value of the potential difference between neighbouring valleys will be some fraction of M_p^4, where M_p is the Planck mass. The potential barriers between minima will be of similar height. Thus, if a vacuum is found with cosmological constant of order $10^{-120} M_p^4$, it will be surrounded by much higher hills and other valleys.

Next consider two large regions of space, each of which has the scalars in some local minimum, the two minima being different. If the local minima are landscape neighbours, then the two regions of space will be separated by a domain wall. Inside the domain wall the scalars go over a 'mountain pass'. The interior of the regions are vacuum-like with cosmological constants. The domain wall, which can also be called a membrane, has additional energy in the form of a membrane tension. Thus there will be configurations of string theory which are not globally described by a single vacuum but instead consist of many domains separated by domain walls. Accordingly, the landscape in field space is reflected in a complicated terrain in real space.

There are scalar fields that are not usually thought of as moduli, but – once we leave the flat plain – I do not think there is any fundamental difference. These are the 4-form field strengths, first introduced in the context of the cosmological constant by Brown and Teitelboim [3]. A simple analogy exists to help visualize these fields and their potential. One can think of $1+1$-dimensional electrodynamics[3] with an electric field E and massive electrons. The electric field is constant in any region of space where there are no charges. The field energy is proportional to the square of the field strength. The electric field jumps by a quantized unit whenever an electron is passed. Going in one direction, say along the positive x axis, the field makes a

---

2 By 'altitude' I am referring to the value of V.
3 In $1+1$ dimensions there is no magnetic field and the electric field is a 2-form, aka a scalar density.

positive unit jump when an electron is passed and a negative jump when a positron is passed. In this model, different vacua are represented by different quantized values of the electric field, while the electrons and positrons are the domain walls. This model is not fundamentally different from the case with scalar fields and a potential. In fact, by bosonizing the theory, it can be expressed as a scalar field theory with a potential

$$V(\phi) = c\phi^2 + \mu \cos \phi. \tag{16.1}$$

If μ is not too small, there are many minima representing the different possible 2-form field strengths, each with a different energy.

In $3+1$ dimensions the corresponding construction requires a 4-form field strength F whose energy is also proportional to F^2. This energy appears in the gravitational field equations as a positive contribution to the cosmological constant. The analogue of the charged electrons are membranes which appear in string theory and function as domain walls to separate vacua with different F. This theory can also be written in terms of a scalar field with a potential similar to Eq. (16.1). Henceforth I will include such fields, along with moduli, as coordinates of the landscape.

Let us now consider a typical compactification of M-theory from eleven to four dimensions. The simplest example is obtained by choosing for the compact directions a 7-torus. The torus has a number of moduli representing the sizes and angles between the seven 1-cycles. The 4-form fields have as their origin a 7-form field strength, which is one of the fundamental fields of M-theory.[4] The 7-form fields have seven anti-symmetrized indices. These non-vanishing 7-forms can be configured so that three of the indices are identified with compact dimensions and the remainder are identified with uncompactified spacetime. This can be done in $(7 \times 6 \times 5)/(1 \times 2 \times 3) = 35$ ways, which means there are that many distinct 4-form fields in the uncompactified non-compact space. More generally, in the kinds of compact manifolds invoked in string theory to try to reproduce Standard Model physics, there can be hundreds of independent ways of 'wrapping' three compact directions with flux, thus producing hundreds of $3+1$-dimensional 4-form fields. As in the case of $1+1$-dimensional quantum electrodynamics, the field strengths are quantized, each in integer multiples of a basic field unit. A vacuum is specified by a set of integers $n_1, n_2, \ldots, n_N$, where N can be as large as several hundred. The energy density of the 4-form fields has the

---

[4] The fact that we have the number 7 appearing in two ways, as the number of compact dimensions and as the number of indices of the field strength, is accidental.

following form:

$$\epsilon = \sum_{i=1}^{N} c_i n_i^2, \qquad (16.2)$$

where the constants c_i depend on the details of the compact space.

The analogue of the electrons and positrons of the $1+1$-dimensional example are branes. The 11-dimensional M-theory has 5-branes which fill five spatial directions and time. By wrapping 5-branes the same way, the fluxes of the 4-forms are wrapped on internal 3-cycles, leaving 2-dimensional membranes in $3+1$ dimensions. These are the domain walls, which separate different values of field strength. There are N types of domain wall, each allowing a unit jump of one of the 4-forms.

Bousso and Polchinski [1] begin by assuming they have located some deep minimum of the field potential at some point Φ_0. The value of the potential is supposed to be very negative at this point, corresponding to a negative cosmological constant, λ_0, of order the Planck scale. Also the 4-forms are assumed to vanish at this point. They then ask what kind of vacua they can obtain by discretely increasing the 4-forms. The answer depends to some degree on the compactification radii of the internal space but – with modest parameters – it is not hard to get such a huge number of vacua that it is statistically likely to have one with cosmological constant $\lambda \sim 10^{-120} M_{\rm p}^4$.

To see how this works, we write the cosmological constant as the sum of the cosmological constant for vanishing 4-form and the contribution of the 4-forms themselves:

$$\lambda = \lambda_0 + \sum_{i=1}^{N} c_i n_i^2. \qquad (16.3)$$

With one hundred terms and modestly small values for the c_n, it is highly likely to find a value of λ in the observed range. Note that no fine-tuning is required, only a very large number of ways to make the vacuum energy. The problem with this proposal was clearly recognized by the authors [1]. The starting point is so far from the supermoduli-space that none of the usual tools of approximate supersymmetry are available to control the approximation. The example was intended only as a model of what might happen because of the large number of possibilities.

More recently, Kachru and colleagues [2] have improved the situation by finding an example which is more under control. These authors subtly use the various ingredients of string theory – including fluxes, branes, anti-branes and instantons – to construct a rather tractable example with a small positive cosmological constant. In addition to arguing that string theory has many vacua with positive cosmological constant, the argument in ref. [2] tends to dispel the idea that vacua not on the supermoduli-space must have vanishing cosmological constant. In other words, there is no evidence in string theory that a hoped for – but unknown – mechanism will automatically force the cosmological constant to zero. It seems very likely that all of the non-supersymmetric vacua have finite λ.

The vacua in ref. [2] are not at all simple. They are jury-rigged Rube–Goldberg contraptions that could hardly have fundamental significance. But in an anthropic theory, simplicity and elegance are not considerations. The only criterion for choosing a vacuum is utility, i.e. does it have the necessary elements, such as galaxy formation and complex chemistry, that are needed for life. That, together with a cosmology that guarantees a high probability that at least one large patch of space will form with that vacuum structure, is all we need.

16.2 The trouble with de Sitter space

The classical vacuum solution of Einstein's equations with a positive cosmological constant is de Sitter space. It is doubtful that it has a precise meaning in a quantum theory such as string theory [4–6]. I want to review some of the reasons for thinking that de Sitter space is at best a meta-stable state.

It is important to recognize that there are two very different ways to think about de Sitter space. The first is to take a global view of the spacetime. The global geometry is described by the metric

$$\mathrm{d}s^2 = R^2 \left\{ \mathrm{d}t^2 - (\cosh t)^2 \mathrm{d}^2 \Omega_3 \right\}, \quad (16.4)$$

where $\mathrm{d}^2 \Omega_d$ is the metric for a unit d-sphere and R is related to the cosmological constant:

$$R = (\lambda G)^{-1/2}. \quad (16.5)$$

Viewing de Sitter space globally would make sense if it were a system that could be studied from the outside by a 'meta-observer'. Naïvely, the meta-observer would make use of a (time-dependent) Hamiltonian to evolve the system from one time to another. An alternative description would use a Wheeler–De Witt formalism to define a wave-function of the Universe on global spacelike slices.

The other way of describing the space is the *causal patch* or 'hot tin can' description. The relevant metric is given by

$$ds^2 = R^2 \left\{ (1-r^2)dt^2 - \frac{1}{(1-r^2)}dr^2 - r^2\, d^2\Omega_2 \right\}. \tag{16.6}$$

In this form, the metric is static and has a form similar to that of a black hole. In fact, the geometry has an horizon at $r=1$. The static patch does not cover the entire global de Sitter space, but is analogous to the region outside a black hole horizon. It is the region which can receive signals from, and send signals to, an observer located at $r=0$. To such an observer, de Sitter space appears to be a spherical cavity bounded by an horizon a finite distance away.

Experience with black holes has taught us to be very wary of global descriptions when horizons are involved. In a black hole geometry there is no global conventional quantum description of both sides of the horizon. This suggests that a conventional quantum description of de Sitter space only makes sense within a given observer's causal patch. The descriptions in different causal patches are complementary [7,8] but cannot be put together into a global description without somehow modifying the rules of quantum mechanics. As in the black hole case, an horizon implies a thermal behaviour with a temperature and an entropy. These are given by

$$T = \frac{1}{2\pi R}; \quad S = \frac{\pi R^2}{G}. \tag{16.7}$$

For the rest of this section, I will be assuming the causal patch description of some particular observer.

If the observed 'dark energy' in the Universe really is a small positive cosmological constant, the ultimate future of the Universe will be eternal de Sitter space. This would mean not that the future comprises totally empty space but that the world will have all the features of an isolated finite thermal cavity with finite temperature and entropy. Thermal equilibrium for such a system is not completely featureless. On short time-scales not much can be expected to happen, but on very long time-scales everything happens. A famous example involves a gas of molecules in a sealed room. Imagine that we start all the molecules in one corner of the room. In a relatively short time the gas will spread out to fill the room and come to thermal equilibrium. During the approach to equilibrium interesting dissipative structures, such as droplets, eddies and vortices, form and then dissipate. The usual assumption is that nothing happens after that. The entropy has reached its maximum value and the second law forbids any further interesting history.

But on a sufficiently long time-scale, large fluctuations will occur. In fact, the phase point will return over and over to the neighbourhood of any point in phase space, including the original starting point. These *Poincaré recurrences* generally occur on a time-scale with an exponential dependence on the thermal entropy of the system. Thus we define the Poincaré recurrence time to be

$$T_\mathrm{r} = \exp S \qquad (16.8)$$

in units of the Planck time. On such a long time-scale the second law of thermodynamics will repeatedly be violated by large-scale fluctuations. Thus, even a pure de Sitter space would have an interesting cosmology of sorts. The causal patch of any observer would undergo Poincaré recurrences in which it would endlessly fluctuate back to a state similar to its starting point, but each time it would be slightly different.

The trouble with such a cosmology is that it relies on very rare 'miracles' to start it off each time. But there are other miracles which could occur and lead to anthropically acceptable worlds with a vastly larger probability than our world. Roughly speaking, the relative probability of a fluctuation leading to a given configuration is proportional to the exponential of its entropy. An example of a configuration far more likely than our own would be a world in which everything was just like our universe except that the temperature of the cosmic microwave background was 10 K instead of 3 K. When I say 'everything is the same', I am including such details as the abundance of the elements.

Ordinarily such a cosmology would be ruled out on the grounds that it would take a huge miracle for the helium and deuterium to survive the bombardment by the extra photons implied by the higher temperature. That is correct, a fantastic miracle would be required, but such miracles would occur far more frequently than the ultimate miracle of returning to the starting point. This can be argued just from the fact that a model at 10 K has a good deal more entropy than one at 3 K. In a world based on recurrences, it would be overwhelmingly unlikely that cosmology could be traced back to something like the inflationary era without a miraculous reversal of the second law along the way. Thus we are forced to conclude that the sealed 'tin can' model must be incorrect, at least for time-scales as long as the recurrence time.

Another difficulty with an eternal de Sitter space involves a mathematical conflict between the symmetry of de Sitter space and the finiteness of the entropy [4]. Basically, the argument is that the finiteness of the de Sitter space entropy indicates that the spectrum of energy is discrete. It is possible

16 The anthropic landscape of string theory

to prove that the symmetry algebra of de Sitter space cannot be realized in a way that is consistent with the discreteness of this spectrum. In fact, this problem is not independent of the issues of recurrences. The discreteness of the spectrum means that there is a typical energy spacing of order

$$\Delta E \sim \exp\left(-S\right). \tag{16.9}$$

The discreteness of the spectrum can only manifest itself on time-scales of order $(\Delta E)^{-1}$, which is simply the recurrence time. Thus there are problems with realizing the full symmetries of de Sitter space for times as long as T_r.

Finally, another difficulty for eternal de Sitter space is that it does not fit at all well with string theory. Generally, the only objects in string theory which are rigorously defined are S-matrix elements. Such an S matrix cannot exist in a thermal background. Part of the problem is again the recurrences which undermine the existence of asymptotic states. Unfortunately, there are no known observables in de Sitter space which can substitute for S-matrix elements. The unavoidable implication of these issues is that eternal de Sitter space is an impossibility in a properly defined quantum theory of gravity.

16.3 de Sitter space is unstable

In ref. [2] a particular string theory vacuum with positive λ was studied. One of the many interesting things that the authors found was that the vacuum is unstable with respect to tunnelling to other vacua. In particular, the vacuum can tunnel back to the supermoduli-space with vanishing cosmological constant. Using instanton methods, the authors calculated that the lifetime of the vacuum is less than the Poincaré recurrence time. This is no accident. To see why it must always be so, let us consider the effective potential that the authors of ref. [2] derived. The only relevant modulus is the overall size of the compact manifold Φ. The potential is shown in Fig. 16.1. The de Sitter vacuum occurs at the point $\Phi = \Phi_0$. However, the absolute minimum of the potential occurs not at Φ_0 but at $\Phi = \infty$. At this point the vacuum energy is exactly zero and the vacuum is one of the 10-dimensional vacua of the supermoduli-space. As was noted long ago by Dine and Seiberg [9], there are always runaway solutions like this in string theory. The potential on the supermoduli-space is zero and so it is always possible to lower the energy by tunnelling to a point on the supermoduli-space.

Suppose we are stuck in the potential well at Φ_0. The vacuum of the causal patch has a finite entropy and fluctuates up and down the walls of the potential. One might think that fluctuations up the sides of the potential are

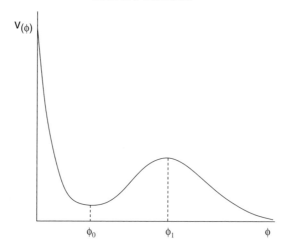

Fig. 16.1. The effective potential derived in ref. [2].

Boltzmann-suppressed. In a usual thermal system there are two things that suppress fluctuations. The first is the Boltzmann suppression by a factor $\exp(-\beta E)$ and the second is entropy suppression by a factor $\exp(S_f - S)$, where S is the thermal entropy and S_f is the entropy characterizing the fluctuation, which is generally smaller than S. However, in a gravitational theory in which space is bounded (as in the static patch), the total energy is always zero, at least classically. Hence the only suppression is entropic. The phase point wanders around in phase space, spending a time in each region proportional to its phase space volume, i.e. $\exp(-S_f)$. Furthermore the typical time-scale for such a fluctuation to take place is of order

$$T_f \sim \exp(S - S_f). \qquad (16.10)$$

Now consider a fluctuation which brings the field ϕ to the top of the local maximum at $\phi = \phi_1$ in the entire causal patch. The entropy at the top of the potential is given in terms of the cosmological constant at the top. It is obviously positive and less than the entropy at ϕ_0. Thus the time for the field to fluctuate to ϕ_1 (over the whole causal patch) is strictly less than the recurrence time $T_r = \exp S$. But once the field gets to the top, there is no obstruction to its rolling down the other side to infinity. It follows that a de Sitter vacuum of string theory is never longer lived than T_r and furthermore we end up at a supersymmetric point of vanishing cosmological constant.

There are other possibilities. If the cosmological constant is not very small, it may tunnel over the nearest mountain pass to a neighbouring valley of smaller positive cosmological constant. This will also take place on a time-scale which is too short to allow recurrences. By the same argument, it will not stay in the new vacuum indefinitely. It may find a vacuum with yet smaller cosmological constant to tunnel to. Eventually it will have to make a transition out of the space of vacua with positive cosmological constant.[5]

16.4 Bubble cosmology

To make use of the enormous diversity of environments that string theory is likely to bring with it, we need a dynamical cosmology which, with high probability, will populate one or more regions of space with an anthropically favourable vacuum. There is a natural candidate for such a cosmology that I will explain from the global perspective.

For simplicity, let us temporarily assume that there are only two vacua: one with positive cosmological constant λ and the other with vanishing cosmological constant. Without worrying how it happened, we suppose that some region of the Universe has fallen into the minimum with positive cosmological constant. From the global perspective it is inflating and new Hubble volumes are constantly being produced by the expansion. Pick a timelike observer who looks around and sees a static region bounded by a horizon. The observer will eventually observe a transition in which the entire observable region slides over the mountain pass and settles into the region of vanishing λ. The observer sees the horizon-boundary quickly recede, leaving in its wake an infinite open Friedmann–Robertson–Walker (FRW) universe with negative spatial curvature. The final geometry has light-like and time-like future infinities similar to those in flat space.

It is helpful to draw some Penrose diagrams to illustrate the history. For this purpose we turn to the global point of view. Figure 16.2 shows the Penrose diagram for a pure de Sitter space. It also shows two observers whose causal patches overlap for some period of time. In Fig. 16.3 the same geometry is shown, except that the formation of a bubble of $\lambda=0$ vacuum is also depicted. The bubble is created at point a and expands with a speed approaching that of light. Eventually the growing bubble intersects the infinite future of the de Sitter space but the geometry inside the bubble continues and forms a future null infinity. Note that even though the observer's final world is infinite, his past light-cone does not include the whole global

---

5 Tunnelling to vacua with negative cosmological constant may or may not be possible. However, such a transition will eventually lead to a crunch singularity. Whether the system survives the crunch is not known. It should be noted that transitions to negative cosmological constant are suppressed and can even be forbidden, depending on magnitudes of the vacuum energies and the domain wall tension. I will assume that such transitions do not occur.

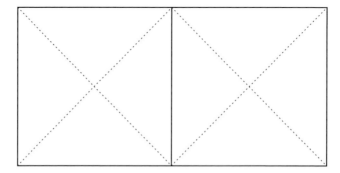

Fig. 16.2. Penrose diagram for pure de Sitter space.

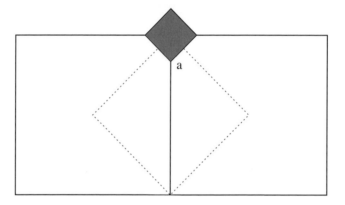

Fig. 16.3. Penrose diagram for a $\lambda = 0$ bubble in de Sitter space.

spacetime. In fact, his causal diamond is not much bigger than it would have been if the space had never decayed. The region outside the bubble is still inflating and disappearing out of causal contact with the observer. From the causal patch viewpoint, the entire world has been swallowed by the bubble.

Now let us take the more global view. The bubble does not swallow the entire global space but leaves part of the space still inflating. Inevitably bubbles will form in this region. In fact, if we follow the worldline of any observer, it will eventually be swallowed by a bubble of $\lambda = 0$ vacuum. The line representing the remote future in Fig. 16.3 is replaced by a jagged fractal as in Fig. 16.4. Any observer eventually ends up at the top of the diagram in one of infinitely many time-like infinities. This process, leading to infinitely many disconnected bubbles, is essentially similar to the process of eternal inflation envisioned by Linde [10].

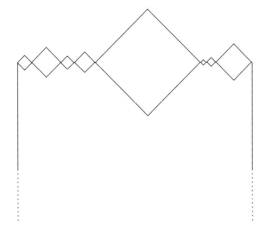

Fig. 16.4. Penrose diagram for many $\lambda = 0$ bubbles in de Sitter space

The real landscape does not comprise only two vacua. If an observer starts with a large value of the cosmological constant, there will be many ways for the causal patch to descend to the supermoduli-space. From the global viewpoint, bubbles will form in neighbouring valleys with somewhat smaller cosmological constant. Since each bubble has a positive cosmological constant, it will be inflating, but the space between bubbles is inflating faster, so the bubbles go out of causal contact with one another. Each bubble evolves in isolation from all the others. Furthermore, in a time too short for recurrences, bubbles will nucleate within bubbles. Following a single observer within his own causal patch, the cosmological constant decreases in a series of events until the causal patch finds itself in the supermoduli-space. Each observer will see a series of vacua descending to the supermoduli-space, and the chance that an observer passes through an anthropically acceptable vacuum is most likely very small. On the other hand, the global space contains an infinite number of such histories, and some of them will be acceptable.

The only problem with the above cosmology is that it is formulated in global coordinates. From the viewpoint of any causal patch, all but one of the bubbles is outside the horizon. As I have emphasized, the application of the ordinary rules of quantum mechanics only makes sense within the horizon of an observer. We do not know the rules for putting together the various patches into one comprehensive global description and, until we do, there cannot be any firm basis for this kind of anthropic cosmology. Nevertheless, the picture is tempting.

16.5 Cosmology as a resonance

The idea of scalar fields and potentials is approximate once we leave the supermoduli-space, as is the notion of a stable de Sitter vacuum. The problem is familiar. How do we make precise sense of an unstable state in quantum mechanics? In ordinary quantum mechanics the clearest situation is when we can think of the unstable state as a resonance in a set of scattering amplitudes. The parameters of a resonance, i.e. its width and mass, are well defined and do not depend on the exact way the resonance was formed. Thus, even black holes have precise meaning as resonant poles in the S-matrix. Normally we cannot compute the scattering amplitudes that describe the formation and evaporation of a black hole, but it is comforting that an exact criterion exists.

In the case of a black hole the density of levels is enormous, being proportional to the exponential of the entropy. The spacing between levels is therefore exponentially small. On the other hand, the width of each level is not very small. The lifetime of a state is the time it takes to emit a single quantum of radiation, and this is proportional to the Schwarzschild radius. Therefore the levels are broadened by much more than their spacing. The usual resonance formulae are not applicable, but the precise definition of the unstable state as a pole in the scattering amplitude is. I think the same things can be said about the unstable de Sitter vacua, but it can only be understood by returning to the 'causal patch' way of thinking. Therefore let us focus on the causal patch of one observer. We have discussed the observer's future history and found that it always ends in an infinite expanding supersymmetric open FRW universe. Such a universe has the usual kind of asymptotic future, consisting of time-like and light-like infinities. There is no temperature in the remote future and the geometry permits particles to separate and propagate freely, just as in flat spacetime.

Now let us consider the observer's past history. The same argument which says that the observer will eventually make a transition to $\lambda = 0$ in the far future can be run backward. The observer could only have reached the de Sitter vacuum by the time-reversed history and so must have originated from a collapsing open universe. The entire history is shown in Fig. 16.5. The history may seem paradoxical, since it requires the second law of thermodynamics to be violated in the past. A similar paradox arises in a more familiar setting. Let us return to the sealed room filled with gas molecules, except that now one of the walls has a small hole that lets the gas escape to unbounded space. Suppose we find the gas filling the room in thermal equilibrium at some time. If we run the system forward, we will eventually

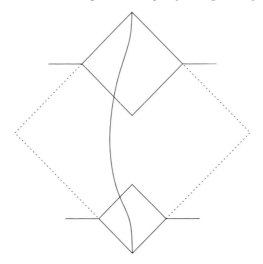

Fig. 16.5. de Sitter space as unstable resonance connecting asymptotic $\lambda = 0$ vacua.

find that all the molecules have escaped and are on their way out, never to return. But it is also true that, if we run the equations of motion backwards, we will eventually find all the molecules outside the room moving away. Thus the only way the starting configuration could have occurred is if the original molecules were converging from infinity toward the small hole in the wall.

If we are studying the system quantum mechanically, the metastable configuration with all the molecules in the room would be an unstable *resonance* in a scattering matrix describing the many-body scatterings of a system of molecules with the walls of the room. Indeed, the energy levels describing the molecules trapped inside the room are complex due to the finite lifetime of the configuration.

This suggests a view of the intermediate de Sitter space in Fig. 16.5 as an unstable resonance in the scattering matrix connecting states in the asymptotic $\lambda = 0$ vacua. In fact, we can estimate the width of the states. Since the lifetime of de Sitter space is always longer than the recurrence time, generally by a huge factor, the width γ satisfies $\gamma \gg \exp(-S)$. On the other hand, the spacing between levels, ΔE, is of order $\exp(-S)$. Therefore $\gamma \gg \Delta E$, so that the levels are very broad and overlapping, as for the black hole. No perfectly precise definition exists in string theory for the moduli fields or their potential when we go away from the supermoduli-space. The only precise definition of the de Sitter vacua seems to be as complex poles in some new sector of the scattering matrix between states on the supermoduli-space.

Knowing that a black hole is a resonance in a scattering amplitude does not tell us much about the way real black holes form. Most of the possibilities for black hole formation are just the time-reverse of the ways in which it evaporates. In other words, the overwhelming number of initial states that can lead to a black hole consist of thermal radiation. Real black holes in our universe form from stellar collapse, which is just one channel in a huge collection of S-matrix 'in states'. In the same way, the fact that cosmological states may be thought of in a scattering framework does not itself shed much light on the original creation process.

16.6 Conclusion

Vacua come in two varieties: supersymmetric and non-supersymmetric. Most likely the latter do not have vanishing cosmological constant, but it is plausible that there are so many of them that they practically form a continuum. Some tiny fraction have a cosmological constant in the observed range. With nothing favouring one vacuum over another, the Anthropic Principle comes to the fore, whether or not we like the idea. String theory provides a framework in which this can be studied in a rigorous way. Progress can certainly be made in exploring the landscape. The project is in its infancy, but in time we should know just how rich it is. We can argue the philosophical merits of the Anthropic Principle, but we cannot argue with quantitative information about the number of vacua with each particular property, such as the cosmological constant, Higgs mass or fine structure constant. That information is there for us to extract.

Counting the vacua is important but not sufficient. A greater understanding of cosmological evolution is essential to determining if the large number of possibilities are realized as actualities. The vacua in string theory with $\lambda > 0$ are not stable and decay on a time-scale smaller than the recurrence time. This is very general and also very fortunate, since there are serious problems with stable de Sitter space. The instability also allows the Universe to sample all or a large part of the landscape by means of bubble formation. In such a world the probability that some region of space has suitable conditions for life to exist can be large.

The bubble universe based on Linde's eternal inflation seems promising, but it is unclear how to think about it with precision. There are real conceptual problems having to do with the global view of spacetime. The main problem is to reconcile two pictures: the causal patch picture and the global picture. String theory has provided a testing ground for some important relevant ideas, such as black hole complementarity [7, 8] and the

Holographic Principle [11, 12]. Complementarity requires the observer's side of the horizon to have a self-contained conventional quantum description. It also prohibits a conventional quantum description that covers the interior and exterior simultaneously. Any attempt to describe both sides as a single quantum system will come into conflict with one of three sacred principles [13]. The first is the Equivalence Principle, which says that a freely falling observer passes the horizon without incident. The second says that experiments performed outside a black hole should be consistent with the rules of quantum mechanics as set down in Dirac's textbook. No loss of quantum information should take place and the time evolution should be unitary. Finally, the rules of quantum mechanics forbid information duplication. This means that we cannot resolve the so-called information paradox by creating two copies (quantum xeroxing) of every bit as it falls through the horizon – at least not within the formalism of conventional quantum mechanics. The Complementarity and Holographic Principles have been convincingly confirmed by the modern methods of string theory (see, for example, ref. [14]). The inevitable conclusion is that a global description of geometries with horizons, if it exists at all, will not be based on the standard quantum rules.

Why is this important for cosmology? The point is that the eternal inflationary production of an infinity of bubbles takes place behind the horizon of any given observer. It is not something that has a description within one causal patch. If it makes sense, a global description is needed. However, if cosmic event horizons are at all like black hole horizons, then any global description will involve wholly new elements. If I were to make a wild guess about which rule of quantum mechanics has to be given up in a global description of either black holes or cosmology, I would guess it is the Quantum Xerox Principle [13]. I would look for a theory which formally allows quantum duplication but cleverly prevents any observer from witnessing it. Perhaps then the replication of bubbles can be sensibly described.

Progress may also be possible in sharpening the exact mathematical meaning of the de Sitter vacua. Away from the supermoduli-space, the concept of a local field and the effective potential is at best approximate in string theory. The fact that the vacua are false meta-stable states makes it even more problematic to be precise. In ordinary quantum mechanics the best mathematical definition of an unstable state is as a resonance in the amplitude for scattering between very precisely defined asymptotic states. Each metastable state corresponds to a pole whose real and imaginary parts define the energy and inverse lifetime of the state.

I have argued that each causal patch begins and ends with an asymptotic 'roll' toward the supermoduli-space. The final states have the boundary conditions of an FRW open universe and the initial states are time-reversals of these. This means we may be able to define some kind of S-matrix connecting initial and final asymptotic states. The various intermediate meta-stable de Sitter phases would be exactly defined as resonances in this amplitude.

At first this proposal sounds foolish. In General Relativity, initial and final states are very different. Black holes make sense. White holes do not. Ordinary things fall into black holes and thermal radiation comes out. The opposite never happens. But this is deceptive. Our experience with string theory has made it clear that the fundamental microphysical input is completely reversible and that black holes are most rigorously defined in terms of resonances in scattering amplitudes.[6] Of course, knowing that a black hole is an intermediate state in a tremendously complicated scattering amplitude does not really tell us much about how real black holes form. For that, we need to know about stellar collapse and the like. But it does provide an exact mathematical definition of the states that comprise the black hole ensemble.

To further illustrate the point, let me tell a story. Two future astronauts in the deep empty reaches of outer space discover a sealed capsule. On further inspection, they find a tiny pinhole in the capsule from which air is slowly leaking out. One says to the other, 'Aha! We have discovered an eternal air tank. It must have been here forever.' The other says, 'No, you fool. If it were here forever, the air would have leaked out an infinitely long time ago.' So, the first one thinks and says, 'Yes, you are right. Let's think. If we wait long enough, all the air will be streaming outward in an asymptotic final state. That is clear. But because of microreversibility, it is equally clear that – if we go far into the past – all the air must have been doing the reverse. In fact, the quantum states with air in the capsule are just intermediate resonances in the scattering of a collection of air molecules with the empty capsule.' The second astronaut looks at the first as if he were nuts. 'Don't be a dope,' he says. 'That's just too unlikely. I guess someone else was here not so long ago and filled it up.'

Both of them can be right. The quantum states of air in a tank are mathematical resonances in a scattering matrix. And it may also be true that the laws of an isolated system of gas and tank may have been temporarily interfered with by another presence. Or we might say that the scattering

---

[6] The one exception is a black hole in anti-de Sitter space, which is stable.

states need to include not only air and tank but also astronauts and their apparatus. It is in this sense that I propose that de Sitter space can be mathematically defined in terms of singularities in some kind of generalized S matrix. But in so doing, I am not really telling you much about how it all started.

From the causal patch view-point, the evolutionary endpoint seems to be an approach to some point on the supermoduli-space. After the last tunnelling, the Universe enters a final open FRW expansion toward some flat supersymmetric solution. This is not to be thought of as a unique quantum state but as a large set of states with similar evolution. Running the argument backward (assuming microscopic reversibility), we expect the initial state to be the time-reversal of one of the many future endpoints. We might even hope for a scattering matrix connecting initial and final states. de Sitter minima would be an enormously large density of complex poles in the amplitude.

One last point: the final and initial states do not have to be 4-dimensional. In fact, in the example given in ref. [2], the modulus describing the overall size of the compact space rolls to infinity, thus creating a 10- or possibly 11-dimensional universe.

Acknowledgements

I am indebted to Simeon Hellerman, Ben Freivogel, Renata Kallosh, Shamit Kachru, Eva Silverstein and Steve Shenker. Special thanks go to Matt Kleban and Andrei Linde for explaining many things to me.

References

[1] R. Bousso and J. Polchinski. Quantization of four-form fluxes and dynamical neutralization of the cosmological constant. *JHEP*, **0006** (2000), 006 [hep-th/0004134].
[2] S. Kachru, R. Kallosh, A. Linde and S. P. Trivedi. de Sitter vacua in string theory. *Phys. Rev.* **D 68** (2003), 046005 [hep-th/0301240].
[3] J. D. Brown and C. Teitelboim. Neutralization of the cosmological constant by membrane creation. *Nucl. Phys.* **B 297** (1988), 787.
[4] N. Goheer, M. Kleban and L. Susskind. The trouble with de Sitter space. *JHEP*, **0307** (2003), 056 [hep-th/0212209].
[5] L. Dyson, J. Lindesay and L. Susskind. Is there really a de Sitter/CFT duality? *JHEP* **0208** (2002), 045 [hep-th/0202163].
[6] L. Dyson, M. Kleban and L. Susskind. Disturbing implications of a cosmological constant. *JHEP*, **0210** (2002), 011 [hep-th/0208013].
[7] L. Susskind, L. Thorlacius and J. Uglum. The stretched horizon and black hole complementarity. *Phys. Rev.* **D 48** (1993), 3743 [hep-th/9306069].

[8] C. R. Stephens, G. 't Hooft and B. F. Whiting. Black hole evaporation without information loss. *Class. Quant. Grav.* **11** (1994), 621 [gr-qc/9310006].

[9] M. Dine and N. Seiberg. Comments on higher derivative operators in some SUSY field theories. *Phys. Lett.* **B 409** (1997), 239.

[10] A. D. Linde. Eternally existing self-reproducing chaotic inflationary universe. *Phys. Lett.* **B 175** (1986), 395.

[11] G. 't Hooft. Dimensional reduction in quantum gravity. In *Salamfest 1993* (1993), pp. 284–96 [gr-qc/9310026].

[12] L. Susskind. The world as a hologram. *J. Math. Phys.* **36** (1995), 6377 [hep-th/9409089].

[13] L. Susskind. Twenty years of debate with Stephen. In *The Future of Theoretical Physics*, eds. G. W. Gibbons, E. P. S. Shellard and S. J. Ranken (Cambridge: Cambridge University Press, 2003), pp. 330–47 [hep-th/0204027].

[14] J. M. Maldacena. Eternal black holes in AdS. *JHEP*, **0304** (2001), 021 [hep-th/0106112].

17
Cosmology and the many worlds interpretation of quantum mechanics

Viatcheslav Mukhanov

Sektion Physik, Ludwig-Maximilians-Universität

17.1 Introduction

Although the mathematical structure of quantum mechanics was understood within a few years after it was invented, numerous quantum paradoxes still disturb 'simple-minded' physicists. Most of them, as 'naïve realists', would probably never take Bohr's own over-philosophical and over-complicated treatment of these paradoxes seriously if they realized the philosophical consequences of the Copenhagen interpretation. To make my meaning clearer, let me quote Bohr's answer to Professor Hoffding's question regarding the double-slit experiment [1]. Bohr was asked: 'What can the electron be said to *be* in its travel from the point of entry to the point of detection?' And he replied: '*To be? To be?* What does it *mean* to be?' However, if one questions the existence of microscopic constituents of macroscopic bodies, then the next logical step would be to question the existence of the macroscopic bodies and even ourselves.

Needless to say, very few (if any) of us, when making experiments or analyzing their results, address the question of what it means 'to be' every time. Even in the context of elementary particles, probably nobody doubts that the particles exist and somehow travel from the point of entry to the point of detection. Moreover, within the accuracy allowed by the uncertainty relation, these particles can be localized and described just as well as macroscopic 'classical' objects.

If this intuitive point of view is correct, then everything is in perfect agreement with what we used to think about the world existing 'out there' and independently of us. Our feeling is that this world can be well described by physical laws which (within a limited accuracy) are in 'one-to-one correspondence' with reality.

Universe or Multiverse?, ed. Bernard Carr. Published by Cambridge University Press.
© Cambridge University Press 2007.

However, this is not what Bohr's interpretation of quantum mechanics tells us. According to Bohr, '... any talk about what the photon is doing between the point of production and the point of reception is ... simply mere talk' [1]. The observer becomes an active 'player in the game', and the physical laws just serve the observer's needs, simply relating the outcomes of the measurements without addressing the question of what the world looks like 'out there'. In this case, quantum mechanics does not make much sense in the absence of observers.

17.2 The Copenhagen picture

In fact, Bohr's interpretation is more than just an interpretation of the equations of quantum mechanics. It puts limits on the applicability of these equations and, at some point, replaces the unitary evolution by a mysterious collapse of the wave-function. One has to stress that this collapse is not derivable – even in principle – from the equations of quantum mechanics. There is no definite quantitative answer as to when the collapse should take place. One usually says that the reduction of the wave-function happens at some point during the interaction of the quantum system with the measuring device. This interaction point can be moved arbitrarily unless it is possible to observe the interference of the 'classically described' macroscopic objects [2]. (Observing the interference is an incredibly difficult enterprise for macroscopic objects because of decoherence.) The 'size of the object' is not a good criterion for telling us when the Schrödinger equation should be replaced by 'classical laws'.

This last statement immediately provokes the question: should not the classical equations just follow from quantum mechanical equations in the limit when the Planck constant $\hbar$ goes to zero, or equivalently when we apply quantum mechanics to macroscopic objects? There are good reasons to believe that this should be the case. In fact, quantum theory is usually constructed by replacing numbers in classical equations by operators which have to satisfy appropriate commutation relations to fulfil the uncertainty relations. In the limit $\hbar \to 0$, all these operators should commute and one could anticipate that the expectation values of operators should satisfy the classical equations to within an accuracy proportional to $\hbar$. However, this is not the case for all admissible quantum states, even if the objects under consideration are macroscopic ('classical'). If one assumes that a macroscopic system can be well described by quantum mechanics, then its generic quantum state is a superposition of various macroscopically different states.

17 Cosmology and the many worlds interpretation

The most famous example of such a state is Schrödinger's cat. It is clear that any operator characterizing the state of the cat produces a nonsensical result if the operator is being averaged over a superposition of 'the dead and alive cat'. If we describe measurements using the quantum mechanical equations, then again the final state of the apparatus is generically a superposition of macroscopically different states. This superposition is a complete analogue of the 'Schrödinger cat' one. The macroscopic superpositions occur commonly if one universally applies the Schrödinger equation. After amplification, the microscopical superpositions generically evolve into quantum states, which, if naïvely 'mapped onto reality', should correspond to 'schizophrenic states' of macroscopic objects. Nobody has ever observed such states. I would like to stress here that the classical limit for the superposition of macroscopic states does not exist even when the Planck constant $\hbar$ goes to zero. Once created, these states survive in the limit $\hbar \to 0$.

Bohr changed the rules of quantum mechanics and put bounds on the applicability of the Schrödinger equation. The reduction postulate introduced by Bohr was mainly needed to get rid of macroscopic superpositions. At first glance, it looks like a simple modification of the equations of motion without far-reaching philosophical consequences. However, these consequences are much deeper than one could expect. For the reduction postulate forces us to change the old classical concept of reality. I do not mean here the rather trivial point that, due to the uncertainty principle, particles in quantum mechanics have no trajectories, or the less trivial point expressed by Einstein as 'God playing dice'. To make the interpretation at least *look* self-consistent, one has to abandon the notion of reality as most of us (even many of those who claim that they agree with Bohr's interpretation) used to understand it. One has to accept that physics does not describe the world 'out there' and its purpose is only to bring some order to our perceptions of the world. The observer (or a 'classical device', as some prefer to say) becomes an inevitable ingredient of any physical theory, playing a central role in the interpretation of the outcome of the equations.

This leads to a complicated philosophical scheme, known as the Copenhagen interpretation, which goes far beyond the 'practical' reduction postulate and statistical interpretation of the wave-function. This scheme is so confusing that it would probably be fair to say that there are many different interpretations of the Copenhagen interpretation. Things become even more dramatic when one applies quantum theory to the Universe. Then the logical implementation of the Copenhagen philosophy looks like a joke.

17.3 Everett's picture

Is there any alternative to the Copenhagen interpretation which would allow us to return to the 'old classical scheme' (perhaps in a modified form)? The answer to this question was given by Everett thirty years after quantum mechanics was invented. This answer is simple but non-trivial. To obtain it, one just has to *interpret* what quantum mechanics really tells us without trying to change its rules to suit our prejudice. As a first step, one has to admit the existence of macroscopic superpositions as a reality. Then how can this be reconciled with the fact that the macroscopic objects we observe are not in 'schizophrenic' states? The only way to avoid a contradiction is to take the next step and admit that quantum mechanics actually always describes an ensemble of many *real universes*. Therefore the discovery of quantum mechanics was in fact the discovery which gave a solid scientific basis to the 'Multiverse versus Universe' debate.

In Everett's interpretation, the different superpositions of the macroscopic states correspond to different ensembles of the universes which they describe. We are back to the 'classical idea' of a one-to-one correspondence between reality and the mathematical symbols which are used to describe it. We also return to the idea that physics describes not only 'our knowledge and perceptions' but also the world 'out there', which *existed* and *will exist* without any observers. However, it is not a single world anymore – instead it is many universes. If we admit this, we also have to accept the 'crazy consequence', namely, the existence of many copies of ourselves which are as real as we are. This is the 'price' one has to pay for the *interpretation* of the theory as it is and for the simple resolution of quantum paradoxes. Since there is nothing wrong any more with the universality of quantum theory, it can be also applied to the whole Universe. Measurements can be completely described by quantum equations and the statistical interpretation can be derived within quantum mechanics (i.e. it does not need to be postulated separately) [3, 4].

According to Everett's interpretation, classical mechanics can always be derived as a limiting case of quantum theory ($\hbar \to 0$). One just needs to expand the wave-function in a preferable basis and then identify the various components of this basis with different quasi-classical universes. In turn, the preferable basis is determined by the requirement that the appropriate expectation values of the 'macroscopic operators' for the state vectors of this basis satisfy the classical equations of motion [5]. The quasi-classical corrections to the equations are proportional to the Planck constant. These corrections characterize 'the strength of quantum interactions' (interference) of different universes. Of course, due to decoherence, two very different universes 'interact' very little and it is *practically impossible* to verify their existence. However, 'practically impossible' does not mean *impossible in principle*. For instance, the existence of our copies in other universes, which at present strongly decohere with 'us', would influence 'our future' on a Poincaré timescale. And if something can *in principle* influence our reality, then it means that this something is as real as 'our reality'. The universes which are just different at microscopic level 'communicate' (interfere) too strongly to speak about well defined 'different' universes. In this case the 'multiverse' becomes a more complex concept. Nevertheless, the existence of this multiverse allows us to understand the interference phenomena.

17.4 Quantum cosmology

Let me turn now to quantum cosmology. According to inflationary models, the primordial inhomogeneities responsible for the large-scale structure originated from the initial vacuum fluctuations, amplified during a stage of accelerated expansion [6]. The initial (vacuum) state of the quantum fluctuations is translationally invariant (i.e. there are no preferred points), and the quantum state remains translationally invariant as the fluctuations are excited. How, in this case, can one understand the fact that the translational invariance is finally broken and that we observe galaxies in certain locations on the sky? One could ask, for instance, why our galaxy is located in the particular place where we observe it and not a few Mpc away. What determined the choice of this place?

If one considers the wave-functional describing the perturbations, then this functional remains translationally invariant even after the occupation numbers in every mode have grown tremendously. After inflation, this wave-functional is a superposition of different components, each one corresponding to macroscopically different universes. One of these universes looks like the one we observe today. In another one, the 'galaxy' similar to our own is shifted to a position 10 Mpc away, and so on. Every 'non-schizophrenic' component of the wave-function describes a state with broken translational invariance. However, their sum remains translationally invariant. If one assumes that the Schrödinger equation is universal and applies it also to observers, then nobody can make a selection of a particular 'real' component from the superposition. Therefore all terms in this superposition come with the 'same rights of existence'. Since we know that one of the components surely corresponds to the reality describing our world, we have to admit that the others also correspond to some reality. Thus we arrive at an ensemble of many *real*, classically different universes, with statistical properties characterized by the correlation functions which can be calculated as usual.

The situation with the wave-function of the whole Universe is very similar. The natural boundary conditions for the wave-function of the Universe [7] always predict states which are superpositions of macroscopically different universes. These universes are highly decoherent, do not interfere with each other, and differ even more than in the case of cosmological perturbations. The wave-function localized near a particular classical trajectory can in principle also be constructed, but only as a result of substantial effort and it does not look very natural. According to Everett's interpretation, all universes are real. One could expect that a theory of quantum gravity will finally allow us to introduce a natural measure to characterize how frequently a universe of a certain type occurs in the ensemble. It is very likely that, according to this (as yet unknown) measure, the ratio of the number of 'big universes' to the number of 'small universes' in the ensemble will be much larger for models with an inflationary stage than for models without such a stage. If the measure is ever constructed, it will allow us to put the statement that 'inflation solves the problem of initial conditions' on a solid mathematical ground. At present, this statement is justified only by our feeling and intuition.

17.5 Conclusions

Probably the main reason why Everett's interpretation has not been taken very seriously by the scientific community up to now is because it predicts the reality of our 'copies'. On the other hand, as we know, the theory of inflation leads to a picture of an eternal self-reproducing universe. It is clear that, in such a universe, one can always find an unlimited number of worlds which are very similar to our world and filled by our copies. Unlike 'quantum copies', these 'eternal universe copies' are less dangerous and more useless, since they will never communicate with us. However, if we seriously believe this picture, then the most important psychological reason to reject the Everett interpretation is gone. Since the 'many worlds' interpretation is the *only* logical interpretation of quantum mechanics, I do not see any reason to substitute it with a 'many *words* interpretation' merely to suit our prejudice. Everett's interpretation is crazy enough to be true, and therefore surely deserves very serious attention.

References

[1] J. A. Wheeler. In *Physical Origins of Time Asymmetry*, eds. J. J. Halliwell, J. Perez-Mercader and W. H. Zurek (Cambridge: Cambridge University Press, 1994), p. 1.
[2] J. von Neumann. In *Mathematical Foundations of Quantum Mechanics* (Princeton: Princeton University Press, 1955).
[3] H. Everett. *Rev. Mod. Phys.* **29** (1957), 454.
[4] H. Everett. In *The Many-Worlds Interpretation of Quantum Mechanics*, eds. B. S. DeWitt and N. Graham (Princeton: Princeton University Press, 1973), p. 3.
[5] V. Mukhanov. *Phys. Lett.* **127A** (1988), 251.
[6] V. Mukhanov and G. Chibisov. *JETP Lett.* **33** (1981), 533.
[7] J. B. Hartle and S. W. Hawking. *Phys. Rev.* **B 28** (1983), 2620.
[8] A. Linde. *Lett. Nuovo Cim.* **39** (1984), 401.
[9] A. Vilenkin. *Phys. Rev.* **D 30** (1984), 509.

18

Anthropic reasoning and quantum cosmology[1]

James B. Hartle

Department of Physics, University of California, Santa Barbara

18.1 Introduction

If the Universe is a quantum mechanical system, then it has a quantum state. This state provides the initial condition for cosmology. A theory of this state is an essential part of any final theory summarizing the regularities exhibited universally by all physical systems and is the objective of the subject of quantum cosmology. This chapter is concerned with the role that the state of the Universe plays in anthropic reasoning – the process of explaining features of the Universe from our existence in it [1]. The thesis will be that anthropic reasoning in a quantum mechanical context depends crucially on assumptions about the Universe's quantum state.

18.2 A model quantum universe

Every prediction in a quantum mechanical universe depends on its state, if only very weakly. Quantum mechanics predicts probabilities for alternative possibilities, most generally the probabilities for alternative histories of the Universe. The computation of these probabilities requires both a theory of the quantum state as well as the theory of the dynamics specifying its evolution.

To make this idea concrete while keeping the discussion manageable, we consider a model quantum universe. The details of this model are not essential to the subsequent discussion of anthropic reasoning but help to fix the notation for probabilities and provide a specific example of what they mean. Particles and fields move in a large – perhaps expanding – box, say, presently 20 000 Mpc on a side. Quantum gravity is neglected – an excellent

---

[1] This is a reworking of an article, reproduced with permission, from *The New Cosmology: Proceedings of the Conference on Strings and Cosmology*, eds. R. Allen, D. Nanopoulos and C. Pope, AIP Conference Proceedings 743 (New York: American Institute of Physics, 2004).

Universe or Multiverse?, ed. Bernard Carr. Published by Cambridge University Press.
© Cambridge University Press 2007.

approximation for accessible alternatives in our Universe later than 10^{-43} s after the big bang. Spacetime geometry is thus fixed with a well defined notion of time, and the usual quantum apparatus of Hilbert space states and their unitary evolution governed by a Hamiltonian can be applied.[2]

The Hamiltonian H and the state $|\Psi\rangle$ in the Heisenberg picture are the assumed theoretical inputs to the prediction of quantum mechanical probabilities. Alternative possibilities at one moment of time t can be reduced to yes/no alternatives represented by an exhaustive set of orthogonal projection operators $\{P_\alpha(t)\}$ ($\alpha = 1, 2, \ldots$) in this Heisenberg picture. The operators representing the same alternatives at different times are connected by:

$$P_\alpha(t) = e^{iHt/\hbar} P_\alpha(0) e^{-iHt/\hbar}. \tag{18.1}$$

For instance, the Ps could be projections onto an exhaustive set of exclusive ranges of the centre-of-mass position of the Earth, labelled by α. The probabilities $p(\alpha)$ that the Earth is located in one or another of these regions at time t is given by

$$p(\alpha|H,\Psi) = \|P_\alpha(t)|\Psi\rangle\|^2. \tag{18.2}$$

The probabilities for the Earth's location at a different time is given by the same formula with different Ps computed from the Hamiltonian by Eq. (18.1). The notation $p(\alpha|H,\Psi)$ departs from usual conventions (e.g. ref. [2]) to indicate explicitly that all probabilities are conditioned on the theory of the Hamiltonian H and quantum state $|\Psi\rangle$.

Most generally, quantum theory predicts the probabilities of sequences of alternatives at a series of times – that is *histories*. An example is a sequence of ranges of centre-of-mass position of the Earth at a series of times giving a coarse-grained description of its orbit. Sequences of sets of alternatives $\{P^k_{\alpha_k}(t_k)\}$ at a series of times t_k ($k = 1, \ldots, n$) specify a set of alternative histories of the model. An individual history α in the set corresponds to a particular sequence of alternatives $\alpha \equiv (\alpha_1, \alpha_2, \ldots, \alpha_n)$ and is represented by the corresponding chain of projection operators C_α:

$$C_\alpha \equiv P^n_{\alpha_n}(t_n) \cdots P^1_{\alpha_1}(t_1), \quad \alpha \equiv (\alpha_1, \ldots, \alpha_n). \tag{18.3}$$

The probabilities of the histories in the set are given by

$$p(\alpha|H,\Psi) \equiv p(\alpha_n, \ldots, \alpha_1|H,\Psi) = \|C_\alpha|\Psi\rangle\|^2 \tag{18.4}$$

---

[2] For a more detailed discussion of this model in the notation used here, see ref. [2]. For a quantum framework when spacetime geometry is not fixed, see, e.g., ref. [3].

provided the set decoheres, i.e. provided the branch state vectors $C_\alpha|\Psi\rangle$ are mutually orthogonal. Decoherence ensures the consistency of the probabilities given by Eq. (18.4) with the usual rules of probability theory.[3]

To use either Eq. (18.2) or (18.4) to make predictions, a theory of *both* H and $|\Psi\rangle$ is needed. No state means no predictions.

18.3 What is predicted?

'If you know the wave-function of the Universe, why aren't you rich?' This question was once put to me by my colleague Murray Gell-Mann. The answer is that there are unlikely to be any alternatives relevant to making money that are predicted as sure bets, conditioned just on the Hamiltonian and quantum state alone. A probability $p(\text{rise}|H, \Psi)$ for the stock market to rise tomorrow could be predicted from H and $|\Psi\rangle$ through Eq. (18.2) in principle. But it seems likely that the result would be a useless $p(\text{rise}|H, \Psi) \approx 1/2$, conditioned just on the 'no boundary' wave-function [7] and M-theory.

It is plausible that this is the generic situation. To be manageable and discoverable, the theories of dynamics and the quantum state must be short – describable in terms of a few fundamental equations and the explanations of the symbols they contain. It is therefore unlikely that H and $|\Psi\rangle$ contain enough information to determine most of the interesting complexity of the present Universe with significant probability [8, 9]. We *hope* that the Hamiltonian and quantum state are sufficient conditions to predict certain large-scale features of the Universe with significant probability. Approximately classical spacetime, the number of large spatial dimensions, the approximate homogeneity and isotropy on scales above several hundred Mpc, and the spectrum of density fluctuations that were the input to inflation are some examples of these. But even a simple feature, such as the time the Sun will rise tomorrow, will not be usefully predicted by our present theories of dynamics and the quantum state *alone*.

The time of sunrise *does* become predictable with high probability if a few previous positions and orientations of the Earth in its orbit are supplied in addition to H and $|\Psi\rangle$. That is a particular case of a *conditional probability* of the form

$$p(\alpha|\beta, H, \Psi) = \frac{p(\alpha, \beta|H, \Psi)}{p(\beta|H, \Psi)} \quad (18.5)$$

---

[3] For a short introduction to decoherence, see ref. [2] or any of the classic expositions of decoherent (consistent) histories quantum theory [4–6].

for alternatives α (e.g. the times of sunrise), given H and $|\Psi\rangle$ and further alternatives β (e.g. a few earlier positions and orientations of the Earth). The joint probabilities on the right-hand-side of Eq. (18.5) are computed using Eq. (18.4), as described in Section 18.2.

Conditioning probabilities on specific information can weaken their dependence on H and $|\Psi\rangle$ but does not eliminate it. That is because any specific information available to us as human observers (such as a few positions of the Earth) is but a small part of that needed to specify the state of the Universe. The chains of the form given in Eqs. (18.3) that define a few previous positions of the Earth in Eqs. (18.4) and (18.5) involve projections P_γ that define a previous position to a certain accuracy. These span a very large subspace of the Hilbert space, so that $P_\gamma|\Psi\rangle$ depends strongly on $|\Psi\rangle$. For example, to extrapolate present data on the Earth to its position 24 hours from now requires that the probability be high that it moves on a classical orbit in that time and that the probability be low that it is destroyed by a neutron star now racing across the Galaxy at near light speed. Both of these probabilities depend crucially on the nature of the quantum state [10].

Many useful predictions in physics are of conditional probabilities of the kind discussed in this section. We next turn to the question of whether we should be part of the conditions.

18.4 Anthropic reasoning – less is more

18.4.1 Anthropic probabilities

In calculating the conditional probabilities for predicting some of *our* observations given others, there can be no objection of principle to including a description of 'us' as part of the conditions:

$$p(\alpha|\beta, \text{'us'}, H, \Psi). \tag{18.6}$$

Drawing inferences using such probabilities is called *anthropic reasoning*. The motivation is the idea that probabilities for certain features of the Universe might be sensitive to this inclusion.

The utility of anthropic reasoning depends on how sensitive probabilities like Eq. (18.6) are to the inclusion of 'us'. To make this concrete, consider the probabilities for a hypothetical cosmological parameter we will call Λ. We will assume that H and $|\Psi\rangle$ imply that Λ is constant over the visible Universe, but only supply probabilities for the various constant values it might take through Eq. (18.4). We seek to compare $p(\Lambda|H,\Psi)$ with $p(\Lambda|\text{'us'}, H, \Psi)$. In principle, *both* are calculable from Eqs. (18.4) and (18.5). Figure 18.1 shows three possible ways in which they might be related.

- $p(\Lambda|H,\Psi)$ is peaked around one value, as in Fig. 18.1(a). The parameter Λ is determined either by H or $|\Psi\rangle$ or by both.[4] Anthropic reasoning is not necessary; the parameter is already determined by fundamental physics.
- $p(\Lambda|H,\Psi)$ is distributed and $p(\Lambda|\text{'us'},H,\Psi)$ is also distributed, as in Fig. 18.1(b). Anthropic reasoning is inconclusive. One might as well measure the value of Λ and use this as a condition for making further predictions,[5] i.e. work with probabilities of the form $p(\alpha|\Lambda,H,\Psi)$.
- $p(\Lambda|H,\Psi)$ is distributed but $p(\Lambda|\text{'us'},H,\Psi)$ is peaked, as in Fig. 18.1(c). Anthropic reasoning helps to explain the value of Λ.

The important point to emphasize is that a *theoretical hypothesis for H and $|\Psi\rangle$ is needed to carry out anthropic reasoning*. Put differently, a theoretical context is needed to decide whether a parameter like Λ *can* vary, and to find out *how* it varies, before using anthropic reasoning to restrict its range. The Hamiltonian and quantum state provide this context. In Section 18.5, we will consider the situation where the state is imperfectly known.

18.4.2 Less is more

While there can be no objections in principle to including 'us' as a condition for the probabilities of our observations, there are formidable obstacles in practice.

- We are complex physical systems requiring an extensive environment and a long evolutionary history, whose description in terms of the fundamental variables of H and $|\Psi\rangle$ may be uncertain, long and complicated.
- The complexity of the description of a condition including 'us' may make the calculation of the probabilities long or impossible as a practical matter.

In practice, therefore, the anthropic probabilities given by Eq. (18.6) can only be estimated or guessed. Theoretical uncertainty in the results is thereby introduced.

The objectivity striven for in physics consists, at least in part, in using probabilities that are not too sensitive to 'us'. We would not have science if anthropic probabilities for observation depended significantly on which individual human being was part of the conditions. The existence of schizophrenic delusions shows that this is possible, so that the notion of 'us' should be restricted to exclude such cases.

---
4 As, for example, in the as yet inconclusive discussions of baby universes [11,12].
5 As stressed by Hawking and Hertog [13].

280 *James B. Hartle*

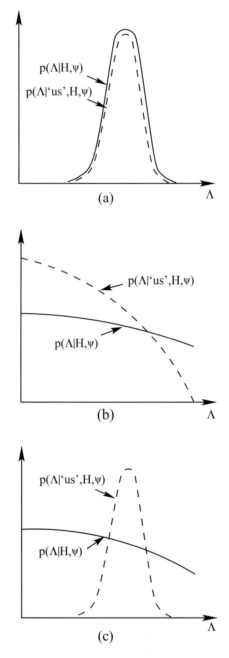

Fig. 18.1. Some possible behaviours for probabilities for the value of a cosmological parameter Λ with and without the condition 'us'. (a) The value of Λ is fixed by H and $|\Psi\rangle$, so anthropic reasoning is not needed. (b) Anthropic probabilities are distributed, so anthropic reasoning is useless in fixing Λ. (c) Anthropic reasoning is useful.

For these reasons, it is prudent to condition probabilities, not on a detailed description of 'us', but on the weakest condition consistent with 'us' that plausibly provides useful results such as those illustrated in Fig. 18.1(c). A short list of conditions of roughly decreasing complexity might include:

- human beings;
- carbon-based life;
- information gathering and utilizing systems;
- at least one galaxy;
- a universe older than 5 Gy;
- no condition at all.

For example, the probabilities used to bound the cosmological constant Λ make use of the fourth and fifth items on this list, under the assumption that including earlier ones will not greatly affect the anthropically allowed range for Λ [1]. (For recent reviews with references to earlier literature, see refs. [14] and [15].) To move down in the above list of conditions is to move in the direction of increasing theoretical certainty and decreasing computational complexity. With anthropic reasoning, less is more.

18.5 Ignorance is *not* bliss

The quantum state of a single isolated subsystem generally cannot be determined from a measurement carried out on it. That is because the outcomes of measurements are distributed probabilistically and the outcome of a single trial does not determine the distribution. Neither can the state be determined from a series of measurements, because measurements disturb the state of the subsystem. The Hamiltonian cannot be inferred from a sequence of measurements on one subsystem for similar reasons. In the same way, we cannot generally determine either the Hamiltonian or the quantum state of the Universe from our observations of it. Rather, these two parts of a final theory are theoretical proposals, inferred from partial data to be sure, but incorporating theoretical assumptions of simplicity, beauty, coherence, mathematical precision, etc. To test these proposals, we search among the conditional probabilities they imply for predictions of observations yet to be made with probabilities very near unity. When such predictions occur, we consider them as successes of the theory; when they do not, we reject the theory and propose another one.

Do we need a theory of the quantum state? To analyze this question, let us consider various degrees of theoretical uncertainty about it.

18.5.1 Total ignorance

In the 'box' cosmology model of Section 18.2, theoretical uncertainty about the quantum state can be represented by a density matrix ρ that specifies probabilities for its eigenstates to be $|\Psi\rangle$. Total ignorance of the quantum state is represented by ρ being proportional to the unit matrix. To illustrate this and the subsequent discussion, assume for the moment that the dimension of the Hilbert space is very large but finite. Then total ignorance of the quantum state is represented by

$$\rho_{\text{tot ign}} = \frac{I}{\text{Tr}(I)}, \qquad (18.7)$$

which assigns equal probability to any member of any complete set of orthogonal states.

The density matrix given by Eq. (18.7) predicts thermal equilibrium, infinite temperature, infinitely large field fluctuations and maximum entropy [9]. In short, its predictions are inconsistent with observations. This is a more precise way of saying that every useful prediction depends in some way on a theory of the quantum state. Ignorance is not bliss.

18.5.2 What we know

A more refined approach to avoiding theories of the quantum state is to assume that it is unknown except for reproducing our present observations of the Universe. The relevant density matrix is given by

$$\rho_{\text{obs}} = \frac{P_{\text{obs}}}{\text{Tr}(P_{\text{obs}})}, \qquad (18.8)$$

where P_{obs} is the projection on our current observations – 'what we know'. Observations in this context mean what we directly observe and record here on Earth and not the inferences we draw from this data about the larger Universe. That is because those inferences are based on assumptions about the very quantum state that Eq. (18.8) aims to ignore. For instance, we observed nebulae long before we understood what they were or where they are. The inference that nebulae are distant clusters of stars and gas relies on assumptions about how the Universe is structured on very large scales that are, in effect, weak assumptions on the quantum state.

Even if we made the overly generous assumption that we had somehow directly observed and recorded every detail of the volume 1 km above the surface of the Earth, say at 1 mm resolution, that is still a tiny fraction ($\sim 10^{-60}$) of the volume inside the present cosmological horizon. The projection operator $P_{\rm obs}$ therefore defines a very large subspace of Hilbert space. We can expect that the entropy of the density matrix Eq. (18.8) will therefore be near maximal, close to that of Eq. (18.7), and that its predictions will be similarly inconsistent with further observations.

In the context of anthropic reasoning, these results show that conditioning probabilities on 'us' alone is not enough to make useful predictions. Rather, in addition a theory of H and $|\Psi\rangle$ is required, as described in Section 18.4.

18.6 A final theory

Let us hope that one day we will have a unified theory based on a *principle* that will specify *both* quantum dynamics (H) and a unique quantum state of the Universe ($|\Psi\rangle$). That would truly be a final theory and a proper context for anthropic reasoning.

Acknowledgements

Appreciation is expressed to the Mitchell Institute for hospitality and to the National Science Foundation for partial support under grant NSF-PHY02-44764.

References

[1] J. Barrow and F. Tipler. *The Anthropic Cosmological Principle* (Oxford: Oxford University Press, 1986).
[2] J. B. Hartle. The quantum mechanics of closed systems. In *Directions in General Relativity, Volume 1: A Symposium and Collection of Essays in Honor of Professor Charles W. Misner's 60th Birthday*, eds. B.-L. Hu, M. P. Ryan and C. V. Vishveshwara (Cambridge: Cambridge University Press, 1993) [gr-qc/9210006].
[3] J. B. Hartle. Spacetime quantum mechanics and the quantum mechanics of spacetime. In *Gravitation and Quantizations, Proceedings of 1992 Les Houches Summer School*, eds. B Julia and J. Zinn-Justin (Amsterdam: North Holland, 1995) [gr-qc/9508023].
[4] R. B. Griffiths. *Consistent Quantum Theory* (Cambridge: Cambridge University Press, 2002).
[5] R. Omnès. *The Interpretation of Quantum Mechanics* (Princeton, NJ: Princeton University Press, 1994).
[6] M. Gell-Mann. *The Quark and the Jaguar* (San Francisco: W. Freeman, 1994).
[7] S. W. Hawking. The quantum state of the universe. *Nucl. Phys.* **B 239** (1984), 257.
[8] J. B. Hartle. Scientific knowledge from the perspective of quantum cosmology. In *Boundaries and Barriers, On the Limits to Scientific*

Knowledge, eds. J. L. Casti and A. Karlqvist (Reading, MA: Addison-Wesley, 1996) [gr-qc/9601046].

[9] J. B. Hartle. The state of the universe. In *The Future of Theoretical Physics and Cosmology: Stephen Hawking 60th Birthday Symposium*, eds. G. W. Gibbons, E. P. S. Shellard, and S. J. Ranken (Cambridge: Cambridge University Press, 2003) [gr-qc/0209046].

[10] J. B. Hartle. Quasi-classical domains in a quantum universe. In *Proceedings of the Cornelius Lanczos International Centenary Conference*, North Carolina State University, December 1992, eds. J. D. Brown, M. T. Chu, D. C. Ellison and R. J. Plemmons (Philadelphia: SIAM, 1994) [gr-qc/9404017].

[11] A. Strominger. Baby universes. In *Quantum Cosmology and Baby Universes, Proceedings of 1989 Jerusalem Winter School for Theoretical Physics*, eds. S. Coleman, J. B. Hartle, T. Piran and S. Weinberg (Singapore: World Scientific, 1991).

[12] S. W. Hawking. The cosmological constant is probably zero. *Phys. Lett.* **B 134**, 403 (1984).

[13] S. W. Hawking and T. Hertog. Populating the landscape: A top down approach. *Phys. Rev.* **D 73** (2006), 123527 [hep-th/0602091].

[14] S. Weinberg. The cosmological constant problems. In *Sources and Detection of Dark Matter and Dark Energy in the Universe*, ed. D. Cline (New York: Springer-Verlag, 2001) [astro-ph/0005265].

[15] A. Vilenkin. Cosmological constant problems and their solutions. In *Astronomy, Cosmology and Fundamental Physics, Proceedings of ESO-CERN-ESA Symposium*, Garching, 4–7 March, 2002, eds. P. A. Shaver, L. DiLella and A. Giménez (Berlin: Springer, 2003), p. 70 [hep-th/0106083].

19
Micro-anthropic principle for quantum theory

Brandon Carter
Observatoire de Paris, Meudon

19.1 Introduction

As a prescription for ascribing *a priori* probability weightings to the eventuality of finding oneself in the position of particular conceivable observers, the anthropic principle was originally developed for application to problems of cosmology [1] and biology [2]. The purpose of this chapter is to provide a self-contained introductory account of the motivation and reasoning underlying the recent development [3] of a more refined version of the anthropic principle that is needed for the provision of a coherent interpretation of quantum theory.

In order to describe ordinary laboratory applications, it is commonly convenient, and entirely adequate, to use a 'Copenhagen' type representation, in which a Hilbert state vector undergoes 'collapse' when an observation is made. However, from a broader perspective it is rather generally recognized that such a collapse cannot correspond to any actual physical process.

A leading school of thought on this subject was founded by Everett [4], who maintained the principle of the physical reality of the Hilbert state, and deduced that – in view of the agreement that no physical collapse process occurs – none of the ensuing branch channels can be 'more real than the rest'. This was despite the paradox posed by the necessity that they be characterized by *different* (my italics) 'weightings', the nature of which was never satisfactorily explained. This intellectual flaw in the Everett doctrine was commonly overlooked, not so much by its adherents, who were seriously concerned about it [5], as by its opponents, who were upset by its revolutionary 'multi-universe' implications.

The main alternative line of development was based on the (widely accepted) principle – which will be adopted as the starting point for the present work – that neither the specialized pure Hilbert space vector, nor

Universe or Multiverse?, ed. Bernard Carr. Published by Cambridge University Press.
© Cambridge University Press 2007.

the von Neumann probability operator that replaces it in more general circumstances, is of an objective physical nature. Rather, they are merely mathematical prediction tools of an entirely subjective nature, as also is the collapse to which they are subjected if and when the relevant information becomes available. However, this approach also came up against a paradox, which was exemplified by the parable of 'Wigner's friend' [6] (who, in the more detailed discussion below, I shall suppose to have been Schrödinger, the owner of the legendary cat). The problem – which became particularly acute in the context of cosmology – was how independent observers (such as Wigner and Schrödinger) can be dealt with objectively, on the same footing, by a probabilistic theory of an intrinsically subjective nature.

The longstanding problem of reconciling objectivity with subjectivity is solved here by the anthropic abstraction, which distinguishes a material observer (such as Wigner) from an abstract perceptor who may or may not perceive himself to be Wigner. The probability of such a perceptor must be attributed to some appropriate micro-anthropic principle of the kind that will be presented below [3].

19.2 Eventualities and observables

Although their ultimate purpose is to account for (and even predict) *events*, i.e. things that actually happen, physical (and other) theories are mainly concerned with what I shall refer to as *eventualities*, meaning things that may or may not actually happen.

Eventualities are subject to partial ordering, as expressible by a statement of the form $e_1 \subset e_2$, which is to be understood as meaning that if an eventuality e_1 happens as an actual event, then so does e_2. On the understanding that the concept of eventuality formally includes the special case of the *null* eventuality, $\emptyset$, which by definition never happens, it can be assumed that any pair of eventualities, e_1 and e_2, will define a corresponding combined eventuality $e_1 \cap e_2$, whose occurrence as an actual event implies and is implied by the occurrence, both at once, of e_1 and e_2, so that we always have $e_1 \cap e_2 \subset e_1$. In particular, the condition for e_1 to be incompatible with e_2 will be expressible as $e_1 \cap e_2 = \emptyset$.

The kinds of (classical and quantum) theory that I know about are all *additive* in the sense that, for each pair of eventualities e_1 and e_2, there will be a well defined sum $e_1 \oplus e_2$ that is an admissable eventuality such that $e_1 \cap (e_1 \oplus e_2) = e_1$. In such a case, it is commonly useful to introduce a corresponding concept of *complementarity*, whereby a set $\{e\}$ of eventualities

$e_1, \ldots, e_\mathcal{N}$ will be describable as complementary if the sum $s = e_1 \oplus \cdots \oplus e_\mathcal{N}$ is an event that must necessarily happen.

An important related concept – on which discussions of quantum theory are commonly based (though it is less fundamental than that of an eventuality) – is that of an *observable*. This term is used to describe a set $\{e\}$ of non-null eventualities that is subject to a condition not just of complementarity, but also of what may be termed mutual *exclusivity*. An awkward feature of this concept (one of the reasons why I prefer to attribute the primary role to eventualities rather than to observables) is that it is difficult to formulate in a manner that transcends the technicalities of the particular kind of theory under consideration.

For a theory that is classical, in a sense whose meaning will be clarified in Section 19.3, a pair of eventualities e_1 and e_2 can be considered mutually exclusive if they satisfy the incompatibility condition $e_1 \cap e_2 = \emptyset$. However, for a quantum theory such incompatibility is merely necessary, not sufficient, for exclusivity in the strong sense (defined below) that is required for observability.

19.3 The classical paradigm

Some of the simplest and most commonly used theories are of the kind describable as *deterministic*, which means that they consist of rules whereby appropriate input data (such as initial conditions) can be use to single out a restricted subclass of events that actually happen within a much broader class of conceivable eventualities. However, a much more widely applicable category of theories are those that are *probabilistic*. Instead of providing rules that clearly distinguish events that happen from other eventualities that do not, such theories merely provide prescriptions for ascribing what is usually called a *probability* (but what some people prefer to call a *propensity*) – meaning a real number P in the range $0 \leq P \leq 1$ – to each of the relevant eventualities, in a manner that must naturally be consistent with the partial ordering. This means that $e_1 \supset e_2$ implies $P\{e_1\} \geq P\{e_2\}$ and, in particular, $P\{\emptyset\} = 0$. The category of probabilistic theories evidently includes deterministic theories as the special case for which the range of probabilities is restricted to two extreme values: $P = 1$ characterizing events and $P = 0$ characterizing other eventualities that do not actually happen.

A particularly important subcategory of probabilistic theories is that which includes the *classical* ones. For the description of a system A in a classical theory, the admissible eventualities will be identifiable as subsets of a corresponding set $I\{A\}$ that is endowed with an ordinary probability measure, whose restriction to a subset $e \subset I$ gives the corresponding probability $P\{e\}$, while the complete set I can be interpreted as representing an eventuality that is certain, meaning that $P\{I\} = 1$. Such a theory will automatically be endowed with an additive structure, whereby any pair of eventualities e_1 and e_2 will not only have a combination given by the intersection $e_1 \cap e_2$, but will also have a well defined sum that is defined as the corresponding union $e_1 \oplus e_2 = e_1 \cup e_2$. Unlike in a quantum theory, its probability will then be given by $P\{e_1 \oplus e_2\} = P\{e_1\} + P\{e_2\} - P\{e_1 \cap e_2\}$.

The simplest example of a classical theory is the system consisting of a tossed coin, which can be described in terms of a total of four eventualities. Two of these eventualities are the independent possibilities e_1 for the tail to turn up and e_2 for the head to turn up. The other two (trivial) eventualities are their sum $I = e_1 \oplus e_2$, representing the certain event of something turning up, and the null eventuality $\emptyset = e_1 \cap e_2$, representing the impossible case of nothing turning up. The latter eventualities must always be characterized by $P\{I\} = 1$ and $P\{\emptyset\} = 0$. The non-trivial part of the probability distribution will be given in the unbiased version of the theory by $P\{e_1\} = P\{e_2\} = 1/2$ but could be different in biased versions. In such a (biased or unbiased) theory, the only non-trivial observable consists of the complementary pair of alternatives $\{e\} = \{e_1, e_2\}$, but of course there is also the trivial observable consisting just of I.

19.4 The Dirac–von Neumann paradigm

As in a classical theory, the admissible eventualities in a quantum theory for the description of a system A will be identifiable with subsets of a corresponding set $I\{A\}$. The essential new feature distinguishing a quantum theory is that I is endowed with a Hilbert space structure, and that the *admissible eventualities* are identifiable not with arbitrary subsets, but only with those that are *Hilbert subspaces*.

If e_1 and e_2 are the Hilbert subspaces representing a pair of admissible eventualities, their intersection $e_1 \cap e_2$ will also be a Hilbert subspace, representing the corresponding conjoint eventuality, but their union $e_1 \cup e_2$ will in general not have the structure of a Hilbert subspace and thus (unlike the classical case) will not represent an admissible eventuality. The eventualities of a quantum theory do, nevertheless, have an additive structure that is naturally induced by the Hilbert space structure: the sum $e_1 \oplus e_2$ is defined to be the Hilbert subspace that is *spanned* by the separate Hilbert subspaces e_1 and e_2. What this means, using the standard notation scheme originally developed by Dirac [7], is that the state $|\Psi\rangle \in e_1 \oplus e_2$ if and only if $|\Psi\rangle$ is a Hilbert space vector having the form $|\Psi\rangle = |\Psi_1\rangle + |\Psi_2\rangle$ for some pair of

Hilbert space vectors such that $|\Psi_1\rangle \in e_1$ and $|\Psi_2\rangle \in e_2$. In the particular case for which every such pair of vectors satisfies the orthogonality condition $\langle\Psi_1|\Psi_2\rangle = 0$, the corresponding subspaces $e_1 \subset I$ and $e_2 \subset I$ will be describable as mutually orthogonal.

Orthogonality in this sense is what characterizes the kind of exclusivity required for the definition of what is generally known as an observable in the context of quantum theory. Thus an *observable* in a quantum theory for the system A (i.e. a qualitative observable, as distinct from a quantitative observable of the related kind to be discussed below) can be formally defined to consist of a complete set $\{e\}$ of mutually orthogonal Hilbert subspaces $e_1, \ldots, e_\mathcal{N}$, where the condition of completeness means that they span the entire Hilbert space $I\{A\}$, i.e. that $e_1 \oplus \cdots \oplus e_\mathcal{N} = I$.

For any particular eventuality, the corresponding subspace $e \subset I$ will determine and be determined by an associated Hilbert space projection operator $\mathbf{e} = \mathbf{e}^2$, defined so as to be automatically Hermitian by the conditions $\mathbf{e}|\Psi\rangle = |\Psi\rangle$ whenever $|\Psi\rangle$ lies in e and $\mathbf{e}|\Psi\rangle = 0$ whenever $|\Psi\rangle$ is orthogonal to the subspace e. The condition for a set $\{e\}$ of eventualities $\{e_i\}$ ($i = 1, \ldots, n$) to constitute an observable is thus expressible as the condition that the corresponding operators should satisfy the orthogonality requirement $\mathbf{e}_i \mathbf{e}_j = 0$ for $i \neq j$ and the completeness condition $\sum_i \mathbf{e}_i = \mathbf{I}$, where $\mathbf{I}$ is the unit operator on I.

In the earliest versions of quantum theory, it was postulated that the relevant probabilities would be given just by the specification of a single state vector $|\Psi\rangle \in I\{A\}$, subject to the normalization condition $\langle\Psi|\Psi\rangle = 1$, according to a prescription expressible in the following familiar form:

$$P_{[\mathcal{O}]}\{e_i\} = \langle\Psi|\mathbf{e}_i|\Psi\rangle. \tag{19.1}$$

This is just a conditional probability, subject to the requirement that the relevant observation, $\mathcal{O}_e$, be actually carried out.

Soon after the original development of this Dirac–Heisenberg paradigm, it was recognized that a prescription of the simple form in Eq. (19.1) is too restrictive for typical cases in which the system A may interact with another (internal or external) system B. The extended system $\widehat{A}$, consisting of the combination of A and B, will be characterized by a Hilbert space $\widehat{I} = I\{\widehat{A}\}$ which is constructed as the tensor product of $I\{A\}$ and $I\{B\}$. This means that a state vector $|\widehat{\Psi}\rangle \in \widehat{I}$ for the extended system will be expressible in terms of a basis of vectors $|\Phi_a\rangle \in I\{B\}$ satisfying the orthonormality condition $\langle\Phi_a|\Phi_b\rangle = \delta_{ab}$. It must therefore have the form

$$|\widehat{\Psi}\rangle = \sum_a |\Phi_a\rangle|\Psi_a\rangle, \tag{19.2}$$

for some corresponding set of vectors $|\Psi_a\rangle \in I\{A\}$ that will not in general be orthonormal but must satisfy the condition $\sum_a \langle \Psi_a | \Psi_a \rangle = 1$ in order for the unit normalization condition $\langle \widehat{\Psi} | \widehat{\Psi} \rangle = 1$ to be satisfied. If e_i is a subspace of dimension $\mathcal{R}_i$ within the original Hilbert space $I\{A\}$ of dimension $\mathcal{N}\{A\}$, then it will determine a corresponding subspace $\widehat{e}_i$ of dimension $\mathcal{R}_i \mathcal{N}\{B\}$ in the tensor product Hilbert space $\widehat{I}$, where $\mathcal{N}\{B\}$ is the dimension of $I\{B\}$. Within the original Hilbert space $I = I\{A\}$, the corresponding projection operator will have rank given by its trace, $\mathcal{R}_i = \text{tr}\{\mathbf{e}_i\}$, while the corresponding operator $\widehat{\mathbf{e}}_i$ of projection onto $\widehat{e}_i$ in $\widehat{I}$ will have rank $\mathcal{R}_i \mathcal{N}\{B\}$. According to the natural extension of Eq. (19.1), a unit state vector $|\widehat{\Psi}\rangle$ in $\widehat{I}$ will specify a (conditional) probability distribution given by

$$P_{[\mathcal{O}]}\{e_i\} = \langle \widehat{\Psi} | \widehat{\mathbf{e}}_i | \widehat{\Psi} \rangle. \qquad (19.3)$$

In order to express such a prescription within the simpler framework of the original Hilbert space $I\{A\}$ of the subsystem A with which we are particularly concerned, it is necessary to use a prescription of the kind whose development was attributed by Dirac to von Neumann. In the Dirac–von Neumann paradigm, instead of being specified by just a single state vector $|\Psi\rangle$, the (conditional) probability distribution (for the outcome of an observation $\mathcal{O}_e$ if actually performed) is specified by a Hermitian probability density operator $\mathbf{P}$ with unit trace on I according to the prescription

$$P_{[\mathcal{O}]}\{e_i\} = \text{tr}\{\mathbf{P}\, e_i\}. \qquad (19.4)$$

This prescription is compatible with the original pure state paradigm, as specified by a single vector satisfying the unit normalization $\langle \Psi | \Psi \rangle = 1$, according to Eq. (19.1). The effect of this can be seen to be the same as taking $\mathbf{P} = |\Psi\rangle\langle\Psi|$ in the general formula Eq. (19.4). The advantage of the von Neumann type formulation in Eq. (19.4) is that it can also express the result of the more general prescription given in Eq. (19.3), whose effect can be seen to be the same as that of taking

$$\mathbf{P} = \sum_a |\Psi_a\rangle\langle\Psi_a|, \qquad (19.5)$$

where the (generally non-orthonormal) set of vectors $|\Psi_a\rangle$ is specified by the decomposition in Eq. (19.2).

Many authors – particularly those influenced by the Everett doctrine [4] – have continued to hanker after the original Heisenberg-type paradigm, meaning the supposition that the probabilities should ultimately be determined by a pure state in a very large all-embracing Hilbert space, characterizing the universe as a whole. Such authors – including Hawking [8] – have been inclined to regard the use of a von Neumann operator as a rather unsatisfactory approximation device that may be made necessary by our ignorance due to the regrettable loss of some of the relevant information in, for example, a black hole. However, my own attitude is like that of the distrustful insurance agent, who doubts whether what was alleged to have been lost was ever actually possessed. I personally see no reason why – to encompass more and more detailed microstructure and more and more extended macrostructure – the process of construction of successively larger and larger Hilbert spaces should ever come to an end. In other words, the search [9] for a single, ultimate, all-embracing 'wave-function of the universe', or even an ultimate, all-embracing von Neumann operator, may be like the pursuit of the proverbially elusive 'Will o' the wisp'. It seems more reasonable to accept that any system sufficiently simple to be amenable to our mathematical analysis can only be a model of an incomplete sub-component of something larger, and that it is therefore unreasonable to demand that it be describable by a pure state rather than a more general von Neumann operator. However that may be, advocates of the Everett doctrine would agree that there can in any case be no harm in working throughout in terms of the von Neumann paradigm, as will be done here, because it includes the more restricted Heisenberg-type pure state paradigm as a special case.

Before continuing, it should be remarked that the term *observable* has been used here to designate what – in a more pedantically explicit terminology – would be called a *qualitative observable*, in order to distinguish it from the *quantitative observables* that are definable as functions thereof. Thus, any qualitative observable $\{e\}$ determines and is determined by a corresponding equivalence class of quantitative variables, in which any particular member E is determined by a corresponding non-degenerate real-valued function E_i with the index labelling the admissible alternatives e_i for $\{e\}$. The condition for non-degeneracy of the function is to be understood as meaning that $E_i \neq E_j$ whenever $i \neq j$. In a quantum theory for a system characterized by a Hilbert space $I\{A\}$, such a quantitative observable will be identifiable with a corresponding Hermitian operator $\mathbf{E}$ whose eigenspaces are the Hilbert subspaces $e_i \subset I\{A\}$, while the corresponding eigenvalues are the real numbers E_i. One therefore has

$$\mathbf{E}|\Psi\rangle = E_i|\Psi\rangle \Leftrightarrow |\Psi\rangle \in e_i. \tag{19.6}$$

Such a quantitative variable E will have a mean (expectation) value $\langle E \rangle$ given by

$$\langle E \rangle = \mathrm{tr}\{\mathbf{P}\mathbf{E}\}, \qquad (19.7)$$

in which the operator $\mathbf{E}$ will be expressible in terms of the relevant projection operators $\mathbf{e}_i$ in the explicit form

$$\mathbf{E} = \sum_i E_i \mathbf{e}_i. \qquad (19.8)$$

The simplest illustration is provided by the familiar Stern–Gerlach example, for which the observable $\mathbf{E}$ represents the spin energy of an electron (with respect to its own rest frame) in a uniform magnetic field. For this application, the relevant Hilbert space I has only two (complex) dimensions, being spanned by a subspace e_1 representing the eventuality that the spin be aligned with the magnetic field, and a subspace e_2 representing the eventuality that it be aligned in the opposite direction. Other eventualities, corresponding to alignment in other directions, will not be characterized by well defined energy values. The quantum analogue of the unbiased coin toss theory (considered in Section 19.3) is the unbiased spin theory, specified by adopting the isotropic probability distribution given (as the high temperature limit of an ordinary thermal distribution) by $\mathbf{P} = \mathbf{I}/2$.

19.5 Sensors and conditional probabilities

Having thus completed a brief overview of the basic quantum mechanical principles that are generally accepted as a matter of consensus, it is now necessary to approach the much more controversial issue of how these rather abstract principles should be interpreted in practice. In particular, how do we relate what might be observable in principle – in the abstract sense of the term as used above – to what is actually observed in the ordinary sense of the word, i.e. the recognition of the actual occurrence of an eventuality in some particular system under consideration?

The first, relatively uncontroversial, point that needs to be made is that the notion of an actual observation of an eventuality in a generic system is usually taken to involve an interaction with a specialized kind of system that I shall refer to as a *sensor*. This might consist of an artificial measuring apparatus of a simple and easily understandable kind, such as a Stern–Gerlach spin-orientation detector, or of something more mysterious, such as the brain of Schrödinger's famous cat.

In order for an observable $\{f\}$ of a system B to be (exactly or approximately) observable, i.e. for the recognition of the actual occurrence of a particular eventuality $f_j \in \{f\}$ to be feasible in practice, it is generally considered to be necessary not just that $\{f\}$ should be observable in the abstract sense formulated above, but also – more particularly – that it should be adequately correlated with a corresponding sensor observable $\{e\}$ in an appropriate sensor system A. The subsets $\widehat{f}_j = f_j \otimes I\{A\}$ and $\widehat{e}_i = e_i \otimes I\{B\}$ in the tensor product space $\widehat{I} = I\{A\} \otimes I\{B\}$ of the combined system will naturally give rise to a conjoint observable $\{c\} = \{e\} \otimes \{f\}$, whose eventualities $\{c_{ij}\}$ are given by the intersection subspaces $\widehat{c}_{ij} = \widehat{e}_i \cap \widehat{f}_j$. The probabilities of these conjoint eventualities will evidently form a matrix with elements given by

$$P_{ij} = P_{[\mathcal{O}]}\{c_{ij}\}. \tag{19.9}$$

The first prerequisite for the desired correlation of $\{e\}$ and $\{f\}$ is that they have the same channel number, $\mathcal{N}_e = \mathcal{N}_f$, i.e. the same number of alternative eventualities, so that the matrix P_{ij} will be square. The final requirement for them to be adequately correlated is that (for a suitable index ordering) the matrix should be more or less diagonal, i.e. that for $i \neq j$ the probability P_{ij} should be zero or very small. (There is an extensive literature on the decoherence processes by which such diagonalization can be brought about [10].)

The conditions of the preceding paragraph are applicable both to classical and quantum systems. In the particular case of ordinary quantum systems, the observables $\{e\}$ and $\{f\}$ will give rise (in the extended Hilbert space $\widehat{I}$) to corresponding sets of projection operators $\widehat{\mathbf{e}}_i$ and $\widehat{\mathbf{f}}_i$ that will automatically commute, $[\widehat{\mathbf{e}}_i, \widehat{\mathbf{f}}_j] = 0$, and whose products,

$$\widehat{\mathbf{c}}_{ij} = \widehat{\mathbf{e}}_i \widehat{\mathbf{f}}_j = \widehat{\mathbf{f}}_i \widehat{\mathbf{e}}_j, \tag{19.10}$$

will be the projection operators specified by the corresponding subspaces $\widehat{e}_i \cap \widehat{f}_j$. This means that (whether it is sufficiently diagonal or not) the probability matrix in Eq. (19.9) will be obtainable from the von Neumann operator $\widehat{\mathbf{P}}$ on $\widehat{I}$ in the form

$$P_{ij} = \mathrm{tr}\{\widehat{\mathbf{P}}\,\widehat{\mathbf{e}}_i \widehat{\mathbf{f}}_j\}. \tag{19.11}$$

It should be remarked that the relation described in the preceding paragraphs is *reflexive*, in the sense that if an observable $\{f\}$ of B is observable by A, then the corresponding observable $\{e\}$ of A will be similarly observable by B. A graphic illustration is provided by the gedanken experiment in which Schrödinger put his cat in a box that was equipped with an anaesthetizing mechanism triggered by a Stern–Gerlach detector. (Schrödinger originally envisaged a lethal mechanism, but that would have conflicted with the Popperian desideratum of repeatability of the experiment.) One way of describing this is to take the detector to be the sensor A, whose reading will tell us about the state of the cat, considered as system B. However, by opening the box, one can see directly whether the cat is still awake, thereby using it as a sensor, A, that will tell us whether the spin measured by the detector, now considered as system B, was up or down. If one also reads the detector as well as opening the box, one can check the validity of the theory: an inconsistency might remind us of the likelihood for the cat to fall asleep spontaneously, with the implication that resort to a less satisfactory probability distribution, with non-vanishing off-diagonal elements, might be more realistic for subsequent repetition of the experiment.

It is commonly convenient to rewrite the expression for a joint probability of the kind given in Eq. (19.9) in terms of the corresponding conditional probability $P_{[i]}\{f_j\}$ for f_j given e_i in the form

$$P_{ij} = P_{[\mathcal{O}]}\{e_i\}P_{[i]}\{f_j\}. \tag{19.12}$$

In the quantum context we are concerned with here, it can be seen that such a conditional probablity for f_i will be given by the prescription whose form is analogous to that of Eq. (19.4), namely

$$P_{[i]}\{f_j\} = \mathrm{tr}\{\widehat{\mathbf{P}}_{[i]}\widehat{\mathbf{f}}_j\}. \tag{19.13}$$

Here, $\widehat{\mathbf{P}}_{[i]}$ is the reduced probability operator associated with the subspace $\widehat{e}_i$, given in terms of the original (unreduced) probability operator $\widehat{\mathbf{P}}$ (on the extended space $\widehat{I}$) by

$$\widehat{\mathbf{P}}_{[i]} = P_i^{-1}\widehat{\mathbf{e}}_i\widehat{\mathbf{P}}\,\widehat{\mathbf{e}}_i. \tag{19.14}$$

This formula automatically ensures that the reduced probability operator has the properties required for qualification as a von Neumann density in its own right, meaning that it is Hermitian with unit trace, i.e.

$$\mathrm{tr}\{\widehat{\mathbf{P}}_{[i]}\} = 1. \tag{19.15}$$

The desideratum that $\{e\}$ should provide an approximate observation of $\{f\}$ is equivalent to the more restrictive requirement that the reduced probability operators should satisfy an approximation of the form $\mathrm{tr}\{\widehat{\mathbf{P}}_{[i]}\widehat{\mathbf{f}}_j\} \approx \delta_{ij}$.

19.6 The subjective nature of a probability operator

The consensus about what is meant (in the abstract sense) in quantum theory by a qualitative or quantitative observable and by a suitably adapted sensor does not extend to the question of what is meant by the occurrence of an actual observation. There is, however, a rather general understanding that it is something that can be performed only by sensors of privileged class for which the title of *observer* is reserved. It would be rather generally agreed, in the context of the example referred to above, that this class would include Schrödinger himself but not his (gedanken) Stern–Gerlach detector. What is more litigious is the status of the cat; would its own discovery that it was still awake count as an actual observation?

Such awkward questions are particularly crucial in the context of what is commonly referred to as the naïve Copenhagen interpretation ('naïve' to distinguish it from other purportedly more sophisticated variants), according to which the von Neumann operator – or the state vector in the pure case – has the status of an objective physical entity that undergoes a (non-unitary) collapse

$$\mathbf{P} \mapsto \mathbf{P}_{[i]} \tag{19.16}$$

to the relevant reduced operator (or reduced state vector in the pure case). This reduced operator is constructed according to the procedure given by Eq. (19.14), when the outcome e_i is actually observed for an observable $\{e\}$.

The problem with this naïve Copenhagen doctrine is how to give a coherent prescription for deciding just when this collapse is supposed to occur. A relativity theorist would object at the outset that a question about when something occurs implicitly refers to the concept of time, a concept that is ultimately elusive and at best dependent on a subjectively arbitrary choice of reference system. However, there also is a more basic problem that arises, even when a reasonably unambiguous Newtonian-type temporal description is available as a good approximation. This would be the case for the cat experiment, even if not in other (e.g. cosmological) contexts.

This more basic problem [6] involves what is known as 'Wigner's friend'. Let us suppose that the friend in question was Schrödinger himself, and that Wigner was interested in the fate of the cat. Wigner would have had no direct access to the Stern–Gerlach detector, but would have been able to telephone Schrödinger to ask what had happened, thus using Schrödinger himself as the sensor, which prior to the opening of the box would have been in a mixed state. As far as Wigner was concerned, the relevant collapse process given by Eq. (19.16) would not have been applicable until the time of the telephone call, whereas from Schrödinger's point of view it would have occurred at the earlier time when the box was opened, while the cat itself would have known even sooner if it had not been put to sleep. One might resolve the discrepancy between Schrödinger's point of view and that of the cat by taking the line (which might be that of a theologian such as Polkinghorne [11]) that the subhuman status of the cat disqualifies it from membership of the privileged class of genuine 'observers'. However, no such specious evasion of the issue is available for the discrepancy between Wigner's point of view and that of Schrödinger, whose equivalent status cannot be so easily denied.

The implication of this well known example is that the naïve Copenhagen interpretation cannot be coherently applied to cases in which several independent (human or other qualified) observers are involved. This means that it can ultimately be acceptable only to a (deliberate or subconscious) solipsist.

The obvious conclusion to be drawn from this is that a probability operator (or state vector in the pure case) should not be thought of as an objective physical entity, and that – as would be agreed even by followers of the Everett doctrine, who refuse such subjectivity – its collapse should not be thought of as a physical process, but as a mathematical step whose application will be appropriate whenever the necessary information, namely the observation of the particular eventuality e_i, becomes available. The operator collapse process described by Eq. (19.16) is thus merely the quantum analogue of the ordinary Bayesian reduction process $P \mapsto P_{[i]}$ for an ordinary classical probability distribution, whereby its *a priori* value is to be replaced by the corresponding *a posteriori* – i.e. conditional – value when the relevant information is supplied. Like the classical probability distributions P and $P_{[i]}$, the corresponding *a priori* and *a posteriori* von Neumann operators $\mathbf{P}$ and $\mathbf{P}_{[i]}$ should be considered to have a status that is not objective but intrinsically subjective.

A corollary of the foregoing conclusion is the anticipation that observers with different personal historical backgrounds should use *different* von Neumann operators, the *a priori* ones being very different and the *a posteriori* less so. However, there will tend to be *a posteriori* agreement when observational information is shared. In discussions of their different opinions about what is appropriate in cosmological contexts, authors such as Hawking and Vilenkin [9] tend to use the definite article for what they call 'the' state of the universe, but the reasoning I am developing here would suggest that such definiteness is unjustifiable, and that the most that is reasonable would be to propose 'an' *a priori* probability operator.

19.7 Everett's concept of branch-channels

Having recognized the incoherence of the naïve Copenhagen interpretation, a newer school of thought, founded by Everett [4], has emphasized – correctly according to the reasoning I am developing here – that there is no physical process of collapse of the probability operator. What is not so clearly correct, or even meaningful, is Everett's concomitant conclusion that all the ensuing 'branches of the universe' remain equally real.

Before the validity of this doctrine can be discussed, it is necessary to explain what is meant by the branches – or more precisely branch-channels – in question. The origin of the idea dates back to before von Neumann, when it was assumed that the relevant probability distribution would be provided by a pure state, as specified by a unit Hilbert space vector that could be represented as a sum, $|\Psi\rangle = \sum_i |\Psi_i\rangle$, of eigenvectors $|\Psi_i\rangle \in e_i$ of the observable $\{e\}$ under consideration. The observation process was commonly described as having a first step consisting of splitting $|\Psi\rangle$ into the set of alternative projections,

$$|\Psi_i\rangle = \mathbf{e}_i |\Psi\rangle, \quad (19.17)$$

onto the relevant eigenspaces. These were described (rather misleadingly) as branches.

According to the naïve Copenhagen doctrine, the observation process would be completed by a second step, consisting of a collapse, whereby the set would be replaced by a single appropriately renormalized branch vector, $P_i^{-1/2}|\Psi_i\rangle$, that would turn up with the corresponding conditional probability $P_i = \langle \Psi_i | \Psi_i \rangle$. On the other hand, the Everett doctrine denied the occurrence of the collapse as a physical process, with the implication that the system would be subsequently describable as being in a mixed state, for which the corresponding von Neumann operator would have the form [12]

$$\mathbf{P} = \sum_i |\Psi_i\rangle\langle\Psi_i|. \quad (19.18)$$

This represents what I shall call the *provisional* probability operator, in order to distinguish it from the relevant (pure) *a priori* probability operator, given by

$$\mathbf{P}_{(0)} = |\Psi\rangle\langle\Psi_i|, \quad (19.19)$$

and whichever *a posteriori* probability operator,

$$\mathbf{P}_{[i]} = P_i^{-1}|\Psi_i\rangle\langle\Psi_i|, \quad (19.20)$$

may turn out to apply.

If the presumption that the system was initially in a pure state is replaced by the more general supposition that it was in an initial state describable by an *a priori* probability operator, $\mathbf{P}_{(0)}$ say, consisting of an arbitrary sum of pure state operators, then – by considering the effect on each member of such a sum – it can be seen that the effect of the first step of the observation process described above will be to provide a provisional probability operator given no longer by the simple formula Eq. (19.18) but by the more general prescription

$$\mathbf{P} = \sum_i P_i \, \mathbf{P}_{[i]}. \tag{19.21}$$

This is known as Luder's rule [13], in which the operators $\mathbf{P}_{[i]}$ are the *a posteriori* probabilities for the relevant output channels, i.e. the relevant eventualities e_i, which are what Everett referred to as 'branches'. In accordance with Eq. (19.14), these *a posteriori* probability operators, and the corresponding probabilities, are given in terms of the *a priori* probability operator $\mathbf{P}_{(0)}$ by

$$\mathbf{P}_{[i]} = P_i^{-1} \mathbf{e}_i \mathbf{P}_{(0)} \mathbf{e}_i, \qquad P_i = \mathrm{tr}\{\mathbf{P}_{(0)} \mathbf{e}_i\}, \tag{19.22}$$

and it is to be noted that they are also recoverable, using expressions of the same form,

$$\mathbf{P}_{[i]} = P_i^{-1} \mathbf{e}_i \mathbf{P} \mathbf{e}_i, \qquad P_i = \mathrm{tr}\{\mathbf{P} \mathbf{e}_i\}, \tag{19.23}$$

from the ensuing provisional probability operator given in Eq. (19.21).

19.8 The deficiency of the Everett interpretation

Before exposing the essential deficiency of the Everett doctrine, I would like to rectify an additional misconception to which it has given rise. In its usual presentation, the use of the term 'branch' is motivated by the notion that the number of relevant channels increases whenever an observation is made. It is important to recognize that this idea – of perpetual multiplication of the relevant number of branch-channels – is, as a general rule, misguided. It is based on the – rarely realistic – presumption that the *a priori* state of the system under consideration is pure, consisting just of a single branch-channel, whereas in a generic case (for the reasons discussed above) $\mathbf{P}_{(0)}$ will already be mixed, involving as many branch-channels as $\mathbf{P}$, so no actual increase occurs. It is thus more appropriate as a metaphor to speak of channels rather than branches, which is why I have chosen, as a compromise, to use the term branch-channel.

In any case, even if the initial state really were pure, the commonly accepted idea that – as more and more information is obtained by successive observations – the number of branches will go on increasing is also unrealistic for another reason, which is that a given finite system cannot continue to acquire information without limit. After a certain amount has been acquired, the system will saturate, so that further information will be able to be taken into account only by a (Landauer-type [14]) process involving the erasure of previously recorded information in order to release the necessary memory space. The number of channels available for useful observation can at most be a small fraction of the number of dimensions needed for a complete physical representation of the sensor, which in practice (if he, she or it is a system constituted from a finite number of molecules with a finite total energy in a finite volume) will, of course, be limited.

Bounded though it must be, the number of branch-channels – meaning the number of eventualities that may be observationally distinguished – in a given (human or other) sensor system can indeed be very large. It is this consideration that has exposed the Everett proposal [4] that all the branches are 'actual, none any more real than the rest' to the criticism [15] that it entails a 'bloated ontology'. However, as I have previously remarked [16], as far as the scientific desideratum of Ockham's razor (meaning economy of formulation) is concerned, it does not matter how extensive or otherwise the ensuing 'ontology' may be.

A more serious reason for dissatisfaction with the Everett doctrine of quantum theory is its failure to apply its own declared rules in a coherent manner, which has made the question of the interpretation of this 'interpretation' the subject of much discussion [5, 17]. The assertion that the branches are 'actual' seems to imply their ontological reality, but Everett's categorical denial that any one is 'more real than the rest' is followed by the Orwellian admission [4] in a subsequent paragraph that 'in order to obtain quantitative results' the branches must be given 'some sort of quantitative measure (weighting)'.

The aim of the Everett programme, as expressed by DeWitt [17], is to construct a theory 'in which it makes sense to talk about the state vector of the whole Universe. This vector never collapses, and hence the Universe as a whole is deterministic'. The troublesome problem [18–20] is how to use such an ultimately deterministic model to obtain the probabilistic predictions that work so well in local applications of quantum theory. As Graham [5] puts it 'Everett attempts to escape from this dilemma by introducing a numerical weight for each world'. The work of Graham and of Hartle [21] has shown that Everett's 'weighting' scheme does successfully reproduce the usual probabilistic predictions, so much so that the distinction between the terms 'weighting' and 'probability' can be seen to be merely semantic.

Changing its name to 'weighting' (or 'propensity', which is another traditionally favoured alternative) does not solve the problem of interpreting the meaning of the 'probability' that is involved.

It is clear that Everett and his followers have so far failed to achieve their declared objective. Their bold attempt to solve the – originally local – interpretation problem by reintroducing determinism at a global level has been helpful for providing a deeper understanding of many of the issues involved, but the question of how much 'reality' should be attributed to the probabilistically 'weighted' branch-channels has nevertheless remained unsolved until now.

My purpose here is to present a recent clarification [3], whereby this issue is not so much decided as transcended. This is in conformity with the precept that questions of ontology are of a theological nature that is beyond the scope of ordinary science (whose modest ambition is to account for appearances, and not for ultimate reality, whatever that may mean). The anthropic approach described below provides a framework in which an intellectually coherent interpretation can be provided in a manner that leaves plenty of scope for adjustment, and that is compatible not only with an (unbloated) 'oriental' option, in which hardly any of the relevant branches need be considered to be 'real', but also with a (scientifically indistinguishable but theologically very different) 'occidental' option, in which they might all be describable as 'actual'.

19.9 The side issue of the provisional distribution

Whereas zealous adherents of the Everett doctrine – and *a fortiori* of the naïve version of the Copenhagen interpretation that was discussed above – would have it that some sort of objective reality can be attributed to the state vector on a sufficiently large scale, and hence to the probability operator that would be relevant on a more local scale, most other schools of thought, including less naïve versions of the dualistic Copenhagen interpretation, would concur with the supposition adopted here to the effect that such entities are essentially of a subjective nature. This contrasts with the status of the Hilbert space operator algebra of eventualities and observables, which have a more objectively well defined nature. According to this principle, the amplitudes (and corresponding 'weightings') of Everett-type branches should be considered as ultimately subjective, whereas the branches themselves can be considered to be objective. This does not, of course, entail that such mathematical structures are 'real' in any ontological sense.

Before leaving the subject of the 'branching' process (misnamed because the number of branches involved in the description of a subsystem need not increase, and might even decrease, when an interaction occurs), it is worth commenting further on the nature of the process whereby an *a priori* probability operator $\mathbf{P}_{(0)}$ is replaced by the corresponding provisional probability operator $\mathbf{P}$ given by Eqs. (19.21) and (19.22). The original discussions of this process were formulated in terms of what Dirac [7] referred to as the Schrödinger picture, wherein states are considered to have a time dependence in which the evolution from an initial time $t_{(0)}$ to a later time t is given by an operator transformation $\mathbf{P}_{(0)} \mapsto \mathbf{P}$ that will be given, in the special case of a pure state for an isolated system, by a corresponding vector transformation $|\Psi_{(0)}\rangle \mapsto |\Psi\rangle$. In the special case of an isolated system, such a transformation will be given by a unitary operator $\mathbf{U}$ (that is continuously generated by some Hermitian Hamiltonian) according to prescriptions of the standard form $|\Psi\rangle = \mathbf{U}|\Psi_{(0)}\rangle$ and $\mathbf{P} = \mathbf{U}\mathbf{P}_{(0)}\mathbf{U}^{-1}$. However, the transformation will in general be of a less simple (non-unitary) type when interaction with an external system is involved. The idea, as discussed by von Neumann, was that the preparation of an actual experimental observation should involve an arrangement whereby a transformation of this latter (non-unitary) type produced a provisional probability operator $\mathbf{P}$ of the required form. This is given by the Luder formula, Eq. (19.21).

As originally pointed out by Dirac [7], a representation in terms of such a Schrödinger picture can be translated into an equivalent representation in terms of the kind of Heisenberg picture that has been implicitly adopted throughout the present discussion. In this kind of representation, the relevant state vector $|\Psi\rangle$ or probability operator $\mathbf{P}$ is considered to be time-independent, and the effect of Schrödinger-type time translations is allowed for by corresponding transformations of the relevant observables and their constituent eventualities. In the special case of an isolated system, these transformations will be of the standard unitary type, so that, for example, if $\mathbf{e}_{(0)}$ is the projection operator corresponding to some particular eventuality at a time $t_{(0)}$, then the corresponding time-transposed eventuality at a later time t will be given by

$$\mathbf{e} = \mathbf{U}^{-1}\mathbf{e}_{(0)}\mathbf{U}. \tag{19.24}$$

The essential advantage of using a picture of this kind is that there is no impediment to its extension to (general relativistic and other) applications, for which no globally well defined Newtonian-type time parametrization may be available, so that the concept of a time translation relation of the form $\mathbf{e}_{(0)} \mapsto \mathbf{e}$ might make sense only for very particular locally related eventualities.

As seen from this Heisenberg (as opposed to Schrödinger) point of view, the process of preparation of an experimental observation in the manner prescribed by von Neumann should be thought of not as the replacement of an *a priori* probability operator $\mathbf{P}_{(0)}$ by a different provisional probability operator $\mathbf{P}$, but as the replacement of an initially envisaged (perhaps maladapted) observable $\{\mathbf{e}_{(0)}\}$ by an appropriately adjusted observable $\{\mathbf{e}\}$, with respect to which the probability distribution already has the required Luderian form.

From this point of view, there is no need to bother about any distinction between *a priori* and provisional probability operators (which – in view of the possibility of using Eq. (19.23) instead of Eq. (19.22) – were, in any case, equivalent for the practical observational purpose under consideration). What matters for the purpose of making what von Neumann would consider to be a satisfactory observation is the choice of a suitably adjusted observable $\mathbf{e}$. However, the main point I wish to emphasize at this stage is that – although it may be of technical interest in particular applications – the importance of the issue of obtaining a satisfactory observation in the sense specified by Luder's rule has been greatly exaggerated, in so far as its relevance to the ultimate interpretation of the observations process is concerned. To start with, there is the consideration that the Luderian desideratum is obtainable not only by the non-trivial process described above, whereby $\{\mathbf{e}\}$ is adjusted to a previously chosen probability operator, but also by the trivial process whereby the subjective *a priori* choice of $\mathbf{P}$ is adjusted *ad hoc* to fit a prescribed observable $\{\mathbf{e}\}$, an adjustment that in no way diminishes the credibility of its implications, as can be seen from the equivalence of the prescriptions given in Eqs. (19.22) and (19.23).

A more fundamental reason why the question of the Luderian transition is irrelevant is that, when an observation has been actually carried out (not merely planned), one will be left just with a single confirmed eventuality e_i. Such a single eventuality might be incorporated with others to constitute a complete observable set (spanning the entire Hilbert space) in many different ways, whose substitution in the Luder formula, Eq. (19.21), would provide many different results. Nevertheless, however that might be, and regardless of any distinction that may or may not have been made between an *a priori* probability distribution $\mathbf{P}_{(0)}$ and a provisional probability $\mathbf{P}$, one will be left with an unambiguously specified *a posteriori* probability distribution $\mathbf{P}_{[i]}$, which is all that matters for the purpose of subsequent predictions one may wish to make.

The upshot is that someone (such as Wigner) concerned about Schrödinger's cat should use the *a posteriori* distribution when the relevant information has become available, and until then should just continue to use the ordinary *a priori* distribution. One should avoid getting sidetracked (as so many of Everett's followers have been) by intermediate Luderian technicalities, whose analysis is of little relevance to the two outstanding issues that remain. In addition to the question of interpretation, which will be addressed from an anthropic point of view below, the other outstanding issue is the usual practical Bayesian dilemma of how to decide quantitatively what *a priori* distribution should be used in a particular context – something that can sometimes be resolved just by symmetry considerations (as in the coin-tossing example described above).

19.10 Perceptions and perceptibles

An important idea that was latent in much of the preceding discussion is that some privileged eventualities and observables are more naturally significant than others. In the discussion of Luder's rule, it was remarked that this can be interpreted as selecting a privileged class of observables, but I should emphasize before continuing that privilege of that kind is not what I am concerned with here, because it is ultimately dependent on an arbitrary subjective choice of the relevant *a priori* probability distribution. The kind of privilege I am concerned with is something that depends on the essential nature of the system under consideration, in a manner that is independent of the choice of the probability distribution. This is something that could be said about Bohm's idea [13] of privileging position with respect to its dynamical conjugate, namely momentum, but that particular choice is something that would not seem very natural to the numerous physicists whose mental life is based in Fourier space.

The kind of privilege that seems to me more relevant for the interpretation question is something that would be rather generally recognized as being imposed by the circumstances in particular cases. It is exemplified most simply by the existence of a privileged choice (determined by the background magnetic field) for the the particular spin eventualities characterized as 'up' and 'down' in the Stern–Gerlach experiment discussed above. It is also exemplified by many familiar kinds of apparatus, such as can be found in scientific laboratories, and increasingly in ordinary homes, whose output is typically presented in terms of what – at the highest resolution – usually turns out to consist of simple integer valued observables, such as the alternative eventualities in the range from 0 to 9 for a digit in a counter output, or the binary alternatives for a particular pixel on a screen to be 'on' or 'off'. It is mathematically possible to use other bases for a Hilbert space description of such systems, for example by working with eventualities defined as linear superpositions of 'on' and 'off' states of screen pixels, but that is evidently not the kind of treatment for which such an apparatus was intended by its designer.

Although the degree of complexity of the systems involved is very different, it seems to me that there is a rather strong analogy between the special role of the 'on' and 'off' states for a pixel on a screen and the 'awake' and 'sleeping' states of Schrödinger's cat. The privileged status of the particular eventualities in question can be accounted for as the result of a process of design that is attributable in the first case, not just to an individual engineer, but to the collective activity of a scientific community, while it is attributable in the second case to a very long history of biological evolution by Darwinian selection. Having said this about the cat, the next thing to be said is that the same applies to Schrödinger and Wigner, for whom the relevant privileged eventualities are states of mind corresponding to the realization of whether or not the cat is awake.

Whatever doubts we may have about the status of the cat, we must recognize that Schrödinger and Wigner are closely analogous to ourselves (i.e. the author and presumed readers of this essay), which means that insight into the working of their minds can be obtained from our own experience. The only eventualities about whose reality we can be sure are the conscious perceptions in our own minds (of which some, namely those occurring in dreams, are evidently uncorrelated with anything outside). These correspond to the 'mind states' whose essential role has been recognized by authors such as Donald [22], Lockwood [23] and particularly Page [24], whose line of approach is followed here. It seems reasonable to postulate the validity of Page's principle, according to which conscious perceptions are the only eventualities that can be considered actually to happen. It also seems reasonable to make the concomitant postulate that these perceptions must belong to some restricted class of privileged eventualities of the kind discussed in the preceding paragraph. I shall refer to the eventualities of this subclass as perceptibles.

In his 'sensible quantum theory' [24], Page has attributed a privileged role to a class of observables that he refers to as 'awareness operators', which I interpret to mean observables whose individual constituent eventualities are the perceptibles introduced in the previous paragraph. Page has used these particular operators to develop a refined version of the Everett interpretation, in which the branches – or as I would prefer to say, channels – that matter are specified with respect to these awareness operators. Thus, whereas Everett's original version might attribute 'actuality' to branches defined with respect to observables of a rather arbitrary kind, Page's more refined version would attribute 'actuality' only to branches of an appropriately restricted kind, namely the channels that are specified by perceptibles. Having thus provided a much clearer idea of which channels are actually needed, Page was still left with the problem of interpreting what, following Everett's evasive example, he referred to as their 'weighting'. The point at which Everett stumbled was in trying to reconcile his recognition that the weighting was needed with his preceding claim that all the branches were equally real. Page came up against the same problem with respect to the claim to the effect that all the perceptibles are actually perceived.

19.11 The anthropic abstraction

A corresponding paradox is reached from a rather different angle in the approach I am developing here, which is in agreement with that of Page [24] in so far as the special role of perceptions is concerned, but differs in affirming that the weighting in question must be considered to have an essentially subjective and probabilistic nature. The intrinsically probabilistic nature of models of the kind advocated here raises the problem of what it can mean to attach a probability to the actuality of an eventuality in the mind of someone else if the only events one can actually observe are those occurring in one's own mind.

Before presenting what I think is the only acceptable way of dealing with this paradoxical problem, I would mention two less satisfactory ways of resolving the issue that have been suggested in the past. The first way is that of the solipsist, who would deny the existence of any conscious perceptions other than his (or her) own [20], with the implication that the apparent analogy between oneself and others (such as Schrödinger) is merely a superficial illusion. The second way –which (unlike that of the solipsist) has been followed up by many physicists, starting with de Broglie – is to revert to a deterministic description of the world, providing a theoretically well defined answer to the question of what really happens by denying the (experimentally well established) validity of the essentially probabilistic description provided by orthodox quantum theory. Neither the first nor the second of these ways can be said to resolve the paradox; they merely evade the issue by dropping one or other of the essential (experimentally motivated) elements of the problem, which is that of providing an inherently probabilistic treatment of perceived reality that respects the apparent symmetry between different people.

A historical analogy is provided by the incompatibility between Maxwellian electromagnetism and Newtonian gravity, which was ultimately resolved by their unification in Einstein's General Relativity. The problem to be dealt with here is that of reconciling subjective probability with objective reality. The only way that I know of solving this problem in a satisfactory manner is the anthropic approach, which faces the issue head on [3] without denying the validity of the considerations that lead to the paradox.

It is worth emphasizing, by the way, that the problem is not specifically a problem of quantum theory, but also arises in probabilistic versions of classical theory, as was recognized, I suspect, by many of those who were hostile to anything associated with the name of Bayes. The importance in this context of the quantum revolution is that it changed the status of Bayesian theorists from that of radicals (because they were willing to abandon determinism) to that of reactionaries (because they continued to use old fashioned Boolean logic).

The situation, as I understand it, is as follows. Suppose that to describe a system that includes ourselves (but, for the sake of finiteness, perhaps not the whole of the Universe) we have set up some (classical or quantum) theory that provides probabilities for an extensive class of eventualities. This class includes a specially privileged subclass of eventualities that I shall refer to as perceptibles, which are the only ones that can be actually observed as conscious perceptions. The set of such perceptions (not just yours and mine, but also those of everyone else) can be described as objective, and it is the only thing in the theory that can be considered to be real

We thus have an objective model attributing probabilities to perceptibles. But what sense can it make to attribute a probability to an observation you cannot make? If you are Wigner, what sense can it make – even in a classical theory – to use an objective distribution attributing probability to something that can only be known by Schrödinger? The contradiction arises when Schrödinger makes the Bayesian transition to the relevant *a posteriori* distribution, while Wigner continues, for the time being, to use the *a priori* distribution. How in these conditions can either of these distributions be considered to be objective?

The resolution to this paradox is provided by what may be called the anthropic abstraction (so called because it underlies that which I designated – perhaps inaptly – as the anthropic principle [25]). The paradox that arises in this case (as in many others) can be attributed to an unnecessary assumption that has been consciously or subconsciously taken for granted. The unnecessary assumption is that of knowing in advance who one is. The *anthropic abstraction* consists in refraining from assuming in advance that one has the identity of some particular sensorial observer in the model, so that one's status *a priori* is that of what I shall refer to as an abstract *perceptor*. It is not until the actual happening of the perception that one can know whether one is Schrödinger, Wigner or whoever else may be included in the model.

It is, of course, to be understood that the perceptible eventualities that are involved in this anthropic approach cannot just be of the elementary type exemplified by the observation that someone else is awake, but that they need to include eventualities of the more complicated kind known as *consistent histories* [26]. The sort of eventuality that needs to be envisaged is not simply that of finding oneself to be Schrödinger, but that of finding oneself to be Schrödinger at a particular instant in his life, with all the memories he would have had at that moment.

The use (which I see no satisfactory way of avoiding without reverting to determinism) of the anthropic abstraction entails the need to adopt some kind of anthropic principle, by which I mean some kind of prescription for attributing appropriate probabilities to the relevant perceptible eventualities. The rather crude kind of anthropic principle that I have put forward on previous occasions [2] was concerned with the attribution of probability to entire observer systems (such as those associated with the names of Schrödinger or Wigner) without getting into the details of particular moments in their lives. For the applications I was then considering, it was sufficient to use a crude statistical treatment attributing equal weight to all terrestrial or extraterrestrial observers who can be considered to be sufficiently like ourselves to be describable as 'anthropic'. However – as several authors have already remarked [24, 25, 27] – the more detailed applications I have been considering here (particularly those involving quantum effects) require the use of a more refined kind of anthropic principle [3], which will distinguish not just between anthropic individuals, but also between different instants in the lives of such individuals.

The question that naturally arises at this point is whether it can suffice to use just the probability weightings that are directly provided by orthodox quantum theory (such as has been discussed above), in conjunction with some prescription for deciding which of the many mathematically defined eventualities in the model should be considered to have the privileged status of perceptibility.

19.12 Uniqueness of the perceptor?

In the subsequent subsections I shall address the scientifically important question of the attribution of the required anthropic probability. However, before doing so, I would like to digress by mentioning another question of a less scientific nature that might become a subject of philosophical discussion in the future.

This is the question of the nature of what I have referred to as a perceptor, whose actual perceptions are the only entities within the model that are considered to be real (which is not to deny the reality, in some theological sense, of other entities beyond the scope of the model). The perceptor acquires an *a posteriori* identity (of an ephemeral nature) as a material observer (such as Schrödinger) on the occasion of an actual perception, but what about the immaterial identity the perceptor might have *a priori*?

Is the perceptor unique? The notion that all anthropic observers might just be avatars of a single perceptor will not seem strange to anyone familiar with oriental (Hindu or Buddhist) religious tradition. (A scientific analogy that comes to mind is Feynman's idea that the Universe is inhabited only by a single electron, which is able to follow all the worldlines that we usually attribute to distinct electrons by also following – but in a time-reversed sense – the worldlines that we attribute to positrons.) The obvious Wheelerian epithet for the succinct encapsulation of this idea – namely that we all share the same abstract identity – is *solipsism without solipsism*.

The postulate of a unique perceptor has the advantage of being particularly economical in the sense required by Ockham's razor. Nevertheless, in the framework of the occidental (Judaeo–Christian–Islamic) religious tradition, it might seem more natural to suppose that there are many distinct perceptors. What is not permissible, however tempting it may seem, is to suppose that distinct perceptors are correlated with distinct anthropic observers, such as Schrödinger and Wigner; the essence of the anthropic abstraction is that a perceptor has the potential for actualization in any observer state that has a non-zero probability amplitude. The only way you, as a material observer, can claim an exclusive monopoly of the potential for actualization of your own perceptor is by adopting an *a priori* probability distribution that attributes no weight to anyone other than yourself, in other words by adopting the (unacceptable) autocentric attitude describable as solipsism *with* solipsism.

For someone whose objection to the Everett doctrine was based not on its failure to follow its own declared rules, but on the ontological bloating [15] implicit in the many-universes doctrine, the present idea that one might adopt a many-perceptor doctrine might be felt to be even worse. Whereas the number of Everett branch-channels is restricted, as I have remarked above, by the limited information content for any finite system, there is no limitation at all on the number of distinct perceptors that might be conceived to exist, and that might all have a chance of undergoing the experience of being Schrödinger at some moment in their lives.

The idea that there might be an unlimited number of distinct perceptors may be abhorrent to anyone for whom ontological economy is a desideratum, but on the other hand it might be extremely attractive to those who still hanker after determinism. Indeed, for those who consider that – in order to be meaningful – the concept of probability must be defined in terms of frequencies of the outcome of many identical performances of the same experiment, the many-perceptor doctrine can provide what is desired. If the number of perceptors is vastly larger than the number of anthropic observers in the model, then each observer state (even those that are relatively improbable) would actually be perceived by a large number (albeit a small fraction) of the perceptors. This would provide the desired frequency interpretation for the probability distribution. By using the anthropic abstraction in this ontologically uninhibited manner, it is at last possible to deliver what the Everett programme sought, which may be epitomized as *probability without probability*.

Multiplication of the number of sensors is not the only way of obtaining probability without probability, if that is what is desired. Another number whose magnification can achieve the same result is the number of perceptions that each particular perceptor is allowed to make. The supposition that there are a large number of perceptors, each allowed to make only a small number of perceptions or possibly only one, is ontologically equivalent to the supposition that there is just a single perceptor who is allowed to make a large number of perceptions. As far as ontology is concerned, all that counts is the total number of perceptions.

Whether – as in the oriental version of the anthropic interpretation – there is a unique perceptor, or whether – as in the occidental version – the number of perceptors is large (even compared with the number of anthropic observers) – is an issue that belongs to the realm of theology rather than science. The same can be said about the (more ontologically relevant) number of total perceptions, which may seem important to those who believe in probability only when formulated in terms of frequencies, but which in no way affects the way the theory is actually applied in practice. All that matters for scientific purposes is the relative probablity distribution for the perceptions, which will now be discussed.

19.13 Anthropic weighting: the proper ansatz?

On the basis of the discussion in the preceding sections, it seems reasonable to suppose that, from the point of view of a perceptor, the 'net' probability P of a particular perception e_i within a particular subsystem (representing the part of the Universe under consideration) should be given by an expression of the form

$$P\{e_i\} = \mathcal{P}_e P_{[\mathcal{O}]}\{e_i\}. \tag{19.25}$$

Here, $P_{[\mathcal{O}]}\{e_i\}$ is the ordinary 'gross' classical or quantum mechanical probability (as calculated in the manner described above) for the particular perceptible eventuality e_i to occur on the occasion when the relevant Page-type awareness observable $\{e\}$ is actually observed, while $\mathcal{P}_e$ is the anthropic factor giving the probability for the perception to belong to that particular observable set. A sensor of the familiar macroscopic but localized kind – exemplified by an ordinary computer or a human observer – will be characterizable by a fairly well defined worldline with a proper time parametrization τ, in terms of which the anthropic probability factor will be expressible as

$$\mathcal{P}_e = \dot{\mathcal{P}} \Delta_e \tau, \tag{19.26}$$

where $\Delta_e \tau$ is the relevant proper time duration and $\dot{\mathcal{P}}$ is a corresponding probability rate, whose integral,

$$\mathcal{P} = \int \dot{\mathcal{P}} \, d\tau, \tag{19.27}$$

will be interpretable as the total probability for the perception to occur at some stage in the life of that particular observer.

Whereas the conditional probability – designated by P in Eq. (19.25) – is of the ordinary kind that is provided by the relevant classical or quantum physical theory for the system under consideration, the anthropic probability factor – designated by $\mathcal{P}$ (which is also conditional in so much as it is restricted to that particular system within the Universe) – can only be provided by what I call an *anthropic principle*.

In my earlier discussions of applications that were not concerned with discrimination between individuals, but with averages over entire populations [2], it was good enough to suppose that – provided they were sufficiently similar to ourselves (motivating the rather debatable choice of the term anthropic) – the relevant total probability $\mathcal{P}$ per observer could be taken to be the *same* for each one. This is in accordance with what Vilenkin has referred to as a postulate of *mediocrity* and what I would refer to as a

19 Micro-anthropic principle for quantum theory

postulate of *approximate symmetry*. The application of such a mediocrity postulate in the present context gave rise to what I call the *weak anthropic principle*, whose purport is that the anthropic probability factor should take a fixed value:

$$\mathcal{P} = \frac{1}{N}, \tag{19.28}$$

where N is the number of anthropic observers that come into existence within the system under consideration. So, if the system were scaled up to include a larger chunk of the Universe, with a larger population number N, then the value of $\mathcal{P}$ would be correspondingly scaled down.

The ordinary (weak) anthropic principle formulated in the preceding paragraph will evidently not be enough for more detailed purposes, such as comparison of the probability of finding oneself to be someone very short lived (as in the case of a child that dies in infancy) with that of finding oneself to be someone more long lived (as in the case of a normal adult). For such a purpose, the most naïve possibility is to adopt the ansatz which I would call the *proper* anthropic principle, meaning the postulate of a fixed universal value for the anthropic probability rate $\dot{\mathcal{P}}$. This would be given numerically by

$$\dot{\mathcal{P}} = \frac{1}{\langle \tau \rangle N}, \tag{19.29}$$

where $\langle \tau \rangle$ is the average total proper lifetime of an anthropic observer in the system. In so far as the total probability over the total lifetime τ of an observer is concerned, adoption of the proper anthropic principle, represented by Eq. (19.29), evidently entails that Eq. (19.28) should be replaced by

$$\mathcal{P} = \frac{\tau}{\langle \tau \rangle N}. \tag{19.30}$$

The foregoing refinement of the original anthropic principle, represented by Eq. (19.28), should be good enough for a wide range of applications. However, for the purpose of comparing observers of very different kinds (for which the qualification anthropic might not be so appropriate), such as extraterrestrials and cats, not to mention babies in our own species, the plausibility of Eq. (19.29) is much less obvious.

19.14 Micro-anthropic principle: the entropic ansatz

A hint toward a more plausible (though not so easily applicable) alternative is discernible in the response to the eschatological problem posed by Islam [28] that was provided by Dyson, who suggested [29] that what really matters is not the proper time duration of an interval, but how much information is effectively processed therein. There is, of course, room for discussion about how to quantify what is effectively processed (as opposed to what is merely stored in a memory), even in the case of an ordinary computer and hence much more so in the case of a feline or human mind. Estimating that the duration of a human 'moment of consciousness', which presumably corresponds to what is denoted here by $\Delta_e\tau$, has the same order of magnitude as supposed in the more recent work of Page [24], namely a significant fraction of a second, Dyson deduced (from the fact that the heat production of an entire human body is typically about 200 W at a temperature of 300 K) that the corresponding entropy production Q_e is of the order of 10^{23} bits. However, experience with the analogous problem for computers indicates that the amount of information S_e that can be judged to have been effectively processed by the mind itself during the corresponding period of perception – and the associated Landauer entropy production [14] – must have a vastly smaller value $S_e \ll Q_e$ that is not so easy to evaluate.

A plausible prescription for the evaluation of the processed information S_e will, however, be available if we have a sufficiently detailed (quantum not just classical) theory to characterize the Hilbert space projection operator $\mathbf{e}_i$ corresponding to a particular perception e_i under consideration. If we suppose that this particular perception belongs to a complete set of eventualities having the same rank (i.e. subspace dimension) $\mathcal{R}_e = \mathrm{tr}\{\mathbf{e}_i\}$ constituting an observable $\{\mathbf{e}\}$ in a Hilbert space of dimension $\mathcal{N} = \mathrm{tr}\{\mathbf{I}\}$, so that the corresponding number of Everett-type branch channels is given by $\mathcal{N}_e = \mathcal{N}/\mathcal{R}_e$, then the associated information capacity will be given by

$$S_e = \log\{\mathcal{N}_e\} = \log\{\mathrm{tr}\{\mathbf{I}\}\} - \log\{\mathrm{tr}\{\mathbf{e}_i\}\}, \qquad (19.31)$$

using a logarithm with base 2 if one wants to use Shannon's bit units, or using a natural logarithm if one wants to use the entropy units preferred by physicists. This information capacity represents the maximum amount of information that can be given – for a probability distribution P_i ($i = 1, \ldots, \mathcal{N}_e$) – by Shannon's formula $S = -\sum_i P_i \log\{P_i\}$.

What I would propose is that Eq. (19.31) be used as an estimate of the amount of information that can be considered to be processed during the perception e_i, and that the corresponding anthropic probability should be postulated to be proportional to this, i.e. the required factor in Eq. (19.25) should be taken to be given by

$$\mathcal{P}_e = \alpha S_e, \qquad (19.32)$$

where α is a fixed proportionality factor that is chosen so as to ensure the usual requirement that the total probability (over all the relevant worldlines) should add up to unity. According to this *micro-anthropic* principle – which might appropriately be described by the term *entropic* principle – the probability rate will not have a fixed value (as was postulated by the proper anthropic principle formulated above) but will be given by

$$\dot{\mathcal{P}} = \alpha \frac{S_e}{\Delta_e \tau}. \tag{19.33}$$

The advantage of using the term 'entropic principle' for this ansatz is that it emphasizes its virtue of being applicable in principle not just to observers qualifiable as anthropic, in the sense of being sufficiently similar to ordinary adult humans, but also to very different kinds, ranging from such familiar examples as babies and cats to the highly exotic extraterrestrial observers whose survival at extremely low temperatures was envisaged by Dyson [29]. A rather obvious application of this entropic principle is its use as evidence against Dyson's conjecture that civilizations constituted by observers capable of surviving at the extremely low temperatures predicted for a non-compact universe in the distant future would be able to survive indefinitely with respect to not just proper time, but also any relevant information-processing measure. If this conjecture were correct, it would mean that the probability measure defined according to Eq. (19.33) by the entropic principle would diverge toward the future. The contrary prediction by Islam [28] that 'it is unlikely that civilization in any form can survive indefinitely' is therefore overwhelmingly favoured by the fact that we do not observe ourselves to be incarnated in asymptotically viable low temperature life-forms (if such can exist) but in carbon-based life-forms adapted to (cosmologically ephemeral) conditions of moderate temperature.

It must be emphasized that the preceding argument against the likelihood of long-term survival is entirely dependent on the acceptance of the kind of *a priori* probability distribution proposed (as a matter of choice, not merely as a tautology) by the anthropic principle and its entropic extension. Dyson's writings in this and other analogous contexts – notably that of the prospects for our own terrestrial civilization [27] – give the impression that he personally prefers an *a priori* probability distribution of the traditional kind, based on what I would refer to as an autocentric (or pre-ordination) principle, to the effect that the attribution of non-zero weighting should be restricted retroactively to wherever one already finds oneself to be. Although it may be logically admissible as an alternative to principles of the anthropic

Table 19.1.

	e_1: C awake	e_2: C asleep	M: gross → net
e_3: M awake, sees C awake	$P_{31} = 1/4$	$P_{32} = 0$	$1/4 \to 1/6$
e_4: M awake, sees C asleep	$P_{41} = 0$	$P_{42} = 1/4$	$1/4 \to 1/6$
e_5: M dreams C awake	$P_{51} = 1/8$	$P_{53} = 1/8$	$1/4 \to 1/6$
e_6: M dreams C asleep	$P_{61} = 1/8$	$P_{62} = 1/8$	$1/4 \to 1/6$
C: gross → net	$1/2 \to 1/6$	$1/2 \to 1/6$	

kind, I would maintain that such an autocentric attitude is scientifically unreasonable, in so much as it violates the desideratum that comparable observers be treated objectively on the same footing. By adopting such an attitude [30], Dyson implicitly assumes for himself a privileged position to which other observers (such as Wigner and Schrödinger) are not admitted.

Before leaving the subject of logically admissible (even if not scientifically reasonable) alternatives to principles of the anthropic kind, I would mention a conceptually possible alternative that is quite the opposite of the autocentric deviation described in the previous paragraph. Instead of prescribing an *a priori* probability distribution with weighting restricted to material observers as in the anthropic case (or to a single privileged observer in the autocentric case), one might go so far as to envisage the attribution of non-zero weighting even to situations where no material observer is present at all. Such an unreasonably overextended weighting (as exemplified by the kind of ubiquity principle that was implicit in Dirac's original argument in favour of his now discredited theory [2, 31] of varying gravitational coupling) might be logical if one could imagine oneself as some sort of disembodied spirit, but (as Dirac's example shows) it cannot be trusted for scientific purposes.

As a toy example to illustrate the application of this micro-anthropic principle, consider a gedanken experiment in which Schrödinger's cat, C, has an equal chance of being awake or dreaming, as also does its master, M, who – if awake – can see whether the cat is too, but – if asleep – has equal chance of dreaming that the cat is awake or asleep, whether or not it actually is. The cat is unconcerned about its master, and so has only two relevant mind states, e_1 or e_2 (awake or dreaming) with entropy $\mathcal{S} = \log 2 = 1$. Schrödinger has four relevant mind states, e_3, e_4, e_5 or e_6 with $\mathcal{S} = \log 4 = 2$, so his net probability is 2/3, while the cat's is 1/3. The conditional 'gross' probability P and absolute 'net' probability P for the relevant eventualities are given in Table 19.1.

19.15 Local application

Whereas the term *entropic* principle has the advantage of avoiding any risk of misunderstanding that the range of applicability of the ansatz in Eq. (19.32) extends beyond observers of narrowly anthropic type, the alternative term *micro-anthropic* principle has the advantage of advertising the applicability of the principle (as of its *proper* predecessor defined by Eq. (19.29)) not just to the entire life of an observer, but to particular parts thereof. The question of whether one is more likely to find oneself nearer the beginning or end of one's life was raised in an epilogue by Leslie [27]. He suggested, on the basis of Everett's own (confusing) presentation of his doctrine [4], that the continual multiplication of the number $\mathcal{N}_e$ of relevant branches entailed a probability distribution that would be heavily biased towards the last moments of life. This assumes that the dogma that all the branch-channels are equally 'real' implies that the corresponding anthropic probability factor should be given by $\mathcal{P}_e \propto \mathcal{N}_e$, rather than by an expression of the entropic form $\mathcal{P}_e \propto \log\{\mathcal{N}_e\}$ that has been advocated here. Having safely survived, and thereby invalidated this alarming prediction, Leslie arrived at the observational conclusion – as argued on purely theoretical grounds at the beginning of this chapter – that this particular interpretation of the Everett doctrine is untenable.

According to the present analysis, the correct answer to Leslie's question is as follows. To start with, it is necessary to reject not only Everett's claim that the relevant branches are 'real' (which might be interpreted as meaning $\mathcal{P}_e \propto \mathcal{N}_e$), but also his attribution to them of an ordinary non-anthropic quantum probability weighting (which might be interpreted as implying the choice of a constant value for $\mathcal{P}_e$). This contradiction between Everett's preaching and practice is resolved in the present approach by what is interpretable as a compromise, according to which the appropriate formula has the logarithmic form $\mathcal{P}_e \propto \log\{\mathcal{N}_e\}$.

The replacement of a linear by a logarithmic dependence merely moderates, but does not avoid, the unrealistic implication that the probability distribution would strongly disfavour the earlier stages of a lifetime if Everett's branching metaphor were to be taken literally. It is therefore obvious that this aspect of what is commonly understood to be Everett's interpretation is also misleading and, as remarked above, it is very easy to see why. The idea of a rapidly increasing number of relevant branch-channels is something that may make sense in the case when, for example, one has just taken delivery of a new computer with entirely empty memory banks, but it will soon cease to be valid when saturation sets in, so that erasure becomes necessary to release occupied space by converting the relevant information into Landauer entropy [14]. Concerning the human case, parents and primary school teachers know that even small children do a lot of forgetting as well as learning, while – as adulthood progresses – the ratio of what is learnt to what is forgotten goes on decreasing, so that it may ultimately become quite small as senility sets in. This means that the relevant number $\mathcal{N}_e$ of Everett branch-channels should normally reach a maximum – not a peak but a plateau – in midlife. It is to be understood that this statement refers to a smoothed average over diurnal variations, because the number of channels involved in conscious perception presumably undergoes considerable reduction during sleep, particularly during deep dreamless phases.

For practical probabilistic purposes, it is only relative values that matter. The intrinsically interesting question of the absolute height of the plateau is beyond the scope of the present investigation, but it is evident from physical considerations that $\mathcal{N}_e$ cannot be nearly as large as the (admittedly gigantic) value of $\exp\{Q_e\}$, where Q_e is the Dyson entropy number discussed above. This exceeds the corresponding Landauer entropy $S_e = \log\{\mathcal{N}_e\}$ (representing the amount of useful information processed during the perception [14]) by an enormous thermodynamical waste factor $W_e = Q_e/S_e \gg 1$. (In the days before valves were replaced by transistors, the relevant waste factors for computers were far worse even than those of their biological analogues, but the spectacular progress of engineering techniques in recent years has brought about an amazing rate of improvement.)

It is a noteworthy coincidence that Dyson's evaluation of Q_e in the human case [29] gave a value of the same order as the Avogadro number, which is interpretable as the number of molecules in a fraction of the order of 10^{-3} of the mass of a human body. If it is supposed that this fraction is comparable with the fraction of molecules that are metabolically active, then it can be deduced that the corresponding metabolic turnover time must be comparable to the mental time interval $\Delta_e\tau$, of the order of a fraction of a second, that was used in Dyson's evaluation.

This observation – that the estimated duration $\Delta_e\tau$ of a conscious perception is roughly comparable with a timescale characterizing metabolic processes throughout the body – may offer a significant clue as to the nature of the (still largely mysterious) mental processes involved. One thing that is clear is that the relevant value of $\Delta_e\tau$ can undergo considerable variation – lengthening, for example, in states of hibernation. In so far as the solution to the problem posed by Leslie is concerned, what is relevant is the age dependence of $\Delta_e\tau$. My impression, with which I think most people would agree, is that the typical duration of a moment of consciousness is relatively

19 Micro-anthropic principle for quantum theory

Fig. 19.1. Some of the participants at the Clifford Centennial meeting organized by John Wheeler at Princeton in February 1970, assembling many of the people whose thoughts contributed to the synthesis presented here. In the front on the left are Bob Dicke with Eugene Wigner, in the centre are Stephen Hawking with the author, and behind them are Bryce DeWitt with Freeman Dyson.

short in early childhood and that, on average – modulo diurnal fluctuations through states of shallow or deep sleep – it increases monotonically throughout life. According to Eq. (19.33), this means that the maximum of the anthropic probability distribution need not coincide with the summit of the midlife plateau where the relevant branch-channel number $\mathcal{N}_e$ and its logarithm S_e are highest, but may actually occur at a more youthful stage.

19.16 Conclusions

The question of unicity or multiplicity (as posed by the title of this volume) arises at several levels in the approach developed here. One level (as discussed in Section 19.12) concerns the number of distinct perceptions involved; by postulating that this number is unlimited, the relevant probablity weightings can be specified in terms of relative frequencies. One thereby obtains an interpretation of quantum mechanics that is compatible with Einstein's *desideratum* that 'God does not play dice', in the sense that uncertainty is no longer involved at an objective global level, but arises only at the subjective level of particular perceptions. It could therefore be said that *we* play dice, but God does not!

A deeper level concerns the plurality of the 'we' in the preceding statement; the basic question (also discussed in Section 19.12) is that of the number of distinct perceptors which can most economically be postulated to be just one (in the sense of Ockham). This issue is of no consequence in so far as purely scientific purposes are concerned, but it does have obvious ethical implications. The injunction to 'love one's neighbour as oneself' acquires a new significance when one recognizes that the 'neighbour' may be another incarnation of 'oneself'.

From a purely scientific point of view, the most interesting level (as discussed in Section 19.10) concerns the relevant numbers of Everett-type branches, which are determined by the solution to the problem of which eventualities should actually be characterized as perceptible. Whereas upper limits are obtainable in the manner suggested by Dyson (as discussed in Section 19.13), it is not so easy to see how to obtain lower limits on what should be considered perceptible.

References

[1] B. Carter. Large number coincidences and the anthropic principle in cosmology. In *Confrontations of Cosmological Theories with Observational Data*, ed. M. Longair (Dordrecht: Reidel, 1974), pp. 291–298.

[2] B. Carter. The anthropic principle and its implications for biological evolution. *Phil. Trans. Roy. Soc.* **A 310** (1983), 347.

[3] B. Carter. Anthropic interpretation of quantum theory. *Int. J. Theor. Phys.* **43** (2004), 721 [hep-th/0403008].

[4] H. Everett. Relative state formulations of quantum theory. *Rev. Mod. Phys.* **29** (1957), 454.

[5] N. Graham. The measurement of relative frequency. In *The Many Worlds Interpretation of Quantum Mechanics*, eds. B. S. DeWitt and N. Graham (Princeton, NJ: Princeton University Press, 1973), pp. 229–253.

[6] E. P. Wigner. On hidden variables and quantum mechanical probabilities. *Am. J. Phys.* **38** (1970), 1005.

[7] P. A. M. Dirac. *Quantum Mechanics* (Oxford: Oxford University Press, 1958).

[8] S. W. Hawking. The unpredictability of quantum gravity. *Commun. Math. Phys.* **87** (1982), 395.

[9] A. Vilenkin. Quantum cosmology and eternal inflation. In *The Future of Theoretical Physics*, eds. G. W. Gibbons, E. P. Shellard and S. J. Ranken (Cambridge: Cambridge University Press, 2003), pp. 649–663 [gr-qc/0204061].

[10] W. H. Zurek. Pointer basis of quantum apparatus: into what mixture does the wave function collapse? *Phys. Rev.* **D 24** (1981), 1516.

[11] J. C. Polkinghorne. *One World: The Interaction of Science and Theology* (Princeton, NJ: Princeton University Press, 1992).

[12] J. von Neumann. *Mathematical Foundations of Quantum Mechanics* (Princeton, NJ: Princeton University Press, 1955).

[13] J. Bub. *Interpreting The Quantum World* (Cambridge: Cambridge University Press, 1997).
[14] V. Vedral. Landauer's erasure, error correction, and entanglement. *Proc. Roy. Soc. Lond.* **A 456** (2000), 969.
[15] J. Leslie. Cosmology, probability, and the need to explain life. In *Scientific Understanding*, eds. N. Rescher (Lanham and London: University Press of America, 1983).
[16] B. Carter. The anthropic principle and the ultra-Darwinian synthesis. In *The Anthropic Principle*, eds. F. Bertola and U. Curi (Cambridge: Cambridge University Press, 1993), pp. 33–63.
[17] B. S. DeWitt. The many universes interpretation of quantum mechanics. In *The Many Worlds Interpretation of Quantum Mechanics*, eds. B. S. De Witt and N. Graham (Princeton, NJ: Princeton University Press, 1973).
[18] D. Deutsch. Quantum theory of probability and decisions. *Proc. Roy. Soc. Lond.* **A 455** (1999), 3129 [quant-ph/9906015].
[19] D. Wallace. Quantum probability and decision theory revisited. *Stud. Hist. Phil. Mod. Phys.* **34** (2003), 87 [quant-ph/0107144].
[20] H. Greaves. Understanding Deutsch's probability in a deterministic universe. *Stud. Hist. Phil. Mod. Phys.* **35** (2004), 423 [quant-ph/0312136].
[21] J. Hartle. Quantum mechanics of individual systems. *Am. J. Phys.* **36** (1968), 704.
[22] M. J. Donald. A mathematical characterisation of the physical structure of observers. *Found. Phys.* **22** (1995), 1111.
[23] M. Lockwood. Many minds' interpretations of quantum mechanics. *Brit. J. Phil. Sci.* **47** (1996), 159.
[24] D. Page. Sensible quantum mechanics: are probabilities only in the mind? *Int. J. Mod. Phys.* **D 5** (1996), 583 [gr-qc/9507024].
[25] N. Bostrom. *Anthropic Bias: Observation Selection Effects in Science and Philosophy* (New York: Routledge, 2002).
[26] M. Gell-Man and J. B. Hartle. Quantum mechanics in the light of quantum cosmology. In *Complexity, Entropy, and the Physics of Information*, ed. W. H. Zurek (Redwood City: Addison Wesley, 1991), pp. 425–458.
[27] J. Leslie. *The End of the World* (New York: Routledge, 1996).
[28] J. Islam. Possible ultimate fate of the universe. *Quart. J. Roy. Astron. Soc.* **18** (1977), 3.
[29] F. J. Dyson. Time without end: physics and biology in an open system. *Rev. Mod. Phys.* **51** (1979), 447.
[30] F. J. Dyson. Reality bites. *Nature* **380** (1996), 296.
[31] R. H. Dicke. Dirac's cosmology and Mach's principle. *Nature* **192** (1960), 440.

Part IV

More general philosophical issues

20
Scientific alternatives to the anthropic principle

Lee Smolin
Perimeter Institute for Theoretical Physics, Waterloo

20.1 Introduction

I have chosen a deliberately provocative title, in order to communicate a sense of frustration I have felt for many years about how otherwise sensible people, some of whom are among the scientists I most respect and admire, espouse an approach to cosmological problems – the Anthropic Principle (AP) – that is easily seen to be unscientific. By calling it unscientific I mean something very specific, which is that it lacks a property necessary for any scientific hypothesis – that it be *falsifiable*. According to Popper [1–4], a theory is falsifiable if one can derive from it unambiguous predictions for practical experiments, such that – were contrary results seen – at least one premise of the theory would have been proven not to true. This introduction will outline my argument in a few paragraphs. I will then develop the points in detail in subsequent sections.

While the notion of falsifiability has been challenged and qualified by philosophers since Popper, such as Kuhn, Feyerabend and others,[1] few philosophers of science or working scientists would be able to take seriously a fundamental theory of physics that had no possibility of being disproved by an experiment. This point is so basic to how science works that it is perhaps worthwhile taking a moment to review its rationale.

Few scientists will disagree that an approach can be considered 'scientific' only to the extent that it requires experts who are initially in disagreement about the status of a theory to resolve their disagreements – to the fullest extent possible – by rational argument from common evidence. As Popper emphasizes, science is the only approach to knowledge whose historical record shows repeatedly that consensus was reached among well trained

---

[1] I will not discuss here the history and present status of the notion of falsifiability. My own views on the methodology of science are discussed elsewhere [5].

Universe or Multiverse?, ed. Bernard Carr. Published by Cambridge University Press.
© Cambridge University Press 2007.

people in this way. But Popper's key point is that this has only been possible because proposed theories have been required to be falsifiable. The reason is that the situation is asymmetric; confirmation of a prediction of a theory does not show that the theory is true, but falsification of a prediction can show it is false.

If a theory is not falsifiable, experts may find themselves in permanent disagreement about it, with no possible resolution of their differences by rational consideration of the evidence. The point is that, to be part of science, X-theorists have to do more than convince other X-theorists that X is true. They have to convince all other hitherto sceptical scientists. If they do not aspire to do this, then – by Popper's definition – they are not doing science. Hence to prevent the progress of science from grounding to a halt, i.e. to preserve what makes science generally successful, scientists have an ethical imperative to consider only falsifiable theories as possible explanations of natural phenomena.

There are several versions of the AP [6–9].[2] There is, of course, the explicitly theological version, which is, by definition, outside of science. I have no reason to quarrel with that here. I also have no argument against straightforward consideration of selection effects, so long as the conditions invoked are known independently and are not part of a speculative theory that is otherwise unsupported by evidence. I will discuss this in some detail below, but – put briefly – there is a vast logical difference between taking into account a known fact (e.g. that most of the galaxy is empty space) and arguing from a speculative and unproven premise (e.g. that there is a large ensemble of unseen universes).

In recent discussions, the version of the AP that is usually proposed as a scientific idea is based on the following two premises.

(A) There exists (in the same sense that chairs, tables and our universe exist) a very large ensemble of 'universes', $\mathcal{M}$, which are completely or almost completely causally disjoint regions of spacetime, within which the parameters of the standard model of physics and cosmology differ. To the extent that they are causally disjoint, we have no ability to make observations in universes other than our own.

(B) The distribution of parameters in $\mathcal{M}$ is random (with some measure) and the parameters that govern our universe are rare.

This is the form of the AP most invoked in discussions related to inflationary cosmology and string theory, and it is the one I will critique in this chapter.

---

2 My understanding of the logical status of the different versions of the Anthropic Principle was much improved by refs. [10]–[12].

Here is the basic argument for why a theory based on **A** and **B** is not falsifiable. If such a theory applies to nature, it follows that our universe is a member of the ensemble $\mathcal{M}$. Thus, we can assume that *whatever properties our universe is known to have, or is discovered to have in the future, at least one member of $\mathcal{M}$ has those properties. Therefore, no experiment, present or future, could contradict **A** and **B**.* Moreover, since by **B** we already assume that there are properties of our universe that are improbable in $\mathcal{M}$, it is impossible to make even a statistical prediction that, were it not borne out, would contradict **A** and **B**.

There are a number of claims in the literature of predictions made from **A** and **B**. By the logic just outlined, these must all be spurious. We will examine the major claims of this kind and demonstrate that they are fallacious. This does not mean that the conclusions are wrong. As we shall see, there are cases in which the part of the argument that is logically related to the conclusion has nothing to do with **A** and **B** but instead relies only on observed facts about our universe. In these cases, the only parts of the argument that are wrong are the parts that fallaciously attribute the conclusion to a version of the AP.

But what if **A** is true? Will it be possible to do science in such a universe? Given what was just said, it is easy to see how a theory could be constructed so as to still be falsifiable. To do this, **B** must be replaced by the following.

(**B′**) It is possible, nevertheless, to posit a mechanism $\mathcal{X}$ by which the ensemble $\mathcal{M}$ was constructed, on the basis of which one can show that almost every universe in $\mathcal{M}$ has a property $\mathcal{W}$ with the following characteristics:[3] (i) $\mathcal{W}$ does not follow from any known law of nature or observation, so it is consistent with everything we know that $\mathcal{W}$ could be false in our universe; (ii) there is a practical experiment that could show that $\mathcal{W}$ is not true in our universe.

If these conditions are satisfied, then an observation that $\mathcal{W}$ is false in our universe disproves **A** and **B′**. Since, by assumption, the experiment can be done, the theory based on these postulates is falsifiable.

Note that what would be falsified is only the specific **B′** dependent on a particular mechanism $\mathcal{X}$. Since $\mathcal{X}$, by generating the ensemble, will imply **A**, what is falsifiable is the postulate that the mechanism $\mathcal{X}$ acts in nature. Conversely, a mechanism that generates a random ensemble – as described by **B** rather than **B′** – cannot be falsified, as I will demonstrate below.

---

[3] As discussed in Section 20.5, because of the issue of selection effects related to the existence of life, this can be weakened to *almost every universe in $\mathcal{E}$ that contains life also has property $\mathcal{W}$*.

Someone might argue that it is logically possible that **A** and **B** are true and that, if so, this would be bad only for those of us who insist on doing science the old-fashioned way. If an otherwise attractive theory points in the direction of **A** and **B**, then we should simply accept this and abandon what may be outmoded ideas about 'how science works'.

If this is the case, then it will always be true that basic questions about our universe cannot be answered by any scientific theory (that is, by a theory that could be rationally argued on the basis of shared evidence to be true). But the fact that it is a possibility does not mean we should worry unduly about it turning out to be true. This is not the only hypothesis about the world that, if true, means that science must remain forever incomplete.

Others argue that it is sufficient to do science with one-way predictions of the following form: 'Our theory has many solutions S_i. One of them, S_1, gives rise to a prediction X. If X is found, that will confirm the combination of our theory and the particular solution S_1. But belief in the theory is not diminished if X is not found, for there are a large number of solutions that do not predict X.'

One problem with this is that it can easily lead to a situation in which the scientific community is indefinitely split into groups that disagree on the likelihood that the theory is true, with no possible resolution. Indeed, it is plausible that this is already the case with string theory, which appears so far unfalsifiable but makes claims of this form. A second problem is that, even if X were found, another theory could be invented that also had a solution that predicted X. If neither were falsifiable, there would be no possibility of deciding which one was true.

Thus, so long as we prefer a science based on what can be rationally argued from shared evidence, there is an ethical imperative to examine only hypotheses that lead to falsifiable theories. If none is available, our job must be to invent some. So long as there are falsifiable – and not yet falsified – theories that account for the phenomena in question, the history of science teaches us to prefer them to their non-falsifiable rivals. Otherwise the process of science stops and further increases in knowledge are ruled out. There are many occasions in the history of science when this might have happened; we know more than people who espoused Ptolemy's astronomy or Lysenko's biology or Mach's dismissal of atoms as forever unobservable, because at least some scientists preferred to go on examining falsifiable theories.

To deflate the temptation to proceed with non-falsifiable theories, it therefore suffices to demonstrate that falsifiable alternatives exist. In this chapter I review one falsifiable alternative to the AP, which is Cosmological

Natural Selection [13–17]. As it is falsifiable, it may very well be wrong. In Section 20.6, I will review this theory in light of developments made since it was first proposed. I will show that, in spite of several claims to the contrary, it has yet to be falsified. However, it remains falsifiable as it makes at least one prediction for a property $\mathcal{W}$ of the kind described in **B'**. But whether it is right or wrong, the fact that a falsifiable theory exists is sufficient to show that the problems that motivate the AP might be genuinely solved by a falsifiable theory.

But if the AP cannot provide a scientific explanation, what are we to make of the claim that our universe is friendly to life? It is essential here to distinguish the different versions of the AP from what I would like to call 'the anthropic observation'.

> **The anthropic observation** Our universe is much more complex than most universes with the same laws but different values of the parameters of those laws. In particular, it has complex astrophysics, including galaxies and long-lived stars, and complex chemistry, including carbon chemistry. These necessary conditions for life are present in our universe as a consequence of the complexity which is made possible by the special values of the parameters.

I will describe this more specifically below. There is good evidence that the anthropic observation is true [6–9, 17] and why it is true is a puzzle that science must solve. However, to achieve this, it does not suffice just to restate what is to be explained as a principle, especially if the resulting theory is not falsifiable. One must discover a reason why it is true that has nothing to do with our own existence. Whether Cosmological Natural Selection is right or wrong, it does provide a genuine explanation for the anthropic observation. This is that the conditions for life, such as carbon chemistry and long-lived stars, serve another purpose, in that they contribute to the reproduction of the universe itself.

20.2 The problem of undetermined parameters

The second half of the twentieth century saw a great deal of progress in our understanding of elementary particle physics and cosmology. In both areas, Standard Models were established, which passed numerous experimental tests. In particle physics, the standard model – described in the mid 1970s – is based on two key insights. The first concerns the unification of the fundamental forces; the second considers why that unification does not prevent the various particles and forces from having different properties. The unifying principle is that all forces are described in terms of gauge fields, based on making symmetries local. However, the symmetries between particles and among forces can be broken naturally when those gauge fields are coupled to matter fields. The standard model of cosmology took longer to establish, but is also based on the behaviour of matter fields when the symmetry breaks. In particular, this leads to the existence of a non-zero vacuum energy, which can both drive the inflation of the early universe and accelerate the expansion today.

In each case, however, there is a catch. The interactions of the gauge fields with each other and with gravity are determined completely by basic symmetries, whose description allows a very small number of parameters. However, the dynamics of the matter fields needed to realize the symmetry-breaking spontaneously and dynamically is arbitrary and requires a large number of parameters. This is because the easiest matter fields to work with are scalar fields, and no transformation properties constrain the form of their self-interactions.

The result is that the standard model of particle physics has more than twenty adjustable parameters. These include the masses of all the basic stable elementary particles (the proton, neutron, electron, muon, neutrino) and also the coupling constants and mixing angles associated with the various interactions. These are not determined by any principle or mechanism we know; they must be specified by hand to bring the theory into agreement with experiment. Similarly, the Standard Model of cosmology has about fifteen parameters.

One of the biggest mysteries of modern science, therefore, is how these thirty-five or so parameters are determined. There are two especially puzzling aspects to this problem. The first is the *naturality problem*. Many of the parameters, when expressed in terms of dimensionless ratios, are extremely tiny or extremely large. In Planck units, the proton and neutron masses are around 10^{-19}, the cosmological constant is 10^{-120}, the coupling constant for the self-interaction of the field responsible for inflation cannot be larger than 10^{-11}, etc.

The second puzzling aspect is the *complexity problem*. Our universe has an array of complex and non-equilibrium structures, spread out over a huge range of scales from clusters of galaxies to living cells. It is not too hard to see that this remarkable circumstance depends on the parameters being fine-tuned to lie within narrow windows. Were the neutron heavier by only 1%, the proton light by the same amount, the electron twice as massive, its electric charge 20% stronger or the neutrino as massive as the electron, there would be no stable nuclei, no stars and no chemistry. The universe would be just hydrogen gas. The anthropic observation stated in the introduction is one way to state the complexity problem.

Despite all the progress in gauge theories, quantum gravity and string theory, not one of these problems has been solved. Not one mass or coupling constant of any particle considered now to be elementary has ever been explained by fundamental theory.

20.3 The failure of unification to solve the problem

For many decades there has been a consensus on how to solve the problems of the undetermined parameters: *unify the different forces and particles by increasing the symmetry of the theory and the number of parameters will decrease.* The expectation that unification reduces the number of parameters in a theory is partly due to historical experience. In several cases, unification has been accomplished by the discovery of a symmetry principle which relates things heretofore unconnected and reduces the number of parameters. This worked, for example, when Newton unified the theory of planetary orbits and when Maxwell showed that light was a consequence of the unification of electricity and magnetism. There is also the following philosophical argument: unification operates in the service of reductionism and this aims to provide a fundamental theory which will answer all possible questions and so cannot have free parameters.

Whatever the arguments for it, the correlation between unification and reduction in the number of parameters has not worked recently. Indeed, the last few times it was tried, it went the other way. One can reduce the number of parameters slightly by unifying all elementary particle physics in one Grand Unified Theory. However, one does not eliminate most of the freedom, because the values of the observed fundamental parameters are traded for the Higgs vacuum expectation values, which are not determined by any symmetry and remain free.

The grand unified theories had two problems. The first is that the simplest version of them, based on the group $SU(5)$, was falsified. It predicted that protons would decay with some minimum rate. The experiment was performed in the 1980s and protons were seen not to decay at that rate. This was the last time there was a significant experimental test of a new theoretical idea about elementary particles.

One can consider more complicated grand unified models, in which protons do not decay or the decay rate is much smaller. But all such models suffer from the second problem – their lack of naturality. They require two Higgs scales, one at around 1 TeV and the other at around 10^{15} TeV. But quantum corrections tend to pull the two scales closer to each other; to keep their ratio so large requires fine-tuning of the coupling constants of the theory to roughly one part in 10^{15}. To solve this problem, supersymmetry was proposed to relate bosons to fermions. One might think this would reduce the number of free parameters, but it goes the other way. The *simplest* supersymmetric extension of the Standard Model has 125 parameters.

Supersymmetry is a beautiful idea, and it was hard not to get very excited about it when it was first introduced. But so far it has to be counted as a disappointment. Had the addition of supersymmetry to what we know led to unique predictions (e.g. for what will be seen at the LHC), that would have been very compelling. The reality has turned out to be quite different. The problem is that, while supersymmetry is not precisely unfalsifiable, it is difficult to falsify, as many negative results can be – and have been – dealt with by changing the parameters of the theory. Supersymmetry would be completely convincing if there were even one pair, out of all the observed fundamental particles, that could be made into superpartners. Unfortunately, this is not the case, and one has to invent superpartners for each one of the presently observed particles.

This introduces a huge amount of arbitrariness. The current situation is that the minimal supersymmetric Standard Model has so much freedom, coming from its 125-dimensional parameter space, that – depending on which region of the parameter space one chooses – there are at least a dozen scenarios that could be probed by the upcoming LHC experiments [18]. Almost any result seen by the LHC could be – and probably will be – promoted as evidence for supersymmetry, whether or not it actually is. To test whether or not particle physics is supersymmetric will take much longer, as it will require measuring enough amplitudes to see if they are related by supersymmetry.

Another possible solution is to unify further the theory by coupling to gravity. There are two well developed approaches to quantum gravity – one non-perturbative, which means it makes no use of a background of classical spacetime, and the other perturbative, which describes small excitations of a classical spacetime. The latter includes loop quantum gravity, spin foam models, dynamical triangulations and others. In recent years, much progress has been made in these directions. Indeed, this has led to the realization that many models of this kind have emergent matter degrees of freedom [19, 20], whose properties are already fixed by the dynamics of the quantum spacetime. There is even a large class of models whose excitations include chiral states that match the properties of the Standard Model fermions [21, 22]. These recent results suggest that at the non-perturbative level, quantum gravity theories are automatically unified with matter, are highly constrained and hence highly predictive and falsifiable.

The perturbative approach, which is string theory, makes very strong assumptions about how the string is to be quantized,[4] and it also makes two physical assumptions: (1) that no matter on how small a scale one looks, spacetime looks classical with small quantum excitations; (2) Lorentz invariance is a good symmetry up to infinite energies and boosts. It is not certain that both these assumptions can be realized consistently. After many years, there are only proofs of consistency and finiteness of perturbative string theory to second (non-trivial) order in perturbation theory, and attempts to go further have not so far succeeded.[5] But these results indicate that the assumptions mentioned previously do put some constraints on particle physics. Supersymmetry is required, and the dimension of spacetime must be ten.

To the order of perturbation theory for which it is known to be consistent, string theory unifies all the interactions, including gravity. It was therefore originally hoped that it would be unique. These hopes were quickly dashed, and indeed the number of string theories for which there is evidence has been growing exponentially as string theorists have developed better techniques to construct them. Originally there were five consistent supersymmetric string theories in ten dimensions. But, the fact that the number of observed dimensions is four led to the hypothesis that the extra six dimensions are curled up very small or otherwise hidden from large-scale observations. Unfortunately, the number of ways to do this is quite large, at least 10^5. In recent years, evidence has been found for many more string theories, which incorporate non-perturbative structures of various dimensions, called 'branes'.

A key problem has been constructing string theories that agree with the astronomical evidence that the vacuum energy (or cosmological constant) is positive. The problem is that a positive cosmological constant is not consistent with supersymmetry. But supersymmetry appears to be necessary to cancel dramatic instabilities related to the existence of tachyons in the spectrum of string theories.

A few years ago, dramatic progress was made on this problem by Kachru and collaborators [25].[6] They found a way round the problem by wrapping flux around cycles of the compactified 6-manifold and thereby discovered evidence for the existence of string theories with positive vacuum energy.

---

[4] A recent paper by Thiemann [23] suggests that, with different technical assumptions, there are consistent string theories in any dimension without supersymmetry.
[5] For details of precisely what has and has not been proven regarding string theory, loop quantum gravity and other approaches to quantum gravity, see ref. [24].
[6] This built on earlier work by Giddings and colleagues [26], Bousso and Polchinski [27] and others.

This evidence is very weak – e.g. they are unable to construct propagation amplitudes even for free, non-interacting strings – but they are able to argue that there are consistent string theories with the desired characteristics. Their low-energy behaviour should be captured by solutions to classical supergravity, coupled to the patterns of the branes in question. They then construct the low-energy, classical, supergravity description.

Of course, the logic here is backwards. Had they been able to show that the required supergravity solutions do not exist, they would have ruled out the corresponding string theories. But the existence of a good low-energy limit is not a sufficient condition for a theory to exist. So, on logical grounds, the evidence for string theories with positive cosmological constant is very weak. However, if one takes the existence of these theories seriously, there is a disturbing consequence: the evidence suggests that the number of distinct theories is vast, of the order of 10^{100} to 10^{500} [28–30]. Each of these theories is consistent with the macroscopic world being 4-dimensional and with the existence of a positive and small vacuum energy. But they disagree about everything else; in particular, they imply different versions of particle physics, with different gauge groups, spectra of fermions and scalars and different parameters.

The fact that there are so many different ways to unify gauge fields, fermions and gravity consistently makes it likely[7] that *string theory will never make any new, testable predictions about elementary particles.*[8] Of course, a very small proportion of the theories will be consistent with current particle physics data. But even if this is only one in 10^{450}, there will still be 10^{50} viable theories. Although some of these may be disproved by some future experiment at higher energy, the number is so vast that it appears likely that, whatever is found, there will be many versions of string theory that agree with it.

20.4 Mechanisms for production of universes

Whatever the fate of the positive vacuum energy solutions of string theory, one thing is clear. At least up till now, the hope that unification would

---

[7] It should be noted that, while some string theorists have argued that this situation calls for some version of the AP [28], others have sought ways to pull falsifiable predictions from the theory [29, 30].

[8] It may be claimed that string theory makes a small number of correct *postdictions*, e.g. that there are fermions, gauge fields, gravitational fields and no more than ten spacetime dimensions. But this is not itself a strong argument for string theory, as there are other approaches to unifying gravity with quantum theory and the Standard Model for which non-trivial properties have also been proven [24]. So there is no evidence that string theory is the unique theory that unifies gravity with the Standard Model.

lead to a unique theory has failed dramatically. So it seems unlikely that the problem of accounting for the values of the parameters of the standard models of particle physics and cosmology will be solved by restrictions coming from the consistency of a unified theory.

The rest of this chapter is devoted to alternative explanations of the choice of parameters. All alternatives I am aware of involve the postulate **A** given in Section 20.1. They also require the further postulate **C**.

(**C**) There are many possible consistent phenomenological descriptions of particle physics at scales much less than the Planck energy. These may correspond to different phases of the vacuum or different theories altogether.

As a result, fundamental physics is assumed to give us, not a single theory, but a space of theories, $\mathcal{L}$, which has been called the *landscape* [14–17].[9] As in biology, we distinguish the space of genotypes from the space of phenotypes, i.e. we distinguish $\mathcal{L}$ from the space of the parameters of the Standard Model $\mathcal{P}$. All multiverse theories then make some version of the 'multiverse hypothesis'.

Multiverse hypothesis. Assuming **A** and **C**, the whole of reality – which we call the multiverse – consists of many different regions of spacetime, within which phenomena are governed by different phenomenological descriptions. For simplicity, we call these 'universes'.

The multiverse is then described by probability distributions ρ_L in $\mathcal{L}$ and ρ_P in $\mathcal{P}$. These describe the population of universes within the ensemble. Multiverse theories can be classified by their answers to the following questions.

(i) How is the ensemble of universes generated?
(ii) What mechanism produces the probability distribution ρ_P?
(iii) What methodology is used to produce predictions for our universe from the ensemble of universes?

We are interested here only in those multiverse theories that make falsifiable predictions. To do this, the ensemble of universes cannot be arbitrarily specified. Otherwise it could be adjusted to agree with any observations by making a typical universe agree with whatever is observed about ours. To have empirical content, the ensemble of universes must be generated by

---

[9] It should be mentioned that the word 'landscape' was chosen in refs. [14]–[17] to make the transition to the concept of *fitness landscape* – well known in evolutionary theory – more transparent.

some dynamical mechanism which is a consequence of general laws. The properties of the ensemble are then determined by laws that have other consequences which can – at least in principle – be checked independently. Two mechanisms for the generation of universes have been studied: eternal inflation and bouncing black hole singularities. We will describe each of these and then contrast their properties.

20.4.1 Eternal inflation

The inflation hypothesis provides a plausible explanation of several observed features of our universe, such as its homogeneity and uniformity [31–34]. The basic idea is that, at very early times, the energy density is dominated by a large vacuum energy, possibly coming from the vacuum expectation value of a scalar field. As the universe expands exponentially due to this vacuum energy, the vacuum expectation value also evolves in its potential. Inflation comes to an end when a local minimum of the potential is reached, converting vacuum energy into thermal energy that is presumed to become the observed cosmic microwave background.

The model appears to be consistent, assuming all scales involved are less than the Planck scale, and has made predictions which have been confirmed.[10] But there are problems. Some have to do with the initial conditions necessary for inflation. It has been shown that a region of spacetime will begin to inflate if the vacuum energy dominates other sources and if the matter and gravitational fields are homogeneous to good approximation over that region. Of course, we do not know the initial conditions for our universe, and we have observed nothing so far of the conditions prior to inflation. But on several plausible hypotheses about the initial state, the conditions required for a region to begin inflating are improbable. For example, the existence of inflation and the smallness of the associated density fluctuations, $\delta \rho / \rho$, requires that the self-coupling of the inflaton be small.

However, once the conditions necessary for inflation are met, it appears likely in some models that inflation does not happen just once. Because of quantum fluctuations, the scalar field will sometimes fluctuate 'up' the potential, so that even after inflation has ended in one region, it will continue in others. This can lead to the scenario known as *eternal inflation* [41–45], in which there are always regions which continue to inflate. There is then a competition between the classical force from the potential, causing the expectation value to decrease or 'roll' towards a local minimum, and the

---

[10] However, it should be noted that other theories make predictions so far indistinguishable from those of inflation [35–40].

quantum fluctuations, which can lead it to increase locally. Given plausible – but not necessary – assumptions, this can result in the creation of a large, or even infinite, number of regions which locally resemble ordinary FRW universes.

20.4.2 Bouncing black hole singularities

A second mechanism for generating new universes is through the formation of black holes. It is known that a collapsing star, such as the remnant of a supernova, will form either a neutron star or a black hole, depending on its mass. There is an upper mass limit for a stable neutron star. Remnants of supernovae larger than this have nothing to restrain them from collapsing to the point at which an event horizon is formed. Rough estimates of this upper mass limit are between 1.5 and $2.5 M_\odot$.

According to the singularity theorem of Penrose, proved on general assumptions, classical general relativity predicts the formation of a singularity at which the curvature of spacetime becomes infinite and spacetime ends. No trajectory of a particle or photon can be continued past the singularity to the future.

This result, however, may be modified by quantum effects. Before the singularity is reached, densities and curvatures reach the Planck scale and quantum gravity dictates the dynamics. As early as the 1960s, pioneers of this field, such as John Wheeler and Bryce DeWitt, conjectured that the effects of quantum gravity would reverse the collapse, removing the singularity and causing the matter that was collapsing to expand [46]. Time then does not end and there is a region of spacetime to the future of where the singularity would have been. The result is the creation of a new expanding region of spacetime, which may grow and become, for all practical purposes, a new universe. This region is inaccessible from the region where the black hole originally formed. The horizon is still there, which means that no light can escape from the new region to the previous universe. Unless the black hole evaporates, the causal structure implies that every event in the new region is to the future of every event in the region of spacetime where the black hole formed.

The transition by which collapse to a singularity is replaced by a new expanding region of spacetime is called a 'bounce'. One can then hypothesize that our own big bang is the outcome of a collapse in a previous universe and that every black hole in our universe is giving rise to a new universe.

The conjecture that singularities in classical general relativity are replaced by bounces has been investigated and confirmed in many semi-classical calculations (see, for example, ref. [48]). It is also suggested by some calculations in string theory [49,50]. In recent years, quantum gravity theory has been developed to the point where the conjecture can be investigated exactly. It has been shown that cosmological singularities do bounce [51–55]. Assuming that the theory of quantum gravity is correct, this means that the big bang in our past could not have been the first moment of time; there must have been something before that. Recent results from models of black hole interiors also strongly support the conjecture that black hole singularities are replaced by bounces [56–58]. These results strengthen the conjecture that quantum gravity effects replace black hole singularities with the birth of new universes.

20.4.3 Comparison of universe-generation mechanisms

In comparing these mechanisms for the reproduction of universes, several issues should be borne in mind.

How reliable is the evidence for the mode of production of universes?

We know that our universe contains black holes. There is observational evidence that many galaxies have large black holes at their centres. They are also believed to form from some supernova explosions, and there may be around 10^{18} such black holes in the observable universe. A number of candidates for such stellar black holes have been found, and the evidence so far, e.g. from studying X-rays from their accretion disks, supports their identification as black holes.

There has been speculation that many black holes may have been created by strong inhomogeneities in the early universe. However, the simplest theories of inflation predict that inhomogeneities were not strong enough to create many such primordial black holes. In any case, were they there, one would expect to see signals of their final evaporation, and no such signals have been detected. Thus, it is likely that the population of black holes is dominated, numerically, by supernova remnants.

We also have reasonable, if not yet compelling, theoretical evidence that black holes bounce [48–50], as well as exact quantum calculation results showing that there was something to the past of our big bang [51–55]. Thus there is plausible evidence that our universe is creating new universes through the mechanism of black hole production and that our own universe was created by such a process.

By contrast, the formation process for new universes in eternal inflation cannot be observed, since it takes place outside of our past horizon. The existence of the process depends entirely on believing in particular inflationary models that lead to eternal inflation. While many do, it is also possible to invent inflationary models that do not. For although there is evidence for inflation in general, several predictions of inflation having been confirmed, the observations do not yet discriminate between models that do and do not predict eternal inflation.

Also, some of the calculations backing up the eternal inflation scenario use very rough methods, based on imprecise theories employing semi-classical estimates for 'the wave-function of the universe'. This is a speculative extension of quantum theory to cosmology which has not been put on firm ground, conceptually or mathematically. Very recently, progress has been made in quantum cosmology which allows precise predictions to be made from a rigorous framework [51–55]. However, while inflation has been studied with these methods, so far the results do not address the conjectures that underlie eternal inflation.

Other approaches to eternal inflation [41] rely only on quantum field theory in curved spacetime. This is better understood, but there are still open questions about its applicability for cosmology. As a consequence, eternal inflation can be considered an interesting speculation, but it is supported by neither observation nor firm mathematical results within a well defined theory of quantum gravity.

What physics is involved in the mechanism of reproduction of universes?

The physical scale governing the birth of universes in eternal inflation is the scale of the inflaton potential in the regime where nucleation of new inflating regions takes place. This is at least the grand unified scale $\sim 10^{15}$ GeV and could be as large as the Planck scale. We have theories about the physics at this scale, but so far no predictions made by these theories have been confirmed experimentally. In fact, the only relevant experimental evidence, coming from proton decay experiments, falsified the simplest grand unified theories.

By contrast, the physical scale that governs black hole production is that of ordinary physics and chemistry. How many stars are massive enough to form supernovae is determined by ordinary chemical processes that govern the formation and cooling of giant molecular clouds. We know the physics of stars and supernovae reasonably well, and knowledge is improving all the time due to progress in theory, observation and experiment. Thus, we understand the physics that controls how many universes are created through black hole formation and we have speculations – but no detailed understanding – of the processes that govern the creation of new universes in eternal inflation.

What is the structure of the multiverse predicted by each theory?

A multiverse formed by black hole bouncing looks like a family tree. Each universe has an ancestor, which is another universe. Our universe has at least 10^{18} children; if they are like ours, they each have roughly the same number of children. The structure of a multiverse formed by eternal inflation is much simpler. Each universe has the same ancestor, which is the primordial vacuum. Universes themselves have no descendants.

20.5 Varieties of anthropic reasoning

Just as there are two modes of production of universes, there are two modes of explanation by which people have tried to draw physical predictions from multiverse models. These are the Anthropic Principle (AP) and Cosmological Natural Selection (CNS). There are actually several different anthropic principles and several different ways that people draw conclusions from them. We discuss the major ones in this section, including the arguments of Dicke, Hoyle and Weinberg that are usually cited as successes of the AP.[11] We consider CNS in Section 20.6.

20.5.1 The theological anthropic principle

It is not surprising that some theologians and scientists take the *complexity* problem as evidence that our universe was created by a benevolent God. They argue that if the best efforts of science lead to an understanding of the laws of nature within which there is choice, and if the choices that lead to a universe with intelligent life are extremely improbable, the very fact that such an improbable choice was made is evidence for intention. This is the old argument from design, recycled from controversies over evolution theory. It should be admitted that it does have force; the discovery of a craft as complex as an airbus on a newly discovered planet would be good evidence for intelligent life there. But this argument has force only so long as there are no plausible alternative explanations for how the choice might have been made. In the case of biology, natural selection provides a falsifiable and so far successful explanation, which renders unnecessary the argument from design.

---

11 I do not use the traditional nomenclature of 'weak' and 'strong' anthropic principles, as these terms have been used in different ways by different authors.

We can learn from the long history of the controversy in biology what tests a proposed explanation must satisfy if it is to be more convincing than the argument for intentional creation of a biofriendly universe.

- There must be a physical mechanism which converts the improbable to the probable, i.e. that raises the probability that a universe such as ours was chosen from infinitesimal to order unity.
- That mechanism must be falsifiable. It must be built from processes or components which can be examined empirically and be seen to function as hypothesized, either by being created in a laboratory or by occurring in nature in our observable universe.

We will see below that these tests are not satisfied by the different versions of the AP used in physics and cosmology. We will then propose a way of reasoning about multiverses that is not anthropic but does satisfy these tests.

20.5.2 Selection effects within one universe

The first anthropic arguments in cosmology were based on the use of *selection effects* within our observable universe. A selection effect is an effect due to the conditions of observation, which must be applied to a set of observations before they can be interpreted properly. A classical example is the following. Early humans observed that all around them was land and water, and above them was sky. From this, they deduced that our universe consists of a vast continent of land, surrounded by water and covered by sky. They were wrong; the reason was that they forgot to take into account the fact that the conditions they observed were necessary for them to exist as intelligent mammals. We now know that our universe is vastly bigger than they imagined and that most of it is filled with nothing but a very dilute gas and radiation. If we picked a point randomly, it would be very unlikely to be on the surface of a planet. But the conditions necessary for our evolution turn the improbable into the probable.

Dicke [59] used this logic to debunk Dirac's 'law of large numbers'. Dirac observed a coincidence [60] between the age of our universe and the proton mass in Planck units (the former is roughly the inverse cube of the latter). He argued that this requires an explanation and proposed one in which the gravitational constant G (and hence the Planck unit) would change in time. Dicke pointed out that the coincidence could be explained by our own existence, without invoking such a variation. Intelligent life requires billions of years of evolution on the surface of a planet near a stable long-lived star, and he was able to argue that the physics of stars implies that these conditions would only hold at an era where Dirac's coincidence was observed. Indeed, there is still no evidence for the variation of G postulated by Dirac. This argument is logically sound, but is quite different from the other types of explanation discussed below.

20.5.3 False uses of the Anthropic Principle

There are other successful arguments which have been called 'anthropic', although they have nothing to do with selection effects or the existence of life. An illustrative example is Hoyle's prediction of a certain resonance in the nuclei of carbon [61]. Hoyle argued that for life to exist there must be carbon. Carbon is indeed plentiful in our universe and must have been made either during the big bang or in stars, as these are the only ways to synthesize copious amounts of chemical elements. Detailed studies show that it could not have been made in the big bang, so it must have been made in stars. Hoyle argued that carbon could only be formed in stars if there were a certain resonant state in carbon nuclei. He communicated this prediction to a group of experimentalists who went on to find this resonant state.

The success of Hoyle's prediction is sometimes used as support for the effectiveness of the AP. However, it has nothing whatsoever to do with the existence of life because the first step of his argument is unnecessary. The fact that we – or other living things – are made of carbon is totally unnecessary to the argument. Indeed, were there intelligent life-forms which evolved without carbon chemistry, they could just as easily make Hoyle's argument.

To be clear why Hoyle's argument does not employ the AP, let us examine its logical schema and then ask which step we would have to question were the prediction falsified. The key steps in the argument are as follows.

(i) X is necessary for life to exist.
(ii) X is true about our universe.
(iii) Using the laws of physics, as presently understood, together with other observed facts Y, we deduce that if X is true of our universe, then so is Z.
(iv) We therefore predict that Z is true.

In Hoyle's case, X is the statement that our universe is full of carbon, Y is the claim that this could only be made in stars, and Z is the existence of a certain resonance in carbon.

It is clear that the prediction of Z at step (iii) in no way depends on step (i). To see this, ask how we would react if Z were found not to be true. Our only option would be to question either Y or the deduction from the presently known laws of physics of Z. We might conclude that the deduction was wrong, for example, if we made a mistake in a calculation. If no such option worked, we might have to conclude that the laws of physics have to be modified. But we would never question (i), because – while true – it plays no role in the logic of the argument leading to the prediction for Z.

There are other examples of this kind of mistaken reasoning, in which an argument promoted as 'anthropic' actually has nothing to do with the existence of life, but is instead a straightforward deduction from observed facts.

20.5.4 Selection effects within a multiverse

More recent arguments termed 'anthropic' are made within the context of multiverse scenarios. It is tricky to pull falsifiable predictions from such scenarios, because (so far) we have only observed one member of the ensemble. But it is not impossible, as I will show shortly.

First, however, we have to dispose of mistaken uses of multiverse selection effects. These are arguments in which point (i) in the above schema for Hoyle's argument is replaced as follows.

(i)′ We live in one member of a multiverse in which the laws of physics vary. X is necessary for life, so by a selection effect we must live in a universe in which X is true.

The other steps in Hoyle's argument remain the same. The substitution of (i)′ for (i) has not changed the logic of the argument; (i)′ is as irrelevant for the argument as (i) was, because (ii) still does the real logical work. Furthermore, if the prediction Z were falsified, we would not question (i)′. Rather, the problem would be to understand why, *in one universe*, X is true without Z being true. This problem must be solved within one universe and is independent of whether or not the universe we live in is part of a multiverse. Had the carbon resonance lines which Hoyle predicted not been found, he would neither have questioned the existence of life nor regarded the result as relevant to the number of universes. Instead, given that carbon is plentiful, he would have examined all the steps in the argument, looking for a loophole. This might have involved exotic new sources of carbon, such as collisions of neutron stars.

Hence, to pull a genuinely falsifiable prediction from a multiverse theory, which genuinely depends intrinsically on the hypothesis that our universe is part of a multiverse, the logic must be different from the schema just given. One way to do this is to fix a multiverse theory $\mathcal{T}$ which gives rise to an ensemble of universes $\mathcal{M}$. We are interested in predictions concerning some set of properties p_i, where i labels the property. The theory may give some *a priori* probability $\rho_\mathcal{M}(i)$ that a universe picked randomly from the ensemble will have property i.

To make the argument precise, we will need to refer to another ensemble which consists of randomly generated universes $\mathcal{R}$. This is produced by taking properties allowed to vary within the theory and selecting their values randomly, according to some measure on the parameter space of the theory. By 'random' we mean that the measure chosen is unbiased with respect to the choice of hypothesis for the physical mechanism that might have produced the ensemble. For example, if we are interested in string theory, we randomly pick universes with different string vacua. The difference between $\mathcal{R}$ and $\mathcal{M}$ is that the former is picked randomly from the physically possible universes, whereas the latter is generated dynamically, by a mechanism prescribed in theory $\mathcal{T}$.

Before comparing this with our universe, we should take into account that we may not live in a typical member of the ensemble $\mathcal{M}$. There will be a sub-ensemble $\mathcal{LM} \subset \mathcal{M}$ of universes that have the conditions for intelligent life to exist. Depending on the theory, the probability for a random universe in $\mathcal{M}$ to also be in $\mathcal{LM}$ may be very small or close to unity. But we already know that, if the theory is true, we are in a universe in $\mathcal{LM}$. So we should compute $\rho_\mathcal{L}(i)$, the probability that a universe randomly picked in the sub-ensemble $\mathcal{L}$ contains property p_i. Similarly, there will be a sub-ensemble $\mathcal{LR} \subset \mathcal{R}$ of those universes within the random ensemble which contain life.

The theory, then, can only make a falsifiable prediction if some restrictions are satisfied. It is no good considering properties that depend on the conditions necessary for life, for they will always be satisfied in a universe where life exists. To find a falsifiable prediction, the following must hold.

- There is a property $\mathcal{B}$ which is independent of the existence of life, i.e. it must be physically and logically possible that universes exist which have life but do not have $\mathcal{B}$. To make this meaningful, we must refer to the ensemble $\mathcal{R}$ of random universes. The probability of a universe in $\mathcal{LR}$ having property $\mathcal{B}$ must be small.
- Within the ensemble $\mathcal{M}$ generated by theory $\mathcal{T}$, there must be a strong correlation between universes with life and those with property $\mathcal{B}$.

- The argument will have force if the property $\mathcal{B}$ has not yet been looked for, so that this property is a genuine prediction of the theory $\mathcal{T}$, vulnerable to falsification at the time the prediction is made.

Under these conditions, we can now proceed to do real science with a multiverse theory. We make the assumption that our universe is a typical member of the ensemble $\mathcal{L}$. We then look for property $\mathcal{B}$. The theory is falsifiable because, if property $\mathcal{B}$ is not seen in our universe, then we know that theory $\mathcal{T}$ which gave rise to the ensemble $\mathcal{L}$ is false. If, however, $\mathcal{B}$ is found, then the evidence favours the ensemble $\mathcal{M}$ produced by the theory over the random ensemble $\mathcal{R}$.

We can draw a very important conclusion from this. To make a falsifiable prediction, a theory must produce an ensemble $\mathcal{M}$ that differs from a random ensemble $\mathcal{R}$. There must be properties that are improbable in $\mathcal{LR}$ and probable in $\mathcal{LM}$. If the two ensembles are identical, and if there is a high probability that a universe with life in $\mathcal{M}$ has property $\mathcal{B}$, this is also true of a universe with life in a randomly generated ensemble. There are two problems with this. First, the particular hypotheses that make up the theory $\mathcal{T}$ are not being tested, for they are empirically equivalent to a random number generator. Second, and more importantly, without the random ensemble, we cannot give meaning to the necessary condition that $\mathcal{B}$ is uncorrelated with the conditions necessary for life. For the observation of $\mathcal{B}$ to be able to falsify $\mathcal{T}$, it must be possible that there exist ensembles in which the probability of $\mathcal{B}$ in universes with life is low. The operational meaning of this is that they are uncorrelated in an ensemble of randomly generated universes.

To put this more strongly, suppose that a theory $\mathcal{T}$ generates an ensemble whose living sub-ensemble $\mathcal{LM}$ is identical to the living sub-ensemble of the random ensemble. Assume this theory predicts a property $\mathcal{B}$ which has probability close to unity in $\mathcal{LM}$. If $\mathcal{B}$ is observed, that does not provide evidence for $\mathcal{T}$, because there is already a complete correlation between $\mathcal{B}$ and life in the ensemble of randomly generated universes.

The conclusion is that no multiverse theory that produces an ensemble identical to $\mathcal{R}$ can give falsifiable predictions. Genuine falsifiable predictions can only be made by a theory whose ensemble $\mathcal{M}$ differs from $\mathcal{R}$. Furthermore, to give a genuine prediction, there must be a property $\mathcal{B}$, not yet observed but observable with present technology, which is probable in $\mathcal{LM}$ but improbable in $\mathcal{LR}$.

A very important consequence of this follows from the following observation: *properties of the ensemble $\mathcal{M}$ generated by a mechanism in a theory $\mathcal{T}$ will be random if that property concerns physics on a scale many orders* *of magnitude different from the scale of the mechanism of production of universes defined by* $\mathcal{T}$. One reason is familiar from statistical mechanics. Ensembles tend to be randomized in observables that are not controlled in their definition. For example, for a gas in a room, the properties of individual atoms are randomized, subject only to their random values being related to the temperature and density in the room.

A second reason has to do with a general property of local field theories, namely the decoupling of scales. In renormalizable field theories, including those of the standard model, there is only weak coupling between modes of the field at very different scales. We can see this applied to eternal inflation models. The mechanism for generating universes involves quantum fluctuations in the presence of a vacuum condensate with energy between 10^{15} and 10^{20} GeV. Properties of the vacuum that influence physics at those scales will play a role in determining the ensemble of universes created. If we consider the space of possible theories (perhaps string vacua), these will be preferentially selected by properties that strongly influence the probability for a quantum fluctuation in this environment to be uniform. These will include coupling constants for interactions manifest on that scale and vacuum expectation values for Higgs fields on that scale. But the exact values of masses or couplings many orders of magnitude lighter are not going to show up.

What will matter is the total number of degrees of freedom, but all particles so far observed are many orders of magnitude lighter and may be treated as massless from the point of view of the physics of the creation of our universe. Hence, changes in the proton–neutron mass difference, or the electron–proton mass ratio, are not going to have a significant influence on the probability for universe creation. The result is that these properties will be randomized in the ensemble $\mathcal{M}$ created by eternal inflation.

As a result, it is reasonable to expect that any standard model parameters that govern low-energy (but not grand unification) physics will have the same distribution in $\mathcal{M}$ as in the random ensemble $\mathcal{R}$. These include the masses of the quarks, leptons and neutrinos, and the scale of electroweak symmetry-breaking (i.e. the weak interactions). It follows that *eternal inflation will not be able to make any falsifiable predictions about any low-energy parameters of the standard model of particle physics. Consequently, no solution to the complexity problem can come from eternal inflation, since that involves the values of these parameters.*

Eternal inflation may be able to make some predictions, but only those restricted to parameters that govern physics at grand unified scales. However, there are claims that eternal inflation – in conjunction with another principle – does lead to predictions about the cosmological constant [62–66] and we examine these next.

20.5.5 The Principle of Mediocrity

A variant of selection effects applied to a multiverse is the 'Principle of Mediocrity' (PM). This is defined by Garriga and Vilenkin [63] as requiring that 'our civilization is typical in the ensemble of all civilizations in the universe'. This means that we weigh the ensemble $\mathcal{M}$ by the number of civilizations in each universe. It follows that all universes outside of $\mathcal{LM}$ have zero weight and that universes with more civilizations are weighed more heavily.

This principle adds several layers of presently untestable assumptions to the analysis. We know nothing reliable about the conditions that generate civilizations. While we can speculate, our genuine knowledge about this is unlikely to improve in the near future. If we conjecture that the number of civilizations will be proportional to the number of spiral galaxies, we can provisionally take the PM to mean that we weigh our ensemble with the number of spiral galaxies in each universe. Alternatively, we can postulate that the number of civilizations is proportional to the fraction of baryons that end up in galaxies [67,68].

Garriga and Vilenkin then argue that certain predictions can be drawn concerning properties of the vacuum energy [63]. We note that, in conformity with the above argument, no predictions are drawn concerning properties that have to do with the parameters of low-energy physics and are uncorrelated in a random ensemble with the existence of life. Still, it is good that people put predictions on the table and we should take them seriously. To do so, we must ask what exactly would be falsified if one or more of their predictions were found to disagree with observation. The argument depends on properties of the eternal inflation theory, some rough guesses about the wave-function of the universe and how to reason with it, and some rationale about the effects of vacuum energy on the creation and evolution of galaxies.

The PM can only have force if it is more stable than the other parts of the argument leading to the predictions. Otherwise a falsification of the prediction may teach us only that the PM is unreliable. To be useful, a methodological principle must be reliable enough that it can be taken as firm and used as part of an argument to negate any hypothesis about physics.

So, is the PM on firmer ground than quantum cosmology or the theory of galaxy formation? I know of no *a priori* argument for the PM. If the multiverse is real, we may indeed live in a universe with the maximal number of civilizations. But it could just as easily be false. There is no reason why we may not live in a universe which is atypical, in that it has some civilizations, but many fewer than other members of the ensemble. Thus, while we can argue for taking into account selection effects coming from the fact that we are in a universe hospitable to life, the PM is on much less firm ground.

The PM is sometimes supported by referring to ensembles within which we are typical individuals. Indeed, there are many ensembles within which this is the case. The problem is that there are also many ensembles with respect to which we are atypical. The PM has little force in human affairs, because – without further specification – it is vacuous, as we are both typical and atypical, depending on what ensembles we are compared against. To see this, let us ask some questions about how typical we are.

(i) Do we live in the universe with the largest number of civilizations?
(ii) Do we live in the universe with the largest number of intelligent beings?
(iii) Do we live in the universe with the largest number of conscious minds?
(iv) Do we live on the planet with the largest number of intelligent beings?
(v) Do we live in the most populous city on my planet?
(vi) Do we live in the most populous country on my planet?
(vii) Are we members of the largest ethnic group on my planet?
(viii) Do we have a typical level of wealth or income on my planet?
(ix) Do we live at a time when more people are alive than at any other?

The answer to questions (i) to (iv) is that there is no way of telling with either present data or any conceivable future data. In my particular case, the answers to questions (v) to (viii) are no, but I know people who can answer yes to one or more of them. Question (ix) is ambiguous. If the ensemble includes all times in the past, the answer is probably yes. If it includes all times in future as well, it is impossible to know the answer.

Given how often any individual fails to be typical in ensembles we know about, it seems to me we are on equally weak ground reasoning from any assertion of answers to (i) to (iv) as we would be reasoning from (v) to (ix). I conclude that the PM is too ambiguous to be useful. It must be supplemented by a specification of the ensemble. When that is done, we can test it, but it is still found to be unreliable. Thus it must be even less reliable in situations where it cannot be tested.

The well known 'doomsday argument' [69–71] illustrates the perils of the use of the PM. Someone begins it by stating 'I am a typical human being'. They may support that by noting the existence of some ensembles within which they are typical. Then they introduce a new ensemble $\mathcal{H}$, consisting of *all human beings who will ever live*. They next assert that, since they are generally typical, they should be typical in that ensemble. They then draw the drastic deduction (which we call $\mathcal{C}$) that *roughly the same number of human beings will live after them as before*. Given that the population has been growing exponentially for a long time, this leads to the conclusion that the population should begin to fall drastically within their lifetime.

There are more details, but we do not need them to see the ways in which the argument is fallacious.[12] The ensemble $\mathcal{H}$ contains an unknown number of human beings, who may live in the future. There is no way, given any information we have at present, to determine if we (living now) are in any way typical or untypical members of $\mathcal{H}$. There is simply no point in guessing. Whether we who have lived so far constitute most of $\mathcal{H}$, an infinitesimal fraction of $\mathcal{H}$ or something in between, depends on events that will take place in the future, most of which we are unable to control, let alone predict.[13] So it is simply impossible with current knowledge to deduce the truth value of $\mathcal{C}$.

However, we can still look to the past. The population has been growing exponentially for at least 10 000 years. Any person living in the last 10 000 years would have had just as much rational basis for following the reasoning from 'I am a typical human being' to conclusion $\mathcal{C}$ as we have. Other facts, such as the existence of weapons of mass destruction or global warming, are irrelevant, as they are not used to support $\mathcal{C}$. (The whole point of the argument is supposed to be that it is independent of facts such as these.)

But would a person have been correct to use this argument to conclude $\mathcal{C}$ a 1000 years ago? Clearly not; they would have been wrong because already many more people have lived since them than had lived before. But $\mathcal{C}$ is supposed to be a consequence of the PM. The conclusion is that there are two cases of individuals to which the PM may be applied. There is a class of

---

12 Another criticism of the argument, from F. Markopoulou (personal communication), is that even to state that a person is typical in the ensemble $\mathcal{H}$ with respect to a given property is to assume that there is a normalizable probability distribution for that property in $\mathcal{H}$. If the property is birth order, then the normalizability of the probability distribution already implies that the population must decrease at some point in the future. Thus, the argument assumes what it claims to demonstrate. The only open issue is when this decrease occurs, but, as we see, this cannot in any case be determined by the argument.

13 It would take us too far afield to analyze why such a fallacious argument is so attractive. It has something to do with the fallacy that every statement that will, at the end of time, have a truth value, has a truth value now. The statement 'I am a typical member of the ensemble $\mathcal{H}$' is one that can only be given a truth value by someone in the unhappy situation of knowing they are the last of us, and they would thus judge it false. No one for whom the statement is true could possibly have enough information to ascribe to it a truth value, for the simple reason that to do so would require knowledge of the future. Thus, logic in this case cannot be Boolean because different observers, at different times, can only make partial judgments as to the truth values of propositions that concern themselves. A more adequate logic is that given by Heyting, which is intimately related to the causal relations amongst events in time [72].

individuals to whom the truth value of $\mathcal{C}$ – and hence of the PM – cannot be checked. Then there is a class of individuals about whom the truth value of $\mathcal{C}$ can be determined. In each and every one of these cases, $\mathcal{C}$ is false. Thus, in every case in which there is an independent check of the consequences of the PM, it turns out to be false. Hence, it is either false or undetermined. Hence there is no evidence for its truth.

20.5.6 Weinberg's argument for the cosmological constant

Recently it has been claimed that the AP, and more specifically the PM, lead to a successful prediction. This is Weinberg's prediction for the value of the cosmological constant Λ, first made in ref. [65] and then elaborated in refs. [66–68]. It is important to note that Weinberg and collaborators make two separate arguments. The first is the following: assuming **A** and **B**, we cannot find ourselves in a universe with too positive a Λ else galaxies would never have formed. The upper limit for Λ predicted by this argument – with all other constants of nature fixed – is about 200 times the present matter density [67] (baryons plus dark matter), which yields roughly $\Omega_\Lambda < 100$. This is about two orders of magnitude larger than the present observed value.

In their second argument, Weinberg and his collaborators attempt to improve this estimate by evoking the PM in the form just discussed. They find that the probability of finding Ω_Λ less than 0.7 is either 5% or 12%, depending on technical assumptions made. Thus one can conclude that, while the actual observed value is somewhat low compared to the mean, it is not unreasonable to argue that the observed value is consistent with the result of the analysis based on the PM.

It might be argued that there is something wrong with my case against the PM; since Weinberg's first paper preceded the supernova and CMB measurements of Λ, his use of the PM has to count as a successful prediction. Indeed, this is perhaps the only successful new prediction in fundamental physics for the value of a physical parameter in decades.[14]

To reply, let us first distinguish the two arguments. One can reasonably conclude that Weinberg's first argument is, in part, correct. Were $\Omega_\Lambda > 100$ (with all other constants of nature fixed), we would have a problem understanding why we live in a universe filled with galaxies. However, this is false use of the AP, of the kind discussed in Section 20.5.4, because the problem

---

14 It is sometimes stated that Weinberg made the only correct prediction for the order of magnitude of the cosmological constant, but a correct prediction was also made by Sorkin and colleagues [73], based on the causal set approach to quantum gravity.

has nothing to do with our own existence. Just as Hoyle's argument has nothing to do with life, but is only based on the observed fact that carbon is plentiful, so Weinberg's first argument only has to do with the fact that galaxies are plentiful. It could be made by a robot or disembodied spirit. Were $\Omega_\Lambda > 100$ (with all other constants fixed), there would be a contradiction between present models of structure formation, which are based on established physical principles, and the observation that our universe is filled with galaxies.

The first argument of Weinberg is sometimes presented as an example of the success of a selection principle within a multiverse. But it then follows the schema given at the beginning of Section 20.5.3, with X now being the existence of many galaxies and Z the requirement that $\Omega_\Lambda < 100$. Were $\Omega_\Lambda = 1000$, the problem would not be with the multiverse hypothesis. Rather, it would be to explain how galaxy formation happened *in a single universe* despite such a large Λ. Therefore, Weinberg's first argument is partly valid, but the part that is correct is just a rational deduction from the fact that there are galaxies. The existence of life and selection effects in a multiverse are completely irrelevant to the argument, as they can be removed from the argument without its logical force being in any way diminished.

Before considering the second argument, there is a caveat to deal with, which is the restriction to a class of universes in which all the other constants of nature are fixed. As pointed out by Rees [74, 75], Tegmark and Rees [76] and Graesser and colleagues [77], it is difficult to justify any claim to make a valid prediction based on this restricted assumption. One should instead consider ensembles in which other cosmological parameters are allowed to vary. When one does this, the constraint on Λ from Weinberg's first argument is considerably weakened. The above authors show this by considering the case of the magnitude of the density fluctuations, usually denoted Q, which is observed to be about 10^{-5}. One can argue that, holding Λ fixed, Q cannot be much more than an order of magnitude larger, for similar reasons. But, as they show, one can have stars and galaxies in a universe in which *both* Q and Λ are raised by several orders of magnitude from their present values (see Fig. 2 of ref. [74]). For example, were $\Omega_\Lambda > 100$, one option to explain the existence of galaxies in our universe would be to raise Q.

From the calculations of refs. [74–77], we conclude that, if we make no other assumptions, the pair (Q, Λ) are each about two orders of magnitude smaller than their most likely values, based only on the existence of stars and galaxies. This is non-trivial, since they are many orders of magnitude away from their natural values. But it still leaves a great deal to be explained.

Let us now turn to Weinberg's second argument, in which he employs the PM. I argued above that the PM provides an unreliable basis for deductions, because the results obtained depend strongly on which ensemble is considered to be typical. It is easy to see that Weinberg's second argument supports this conclusion. Were this argument reliable, it would be robust under reasonable changes of the ensemble considered. But, as shown by Graesser and colleagues [77] and earlier authors [63, 64], if the ensemble is taken to be universes in which Λ varies but all other constants are held fixed, then an application of the PM leads to the conclusion that the probability of Λ being as small as observed is around 10%. But if we consider an ensemble in which Q as well as Λ varies, the probability comes down to order 10^{-4}, with the precise estimate depending on various assumptions made (see Table 1 of ref. [77]).

We draw two conclusions from this. First, the PM is unreliable because the conclusions drawn from it are ambiguous in that they depend strongly on the ensemble considered. Second, if taken seriously, it nevertheless leads to the conclusion that the probability of the observed value of Λ is of order 10^{-4}. This is because, in all modern cosmological theories, Q depends on the parameters of the inflation potential, such as its mass and self-coupling. In any fundamental model of particle physics in which parameters vary, these would certainly be among the parameters expected to do so.

Thus, Weinberg's two arguments illustrate the conclusions we reached earlier. The AP itself, in the form of **A** and **B**, makes no predictions. Arguments that it has led to predictions are false; the effective part of the argument is never the existence of life or intelligent observers but only observed facts about our universe. The PM cannot help, because it is easily shown to be unreliable. Any argument that it leads to a correct conclusion can be easily turned into an argument for an incorrect conclusion by reasonable changes in the definition of the ensemble in which we are assumed to be typical.

20.5.7 Aguirre's argument against the Anthropic Principle

We now mention one final argument against the AP, given by Aguirre [78]. He points out that intelligent life would be possible in universes with parameters very different from our own. He gives the particular example of a cold big bang model. This is a class of models which disagree with observations but still have galaxies, carbon and long-lived stars. This is sufficient, because it follows that any argument that incorporates a version of the AP would have to explain why we do not live in a cold big bang universe.

Given that cold big bang universes share the property of our universe of having abundant formation of galaxies and stars, none of the versions of the AP can do this. Thus, either we leave unexplained why we do not live in a cold big bang universe, or we have to find an explanation other than the AP for the parameters of our universe.

20.6 Cosmological Natural Selection

I believe that I have demonstrated conclusively that the version of the AP described by **A** and **B** is never going to give falsifiable predictions for the parameters of physics and cosmology. For the few times an argument called 'anthropic' has led to a successful prediction, as in the case of Hoyle's argument and Weinberg's first argument, examination shows that it rests entirely on a straightforward deduction from an observed fact about our universe.

Thus, if we are to understand the choices of parameters in the context of a falsifiable theory, we need an alternative approach. One alternative to deriving predictions from a multiverse theory is patterned on the successful model of natural selection in biology. This was also originally motivated by asking the question of how science can explain improbable complexity. To my knowledge, only in biology do we successfully explain why some parameters – in this case the genes of all the species in the biosphere – come to be set to very improbable values, with the consequence that the system is vastly more complex and stable than it would be for random values. The intention is then not to indulge in some mysticism about 'living universes', but merely to borrow a successful methodology from the only area of science that has successfully solved a problem similar to the one we face.[15]

The methodology of natural selection, applied to multiverse theories, is described by three hypotheses.

(i) A physical process produces a multiverse with long chains of descendants.
(ii) For the space $\mathcal{P}$ of dimensionless parameters of the standard model of physics, there is a fitness function $F(p)$ on $\mathcal{P}$ which is equal to the average number of descendants of a universe with parameters p.
(iii) The dimensionless parameters p_{new} of each new universe differ, on average, by *small* random amounts from those of its immediate ancestor

---

[15] Other approaches to cosmology which employ phenomena analogous to biological evolution have been proposed, for example, by Davies [79], Gribbin [80], Kauffman [81] and Nambu [82]. We note that Linde sometimes employs the term 'Darwinian' to describe eternal inflation [83–89]. However, because each universe in eternal inflation has the same ancestor, there is no inheritance and no modification of parameters analogous to the case of biology.

(i.e. small compared with the change that would be required to change $F(p)$ significantly).

Their conjunction leads to a predictive theory, because – using standard arguments from population biology – after many iterations from a large set of random starts, the population of universes, given by a distribution $\rho(p)$, is peaked around local extrema of $F(p)$. With more detailed assumptions, more can be deduced, but this is sufficient to lead to observational tests of these hypotheses. This implies the following prediction.

($\mathcal{S}$) If p is changed from its present value in any direction in $\mathcal{P}$, the first significant changes in $F(p)$ encountered must be to decrease $F(p)$.

The point is that the process defined by the three hypotheses drives the probability distribution $\rho(p)$ to the local maxima of the fitness function and keeps it there. This is much more predictive than the AP, because the resulting probability distribution would then be much more structured and very far from random. If, in addition, the physics that determines the fitness function is well understood, detailed tests of the general prediction $\mathcal{S}$ become possible, as we will now see.

20.6.1 Predictions of Cosmological Natural Selection

It is important to emphasize that the process of natural selection is very different from a random sprinkling of universes in the parameter space $\mathcal{P}$, which would produce only a uniform distribution. To achieve a distribution peaked around the local maxima of a fitness function requires two conditions. The change in each generation must be small, so that the distribution can 'climb the hills' in $F(p)$ rather than jump around randomly, and so that it can stay in the small volumes of $\mathcal{P}$ where $F(p)$ is large and not diffuse away. It requires many steps to reach local maxima from random starts, which implies that long chains of descendants are needed.

As a result, of the two mechanisms for universe production studied so far, only black hole bouncing fits the conditions necessary for natural selection. This is also fortunate, because the physics that goes into the fitness function is well understood in this case, at least in the neighbourhood of the parameters of our universe. The physical processes that strongly influence the number of black holes produced are nucleosynthesis, galaxy formation, star formation, stellar dynamics, supernova explosions and the formation and stability of neutron stars. All of these processes, except perhaps galaxy formation, are understood in some detail, and in several of them our theories make precise predictions which have been tested. We are then on reasonably firm ground asking what happens to each of these processes when we make small changes in the parameters. Thus, for the rest of this chapter, CNS will be taken to mean the process of reproduction of universes through black hole bounces, supplemented by the above hypotheses.

The hypothesis that the parameters p change by small random amounts should be ultimately grounded in fundamental physics. We note that this is compatible with string theory, in the sense that a great many string vacua likely populate the space of low-energy parameters. It is plausible that when a region of the universe is squeezed to the Planck density and heated to the Planck temperature, phase transitions may occur, leading to jumps from one string vacua to another. But so far there have been no detailed studies of these processes which would have checked the hypothesis that the change in each generation is small. One study of a bouncing cosmology in quantum gravity also lends support to the hypothesis that the parameters change in each bounce [90].

20.6.2 Successes of the theory

Details of the arguments for CNS, as well as references to the astrophysical literature on which the arguments are founded, can be found in [13–17]. Here I will only summarize the conclusions. The crucial conditions necessary for forming many black holes as the result of massive star formation are as follows.

- There should be a few light stable nuclei, at least up to helium, so that gravitational collapse leads to long-lived stable stars.
- Carbon and oxygen nuclei should be stable, so that giant molecular clouds form and cool efficiently, giving rise to the efficient formation of stars massive enough to give rise to black holes.
- The number of massive stars is increased by feedback processes, whereby massive star formation catalyzes more massive star formation. This is called 'self-propagated star formation', and there is good evidence that it makes a significant contribution to the number of massive stars produced. This requires a separation between the timescale required for star formation and the lifetime of massive stars, which implies – among other things – that nucleosynthesis should not proceed so far that the universe is dominated by long-lived hydrogen-burning stars.
- Feedback processes involved in star formation also require that supernovae should eject enough energy and material to catalyze the formation

of massive stars, but not so much that there are not many supernova remnants over the upper mass limit for stable neutron stars.
- The parameters governing nuclear physics should be tuned, as much as possible consistent with the foregoing, so that the upper mass limit for neutron stars is as low as possible.

The study of the first four conditions leads to the conclusion that the number of black holes produced in galaxies will be decreased by almost any change in low-energy parameters: a reversal of the sign of $\Delta m = m_n - m_p$ will result in a universe dominated by a gas of neutrons; a small increase in Δm compared to m_n will destabilize helium and carbon; an increase in m_e of order m_e will destabilize helium and carbon; an increase in m_ν of order m_e will destabilize helium and carbon; a small increase in α will destabilize all nuclei; a small decrease in α_S, the strong coupling constant, will destabilize all nuclei; an increase or decrease in G_F, the Fermi constant, of order unity will decrease the energy output of supernovae, and one sign will lead to a universe dominated by helium. Thus, the CNS hypothesis explains the values of all the parameters that determine low-energy physics and chemistry: the masses of the proton, neutron, electron and neutrino and the strengths of the strong, weak and electromagnetic interactions.

However, explanation is different from prediction. These cannot be considered independent predictions of the theory, because the existence of carbon and oxygen, plus long-lived stars, are also conditions of our own existence. Hence selection effects prevent us from claiming these as unique predictions of CNS. If the theory is to make falsifiable tests, it must involve changes of parameters that do not affect the conditions necessary for our own existence. There are such tests, and they will be described shortly. Before discussing them, however, we should address several criticisms that have been made.

20.6.3 Previous criticisms

Several arguments have been made that $\mathcal{S}$ is contradicted by present observation [91–93]. These are found to depend either on confusion about the hypothesis itself or on too simple assumptions about star formation. For example, it was argued in ref. [91] that star formation would proceed to more massive stars were the universe to consist only of neutrons, because there would be no nuclear processes to impede direct collapse to black holes. This kind of argument ignores the fact that the formation of stars massive enough to become black holes requires efficient cooling of giant molecular clouds. The cooling processes that appear to be dominant require carbon and oxygen, both for formation of CO, whose vibrational modes are the most efficient mechanism of cooling, and because dust grains consisting of carbon and ice provide efficient shielding of star-forming regions from starlight. But even processes cooling molecular clouds to 5–20 K are not enough; formation of massive stars appears to require that the cores of the cold clouds are disturbed by shock-waves, which come from ionized regions around other massive stars and supernovae. For these reasons, our universe appears to produce many more black holes than would a universe consisting of just neutrons.[16]

Vilenkin (personal communication) has raised an issue concerning the cosmological constant. Were Λ (the vacuum energy) raised from its present value, he notes that galaxy formation would not have taken place at all. One might add that, with even a slight increase, galaxy formation would have been cut off, leading only to small galaxies unable to sustain the process of self-propagated star formation that is apparently necessary for copious formation of massive stars. This counts as a success of the theory.

On the other hand, were Λ smaller than its present value, there might be somewhat increased massive star formation, due to the fact that, at the present time, the large spiral galaxies are continuing to accrete matter through several processes. These include the accretion of intergalactic gas onto the disks of galaxies and the possible flow of gas from large gaseous disks that the visible spiral galaxies may be embedded in. It is difficult to estimate exactly how much the mass of spiral galaxies would be increased by this process, but Vilenkin claims it could be as much as 10–20%.

However, lowering Λ would also increase the number of mergers of spiral galaxies and the number of absorptions of dwarf galaxies by spirals. These mergers and absorptions are believed to convert spiral galaxies to elliptical galaxies by destroying the stellar disk and heating the gas. The result is to cut off the formation of massive stars, leaving much gas unconverted to stars.

There is then a competition between two effects. Raise Λ and galaxies do not form or do not grow large enough to support disks and hence massive star formation. Decrease Λ and the dominant effect may be to cut off massive star formation, due to increased mergers and absorptions converting spiral to elliptical galaxies. One can conjecture that the present value of Λ maximizes the formation of black holes.

It has also been claimed that $\mathcal{S}$ is untestable with present knowledge [93, 94]. In the following, I will show that these claims are also false, by explaining

---

16 For details, see the appendix of ref. [17], which addresses the objections published in refs. [91–93] and elsewhere.

why a single observation of an astrophysical object that very well might exist – a heavy neutron star – would refute $\mathcal{S}$. After this, I describe two more kinds of observations that could refute $\mathcal{S}$ in the near future. These involve more accurate observations of the spectrum of fluctuations in the cosmic microwave background (CMB) and the initial mass function for star formation in the absence of carbon.

20.6.4 Why a single heavy pulsar would refute $\mathcal{S}$

Bethe and Brown [95] have hypothesized that neutron star cores contain a condensate of K^- mesons. Their calculations show that there is a critical value μ_c for the strange quark mass μ such that, for $\mu < \mu_c$, neutron star cores consist of approximately equal numbers of protons and neutrons with the charge balanced by a condensate of K^- mesons. The reason is that in nuclear matter the effective mass of the K^- is renormalized downward by an amount depending on the density ρ. Given a choice of the strange quark mass, let $\rho_0(\mu)$ be the density where the renormalized kaon mass is less than the electron mass. Then μ_c is the value of μ where $\rho_0(\mu)$ becomes less than the density ρ_e at which the electrons react with the protons to form neutrons. In either case, one neutrino per electron is produced, leading to a supernova.

Bethe and Brown and collaborators [95–98] claim that $\mu < \mu_c$ but their calculations involve approximations, such as chiral dynamics, so are insufficiently accurate to exclude $\mu_c > \mu$. However, as μ decreases, the accuracy of the calculations increases as μ^{-2}, so, even if we are not sure that $\mu < \mu_c$, we can be reasonably confident of the existence of a critical value μ_c. We may then reason as follows. If $\mu < \mu_c$, the upper mass limit is low, approximately $1.5 M_\odot$. If $\mu > \mu_c$, neutron stars have the conventional equations of state and the upper mass limit is higher, almost certainly above $2 M_\odot$ [99]. Therefore, a single observation of a neutron star whose mass was sufficiently high could show $\mu > \mu_c$, refuting Bethe and Brown's claim. A mass of $2.5 M_\odot$ would certainly suffice, although any value higher than $1.5 M_\odot$ would be troubling if one believed Bethe and Brown's upper limit. Furthermore, this would refute $\mathcal{S}$ because a decrease in μ would then lead to a world with a lower upper mass limit for neutron stars, and therefore more black holes. All well measured neutron star masses are currently from binary pulsar data and are below $1.5 M_\odot$ [100, 101]; other methods yield less precise estimates [102].

We note that this argument is independent of any issue of selection effects associated with 'anthropic reasoning', because the value of the strange quark mass μ may be varied within a large range before it produces a significant effect on chemistry. Sceptics might reply that, were $\mathcal{S}$ so refuted, it could be modified to a new unrefuted $\mathcal{S}'$ by the additional hypothesis that μ is not an independent parameter and cannot be varied without also, say, changing the proton–neutron mass difference, leading to large effects in star formation. Of course, most theories can be saved by the proliferation of *ad hoc* hypotheses, but science tends to reject hypotheses that require special fixes. There are occasions where such a fix is warranted. The present case would only be among them if there were a preferred fundamental theory, which had strong independent experimental support, in which μ was not an independent parameter but could not be changed without altering the values of parameters that strongly affect star formation and evolution.

20.6.5 How observations of the CMB could refute $\mathcal{S}$

It might be observed that there would have been many more primordial black holes (PBHs) if the spectrum of primordial fluctuations, $f(n)$, were tilted to increase their amplitude on small scales [103]. This does not refute $\mathcal{S}$ directly unless the inflationary model has a parameter that can be varied to achieve this tilt. The Standard Model does not, but it is reasonable to examine whether some plausible extension of it, $\mathcal{E}$, might do so. One plausible extension is to add a field that could serve as the inflaton. The spectrum of primordial fluctuations may then be predicted as a function of the parameters of $\mathcal{E}$. Thus $\mathcal{S}$ is refuted if (a) some model $\mathcal{E}$ of inflation is observationally confirmed and (b) that particular model has some parameter, p_{inf}, that can be modified to *increase the total number of PBHs produced*. Given the accuracy expected for observations of the CMB from the WMAP and PLANCK satellites, there is a realistic possibility that these will distinguish between different hypotheses $\mathcal{E}$ and measure the values of their parameters.

In the standard 'new' inflationary scenario [34], there is no parameter that fulfils the function required of p_{inf}. There is the inflaton coupling, λ, and it is true that the amplitude of $f(n)$ is proportional to λ, so that the number of PBHs can be increased by increasing this. However, the size of the region that inflates, R, scales as $e^{\lambda^{-1/2}}$. This means that λ should be at the lower limit of the range of values for which galaxy formation occurs. An exponentially larger universe, which produces black holes only through supernova remnants, still has vastly more black holes than an exponentially smaller universe with many primordial black holes. In fact, if the observations confirm the new inflationary scenario, $\mathcal{S}$ is refuted if λ is not tuned to the value that maximizes the total production of black holes in the inflated region [13]. Because of the exponential decrease in R with increasing λ, this is likely to be close to the *smallest* value that leads to appreciable black hole production. This should correspond to the smallest λ that allows prolific formation of galaxies [13].

This seems consistent with the actual situation, in which there appears to have been little PBH production. Therefore, given that $Q \equiv \delta\rho/\rho \approx 10^{-5}$, the primary mode of production of black holes seems to be through massive star production in galaxies that do not form until rather late. However, there are non-standard models of inflation that have parameters p_{inf} that can be varied in a manner that tilts $f(n)$ so that more PBHs are created without decreasing R [104–109]. If future CMB observations show that standard new inflation is ruled out, so that only models with such a parameter p_{inf} are allowed, then $\mathcal{S}$ will be refuted. This is a weaker argument than the first one, but – given the scope for increased accuracy of the CMB measurements – such a refutation is plausible.

20.6.6 How early star formation could refute $\mathcal{S}$

As shown in refs. [6, 13, 94], there are several directions in $\mathcal{P}$ which lead to universes that contain no stable nuclear bound states. It is argued in refs. [13, 17] that this leads to a strong decrease in $F(p)$, because the gravitational collapse of objects more massive than the upper mass limit of neutron stars seems to depend on the cooling mechanisms in giant molecular clouds, which are dominated by radiation from CO. In a universe without nuclear bound states, the upper mass limit for stable collapsed objects is unlikely to decrease dramatically (as the dominant factor ensuring stability is Fermi statistics), while collapsed objects larger than the upper mass limit are likely to be less common without CO cooling.

In the absence of bound states, the main cooling mechanism involves molecular hydrogen [110], but there are two reasons to suppose this would not lead to plentiful collapse of massive objects in a world with nuclear bound states. The first is that there would be no dust grains, and these appear to be the primary catalysts for the binding of molecular hydrogen. The second is that molecular hydrogen is, anyway, a less efficient coolant than CO [110]. Given present uncertainties in star formation processes, this is a weaker argument than the first. However, since these uncertainties are unlikely to be reduced in the near future, let us ask whether this argument could be refuted by any possible observations.

In the present universe the collapse of massive objects is dominated by processes that involve nuclear bound states, but we have a laboratory for the collapse of objects in the absence of nuclear bound states – our universe before enrichment with metals. Indeed, we know that massive objects must have collapsed then, otherwise carbon, oxygen and other elements would never have been produced. But, given that CO acts as a catalyst for the formation of heavy elements and that the dust formed from heavy elements produced in stars is a catalyst for molecular binding, there is an instability whereby any chance formation of massive objects leads in a few million years to both an enrichment of the surrounding medium and the production of significant quantities of dust. These greatly increase the probability of forming additional massive objects, so the initial rate of formation of heavy objects in the absence of enrichment does not have to be very high to explain how our universe first became enriched.

This shows that the collapse of some heavy objects before enrichment does not refute the argument that the number of black holes produced in a universe without nuclear bound states would be much less than at present. But nor does it establish the argument. It is still consistent with present knowledge that the production of massive objects in the absence of heavy elements proceeds efficiently under the right conditions, so that there may have been a great deal of early star formation uncatalyzed by any process involving heavy elements. This could lead to a refutation of $\mathcal{S}$ because, in a world without nuclear bound states, many more massive collapsed objects would become black holes than do in our universe, where the collapse is delayed by stellar nucleosythesis.

The question is then whether a combination of observation and theory could disentangle the strong catalytic effects of heavy elements, leading to a strong positive feedback in massive star formation, from the initial rate of massive star formation without heavy elements. Although models of star formation, with and without heavy elements are not sufficiently developed to distinguish the two contributions at early times, it is likely that this will become possible as our star formation models improve. If so, future observations may yield enough information about early star formation to distinguish the two effects. If the conclusion is that the number of black holes formed is greater in a world without nuclear bound states than in our own, then $\mathcal{S}$ would be refuted.

20.7 Conclusions

This chapter was written with the hope of contributing to a debate about the possible role of the AP in physics and cosmology. Having carefully considered the arguments and engaged several proponents in conversation and correspondence, it seems to me incontrovertible not only that the AP is unscientific, but also that its role is negative. To the extent that it is espoused to justify continued interest in unfalsifiable theories, it may play a destructive role in the progress of science. The main points of my argument are as follows.

- No theory can be a candidate for a physical theory that does not make falsifiable predictions. To violate this maxim is to risk the development of a situation in which the scientific community splits into groups divided by different unverifiable faiths, because there is no possibility of killing popular theories by rational argument from shared evidence.
- The version of the AP described by **A** and **B** cannot lead to falsifiable theories.
- Claimed successful predictions of anthropic reasoning involve either uncontroversial use of selection effects within one universe, as in the argument of Dicke, or simple deductions from observed facts, with life or ensembles of universes playing no role in the prediction, as in the arguments of Hoyle and Weinberg. There are no successful anthropic predictions that do not fall into these two classes. Hence all claims for the success of the AP are false.
- The Principle of Mediocrity is ambiguous, because a reasonable change in the definition of the ensemble in which we or our civilization are taken to be typical can often turn an argument for a correct conclusion into an argument for an incorrect one. In specific applications, it is often unreliable. When the claims made can be tested, as in the doomsday argument, it often leads to false conclusions.
- Eternal inflation cannot lead to an explanation of the low-energy parameters of the Standard Model and thus to a resolution of why these parameters allow stars and organic chemistry, because these parameters play no role in the mechanism that generates the probability distribution for universes created by eternal inflation.
- It is possible to derive falsifiable predictions from a multiverse theory if the following conditions are satisfied: (1) the ensemble of universes generated differs strongly from a random ensemble, constructed from an unbiased measure; (2) almost all members of the ensemble have a property W that is not a consequence of either the known laws of physics or a requirement for the existence of life; (3) it must be possible to establish whether W is true or not in our universe by a practical experiment.
- There is at least one example of a falsifiable theory satisfying these conditions, which is CNS. Among the properties W that make the theory

falsifiable is that the upper mass limit of neutron stars is less than $1.6 M_\odot$. This and other predictions of CNS have yet to be falsified, but they could be by observations in progress.

It must then be considered unacceptable for any fundamental theory of physics to rely on the AP in order to make contact with observations. When such claims are made, as they have been recently for string theory [31–33], this can only be considered as a sign that the theory is in deep trouble and at great risk of venturing outside the bounds of science.

There are, of course, alternatives. String theory might be shown to imply the conditions necessary for CNS, in which case it would yield falsifiable predictions. Or another mechanism for the selection of parameters might turn out to lead to falsifiable predictions. What is clear is that some falsifiable version of the theory must be found. Otherwise the theory cannot be considered scientific, because there will be no way to establish its truth or falsity by a means which allows consensus to be established by rational argument from shared evidence.

Acknowledgements

I am grateful to numerous people for conversation and correspondence on these topics, including Anthony Aguirre, Stephon Alexander, Tom Banks, John Barrow, Michael Dine, George Ellis, Ted Jacobson, Joao Magueijo, Fotini Markopoulou, John Moffat, Slava Mukhanov, Holger Nielsen, Joe Polchinski, Martin Rees, Sherrie Roush, Alex Vilenkin, Peter Woit, and colleagues at Perimeter Institute. I am also grateful to Michael Douglas, Jaume Garriga, Leonard Susskind, Alex Vilenkin and Steven Weinberg for helpful clarifications and references. This research was supported partly by the Jesse Phillips Foundation. Finally, I am grateful to the Templeton Foundation and the organizers of the 2001 Cambridge Conference for including anthro-sceptics like myself in the debate, in the true spirit of science.

References

[1] K. Popper. *Conjectures and Refutations* (London: Routledge and Keagan Paul, 1963), pp. 33–39.
[2] K. Popper. In *Readings in the Philosophy of Science*, ed. T. Schick (Mountain View, CA: Mayfield Publishing Company, 2000), pp. 9–13.
[3] K. Popper. *The Open Society and its Enemies* (Princeton, NJ: Princeton University Press, 1971).
[4] K. Popper. *The Logic of Scientific Discovery* (London: Routledge, 1992).
[5] L. Smolin. *The Trouble with Physics* (Boston, MA: Houghton Mifflin, 2006).

[6] B. J. Carr and M. J. Rees. The anthropic principle and the structure of the physical world. *Nature* **278** (1979), 605.

[7] B. Carter. The significance of numerical coincidences in nature. Cambridge University preprint (1967).

[8] B. Carter. Number coincidences and the anthropic principle in cosmology. In *Confrontation of Cosmological Theories with Observational Data, IAU Symposium No. 63*, ed. M. Longair (Dordrecht: Reidel, 1974), p. 291.

[9] J. D. Barrow and F. J. Tipler *The Anthropic Cosmological Principle* (Oxford: Oxford University Press, 1986).

[10] S. Roush. Copernicus, Kant, and the anthropic cosmological principles. *Stud. Hist. Phil. Mod. Phys.* **34** (2003), 5–35.

[11] S. Roush. Anthropic principle. In *The Philosophy of Science: An Encyclopedia*, eds. J. Pfeifer and S. Sarkar (London: Routledge, 2003).

[12] A. Feoli and S. Rampone. Is the strong anthropic principle too weak? *Nuovo Cim.* **B 114** (1999), 281.

[13] L. Smolin. Did the universe evolve? *Class. Quant. Grav.* **9** (1992), 173.

[14] L. Smolin (1994) On the fate of black hole singularities and the parameters of the standard model. [gr-qc/9404011].

[15] L. Smolin. Cosmology as a problem in critical phenomena. In *Proceedings of the Guanajuato Conference on Complex Systems and Binary Networks*, eds. R. Lopez-Pena, R. Capovilla, R. Garcia-Pelayo, H. Waalebroeck and F. Zertuche (Berlin: Springer, 1995) [gr-qc/9505022].

[16] L. Smolin. Experimental signatures of quantum gravity. In *Proceedings of the Fourth Drexel Conference on Quantum Nonintegrability*, eds. B. L. Hu and D. H. Feng (Boston: International Publishers, 1996) [gr-qc/9503027].

[17] L. Smolin. *The Life of the Cosmos* (New York: Oxford University Press, 1997).

[18] P. Binetruy, G. L. Kane, B. D. Nelson, L-T. Wang and T. T. Wang. Relating incomplete data and incomplete theory. *Phys. Rev.* **D 70** (2004), 095006 [hep-ph/0312248].

[19] F. Markopoulou. Towards gravity from the quantum (2006) [hep-th/0604120].

[20] D. W. Kribs and F. Markopoulou. Geometry from quantum particles (2005) [gr-qc/0510052].

[21] S. O. Bilson-Thompson, F. Markopoulou and L. Smolin. Quantum gravity and the standard model (2006) [hep-th/0603022].

[22] S.O. Bilson-Thompson. A topological model of composite preons. (2005) [hep-ph/0503213].

[23] T. Thiemann. The LQG – string:loop quantum gravity quantization of string theory I. Flat target space. *Class. Quant. Grav.* **23** (2006), 1923 [hep-th/0401172].

[24] L. Smolin. How far are we from the quantum theory of gravity? (2003) [hep-th/0303185].

[25] S. Kachru, R. Kallosh, A. Linde and S. P. Trivedi. de Sitter vacua in string theory. *Phys. Rev.* **D 68** (2003), 046005 [hep-th/0301240].

[26] S. B. Giddings, S. Kachru and J. Polchinski. Hierarchies from fluxes in string compactifications. *Phys. Rev.* **D 66** (2002), 106006 [hep-th/0105097].

[27] R. Bousso and J. Polchinski. Quantization of four-form fluxes and dynamical neutralization of the cosmological constant. *JHEP*, **0006** (2000), 006 [hep-th/0004134].

[28] L. Susskind. This volume (2007) [hep-th/0302219].

[29] M. Douglas. The statistics of string/M theory vacua. *JHEP*, **0305** (2003), 046 [hep-th/0303194].

[30] M. Douglas. Statistical analysis of the supersymmetry breaking scale (2004) [hep-th/0405279].

[31] A. H. Guth. The inflationary universe: A possible solution to the horizon and flatness problems. *Phys. Rev.* **D 23** (1981), 347.

[32] A. D. Linde. A new inflationary universe scenario: A possible solution of the horizon, flatness, homogeneity, isotropy and primordial monopole problems. *Phys. Lett.* **B 108** (1982), 389.

[33] A. Albrecht and P. J. Steinhardt. Cosmology for grand unified theories with radiatively induced symmetry breaking. *Phys. Rev. Lett.* **48** (1982), 1220.

[34] A. D. Linde. *Particle Physics and Inflatioary Cosmology* (Chur, Switzerland: Harwood Academic, 1990).

[35] P. J. Steinhardt and N. Turok. A cyclic model of the universe. *Science*, **296** (2002), 1436.

[36] P. J. Steinhardt and N. Turok. Cosmic evolution in a cyclic universe. *Phys. Rev.* **D 65** (2002), 126003 [hep-th/0111098].

[37] J. W. Moffat. Superluminary universe: A possible solution to the initial value problem in cosmology. *Int. J. Mod. Phys.* **D 2** (1993), 351 [gr-qc/9211020].

[38] A. Albrecht and J. Magueijo. A time varying speed of light as a solution to cosmological puzzles. *Phys. Rev.* **D 59** (1999), 043516 [astro-ph/9811018].

[39] J. D. Barrow. Cosmologies with varying light speed. *Phys. Rev.* **D 59** (1999), 043515.

[40] J. Magueijo. New varying speed of light theories. *Rep. Prog. Phys.* **66** (2003), 2025 [astro-ph/0305457].

[41] A. Vilenkin. The birth of inflationary universes. *Phys. Rev.* **D 27** (1983), 2848.

[42] A. Linde. The inflationary universe. *Rep. Prog. Phys.* **47** (1984), 925.

[43] A. Linde, D. Linde and A. Mezhlumian. From the big bang theory to the theory of a stationary universe. *Phys. Rev.* **D 49** (1994), 1783 [gr-qc/9306035].

[44] J. Garcia-Bellido and A. Linde. Stationarity of inflation and predictions of quantum cosmology. *Phys. Rev.* **D 51** (1995), 429 [hep-th/9408023].

[45] J. Garriga and A. Vilenkin. A prescription for probabilities in eternal inflation. *Phys. Rev.* **D 64** (2001), 023507 [gr-qc/0102090].

[46] J. A. Wheeler. Beyond the end of time. In *Black Holes, Gravitational Waves and Cosmology: An Introduction to Current Research*, eds. M. Rees, R. Ruffini and J.A. Wheeler (New York: Gordon and Breach).

[47] C. Misner, K. Thorne and J. A. Wheeler. *Gravitation* (San Francisco: Freeman, 1974).

[48] V. P. Frolov, M. A. Markov and M. A. Mukhanov. Through a black hole into a new universe? *Phys. Lett.* **B 216** (1989), 272.

[49] A. Lawrence and E. Martinec. String field theory in curved space-time and the resolution of spacelike singularities. *Class. Quant. Grav.* **13** (1996), 63 [hep-th/9509149].

[50] E. Martinec. Spacelike singularities in string theory. *Class. Quant. Grav.* **12** (1995), 941 [hep-th/9412074].

[51] M. Bojowald. Isotropic loop quantum cosmology. *Class. Quant. Grav.* **19** (2002), 2717 [gr-qc/0202077].

[52] M. Bojowald. Inflation from quantum geometry. *Phys. Rev. Lett.* **89** (2002), 261301 [gr-qc/0206054].

[53] M. Bojowald. The semi-classical limit of loop quantum cosmology. *Class. Quant. Grav.* **18** (2001), L109 [gr-qc/0105113].

[54] M. Bojowald. Dynamical initial conditions in quantum cosmology. *Phys. Rev. Lett.* **87** (2001), 121301 [gr-qc/0104072].

[55] S. Tsujikawa, P. Singh and R. Maartens. Loop quantum gravity effects on inflation and the CMB. *Class. Quant. Grav.* **21** (2003), 5767 [astro-ph/0311015].

[56] V. Husain and O. Winkler. Quantum resolution of black hole singularities. *Class. Quant. Grav.* **22** (2005), L127 [gr- qc/0410125].

[57] L. Modesto. Disappearance of black hole singularity in quantum gravity. *Phys. Rev.* **D 70** (2004), 124009 [gr-qc/0407097].

[58] A. Ashtekar and M. Bojowald. Quantum geometry and the Schwarzschild singularity. *Class. Quant. Grav.* **23** (2006), 391 [gr-qc/0509075].

[59] R. H. Dicke. Principle of equivalence and weak interactions. *Rev. Mod. Phys.* **29** (1957), 355.

[60] P. A. M Dirac. The cosmological constants. *Nature*, **139** (1937), 323.

[61] F. Hoyle. On the fragmentaion of gas clouds into galaxies and stars. *Astrophys. J.* **118** (1953), 513.

[62] A.Vilenkin. Predictions from quantum cosmology. *Phys. Rev. Lett.* **74** (1995), 846 [gr-qc/9406010].

[63] J. Garriga and A. Vilenkin. Testable anthropic predictions for dark energy. *Phys. Rev.* **D 67** (2003), 043503 [astro-ph/0210358].

[64] J. Garriga, M. Livio and A. Vilenkin. The cosmological constant and the time of its dominance. *Phys. Rev.* **D 61** (2000), 023503 [astro-ph/9906210].

[65] S. Weinberg. Anthropic bound on the cosmological constant. *Phys. Rev. Lett.* **59** (1987), 2067.

[66] H. Martel, P. Shapiro and S. Weinberg. Likely values of the cosmological constant. *Astrophys. J.* **492** (1998), 29 [astro-ph/9701099].

[67] S. Weinberg. A priori probability distribution of the cosmological constant. *Phys. Rev.* **D 61** (2000), 103505 [astro-ph/0002387].

[68] S. Weinberg. The cosmological constant problems. In *Sources and Detection of Dark Matter and Dark Energy in the Universe*, ed. D. Cline (New York: Springer-Verlag, 2000) [astro-ph/0005265].

[69] H. Nielsen. Random dynamics and relations between the number of fermion generations and the fine structure constants. *Act. Phys. Polon.* **B 20** (1989), 427.

[70] J. R. Gott III. Implications of the Copernican principle for our future prospects. *Nature*, **363** (1993), 315.

[71] J. Leslie. *The End of the World. The Ethics and Science of Human Extinction* (London: Routledge, 1996).

[72] F. Markopoulou. The internal description of a causal set: What the universe looks like from the inside. *Commun. Math. Phys.* **211** (2000), 559 [gr-qc/9811053].

[73] M. Ahmed, S. Dodelson, P. B. Greene and R. Sorkin. Ever-present lambda. *Phys. Rev.* **D 69** (2004), 103523 [astro-ph/0209274].

[74] M. J. Rees. Anthropic reasoning. Clues to a fundamental theory. *Complexity*, **3** (1997), 17.

[75] M. J. Rees. Numerical coincidences and 'tuning' in cosmology. In *Fred Hoyle's Universe*, eds. C. Wickramasinghe, G. Burbidge and J. V. Narlikar (Dordrecht: Kluwer, 2003), p. 87.

[76] M. Tegmark and M. J. Rees. Why is the CMB fluctuation level 10^{-5}? *Astrophys. J.* **499** (1998), 526 [astro-ph/9709058].

[77] M. L. Graesser, S. D. H. Hsu, A. Jenkins and M. B. Wise. Anthropic distribution for cosmological constant and primordial density perturbations. *Phys. Lett.* **B 600** (2004), 15 [hep-th/0407174].

[78] A. Aguirre. The cold big-bang cosmology as a counter-example to several anthropic arguments. *Phys. Rev.* **D 64** (2001), 083508 [astro-ph/0106143].

[79] P. Davies. *The Mind of God* (New York: Simon and Schuster, 1991).

[80] J. Gribbin. *In the Beginning: After COBE and Before the Big Bang* (New York: Little Brown, 1993).

[81] S. Kauffman. *Investigations into Autonomous Agents* (Oxford: Oxford University Press, 2002).

[82] Y. Nambu. Distribution of particle physics. *Prog. Theor. Phys. Suppl.* **85** (1985), 104.

[83] A. Linde. Particle physics and inflationary cosmology. *Phys. Today*, **40** (1987), 61.

[84] A. Linde. *Particle Physics and Inflationary Cosmology* (Chur, Switzerland: Harwood, 1990).

[85] A. Linde. Universe multiplication and the cosmological constant problem. *Phys. Lett.* **B 200** (1988), 272.

[86] A. Linde. Life after inflation and the cosmological constant problem. *Phys. Lett.* **B 227** (1989), 352.

[87] A. Linde. Extended chaotic inflation and spatial variations of the gravitational constant. *Phys. Lett.* **B 238** (1990), 1680.

[88] J. Garcia-Bellido, A. Linde and D. Linde. Fluctuations of the gravitational constant in the inflationary Brans-Dicke cosmology. *Phys. Rev.* **D 50** (1994), 730 [astro-py/9312039].

[89] J. Garcia-Bellido and A. Linde. Stationary solutions in Brans-Dicke stochastic inflationary cosmology. *Phys. Rev.* **D 52** (1995), 6780 [gr-qc/9504022].

[90] R. Gambini and J. Pullin. Discrete quantum gravity: A mechanism for selecting the value of fundamental constants. *Int. J. Mod. Phys.* **D 12** (2003), 1775 [gr-qc/0306095].

[91] T. Rothman and G. F. R. Ellis. Smolin's natural selection hypothesis. *Quart. J. Roy. Astron. Soc.* **34** (1993), 201.

[92] E. R. Harrison. The natural selection of universes containing intelligent life. *Quart. J. Roy. Astron. Soc.* **36** (1995), 193.

[93] J. Silk. *Science*, **227** (1997), 644.

[94] M. Rees. In *Before the Beginning* (Reading, MA: Addison Wesley, 1997).

[95] G. E. Brown and H. A. Bethe. A scenario for a large number of low-mass black holes in the Galaxy. *Astrophys. J.* **423** (1994), 659.

[96] G. E. Brown and J. C. Weingartner. Accretion onto and radiation from the compact object formed in SN 1987A. *Astrophys. J.* **436** (1994), 843 .

[97] G. E. Brown. Kaon condensation in dense matter. *Nucl. Phys.* **A 574** (1994), 217.

[98] H. A. Bethe and G. E. Brown. Observational constraints on the maximum neutron star mass. *Astrophys. J. Lett.* **445** (1995), L129.

[99] G. B. Cook, S. L. Shapiro and S.A. Teukolsky. Rapidly rotating neutron stars in general relativity. Realistic equations of state. *Astrophys. J.* **424** (1994), 823.

[100] S. E. Thorsett, Z. Arzoumanian, M. M. McKinnon and J. H. Taylor. The masses of two binary neutron star systems. *Astrophys. J. Lett.* **405** (1993), L29.

[101] D. J. Nice, R. W. Sayer and J. H. Taylor. PSR J1518+4904: A mildly relativistic binary pulsar system. *Astrophys. J.* **466** (1996), L87.

[102] J. Casares, P. Charles and E. Kuulkers. The mass of the neutron star in Cyg X-2 (V1341 Cyg). *Astrophys. J. Lett.* **493** (1997), L39.

[103] B. J. Carr, J. H. Gilbert and J. E. Lidsey. Black hole relics from inflation: Limits on blue perturbation spectra. *Phys. Rev.* **D 50** (1994), 4853.

[104] M. Bucher, A. S. Goldhaber and N. Turok. An open universe from inflation. *Phys. Rev.* **D 52** (1995), 3314–3337 [hep-ph/9411206, PUPT-94-1507].

[105] M. Bucher and N. Turok. Open inflation with arbitrary false vacuum mass. *Phys. Rev.* **D 52** (1995), 5538–5548 [hep-ph 9503393, PUPT-95-1518].

[106] J. R. Gott. Creation of open universes from de Sitter space. *Nature*, **295** (1982), 304.

[107] J. Garcia-Bellido and A. Linde. Tilted hybrid inflation. *Phys. Lett.* **B 398** (1997), 18 [astro-ph/9612141].

[108] J. Garcia-Bellido and A. Linde. Open hybrid inflation. *Phys. Rev.* **D 55** (1997), 7480–7488 [astro-ph/9701173].

[109] D. H. Lyth and A. Riotto. Particle physics models of inflation and the cosmological density perturbation. *Phys. Rep.* **314** (1999), 1 [hep-ph/9807278].

[110] M. Tegmark, J. Silk, M. Rees, A. Blanchard, T. Abel and F. Palla. How small were the first cosmological objects? *Astrophys. J.* **474** (1997), 1 [astro-ph/9603007].

21

Making predictions in a multiverse: conundrums, dangers, coincidences

Anthony Aguirre
Department of Physics, University of California, Santa Cruz

21.1 Introduction

The standard models of particle physics and cosmology are both rife with numerical parameters that must have values fixed by hand to explain the observed world. The world would be a radically different place if some of these constants took a different value. In particular, it has been argued that if any one of six (or perhaps a few more) numbers did not have rather particular values, then life as we know it would not be possible [1]; atoms would not exist, or no gravitationally bound structures would form in the Universe, or some other calamity would occur that would appear to make the alternative universe a very dull and lifeless place. How, then, did we get so lucky as to be here?

This question is an interesting one because all of the possible answers to it that I have encountered or devised entail very interesting conclusions. An essentially exhaustive list of such answers follows.

(i) We just got very lucky. All of the numbers could have been very different, in which case the Universe would have been barren, but they just happened by pure chance to take values in the tiny part of parameter space that would allow life. We owe our existence to one very, very, very lucky roll of the dice.

(ii) We were not particularly lucky. Almost any set of parameters would have been fine, because life would find a way to arise in nearly any type of universe. This is quite interesting because it implies (at least theoretically) the existence of life-forms radically different from our own, for example existing in universes with no atoms or bound structures, or overrun with black holes, etc.

(iii) The Universe was specifically designed for life. The choice of constants only happened once, but their values were determined in some way by

Universe or Multiverse?, ed. Bernard Carr. Published by Cambridge University Press.
© Cambridge University Press 2007.

the need for us to arise. This might involve divine agency or some radical form of Wheeler's 'self-creating Universe' or super-advanced beings that travel back in time to set the constants at the beginning of the Universe, etc. However one feels about this possibility, one must admit that it would be interesting if true.

(iv) We did not have to get lucky, because there are many universes with different sets of constants, i.e. the dice were rolled many, many times. We are necessarily in one of the universes that allows life, just as we necessarily reside on a planet that supports life, even when most others may not. This is interesting because it means that there are other very different universes coexisting with ours in a 'multiverse'.

These four answers – luck, *élan vital*, design and multiverse – will appeal at different levels to different people. But I think it is hard to argue that the multiverse is necessarily less reasonable than the alternatives. Moreover, as discussed at length elsewhere in this volume, there are independent reasons to believe – on the basis of inflation, quantum cosmology and string/M-theory – that there might naturally be many regions outside the observable one, governed by different sets of low-energy physics. I am not aware of any independent scientific argument for the other three possible explanations.

Whether they are contemplated as an answer to the 'why are we lucky' question, or because they are forced upon us from other considerations, multiverses come at a high price. Even if we have in hand a physical theory and cosmological model that lead to a multiverse, how do we test it? If there are many sets of constants, which ones do we compare to those we observe? In Section 21.2, I will outline what I think a sound prediction in a multiverse would look like. As will become clear, this requires many ingredients and there are serious difficulties in generating some of these ingredients, even with a full theory in hand. For this reason, many short-cuts have been devised to try to make predictions more easily. In Section 21.3, I will describe a number of these and show the cost that this convenience entails. Finally, in Section 21.4, I will focus on the interesting question of whether the anthropic approach to cosmology might lead to any *general* conclusions about how the study of cosmology will look in coming years.

21.2 Making predictions in a multiverse

Imagine that we have a candidate physical theory and a set of cosmological boundary conditions (hereafter denoted $\mathcal{T}$) that predict an ensemble of physically realized systems, each of which is approximately homogeneous in some coordinates and can be characterized by a set of parameters (i.e. the constants appearing in the standard models of particle physics and cosmology). I assume here that the laws of physics themselves retain the same form. Let us denote each such system a 'universe' and the ensemble a 'multiverse'. Given that we can observe only one of these universes, what conclusions can we draw regarding the correctness of $\mathcal{T}$, and how?

One possibility would be that there is a parameter for which *no* universes in the ensemble have the observed value. In this case, $\mathcal{T}$ would be ruled out. (Note that any $\mathcal{T}$ in which at least one parameter has an excluded range in *all* universes is thus rigorously falsifiable, which is a desirable feature for any theory.) Or perhaps some parameter takes only one value in all universes, and this value matches the observed one. This would obviously be a significant accomplishment of the theory. Both possibilities are good, as far as they go, and seem completely uncontroversial. But they do not go far enough. What if our observed parameter values appear in some but not all of the universes? Could we still rule out the theory if those values are incredibly rare, or gain confidence if they are extremely common?

I find it hard to see why not. If some theory predicts outcome A of some experiment with probability $p = 0.99\,999\,999$ and outcome B with probability $1 - p$, I think we would be reluctant to accept the theory if a single experiment were performed and showed outcome B, *even if we did not get to repeat the experiment*. In fact, it seems consistent with all normal scientific methodology to rule out the theory at $99.999\,999\%$ confidence – the problem is just that we will not be able to *increase* this confidence without repeating our measurements. This seems to be exactly analogous to the multiverse situation *if* we can compute, given our $\mathcal{T}$, the *probability* that we should observe a given value for some observable.

Can we compute this probability distribution in a multiverse? Perhaps. I will argue that to do so, in a sensible way, we would need seven successive ingredients.

(i) First, of course, we require a multiverse: an ensemble of regions, each of which would be considered a universe to observers inside it (i.e. its properties would be uniform for as far as those observers could see), but each of which may have different properties.

(ii) Next we need to isolate the set of parameters characterizing the different universes. This might be the set of twenty or so free parameters in the Standard Model of particle physics (see refs. [2, 3] and references therein), plus a dozen or so cosmological parameters [4, 5]. There might be additional parameters that become important in other universes, or differences (such as different forms of the physical laws) that cannot be characterized by differences in a finite set of parameters. But, for simplicity, let us assume that some set of numbers α_i (where $i = 1, \ldots, N$) fully specifies each universe.

(iii) Given our parameters, we need some *measure* with which to calculate their multi-dimensional probability distribution $P(\alpha_i)$. We might, for example, 'count each universe equally' to obtain the probability $P_{\rm U}(\alpha_i)$, defined to be the chance that a randomly chosen universe from the ensemble would have the parameter values α_i.[1] This can be tricky, however, because it depends on how we delineate the universes. Suppose that $\alpha_1 = a$ universes happen to be 10^{10} times larger than $\alpha_1 = b$ universes. What would then prevent us from 'splitting' each $\alpha_1 = a$ universe into 10 or 100 or 10^{10} universes, thus radically changing the relative probability of $\alpha_1 = a$ versus $\alpha_1 = \beta$? These considerations might lead us to take a different measure, such as volume. We could then define $P_{\rm V}(\alpha_i)$ as the chance that a random point in space resides in a universe with parameter values α_i. But in an expanding universe volume increases, so this would depend on the time at which we chose to evaluate the volume in each universe. We might then consider some 'counting' object that endures, say a baryon (which is relatively stable), and define $P_{\rm B}(\alpha_i)$, the chance that a randomly chosen baryon resides in a universe with parameter values α_i. But now we have excluded from consideration universes with no baryons. Do we want to do that? This will be addressed in step (v). For now, note only that it is not entirely clear, even in principle, which measure we should place over our multiverse. We can call this the 'measure problem'.

(iv) Once we have chosen a measure object M, we still need actually to compute $P_{\rm M}(\alpha_i)$, and this may be far from easy. For example, in computing $P_{\rm V}$, some universes may have infinite volume. In this case, values of α_i leading to universes with finite volume will have zero probability. How, though, do we compare two infinite volumes? The difficulty can be seen by considering how we would count the fraction of red and blue marbles in an infinite box. We could pick one red, then one blue, and find a 50:50 split. But we could also repeatedly pick one red, then two blue, or five red, then one blue. We could do this forever and so obtain any ratio we like. What we would like to do is just 'grab a bunch of marbles at random' and count the ratio. However, in the multiverse

---

[1] Note that this is really shorthand for $({\rm d}\mathcal{P}/{\rm d}\alpha_1 \cdots {\rm d}\alpha_N)({\rm d}\alpha_1 \cdots {\rm d}\alpha_N)$, the probability that α_i are all within the interval $[\alpha_i, \alpha_i + {\rm d}\alpha_i]$, so $\mathcal{P}(\alpha_i)$ is a cumulative probability distribution.

case, it is not so clear how to pick this random ordering of marbles. This difficulty, which might be termed the 'ordering problem' [5], has been discussed in the context of eternal inflation [6–10] and a number of plausible prescriptions have been proposed. But there does not seem to be any generic solution or convincing way to prove that one method is correct.

(v) If we have managed to calculate $P_M(\alpha_i)$, do we have a prediction? At least we have an answer to the question 'Given that I am (or can associate myself with) a randomly chosen M-object, which sort of universe am I in?' But this is *not* necessarily the same as the more general question 'What sort of universe am I in?' First, different M-objects will generally give different probabilities, so they cannot all be the answer to the same question. Second, we may not be all that closely associated with our M-object (which was chosen mainly to provide *some* way to compute probabilities) because it does not take into account important requirements for our existence. For example, if M were volume, I could ask 'What should I observe, given that I am at a random point in space?' However, we are *not* at a random point in space (which would on average have a density of 10^{-29} g cm^{-3}) but at one of the very rare points with density ~ 1 g cm^{-3}. The reason for this improbable situation is obviously 'anthropic' – we just do not worry about it because we can observe many other regions at the proper density. (If we could not see such regions, we might be more reluctant to accept a cosmological model with such a low average density.) Finally, it might be argued that the question we have answered through our calculation is not specific enough, because we know a lot more about the Universe than its volume or baryon content. We might, instead, ask: 'Given that I am in a universe with the properties we have already observed, what should I observe in the future?'

As discussed at length in ref. [11], these different questions can be usefully thought of as arising from different choices of *conditionalization*. The probabilities $P_M(\alpha_i)$ are conditioned on as little as possible, whereas the anthropic question – 'Given that I am a randomly chosen *observer*, what should I measure?' – specifies probabilities conditioned on the existence of an 'observer'. The approach of 'Given what I know now, what will I see?' specifies probabilities conditioned on being in a universe with all of the properties that we have already observed. These are three genuinely different approaches to making predictions in a multiverse that may be termed, respectively, 'bottom-up', 'anthropic' and 'top-down'.

Let us denote by O the conditionalization object used to specify these conditional probabilities. In bottom-up reasoning it would be the same as the M-object; in the anthropic approach it would be an observer; and in the top-down approach it could be a universe with the currently known properties of our own. It can be seen that they cover a spectrum, from the weakest conditionalization (bottom-up) to the most stringent (top-down). Like our initial M-object, choosing a conditionalization is unavoidable and important, and there is no obviously correct choice to make. (See refs. [12] and [13] for similar conditionalization schemes.)

(vi) Having decided on a conditionalization object O, the next step is to compute the number $N_{O,M}(\alpha_i)$ of O-objects per M-object, for each set of values of the parameters α_i. For example, if we have chosen to condition on observers, but have used baryons to define our probabilities, then we need to calculate the number of observers per baryon as a function of cosmological parameters. We can then calculate $P_O(\alpha_i) \equiv P_M(\alpha_i) N_{O,M}(\alpha_i)$, i.e. the probability that a randomly chosen O-object (observer) resides in a universe with parameters α_i. There are possible pitfalls in doing this. First, if $N_{O,M}$ is infinite, then the procedure clearly breaks down because P_O then becomes undefined. This is why the choice of the M-object should require as little as possible and hence be associated with the minimal-conditionalization bottom-up approach. This difficulty will occur generically if the existence of an O-object does not necessarily entail the existence of an M-object. For example, if the M-object were a baryon but the O-object were a bit of volume, then N would be infinite for α_i corresponding to universes with no baryons. The problem arises because baryons require volume but volume does not require baryons. This seems straightforward but becomes much murkier when we consider the *second* difficulty in calculating N, which is that we may not be able to define precisely what an O-object is or what it takes to make one. If we say that the O-object is an observer, what exactly does that mean? A human? A carbon-based life-form? Can observers exist without water? Without heavy elements? Without baryons? Without volume? It seems hard to say, so we are forced to choose some proxy for an observer, for example a galaxy or a star with planet, etc. But our probabilities will perforce depend on the chosen proxy, and this must be kept in mind.

It is worth noting a small bit of good news here. If we do manage to compute N_{O,M_1} consistently for some measure object M_1, then insofar as we want to condition our probabilities on O-objects, we have solved the measure problem. If we could consistently calculate N_{O,M_2} for a different measure object M_2, then we should obtain the same result for N_O, i.e. $N_{O,M_1} P_{M_1} = N_{O,M_2} P_{M_2}$. Thus, our choice of M_1 (rather than M_2) becomes unimportant.

(vii) The final step in making predictions is to assume that the probability that we will measure some set of α_i is given by the probability that a randomly chosen O-object will do so. This assumption really entails two others: first, that we are somehow directly associated with O-objects; second, that we have not – simply by bad luck – observed highly improbable values of the parameters. The assumption that we are *typical* observers has been termed the 'principle of mediocrity' [14]. One may argue about this assumption, but *some* assumption is necessary if we are to connect our computed probabilities to observations, and it is difficult to see what alternative assumption would be more reasonable.

The result of all this work would be the probability $P_O(\alpha_i)$ that a randomly selected O-object (out of all of the O-objects that exist in the multiverse) would reside in a universe governed by parameters α_i, along with a reason to believe that this same probability distribution should govern what we observe. We can then make observations or consider some already made ones. If the observations are highly improbable according to our predictions, we can rule out the candidate $\mathcal{T}$ at some confidence that depends on how improbable our observations were. Apart from the manifest and grave difficulties involved in actually completing the seven steps listed above in a convincing way, I think the only real criticism that can be levelled at this approach is that, unless $P = 0$ for our observed parameters, there will always be the chance that the $\mathcal{T}$ was correct and we measured an unlikely result. Usually, we can rid ourselves of this problem by repeating our experiments to make P as small as we like (at least in principle), but here we do not have that option – once we have 'used up' the measurement of all the parameters required to describe the Universe (which appears to be surprisingly few according to current theories), we are done.

21.3 Easing predictions in a multiverse: a bestiary of shortcuts

Although the idea of a multiverse has been around for a while, no one has ever really come close to making the sort of calculation outlined in the previous section.[2] Instead, those wishing to make predictions in a multiverse context have made strong assumptions about which parameters α_i actually vary

---

[2] The most ambitious attempt is probably the recent one by Tegmark [3,5].

across the ensemble, about the choice of O-object, and about the quantities $P_M(\alpha_i)$ and $N_{O,M}(\alpha_i)$ that go into predicting the measurement probabilities. Some of these short-cuts aim simply to make a calculation tractable; others are efforts to avoid anthropic considerations or, alternatively, to use anthropic considerations to avoid other difficulties.

I would not have listed any ingredients that I thought could be omitted from a sound calculation, so all of these short-cuts are necessarily incomplete (some, in my opinion, disastrously so). But by listing and discussing them, I hope to give a flavour of what sort of anthropic arguments have been made in the literature and where they may potentially go astray.

21.3.1 Anthropic arguments only allow one set of parameters

This assumption underlies a sort of anthropic reasoning that has earned the anthropic principle a lot of ill will. It goes something like this: 'Let's assume that lots of universes, governed by lots of different parameter values, exist. Then, since only universes with parameter values almost exactly the same as ours allow life, we must be in one of those, and we should not find it strange if our parameter values seem special.' In the conventions I have described, this is essentially equivalent to setting O-objects to be observers and then hoping that the 'hospitality factor' $N_{O,M}(\alpha_i)$ is very narrowly peaked around one particular set of parameters. In this case, the *a priori* probabilities P_M are almost irrelevant because the shape of $N_{O,M}$ will pick out just one set of parameters. Because our observed values α_i^{obs} definitely allow observers, the allowed set must then be very near α_i^{obs}.

There are three problems with this type of reasoning. First, it is rather circular: it entails picking the O-object to be an observer, but then quickly substituting it with a 'universe just like ours', on the grounds that such universes will definitely support life.[3] Thus we conclude that the universe should be very much like the observed one. The way to avoid this silliness is to allow at least the possibility that there are life-supporting universes with $\alpha_i \neq \alpha_i^{obs}$, i.e. to discard the unproven *assumption* that $N_{O,M}$ has a single, dominant, narrow peak.

The second problem is that, if $N_{O,M}$ were *so* narrowly peaked as to render P_M irrelevant, then we would be in serious trouble as theorists, because we would lose any ability to distinguish between candidates for our fundamental theory. Unless our observed universe is *impossible* in the theory, then the

---

3 One can also argue that, if there were other, more common, universes that supported life, we ought to be in them; since we are not, we should assume that almost all life-supporting universes are like our own. But this argument is also circular, since it assumes the anthropic argument works.

anthropic factor would force the predictions of the theory to match our observations. As discussed in Section 21.3.2, this is not good.

The third problem with $N_{O,M}$ being an extremely peaked function is that it does not appear to be true! As discussed below, for any reasonable surrogate for observers (for example galaxies like our own or stars with heavy elements, etc.), calculations using our current understanding of galaxy and structure formation seem to indicate that the region of parameter space in which there can be many of those objects may be small compared to the full parameter space but much larger than the region compatible with our observations.

21.3.2 Just look for zero-probability regions in parameter space

As mentioned in Section 21.2, one (relatively) easy thing to do with a multiverse theory is to work out which parameter combinations cannot occur in *any* universe. If the combination we actually observe is one of these, then the theory is ruled out. This is unobjectionable, but a rather weak way to test a theory because, if we are given two theories that are *not* ruled out, then we have no way of judging one to be better, even if the parameter values we observe are generic in one and absurdly rare in the other.[4]

This is not how science usually works. For example, suppose our theory is that a certain coin-tossing process is unbiased. If our only way to test this theory were to look for experimental outcomes that are impossible, then the theory would unfalsifiable; we would have to accept it for *any* coin we are confronted with, because *no* sequence of tosses would be impossible! Even if 10 000 tosses in a row all came up heads, we would have no grounds for doubting our theory because, while getting heads 10 000 times in a row on a fair coin is absurdly improbable, it is not impossible. Nor would we have reason to prefer the (seemingly much better) 'nearly every toss comes out heads' theory. Clearly this is a situation we would like to improve on, as much in universes as in coin tosses.

21.3.3 Let us look for overwhelmingly more probable values

One possible improvement would be employ the 'bottom-up' reasoning described in Section 21.2 and assume that we will observe a 'typical' set of parameters in the ensemble. This amounts to using the *a priori* (or

---

[4] Amusingly, in terms of testing $\mathcal{T}$, the approach which makes *no* assumptions about $N_{O,M}$ is equivalent to the approach just described of making the very strong assumption that $N_{O,M}$ allows only one specific set of parameter values, because – in either case – a theory can only be ruled out if our observed values are impossible in that theory.

'prior') probabilities P_M for some choice of M-object, such as universes, and just ignoring the conditionalization factor $N_{O,M}$. There are two possible justifications for this. First, we might simply want to avoid any sort of anthropic issues on principle. Second, we might hope that some parameter values are much more common than others, to the extent that the $N_{O,M}$ factor becomes irrelevant, i.e. that P_M (rather than $N_{O,M}$) is very strongly peaked around some particular parameter values.

The problem with this approach is the 'measure problem' previously discussed: there is an implicit choice of basing probabilities on universes rather than on volume elements or baryons. Each of these measures has problems – for example, it seems that probabilities based on 'universes' depend on how the universes are delineated, which can be ambiguous. Moreover, there seems to be no reason to believe that predictions made using any two measures should agree particularly well. For example, as discussed elsewhere in this volume [15], in the string theory 'landscape' there are many possible parameter sets, depending on which metastable minimum one chooses in a potential that depends in turn on a number of fluxes that can take a large range of discrete values. Imagine that exponentially many more minima lead to $\alpha_1 = a$ than to $\alpha_1 = b$. Should we expect to observe $\alpha_1 = a$? Not necessarily, because the relative number of a and b universes that *actually come into existence* may easily differ exponentially from the relative number of a and b minima. (This seems likely to me in an eternal-inflation context, where the relative number of universes could depend on exponentially suppressed tunnellings between vacua.) Even worse, these may in turn differ exponentially (or even by an infinite factor) from the relative numbers of baryons or relative volumes.

In short, while we are free to use bottom-up reasoning with any choice of measure object we like, we are not free to assert that other choices would give similar predictions, or that conditionalization can be rendered irrelevant. So we had better have a good reason for our choice.

21.3.4 Let us fix some parameters as observed and predict others

Another way in which one might hope to circumvent anthropic issues is to condition the probabilities on some or all observations that have already been made. In this 'top-down' (or perhaps 'pragmatic') approach we ask: 'Given everything that has been observed so far, what will we observe in some future measurement?' This has a certain appeal, as this is often done in experimental science; we do not try to *predict* what our laboratory will look like, just what will happen given that the laboratory is in a particular state at a given time. In the conventions of Section 21.2, the approach could consist of choosing the O-object to be universes with parameters agreeing with the measured values. While appealing, this approach suffers some deficiencies.

- It still does not completely avoid the measure problem, because even once we have limited our consideration to universes that match our current observations, we must still choose a measure with which to calculate the probabilities for the remaining ones.
- Through our conditioning, we may accept theories for which our parameter values are wildly improbable, without supplying any justification as to why we observe such improbable values. This is rather strange. Imagine that I have a theory in which the cosmological constant Λ is (with very high probability) much higher than we observe and the mass of the particle providing the dark matter m_{DM} almost certainly exceeds $1000\,\text{GeV}$. I condition on our observed Λ by simply accepting that I am in an unusual universe. Now, if I measured $m_{DM} = 1\,\text{GeV}$, I would like to say my theory is ruled out. However, according to top-down reasoning, I should have already ruled it out if I had done my calculation in 1997, before Λ was measured. And someone who invented the very same theory next week – but had not been told that I have already ruled it out – would *not* rule it out, but instead just take the low value of m_{DM} (along with the observed Λ) as part of the conditionalization!
- If we condition on everything we have observed, we obviously give up the possibility of *explaining* anything through our theory.

These two issues motivate variations on the top-down approach in which only some current observations are conditioned on. Two possibilities are as follows.

(i) We might *start* by conditioning on all observations, then progressively condition on less and less and try to 'predict' the things we have decided not to condition on (as well, of course, as any *new* observations) [16,17]. The more we can predict, the better our theory is. The problem is that either: (a) we will get to the point where we are conditioning on as little as possible (the bottom-up approach) and hence the whole conditionalization process will have been a waste of time; or (b) we will still have to condition on some things and admit that either these have an anthropic explanation or we just choose to condition on them (leading to the peculiar issues discussed above).

(ii) We might choose at the outset to condition on things that we think may be fixed anthropically (without trying to generate this explanation), then try to predict the others [18]. This is nice in being relatively easy and in providing a justification for the conditionalization. It suffers from three problems: (a) we have to guess which parameters are anthropically important and which are not; (b) even if a parameter is anthropically unimportant, it may be strongly correlated in P_M with one that is; and (c) we still face the measure problem, which we cannot avoid by conditioning on observers, because we are avoiding anthropic considerations.

21.3.5 Let us assume that just one parameter varies

Most of the 'short-cuts' discussed so far have been attempts to avoid anthropic considerations. But we may, instead, consider how we might try to formulate an anthropic prediction (or explanation) for some observable, without going through the full calculation outlined in Section 21.2. The way of doing this that has been employed in the literature (largely in the efforts of Vilenkin and collaborators [19–21]) is as follows.

First, one fixes all but one of the parameters to the observed values. This is done for tractability and/or because one hopes that they will have non-anthropic explanations. Let us call the parameter that is allowed to vary across the ensemble α.

Second, an O-object is chosen such that – *given that only α varies* – the number $N_{O,M}$ of these objects (per baryon or comoving volume element) in a given universe is hopefully calculable and proportional to the number of observers. For example, if only Λ varies across the ensemble, galaxies might make reasonable O-objects because a moderately different Λ will probably not change the number of observers per galaxy, but *will* change the number of galaxies in a way that can be computed using fairly well understood theories of galaxy and structure formation (e.g. see refs. [5], [20] and [22]–[26]).

Third, it is assumed that $P_M(\alpha)$ is either flat or a simple power law, without any complicated structure. This can be done just for simplicity, but it is often argued to be natural [27–29]. The flavour of this argument is as follows. If P_M is to have an interesting structure over the relatively small range in which observers are abundant, there must be a parameter of order the observed α in the expression for P_M. But it is precisely the absence of this parameter that motivated the anthropic approach. For example, if the expression for $P_M(\Lambda)$ contained the energy scale ~ 0.01 eV corresponding to the observed Λ, the origin of that energy scale would probably be more interesting than our anthropic argument, as it would provide the basis for a (non-anthropic) solution to the cosmological constant problem!

Under these (fairly strong) assumptions, we can then actually calculate $P_O(\alpha)$ and see whether or not the observed value is reasonably probable given this predicted distribution. For example, when Λ alone is varied, a randomly chosen galaxy is predicted to lie in a universe with Λ comparable to (but somewhat larger than) the value we see [23].[5]

I consider this sort of reasoning respectable, *given the assumptions made*. In particular, the anthropic argument in which *only* Λ varies is a relatively clean one. But there are a number of pitfalls when it is applied to parameters other than Λ or when one allows multiple parameters to vary simultaneously.

- Assuming that the abundance of observers is proportional to that of galaxies only makes sense if the number of galaxies – and not their properties – changes as α varies. However, changing nearly any cosmological parameter will change the properties of typical galaxies. For example, increasing Λ will decrease the number of galaxies, but will also make them smaller on average, because a high Λ squelches structure formation at the late times when massive galaxies form. Similarly, increasing the amplitude of primordial perturbations would lead to smaller, denser – but more numerous – galaxies, as would increasing the ratio of dark matter to baryons. In these cases, we must specify in more detail what properties an observer-supporting galaxy should have, and this is very difficult without falling into the circular argument that only galaxies like ours support life. Finally, this sort of strategy seems unlikely to work if we try to change *non*-cosmological parameters, as this could lead to radically different physics and the necessity of thinking *very* hard about what sort of observers there might be.
- The predicted probability distribution clearly depends on P_M, and the assumption that P_M is flat or a simple power law can break down. This can happen even for Λ [23, 30], but perhaps more naturally for other parameters – such as the dark matter density – for which particle physics models can already yield sensible values. Moreover, this breakdown is much more probable if (as discussed below and contrary to the assumption made above) the hospitality factor $N_{O,M}(\alpha)$ is significant over many orders of magnitude in α.
- Calculations of the hospitality factor $N_{O,M}(\alpha)$ can go awry if α is changed more than a little. For example, a neutrino mass slightly larger than

---

[5] This is for a 'flat' probability distribution $dP_M/d\lambda \propto \lambda^\alpha$ with $\alpha = 0$. For $\alpha > 0$, higher values would be predicted; for $\alpha < 0$, lower values would be.

observed would suppress galaxy formation by erasing small-scale structure. But neutrinos with a large ($\gtrsim 100$ eV) mass would act as dark matter and lead to strong halo formation. Whether these galaxies would be hospitable is questionable, since they would be very baryon-poor, but the point is that the physics becomes *qualitatively* different. As another example, a lower photon/baryon ratio (n_γ/n_b) would lead to earlier, denser galaxies. But a *much* smaller value would lead to qualitatively different structure formation, as well as the primordial generation of heavy elements [4]; these changes are very dangerous because, over orders of magnitude in α, $P_M(\alpha)$ will tend to change by many orders of magnitude. Thus, even if these alternative universes only have a few observers in them, they may dominate P_O and hence qualitatively change the predictions.

- Along the same lines, but perhaps even more pernicious, when multiple parameters are varied simultaneously, the effects of some variations can offset the effect of others so that universes quite different from ours can support many of our chosen O-objects. For example, increasing Λ cuts off galaxy formation at a given cosmic density, but raising Q (the amplitude of the presumably scale-invariant perturbations on the horizon scale) causes galaxies to form earlier, thus nullifying the effect of Λ. This can be seen in the calculations of refs. [24] and [31] and is discussed explicitly in refs. [3]–[5] and [26]. Many such degeneracies exist, because raising Λ, n_γ/n_b or the neutrino mass all decrease the efficiency of structure formation, while raising the density $\rho_{\rm DM}$ of dark matter relative to the density ρ_b of baryons, or raising Q, would increase the efficiency. As an extreme case, it was shown in ref. [4] that, if Q and n_γ/n_b are allowed to vary with Λ, then universes in which Λ is 10^{17} times our observed value could arguably support observers! Including more cosmological or non-cosmological parameters can only make this problem worse.

These problems indicate that, while anthropic arguments concerning Λ in the literature are relatively 'clean', it is unclear whether other parameters (taken individually) will work as nicely. More importantly, a number of issues arise when several parameters are allowed to vary at once, and there does not seem to be any reason to believe that success in explaining one parameter anthropically will persist when additional parameters are allowed to vary. It may do so in some cases – for example, allowing neutrino masses to vary in addition to Λ does not appear to spoil the anthropic explanation of a small but non-zero Λ [25]. On the other hand, allowing Q to vary does unless $P_M(Q)$ is strongly peaked at small values [26]. I suspect that allowing $\rho_{\rm DM}/\rho_b$ or n_γ/n_b to vary along with Λ would have a similar effect.

21.3.6 So what should we do?

For those serious about making predictions in a multiverse, I would propose that rather than working to generate additional incomplete anthropic arguments by taking short-cuts, a much better job must be done with each of the individual ingredients. For example, our understanding of galaxy formation is sufficiently strong that the multi-dimensional hospitality factor $N_{O,M}(\alpha_i)$ could probably be computed for α_i within a few orders of magnitude of the observed values, if we take the O-objects to be galaxies with properties in some range. Second, despite some nice previous work, I think the problem of how to compute P_M in eternal inflation is an open one. Finally, the currently popular string/M-theory landscape cannot hope to say much about P_M until its place in cosmology is understood – in particular, we need a better understanding both of the statistical distribution of field values that result from evolution in a given potential and of how transitions between vacua with different flux values occur and exactly what is transitioning.

21.4 Are there general predictions of anthropic reasoning?

The preceding sections might suggest that it will be a huge project to compute sound predictions of cosmological and physical parameters from a multiverse theory in which they vary. It may indeed be a long time before any such calculation is believable. It is therefore worth asking if there is any way nature might indicate whether the anthropic approach is sensible, i.e. does it make any *general* predictions, even without the full calculation of P_O? Interestingly, I think the answer might be yes; I am aware of two such general (though somewhat vague) predictions of the anthropic approach.

To understand the first, assume that only one parameter, α, varies and consider $p(\log \alpha) = \alpha P_M(\alpha)$, the probability distribution in $\log \alpha$, given by some theory $\mathcal{T}$. For $\log \alpha$ near the observed value $\log \alpha^{\text{obs}}$, p can either rise, fall or remain approximately constant with increasing $\log \alpha$. In the first two cases, $\mathcal{T}$ would predict that we should see a value of α that is, respectively, higher or lower than we actually do *if* no anthropic conditionalization $N_{O,M}$ is applied. Now suppose we somehow compute $N_{O,M}(\alpha)$ and find that it falls off quickly for values of α much smaller or larger than we observe, i.e. that only a range $\alpha_{\min} \lesssim \alpha \lesssim \alpha_{\max}$ is 'anthropically acceptable'. Then we have an anthropic argument explaining α^{obs}, because this fall-off means that P_O will only be significant near α^{obs}. But now note that *within* the anthropically acceptable range, P_O will be peaked near $\alpha_{\max}$ if p is increasing with α, or near $\alpha_{\min}$ if p is decreasing with α. That is, we should expect α^{obs} at one edge of the anthropically acceptable range. This idea has been called the 'principle of living dangerously' [32]. It asserts that, for a parameter that is anthropically determined, we should expect that a calculation of $N_{O,M}$ would reveal that observers would be strongly suppressed either for α slightly larger or slightly smaller than α^{obs}, depending on whether p is rising or falling.

Now this is not a very specific prediction; exactly where we would expect α^{obs} to lie depends on both the steepness of $p(\log \alpha)$ and the sharpness of the cut-off in $N_{O,M}$ for α outside the anthropically acceptable range. And it would not apply to anthropically determined parameters in all possible cases. (For example, if p were flat near α^{obs} but very high at $\alpha \gg \alpha^{\text{obs}}$, anthropic effects would be required to explain why we do not observe the very high value; but any region within the anthropically acceptable range would be equally probable, so we would not expect to be living on the edge.) Despite these caveats, this is a prediction of sorts, because the naïve expectation would probably be for our observation to place us somewhere in the interior of the region of parameter space that is hospitable to life rather than at the edge.

A second sort of general prediction of anthropic reasoning is connected to what might be called 'cosmic coincidences'. For example, many cosmologists have asked why the current density in vacuum energy, dark matter, baryons and neutrinos are all within a couple of orders of magnitude of each other – making the Universe a much more complicated place than it might be. Conventionally, it has been assumed that these are just coincidences which follow directly from fundamental physics that we do not understand. But if the anthropic approach to cosmology is correct (that is, if it is the real answer to the question of why these densities take the particular values they do), then the explanation is quite different; the densities are bound together by the necessity of the existence of observers, because only certain combinations will do.

More explicitly, suppose several cosmological parameters are governed by completely unrelated physics, so that their individual prior probabilities P_M simply multiply to yield the multi-dimensional probability distribution. For example, we might have $P_M(\Lambda, \Omega_{\text{DM}}/\Omega_b, Q) = P_M(\Lambda) P_M(\Omega_{\text{DM}}/\Omega_b) P_M(Q)$. But even if P factors like this, the hospitality factor $N_{O,M}$ will almost certainly not do so; if galaxies are O-objects, the number of galaxies formed at a given Λ will depend on the other two parameters and only certain combinations will give a significant number of observers. Thus, $P_O = N_{O,M} P_M$ will likewise have correlations between the different parameters that lead to only particular combinations (e.g. those with $\Omega_{\text{DM}}/\Omega_b \sim 1\text{--}10$ for a given Q and Λ) having high probabilities. The cosmic coincidences would be explained in this way.

This anthropic explanation of coincidences, however, should not only apply to things that we have already observed. If it is correct, then it should apply also to *future* observations; that is, we should expect to uncover yet more bizarre coincidences between quantities that seem to follow from quite unrelated physics.

How might this actually happen? Consider dark matter. We know fairly precisely how much dark matter there is in the Universe and what its basic properties are. But we have no idea what it is, and many possible candidates have been proposed. In fact, we have no *observational* reason to believe that dark matter is one substance at all; in principle, it could be equal parts axions, supersymmetric particles and primordial black holes. The reason most cosmologists do not expect this is that it would be a strange coincidence if three substances involving quite independent physics all had roughly the same density. But, of course, this would be just like the suprising-but-true coincidences that already hold in cosmology.

In the anthropic approach, the comparability of these densities could be quite natural [11]. To see why, imagine that there are two *completely independent* types of dark matter permeating the ensemble; in each universe, they have particular densities ρ_1 and ρ_2 out of a wide range of possibilities, so that the densities in a randomly chosen universe (or around a randomly chosen baryon, etc.) will be given probabilistically by $P_M(\rho_1)P_M(\rho_2)$. Under these assumptions, there is no reason to expect that we should observe $\rho_1 \sim \rho_2$ based just on these *a priori* probabilities. However, now suppose that $N_{O,M}$ picks out a particular narrow range of *total* dark matter density as anthropically acceptable – that is, $N_{O,M}(\rho_1 + \rho_2)$ is narrowly peaked about some ρ_{anth}. In this case, the peak of the probability distribution $P_O(\rho_1, \rho_2)$ will occur where $P_M(\rho_1)P_M(\rho_2)$ is maximized *subject to the condition* that $\rho_1 + \rho_2 \simeq \rho_{\text{anth}}$. For simplicity, let both prior probabilities be power laws: $P_M(\rho_1) \propto \rho_1^\alpha$ and $P_M(\rho_2) \propto \rho_2^\beta$. Then it is not hard to show that, if $\alpha \geq 0$ and $\beta \geq 0$, then the maximum probability will occur when $\rho_1/\rho_2 = \alpha/\beta$. That is, the two components are likely to have similar densities unless the *power law indices* of their probability distributions differ by orders of magnitude.[6] Of course, there are many ways in which this coincidence could fail to occur (e.g. negative power law indices or correlated probabilities), but the point is that there is a quite natural set of circumstances in which the

---

6 Extremely high power law indices are uncomfortable in the anthropic approach because they would lead to P_O being peaked where $N_{O,M}$ is declining, i.e. we should be living *outside* the anthropically comfortable range – not just dangerously but recklessly.

components are coincident, even though the *fundamental* physics is completely unrelated.

21.5 Conclusions

The preceding sections should have convinced the reader that there are good reasons for scientists to be very worried if we live in a multiverse; in order to test a multiverse theory in a sound manner, we must perform a fiendishly difficult calculation of $P_O(\alpha_i)$, the probability that an O-object will reside in a universe characterized by parameters α_i. And, because of the shortcomings of the short-cuts one may (and presently must) take in doing this, almost any particular multiverse prediction is going to be easy to criticize. Much worse, we face an unavoidable and important choice in what O should be: a possible universe, an existing universe, a universe matching current observations, a bit of volume, a baryon, a galaxy, an 'observer' etc.

I find it disturbingly plausible that observers really are the correct conditionalization object, that their use as such is the correct answer to the measure problem, and that anthropic effects are the real explanation for the values of some parameters (just as for the local density that we observe). Many cosmologists appear to believe that taking the necessity of observers into account is shoddy thinking, employed only because it is the *easy* way out of solving problems the 'right' way. But the arguments of this chapter suggest that the truth may well be exactly the opposite; the anthropic approach may be the right thing to do in principle, but nearly impossible in practice.

Even if we cannot calculate P_O in the foreseeable future, however, cosmology in a multiverse may not be completely devoid of predictive power. For example, if anthropic effects are at work, they should leave certain clues. First, if we could determine the region of parameter space hospitable to observers, we should find that we are living on the outskirts of the habitable region rather than somewhere in its middle. Second, if the anthropic effects are the explanation of the parameter values – and coincidences between them – that we see, then it ought to predict that new coincidences will be observed in future observations.

If, in the next few decades, dark matter is resolved into several equally important components, dark energy is found to be three independent substances, and several other 'cosmic coincidences' are observed, even some of the most die-hard sceptics might accede that the anthropic approach may have validity – why else would the Universe be so baroque? On the other hand, if the specification of the basic cosmological constituents is essentially complete, and the associated parameters are in the middle of a relatively large region of parameter space that might arguably support observers, then I think the anthropic approach would lose almost all its appeal. We would be forced to ask why the universe is not much weirder.

References

[1] M. Rees. *Just Six Numbers* (New York: Basic Books, 2000).
[2] C. J. Hogan. Why the universe is just so. *Rev. Mod. Phys.* **72** (2000), 1149 [astro-ph/9909295].
[3] M. Tegmark, A. Aguirre, M. J. Rees and F. Wilczek. Dimensionless constants, cosmology and other dark matters. *Phys. Rev.* **D 73** (2006), 23505 [astro-ph/0511774].
[4] A. Aguirre. The cold big-bang cosmology as a counter-example to several anthropic arguments. *Phys. Rev.* **D 64** (2001), 83508.
[5] M. Tegmark. What does inflation really predict? *J. Cosmol. Astropart. Phys.* **0504** (2005), 001 [astro-ph/0410281].
[6] A. D. Linde and A. Mezhlumian. On regularization scheme dependence of predictions in inflationary cosmology. *Phys. Rev.* **D 53** (1996), 4267 [gr-qc/9511058].
[7] V. Vanchurin, A. Vilenkin and S. Winitzki. Predictability crisis in inflationary cosmology and its resolution. *Phys. Rev.* **D 61** (2000), 083507 [gr-qc/9905097].
[8] A. H. Guth. Inflation and eternal inflation. *Phys. Rep.* **333** (2000), 555 [astro-ph/0002156].
[9] J. Garriga and A. Vilenkin. A prescription for probabilities in eternal inflation. *Phys. Rev.* **D 64** (2001), 023507 [gr-qc/0102090].
[10] J. Garriga, D. Schwartz-Perlov, A. Vilenkin and S. Winitzki. Probabilities in the inflationary multiverse. *J. Cosmol. Astropart. Phys.* **0601** (2006), 017 [hep-th/0509184].
[11] A. Aguirre and M. Tegmark. Multiple universes, cosmic coincidences, and other dark matters. *J. Cosmol. Astropart. Phys.* **0501** (2005), 003 [hep-th/0409072].
[12] N. Bostrum. *Anthropic Bias: Observation Selection Effects in Science and Philosophy* (New York: Routledge, 2002).
[13] J. B. Hartle. Anthropic reasoning and quantum cosmology. *AIP Conf. Proc,* **743** (2005), 298 [gr-qc/0406104].
[14] A. Vilenkin. Predictions from quantum cosmology. *Phys. Rev. Lett.* **74** (1995), 846 [gr-qc/9406010].
[15] L. Susskind. This volume (2007) [hep-th/0302219].
[16] S. W. Hawking and T. Hertog. Why does inflation start at the top of the hill? *Phys. Rev.* **D 66** (2002), 123509 [hep-th/0204212].
[17] A. Albrecht. Cosmic inflation and the arrow of time. In *Science and Ultimate Reality*, eds. J. D. Barrow, P. C. W. Davies and C. L. Harper (Cambridge: Cambridge University Press, 2004), pp. 363–401 [astro-ph/0210527].
[18] M. Dine. Supersymmetry, naturalness and the landscape. In *Pascos 2004, Themes in Unification* (Singapore: World Scientific, 2004), pp. 249–63 [hep-th/0410201].
[19] J. Garriga and A. Vilenkin. Anthropic prediction for Lambda and the Q catastrophe. *Prog. Theor. Phys. Suppl.* **163** (2005), 245 [hep-th/0508005].

[20] M. Tegmark, A. Vilenkin and L. Pogosian. Anthropic predictions for neutrino masses. *Phys. Rev.* **D 71** (2005), 103523 [astro-ph/0304536].
[21] J. Garriga and A. Vilenkin. Testable anthropic predictions for dark energy. *Phys. Rev.* **D 67** (2003), 043503 [astro-ph/0210358].
[22] J. Garriga, M. Livio and A. Vilenkin. The cosmological constant and the time of its dominance. *Phys. Rev.* **D 61** (2000), 023503 [astro-ph/9906210].
[23] J. Garriga and A. Vilenkin. On likely values of the cosmological constant. *Phys. Rev.* **D 61** (2000), 083502 [astro-ph/9908115].
[24] J. Garriga and A. Vilenkin. Solutions to the cosmological constant problems. *Phys. Rev.* **D 64** (2001), 023517 [hep-th/0011262].
[25] L. Pogosian, A. Vilenkin and M. Tegmark, Anthropic predictions for vacuum energy and neutrino masses. *J. Cosmol. Astropart. Phys.* **0407** (2004), 005 [astro-ph/0404497].
[26] M. L. Graesser, S. D. H. Hsu, A. Jenkins and M. B. Wise. Anthropic distribution for cosmological constant and primordial density perturbations. *Phys. Lett.* **B 600** (2004), 15 [hep-th/0407174].
[27] A. Vilenkin. Quantum cosmology and the constants of nature. In *Tokyo 1995, Cosmological Constant and the Evolution of the Universe* (Tokyo: Universal Academy Press, 1995), pp. 161–8 [gr-qc/9512031].
[28] S. Weinberg. A priori probability distribution of the cosmological constant. *Phys. Rev.* **D 61** (2000), 103505 [astro-ph/0002387].
[29] L. Smolin. This volume (2007) [hep-th/0407213].
[30] D. Schwartz-Perlov and A. Vilenkin. Probabilities in the Bousso–Polchinski multiverse. *J. Cosmol. Astropart. Phys.* **6** (2006), 010 [hep-th/0601162].
[31] M. Tegmark and M. J. Rees. Why is the CMB fluctuation level 10^{-5}? *Astrophys. J.* **499** (1998), 526 [astro-ph/9709058].
[32] S. Dimopoulos and S. Thomas. Discretuum versus continuum dark energy. *Phys. Lett.* **B 573** (2003), 13 [hep-th/0307004].

22
Multiverses: description, uniqueness and testing

George Ellis
Department of Applied Mathematics, University of Cape Town

22.1 Introduction

The idea of a multiverse – an ensemble of universes or expanding domains like the one we see around us – has recently received increasing attention in cosmology. It has been conceived of as occurring either in separate places or times in the same overall encompassing universe (as in chaotic inflation) or through splitting of the quantum wave-function (as in the Everett interpretation of quantum mechanics) or as a set of totally disjoint universes, with no causal connection whatsoever. Physical properties may be different in the different universes or in different expanding domains within a single universe.

In this context, definitions are important. Some workers refer to the separate expanding regions in chaotic inflation as 'universes', even though they have a common causal origin and are all part of the same single spacetime. In keeping with long established use, I prefer to use the word 'universe' to refer to the single unique connected spacetime of which our observed region (centred on our galaxy and bounded by our visual horizon in the past) is a part. I will describe situations such as chaotic inflation – with many expanding domains – as a 'multi-domain universe'. Then we can reserve the term 'multiverse' for a collection of genuinely disconnected spacetimes, which are not causally related. When the discussion pertains to both disjoint collections of universes and different domains of a multi-domain universe, I will refer to an 'ensemble of universe domains' or 'ensemble' for short. There are basically three motivations for proposing an ensemble.

Generating mechanisms

It has been claimed that a multi-domain universe is the inevitable outcome of the physical processes that generated our own expanding region from a primordial quantum configuration; they would therefore have generated

Universe or Multiverse?, ed. Bernard Carr. Published by Cambridge University Press.
© Cambridge University Press 2007.

many other such regions. This was first modelled in a specific way by Vilenkin [1] and then developed by Linde [2,3] in his chaotic cosmology scenario. Since then, many others – including Sciama [4], Leslie [5], Deutsch [6], Tegmark [7,8], Smolin [9], Lewis [10], Weinberg [11], Rees [12,13] and Davies [14] – have discussed ways in which an ensemble of universe domains might originate physically.

Universality

The existence of an ensemble can be seen as the result of an overall philosophical stance underlying physics: the idea that 'everything that can happen does happen' [4,8,15]. This is a logical conclusion of the Feynman path integral approach to quantum theory, viewed as a basic underlying physical principle that provides a foundation for quantum physics and, in some sense, a fundamental approach to the nature of existence. Clearly this implies that we consider all possible alternative physics, as well as all possible alternative spacetime geometries.

Fine-tuning and anthropic issues

An ensemble has been proposed as an explanation for the way that our universe appears to be anthropically fine-tuned for the existence of life and the appearance of consciousness. It is now clear that, if any of a number of parameters which characterize the observed universe – including both fundamental constants and initial conditions – were slightly different, no complexity of any sort would come into existence and hence no life would appear and no Darwinian evolution would take place [16,17]. For example, Rees [18] suggests there are just six numbers that must be fine-tuned in order that life can exist:

(1) N = ratio of electrical and gravitational forces between protons = 10^{36};
(2) E = nuclear binding energy as a fraction of rest mass energy = 0.007;
(3) Ω = amount of matter in universe in units of critical density = 0.3;
(4) Λ = cosmological constant in units of critical density = 0.7;
(5) Q = amplitude of density fluctuations for cosmic structures = 10^{-5};
(6) D = number of spatial dimensions = 3.

A multiverse seems to be the only scientific way of explaining the precise adjustment of all these parameters simultaneously, so that complexity and life eventually emerged. The existence of a sufficiently large collection of expanding domains, covering the full range of possible combinations of parameter values, ensures that in some of them life would arise. In most of the domains, it will not do so because conditions will be wrong. But in a few of them conditions will happen to work out right. So, although there is an incredibly small probability of a domain existing that will allow life, if there exist enough domains, it becomes essentially inevitable that somewhere the right mix of circumstances will occur. If physical cosmogonic processes naturally produced such a variety of expanding domains, we necessarily find ourselves in one in which all the many conditions for life have been fulfilled.

This is analogous to the way in which we look upon the special character of the Solar System. We do not agonize about how initial conditions for the Earth and Solar System have allowed life to emerge. We realize that the Solar System is one of hundreds of billions of planetary systems in the Milky Way and accept that (though the probability of any one of them being bio-friendly is very low) at least a few will naturally be so. No direct fine-tuning is required, provided we take for granted both the nature of the laws of physics and the specific initial conditions in the universe. The processes of star formation throughout our galaxy naturally lead to the generation of the full range of possible stellar systems and planets.

An important point is that, in order for an ensemble with varied properties to explain fine-tuning, it must be an *actually existing ensemble* and not a potential or hypothetical one.[1] This is essential for any such anthropic argument.

Combinations of the above

One may finally note that these motivations are not necessarily in conflict with each other; one might, for example, attempt to propose a generating mechanism based on the ideas of universality that will also provide an anthropic explanation.

22.2 Describing multiverses

In considering how multiverses should be defined, it is important to note the key distinction between the collection of all possible universes and ensembles of really existing universes [20, 21]. We need first to describe the space of possibilities, characterizing the kinds of universes or expanding domains that can exist in any of the ensembles envisaged, and then to specify a distribution function on that space, characterizing those that actually exist in a specific realized ensemble.

---

1 An example of a paper that apparently only considers hypothetical ensembles is the contribution by Bjorken to this volume [19]. Although he talks about 'constructing ensembles', we are regrettably unable to do this.

22.2.1 Spaces of possibilities

The 'possibility space' M is the set of all possible universes m, each of which can be described by a set of states s in a state space S. Each universe will be characterized by a set of distinguishing parameters p, which are coordinates in S. The set of all possible parameters p form a parameter space P. One of the issues that arises is the 'equivalence problem' – the same universe m will in general be represented by a variety of different parameters. One can either factor out these multiple representations, going to the corresponding quotient space where each universe is represented just once, or try to identify the different representations of the same space in a naturally occurring parameter space. The latter is the better option, because the quotient space does not have a good manifold structure.

Each universe m in M will evolve from its initial state to the final state according to the dynamics operative, with some or all of its parameters varying as it does so. Thus, each such path in S (in degenerate cases, a point) is representative of one of the universes in M. If an ensemble contains numerous different FLRW-like[2] domains within a single overall connected spacetime, we can characterize the properties of each of these expanding domains by such a description.

The very description of the space M of possibilities is based on an assumed set of laws of behaviour – either laws of physics or meta-laws that determine the laws of physics. Without this, we have no basis for setting up a description of S. Indeed, these regularities are characterized by the parameters p used to describe these spaces. These parameters belong to various categories, which we now list. We denote them as $p_j(i)$, where j indicates the category and i denotes the parameter itself.

Physics parameters

- $p_1(i)$ are the basic physics parameters within each universe, excluding gravity. These characterize the non-gravitational laws of physics and related constants (e.g. the fine-structure constant) and parameters describing basic particle properties (mass, charge, spin, etc.) They should be dimensionless, or one may be describing the same physics in other units.
- $p_2(i)$ are the parameters describing the cosmological (gravitational) dynamics, e.g. $p_2(1) = 1$ indicates that Einstein gravity dominates, $p_2(1) = 2$ indicates that Brans–Dicke theory dominates, $p_2(1) = 3$ indicates that electromagnetism dominates, etc. Associated with each choice

---
2 FLRW = Friedmann–Lemaître–Robertson–Walker.

are the relevant parameter values, e.g. $p_2(2) = G$, $p_2(3) = \Lambda$ and $p_2(4) = \omega$ in the Brans–Dicke case. If gravity can be derived from more fundamental physics in some unified theory, these will be related to the parameters $p_1(i)$. For example, Λ may be determined from quantum field theory and basic matter parameters.

Cosmological parameters

- $p_3(i)$ are the cosmological parameters characterizing the matter content of a universe. These encode the presence of radiation, baryonic matter, dark matter, neutrinos, scalar fields, etc. In each case, we must specify the relevant equations of state and the auxiliary functions needed to determine the physical behaviour of matter (e.g. a barotropic equation of state for a fluid or the potential for a scalar field). These are characterizations of the macro-states of matter arising out of fundamental physics, so the possibilities here will be related to the parameters in $p_1(i)$.
- $p_4(i)$ are the geometrical parameters characterizing the spacetime geometry of the cosmological solutions envisaged, e.g. in the standard FLRW models, the scale factor $a(t)$, Hubble parameter $H(t)$ and spatial curvature k. These parameters will be related to the amount of each type of matter, characterized by $p_3(i)$ through the gravitational equations specified in $p_2(i)$. For example, Einstein's field equations relate the Hubble parameter and density parameters to the spatial curvature. One of the key issues is how the FLRW models, providing very good representations of the region we actually observe, relate to more generic models. This cannot be investigated within the context of FLRW models alone, since these are extremely special geometrically. More general models will contain anisotropy and inhomogeneity, so we must include spacetime functions characterizing the shear, vorticity and spatial density variations, as well as their dynamic behaviour [22, 23]. We also have to characterize all possible global spacetime connectivities, e.g. the many possible topologies of the space sections in FLRW universes.

Parameters relating to existence of life

- $p_5(i)$ are the parameters related to the functional emergence of complexity in the hierarchy of structure, e.g. the existence of chemically complex molecules which provide the foundations for life. Thus $p_5(1)$ might be the number of different types of atoms (as specified in the periodic table), $p_5(2)$ the number of different states of matter (crystal, glass, liquid, gas, plasma) and $p_5(3)$ the number of different types of molecules, characterized in a suitable way. These are properties emerging out of the

fundamental physics in operation, and so are related to the parameter set $p_1(i)$ [24]. Since we are considering physics quite different from that experienced on Earth, we cannot take any of these properties for granted.
- $p_6(i)$ are the biologically relevant parameters related specifically to the functional emergence of life and consciousness. For example, $p_6(1)$ might characterize the possibility of supra-molecular chemistry and $p_6(2)$ the possibility of living cells. This builds on the complexity allowed by $p_5(i)$ and again relates to the parameter set $p_1(i)$. However, when considering altered physics, we do not know what the necessary or sufficient conditions for intelligent life are. Thus we cannot at present fully characterize these parameters.

Choices have to be made for all these parameter sets on some basis, be it physical or philosophical. Major issues arise and these determine the nature of the ensembles envisaged. In particular, one must first address the following fundamental issue.

Issue (1) What determines the possibility space M? What range of possibilities will be contemplated? Where does this structure come from?

What is the meta-cause that delimits this set of possibilities? Whatever choice we make, the question will arise of why we only consider this range of options. Have we really considered the full range of possibilities? What is it that determines that range?

22.2.2 Realized multiverses

Given a description of the set of possibilities, the issue then is which of them occur in reality. This is described by a distribution function and a measure.

The distribution function

To describe a physically realized multiverse or ensemble, we need to define a distribution function $f(m)$ on M, specifying how many times each type of possible universe m is realized in the ensemble. The class of models comprising the ensemble is determined by all the parameters held constant across the ensemble; we call these 'class parameters'. The set of models occurring within the class is then determined by the parameters allowed to vary; we call these 'member parameters'. If a realized universe contains numerous different FLRW-like expanding domains within a single spacetime, each evolving separately from the others, we can characterize the ensemble by a distribution function for the FLRW parameters of these causally separated domains.

A measure

For continuous parameters, as a counterpoint to the distribution function, we need a volume element $\pi = \Pi_{i,j} m_{ij}(m) \mathrm{d}p_j(i)$ on the parameter space, characterised by weights $m_{ij}(m)$. The number of universes corresponding to the set of parameter increments $\mathrm{d}p_j(i)$ will then be given by $\mathrm{d}N = f(m)\pi$. We can also find the average value of any physical quantity from f and π. However, the obvious choices for the measure may not be the best ones. Indeed, what looks best is highly coordinate-dependent and can be changed arbitrarily by choosing new coordinates. Furthermore, the resulting integrals may often diverge. How to choose an optimal measure remains an open question, even in the FLRW case [25], and the use of ensembles is not well defined until this is clarified. This is a mathematical issue, but it has important physical implications.

Thus, a realized ensemble E of universes is characterized by a possibility space M, together with a measure π and distribution function $f(m)$ on M. Another major issue arises here, and this determines the nature of the ensembles claimed actually to exist.

> *Issue (2)* What determines the distribution function $f(m)$? What is the meta-cause that delimits the set of realizations out of the set of possibilities?

The answer to this has to be different from the answer to Issue 1, because here we are describing the contingency of selection of a subset of possibilities for realization from the total set of possibilities – determination of the latter being what is considered in Issue 1. Whatever choice we make, the following question will arise: Why has this particular set of universes or expanding domains come into being? What kind of meta-cause could be in operation here? Some choices will correspond to an actually existing multiverse and others to one or more multi-domain universes.

22.3 Issues arising

Several major physical and philosophical issues arise in considering ensembles of universes: (1) their non-uniqueness; (2) the emergence of self-consciousness; (3) whether fine-tuning or generic primordial conditions is more likely; (4) the question of realized infinities; (5) the source of regularities. We now consider these in turn.

22.3.1 Non-uniqueness

The first important point is that a multiverse or ensemble is not a unique concept; there can be many quite different realizations. Saying that everything that can happen does happen will not specify a unique multiverse for two reasons.

Non-uniqueness in realized models

Given the space of possibilities, realized multiverses are by no means unique. Their description requires the existence of a well defined and physically motivated distribution function $f(m)$ on the space of all possible universes. The mere fact that different distribution functions are possible shows that the concept of an ensemble is not unique. Hence one has to ask: How does the choice of these functions originate and what is their rationale?

These questions can be partially answered scientifically. A really existing ensemble of universes or domains may arise from the operation of a generating process, which adequately explains the origin of its member and their ranges of characteristics and parameter distribution from a more fundamental potential or primordial quantum configuration. That is, there may be a specific generating process which determines $f(m)$.

However, when it comes to the question of what is responsible for the operation of a specific generating process rather than some other one which would generate a different ensemble, an adequate answer cannot be given scientifically. This is the question of why the primordial dynamics leading to the given existing ensemble of universes is of one type rather than another. Even if we could establish $f(m)$ in detail, it is difficult to imagine how we would scientifically explain why one generating process (operating prior to the existence of the universe) was instantiated rather than some other one. The only possible answer comes from philosophical considerations. *A priori*, there is a complete arbitrariness in the resultant models because the generating process is not uniquely determined by any provable principle.

Non-uniqueness of possible models

The 'space of all possible universes' is not an easily delimited or unique concept, but the choice of what is included here determines what kind of multiverse can exist. How wide a variation of properties are we prepared to consider in our class of multiverses? Are we prepared to consider: universes with quite different physics (e.g. not coming from a variational principle and not based on quantum field theory); universes with different kinds of logic and perhaps alternative forms of mathematics; universes allowing magic, such as envisaged in the Harry Potter novels by J. K. Rowling?

If we are not prepared to allow such situations, what is the rationale for this? If the underlying principle is that 'all that can happen does happen', then there should conceivably be universes of all these kinds, as well as theistic and non-theistic ones, universes based in beauty and ethereal vibrations rather than physics, etc. Science fiction provides a fruitful source of such ideas. We surely cannot even conceive of all the alternatives, much less write them down systematically. In many of these classes of universes, no processes generating expanding domains would be possible, because the requisite kind of physics would not be realized. If any such possibilities are to be excluded, we have to be given both a meta-principle excluding them and a justification as to why that meta-principle should apply in the ensemble. Only philosophy can justify such a choice.

Implications of non-uniqueness

We conclude that the idea of a multiverse or ensemble is not a unique concept but a whole package of differing possibilities. This has several implications. First, it is not necessarily true that life can exist in any universe in an ensemble; this may or may not be true. Second, in considering experimental or observational tests for the existence of an ensemble, one needs to be told *which specific kind of ensemble is claimed to exist* and how to test that claim. How can we observationally or experimentally exclude specific classes of multiverse, e.g. those in which there may be any number of copies of our own universe. Thus one question would be: How do you determine how many copies of our own universe exist in any claimed ensemble?

22.3.2 The emergence of consciousness

Related to this question is the fact that we do not know the sufficient conditions for the emergence of consciousness. Given the kind of physics which holds here and now, necessary conditions for life as we know it depend on suitable choices of parameters in fundamental physics [7, 24]. However, we would have no handle on how to deal with this issue if physics were substantially different. Indeed, we do not even know what the necessary conditions are within the broad kind of physics that holds here and now. Hence we cannot properly characterize the domains in any ensemble where intelligent life may be able to exist. We have a good idea of many *necessary* conditions, but we do not know the *sufficient* conditions.

22.3.3 Special or general

An ensemble may include singular and non-singular, as well as special and general, universes. Which is more fundamental and so more likely to occur: fine-tuning to produce the very special initial conditions for our observed universe[3] or generic primordial conditions? While the introduction of ensembles has in effect been driven by the second view, they can equally well fulfil the first, the difference between them coming through the choice of distribution function on the space of possibilities. Hence, the mere existence of an ensemble does not by itself support either view. A generating mechanism might spit out numerous copies of one (successful) universe[4] rather than numerous universes with greatly different properties. Whichever is the case will determine the form of the distribution function [25]. Neither possibility can be excluded without some understanding of the creation mechanism. That mechanism cannot be tested in the true multiverse case and may not be testable in practice even in the multi-domain case. Until it is tested, any specific form assigned to it – in particular, the assumption that it creates generality rather than speciality – is metaphysical rather than physical.

This underlines the point that the present philosophical predilection for generality does not necessarily reflect the nature of physical reality. Indeed, philosophical predilection in cosmology has oscillated from assuming the universe is very special (in the sense postulated by the Cosmological Principle [26, 27]) to assuming that it has attained its present nature through operation of standard physics on generic initial conditions [28, 29]. The present tendency to assume that only the latter assumption is allowed is a philosophical supposition rather than a physical requirement. Either assumption may be allowed by appropriate assumptions about the pre-physics that leads to universe generation.

22.3.4 Problems with infinity

It is often claimed that really existing ensembles involve an infinity of universes. However, Hilbert strongly argues that a really existing infinite set is impossible [30]. He points out that the actual existence of the infinite directly or indirectly leads to well recognized unresolvable contradictions in set theory (e.g. the Russell paradox, involving the set of all sets which do not contain themselves) and thus in the definitions and deductive foundations

---

[3] Note that inflation only changes this speciality requirement to some degree; it does not work for very inhomogeneous or anisotropic conditions and, in any case, a general ensemble will have both inflationary and non-inflationary universes.

[4] For example, $f(m) = 10^{100}\delta(m_0)$, where m_0 is the universe domain we inhabit.

of mathematics itself. Hilbert's basic position is that 'the infinite is nowhere to be found in reality, no matter what experiences, observations and knowledge are appealed to'. One can apply this to both individual universes and multiverses. If we assume 'all that can happen does happen', then we run into uncountable infinities of actually existing universes, even in the FLRW case. For example, we have to include all values of the matter and radiation density parameters at a given value of the Hubble parameter, as well as all positive and negative values of the cosmological constant – a triply uncountable set of models. Hilbert would urge us to avoid such catastrophes.

Thus it is important to recognize that infinity is not an actual number which we can ever specify or determine – it is simply the code-word for 'it continues without end'. And something that is not specifiable or determinate in extent is not physically realizable. Whenever infinities emerge in physics, we can be reasonably sure there has been a breakdown in our model. It is plausible that this applies here too.

In addition to these mathematically based considerations, there are philosophical problems with spatial infinities [31], as well as physical arguments against the existence of an infinity in cosmological models. On the one hand, quantum theoretical considerations suggest that spacetime may be discrete at the Planck scale; indeed, some specific quantum gravity models have been shown to incorporate this feature. Not only would this remove the real line as a physics construct, but it could even remove the ultraviolet divergences that otherwise plague field theory – a major bonus. On the other hand, there are problems with putting boundary conditions for physical theories at infinity, and it was for this reason that Einstein preferred to consider cosmological models with compact spatial sections (thus removing the occurrence of spatial infinity). This was a major motivation for his static model, proposed in 1917, which necessarily has compact space sections. Wheeler picked up on this theme and wrote about it extensively [32]. Consequently, the famed textbook *Gravitation* by Misner, Thorne and Wheeler [33] only considered spatially compact, positively curved cosmological models in the main text – those with flat and negative spatial curvatures were relegated to a subsection on 'Other models'.

This theme recurs in present speculations on higher-dimensional theories, where the further dimensions are often (as in the original Kaluza–Klein picture) assumed to be compact. Unless one has some good physical reason for supposing otherwise, one might also expect this to be true of the three dimensions that expanded to a large size, even though this may necessitate 'non-standard' topologies for these spatial sections. Such topologies are commonplace (indeed they are essential) in M-theory. Thus physics also supports the idea that our universe may have compact spatial sections, which avoids infrared divergences as well.

22.3.5 The existence of regularities

Why is there a uniform structure across all universes m in M? Any consistent description of M (e.g. through a prescribed set of parameters P) demands regularities across all its members. But that means the existence of similarities which have a common causal source – a generating mechanism of some kind. Such a mechanism is implied in scenarios such as chaotic inflation, where the ensemble consists of domains that are connected to each other in the past, even though they would be causally disjoint in a non-inflationary model. In this case, the existence of such similarities is not mysterious because all members of the ensemble arise from a common origin through a prescribed generating mechanism.

Consider now a genuine multiverse, where the universes are completely disconnected from each other. Why should there be any regularity at all in the properties of these universes? If there are such regularities and specific resulting properties, this suggests that a mechanism created that family of universes and that a causal link to a higher domain was the source of processes leading to these regularities. This, in turn, means that the individual universes making up the ensemble are not actually independent of each other but just products of a single process, or meta-process, as in the case of chaotic inflation. A common generating mechanism is clearly a causal connection, even if not situated in a single connected spacetime, and some such mechanism is needed if all the universes in an ensemble are to have the same properties, for example being governed by the same physical laws or meta-laws.

As emphasized when we considered the description of ensembles, any multiverse with regular properties that we can characterize systematically is necessarily of this kind. If it did not have regularities of properties across the class of universes included in the ensemble, we could not even describe it, much less calculate any properties or characterize a distribution function.

Thus, in the end, the idea of a completely disconnected multiverse – with regular properties but no common causal mechanism – is not viable. Some pre-realization causal mechanism must necessarily determine the properties of the universes in the ensemble. What are claimed to be totally disjoint universes must in some sense be causally connected, albeit in some pre-physical or meta-physical domain.

22.3.6 Hypothesized mechanisms

Many researchers have invoked processes at or near the Planck era to generate a really existing ensemble of expanding domains, one of which is our own. The earliest explicit proposal of this kind was by Vilenkin [1], and Linde's chaotic inflationary proposal [2,3,34] is another well known example. The scalar field (inflaton) in these scenarios drives inflation and leads to the generation of a large number of causally disconnected regions. Our own observable region is then situated in a much larger universe that is inhomogeneous on the largest scales. No FLRW approximation is possible globally; rather there are many FLRW-like domains with different FLRW parameters in a single fractal universe [35].

In a stochastic approach [1,36,37], probability distributions can be derived for such models from specified inflaton potentials by using the slow-roll approximation. This kind of scenario suggests how overarching physics – or a *law of laws* represented by the inflaton field and its potential, together with the Friedmann and Klein–Gordon equations – can lead to a really existing ensemble of many different FLRW-like regions of a larger universe. Some versions relate this to the huge numbers of vacua of string theory [14,38]. However, these proposals rely on extrapolations of presently known physics to realms far beyond where its reliability is assured. They also employ inflaton potentials which as yet have no connection to known low-energy particle physics. Nor are these proposals directly testable, since we have no astronomical evidence that the other FLRW-like regions exist. There also remains the problem of infinities, which we discussed above: the continual reproduction of different inflating regions in eternal inflation is claimed to lead to an infinite number of domains. In this case, the supposed infinity of really existent FLRW-like domains derives from the assumed initial infinite (flat or open) space sections, and we have already pointed out the problems in assuming that such space sections are actually realized.

The various physical proposals made for generating mechanisms are either based on unproven physics acting in an existent universe or conceived as mechanisms that precede the existence of physics (since they precede the existence of the universe). In either case, the question raised above recurs in a new form: Why *that* physics and why *this* specific distribution function?

22.4 Problems with proof

The issue of testability underlies the question of whether multiverse proposals are really scientific. There are observational barriers preventing direct proof of the existence of any expanding domains other than the one we live in.

Furthermore, the supposed underlying physics is untested and may always be untestable. Consequently, ensemble proposals are not scientific in the usual sense. In order to make them so – e.g. by showing they are based on a theory that has gained credibility by unifying or clarifying some other mysteries – one has to search for extended ways of relating them to observations.

In looking at these issues, what faces us is not just showing that a multiverse exists. If this is a scientific proposition, one should be able to show what *specific* kind of ensemble is involved, and we have seen there are many different kinds. If you cannot show which *particular* one exists, it is doubtful you can show that any exists at all. If you cannot describe the characteristics of a purported object, then what scientific content can this claim have?

22.4.1 Evidence from observations: true multiverses

There can be no direct evidence for the existence of other universes in a true multiverse, as there is no possibility of even an indirect causal connection. The universes are completely disjoint and nothing that happens in one can affect what happens in another. Since there can be no direct or indirect evidence for such systems, what weight does the claim for their existence carry?

Experimental or observational testing requires some kind of causal connection between an object and an experimental apparatus, so that some characteristic of the object affects the output of the apparatus. But in a true multiverse, this is not possible. No scientific apparatus in one universe can be affected in any way by any object in another universe. The implication is that the supposed existence of true multiverses can only be a metaphysical assumption. It cannot be a part of science, because science involves experimental or observational tests to enable correction of wrong theories. However, no such tests are possible here because there is no relevant causal link.

Now, there is nothing wrong with metaphysical argument when it is appropriate. In this context, it involves a well argued case, based on a generalization of tested physical principles, to produce a conclusion that cannot be tested by any observation whatsoever. However, it must not be confused with science proper, and the claim that the existence of unconnected multiverses is scientifically testable cannot be supported within the usually understood concept of science. A belief that is justified by faith, unsupported by direct or indirect evidence, should be clearly identified as such, so that one knows precisely what one is being asked to support. I suggest the claim that properly disjoint multiverses exist is a metaphysical one, which, by its very nature, can never become a scientific one.

In any case, we have already commented that the idea of completely disconnected multiverses is vitiated by the need to have a common description of the properties of the members of a multiverse. The existence of such a commonality, describable by mathematical relations, must imply some reason for its existence. This presumably comes because the universes were all produced by a common mechanism pre-existing the universes and governing all of them.[5] So they are not causally unrelated after all, even though the relevant causal mechanism preceded the universes. In order to remain within the domain of science, it seems reasonable to restrict consideration of universe ensembles to multi-domain ensembles rather than multiverses proper (without any common causal mechanism), because the latter cannot even be described, much less shown in any scientific sense to have specific properties. I propose that they should be excluded from the scientific debate.

22.4.2 Evidence and physics: multi-domain universes

In the case of multi-domain universes, there is no direct evidence for the other regions in the ensemble. Nor can there be, since they lie beyond the visual horizon. We will never (on a realistic timescale) receive any information from them. Most are even beyond the particle horizon, so there is no causal connection with them. Nevertheless, they are the result of a common causal mechanism, whether that occurred in a single universe or in some meta-space before the universe came into existence. Thus, one way to make a reasonable claim for existence of a multiverse would be to show that its existence was a more or less inevitable consequence of well established physical laws and processes.

This is essentially the claim made for chaotic inflation. However, as pointed out above, the problem is that the proposed underlying physics has not been tested and may indeed be untestable. There is no evidence that the postulated physics is true, even in our universe, much less in some pre-existing meta-space that might generate an ensemble of universes. The issue is not just that the inflaton is unidentified and its potential untested – it is that we are assuming quantum field theory remains valid far beyond the domain where it has been tested. Can we have faith in that extreme extrapolation, despite all the unsolved problems related to the foundations of quantum theory, the divergences of quantum field theory and the failure of that theory to provide a satisfactory resolution of the cosmological constant problem?

---

[5] One might claim that the one multiverse proposal free from this objection is Tegmark's extreme version of all that can happen does happen [8,15]; but this suffers from major problems to do with infinity and we cannot even describe this possibility, let alone prove it occurs. Claiming existence of something you cannot even properly characterize has dubious scientific merit.

There are two requirements which must be met for a potentially viable ensemble theory. The first is to provide some credible link between presently known physics and the proposed physics underlying the ensemble – a vast extrapolation. The second is to provide at least indirect evidence that the proposed scalar potentials – or some other overarching cosmic principles – really were functioning in the very early universe. Neither requirement has yet been satisfied by any multi-domain proposal. Despite these problems, several arguments have been provided to suggest that we can indeed verify the multiverse proposal and we now discuss these.

22.4.3 The slippery slope argument

One argument, due to Rees [12,13], encourages us to go down the following slippery slope. First, we believe on good grounds that galaxies exist within the visual horizon but beyond the detection limits of current telescopes. We have not seen them yet, but will do so as detectors improve. Second, nobody seriously doubts that galaxies exist in other similar domains that intersect with ours; we share some galaxies with them, even though we do not see the same ones. We believe conditions there are similar to here and have some evidence that this is the case because we can see part of those domains. Third, by a further extension of this argument, it is reasonable to expect that there exist never-observable regions, with no observational connection to our own big bang domain, but with a joint causal past (as in chaotic inflation). Finally, it is reasonable to take another step and assume galaxies exist even in completely disjoint universes – with no causal connection with ours whatsoever. At each step we can reasonably assume that galaxies will continue to exist, because it is just like the previous step.

The problem is that the later steps assume a continuity of structure for which there is no evidence and whose existence would surely rely on a particular kind of common causal mechanism. But there is no guarantee of such continuity at either the last step (the transition to the multiverse) or the preceding one (the transition to a multi-domain universe).[6] To illustrate this, one could use exactly the same argument to re-assert what used to be taken for granted [26, 27]: that the universe has a *global* Robertson–Walker metric, with spatial homogeneity and isotropy continuing without bound in all directions. But that argument would disprove chaotic inflation, another multiverse contender. Despite this, the proposals are neither provable nor disprovable, because no evidence is available.

6 A discussion of testability in this context is given in ref. [39].

22.4.4 The speciality argument

Another argument is based on anthropic considerations and involves the issue of whether our universe is more finely tuned than our presence requires or merely typical of the subset in which we could have emerged. In the first case, the existence of a multiverse to explain such 'over-tuning' would be refuted, so the idea would be a scientific one.

The main application of this argument so far has been to the cosmological constant Λ. This is a two-stage argument and one version goes as follows [13]. First, naïve quantum physics implies that Λ should be very large. But our presence requires it to be small enough for galaxies and stars to form; Λ must therefore be below this galaxy-forming threshold. This explains the observed low value of Λ as a selection effect in a really existing ensemble of universes. The probability of a universe with small Λ being selected at random from the set of all universes in the ensemble is very small, but, when we add the prior that life exists, it becomes large. In this way, one justifies the existence of a biophilic universe, even though its *a priori* probability is extremely small. On the other hand, we would not expect any universe that allows life to be more fine-tuned than necessary. Present data indicates that Λ is not much below the threshold value, so our universe is not markedly more special that it needs to be as far as Λ is concerned. Consequently, explaining its fine-tuning by assuming a multiverse is acceptable and suggests that the same argument can be applied to other parameters

Is this argument compelling? As Hartle [40] has pointed out, for the first stage to be useful, we need an *a priori* distribution $f(\Lambda)$ for Λ that is very broad, combined with a very narrow set of values that allow life, these being centred well away from the most probable *a priori* values for Λ. This is indeed the case if we suppose that Λ is centred at a very large value Λ_{prob} (as suggested by field theory) but with a long tail to smaller values, and that $p_{\text{life}}(\Lambda)$ is centred at zero and has a very narrow distribution, as implied by astrophysics. Because the biophilic range with $p_{\text{life}}(\Lambda) > 0$ is so narrow, the function $f_{\text{cc}}(\Lambda)$ will not vary much in this range, so it is equally likely to take any value there. Thus a uniform probability assumption should be reasonably well satisfied within the biophilic range of Λ. Furthermore, this result does not depend critically on a Gaussian assumption for either distribution function – it would be true quite generally [11].

As regards the second stage of the argument, because of the uniform probability assumption, there is an equal *a priori* probability of any value for Λ within the biophilic range, so it is not clear why the expected values for existence of galaxies should pile up near the anthropic limit. Indeed, one might expect the probability of galaxy formation to be maximal at the *centre* of the biophilic range; the probability drops to zero at the edges because it vanishes outside. In any case, the probabilities are non-zero throughout the biophilic range, so this gives no justification for ruling out a multiverse with any specific value for Λ there.

An alternative form of the analysis uses more detailed calculations of structure formation. This is based on the 'Principle of Mediocrity', the assumption that our civilization is typical in the ensemble of all civilizations in our universe [41,42]. The key equation used is $dP(\rho_V) = P_*(\rho_V)N(\rho_V)d\rho_V$, where $dP(\rho_V)$ is the probability that intelligent life will occur for a vacuum energy density (i.e. an effective cosmological constant) between ρ_V and $\rho_V + d\rho_V$, $P_*(\rho_V)d\rho_V$ is the *a priori* probability that the vacuum energy density will be in this range, and $N(\rho_V)$ is the fraction of baryons that end up in galaxies (it being assumed that life is then guaranteed). The uniform probability assumption implies that $P_*(\rho_V)$ is nearly constant in the range where $N(\rho_V)$ is non-zero. Martel and colleagues [43] calculated $N(\rho_V)$ and showed that – under a specific set of physical assumptions – 5% to 12% of universes would have cosmological constants smaller than our own. This was stated to be a reasonable level of expectation, so the multiverse assumption is not disproved in this case.

But suppose another reasonable set of physical assumptions leads to 0.01% or 0.0001%, as may happen, for example, if we allow other parameters to vary together with Λ [44]. Should we then reject those assumptions? In a context where we already know extreme improbabilities are involved, we can choose this threshold according to taste; the whole point of the argument is to justify a value of $P_*(\rho_V)$ that is different by 120 orders of magnitude from its maximum value. Furthermore, the mediocrity assumption may or may not be true; its adoption is merely a philosophical presupposition. The physical assumptions used in the specific calculation may be reasonable, but they are not scientifically proven.

In the end, there is no way to avoid philosophical decisions, and these shape the outcome of the enquiry. The probabilistic argument is not decisive, because no impossibility is involved. It may give one grounds for saying an ensemble is not unlikely, given the assumptions made, but that is a far cry from proving that it exists.

22.4.5 The set of possibilities

Turning to the prior question of what determines the space of all possible universes from which a really existing universe or ensemble of domains is drawn (Issue 1 in Section 22.2.1), we find ourselves in even more uncertain waters. Suppose we demand some basic meta-principle which delimits the set of possibilities and so ultimately characterizes all the regularities in this possibility space (e.g. that they all obey the principles of quantum field theory). Where would such a principle originate? How can we possibly test its validity for all possible universes? We simply do not have enough theoretical knowledge to describe and delimit reliably the realm of what is possible, and it is unclear how this situation could ever change. Much less do we have any possibility of scientifically testing the nature of this domain. Any assumptions we make about it will necessarily be made on philosophical grounds. However well justified these may be, other possibilities will remain open. Lacking any observational or experimental restrictions, the theory can take any form we like and may be used to justify any conclusion.

22.4.6 Testability: refutation

Despite the gloomy prognosis given above, there are some specific cases where the existence of a multi-domain scenario *can* be disproved. These are when we live in a 'small universe', where we have already seen all the way around it [45, 46]. In this case, the universe closes up on itself in a single FLRW-like domain, so there can be no further domains that are causally connected to us.

This proposal is observationally testable; indeed it has been suggested that the power spectrum of the microwave background fluctuations (in particular, the lack of power on the largest angular scales) might already support it [47]. The proposal can be further tested by looking for identical circles in the microwave sky or for an alignment of the quadrupole and octopole axes. (A preliminary identification of the first effect has been made by Roukema and colleagues [48]), and evidence for the second effect is also good according to Katz and Weeks [49]). Confirmation of the small universe hypothesis would disprove the usual multi-domain chaotic inflation scenario but not a true multiverse proposal, since that cannot be observationally falsified or verified.

22.5 Explanatory usefulness and existence

We have seen that there are major problems in confirming any ensemble proposal in the usual scientific manner. We can also ask whether the idea is useful. It does provide a plausible kind of explanation of anthropic coincidences (e.g. in the context of the cosmological constant), but ultimately its usefulness as a scientific proposal is dubious. Also any ensemble proposal is not a final explanation; it just pushes the ultimate question back one stage further. For if one assumes the existence of a multiverse, the deeper issue then becomes: Why this multiverse rather than another one? Why an ensemble that allows life rather than one that does not [14]? The only multiverse proposal that *necessarily* admits life is Tegmark's extreme version of 'all that can happen does happen' [8]. But then why should this be the one that exists, with its extraordinary profligacy of infinities? The crucial existential questions recur and the multiverse proposal *per se* cannot answer them.

Given its essentially philosophical nature, a useful question is whether there is a philosophically preferable version of the multiverse proposal. In my view, Smolin's idea of a Darwinian evolutionary process in cosmology [9] is the most radical and satisfactory one, because it introduces the crucial idea of natural selection – the one process that can produce apparent design out of mechanistic interactions – into cosmology. Thus it extends fundamental physics to include central biological principles, making it a much more inclusive view than any of the other proposals, which are all based purely on theoretical physics. However, this proposal is incomplete in several ways [50], so it would be helpful to have its physical basis investigated in more detail.

So finally the question is: Does a multiverse in fact exist? The considerations of this article suggest that, in scientific terms, we simply do not know and probably never will. In the end, belief in a multiverse will always be a matter of faith that the logical arguments proposed give the correct answer in a situation where direct observational proof is unattainable and the supposed underlying physics is untestable.

This situation would change if we were able to point to compelling reasons, based on scientific evidence, for a particular specifiable ensemble, or at least a narrowly defined class of ensembles. One way in which this could be accomplished would be by accumulating evidence that an inflaton potential, capable of generating a particular ensemble of domains, was dominant in the very early universe. Otherwise, there will be no way of ever knowing which (if any) particular ensemble is realized. We can always claim whatever we wish about an ensemble, provided it includes at least one universe that admits life. It may contain only one universe or a vast number of them (but not an infinite number if the above arguments are correct) and the evidence will not allow us to choose. Gardner [51] puts it this way:

> There is not the slightest shred of reliable evidence that there is any universe other than the one we are in. No multiverse theory has so far provided a prediction that can be tested. As far as we can tell, universes are not even as plentiful as two blackberries.

The existence of multiverses is neither established nor scientifically establishable. The concept is justified by philosophy rather than science. They have explanatory power, but the philosophical nature of their justification must be appreciated.

Acknowledgements

I thank Bill Stoeger and Uli Kirchner for their comments.

References

[1] A. Vilenkin. The birth of inflationary universes. *Phys. Rev.* **D 27** (1983), 2848.
[2] A. D. Linde. Chaotic inflation. *Phys. Lett.* **B129** (1983), 177.
[3] A. D. Linde. *Particle Physics and Inflationary Cosmology* (Chur: Harwood Academic Publishers, 1990).
[4] D. W. Sciama. Is the universe unique? In *Die Kosmologie der Gegenwart*, eds. G Borner and J. Ehlers (Munich: Serie Piper, 1993).
[5] J. Leslie. *Universes* (London: Routledge, 1996).
[6] D. Deutsch. *The Fabric of Reality. The Science of Parallel Universes* (London: Penguin 1997).
[7] M. Tegmark. Is the Theory of Everything merely the ultimate ensemble theory? *Ann. Phys.* **270** (1998), 1.
[8] M. Tegmark. Parallel universes. In *Science and Ultimate Reality*, eds. J. D. Barrow, P. C. W. Davies and C. L. Harper (Cambridge: Cambridge University Press, 2004) p. 459.
[9] L. Smolin. *The Life of the Cosmos* (Oxford: Oxford University Press, 1997).
[10] D. K. Lewis. *On the Plurality of Worlds* (Oxford: Blackwell, 2000).
[11] S. W. Weinberg. The cosmological constant problems. In *Sources and Detection of Dark Matter and Dark Energy in the Universe*, ed. D. Cline (Berlin, New York: Springer-Verlag, 2001) [astro-ph/0005265].
[12] M. J. Rees. *Our Cosmic Habitat* (Princeton: Princeton University Press, 2001).
[13] M. J. Rees. Numerical coincidences and 'tuning' in cosmology. In *Fred Hoyle's Universe*, eds. C. Wickramasinghe, G. R. Burbidge and J. V. Narlikar (Dordrecht: Kluwer, 2003), p. 95.
[14] P. C. W. Davies. Multiverse cosmological models. *Mod. Phys. Lett.*, **A19** (2004), 727.
[15] M. Tegmark. Parallel universes. *Sci. Am.* **288**(5) (2003), 41.
[16] J. D. Barrow and F. J. Tipler. *The Anthropic Cosmological Principle* (Oxford: Oxford University Press, 1986).
[17] Y. Y. Balashov. Resource Letter AP-1. The Anthropic Principle. *Am. J. Phys.* **54** (1991), 1069.
[18] M. J. Rees. *Just Six Numbers, The Deep Forces that Shape the Universe* (London: Weidenfeld and Nicholson, 1999).
[19] J. D. Bjorken. This volume (2007) [astro-ph/0404233].
[20] G. F. R. Ellis, U. Kirchner and W. R. Stoeger, Multiverses and physical cosmology. *Mon. Not. Roy. Astron. Soc.* **347** (2003), 921.

[21] W. R. Stoeger, G. F. R. Ellis and U. Kirchner. Multiverses and cosmology. Philosophical issues (2004) [astro-ph/0407329].
[22] J. Wainwright and G. F. R. Ellis (eds.). *The Dynamical Systems Approach to Cosmology* (Cambridge: Cambridge University Press, 1996).
[23] .G. F. R. Ellis and H. van Elst. Cosmological models. In *Theoretical and Observational Cosmology*, ed. M. Lachieze-Ray (Boston: Kluwer, 1999).
[24] C. Hogan. Why the universe is just so. *Rev. Mod. Phys.* **72** (2000), 1149.
[25] U. Kirchner and G. F. R. Ellis. A probability measure for FRW models. *Class. Quant. Grav.* **20** (2003), 1199.
[26] H. Bondi. *Cosmology* (Cambridge: Cambridge University Press, 1960).
[27] S. W. Weinberg. *Gravitation and Cosmology* (New York: Wiley, 1972).
[28] C. W. Misner. The isotropy of the Universe. *Astrophys J.* **151** (1968), 431.
[29] A. H. Guth. Inflationary universe. A possible solution to the horizon and flatness problems. *Phys. Rev.* **D 23** (1981), 347.
[30] D. Hilbert. On the infinite. In *Philosophy of Mathematics*, eds. P. Benacerraf and H. Putnam (Englewood Cliff, NJ: Prentice Hall, 1964), p. 134.
[31] G. F. R. Ellis and G. B. Brundrit. Life in the infinite universe. *Quart. J. Roy. Astron. Soc.* **20** (1979), 37.
[32] J. A. Wheeler. *Einstein's Vision* (Berline: Springer Verlag, 1968).
[33] C. W. Misner, K. S. Thorne and J. A. Wheeler. *Gravitation* (San Francisco: Freeman, 1973).
[34] A. D. Linde. Inflation, quantum cosmology and the anthropic principle. In *Science and Ultimate Reality*, eds. J. D. Barrow, P. C. W. Davies and C. L. Harper. (Cambridge: Cambridge University Press, 2003), p. 426.
[35] A. H. Guth. Eternal inflation. *N.Y. Acad. Sci.* **321** (2001), 66 [astro-ph/0101507].
[36] A. A. Starobinsky. Current trends in field theory, quantum gravity and strings. In *Lecture Notes in Physics*, **246** (1986), 107.
[37] A. D. Linde, D. A. Linde and A. Mezhlumian. From the big bang theory to the theory of a stationary universe. *Phys. Rev.* **D 49** (1994), 1783.
[38] L. Susskind. This volume (2007) [hep-th/0302219].
[39] G. F. R. Ellis. Cosmology and verifiability. *Quart. J. Roy. Astron. Soc.* **16** (1975), 245.
[40] J. Hartle. This volume (2007) [gr-qc/0406104].
[41] A. Vilenkin. Predictions from quantum cosmology. *Phys. Rev. Lett.* **74** (1995), 846.
[42] S. W. Weinberg. *A priori* probability distribution of the cosmological constant. *Phys. Rev.* **D 61** (2000), 103505.
[43] H. Martel, P. Shapiro and S. W. Weinberg. Likely values of the cosmological constant. *Astrophys. J.* **482** (1998), 29.
[44] M. L. Graesser, S. D. Hsu, A. Jenkins and M. B. Wise. Anthropic distribution for cosmological constant and primordial density perturbations. *Phys. Lett.* **B 600** (2004), 15 [hep-th/0407174].
[45] G. F. R. Ellis and G. Schreiber. Observational and dynamic properties of small universes. *Phys. Lett.* **A 115** (1986), 97.
[46] M. Lachieze-Ray and J. P. Luminet. Cosmic topology. *Phys. Rep.* **254** (1995), 135.
[47] J. P. Luminet, J. R. Weeks, A. Riazuelo, R. Lehoucq and J. P. Uzan. Dodecahedral space topology as an explanation for weak wide-angle

temperature correlations in the cosmic microwave background. *Nature*, **425** (2003), 593.
[48] B. F. Roukema, B. Lew, M. Cechowska, A. Masecki and S. Bajtlik. A hint of Poincaré dodecahedral topology in the WMAP First Year Sky Map. *Astron. Astrophys.* **423** (2004), 821 [astro-ph/0402608].
[49] G. Katz and J. Weeks, Polynomial interpretation of multipole vectors. *Phys. Rev.* **D 70** (2004), 063527 [astro-ph/0405631].
[50] T. Rothman and G. F. R. Ellis. Smolin's natural selection hypothesis. *Quart. J. Roy. Astron. Soc.* **34** (1992), 201.
[51] M. Gardner. *Are Universes Thicker than Blackberries?* (New York: W. W. Norton, 2003).

23
Predictions and tests of multiverse theories

Don N. Page
Institute for Theoretical Physics, University of Alberta

23.1 Multiverse explanations for fine-tuning

Many of the physical parameters of the observed part of the Universe, whether constants of nature or cosmological boundary conditions, seem fine-tuned for life and us [1–4]. There are three common explanations for this. One is that there is a 'Fine-Tuner' who providentially selected the physical parameters so that we can be here. Another is that it is just a coincidence that the parameters turned out to have the right values for us to be here. A third is that the observed Universe is only a small part of a much vaster Universe or multiverse or megaverse or holocosm (my own neologism for the whole), and that the physical parameters are not the same everywhere but take values permitting us in our part.

These three explanations are not necessarily mutually exclusive. For example, combining a Fine-Tuner with coincidence but without a multiverse, perhaps the Universe was providentially created by a God who had a preference for a particularly elegant single universe which only coincidentally gave values for the physical parameters that allowed us to exist. Or, for a Fine-Tuner with a multiverse but without coincidences, perhaps God providentially created a multiverse for the purpose of definitely creating us somewhere within it. Or, for coincidence and a multiverse without a Fine-Tuner, if the Universe were not providentially created, it might be a multiverse that has some parts suitable for us just coincidentally. Or, it might even be that all three explanations are mutually true, say if God providentially created a multiverse for reasons other than having us within it, and yet it was a coincidence that this multiverse did contain us.

On the other hand, it seems conceivable (in the sense that I do not see any obvious logical contradiction) that the Universe is determined by some sort of blind necessity that requires both our own existence and a single world with a single set of physical parameters. In this case, the Universe is not providential (in the sense of being foreseen by any God), but nor is our existence coincidental.

Thus, logically, I do not see that we can prove that any combination of the three explanations is either correct or incorrect. However, it does seem a bit implausible that none of these explanations is at least partially correct, and it also seems rather implausible that the large number of fine-tunings that have been noticed are mere coincidences.

I should perhaps at this point put my metaphysical cards on the table and say that – as an evangelical Christian – I do believe the Universe was providentially created by God, and that – as a quantum cosmologist with a sympathy toward the Everett 'many worlds' version of quantum theory – I also strongly suspect that the Universe is a multiverse, with different parts having different values of the physical parameters. It seems plausible to me that – in a quantum theory with no arbitrary collapses of the wave-function – God might prefer an elegant physical theory (perhaps string/M-theory with no adjustable dimensionless parameters) that would lead to a multiverse that nevertheless has been created providentially by God with the purpose of having life and us somewhere within it.

Although personally I have less confidence in string/M-theory than in either providence or the multiverse, nevertheless string/M-theory is very attractive. It does seem to be the best current candidate for a dynamical theory of the Universe (i.e. for its evolution, if not its state), and it does strongly appear to suggest a multiverse. Since string/M-theory has no adjustable dimensionless constants, if it predicted just a single set of parameters, it would seem very surprising if these parameters came out right for our existence. Thus, if string/M-theory – or some alternative with no adjustable dimensionless constants – were correct, it would seem much more plausible that it would lead to a multiverse, with different parts of the Universe having different physical parameters.

Indeed, string theorists [5–15] have argued that string/M-theory leads to an immense multiverse or landscape of different values of physical parameters and 'constants of nature'. It is not yet known whether the range of values can include the physical parameters that allow life, such as those within our part of the Universe, but that does seem at least plausible with the enormous range suggested in the string landscape or 'stringscape'.

One objection that is often raised against the multiverse is that it is unobservable. Of course, this depends on how the multiverse is defined. One definition would be the existence of different parts, where some physical parameters are different, but this just shifts the arbitrariness to the choice of this set of physical parameters. Obviously if some quantity which varies with position (such as energy density) were included in the set of physical parameters, then even what we can see could be considered a multiverse. But if we just include the so-called 'constants of nature', such as the fine structure constant and various other coupling constants and the mass ratios of the various elementary particles, then what we can observe directly seems to consist of a single universe. Indeed it would be rather natural – if *ad hoc* – to define a multiverse with respect to the physical parameters that have no observable variation within the part we can directly see. In this case, the multiverse becomes unobservable, and it becomes an open question whether parts of the Universe we cannot see have different values of these constants. Many would argue that it is a purely metaphysical concept that has no place in science.

However, in science we need not restrict our entities to be observable – we just want the simplest theory, whether using observable or unobservable entities, to explain and predict what is observable. One cannot test scientifically a theory that makes predictions about what is unobservable, but one can test a theory that makes use of unobservable entities to explain and predict the observable ones. Therefore, if we find a multiverse theory that is simpler and more explanatory and predictive of what is observed than the best single-universe theory, then the multiverse theory should be preferred. The success of such a multiverse theory itself would then give credence to the existence of the unobservable multiverse.

Another objection that is often raised against multiverse theories is that, naïvely, they can 'explain' anything and predict nothing, so that they cannot be tested and considered scientific. The idea is that if a multiverse gives all possible physical parameters or other conditions somewhere within the multiverse, then the parameters and conditions we observe will exist somewhere. Hence what we observe is 'explained' at least somewhere. On the other hand, if every possibility exists, then we cannot predict any non-trivial restriction on what might be observed. If a theory makes no non-trivial predictions, then it cannot be tested against observations, and it can hardly be considered scientific.

23.2 Testable multiverse explanations

Sufficiently sophisticated multiverse theories can provide predictions as well as explanations, and hence can be tested against observations scientifically. Unlike single-universe theories, in each of which one can, in principle, predict uniquely the physical parameters, in multiverse theories one usually can make only statistical predictions for ranges of parameters, but this can still be much better than making no prediction at all. However, to make such statistical predictions, the multiverse theory needs to include a *measure* for the different observations that can be made. If it allows all possible observations without putting any measure on them, then one can make no predictions.

Since we have strong evidence that we live in a quantum universe, it would be natural to seek a quantum multiverse theory. If this just includes some quantum states, unitary evolution, path integrals, operators, some operator algebra and the like, one has the bare quantum theory eloquently described by Sidney Coleman,[1] which by itself does not give any measures or probabilities. The Copenhagen version of quantum theory does give these, but at the apparent cost of the collapses of the wave-function at times undetermined by the theory and to states that are random.

Here I shall take essentially an Everett 'many worlds' view that, in actuality, there is no collapse of the wave-function. However, to achieve testability of the quantum theory, I shall assume that there is one aspect of the Copenhagen version that should be added to the bare quantum theory: measures for observations that are expectation values of certain corresponding 'awareness operators'.

In Copenhagen theory, these operators are projection operators, and the measures are the probabilities for the results of the collapse of the wave-function. Here I shall not necessarily require the operators to be projection operators, though – to give the positivity properties of measures – I shall assume they are at least positive operators. Also, I shall not assume that anything really random occurs, such as wave-function collapse, but that there are simply measures for all the different observations that might occur. In testing the theory against one's observation, one can regard that observation as being selected at random (with the theory-given measure) from the set of all possible observations, but ontologically one can assume that all possible observations with non-zero measure really do occur, so that there is never a real physical random choice between them.

For the quantum theory to be fundamental, one would need to specify which observations have measures and what the corresponding operators are whose expectation values give those measures. In my opinion, the most fundamental aspect of a true observation is a conscious perception or awareness of the observation. Therefore, I have developed the framework of 'Sensible Quantum Mechanics' (SQM) [16] or 'Mindless Sensationalism' [17] for

---

[1] S. Coleman. Quantum mechanics with the gloves off. Dirac Memorial Lecture, St. John's College, University of Cambridge, June 1993 (unpublished).

giving the measures for sets of conscious perceptions as expectation values of corresponding positive operators that I call 'awareness operators'. This is only a framework (analogous to the bare quantum theory without the detailed form of the unitary evolution or operator algebra) rather than a detailed theory, since I have no detailed proposal for the sets of possible conscious perceptions or for the corresponding positive operators. Presumably, for human conscious perceptions, these operators are related to states of human brains, so understanding them better would involve brain physics. However, I do not see how they could be deduced purely from an external examination of a brain, since we cannot then know what is being consciously experienced by the brain.

To avoid the complications of brain physics, one might use the observed correlation between external stimuli and conscious experiences to replace the unknown awareness operators acting on brain-states with surrogate operators acting on the correlated external stimuli. Of course, this would not work well for illusory or hallucinatory conscious perceptions, for which the fundamental awareness operators would presumably still work if they were known. However, one might prefer to focus on conscious perceptions that are correlated with external stimuli and hence better fit what is usually meant by observations.

If the awareness operator for a conscious perception is correlated with a single set of external stimuli at a single time, it could be approximately replaced by a single projection operator onto some external system. Alternatively, if it is correlated with a sequence of measurement processes, then it could be approximately replaced by a product of projection operators or a sum of such products, a class operator of the decoherent histories approach to quantum theory [18–20].

Therefore, though I would not regard either the projection operators of Copenhagen quantum theory or the class operators of decoherent histories quantum theory as truly fundamental in the same way that I believe awareness operators are, it might be true that, in certain circumstances, these are reasonable approximations to the fundamental awareness operators. Then one can take their expectation values in the quantum state of the Universe as giving the measure for the corresponding conscious perception.

One example of this replacement would be to calculate the measure for conscious perceptions of a certain value of the Hubble constant. In principle, in SQM this would be the expectation value of a certain awareness operator that presumably acts on suitable brain-states in which the observer is consciously aware of that particular Hubble constant value. But the expectation value of this operator might also be well approximated by that of some suitable operator acting on the logarithm of the expansion rate of the part of the Universe that is observed. Because the latter operator does not involve brain physics, it might be easier to study scientifically and so could be used as a good surrogate for the actual awareness operator.

However, it would presumably not be a good approximation to use the latter operator if its expectation value depended significantly on parts of the Universe where there are no conscious observers; if one wants to use it to mimic the expectation value for the perceptions of conscious observations, one must include a selection effect which restricts the operator to parts of the Universe where there are conscious observers.

To include this selection effect in operators that are external to brains (or whatever directly has the conscious perceptions), so that their expectation values can be good approximations for those of the fundamental awareness operators, is a difficult task, since we do not know the physical requirements for conscious observers. For example, there is nothing within our current understanding of physics that would tell us whether or not some powerful computer is conscious, unless one makes assumptions about what is necessary for consciousness. Also, I know of nothing within our current understanding of physics that would enable us to predict that I am currently conscious of some of my visual sensations but not of my heartbeat, since presumably information about both is being processed by my brain and would be incorporated in a purely physical analysis.

Nevertheless, to formulate a very crude guess for a selection effect for conscious observers, one might make the untested hypothesis that typical observers are like us in requiring suitable complex chemical reactions and perhaps a liquid compound like water. Then one could use the existence of liquid water as a very crude selection effect for observers and attach it onto other projection or class operators used to approximate some conscious perception depending on the external stimuli that are described by the projection or class operators.

Thus one might use projection or class operators to ask the following two questions. Does liquid water exist in part of the Universe? Is that part of the Universe expanding at a suitable logarithmic rate? If the answer to these questions is yes with some measure, then one might expect that there would be a roughly corresponding expectation value for the awareness operator for conscious perceptions of that value of the Hubble constant. This is an extremely crude approximation to what I postulate would objectively exist as the expectation value of the true corresponding awareness operator, but since these awareness operators are, as yet, largely unknown, the crude approximation may be useful during our present ignorance.

One problem with calculating the measure for sets of conscious perceptions as expectation values of corresponding 'awareness operators' is that, naïvely, one might get infinite values. By itself this would not necessarily be a problem, since only ratios of measures are testable as conditional probabilities. However, when the measures themselves are infinite, it is usually ambiguous how to take their ratios.

The problem arises if the awareness operators are sums of positive operators that are each localized within finite spacetime regions (as one would expect if the operators correspond to finite conscious beings). Assume that one such operator in the sum has support within one of N spacetime regions of equal volume within the total spacetime. Then, by translational or diffeomorphism invariance, one would expect the sum of the operators for a particular awareness operator to include a sum over the corresponding operators in each of the N regions. (There would also be a sum over operators that overlap different regions, but we need not consider those for this argument.) This is essentially just the assumption that, if a suitable brain-state for some conscious perception can occur in one of the N spacetime regions, then it can also occur (depending on the quantum state) in any of the other $N-1$ regions. Also, where it occurs in some coordinate system should not affect the content of the conscious perception produced by the corresponding brain-state.

If the conditions for observers with the corresponding conscious perception occur within all N spacetime regions, so that the expectation value of the operator within each region has a positive expectation value bounded from below by a positive number ϵ, then the total awareness operator (a sum of at least the individual positive operators within each of the N regions) will have an expectation value at least as large as $N\epsilon$. This is infinite if the number N of such spacetime regions is infinite. Essentially the argument is that, if the measure for a conscious perception has a strictly positive (but finite) expectation value for each spacetime volume in some region, then for an infinite volume of spacetime where this is true the measure will be infinite. One can regard this as arising from the infinite number of conscious observers that arise in an infinite volume of spacetime with conditions suitable for life and conscious observers.

Since inflation tends to produce a universe that is arbitrarily large (with an infinitely large expectation value for the spatial volume at any fixed time after inflation and hence presumably infinitely many conscious observers), it tends to produce an infinite measure for almost all non-zero sets of conscious observations. There has been a lot of discussion in the literature [21–27] of how to obtain well defined ratios of these infinite measures (or of related quantities, since the discussion is not usually in terms of measures for conscious perceptions), but I think it is fair to say that there is as yet no universally accepted solution.

This is a serious problem that needs to be solved before we can hope to make rigorous testable predictions for an inflationary multiverse. A vague hope is that somehow the dimensionality of the part of the Hilbert space (or quantum state space if it is bigger than the Hilbert space) where conscious observers are supported is finite, so that – for all finite quantum states – the expectation values of all finite positive operators (including the awareness operators) would be finite, thus giving finite measures for all conscious perceptions. But what would limit conscious observers to a finite-dimensional part of the presumably infinite-dimensional quantum state space eludes me.

23.3 Testing multi-observation theories with typicality

If we can find a theory that gives finite measures for sets of observations (perhaps conscious perceptions) or which can be approximated as the expectation values of other positive operators, how can we test it? If the theory predicts a unique observation (at least unique under some condition, such as observing a clock reading to have some value), then one can simply check whether one's observation fits the prediction. This would typically be the case in a classical model of a universe with a single observer who reads a clock that gives monotonically increasing readings (so that there is only a single observation for each clock reading).

Although a classical solipsist might believe this is true for his universe, for most of us the evidence is compelling that there are many observers and hence presumably many different observations even at one value of some classical time. Quantum theory further suggests that there are many possible observations – even for a single observer at a single time.

There is a debate as to whether the observations given by quantum theory are actual or are merely unrealized possibilities. The Copenhagen view seems to imply that – for each value of the time and for each observer – there is only one observation that is actualized (say by collapse of the wavefunction), so that all the other possibilities are unrealized. This seems to come from a naïvely WYSIWYG[2] view of the Universe, so to me it is much simpler to suppose that all possible observations predicted by the quantum theory are actualized, with no ugly collapse of the wave-function to give a single actualized observation for each observer at each time. We are already

---
2 What you see is what you get.

used to the idea of many different times (which are effectively just different branches of the quantum state, at least in the Wheeler–DeWitt approach to quantum gravity) and – except for solipsists – to the idea of many different observers, so why should we not accept the simple prediction from quantum theory of many observations at the same time by the same observer?

In any case, whether in a classical universe or a quantum universe without collapse of the wave-function, each time an observation occurs, there are many observations even at the same time, and so one needs to be able to test this. To do this for a theory that gives measures for all sets of observations, I would propose using the concept of 'typicality' [16], which is a suitable likelihood that one may use to test or compare theories or to calculate their posterior probabilities in a Bayesian analysis after assigning their prior probabilities.

The basic idea is to choose a set of possible observations that each give a single real parameter, such as the Hubble constant or the value of one of the constants of nature. Then we use the measure for sets of observations to get the measure for all ranges of this single real parameter. For simplicity, we normalize the total measure in the set of observations being considered to be unity.

Now we want to test our observation against the theory by calculating the typicality for that observation within the set. For simplicity, I shall call the observation being tested the 'actual' observation, even though the theory would say that all possible observations with non-zero measure are realized as actual observations. To do this, one calculates the total 'left' and 'right' measures for all possible observations in the set under consideration, i.e. the total measure to the left or right of and including the 'actual' observation when they are ordered on the x-axis by the value of the real parameter under consideration. These two measures will add up to unity plus the measure of the 'actual' observation, which is counted in both of the measures.

Next take the smaller of these two measures (if the 'actual' observation is not in the middle of the total measure) as the 'extreme' measure of the 'actual' observation. We then use the normalized measure of the set of observations to calculate the probability that a random observation within the set would give an 'extreme' measure as small as that of the 'actual' observation. This probability is what I call the 'typicality' of the actual observation of the real parameter within the chosen set of possible observations. The typicality is thus the probability that a random observation in the set is at least as extreme as the actual observation. It depends not only on the actual observation, but also on the theory predicting the measure for the sets of observations. This is what is needed to calculate the conditional probability of subsets of observations within the set under consideration.

In the case in which the real parameter takes a continuum of values and there is zero measure for an observation to have precisely any particular value, the left plus right measures add up to unity. Then the extreme measure (the smaller of the left and right measures) will take continuous values from zero to one-half with a uniform probability distribution, so the typicality is twice the extreme measure. In this simple case, the typicality is a random variable with a uniform probability distribution ranging from zero (if the actual parameter is at the extreme left or right) to unity (if the actual value is in the middle of its measure-weighted range, with both left and right measures being one-half).

If the real parameter takes on discrete values, then the situation is more complicated. For example, suppose that the real parameter is k, with possible values $k=-1$ (with measure 0.2), $k=0$ (with measure 0.35) and $k=+1$ (with measure 0.45). Then $k=-1$ has a left measure of 0.2 and a right measure of $0.2+0.35+0.45=1$ for an extreme measure of 0.2; $k=0$ has a left measure of $0.2+0.35=0.55$ and a right measure of $0.35+0.45=0.8$ for an extreme measure of 0.55; and $k=+1$ has a left measure of $0.2+0.35+0.45=1$ and a right measure of 0.45 for an extreme measure of 0.45. Thus the probability of an extreme measure of 0.2 is 0.2 (the probability of $k=-1$); the probability of an extreme measure of 0.45 is 0.45 (the probability of $k=+1$); and the probability of an extreme measure of 0.55 is 0.35 (the probability of $k=0$). The typicality of $k=-1$ is the probability that the extreme measure will be at least as small as 0.2, which is 0.2; the typicality of $k=0$ is the probability that the extreme measure will be at least as small as 0.55, which is $0.2+0.45+0.35=1$, and the typicality of $k=+1$ is the probability that the extreme measure will be at least as small as 0.45, which is $0.2+0.45=0.65$.

Note that only for the most extreme parameter value or values (for which the extreme measure is the smallest possible within the set) is the typicality the same as the normalized measure of the observation giving that value itself. For less extreme parameter values, the typicality is greater than the measure of the observations giving that parameter value. On the other hand, the least extreme parameter value or values (the middle one, for which the 'extreme' measure is the greatest possible within the set) has a typicality of unity. Thus the typicality always attains its upper limit of unity for some member of the set, but the lowest value it attains is the measure of the most extreme observation (which would be zero if the observed parameter formed a continuum with zero measure for any particular value of the parameter).

The typicality is thus a likelihood, given a theory for the measures of sets of values of a real parameter, for a parameter chosen randomly with the probability measure given by the theory, to be at least as extreme as the 'actual' observed parameter. The typicality has the advantage over the probability measure for the actual observed parameter of being a probability that has values up to unity for some possible observation. This differs from the probability measure for the parameter itself, which may have a very small upper limit (e.g. if there is an enormous number of possible discrete values for the parameter) or even a zero upper limit (e.g. if the parameter ranges over a continuum and has a smooth probability density, with no delta functions at any particular values of the parameters).

If one uses the probability measure itself as a likelihood, one cannot directly perform a Bayesian analysis with an observation of a continuous parameter having a smooth probability density, since the resulting likelihood will be zero for all possible observed values of the parameter. One might try to use the probability density instead of the probability itself, but this depends on the coordinatization of the parameter and so gives ambiguous results. For example, one would get a different likelihood for an observed value of the Hubble constant H by using its probability density than one would for H^2.

Another approach that is often used for results that have a large number of possible values is to bin them and then use the total probability for the bin in which the actual observation lies as the likelihood. But again this depends on the bins and so gives ambiguous results. The ambiguity of both the probability density and the binning are avoided if one uses the typicality as I have defined it here.

Admittedly, if there are $N > 1$ parameters being observed, then there are ambiguities even with the typicalities. First, with more than one parameter, one gets more than one typicality. Second, if there are N independent parameters, one can construct N independent combinations of them in arbitrarily many different ways. Both of these problems are related to the issue of how one chooses to test a theory, which has no unique answer.

Once one has made a choice of what set of observations to include and what parameter to determine the typicality for, how do we use the typicality to test a theory? It can be used – like any other likelihood – in the following manner. Let H_n be an hypothesis that gives measures to observations in the set, so that an actual observation O has typicality $T_n(O)$ according to this hypothesis. At the simplest level, one can say that, if $T_n(O)$ is low, then H_n is ruled out at the corresponding level. For example, if $T_n(O) < 0.01$, then one can say that H_n is ruled out at the 99% confidence level.

A better approach would be to assign initial or prior probabilities $P_i(H_n)$ to different hypotheses H_n, labelled by different values of n. Then the typicalities $T_n(O)$ for these different hypotheses would be used as weights to adjust the $P_i(H_n)$ to final or posterior probabilities $P_f(H_n)$ that are given by Bayes' formula:

$$P_f(H_n) = \frac{T_n(O)P_i(H_n)}{\sum_m T_m(O)P_i(H_m)}. \tag{23.1}$$

Apart from the ambiguity of choosing the set of possible observations and the parameter to be observed, and the physics problem of calculating the typicalities $T_m(O)$ assigned by each theory H_m, there is now the new ambiguity of assigning prior probabilities $P_i(H_m)$ to the theories themselves. This appears to be a purely subjective matter, though – in the spirit of Ockham's razor – scientists would generally assign higher prior probabilities to simpler theories. Of course, there are arbitrarily many ways to do that. However, if one just considered an infinite countable set of theories that one could order in increasing order of complexity, from the simplest H_1 to the next simplest H_2 and so on, then one simple assignment of prior probabilities would be

$$P_i(H_m) = 2^{-m}. \tag{23.2}$$

The idea of restricting attention to a countable set of theories seems plausible, since humans could really consider only a finite set of theories, but it could be inappropriate if the ultimate theory of the Universe contained an infinite amount of information, even if merely in the form of a single real coupling constant or some other parameter whose digits are not compressible (i.e. generated by a finite amount of input information). Note that it is considered to be a merit of string/M-theory that there is not even the possibility of having infinite amounts of information in any dimensionless coupling constants, at least in the dynamical equations of the theory, although it is apparently not yet ruled out that the quantum state might have an infinite amount of information. This might apply to the expectation value of the dilaton, although most theorists would also prefer to avoid this possibility of an infinite amount of information in the dilaton.

23.4 Testing the single-universe and multiverse hypotheses

Tegmark [28–30] has classified multiverse hypotheses into Levels 1, 2, 3 and 4. Level 1 comprises regions beyond our cosmic horizon, with the same 'constants of nature' as our own region. Level 2 describes other post-inflation bubbles, perhaps with different 'constants of nature'. Level 3 is the Everett many worlds of quantum theory, with the same features as Level 2. Level 4 includes other mathematical structures, with different fundamental equations of physics as well as different constants of nature.

Levels 1–3 can all come from a single universe if we define a universe to be some quantum state in some quantum state space (e.g. some C*-algebra state). In this case, the quantum state space may be regarded as a set of quantum operators and their algebra, and the quantum state as an assignment of an expectation value to each quantum operator. To obtain measures for observations in the form of conscious perceptions, one must add to this bare quantum theory an assignment of a particular positive operator for each set of conscious perceptions. The resulting 'awareness operators' then form a positive-operator-valued set obeying the appropriate sum rules when one forms unions of disjoint sets of conscious perceptions, so that the resulting expectation values have the properties of a measure on sets of conscious perceptions [16, 17].

Different hypotheses H_m that each specify a single SQM universe would give different quantum state spaces, different operator algebras, different quantum states, different sets of conscious perceptions and/or different sets of awareness operators corresponding to the sets of conscious perceptions. (A quantum state is here defined, in the C*-algebra sense, as the quantum expectation values for all possible quantum operators in the set.) By the SQM rule that the measure for each set of conscious perceptions is the expectation value given by the quantum state for the corresponding awareness operator, a definite SQM theory H_n would give a definite measure for each set of possible conscious perceptions. This would be a theory of a single SQM universe, though that universe could be a multiverse in the senses of Levels 1–3.

Then, by the procedure outlined above, from one's actual observation O, a sufficient intelligence should be able to calculate for each H_m the typicality $T_m(O)$ of that observation. If one has a set of such theories with prior probabilities $P_i(H_m)$, then one can use a Bayesian analysis to calculate the posterior probability $P_f(H_n)$ for any specific theory H_n and thereby test the theory at a statistical or probabilistic level.

But what if there is more than one universe? Tegmark [28–31] has raised the possibility of a multiverse containing different mathematical structures, and it certainly seems logically conceivable that reality may consist of more than one universe in the sense of Levels 1–3. Tegmark discusses a Level 4 multiverse which, as he describes it, includes all mathematical structures. This seems to me logically inconsistent and inconceivable. My argument against Level 4 is that different mathematical structures can be contradictory, and contradictory ones cannot co-exist. For example, one structure could assert that spacetime exists somewhere and another that it does not exist at all. However, these two structures cannot both describe reality.

Now, one could say that different mathematical structures describe different existing universes, so that they each apply to separate parts of reality and cannot be contradictory. But this set of existing universes, and the different mathematical structures with their indexed statements about each of them, then forms a bigger mathematical structure. At the ultimate level, there can be only one world and, if mathematical structures are broad enough to include all possible worlds or at least our own, there must be one unique mathematical structure that describes ultimate reality. So I think it is logical nonsense to talk of Level 4 in the sense of the co-existence of all mathematical structures. However, one might want to consider how to test levels of the multiverse between Levels 1–3 and 4.

One way to extend an SQM universe to a multiverse might be to allow more than one quantum state on the same quantum state space, while keeping the other parts of the structure – such as the awareness operators – the same. Then, if a weight is assigned to each of these different quantum states, one can get the measure for each set of conscious perceptions as the weighted sum of the measures for each quantum state. But this is equivalent to defining a new single quantum state in a new single-universe theory that is the weighted sum of these different quantum states in the original description. That is, the new single quantum state would be defined to give as the expectation value of each operator the weighted sum of the expectation values that the different quantum states would give. (If the quantum state can be described by a density matrix, then the new density matrix would be the weighted sum of the old ones.)

Since the measure for a set of conscious perceptions in an SQM universe is the expectation value given by the quantum state of the awareness operator corresponding to the set of conscious perceptions, one would get the same measure by using the new quantum state as by taking the weighted sum of the measures in the old description in which there are different quantum states.

Another way to get a broader multiverse would be to keep the same quantum state space, quantum operators, operator algebra and set of possible conscious perceptions, but to include different sets of awareness operators in different SQM universes. But again, if one weights the resulting measures for each universe to get a total measure for this multiverse, this would be equivalent to forming a single new set of awareness operators that are each

23 *Predictions and tests of multiverse theories* 425

the weighted sum of the corresponding awareness operators in each of the different universes.

Yet another way to extend the multiverse would be to include universes with separate quantum state spaces, each with its own quantum state and awareness operators. If each of these universes has a weight, then one can again get the total measure for each set of conscious perception by taking the weighted sum of the measure for that set in each universe. This would be equivalent to defining a total quantum state space whose quantum operators were generated by the tensor sum of the operators in each of the original sets of operators that correspond to the original separate quantum state spaces. One could take operators from different original sets as commuting to define the quantum algebra of the new set.

The new single quantum state could then be defined by giving – on any sum of operators from the separate sets of operators – the weighted sum of expectation values that the old quantum states gave. For products of operators from different sets, one could just take the new expectation value to be the product of the weighted old expectation values for the separate operators in each set. The new awareness operators could be defined as the sum of the original awareness operators. Since this would involve only sums from the different sets of operators and not products, the expectation values of the new awareness operators would all be linear in the weights for the original separate universes in the new single quantum state and hence would give in that new single quantum state the same measure as the weighted sum of the original measures.

Each of these three simple-minded ways to attempt to extend the multiverse produces nothing new, at least for the measures of sets of conscious perceptions. Thus a single SQM universe is a fairly broad concept, encompassing a wide variety of ways of generating measures for conscious perceptions. In fact, one could argue that any assignment of measures for conscious perceptions could come from a single SQM universe, since one could just define awareness operators for all sets of conscious perceptions and embed these into a larger set of quantum operators with some algebra. One could then just choose the quantum state to give the desired expectation values for all of the awareness operators.

In principle, one could even choose the algebra of operators to be entirely commuting, so that the resulting quantum theory would be entirely classical, though still possibly giving the Everett many worlds rather than just a single classical world. Thus, even a universe that gives exactly the same measures for conscious perceptions as ours, and hence the same typicalities for all observations, could, in principle, be entirely classical in the sense of being commutative. We cannot prove that a universe is quantum just from our observations.

However, surely such a classical description of our conscious perceptions would involve a more complicated SQM universe than one in which there are non-commuting operators (and presumably even non-commuting awareness operators). Thus, it is on the grounds of simplicity and Ockham's razor that we assign higher probabilities to non-commuting quantum theories that explain our observations, even though the likelihoods for our observations can be precisely the same in a classical theory. In a similar way it might turn out that, although a multiple-SQM-universe theory could be reduced to a single-SQM-universe theory in one of the ways outlined above, the description could be simpler in terms of the former or even in terms of universes that are not SQM.

If we do have a true multiverse of different universes, each of which gives a measure for each set of conscious perceptions, then to get a measure covering the whole of reality, we would need a measure for each of the individual universes. Suppose each universe is described by an hypothesis H_n that assigns a measure $\mu_n(S)$ for each set S of conscious perceptions. When we were considering single universes, we considered different H_m just as theoretical alternative possibilities and discussed assigning subjective prior probabilities $P_i(H_m)$ to them. But when we are considering true multiverses, we need an objective weight $w(H_n)$ for each universe, since each universe with non-zero weight is being considered as actually existing. Therefore the total measure for each set of conscious perceptions from this extended multiverse would be $\mu(S) = \sum_n w(H_n)\mu_n(S)$.

Extending the multiverse to multiple SQM universes (or to any ensemble in which there is a prediction for the measure for all sets of conscious perceptions from each universe) replaces our uncertainty about which H_n is correct with the uncertainty about which $w(H_n)$ is correct. It would replace Tegmark's question [28–31] 'Why these equations?' with 'Why this measure?' We cannot evade some form of this question by invoking ever higher levels of the multiverse, even though this may provide a simpler description of a world.

In the sense that an SQM universe is a single universe, it may still encompass Level 1–3 multiverses. At the true multiverse level, we need not just a single theory H_n for a single universe, but also a meta-theory I for the measure or weight $w(H_n)$ of the single universes within the set of actually existing multiverses. However, since we do not yet know what the correct meta-theory is, just as we do not yet know what the correct theory H_n is for our single universe, we may wish to consider various theoretically possible meta-theories, I_M, labelled by some index M in the same way that n labelled the single universe described by the theory H_n. Then meta-theory I_M says that single universes exist with measures $w_{M,n} \equiv w_M(H_n)$ and so a set of conscious perceptions S would have measure $\mu_M(S) = \sum_n w_{M,n} \mu_n(S)$. From the measure for conscious perceptions, one can follow the procedure outlined in Section 23.3 to get the typicality $T_M(O)$ of an observation O in meta-theory I_M.

For example, if the single universes described by H_n are labelled by the positive integers n in order of increasing complexity, and if the meta-theories I_M are labelled by the positive integers M, one might imagine the following choice for the weights $w_{M,n}$ of the meta-theory I_M to give the universe H_n:

$$w_{2m-1,n} = \frac{1}{m}\left(\frac{m}{m+1}\right)^n, \quad w_{2m,n} = \delta_{mn}. \tag{23.3}$$

Then, for odd M, one gets a geometric distribution of weights over all single universes described by the theories H_n, with the mean of n being $m+1$. However, for even M, one gets a non-zero (unit) weight only for the unique single universe described by the theory H_m. Thus the odd members of this countable sequence of meta-theories do indeed give multiverse theories with various weights, but the even members give single-universe theories.

Just as in a Bayesian analysis for single-universe theories we needed subjective prior probabilities $P_i(H_m)$ for the possible single-universe theories H_m, so now for a Bayesian analysis of multiverse meta-theories I_M, we need subjective prior probabilities $P_i(I_M)$. Again, although these subjective prior probabilities are really arbitrary, we may wish to invoke Ockham's razor for the meta-theories and assign the simpler ones the greater prior probabilities. For example, if we can re-order the I_M in increasing order of complexity by another natural number $N(M)$, one might use the simple subjective prior probability assignment:

$$P_i(I_M) = 2^{-N(M)}. \tag{23.4}$$

This would imply that the simplest meta-theory ($N=1$) is assigned 50% prior probability of being correct, the next simplest ($N=2$) 25%, etc.

For a more *ad hoc* choice, one could take the meta-theory weights given by the hybrid model of Eq. (23.3) for both single and multiple universes and arbitrarily set

$$P_i(I_{2m-1}) = P_i(I_{2m}) = 2^{-m-1}. \tag{23.5}$$

This gives a total prior probability of 1/2 for single-universe (even M) theories and 1/2 for multiple-universe (odd M) theories. This might be viewed as a compromise assignment if one is *a priori* ambivalent about whether a single-universe or multiple-universe theory should be used.

23.5 Conclusions

Even though multiverse theories usually involve unobservable elements, they may give testable predictions for observable elements if they include a well defined measure for observations. One can then analyze them by Bayesian means, using the theory-dependent typicality of the result of observations as a likelihood for the theory, though there is still an inherent ambiguity in assigning prior probabilities to the theories.

One can try to avoid specifying the equations or other properties of an individual universe by assuming that there is an ensemble of different universes, but this replaces the question of the equations with the question of the measure for the different universes in the ensemble. There is no apparent way to avoid having non-trivial content in a testable theory fully describing all of reality.

Acknowledgements

I am very grateful for discussions with other participants at the Templeton conferences on the 'multiverse' theme in Cambridge in 2001 and Stanford in 2003. My ideas have also been sharpened by e-mail discussions with Robert Mann. This research has been supported in part by the Natural Sciences and Engineering Research Council of Canada.

References

[1] B. Carter. In *Confrontation of Cosmological Theory with Observational Data*, ed. M. S. Longair (Dordrecht: Reidel, 1974), pp. 291–298.
[2] B. J. Carr and M. J. Rees. *Nature*, **278** (1979), 605.
[3] J. D. Barrow and F. J. Tipler. *The Anthropic Cosmological Principle* (Oxford: Clarendon Press, 1986).
[4] M. Rees. *Just Six Numbers: The Deep Forces that Shape the Universe* (New York: Basic Books, 2000).
[5] M. J. Duff, B. E. W. Nilsson and C. N. Pope. *Phys. Rep.* **130** (1986), 1.
[6] R. Bousso and J. Polchinski. *JHEP*, **0006** (2000), 006 [hep-th/0004134].
[7] S. Kachru, R. Kallosh, A. Linde and S. P. Trivedi. *Phys. Rev.* **D 68** (2003), 046005 [hep-th/0301240].
[8] L. Susskind. This volume (2007) [hep-th/0302219].

[9] L. Susskind. In *From Fields to Strings: Circumnavigating Theoretical Physics, Ian Kogan Memorial Collection, Vol. 3*, eds. M. Shifman, A. Vainshtein and J. Wheater (Oxford: Oxford University Press, 2005), pp. 1745–1749 [hep-th/0405189].
[10] L. Susskind. *The Cosmic Landscape, String Theory and the Illusion of Intelligent Design* (New York: Little, Brown and Company, 2005).
[11] M. R. Douglas. *JHEP*, **0305** (2003), 046 [hep-th/0303194].
[12] M. R. Douglas. *Compt. Rend. Phys.* **5** (2004), 965 [hep-th/0401004].
[13] M. R. Douglas. (2006) [hep-th/0602266].
[14] T. Banks, M. Dine and E. Gorbatov. *JHEP*, **0408** (2004), 058 [hep-th/0309170].
[15] H. Firouzjahi, S. Sarangi and S. H. H. Tye. *JHEP*, **0409** (2004), 060 [hep-th/0406107].
[16] D. N. Page. *Int. J. Mod. Phys.* **D 5** (1996), 583 [gr-qc/9507024]; also quant-phys/9506010.
[17] D. N. Page. In *Consciousness, New Philosophical Perspectives*, eds. Q. Smith and A. Jokic (Oxford: Clarendon Press, 2003), pp. 468–506 [quant-ph/0108039].
[18] R. B. Griffiths. *J. Stat. Phys.* **36** (1984), 219.
[19] M. Gell-Mann and J. B. Hartle. In *Complexity, Entropy, and the Physics of Information, SFI Studies in the Science of Complexity, Vol. VIII*, ed. W. Zurek (Reading, MA: Addison-Wesley, 1990).
[20] R. Omnès. *Interpretation of Quantum Mechanics* (Princeton, NJ: Princeton University Press, 1994).
[21] A. Linde and A. Mezhlumian. *Phys. Rev.* **D 53** (1996), 4267 [gr-qc/9511058].
[22] A. Vilenkin. *Phys. Rev. Lett.* **81** (1998), 5501 [hep-th/9806185].
[23] A. Vilenkin. *Phys. Rev.* **D 59** (1999), 123506 [gr-qc/9902007].
[24] V. Vanchurin, A. Vilenkin and S. Winitzki. *Phys. Rev.* **D 61** (2000), 083507 [gr-qc/9905097].
[25] J. Garriga and A. Vilenkin. *Phys. Rev.* **D 64** (2001), 043511 [gr-qc/0102010].
[26] J. Garriga and A. Vilenkin. *Phys. Rev.* **D 64** (2001), 023507 [gr-qc/0102090].
[27] J. Garriga and A. Vilenkin. *Phys. Rev.* **D 67** (2003), 043503 [astro-ph/0210358].
[28] M. Tegmark. *Sci. Am.* **288** (5) (2003), 41.
[29] M. Tegmark. *Spektrum Wiss.* (8) (2003), 34.
[30] M. Tegmark. In *Science and Ultimate Reality*, eds. J. D. Barrow, P. C. W. Davies and C. L. Harper (Cambridge: Cambridge University Press, 2003) pp. 459–491 [astro-ph/0302131].
[31] M. Tegmark. *Ann. Phys.* **270** (1998), 1 [gr-qc/9704009].

24

Observation selection theory and cosmological fine-tuning

Nick Bostrom
Faculty of Philosophy, Oxford University

24.1 Introduction

When our measurement instruments sample from only a subspace of the domain that we are seeking to understand, or when they sample with uneven sampling density from the target domain, the resulting data will be affected by a selection effect. If we ignore such selection effects, our conclusions may suffer from selection biases. A classic example of this kind of bias is the election poll taken by the *Literary Digest* in 1936. On the basis of a large survey, the *Digest* predicted that Alf Langdon, the Republican presidential candidate, would win by a large margin. But the actual election resulted in a landslide for the incumbent, Franklin D. Roosevelt. How could such a large sample size produce such a wayward prediction? The *Digest*, it turned out, had harvested the addresses for its survey mainly from telephone books and motor vehicle registries. This introduced a strong selection bias. The poor of the depression era – a group that disproportionally supported Roosevelt – often did not have phones or cars.

The *Literary Digest* suffered a major reputation loss and soon went out of business. It was superseded by a new generation of pollsters, including George Gallup, who not only got the 1936 election right, but also managed to predict what the *Digest*'s prediction would be to within 1%, using a sample size that was only one-thousandth as large. The key to his success lay in his accounting for known selection effects. Statistical techniques are now routinely used to correct for many kinds of selection bias.

Observation selection effects are an especially subtle kind of selection effect, introduced not by limitations in our measurement apparatus, but by the fact that all evidence is preconditioned on the existence of an observer 'having' the evidence and building the instrument in the first place. Only quite recently have observation selection effects become the subject of

Universe or Multiverse?, ed. Bernard Carr. Published by Cambridge University Press.
© Cambridge University Press 2007.

systematic study. Observation selection effects are important in many scientific areas, including cosmology and parts of evolution theory, thermodynamics, the foundations of quantum theory and traffic analysis. There are also interesting applications to the search for extraterrestrial life and questions such as whether we might be living in a computer simulation created by an advanced civilization [1].

Observation selection theory owes a large debt to Brandon Carter, a theoretical physicist who wrote several seminal papers on the subject, the first one published in 1974 [2–4]. Although there were many precursors, one could fairly characterize Carter as the father of observation selection theory – or 'anthropic reasoning' as the field is also known. Carter coined the terms 'weak' and 'strong' anthropic principle, intending them to express injunctions to take observation selection effects into account. But while Carter knew how to apply his principles to good effect, his explanations of the methodology they were meant to embody were less than perfectly clear. The meaning of the anthropic principles was further obscured by some later interpreters, who bestowed them with additional content that had nothing to do with observation selection effects. This contraband content, which was often of a speculative, metaphysical or teleological nature, caused anthropic reasoning to fall into disrepute. Only recently has this trend been reversed.

Since Carter's contributions, considerable effort has been put into working out the applications of anthropic principles, especially as they pertain to cosmological fine-tuning. There have also been many philosophical investigations into the foundations of anthropic reasoning. These investigations have revealed several serious paradoxes, such as the Doomsday argument (which one may or may not regard as paradoxical) [5], the Sleeping Beauty problem [6, 7], and the Adam and Eve thought experiments [8]. It is still controversial what conclusions we should draw from the apparent fine-tuning of our universe, as well as whether and to what extent our universe really is fine-tuned, and even what it means to say that it is fine-tuned.

Developing a theory of observation selection effects that caters to legitimate scientific needs, while sidestepping philosophical paradoxes, is a non-trivial challenge. In my recent book *Anthropic Bias: Observation Selection Effects in Science and Philosophy* [7], I presented the first mathematically explicit general observation selection theory and explored some of its implications. Before sketching some of the basic elements of this theory and discussing how it pertains to the multiverse hypothesis, let us briefly consider some of the difficulties that confront attempts to create a method for dealing with observation selection effects.

24.2 The need for a probabilistic anthropic principle

The anthropic principles that Carter proposed, even setting aside the inadequacies in their formulation, were insufficiently strong for many scientific applications. A particularly serious shortcoming is that they were not probabilistic.

Carter's principles enable us to deal with some straightforward cases. Consider a simple theory that says that there are one hundred universes, and that ninety of these are lifeless and ten contain observers. What does such a theory predict that we should observe? Clearly not a lifeless universe. Since lifeless universes contain no observers, an observation selection effect precludes them from being observed, as enunciated by the strong anthropic principle. So, although the theory claims that the majority of universes are lifeless, it nevertheless predicts that we should observe one of the atypical ones that contain observers.

Now take a slightly more complicated case. Suppose a theory says that there are one hundred universes of the following description:

> ninety type-A universes, which are lifeless;
> nine type-B universes, which contain one million observers each;
> one type-C universe, which contains one billion observers.

What does this theory predict that we should observe? (We need to know the answer to this question in order to determine whether it is confirmed or disconfirmed by our observations.) As before, an obvious observation selection effect precludes type-A universes from being observed, so the theory does not predict that we should observe one of those. But what about type-B and type-C universes? It is logically compatible with the theory that we should be observing a universe of either of these kinds. However, probabilistically it is more likely, conditional on the theory, that we should observe the type-C universe, because that is what the theory says that over 99% of all observers observe. Finding yourself in a type-C universe would, in many cases, tend to confirm such a theory, to at least some degree, compared with other theories that imply that most observers live in type-A or type-B universes.

To obtain this result, we must introduce a probabilistic strengthening of the anthropic principle along the lines of what I have called the 'Self-Sampling Assumption' [7, 9, 10]:

> (SSA) One should reason as if one were a random sample from the set of all observers in one's reference class.[1]

---

[1] Related principles have been explored in, for example, refs. [11]–[16]

With the help of SSA, we can calculate the conditional probabilities of us making a particular observation, given one theory or another, by comparing what fraction of the observers in our reference class would be making such observations according to the competing theories.

What SSA does is enable us to take indexical information into account. Consider the following two evidence statements concerning the cosmic microwave background temperature (CMBT):

> E: an observation of CMBT = 2.7 K is made.
>
> E*: *we* make an observation of CMBT = 2.7 K.

Note that E* implies E, *but not vice versa*: E*, which includes a piece of indexical information, is logically stronger than E. Consequently, it is E* that dictates what we should believe if these different evidence statements lead to different conclusions. This follows from the principle that all relevant information should be taken into account.

Let us examine a case where it is necessary to use E* rather than E [17]. Consider two rival theories about the CMBT at the current cosmic epoch. Let T_1 be the theory we actually hold, claiming that CMBT = 2.7 K. Let T_2 say that CMBT = 3.1 K. Now suppose that our universe is infinitely large and contains an infinite number of stochastic processes of suitable kinds, such as radiating black holes. If for each such random process there is a finite, non-zero probability that it will produce an observer in any particular brain-state (subjectively making an observation e), then, because there are infinitely many independent 'trials', the probability for any given observation e that e will be made by some observer somewhere in our universe is equal to unity. Let B be the proposition that this is the case. We might wonder how we could possibly test a conjunction like T_1&B or T_2&B. For whatever observation e we make, both these conjunctions predict equally well (with probability unity) that e should be made. According to Bayes's theorem, this entails that conditionalizing on e being made will not affect the probability of T_1&B or T_2&B. And yet it is obvious that the observations we have actually made support T_1&B over T_2&B. For, needless to say, it is because of our observations that we believe that CMBT = 2.7 K and not 3.1 K.

This problem is solved by going to the stronger evidence statement E* and applying SSA. For any reasonable choice of reference class, T_1&B implies that a much larger fraction of all observers in that class should observe CMBT = 2.7 K than does T_2&B. (According to T_1&B, all normal observers observe CMBT = 2.7 K, while T_2&B implies that only some exceptional black-hole-emitted observers, or those who suffer from rare illusions, observe

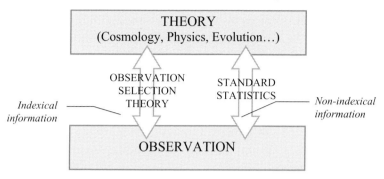

Fig. 24.1. Observation selection theory complements standard statistics and is required for cases where either the evidence or the hypothesis includes indexical information.

CMBT = 2.7 K.) Given these facts, SSA implies

$$P(E^*|T_1 \& B) \gg P(E^*|T_2 \& B). \tag{24.1}$$

From Eq. (24.1), it is then easy to show that our actual evidence E* does indeed give us reason to believe in $T_1 \& B$ rather than $T_2 \& B$. In other words, SSA makes it possible for us to know that CMBT = 2.7 K. This is illustrated in Fig. 24.1.

For the moment, we are setting aside the problem of exactly how the reference class is to be defined. In the above example, any reference class definition satisfying some very weak constraints would do the trick. To keep things simple, we also ignore the problem of how to generalize SSA to deal with infinite domains. Strictly speaking, such an extension, which might involve focusing on densities rather than sets of observers, would be necessary to handle the present example, but it would add complications that would distract from basic principles.

24.3 Challenges for observation selection theory

So far, so good. SSA can derive additional support from various thought experiments, and it can be applied to a number of scientific problems where it yields results that are less obvious but nevertheless valid.

Unfortunately, if we use SSA with the universal reference class, the one consisting of all intelligent observers, we encounter paradoxes. One of these is the notorious Doomsday argument, which purports to show that we have systematically underestimated the probability that our species will become extinct soon. The basic idea behind this argument is that our position in the sequence of all humans that will ever have lived would be much more probable if the total number of humans is, say, 200 billion rather than 200 trillion. Once we take into account this difference in the conditional probability of our observed birth rank, the argument goes, hypotheses that imply that very many humans are yet to be born are seen to be much less probable than we would have thought if we considered only the ordinary evidence (about the risk of germ warfare, nuclear war, meteor strikes, destructive nanotechnology, etc.). The prospects of our descendants ever colonizing the Galaxy would be truly dismal, as this would make our own place in human history radically atypical.

The most common initial reaction to the Doomsday argument is that it must be wrong; moreover, that it is wrong for some obvious reason. Yet when it comes to explaining *why* it is wrong, it turns out that there are almost as many explanations as there are people who disbelieve the Doomsday arguments. And the explanations tend to be mutually inconsistent. On closer inspection, all these objections, which allege some trivial fallacy, turn out to be themselves mistaken [5, 7, 18].

Nevertheless, the Doomsday argument has some backers, and while the way in which it aims to derive its conclusion is definitely counter-intuitive, it may not quite qualify as a paradox. It is therefore useful to consider the following thought experiment [8], which has a structure similar to the Doomsday argument but yields a conclusion that is even harder to accept.

Serpent's Advice. Eve and Adam, the first two humans, knew that if they gratified their flesh, Eve might bear a child, and that if she did, they would both be expelled from Eden and go on to spawn billions of progeny that would fill the Earth with misery. One day a serpent approached the couple and spoke thus: 'Pssssst! If you take each other in carnal embrace, then either Eve will have a child or she won't. If she has a child, you will have been among the first two out of billions of people. Your conditional probability of having such early positions in the human species, given this hypothesis, is extremely small. If, on the other hand, Eve does not become pregnant, then the conditional probability, given this, of you being among the first two humans is equal to one. By Bayes's theorem, the risk that she shall bear a child is less than one in a billion. Therefore, my dear friends, step to it and worry not about the consequences!'

It is easy to verify that, if we apply SSA to the universal reference class, the serpent's mathematics is watertight. Yet surely it would be irrational for Eve to conclude that the risk of her becoming pregnant is negligible.

One can try to revise SSA in various ways or to impose stringent conditions on its applicability. However, it is difficult to find a principle that satisfies all constraints that an observation selection theory ought to satisfy – a principle that both serves to legitimize scientific needs and at the same time is probabilistically coherent and paradox-free. We lack the space here to elaborate on the multitude of such theory constraints. It is easy enough to formulate a theory that passes a few of these tests, but it is hard to find one that survives them all.

24.4 Sketch of a solution

The solution, in my view, begins with the realization that the problem with SSA is not that it is too strong but that it is not strong enough. SSA tells you to take into account one kind of indexical information: information about which observer you are. But you have more indexical information than that. You also know *which temporal segment* of that observer, which 'observer-moment', you currently are. We can formulate a 'Strong Self-Sampling Assumption' that takes this information into account as follows [7]:

> (SSSA) Each observer-moment should reason as if it were randomly selected from the class of all observer-moments in its reference class.

Arguments can be given for why SSSA expresses a correct way of reasoning about a number of cases.

To cut a long story short, we find that the added analytical firepower provided by SSSA makes it possible to relativize the reference class, so that different observer-moments of the same observer may place themselves in different reference classes without that observer being probabilistically incoherent over time. This relativization of the reference class, in turn, makes it possible coherently to reject the serpent's advice to Eve, while still enabling legitimate scientific inferences to go through. Recall, for instance, the case we considered above, about our observation of $CMBT = 2.7$ K, supporting the theory that this is the actual local temperature even when evaluated in the context of a cosmological theory that asserts that all possible human observations are made. This result would be obtained almost independently of how we defined the reference class. So long as the reference class satisfies some very weak constraints, the inference works. This 'robustness' of an inference under different definitions of the reference class turns out to be a hallmark of those applications of anthropic reasoning that are scientifically respectable. By contrast, the applications that yield paradoxes rely on specific definitions of the reference class and collapse when a different reference class is chosen. The serpent's reasoning, for example, works only if we place the observer-moments of Adam and Eve prior to sinning in the same reference class as the observer-moments of those (very different) observers that may come to exist centuries later as a result of the first couple's moral lapse. The very fact that this absurd consequence would follow from selecting such a reference class gives us a good reason to use another reference class instead.

The idea expressed vaguely in SSSA can be formalized into a precise principle that specifies the evidential bearing of a body of evidence e on a hypothesis h. I have dubbed this the 'Observation Equation' [7]:

$$P_\alpha(h|e) = \frac{1}{\gamma} \sum_{\sigma \in \Omega_h \cap \Omega_e} \frac{P_\alpha(w_\sigma)}{|\Omega_\alpha \cap \Omega(w_\sigma)|}. \tag{24.2}$$

Here α is the observer-moment whose subjective probability function is P_α, Ω_h is the class of all possible observer-moments about whom h is true, Ω_e is the class of all possible observer-moments about whom e is true, Ω_α is the class of all observer-moments that α places in the same reference class as herself, w_α is the possible world in which α is located, and γ is a normalization constant. The quantity in the denominator is the cardinality of the intersection of two classes, Ω_α and $\Omega(w_\sigma)$, the latter being the class of all observer-moments that exist in the possible world w_σ.

The Observation Equation can be generalized to allow for different observer-moments within the reference class having different weights $\mu(\sigma)$. This option is of particular relevance in the context of the 'many worlds' version of quantum mechanics, where the weight of an observer-moment would be proportional to the amplitude squared of the branch of the universal wave-function where that observer-moment lives.

The Observation Equation expresses the core of a quite general methodological principle. Two of its features deserve to be highlighted here. The first is that by dividing the terms of the sum by the denominator, we are factoring out the fact that some possible worlds contain more observer-moments than do others. If one omitted this operation, one would, in effect, assign a higher prior probability to possible worlds that contain a greater number of observers (or more long-lived observers). This would be equivalent to accepting the Self-Indication Assumption, which prescribes an *a priori* bias towards worlds that have a greater population. But, although the Self-Indication Assumption has its defenders (see, e.g., ref. [19]), it leads to paradoxical consequences, as shown by the Presumptuous Philosopher thought experiment [7]. In a nutshell, this thought experiment points out that the Self-Indication Assumption implies that we should assign probability one to the cosmos being infinite, even if we had strong empirical evidence that it was finite; and this implication is very hard to accept.

A second feature to highlight is that the only possible observer-moments that are taken into account by an agent are those that the agent places in the same reference class as herself. Observer-moments that are outside this reference class are treated, in a certain sense, as if they were rocks or other lifeless objects. Thus, the question of how to define 'observer' is replaced by the question of how an agent should select an appropriate reference class for a particular application. This reference class will often be a proper subset of intelligent observers or observer-moments.

Bounds can be established on permissible definitions of the reference class. For example, if we reject the serpent's advice, we must not use the universal reference class that places all observer-moments in the same reference class. If we want to conclude on the basis of our evidence that CMBT = 2.7 K, we must not use the minimal reference class that includes only subjectively indistinguishable observer-moments, for such a reference class would block that inference.

It is an open question whether additional constraints can be found that would always guarantee the selection of a unique reference class for all observer-moments, or whether there might instead, as seems quite likely, be an unavoidable element of subjective judgment in the choice of reference class. This latter contingency would parallel the widely acknowledged element of subjectivity inherent in many other kinds of scientific judgments that are made on the basis of limited or ambiguous evidence.

24.5 Implications for cosmological fine-tuning

One immediate implication of observation selection theory for cosmological fine-tuning is that it allays worries that anthropic reasoning is fundamentally unsound and inevitably plagued by paradoxes. It thereby puts the multiverse explanation of fine-tuning on a more secure methodological footing.

A multiverse theory can potentially explain cosmological fine-tuning, provided several conditions are met. To begin with, the theory must assert the existence of an ensemble of physically real universes. The universes in this ensemble would have to differ from one another with respect to the values of the fine-tuned parameters, according to a suitably broad distribution. If observers can exist only in those universes in which the relevant parameters take on the observed fine-tuned values (or if the theory at least implies that a large portion of all observers are likely to live in such universes), then an observation selection effect can be invoked to explain why we observe a fine-tuned universe. Moreover, in order for the explanation to be completely satisfactory, this postulated multiverse should not itself be significantly fine-tuned. Otherwise the explanatory problem would merely have been postponed; for we would then have to ask, how come the *multiverse* is fine-tuned? A multiverse theory meeting these conditions could give a relatively high conditional probability to our observing a fine-tuned universe. It would thereby gain a measure of evidential support from the finding that our universe is fine-tuned. Such a theory could also help *explain* why we find ourselves in a fine-tuned universe, but to do this the theory would also have to meet the ordinary crew of *desiderata* – it would have to be physically plausible, fit the evidence, be relatively simple and non-gerrymandered, and so forth. Determining whether this potential anthropic explanation of fine-tuning actually succeeds requires a lot of detailed work in empirical cosmology.

One may wonder whether these conclusions depend on fine-tuning *per se* or whether they follow directly from the generic methodological injunction that we should, other things being equal, prefer simpler theories with fewer free variables to more complex theories that require a larger number of independent stipulations to fix their parameters (Ockham's razor). In other words, how does the fact that *life would not have existed if the constants of our universe had been slightly different* play a role in making fine-tuning cry out for an explanation and in suggesting a multiverse theory as the remedy?

Observation selection theory helps us answer these questions. It is not just that all single-universe theories in the offing would seem to require delicate handpicking of lots of independent variable values that would make such theories unsatisfactory – the fact that life would not otherwise have existed adds to the support for a multiverse theory. It does so by making the anthropic multiverse explanation possible. A simple multiverse theory could potentially give a high conditional probability to us observing the kind of universe we do, because it says that only that kind of universe – among all the universes in a multiverse – would be observed (or, at least, that it would be observed by a disproportionately large fraction of the observers). The observation selection effect operating on the fact of fine-tuning *concentrates* the conditional probability on us observing a universe like the one we live in. This is illustrated by Fig. 24.2.

Further, observation selection theory enables us to answer the question of how big a multiverse has to be in order to explain our evidence. The upshot is that bigger is not always better [7]. The postulated multiverse would have to be large and varied enough to make it probable that some universe like ours should exist. Once this objective is reached, there is no additional

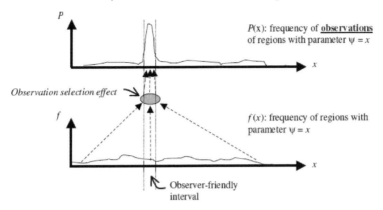

Fig. 24.2. An observation selection acts like a focusing lens, concentrating conditional probability on a small set of observational parameter values.

anthropic ground for thinking that a theory that postulates an even bigger ensemble of universes is therefore, other things equal, more probable. The choice between two multiverse theories that both give a high probability to a fine-tuned universe like ours existing must be made on other grounds, such as simplicity or how well they fit with the rest of physics.

A multiverse would not have to be large enough to make it probable that a universe *exactly* like ours should exist. A multiverse theory that entails such a huge cosmos that one would expect a universe exactly like ours to be included in it does not have an automatic advantage over a more frugal competitor. Such an advantage would have to be earned, for example by being a simpler theory. There is, as we noted earlier, no general reason for assigning a higher probability to theories that entail that there is a greater number of observers in our reference class. Increasing the membership in our reference class might make it more likely that the reference class should contain some observer who is making exactly the observations that we are making, but it would also make it more surprising that we should happen to be that particular observer rather than one of the others in the reference class. The net effect of these two considerations is to cancel each other out. All the observation selection effect does is concentrate conditional probability on the observations represented by the observer-moments in our reference class so that, metaphorically speaking, we can postulate stuff outside the reference class 'for free'. Postulating additional stuff *within* the reference class is not *gratis* in the same way, but would have to be justified on independent grounds.

It is, consequently, in major part an empirical question whether a multiverse theory is more plausible than a single-universe theory, and whether a larger multiverse is more plausible than a smaller one. Anthropic considerations are an essential part of the methodology for addressing these questions, but the answers will depend on the data.

In its current stage of development, observation selection theory falls silent on problems where the solution depends sensitively on the choice of reference class. For example, suppose a theory implies that the overwhelming majority of all observers that exist are of a very different kind from us. Should these radically different observers be in our reference class? If we do place them in our reference class (or, more precisely, if we place their observer-moments in the same reference class as our own current observer-moments), then a theory that implies that the overwhelming majority of all observers are of that different kind would be contra-indicated by our evidence, roughly because – according to that theory – we should have thought it highly unlikely that we should have found ourselves to be the kind of observer that we are rather than a more typical kind of observer. That is to say, such a theory would be disconfirmed compared to an equally simple theory that implied that a much larger fraction of all observers would be of our kind. Yet if we exclude the other kind of observer from our reference class, our evidence would not count against the theory. In a case like this, the choice of reference class makes a difference to our interpretation of our evidence.

Further developments of observation selection theory would be needed to determine whether there is a unique objectively correct way of resolving such cases. In the meantime, it is a virtue of the methodological framework encapsulated by the Observation Equation that it brings this indeterminacy into the open and does not surreptitiously privilege one particular reference class over potentially equally defensible alternatives.

References

[1] N. Bostrom. Are you living in a computer simulation? *Phil. Quart.* **53** (2003), 243.

[2] B. Carter. Large number coincidences and the anthropic principle in cosmology. In *Confrontation of Cosmological Theories with Data*, ed. M.S. Longair (Dordrecht: Reidel, 1974), pp. 291–298.

[3] B. Carter. The anthropic principle and its implications for biological evolution. *Phil. Trans. Roy. Soc.* **A 310** (1983), 347

[4] B. Carter. The anthropic selection principle and the ultra-Darwinian synthesis. In *The Anthropic Principle*, eds. F. Bertola and U. Curi (Cambridge: Cambridge University Press, 1989), pp. 33–63.

[5] N. Bostrom. The doomsday argument is alive and kicking. *Mind*, **108** (1999), 539.

[6] N. Bostrom. Sleeping beauty and self-location: a hybrid model. *Synthese* (2006); in press, see also www.nickbrostrom.com
[7] N. Bostrom. *Anthropic Bias: Observation Selection Effects in Science and Philosophy* (New York: Routledge, 2002).
[8] N. Bostrom. The doomsday argument, Adam & Eve, UN++, and Quantum Joe. *Synthese* **127** (2001), 359.
[9] N. Bostrom. Investigations into the doomsday argument. (1997) http://www.anthropic-principles.com/preprints/inv/investigations.html.
[10] N. Bostrom. Observer-relative chances in anthropic reasoning? *Erkenntnis*, **52** (2000), 93.
[11] R. J. Gott. Implications of the Copernican Principle for our future prospects. *Nature*, **363** (1993), 315.
[12] A. Vilenkin. Predictions from quantum cosmology. *Phys. Rev. Lett.* **74** (1995), 846.
[13] D. N. Page. Sensible quantum mechanics: Are probabilities only in the mind? *Int. J. Mod. Phys.* **D 5** (1996), 583.
[14] D. N Page. Can quantum cosmology give observational consequences of many-worlds quantum theory? In *General Relativity and Relativistic Astrophysics, 8th Canadian Conference*, eds. C. P. Burgess and R. C. Myers (Melville, New York: American Institute of Physics, 1999), pp. 225–232.
[15] M. Tegmark. Is the 'Theory of Everything' merely the ultimate ensemble theory? *Ann. Phys.* **270** (1998), 1 [gr-gc/9704009].
[16] A. Linde and A. Mezhlumian. On regularization scheme dependence of predictions in inflationary cosmology. *Phys. Rev.* **D 53** (1996), 4267.
[17] N. Bostrom. Self-locating belief in big worlds: Cosmology's missing link to observation. *J. Phil.* **99**, (2002), 607623.
[18] J. Leslie. *The End of the World: The Science and Ethics of Human Extinction* (London: Routledge, 1996).
[19] K. Olum. The Doomsday argument and the number of possible observers. *Phil. Quart.* **52** (2002), 164.

25

Are anthropic arguments, involving multiverses and beyond, legitimate?[1]

William R. Stoeger, S. J.
Vatican Observatory Research Group, Steward Observatory, University of Arizona

25.1 Introduction

Though there has been much discussion of the Anthropic Principle (AP) over the last 35 years or so, it is still a very tantalizing and controversial subject, on the boundary between scientific cosmology and philosophy. As new scenarios and theories emerge for describing and explaining the origin of our observable universe, AP considerations inevitably surface. So, a critical review of the meaning and status of the AP – as well as of the directions anthropic arguments are now taking, their legitimacy and the fundamental philosophical issues involved – is perhaps warranted.

The anthropic idea was first introduced in 1961 by Robert Dicke, who noted the comparability of several very large numbers when fundamental physical constants are combined, and suggested that this might be connected with the conditions necessary for the presence of observers [1]. A decade later, Barry Collins and Stephen Hawking, realizing that the initial conditions for our universe seemed to be very special, suggested the following: 'The fact that we have observed the universe to be isotropic is therefore only a consequence of our own existence' [2]. One way of explaining this, they speculated, would be to have an 'infinite set of universes with all possible initial conditions' – thus anticipating the way many cosmologists now interpret the AP.

The following year, Brandon Carter – obviously stimulated by Dicke's seminal suggestion, since he referred to it several times in his paper – introduced the term 'anthropic principle'. His initial formulation of the AP was as follows: 'What we can expect to observe must be restricted by the conditions necessary for our presence as observers' [3]. Subsequently, Carter

---

[1] This is a revised and expanded version of an article, 'The Anthropic Principle revisited', which appeared in *Phil. Sci.* **10** (2003), 9–33. Published here with permission.

Universe or Multiverse?, ed. Bernard Carr. Published by Cambridge University Press.
© Cambridge University Press 2007.

made a distinction between the 'weak' and 'strong' forms of the AP [4]. However, these terms have been used in many different ways, corresponding to similar but inequivalent formulations of the AP, which has led to considerable confusion. (In summarizing the origin and early history of the AP, I have followed ref. [5].)

First, it is important to recognize that the AP is not really a principle. Its fundamental content is that our universe *appears* to be fine-tuned for life and for consciousness – or perhaps, more precisely, for complexity. This appearance of fine-tuning originates from analysis of and reflection upon the results of a very broad range of experimental and theoretical investigations, indicating the extreme sensitivity of our universe's capacity for generating and sustaining complexity to very small changes in the laws and constants of nature, in the properties of the basic constituents of matter and in the initial conditions of our universe (for example the expansion rate or mass–energy density at some early time). A classical compendium on the AP, including a wide range of examples of fine-tuning, is given in ref. [6]. Changing the value of any of a large number of parameters even a little would so change our universe as to preclude the emergence of complexity – and therefore life and consciousness. The different formulations of the AP – and all the controversy which surrounds it – really trace back to the issue of what conclusions can be legitimately drawn from this apparent fine-tuning and what presuppositions are justifiable.

25.2 Weak and strong versions of the AP

From the earliest AP discussions, it was recognized that there are both weak (WAP) and strong (SAP) formulations. The weak versions assert that, since there are observers in our universe, its characteristics, including the values of the fundamental constants and the initial conditions, must be consistent with the presence of such observers (see ref. [5], p. 372). Thus the existence of observers acts *a posteriori* to select values of the fundamental constants and other important parameters. These versions of the AP just specify the conditions which have been fulfilled for complexity and life to arise – they do not explain how or why those conditions have been realized. In fact, some writers describe the WAP as just a selection effect.

Strong versions of the AP go much further: they asssert that our universe – right from the start – *had* to be such that the appearance of observers is inevitable. That is, they purport to account in a basic way for our universe being life-bearing. For instance, one version of the SAP would be as follows: 'The universe must be such as to admit the existence of observers within it at some stage' (see ref. [5], p. 376). Here, the words 'must be' indicate *a priori* necessity – not the consequence of there being observers now [5]. The eventual emergence of observers somehow explains why our universe possessed its initial characteristics – it has these characteristics *in order* that observers will appear.

From this, it is clear that some evidence or justification for the requirement of having observers must be provided. Many – but not all – such formulations do this by incorporating an explicit or implicit finality or purposiveness in our universe, which goes considerably beyond what can be concluded from the natural sciences themselves. Sometimes this is done on philosophical grounds, sometimes on theological ones.

25.3 Two principal versions of the SAP

Over the past decade, two very different – and certainly inequivalent – versions of the SAP have been discussed. The first is essentially the way it was first formulated: the characteristics of our universe are chosen to ensure the appearance of life and observers. But this raises the issue of what or who tailors the laws of nature and the fundamental constants in this way, which immediately goes beyond the domain of science.

Thus, a second version of the SAP has become popular, which – at first sight – keeps it within the realm of the natural sciences. This asserts that our universe – or our domain – is one of a large, actually existing, ensemble of universes or domains, each having different laws, fundamental constants and initial conditions. In fact, a frequent, but much less adequate, specification of this ensemble is that it contains all possible universes. The presupposition here is that there exist universes or domains representing the full range of possibilities [7]. There is then some probability that any one of the these really existing universes will allow the emergence of life and observers.

This, in one sense, *does* explain why our universe is life-bearing, providing the presuppositions can be justified and providing a meaningful probability measure can be defined on the space of the ensemble [7,8]. But this explanation is obviously incomplete. It immediately invites further understanding of the process by which this particular cosmic ensemble emerged and why it contains universes or domains which allow for the emergence of complexity. And if we can substantiate the operation of such a fertile cosmogonic process, then we may certainly want to seek an explanation for its origin, however we have come to understand and model its scientifically accessible underpinnings.

Thus, this formulation of the second version of the SAP clearly manifests its inequivalence with the first version, as well as the extraordinarily strong presuppositions on which it rests. In fact, if we use it to argue from the presence of observers in our universe to the existence of a certain type of ensemble of universes, then it seems to reduce to the WAP. However, the characteristic feature of this version – the existence of at least a large subset of possible domains, some of which are life-bearing – really takes us beyond the WAP. It solves the fine-tuning problem, but does not explain in any *a priori* way why our universe should have observers in it at some stage, much less why the ensemble of all existing universes should include some which admit their emergence.

It is certainly true that, if the ensemble exists, then our universe itself *must* exist. But this is obviously a very weak form of the SAP. The 'necessity' of the existence of our life-bearing universe rests on the presupposed existence of all possible universes, or at least of a large number of universes of a broad range of types. Clearly – to achieve equivalence with the first version of the SAP – we require an adequate explanation for the necessity of the encompassing ensemble – or at least some explanation or justification for its *de facto* realization. Anchoring this version of the SAP really requires some compelling cosmological account of the ensemble, which is by no means unique [8], or – even better – of why it *must* exist. This would make the multiverse version of the SAP equivalent to the first version. However, the multiverse version will never be able to go that far, since it strives to avoid scientifically inaccessible causes and explanations.

Another strong reason for stressing the multiverse version of the SAP is that we now recognize that there are a number of natural ways in which an ensemble of actually existing 'universes' could arise in quantum cosmology: for example, Andrei Linde's chaotic or eternal inflation scenario [9]. However, as we shall see later, such suggestions are not yet very secure. Furthermore, there are serious physical and philosophical issues which need to be resolved before they can be regarded as evidential [7, 8]. Until then, this version of the SAP, despite its popularity, must be relegated to the category of (at best) informed cosmological speculation.

In discussing the multiverse version of the SAP more fully, several points should be emphasized.

- As it now stands, it does not provide either an adequate or complete – let alone an ultimate – scientific explanation. Only strong evidence for – and an adequate description of – the process by which the ensemble emerges can do that. To constitute an ultimate explanation, that process must further be shown to be necessary, an understandable accident that was always a possibility, or intended by some transcendent agent for a specific reason. But the scenarios by which the ensembles may have originated are still very uncertain and *ad hoc*, so it is impossible to envision them as necessary or providing any fundamental or final philosophical explanation.

- Once we grant that the ensemble embracing our universe really does exist, then saying this ensemble explains how our universe is fine-tuned for life does have some meaning and validity – in terms of the probability of any one of the universes being like our own. But this requires that there be some well defined distribution function on the space of all possible universes, with an associated probability measure [7, 8].
- It is very difficult, if not impossible, to define a really existing ensemble of all possible universes in a meaningful way which avoids infinities [7, 8]. Also, in order for the ensemble idea to work, it cannot just be an ensemble of conceptually possible universes – it must really exist. Any power this version possesses relies on the universes or domains having a *bona fide* existence. Possible or potential existence has no *a posteriori* implications and explains nothing (see ref. [5], p. 371).

25.4 Scientific status of ensembles of universes

We have seen that the cosmic ensemble version of the SAP is not as strong as the original version. It is incomplete, requiring understanding of the process generating the ensemble. Furthermore, even with this understanding it cannot provide an ultimate explanation of the fine-tuning. Nevertheless, we can accept it such as it is, acknowledging that it may become more compelling as our understanding of the early universe improves. With this in mind, we shall reflect in more detail on what has come to be known as the 'multiverse' proposal [7].

First, as emphasized in Section 25.3, there are well supported but still preliminary indications that whatever process or event gave birth to our universe or domain also generated a large number of other universes or domains. This is why so many cosmologists and theoretical physicists are taking the idea seriously. Several lines of current research and speculation are probing, accumulating evidence for and attempting to model the primordial emergence of such an ensemble. Besides Linde's chaotic inflationary programme, there are a number of others, including those of Steven Weinberg [12, 13] and Jaume Garriga and Alex Vilenkin [14–16], who have suggested that random quantum fluctuations generated during inflation could have led to a large number of different cosmic regions, each with a different vacuum energy density. All of them would then evolve differently, with significantly different physics perhaps emerging in subsequent (GUT and electro-weak) spontaneous-symmetry-breaking transitions. Recently, superstring theory has provided prospects for generating multiverses. Some versions of it provide 'landscapes' populated by extremely large numbers of vacua, each of which could initiate a separate universe domain [17–20].

Ensembles of universes can also be generated in many other ways: for example, through decoherence from the mixed quantum gravity states which may have characterized the Planck era, or through the re-expansion into different domains of regions which had earlier collapsed to form black holes [21]. In the latter case, Lee Smolin envisions a type of natural selection operating on the resulting ensemble of expanding regions, rendering a significant subset of them bio-friendly. Finally, ensembles of universes can develop from the cosmic branching allowed by the Everett–Wheeler interpretation of quantum theory. In a recent popular article [22], Max Tegmark presents the case for multiverses and describes the different processes through which they may arise. All such scenarios are scientifically plausible. But, if they are to be taken seriously, they must continue to receive support from theoretical and observational advances in early and late universe research.

Even when such multiverse scenarios are better established, their deployment in anthropic arguments requires a proper characterization of the ensembles, with well defined (finite) probability distribution functions and meaningful probability measures [7]. If all these requirements are eventually fulfilled, there still remains the philosophical question of the legitimacy of appealing to ensembles whose existence is not testable. This raises the more fundamental issue of what kinds of testability are appropriate in the natural sciences. What concept of testability, if any, can legitimate reliance on cosmic ensembles for scientific conclusions? It is important in this regard to note that there is a general consensus that the acceptability of any appeal to multiverses depends on there being a testable theory which independently predicts their existence. This requirement is crucial and must be kept in mind in evaluating these theories and in contemplating their use in anthropic arguments.

That understood, are there concepts of testability which would enable multiverses to be scientifically legitimate? I believe that there are. One very compelling approach is that of 'retroduction' or 'abduction', first described in detail by the American philosopher of science C. S. Peirce [23, 24] and more recently emphasized by Ernan McMullin [25–28]. 'Retroduction' is inference from observed consequences of a postulated hypothesis to the explanatory antecedents contained in the hypothesis – that is, it is an inference based on the success or fruitfulness of an hypothesis in accounting for and better understanding a set of phenomena. Scientists construct hypotheses, which are then used to describe and probe the phenomena more profoundly. As they do so, they modify – or even replace – the original hypotheses in order to make them more fruitful and more precise in what they reveal and explain. As McMullin himself emphasizes, the hypotheses may often involve the existence of hidden properties or entities (like multiverses) which are basic to the explanatory power they possess. As the hypotheses become more fruitful in explaining a set of natural phenomena and their inter-relationships, and more central to the research of a given discipline, they become more reliable as accounts of reality. Even if the hidden properties are never directly detected, the success of the hypotheses which rely on them indirectly leads us to affirm that either they – or something very much like them – must exist. We can regard hypotheses as fruitful or successful if they: (1) account for all relevant data (empirical adequacy); (2) provide *long-term* explanatory success and stimulate productive lines of further enquiry (theory fertility); (3) establish the compatibility of previously disparate domains of phenomena or facts (unifying power); and (4) manifest consistency (or correlation) with other established theories [29].

This way of looking at how science works provides us with a criterion for testing theories which imply the existence of a multiverse. If such a theory successfully explains various aspects of what we see and measure in our universe, and continues to provide a secure basis for further cosmological understanding, then that strongly supports the existence of such universes, even though we may never be able to detect them directly. This criterion can be summarized as: Does the multiverse theory lead to greater intelligibity of the reality around us?

25.5 Using anthropic arguments in scientific cosmology

Setting aside for the time being the controversial SAP and multiverse ideas, we now turn our attention to a more modest application of anthropic arguments: their use in deciding purely scientific issues in cosmology. The extreme sensitivity of the character of our universe to slight changes in fundamental constants, the properties of fields and particles, initial conditions, etc. shows that – with enough knowledge – we can determine the values of these parameters on anthropic grounds.

The general form of such arguments is very straightforward. For life to exist in our universe, a given parameter A must be in the range A_1 to A_2. Life exists in our universe; therefore the value of A *is* between A_1 and A_2. However, it is important to recognize that this is a *necessary* but not *sufficient* condition for life. The main idea is that, using such anthropic arguments, cosmology can determine the values of key parameters without directly measuring them. This would be important whenever we did not have the capability of measuring the parameter A.

Three questions arise in considering such arguments: (1) Are they legitimate? (2) Do we need them in cosmology? (3) Do they suffice from a scientific point of view? The first question is easy to answer – the logic of the argument is clearly valid, so anthropic arguments are certainly legitimate. Establishing the major premise requires a great deal of theoretical work, however, and usually involves assumptions about what is essential for the emergence of life and how those essentials can be realized. Furthermore, as discussed below, such arguments demand a more complete understanding of the underlying 'laws of nature' which are at the basis of the parameter constraints.

Moving to the second question, we can say that, in some circumstances, we may 'need' such arguments or at least find them 'useful' until better scientific evidence is available. One of the drawbacks of anthropic arguments is that establishing that a given parameter must have a certain range of values for life normally takes a great deal of scientific investigation. The better and more reliable the underlying scientific theory enabling us to make that determination, the better and more reliable the anthropic arguments we can construct. But often, by the time we have reached that stage, we already know or have a good idea of what range of values a certain parameter has, independently of anthropic arguments. From this we might conclude that, whatever the state of our knowledge, anthropic arguments can always serve as consistency checks on conclusions we have reached by other means.

The answer to the third question – Are anthropic arguments sufficient? – is obviously no from a scientific point of view. The anthropic connection never stands by itself, but reflects a deeper and more fundamental set of relationships in the laws of nature – whether or not we understand them. Those deeper and more fundamental relationships will always be vulnerable to scientific determination or philosophical reflection, at least in principle.

25.6 Undermining anthropic arguments

The difficulty of reliably establishing the ranges of parameter values necessary for life in our universe is illustrated by the work of Anthony Aguirre [30]. He has demonstrated that there are more regions of cosmological parameter space which allow life than we had originally suspected. And some of these regions are isolated from each other.

This is true, for instance, for the cosmological parameter η, which is the ratio of the number density of baryons (protons and neutrons) to the number density of photons. This is a measure of the cosmological entropy density. In our universe, $\eta \approx 10^{-9}$, which indicates that the early universe was very hot. If we found $\eta \approx 1$ instead, our universe would have started out relatively cold. Such universes are referred to as 'cold big bang' models. Aguirre has shown that several classes of such models allow the formation of stars, and hence the production of heavy elements, and would therefore be open to the emergence of life. This set of bio-friendly cosmological models is disconnected in η-parameter space from the hot big bang models.

This unexpected development undermines anthropic arguments somewhat – or at least makes the conclusions we can draw from them less certain. We originally expected anthropic arguments to yield tightly constrained parameter ranges for life. But now, in at least some cases, we find that these ranges are somewhat broader and perhaps even disconnected from one another. We do not know if there are other cases of this sort. But if we are going to rely on such arguments, we have to be sure that we have theoretically explored the full range of cosmological parameter space for isolated bio-friendly islands.

Despite this uncertainty, we can still legitimately assert that: (1) the conditions for life have been fulfilled; and (2) the values of the parameters which characterize our universe must fall within certain relatively narrow ranges for these conditions to be maintained. However, given Aguirre's results, we need to be cautious in asserting precisely what these bio-friendly ranges are.

25.7 The SAP, final theories and alternative universes

Any version of the SAP presupposes that the laws of nature that characterize our universe could have been significantly different in terms of at least one of the following: initial conditions, particle properties (for example, masses), fundamental constants (for example, coupling parameters) and laws of nature (for example, different fundamental interactions). The key point is this: if a 'final theory' or 'a theory of everything' specifies a unique universe – that is, a universe with precise laws, values of the fundamental constants and initial conditions – then there is no need for, or even the possibility of, anthropic arguments. The universe could not have been any different without violating the theoretical consistency 'imposed' by the final theory. However, even if this were the case, we would – from most philosophical points of view – still need a sufficient explanation for the existence of the universe and for its precise order.

The extreme consequence of a final theory that specified a unique universe, accounting for all its characteristics precisely and exhaustively, is difficult to imagine. It is just possible that a final theory could achieve this. However, it seems very unlikely that it would fully determine the conditions for the universe as it exited the Planck or inflationary era, for example its expansion rate at this point and the initial entropy. In other words, to make anthropic arguments unnecessary scientifically or vacuous philosophically, we would need an adequate theory of initial conditions. We would also need a theory to specify the parameter values after spontaneous symmetry-breaking transitions.

An alternative would be a process, or a combination of processes, which renders the universe which emerges from the quantum cosmological womb *insensitive* to initial conditions. If such processes operated in the early universe, there would be no need for us to know or to explain the initial conditions in order to model how our universe evolved to its present form. It would have done so, no matter what the initial conditions, due to the 'smoothing' action of these primordial processes. They would bring the infant universe to the primordial homogeneity, no matter how it 'began'. This attractive suggestion is sometimes referred to as the 'Cosmological Indifference Principle' (see ref. [5], p. 359).

Two proposals for such 'indifference-rendering' are the chaotic cosmology programme of Charles Misner [31] and the (now almost orthodox) inflationary scenarios. In chaotic cosmology, Misner envisaged viscous forces dissipating any initial anisotropies to yield an isotropic expanding universe with very smooth spatial sections. It was eventually realized that such processes cannot accomplish this, but inflation is now invoked to fulfil this function.

As long as inflation can be initiated, it severely attenuates all initial inhomogeneities and anisotropies, ensuring that the resulting domain is nearly flat, causally connected and smooth on very large scales. At the same time, it preserves the low-amplitude quantum fluctuations of the early universe. These gradually develop into galaxies and clusters of galaxies but within a large-scale, nearly homogeneous background. Although it now appears that inflation is also incapable of rendering our universe insensitive to initial conditions – because the onset of inflation itself seems to require very special initial conditions [32,33] – attempts to realize the Cosmological Indifference Principle persist.

In fact, the multiverse idea may itself be interpreted this way (see ref. [5], p. 285). Taken alone, our universe requires very finely tuned initial conditions. Placing it in a really existing ensemble of other universes or domains seems, at first sight, to dispense with the need for that fine-tuning. However, as we have seen, they are certainly not uniquely specified. Accounting for the existence and specific character of our multiverse requires an adequate generating process or principle, which must explain the particular distribution function specifying it. This may itself require fine-tuning.

In comparing the two opposing philosophical perspectives represented by the anthropic and indifference principles, McMullin points out that the first inevitably involves mind and teleology (see ref. [5], pp. 259–367). This always threatens to take us beyond the domain of natural science to philosophy and theology. The indifference preference studiously avoids any direct appeal to such influences, relying instead on the dynamics inherent in and emerging from the mass–energy distribution itself [8].

25.8 The SAP and transcendent explanations

We have considered the legitimacy and scientific potential of anthropic arguments and we have come to a number of conclusions about the philosophical reach of the two versions of the SAP. We summarize these in five statements.

- Leaving aside the issue of an ultimate explanation, as long as the selection of initial conditions and the fundamental constants cannot be explained by some physical process or relationship, or rendered indifferent by one, a 'transcendent' explanation – one that takes us beyond natural science – is needed *if the Principle of Sufficient Reason continues to hold*.[2] This may take the form of a divine creative agent or a really existing multiverse.
- If we do have good evidence, and an adequately specific model for, the multiverse to which our own universe belongs, thus providing some explanation for its bio-friendly characteristics, this would not be a complete – let alone an ultimate – explanation. We would still require an explanation for the existence and bio-friendly character of the multiverse itself (bearing in mind that there is no unique prescription for it) and for the process through which it emerged – as well as a philosophically ultimate explanation.
- If we have a final theory which uniquely specifies all the characteristics of our universe, including the initial conditions, we cannot employ the

---

[2] The Principle of Sufficient Reason, which many philosophers maintain holds in all circumstances, requires that, for every state of affairs, event or outcome, there is an adequate reason or explanation. If, in some fundamental regime (for example quantum cosmology), this were not the case, then we might be able to forego searching for a further, deeper understanding. I personally do not believe this is the case, but it is a possible philosophical stance.

fine-tuning arguments of the SAP either scientifically or philosophically. There would then be only one way in which our universe could exist consistently. This is very unlikely, but we cannot rule it out at present.
- Even if the previous option applies, it would still not eliminate the need for an ultimate explanation or 'cause'. Nor would it invalidate philosophical arguments from contingency for the existence of God. (Here again we would be invoking the Principle of Sufficient Reason.)
- If we have a final theory that still allows some 'play' in the laws of nature, then a theological answer in terms of intentional action by a divine agent or Creator is certainly acceptable, as long as we are allowing ourselves to go beyond the natural sciences and admit a theological or metaphysical frame of reference. Science can neither support nor exclude such a conclusion. It cannot even adjudicate the question. However, in going beyond the sciences, we must avoid putting God in the 'scientific gaps'. Perhaps our final theory is not really final! We should ensure that the divine agent is always a primary or ultimate cause – not one that could conceivably be filled by some unknown secondary or created cause [34].

References

[1] R. H. Dicke. Dirac's cosmology and Mach's principle. *Nature*, **192** (1961), 440.
[2] C. B. Collins and S. W. Hawking. Why is the universe isotropic? *Astrophys. J.* **180** (1973), 317.
[3] B. Carter. Large number coincidences and the anthropic principle in cosmology. In *Confrontation of Cosmological Theory with Astronomical Data*, ed. M. S. Longair (Dordrecht: Reidel, 1974), pp. 291–298.
[4] B. Carter. The anthropic principle and its implications for biological evolution. *Phil. Trans. Roy. Soc. Lond.* **A 310** (1983), 347.
[5] E. McMullin. Indifference principle and anthropic principle in cosmology. *Studies Hist. Phil. Sci.* **24** (1993), 359.
[6] J. D. Barrow and F. J. Tipler. *The Cosmological Anthropic Principle* (Oxford: Oxford University Press, 1986).
[7] G. F. R. Ellis, U. Kirchner and W. R. Stoeger. Multiverses and physical cosmology. *Mon. Not. Roy. Astron. Soc.* **347** (2003), 921 [astro-ph/0305292v3].
[8] W. R. Stoeger, G. F. R. Ellis and U. Kirchner. Multiverses and cosmology: Philosophical issues (2004) [astro-ph/0407329v2].
[9] A. Linde. Chaotic inflation. *Phys. Lett.* **B 129** (1983), 177.
[10] A. Linde. *Particle Physics and Inflationary Cosmology* (Chur: Harwood Academic Publishers, 1990).
[11] A. Linde. Inflation, quantum cosmology and the anthropic principle. In *Science and Ultimate Reality*, eds. J. D. Barrow, P. C. W. Davies and C. L. Harper (Cambridge: Cambridge University Press, 2004), pp. 426–458 [hep-th/0211048].

[12] S. Weinberg. The cosmological constant problem. In *Sources and Detection of Dark Matter and Dark Energy in the Universe*, ed. D. Cline (Berlin: Springer-Verlag, 2001) [astro-ph/0005265].

[13] S. Weinberg. A priori probability distribution of the cosmological constant. *Phys. Rev.* **D 61** (2000), 103505 [astro-ph/0002387].

[14] J. Garriga and A. Vilenkin. Many worlds in one. *Phys. Rev.* **D 64** (2001), 043511.

[15] J. Garriga and A. Vilenkin. A prescription for probabilities in eternal inflation. *Phys. Rev.* **D 64** (2001), 0235074 [gr-qc/0102090].

[16] J. Garriga and A. Vilenkin. Testable anthropic predictions for dark energy. *Phys. Rev.* **D 67** (2003), 043503 [astro-ph/0210358].

[17] S. Kachru, R. Kallosh, A. Linde and S. P. Trivedi. de Sitter vacua in string theory. *Phys. Rev.* **D 68** (2003), 046005.

[18] L. Susskind. This volume (2007) [hep-th/03021219].

[19] D. Freivogel and L. Susskind. Framework for the string theory landscape. *Phys. Rev.* **D 70** (2004), 126007.

[20] B. Freivogel, M. Kelban, M. R. Martinez and L. Susskind. Observational consequences of a landscape. *JHEP*, **0603** (2006), 039 [hep-th/0505232].

[21] L. Smolin. *The Life of the Universe* (Oxford: Oxford University Press, 1999).

[22] M. Tegmark. Parallel Universes, *Sci. Am.* **288** (5) (May 2003), 30.

[23] C. S. Peirce. In *Collected Papers, Vols. 1–6*, eds. C. Hartstone and P. Weiss (Cambridge MA: Harvard University Press, 1931–1935), vol. 1, para. 65; vol. 5, para. 188.

[24] C. S. Peirce. In *Collected Papers, Vols. 7 & 8*, ed. A. Burks (Cambridge, MA: Harvard University Press, 1958), vol. 7, paras 202–207; 218–222.

[25] E. McMullin. *The Inference that Makes Science* (Milwaukee, WI: Marquette University Press, 1992), p. 112.

[26] E. McMullin. Is philosophy relevant to cosmology? *Am. Phil. Quart.* **18** (1981), 177–189. Reprinted in *Physical Cosmology and Philosophy*, ed. J. Leslie (New York: Macmillan, 1990), pp. 29–50.

[27] E. McMullin. Truth and explanatory success. In *Proc. Am. Cath. Phil. Assoc.* **59** (1985), 206.

[28] E. McMullin. Models of scientific inference. *CTNS Bull.* **8** (2) (1988), 1–14.

[29] P. Allen. *A Philosophical Framework within the Science-Theology Dialogue: A Critical Reflection on the Work of E. McMullin*. Ph.D. dissertation, St. Paul University, Ottawa, 2001, p. 113.

[30] A. Aguirre. The cold big-bang cosmology as a counter-example to several anthropic arguments. *Phys. Rev.* **D 64** (2001), 083508 (astro-ph/0106143).

[31] C. W. Misner. The isotropy of the universe. *Astrophys. J.* **151** (1968), 431.

[32] R. Penrose. Difficulties with inflationary cosmology. *Ann. N. Y. Acad. Sci.* **571** (1989), 249.

[33] G. F. R. Ellis, P. McEwan, W. R. Stoeger and P. Dunsby. Causality in inflationary universes with positive spatial curvature. *Gen. Rel. Grav.* **34** (2002), 1461.

[34] W. R. Stoeger. Describing God's action in the world in light of scientific knowledge of reality. In *Chaos and Complexity: Scientific Perspectives on Divine Action*, eds. R. J. Russell, N. Murphy and A. R. Peacocke (Rome: Vatican Observatory Publications and the Center for Theology and the Natural Science, 1995), pp. 239–261.

26

The multiverse hypothesis: a theistic perspective

Robin Collins
Department of Philosophy, Messiah College

26.1 Introduction

In articles published in physics journals, the multiverse hypothesis is strictly regarded from a non-theistic perspective, as a possible explanatory hypothesis for the life-permitting values of the constants of physics. Further, there have been several attempts to make specific predictions with regard to the values of these constants from a multiverse hypothesis, such as the value of the cosmological constant [1–3]. Such approaches reflect the legitimate methodological naturalism of physics. However, in wider-ranging philosophical discussions of the multiverse hypothesis – as found in various books on the topic – the issue arises as to what is the relation between the multiverse hypothesis and much larger philosophical issues, particularly whether reality is ultimately impersonal or personal in nature. In such contexts, the multiverse hypothesis is often presented as *the* atheistic alternative to a theistic explanation – such as that offered by John Polkinghorne [4] – of the purported fine-tuning of the cosmos for intelligent life. In this contribution, I will attempt to explain why, contrary to the impression one often gets, contemporary physics and cosmology are not only compatible with theism, but could arguably be thought to suggest a theistic explanation of the Universe or multiverse. I do not expect necessarily to convince anyone of the theistic point of view, realizing that many factors – both theoretical and personal – underlie our views of the ultimate nature of reality. Further, since there is no contribution to this collection arguing for an atheistic point of view, I will attempt to give voice – and replies – to some of the significant concerns and objections of non-theists.

26.2 Terminology: God and multiple universes

First, we need to clarify what we mean by theism and the multiverse hypothesis. With regard to theism, I take the theistic hypothesis to be the claim that an omnipotent and omniscient being is ultimately responsible for the existence of the Universe. The concept of God I will assume is the standard so-called Anselmian one, according to which God is defined as the greatest possible being, but this is not essential to my argument. It is often claimed that this conception of God is central to all of the world's theistic religious traditions – Islam, Judaism, Christianity and theistic versions of Hinduism. As an alternative to the Anselmian conception, one could simply think of the God hypothesis as the minimal one needed to explain the existence of the Universe via some sort of intelligent agent; that is, the hypothesis that there exists some highly intelligent and very powerful, and at least partly transcendent, agent that is ultimately responsible for the existence of the Universe.

Second, we need to be clear on what we mean by the multiverse hypothesis. There are essentially two kinds of multiverse hypotheses: what could be called the *physical* or *universe-generator* version and the *metaphysical* version. In the universe-generator version, some particular real physical process – such as an inflaton field – is postulated that generates the many universes (or domains), whereas in the metaphysical version the universes are thought to exist on their own, without being generated by any physical process.[1] In this chapter, I will primarily restrict myself to discussing the physical version, but the arguments in Section 26.7 on the beauty of the laws of nature apply to both versions.

26.3 The compatibility between theism and the multiverse hypothesis

In this section, I will argue not only that theism is compatible with the universe-generator version of the multiverse hypothesis, but also that theists might even have reasons for preferring a multiverse over a single universe. Since within the world's theistic traditions, God is considered infinite and infinitely creative, it makes sense that creation would reflect these attributes, and hence that physical reality might be much larger than one universe. Further, it makes sense that an infinitely creative God might create these many universes via some sort of universe-generator, since arguably this would be somewhat more elegant and ingenious than just creating them *ex nihilo*. Nonetheless, the idea that the Universe is infinite, or that there is some multiverse, has not been stressed in Western theology. A large part

---

1 I define a 'universe' as any region of spacetime that is disconnected from other regions in such a way that the parameters of physics in that region could differ significantly from what they are in other regions. The typical form of the metaphysic0al multiverse hypothesis is the claim that all possible mathematical structures are actual [5] or that all possible realities exist [6].

of the reason for this seems to be historical and not intrinsic to the Western theistic conception of God. The highly influential late medieval theology, for instance, was self-consciously based on Aristotelean metaphysics. For Aristotle, however, space was defined in terms of the extension enclosed by a physical object – such as the medieval crystalline spheres in the case of our universe – which were conceived of as necessarily finite. Indeed, many felt that restricting God to creating one universe was contrary to the omnipotence of God. Only with the eventual questioning of Aristotle by thinkers such as Nicholas of Cusa and, later, Giordano Bruno did the positive suggestion emerge that space was infinite, with perhaps an infinity of worlds. Although Bruno was considered a heretic by the Roman Catholic Church at the time for a variety of reasons, he spoke for many theistic thinkers when he declared [7]:

Thus is the excellence of God magnified and the greatness of his kingdom made manifest; he is glorified not in one, but in countless suns; not in a single earth, but in a thousand, I say, in an infinity of worlds.

This belief in a plurality of worlds, many of which are inhabited, was further developed and elaborated by Isaac Newton and many of his contemporaries, such as Gottfried Leibnitz. As Michael Crowe [8] notes, by 1750 belief in the plurality of worlds 'had been championed by an array of authors, including some of the most prominent of the age'.[2]

Indeed, the fact that the multiverse scenario fits well with an idea of an infinitely creative God, and that so many factors in contemporary cosmology and particle physics conspire together to make an inflationary multiverse scenario viable, should give theists good reason to consider a theistic version of it. Added to these reasons is the fact that science has progressively shown that the visible part of the Universe is vastly larger than we once thought, with a current estimate of some 300 billion galaxies with 300 billion stars per galaxy. Thus, it makes sense that this trend will continue and physical reality will be found to be much larger than a single universe.

Of course, one might object that creating a fine-tuned universe by means of a universe-generator would be an inefficient way for God to proceed. But this assumes that God does not have some motive for creation – such as that of expressing His/Her infinite creativity and ingenuity – other than creating a life-permitting cosmos using the least amount of material. But why would one make this assumption unless one already had a pre-existing model of God as something more like a great engineer instead of a great

---

2 A more contemporary Christian writer who has imaginatively developed this theme is the late C. S. Lewis in his fantasy series *Chronicles of Narnia*, in which God is imagined to have created a large number of different realms of being.

artist? Further, an engineer with infinite power and materials available would not necessarily care much about efficiency.

26.4 Understanding the fine-tuning

Our next question is this: Does the multiverse hypothesis undercut the case for some sort of design from fine-tuning, as advocated by various thinkers [9]? I will argue that, at most, it mitigates the case by rendering it less quantitative. First, however, we will need to sketch the fine-tuning argument itself.

Fine-tuning has been widely claimed to provide evidence of, or at least suggest, some sort of divine design of the Universe. Elsewhere, I have attempted to develop this argument in a more principled way [10]. As I develop it, the 'core version' of the argument essentially involves claiming that the existence of intelligent-life-permitting values for the constants of physics is not surprising under theism, but highly surprising under the non-design, non-multiverse hypothesis – that is, the hypothesis that there is only a single universe and that it exists as a brute fact without any further explanation. Further, the reason it seems highly surprising under the non-design, non-multiverse hypothesis is that, for certain constants of physics, the range of intelligent life-permitting values is purportedly small compared with some non-arbitrarily defined comparison range – such as the range of force strengths in nature when discussing the fine-tuning of gravity and other forces.

Using what could be called the 'surprise principle', it follows that the existence of intelligent-life-permitting values for the constants provides evidence in favour of theism over the non-design, non-multiverse hypothesis.[3] Note that no claim is being made here that theism is the best explanation of the constants being intelligent-life-permitting. To judge that a hypothesis is the best explanation of a body of data involves an overall assessment of the hypothesis, not simply how well it explains the particular data in

---

3 The surprise principle can be stated as follows. Let H1 and H2 be two competing non-*ad-hoc* hypotheses; that is, hypotheses that were not constructed merely to account for the data E in question. According to the surprise principle, if a body of data E is less surprising under hypotheses H1 than H2, then the data E provides evidence in favour of H1 over H2. The best way, I believe, of explicating what the notion of surprise is here is in terms of what philosophers call *conditional epistemic probability*, in which case the above principle becomes a version of the *likelihood principle* or *the principle of relevance*, which is a standard principle of probabilistic confirmation theory (see, e.g., ref. [11]). Unlike what D. H. Mellor [12] assumes in his objection to Martin Rees's claim that cosmic fine-tuning supports the multiverse hypothesis, conditional epistemic probability is not a measure of ignorance. Rather, it has to do with relations of support or justification between propositions. As John Maynard Keynes stated in his treatise on probability [13], 'if a knowledge of h justifies a rational belief in a of degree α, we say that there is a 'probability-relation' of degree α between a and h'. Although I think Keynes's account needs further work, I believe it is on the right track. For a recent development of this notion of conditional epistemic probability, see ref. [14].

question. The fact that Johnny's fingerprints are on the murder weapon might significantly support the claim that Johnny committed the murder. Nonetheless, Johnny's committing the murder might not be the best explanation of the fingerprints, since we might have strong, countervailing evidence that he did not commit the murder. Perhaps, for instance, five reliable witnesses saw Johnny at a party at the time of the murder. Similarly, all I claim is that the evidence of fine-tuning supports theism over the non-design, non-multiverse hypothesis. However, to judge whether we should infer that theism is the best explanation of the structure of the Universe – versus simply accepting the Universe as a brute given – involves many factors beyond the evidence of fine-tuning.

One of the key claims in the above argument is that the existence of a universe with intelligent-life-permitting cosmic conditions is not surprising under theism. This claim needs support instead of merely being assumed in an *ad hoc* way. Essentially, the argument is that if God is good – an assumption that is part of classical theism – then it is not surprising that God would create a world with intelligent beings, because the existence of such beings has positive value, at least under the theistic hypothesis.

Philosophers of religion offer a variety of justifications for the claim that God is perfectly good, or at least why ascribing goodness to God is not arbitrary. Here I will simply present two. First, some philosophers appeal to the Anselmian conception of God mentioned above, arguing that a being that is perfectly good is greater than one that is not perfectly good, and hence the characteristic of perfect goodness should be ascribed to God. Second, other philosophers – such as Richard Swinburne [15] – argue that, once grasped, the goodness or beauty of a state of affairs gives any conscious agent a reason to prefer that state of affairs. The idea is that part of grasping that a state of affairs has value – whether moral or aesthetic – is to 'feel' the desirability of the state, and hence have some motivation to bring it about. Under this view, for instance, people only do evil either because they do not grasp the disvalue of doing evil, or because some other influence tempts them to do what they recognize as having ethical disvalue. Since God is perfectly free – that is, God is not subject to countervailing desires in the way we are – God would have no motive to act against the good or beautiful. So, we would naturally expect God to act to bring about states of beauty and goodness. Whether one buys this sort of argument or not, I think that at minimum one has to admit that it is in no way arbitrary or *ad hoc* to hold that God has the desire to bring about states of goodness and beauty.

If this is right, then theism provides a natural connection between the moral and aesthetic realms of value and any such value we might have reason to believe that the Universe is structured to realize. One need not necessarily invoke a personal God to provide this connection, however. John Leslie [16], for instance, proposes what he calls a 'neo-Platonic principle of ethical requiredness', as suggested by Book VI of Plato's *Republic*, which could be thought of as a 'God substitute'. According to this principle, what ought to exist does exist. Further, Leslie claims, this principle exists as a matter of metaphysical necessity, much as many philosophers view the truths of mathematics, such as $2+2=4$. Thus its existence is self-explanatory. Leslie points out, however, that his neo-Platonic principle is compatible with theism, and might even entail theism: since God is a being of supreme value, one could argue that the principle entails that God would exist. Thus, even if we adopt Leslie's hypothesis, this would not necessarily provide an alternative to the theistic explanation of the fine-tuning of the Universe.[4]

26.5 Multiverse-generator needs design

In this section, I will argue that – even if a multiverse-generator exists – the argument for theism from the fine-tuning of the constants for intelligent life is not completely eliminated. The argument essentially goes as follows. The multiverse-generator itself, whether of the inflationary variety or some other type, seems to need to be 'well designed' in order to produce life-sustaining universes. After all, even a mundane item like a bread machine, which only produces loaves of bread instead of universes, must be well designed as an appliance and must have the right ingredients (flour, water and yeast) to produce decent loaves of bread. If this is right, then invoking some sort of multiverse-generator as an explanation of the fine-tuning serves to kick the issue of design up one level, to the question of who designed the multiverse-generator.

The inflationary multiverse scenario, widely considered as the most physically viable, provides a good test case of this line of reasoning. The inflationary multiverse-generator can only produce life-sustaining universes (or regions of spacetime) because it has the following 'components' or 'mechanisms':

(1) *A mechanism to supply the energy needed for the bubble universes.* This mechanism is the hypothesized inflaton field. By imparting a constant

---

4 It should be noted, however, that in Leslie's recent book [17] he argues for a pantheistic conception of God based on this principle and other considerations. In his view, every universe (or reality) whose sum value (both moral and aesthetic) is positive exists as a thought in the mind of God, with our universe itself existing as one such thought. (Like models in a computer simulation, God's thoughts are considered to have substantial structure – and thus substantial existence – in God's mind.) For a short critique of this fascinating book and Leslie's response, see my review essay [18] and ref. [19].

energy density to empty space as space expands, the inflaton field can act 'as a reservoir of unlimited energy' [20] for the bubbles.

(2) *A mechanism to form the bubbles.* This mechanism relates to Einstein's equations of general relativity. Because of their peculiar form, Einstein's equations dictate that space expands at an enormous rate in the presence of a field – like the inflaton – which imparts a constant (and homogeneous) energy density to empty space. This causes both the formation of the bubble universes and the rapid expansion which keeps them from colliding.

(3) *A mechanism to convert the energy of the inflaton field to the normal mass/energy we find in our universe.* This mechanism is Einstein's equivalence of mass and energy, combined with an hypothesized coupling between the inflaton field and normal mass/energy fields we find in our universe.

(4) *A mechanism that allows enough variation in constants of physics among universes.* Currently, the most physically viable candidate for this mechanism is superstring or M-theory. Superstring theory might allow enough variation in the constants of physics among bubble universes to make it reasonably likely that a fine-tuned universe would be produced, but no one knows for sure.[5]

Without all these 'components', the multiverse-generator would almost certainly fail to produce a single life-sustaining universe. If, for example, the universe obeyed Newton's theory of gravity instead of Einstein's, the vacuum energy of the inflaton field would at best create a gravitational attraction causing space to contract rather than expand.

In addition to the four factors listed above, the inflationary multiverse generator can only produce life-sustaining universes because the right background laws are in place. Specifically, the background laws must be such as to allow the conversion of the mass/energy into the material forms required for the sort of stable complexity needed for life. For example, without the principle of quantization, all electrons would be sucked into the atomic nuclei and hence atoms would be impossible; without the Pauli exclusion principle, electrons would occupy the lowest atomic orbit and hence complex and

---

5 See Leonard Susskind's contribution to this volume [21] for the variations allowed by superstring theory. The other leading alternatives to string theory being explored by physicists, such as the currently proposed models for Grand Unified Theories (GUTs), do not appear to allow for enough variation. The simplest and most studied GUT, $SU(5)$, allows for three differing sets of values for the fundamental constants of physics when the other non-$SU(5)$ Higgs fields are neglected [22]. Including all the other Higgs fields, the number of variations increases to perhaps several dozen [23]. Merely to account for the fine-tuning of the cosmological constant, however, which is estimated to be fine-tuned to be at least one part in 10^{53}, would require on the order of 10^{53} variations of the physical constants among universes.

varied atoms would be impossible; without a universally attractive force between all masses, such as gravity, matter would not be able to form sufficiently large material bodies (such as planets) for complex, intelligent life to develop or for long-lived stable energy sources like stars to exist.[6]

In sum, even if an inflationary multiverse-generator exists, it must involve just the right combination of laws, principles and fields for the production of life-permitting universes; if one of the components were missing or different – such as Einstein's equation or Pauli's exclusion principle – it is unlikely that any life-permitting universes could be produced. In the absence of alternative explanations, it follows from the surprise principle that the existence of such a system could be considered to suggest design since it seems very surprising that such a system would have just the right components as a brute fact, but not surprising under the theistic design hypothesis. Thus, it does not seem that one can completely escape the suggestion of design merely by hypothesizing some sort of multiverse-generator.

It must be admitted, however, that if such a multiverse-generator could be verified, the sort of *quantitative* evidence for design based on the fine-tuning of the constants would be eliminated. Whereas the degree of fine-tuning of a particular constant of physics could arguably be assigned a number – such as that corresponding to the ratio of the length of its intelligent-life-permitting range to some non-arbitrarily specified 'theoretically possible' range – we cannot provide a quantitative estimate for the degree of apparent design in the cases mentioned above. All we can say is that if certain seemingly highly specific sorts of laws were not in place, no life-sustaining universes could be generated. Thus, depending on the weight one attaches to such quantitative estimates, the evidence for design would be mitigated, although not completely eliminated.

26.6 Multiverses, design and the beauty of the laws of nature

Next, I want to look at what many consider another powerful suggestion of design from modern physics, that arising from the 'beauty' and 'elegance' of the laws of nature. This suggestion of design bypasses the multiverse objection to the design argument, whether the multiverse hypothesis is of

---

6 Although some of the laws of physics can vary from universe to universe in string theory, these background laws and principles are a result of the structure of string theory and therefore cannot be explained by the inflationary/superstring multiverse hypothesis since they must occur in all universes. Further, since the variation among universes would consist of variation of the masses and types of particles, and the form of the forces between them, complex structures would almost certainly be atom-like, and stable energy sources would almost certainly require aggregates of matter. Thus, the above background laws seem necessary for there to be complex, embodied intelligent observers in any of the many universes generated in this scenario, not merely a universe with our specific types of particles and forces.

the universe-generator or metaphysical variety. The idea that the laws of nature are beautiful and elegant is commonplace in physics, with entire books devoted to the topic. Indeed, Steven Weinberg – who is no friend of theism – devotes an entire chapter of his book *Dreams of a Final Theory* [24] to beauty as a guiding principle in physics. As Weinberg notes, 'mathematical stuctures that confessedly are developed by mathematicians because they seek a sort of beauty are often found later to be extraordinarily valuable by physicists' (see p. 153 of ref. [24]). To develop our argument, however, we need first to address what is meant by beauty. As Weinberg notes, the sort of beauty exemplified by physics is that akin to classical Greek architecture. The highpoint of the classical conception of beauty could be thought of as that of William Hogarth in his 1753 classic *The Analysis of Beauty* [25]. According to Hogarth, simplicity with variety is the defining feature of beauty or elegance, as illustrated by a line drawn around a cone. He went on to claim that simplicity apart from variety, as illustrated by a straight line, is boring rather than elegant or beautiful.

The laws of nature seem to manifest just this sort of simplicity with variety: we inhabit a world that could be characterized as having fundamental simplicity that gives rise to the enormous complexity needed for intelligent life. To see this more clearly, we will need to explicate briefly the character of physical law, as discovered by modern physics. I will do this in terms of various levels.

The physical world can be thought of as ordered into the following, somewhat overlapping, levels. Level 1 consists of observable phenomena. The observable world seems to be a mixture of order and chaos: there is regularity, such as the seasons or the alternation of day and night, but also many unique, unrepeatable events that do not appear to fall into any pattern. Level 2 consists of postulated patterns that exist among the observable phenomena, such as Boyle's law of gases. The formulation of level 2 marks the beginning of science as understood in a broad sense. Level 3 consists of a set of postulated underlying entities and processes hypothesized to obey some fundamental physical laws. Such laws might be further explained by deeper processes and laws, but these will also be considered to inhabit level 3. So, for instance, both Newton's law of gravity and Einstein's equations of general relativity would be considered to be level 3. The laws at level 3, along with a set of initial (or boundary) conditions, are often taken to be sufficient to account for the large-scale structure of the Universe.

Level 4 consists of fundamental principles of physics. Examples of this are the principle of the conservation of energy and the gauge principle (that is, the principle of local phase invariance), the principle of least action, the anti-commutation rules for fermions (which undergird Pauli's exclusion principle) and the correspondence principle of quantum mechanics (which often allows one to write the quantum mechanical equations for a system by substituting quantum operators for certain corresponding variables into a classical equation for the system). These are regulative principles that, when combined with other principles (such as choosing the simplest Lagrangian), are assumed to place tight constraints on the form that the laws of nature can take in the relevant domain. Thus, they often serve as guides to constructing the dynamical equations in a certain domain.

The laws at level 3 and the principles of level 4 are almost entirely cast in terms of mathematical relations. One of the great achievements of science has been the discovery that a deeper order in observable phenomena can be found in mathematics. As has been often pointed out, the pioneers of this achievement – such as Galileo, Kepler, Newton and Einstein – had a tremendous faith in the existence of a mathematical design to nature, although it is well known that Einstein did not think of this in theistic terms but in terms of a general principle of rationality and harmony underlying the Universe. As Morris Kline, one of the most prominent historians of mathematics, points out [26]:

From the time of the Pythagoreans, practically all asserted that nature was designed mathematically... During the time that this doctrine held sway, which was until the latter part of the nineteenth century, the search for mathematical design was identified with the search for truth.

Level 5 consists of the basic mathematical structure of current physics, for example the mathematical framework of quantum mechanics, though there is no clear separation between much of level 5 and level 4. Finally, one might even want to invoke a level 6, which consists of the highest-level guiding metaphysical principles of modern physics – for example, that we should prefer simple laws over complex laws, or that we should seek elegant mathematical explanations for phenomena.

Simplicity with variety is illustrated at all the above levels, except perhaps level 6. For example, although observable phenomena have an incredible variety and much apparent chaos, they can be organized via relatively few simple laws governing postulated unobservable processes and entities. What is more amazing, however, is that these simple laws can in turn be organized under a few higher-level principles (level 4) and form part of a simple and elegant mathematical framework (level 5).

One way of thinking about the way in which the laws fall under these higher-level principles is as a sort of fine-tuning. If one imagines a space of all possible laws, the set of laws and physical phenomena we have are just those that meet the higher-level principles. Of course, in analogy to the case of the fine-tuning of the parameters of physics, there are bound to be other sets of laws that meet some other relatively simple set of higher-level principles. But this does not take away from the fine-tuning of the laws, or the case for design, any more than the fact that there are many possible elegant architectural plans for constructing a house takes away from the design of a particular house. What is important is that the vast majority of variations of these laws end up causing a violation of one of these higher-level principles, as Einstein noted about general relativity. Further, it seems that, in the vast majority of such cases, such variations do not result in new, equally simple higher-level principles being satisfied. It follows, therefore, that these variations almost universally lead to a less elegant and simple set of higher-level physical principles being met. Thus, in terms of the simplicity and elegance of the higher-level principles that are satisfied, our laws of nature appear to be a tiny island surrounded by a vast sea of possible law structures that would produce a far less elegant and simple physics.

As testimony to the above point, consider what Steven Weinberg and other physicists have called the 'inevitability' of the laws of nature (see, e.g., pp. 135–153 and 235–237 of ref. [24]). The inevitability that Weinberg refers to is not the inevitability of logical necessity, but rather the contingent requirement that the laws of nature in some specified domain obey certain general principles. The reason Weinberg refers to this as the 'inevitability' of the laws of nature is that the requirement that these principles be met often severely restricts the possible mathematical forms the laws of nature can take, thus rendering them in some sense inevitable. If we varied the laws by a little bit, these higher-level principles would be violated.

This inevitability of the laws is particularly evident in Einstein's general theory of relativity. As Weinberg notes, 'once you know the general physical principles adopted by Einstein, you understand that there is no other significantly different theory of gravitation to which Einstein could have been led' (p. 135 of ref. [24]). As Einstein himself said, 'To modify it [general relativity] without destroying the whole structure seems to be impossible.'

This inevitability, or near-inevitability, is also illustrated by the gauge principle, the requirement that the dynamical equations expressing the fundamental interactions of nature – the gravitational, strong, weak and electromagnetic forces – be invariant under the appropriate local phase transformation. When combined with the heuristic of choosing the simplest interaction Lagrangian that meets the gauge principle and certain other background constraints, this has served as a powerful guide in constructing the equations governing the forces of nature (e.g. the equations for the forces between quarks). Yet, as Ian Atchison and Anthony Hey point out, there is no compelling logical reason why this principle must hold. Rather, they claim, this principle has been almost universally adopted as a fundamental principle in elementary particle physicists because it is 'so simple, beautiful and powerful (and apparently successful)' [27]. Further, as Alan Guth points out [28], the original 'construction of these [gauge] theories was motivated mainly by their mathematical elegance'. Thus, the gauge principle provides a good example of a contingent principle of great simplicity and elegance that encompasses a wide range of phenomena, namely the interactions between all the particles in the Universe.

26.7 Potential non-theistic explanations of beauty and elegance

Theism offers a natural, non-*ad-hoc*, explanation of why the laws of nature can be encompassed by such higher-level principles. As mentioned above, it has been part of traditional theism that God would be motivated to bring about an aesthetically pleasing universe. Can a non-theistic, non-design view of reality offer an explanation? One reaction to this purported need for a theistic explanation might be to claim that it is no more surprising that such higher-level principles exist than that nature has simple laws. Further, it could be argued that, just as simple laws (along with the initial conditions) determine the various phenomena of nature, so the higher-level principles determine which laws of nature exist. Given that we take the higher-level principles as ontologically primary, it is therefore no surprise that the lower-level laws fall under them; once such principles are given, everything else follows. Accordingly, the only things that could be surprising are that the higher-level principles are mathematically simple or that they exist at all. But why should these be surprising?

One flaw in the above argument is that it does not seem that one can plausibly think of these principles as in themselves having any causal power to dictate the lower-level phenomena or laws. It is easy to be misled at this point, however, into thinking that they do. Because we can often derive (with a few additional assumptions) the lower-level laws from the higher-level principle, it is easy to think that somehow these higher-level principles make the lower-level laws what they are. Rather, the causation or dependence is in the other direction: it is because the laws and phenomena are what they are that these principles hold universally, not the other way around. The principles are therefore not ontologically primary.

An analogy from architecture might help to illustrate this point: insofar as the placement of windows in a building follows higher-level principles, it is not because the principles in themselves have a special power to make the windows have the right positions. Rather, it is because of the position of the windows that the higher-level principles hold. Further, insofar as the higher-level principles could be said to have a causal efficacy to determine the placement of the windows, it is only via the causal powers of intelligent agents, such as the people who constructed the building.

One reason for claiming that these principles have no intrinsic causal powers is that – except for being an intention or thought in some mind, human or transcendent – it is difficult to see how these higher-level principles could be anything over and above the patterns into which the laws and phenomena of nature fall. For example, they do not appear to be reducible to the causal powers of actual entities, as some philosophers claim about the laws of nature.[7] Instead, insofar as entities possess causal powers, the principles describe the arrangement of the causal powers of a diverse class of such entities – for example the fundamental particles – and therefore cannot be the powers of any given entity. And even if this arrangement were the result of the causal power of some single type of entity – for example a superstring – it would still be surprising that the arrangement could be captured and unified by a few simple higher-level mathematical rules. This is analogous to the claim that, even if the fine-tuning of the constants of physics for life were to be explained by some Grand Unified Theory, it would still be amazing that the theory that happened to exist was one that yielded values for the constants that were intelligent-life-permitting.

One could still insist that there is no reason to think that it is surprising that individual laws collectively fall under simple and elegant mathematical rules which are, in turn, expressible in an elegant and incredibly fruitful system of mathematics, such as the complex numbers in the case of quantum mechanics. Perhaps one could argue at this point that it is just an unexplainable fact about the world, and that all explanations must come to an end at some point. Invoking God, it could be argued, merely moves the problem up one level: the theist is still left with postulating the existence of God as an unexplained given. Here the debate can take an even more philosophical turn – with the theist offering reasons for why God is a superior stopping point for explanation and the atheist denying those reasons. All we will say here is that given that theism makes more sense of the existence of

---

7 See, for example, the book by Rom Harré and Edward Madden [29]. Under the conception of laws as expressing causal powers, Einstein's equation of general relativity would be seen as being grounded in the causal powers of matter to bend spacetime.

this fine-tuning for beauty and elegance than atheism, it offers us a reason for believing in theism.

Further, this 'fine-tuning' for simplicity and elegance cannot be explained either by the universe-generator or metaphysical multiverse hypothesis, since there is no reason to think that intelligent life could only arise in a universe with simple, elegant underlying physical principles. Certainly, a somewhat orderly macroscopic world is necessary for intelligent life, but there is no reason to think that this requires a simple and elegant underlying set of physical principles. This is especially clear when one considers how radically different the framework and laws of general relativity and quantum mechanics are from the world of ordinary experience: although the regularities of the everyday world are probably derived from the underlying laws of quantum mechanics and general relativity, they do not reflect the structure of those laws. Indeed, it is this difference in structure between the classical, macroscopic world and the quantum world that has largely given rise to the interpretive problems of quantum mechanics. Thus, there is little reason based on an observation selection effect to expect the sort of macroscopic order necessary for intelligent life to be present in the underlying, microscopic world.

Another alternative is to attempt to explain the simplicity of the world by some sort of metaphysical principle, according to which the world is more likely to be simple than complex. One problem with this view is that there are many, many simpler possible worlds than ours, such as one with a single particle that simply travels in a straight line. The enormous actual complexity of our world thus strongly testifies against this claim.

One way of getting around this problem is to combine this metaphysical principle of simplicity with a metaphysical many-universe hypothesis, according to which all possible mathematical structures are instantiated in some universe or another. One such view is suggested by Max Tegmark [30]. Using this hypothesis, one could explain the actual complexity of the Universe by claiming that only sufficiently complex universes could contain embodied observers, whereas the simplicity of the world would be explained by claiming that simple worlds are given a higher probabilistic weight than complex worlds. As Tegmark suggests, 'One could reward simplicity by weighting each mathematical structure by 2^{-n}, where n is the algorithmic information content measured in bits, defined as the length of the shortest bit string... that would specify it' (see p. 16 of ref. [30]).

Even granting the many-universe hypothesis, however, such a metaphysical principle runs into severe problems. For one thing, simplicity seems to be conceptual-framework relative and so it is difficult to see how there could be any such metaphysical principle. Any mathematical equation, for instance, can be written in a simple form given that one constructs the right mathematical properties. For example, consider the equation $Y = 2x + 4x^2 + 7.1x^5$. Define $F(x)$ by the expression on the right-hand side of the equation. Given that the concept of a function $F(x)$ is part of our mathematical repertoire, we can write the above equation as $Y = F(x)$, which is much simpler than our original way of expressing Y. The only way I can see around this problem is to postulate a set of primitive mathematical properties and then define complexity in terms of the shortest bit string that would specify the mathematical equation using only those primitive properties. Without such a postulate, simplicity will be relative to the repertoire of mathematical properties with which one has to work. An example of this viewpoint-dependence occurs when Newton's law of gravity is translated into the framework of general relativity and vice versa. When this is done, however, their respective simplicity vanishes. As Misner, Thorne and Wheeler point out, expressed in the conceptual framework of general relativity, Newton's gravitational law is extremely complex. On the other hand, if expressed in the Newtonian framework, 'Einstein's field equations (ten of them now!) are horrendously complex' [31]. So the respective simplicity of each is dependent on the conceptual framework in which it is written.

Other problems plague this appeal to a principle of simplicity. For example, the Universe appears to be infinitely complex when one takes into account the complexity of the initial conditions – such as the initial continuous distribution of mass/energy at some chosen surface of constant proper time. Such a distribution would be continuous, and so would take a non-denumerably infinite amount of information to specify. So one's metaphysical principle of simplicity would have to be much more restrictive, such as dictating that some global feature – such as the way the distribution of mass/energy develops with time – can be described using a simple rule, which makes it even less plausible.

Even breaking it into initial conditions plus development in time is metaphysically arbitrary, however, since it is well known that in general relativity there is no single way of deciding which sets of space-like separated points will count as hypersurfaces of constant time. On the other hand, even if there were a preferred set of hypersurfaces – such as that corresponding to those hypersurfaces yielding the same proper time for each element of the cosmic fluid – the temporal development of the mass/energy is describable by many different mathematical systems of equations, as pointed out above. For example, since the quantity of mass/energy at any spacetime point is given by a real number, the complete temporal development of the mass/energy with time is given by an exceedingly complex function over the real numbers that maps mass/energy distributions (or probabilities of such distributions) from some arbitrarily chosen initial hypersurface to all future hypersurfaces. So the 'manifest' temporal development of mass/energy is actually exceedingly complex, at least as expressed in terms of mathematical terms we normally treat as basic.

What is amazing is that, presumably, this mass/energy distribution can be reconstructed using a relatively few simple rules expressed within the complex numbers – that is, the rules of some quantum mechanical theory, such as a (yet undeveloped) theory of quantum gravity. So, the principle of simplicity would require that the weight we attach to a type of universe – that is, what proportion of the space of universes be encompassed by this type – should not be determined by the degree of complexity of the mathematical system a type of universe exemplifies since it exemplifies multiple systems. Rather, it would need to be determined by the simplest mathematical framework in terms of which its temporal development can be expressed. Given all these complications, such a principle seems to lose almost any intuitive appeal.[8]

The alternative to such a mind-independent principle of simplicity, with all its metaphysical baggage, is to claim that simplicity in physics is relative to both our conceptual framework (instead of some primitive set of mathematical properties) and to our way of breaking up the elements of physical reality (e.g. into laws and initial conditions). According to theism, our minds and the world ultimately owe their origin to God, so it makes sense that the Universe would – at some deep level – be reflective of the preferences of the human mind. Thus, the dependence of simplicity and elegance on our conceptual framework does not present a problem for the theist. However, this partially anthropomorphic characterization of simplicity, as occurs in physics, does not fit well with a mind-independent principle of simplicity

---

8 Since within the metaphysical multiverse hypothesis there will exist worlds that have local islands of order that contain conscious observers, but are otherwise extremely chaotic, the only way to have an expectation for what our universe should look like is to consider ourselves to be generic observers. This raises perhaps the most serious objection to Tegmark's proposal: saying that a certain type of world has a measure X in our space of all possible mathematical structures cannot explain anything, or lead to any sort of predictions, unless that measure is tied to what we, considered as generic observers, should rationally expect. But mathematical measures themselves do not generate expectations; one can put an infinite number of different measures over a space, but clearly all of them could not imply an expectation. In order for a measure to have any significance, one must have some well justified connection between that measure and our expectations. I see no way of establishing such a connection, especially with the odd type of principle of weighting that would actually be required, as elaborated above.

that determines the weight given to various universes – or more precisely, sets of universes – as Tegmark proposes.

Another non-theistic response to this apparent fine-tuning for beauty and elegance of the laws of nature is that the idea of beauty is purely subjective, simply the result of our reading into nature anthropomorphic patterns, in the same way as humans have read meaningful patterns – such as the Bear or the Big Dipper – into the random pattern of stars in the night sky. The major problem with this explanation is that it does not account for the surprising success of the criterion of beauty in the physical sciences. We would not expect patterns that are merely subjective to serve as a basis for theories that make highly accurate predictions, such as the successful prediction of quantum electrodynamics – to nine significant figures – of the quantum correction to the gyromagnetic ratio of the electron. Merely subjectively reading a pattern into an otherwise random pattern is predictively useless. The second problem is that there are significant objective aspects of beauty, at least in the classical sense of beauty, that one can clearly demonstrate in the realm of physics (e.g. symmetry).[9]

A second way of discounting the significance of the apparent beauty of the laws of nature, suggested by Steven Weinberg (see p. 158 of ref. [24]), is that after the scientific revolution scientists unconsciously modified their criteria of beauty to fit nature. The problem with this explanation is that we can point to objective features of the underlying world – its symmetry, its simplicity in variety, its inevitability – that clearly fit the general criteria of the so-called *classical* conception of beauty exemplified in Greek architecture. This is a category of great human significance that originated long before the scientific revolution. Further, the mere fact that scientists use the term 'beautiful' instead of some other category to describe the underlying order indicates that they sense a deep congruence between the order of nature and those features normally associated with beauty in other, non-scientific contexts. It is this congruence that Weinberg's evolutionary explanation fails to explain. Of course, evolution could account for why we have the category of beauty – namely, that it was of survival value. But it cannot account for why a category that has often been considered to have transcendent, even religious, value fits the underlying mathematical order of nature which is so far removed from the order of everyday life where selection pressures played a role. From an evolutionary perspective, the outstanding question

---

9 This is not to say, however, that the Universe exemplifies elegant and simple laws apart from some preferred categorization scheme; rather, as argued above, the concept of the laws of nature being beautiful seems partly dependent on non-arbitrary, intersubjectively shared human conceptual categories.

still remains as to why a category selected for survival value – for example, because it helped attract mates – would serve as such a useful guide to the underlying order of nature.

Finally, the form of argument in this case for design has the same form as that in the case of the fine-tuning of the constants for intelligent life, except that this fine-tuning cannot be explained by the multiverse hypothesis. One way of putting the argument is in terms of the 'surprise principle' we invoked in the argument for the fine-tuning of the constants for the emergence of intelligent life. Specifically, as applied to this case, one could argue that the fact that the phenomena and laws of physics are fine-tuned for simplicity with variety is highly surprising under the non-design hypothesis, but not highly surprising under theism. Thus, the existence of such fine-tuned laws provides significant evidence for theism over the non-design hypothesis. Another way one could explicate this argument is as follows. Atheism seems to offer no explanation for the apparent fine-tuning of the laws of nature for beauty and elegance (or simplicity with variety). Theism, on the other hand, seems to offer such a natural explanation: for example, given the classical theistic conception of God as the greatest possible being, and hence a being with a perfect aesthetic sensibility, it is not surprising that such a God would create a world of great subtlety and beauty at the fundamental level. Given the rule of inference that – all else being equal – a natural non-*ad-hoc* explanation of a phenomenon is always better than no explanation at all, it follows that we should prefer the theistic explanation to the claim that the elegance and beauty of the laws of nature is just a brute fact.[10]

26.8 Conclusion

Many scientists feel very uncomfortable, if not hostile, to linking science and religion. As many leading historians have pointed out, however, natural theology and religion were closely linked with scientific practice, and indeed provided much of the inspiration for scientific work, until the late nineteenth century.[11]

---

10 For further elaborations of the fine-tuning arguments mentioned in this chapter, with answers to various objections, see my fine-tuning website at *www.fine-tuning.org*.

11 See, for example, the book by John Brooke and Geoffrey Cantor [32]. For additional work concerning the relation between science and religion from the Middle Ages onward, see the books by Edward Grant [33] and David C. Lindberg [34, 35]. As the evolutionary biologist H. Allen Orr says in his recent review [36] of Richard Dawkins recent book *A Devil's Chaplain: Reflections on Hope, Lies, Science, and Love*:

> The popular impression of long warfare between Church and science – in which an ignorant institution fought to keep a fledgling science from escaping the Dark Ages – is nonsense, little more than Victorian propaganda. The truth, which emerged only from the last [twentieth] century of scholarship, is almost entirely unknown among scientists: the medieval Church was

This unease with a science/religion dialogue extends to an unease with publicly discussing anything metaphysical at all in relation to science, including such topics as the anthropic principle and the multiverse hypothesis. On careful analysis of the overall purpose of doing science, however, I think it becomes clear that scientists should be talking about these issues, and doing so in dialogue with other thinkers, such as philosophers and theologians. To see this, note that science could roughly be divided into those aspects which have practical, technological consequences, and those aspects that are at present of merely theoretical interest. What justifies the highly theoretical branches of science, such as current cosmology? One justification is that we cannot know for sure whether these highly abstract studies will eventually lead to technological innovations that contribute to human flourishing. I do not believe, however, that such a far-off hope is sufficient to maintain the broad public support for these theoretical branches of science necessary to retain both funding for publications and student interest. This is especially true given other pressing needs, such as cancer research, for which the payoff is much more immediate.

The other major purpose of science is to help fulfil the human desire to understand the world, which forms a crucial aspect of what could be called 'human flourishing needs'. These needs consist of meaning, understanding, transcendence, connectedness, belonging, growth and creativity. For many people, these are more important than mere survival needs – such as food and shelter – as demonstrated by the willingness of people throughout history to risk their lives for a cause that gives them purpose. Since these needs must be addressed, our only alternatives are to address them through developing overarching stories of human life and its place in the cosmos that incorporate the best that science has to offer, or by stories lacking such critical input. Bringing our scientific understanding into this larger context, I believe, will serve to sustain and invigorate public interest and support of science, especially in its more theoretical branches.

Addressing these human flourishing needs lies at the core of religion, though often religion has neglected and hampered the need for growth and

a leading patron of science; most theologians studied 'natural philosophy'; and the medieval curriculum was perhaps the most scientific in Western history.

Further, as pointed out by historians Frances Yates [37] and Robert Merton [38], belief in God is largely responsible for the assumption during the scientific revolution and beyond that the Universe is both harmoniously designed and accessible to human reason, an assumption that lies at the foundation of scientific practice. Science was seen as a way of revealing God's handiwork in creation. This was especially true in England. Thus, the current dogma forbidding the mutual positive interaction between science and religion does not appear to be intrinsic to science unless one takes the highly presumptuous view that until the late nineteenth century, scientists did not really understand the nature of science.

creativity and given problematic answers to these other needs. Further, religion has typically relied on revelation and common deep human intuitions about the nature of reality as its starting point in understanding the world, whereas science has relied on sense experience and its own methodology in forming hypotheses based on those experiences. Since these needs are common to human beings, they provide a very broad context for science–religion dialogue without requiring knowledge of some particular religious tradition and its sacred texts. Of course, discussions relating to particular religious traditions can still proceed, but these would not comprise the core of the dialogue.

Finally, although discussions of the existence and nature of God are certainly critically relevant to the issue of meaning and purpose in the cosmos, these issues can also be discussed without invoking God. For example, non-theistic understandings of meaning and purpose in the cosmos have been explored by philosophers Quentin Smith [39] and Milton Munitz [40], biologist Ursula Goodenough [41] and (to some extent) physicist Paul Davies [42]. This article should therefore be understood as a contribution to this larger discussion, not merely as an apology for a theistic perspective. I hope this chapter provides some understanding of why a theist might not only be sympathetic to the multiverse hypothesis, but might even see some of the findings of physics and cosmology as supportive of theism.

References

[1] S. Weinberg. This volume (2007).
[2] N. Bostrom. This volume (2007).
[3] A. Vilenkin. This volume (2007).
[4] J. Polkinghorne. *Belief in God in an Age of Science* (New Haven: Yale University Press, 1998), pp. 1–12.
[5] M. Tegmark. This volume (2007).
[6] D. Lewis. *On the Plurality of Worlds* (New York: Basil Blackwell, 1986).
[7] A. Koyré. *From the Closed World to the Infinite Universe* (Baltimore: Johns Hopkins University Press, 1957), p. 42.
[8] M. Crowe. *The Extraterrestrial Life Debate, 1750–1900: The Idea of the Plurality of Worlds from Kant to Lowell* (Cambridge: Cambridge University Press, 1986), p. 37.
[9] L. Susskind. *The Cosmic Landscape, String Theory and the Illusion of Intelligent Design* (New York: Little, Brown and Company, 2005).
[10] R. Collins. God, design, and fine-tuning. In *God Matters: Readings in the Philosophy of Religion*, eds. R. Martin and C. Bernard (New York: Longman Press, 2002).
[11] A. W. F. Edwards *Likelihood* (Baltimore: Johns Hopkins University Press, 1973).

[12] D. H. Mellor. Too many universes. In *God and Design: The Teleological Argument and Modern Science*, ed. N. Mason (London: Routledge, 2003).
[13] J. M. Keynes. *A Treatise on Probability* (London: Macmillan, 1921), p. 4.
[14] A. Plantinga. *Warrant and Proper Function* (Oxford: Oxford University Press, 1993), chaps. 8 and 9.
[15] R. Swinburne. *The Existence of God* (Oxford: Clarendon Press, 1979), pp. 97–102.
[16] J. Leslie. *Universes* (New York: Routledge, 1989), chap. 8.
[17] J. Leslie. *Infinite Minds: A Philosophical Cosmology* (Oxford: Clarendon Press, 2001).
[18] R. Collins. Theism or pantheism? A review essay on John Leslie's 'Infinite Minds'. *Philosophia Christi*, **5** (2003), 563.
[19] J. Leslie. Pantheism and Platonic creation. A reply to Robin Collins. *Philosophia Christi*, **5** (2003), 575.
[20] J. Peacock. *Cosmological Physics* (Cambridge: Cambridge University Press, 1999), p. 2.
[21] L. Susskind. This volume (2007).
[22] A. Linde. *Particle Physics and Inflationary Cosmology* (Pennsylvania: Harwood Academic Publishers, 1990), p. 33.
[23] A. Linde *Inflation and Quantum Cosmology* (New York: Academic Press, 1990), p. 6.
[24] S. Weinberg. *Dreams of a Final Theory* (New York: Vintage Books, 1992).
[25] W. Hogarth. *The Analysis of Beauty* (London: J. Reeves, 1753).
[26] M. Kline. *Mathematical Thought, from Ancient to Modern Times, Vol. 1* (Oxford: Oxford University Press, 1972), p. 153.
[27] I. Aitchison and A. Hey. *Gauge Theories in Particle Physics, a Practical Introduction*, 2nd edn (Bristol: Adam Hilger, 1989), p. 60.
[28] A. Guth. *The Inflationary Universe, The Quest for a New Theory of Cosmic Origins* (New York: Helix Books, 1997), p. 124.
[29] R. Harré and E. Madden. *Causal Powers, Theory of Natural Necessity* (Oxford: Basil Blackwell, 1975).
[30] M. Tegmark. Parallel universes. In *Science and Ultimate Reality*, eds. J. D. Barrow, P. C. W. Davies and C. L. Harper (Cambridge: Cambridge University Press, 2004), p. 488 [astro-ph/030213].
[31] C. Misner, K. Thorne and J. Wheeler. *Gravitation* (San Francisco: W. H. Freeman & Co, 1973), pp. 302–303.
[32] J. Brooke and G. Cantor. *Reconstructing Nature, the Engagement of Science and Religion* (Oxford: Oxford University Press, 1998).
[33] E. Grant. *The Foundations of Modern Science in the Middle Ages* (Cambridge: Cambridge University Press, 1996).
[34] D. C. Lindberg. *The Beginnings of Western Science* (Chicago: University of Chicago Press, 1992).
[35] D. C. Lindberg. In *God and Nature: Historical Essays on the Encounter Between Christianity and Science*, eds. D. C. Lindberg and R. L. Numbers (Los Angeles, CA: University of California Press, 1986, pp. 19–48).
[36] A. H. Orr. A passion for evolution. *New York Review of Books* **51** (February 26, 2004).
[37] F. Yates. *Giordano Bruno and the Hermetic Tradition* (Chicago, IL: Chicago University Press, 1964).

[38] R. Merton. Science, technology, and society in 17th century England. *Osiris*, **4** (1938), 360.

[39] Q. Smith. *The Felt Meanings of the World, a Metaphysics of Feeling* (West Lafayette, ID: Purdue University Press, 1986).

[40] M. K. Munitz. *Cosmic Understanding, Philosophy and Science of the Universe* (Princeton: Princeton University Press, 1986).

[41] U. Goodenough. *The Sacred Depths of Nature* (New York: Oxford University Press, 1998).

[42] P. Davies. Traveling through time. In *Science and Theology News* (July 1, 2001); www.stnews.org/Commentary-731.htm.

27
Living in a simulated universe

John D. Barrow

Centre for Mathematical Sciences, Cambridge University

27.1 Introduction

A good point of philosophy is to start with something so simple as not to seem worth stating, and to end with something so paradoxical that no one will believe it.
Bertrand Russell

Of late, there has been much interest in multiverses. What sorts could there be? And how might their existence help us to understand those life-supporting features of our own universe that would otherwise appear to be merely very fortuitous coincidences [1, 2]? At root, these questions are not ultimately matters of opinion or idle speculation. The underlying Theory of Everything, if it exists, may require many properties of our universe to have been selected at random, by symmetry-breaking, from a large collection of possibilities, and the vacuum state may be far from unique.

The favoured inflationary cosmological model – that has been so impressively supported by observations of the COBE and WMAP satellites – contains many apparent 'coincidences' that allow our universe to support complexity and life. If we were to consider a 'multiverse' of all possible universes, then our observed universe appears special in many ways. Modern quantum physics even provides ways in which the possible universes that make up the multiverse of all possibilities can actually exist.

Once you take seriously that all possible universes can (or do) exist, then a slippery slope opens up before you. It has long been recognized that technical civilizations, only a little more advanced than ourselves, will have the capability to simulate universes in which self-conscious entities can emerge and communicate with one another [3]. They would have computer power that exceeded ours by a vast factor. Instead of merely simulating their weather or the formation of galaxies, like we do, they would be able to go further and watch the appearance of stars and planetary systems. Then, having coupled

Universe or Multiverse?, ed. Bernard Carr. Published by Cambridge University Press.
© Cambridge University Press 2007.

the rules of biochemistry into their astronomical simulations, they would be able to watch the evolution of life and consciousness – all speeded up to occur on whatever timescale was convenient for them. Just as we watch the life cycles of fruit flies, they would be able to follow the evolution of life, watch civilizations grow and communicate with each other and argue about whether there existed a Great Programmer in the Sky who created their universe and who could intervene at will in defiance of the laws of nature they habitually observed.

Once this capability to simulate universes is achieved, fake universes will proliferate and will soon greatly outnumber the real ones. Thus Nick Bostrom [4,5] and Brian Weatherson [6] have argued that a thinking being here and now is more likely to be in a simulated reality than a real one. Motivated by this alarming conclusion, there have even been suggestions as how best to conduct ourselves if we have a high probability of being simulated beings in a simulated reality. Robin Hanson [7] suggests that you should act so as to increase the chances of continuing to exist in the simulation or of being resimulated in the future:

If you might be living in a simulation, then all else equal you should care less about others, live more for today, make your world look more likely to become rich, expect to and try more to participate in pivotal events, be more entertaining and praiseworthy, and keep the famous people around you happier and more interested in you.

In response, Paul Davies [8,9] has argued that this high probability of living in a simulated reality is a *reductio ad absurdum* for the whole idea that multiverses of all possibilities exist. It would undermine our hopes of acquiring any sure knowledge about our universe.

The multiverse scenario was originally suggested by some cosmologists as a way to avoid the conclusion that our universe was specially designed for life by a Grand Designer. Others saw it as a way to avoid having to say anything more about the problem of fine-tuning at all. However, we see that once conscious observers are allowed to intervene in our universe, rather than being merely lumped into the category of 'observers' who do nothing, we end up with a scenario in which the gods reappear in unlimited numbers in the guise of the simulators who have power of life and death over the simulated realities that they bring into being. The simulators determine the laws, and can change the laws, that govern their worlds. They can engineer anthropic fine-tunings [10]. They can pull the plug on the simulation at any moment; intervene or distance themselves from their simulation; watch as the simulated creatures argue about whether there is a god who controls or intervenes; work miracles or impose their ethical principles upon the simulated reality. All the time they can avoid having even a twinge of conscience about hurting anyone because their toy reality is not real. They can even watch their simulated realities grow to a level of sophistication that allows them to simulate higher-order realities of their own.

27.2 How would we tell if we lived in a simulation?

Faced with these perplexities, do we have any chance of winnowing fake realities from true? What might we expect to see if we made scientific observations from within a simulated reality?

Firstly, the simulators will have been tempted to avoid the complexity of using a consistent set of laws of nature in their worlds when they can simply patch in 'realistic' effects. When the Disney company makes a film that features the reflection of light from the surface of a lake, it does not use the laws of quantum electrodynamics and optics to compute the light scattering. That would require a stupendous amount of computing power and detail. Instead, the simulation of the light scattering is replaced by plausible rules of thumb that are much briefer than the real thing but give a realistic looking result – as long as no one looks too closely. There would be an economic and practical imperative for simulated realities to stay that way if they were purely for entertainment. But such limitations to the complexity of the simulations programming would presumably cause occasional tell-tale problems; perhaps they would even be visible from within.

Even if the simulators were scrupulous about simulating the laws of nature, there would be limits to what they could do. Assuming the simulators, or at least the early generations of them, have a very advanced knowledge of the laws of nature, it is likely they would still have incomplete knowledge of them (some philosophers of science would argue this must always be the case). They may know a lot about the physics and programming needed to simulate a universe, but there will be gaps or, worse still, errors in their knowledge of the laws of nature. These would, of course, be subtle and far from obvious; otherwise our 'advanced' civilization would not be advanced. These lacunae would not prevent simulations being created and running smoothly for long periods of time. But gradually the little flaws would begin to build up.

Eventually, their effects would snowball, and these realities would cease to compute. The only escape is if their creators intervene to patch up the problems one by one as they arise. This is a solution that will be very familiar to the owner of any home computer who receives regular updates in order to protect it against new forms of invasion or repairs gaps that its original creators had not foreseen. The creators of a simulation could offer this type of temporary protection, updating the working laws of nature to include extra things they had learnt since the simulation was initiated.

In this kind of situation, logical contradictions will inevitably arise and the laws in the simulations will appear to break down now and again. The inhabitants of the simulation – especially the simulated scientists – will occasionally be puzzled by the experimental results they obtain. The simulated astronomers might, for instance, make observations that show that their so-called constants of nature are very slowly changing.[1]

It is likely that there could even be sudden glitches in the laws that govern these simulated realities. This is because the simulators would most likely use a technique that has been found to be effective in all other simulations of complex systems: the use of error-correcting codes to put things back on track.

Take our genetic code, for example. If it were left to its own devices, we would not last very long. Errors would accumulate and death and mutation would quickly follow. We are protected from this by the existence of a mechanism for error-correction that identifies and corrects mistakes in genetic coding. Many of our complex computer systems possess the same type of internal spell-checker to guard against error accumulation.

If the simulators used error-correcting computer codes to guard against the fallibility of their simulations as a whole (as well as simulating them on a smaller scale in our genetic code), then every so often a correction would take place to the state or the laws governing the simulation. Mysterious sudden changes would occur that would appear to contravene the very laws of nature that the simulated scientists were in the habit of observing and predicting.

We might also expect that simulated realities would possess a similar level of maximum computational complexity across the board. The simulated creatures should have a similar complexity to the most complex simulated non-living structures – something that Stephen Wolfram [14] (for

---

[1] There is tantalizing evidence that there might be small changes in some of our constants of nature, notably the fine structure constant, by a few parts in a million over the 14 billion year age of our universe. The fundamental perspective provided by string theory leads us to expect that space has more dimensions than the three large ones that we observe directly. This means that the true constants of nature exist in the total number of dimensions and that the 3-dimensional 'constants' that we have defined are merely 'shadows' that do not need to be constant. In fact, if the extra dimensions of space were to change in time in any way, then our 3-dimensional constants would be seen to change at the same rate. For the observational evidence for such changes, see refs. [11]–[13].

quite different reasons, nothing to do with simulated realities) has coined the Principle of Computational Equivalence.

One of the most common worries about distinguishing a simulated reality from a true one from the inside is the suggestion that the simulators would be able to take into account some difference one might think of ahead of time and pre-adjust the simulation to avoid the mismatch. This new simulated reality might then develop its own disparities with true reality, but they would be plugged by another act of predestination. The question is whether this is possible in the limit where the number of these adjustments becomes arbitrarily large. The problem is similar to that first considered by Karl Popper [15] to identify the self-referential limits of computers. The same argument was used in a different context in many publications by the late Donald MacKay [16] as an argument against the possibility of predestination, if this is knowable by those whose futures are being predicted. It is only possible to make a correct prediction of someone's future actions if it is not made known to them [17]. Once it is made known, it is always possible for them to falsify it. Thus, it is not possible for there to be an unconditionally binding prediction of someone's future actions. Clearly the same argument applies to predicting elections:[2] there cannot be a public prediction of the outcome of an election that unconditionally takes into account the effect of the prediction itself on the electorate. This type of uncertainty is irreducible in principle. If the prediction is not made public, it could be 100% correct.

So we suggest that, if we live in a simulated reality, we should expect occasional sudden glitches, small drifts in the supposed constants and laws of nature over time [21], and a dawning realization that the flaws of nature are as important as the laws of nature for our understanding of true reality.

References

[1] M. Tegmark. Parallel universes. *Sci. Am.* (May 2003), pp. 41–51.
[2] M. J. Rees. *Our Cosmic Habitat* (Princeton, NJ: Princeton University Press, 2001).
[3] J. D. Barrow. *Pi in the Sky: Counting, Thinking and Being* (Oxford: Oxford University Press, 1992), chap. 6.
[4] N. Bostrom. Are you living in a computer simulation? *Phil. Quart.* **53** (2003), 243.
[5] N. Bostrom. The simulation argument: Reply to Weatherson. *Phil. Quart.* **55**, (2005), 90.
[6] B. Weatherson. Are you a sim? *Phil. Quart.* **53** (2003), 425.

---

2 Although there is a famous false argument by Herbert Simon claiming the opposite [18]; it is also reprinted in ref. [19]. The fallacy arose because of the illicit use of the fixed point theorem of Brouwer in a situation where the variables are discrete rather than continuous. See ref. [20] for a detailed explanation.

[7] R. Hanson. How to live in a simulation. *J. Evolution and Technology*, **7** (2001) [http://www.transhumanist.com].

[8] P. C. W. Davies. A brief history of the multiverse. *New York Times* (April 12, 2003).

[9] P. C. W. Davies. This volume (2007).

[10] E. R. Harrison. The natural selection of universes containing intelligent life. *Quart. J. Roy. Astron. Soc.* **36** (1995), 193.

[11] J. D. Barrow and J. K. Webb. Inconstant constants. *Sci. Am.* **292**(6) (2005), 56.

[12] J. K. Webb, M. T. Murphy, V. V. Flambaum *et al.* Further evidence for cosmological evolution of the fine structure constant. *Phys. Rev. Lett.* **87** (2001), 091301.

[13] E. Reinhold, R. Buning, U. Hollenstein, A. Ivanchik, P. Petitjean and W. Ulbachs. Indication of a cosmological variation of the proton-electron mass ratio based on laboratory measurement and reanalysis of H_2 spectra. *Phys. Rev. Lett.* **96** (2006), 151101.

[14] S. Wolfram. *A New Kind of Science* (Chicago, IL: Wolfram Inc., 2002).

[15] K. Popper. Indeterminism in quantum physics and in classical physics. *Brit. J. Phil. Sci.* **1** (1950), 117, 173.

[16] D. MacKay. *The Clockwork Image* (London: IVP, 1974), p. 110.

[17] J. D. Barrow. *Impossibility* (Oxford: Oxford University Press, 1988), chap. 8.

[18] H. Simon. Bandwagon and underdog effects in election predictions. *Public Opinion Quart.* **18** (Fall issue, 1954), 245.

[19] S. Brams. *Paradoxes in Politics* (NY: Free Press, 1976), pp. 70–7.

[20] K. Aubert. Spurious mathematical modelling. *The Mathematical Intelligencer*, **6** (1984), 59.

[21] J. D. Barrow. *The Constants of Nature: From Alpha to Omega* (London: Jonathan Cape, 2002).

28
Universes galore: where will it all end?

Paul Davies
Australian Centre for Astrobiology, Macquarie University

28.1 Some sort of philosophy is inescapable

Most scientists concede that there are features of the observed Universe which appear contrived or ingeniously and felicitously arranged in their relationship to the existence of biological organisms in general and intelligent observers in particular. Often these features involve so-called fine-tuning in certain parameters, such as particle masses or coupling constants, or in the cosmic initial conditions, without which life (at least life as we know it) would be either impossible or very improbable. I term this state of affairs bio-friendliness or biophilicity. Examples of such fine-tuning have been thoroughly reviewed elsewhere [1] and in this volume, so I will not list them here.

It is normally remarked that cosmic bio-friendliness has two possible explanations (discounting sheer luck). One is that the Universe has been designed by a pre-existing creator with life in mind. The other, which is often motivated explicitly or implicitly by a reaction to supernatural explanations, is the multiverse. According to the latter explanation, what we call 'the Universe' is but a small component in a vastly larger assemblage of 'universes', or cosmic regions, among which all manner of different physical laws and conditions are somewhere instantiated. Only in those 'Goldilocks' regions where, by accident, the numbers come out just right will observers like ourselves arise and marvel at the ingenious arrangement of things. Thus the reason why we observe a universe so suspiciously contrived for life is because we obviously cannot observe one that is inimical to life. This is the so-called anthropic, or biophilic selection, principle [2].

Before reviewing the pros and cons of the multiverse explanation, I should like to make a general point. All cosmological models are constructed by augmenting the results of observations by some sort of philosophical principle. Two examples from modern scientific cosmology are the principle

Universe or Multiverse?, ed. Bernard Carr. Published by Cambridge University Press.
© Cambridge University Press 2007.

of mediocrity, sometimes known as the Copernican principle, and the biophilic selection principle. The principle of mediocrity states that the portion of the Universe we observe is not special or privileged, but is representative of the whole. Ever since Copernicus demonstrated that Earth does not lie at the centre of the Universe, the principle of mediocrity has been the default assumption; indeed, it is normally referred to as simply the 'cosmological principle'. It underpins the standard Friedmann–Robertson–Walker cosmological models.

In recent years, however, an increasing number of cosmologists have stressed the inherent limitations of the principle of mediocrity. Scientific observations necessarily involve observer selection effects, especially in astronomy. One unavoidable selection effect is that our location in the Universe must be consistent with the existence of observers. In the case of humans at least, observers imply life. (There is no reason why non-living observers could not exist, and indeed we may conjecture that advanced technological communities may create them. However, it is normally assumed that the emergence of life and intelligence is a precursor to the creation of non-living sentient beings, although there is no logical impediment to abiological sentence arising *de novo*.) Stated this way – that the Universe we observe must be consistent with the existence of observers – the biophilic principle seems to be merely a tautology. However, it carries non-trivial meaning when we drop the tacit assumption that the Universe, and the laws of nature, necessarily assume the form that we observe. If the Universe and its laws could have been otherwise, then one explanation for why they are as they are might be that we (the observers) have selected it from a large ensemble of alternatives.

This biophilic selection principle becomes more concrete when combined with the assumption that what we have hitherto regarded as absolute and universal laws of physics are, in fact, more like local by-laws: they are valid in our particular cosmic patch, but they might be different in other regions of space and/or time [3]. This general concept of 'variable laws' has been given explicit expression through certain recent theories of cosmology and particle physics, especially by combining string/M-theory with inflationary cosmology. There is little observational evidence for a domain structure of the Universe within the scale of a Hubble volume, but on a much larger scale there could exist domains in which the coupling constants and particle masses in the Standard Model may be inconsistent with life. It would then be no surprise that we find ourselves located in a (possibly atypical) life-encouraging domain, as we could obviously not be located where life was impossible.

Once it is conceded that the Universe could have been otherwise – that the laws of physics and the cosmic initial conditions did not have to assume the form we observe – then a second philosophical issue arises. The multiverse will contain a set of 'universes' that serve as instantiations for certain laws and initial conditions. What, then, determines the selection of universes on offer? Or to express it more graphically, using Stephen Hawking's words [4]: 'What is it that breathes fire into the equations and makes a universe for them to govern?'

Only two 'natural' states of affairs commend themselves in this regard. The first is that nothing exists; the second is that everything exists. The former we may rule out on observational grounds. So might it be the case that everything that can exist, does exist? That is indeed the hypothesis proposed by some cosmologists, most notably Max Tegmark [5]. At first sight this hypothesis appears extravagant. The problem, however, for those who would reject it is that, if *less* than everything exists, then there must be some rule that divides those things that actually exist from those that are merely possible but are in fact non-existent. One is bound to ask: What would this rule be? Where would it come from? And why *that* rule rather than some other?

28.2 An old-fashioned Cosmic Designer is a poor explanation

Since most of the contributions to this volume are written from a scientific perspective, I shall not dwell at length on why one might feel uncomfortable with the crude idea of a Cosmic Designer who contemplates a 'shopping list' of possible universes, figures out one that will contain life and observers, and then sets to work creating it, discarding the alternatives. The central objection to the hypothesis is its *ad hoc* nature. Unless one already has some other reason to believe in the existence of the Designer, then merely declaring 'God did it!' tells us nothing at all.

It has been argued (see, for example, ref. [6]) that an infinite God is a simpler explanation for existence than just accepting the Universe as a brute fact, and therefore to be preferred on the grounds of Ockham's razor. Dawkins [7] has countered that God must be at least as complex as the system that God creates. But considerable care is needed in using terms like 'simple' and 'complex'. A branch of mathematics called algorithmic complexity theory [8] can be used to provide rigorous definitions of simplicity and complexity. One surprising feature of these definitions is that the whole can sometimes be simpler than its component parts. Thus, God-plus-Universe can be simpler than either God or the Universe in isolation. I shall return to this topic in Section 28.3.5.

A further difficulty with divine selection concerns the notion of free choice. Christian theologians traditionally assert that God is a *necessary* being (see, for example, ref. [9]), i.e. it is logically impossible for God not to exist. If so, we are invited to believe that a necessary being did not necessarily create the Universe as it is (otherwise there is no element of choice and nature is reduced to a subset of the divine being rather than a creation of this being). But can a necessary being act in a manner that is not necessary? On the other hand, if God is regarded as not necessary but contingent, then on what, precisely, is God's existence and nature contingent? If we do not ask, we gain nothing by invoking such a contingent God, whose existence would then have to be accepted as a brute fact. One might as well simply accept a contingent universe as a brute fact, and be done with it. If we do ask, then we accept that reality is larger than God and that an account of the Universe must involve explanatory elements beyond God's being. But if we accept the existence of such explanatory elements, why is there any need to invoke divine elements also?

There is a long tradition of attempts to reconcile a necessary God with a contingent single universe (see, for example, ref. [10]). But one is bound to ask, even if such reconciliation were possible, why God freely chose to make this universe rather than some other. If the choice is purely whimsical, then the Universe is absurd and reasonless once more. On the other hand, if the choice proceeds from God's nature (for example, a good God might make a universe inhabited by sentient beings capable of joy), then one must surely ask: *Why* was God's nature such as to lead to *this* choice of universe rather than some other? This further worry would be addressed, in turn, only by proving – not only that God exists necessarily – but that God's entire nature is also necessary. Such a conclusion would entail proving that, for example, an evil creator capable of making a world full of suffering is not merely undesirable but *logically impossible*.

28.3 Shortcomings of anthropic/multiverse explanations

If theological explanations for cosmic biophilicity are problematic, then multiverse explanations are not without their difficulties too. In what follows, I shall review some of the challenges that have been made to multiverse/anthropic reasoning.

28.3.1 It's not science because it's not testable

It is sometimes objected that, because our observations are limited to a single universe (e.g. a Hubble volume), then the existence of 'other universes' cannot be observed, and so their existence cannot be considered a proper scientific hypothesis. Even taking into account the fact that future observers will see a larger particle horizon, and so have access to a bigger volume of space, most regions of the multiverse (at least in the eternal inflation model) can never be observed, even in principle. While this may indeed preclude direct confirmation of the multiverse hypothesis, it does not rule out the possibility that it may be tested indirectly. Almost all scientists and philosophers accept the general principle that the prediction of unobservable entities is an acceptable scientific hypothesis if those entities stem from a theory that has other testable consequences. At this stage, string/M-theory does not have any clear-cut experimental predictions, but one may imagine that a future elaboration of the theory would produce testable consequences. These theories are not idle speculations, but emerge from carefully considered theoretical models with some empirical justification.

A test of the multiverse hypothesis may be attained by combining it with biophilic selection. This leads to statistical predictions about the observed values of physical parameters [3]. If we inhabit a typical biophilic region of the multiverse, we would expect any biologically relevant adjustable parameters to assume typical values. If one considers a vast parameter space of possible universes, there will be one or more biophilic patches – or subsets – of the space, and a typical biophilic universe would not lie *close to the centre* of such a patch (i.e. it would not be *optimally biophilic*). In other words, there is no *a priori* reason why the laws of physics should be more bio-friendly than is strictly necessary for observers to arise. If, therefore, we discovered that some parameter (such as the amount of dark energy) assumed a value in a tiny subset located deep inside the biophilic parameter range, this would be evidence against it being a random variable that had been anthropically selected.

There is a hidden assumption in the foregoing reasoning, which is that life originates only once in each universe. If life happens many times, then optimally biophilic universes may contain many more observers than *minimally* biophilic universes, and this weighting factor must be taken into account when considering a randomly chosen observer. We must now distinguish between biophilicity in relation to *laws* and biophilicity in relation to *contingency*. Regarding the latter, it is possible that life is indeed 'a damned close-run thing' (to paraphrase Lord Wellington) – a statistical fluke, unique in the observable Universe, arising from a highly improbable molecular accident.

If, however, we discover a second genesis of life – an independent origin on a nearby planet – then this would imply that the Universe is teeming with life and is at least *near*-optimally biophilic in relation to contingency.

But even in the absence of any data concerning multiple geneses, we may still consider biophilicity in relation to the laws of the Universe. At first glance, there is little reason to suppose that the Universe is minimally biophilic in this respect. Take the much-cited example of carbon abundance. The existence of carbon as a long-lived element depends on the ratio of electromagnetic to strong nuclear forces, which determines the stability of the nucleus. But nuclei much heavier than carbon are stable, so the life-giving element lies comfortably within the stability range. The electromagnetic force could be substantially stronger, without threatening the stability of carbon. Now, it is true that, if it were stronger, then the specific nuclear resonance responsible for abundant carbon would be inoperable, but it is not clear how serious this would be. Life could arise in a universe where carbon was merely a trace element, or abundant carbon could occur because of different nuclear resonances. Of course, if it could be shown that other, heavier, elements are essential for life, this objection would disappear. (The prediction that much heavier elements are essential for life could be an interesting prediction of the multiverse theory.)

A simpler example is the amount of dark energy in the Universe. Once again, the observed value is comfortably in the middle of the biologically acceptable parameter range. Theory predicts that the density of dark energy (Λ) should be vastly greater than the observed value, so we might expect in the multiverse explanation that the observed value would be near the top end of the biologically permissible parameter range. But Λ could be an order of magnitude bigger without threatening the existence of galaxies and stars, and hence life [3]. On the face of it, therefore, the observed Universe is not minimally biophilic, and many scientists seem to think it is actually optimally biophilic.

Another consideration concerns the very existence of physical laws. In those versions of the multiverse in which even the appearance of law is attributed to anthropic selection, there is clearly a problem about minimal biophilicity. The multiverse explanation would lead us to expect that we live in a universe that has the minimal degree of order consistent with the existence of observers. Departures from order, or lawfulness, that are not biologically threatening should therefore be permitted. To take a simple example, consider the law of conservation of electric charge. The charge on the electron could happily fluctuate by, say, one part in 10^6 without disrupting biochemistry. In fact, measurement of the anomalous magnetic moment of the electron fixes the electric charge to eleven significant figures – a stability far in excess of that needed to ensure the viability of living organisms. So either the electric charge is fixed by a law of nature, in which case the multiverse cannot be invoked to explain this particular aspect of cosmic order, or there is some deep linkage between the charge on the electron and some aspect of physics upon which the existence of life depends far more sensitively. But it is hard to see what this might be.

28.3.2 Measures of fine-tuning are meaningless

Intuitively we may feel that some physical parameters are remarkably fine-tuned for life, but can this feeling ever be made mathematically precise? The fact that a variation in the strength of the strong nuclear force by only a few per cent may disrupt the biological prospects for the Universe appears to offer a surprisingly narrow window of biophilic values, but what determines the measure on the space of parameters? If the strength of the nuclear force could, in principle, vary over an infinite range, then any finite window, however large, would be infinitesimally improbable if a uniform probability distribution is adopted. Even the simple expedient of switching from a uniform to a logarithmic distribution can have a dramatic change on the degree of improbability of the observed values, and hence the fineness of the fine-tuning. There will always be an element of judgement involved in assessing the significance, or degree of surprise, that attaches to any given example.

28.3.3 Humans are more than mere observers

Most often discussed in relation to anthropic selection is the matter of fine-tuning of certain physical parameters, such as the relative strengths of the fundamental forces of nature. Had these parameters taken on values outside a relatively narrow range, then it is likely that the Universe would go unobserved. Whilst trivially true, this explanation is unacceptably narrow, because it treats humans as mere observers. That is, it is merely necessary for there to exist observers *of some sort* for the argument to work (the term 'anthropic' is an acknowledged misnomer in this respect). Indeed, the application of anthropic reasoning usually ignores even the conditions necessary for intelligent observers to evolve and restricts attention simply to the existence of life.

Humans, however, are more than mere observers. They also have the ability to *understand* the Universe through logical reasoning and the scientific method [11]. This remarkable fact, often taken for granted by scientists, cannot be explained by anthropic/multiverse reasoning. It is perfectly possible for there to exist a universe that permits the existence of observers who nevertheless do not, or cannot, make much sense of nature. Thus cats and dogs surely qualify as observers, but are not, like humans, privy to the deep mathematical rules on which the Universe runs. Moreover, in a general multiverse scenario, the vast majority of universes that permit the existence of observers with the same intellectual prowess as humans will not be comprehensible to those observers. For example, there are many ways that the laws of physics we observe could be more complex without threatening the existence of biology: non-computability of the laws, forces varying with time in a complicated way that leaves chemistry largely unaffected, legions of additional weak forces that do not substantially affect the formation of galaxies, stars and planets, millions of species of neutrinos, etc. In fact, the physics of the Universe is *extremely special*, inasmuch as it is both simple and comprehensible to the human mind.

28.3.4 The blunderbuss objection

It is trivially true that, in an infinite universe, anything that can happen will happen. But this catch-all explanation of a particular feature of the Universe is really no explanation at all. We should like to understand the bio-friendliness of *this* universe. To postulate that all possible universes exist does not advance our understanding at all. A good scientific theory is analogous to a well targeted bullet that selects and explains the object of interest. The multiverse is like a blunder buss – hitting everything in sight.

To put this point into context, imagine that the Universe we observe is divided into Planck-sized three-dimensional cells. Each cell may be assigned a finite set of numbers that determines its state; for example, the amplitudes of all fields at that point. Now imagine that the digits of π are expressed in binary form and used to label the state of each cell in sequence, using as many digits as necessary to specify the field amplitudes to any desired precision. When all cells in the observable Universe have been labelled, a state of the entire Universe is determined. Now imagine that the process is repeated with the further digits of π. Another state is defined. This process may be continued for a stupendous number of steps (Planck times), giving us a 'cosmic history'. Most of the cosmic history will be random noise, lacking even the semblance of causal order. But, by the very definition of randomness, we are assured that, sooner or later, the observed cosmic history will be generated [12]. So too will *all other* cosmic histories: the digits of π

contain all possible worlds. Should we be satisfied, therefore, that we have explained the Universe, together with all its remarkable features, such as biophilicity, by saying merely that it is a manifestation of π? Or perhaps of e, or of almost any real number we like to pick? Clearly not. Saying that our world is buried in the limitless noise of the digits of π does not make π a magic generator of reality. It merely highlights the vacuousness of seeking to appeal to everything in order to explain something in particular.

28.3.5 The multiverse is really an old-fashioned God in disguise

In this section, I shall argue that, in a certain mathematical sense, the most general multiverse models (e.g. Tegmark's Level 4 version) are ontologically equivalent to naïve deism, by which I mean the existence of a Cosmic Designer/Selector who judiciously picks a single real universe from an infinite shopping list of possible but unreal universes. Indeed, I suspect the general multiverse explanation is simply naïve deism dressed up in scientific language. Both appeal to an infinite unknown, invisible and unknowable system. Both require an infinite amount of information to be discarded just to explain the (finite) universe we observe. It would be instructive to quantify and compare the degree of credulity we might attach to various competing multiverse and theological models using algorithmic complexity theory. It seems likely that some versions of both the multiverse and naïve deism would be equivalently complex and, in most cases, infinitely complex. They may employ different terminology but, in essence, both explanations are the same. If I am right, then the multiverse is scarcely an improvement on naïve deism as an explanation for the physical universe. It is basically just a religious conviction rather than a scientific argument.

I will make an even stronger claim. I believe that naïve deism and the general multiverse concept will turn out to be of equivalent complexity because *they are contained within each other*. Consider the most general multiverse theories (Tegmark's Level 4), where even laws are abandoned and anything at all can happen. At least some of these universes will feature miraculous events – water turning into wine, etc. They will also contain thoroughly convincing religious experiences, such as direct revelation of a transcendent being. It follows that a general multiverse set must contain a subset that conforms to traditional religious notions of God and design. It could be countered, however, that this subset is embedded in a much bigger set in which no coherent theological plan is discernible, so that a random observer would be unlikely to encounter a world in which a God was seen to be at work. But this is to ignore the possibility of simulated realities (see Section 28.3.6).

28.3.6 Real versus fake universes

The starting point of all anthropic – multiverse arguments is the existence of *observers*. This raises the question of what constitutes an observer. I shall assume that 'observership' is a product of physical processes, for example electrochemical activity in the brain. It then follows that observers may be created artificially by sufficiently advanced technology. Possibly this merely requires bigger and better computing systems, as argued by proponents of strong AI; possibly it requires a new form of technology, as argued by Roger Penrose [13]. For my purposes, it does not matter. In a multiverse, there will be a subset of universes in which advanced technology like ours emerges, and a sizeable sub-subset will contain at least one technological civilization that reaches the point of simulating consciousness. It is but a small step from simulating consciousness to simulating a community of conscious beings and an entire virtual world for them to inhabit.

This notion has been popularized in *The Matrix* series of science fiction movies. For any given 'real' world, there would be a vast, indeed infinite, number of possible virtual worlds. A randomly selected observer would then be overwhelmingly more likely to experience a virtual simulation than the real thing. Thus there is little reason to suppose that *this* world (the one you and I are observing now) is other than a simulated one [14, 15]. But the denizens of a simulated virtual world stand in the same ontological relationship to the intelligent system that designed and created their world as human beings stand in relation to the traditional Designer/Creator Deity (a fact not lost on science fiction writers from Olaf Stapledon onwards), but with God now in the guise – not of a Grand Architect – but of a Grand Software Engineer. The creator of the virtual worlds is a transcendent designer with the power to create or destroy simulated universes at will, alter the circumstances within them, devise laws, perform miracles, etc. Taken to its logical extreme, the multiverse explanation is a convincing argument for the existence of (a rather old-fashioned form of) God! This is certainly ironical, since it was partly to do away with such a God that the multiverse was originally invoked.

Worse still, there is no end to the hierarchy of levels in which worlds and designers can be embedded. If the Church–Turing thesis is accepted, then simulated systems are every bit as good as the original real universe at simulating their own conscious sub-systems, sub-sub-systems, and so on *ad infinitum*: gods and worlds, creators and creatures, in an infinite regress, embedded within each other. We confront something more bewildering than an infinite tower of virtual turtles: a turtle fractal of virtual observers, gods and universes in limitlessly complex inter-relationships. If *this* is the ultimate reality, there would seem to be little point in pursuing scientific inquiry at all into such matters. Indeed, to take such a view is as pointless as solipsism. My point is that to follow the multiverse theory to its logical extreme means effectively abandoning the notion of a rationally ordered real world altogether, in favour of an infinitely complex charade, where the very notion of 'explanation' is meaningless.

This is the 'slippery slope' referred to by Rees [16]. At one end of the slope is the perfectly unobjectionable idea that there may be regions beyond a Hubble distance that possess, say, a lower average matter density or slightly less dark energy. At the bottom of the slope is the 'fantasy-verse' of arbitrary virtual realities, whimsically generated by a pseudo-Deity designer.

28.3.7 Multiverses merely shift the problem up one level

Multiverse proponents are often vague about how the parameter values are chosen across the defined ensemble. If there is a 'law of laws' describing how parameter values are assigned as one slips from one universe to the next, then we have only shifted the problem of cosmic biophilicity up one level. Why? First, because we need to explain where the law of laws comes from. But there is a second problem. Each law of laws specifies a different version of the multiverse, and not all multiverses are bound to contain at least one biophilic universe. In fact, on the face of it, most multiverses would *not* contain even one component universe in which all the parameter values were suitable for life. To see this, note that each parameter will have a small range of values – envisage it as a highlighted segment on a line – consistent with biology. Only in universes where all the relevant highlighted segments intersect in a single patch (i.e. all biophilic values are instantiated together) will biology be possible. If the several parameters vary *independently* between universes, each according to some rule, then for most sets of rules the highlighted segments will not concur. So we must not only explain why there is any law of laws; we must also explain why the actual law of laws (i.e. the actual multiverse) happens to be one that intersects the requisite patch of parameter space that permits life.

Often it is asserted that there is no law of laws, only randomness. Thus in Smolin's version of the multiverse, gravitational collapse events 'reprocess' the existing laws with small random variations [17]. In this case, given an infinite multiverse, randomness would ensure that at least one biophilic universe exists with a finite (albeit minute) probability. (That is, there will always be a patch of parameter space somewhere with all highlighted segments intersecting.) Plausible though this is, the assumption of randomness is not without its problems. Without a proper measure over the parameter space, probabilities cannot be properly defined. There is a danger of predicting meaningless or paradoxical results. There is also a danger in some multiverse models that the biophilic target universes may form only a set of measure zero in the parameter space, and thus be only infinitesimally probable. Furthermore, in some models, various randomness measures may be inconsistent with the underlying physics. For example, in the model of a single spatially infinite Universe in which different supra-Hubble regions possess different total matter densities, it is inconsistent to apply the rule that any value of the density may be chosen randomly in the interval $[0, \rho]$, where ρ is some arbitrarily large density (e.g. the Planck density). The reason is that for all densities above a critical value (very low compared with the Planck density), the Universe is spatially finite, and so inconsistent with the assumption of an infinite number of finite spatial regions [18].

The need to rule out these 'no-go' zones of the parameter space imposes restrictions on the properties of the multiverse that are tantamount to the application of an additional overarching biophilic principle. There would seem to be little point in invoking an infinity of universes only then to impose biophilic restrictions at the multiverse level. It would be simpler to postulate a single universe with a biophilic principle.

28.4 The third way

Considerations of anthropic fine-tuning seek to explain the appearance of an otherwise puzzling link between the universe on one hand and life on the other. Why should there be a connection? What does the Universe know about life? What do the laws of physics care about consciousness?

The most obvious way to establish a link between life and cosmos is to postulate a 'life principle' (or, extending this to encompass observers, a 'mind principle'). Indeed, many scientists have suggested just such a thing. It is often claimed by astrobiologists that life is 'written into the laws of physics' or 'built into the nature of the Universe' [19]. Thus Sydney Fox, in his theory of biogenesis, claimed that the laws of physics and chemistry were rigged in favour of those reactions that lead to life [20]. Others, such as Christian de Duve [21] and Stuart Kauffman [22], have hinted that somehow chemistry favours life and can fast-track matter and energy to the living state.

John Wheeler, in his 'participatory universe' principle, has even claimed something along those lines for mind [23].

Is there any evidence for such a principle? The laws of physics, as we now understand them, do not offer much promise in this regard. The reason is not hard to find. Life is incredibly complex but the laws of physics are, in the algorithmic sense, simple. So life cannot be contained in the laws of physics. Contrast this with another state of matter: crystals. The structures of crystals are determined by the symmetries of the electromagnetic force, and so they *are* built into the laws of physics. Basic geometry underlies them. Given the laws of physics, the structure of, say, common salt crystals may be deduced from purely geometrical considerations. Crystals are simple and have low information content, concordant with the low information content of the laws of physics. But one could not predict the structure of, say, a bacterium, nor even its genome sequence, from the laws of physics, because the genome has very high information content. It was for that sound mathematical reason that Jacques Monod declared 'we are alone' and 'the Universe is not pregnant with life'. In his opinion, life is just a stupendously improbable accident [24].

The root cause of the difficulty goes back at least to the time of Newton and the deep dualism that pervades all of science: the dualism between eternal universal laws and time-dependent contingent states. Because laws are general, simple, low in information content and unchanging with time, most specific states of matter cannot be built into them. States of matter are generally local, special, complex, high in information content and time-dependent. So the very structure of traditional scientific explanation precludes our finding a direct link between the underlying laws of the Universe (as we at present understand them) and the emergence of an exceedingly specific and peculiar state of matter such as 'life' – still less an even more specific and peculiar state such as 'mind'. Therefore, if we wish to postulate such a link, then the traditional dualism of laws and states must go.

Aristotle did not make a sharp distinction between laws and states. By introducing different categories of causation, and specifically by including final causes, he could speculate on how the Universe might develop in a directed manner toward certain special states. For Aristotle, life was indeed built into the nature of the Universe through final causation. Such goal-directed or purposeful influences in nature are termed *teleological* by philosophers.

The assumption of a link between laws and product states such as life inevitably amounts to slipping an element of teleology into physics. This is very unfashionable, but I believe it is unavoidable if we are to take life and mind seriously as *fundamental* rather than *incidental* features of the Universe. And the bio-friendliness of the Universe suggests that they *are* fundamental. We need not be as crude as Aristotle, by nailing down the final state in advance and constraining the Universe to generate it; de Duve, for example, has suggested in the context of biological evolution that the general trend (e.g. from simple to complex, from mindless to mental) is law-like, although the specific details are contingent [21]. In my essay 'The physics of downward causation' [25], I have suggested that such a felicitous mix of law and chance might be generalized to cosmology, producing directional evolution from simple through complex states, to life and mind.

Obviously these are just words, whereas what is required are concrete mathematical models. To investigate the basic ideas, I have developed some cellular automaton models with the help of Neil Rabinowitz. Recall that, in a conventional cellular automaton system, one starts with a 1-dimensional array of cells, or pixels, each of which can be in one of two states: filled or unfilled ('on' or 'off'). An update rule is specified that determines whether a given pixel remains on or off, is switched from on to off or vice versa. This rule is based on the state of the near neighbours, and there are 256 possible simple local rules [26]. This system thus mimics the physics of a causally closed system subject to local dynamical laws. An initial state is specified, for example a random scatter of filled cells, and the array is evolved forward in discrete time steps. A variety of interesting behaviour results. Crucially, the conventional automaton retains the ancient dynamical dualism: the update rules are always independent of the states.

As a first departure from the conventional prescription, we decided to start with a random input state and tried switching between two different rules either randomly or periodically. The results of one interesting case are shown in Figs. 28.1–28.3. This features two automata, designated 87 and 90 according to Wolfram's classification scheme [26]. Applied on its own, rule 87 leads to structured, but relatively dull, quasi-periodic spatial structures that move across the array at uniform speed (Fig. 28.1). Rule 90 merely perpetuates the random noise (Fig. 28.2). Thus, individually, rules 87 and 90 do not lead to interesting dynamical behaviour. However, when the rules are interspersed, the story is very different. Figure 28.3 shows the outcome when rule 90 is applied and interrupted every seven steps by rule 87. The upshot is the evolution of a form of organized complexity from disorganized, or random, input. Although there is nothing explicitly teleological in the set-up, a form of directionality – order out of chaos – is discerned. With a bit of experimentation, this rule-interspersion technique can be used to combine order and chaos in a suggestively creative manner, getting 'the

28 Universes galore: where will it all end?   501

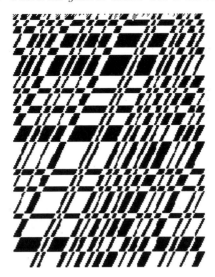

Fig. 28.1. Rule 87 cellular automaton with random initial state. Time runs downward.

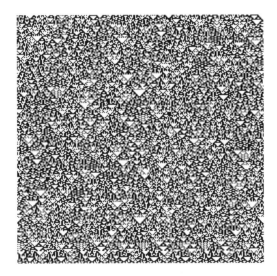

Fig. 28.2. Rule 90 cellular automaton with random initial state.

best of both worlds' – the unpredictability and novelty of chaos with the coherence of order. Our results are reminiscent of Parrondo's games [27], in which two games of chance, each of which when played individually have an expectation of loss, when combined can lead to an expectation of gain. Parrondo's games show that, counter-intuitively, two losses can make a win. Figures 28.1–28.3 appear to be a cellular automaton analogue.

Fig. 28.3. Rule 90 interrupted every seven steps by rule 87 cellular automaton with random initial state.

To incorporate fully my 'third way' idea, we must alter the automaton rules so that they depend explicitly on some aspect of the state. To take a very simple example of state-dependent laws, the rule may be chosen to be A if the total number of filled pixels is even and B if it is odd. Alternatively, some statistical measure, such as the entropy or complexity (defined by some prescription) may be used as the discriminator of the rules. Whatever choice is made, the behaviour of a group of pixels now depends not only on the state of the neighbouring pixels, in analogy with conventional physical laws, but on the *global state too*. This is therefore an explicit form of *top-down*, or whole–part, causation [25]. Although our work is at a preliminary stage, the hope is that simple mathematical models might capture the elusive notion that certain complex states are favoured by acting as attractors in the product-space of states and laws. This idea could be placed in a restricted multiverse context by considering how some universes, or regions thereof, generate their bio-friendly laws in an evolutionary sense, and thus become observed. So biology does not actually select a pre-ordained universe; rather, physics and biology co-evolve under the action of a (precise) principle operating at the multiverse level, in such a manner that teleological behaviour emerges. So this is a theory in which life and mind, goal and purpose, arise in a law-like manner from a dynamic universe (or multiverse). The key feature is that there is *a causal link between laws and product states* (in contrast to Darwinian evolution, where mutations and selection events form causally disjoint chains). Thus life is neither a statistical fluke in an indifferently random set of laws/universes, nor is the Universe designed in an *ad hoc* way for life. Instead, life and mind, laws and universes, are common products of an overarching principle.

If I were to pick a symbol to characterize this set of still rather woolly ideas, it is that of a self-consistent, self-supporting loop. It has some elements in common with Wheeler's idea of a loop in which nature and observer are mutually enfolded [23]. I have described it as a 'turtle loop' in the context of the famous 'tower of turtles' metaphor [28].

As a final illustration of an implicit loop, consider the fact that the mathematics describing the underlying laws of physics is a product of the human mind. The mental realm occupies a conceptually higher level than the physical realm of particles and fields to which this mathematics applies. Why should something created at this higher level apply so famously well [29] to the physical realm? Why should 'software' apply to 'hardware'? More specifically, the concept of what constitutes a computable function (software) is based on the idea of a classical Turing machine (hardware). As stressed by David Deutsch [30], the existence of such a physical device depends on the specific nature of the laws of physics. Thus the concept of computability depends on what the physics of the particular world allows to be computed. So the laws of the Universe permit the existence of physical systems (human beings, Turing machines) that can output the mathematics of those very same laws. This remarkable self-consistent loop is by no means guaranteed [28]. (It also constitutes a further example of why human beings are more than mere observers, which I considered in Section 28.3.3. Human beings are also 'computers'.) There could be many universes with computable laws that do not admit physical systems which can actually output the computable functions describing those laws. Or there could be universes with non-computable laws [31]. Since there is an intimate connection [12] between Turing machines and self-reproducing machines (i.e. life), we glimpse a link between life and laws.

References

[1] J. D. Barrow and F. J. Tipler. *The Anthropic Cosmological Principle* (Oxford: Oxford University Press, 1986).
[2] B. Carter. Large number coincidences and the anthropic principle in cosmology. In *Confrontation of Cosmological Theory with Observational Data*, ed. M. Longair (Dordrecht: Reidel, 1974), pp. 291–298.

[3] M. J. Rees. Numerical coincidences and tuning in cosmology. *Astrophys. Space Sci.* **285** (2003), 375.
[4] S. W. Hawking. *A Brief History of Time* (New York: Bantam, 1988), p. 174.
[5] M. Tegmark. Parallel universes. In *Science and Ultimate Reality*, eds. J. D. Barrow, P. C. W. Davies and C. Harper (Cambridge: Cambridge University Press, 2003, p. 459).
[6] R. Swinburne. *The Coherence of Theism*, revised edn (Oxford: Clarendon Library of Logic and Philosophy, Clarendon Press, 1993).
[7] R. Dawkins. *The Blind Watchmaker* (New York: Norton, 1986).
[8] M. Li and P. Vitanyi. *An Introduction to Kolmogorov Complexity and its Applications* (New York: Springer-Verlag, 1997).
[9] J. Haught. *What is God?* (New York: Paulist Press, 1986).
[10] K. Ward. *God, Chance and Necessity* (Oxford: Oneworld, 1996).
[11] P. C. W. Davies. *The Mind of God* (London and New York: Simon & Schuster, 1992).
[12] W. Poundstone. *The Recursive Universe* (Oxford: Oxford University Press, 1985).
[13] R. Penrose. *The Emperor's New Mind* (Oxford: Oxford University Press, 1989).
[14] N. Bostrom. Are you living in a computer simulation? *Phil. Quart.* **53** (2003), 243.
[15] M. Brooks. Life's a sim and then you're deleted. *New Scientist* (27 July 2002), p. 48.
[16] M. J. Rees. *Our Cosmic Habitat* (Princeton: Princeton University Press, 2001).
[17] L. Smolin. Did the universe evolve? *Class. Quant. Grav.* **9** (1995), 173.
[18] R. Holder. *God, the Multiverse and Everything: Modern Cosmology and the Argument from Design* (Aldershot: Aldgate, 2004).
[19] R. Shapiro. *Origins: A Skeptic's Guide to the Creation of Life on Earth* (New York: Summit Books, 1986).
[20] S. Fox. Prebiotic roots of informed protein synthesis. In *The Roots of Modern Biochemistry*, eds. H. Kleinhauf, H. von Doren and L. Jaenicke (Berlin: De Gruyter, 1988), p. 897.
[21] C. de Duve. *Vital Dust* (New York: Basic Books, 1995).
[22] S. Kauffman. *At Home in the Universe* (Oxford: Oxford University Press, 1995).
[23] J. A. Wheeler. Beyond the black hole. In *Some Strangeness in the Proportion*, ed. H. Woolf (Reading, MA: Addison-Wesley, 1980), p. 341.
[24] J. Monod. *Chance and Necessity*, trans. A. Wainhouse (London: Collins, 1971), p. 145.
[25] P. C. W. Davies. The physics of downward causation. In *The Re-emergence of Emergence*, eds. P. Clayton and P. C. W. Davies (Oxford: Oxford University Press, 2006).
[26] S. Wolfram. *A New Kind of Science* (Champaign, IL: Wolfram Media Inc., 2002).
[27] J. M. R. Parrondo. Reversible ratchets as Brownian particles in an adiabatically changing periodic potential. *Phys. Rev.* **E 57** (1998), 7297.
[28] P. C. W. Davies. Why is the physical world so comprehensible? In *Complexity, Entropy and the Physics of Information*, SFI Studies in the

Sciences of Complexity, vol. VIII, ed. W. H. Zurek (New York: Addison-Wesley, 1990), p. 61.
[29] E. Wigner. The unreasonable effectiveness of mathematics in the natural sciences. *Commun. Pure Appl. Math.* **13** (1960), 1.
[30] D. Deutsch. Quantum theory, the Church–Turing principle and the universal quantum computer. *Proc. Roy. Soc.* **A 400** (1985), 97.
[31] J. Hartle. Excess baggage. In *Particle Physics and the Universe: Essays in Honor of Gell-Mann*, ed. J. Schwarz (Cambridge: Cambridge University Press, 1991).

Index

Italics indicate a prime reference.

absorption lines 14, 84, 355
acceleration, cosmic 6, 82, 105, 166, 193, 195, 209
accelerators 14
accretion disks 336, 355
Adam and Eve 432, 437
Agrawal, V. 144
Aguirre, Anthony 31, 81, 84, 176, 350, *367*, 452, 453
Alpher, Ralph 15
Andromeda Galaxy (M31) 9
Anselm 460, 463
anthropic
  coincidences 78, 85, 405
  predictions 17, 20, 163, 164, 168, 171, 177
  reasoning 15, 20–3, 31–3, 39, 43, 45, 50, 53, 65, 144, 145, 151, 152, 157, 161, 191, 209, 227, 275, 278-83, 238, 356, 360, 374, 381, 382, 432, 437, 439, 490, 493
  selection 6, 66–8, 81, 85, 163, 173, 221, 226, 492, 493
  weighting 24, 52, 285, 310, 313, 314, 474, 491
Anthropic Principle (AP)
  micro version of 23, 285, 286, 311, 313
  strong version of 3, 4, 26, 78, 83, 144, 433
  weak version of 78, 311, 315
Anti de Sitter (AdS) space 202–7, 264
Anti de Sitter/Conformal Field Theory (AdS/CFT) correspondence 101, 113
antimatter 13, 69
Aristotle, *see* paradigm, Aristotelian
Arkani-Hamed, N. 34, 144
Arkani-Hamed, Dimopoulos and Kachru (ADK) 36, 37
Ashok, S.K. 31
Atchison, Ian 470
atheism 472, 476
atoms, *see* periodic table
atomic nuclei, *see* nuclei
awareness 67, 117, 310, *see also* consciousness
awareness operators 304, 305, 414–18, 423–6

axion 20, 51, 52, 71, 82, 140-4, 153–62, 201, 205, 237, 238, 383
baby universe, *see* universe
Banks, Tom 187
Barrow, John 27, 80, 118, 198, *481*
baryons 37, 67–71, 82, 141–3, 157, 158, 176, 231, 345, 348, 370, 372, 376, 379, 382, 404, 453
  density of 72, 82, 141, 142, 153, 156, 158, 453
baryon, synthesis of 82
baryon-to-photon ratio 69, 70, 71, 134, 176, 184
  see also photon-to-baryon ratio
Bayes, Thomas 24, 306, 422, 434, 436
Bayesian prior/statistics 26, 67, 68, 296, 303, 306, 419, 421, 423, 427, 428
beauty 145, 234, 396, 463, 470, 475
  God and 470, 471, 472, 476
  mathematics and 14, 27
  nature and 234, 247, 281, 460, 466, 467, 475
  see also Sleeping Beauty
beryllium 79, 80
Bethe, Hans 356
Beyond the Standard Model (BSM) 52, 184, 211, 236, 245
big bang
  cold 84, 350, 351, 433, 453
  hot 15, 130, 142, 453
  see also pre-big-bang
big crunch 174
biochemistry 31, 223, 482, 492
biophilicity 58, 403, 404, 487, 490–8
biophilic selection principle 488, 491
Bjorken, James 20, 21, *181*, 389
Black Cloud 69
black-body (thermal) radiation 9, 183, 185, 262, 264, 334
black hole 58, 59, 63, 71, 107, 253, 260-4, 291, 352, 353, 356, 367
  baby universe and 107, 139, 140, 189

black hole (*cont.*)
  bouncing 334–8, 352
  complementarity 262
  evaporation of 113, 118, 260, 336
  event horizon of 100, 253, 335
  formation/production of 25, 84, 260, 262, 335–7, 353–9, 450, 497
  interior of 102, 189
  Hawking radiation from 138, 264, 434
  Schwarzschild radius of 260
  see also primordial black hole
Bohr, Niels 267–9
Bohr radius 223
Boltzmann, Ludwig 24, 67, 256
Boolean logic 306, 347
Bose–Einstein condensate 155
boson 330
  weak 216, 233
  Goldstone 248
Bostrom, Nick 26, *431*, 482
bottom-up 91, 93, 95, 97, 371–2, 375–7
Boyle's law 467
bouncing universe, *see* universe
Bousso, Raphael 30, 192, 251
Brahe, Tycho 8
brain-states 415, 417, 434
branches, *see* quantum
branes/membranes 6, 13, 208, 229, 244, 250, 251, 331, 332
  see also D-branes
Brans–Dicke (BD) 134, 390, 391
Brooke, John 376
Brouwer, L. 485
Brown, G.E. 356
Brown, J.D. 249
Brownlee, D. 182
Bruno, Giordano 102, 461
bulk (higher-dimensional) 6, 13, 85, 212
bubble
  inflationary 4, 19, 92, 101, 105–7, 169, 258, 263, 423, 465
  nucleation of 38, 166, 174, 240, 241, 259, 337
  vacuum 166, 259
  see also universe

Cabibbo–Kobayashi–Maskawa (CKM) matrix 52, 222, 223
Calabi-Yau (CY) manifold 10, 48, 91, 203, 204
Cambridge 17, 18, 29, 41
Cantor, Geoffrey 476
carbon 113, 327, 349–56, 359
  life based on 281, 313, 372
  synthesis in stars 34, 36, 57, 77, 80, 82, 144, 338–41, 349, 351, 360
  see also triple-alpha reaction
Carr, Bernard *3*, 77
Carter, Brandon 3, 18, 19, 23, 24, 85, 86, *285*, 432, 433, 445
cat, *see* Schrödinger

causation 470, 499, 500, 502
cellular
  automata 27, 500–2
  reaction network 223
chaos 467, 468, 500, 501
  see also universe, chaotic model
charge parity (CP) 20
  violation of 22, 82, 94, 140, 231, 236–9, 242, 244
chemistry 46, 50, 58, 69, 80, 168, 186, 223, 224, 252, 327, 328, 337, 340, 354, 357, 360, 392, 494, 498
chirality, *see* symmetry
Christianity 308, 412, 460, 461, 490, 498
Church (Catholic) 39, 461, 476
Church-Turing 496
civilization 37, 172, 313, 345, 346, 360, 404, 432, 481–3, 496
closed universe, *see* universe
coincidence problem 171, 195, 199–201
cold big bang, *see* big bang
cold dark matter (CDM) 42
Coleman, Sidney 38, 416
Collins, Barry 445
Collins, Robin 16, 27, *459*
collisions
  of branes 7
  of neutron stars 341
Coma cluster 70
compactification, *see* dimensions
complementarity 65, 263, 286
  see also black hole
complexity 18, 50, 60, 61, 69, 86, 109, 120, 122, 123, 237, 277, 279, 281, 304, 327–9, 338, 344, 351, 388, 391, 392, 422, 427, 446, 465, 467, 472–4, 481, 483, 484, 489, 495, 500, 502
  see also Pyramid of Complexity
Complexity Principle, *see* Principle of Complexity
Compton cooling 70
computer simulation, *see* simulation
Comte, August 14
concordance (cosmological) 19, 100
consciousness 13, 61, 77, 83, 85, 88, 112, 140, 312, 316, 388, 392–5, 416, 446, 482, 496, 498
  see also self-aware substructure
conservation laws 34, 69, 467, 492
constants
  dimensionless 106, 412
  renormalization of 187, 242
  scanning of 33–7
  variation of 40, 46, 48, 52, 83–5, 96, 98, 152, 157, 161, 169, 186, 188, 228, 239, 240, 339, 340, 360, 394, 413, 465, 466, 469, 493, 497
  see also coupling constants
  see also fine-structure constants
  see also fine-tuning

contingency 5, 65, 83, 156, 157, 393, 439, 456, 469, 470, 490, 492, 499, 502
continuum 212, 215, 221, 247, 248, 262, 420, 421 *see also* discretum
Copenhagen interpretation 5, 24, 267–70, 285, 295–7, 300, 414, 415, 418
Copernicus, Nicolaus 7, 49, 488
Cosmic Background Explorer (COBE) satellite 9, 481
cosmic background radiation 9, 13, 81, 102, 121
cosmic microwave background (CMB) 15, 32, 44, 69–71, 80, 93, 96–8, 101–5, 121, 170, 227, 254, 334, 356, 405, 434
cosmic microwave background temperature (CMBT) 434, 435, 437, 439
cosmocentric view 8, 10
cosmological concordance *see* concordance
cosmological constant (CC) 211–14
    problem (CCP) 134, 135, 136, 138, 165, 166, 202, 209, 211–15, 217, 379, 401
    see also dark energy
    see also Einstein
    see also gravity, repulsive
Cosmological Indifference Principle, *see* Indifference Principle
Cosmological Natural Selection (CNS) 25, 27, 107, 327, 338, 350–4, 360, 361, 450
cosmological nucleosynthesis, *see* nucleosynthesis
Cosmological Principle 8, 46, 127
cosmology
    see big bang
    see bottom-up
    see chaos
    see pre-big-bang
    see quantum
    see top-down
coupling constants 134, 145, 344, 413, 422, 487
    dimensionless 35, 36, 184, 185, 235
    in Standard Model 21, 108, 223, 231, 234, 328
    relationships between 78
    fine-tuning of 330
    running of 159
    variation in 8, 54, 78, 83, 488
Creator 3, 14, 16, 26, 39, 60, 456, 483, 484, 487, 490, 496, 497
    see also God
critical density, *see* density
Crowe, Michael 461
Curtis, Heber 9
curvature of space 38, 59, 85, 102, 105, 106, 161, 202, 257, 335, 391, 397
cyclic model, *see* universe

dark energy 10, 20, 21, 47, 53, 157, 161, 162, 166, 188, 191–4, 199, 211, 214, 221
    cosmological constant and 68, 69, 135, 215, 253

density/scale of 53, 156, 182, 184, 185, 193, 491, 492, 497
equation of state of 215
models of 194, 200, 201
see also quintessence
dark matter
    see axion, cold dark matter
Darwin, Charles 39, 40, 60
Darwinian selection 131, 229, 304, 357, 388, 406, 503
    see also evolution theory, *see also* natural selection
Davies, Paul 14, 27, 65, 86, 351, 388, 478, 482, *487*
Dawkins, Richard 476, 489
D-branes, 202–7, 212, 221
    see also branes/membranes
decay
    of beryllium 80
    of neutron 34, 238
    of nuclei 12
    of pseudo-LSP 160
    see also proton, vacuum
de Broglie, Louis 305
decoherence, 101, 112, 113, 121, 268, 271, 277, 293, 415, 450
de Cusa, Nicholas 8, 461
de Duve, Christian 498, 500
deism 495
density
    critical value of 78, 388
    fluctuations in 46, 53, 69, 70, 78, 81, 85, 131, 134, 136, 139, 142, 155, 161, 166, 167, 171, 175, 271, 336, 363, 379, 380, 388, 454
    matrix 282, 283, 424
    parameter 32, 80, 98, 193, 391, 397, *see also* Omega
    see also baryon density
design
    cosmic 16, 489
de Sitter (dS) space 21, 67, 92, 103, 134, 138, 182–3, 189, 191, 202, 252–65
deuterium 110, 238, 254
Deutsch, David 388, 503
DeWitt, Bryce 317
    see also Wheeler–DeWitt
dice 111, 269, 317
Dicke, Robert 77, 78, 81, 317, 338, 339, 360, 445
    see also Brans–Dicke
dimensions
    compactification of 19, 22, 30, 48, 96, 134, 192, 193, 200–4, 208, 228, 250, 251
    extra/higher 6, 10, 13, 14, 22, 64, 229, 236, 240, 484
    large 95, 97
Dimopoulos, Savas 21, 34, *211*
    see also Arkani-Hamed, Dimopoulos and Kachru (ADK)
Dine, Michael 33, 256

Dirac, Paul 24, 263, 288–90, 301,314
    see also law of large numbers
Dirac–Heisenberg, see paradigm
Dirac–von Neumann, see paradigm
discretum 212, 214, 215
Distler, J. 36
DNA (double helix) 11, 223, 229
Dobzhansky, Theodosius 229
domain
    expanding 387–9, 392, 395, 399
    wall 249–51, 257
Donald, M.J. 305
Donoghue, John 22, *231*
Doomsday argument 346, 360, 432, 435, 436
double-slit experiment 267
Douglas, Mike 31, 33, 36, 192, 208
Duff, Mike 30
dwarf galaxies 72, 142, 172, 173, 355
Dyson, Freeman 15, 311–14, 316, 318

early universe 19, 20, 35, 69, 71, 79, 80, 142, 166, 171, 191, 195, 328, 336, 402, 406, 449, 453
    phase transitions in 38, 221, 353
Earth 7, 8, 29, 31, 40, 47, 86, 88, 108, 139, 152, 181, 182, 276–8, 282, 389, 392, 436, 461, 488
Efstathiou, George 81, 170, 201
Einstein, Albert 6, 8, 29–31, 48, 50, 64, 65, 100, 102, 118, 468
    field equations of 123, 130, 252, 391, 465, 466, 473
    cosmological constant of 165
    mass-energy relation of 165, 465
    see also static model
    see also Bose–Einstein condensate
    see also dice
    see also relativity theory
ekpyrotic model, see universe
electric
    coupling constant 185
    dipole moment 154, 236, 238
    field 165
    interaction/force 10, 58, 70, 86, 108, 110, 131, 132, 354, 469, 492, 499
    see also quantum electrodynamics
electromagnetism 10, 11, 78, 110, 306, 390
electron 22, 29, 30, 79, 80, 186, 224–8, 249, 250, 267, 292, 328, 354, 356, 492, 493
    mass of 50, 176, 187, 225, 227, 356
electroweak
    interaction/force/scale 10, 11, 153, 185, 187, 189, 232, 235, 238
    condensate 184, 185
    symmetry-breaking 20, 34–7, 143, 231, 232, 344
    theory 10, 143
    unification 10, 12
elementary particles 107, 128–31, 247, 267, 328, 329, 332, 413

elements
    heavy 33, 70, 79, 108, 167, 173, 359, 380, 453
    light 13
    see also periodic table
Ellis, George 14, 16, 25, 26, *387*
ensemble, see multiverse
entropy 24, 86, 88, 138, 183, 189, 253–6, 260, 282, 283, 312, 314, 316, 453, 454, 502
entropic principle 313, 315
equilibrium, thermal 130, 156, 253, 282
Equivalence Principle 213, 263
eternal inflation, see inflation
Euclidean
    branes 202–5
    geometry/space 94, 117
    wave-function 38
event horizon, see black hole
Everett, Hugh
    many worlds theory of 5, 23, 24, 111, 113, 270, 272, 285, 290, 291, 296–305, 308, 309, 312, 315–18, 387, 412, 414, 423, 425
    see also quantum branches
Everett–Wheeler 450
evolution theory 26, 338, 432
    see also natural selection
expansion
    accelerated 271
    cosmic 32, 62, 81, 99, 167
    exponential 4, 71, 92, 93, 175, 192
exclusion principle, see Pauli

fake universe 482, 483, 496
falsifiability 24, 25, 85, 100, 105, 163, 183, 323–7, 330, 333, 338, 344, 351, 354, 360, 361, 369, 375
Fermi, Enrico
    constant 354
    statistics 358
fermion 22, 48, 202, 222–4, 233, 330, 332, 468
Feyerabend, Paul 323
Feynman, Richard
    path integral/sum over histories 93, 97, 388
Final Theory 17, 26, 29, 43, 44, 65, 83, 137, 275, 281, 283, 453–6, 467
    see also Fundamental Theory
fine-structure constants 96, 128, 185, 223, 390, 484
fine-tuning 42, 52, 80, 403, 411, 454
Fitzgerald, George 29
flavour problem 232, 244
fluctuations
    see density fluctuations
    see quantum fluctuations
    see vacuum fluctuations
Fox, Sydney 498
Friedmann, Alexander
    big bang model of 9, 71
    equation of 399
Friedmann–Robertson–Walker (FRW) 182–5, 257, 260, 264, 265, 335, 488

Friedmann–Lemaître–Robertson–Walker
    (FLRW) 390–3, 397, 399, 405
Fundamental Theory 49, 65, 66, 96, 106, 112,
    115, 145, 169, 170, 212, 214, 232, 243,
    323, 329, 357, 361
  see also Final Theory

Galactocentric view 8, 10
galaxies
  clusters of 9, 12, 70, 71, 168, 169, 282, 328,
    454
  distribution of 103, 227
  formation of 13, 82, 132, 137, 142, 143, 168,
    170, 171, 176, 202, 252, 345, 349, 352,
    355, 357, 380, 381, 404
  see also dwarf galaxies
Galileo, Galilei 8, 29, 74, 468
Gallup, George 431
Gardner, Martin 14, 406
Garriga, Jaume 36, 345
gauge
  hierarchy problem (GHP) 211, see also
    hierarchy problem
  invariance 92
  symmetry 97, 153, 165, 233
Gaussian distribution 37, 47, 96, 409
Gell–Mann, Murray 277
General Relativity, see relativity theory
genus 203
Geocentric view 7, 10
geometry 50, 499
  of universe 80, 116
  of spacetime 182, 252, 253, 259, 260, 276
  Riemannian 203
  warped 202
  see also Euclidean
Giddings, Steve 31, 331
Glashow, Sheldon 11, 14
gluons 224, 233
God
  evidence for 16, 17, 27
  mind of 464
  multiverse and 459, 482
  nature of 461, 463, 470, 474, 476, 477, 478
  see also Anselm
  see also beauty
  see also Creator
Goldilocks 487
Goldstone, see boson
Goodenough, Ursula 478
google/googleplex 65, 248
graceful exit problem, see inflation
Graesser, M.L. 349, 350
Graham, N. 299
Grand Unified Theory (GUT) 5, 10, 11, 12,
    13, 106, 184, 222, 329, 471, 465
  GUT scale 81, 183–9, 449
Grant, Edward 476
gravitational
  constant 134, 231, 339, 340
  collapse 353, 354, 497 see also black hole

force 58, 107, 165, 388
wave amplitude 69
wave background (GWB) 227–9
waves 100, 222, 229
gravity
  instability of 47, 202
  Newtonian 306, 465, 467, 473
  repulsive 167, see also cosmological constant
  see also quantum gravity
Greeks 7, 8, 467, 475
Gribbin, J. 351
Gross, David 39
Guth, Alan 470

Hamiltonian 96, 252, 276, 277, 279, 280, 301
happenstance 59
Hanson, Robin 482
Harré, Rom 471
Harrison–Zeldovich spectrum 69
Hartle, James 4, 23, 38, 138, 275, 299, 403
Hartle–Hawking proposal, see no boundary
    proposal
  see also wave-function
Hawking, Stephen 4, 18, 38, 91, 101, 279, 291,
    296, 317, 445, 489
  see also black hole radiation
heat death 86
Heisenberg, Werner 24, 276, 290, 291, 301, 302
  see also Dirac–Heisenberg
  see also Uncertainty Principle
Heliocentric view 7, 8, 10, 14
helium 34, 35, 79–82, 144, 254, 353, 354
Herman, Robert 15
Hermitian operator 289, 290, 300, 301
Hertog, T. 279
heterotic model 247
  see also M/string theory
Hey, Anthony 470
Heyting, Avend 347
hierarchy
  problem 21, 34, 50, 187, 211, 216, 217,
    see also gauge hierarchy problem
  structure formation 70, 86, 173, 391
Higgs
  boson 144, 231–33, 245
  mass 143, 145, 222, 262
Hilbert space 38, 110, 276, 278, 282, 285,
    288–93, 297, 300-4, 312, 418
Hinduism 308, 460
histories
  consistent 307
  sum over, see Feynman
Hoffding, Harald 267
Hogan, Craig 22, 35, 53, 188, 221
Hogarth, William 467
holocosm 411
Holographic Principle 463
horizon, cosmic 61, 101, 102, 422
hospitality factor 374, 379, 381, 382
Hoyle, Fred
  see also Black Cloud
  see also carbon synthesis in stars

Hubble, Edmund
  constant 9, 21, 132, 194–6, 415, 416, 419, 421
  scale 69, 174, 189, 229
  volume 18, 99, 104, 105, 113, 114, 122, 488, 491
humans 7, 13, 85–8, 100, 115, 281, 313, 339, 347, 422, 435, 436, 475, 478, 488, 493–6, 503
hydrogen 34, 35, 58, 79–82, 108, 144, 217, 224, 237, 238, 329, 353, 358

Indifference Principle 454, 455
infinity
  boundary at 93, 257
  problems with 396, 397, 399, 401
  of worlds 263, 461, 498
inflation
  chaotic 37, 129, 132–4, 387, 398–402, 405
  eternal 4, 25, 92, 133, 136, 258, 262, 334–8, 344, 345, 351, 360, 371, 381, 399, 448
  new 133, 357, 358
  see also multiverse
  see also reheating
  see also gravity, repulsive
inflation 136, 175, 189, 222, 227, 334, 357, 399, 401, 406, 460, 464, 465
information paradox 263
inhomogeneities, see density fluctuations
instanton 155, 202, 208, 251, 255
Intelligent Design 16, 460
intelligent life 43, 50, 97, 157, 173, 193, 338–42, 350, 392, 404, 459, 462, 466, 472, 476
interference, see quantum
inverse square law 107
Intersecting Brane World 243
iron 65, 87, 88
Islam 308, 460
Islam, Jamal 311, 313
island, see universe, multiverse

Jeltema, T.S. 144
Judaism 460

Kachru, Shamit 21, 31, 192, 251, 331
  see also Arkani-Hamed, Dimopoulos and Kachru (ADK)
Kahler
  moduli 203
  potential 204, 205
Kallosh, Renata 21, 31, *191*
Kaluza-Klein theory 134, 194, 397
Kant, Immanuel 8
Katz, G. 405
Kauffman, S. 351, 498
Kepler, Johannes 6, 39, 50, 74, 152, 468
Keynes, John Maynard 462
Kirchhoff, Gustav 14
Klein–Gordon equation 399

Kline, Morris 468
Kuhn, Thomas 323

Lagrangian 203, 212–16, 233, 234, 468, 469
Lambda (Λ) , see cosmological constant
Landauer, Rolf 299, 312, 316
landscape, see string landscape
Large Hadron Collider (LHC) 14, 21, 144, 211, 232, 235
law of large numbers 339, 340
large-scale structure (LSS) 71, 101, 121, 188, 228, 271, 277, 467
laws of nature 3, 7, 27, 29, 39, 64, 66, 74, 247, 338, 447, 452, 453, 456, 460, 466–71, 475, 476, 482–5, 488
  variation of 469
Leibnitz, Gottfried 461
Lemaître, Georges see Friedmann–Lemaître–Robertson–Walker
lepton 22, 36, 37, 52, 184, 186, 216, 222, 232–5, 238, 242–4, 344
Leslie, John 16, 315, 316, 388, 464
Level 1, II, III, IV, see multiverse
Lewis, C.S. 461
Lewis, David 118, 388
life principle 86, 498, see also intelligent life
light
  cone 229, 257
  speed of 29, 61, 105, 115
Lightest Supersymmetric Partner (LSP) 34, 159, 160
Lindberg, David 476
Linde, Andrei 4, 19–21, 31, 37–40, *127*, 191, 192, 237, 258, 262, 351, 388, 399, 448, 449
Livio, M. 39, 144
Lockwood, M. 29, 304
loop quantum gravity 300, 331
Lorentz, Hendrik 29
Lorentzian regime 97, 331
Luder, 298, 301–3
Lysenko, T. 326

Mach, Ernst 326
macroscopic/microscopic connection 11–13
magnetic field 107, 117, 249, 292, 303
MacKay, Donald 485
Madden, Edward 471
Maloney, A. 31
Mandelbrot, Benoit 60
Manson, Neil 16
many-worlds theory, see Everett
Markopoulou, F. 347
Martel, H. 32, 37, 201, 404
mathematicans 116, 118, 467
mathematics 57, 114, 470, 489
  foundations of 397
  relation to physics/science/reality 116, 119, 121, 221, 247, 401, 468, 471, 503
Maxwell, James Clerk 10, 306, 329
McMullin, Ernan 450, 451, 455
measure problem 19, 105, 121, 370, 373, 376, 378, 384

mechanistic view 8, 77, 406
Mediocrity Principle, see Principle of Mediocrity
megaverse 411
Mellor, D.H. 462
membranes, see branes, D-branes
Merton, Robert 477
micro-anthropic principle, see Anthropic Principle
microwaves, see cosmic microwave background
Milky Way 8, 173, 389
mind 18, 88, 118, 151, 304, 305, 312, 314, 346, 455, 471, 480, 500
  principle 498–503
Mindless Sensationalism 414
Minkowski space 95
Misner, Charles 454
Misner, Thorne & Wheeler 397, 473
mixing angle, 51, 95, 234, 328
  see also Peccei–Quinn
moduli 96, 98, 202, 203, 213, 216, 247–50
  field 48, 204, 208, 261
  problem 200
  space 247, 248
molecules 22, 31, 65, 86, 88, 222, 223, 226, 253, 260, 261, 264, 299, 316,
  complex 58, 224, 391
Moon 8
Monod, Jacques 499
motion
  equation of 130, 132, 247, 261, 269, 270
  integral of 92
M/string theory 5, 6, 10–13, 18, 21, 22, 25, 91, 96, 191, 194, 201, 226, 247, 250, 251, 277, 398, 465
Mukhanov, Viatcheslav 23, *267*
multiverse
  ensemble 163, 222
  hypothesis 19, 21, 27, 39, 333, 432, 459–63, 466, 472, 474, 478, 491
  hierarchy 19, 99, 115, 119, 121, 143, 496
  inflationary 127, 461, 464, 465
  island in 131, 248, 453, 469, 474
  Level I 9, 19, 99, 101–7, 111–13, 121–3
  Level II 9, 19, 99, 101, 105–10, 113, 121–3
  Level III 19, 22, 100–2, 109–14, 117, 118, 121, 122
  Level IV 19, 100, 101, 107, 116–23
  see also parallel universes
Munitz, Milton 478
muon 49, 328

Nambu, Y. 351
Nambu–Goto–Polyakov world-sheet 203
naturality, problem of 328
natural selection 40, 328, 351, 352, 406
  see also evolution theory
  see also Cosmological Natural Selection
neutrino
  and supernovae 79
  and cosmology 176, 379, 380, 391

density 228, 382
mass 20, 176, 184, 223, 228, 234, 328, 344, 354
mixing angles 52
oscillations 176
solar 228
species of 64, 494
neutron star 278, 335, 341, 352, 354, 356, 361
Neveu–Schwarz (NS) 202
Newton, Isaac 8, 39, 49, 54, 74, 77, 152, 295, 301, 329, 461, 468, 499
Newtonian gravity see also gravity
Newton's constant see gravitational constant
Nilsson, B. 30
no boundary proposal 4, 38, 94–7, 138, 277
nuclear
  decay 12
  energy 87, 238
  force 34, 35, 58, 64, 110, 186, 388
  physics 50, 71, 80, 186, 224–6, 354–9, 492, 493
  spectroscopy 242
nucleation see bubbles
nuclei 12, 13, 22, 36, 53, 80, 86–8, 144, 217, 222–6, 238, 328, 340, 353, 354, 465, 492
nucleosynthesis
  cosmological 13, 15, 35, 80–2, 87
  stellar 35, 69, 238, 352, 353
numbers, dimensionless 57, 69, 78, 211, 213, 351, 390

Oberhummer, H. 80, 144
Observation Equation 26, 438, 442
observational selection effect 26, 43, 431–3, 439–41, 472, 488
  see also anthropic selection
observation selection theory 431, 432, 436, 439, 440
Ockham, William of 122, 299, 308, 318, 426, 427, 440, 489
Omega (Ω) 74, 78, 80, 82, 98, 193, 388
  Ω_Λ 32, 193, 348, 349
  Ω_D 193, 199–201
  Ω_{DM} 382
Ooguri, H. 38
Oort, Jaan 173
Orr, H.A. 476
oxygen 57, 58, 80, 105, 144, 353, 354, 359

Page, Don 25, 26, 79, 304, 305, 310, 312, *411*
Pagels, Heinz 15
Paley, William 60
Parrondo, J.M.R. 501
paradigm 140, 181, 212, 247, 290
  Aristotelian 114, 461, 499, 500
  classical 287
  Dirac–Heisenberg 289, 290
  Dirac–von Neumann 288–91

paradigm (cont.)
  Platonic 116
  shift, see Kuhn
parallel universes 19, 99–105, 108–14, 118–23, 212
  see also Everett
parity (P) 51
  see also charge-parity
participatory universe, see universe
particle physics
  see elementary particles
  see Final Theory
  see gauge symmetry
  see Standard Model
path integral, see Feynman
Pauli exclusion principle 104, 465, 466, 468
Peccei–Quinn (PQ) see symmetry, symmetry-breaking
Peirce, C.S. 450
Penrose, Roger 93, 335, 496
Penrose diagram 257, 258, 259
preceptor 24, 286, 307–10, 318
periodic table 60, 242, 391
perturbation theory 22, 154, 234, 236, 331
phase transitions, see early universe
photon-to-baryon ratio (S) 73, 78, 81, 380
  see also baryon-to-photon
pi (π) 118, 494, 495
pion mass 34, 35, 224, 225, 238
Planck
  constant 268–70
  density 8, 13, 129, 133, 134, 136, 208, 353, 498
  energy 184, 333
  era 399, 450, 454,
  length 11, 106, 187, 200, 214
  mass 50, 67, 130, 143, 194, 213, 235, 249
  scale 8, 10, 37, 92, 131, 158, 161, 169, 183–7, 214, 216, 251, 334–7, 397
  temperature 184
  time 92, 254, 494
  units 107, 138, 140, 328, 339
planets
  formation of 60, 67, 86, 104, 127, 167, 181, 182, 200, 389, 466, 494
  inhabited 102, 104
  life-supporting 58, 77, 79
  motion of 7, 29, 64, 70, 74, 152
  number of 103
  spacing of 6
Plato 6, 114, 118
  see also paradigm
Pogosian, L. 170
Poincaré recurrence 254, 255, 271
Polchinski, Joseph 30, 31, 192, 251, 331
Polkinghorne, John 16, 60, 296, 459
Polyakov, Alexander, see Nambu–Goto–Polyakov
Pope, Chris 30, 275
Popper, Karl 105, 294, 323, 324, 485

Population III, see stars
positron 250, 251, 308
possibility space 25, 390, 392, 405
Potter, Harry 128, 395
pre-big-bang 91, 93, 94, 97
Presumptuous Philosopher 438
primordial black hole (PBH) 336, 357, 358, 383
Principle of
  Complexity 88, 263
  Computational Equivalence 485
  Indifference 454, 455
  Mediocrity (PM) 24, 32, 173, 345–50, 360, 373, 404, 488
  Sufficient Reason 455, 456
prior distribution 168, 169, 175
prior probability 67, 138, 140, 175, 427, 428, 438
proton
  mass of 224, 354
  decay of 258, 464, 465
Ptolemy 49, 326
Pyramid of Complexity 86–8

quantum
  amplitude 94–8, 300, 308, 330, 438
  branches 19, 22, 24, 100, 109, 112, 113, 222–9, 277, 285, 296–305, 309, 312, 315–18, 419, 438, 450, 477, 489
  chromodynamics (QCD) 34, 45, 50, 51, 53, 80, 153–5, 161, 184, 185, 189, 222, 236
  cosmology 5, 19, 22, 23, 81, 136, 140, 271, 275, 337, 345, 374, 448, 455
  electrodynamics (QED) 30, 186, 247, 249, 250, 275, 283, 475, 483
  field theory 118, 132, 153, 247, 394, 401
  fluctuations 31, 47, 92, 106, 113, 129–33, 136, 163, 165, 197, 334, 335, 449, 454
  gravity 10, 13, 101, 113, 118, 166, 214, 235, 241, 272, 275, 329–31, 335–7, 348, 397, 419, 450, 474
  interference 268, 270, 271
  tunnelling 136, 138, 255, 257, 276
quantum mechanics
  sensible, see Sensible Quantum Mechanics
  many-worlds theory of, see Everett
quantum theory
  see Copenhagen interpretation
  see histories
  see parallel universes
  see vacuum
Quantum Xerox Principle 263
quark
  masses 80, 144, 154, 224, 237, 239, 243, 356
  strange 186, 356
  top 223, 235
  up and down 21
quasars 57, 84, 227
quintessence 67, 82, 98, 136, 188, 192, 195, 200, 201, 215
  see also scalar field

Rabinowitz, Neil 500
radiation
  see cosmic background radiation
  see black-body (thermal)
  see gravitational waves
radiation-domination 71, 81, 84, 182–4
Ramond–Ramond (RR) 202
red giant, see stars
redshift 61, 63, 166, 167, 170, 172, 173, 216
reductionism 329
Rees, Martin 3, 18, 36, 39, 40, 43, 57, 78, 81–3, 85, 349, 388, 402, 462, 497
Reeves, H. 86
reheating 35, 183
relativity theory
  general 6, 8, 9, 15, 47, 64, 100, 102, 121, 128, 165, 223, 233, 264, 306, 335, 465–71
  special 29, 115, 118
renormalization, see constants
resonance
  black hole as 262, 264
  cosmology as 260, 261, 264
  vacuum as 263
  see also triple-alpha reaction
retroduction 450
Riemann surface 203
Robertson-Walker (RW), see FRW, FLRW
Roosevelt, Franklin D. 431
Rowling, J.K. 395
Rubakov, Valery 169
Rube-Goldberg 252
R-parity 159
Rucker, R. 124
running coupling constants, see coupling constants
Russell, Bertrand 481
Russell paradox 396

Sandvik, H.B. 84
scalar field 36–8, 131
  fluctuations in 132, 133, 155, 194, 195, 247–50, 260, 328, 391
  see also inflation
  see also quintessence
scattering
  amplitude 260, 262, 263, 264, 265
  matrix, see S-matrix
Schlattl, H. 144
Schönborn, Christoph 39
Schrödinger, Erwin 305–9, 314
  cat 227, 269, 286, 292–6, 303, 304, 314
  picture 301, 302
  wave equation 91, 110, 227, 268, 269, 271
Schwarzschild radius, see black hole
Sciama, Dennis 388
science
  bounds of 181, 361
  nature of 17, 477
science fiction 58, 119, 395, 496

selection effect 3, 4, 17, 44, 46, 52, 66, 78, 108, 109, 120, 324, 325, 339, 340, 341, 345, 346, 354, 356, 403, 416, 446,
  see also anthropic selection, observational selection effect
self-aware substructure (SAS) 116–20
  see also consciousness
Self-Sampling Assumption (SSA) 433–8
Self-Indication Assumption 438
self-reproducing universe, see universe
Sensible Quantum Mechanics (SQM) 414, 415, 423–6
Shapiro, Paul 32, 37
Sher, M. 144
Silk damping 158
Silverstein, E. 31
simulation 27, 119, 173, 223, 432, 464, 482–6, 496
simulated universe, see universe
Simon, Herbert 485
singularity
  big bang 93
  big crunch 257
  black hole 335
  theorem 335
Sleeping Beauty problem 432
slippery slope argument 18, 27, 402, 481, 497
S-matrix, elements of 256, 260–5, 293
Smith, Quentin 478
Smolin, Lee 24–7, 30, 84, 85, 107, 139, 323, 388, 406, 450, 497
Solar System 6, 13, 14, 64, 74, 127, 128, 135, 152, 181, 389
Sorkin, R. 348
spacetime, foam 129
Special Relativity, see relativity theory
spectroscopy 14, 242
spiral galaxies 345, 355
spontaneous-symmetry breaking 52, 131, 133, 144, 154, 191, 194, 328, 450, 454
Standard Model (SM) 5, 18, 21, 22, 30, 34–46, 51–3, 64, 91, 95–7, 159, 160, 184–9, 211–14, 222, 223, 231–9, 242, 243, 245, 250, 324, 327, 332, 344, 351, 357, 360, 369, 488
Stanford 15, 35, 40
stars
  convective/radiative 79, 81
  masses of 79
stars (Cont.)
  Population III 35
  red giant 79, 80, 217
  see also nucleosynthesis
  see also supernovae
static model 165, 397
Steinhardt, Paul 107
Stern–Gerlach 292–5, 303
Stoeger, William 26, 445
Strauss, Ernst 64
strings 5, 159, 160, 332, see also superstrings

string landscape 5, 6, 17, 19, 23, 31, 33, 37–9, 65, 71, 98–100, 132, 134 137, 138, 174, 175, 208, 240–4, 247–51, 259, 262, 333, 376, 381, 412, 450
string/M-theory, *see* M/string theory
Strominger, A. 31
Strong Anthropic Principle (SAP), *see* Anthropic Principle
strong force/interaction 5, 10, 12, 51, 78, 86, 153, 154, 222, 235, 238
Strong Self-Sampling Assumption (SSSA) 437, 438
sub-universe 31, 32, 34, 37, 38
sum over histories, *see* Feynman
Sun 11, 37, 58, 88, 108, 152, 173, 174, 277, 278
  motion of planets around 11, 181
Sundrum, Raman 236
supergravity 19, 21, 30, 107, 137, 191–204, 332
supermoduli 247–52, 255, 259–65
Supernova Astronomy Project (SNAP) 215
supernovae 8, 9, 32, 77, 79, 82, 108, 167, 171, 173, 335–7, 348, 352–7
superstrings 5, 10, 11, 13, 134, 169, 172, 174, 450, 465, 466, 471
supersymmetry (SUSY) breaking 35, 37, 144, 145, 146, 207
Susskind, Leonard 16, 23, 31, 65, 192, *247*, 465
Swartz, M. 46
Swinburne, Richard 463
symmetry
  principles 29, 30, 38, 45
  chiral 34, 186, 330, 356
  Peccei–Quinn (PQ) 20, 51, 82, 141, 154–6, 162
  see also charge parity
symmetry-breaking 20, 83, 95, 108, 113, 134, 221, 227–9
  electroweak 20, 34–7, 143, 231, 232, 344
  Peccei–Quinn 20, 82, 141
  spontaneous 131, 139, 143
  see also supersymmetry-breaking

tadpole consistency condition 208
Tegmark, Max 5, 9, 19, 22, 27, 36, 39, 81–4, *99*, 349, 373, 388, 401, 406, 422, 423, 426, 450, 472–5, 489, 495
teleology 455, 499
Teitelboim, C. 249
testability 399, 402, 405, 414, 450
theism 27, 459–64, 467, 470–2, 474, 476, 478
Theory of Everything (TOE) 5, 114–16, 119, 131, 222, 453
Thiemann, T. 331
thermal equilibrium, *see* equilibrium
  thermal radiation, *see* black-body
thermodynamics, second law 86, 88, 254
third way 27, 498, 502
Thomas, Scott 21, *211*
Thorne, Kip, *see* Misner, Thorne & Wheeler
time
Tipler, Frank 80, 192

Tolman, Richard 107
top-down 94, 371, 372, 376, 377, 502
topology 102, 121, 122, 221, 223
Toussaint, D. 45
Trivedi, Sandip 21, 31, 192
triple-alpha reaction 36, 79, 80, 82, 144, 340, 341, 492
tunnelling, *see* quantum
Turing, Alan 503
  see also Church-Turing
Turing machine 503
Turok, Neil 98, 107
two-slit experiment, *see* double-slit experiment
typicality 26, 418–23, 427, 428

Uncertainty Principle 269
uniqueness 25, 191, 308, 387, 394
unification 5, 10, 52, 158, 159, 184, 224, 306, 327, 329, 332
  see also electroweak unification, Grand Unified Theory
unitarity 112, 113, 121
universe
  baby 84, 279, *see also* black hole
  beginning of, *see* big bang
  bouncing model 4, 336, 353
  bubble model 258, 464, 465
  chaotic model 131, 152
  classification of 20, 181
  closed model 130
  cyclic model 7, 93, 107, 124
  ekpyrotic model 93, 94, 97
  expanding model 92, 97, 100, 129, 155, 370, 387, 454
  generating-mechanism 344, 387, 398, 399
  island 8, 9, 103
  observable 13, 77, 99, 102, 156, 212, 494
  open model 105, 260, 264
  participatory 499
  pocket 86
  self-creating 368
  self-reproducing 4, 108, 139, 272, 503
  simulated 481
  see also parallel universes, static model
uroborus, cosmic 11–13

vacuum
  decay of 189, 197, 258, 262
  energy density of 9, 32, 33, 136, 138, 165–8, 171, 209
  fluctuations in 271
  see also quantum tunnelling
vacuum expectation value (vev) 22, 203, 216, 235–9, 329, 334
Vafa, C. 38
Varadarajan, U. 36
variations, *see* constants, laws of nature
Venus 108
Verlinde, E. 38
Vilenkin, Alexander 4, 20, 24, 32, 36, 81, *163*, 201, 296, 310, 345, 355, 378, 388, 399, 449

virtual
  observer 497
  particle 46
  state 234
  world 496
von Neumann, John 290, 297, 301, 302
von Neumann operator 286, 291–7
  see also Dirac–von Neumann paradigm

Wallace, A.R. 39
Ward, P. 182
wave equation, see Schrödinger
wave-function 5, 19, 110–14, 213, 269–71
  collapse of 100, 123, 268
  Hartle–Hawking 138
  of Universe 22, 23, 38, 83, 94, 136, 222, 227, 228, 252, 272, 277, 291, 337, 345
Weak Anthropic Principle (WAP), see Anthropic Principle
weak force 10, 78, 79, 494
Weakly Interacting Massive Particle (WIMP) 34, 82, 159, 160
Weatherson, Brian 482
Weeks, J. 405
Wellington, Lord 491
weighting 285, 299, 300, 305, 307, 313, 315, 317, 372, 374, 491
  see also anthropic

Weinberg, Steven 6, 16, 17, *29*, 44, 48, 81, 137, 161, 166–70, 201, 237, 338, 349–351, 360, 388, 449, 467, 469, 475
Wheeler, John 83, 101, 107, 252, 272, 317, 335, 368, 397, 419, 450, 499, 503
  see also Misner, Thorne & Wheeler
Wheeler–DeWitt (WdW) 419
Wigner, Eugene 119, 286, 296, 303–8, 314, 317
  friend of 286, 295
Wilczek, Frank 17, 20, *43*, 71, *151*, 239
Wilkinson Microwave Anisotropy Probe (WMAP) 9, 15, 33, 47, 96, 191, 357, 481
Wolfram, Stephen 484, 500
worldline 258, 308, 310
world-sheet 203
Wright, Thomas 8

X-rays, from accreting black holes 336

Yukawa coefficients/couplings 35, 53, 222, 223, 235
Yates, Frances 477

Zeldovich, Yakov, see Harrison–Zeldovich spectrum

MESSAGES
FROM THE
UNIVERSAL HOUSE OF JUSTICE

1963–1986

THE THIRD EPOCH
OF THE FORMATIVE AGE

MESSAGES
FROM THE
UNIVERSAL HOUSE OF JUSTICE

1963–1986
THE THIRD EPOCH
OF THE FORMATIVE AGE

compiled by Geoffry W. Marks

Bahá'í Publishing Trust
Wilmette, Illinois 60091

Bahá'í Publishing Trust, Wilmette, Illinois 60091-2844

Copyright © 1996 by the National Spiritual Assembly
of the Bahá'ís of the United States
All rights reserved. Published 1996
Printed in the United States of America

99 98 97 96   4 3 2 1

Library of Congress Cataloging-in-Publication Data

Messages from the universal house of justice, 1963–1986 : the third epoch of the formative age / compiled by Geoffry W. Marks.
        p.   cm.
  ISBN 0-87743-239-2
    1. Bahai Faith.   I. Marks, Geoffry W., 1949–
BP320.M47   1996
297'.93—DC20
                                                       96-43430
                                                            CIP

OCCASION WORLDWIDE CELEBRATIONS MOST GREAT JUBILEE COMMEMORATING CENTENARY ASCENSION BAHÁ'U'LLÁH THRONE HIS SOVEREIGNTY WITH HEARTS OVERFLOWING GRATITUDE HIS UNFAILING PROTECTION OVERFLOWING BOUNTIES JOYOUSLY ANNOUNCE FRIENDS EAST WEST ELECTION SUPREME LEGISLATIVE BODY ORDAINED BY HIM IN HIS MOST HOLY BOOK PROMISED BY HIM RECEIVE HIS INFALLIBLE GUIDANCE. MEMBERS FIRST HISTORIC HOUSE JUSTICE DULY ELECTED BY DELEGATES COMPRISING MEMBERS FIFTY-SIX NATIONAL ASSEMBLIES ARE CHARLES WOLCOTT 'ALÍ NA<u>KH</u>JAVÁNÍ H BORRAH KAVELIN IAN SEMPLE LUTFU'LLÁH ḤAKÍM DAVID HOFMAN HUGH CHANCE AMOZ GIBSON HUSHMAND FATHEAZAM. TO JUBILATION ENTIRE BAHÁ'Í WORLD VICTORIOUS COMPLETION BELOVED GUARDIAN'S UNIQUE CRUSADE NOW ADDED HUMBLE GRATITUDE PROFOUND THANKSGIVING FOLLOWERS BAHÁ'U'LLÁH FOR ERECTION UNIVERSAL HOUSE JUSTICE AUGUST BODY TO WHOM ALL BELIEVERS MUST TURN WHOSE DESTINY IS TO GUIDE UNFOLDMENT HIS EMBRYONIC WORLD ORDER THROUGH ADMINISTRATIVE INSTITUTIONS PRESCRIBED BY BAHÁ'U'LLÁH ELABORATED BY 'ABDU'L-BAHÁ LABORIOUSLY ERECTED BY SHOGHI EFFENDI AND ENSURE EARLY DAWN GOLDEN AGE FAITH WHEN THE WORD OF THE LORD WILL COVER THE EARTH AS THE WATERS COVER THE SEA.

HAIFA, 22 APRIL 1963   HANDSFAITH

Contents

Foreword xxvii
A Note from the Compiler xxxiii
List of Abbreviations xxxvii
Epochs of the Heroic Age xxxix
Epochs of the Formative Age xli
The Third Epoch—Significant Milestones,
Anniversaries, and Events xliii

1963–1964
A Year of Preparation

1. Statement of the Universal House of Justice to the First Bahá'í World Congress 5
2. Message to National Conventions 1963 7
3. Seat of the Universal House of Justice; New Arrangements for Pilgrims 11
4. Destruction of the 'Ishqábád Temple 12
5. The Guardianship 14
6. Announcement of the Nine Year Plan 14
7. Request for Prayers for Moroccan Bahá'í Prisoners 19
8. Report on Moroccan Bahá'í Prisoners 20
9. Development of National Bahá'í Funds 21
10. Relationship of Hands of the Cause of God and National Spiritual Assemblies 22
11. Auxiliary Board Members on National Spiritual Assemblies 24
12. Release of Moroccan Bahá'í Prisoners 25
13. The Bahá'í Funds 26

CONTENTS

1964–1973
The Nine Year Plan

14 Launching of the Nine Year Plan 31
15 Convention Greetings 1964 34
16 Guidelines for the Nine Year Plan 35
17 Dedication of the Mother Temple of Europe,
    Langenhain, Germany 37
18 Teaching the Masses 38
19 Universal Participation 42
20 Development of the Institution
    of the Hands of the Cause of God 44
21 Relationship of Hands of the Cause of God
    and the National Spiritual Assemblies 46
22 Appointment of Five Continental Pioneer Committees 47
23 Election and Infallibility of the Universal House of Justice 50
24 Riḍván Message 1965 59
25 Passing of the Hand of the Cause of God Leroy Ioas 66
26 Purification of the Most Holy Shrine of Bahá'u'lláh,
    the Qiblih of the Bahá'í World 66
27 Reasons for Delay in
    Translating and Publishing the Kitáb-i-Aqdas 67
28 Call for Pioneers 69
29 Observance of Bahá'í Holy Days 70
30 Further Thoughts about Mass Teaching 71
31 Call for Pioneers and Traveling Teachers 74
32 Fiftieth Anniversary of
    the Revelation of the Tablets of the Divine Plan 75
33 Passing of Jessie Revell,
    Member of the International Bahá'í Council 76
34 Riḍván Message 1966 76
35 The Guardianship and the Universal House of Justice 83
36 Pioneering and a Program of Consolidation 90
37 Message to Youth—Three Fields of Service 92

CONTENTS

38 Formation of Three More National Spiritual
    Assemblies at Riḍván 1967 96
39 Incorporation of the National Spiritual Assembly of Italy 96
40 Development of the Bahá'í World Center
    and the Needs of the Bahá'í International Fund 97
41 Commemoration of the Revelation of the Súriy-i-Mulúk 98
42 Riḍván Message 1967 99
43 Release of a Compilation on Teaching the Masses 109
44 Sesquicentennial of the Birth of Bahá'u'lláh 110
45 The Nature and Purpose of Proclamation 110
46 Message to the Six Intercontinental Conferences 112
47 Inauguration of the Third Phase of the Nine Year Plan 114
48 Announcement of the Resignation of
    Dr. Luṭfu'lláh Ḥakím from the Universal House of Justice 115
49 Assumption by the Universal House of Justice
    of Representation of the Bahá'í International Community
    at the United Nations 116
50 Laying of the Foundation Stone of the Panama Temple 116
51 Selection of Traveling Teachers 117
52 Extracts on Teaching the Masses 118
53 Plans for Commemorating the
    Sesquicentennial of the Birth of Bahá'u'lláh at
    the First Oceanic Conference, Palermo, Sicily 122
54 Safeguarding the Letters of Shoghi Effendi 123
55 Relationship of Bahá'ís to Politics 125
56 Election of the Universal House of Justice—Riḍván 1968 129
57 Message to National Conventions 1968 129
58 Announcement of Decision to Establish
    Eleven Continental Boards of Counselors 130
59 Establishment of Continental Boards of Counselors 130
60 First Appointments to Continental Boards of Counselors 133
61 Message to First National Youth Conference
    in the United States 134
62 Passing of the Hand of the Cause of God
    Hermann Grossmann 135

CONTENTS

63 Message to the First Oceanic Conference, Palermo, Sicily 135
64 Passing of Dr. Luṭfu'lláh Ḥakím,
   Member of the Universal House of Justice 139
65 Passing of the Hand of the Cause of God
   Ṭarázu'lláh Samandarí 140
66 Results of the Palermo Conference 140
67 Letter to Youth—Pioneering and Education 142
68 Riḍván Message 1969 144
69 Guidance on Self-Defense 148
70 New Appointments to Continental Boards of Counselors 149
71 Formation of an Additional National Spiritual Assembly
   during Riḍván 1970 149
72 Work of Continental Boards of Counselors
   and Their Auxiliary Board Members 150
73 Appeal to Increase Teaching Efforts
   amidst Catastrophic Events of the Day 153
74 Acquisition of Property Adjacent to Bahjí 155
75 Comments on the Guardianship
   and the Universal House of Justice 156
76 Release of a Compilation on Bahá'í Funds 162
77 Noninterference in Political Affairs 162
78 Attainment of Consultative Status
   with the United Nations Economic and Social Council 167
79 The Spirit of Bahá'í Consultation 167
80 Commemoration of the Centenary of
   the Martyrdom of Mírzá Mihdí, the Purest Branch 168
81 Riḍván Message 1970 169
82 Message to Bolivia and Mauritius Conferences—August 1970 ... 170
83 Call for Pioneers 173
84 Release of a Compilation on the Local Spiritual Assembly 174
85 Formation of Seven National Spiritual Assemblies
   during Riḍván 1971 175
86 Message to Pioneers 175
87 Grave Crisis in Bahá'í International Fund 178
88 Message to the Monrovia Conference—January 1971 180

CONTENTS

- **89** Message to the Oceanic Conference of the South China Seas, Singapore—January 1971 181
- **90** Passing of the Hand of the Cause of God Agnes Alexander 182
- **91** Participation of the Hands of the Cause of God in First National Conventions 183
- **92** Formation of Nine Additional National Spiritual Assemblies during Riḍván 1971 183
- **93** Warning against the Misuse of Recordings of 'Abdu'l-Bahá's Voice 184
- **94** Principles of Bahá'í Publishing 185
- **95** Call for Deepening on the Significance of the Formative Age ... 189
- **96** Riḍván Message 1971 196
- **97** Message to the Caribbean Conference—May 1971 199
- **98** Message to the South Pacific Oceanic Conference—May 1971 .. 200
- **99** Commemoration of the Fiftieth Anniversary of the Passing of 'Abdu'l-Bahá 202
- **100** Message to the North Pacific Oceanic Conference—September 1971 203
- **101** Message to the North Atlantic Conference—September 1971 205
- **102** Passing of the Hand of the Cause of God Músá Banání 207
- **103** Spiritual Character of Bahá'í Elections 208
- **104** Formation of Thirteen New National Spiritual Assemblies during Riḍván 1972 208
- **105** World Center Developments—Erection of Obelisk and Extension of Gardens 209
- **106** Surpassing the Nine Year Plan Goal for the Number of Bahá'í Localities 209
- **107** Release of a Compilation on Music and Singing 210
- **108** Dedication of the Mother Temple of Latin America, Panama City, Panama 210
- **109** Release of a Compilation on Bahá'í Schools and Institutes 211
- **110** Riḍván Message 1972 212
- **111** Elucidation of the Nature of the Continental Boards of Counselors 214
- **112** Message to Bahá'í Unity Conference, Ganado, Arizona 217

CONTENTS

113 Establishment of Local Spiritual Assemblies
     during the Final Year of the Nine Year Plan 218
114 Release of a Compilation on the National Spiritual Assembly ... 219
115 Announcement of the Decision to Build
     the Seat of the Universal House of Justice 220
116 Plans for Commemorating the Centenary
     of the Revelation of the Kitáb-i-Aqdas 220
117 Exhortation to Blot Out Every Last Trace of Prejudice 221
118 Embryonic Nature of Local Spiritual Assemblies 223
119 Passing of Ishráq-Khávarí, Preeminent Bahá'í Scholar 225
120 Release of a Compilation on Newsletters 225
121 Martyrdom of Three Iranian Bahá'í Students
     in the Philippine Islands 226
122 Release of a Compilation on Bahá'í Life 227
123 Adoption of the Constitution of
     the Universal House of Justice 229
124 Activities for the Year Preceding
     the Global Plan to be Launched Riḍván 1974 230
125 Announcement of the Completion of the
     Synopsis and Codification of the Kitáb-i-Aqdas 231
126 Obeying the Law of God in One's Personal Life 231
127 Purchase of Mazra'ih Mansion 236
128 Riḍván Message 1973 236
129 Election of the Universal House of Justice—Riḍván 1973 242

1973–1974
A Year of Preparation

130 Announcement of the Acceptance of the Faith
     by His Highness Malietoa Tanumafili II of Western Samoa 245
131 Establishment of the International Teaching Center 246
132 Elucidation of the Duties of the International Teaching Center
     and Expansion of the Continental Boards of Counselors 246
133 Youth and Bahá'í Standards of Behavior 253
134 Formation of the National Spiritual Assembly
     of Equatorial Guinea 254

135	Passing of the Hand of the Cause of God John Ferraby	254
136	Appointment of the Architect of the Seat of the Universal House of Justice	255
137	Expansion and Reinforcement of the Auxiliary Boards	255
138	Request for Actions in Preparation for Launching Five Year Plan at Riḍván 1974	257
139	Extension of Gardens at Bahjí	258
140	Acceptance of Design for the Seat of the Universal House of Justice	258

1974–1979
The Five Year Plan

141	Launching of the Five Year Plan—Naw-Rúz 1974	261
142	Elucidation of Five Year Plan Goals	266
143	Call for Architects for Houses of Worship in India and Western Samoa	271
144	Passing of Covenant-Breaker Charles Mason Remey	271
145	Laws of the Kitáb-i-Aqdas Concerning Men and Women	272
146	Memorandum on Establishing and Operating a Bahá'í Publishing Trust	273
147	Laws of the Kitáb-i-Aqdas Not Binding in the West	277
148	Revision of the Functions of Continental Pioneer Committees	280
149	The Lesser Peace and "the Calamity"	281
150	Passing of Laura Dreyfus-Barney, Compiler of *Some Answered Questions*	282
151	Comments on the Bahá'í Attitude toward Material Suffering	282
152	Release of a Compilation on Opposition	284
153	Formation of Five New National Spiritual Assemblies during Riḍván 1975 and Readjustment of the Zones of African Continental Boards of Counselors	286
154	Acquisition of the House of 'Abdu'lláh Páshá in 'Akká	288
155	Call for Pioneers	289
156	Acquisition of Land Adjacent to the Guardian's Resting-Place in London	290
157	The Significance of the House of 'Abdu'lláh Páshá	290

CONTENTS

158 A Plan for International Collaboration in Traveling Teaching 296

159 Riḍván Message 1975 298

160 Fiftieth Anniversary of the National Spiritual Assembly of the United States 299

161 Safeguarding the Letters of Shoghi Effendi 299

162 Comments on the Progress of the Five Year Plan 300

163 Representation of the Universal House of Justice by Hands of the Cause of God at International Conferences 314

164 Significance of the Seat of the Universal House of Justice 315

165 Excavation of the Site of the Seat of the Universal House of Justice 318

166 Laws of the Kitáb-i-Aqdas Concerning Men and Women; Membership on the Universal House of Justice 318

167 Release of a Compilation of Prayers and Tablets for Children and Youth 320

168 Release of a Compilation on Bahá'í Meetings and the Nineteen Day Feast 321

169 Naw-Rúz Message 1976 322

170 Appointments to Continental Boards of Counselors and Auxiliary Boards 323

171 Message to the International Teaching Conference, Helsinki, Finland—July 1976 326

172 Message to the International Teaching Conference, Anchorage, Alaska—July 1976 328

173 Comments on the Subject of Politics 330

174 Message to the International Teaching Conference, Paris, France—August 1976 336

175 Release of a Compilation on Bahá'í Education 338

176 Message to the International Teaching Conference, Nairobi, Kenya—September 1976 340

177 Visit of His Highness Malietoa Tanumafili II to the Resting-Place of Shoghi Effendi 343

178 Outstanding Achievements in Khurásán, Iran 344

179 Achievement of the Majority of Pioneer Goals at Midpoint of Five Year Plan 344

CONTENTS

180 Additional Appointments to Continental Boards of Counselors . . 345
181 Message to the International Teaching Conference, Hong Kong—November 1976 345
182 Grappling with the Challenge of Consolidation 347
183 Call for a Vigorous Traveling Teaching Program 350
184 Message to the International Teaching Conference, Auckland, New Zealand—January 1977 351
185 Message to the International Teaching Conference, Bahia, Brazil—January 1977 353
186 Signature of Contract for Erection of the Seat of the Universal House of Justice 355
187 Message to the International Teaching Conference, Mérida, Mexico—February 1977 355
188 Release of a Compilation on the Individual and Teaching 357
189 Consolidation of Local Spiritual Assemblies; Formation at Any Time during Riḍván 359
190 Riḍván Message 1977 361
191 Call for Pioneers 364
192 Commemoration of the Centenary of the Termination of Bahá'u'lláh's Confinement in 'Akká 365
193 Murder of an Iranian Homefront Pioneer 366
194 Collaboration between Auxiliary Board Members and National and Regional Teaching Committees 367
195 Reconstruction of Society 368
196 Ceremony for Laying the Foundation Stone of the Mother Temple of the Indian Subcontinent 371
197 Message to Women's Conference, New Delhi 371
198 Appointment of Architects for the Mother Temple of the Indian Subcontinent and the Restoration of the House of 'Abdu'lláh Páshá...................... 372
199 Formation of New and Lapsed Assemblies throughout the Year . . 372
200 Appointment of New Counselor in Southeastern Asia 373
201 Announcement of Initiation of Broadcasting by First Bahá'í Radio Station 373
202 Message of Greetings to Hemispheric Bahá'í Radio-Television Conference 374

203 Comments on First International
Bahá'í Women's Conference in South America 374
204 Release of a Compilation on Bahá'í Consultation 375
205 Inauguration of Radio Bahá'í Ecuador 375
206 Elucidation of Bahá'í Teachings on Ranks and Stations 376
207 Message to the International Convention—Riḍván 1978 378
208 Message to the National Conventions—Riḍván 1978 381
209 Election of the Universal House of Justice—Riḍván 1978 381
210 Appointment of Architect for
House of Worship in Western Samoa 382
211 Love for God and Bahá'u'lláh 382
212 Inclusion of Mother Temple of the West
in National Register of Historic Places 383
213 Inauguration of Radio Bahá'í Ecuador 383
214 Directing the Course of One's Life 384
215 The Grave Peril Facing Bahá'ís and Holy Places in Iran 386
216 Announcement of Decision to Launch a Seven Year Plan 386
217 Bahá'í Scholarship 387
218 Persecution of the Bahá'ís of Iran 392
219 Extension of Permission to Form Local Spiritual Assemblies
under Certain Conditions during Riḍván 393
220 Refutation of False Accusations against Iranian Bahá'ís 394

1979–1986
The Seven Year Plan

221 Launching of the Seven Year Plan—Naw-Rúz 1979 401
222 Elucidation of Seven Year Plan Goals 407
223 Member of the Universal House of Justice
to Address Significant Meetings in North America and Europe . . 412
224 Living a Chaste and Holy Life 413
225 Seizure of the House of the Báb 414
226 Call for Prayers for the Bahá'ís of Iran 415
227 The Increasing Dangers Facing Iranian Bahá'ís 416

CONTENTS

228	Advice on Use of Pioneers and Traveling Teachers	420
229	The Setting of Counselors' Terms of Service	421
230	New Member of International Teaching Center	421
231	Conversion of Iranian National Bahá'í Headquarters into an Islamic University	421
232	Call for Pioneers	422
233	Release of Bahá'í Prisoners in Baghdad	422
234	Effects of Bahá'í Actions on the Situation in Iran	423
235	Demolition of the House of the Báb	424
236	Results of Significant Meetings Addressed by Member of the Universal House of Justice	425
237	Assassination of the Hand of the Cause of God Enoch Olinga	426
238	Omission of Bahá'ís as a Recognized Religious Minority from Iranian Constitution	426
239	Refutation of Accusations against Iranian Bahá'ís	427
240	Release of a Compilation on Inspiring the Heart	430
241	Demolition of the House of the Báb and Adjacent Bahá'í Properties	431
242	Excavation of Temple Site in Western Samoa	431
243	Passing of the Hand of the Cause of God Raḥmatu'lláh Muhájir	431
244	Release of a Compilation on Divorce	432
245	Passing of Inparaju Chinniah, Continental Counselor	433
246	Message to Iranian Bahá'ís throughout the World	433
247	Passing of the Hand of the Cause of God Hasan M. Balyuzi	442
248	Naw-Rúz Message 1980	442
249	Release of a Compilation on the Importance of Prayer, Meditation, and the Devotional Attitude	446
250	Contract for Construction of the Indian Temple	447
251	Comments on the Kitáb-i-Aqdas	447
252	Our Attitude and Actions toward the Impending Catastrophe	451
253	Ominous Increase in Pressures on Iranian Bahá'ís	452
254	Execution of Two Members of the Tabríz Spiritual Assembly	453

CONTENTS

255	Vigorous Steps Taken in Defense of the Bahá'ís of Iran	454
256	Passing of the Hand of the Cause of God Adelbert Mühlschlegel	454
257	Commencement of Construction of Indian Temple	455
258	Translation of the Long Healing Prayer and the Fire Tablet	455
259	Arrest of National Spiritual Assembly of Iran and Two Auxiliary Board Members	456
260	Another False Accusation against the Iranian Bahá'ís	457
261	Martyrdom of Seven Bahá'ís in Yazd	457
262	Defense of the Bahá'ís of Iran in *Le Monde*	458
263	Worldwide Response to Iranian Persecutions	461
264	Property Acquisitions in the Holy Land	462
265	Passing of Leonora Stirling Armstrong, Spiritual Mother of South America	463
266	Attendance of the Members of a Spiritual Assembly at Its Meetings	463
267	Further Development of the Continental Boards of Counselors	464
268	Martyrdom of Riḍá Fírúzí in Tabríz; Uncertainty of Whereabouts of National Assembly Members	467
269	Passing of the Hand of the Cause of God Abu'l-Qasim Faizi	467
270	Immolation of Bahá'í Couple in Núk Village	468
271	Message to Counselors' Convocations on Five Continents	469
272	Relationship between Husband and Wife	470
273	Assassination of Dr. Manúchihr Ḥakím, Chairman of Iranian National Spiritual Assembly	474
274	Events Related to the Execution of Dr. Manúchihr Ḥakím	474
275	Second Phase of the Seven Year Plan	476
276	Return of the Hand of the Cause of God William Sears to Africa	480
277	Execution of Two Bahá'ís in Shiraz	480
278	Attendance of Bahá'ís from Rural Areas at International Conferences	481
279	Deployment of Pioneers and Traveling Teachers	481
280	Comments on Aspects of the Seven Year Plan	484

CONTENTS

281 Execution of Three Bahá'ís in Shiraz 493
282 Imminent Obliteration of the Site of the House of the Báb 494
283 Establishment of European Branch Office
    of Bahá'í International Community 494
284 Bahá'í International Fund—Recent Victories
    and Immediate Challenges 495
285 Execution of Seven Members of
    the Spiritual Assembly of Hamadan 497
286 Execution of Three Bahá'ís in Tehran 497
287 Execution of Four Bahá'ís in Iran 497
288 Announcement of Representatives to the
    International Conferences 498
289 Tests of Bahá'í Community Life—A Balanced Perspective 498
290 Execution of Two Bahá'ís in Káshmar 501
291 Message to First Women's Conference, United Kingdom 501
292 Execution of Auxiliary Board Member and Seven Assembly
    Members in Tabríz; Abduction of Two Teenage Girls 502
293 A New Phase of Persecution in Iran 503
294 Release of a Compilation on the Assistance of God 504
295 Release of a Compilation of Prayers and
    Passages from the Bahá'í Writings 504
296 Execution of Six Bahá'ís in Tehran and Dáryún 505
297 The Worsening Situation in Yazd 506
298 Further Deterioration of the Situation in Iran 507
299 Voice of America Report on Campaign
    of Arrests against Iranian Bahá'ís 507
300 Further Acts of Persecution in Iran 508
301 Grave Developments in Iran 508
302 Call for Pioneers 509
303 Release of a Compilation on Excellence in All Things 509
304 Inauguration of Radio Bahá'í Peru 510
305 Demolition of the House of Bahá'u'lláh in Tákur;
    Seizure of Cemetery in Tehran 510
306 Arrest of Eight Members of the Iranian
    National Spiritual Assembly 511

CONTENTS

307 Execution of Eight Members of the
    Iranian National Spiritual Assembly 511
308 Proselytizing, Development, and the Covenant 513
309 Iranian Supreme Court's Denial of
    Execution of Iranian National Spiritual Assembly Members 519
310 Appeal by Bahá'í International Community to Iranian Leaders .. 520
311 Secret Execution of Six
    Members of Tehran Assembly and Hostess 521
312 Using *A Cry from the Heart*
    to Explain the Tragic Events in Iran 521
313 Plans for Commemorating the Fiftieth Anniversary
    of the Ascension of Bahíyyih Khánum 522
314 Golden Opportunity Offered by Reign of Relentless Terror;
    Actions Taken to Defend the Bahá'ís of Iran 525
315 Call for a Day of Prayer for the Bahá'ís of Iran;
    Inviting Sympathetic Organizations and Individuals 532
316 Release of a Compilation on Family Life 533
317 Mounting Cruelties and the Heightened
    Steadfastness of the Iranian Bahá'ís 534
318 Relocation of Manila Conference 535
319 Transfer of Manila Conference to Canberra, Australia 535
320 Representation of the Universal House of Justice at
    Lagos Conference by Hand of the Cause of God John Robarts .. 536
321 Riḍván Message 1982 536
322 Further Developments of the Situation in Iran 539
323 Six Martyrdoms, More Arrests Resulting from
    Fresh Outburst of Persecutions 540
324 Execution of Two Members
    of the Spiritual Assembly of Urúmíyyih 541
325 Appointment of New Counselor in Asia 541
326 Passing of Amoz Gibson,
    Member of the Universal House of Justice 541
327 Call for Ballots to Elect One Member
    of the Universal House of Justice 542
328 Execution of Two Members of
    the Kháníábád Spiritual Assembly 543

329	Message to the International Conference in Dublin, Ireland—2 June 1982	543
330	Comments on the Station of 'Abdu'l-Bahá and the Language Used by the Central Figures	545
331	Martyrdom of Four Bahá'ís in Qazvín and Surrounding Areas	548
332	Election of Glenford Mitchell to the Universal House of Justice	548
333	Guidelines Concerning the Registration of Children and Youth	549
334	Commemoration at the World Center of the Fiftieth Anniversary of the Passing of the Greatest Holy Leaf and Inauguration of the Seat of the Universal House of Justice	552
335	Execution of 'Abbás-'Alí Ṣádiqípúr in Shiraz	553
336	Elucidation of the Principle of Confidentiality	553
337	Message to the International Conference in Quito, Ecuador—6–8 August 1982	555
338	Message to the International Conference in Lagos, Nigeria—19–22 August 1982	557
339	Execution of 'Alí Na'ímíyán in Urúmíyyih	559
340	Message to the International Conference in Montreal, Canada—3–5 September 1982	559
341	Message to the International Conference in Canberra, Australia—2–5 September 1982	563
342	Comments on the First Danish Bahá'í Women's Conference	565
343	Guidelines for Behavior of Children at Feast	566
344	Intensification of Oppression of Iranian Bahá'ís	566
345	Execution of Dr. Ḍíyá'u'lláh Aḥrárí in Shiraz	567
346	Passing of the Hand of the Cause of God Paul Haney	567
347	Imprisonment of More Than Eighty Bahá'ís in Shiraz	568
348	Establishment of First Bahá'í Radio Station in North America	568
349	Torture of Bahá'ís Imprisoned in Shiraz	569
350	Call for Another Day of Prayer	569
351	Execution of Hidáyat Síyávushí in Shiraz	570
352	Prayer of 'Abdu'l-Bahá for Martyred Friends and Relatives	570
353	Release of a Compilation on the Importance of Deepening	571
354	Occupation of the Seat of the Universal House of Justice	571

355	Execution of Two Bahá'ís in Shiraz	572
356	Execution of Mrs. Ṭúbá Zá'irpúr in Shiraz	573
357	Message to Youth Conferences in Costa Rica and Honduras	573
358	Riḍván Message 1983	574
359	Election of the Universal House of Justice—Riḍván 1983	577
360	Execution of Two Bahá'ís in Isfahan	577
361	Further Evolution of the International Teaching Center	577
362	Statement by the President of the United States about the Bahá'ís of Iran	580
363	Execution of Six Bahá'ís in Shiraz	580
364	Execution by Hanging of Ten Bahá'í Women in Shiraz	581
365	Message to Youth—Summons to Vindicate Martyred Youth	582
366	Appointment of Continental Counselors	583
367	Execution of Suhayl Húshmand in Shiraz	583
368	Attacks against Bahá'í Villagers in Mázindarán; Kidnapping of Two Bahá'ís in Tehran	584
369	Message to the European Youth Conference in Innsbruck	584
370	Continuing Harassments and Arrests in Iran	586
371	Formation of Nine New National Spiritual Assemblies at Riḍván	586
372	Intensification of Waves of Arrests in Iran	587
373	Arrest of More Bahá'ís in Iran	587
374	Passing of Sylvia Ioas, Member of the International Bahá'í Council	587
375	Requisites for Spiritual Growth	588
376	Preparations for International Youth Year	590
377	Banning of Bahá'í Administration in Iran	591
378	Open Letter to Iranian Authorities from the National Spiritual Assembly of Iran	592
379	Social and Economic Development— New Field of Bahá'í Service	601
380	Passing of Continental Counselor Raúl Pavón	604
381	Continued Pressure against the Bahá'ís of Iran	605

CONTENTS

382	Appointment of Continental Counselors	606
383	Response of Individuals to Persecution of Iranian Bahá'ís	606
384	Service in Voluntary Nonsectarian Organizations	611
385	Acute Needs of the International Funds	612
386	Message to Youth—International Youth Year	614
387	Death of Three More Iranian Bahá'ís and Arrest of 250 More	617
388	Passing of Ethel Revell, Member of the International Bahá'í Council	618
389	Message to National Bahá'í Women's Conference, Kenya	618
390	Persecution of Bahá'ís in Morocco	619
391	Inaugural Broadcast of Radio Bahá'í Bolivia	619
392	Insidious Turn of Persecution of Iranian Bahá'ís	620
393	Execution of Three by Firing Squad	620
394	Riḍván Message 1984	621
395	Use of Torture to Extract False Confessions from Iranian Bahá'ís	625
396	Summary of the Situation in Iran	627
397	The Development of the Soul and the Reconstruction of Society	629
398	Message to Bahá'í Women's Conference in Cotonou, Benin	631
399	Execution of Nuṣratu'lláh Vaḥdat in Mashhad	632
400	Execution of Iḥsánu'lláh Kathírí in Tehran	632
401	Results of Bahá'í Women's Conference in Cotonou, Benin	632
402	Relationship between Husband and Wife—Further Comments	633
403	Dedication of the Mother Temple of the Pacific Islands, Apia, Western Samoa	635
404	Introduction of the Law of Ḥuqúqu'lláh to the West	637
405	Understanding the Laws of the Faith	638
406	Execution of Manúchihr Rúḥí	639
407	Roles of Parents within the Bahá'í Family	640
408	Message to Youth Conference, London, Ontario	641
409	Safeguarding the Letters of Shoghi Effendi	642
410	Call for Pioneers	643

CONTENTS

411 Martyrdom of Iranian National Spiritual Assembly Member Shápúr Markazí; Death of Amínu'lláh Qurbánpúr in Prison 644

412 The Universal House of Justice's Power of Elucidation 645

413 Execution of Member of Disbanded Iranian National Spiritual Assembly and of Bahá'ís from Karaj and Mashhad 647

414 Passing of the Hand of the Cause of God Shu'á'u'lláh 'Alá'í 647

415 Invitation to Participate in the Formulation of the Next Plan ... 648

416 Execution of Six Bahá'ís in Tehran 649

417 Execution of Dr. Farhád Aṣdaqí after Four Months of Torture; Death of Two Bahá'ís in Tabríz after More Than Two Years' Imprisonment 649

418 Adoption of Another Oppressive Measure by the Iranian Government 650

419 Renewed Call for Support of International Funds 650

420 International Year of Peace 652

421 Execution of Rúḥu'lláh Ḥaṣúrí in Yazd; Confirmation of Death of Rustam Varjávandí in Prison 655

422 Elucidation of the Lesser Peace and the Supreme Tribunal 655

423 Passing of Lloyd Gardner, Continental Counselor 660

424 Execution of Rúḥu'lláh Bahrámsháhí in Yazd and Nuṣratu'lláh Subḥání in Tehran 660

425 Human Suffering and the Reconstruction of Society 661

426 Responsibilities of Youth at the Age of Maturity 665

427 Riḍván Message 1985 666

428 Message to Youth—Scaling the Heights of Excellence 668

429 Announcement of Completion of a Statement on Peace Addressed to the Peoples of the World 669

430 Release of a Compilation on the Law of Ḥuqúqu'lláh 670

431 Message to International Youth Conference, Columbus, Ohio .. 670

432 Inauguration of Public Information Office at the World Center 671

433 Election of Delegates to National Conventions 671

434 Dr. Martin Luther King, Jr., and Bahá'í Peace Activities 674

435 Development of Local and National Bahá'í Funds 676

436 Release of a Compilation on Peace 679

437	Execution of Two More Iranian Bahá'ís; Bahá'í Students Pressured	680
438	The Promise of World Peace	681
439	Appointment of Continental Boards of Counselors	696
440	Reinterment of Bahá'u'lláh's Faithful Half-Brother	698
441	Execution of 'Azízu'lláh Ashjárí by Firing Squad	699
442	Presentation of Peace Statement to the Secretary-General of the United Nations	700
443	Presentation of Peace Statement to the President of the United States	701
444	Adoption of Resolution in Support of Iranian Bahá'ís by the United Nations General Assembly—Affirmation of Emergence from Obscurity	701
445	Message to Conference of Counselors in the Holy Land	702
446	Release of a Compilation on Women	704
447	Report from Conference in the Holy Land; the Inception of the Fourth Epoch of the Formative Age	704
448	Women—Their Role in Society and the Establishment of Peace; Membership on the Universal House of Justice	707
449	Message to National Youth Conference, Adelaide, Australia	709
450	Inaugural Broadcast of Radio Bahá'í Panama	710
451	Epochs of the Formative Age	710
452	Plans for the Dedication of the Mother Temple of the Indian Subcontinent, New Delhi	716
453	The Six Year Plan—Major Objectives and Setting National Goals	717
454	Passing of Angus Cowan, Continental Counselor	723
455	Women—Reasons for Exemption from Acts of Worship	724
456	Riḍván Message 1986	724
	Glossary	729
	Bibliography	759
	Index of Names	765
	General Index	773

Foreword

On 21 April 1963, one hundred years after Bahá'u'lláh's public declaration of His world-redeeming mission, the first International Bahá'í Convention was held on the majestic slopes of Mount Carmel in Haifa, Israel. The purpose of this momentous gathering was to elect, for the first time, the Universal House of Justice, the supreme governing and legislative body of the Bahá'í Faith.

The election of the Universal House of Justice was greeted with great joy by the worldwide Bahá'í community, for it ensured the continuation of divine guidance until the advent of the next Manifestation of God, thereby fulfilling the prophecy that there would come a "Day which shall not be followed by night."

The continuation of divine guidance—the primary theme of this book—is a unique feature of the Bahá'í Faith. Unlike any of the Manifestations of God Who preceded Him, Bahá'u'lláh, the Founder of the Bahá'í Faith, made a Covenant with His followers to direct and channel the forces released by His Revelation, guaranteeing the continuity of infallible guidance after His death through institutions to which all of His followers must turn. In His will and testament Bahá'u'lláh designated His eldest son, 'Abdu'l-Bahá, His successor, the authoritative Interpreter of His writings, and the Center of His Covenant. In His own will and testament 'Abdu'l-Bahá perpetuated the Covenant through the Administrative Order ordained in Bahá'u'lláh's writings. 'Abdu'l-Bahá appointed as His twin successors the institutions of the Guardianship and the Universal House of Justice. He named His eldest grandson, Shoghi Effendi, the Guardian of the Cause of God and expounder of the Word of God, while giving the Universal House of Justice the role of legislating on matters not explicitly revealed in the writings of Bahá'u'lláh and 'Abdu'l-Bahá or dealt with by Shoghi Effendi.

Writing about the Universal House of Justice, Bahá'u'lláh explained that, "Inasmuch as for each day there is a new problem and for every problem an expedient solution, such affairs should be referred to the Ministers of the House of Justice that they may act according to the needs and requirements of the time. . . . It is incumbent upon all to be obedient unto them." Moreover, 'Abdu'l-Bahá asserts in His Will and Testament that "Unto the Most Holy Book every one must turn, and all that is not expressly recorded therein must be referred to the Universal House of Justice. That which this body, whether unanimously or by a majority doth carry, that is verily the Truth and the Purpose of God Himself."

Thus the fundamental purpose of the Universal House of Justice is to ensure the continuity of the divine guidance that flows from the Source of the Faith, to safeguard the unity of its followers, and to maintain the integrity and flexibility of its teachings. Its origin, its authority, its duties, and its sphere of action are all derived from the writings of Bahá'u'lláh. His writings, along with those of 'Abdu'l-Bahá and of Shoghi Effendi constitute its bedrock foundation and its binding terms of reference.

Messages from the Universal House of Justice, 1963–1986: The Third Epoch of the Formative Age brings together letters, cables, telexes, and electronic mail messages sent to Bahá'í institutions and individuals during a twenty-three-year period that the Universal House of Justice has called the Third Epoch of the Formative Age of the Faith. The Formative Age stands between the Heroic Age (1844–1921), which is associated with the ministries of the Báb (the Prophet-Herald of the Bahá'í Faith), Bahá'u'lláh, and 'Abdu'l-Bahá, and the Golden Age, which will witness the dawning of the Most Great Peace, the emergence of a commonwealth of all the nations of the world, and the advent of the Kingdom of God on earth.

The First Epoch of the Formative Age (1921–44/46) saw the birth of the Bahá'í Administrative Order and the primary stages in the erection of its framework. This epoch was characterized by its focus on forming local and national institutions in all the continents, thereby laying the foundation needed to support future systematic teaching activities. The First Epoch also witnessed the initiation of the first Seven Year Plan (1937–44), at the direction of Shoghi Effendi, by the Bahá'ís of the United States and Canada. This plan, inspired by 'Abdu'l-Bahá's Tablets of the Divine Plan, constituted the Bahá'í community's first systematic teaching campaign and began the initial stage of carrying out 'Abdu'l-Bahá's Divine Plan in the Western Hemisphere.

The Second Epoch of the Formative Age (1946–63) witnessed the further development and maturation of the Administrative Order and the execution of a series of teaching plans that carried the Faith beyond the confines of the Western Hemisphere and Europe. Its two most distinguishing characteristics were the rise and steady consolidation of the World Center of the Faith in Haifa, including the appointment of the Hands of the Cause of God, the introduction of the Auxiliary Boards, and the establishment of the International Bahá'í Council, and the simultaneous and often spontaneous prosecution of Bahá'í national plans in both the East and the West. Shoghi Effendi now used the internally consolidated and administratively experienced National Spiritual Assemblies to launch the Ten Year World Crusade (1953–63) in which he summoned the twelve existing National Spiritual Assemblies to participate in one concentrated, united effort to spread the Faith more widely throughout the world. This enterprise resulted in the opening of 131 countries to the Faith, the formation of forty-four National Spiritual Assemblies, and a vast increase in the ethnic diversity of the Bahá'í community.

However, the victorious completion of the Ten Year World Crusade came with a heavy price. In 1957, midway through the Crusade, Shoghi Effendi died unexpectedly of a heart attack. The Bahá'ís were devastated. Not only were they deprived of their Guardian, but they were also deprived of their source of divine guidance. The Guardianship was a hereditary office, and Shoghi Effendi had left no heir.

How would the unity of the Faith be preserved? How could the conflict and contention that has divided other religions be prevented? Clearly, unless the Bahá'í Faith could safeguard its unity, it could never hope to unify a sorely divided world. Into this chasm of doubt and despair stepped the Hands of the Cause of God, whom Shoghi Effendi had appointed and to whom he had referred as the "Chief Stewards of Bahá'u'lláh's embryonic World Commonwealth who have been invested by the unerring Pen of the Center of His Covenant with the dual function of guarding over the security, and of insuring the propagation, of His Father's Faith." Through their heroic stewardship they guided the community along the course Shoghi Effendi had set. By the end of the nearly six-year interregnum they had not only preserved the unity of the Faith but also inspired a surge of sacrificial service leading to the Ten Year Crusade's victorious conclusion and crowning achievement, the culminating event of the Second Epoch—the election of the Universal House of Justice.

The Universal House of Justice proclaimed in a letter dated October 1963 that the Third Epoch of the Formative Age had begun. Outlining what would become the distinctive characteristics of the new epoch, it said the Bahá'í Faith "must now grow rapidly in size, increase its spiritual cohesion and executive ability, develop its institutions and extend its influence into all strata of society." Moreover,

> its members, must, by constant study of the life-giving Word, and by dedicated service, deepen in spiritual understanding and show to the world a mature, responsible, fundamentally assured and happy way of life, far removed from the passions, prejudices and distractions of present day society.

The Third Epoch called the Bahá'ís of the world to a yet more mature level of functioning that would be consistent with expected vast increases in the community's size and diversity, its emergence as a model to humankind, and the extension of its influence in the world at large.

The nature of individual and community maturity for which the Universal House of Justice called is evident as one reads the book. Through the release of translations and compilations of Bahá'í writings and through a voluminous correspondence in which perplexing questions about various personal and social problems are discussed, the Universal House of Justice guides individuals and communities in their efforts to live a Bahá'í life. Matters of spiritual growth, sexual morality, relations between husband and wife, family life, prejudice, politics, possession of firearms, ranks and stations in the Faith, the tests of Bahá'í community life, and the suffering of humanity are a few of the topics of individual and community life that are addressed.

Numerous accounts of the vicious persecution of the Iranian Bahá'ís and of their efforts and the efforts of the Bahá'í communities around the world to help them testify to the spiritual and administrative maturation of the worldwide Bahá'í community.

The increasing maturity of the worldwide Bahá'í community is further evidenced as the Universal House of Justice guides the Bahá'í community to progressively more mature levels of functioning through three major world teaching plans conducted during the Third Epoch of the Formative Age: the Nine Year Plan (1964–73), the Five Year Plan (1974–79), and the Seven Year Plan (1979–86). Among the results and achievements won during these three plans are the Faith's gradual emergence from obscurity, its assumption by the Universal House of Justice of the role of a promoter of world peace, the initiation of activities intended to foster social and economic development, the establishment of the Continental Boards of Counselors and the International Teaching Center, the introduction of assistants to the Auxiliary Boards, and—one of the crowning events of the Third Epoch—the construction and occupation of the seat of the Universal House of Justice at the World Center of the Faith.

The publication of *Messages from the Universal House of Justice, 1963–1986: The Third Epoch of the Formative Age* comes ten years after the inauguration of the Fourth Epoch. Although the Bahá'í Faith is well into this new stage of development, and new patterns of growth and change are evident, yet the study of the messages contained in this volume yields a treasure of insight, knowledge, and understanding of the principles guiding the unfoldment of the Bahá'í Faith and the efforts of individuals and communities as they strive to adopt the pattern of life prescribed by Bahá'u'lláh. Moreover, the inclusion of virtually every major message of the Third Epoch enables one to survey significant developments in the organic growth of the Bahá'í world community, to follow its increasing maturity and acquisition of new powers and capacities, to perceive a continuity in its development, and to detect patterns in the interplay between the forces of light and darkness, between crisis and victory. Individuals, communities, and institutions will find the book an important aid in making decisions as they apply themselves to meeting current challenges and will derive continual inspiration and encouragement from the majestic pronouncements, authoritative elucidations, and ennobling, luminous, and loving counsels of the Universal House of Justice. The messages themselves are sufficient proof that the Covenant of Bahá'u'lláh is unbroken, that the channel of divine guidance remains open, and that "the Day which shall not be followed by night" has at last dawned upon the world.

NATIONAL SPIRITUAL ASSEMBLY OF THE
BAHÁ'ÍS OF THE UNITED STATES

A Note from the Compiler

Messages from the Universal House of Justice, 1963–1986: The Third Epoch of the Formative Age brings under one cover 456 letters, cables, telexes, and electronic mail messages from the Universal House of Justice to Bahá'í institutions, to Bahá'ís gathered at special events, to individuals, or—in one case—to the peoples of the world.

A volume long overdue, it enlarges upon and replaces two earlier compilations of messages: *Wellspring of Guidance: Messages, 1963–1968*, published in 1969, and *Messages from the Universal House of Justice, 1968–1973*, published in 1976. In preparing the new volume several factors where considered. First was the need to make available the messages contained in the first two compilations and to add thirty-six messages from the 1963 through 1973 period that had not been included. Second was the need to publish messages received after 1973. Thus it was decided to publish a new volume that would encompass and expand upon the first two books.

The time frame of the new volume became clear in January 1986 when the Universal House of Justice announced the closing of the Third Epoch (1963–1986) and the inception of the Fourth Epoch, signaled by new developments in the maturation of the institutions of the Faith. Hence *Messages from the Universal House of Justice, 1963–1986* enables the reader to survey the full sweep of the Third Epoch, beginning with the first message issued by the Universal House of Justice after its establishment in 1963 and ending with the annual Riḍván message of 1986. The messages record the progress of the Faith through the Nine, Five, and Seven Year Plans, chronicling extraordinary advances in proclamation, expansion, and consolidation. They reach a climax in the last two years of the Seven Year Plan with the Faith's emergence from obscurity, the release of a statement on peace addressed to the peoples of the world, and the inception of the Fourth Epoch of the Formative Age. These and countless other achievements stand in marked contrast to the contemporaneous persecution of the Bahá'ís of Iran, throwing into sharp relief the interplay between crisis and victory. The dynamic interaction between these two forces is a chief characteristic of the unfoldment of the Bahá'í Faith and a prominent theme in this volume.

Messages from the Universal House of Justice, 1963–1986 includes numerous features that have been introduced for the reader's convenience. A table of epochs of the Heroic and Formative Ages of the Bahá'í Faith and a list of significant milestones, anniversaries, and events in the Third Epoch offer a brief overview of recent Bahá'í history. Each message has been given a heading and assigned a number, and the paragraphs or other divisions of text within the messages have been numbered to facilitate reference. An extensive index is included, its locators corresponding to the numbering system used in the text. Subheadings have been added to many longer messages; spellings have been Americanized; and, for consistency, dates have been given in a day-month-year format. Addressees, salutations, and complimentary closings have been retained, except for a few cases in which confidentiality required their omission. Because the compilation of messages covers almost a quarter of a century, footnotes have been added to explain historical details, allusions, technical terms, sources of most quotations, cross references to other messages on the same topic, and references to further details in the glossary or in other books. To help keep the notes as brief as possible, the titles of many Bahá'í books were abbreviated. A list of abbreviations and the titles they stand for appears at the front of the book. A glossary has been included to give background information and define technical Bahá'í terms. A bibliography provides information about books and compilations referred to in the text and footnotes.

No changes have been made to the texts of the messages without explicit permission from the Universal House of Justice. Extracts from the writings of Bahá'u'lláh, the Báb, and 'Abdu'l-Bahá that are quoted in the messages have been checked against authenticated sources and authorized translations and have been updated accordingly where appropriate. Noticeable changes in the translations of such extracts are footnoted. The dates of certain cables may differ slightly from those of the same cables published elsewhere because cables of general interest that were addressed to a specific National Spiritual Assembly were usually conveyed to other National Spiritual Assemblies by mail a few days later. In such cases the date of the original cable is used.

Occasionally the reader will encounter references to enclosures or attachments, such as compilations and lists of pioneer assignments given to National Spiritual Assemblies. Such enclosures have been omitted because of space limitations. Most of the compilations have already been published, and, wherever possible, a reference to the published work is given.

We would like to express our deepest appreciation to the Universal House of Justice for its assistance in making available many messages that were previously unpublished and particularly for its direction in all phases of the production of this volume. Thanks are due to Ivor Stoakley, Diane Taherzadeh, and Lewis Walker for their assistance with footnotes and glossary entries; to John Walbridge for helpful comments on the glossary; to Dr. Manuchehr Derakhshani for assistance with Persian and Arabic transliterations; to Dr. Betty

J. Fisher and Terry J. Cassiday of the Bahá'í Publishing Trust whose insights, scholarship, hard work, and close attention to detail have been invaluable; and to my wife, Amy, for her unwavering encouragement and support. We hope this book will prove to be a useful and convenient source of guidance and inspiration for both individuals and institutions the world over as they strive to serve the Cause of Bahá'u'lláh and carry the healing balm of His teachings to all humanity.

GEOFFRY W. MARKS

List of Abbreviations

ABL	*'Abdu'l-Bahá in London*
ADJ	*The Advent of Divine Justice*, 1990 ed.
BA	*Bahá'í Administration*
BN	*Bahá'í News*
BNE	*Bahá'u'lláh and the New Era*
BP	*Bahá'í Prayers*, 1991 U.S. ed.
BW 9	*The Bahá'í World, Volume IX*
BW 13	*The Bahá'í World, Volume XIII*
BW 14	*The Bahá'í World, Volume XIV*
BW 15	*The Bahá'í World, Volume XV*
BW 16	*The Bahá'í World, Volume XVI*
BW 17	*The Bahá'í World, Volume XVII*
BW 18	*The Bahá'í World, Volume XVIII*
BW 19	*The Bahá'í World, Volume XIX*
CC	*The Compilation of Compilations*
CF	*Citadel of Faith*
CUHJ	*Constitution of the Universal House of Justice*
DB	*The Dawn-Breakers*
DND	*Dawn of a New Day*
ESW	*Epistle to the Son of the Wolf*
FSTS	*From Strength to Strength*
GPB	*God Passes By*
GWB	*Gleanings from the Writings of Bahá'u'lláh*
KA	*The Kitáb-i-Aqdas*
KI	*The Kitáb-i-Íqán*
LFG	*Letters from the Guardian to Australia and New Zealand*
LG	*Lights of Guidance*
MA	*Messages to America*
MBW	*Messages to the Bahá'í World*

ABBREVIATIONS

MC	*Messages to Canada*
MF	*Memorials of the Faithful*
PB	*The Proclamation of Bahá'u'lláh*
PDIC	*The Promised Day Is Come*, 1996 ed.
PM	*Prayers and Meditations*
PP	*The Priceless Pearl*
PT	*Paris Talks*
PUP	*The Promulgation of Universal Peace*, 1982 ed.
SAQ	*Some Answered Questions*, 1984 ed.
SBS	*Selections from Bahá'í Scriptures*
SC	*A Synopsis and Codification of the Kitáb-i-Aqdas*
SDC	*The Secret of Divine Civilization*
SWAB	*Selections from the Writings of 'Abdu'l-Bahá*
TABA	*Tablets of Abdul-Baha Abbas*
TB	*Tablets of Bahá'u'lláh Revealed after the Kitáb-i-Aqdas*
TDP	*Tablets of the Divine Plan*, 1992 ed.
TN	*A Traveler's Narrative*
UD	*Unfolding Destiny*
WOB	*The World Order of Bahá'u'lláh*
WT	*Will and Testament of 'Abdu'l-Bahá*

Epochs of the Heroic Age[1]

1844–1853
The First Epoch

The Bábí Dispensation

1853–1892
The Second Epoch

The Ministry of Bahá'u'lláh

1892–1921
The Third Epoch

The Ministry of 'Abdu'l-Bahá

---

1. Based on the attachment to the letter dated 5 February 1986 from the Universal House of Justice to all National Spiritual Assemblies. See message no. 451.

Epochs of the Formative Age[1]

1921–1944/46
The First Epoch

1921–1936	Period of Preparation
1937–1944	The first Seven Year Plan of the Bahá'ís of the United States and Canada
1944–1946	Consolidation of victories

1946–1963
The Second Epoch

1946–1953	The second Seven Year Plan of the Bahá'ís of the United States and plans of varying duration pursued by nine other regional and national Bahá'í communities (Canada; Central America and South America; Australia and New Zealand; India, Pakistan, and Burma; the British Isles; Germany and Austria; Persia; Egypt; Iraq)[2]
1953–1963	The Ten Year World Crusade

1963–1986
The Third Epoch

1963–1964	Year of Preparation
1964–1973	Nine Year Plan
1973–1974	Year of Preparation
1974–1979	Five Year Plan
1979–1986	Seven Year Plan

---

1. Based on the attachment to the letter dated 5 February 1986 from the Universal House of Justice to all National Spiritual Assemblies. See message no. 451.

2. See FSTS, pp. 33–38.

1986–
The Fourth Epoch

1986–1992	Six Year Plan
1992–1993	Holy Year Commemorating the Ascension of Bahá'u'lláh and the Inauguration of His Covenant
1993–1996	Three Year Plan
1996–2000	Four Year Plan

The Third Epoch
Significant Milestones, Anniversaries, and Events

Milestones

Riḍván 1963	First election of the Universal House of Justice
April 28–May 2 1963	First Bahá'í World Congress commemorating Bahá'u'lláh's declaration of His mission and the victorious consummation of the Ten Year World Crusade
Riḍván 1964	Inauguration of the Third Epoch with the launching of the Nine Year Plan
Riḍván 1973	End of the Nine Year Plan
Naw-Rúz 1974	Beginning of the Five Year Plan
Riḍván 1979	End of the Five Year Plan Beginning of the Seven Year Plan
January 1986	Inauguration of the Fourth Epoch
Riḍván 1986	End of the Seven Year Plan Beginning of the Six Year Plan

Significant Anniversaries

26 March 1966	Fiftieth anniversary of the revelation of the first Tablets of the Divine Plan
April 1963	Centenary of Bahá'u'lláh's public declaration of His Mission
September/October 1967	Celebration of the centenary of Bahá'u'lláh's proclamation to the Kings
27 September 1967	Celebration of the centenary of the revelation of the Súriy-i-Mulúk
12 November 1967	Sesquicentennial of the Birth of Bahá'u'lláh
31 August 1968	Centenary of the arrival of Bahá'u'lláh in the Holy Land
23 June 1968	Centenary of the passing of Mírzá Mihdí, the Purest Branch

28 November 1971	Fiftieth anniversary of the passing of 'Abdu'l-Bahá
1973	Centenary of the revelation of the Kitáb-i-Aqdas
June 1977	Centenary of the termination of Bahá'u'lláh's confinement in 'Akká
July 1982	Fiftieth anniversary of the passing of the Greatest Holy Leaf
4 November 1982	Twenty-fifth anniversary of the passing of Shoghi Effendi

Significant Events

Riḍván 1963	Convening of 1st International Bahá'í Convention, Haifa, and election of the Universal House of Justice
Riḍván 1964	Launching of the Nine Year Plan
July 1964	Dedication of the Mother Temple of Europe, Langenhain, near Frankfurt am Main, Germany
November 1965	Purification of the Most Holy Tomb
October 1967	Opening of a period of proclamation inaugurated by the presentation by the Universal House of Justice to 140 heads of state of a special edition of *The Proclamation of Bahá'u'lláh*
	Commemoration of the centenary of Bahá'u'lláh's proclamation to the kings with the holding of six intercontinental conferences in Panama City, Panama; Wilmette, Illinois, U.S.A.; Sydney, Australia; Kampala, Uganda; Frankfurt am Main, Germany; and New Delhi, India
	Assumption by the Universal House of Justice of representation of the Bahá'í International Community at the United Nations
October 1967	Laying of the foundation stone of the Mother Temple of Latin America, Panama City, Panama
February 1968	Embracing of the Faith by His Highness Malietoa Tanumafili II of Western Samoa
Riḍván 1968	Convening of the 2nd International Bahá'í Convention, Haifa, and election of the Universal House of Justice
June 1968	Establishment of the Continental Boards of Counselors
August 1968	Commemoration of the centenary of Bahá'u'lláh's arrival in the Holy Land with the holding of First Oceanic Conference, Palermo, Sicily

THE THIRD EPOCH • MILESTONES, ANNIVERSARIES, AND EVENTS

February 1970	Bahá'í International Community's attainment of consultative status with the United Nations Economic and Social Council
August 1970– September 1971	Holding of eight oceanic and continental conferences in La Paz, Bolivia; Rose Hill, Mauritius; Monrovia, Liberia; Djakarta, Indonesia; Suva, Fiji; Kingston, Jamaica; Sapporo, Japan; Reykjavik, Iceland
April 28–May 2 1972	Dedication of the Mother Temple of Latin America, Panama City, Panama
June 1972	Announcement of the decision to build the Seat of the Universal House of Justice
September 1972	Murder of three Iranian Bahá'í students in the Philippine Islands
November 1972	Adoption of the Constitution of the Universal House of Justice
March 1973	Purchase of the Mazra'ih Mansion
Riḍván 1973	Convening of the 3rd International Bahá'í Convention, Haifa, and election of the Universal House of Justice
	Publication of *A Synopsis and Codification of the Laws and Ordinances of the Kitáb-i-Aqdas*
May 1973	Announcement of the embracing of the Bahá'í Faith by His Highness Malietoa Tanumafili II of Western Samoa
June 1973	Establishment of the International Teaching Center
February 1974	Acceptance of the design for the seat of the Universal House of Justice
Riḍván 1974	Launching of the Five Year Plan
January 1975	Acquisition of the House of 'Abdu'lláh Páshá in 'Akká
April 1975	Excavation of the site of the seat of the Universal House of Justice
July 1976– January 1977	Holding of eight international teaching conferences in Helsinki, Finland; Anchorage, Alaska; Paris, France; Nairobi, Kenya; Hong Kong; Auckland, New Zealand; Bahia, Brazil; and Mérida, Mexico.
September 1976	Visit of His Highness Malietoa Tanumafili II to the resting-place of Shoghi Effendi in London, England
May 1977	Murder of Rúḥu'lláh Taymúrí-Muqaddam, an Iranian homefront pioneer, in his home by fanatics

October 1977	Laying of foundation stone for Mother Temple of Indian subcontinent, New Delhi, India
December 1977	Initiation of broadcasting by the first Bahá'í radio station, Otavalo, Ecuador
Riḍván 1978	Convening of the 4th International Bahá'í Convention, Haifa, and election of the Universal House of Justice
August 1978	Inauguration of Radio Bahá'í Ecuador
December 1978–January 1979	Intensification of the persecution of the Bahá'ís in Iran
January 1979	Laying of the foundation stone of the Mother Temple of the Pacific Islands
Naw-Rúz 1979	Launching of the Seven Year Plan
June 1979	Fixing of Counselors' term of service at five years
June–July 1979	Numerous appeals made by Local Spiritual Assemblies around the world to Ayatollah Khomeini, asking him to stop the persecution of the Bahá'ís of Iran
June–August 1979	Addresses by H. Borrah Kavelin, representative of the Universal House of Justice, at a series of significant meetings in North America and Europe about the importance of redeeming the Iranian Bahá'ís' sacrifices through service to the Faith
September 1979	Demolition of the House of the Báb in Shiraz
November 1979	Arrest of a member of the National Spiritual Assembly of the Bahá'ís of Iran
December 1979	Commencement of excavation of the site of the Mother Temple of the Pacific Islands, Apia, Western Samoa
July 1980	Commencement of construction of Mother Temple of the Indian Subcontinent, New Delhi, India
August 1980	Arrest of all nine members of the National Spiritual Assembly of Iran and two Auxiliary Board members
November 1980	Appointment of a new contingent of Continental Counselors to a five-year term of service
January 1981	Execution of chairman of National Spiritual Assembly of Iran
May 1981	Establishment of a European Branch Office of the Bahá'í International Community
August 1981	Increase in the severity of the persecution of the Bahá'ís of Iran

November 1981	Arrest of six members of National Spiritual Assembly of Iran
	Inauguration of Radio Bahá'í Peru
December 1981	Demolition of the House of Bahá'u'lláh in Tákur
	Arrest and execution of eight members of the National Spiritual Assembly of Iran
May–September 1982	Holding of five international conferences in Lagos, Nigeria; Montreal, Canada; Quito, Ecuador; Dublin, Ireland; and Canberra, Australia
July 1982	By-election to replace one member of the Universal House of Justice
December 1982	Establishment of first Bahá'í radio station in North America
January 1983	Occupation by the Universal House of Justice of its permanent seat
April 1983	Opening of the House of 'Abdu'lláh Páshá to pilgrims
Riḍván 1983	Convening of the 5th International Bahá'í Convention, Haifa, and election of the Universal House of Justice
May 1983	Increase in the responsibilities of the International Teaching Center
	Public defense of the Bahá'ís of Iran by the President of the United States
June 1983	Execution of seventeen Bahá'ís in Shiraz, including ten women
	Call for the youth of the world to vindicate the martyred youth of Shiraz
August 1983	Banning of Bahá'í administration in Iran
September 1983	Sending of an open letter from the National Spiritual Assembly of Iran to the authorities responding to the false charges made against the Bahá'í community
October 1983	Call for Bahá'í involvement in social and economic development activities
April 1984	Attempt to achieve false confessions from the Bahá'ís of Iran by use of torture
	Inauguration of Radio Bahá'í Bolivia
August 1984	Dedication of the Mother Temple of the Pacific Islands, Apia, Western Samoa

July 1985	Release of compilation on the Law of Ḥuqúqu'lláh
	Establishment of Public Information Office at the Bahá'í World Center
October 1985	Release by the Universal House of Justice of a statement on peace addressed to the peoples of the world
	Appointment of a new and expanded contingent of members of the Continental Boards of Counselors
November 1985	Presentation of the peace statement to the Secretary General of the United Nations
December 1985	Emergence of the Bahá'í Faith from obscurity signaled by adoption of a resolution by the United Nations General Assembly supporting the Iranian Bahá'í community
December 1985–January 1986	Conference of the Continental Boards of Counselors in the Holy Land with Hands of the Cause of God, members of the Universal House of Justice, and the International Teaching Center
January 1986	Announcement of the beginning of the Fourth Epoch of the Formative Age by the Universal House of Justice in a letter to the Bahá'ís of the world
	Inauguration of Radio Bahá'í Panama
February 1986	Setting of date for dedication of Mother Temple of the Indian Subcontinent, New Delhi, India
	Invitation from the Universal House of Justice to National Spiritual Assemblies and Continental Counselors to consult about national goals within the framework of the major objectives of the Six Year Plan set by the Universal House of Justice
Riḍván 1986	Review by the Universal House of Justice of achievements of the Seven Year Plan, comments on the Faith's emergence from obscurity and the maturation of the institutions of the Cause and announcement of World Center goals of publishing an English translation of the Kitáb-i-Aqdas, education of the Bahá'ís about the law of Ḥuqúqu'lláh, planning for the completion of the remaining buildings on the Arc, and broadening the basis of the international relations of the Faith
	Launching of the Six Year Plan

MESSAGES
FROM THE
UNIVERSAL HOUSE OF JUSTICE

1963–1986

1963–1964
A Year of Preparation

1

Statement of the Universal House of Justice to the First Bahá'í World Congress[1]

30 April 1963

"All praise, O my God, be to Thee Who art the Source of all glory and majesty, of greatness and honor, of sovereignty and dominion, of loftiness and grace, of awe and power. Whomsoever Thou willest Thou causest to draw nigh unto the Most Great Ocean, and on whomsoever Thou desirest Thou conferrest the honor of recognizing Thy Most Ancient Name. Of all who are in heaven and on earth, none can withstand the operation of Thy sovereign Will. From all eternity Thou didst rule the entire creation, and Thou wilt continue for evermore to exercise Thy dominion over all created things. There is none other God but Thee, the Almighty, the Most Exalted, the All-Powerful, the All-Wise."[2]

Beloved friends: On this glorious occasion, the celebration of the Most Great Jubilee, we raise our grateful thanks to Bahá'u'lláh for all His bounties showered upon the friends throughout the world. This historic moment marks at one and the same time the fulfillment of Daniel's prophecy, the Hundredth Anniversary of the Declaration of the Promised One of all ages, the termination of the first epoch of the Divine Plan of 'Abdu'l-Bahá designed to establish the Faith of God in all the world, and the successful conclusion of our beloved Guardian's world-encircling Crusade, enabling his lovers and loved ones everywhere to lay this glorious harvest of victory in his name at the feet of the Blessed Beauty.[3] This Most Great Jubilee is the crowning victory of the lifework of Shoghi Effendi, Guardian of the Cause of God. He it was, and he alone, who unfolded the potentialities of the widely scattered, numerically small, and largely unorganized Bahá'í community which had been called into

---

1-1. The statement was presented by Mr. David Hofman, a member of the Universal House of Justice, to the Bahá'ís assembled in Royal Albert Hall, London, 28 April–2 May 1963, at the Bahá'í World Congress, which commemorated the Most Great Jubilee, the one hundredth anniversary of Bahá'u'lláh's declaration of His mission, and also marked the victorious completion of the Ten Year Crusade launched by Shoghi Effendi in April 1953.

1-2. Bahá'u'lláh, in BP, p. 120.

1-3. 'Abdu'l-Bahá explained that Daniel's prophecy ("Blessed is he who cometh unto the thousand three hundred and thirty five days," Dan. 12:12) points to one century after Bahá'u'lláh's appearance and declaration of His mission in 1863 and refers to the spread of the teachings of the Bahá'í Faith across the world.

The first epoch of the Divine Plan, 'Abdu'l-Bahá's program of action for the worldwide propagation of the Faith, began in 1937 and ended in 1963, with the conclusion of the Ten Year Crusade (1953–63). For further information, see glossary entries on Divine Plan; Crusade, Ten Year World; Epochs; Plans; Tablets of the Divine Plan.

"Blessed Beauty" refers to Bahá'u'lláh.

being during the Heroic Age of the Faith.[4] He it was who unfolded the grand design of God's Holy Cause, set in motion the great plans of teaching already outlined by 'Abdu'l-Bahá, established the institutions and greatly extended the endowments at the World Center, and raised the Temples of America, Africa, Australasia and Europe, developed the Administrative Order of the Cause throughout the world, and set the Ark of the Cause true on its course. He appointed the Hands of the Cause of God.

1.3 The paeans of joy and gratitude, of love and adoration which we now raise to the throne of Bahá'u'lláh would be inadequate, and the celebrations of this Most Great Jubilee in which, as promised by our beloved Guardian, we are now engaged, would be marred were no tribute paid at this time to the Hands of the Cause of God. For they share the victory with their beloved commander, he who raised them up and appointed them. They kept the ship on its course and brought it safe to port. The Universal House of Justice, with pride and love, recalls on this supreme occasion its profound admiration for the heroic work which they have accomplished. We do not wish to dwell on the appalling dangers which faced the infant Cause when it was suddenly deprived of our beloved Shoghi Effendi, but rather to acknowledge with all the love and gratitude of our hearts the reality of the sacrifice, the labor, the self-discipline, the superb stewardship of the Hands of the Cause of God. We can think of no more fitting words to express our tribute to these dearly loved and valiant souls than to recall the Words of Bahá'u'lláh Himself: "Light and glory, greeting and praise be upon the Hands of His Cause, through whom the light of fortitude hath shone forth and the truth hath been established that the authority to choose rests with God, the Powerful, the Mighty, the Unconstrained, through whom the ocean of bounty hath surged and the fragrance of the gracious favors of God, the Lord of mankind, hath been diffused."[5]

1.4 The members of the Universal House of Justice, all being in Haifa at the time of the election, were able to visit the Holy Shrines of Bahá'u'lláh, the Báb and of 'Abdu'l-Bahá where they prostrated themselves at the Sacred Thresholds and humbly sought strength and assistance in the mighty task before them. Later in London they have paid homage at the resting-place of Shoghi Effendi, the blessed and sacred bough of the Tree of Holiness.

1.5 As soon as the House of Justice is able to organize its work and deploy its forces it will examine carefully all the conditions of the Cause of God, and

---

1-4. The Heroic Age, the first of three Ages of the Bahá'í Dispensation, began in 1844 with the Declaration of the Báb, ended in 1921 with the passing of 'Abdu'l-Bahá, and spanned the ministries of the Báb, Bahá'u'lláh, and 'Abdu'l-Bahá. It was followed by the Formative Age, which in turn will be followed by the Golden Age. For a fuller discussion of the Ages of the Bahá'í Dispensation, see the glossary.
1-5. TB, p. 83.

communications will be made to the friends. At this time we call upon the believers everywhere to follow up vigorously the opportunities opened up by the World Crusade. Consolidation and deepening must go hand in hand with an eager extension of the teaching work so that the onward march of the Cause may continue unabated in preparation for future plans. Now that the attention of the public is becoming more and more drawn to the Cause of God the friends must brace themselves and prepare their institutions to sustain the gaze of the world, whether it be friendly or hostile, eager or idle.

The Universal House of Justice greets you all lovingly and joyfully at this time, and asks you to pray fervently for its speedy development and the spiritual strengthening of its members. 1.6

2
Message to National Conventions 1963
7 May 1963

To the annual National Conventions of the Bahá'í World

Beloved Friends,

The marvelous happenings which have transpired during and immediately after the twelve days of Riḍván attest the greatness of the Cause of God, and fill every Bahá'í heart to overflowing with joy and gratitude.[1] It was in obedience to the summons of the Lord of Hosts Himself that the elected representatives of the fifty-six national and regional communities of the Bahá'í world were called to elect, in the shadow of God's Holy Mountain and in the house of the Center of His Covenant, the members of the Universal House of Justice.[2] It was the Sign of God on Earth, the Dayspring of Divine Guid- 2.1

2-1. The happenings referred to were the First International Bahá'í Convention held 21–23 April 1963 at the Bahá'í World Center, during which the Universal House of Justice was first elected, and the first Bahá'í World Congress, held in London 28 April through 2 May 1963. Immediately following the World Congress the Universal House of Justice met for several days in the council room of the National Spiritual Assembly of the Bahá'ís of the British Isles at 27 Rutland Gate, dealing with essential matters concerning its establishment and initial functioning. At the same time, the Hands of the Cause of God were meeting in conclave in another room of the same Ḥaẓíratu'l-Quds, preparing a first draft of the next teaching plan for submission to the Universal House of Justice and arranging the future disposition of their own work. For the message issued by the Hands of the Cause of God announcing the results of the election of the Universal House of Justice, see the epigraph.

2-2. Bahá'u'lláh wrote, in *Tablets of Bahá'u'lláh* (pp. 27, 68): "Inasmuch as for each day there is a new problem and for every problem an expedient solution, such affairs should be referred to the Ministers of the House of Justice that they may act according to the needs and requirements of the time.... It is incumbent upon all to be obedient unto them." "It is incumbent upon the Trustees of the House of Justice to take counsel together regarding those things which have not outwardly been revealed in the Book, and to enforce that which is agreeable to them.

ance, the Guardian of the Cause of God, who gathered more than six thousand Bahá'ís from all parts of the earth to the celebration of the Most Great Jubilee in London.

Events of Spiritual and Administrative Significance

2.2 The first of these historic occasions was marked by events of extreme spiritual and administrative significance at the World Center of the Faith. The daily visits of large groups of believers, of many varying backgrounds, to the sacred Shrines in the twin holy cities;[3] the holding of the First International Bahá'í Convention and the successful accomplishment of its main task; the celebration of the Riḍván Feast by some three hundred believers in the company of the Hands of the Cause of God in the precincts of the Ḥaram-i-Aqdas,[4] are events of unique character and untold significance in the history of our beloved Faith.

2.3 The celebration of the Most Great Jubilee in London must be described elsewhere. Suffice it to say now that this greatest gathering of Bahá'ís ever held in one place was permeated by a spirit of such bliss as could only have come from the outpourings of the Abhá Kingdom. The review of the progress of the Cause, the presentation of believers from the new races and countries of the world brought within the pale of the Faith during the Beloved Guardian's Ten Year Crusade, of the Knights of Bahá'u'lláh, those valiant souls who carried the banner of Bahá'u'lláh to the unopened and often inhospitable regions of the earth, the spontaneous outbursts of singing of "Alláh-u-Abhá," the informal gatherings, the constant greetings of Bahá'u'lláh's warriors known to each other only by name and service, the youth gatherings, the unprecedented publicity in the press, on radio and television, the daily stream of visitors to the beloved Guardian's resting-place, the radiant faces and heightened awareness of the true and real brotherhood of the human race within the Kingdom of the Everlasting Father, are among the outstanding events of this supreme occasion, the crowning victory of the lifework of Shoghi Effendi.

Reaffirmation of Tribute to the Hands of the Cause of God

2.4 The Universal House of Justice wishes to reaffirm at this time the tribute which it felt moved to pay to the Hands of the Cause of God at the World

---

God will verily inspire them with whatsoever He willeth, and He, verily, is the Provider, the Omniscient." 'Abdu'l-Bahá, in His Will and Testament (p. 19), wrote: "Unto the Most Holy Book every one must turn, and all that is not expressly recorded therein must be referred to the Universal House of Justice." The election of the Universal House of Justice was held in the House of 'Abdu'l-Bahá in Haifa, 7 Haparsim Street, at the foot of Mount Carmel, on 21 April 1963. The results of the election were announced to the delegates the following day, and the Hands of the Cause of God announced the results in a cable to the Bahá'í world (see epigraph).

2-3. Haifa and 'Akká.

2-4. Ḥaram-i-Aqdas (the Most Holy Court) is a designation Shoghi Effendi gave to the northwestern quadrant of the garden surrounding the Shrine of Bahá'u'lláh in Bahjí.

Congress, those precious souls who have brought the Cause safely to victory in the name of Shoghi Effendi. We wish also to remember the devoted work of their Auxiliary Board members, as well as the services of the Knights of Bahá'u'lláh, of the army of pioneers, the members of the National and Regional Spiritual Assemblies, the services and prayers and sacrifices of the believers everywhere, all of which in the sum total have attracted such bounties and favors from Bahá'u'lláh.

The Universal House of Justice, in several sessions held in the Holy Land and in London, has been able to initiate its work and to make arrangements for the establishment of the Institution in Haifa. It has no officers, and henceforth its communications to the Bahá'í World will be signed *Universal House of Justice* over an embossed seal.

Second Epoch of 'Abdu'l-Bahá's Divine Plan

The Cause of God, launched on the sea of the Divine Plan of 'Abdu'l-Bahá, has achieved, under the superb leadership of its beloved Guardian, a spread throughout the world and a momentum which must now carry it forward on the next stage of its world-redeeming mission, the second epoch of the Divine Plan.⁵ The Universal House of Justice, in close consultation with the Hands of the Cause, is examining the vast range of Bahá'í activity and growth in order to prepare a detailed plan of expansion for the whole Bahá'í community, to be launched at Riḍván 1964. But there are some objectives to be achieved at once.

Immediate Objectives

The consolidation of the goals and new communities of the Bahá'í world is an urgent and immediate task facing the fifty-six National Spiritual Assemblies, and an essential preparation for the launching of the new plans. Pioneers must be maintained at their posts and all the Local Spiritual Assemblies strengthened through a firm establishment of Bahá'í community life and an active teaching program. Those National Spiritual Assemblies which rest on the basis of a small number of Local Spiritual Assemblies must make great efforts to insure that this number will be increased at Riḍván 1964. Pioneers ready to go to consolidation areas, as well as those eager to open new territories, should make their offers through their National Spiritual Assembly.

The great work of teaching must be extended, not only in those areas where mass conversion is beginning, but everywhere. The high intensity of teaching activity reached at the end of the World Crusade, far from slackening, must now be increased as the friends everywhere draw on the vast spiritual powers released as a result of the celebration of the Most Great Jubilee and the emergence of the Universal House of Justice.

2-5. For information on the epochs of the Divine Plan, see message no. 451.

2.9 The Ten Year Crusade witnessed the completion of the structure of the Mother Temple of Europe.[6] It is now imperative to complete, without delay, the interior decoration, to install utilities and lay access roads, to landscape grounds and to construct the caretaker's house. This work will cost not less than $210,000.00, but if delayed it will cost considerably more. The House of Justice calls upon the National Spiritual Assemblies to allocate substantial budgets for the immediate completion of this work.

Preview of the New Plan

2.10 The plan to be embarked upon next Riḍván,[7] the details of which will be announced during the coming year, will include such projects as the extension and embellishment of the endowments at the World Center; collation of the Writings of Bahá'u'lláh, 'Abdu'l-Bahá and Shoghi Effendi; continual reinforcement of the ties binding the Bahá'í World to the United Nations;[8] formation of many more National Spiritual Assemblies, both by division of existing Regional Spiritual Assemblies and the development of new Bahá'í communities, together with the purchase of national Ḥaẓíratu'l-Quds, Temple sites and national endowments; the opening of new territories to the Faith; detailed plans for National Spiritual Assemblies involving, in some areas, consolidation goals, in others the multiplication of Bahá'í institutes and schools, in others a great enrichment of Bahá'í literature, and in all a vast increase in the number of Bahá'ís, and the holding of oceanic and intercontinental conferences.

2.11 All such expansion and development of the Faith will be dependent upon the Bahá'í Fund. The Universal House of Justice calls the attention of every believer to this vital and pressing matter, and asks the National Spiritual Assemblies to pay special attention to the principle of universal participation, so that every single follower of Bahá'u'lláh may make his offering, however small or great, and thereby identify himself with the work of the Cause everywhere. It is our hope that a constant flow of contributions to the International Fund will make it possible to build up sufficient reserves for the launching of the new plan in 1964.

2.12 Beloved friends, we enter the second epoch of the Divine Plan blessed beyond compare, riding the crest of a great wave of victory produced for us by our beloved Guardian. The Cause of God is now firmly rooted in the

---

2-6. The Mother Temple of Europe is the Bahá'í House of Worship in Langenhain, near Frankfurt am Main, Germany. Its cornerstone was laid on 20 November 1960; construction was completed in the spring of 1963. It was dedicated in July 1964 (see message no. 17).
2-7. The Nine Year Plan, 1964–73.
2-8. For more details on the development of the relationship between the Bahá'í community and the United Nations, see messages no. 49, 78, 263, 283.

world. Forward then, confident in the power and protection of the Lord of Hosts, who will, through storm and trial, toil and jubilee, use His devoted followers to bring to a despairing humanity, the life-giving waters of His supreme Revelation.

THE UNIVERSAL HOUSE OF JUSTICE

3
Seat of the Universal House of Justice; New Arrangements for Pilgrims
16 JUNE 1963

To all National Spiritual Assemblies

Dear Friends,

3.1 The Universal House of Justice has been deeply moved and its hopes have been raised high by the many messages of love, devotion and eager anticipation which have been received from National Conventions and National Spiritual Assemblies.

3.2 Two decisions have been taken by the Universal House of Justice involving a further development of the institutions at the World Center. The former offices of the International Bahá'í Council at 10 Haparsim Street being inadequate for the far greater volume of work facing the Universal House of Justice, it has been decided to take over the whole of this building (until now called the Western Pilgrim House) as the seat, for the present time, of the Universal House of Justice.[1]

3.3 This decision made it necessary to find other accommodation for the western pilgrims and led directly to the second decision. After careful consideration of the alternatives the House of Justice has decided that the time has come to take the significant step, anticipated by our beloved Guardian, of housing all pilgrims in one place.[2] It was found possible, by slight alterations, to accommodate all pilgrims, without lessening the number, in the former Eastern Pilgrim House and its adjacent buildings. We have therefore established one Pilgrim House, at the Bahá'í Gardens on Mount Carmel. The friends should note that this is where they should go on arrival.

3-1. The International Bahá'í Council, a body first appointed by Shoghi Effendi in 1951 and elected in 1961 after his passing by members of National Spiritual Assemblies, ceased to exist upon the election of the Universal House of Justice in 1963. The Universal House of Justice made 10 Haparsim Street its seat until 1 February 1983; the building now houses the International Teaching Center. For more information on the International Bahá'í Council, see the glossary.

3-2. In April 1969, during the Nine Year Plan, the Universal House of Justice took a new action, opening the door of pilgrimage to a greater number of believers and instructing the National Spiritual Assemblies in procedures to be followed. See the Riḍván 126 (1969) message, no. 68.

3.4 All friends whose pilgrimages have been confirmed for 1963–64 are therefore expected. There are still vacancies after December 1963, but only a very few before that date.

3.5 We have asked the Hands of the Cause residing in the Holy Land to continue to be responsible for the program of the pilgrims while they are here, but letters requesting permission to come should be addressed to the Universal House of Justice.

With loving greetings,
In His Service,
THE UNIVERSAL HOUSE OF JUSTICE

4
Destruction of the 'Ishqábád Temple
25 AUGUST 1963

To the National Spiritual Assemblies of the Bahá'í World

Dear Bahá'í Friends,

4.1 The whole Bahá'í World will be grief-stricken at the news of the sad fate which has overtaken the Mashriqu'l-Adhkár in 'Ishqábád, the first Temple raised to the glory of Bahá'u'lláh.¹ Due to its unsafe condition, resulting from earthquakes, the building has been entirely demolished and the site cleared.

4.2 The building of this edifice, the only structure of its kind to be raised and completed in the lifetime of 'Abdu'l-Bahá, was described by the beloved Guardian as "a lasting witness to the fervor and the self-sacrifice of the Oriental believers." This "enterprise," the Guardian further wrote, "must rank not only as the first major undertaking launched through the concerted efforts of His followers in the Heroic Age of His Faith, but as one of the most brilliant and enduring achievements in the history of the first Bahá'í century."²

4.3 The Bahá'í center in 'Ishqábád was founded in the days of Bahá'u'lláh. Already during His lifetime preliminary steps had been adopted by the friends of that community to build, in accordance with the provisions of the Most Holy Book, a Mashriqu'l-Adhkár.³

4.4 However, the project had to be postponed until 1902, at the end of the first decade of the Ministry of 'Abdu'l-Bahá, when He initiated its construction, called on the friends in the East to offer their contributions towards the

---

4-1. The temple was located in Turkmenistan, near the Iranian border, north of the Iranian province of Khurásán.
4-2. GPB, p. 300. The first Bahá'í century ended in 1944.
4-3. The Most Holy Book is the Kitáb-i-Aqdas, Bahá'u'lláh's book of laws.

fulfillment of this goal, and personally encouraged and directed its development at every stage. The Báb's cousin, the venerable Ḥájí Mírzá Muḥammad-Taqí, the Vakílu'd-Dawlih, offered his total wealth towards this meritorious enterprise, established his residence in that city, and personally supervised its construction.

The laying of the cornerstone of this edifice at a ceremony attended by the delegate of the Czar—the Governor-General of Turkistan—and the initial steps taken to raise this first House of Worship of the Bahá'í World, inspired the friends in America, who, in 1903, eager to demonstrate the quality of their faith, petitioned 'Abdu'l-Bahá for permission to erect the first Ma_sh_riqu'l-Adhkár of the West. 4.5

In addition to the Temple itself, two schools, one for boys and one for girls, and a pilgrim house were built. The local community, and the activities of the friends throughout the provinces of Turkistan expanded and developed in stature until 1928, when the law expropriating religious edifices was applied to this Temple. However, under the terms of two five-year leases, the Bahá'í community was permitted to continue to use the building as a house of worship. In 1938 the Temple was completely expropriated and converted into an art gallery. 4.6

In 1948 violent earthquakes shook the whole town causing devastation and ruin. The building was seriously damaged. The only section which remained relatively secure was the central rotunda. Heavy yearly rains further weakened the structure to such a degree as to endanger the safety of houses in the vicinity. It was at this point that the authorities decided to demolish the remaining edifice and clear the site. 4.7

A reliable report, recently received, indicates that had the Temple been restored to us at this point, we should have had no option but to raze the building ourselves. 4.8

Please share this news with the friends but we do not wish National Assemblies, Local Assemblies or individual believers to take any action. 4.9

With loving greetings,
THE UNIVERSAL HOUSE OF JUSTICE

5

The Guardianship
6 October 1963

To all National Spiritual Assemblies

Beloved Friends,

5.1 We wish to share with you the text of the following resolution:

> After prayerful and careful study of the Holy Texts bearing upon the question of the appointment of the successor to Shoghi Effendi as Guardian of the Cause of God, and after prolonged consideration of the views of the Hands of the Cause of God residing in the Holy Land, the Universal House of Justice finds that there is no way to appoint or to legislate to make it possible to appoint a second Guardian to succeed Shoghi Effendi.

5.2 Please share this message with the friends in your jurisdiction.

With loving greetings,
THE UNIVERSAL HOUSE OF JUSTICE

6

Announcement of the Nine Year Plan
October 1963

To the Followers of Bahá'u'lláh throughout the World

Fellow believers in the Cause of God,

6.1 Six years ago when nearing the midway point of the Ten Year Crusade, the Bahá'í World found itself abruptly deprived of the guiding hand of its beloved Guardian. The anguish which then seized our hearts, far from paralyzing the progress of the Cause, stiffened our resolve and fired our zeal to complete the tasks which God, through His Chosen Branch, had laid upon us.[1] The august institution of the Hands of the Cause of God which he had, but recently, in compliance with the instructions of the Master's Will, raised up, kept the people of this Cause faithfully to the path which had been shown to us by the pen of divine guidance, and brought us not only to the triumphal conclusion of that Crusade but to the culminating point of the construction of the framework of Bahá'u'lláh's World Order.[2]

6-1. The Chosen Branch is Shoghi Effendi, Bahá'u'lláh's great-grandson.
6-2. 'Abdu'l-Bahá included in His Will and Testament a provision for the appointment of Hands of the Cause of God (see WT, pp. 12–13).

In March 1930 Shoghi Effendi wrote that Bahá'u'lláh and 'Abdu'l-Bahá had "in unequivocal and emphatic language, appointed those twin institutions of the House of Justice and of the Guardianship as their chosen Successors, destined to apply the principles, promulgate the laws, protect the institutions, adapt loyally and intelligently the Faith to the requirements of progressive society, and consummate the incorruptible inheritance which the Founders of the Faith have bequeathed to the world." After long and prayerful consultation, the House of Justice, as the friends have already been informed, found that there is no way in which it can legislate for a second Guardian to succeed Shoghi Effendi.³ The Universal House of Justice has therefore begun, in humble obedience to the Will of God, and strengthened by daily prayer in the Holy Shrines, to undertake the heavy tasks laid upon it. In the words of our beloved Guardian it "will guide, organize and unify the affairs of the Movement throughout the world" and "will have to consider afresh the whole situation, and lay down the principle which shall direct, so long as it deems advisable, the affairs of the Cause."⁴ 6.2

The Covenant of Bahá'u'lláh

The Covenant of Bahá'u'lláh is unbroken, its all-encompassing power inviolate. The two unique features which distinguish it from all religious covenants of the past are unchanged and operative. The revealed Word, in its original purity, amplified by the divinely guided interpretations of 'Abdu'l-Bahá and Shoghi Effendi, remains immutable, unadulterated by any man-made creeds or dogmas, unwarrantable inferences or unauthorized interpretations. The channel of divine guidance, providing flexibility in all the affairs of mankind, remains open through that Institution which was founded by Bahá'u'lláh and endowed by Him with supreme authority and unfailing guidance, and of which the Master wrote: "Unto this body all things must be referred."⁵ How clearly we can see the truth of Bahá'u'lláh's assertion: "The Hand of Omnipotence hath established His Revelation upon an unassailable, an enduring foundation. Storms of human strife are powerless to undermine its basis, nor will men's fanciful theories succeed in damaging its structure."⁶ 6.3

Responsibility of the Institutions

As the significance of the Cause of God continues in the years ahead to become more clearly apparent to the eyes of men, a great responsibility to watch over its security rests upon all of its institutions. The Institution of the Hands of the Cause of God, charged in the sacred Texts with the specific 6.4

6-3. See message no. 6.
6-4. BA, pp. 39, 41.
6-5. WT, p. 14.
6-6. Quoted in WOB, p. 109.

duties of protecting and propagating the Faith, has a particularly vital responsibility to discharge. In their capacity as protectors of the Faith, the Hands will continue to take action to expel Covenant-breakers and to reinstate those who sincerely repent, subject in each instance to the approval of the Universal House of Justice.7 Exercising their function of propagating the Faith, the Hands of the Cause will inspire, advise and assist the National Spiritual Assemblies in the work as they did in the time of our beloved Shoghi Effendi, assisted by the members of their Auxiliary Boards who will continue to fulfill those functions outlined for them by him.

6.5 We stand now upon the threshold of the second epoch of 'Abdu'l-Bahá's Divine Plan, with the outposts of the Cause established in the remotest corners of the earth, and having already witnessed the beginnings of that entry into the Faith by troops promised by the Master Himself. The foundation of the Kingdom has been securely laid, the framework has been raised. The friends must now consolidate these achievements, safeguard their institutions and gather the peoples and kindreds of the world into the Ark which the Hand of God has built.

Second Global Plan

6.6 Next Riḍván will be launched the second of those world-encircling enterprises destined in the course of time to carry the Word of God to every human soul. The Standard-Bearers of this Nine Year Plan are the Hands of the Cause of God. The responsibility for directing the work will rest upon the shoulders of the National Spiritual Assemblies, the generals of the Army of Light, under the guidance of the Universal House of Justice.

6.7 As the first step inaugurating this great undertaking we rejoice to announce the formation next Riḍván of nineteen National Spiritual Assemblies, resulting in the dissolution of six of the existing Regional National Spiritual Assemblies, and bringing the total number of these pillars of the Universal House of Justice to sixty-nine. The National and Regional National Assemblies now to be formed are:

1. The National Spiritual Assembly of the Bahá'ís of North West Africa, with its seat in Tunis, comprising Tunisia, Algeria, Morocco, Spanish Sahara, Rio de Oro, Mauritania, the Canary Is., and Madeira.
2. The National Spiritual Assembly of the Bahá'ís of West Africa, with its seat in Monrovia, comprising Liberia, Sénégal, Gambia, Portuguese

---

6-7. To maintain the unity and incorruptibility of the Faith, the Covenant of Bahá'u'lláh established a Center of authority to which all are to turn. This Center has been, successively, 'Abdu'l-Bahá, Who is uniquely the Center of the Covenant; Shoghi Effendi, the Guardian of the Faith; and the Universal House of Justice, which is the apex of its Administrative Order. A Bahá'í who turns against and defies this Center breaks the Covenant and, if he is adamant in his disobedience, is expelled from the Faith as a Covenant-breaker.

Guinea, Guinée, Sierra Leone, Mali, Upper Volta, Ivory Coast and Cape Verde Is.
3. The National Spiritual Assembly of the Bahá'ís of West Central Africa, with its seat in Victoria, comprising Cameroon, Spanish Guinea, St. Thomas I., Fernando Po I., Corisco I., Nigeria, Niger, Dahomey, Togo, and Ghana.
4. The National Spiritual Assembly of the Bahá'ís of Uganda and Central Africa, with its seat in Kampala, comprising Uganda, Rwanda, Burundi, the Republic of the Congo (Ex-Belgian), the Congo Republic (Ex-French), Central African Republic, Gabon and Chad.
5. The National Spiritual Assembly of the Bahá'ís of Kenya, with its seat in Nairobi.
6. The National Spiritual Assembly of the Bahá'ís of Tanganyika and Zanzibar, with its seat in Dar-es-Salaam, comprising Tanganyika, Zanzibar, Mafia I., and Pemba I.
7. The National Spiritual Assembly of the Bahá'ís of South Central Africa, with its seat in Salisbury, comprising Nyasaland, Northern Rhodesia, Southern Rhodesia and Bechuanaland.
8. The National Spiritual Assembly of the Bahá'ís of South and West Africa, with its seat in Johannesburg, comprising Angola, Southwest Africa, South Africa, Zululand, Swaziland, Basutoland, Mozambique and St. Helena.
9. The National Spiritual Assembly of the Bahá'ís of the Indian Ocean, with its seat in Port Louis, comprising Mauritius, the Malagasy Republic, Réunion I., Seychelles Is., Comoro Is., and the Chagos Archipelago.
10. The National Spiritual Assembly of the Bahá'ís of the Hawaiian Islands, with its seat in Honolulu.
11. The National Spiritual Assembly of the Bahá'ís of the South Pacific Ocean, with its seat in Suva, comprising the Gilbert and Ellice Is., Nauru I., Fiji, Samoa Is., Tonga Is., and Cook Is.
12. The National Spiritual Assembly of the Bahá'ís of the South West Pacific Ocean, with its seat in Honiara, comprising the Solomon Is., New Hebrides Is., New Caledonia and Loyalty Is.
13. The National Spiritual Assembly of the Bahá'ís of North East Asia, with its seat in Tokyo, comprising Japan, Formosa, Hong Kong and Macau.
14. The National Spiritual Assembly of the Bahá'ís of Korea, with its seat in Seoul.
15. The National Spiritual Assembly of the Bahá'ís of Malaysia, with its seat in Kuala Lumpur, comprising Malaya, Singapore, Brunei, Sabah and Sarawak.
16. The National Spiritual Assembly of the Bahá'ís of Indonesia, with its seat in Djakarta, comprising Indonesia, the Mentawai Is., Portuguese Timor and West Irian.

17. The National Spiritual Assembly of the Bahá'ís of Vietnam, with its seat in Saigon, and having jurisdiction over the Bahá'ís of Cambodia.
18. The National Spiritual Assembly of the Bahá'ís of Thailand, with its seat in Bangkok, and having jurisdiction over the Bahá'ís of Laos.
19. The National Spiritual Assembly of the Bahá'ís of the Philippines, with its seat in Manila.

6.8 The detailed goals of the Plan, which will include sixty-nine national plans, have yet to be announced, but they must be such as to develop still further the World Center of the Faith and the work of its institutions; to consolidate those territories which have already been opened to the Faith; to bring God's healing Message to many more of the peoples and territories of the world including all the unopened territories of the Ten Year Crusade and all the remaining independent states of the planet; and to achieve worldwide proclamation of the Faith to mark the Centenary of Bahá'u'lláh's Proclamation to the kings and rulers in 1867–1868.[8]

6.9 In the spring of 1968 the next election for the Universal House of Justice will take place.

Third Epoch of the Formative Age

6.10 Beloved friends, the Cause of God, guarded and nurtured since its inception by God's Messengers, by the Center of His Covenant and by His Sign on earth, now enters a new epoch, the third of the Formative Age.[9] It must now grow rapidly in size, increase its spiritual cohesion and executive ability, develop its institutions and extend its influence into all strata of society. We, its members, must, by constant study of the life-giving Word, and by dedicated service, deepen in spiritual understanding and show to the world a mature, responsible, fundamentally assured and happy way of life, far removed from the passions, prejudices and distractions of present day society. Relying upon God alone, we can promote His Cause and establish His Kingdom on earth. Only thus can we prove our love for Those Who brought this new day into being. Only thus can we prove the truth of Their Divine Mission and demonstrate how valid was Their Sacrifice.

THE UNIVERSAL HOUSE OF JUSTICE

6-8. For details on Bahá'u'lláh's proclamation to the kings and rulers of the world and a description of the events held throughout the Bahá'í world to commemorate its anniversary, see letters dated Riḍván 1965, 17 March 1967, and October 1967 (messages no. 24, 41, and 46). For the Tablets Bahá'u'lláh sent to the kings and rulers of the world, see *Proclamation of Bahá'u'lláh*.

6-9. For information on the Formative Age, see entries on Ages and Epochs in the glossary.

7
Request for Prayers for Moroccan Bahá'í Prisoners
17 October 1963

To all National Spiritual Assemblies

Dear Friends,

7.1 Our Bahá'í friends who were tried at Nador, Morocco, last year and who received sentences of death or life imprisonment are still being held in prison.¹ Their appeal which has been pending for nearly a year still has not been heard. However, we are pleased to report that another Bahá'í prisoner, Mohammad Manaan, of Tangier was recently provisionally released.

7.2 We are calling for believers throughout the world to pray at the Feast of Qawl (November 23rd) that early favorable action may be taken in the cases of these devoted and steadfast friends whose long suffering on behalf of our Faith has been an inspiration to Bahá'ís everywhere.

7.3 The friends at the World Center will join you on that day by saying prayers at the Holy Shrines for our Moroccan friends.

With loving greetings,
THE UNIVERSAL HOUSE OF JUSTICE

7-1. In April 1962 fourteen Moroccan Bahá'í men were arrested in Nador. On 31 October 1962, after more than six months of imprisonment, they were tried before the Criminal Court of Nador on charges of rebellion and disorder, attacks on public security, constitution of an association of criminals, and attacks on religious faith. On 10 December 1962 they were found guilty and sentenced. Their cases were then appealed to the Supreme Court of Morocco, which overruled the lower court's verdict and released the men. A worldwide campaign to publicize the plight of the Bahá'í prisoners focused public opinion, favorable to the Bahá'ís, on the Moroccan authorities. For further information, see BW 13:288–89, and messages no. 8 and 12.

8
Report on Moroccan Bahá'í Prisoners
15 November 1963

To the National Spiritual Assemblies of the
British Isles, France, and the United States

Dear Friends,

8.1 The case of the Bahá'í prisoners in Morocco has passed through three stages:

(a) The arrest, detention and trial which began in April, 1962 and ended on December 15, 1962 when they appeared in court and received their respective sentences.

(b) The second stage began when the verdict was known and widely publicized, attracting the attention of many influential people both within and without Morocco resulting in widespread sympathy on behalf of the prisoners and many petitions to the Moroccan government.

(c) The third stage has been characterized by relative quiet and lack of publicity while all are waiting for the appeal to be heard. We are now in this stage, the most difficult of all.

8.2 The Universal House of Justice wishes to share with you a summary of the present situation as it appears from reports received from many sources, and to outline the policies which it believes should govern our immediate actions.

8.3 The prisoners, except for one who will be referred to later, are still in prison awaiting their unknown fate with great firmness and patience. For more than a year and a half they have withstood all manner of harassment and difficulty. We are informed that the three sentenced to death have been kept in solitary confinement for twenty-three hours of every day. Because of recent political developments it has become more and more difficult for Bahá'í friends to meet these dear souls in prison, but when it has been possible the Bahá'ís and non-believers alike are astonished that the prisoners are able to demonstrate such exemplary steadfastness and patience.

8.4 It is still not known when the appeal will be heard. At first it was promised that it would be among the first cases taken up after the courts reconvened in October. Again, some felt that it would be delayed until the opening of parliament in November. Now it is thought that the appeal will be heard very soon because it has been presented to the Court in such a way that they cannot delay it longer than one month. A favorable result is also anticipated.

8.5 The prisoner in Tangier, Mr. Ma'anan, has been provisionally released. His release followed a finding by the magistrate of that city that he could not find any convincing evidence in the file proving the charges regarding political subversion and incitement against Islam were justified. In the view of the magistrate the prisoner is guilty of no other crime than that of being a Bahá'í. Nevertheless, the prosecution appealed from this finding, and it was only after later efforts on the part of the legal committee in Morocco that Mr. Ma'anan was released.

8.6 Recently there has been a proposal for a change in the law which would provide a penalty of up to three years in prison for persons who are convicted of enticing Muslims to change their faith. While this law cannot be applied retroactively to the Bahá'í prisoners, it is felt that this change in the law together with the result in the Tangier case will have a favorable effect on the other cases since it indicated the injustice of the convictions and the harshness of the penalties meted out.

8.7 We have had many excellent suggestions from all of you as to courses of action which might be taken to secure the release of our fellow Bahá'ís. All have been carefully considered and weighed together in arriving at our present course of action. . . .

8.8 If the appeal should be further delayed, we must reconsider our present policy and do whatever is reasonably required to bring about an early hearing of the appeal. In the meantime the entire Bahá'í world is offering prayers on behalf of the prisoners at the Feast of Qawl.

8.9 This letter is for your own information and is not to be circulated.

8.10 With grateful appreciation and with warm Bahá'í love and greetings,

THE UNIVERSAL HOUSE OF JUSTICE

9
Development of National Bahá'í Funds
15 NOVEMBER 1963

To all National Spiritual Assemblies

Dear Friends,

9.1 While it is understood that many communities do not have the resources to fully sustain their administrative and teaching work, it is important that each National Assembly make every effort to work out a program and budget which will enable it to become self-sustaining as rapidly as possible.

9.2 To accomplish this, the friends should be made to appreciate the bounties which come from regular and systematic contribution to the National Fund. The amount of the contribution is not as important as universal participation. In a letter to the National Spiritual Assembly of Central and East Africa dated August 8, 1957 the beloved Guardian said:

9.2a   The institution of the National Fund, so vital and essential for the uninterrupted progress of these activities must, in particular, be assured of the wholehearted, the ever-increasing and universal support of the mass of believers, for whose welfare, and in whose name, these beneficent activities have been initiated and have been conducted. All, no matter how modest their resources, must participate. . . .

9.3   National Assemblies needing supplemental assistance for budgets after Riḍván, 1964 should direct their appeal to the Universal House of Justice. They should not appeal to another National Assembly except in specific cases authorized by the Universal House of Justice.

9.4   We shall pray that the bounties of the Almighty will descend upon the growing Administrative Order of His promised Kingdom on Earth so that we may be enabled to take His healing message to all peoples everywhere.

With loving greetings,
THE UNIVERSAL HOUSE OF JUSTICE

10
Relationship of Hands of the Cause of God and National Spiritual Assemblies
19 NOVEMBER 1963

To all National Spiritual Assemblies

Dear Friends,

10.1   The relationship of the Continental Hands[1] and their respective National Assemblies was clearly delineated in the Message of the beloved Guardian dated June 4, 1957:

10.1a   CALL UPON HANDS NATIONAL ASSEMBLIES EACH CONTINENT SEPARATELY ESTABLISH HENCEFORTH DIRECT CONTACT DELIBERATE WHENEVER FEASIBLE AS FREQUENTLY POSSIBLE EXCHANGE REPORTS TO BE SUBMITTED THEIR RESPECTIVE AUXILIARY BOARDS NATIONAL COMMITTEES EXERCISE UNRELAXING VIGILANCE CARRY OUT UNFLINCHINGLY SACRED INESCAPABLE DUTIES. SECURITY PRECIOUS FAITH PRESERVATION SPIRITUAL HEALTH BAHÁ'Í COMMUNITIES VITALITY FAITH ITS INDIVIDUAL MEMBERS PROPER FUNCTIONING ITS LABORIOUSLY ERECTED INSTITUTIONS FRUITION ITS WORLDWIDE ENTERPRISES FULFILLMENT ITS ULTIMATE DESTINY ALL DIRECTLY DEPENDENT BEFITTING DISCHARGE

10-1. Hands of the Cause of God to whom continental responsibilities are assigned as distinct from those whose responsibility it is to serve in the Holy Land.

WEIGHTY RESPONSIBILITIES NOW RESTING MEMBERS THESE TWO INSTITUTIONS OCCUPYING WITH UNIVERSAL HOUSE JUSTICE NEXT INSTITUTION GUARDIANSHIP FOREMOST RANK DIVINELY ORDAINED ADMINISTRATIVE HIERARCHY WORLD ORDER BAHÁ'U'LLÁH.

10.2 These highly significant words of the beloved Guardian clearly foreshadow the tremendous importance of the joint responsibility which he placed upon the Hands and the National Assemblies in protecting the believers and fostering the worldwide development of the Faith.

10.3 We stress the necessity of doing everything possible to expedite and facilitate this free play of consultation between the Hands and the National Assemblies within the spirit and framework of these instructions of Shoghi Effendi. The two institutions have joint and complementary functions which can be discharged successfully only if the greatest degree of understanding and cooperation exists between them.

10.4 We are confident that this relationship will be further strengthened and developed in the manner envisioned by the Guardian, and that it will contribute in large measure to the successful achievement of the goals of the Nine Year Plan which is to be inaugurated this coming Riḍván.

> With loving Bahá'í greetings,
> THE UNIVERSAL HOUSE OF JUSTICE

11
Auxiliary Board Members on National Spiritual Assemblies
25 November 1963

To all National Spiritual Assemblies

Beloved Friends,

11.1 The Hands of the Cause at their London Conclave last May decided that Auxiliary Board members should be free of administrative responsibilities in order to devote their full energies to their work as Board members.[1] Those Board members serving on National Assemblies were therefore requested to decide, before Riḍván 1964, in which capacity they could best serve the Cause.

11.2 In view of this request made by the Hands of the Cause to their Auxiliary Board members, the Universal House of Justice has reached the following decisions:

11.2a 1. National Assemblies in whose areas of jurisdiction Board members reside, should point out to the delegates at Convention that whilst teaching and administrative duties are not mutually exclusive, it is desirable that Auxiliary Board members, whether for teaching or protection, be left free to concentrate on the work allotted to them by the Hands of the Cause in each Continent. The following extract from the Guardian's letter, written through his secretary, could be shared with the delegates for their guidance when casting their votes:

11.2b "The teachers of the Cause can surely become members of any Assembly or committee. There should be no incapacity attached to them. But Shoghi Effendi would just prefer to see them devote all their time to teaching and leave the administrative functions for those who cannot serve as teachers." (*Bahá'í News*, October 1932)

11.2c 2. Should Board members still be elected on National Assemblies after the above explanation, it should rest with each Board member to decide

---

11-1. After Shoghi Effendi's passing in 1957, the functions of preserving the unity of the Bahá'í community, guiding it to the victorious conclusion of the Ten Year Crusade, and calling the election of the Universal House of Justice fell to the Hands of the Cause of God in their capacity as "the Chief Stewards of Bahá'u'lláh's embryonic World Commonwealth" (MBW, p. 127). In order to carry out these vital functions, the Hands of the Cause of God met periodically in conclaves at the World Center to confer on the affairs of the Cause. After the Most Great Jubilee, the one hundredth anniversary of Bahá'u'lláh's declaration of His mission, the Hands of the Cause of God met in London to discuss a number of matters, among those being the evolving role of the members of the Auxiliary Board in light of the increasing maturity of the Bahá'í Administrative Order, as evidenced by the formation of the Universal House of Justice at Riḍván 1963 and the existence of fifty-six National Spiritual Assemblies. The Hands of the Cause of God submitted their views to the Universal House of Justice, which made the decisions enumerated in this letter.

which of the two functions he feels best suited to perform. The Universal House of Justice, therefore, approves the request made by the Hands of the Cause to their Board members that they choose between the two functions.

3. In the event of the resignation of Auxiliary Board members from National Assemblies, this should be considered as good reason for resignation. If, however, the National Assembly, in view of special conditions, should consider such resignations to be detrimental to the interests of the Faith of that National Community, and the Board member should insist upon resigning on the grounds of his membership on the Board, the matter should be at once referred to the Universal House of Justice by the National Spiritual Assembly for examination and final decision. Pending such a decision, the Board member should continue his membership on the National Assembly, and explain his position to the Hands of the Cause in his continent. 11.2d

4. Measures will have to be taken by National Assemblies according to normal procedure to fill any vacancies created in this way. If the vacancy is recognized while Convention is still in session, a by-election could be arranged before Convention disbands. 11.2e

With loving Bahá'í greetings,
THE UNIVERSAL HOUSE OF JUSTICE

12
Release of Moroccan Bahá'í Prisoners
18 DECEMBER 1963

To all National Spiritual Assemblies

Dear Friends,

Following twenty months of close confinement, and one year after hearing sentences of death and imprisonment imposed on them, our brave and steadfast brother believers in Morocco were set free on Friday, December 13, 1963. The long-sought objective has been obtained. 12.1

As we prayed for their release during the Feast of Qawl, now let the entire Bahá'í world join in prayers of thanksgiving for this joyful outcome and that the Blessed Beauty enabled these devoted friends to remain firm in their faith through this trying ordeal.[1] 12.2

12-1. See message dated 17 October 1963 (no. 7) requesting prayers worldwide at the Feast of Qawl (Speech), 23 November.

12.3 In view of the situation which continues to be delicate in Morocco, you are urged not to seek publicity in this matter. If you are contacted by the press, make no statement beyond expressing thankfulness for the just decision.

12.4 Each National Assembly is requested, wherever possible, to express gratification to King Hassan II of Morocco for this just decision of the Supreme Court. This should be done in writing through the Moroccan Embassy, or Consulate, serving your area.

12.5 We also request that wherever you consider it advisable you write letters of appreciation to individuals and organizations within your jurisdiction who offered assistance in connection with this case.

With warm and loving Bahá'í greetings,
THE UNIVERSAL HOUSE OF JUSTICE

13
The Bahá'í Funds
18 DECEMBER 1963

To the Bahá'ís of East and West

Dear Friends,

13.1 With the rapid approach of the launching of the Nine Year Plan, the Universal House of Justice feels that it is timely to lay clearly before the Bahá'ís of all countries, the needs of the Fund at all its levels: local, national, continental and international.

13.2 The continual expansion of the Faith and the diversification of the activities of Bahá'í communities make it more and more necessary for every believer to ponder carefully his responsibilities and contribute as much and as regularly as he or she can. Contributing to the Fund is a service that every believer can render, be he poor or wealthy; for this is a spiritual responsibility in which the amount given is not important. It is the degree of the sacrifice of the giver, the love with which he makes his gift, and the unity of all the friends in this service which bring spiritual confirmations. As the beloved Guardian wrote in August 1957: "All, no matter how modest their resources, must participate. Upon the degree of self-sacrifice involved in these individual contributions will directly depend the efficacy and the spiritual influence which these nascent administrative institutions, called into being through the power of Bahá'u'lláh, and by virtue of the Design conceived by the Center of His Covenant, will exert."[1]

13-1. The nascent administrative institutions referred to are the Local and National Spiritual Assemblies. Bahá'u'lláh ordained the institution of the Local Spiritual Assembly in the Kitáb-i-

Not only the individual's responsibility to contribute is important at this 13.3
time, but also the uses to which the fund is put and the areas in which it is
expended.

Much of the present rapid expansion of the Faith is taking place in areas 13.4
of great poverty where the believers, however much they sacrifice, cannot produce sufficient funds to sustain the work. It is these very areas which are the most fruitful in teaching, and a sum of money spent here will produce ten times—even a hundred times—the results obtainable in other parts of the world. Yet in the past months the Universal House of Justice has had to refuse a number of appeals for assistance from such areas because there just was not enough money in the International Fund.

It should therefore be the aim of every local and national community to 13.5
become not only self-supporting, but to expend its funds with such wisdom and economy as to be able to contribute substantially to the Bahá'í International Fund, thus enabling the House of Justice to aid the work in fruitful but impoverished areas, to assist new National Assemblies to start their work, to contribute to major international undertakings of the Nine Year Plan such as Oceanic Conferences, and to carry forward the work of beautifying the land surrounding the Holy Shrines at the World Center of the Faith.

Nor should the believers, individually or in their Assemblies, forget the 13.6
vitally important Continental Funds which provide for the work of the Hands of the Cause of God and their Auxiliary Boards. This divine institution, so assiduously fostered by the Guardian, and which has already played a unique role in the history of the Faith, is destined to render increasingly important services in the years to come.

In the midst of a civilization torn by strifes and enfeebled by materialism, 13.7
the people of Bahá are building a new world. We face at this time opportunities and responsibilities of vast magnitude and great urgency. Let each believer in his inmost heart resolve not to be seduced by the ephemeral allurements of the society around him, nor to be drawn into its feuds and short-lived enthusiasms, but instead to transfer all he can from the old world to that new one which is the vision of his longing and will be the fruit of his labors.

With loving greetings,
THE UNIVERSAL HOUSE OF JUSTICE

Aqdas (KA ¶30, n 49.); 'Abdu'l-Bahá established the institution of the National Spiritual Assembly (see WT, pp. 14–15).

The design conceived by 'Abdu'l-Bahá is embodied in His Will and Testament, the Charter of the New World Order.

1964–1973

The Nine Year Plan

14
Launching of the Nine Year Plan
APRIL 1964

To the Bahá'ís of the World
Dearly loved Friends,

14.1 The divinely propelled process, described in such awe-inspiring words by our beloved Guardian, which began six thousand years ago at the dawn of the Adamic cycle and which is destined to culminate in "the stage at which the light of God's triumphant Faith shining in all its power and glory will have suffused and enveloped the entire planet," is now entering its tenth and last part.[1]

14.2 The Ten Year Crusade, so recently consummated in a blaze of victory and rejoicing, constituted the entire ninth part of this process. It saw the Cause of God leap forward in one mighty decade-long effort to the point at which the foundations of its Administrative Order were laid throughout the world, thus preparing the way for that awakening of the masses which must characterize the future progress of the Faith.

14.3 From the beginning of this Dispensation the most urgent summons of the Word of God, voiced successively by the Báb and Bahá'u'lláh, has been to teach the Cause. 'Abdu'l-Bahá, in His own words, "spent His days and nights in promoting the Cause and urging the peoples to service."[2] Shoghi Effendi, discharging the sacred mission laid upon him, raised the Administrative Order of the Faith, already enshrined within the Sacred Writings, and forged it into a teaching instrument to accomplish through a succession of plans, national, international, and global, the entire Divine Plan of 'Abdu'l-Bahá, and he clearly foresaw in the "tremendously long" tenth part of the process already referred to, a series of plans to be launched by the Universal House of Justice, extending over "successive epochs of both the Formative and Golden Ages of the Faith."[3]

14-1. MBW, p. 155. The Bahá'í teachings assert that Adam was a Prophet of God and that He inaugurated the Adamic, or prophetic, cycle that ended with the advent of the Báb and that is followed by the cycle of fulfillment Bahá'u'lláh inaugurated. Shoghi Effendi explains that "The Faith of Bahá'u'lláh should indeed be regarded, if we wish to be faithful to the tremendous implications of its message, as the culmination of a cycle, the final stage in a series of successive, of preliminary and progressive revelations. These, beginning with Adam and ending with the Báb, have paved the way and anticipated with an ever-increasing emphasis the advent of that Day of Days in which He Who is the Promise of All Ages should be made manifest" (WOB, p. 103).

For a full description of the ten-part, divinely propelled process and for a discussion of the nature of cycles, see entries on Ten-Part Process and Cycles in the glossary.

14-2. WT, p. 10.
14-3. MBW, pp. 153, 155.

14.4 The first of these plans is now before us. Opening at Riḍván 1964, while the memories of the glorious Jubilee of 1963 still surge within our hearts, it must, during its nine-year course, witness a huge expansion of the Cause of God and universal participation by all believers in the life of that Cause.

World Center Goals

14.5 At the World Center of the Faith the tasks of the Plan include publication of a synopsis and codification of the Kitáb-i-Aqdas, the Most Holy Book; formulation of the Constitution of the Universal House of Justice; development of the Institution of the Hands of the Cause of God, in consultation with the body of the Hands of the Cause, with a view to the extension into the future of its appointed functions of protection and propagation; continued collation and classification of the Bahá'í Sacred Scriptures as well as of the writings of Shoghi Effendi; continued efforts directed towards the emancipation of the Faith from the fetters of religious orthodoxy and its recognition as an independent religion; the preparation of a plan for the befitting development and beautification of the entire area of Bahá'í property surrounding the Holy Shrines; extension of the existing gardens on Mount Carmel; development of the relationship between the Bahá'í Community and the United Nations; the holding of Oceanic and Intercontinental Conferences; the coordination of worldwide plans to commemorate, in 1967/68, the centenary of Bahá'u'lláh's Proclamation to the kings and rulers which centered round His revelation of the Súriy-i-Mulúk in Adrianople.[4]

International Goals

14.6 In the world community the Plan involves the opening of seventy virgin territories and the resettlement of twenty-four; the raising of the number of National Spiritual Assemblies, the pillars sustaining the Universal House of Justice, to one hundred and eight, nine times the number which embarked on the first historic World Crusade in 1953; increasing the number of Local Spiritual Assemblies to over thirteen thousand seven hundred, scattered throughout the territories and islands of the world, at least one thousand seven hundred of them to be incorporated; the raising of the number of localities where Bahá'ís reside to over fifty-four thousand; the building of two more Mashriqu'l-Adhkárs, one in Asia and one in Latin America;[5] the acquisition

---

14-4. Tablet to the Kings, referred to by Shoghi Effendi as "the most momentous Tablet revealed by Bahá'u'lláh" (GPB, p. 171). For passages of the Súriy-i-Mulúk that have been translated into English, see PB, pp. 7–12, 47–54, and 102–03. For details on the celebration of the centenary, see messages dated October 1963, Riḍván 1965, 17 March 1967, and October 1967 (nos. 6, 24, 41, and 46).

14-5. The Mashriqu'l-Adhkár of Asia was to be located in Tehran, Iran. Antagonism toward the Bahá'í Faith caused the plans for its construction to be held in abeyance. The Mashriqu'l-Adhkár of Latin America is in Panama and was dedicated in 1972.

of thirty-two Teaching Institutes, fifty-two national Ḥaẓíratu'l-Quds, fifty-four national Endowments, and sites for sixty-two future Temples; wide extension of recognition by civil authorities of the Bahá'í Holy Days and Bahá'í Marriage Certificates; the translation of literature into one hundred and thirty-three more languages, and its enrichment in major languages into which translations have already been made; the establishment of four new Bahá'í Publishing Trusts, and a vast increase in the financial resources of the Faith.

The Role of the Individual

14.7 The healthy development of the Cause requires that this great expansion be accompanied by the dedicated effort of every believer in teaching, in living the Bahá'í life, in contributing to the Fund, and particularly in the persistent effort to understand more and more the significance of Bahá'u'lláh's Revelation. In the words of our beloved Guardian, "One thing and only one thing will unfailingly and alone secure the undoubted triumph of this sacred Cause, namely the extent to which our own inner life and private character mirror forth in their manifold aspects the splendor of those eternal principles proclaimed by Bahá'u'lláh."[6]

Twin Objectives of the Plan

14.8 Expansion and universal participation are the twin objectives of this initial phase of the second epoch of the Divine Plan, and all the goals assigned to the sixty-nine National Communities are contributory to them. The process of cooperation between National Spiritual Assemblies, already initiated by the beloved Guardian, will, during the course of this Plan, apply to over two hundred specific projects and will further strengthen this process which may well assume great importance in future stages of the Formative Age.

14.9 Once more, dear friends, we enter the battle but with an incomparably greater array than that which embarked upon the World Crusade in 1953. To that small force of twelve national communities, now veteran campaigners, have been added fifty-seven new legions, each under the generalship of a National Spiritual Assembly, each destined to become a veteran of this and future campaigns. That Crusade began with slightly more than six hundred Local Spiritual Assemblies, the greater part of which were situated in Persia, North America and Europe; the home fronts now comprise nearly four thousand six hundred Local Spiritual Assemblies scattered throughout the continents and islands of the world. We begin this Plan with a tremendous momentum, exemplified by the addition, since last Riḍván, of over four thousand new centers and thirteen National Spiritual Assemblies, and by the beginning, in several countries, of that entry by troops into the Cause of God prophesied by 'Abdu'l-Bahá and so eagerly anticipated by Him.

14-6. BA, p. 66.

Standard-Bearers of the Plan

14.10 The Standard-Bearers of this Nine Year Plan are those same divinely appointed, tried and victorious souls who bore the standard of the World Crusade, the Hands of the Cause of God, whose advice and consultation have been invaluable in the working out of this Nine Year Plan. Supported by their "DEPUTIES, ASSISTANTS [and] ADVISERS,"[7] the members of the Auxiliary Boards, they will inspire and protect the army of God, lead through every breach to the limit of available resources and sustain those communities struggling over intractable or stony ground, so that by 1973 the celebrations befitting the centenary of the Revelation of the Most Holy Book may be undertaken by a victorious, firmly established, organically united world community, dedicated to the service of God and the final triumph of His Cause.

14.11 Therefore let each of the sixty-nine communities seize its tasks, at once consider how best to accomplish them within the allotted span, raise its band of pioneers, consecrate itself to unremitting labor and set out on its mission. Now is the golden opportunity. For whatever convulsions the waywardness of a godless and materialistic age may yet precipitate in the world, however grievous may be the effects of the rolling up of the present order on the plans and efforts of the Community of the Most Great Name, we must seize the opportunities of the hour and go forward confident that all things are within His mighty grasp and that, if we but play our part, total and unconditional victory will inevitably be ours.

THE UNIVERSAL HOUSE OF JUSTICE

15
Convention Greetings 1964
19 APRIL 1964

To the National Spiritual Assembly of the Bahá'ís of the United States

15.1 OCCASION RIḌVÁN FESTIVAL EXTEND LOVING GREETINGS HANDS DELEGATES FRIENDS PRESENT HISTORIC CONVENTIONS LAUNCHING FIRST ENTERPRISE SECOND EPOCH UNFOLDMENT 'ABDU'L-BAHÁ'S DIVINE PLAN. HAIL FORMATION THIRTEEN ADDITIONAL NATIONAL SPIRITUAL ASSEMBLIES TESTIFYING RESISTLESS DEVELOPMENT WORLD ORDER BAHÁ'U'LLÁH. CONVEY ALL FRIENDS JOYFUL NEWS DEDICATION MOTHER TEMPLE EUROPE JULY FOURTH. ASSURE ARDENT PRAYERS VICTORIES BOUNTIFUL FAVORS COURSE PLAN NOW LAUNCHED.

UNIVERSAL HOUSE OF JUSTICE

14-7. Shoghi Effendi, MBW, p. 59.

16
Guidelines for the Nine Year Plan
14 May 1964

To all National Spiritual Assemblies

Dear Friends,

16.1 We have been receiving with deep joy and gratification reports of the Conventions held throughout the Bahá'í world during the Riḍván period just concluded and the national and regional elections resulting in the formation of sixty-nine National Spiritual Assemblies.

16.2 The announcement of the goals of the Nine Year Plan has received universally a warm response. Individuals, Conventions and Assemblies have pledged loyalty and given assurance of the execution of the assigned goals with determination and dedication. In two cases a goal has already been declared accomplished.

16.3 If your Assembly has been given what appears to be a heavy task, particularly in terms of homefront goals, you should consider how this can be phased so that your teaching program is conducted systematically. When you do work out such an internal plan we would appreciate receiving a copy.

16.4 We feel that at this time certain guidelines could be given to the National Spiritual Assemblies to assist them in the understanding and attainment of the goals assigned to them.

16.5 **New and Unsettled Territories:** this aspect of the Plan is of vital importance. It involves the establishment of points of light in new territories, the further diffusion of the radiance of the Faith of God and is directly linked with the prestige and international status of the Faith. It is essential that the enthusiasm generated at the Conventions be capitalized upon and the pioneers who have offered their services are directed wisely and energetically to the posts to be opened and resettled.

16.6 **Acquisition of Properties:** as prices generally are soaring everywhere it is important that properties called for in the Plan be purchased as soon as funds are made available. Your National Assembly is no doubt conscious of the fact that wherever a Ḥaẓíratu'l-Quds has been allocated it should be a modest structure, acquired in a dignified location within the civil limits of the town or city named in the Plan. However, a Temple site could be near the city, as it is not essential that it be located within the civil limits of the city.

16.7 **National Endowments:** the object here is not to buy an extensive property. A token piece of land, possibly donated by a believer, is all that is required for the moment.

16.8 **Teaching Institute:** this is essentially an activity aimed at deepening the knowledge of the friends to prepare them for active participation in the teaching work. In some countries it may continue to be an activity conducted either in local Bahá'í Centers or possibly housed in hired quarters, like most Summer Schools. However, in other countries, and particularly in mass teaching areas, it may have to be a modest structure acquired or erected in the rural areas where the majority of the believers reside rather than in capital cities, to obviate transportation expenses for those attending.

16.9 **Recognition of the Faith:** such goals as national and local incorporations, as well as applications to obtain recognition of the Bahá'í Marriage Certificate and Bahá'í Holy Days, should be embarked upon only when circumstances are propitious and there is reasonable assurance that the goal can be attained.

16.10 **New Languages:** the translation of Bahá'í literature into new languages should be carefully coordinated with the settlement of pioneers in areas using such languages. Therefore, this is a goal which must be pursued vigorously and without delay so that the teaching work may proceed effectively in these areas.

16.11 If in the pursuit of any of the above or other goals your Assembly should experience any difficulty or need any clarification, you are welcome to write to us at once.

16.12 We assure you of our constant prayers at the Holy Shrines for the befitting discharge of your responsibilities under the Nine Year Plan.

With loving Bahá'í greetings,
THE UNIVERSAL HOUSE OF JUSTICE

17
Dedication of the Mother Temple of Europe, Langenhain, Germany
JULY 1964

To the beloved of God gathered in the European Teaching Conference called on the occasion of the dedication of the Mother Temple of Europe

Dear Bahá'í Friends,

17.1 We have just witnessed the dedication of the Mother Temple of Europe—a project of untold significance and tremendous potential for the spread of the light of God's Faith in that Continent. One of the major achievements called for by our beloved Guardian at the outset of the Ten Year Crusade, this Mashriqu'l-Adhkár was triumphantly raised during its closing years as the fruit of long and arduous labors in the face of determined opposition and upon the sacrificial gifts of believers from all parts of the world. Now dedicated in the opening months of the Nine Year Plan, it forms a striking link between these two great crusades, demonstrating afresh the organic progress of the Cause whereby the efforts exerted in one period bear fruit in the next, which in turn endow the Bahá'í Community with new and greater capacities for the winning of still greater victories.

17.2 You are now gathered in this Conference to deliberate on ways and means of accomplishing the goals which are set before you. Let every believer, as he considers in detail these various goals, bear in mind four supreme objectives: to carry the Message of Bahá'u'lláh to every stratum of society, not only in the towns and cities but also in the villages and country districts where the virus of materialism has had much less effect on the lives of men; to take urgent, wise and well-considered steps to spread the Faith to those countries of Eastern Europe in which it has not yet become established; to reinforce strongly the heroic band of pioneers in the islands of the Mediterranean and the North Sea—islands which are to play such an important role in the awakening of the entire continent—as well as to prosecute energetically the goals you are called upon to achieve in other continents and oceans; and to foster the cooperation between National Communities and between National Spiritual Assemblies and the Hands of the Cause of God which has contributed so markedly to the work of the Faith on that Continent and is so essential for its future development.

17.3 Above all let every European Bahá'í have ever-present in his mind that these are the five years during which Bahá'u'lláh sojourned on the soil of that Con-

17.4 tinent a century ago.¹ Let him resolve so to deepen his knowledge of the Faith and so to increase his standards of self-sacrifice and dedication to the Cause as to play his part in building a Community which will be worthy of this supreme bounty and which will be a beacon light to the peoples of this fear-wracked world.

17.4 In 1953 Shoghi Effendi wrote that the Continent of Europe had "at last at this critical hour—this great turning point in its fortunes—entered upon what may well be regarded as the opening phase of a great spiritual revival that bids fair to eclipse any period in its spiritual history."² Those who have been privileged to witness the extraordinary strengthening and consolidation of the Cause in Europe during the course of the last eleven years are well aware of the reservoir of spiritual potential that has been building up and the transformation of the life of the European Bahá'í Community that has ensued. May the completion and dedication of the Mashriqu'l-Adhkár be the signal for the unleashing of this potential, bringing about on the European mainland and in the islands around its shores a quickening of the process of individual conversion comparable to those events which have transpired with such astonishing suddenness in other continents of the globe.

THE UNIVERSAL HOUSE OF JUSTICE

18
Teaching the Masses
13 JULY 1964

To all National Spiritual Assemblies

Dear Bahá'í Friends,

18.1 When the masses of mankind are awakened and enter the Faith of God, a new process is set in motion and the growth of a new civilization begins. Witness the emergence of Christianity and of Islam. These masses are the rank and file, steeped in traditions of their own, but receptive to the new Word of God, by which, when they truly respond to it, they become so influenced as to transform those who come in contact with them.

18.2 God's standards are different from those of men. According to men's standards, the acceptance of any cause by people of distinction, of recognized fame and status, determines the value and greatness of that cause. But, in the words

---

17-1. Bahá'u'lláh was exiled from Constantinople (now Istanbul) to Adrianople (now Edirne, Turkey), where He remained from 12 December 1863 through 12 August 1868. Edirne lies west of the Bosphorus, near Turkey's border with Greece.

17-2. MBW, p. 161.

of Bahá'u'lláh: "The summons and the message which We gave were never intended to reach or to benefit one land or one people only. Mankind in its entirety must firmly adhere to whatsoever hath been revealed and vouchsafed unto it." Or again, "He hath endowed every soul with the capacity to recognize the signs of God. How could He, otherwise, have fulfilled His testimony unto men, if ye be of them that ponder His Cause in their hearts."[1] In countries where teaching the masses has succeeded, the Bahá'ís have poured out their time and effort in village areas to the same extent as they had formerly done in cities and towns. The results indicate how unwise it is to solely concentrate on one section of the population. Each National Assembly therefore should so balance its resources and harmonize its efforts that the Faith of God is taught not only to those who are readily accessible but to all sections of society, however remote they may be.

18.3 The unsophisticated people of the world—and they form the large majority of its population—have the same right to know of the Cause of God as others. When the friends are teaching the Word of God they should be careful to give the Message in the same simplicity as it is enunciated in our Teachings. In their contacts they must show genuine and divine love. The heart of an unlettered soul is extremely sensitive; any trace of prejudice on the part of the pioneer or teacher is immediately sensed.

18.4 When teaching among the masses, the friends should be careful not to emphasize the charitable and humanitarian aspects of the Faith as a means to win recruits. Experience has shown that when facilities such as schools, dispensaries, hospitals, or even clothes and food are offered to the people being taught, many complications arise. The prime motive should always be the response of man to God's message, and the recognition of His Messenger. Those who declare themselves as Bahá'ís should become enchanted with the beauty of the Teachings; and touched by the love of Bahá'u'lláh. The declarants need not know all the proofs, history, laws, and principles of the Faith, but in the process of declaring themselves they must, in addition to catching the spark of faith, become basically informed about the Central Figures of the Faith, as well as the existence of laws they must follow and an administration they must obey.

18.5 After declaration, the new believers must not be left to their own devices. Through correspondence and dispatch of visitors, through conferences and training courses, these friends must be patiently strengthened and lovingly helped to develop into full Bahá'í maturity. The beloved Guardian referring to the duties of Bahá'í Assemblies in assisting the newly declared believer has written: ". . . the members of each and every Assembly should endeavor, by their patience, their love, their tact and wisdom, to nurse, subsequent to his

---

18-1. GWB, pp. 96, 105–06.

admission, the newcomer into Bahá'í maturity, and win him over gradually to the unreserved acceptance of whatever has been ordained in the teachings."[2]

Expansion and Consolidation—Simultaneous Processes

18.6 Expansion and consolidation are twin processes that must go hand in hand. The friends must not stop expansion in the name of consolidation. Deepening the newly enrolled believers generates tremendous stimulus which results in further expansion. The enrollment of new believers, on the other hand, creates a new spirit in the community and provides additional potential manpower that will reinforce the consolidation work.

18.7 We would like to share with you some of the methods used by National Assemblies in various continents that have proved useful in teaching the masses, and attach a list. Certain of these may be valuable in your area, in addition to any methods you may yourself devise.

18.8 We are fervently praying that all National and Local Spiritual Assemblies, supported by the individual believers, will achieve outstanding success in the fulfillment of this glorious objective.

18.9 Please share this communication with all the friends.

With loving Bahá'í greetings,
The Universal House of Justice

Teaching the Masses
Annex

18.10 1. Materials are sent at once to the new believers. In some places this material is in the form of printed cards, mainly in color, portraying a Bahá'í theme or principle. This helps the new believer to know that his declaration has been accepted and to feel that he now belongs to the new Faith.

18.11 2. Training courses of about 2 weeks duration are held. To facilitate attendance and reduce cost, a number of villages are grouped together as one zone in which the course is held. The students to the courses are usually selected, so that the more capable participate, and teaching is facilitated. Transportation expenses, feeding and accommodation are provided, if it is found that the participants are unable to cover such expenses themselves. The material to be taught is prepared ahead of time, presented in simple language, and translated into the vernacular. After the course, the more promising students are picked out, and with their consent, are requested to undertake teaching projects for a limited period. It is sometimes found that long-term projects are also useful. These projects generally are carefully planned as to their duration, places to be visited, and material to be taught. If the traveling teachers are not able to cover their expenses, trav-

---

18-2. MA, p. 11.

eling and living expenses are provided by the Fund for the execution of a given and temporary teaching project.
3. Shorter training courses in the form of conferences over a long weekend are held.   18.12
4. These activities—training courses and conferences—are repeated as frequently as possible and are not dependent upon the acquisition of Teaching Institutes. In the absence of such Institutes, these courses and conferences are normally held in Bahá'í homes or hired quarters, such as schools, etc. In order to facilitate the physical catering and accommodation of the participants they are sometimes asked to come to the course with their eating utensils and bedding.   18.13
5. In the visits made to the villages, the visiting teacher meets with the Local Communities to give them basic Bahá'í knowledge, such as living the Bahá'í life, the importance of teaching, prayer, fasting, Nineteen Day Feasts, Bahá'í elections, and contributions to the Fund. The question of contributions to the Fund is of utmost importance, so that the new believers may quickly feel themselves to be responsible members of the Community. Each National Assembly must find ways and means to stimulate the offering of contributions, in cash or kind, to make it easy for the friends to contribute and to give proper receipts to the donors.   18.14

These are but suggestions based on experience which may help you in your efforts to teach and deepen the spiritually starved multitudes in your area.   18.15

In the course of carrying out such a tremendous spiritual campaign among the masses, disappointments will well be encountered. We tabulate a few instances that have been brought to our notice: 18.16

a) Visiting pioneers or teachers may find in some places newly enrolled believers not so enthusiastic about their religion as expected, or not adjusting to standards of Bahá'í life, or they may find them thinking of material benefits they may hope to derive from their new membership. We should always remember that the process of nursing the believer into full spiritual maturity is slow, and needs loving education and patience.

b) Some teaching committees, in their eagerness to obtain results, place undue emphasis on obtaining a great number of declarations to the detriment of the quality of teaching.

c) Some traveling teachers, in their desire to show the result of their services, may not scrupulously teach their contacts, and in some rare cases, if, God forbid, they are insincere, may even give false reports.

Such irregularities have happened and can be repeated, but must not be a source of discouragement. By sending a team of teachers to an area, or by sending at intervals other teachers to those areas, and through correspondence and reports, such situations can be detected and immediately adjusted. The   18.17

administration of the Faith must at all times keep in close touch with the teaching work.

18.18 To sum up:
1. Teaching the waiting masses is a reality facing each National Assembly.
2. The friends must teach with conviction, determination, genuine love, lack of prejudice, and in a simple language addressed to the heart.
3. Teaching must be followed up by training courses, conferences, and regular visits to deepen the believers in their knowledge of the Teachings.
4. The close touch of the National Office or Teaching Committees with the work is most essential, so that through reports and correspondence not only is information obtained and verified, but stimulation and encouragement is given.
5. Expansion and consolidation go hand in hand.

19
Universal Participation
SEPTEMBER 1964

To the Bahá'ís of the World

Dearly loved Friends,

19.1 In our message to you of April, 1964, announcing the Nine Year Plan, we called attention to two major themes of that Plan, namely "a huge expansion of the Cause of God and universal participation by all believers in the life of that Cause."

19.2 The enthusiastic vigor with which the believers throughout the world, under the devoted guidance of their National Spiritual Assemblies, have arisen to meet the challenge of the Plan, augurs well for the huge expansion called for. We now ask you to bend your efforts and thoughts, with equal enthusiasm, to the requirements of universal participation.

19.3 In that same message we indicated the meaning of universal participation: "the dedicated effort of every believer in teaching, in living the Bahá'í life, in contributing to the Fund, and particularly in the persistent effort to understand more and more the significance of Bahá'u'lláh's Revelation. In the words of our beloved Guardian, 'One thing and only one thing will unfailingly and alone secure the undoubted triumph of this sacred Cause, namely the extent to which our own inner life and private character mirror forth in their manifold aspects the splendor of those eternal principles proclaimed by Bahá'u'lláh.'"[1]

19.4 "Regard the world as the human body," wrote Bahá'u'lláh to Queen Vic-

---

19-1. BA, p. 66.

toria.[2] We can surely regard the Bahá'í world, the army of God, in the same way. In the human body, every cell, every organ, every nerve has its part to play. When all do so the body is healthy, vigorous, radiant, ready for every call made upon it. No cell, however humble, lives apart from the body, whether in serving it or receiving from it. This is true of the body of mankind in which God "hast endowed each and all with talents and faculties,"[3] and is supremely true of the body of the Bahá'í World Community, for this body is already an organism, united in its aspirations, unified in its methods, seeking assistance and confirmation from the same Source, and illumined with the conscious knowledge of its unity. Therefore, in this organic, divinely guided, blessed and illumined body the participation of every believer is of the utmost importance, and is a source of power and vitality as yet unknown to us. For extensive and deep as has been the sharing in the glorious work of the Cause, who would claim that every single believer has succeeded in finding his or her fullest satisfaction in the life of the Cause? The Bahá'í World Community, growing like a healthy new body, develops new cells, new organs, new functions and powers as it presses on to its maturity, when every soul, living for the Cause of God, will receive from that Cause, health, assurance and the overflowing bounties of Bahá'u'lláh which are diffused through His divinely ordained order.

19.5 In addition to teaching every believer can pray. Every believer can strive to make his "own inner life and private character mirror forth in their manifold aspects the splendor of those eternal principles proclaimed by Bahá'u'lláh."[4] Every believer can contribute to the Fund. Not all believers can give public talks, not all are called upon to serve on administrative institutions. But all can pray, fight their own spiritual battles, and contribute to the Fund. If every believer will carry out these sacred duties, we shall be astonished at the accession of power which will result to the whole body, and which in its turn will give rise to further growth and the showering of greater blessings on all of us.

19.6 The real secret of universal participation lies in the Master's oft expressed wish that the friends should love each other, constantly encourage each other, work together, be as one soul in one body, and in so doing become a true, organic, healthy body animated and illumined by the spirit. In such a body all will receive spiritual health and vitality from the organism itself, and the most perfect flowers and fruits will be brought forth.

19.7 Our prayers for the happiness and success of the friends everywhere are constantly offered at the Holy Shrines.

With loving Bahá'í greetings,
THE UNIVERSAL HOUSE OF JUSTICE

19-2. ESW, p. 62.
19-3. 'Abdu'l-Bahá, in BP, p. 103.
19-4. BA, p. 66.

20
Development of the Institution of the Hands of the Cause of God
NOVEMBER 1964

To the Bahá'ís of the World

Beloved Friends,

20.1 Once again the World Center of our Faith has been the scene of historic events, affecting profoundly the immediate prosecution of the Nine Year Plan and the future development of the World Order of Bahá'u'lláh. The occasion was the gathering in the Holy Land, for a period of fourteen days, of the Hands of the Cause of God to discuss their vital responsibilities, and particularly as Standard-Bearers of the Nine Year Plan.

20.2 The Universal House of Justice took advantage of this opportunity not only to receive the advice, opinions and views of the Hands on the progress of the Nine Year Plan but to consult them on the highly important goal announced at Riḍván 1964 under World Center Goals as "Development of the Institution of the Hands of the Cause of God, in consultation with the body of the Hands of the Cause, with a view to the extension into the future of its appointed functions of protection and propagation."

20.3 It was apparent that the elucidation of this vital goal, affecting as it does the relationship of the Hands of the Cause of God to all other institutions of the Cause, was imperative to the prosecution of the all-important teaching work and the development of the Bahá'í World Order.

20.4 Accordingly, the Universal House of Justice gave its full attention to this matter and, after study of the sacred texts and hearing the views of the Hands of the Cause themselves, has arrived at the following decisions:

20.4a There is no way to appoint, or to legislate to make it possible to appoint, Hands of the Cause of God.

20.4b Responsibility for decisions on matters of general policy affecting the Institution of the Hands of the Cause, which was formerly exercised by the beloved Guardian, now devolves upon the Universal House of Justice as the supreme and central institution of the Faith to which all must turn.

20.5 It is with great joy that we are able to share with you the initial steps now taken to attain the goal.

Continental Zones

20.6 The assignment of the Hands to various continents remains unchanged but, in order to expedite the work, the continents of Asia and the Western Hemisphere will each be divided into zones for the day-to-day work of the

Hands, one or more Hands being responsible for each zone. Asia will consist of two zones: the Middle East comprising the countries from and including Pakistan westwards and also Asiatic U.S.S.R.; and South and East Asia comprising the remainder of the continent. The Western Hemisphere will consist of three zones: North America, Central America (including Mexico) and the Antilles, and South America. The Hawaiian Islands will be in the Australasian continental area, as listed in the recently issued statistical summary.

Increases in Membership of Auxiliary Boards

20.7 The number of members of the Auxiliary Boards for the propagation of the Faith will be increased in every continent, raising the total number of Auxiliary Board members in Africa from eighteen to twenty-seven; in Asia from fourteen to thirty-six; in Australasia from four to nine; in Europe from eighteen to twenty-seven; and in the Western Hemisphere from eighteen to thirty-six.

20.8 The Hands of the Cause in each continent are called upon to appoint one or more members of their Auxiliary Boards to act in an executive capacity on behalf of and in the name of each Hand, thereby assisting him in carrying out his work.

Freeing Hands of the Cause from Elected or Appointed Positions

20.9 The exalted rank and specific functions of the Hands of the Cause of God make it inappropriate for them to be elected or appointed to administrative institutions, or to be elected as delegates to national conventions. Furthermore, it is their desire and the desire of the House of Justice that they be free to devote their entire energies to the vitally important duties conferred upon them in the Holy Writings. The importance of close collaboration between the Hands of the Cause and National Spiritual Assemblies cannot be overstressed, and a separate communication is being addressed to National Assemblies on this subject, supplementing guidance given in earlier letters.[1]

20.10 We anticipate announcing at Riḍván 1965 plans for Oceanic and Intercontinental Conferences, an overall plan for worldwide proclamation of the Faith during 1967–68, the centenary year of the revelation of the Súriy-i-Mulúk,[2] involving cooperation of National and Local Assemblies throughout the world, and conditions of entry for a competition for the design of the Mashriqu'l-Adhkár of Panama.

20.11 Teaching the masses is the greatest challenge now facing the followers of Bahá'u'lláh. No work is more important than that of carrying His Message with utmost speed to the bewildered and thirsting peoples of a spiritually

---

20-1. See messages dated October 1963 (no. 6) and 19 November 1963 (no. 10). See also the message dated November 1964 that follows.
20-2. The Súriy-i-Mulúk is Bahá'u'lláh's Tablet to the Kings, portions of which are published in *Proclamation of Bahá'u'lláh*, pp. 7–12, 47–54, and 102–03.

parched world. Now, as the Hands return to their various continents, reinforced by a wider and more efficient organization of their work, we are confident that the whole Bahá'í world will, with rising enthusiasm and ever-increasing success, press forward with the teaching work, greatly increase the flow of pioneers, more widely participate in the financial support of the work of the Cause and add rapidly to the list of goals already accomplished.

With loving Bahá'í greetings,
THE UNIVERSAL HOUSE OF JUSTICE

21
Relationship of Hands of the Cause of God and the National Spiritual Assemblies
NOVEMBER 1964

To National Spiritual Assemblies of the Bahá'í World

Dear Bahá'í Friends,

21.1 The gathering in the Holy Land of the Hands of the Cause of God has been an occasion of vital significance to the Faith. Please share the enclosed general message, as soon as possible, with all believers under your jurisdiction.[1]

21.2 We now wish to elaborate the recent decisions as they affect the relationship between the Institution of the Hands of the Cause and yourselves, the National Spiritual Assemblies of the world.

21.3 It is of the utmost importance that the Hands of the Cause and National Spiritual Assemblies be fully informed of the situation of the Cause in the areas for which they are responsible. We ask you therefore to work out with the Hands in your continent more efficient and easier methods of communication. The sharing of National Assembly minutes with the Hands of the Cause is entirely a matter for each National Spiritual Assembly to decide, but it is vitally important for you to regularly provide the Hands of the Cause with all information which is necessary to their work, including copies of pertinent committee reports.

21.4 The Hands of the Cause are preparing a schedule of proposed meetings with National Assemblies and will also be inviting members of National Assemblies to meet them in conferences with their Board members from time to time, a form of consultation which has been found most effective wherever it has been practiced.

21.5 Members of Auxiliary Boards should be freed from administrative responsibilities including serving on Committees and as delegates to conventions. In

---

21-1. See message no. 20.

the event of any member of a National Assembly accepting appointment to a Board, the National Assembly should accept this as valid reason for that member's resignation from the Assembly; should a Board member be elected to a National Assembly, he must choose on which body he will serve.

21.6 We ask each National Assembly to extend a warm and cordial invitation to the Hands of its continent to attend its national convention. All Hands of the Cause present should be given the freedom of the convention. If no continental Hands can attend a convention they may appoint one or two Board members to act as special deputies for that convention, who will, of course, be warmly welcomed and given the courtesy of taking part in convention as deputies of the Hands.

21.7 The increase in the numbers of Board members will inevitably be reflected in an increase in the needs of the Continental Funds. This is a matter for discussion with the Hands in your continent, and we feel sure that you will do your utmost to meet the new requirements, bearing in mind the importance which the beloved Guardian attached to direct contributions to these Funds by National and Local Spiritual Assemblies, as well as by individual believers.

21.8 The fostering of this important relationship between the exalted body of the Hands of the Cause and the National Spiritual Assemblies of the Bahá'í world will inevitably strengthen the foundation and functioning of the Cause of God and enable its embryonic world order to grow as a healthy tree under whose shade all mankind will eventually find security and peace.

With loving Bahá'í greetings,
THE UNIVERSAL HOUSE OF JUSTICE

22

Appointment of Five Continental Pioneer Committees
8 FEBRUARY 1965

To the National Spiritual Assemblies of Australia, the British Isles, Germany, Persia and the United States

Dear Bahá'í Friends,

22.1 To intensify the prosecution of the Nine Year Plan we call upon your Assemblies to appoint Pioneer Committees respectively for Australasia, Africa, Europe, Asia and the Americas for raising up and deploying the huge army of pioneers necessary to win the goals of the Plan.

22.2 The importance of this step is emphasized by the need, during the coming Bahá'í year, to settle no less than four hundred and two pioneers in teaching and consolidation areas—twenty-eight for Australasia, seventy-one for Africa, sixty-seven for Europe, one hundred and sixty-eight for Asia, and sixty-eight for the Americas.

22.3 These Pioneer Committees, to be appointed by and be responsible to your respective National Assemblies, will have functions set forth in the enclosed outline.[1] Since they are essentially service committees, a membership of three will suffice. The secretary should be a competent and knowledgeable Bahá'í having time, ability and facilities for carrying on a volume of correspondence. All members of the committee should have organizing ability as well as an aptitude for dealing with problems in a warm and loving Bahá'í way.

22.4 The Pioneer Committees will in no way infringe upon or substitute for existing committees which your Assemblies may have already appointed to deal with the teaching and pioneer requirements of either internal or external goals assigned to your Assembly. Rather they will supplement the work of existing committees of all National Assemblies in the respective continental areas, provide an effective means of exchange of vital information, and assist in the processing of pioneer applications and the transfer of pioneers to goal areas.

22.5 The Pioneer Committees should be appointed immediately, and, as soon as acceptances have been received and the committees organized, you should send us the names of the committee members and the address of the committee for inclusion in a dossier to be sent to all National Assemblies as an addendum to the Riḍván message.

22.6 The Pioneer Committees should be prepared to assume their full responsibilities and functions at Riḍván so that no time will be lost in the rapid transfer of pioneers to their posts, thus avoiding any possible dampening of their spirits.

22.7 In the meantime, we will be forwarding additional information and instructions which will enable the committees to swing into action at Convention time. Since announcement of the appointment of these committees will be a part of the Riḍván message, your Assembly should wait until the Convention to inform the friends of these plans.

22.8 Time is of the essence. It is imperative that we have the names of Pioneer Committee members and the address of the committee within thirty days of your receipt of these instructions.

22.9 Assuring you of prayers at the Holy Shrines.

Deepest love,
THE UNIVERSAL HOUSE OF JUSTICE

22-1. The Universal House of Justice later assumed responsibility for appointing the members of Continental Pioneer Committees and transferred responsibility for directing their work to the International Teaching Center (see message dated 19 May 1983, no. 361). For other letters on the evolving duties of Continental Pioneer Committees, see messages dated Riḍván 1965, 18 March 1966, Riḍván 1966, and 22 July 1974 (nos. 24, 31, 34, and 148).

Pioneer Committees

Responsibilities and Functions

I. To assist National Spiritual Assemblies and their relevant committees in the following respects: 22.10
   a. Information on availability and qualifications of pioneers.
   b. Act as a clearing house for information both continentally and intercontinentally.
   c. Determine travel and subsistence budgets needed.
   d. Supply information on needs for pioneers in various localities within their respective continental areas, this information to be furnished initially by the Universal House of Justice, and any later needs to be cleared through the Universal House of Justice.
   e. Supply information on the types of pioneering service needed in specific localities, such as: for teaching in mass conversion areas, or for deepening in administration; also special qualifications for pioneers to specific localities (e.g., language, ethnic background, work opportunities, etc.), and useful information about the territory itself (e.g., climate, geography, living conditions, visas and governmental regulations).

II. To assist prospective pioneers in the following respects: 22.11
   a. Place them in contact with the National Assemblies and/or national committees responsible for settling or consolidating specific localities in which pioneers would like to serve or are qualified to serve.
   b. Furnish useful information (such as that set forth in items I. d and e above) which will enable the prospective pioneers to determine where, when and how they can best volunteer their services.
   c. Work out, in consultation with the relevant National Assembly and the prospective pioneer, such assistance budgets as may be necessary.

The Committee will obtain the information called for above from the various National Assemblies (or their committees) in their continental areas, and there should be an exchange of vital information between the five Pioneer Committees. 22.12

Pioneer Committees will in no case assume direct responsibility for filling goals. Their function is in the nature of secondary assistance to the responsible or assisting National Assembly (or national committee). If a National Assembly can fill a goal without assistance, it need not consult a Pioneer Committee, but should keep the relevant Pioneer Committee informed as to the status of pioneer goals. 22.13

Pioneer Committees will clear with the prospective pioneer's own National Assembly before offering his services. 22.14

22.15 Foremost is the Pioneer Committee's responsibility to expedite and facilitate the transfer of pioneers from the place where they are to the goal areas where they will serve. Only such procedures as will be useful to this purpose should be adopted.

23
Election and Infallibility of the Universal House of Justice
9 March 1965

The National Spiritual Assembly of the Bahá'ís of the Netherlands

Dear Bahá'í Friends,

23.1 We are glad that you have brought to our attention the questions perplexing some of the believers. It is much better for these questions to be put freely and openly than to have them, unexpressed, burdening the hearts of devoted believers. Once one grasps certain basic principles of the Revelation of Bahá'u'lláh such uncertainties are easily dispelled. This is not to say that the Cause of God contains no mysteries. Mysteries there are indeed, but they are not of a kind to shake one's faith once the essential tenets of the Cause and the indisputable facts of any situation are clearly understood.

23.2 The questions put by the various believers fall into three groups. The first group centers upon the following queries: Why were steps taken to elect a Universal House of Justice with the foreknowledge that there would be no Guardian? Was the time ripe for such an action? Could not the International Bahá'í Council have carried on the work?

The Election of the Universal House of Justice

23.3 At the time of our beloved Shoghi Effendi's death it was evident, from the circumstances and from the explicit requirements of the Holy Texts, that it had been impossible for him to appoint a successor in accordance with the provisions of the Will and Testament of 'Abdu'l-Bahá. This situation, in which the Guardian died without being able to appoint a successor, presented an obscure question not covered by the explicit Holy Text, and had to be referred to the Universal House of Justice. The friends should clearly understand that before the election of the Universal House of Justice there was no knowledge that there would be no Guardian. There could not have been any such foreknowledge, whatever opinions individual believers may have held. Neither the Hands of the Cause of God, nor the International Bahá'í Council, nor any other existing body could make a decision upon this all-important matter.¹ Only the House of Justice had authority to pronounce upon it. This was one urgent reason for calling the election of the Universal House of Justice as soon as possible.

23.4 Following the passing of Shoghi Effendi the international administration of the Faith was carried on by the Hands of the Cause of God with the complete agreement and loyalty of the National Spiritual Assemblies and the body of the believers. This was in accordance with the Guardian's designation of the Hands as the "Chief Stewards of Bahá'u'lláh's embryonic World Commonwealth."²

23.5 From the very outset of their custodianship of the Cause of God the Hands realized that since they had no certainty of divine guidance such as is incontrovertibly assured to the Guardian and to the Universal House of Justice, their one safe course was to follow with undeviating firmness the instructions and policies of Shoghi Effendi. The entire history of religion shows no comparable record of such strict self-discipline, such absolute loyalty and such complete self-abnegation by the leaders of a religion finding themselves suddenly deprived of their divinely inspired guide. The debt of gratitude which mankind for generations, nay, ages to come, owes to this handful of grief-stricken, steadfast, heroic souls is beyond estimation.

23.6 The Guardian had given the Bahá'í world explicit and detailed plans covering the period until Riḍván 1963, the end of the Ten Year Crusade. From that point onward, unless the Faith were to be endangered, further divine guidance was essential. This was the second pressing reason for the calling of the election of the Universal House of Justice. The rightness of the time was further confirmed by references in Shoghi Effendi's letters to the Ten Year Crusade's being followed by other plans under the direction of the Universal House of Justice. One such reference is the following passage from a letter addressed to the National Spiritual Assembly of the British Isles on 25th February 1951, concerning its Two Year Plan which immediately preceded the Ten Year Crusade:

> 23.6a On the success of this enterprise, unprecedented in its scope, unique in its character and immense in its spiritual potentialities, must depend the initiation, at a later period in the Formative Age of the Faith, of undertakings embracing within their range all National Assemblies functioning throughout the Bahá'í world—undertakings constituting in

---

23-1. The International Bahá'í Council, first appointed by Shoghi Effendi in 1951 and, after his passing, elected in 1961 by members of National Spiritual Assemblies, was a precursor of the Universal House of Justice. It ceased to exist upon the election of the Universal House of Justice in 1963.

23-2. MBW, p. 127.

themselves a prelude to the launching of worldwide enterprises destined to be embarked upon, in future epochs of that same Age, by the Universal House of Justice, that will symbolize the unity and coordinate and unify the activities of these National Assemblies.³

23.7 Having been in charge of the Cause of God for six years, the Hands, with absolute faith in the Holy Writings, called upon the believers to elect the Universal House of Justice, and even went so far as to ask that they themselves be not voted for. The sole, sad instance of anyone succumbing to the allurements of power was the pitiful attempt of Charles Mason Remey to usurp the Guardianship.⁴

Principles Governing the Election of the House of Justice

23.8 The following excerpts from a Tablet of 'Abdu'l-Bahá state clearly and emphatically the principles with which the friends are already familiar from the Will and Testament of the Master and the various letters of Shoghi Effendi, and explain the basis for the election of the Universal House of Justice. This Tablet was sent to Persia by the beloved Guardian himself, in the early years of his ministry, for circulation among the believers.

23.9 . . . for 'Abdu'l-Bahá is in a tempest of dangers and infinitely abhors differences of opinion . . . Praise be to God, there are no grounds for differences.

23.10 The Báb, the Exalted One, is the Morn of Truth, the splendor of Whose light shineth through all regions. He is also the Harbinger of the Most Great Light, the Abhá Luminary. The Blessed Beauty is the One promised by the sacred books of the past, the revelation of the Source of light that shone upon Mount Sinai, Whose fire glowed in the midst of the Burning Bush. We are, one and all, servants of Their threshold, and stand each as a lowly keeper at Their door.

23.11 My purpose is this, that ere the expiration of a thousand years, no one has the right to utter a single word, even to claim the station of Guardianship. The Most Holy Book is the Book to which all peoples shall refer, and in it the Laws of God have been revealed. Laws not men-

---

23-3. UD, p. 261.
23-4. Charles Mason Remey, a prominent early American believer who was much loved by 'Abdu'l-Bahá, traveled widely in service to the Faith. In January 1951 he was appointed by Shoghi Effendi to be President of the International Bahá'í Council and, the following year, was appointed a Hand of the Cause of God to serve in the Holy Land. After the passing of Shoghi Effendi on 4 November 1957, Charles Mason Remey served as one of the nine Custodian Hands designated by the body of the Hands of the Cause to administer the Faith from Haifa. In April 1960 he proclaimed himself to be the Second Guardian of the Cause (see BW 13:353 n). This attempt to usurp the Guardianship resulted in his expulsion from the Faith as a Covenant-breaker. For the announcement of his passing, see the message dated 5 April 1974 (no. 144).

tioned in the Book should be referred to the decision of the Universal House of Justice. There will be no grounds for difference . . . Beware, beware lest anyone create a rift or stir up sedition. Should there be differences of opinion, the Supreme House of Justice would immediately resolve the problems. Whatever will be its decision, by majority vote, shall be the real truth, inasmuch as that House is under the protection, unerring guidance and care of the one true Lord. He shall guard it from error and will protect it under the wing of His sanctity and infallibility. He who opposes it is cast out and will eventually be of the defeated.

23.12 The Supreme House of Justice should be elected according to the system followed in the election of the parliaments of Europe. And when the countries would be guided, the Houses of Justice of the various countries would elect the Supreme House of Justice.

23.13 At whatever time all the beloved of God in each country appoint their delegates, and these in turn elect their representatives, and these representatives elect a body, that body shall be regarded as the Supreme House of Justice.

23.14 The establishment of that House is not dependent upon the conversion of all the nations of the world. For example, if conditions were favorable and no disturbances would be caused, the friends in Persia would elect their representatives, and likewise the friends in America, in India, and other areas would also elect their representatives, and these would elect a House of Justice. That House of Justice would be the Supreme House of Justice. That is all.
(*Makátíb-i-'Abdu'l-Bahá*, Vol. III, pp. 500–501)

23.15 The friends should realize that there is nothing in the Texts to indicate that the election of the Universal House of Justice could be called only by the Guardian. On the contrary, 'Abdu'l-Bahá envisaged the calling of its election in His own lifetime. At a time described by the Guardian as "the darkest moments of His [the Master's] life, under 'Abdu'l-Ḥamíd's regime, when He stood ready to be deported to the most inhospitable regions of Northern Africa," and when even His life was threatened, 'Abdu'l-Bahá wrote to Ḥájí Mírzá Táqí Afnán, the cousin of the Báb and chief builder of the 'Ishqábád Temple, commanding him to arrange for the election of the Universal House of Justice should the threats against the Master materialize.[5] The second part of the Master's Will is also relevant to such a situation and should be studied by the friends.

23-5. WT, p. 20; WOB, p. 17.

The Authority of the Universal House of Justice

23.16   The second series of problems vexing some of the friends centers on the question of the infallibility of the Universal House of Justice and its ability to function without the presence of the Guardian. Particular difficulty has been experienced in understanding the implications of the following statement by the beloved Guardian:

23.16a   Divorced from the institution of the Guardianship the World Order of Bahá'u'lláh would be mutilated and permanently deprived of that hereditary principle which, as 'Abdu'l-Bahá has written, has been invariably upheld by the Law of God. "In all the Divine Dispensations," He states, in a Tablet addressed to a follower of the Faith in Persia, "the eldest son hath been given extraordinary distinctions. Even the station of prophethood hath been his birthright." Without such an institution the integrity of the Faith would be imperiled, and the stability of the entire fabric would be gravely endangered. Its prestige would suffer, the means required to enable it to take a long, an uninterrupted view over a series of generations would be completely lacking, and the necessary guidance to define the sphere of the legislative action of its elected representatives would be totally withdrawn.

("The Dispensation of Bahá'u'lláh," *The World Order of Bahá'u'lláh*, p. 148)

23.17   Let the friends who wish for a clearer understanding of this passage at the present time consider it in the light of the many other texts which deal with the same subject, for example the following passages gleaned from the letters of Shoghi Effendi:

23.17a   They have also, in unequivocal and emphatic language, appointed those twin institutions of the House of Justice and of the Guardianship as their chosen Successors, destined to apply the principles, promulgate the laws, protect the institutions, adapt loyally and intelligently the Faith to the requirements of progressive society, and consummate the incorruptible inheritance which the Founders of the Faith have bequeathed to the world.

(Letter dated 21 March 1930, *The World Order of Bahá'u'lláh*, p. 20)

23.17b   It must be also clearly understood by every believer that the institution of Guardianship does not under any circumstances abrogate, or even in the slightest degree detract from, the powers granted to the Universal House of Justice by Bahá'u'lláh in the "Kitáb'l-Aqdas," and repeatedly and solemnly confirmed by 'Abdu'l-Bahá in His Will. It does not constitute in any manner a contradiction to the Will and Writings of Bahá'u'lláh, nor does it nullify any of His revealed instructions. It

enhances the prestige of that exalted assembly, stabilizes its supreme position, safeguards its unity, assures the continuity of its labors, without presuming in the slightest to infringe upon the inviolability of its clearly defined sphere of jurisdiction. We stand indeed too close to so monumental a document to claim for ourselves a complete understanding of all its implications, or to presume to have grasped the manifold mysteries it undoubtedly contains. . . .
(Letter dated 27 February 1929, *The World Order of Bahá'u'lláh*, p. 8)

23.17c From these statements it is made indubitably clear and evident that the Guardian of the Faith has been made the Interpreter of the Word and that the Universal House of Justice has been invested with the function of legislating on matters not expressly revealed in the teachings. The interpretation of the Guardian, functioning within his own sphere, is as authoritative and binding as the enactments of the International House of Justice, whose exclusive right and prerogative is to pronounce upon and deliver the final judgment on such laws and ordinances as Bahá'u'lláh has not expressly revealed. Neither can, nor will ever, infringe upon the sacred and prescribed domain of the other. Neither will seek to curtail the specific and undoubted authority with which both have been divinely invested.
("The Dispensation of Bahá'u'lláh," *The World Order of Bahá'u'lláh*, pp. 149–50)

23.17d Each exercises, within the limitations imposed upon it, its powers, its authority, its rights and prerogatives. These are neither contradictory, nor detract in the slightest degree from the position which each of these institutions occupies.
("The Dispensation of Bahá'u'lláh," *The World Order of Bahá'u'lláh*, p. 148)

23.17e Though the Guardian of the Faith has been made the permanent head of so august a body he can never, even temporarily, assume the right of exclusive legislation. He cannot override the decision of the majority of his fellow-members . . .
("The Dispensation of Bahá'u'lláh," *The World Order of Bahá'u'lláh*, p. 150)

23.18 Above all, let the hearts of the friends be assured by these words of Bahá'u'lláh:

23.18a The Hand of Omnipotence hath established His Revelation upon an unassailable, an enduring foundation. Storms of human strife are powerless to undermine its basis, nor will men's fanciful theories succeed in damaging its structure.
(*The World Order of Bahá'u'lláh*, p. 109)

and these of 'Abdu'l-Bahá:

23.18b Verily, God effecteth that which He pleaseth; naught can annul His Covenant; naught can obstruct His favor nor oppose His Cause! He doeth with His will that which pleaseth Him and He is powerful over all things! . . .
(*Tablets of Abdul-Baha Abbas*, Vol. III, p. 598)

23.19 It should be understood by the friends that before legislating upon any matter the Universal House of Justice studies carefully and exhaustively both the Sacred Texts and the Writings of Shoghi Effendi on the subject. The interpretations written by the beloved Guardian cover a vast range of subjects and are equally as binding as the Text itself.

Interpretations of the Guardian and Elucidations of the Universal House of Justice

23.20 There is a profound difference between the interpretations of the Guardian and the elucidations of the House of Justice in exercise of its function to "deliberate upon all problems which have caused difference, questions that are obscure and matters that are not expressly recorded in the Book." The Guardian reveals what the Scripture means; his interpretation is a statement of truth which cannot be varied. Upon the Universal House of Justice, in the words of the Guardian, "has been conferred the exclusive right of legislating on matters not expressly revealed in the Bahá'í writings."[6] Its pronouncements, which are susceptible of amendment or abrogation by the House of Justice itself, serve to supplement and apply the Law of God. Although not invested with the function of interpretation, the House of Justice is in a position to do everything necessary to establish the World Order of Bahá'u'lláh on this earth. Unity of doctrine is maintained by the existence of the authentic texts of Scripture and the voluminous interpretations of 'Abdu'l-Bahá and Shoghi Effendi, together with the absolute prohibition against anyone propounding "authoritative" or "inspired" interpretations or usurping the function of Guardian. Unity of administration is assured by the authority of the Universal House of Justice.

23.21 "Such," in the words of Shoghi Effendi, "is the immutability of His revealed Word. Such is the elasticity which characterizes the functions of His appointed ministers. The first preserves the identity of His Faith, and guards the integrity of His law. The second enables it, even as a living organism, to expand and adapt itself to the needs and requirements of an ever-changing society."
(Letter dated 21 March 1930, *The World Order of Bahá'u'lláh*, p. 23)

23-6. WOB, p. 153.

Every true believer, if he is to deepen in his understanding of the Cause of Bahá'u'lláh, must needs combine profound faith in the unfailing efficacy of His Message and His Covenant, with the humility of recognizing that no one of this generation can claim to have embraced the vastness of His Cause nor to have comprehended the manifold mysteries and potentialities it contains. The words of Shoghi Effendi bear ample testimony to this fact: 23.22

> How vast is the Revelation of Bahá'u'lláh! How great the magnitude of His blessings showered upon humanity in this day! And yet, how poor, how inadequate our conception of their significance and glory! This generation stands too close to so colossal a Revelation to appreciate, in their full measure, the infinite possibilities of His Faith, the unprecedented character of His Cause, and the mysterious dispensations of His Providence. 23.22a
> (Letter dated 21 March 1930, *The World Order of Bahá'u'lláh*, p. 24)

> We are called upon by our beloved Master in His Will and Testament not only to adopt it [Bahá'u'lláh's new world order] unreservedly, but to unveil its merit to all the world. To attempt to estimate its full value, and grasp its exact significance after so short a time since its inception would be premature and presumptuous on our part. We must trust to time, and the guidance of God's Universal House of Justice, to obtain a clearer and fuller understanding of its provisions and implications.... 23.22b
> (Letter dated 23 February 1924, published in *Bahá'í Administration*, p. 62)

> As to the order and the management of the spiritual affairs of the friends, that which is very important now is the consolidation of the Spiritual Assemblies in every center, because on these fortified and unshakable foundations, God's Supreme House of Justice shall be erected and firmly established in the days to come. When this most great Edifice shall be reared on such an immovable foundation, God's purpose, wisdom, universal truths, mysteries and realities of the Kingdom, which the mystic revelation of Bahá'u'lláh has deposited within the Will and Testament of 'Abdu'l-Bahá, shall gradually be revealed and made manifest. 23.22c
> (Letter dated 19 December 1923—translated from the Persian)

Statements such as these indicate that the full meaning of the Will and Testament of 'Abdu'l-Bahá, as well as an understanding of the implications of the World Order ushered in by that remarkable document can be revealed only gradually to men's eyes, and after the Universal House of Justice has come into being. The friends are called upon to trust to time and to await the guidance of the Universal House of Justice, which, as circumstances require, will make pronouncements that will resolve and clarify obscure matters. 23.23

The Authority to Expel Members of the House of Justice

23.24 The third group of queries raised by the friends concerns details of functioning of the Universal House of Justice in the absence of the Guardian, particularly the matter of expulsion of members of the House of Justice. Such questions will be clarified in the Constitution of the House of Justice, the formulation of which is a goal of the Nine Year Plan. Meanwhile the friends are informed that any member committing a "sin injurious to the common weal," may be expelled from membership of the House of Justice by a majority vote of the House itself.[7] Should any member, God forbid, be guilty of breaking the Covenant, the matter would be investigated by the Hands of the Cause of God, and the Covenant-breaker would be expelled by decision of the Hands of the Cause of God residing in the Holy Land, subject to the approval of the House of Justice, as in the case of any other believer. The decision of the Hands in such a case would be announced to the Bahá'í world by the Universal House of Justice.

23.25 We are certain that when you share this letter with the friends and they have these quotations from the Scriptures and the Writings of the Guardian drawn to their attention, their doubts and misgivings will be dispelled and they will be able to devote their every effort to spreading the Message of Bahá'u'lláh, serenely confident in the power of His Covenant to overcome whatever tests an inscrutable Providence may shower upon it, thus demonstrating its ability to redeem a travailing world and to upraise the Standard of the Kingdom of God on earth.

With loving greetings,
THE UNIVERSAL HOUSE OF JUSTICE

23-7. WT, p. 14. See also CUHJ, p. 12.

24
Riḍván Message 1965
RIḌVÁN 1965

The Bahá'ís of the World

Dearly loved Friends,

The tide of victory which carried the Bahá'í World community to the celebrations of the Most Great Jubilee is still rising. A ceaseless shower of divine confirmation rains upon our efforts, its evidences apparent in the many noteworthy achievements of the few brief months since the launching of the Nine Year Plan. The most spectacular of these is the increase in the number of centers where Bahá'ís reside from fifteen thousand one hundred and sixty-eight at Riḍván 1964 to twenty-one thousand and six at the present time, an increase of nearly six thousand in one year. No less remarkable is the progress of the teaching work in India where the number of believers now exceeds a hundred and forty thousand, an increase of more than thirty thousand since Riḍván 1964. Pioneers are moving to those few remaining territories of the earth as yet unillumined by the light of God's new Revelation; "the vast increase" in the size of the Cause, called for at the launching of the Plan, appears to be developing, while in country after country the institutions and endowments of the Faith are being steadily and firmly established. 24.1

World Center Goals

During the past twelve months the goals assigned to the World Center have been actively pursued. Basic decisions and actions to implement the goal of "Development of the Institution of the Hands of the Cause of God, with a view to extension into the future of its appointed functions of protection and propagation," have already been conveyed to the friends. Following their meeting in the Holy Land last October, the members of this august body, the Standard-Bearers of this Nine Year Plan as well as of the beloved Guardian's Ten Year Crusade, already laden with honors and services, have arisen with renewed and matchless vigor to rouse the spirits of the friends to meet the supreme teaching challenge, to lend their counsel and assistance to the administrative bodies, and to diffuse the divine fragrances and love of God through all the world. The increase in the numbers of Board members and the new executive arrangements will, it is confidently anticipated, enable the beloved Hands to discharge their important duties with even greater effectiveness and give them more time to travel and teach. 24.2

A preliminary survey of the conditions affecting the construction of the first Mashriqu'l-Adhkár of Latin America, one of the two edifices to be erected during the Plan, has already been undertaken, and we now invite Bahá'í and 24.3

non-Bahá'í architects to submit designs for the Panama Temple.¹ The terms and conditions of the submission, and the specifications of the structure, may be obtained from the National Spiritual Assembly of Panama, whose choice of design will be subject to the ultimate approval of the Universal House of Justice. It is our hope that the construction of this sacred House of Worship, in a location accorded such special significance by both the Master and the Guardian, will be speedily accomplished, so that its beacon of spiritual light may radiate to all the Americas.

Worldwide Achievements

24.4 During the past twelve months the following new territories have been opened to the Faith: in the continent of Africa, Gabon, Ifni, Mali, Mauritania, Rodrigues Island and Upper Volta; in the continent of America, Aruba Island, Cozumel Island, Guadeloupe, Las Mujeres Island, Prince of Wales Island and St. Vincent; in the continent of Asia, the Ryukyu Islands; in the continent of Australasia, the Line Islands; in the continent of Europe, the Isle of Wight, the East and West Frisian Islands. The following territories have been reopened: in the continent of Africa, Mafia Island; in the continent of America, Antigua, French Guiana and Martinique; West Irian in the continent of Asia; and Admiralty Islands in Australasia. National Ḥaẓíratu'l-Quds have been acquired in nine places, the seats of National Spiritual Assemblies, and land has been acquired in two others on which to build this institution. Six National Spiritual Assemblies have become incorporated and the Faith has been recognized in Cambodia, a country destined to have its own National Spiritual Assembly during the Nine Year Plan. National Endowments have been acquired in eight countries; six Teaching Institutes have been established, and land has been acquired for six others; a Bahá'í Publishing Trust for the provision of literature in the French language has been established in Brussels; Bahá'í Holy Days have been recognized in three territories; Bahá'í literature has been published in the following eleven new languages: Ibibio-Efik in the continent of Africa, Aguacateca, Athabascan, Cariña and Motilon-Yukpa in the continent of America, Kenyah, Melanau and Temiar in the continent of Asia, and Ghari, Marshallese and Motua in Australasia. The progress of the Cause in Borneo makes possible the achievement of a goal supplementary to the Plan, namely the establishment at Riḍván 1966 of the National Spiritual Assembly of the Bahá'ís of Brunei.

Two Conditions of the Bahá'í World Community

24.5 The passage of the first year of the Plan discloses two conditions in the Bahá'í World community. The first, within the Faith itself, is its capacity to

---

24-1. The other edifice that was to be erected during the Nine Year Plan was the Mashriqu'l-Adhkár planned for Tehran, Iran. Due to antagonism toward the Bahá'í Faith in that country, plans for its construction were held in abeyance.

accomplish all and any definitive goals assigned to it, goals such as the purchasing of Ḥaẓíratu'l-Quds, Temple Sites, Endowments, or the incorporation of Spiritual Assemblies; such objective and highly important goals as these, by which the Cause is established physically, legally and socially in the world, are now taken in its stride by the Administrative Order. It should be noted, moreover, that the accomplishment of many goals of this type, involves inter-Assembly cooperation, an international activity vital to the development of world order.

The second condition apparent after the passage of the first year of the Plan, involves the relationship of the Cause to humanity. Almost universally there is a sense of an impending breakthrough in large-scale conversion. Reports of the Hands of the Cause and of Board members constantly mention it; many National Spiritual Assemblies believe that they have reached the shores of this ocean. And, indeed, entry into the Cause by troops has been a fact in some areas for a number of years. But greater things are ahead. The teaching of the Faith must enkindle a world-encircling fire in whose light the Cause and the world—protagonists of the greatest drama in human history—are clearly illumined. Destiny is carrying us to this climax; we must gird ourselves for heroism. 24.6

Four Immediate Tasks

Four challenging and immediate tasks present themselves. The first is to raise and dispatch, during the coming year, no less than four hundred and sixty pioneers who will open the fifty-four remaining virgin territories of the Plan, resettle the eighteen unoccupied ones, reinforce areas where the numbers and cohesion of the Bahá'í communities are at present inadequate to launch effective teaching plans, and support and extend the work in the areas of mass teaching. Let every believer consider this challenge, be he, in the words of the beloved Guardian, "in active service or not, of either sex, young as well as old, rich or poor, whether veteran or newly enrolled . . ." 24.7

To assist the pioneer efforts of the friends and their transfer to their posts during the next twelve months we announce the formation of five Continental Pioneer Committees, namely: Pioneer Committee for Africa appointed by the National Spiritual Assembly of the Bahá'ís of the British Isles; Pioneer Committee for the Americas appointed by the National Spiritual Assembly of the Bahá'ís of the United States; Pioneer Committee for Asia appointed by the National Spiritual Assembly of the Bahá'ís of Persia; Pioneer Committee for Australasia appointed by the National Spiritual Assembly of the Bahá'ís of Australia; Pioneer Committee for Europe appointed by the National Spiritual Assembly of the Bahá'ís of Germany. 24.8

These Committees will in no way infringe the responsibilities of other Pioneer Committees, or of National Spiritual Assemblies, who are in charge of the teaching work, and under whose jurisdiction they will function. They are 24.9

established to facilitate and assist the work of these national bodies by providing effective exchange of vital information, both continentally and intercontinentally, by assisting in the routing of pioneer offers and in the transfer of pioneers to their posts.

24.10 A careful estimate has been made of the pioneer needs of every area during the next twelve months and the result, including those for the seventy-two areas mentioned above, is a call for four hundred and sixty-one pioneers; eighty-six for Africa, ninety-six for the Americas, one hundred and ninety-one for Asia, twenty-nine for Australasia, and fifty-nine for Europe. Each National Spiritual Assembly has been consulted as to its pioneer needs and these have been made known to all National Spiritual Assemblies as well as to the five Continental Pioneer Committees, who will be kept currently informed of progress by the National Spiritual Assemblies. The friends, therefore, are urged to consult their National Spiritual Assemblies for information about pioneer needs and responsibilities both of their own communities and in general.

24.11 For the first time in Bahá'í history, an International Deputization Fund has been established at the World Center under the administration of the Universal House of Justice. From it supplementary support will be given to specific pioneering projects when other funds are not available. All friends, and particularly those who are unable to respond to the pioneer call are invited to support this Fund, mindful of the injunction of Bahá'u'lláh, "Center your energies in the propagation of the Faith of God. Whoso is worthy of so high a calling, let him arise and promote it. Whoso is unable, it is his duty to appoint him who will, in his stead, proclaim this Revelation, Whose power hath caused the foundations of the mightiest structures to quake, every mountain to be crushed into dust, and every soul to be dumbfounded."[2]

24.12 The second challenge facing us is to raise the intensity of teaching to a pitch never before attained, in order to realize that "vast increase" called for in the Plan. Universal participation and constant action will win this goal. Every believer has a part to play, and is capable of playing it, for every soul meets others, and, as promised by Bahá'u'lláh, "Whosoever ariseth to aid Our Cause God will render him victorious . . ."[3] The confusion of the world is not diminishing, rather does it increase with each passing day, and men and women are losing faith in human remedies. Realization is at last dawning that "There is no place to flee to" save God.[4] Now is the golden opportunity; people are willing, in many places eager, to listen to the divine remedy.

24.13 The third challenge is to acquire as rapidly as possible all the remaining National Ḥaẓíratu'l-Quds, Temple Sites, National Endowments and Teaching Institutes called for in the Plan. The speedy conclusion of these projects will

---

24-2. GWB, pp. 196–97.
24-3. Quoted in MBW, p. 101.
24-4. GWB, p. 203.

save tremendous expense later and endow the Faith with increasingly valuable properties. These basic possessions are the embryos of mighty institutions of the future, but it is this generation, which, for its own protection and as its gift to posterity, must acquire them. We call upon the National Spiritual Assemblies charged with responsibility in this field to accord it high priority. A further, but equally important consideration, is, that the achievement of this goal in the early years of the Plan will liberate the energies and resources of the growing world community for a concentrated, resolute and relentless pursuit in its later stages of great victories whose foundations are now being laid.

The Centenary of Bahá'u'lláh's Proclamation to the Kings

24.14 The fourth challenge is to prepare national and local plans for the befitting celebration of the centenary of Bahá'u'lláh's proclamation of His Message in September/October, 1867, to the kings and rulers of the world, celebrations to be followed during the remainder of the Nine Year Plan by a sustained and well-planned program of proclamation of that same Message to the generality of mankind.

24.15 A review of the historic proclamation by Bahá'u'lláh, as described by Shoghi Effendi in *God Passes By*, reveals that its "opening notes" were "sounded during the latter part of Bahá'u'lláh's banishment to Adrianople," and that, six years later, it "closed during the early years of His incarceration in the prison-fortress of 'Akká." These "opening notes" were the mighty and awe-inspiring words addressed by Him to the kings and rulers collectively in the Súriy-i-Mulúk, "the most momentous Tablet revealed by Bahá'u'lláh." It was penned some time during the months of September and October, 1867, and was followed by "Tablets unnumbered . . . in which the implications of His newly-asserted claims were fully expounded." "Kings and emperors, severally and collectively; the chief magistrates of the Republics of the American continent; ministers and ambassadors; the Sovereign Pontiff himself; the Vicar of the Prophet of Islám; the royal Trustee of the Kingdom of the Hidden Imám; the monarchs of Christendom, its patriarchs, archbishops, bishops, priests and monks; the recognized leaders of both the Sunní and Shí'ah sacerdotal orders; the high priests of the Zoroastrian religion; the philosophers, the ecclesiastical leaders, the wise men and the inhabitants of Constantinople—that proud seat of both the Sultanate and the Caliphate; the entire company of the professed adherents of the Zoroastrian, the Jewish, the Christian and Muslim Faiths; the people of the Bayán; the wise men of the world, its men of letters, its poets, its mystics, its tradesmen, the elected representatives of its peoples; His own countrymen"; all were "brought directly within the purview of the exhortations, the warnings, the appeals, the declarations and the prophecies which constitute the theme of His momentous summons to the leaders of mankind . . ." "Unique and stupendous as was this proclamation, it proved to be but a prelude to a still mightier revelation of the creative power of its Author, and to what may well rank as the most signal act of His ministry—the promulgation of the Kitáb-i-Aqdas."[5] In this, the Most Holy Book, revealed in 1873, Bahá'u'lláh not only once more announces to the kings of the earth collectively that "He Who is the King of Kings hath appeared" but addresses reigning sovereigns distinctively by name and proclaims to the "Rulers of America and the Presidents of the Republics therein" that "the Promised One hath appeared."[6] Such was the proclamation of Bahá'u'lláh to mankind. As He Himself testified, "Never since the beginning of the world hath the Message been so openly proclaimed."[7]

24.16 The celebration of this fate-laden centenary period will open with a visit, in September 1967, on the Feast of Mashíyyat, by a few appointed representatives of the Bahá'í World to the site of the house in Adrianople, where the historic Súriy-i-Mulúk was revealed.

Six Intercontinental Conferences

24.17 Immediately following this joyful and pious act, six Intercontinental Conferences will be simultaneously held during the month of October in Panama City, Wilmette, Sydney, Kampala, Frankfurt, and New Delhi. The host and convener of each Conference will be the National Spiritual Assembly in whose area it takes place. The following Hands of the Cause of God will represent the Universal House of Justice at these Conferences: Panama City—Amatu'l-Bahá Rúḥíyyih Khánum, who will, on that occasion, lay the foundation stone of the Temple; Wilmette—Leroy Ioas; Sydney—Ugo Giachery; Kampala—'Alí-Akbar Furútan; Frankfurt—Paul Haney; New Delhi—Abu'l-Qásim Faizi.

24.18 All National Spiritual Assemblies are called upon to arrange befitting observances, on a national and local scale, of the opening of the centenary period during September/October, 1967, and between the above Conferences and Riḍván 1968, at which time the second International Convention for the election of the Universal House of Justice will be held at the World Center.

24.19 The successful carrying out of all these plans will constitute a befitting commemoration, commensurate with the resources of the Bahá'í World community, of the sacred event they recall.

A Period of Proclamation

24.20 These six Conferences, like the epoch-making event whose centenary they commemorate, will sound the "opening notes" of a period of proclamation of the Cause of God extending through the remaining years of the Nine Year Plan to the centenary, in 1973, of the Revelation of the Kitáb-i-Aqdas, an

---

24-5. See GPB, pp. 212, 171, 212, 213.
24-6. GWB, p. 211; PB, p. 63; KA ¶88.
24-7. Quoted in GPB, p. 212.

activity which calls for the ardent and imaginative study of all National and Local Spiritual Assemblies throughout the world.

24.21 The international scene will witness the holding of Oceanic Conferences forecast by Shoghi Effendi. The first one will be held during August 1968 on an island in the Mediterranean Sea to commemorate Bahá'u'lláh's voyage upon that sea, a hundred years before, from Gallipoli in Turkey to the Most Great Prison in 'Akká. In the subsequent years of the Nine Year Plan, others will be held in the Atlantic Ocean, in the Caribbean Sea, the Pacific Ocean, and the Indian Ocean.

24.22 In calling upon all National Spiritual Assemblies to consider now the appointment of National Proclamation Committees charged with laying feasible and effective plans for the proclamation of the Faith throughout the entire centenary period, we can do no better than call attention to the following passage from a letter written by our beloved Guardian in connection with the celebrations of the centenary of the birth of the Bahá'í Era:

> 24.22a An unprecedented, a carefully conceived, efficiently co-ordinated, nation-wide campaign, aiming at the proclamation of the Message of Bahá'u'lláh, through speeches, articles in the press, and radio broadcasts, should be promptly initiated and vigorously prosecuted. The universality of the Faith, its aims and purposes, episodes in its dramatic history, testimonials to its transforming power, and the character and distinguishing features of its World Order should be emphasized and explained to the general public, and particularly to eminent friends and leaders sympathetic to its cause, who should be approached and invited to participate in the celebrations. Lectures, conferences, banquets, special publications should, to whatever extent is practicable and according to the resources at the disposal of the believers, proclaim the character of this joyous Festival.[8]

Gathering Momentum of the Process Launched in 1953

24.23 The majestic process launched by our beloved Guardian in 1953, when he called the widely scattered, obscure Bahá'í World community to embark upon that first, glorious, world-encompassing crusade, is gathering momentum, and posterity may well gaze with awe upon the development, by so small a fraction of the human race and in a world entangled in opposition, enmity and disruption, of the very pattern and sinews of world order. This divinely propelled and long-promised development must continue its historic course until its final consummation in the glories and splendors of the World Order of Bahá'u'lláh, the Kingdom of God on earth.

THE UNIVERSAL HOUSE OF JUSTICE

24-8. MA, p. 62.

25
Passing of the Hand of the Cause of God Leroy Ioas
22 July 1965

To the National Spiritual Assembly of the United States

25.1 GRIEVE ANNOUNCE PASSING OUTSTANDING HAND CAUSE LEROY IOAS. HIS LONG SERVICE BAHÁ'Í COMMUNITY UNITED STATES CROWNED ELEVATION RANK HAND FAITH PAVING WAY HISTORIC DISTINGUISHED SERVICES HOLY LAND. APPOINTMENT FIRST SECRETARY GENERAL INTERNATIONAL BAHÁ'Í COUNCIL PERSONAL REPRESENTATIVE GUARDIAN FAITH TWO INTERCONTINENTAL CONFERENCES ASSOCIATION HIS NAME BY BELOVED GUARDIAN OCTAGON DOOR BÁB'S SHRINE TRIBUTE SUPERVISORY WORK DRUM DOME THAT HOLY SEPULCHER NOTABLE PART ERECTION INTERNATIONAL ARCHIVES BUILDING ALL ENSURE HIS NAME IMMORTAL ANNALS FAITH. LAID TO REST BAHÁ'Í CEMETERY CLOSE FELLOW HANDS ADVISE HOLD BEFITTING MEMORIAL SERVICES. . . .[1]

Universal House of Justice

26
Purification of the Most Holy Shrine of Bahá'u'lláh, the Qiblih of the Bahá'í World
11 November 1965

To all National Spiritual Assemblies

26.1 ANNOUNCE BAHÁ'Í WORLD REMOVAL FROM IMMEDIATE PRECINCTS HOLY SHRINE BAHÁ'U'LLÁH REMAINS MÍRZÁ ḌÍYÁ'U'LLÁH YOUNGER BROTHER MÍRZÁ MUḤAMMAD-'ALÍ HIS ACCOMPLICE IN EFFORTS SUBVERT FOUNDATIONS COVENANT GOD SOON AFTER ASCENSION BAHÁ'U'LLÁH. THIS FINAL STEP IN PROCESS PURIFICATION SACRED INTERNATIONAL ENDOWMENTS FAITH IN BAHJÍ FROM PAST CONTAMINATION WAS PROVIDENTIALLY UNDERTAKEN UPON REQUEST FAMILY OLD COVENANT-BREAKERS A PROCESS WHOSE INITIAL STAGE WAS FULFILLED BY 'ABDU'L-BAHÁ WHICH GATHERED MOMENTUM EARLY YEARS BELOVED GUARDIAN'S MINISTRY THROUGH EVACUATION MANSION ATTAINED CLIMAX THROUGH PURIFICATION ḤARAM-I-AQDAS AND NOW CONSUMMATED THROUGH CLEANSING INNER SANCTUARY MOST HALLOWED SHRINE QIBLIH BAHÁ'Í WORLD PRESAGING EVENTUAL CONSTRUCTION BEFITTING MAUSOLEUM AS ANTICIPATED BELOVED SIGN GOD ON EARTH.

Universal House of Justice

25-1. For an account of the life and services of Leroy Ioas, see BW 14:291–300. For an explanation of the International Bahá'í Council, see the glossary.

27
Reasons for Delay in Translating and Publishing the Kitáb-i-Aqdas
6 December 1965

The National Spiritual Assembly of the Bahá'ís of the United States
Dear Bahá'í Friends,

We have received a number of inquiries as to the translation and publication of the Kitáb-i-Aqdas from friends who are unable to read it in its original form. We feel the following extract from a letter written on behalf of the beloved Guardian by his secretary dated December 27, 1941, addressed to the National Spiritual Assembly of India and Burma clarifies this question: 27.1

> The reason it [the Kitáb-i-Aqdas] is not circulated amongst all the Bahá'ís is, first, because the Cause is not yet ready or sufficiently matured to put all the provisions of the Aqdas into effect and, second, because it is a book which requires to be supplemented by detailed explanations and to be translated into other languages by a competent body of experts. The provisions of the Aqdas are gradually, according to the progress of the Cause, being put into effect already, both in the East and the West. . . . 27.1a

As is well known, the beloved Guardian has already given in *God Passes By,* pp. 214–15, a summary of the contents of this Most Holy Book, and included the codification of all the laws of the Kitáb-i-Aqdas as one of the objectives of the Ten Year Crusade. It is the intention of the Universal House of Justice to achieve this objective by publishing a synopsis and codification of these laws during the current Nine Year Plan. 27.2

Much of the Kitáb-i-Aqdas has already been translated by the beloved Guardian and has been given to the friends in the West, although not designated, in every case, as coming from the Most Holy Book.[1] We give you below a list of such references for your guidance: 27.3

Gleanings from the Writings of Bahá'u'lláh	Sections XXXVII, LVI, LXX, LXXI, LXXII, XCVIII, CV, CLV, CLIX, and CLXV
The Promised Day Is Come	pp. 26 (1st para.), 36–37 (until the end of 2nd para.), 40 (2nd para.), and 84–85 (until the end of 1st para.)

27-1. In 1973, the last year of the Nine Year Plan, the Universal House of Justice published *A Synopsis and Codification of the Kitáb-i-Aqdas.* The volume includes all of the extracts in the list of references that follows. The Bahá'í World Center published a copiously annotated English translation of the Kitáb-i-Aqdas and related texts in 1992.

The Challenging Requirements of the Present Hour[2]	pp. 16–17 (until the end of 1st para.)
Bahá'í Administration	p. 21 (1st para.)
The World Order of Bahá'u'lláh	p. 134 (2nd para.)
The Bahá'í Community (1963 edition)	p. 4 (2nd & 3rd paras.)
Star of the West, Vol. XIV	pp. 112–14

27.4 The two reasons given by the Guardian in the extract of the letter quoted above need further amplification:

27.4a 1. As regards the first reason, regarding the timeliness of putting into effect all the provisions of the Kitáb-i-Aqdas, it must be borne in mind that the beloved Guardian further stated:

> ... the Laws revealed by Bahá'u'lláh in the Aqdas are, whenever practicable and not in direct conflict with the Civil Law of the land, absolutely binding on every believer or Bahá'í institution whether in the East or in the West. Certain laws, such as fasting, obligatory prayers, the consent of the parents before marriage, avoidance of alcoholic drinks, monogamy, should be regarded by all believers as universally and vitally applicable at the present time. Others have been formulated in anticipation of a state of society destined to emerge from the chaotic conditions that prevail today. When the Aqdas is published this matter will be further explained and elucidated. What has not been formulated in the Aqdas, in addition to matters of detail and of secondary importance arising out of the application of the Laws already formulated by Bahá'u'lláh, will have to be enacted by the Universal House of Justice. ...
>
> (*Bahá'í News*, October 1935)

The Guardian has further written:

> It should be noted in this connection that this Administrative Order is fundamentally different from anything that any Prophet has previously established, inasmuch as Bahá'u'lláh has Himself revealed its principles, established its institutions, appointed the person to interpret His Word and conferred the necessary authority on the body designed to supplement and apply His legislative ordinances. Therein lies the secret of its strength, its fundamental distinction, and the guarantee against disintegration and schism. ...
>
> (*The World Order of Bahá'u'lláh*, p. 145)

27-2. This message of Shoghi Effendi was later published in CF, pp. 4–38; the passage from the Kitáb-i-Aqdas appears on pp. 18–19 (see also PB, p. 63).

2. As to the second reason given by the beloved Guardian in the extract referred to above, it must be noted that the supplementary material to go with the publication of the laws of the Kitáb-i-Aqdas may well include the following items, all of which require careful research and translation:

    a. The Annex to the Kitáb-i-Aqdas, the Questions and Answers. (*God Passes By*, p. 219)

    b. Tablets of Bahá'u'lláh in "elaboration and elucidation of some of the laws He [Bahá'u'lláh] had already laid down." (*God Passes By*, p. 216)

    c. Tablets of Bahá'u'lláh establishing "subsidiary ordinances designed to supplement the provisions of His Most Holy Book." (*God Passes By*, p. 216)

    d. The Letters and Writings of 'Abdu'l-Bahá and Shoghi Effendi in interpretation of the laws and ordinances of the Kitáb-i-Aqdas.

    e. Other explanations and footnotes that may be required in elucidation of the provisions of that Book.

We hope the foregoing will clarify the matter for the friends. . . .

With loving Bahá'í greetings,
THE UNIVERSAL HOUSE OF JUSTICE

28
Call for Pioneers
10 December 1965

To all National Spiritual Assemblies

ANNOUNCE ALL BELIEVERS REJOICE RESPONSE BAHÁ'Í WORLD PIONEER CALL RAISED RIḌVÁN MESSAGE REQUIRING 460 PIONEERS COURSE CURRENT YEAR. THUS FAR 93 SETTLED POSTS INCLUDING 15 VIRGIN TERRITORIES ST. ANDRES ISLAND PROVIDENCIA ISLAND MARMARA ISLAND CHAD NIGER CAYMAN ISLANDS TURKS AND CAICOS ISCHIA GOTLAND ALASKA PENINSULA BARBUDA ST. KITTS-NEVIS INNER HEBRIDES BORNHOLM CAPRI 35 ADDITIONAL SETTLED SAME GOALS 167 MORE ARISEN AND IN PROCESS SETTLING TOTALING 295 SOULS RESPONDED CALL. FURTHER 200 BELIEVERS NEEDED NEXT FOUR SWIFTLY PASSING MONTHS FILL REMAINING GOALS. FATE PIONEER PLAN HANGING BALANCE PRAYING FERVENTLY HOLY SHRINES REQUIRED NUMBER HEROIC SOULS PROMPTLY ARISE MEET CHALLENGE CRITICAL HOUR URGE NATIONAL ASSEMBLIES NEEDING FUNDS EXECUTE ASSIGNMENTS APPLY IMMEDIATELY INTERNATIONAL DEPUTIZATION FUND. IMPERATIVE SETTLE ALL TERRITORIES ANNOUNCED RIḌVÁN EXCEPT THOSE DEPENDENT FAVORABLE CIRCUMSTANCES. VIRGIN AND RESETTLEMENT TERRITORIES PRIORITY. CONFIDENT SPIRIT DEVOTION FRIENDS GLORIOUS FAITH ENSURE BRILLIANT VICTORY THIS PRIMARY OBJECTIVE SO VITAL NINE YEAR PLAN.

UNIVERSAL HOUSE OF JUSTICE

29
Observance of Bahá'í Holy Days
28 January 1966

To National Spiritual Assemblies

Dear Bahá'í Friends,

29.1 From time to time questions have arisen about the application of the law of the Kitáb-i-Aqdas on the observance of Bahá'í Holy Days. As you know, the recognition of Bahá'í Holy Days in at least ninety-five countries of the world is an important and highly significant objective of the Nine Year Plan, and is directly linked with the recognition of the Faith of Bahá'u'lláh by the civil authorities as an independent religion enjoying its own rights and privileges.

29.2 The attainment of this objective will be facilitated and enhanced if the friends, motivated by their own realization of the importance of the laws of Bahá'u'lláh, are obedient to them. For the guidance of believers we repeat the instructions of the beloved Guardian:

29.2a He wishes also to stress the fact that, according to our Bahá'í laws, work is forbidden on our nine Holy Days. Believers who have independent businesses or shops should refrain from working on these days. Those who are in government employ should, on religious grounds, make an effort to be excused from work; all believers, whoever their employers, should do likewise. If the government, or other employers, refuse to grant them these days off, they are not required to forfeit their employment, but they should make every effort to have the independent status of their Faith recognized and their right to hold their own religious Holy Days acknowledged.
(From letter written on behalf of the Guardian to the American National Spiritual Assembly, dated 7 July 1947—*Bahá'í News*, No. 198, page 3)

29.2b This distinction between institutions that are under full or partial Bahá'í control is of a fundamental importance. Institutions that are entirely managed by Bahá'ís are, for reasons that are only too obvious, under the obligation of enforcing all the laws and ordinances of the Faith, especially those whose observance constitutes a matter of conscience. There is no reason, no justification whatever, that they should act otherwise . . . The point which should be always remembered is that the issue in question is essentially a matter of conscience, and as such is of a binding effect upon all believers. . . .
(From letter written on behalf of the Guardian to the American National Spiritual Assembly, dated 2 October 1935—*Bahá'í News*, No. 97, page 9)

In addition, steps should be taken to have Bahá'í children excused, on religious grounds, from attending school on Bahá'í Holy Days wherever possible. The Guardian has said: 29.3

> Regarding children: at fifteen a Bahá'í is of age as far as keeping the laws of the Aqdas is concerned—prayer, fasting, etc. But children under fifteen should certainly observe the Bahá'í Holy Days, and not go to school, if this can be arranged, on these nine days.
> (From letter written on behalf of the Guardian to the American National Spiritual Assembly, dated 25 October 1947) 29.3a

National Assemblies should give this subject their careful consideration, and should provide ways and means for bringing this matter to the attention of the believers under their jurisdiction so that, as a matter of conscience, the mass of believers will uphold these laws and observe them. 29.4

With loving Bahá'í greetings,
THE UNIVERSAL HOUSE OF JUSTICE

30
Further Thoughts about Mass Teaching
2 FEBRUARY 1966

To all National Spiritual Assemblies Engaged in Mass Teaching Work

Dear Bahá'í Friends,

Since writing to the National Spiritual Assemblies of the world regarding the importance of teaching the masses, we have received reports from all over the world indicating the steady increase in the number of believers, the concentration of the friends on the more receptive areas, however remote these may have been, and the opening up of new and challenging fields for expansion and service. In this letter we wish once again to stress the importance of this subject, share with you our thoughts regarding the supreme need to preserve the victories you have already won and the necessity to pursue the vital work in which you are engaged and to which the eyes of your sister communities in East and West are turned with admiration. 30.1

It has been due to the splendid victories in large-scale conversion that the Faith of Bahá'u'lláh has entered a new phase in its development and establishment throughout the world. It is imperative, therefore, that the process of teaching the masses be not only maintained but accelerated. The teaching committee structure that each National Assembly may adopt to ensure best results in the extension of its teaching work is a matter left entirely to its dis- 30.2

cretion, but an efficient teaching structure there must be, so that the tasks are carried out with dispatch and in accordance with the administrative principles of our Faith. From among the believers native to each country, competent traveling teachers must be selected and teaching projects worked out. In the words of our beloved Guardian, commenting upon the teaching work in Latin America: "Strong and sustained support should be given to the vitally needed and highly meritorious activities started by the native . . . traveling teachers, . . . who, as the mighty task progresses, must increasingly bear the brunt of responsibility for the propagation of the Faith in their homelands."[1]

30.3 While this vital teaching work is progressing each National Assembly must ever bear in mind that expansion and consolidation are inseparable processes that must go hand in hand. The interdependence of these processes is best elucidated in the following passage from the writings of the beloved Guardian: "Every outward thrust into new fields, every multiplication of Bahá'í institutions, must be paralleled by a deeper thrust of the roots which sustain the spiritual life of the community and ensure its sound development. From this vital, this ever-present need attention must, at no time, be diverted; nor must it be, under any circumstances, neglected, or subordinated to the no less vital and urgent task of ensuring the outer expansion of Bahá'í administrative institutions. That this community . . . may maintain a proper balance between these two essential aspects of its development . . . is the ardent hope of my heart."[2] To ensure that the spiritual life of the individual believer is continuously enriched, that local communities are becoming increasingly conscious of their collective duties, and that the institutions of an evolving administration are operating efficiently, is, therefore, as important as expanding into new fields and bringing in the multitudes under the shadow of the Cause.

30.4 These objectives can only be attained when each National Spiritual Assembly makes proper arrangements for all the friends to be deepened in the knowledge of the Faith. The National Spiritual Assemblies in consultation with the Hands of the Cause, who are the Standard-Bearers of the Nine Year Plan, should avail themselves of the assistance of Auxiliary Board members, who, together with the traveling teachers selected by the Assembly or its Teaching Committees, should be continuously encouraged to conduct deepening courses at Teaching Institutes and to make regular visits to Local Spiritual Assemblies. The visitors, whether Board members or traveling teachers should meet on such occasions not only with the Local Assembly but, of course, with the local community members, collectively at general meetings and even, if necessary, individually in their homes.

30-1. CF, p. 15.
30-2. LFG, p. 76.

The subjects to be discussed at such meetings with the Local Assembly and 30.5
the friends should include among others the following points:

1. the extent of the spread and stature of the Faith today;

2. the importance of the daily obligatory prayers (at least the short prayer);

3. the need to educate Bahá'í children in the Teachings of the Faith and encourage them to memorize some of the prayers;

4. the stimulation of youth to participate in community life by giving talks, etc. and having their own activities, if possible;

5. the necessity to abide by the laws of marriage, namely, the need to have a Bahá'í ceremony, to obtain the consent of parents, to observe monogamy; faithfulness after marriage; likewise the importance of abstinence from all intoxicating drinks and drugs;

6. the local Fund and the need for the friends to understand that the voluntary act of contributing to the Fund is both a privilege and a spiritual obligation. There should also be discussion of various methods that could be followed by the friends to facilitate their contributions and the ways open to the Local Assembly to utilize its local Fund to serve the interests of its community and the Cause;

7. the importance of the Nineteen Day Feast and the fact that it should be a joyful occasion and rallying point of the entire community;

8. the manner of election with as many workshops as required, including teaching of simple methods of balloting for illiterates, such as having one central home as the place for balloting and arranging for one literate person, if only a child, to be present at that home during the whole day, if necessary;

9. last but not least, the all-important teaching work, both in the locality and its neighboring centers, as well as the need to continuously deepen the friends in the essentials of the Faith. The friends should be made to realize that in teaching the Faith to others they should not only aim at assisting the seeking soul to join the Faith, but also at making him a teacher of the Faith and its active supporter.

All the above points should, of course, be stressed within the framework 30.6
of the importance of the Local Spiritual Assembly, which should be encouraged to vigorously direct its attention to these vital functions and become the very heart of the community life of its own locality, even if its meetings should become burdened with the problems of the community. The local friends should understand the importance of the law of consultation and realize that

it is to the Local Spiritual Assembly that they should turn, abide by its decisions, support its projects, cooperate wholeheartedly with it in its task to promote the interests of the Cause, and seek its advice and guidance in the solution of personal problems and the adjudication of disputes, should any arise amongst the members of the community.

30.7 As the Universal House of Justice intends to have on file a full record of the progress of the teaching work in large-scale conversion areas, we request you to send us any published material, such as forms, cards, pamphlets, pictures, audiovisual aids, deepening booklets, etc. that you are currently using, with adequate explanations by your Assembly as to how they are being used, and any comments you may wish to make about their usefulness. Your National Assembly should also feel free to share with us your problems and needs as well as any recommendations you may have. We are looking forward to receiving a prompt reply to this letter, as we feel that an early evaluation of the methods used in various fields of teaching is vital and essential at this time.

With loving Bahá'í greetings,
THE UNIVERSAL HOUSE OF JUSTICE

31
Call for Pioneers and Traveling Teachers
18 MARCH 1966

To all National Spiritual Assemblies

Dear Bahá'í Friends,

31.1 The message of the Universal House of Justice to the Bahá'í world this Riḍván will raise a call for volunteers to engage in traveling teaching in all parts of the world and for whatever periods of time are possible. The purpose is to develop a band of international teachers who will, by the very fact of being visitors from other countries, stimulate interest on the various homefronts which they visit.

31.2 The Continental Pioneer Committees for Africa, the Americas, Asia, Australasia and Europe have been given the additional function of assisting National Assemblies to make the most efficient use of any who may volunteer to travel and teach in countries other than their own.[1] The procedures to be followed are attached to this letter. As you will note we are extending the use of the International Deputization Fund to deserving international teaching projects which cannot otherwise be financed.

31-1. See message no. 22.

This advanced information is given to National Assemblies so that they may be prepared to accept and utilize the offers of traveling teachers as soon as they arise. 31.3

It is requested that the subject of traveling teachers be placed on the Agenda for consultation at the National Convention, and that the friends be given every encouragement to respond to this call. 31.4

Since the announcement of this plan is a part of the Riḍván message, your Assembly should wait until the Convention before announcing it to the friends. 31.5

With loving Baháʾí greetings,
THE UNIVERSAL HOUSE OF JUSTICE

P.S. The locale of the Pioneer Committee for Africa is being changed to Uganda effective at Riḍván. For the time being please contact the Committee in care of that National Assembly. 31.6

32
Fiftieth Anniversary of the Revelation of the Tablets of the Divine Plan
22 MARCH 1966

To the National Spiritual Assembly of the Baháʾís of the United States

JOYOUSLY HAIL FIFTIETH ANNIVERSARY REVELATION FIRST OF TABLETS DIVINE PLAN CHARTER PROPAGATION FAITH THROUGHOUT WORLD.[1] PRAYING SHRINES OBSERVANCE OCCASION MAY BE SOURCE RENEWED ENTHUSIASM DEDICATION FRIENDS ACCOMPLISH GOALS WIN FRESH LAURELS SHARE NATIONAL SPIRITUAL ASSEMBLIES CANADA ALASKA HAWAII. 32.1

THE UNIVERSAL HOUSE OF JUSTICE

32-1. The Tablets of the Divine Plan are fourteen Tablets ʿAbdu'l-Bahá revealed in 1916 and 1917 to the Baháʾís of the United States and Canada. In them He conveys His mandate for the transmission of the Faith of Baháʾuʾlláh throughout the world.

33

Passing of Jessie Revell, Member of the International Bahá'í Council

15 APRIL 1966

To all National Spiritual Assemblies

33.1 WITH PROFOUND GRIEF ANNOUNCE PASSING JESSIE REVELL HER TIRELESS STEADFAST DEVOTION FAITH SINCE BEFORE MASTER'S VISIT AMERICAN CONTINENT EARNED LOVE TRUST ADMIRATION SHOGHI EFFENDI CROWNED BY APPOINTMENT INTERNATIONAL BAHÁ'Í COUNCIL DISTINGUISHED BY SERVICE TREASURER BOTH APPOINTED ELECTED COUNCILS.[1] URGE NATIONAL ASSEMBLIES HOLD MEMORIAL GATHERINGS TRIBUTE UNFORGETTABLE EXEMPLARY SERVICES FAITH. . . .

THE UNIVERSAL HOUSE OF JUSTICE

34

Riḍván Message 1966

RIḌVÁN 1966

The Bahá'ís of the World

Dearly loved Friends,

34.1 The Fiftieth Anniversary of the revelation by 'Abdu'l-Bahá, in March and April 1916, of the first Tablets of the Divine Plan, has witnessed the conclusion of a feat of pioneering unparalleled in the annals of the Cause. A year ago the call was raised for four-hundred-and-sixty-one pioneers to leave their homes within twelve months and scatter throughout the planet to broaden and strengthen the foundations of the world community of Bahá'u'lláh. There is every hope that with the exception of thirty-four posts whose settlement is dependent upon favorable circumstances all the pioneer goals will be filled by Riḍván or their settlement will be assured by firm commitments. The gratitude and admiration of the entire Bahá'í world go out to this noble band of dedicated believers who have so gloriously responded to the call. These pioneers, who have arisen for the specified goals, have been reinforced by a further forty-five believers who have settled in the goal territories, while sixty-nine more have left their homes to reside in twenty-six other countries already opened to the Faith. All told, in the course of the year, five-hundred-and-five Bahá'ís have arisen to pioneer beyond their homelands, the largest number ever to do so in any one year in the entire history of the Cause.

33-1. For an account of the life and services of Jessie Revell, see BW 14:300–03. For an explanation of the International Bahá'í Council, see the glossary.

This is a resounding victory, and in the light of the Master's statement in 34.2
the first of the Tablets of the Divine Plan, "It has often happened that one
blessed soul has become the cause of the guidance of a nation," of wonderful portent for the future.¹ Its immediate results are the opening of twenty-four new territories to the Faith, the resettlement of four others, and the consolidation of ninety-three more. The newly opened territories are: Chad and Niger in Africa; Alaskan Peninsula, Barbuda, Cayman Islands, Chiloé Island, Providencia Island, Quintana Roo Territory, Saba, St. Andrés Island, St. Eustatius, St. Kitts-Nevis, St. Lawrence Island, Tierra del Fuego, and Turks and Caicos Islands in the Americas; Laccadive Islands and Marmara Island in Asia; Niue Island in Australasia; and Bornholm, Capri, Elba, Gotland, Inner Hebrides, and Ischia in Europe.

The resettled territories are: Corisco Island and Spanish Guinea in Africa 34.3
and Maldive Islands and Nicobar Islands in Asia.

As announced last Riḍván, the first Convention of the Bahá'ís of Brunei 34.4
will be held this year, during the second weekend of the Riḍván period, when
the first National Spiritual Assembly of the Bahá'ís of Brunei will be elected.
Hand of the Cause Collis Featherstone will represent the World Center of the
Faith on this historic occasion.

Formation of Nine National Assemblies at Riḍván 1967

A further result of the confirmations which have rewarded the tremendous 34.5
teaching effort of the past two years is the call now made by the House of
Justice for the formation at Riḍván 1967 of the following nine National Spiritual Assemblies: in Africa—the National Spiritual Assembly of Algeria and
Tunisia with its seat in Algiers; the National Spiritual Assembly of Cameroon
Republic with its seat in Victoria and with Spanish Guinea, Fernando Po,
Corisco and São Tomé and Príncipe Islands assigned to it; the National Spiritual Assembly of Swaziland, Mozambique and Basutoland with its seat in
Mbabane; the National Spiritual Assembly of Zambia with its seat in Lusaka.
In the Americas—the National Spiritual Assembly of the Leeward, Windward
and Virgin Islands with its seat in Charlotte Amalie. In Asia—the National
Spiritual Assembly of Cambodia with its seat in Phnom Penh; the National
Spiritual Assembly of Eastern and Southern Arabia with its seat in Bahrayn;
the National Spiritual Assembly of Taiwan with its seat in Taipei. In Australasia—the National Spiritual Assembly of the Gilbert and Ellice Islands
with its seat in Tarawa. These nine new National Spiritual Assemblies constituting, together with the new National Spiritual Assembly of Brunei, ten additional pillars of the Universal House of Justice, will bring to seventy-nine the
number which will take part during Riḍván 1968 in the second International
Convention for the election of that Institution.

34-1. TDP 10.3.

Tribute to the Hands of the Cause of God

34.6 This momentous year cannot be allowed to pass without mention of the tireless and dedicated services of the beloved Hands of the Cause, the Standard-Bearers of the Nine Year Plan, and the able support rendered them by their Auxiliary Boards. The special missions which they have discharged on behalf of the Universal House of Justice, the teaching tours they have undertaken, the conferences they have organized, their constant work at the World Center, and above all their never-ending encouragement of the friends and watchfulness over the welfare of the Cause of God, have given distinction and effective leadership to the work of the entire community. The grievous loss which they sustained in the passing of Hand of the Cause Leroy Ioas is shared by the whole Bahá'í world.

Threefold Purpose of the Intercontinental Conferences

34.7 The splendid achievements in the pioneering and teaching fields, together with the enthusiastic attention given to the preparation of plans for the befitting celebration of the centenary of Bahá'u'lláh's proclamation of His Message to the kings and rulers of the world, have sealed with success the first, and opened the way for the second phase of the Nine Year Plan, a phase in which the Bahá'í world must prepare and arm itself for the third phase, beginning in October 1967 when the six intercontinental conferences will sound the "opening notes" of a period of proclamation of the Cause of God extending through the remaining years of the Nine Year Plan to the centenary, in 1973, of the revelation of the Kitáb-i-Aqdas.[2] The threefold purpose of these conferences is to commemorate the centenary of the opening of Bahá'u'lláh's Own proclamation of His Mission, to proclaim the Divine Message, and to deliberate upon the tasks of the remaining years of the Nine Year Plan.

Five Tasks of the Second Phase of the Nine Year Plan

34.8 Five specific tasks face the Bahá'í world as it enters this second phase of the Plan:

The first is to complete the settlement of the pioneers, and the dispatch of others wherever needed.

The second is intensive preparation for the third phase of the Plan through development of new teaching measures and expansion of the various Bahá'í funds at international, national and local levels.

The third is acceleration of the provision of Bahá'í literature, particularly its translation and publication in those languages in which, as yet, none has been published or the supply is inadequate.

34-2. GPB, p. 212.

The fourth is the acquisition of the remaining national Ḥaẓíratu'l-Quds, Temple sites, national endowments and teaching institutes called for in the Plan, before the developing inflation now affecting nearly the whole world adds too greatly to the financial burden of acquiring these properties.

The fifth is development of the Panama Temple Fund. The Universal House of Justice is initiating this Fund with a contribution of $25,000, and now calls upon the believers and Bahá'í communities to contribute liberally and continuously until the funds for the completion of this historic structure are assured. Such contributions should be sent directly to the National Spiritual Assembly of Panama. More than fifty designs have been received, and the House of Justice is now considering the recommendations of the National Assembly. The choice will be announced and the friends will be kept fully informed of the progress of this highly significant and inspiring project.

Our Challenge—To Raise the Intensity of Teaching

34.9 Every individual follower of Bahá'u'lláh, as well as the institutions of the Faith, at local, national, continental and world levels, must now meet the challenge to raise the intensity of teaching to a pitch never before attained, in order to realize that vast increase called for in the Plan. For those believers living in countries where they have freedom to teach their Faith, this challenge is the more sharply pointed by the oppressive measures imposed on the Faith elsewhere. In Persia the believers are denied their elementary rights and the Faith is still largely proscribed. In Iraq the national and one local Ḥaẓíratu'l-Quds have been seized and the activities of the friends severely restricted. In Egypt Bahá'í properties are still confiscated and recently several believers were imprisoned for a period, and are now awaiting trial. New oppression has broken out in Indonesia where the national Ḥaẓíratu'l-Quds has been seized and organized activities of the believers have been forbidden. In yet other countries the believers are subject to restrictions and surveillance. The friends in all cases are steadfast and confident, looking forward to their emancipation and the eventual triumph of the Cause.

34.10 The challenge to the local and national administrative institutions of the Faith is to organize and promote the teaching work through systematic plans, involving not only the regular fireside meetings in the homes of the believers, the public meetings, receptions and conferences, the weekend, summer and winter schools, the youth conferences and activities, all of which are so vigorously upheld at present, but in addition through a constant stream of visiting teachers to every locality. The forces released by this latter process have been extolled by Bahá'u'lláh in these words:

34.10a The movement itself from place to place, when undertaken for the sake of God, hath always exerted, and can now exert, its influence in the world. In the Books of old the station of them that have voyaged far and near in order to guide the servants of God hath been set forth and written down.³

while 'Abdu'l-Bahá in the Tablets of the Divine Plan, says:

34.10b Teachers must continually travel to all parts of the continent, nay, rather, to all parts of the world . . .⁴

34.11 Such plans must be initiated and developed now, during this period of preparation, so that they may be fully operative by the beginning of the proclamation period from which time they must be relentlessly pursued until the end of the Plan.

34.12 The Universal House of Justice attaches such importance to this principle of traveling teaching that it has decided to develop it internationally, and now calls for volunteers to offer their services in this field. By their visits to lands other than their own, these friends will lend a tremendous stimulus to the proclamation and teaching of the Cause in all continents. It is hoped that such projects will be self-supporting, since the International Deputization Fund will still be needed for pioneering. However, when a proposal which is considered to be of special benefit to the Faith cannot be financed by the individual or the receiving National Assemblies, the House of Justice will consider a request for assistance from the Deputization Fund. Offers, which may be for any period, should be made to one's own National Spiritual Assembly or to the Continental Pioneer Committees, which have been given the additional task of assisting National Assemblies to implement and coordinate this new enterprise. Let those who arise recall the Master's injunction to "travel like 'Abdu'l-Bahá . . . sanctified and free from every attachment and in the utmost severance."⁵

Consolidation—Coequal with Expansion

34.13 Simultaneous and coequal with this vast, ordered and ever-growing teaching effort, the work of consolidation must go hand in hand. In fact these two processes must be regarded as inseparable parts of the expansion of the Faith. While the work of teaching inevitably goes first, to pursue it alone without consolidation would leave the community unprepared to receive the masses who must sooner or later respond to the life-giving message of the Cause. The guidance of our beloved Guardian in this vital matter is, as ever, clear and

---

34-3. Quoted in ADJ, p. 84.
34-4. TDP 8.11.
34-5. TDP 8.11.

unambiguous: "Every outward thrust into new fields, every multiplication of Bahá'í institutions, must be paralleled by a deeper thrust of the roots which sustain the spiritual life of the community and ensure its sound development. From this vital, this ever-present need attention must, at no time, be diverted; nor must it be, under any circumstances, neglected, or subordinated to the no less vital and urgent task of ensuring the outer expansion of Bahá'í administrative institutions."[6] A proper balance between these two essential aspects of its development must, from now on, as we enter the era of large-scale conversion, be maintained by the Bahá'í Community. Consolidation must comprise not only the establishment of Bahá'í administrative institutions, but a true deepening in the fundamental verities of the Cause and in its spiritual principles, understanding of its prime purpose in the establishment of the unity of mankind, instruction in its standards of behavior in all aspects of private and public life, in the particular practice of Bahá'í life in such things as daily prayer, education of children, observance of the laws of Bahá'í marriage, abstention from politics, the obligation to contribute to the Fund, the importance of the Nineteen Day Feast and opportunity to acquire a sound knowledge of the present-day practice of Bahá'í administration.

The Urgent Need for an Increased Flow of Funds

34.14 The onward march of the Faith requires, and is indeed dependent upon, a very great increase in contributions to the various funds. All the goals assigned to the World Center of the Faith, and particularly those dealing with the development and beautification of the properties surrounding the Holy Shrines and the extension of the gardens on Mount Carmel entail heavy expenditures. The building of the two Temples called for in the Plan will require further large sums,[7] and the worldwide process of teaching and consolidation now to be intensified must be sustained by a greatly increased and uninterrupted flow of funds. The International Deputization Fund must be maintained and expanded, not only for further pioneering needs, but in order to assist and develop the traveling teacher program now called for. Since only those who have openly proclaimed their recognition of Bahá'u'lláh are permitted to contribute financially to the establishment of His World Order, it is apparent that more, much more is required from the few now so privileged. Our responsibilities in this field are very great, commensurate indeed with the bounty of being the bearers of the Name of God in this day.

34-6. LFG, p. 76.

34-7. The two Temples referred to are the Houses of Worship that were to be built in Asia and Latin America. Plans for the construction of a Temple in Asia, to be located in Iran, had to be held in abeyance, due to antagonism toward the Faith in that country. The Temple for Latin America, completed in 1986, is in Panama.

The Individual's Challenge

34.15 The challenge to the individual Bahá'í in every field of service, but above all in teaching the Cause of God is never-ending. With every fresh affliction visited upon mankind our inescapable duty becomes more apparent, nor should we ever forget that if we neglect this duty, "others" in the words of Shoghi Effendi, "will be called upon to take up our task as ministers to the crying needs of this afflicted world."[8] Now, it seems, we may well be entering an era of the longed-for expansion of our beloved Faith. Mankind's growing hunger for spiritual truth is our opportunity. While reaching forth to grasp it we would do well to ponder the following words of Bahá'u'lláh:

34.15a Your behavior towards your neighbor should be such as to manifest clearly the signs of the one true God, for ye are the first among men to be recreated by His Spirit, the first to adore and bow the knee before Him, the first to circle round His throne of glory.[9]

34.16 As humanity plunges deeper into that condition of which Bahá'u'lláh wrote, "to disclose it now would not be meet and seemly," so must the believers increasingly stand out as assured, orientated and fundamentally happy beings, conforming to a standard which, in direct contrast to the ignoble and amoral attitudes of modern society, is the source of their honor, strength and maturity.[10] It is this marked contrast between the vigor, unity and discipline of the Bahá'í community on the one hand, and the increasing confusion, despair and feverish tempo of a doomed society on the other, which, during the turbulent years ahead will draw the eyes of humanity to the sanctuary of Bahá'u'lláh's world-redeeming Faith.

34.17 The constant progress of the Cause of God is a source of joy to us all and a stimulus to further action. But not ordinary action. Heroic deeds are now called for such as are performed only by divinely sustained and detached souls. 'Abdu'l-Bahá, the Commander of the hosts of the Lord, in one of the Tablets of the Divine Plan, uttered this cry: "O that I could travel, even though on foot and in the utmost poverty, to these regions and, raising the call of 'Yá Bahá'u'l-Abhá' in cities, villages, mountains, deserts and oceans, promote the Divine teachings! This, alas, I cannot do. How intensely I deplore it." And He concluded with this heart-shaking appeal, "Please God, ye may achieve it."[11]

THE UNIVERSAL HOUSE OF JUSTICE

34-8. BA, p. 66.
34-9. GWB, p. 316–17.
34-10. GWB, p. 118.
34-11. TDP 7.8.

35
The Guardianship and the Universal House of Justice
27 May 1966

To an individual Bahá'í[1]

Dear Bahá'í Friend,

... You query the timing of the election of the Universal House of Justice in view of the Guardian's statement: "... given favorable circumstances, under which the Bahá'ís of Persia and of the adjoining countries under Soviet rule, may be enabled to elect their national representatives ... the only remaining obstacle in the way of the definite formation of the International House of Justice will have been removed."[2] On 19th April 1947 the Guardian, in a letter written on his behalf by his secretary, replied to the inquiry of an individual believer about this passage: "At the time he referred to Russia there were Bahá'ís there, now the Community has practically ceased to exist; therefore the formation of the International House of Justice cannot depend on a Russian National Spiritual Assembly. But other strong National Spiritual Assemblies will have to be built up before it can be established." 35.1

The Provisions of 'Abdu'l-Bahá's Will

You suggest the possibility that, for the good of the Cause, certain information concerning the succession to Shoghi Effendi is being withheld from the believers. We assure you that nothing whatsoever is being withheld from the friends for whatever reason. There is no doubt at all that in the Will and Testament of 'Abdu'l-Bahá Shoghi Effendi was the authority designated to appoint his successor, but he had no children and all the surviving Aghṣán had broken the Covenant.[3] Thus, as the Hands of the Cause stated in 1957, it is clear that there was no one he could have appointed in accordance with the provisions of the Will. To have made an appointment outside the clear and specific provisions of the Master's Will and Testament would obviously have been an impossible and unthinkable course of action for the Guardian, the divinely appointed upholder and defender of the Covenant. Moreover, that same Will had provided a clear means for the confirmation of the 35.2

35-1. Passages from a letter written by the Universal House of Justice on 27 May 1966 in response to questions asked by an individual believer on the relationship between the Guardianship and the Universal House of Justice.

35-2. WOB, p. 7.

35-3. Aghṣán (Branches) are the sons and male descendants of Bahá'u'lláh. In His Will and Testament 'Abdu'l-Bahá wrote, "should the firstborn of the Guardian of the Cause of God not manifest in himself the truth of the words: 'The child is the secret essence of its sire,' that is, should he not inherit of the spiritual within him (the Guardian of the Cause of God) and his glorious lineage not be matched with a goodly character, then must he (the Guardian of the Cause of God) choose another branch [Aghṣán] to succeed him" (WT, p. 12).

Guardian's appointment of his successor, as you are aware. The nine Hands to be elected by the body of the Hands were to give their assent by secret ballot to the Guardian's choice. In 1957 the entire body of the Hands, after fully investigating the matter, announced that Shoghi Effendi had appointed no successor and left no will. This is documented and established.

A Sign of Infallible Guidance

35.3 The fact that Shoghi Effendi did not leave a will cannot be adduced as evidence of his failure to obey Bahá'u'lláh—rather should we acknowledge that in his very silence there is a wisdom and a sign of his infallible guidance. We should ponder deeply the writings that we have, and seek to understand the multitudinous significances that they contain. Do not forget that Shoghi Effendi said two things were necessary for a growing understanding of the World Order of Bahá'u'lláh: the passage of time and the guidance of the Universal House of Justice.

The Infallibility of the Universal House of Justice

35.4 The infallibility of the Universal House of Justice, operating within its ordained sphere, has not been made dependent upon the presence in its membership of the Guardian of the Cause. Although in the realm of interpretation the Guardian's pronouncements are always binding, in the area of the Guardian's participation in legislation it is always the decision of the House itself which must prevail. This is supported by the words of the Guardian: "The interpretation of the Guardian, functioning within his own sphere, is as authoritative and binding as the enactments of the International House of Justice, whose exclusive right and prerogative is to pronounce upon and deliver the final judgment on such laws and ordinances as Bahá'u'lláh has not expressly revealed. Neither can, nor will ever, infringe upon the sacred and prescribed domain of the other. Neither will seek to curtail the specific and undoubted authority with which both have been divinely invested.

35.5 "Though the Guardian of the Faith has been made the permanent head of so august a body he can never, even temporarily, assume the right of exclusive legislation. He cannot override the decision of the majority of his fellow-members, but is bound to insist upon a reconsideration by them of any enactment he conscientiously believes to conflict with the meaning and to depart from the spirit of Bahá'u'lláh's revealed utterances."[4]

35.6 However, quite apart from his function as a member and sacred head for life of the Universal House of Justice, the Guardian, functioning within his own sphere, had the right and duty "to define the sphere of the legislative action" of the Universal House of Justice.[5] In other words, he had the author-

---

35-4. WOB, p. 150
35-5. WOB, p. 148.

ity to state whether a matter was or was not already covered by the Sacred Texts and therefore whether it was within the authority of the Universal House of Justice to legislate upon it. No other person, apart from the Guardian, has the right or authority to make such definitions. The question therefore arises: In the absence of the Guardian, is the Universal House of Justice in danger of straying outside its proper sphere and thus falling into error? Here we must remember three things: First, Shoghi Effendi, during the thirty-six years of his Guardianship, has already made innumerable such definitions, supplementing those made by 'Abdu'l-Bahá and by Bahá'u'lláh Himself. As already announced to the friends, a careful study of the Writings and interpretations on any subject on which the House of Justice proposes to legislate always precedes its act of legislation. Second, the Universal House of Justice, itself assured of divine guidance, is well aware of the absence of the Guardian and will approach all matters of legislation only when certain of its sphere of jurisdiction, a sphere which the Guardian has confidently described as "clearly defined." Third, we must not forget the Guardian's written statement about these two Institutions: "Neither can, nor will ever, infringe upon the sacred and prescribed domain of the other."[6]

35.7 As regards the need to have deductions made from the Writings to help in the formulation of the enactments of the House of Justice, there is the following text from the pen of 'Abdu'l-Bahá:

35.7a Those matters of major importance which constitute the foundation of the Law of God are explicitly recorded in the Text, but subsidiary laws are left to the House of Justice. The wisdom of this is that the times never remain the same, for change is a necessary quality and an essential attribute of this world, and of time and place. Therefore the House of Justice will take action accordingly.

35.7b Let it not be imagined that the House of Justice will take any decision according to its own concepts and opinions. God forbid! The Supreme House of Justice will take decisions and establish laws through the inspiration and confirmation of the Holy Spirit, because it is in the safekeeping and under the shelter and protection of the Ancient Beauty, and obedience to its decisions is a bounden and essential duty and an absolute obligation, and there is no escape for anyone.

35.7c Say, O people: Verily the Supreme House of Justice is under the wings of your Lord, the Compassionate, the All-Merciful, that is, under His protection, His care, and His shelter; for He has commanded the firm believers to obey that blessed, sanctified and all-subduing body, whose sovereignty is divinely ordained and of the Kingdom of Heaven and whose laws are inspired and spiritual.

35-6. WOB, p. 148, 150.

35.7d Briefly, this is the wisdom of referring the laws of society to the House of Justice. In the religion of Islam, similarly, not every ordinance was explicitly revealed; nay not a tenth part of a tenth part was included in the Text; although all matters of major importance were specifically referred to, there were undoubtedly thousands of laws which were unspecified. These were devised by the divines of a later age according to the laws of Islamic jurisprudence, and individual divines made conflicting deductions from the original revealed ordinances. All these were enforced. Today this process of deduction is the right of the body of the House of Justice, and the deductions and conclusions of individual learned men have no authority, unless they are endorsed by the House of Justice. The difference is precisely this, that from the conclusions and endorsements of the body of the House of Justice whose members are elected by and known to the worldwide Bahá'í community, no differences will arise; whereas the conclusions of individual divines and scholars would definitely lead to differences, and result in schism, division, and dispersion. The oneness of the Word would be destroyed, the unity of the Faith would disappear, and the edifice of the Faith of God would be shaken.[7]

The Continuity of Authority

35.8 In the Order of Bahá'u'lláh there are certain functions which are reserved to certain institutions, and others which are shared in common, even though they may be more in the special province of one or the other. For example, although the Hands of the Cause of God have the specific functions of protection and propagation, and are specialized for these functions, it is also the duty of the Universal House of Justice and the Spiritual Assemblies to protect and teach the Cause—indeed teaching is a sacred obligation placed upon every believer by Bahá'u'lláh. Similarly, although after the Master authoritative interpretation was exclusively vested in the Guardian, and although legislation is exclusively the function of the Universal House of Justice, these two Institutions are, in Shoghi Effendi's words, "complementary in their aim and purpose." "Their common, their fundamental object is to ensure the continuity of that divinely appointed authority which flows from the Source of our Faith, to safeguard the unity of its followers and to maintain the integrity and flexibility of its teachings."[8] Whereas the Universal House of Justice cannot undertake any function which exclusively appertained to the Guardian, it must continue to pursue the object which it shares in common with the Guardianship.

35-7. Rahíq-i-Makhtúm 1:302–04; BN, no. 426 (Sept. 1966): 2.
35-8. WOB, p. 148.

The Principle of Inseparability

35.9 As you point out with many quotations, Shoghi Effendi repeatedly stressed the inseparability of these two institutions. Whereas he obviously envisaged their functioning together, it cannot logically be deduced from this that one is unable to function in the absence of the other. During the whole thirty-six years of his Guardianship Shoghi Effendi functioned without the Universal House of Justice. Now the Universal House of Justice must function without the Guardian, but the principle of inseparability remains. The Guardianship does not lose its significance nor position in the Order of Bahá'u'lláh merely because there is no living Guardian. We must guard against two extremes: one is to argue that because there is no Guardian all that was written about the Guardianship and its position in the Bahá'í World Order is a dead letter and was unimportant; the other is to be so overwhelmed by the significance of the Guardianship as to underestimate the strength of the Covenant, or to be tempted to compromise with the clear texts in order to find somehow, in some way, a "Guardian."

Our Part—Fidelity, Integrity, and Faith

35.10 Service to the Cause of God requires absolute fidelity and integrity and unwavering faith in Him. No good but only evil can come from taking the responsibility for the future of God's Cause into our own hands and trying to force it into ways that we wish it to go regardless of the clear texts and our own limitations. It is His Cause. He has promised that its light will not fail. Our part is to cling tenaciously to the revealed Word and to the Institutions that He has created to preserve His Covenant.

35.11 It is precisely in this connection that the believers must recognize the importance of intellectual honesty and humility. In past dispensations many errors arose because the believers in God's Revelation were overanxious to encompass the Divine Message within the framework of their limited understanding, to define doctrines where definition was beyond their power, to explain mysteries which only the wisdom and experience of a later age would make comprehensible, to argue that something was true because it appeared desirable and necessary. Such compromises with essential truth, such intellectual pride, we must scrupulously avoid.

35.12 If some of the statements of the Universal House of Justice are not detailed the friends should realize that the cause of this is not secretiveness, but rather the determination of this body to refrain from interpreting the teachings and to preserve the truth of the Guardian's statement that "Leaders of religion, exponents of political theories, governors of human institutions . . . need have no doubt or anxiety regarding the nature, the origin, or validity of the institutions which the adherents of the Faith are building up throughout the world. For these lie embedded in the teachings themselves, unadulterated and unobscured by unwarrantable inferences, or unauthorized interpretations of His Word."⁹

Authoritative Interpretation and Individual Understanding

35.13 A clear distinction is made in our Faith between authoritative interpretation and the interpretation or understanding that each individual arrives at for himself from his study of its teachings. While the former is confined to the Guardian, the latter, according to the guidance given to us by the Guardian himself, should by no means be suppressed. In fact such individual interpretation is considered the fruit of man's rational power and conducive to a better understanding of the teachings, provided that no disputes or arguments arise among the friends and the individual himself understands and makes it clear that his views are merely his own. Individual interpretations continually change as one grows in comprehension of the teachings. As Shoghi Effendi explained: "To deepen in the Cause means to read the writings of Bahá'u'lláh and the Master so thoroughly as to be able to give it to others in its pure form. There are many who have some superficial idea of what the Cause stands for. They, therefore, present it together with all sorts of ideas that are their own. As the Cause is still in its early days we must be most careful lest we fall under this error and injure the Movement we so much adore. There is no limit to the study of the Cause. The more we read the Writings, the more truths we can find in them and the more we will see that our previous notions were erroneous."¹⁰ So, although individual insights can be enlightening and helpful, they can also be misleading. The friends must therefore learn to listen to the views of others without being overawed or allowing their faith to be shaken, and to express their own views without pressing them on their fellow Bahá'ís.

The Covenant—The Cord to which All Must Cling

35.14 The Cause of God is organic, growing and developing like a living being. Time and again it has faced crises which have perplexed the believers, but each time the Cause, impelled by the immutable purpose of God, overcame the crisis and went on to greater heights.

35.15 However great may be our inability to understand the mystery and the implications of the passing of Shoghi Effendi, the strong cord to which all must cling with assurance is the Covenant. The emphatic and vigorous language of 'Abdu'l-Bahá's Will and Testament is at this time, as at the time of His own passing, the safeguard of the Cause:

35-9. WOB, p. 24.
35-10. Written by the Guardian's secretary on his behalf to an individual believer, on 25 August 1926.

Unto the Most Holy Book every one must turn and all that is not expressly recorded therein must be referred to the Universal House of Justice. That which this body, whether unanimously or by a majority doth carry, that is verily the truth and the purpose of God Himself. Whoso doth deviate therefrom is verily of them that love discord, hath shown forth malice and turned away from the Lord of the Covenant. . . ." And again: ". . . All must seek guidance and turn unto the Center of the Cause and the House of Justice. And he that turneth unto whatsoever else is indeed in grievous error."[11]   35.16

The Universal House of Justice: Recipient of Divine Guidance

The Universal House of Justice, which the Guardian said would be regarded by posterity as "the last refuge of a tottering civilization," is now, in the absence of the Guardian, the sole infallibly guided institution in the world to which all must turn, and on it rests the responsibility for ensuring the unity and progress of the Cause of God in accordance with the revealed Word.[12] There are statements from the Master and the Guardian indicating that the Universal House of Justice, in addition to being the Highest Legislative Body of the Faith, is also the body to which all must turn, and is the "apex" of the Bahá'í Administrative Order, as well as the "supreme organ of the Bahá'í Commonwealth."[13] The Guardian has in his writings specified for the House of Justice such fundamental functions as the formulation of future worldwide teaching plans, the conduct of the administrative affairs of the Faith, and the guidance, organization and unification of the affairs of the Cause throughout the world. Furthermore in *God Passes By* the Guardian makes the following statement: "the Kitáb-i-Aqdas . . . not only preserves for posterity the basic laws and ordinances on which the fabric of His future World Order must rest, but ordains, in addition to the function of interpretation which it confers upon His Successor, the necessary institutions through which the integrity and unity of His Faith can alone be safeguarded."[14] He has also, in "The Dispensation of Bahá'u'lláh," written that the members of the Universal House of Justice "and not the body of those who either directly or indirectly elect them, have thus been made the recipients of the divine guidance which is at once the lifeblood and ultimate safeguard of this Revelation."[15]   35.17

As the Universal House of Justice has already announced, it cannot legislate to make possible the appointment of a successor to Shoghi Effendi, nor can it legislate to make possible the appointment of any more Hands of the Cause, but it must do everything within its power to ensure the performance   35.18

35-11. WT, pp. 19–20, 26.
35-12. WOB, p. 89.
35-13. GPB, p. 332; WOB, p. 7.
35-14. GPB, pp. 213–14.
35-15. WOB, p. 153.

of all those functions which it shares with these two mighty Institutions. It must make provision for the proper discharge in future of the functions of protection and propagation, which the administrative bodies share with the Guardianship and the Hands of the Cause; it must, in the absence of the Guardian, receive and disburse the Ḥuqúqu'lláh, in accordance with the following statement of 'Abdu'l-Bahá: "Disposition of the Ḥuqúq, wholly or partly, is permissible, but this should be done by permission of the authority in the Cause to whom all must turn";[16] it must make provision in its Constitution for the removal of any of its members who commits a sin "injurious to the common weal."[17] Above all, it must, with perfect faith in Bahá'u'lláh, proclaim His Cause and enforce His Law so that the Most Great Peace shall be firmly established in this world and the foundation of the Kingdom of God on earth shall be accomplished.

With loving Bahá'í greetings,
THE UNIVERSAL HOUSE OF JUSTICE

36
Pioneering and a Program of Consolidation
5 JUNE 1966

To all National Spiritual Assemblies

Dear Bahá'í Friends,

36.1 The historic and dazzling accomplishment last year of a feat in pioneering unparalleled in the annals of our Faith should not blind us either to the need of filling the remaining gaps which are still open, or to the ever-present necessity of reinforcing the settled territories with a well-conceived program of consolidation.

36.2 The tasks that lie ahead of us in this particular field of Bahá'í activity are as follows:

1. As pointed out in the Riḍván Message, the settlement of the minimum number of pioneers called for under last year's goals should be completed. A recent letter has been sent to all National Assemblies responsible for supplying manpower, calling on them to find and dispatch as soon as possible the pioneers to the few remaining territories for which no offers have as yet been received and to expedite the completion of those projects which are in process.

35-16. See CC 1:512. For information on Ḥuqúqu'lláh, see the glossary.
35-17. WT, p. 14. See also CUHJ, p. 12.

2. The pioneers who have already settled or are settling in their posts, particularly in virgin and unoccupied territories must be reminded that their movement to their goals is far from being a short stay designed to class a particular territory or island as opened, or label it as having received one or more pioneers, even if, in some cases, new believers native to the land have been enrolled. It is basically and clearly intended to establish the Faith of God securely and firmly in the hearts of people of the area and to ensure that its divinely ordained institutions are understood, adopted and operated by them. The perseverance of the pioneers in their posts, however great the sacrifices involved, is an act of devoted service, which, as attested by our teachings, will have an assured reward in both worlds. The admonitions of the Guardian on this subject are too numerous to cite and amply demonstrate the vital nature of this clear policy.

3. The pioneers and settlers, as well as the National Assemblies responsible for the administration of the Faith in areas assigned to them, should ever bear in mind that in the initial stages of the establishment of the Faith in any territory, the obscurity surrounding the work of the pioneer or the local Bahá'ís is in itself a protection to the Faith. Patience, tact and wisdom should be exercised. Public attention should not be attracted to the Faith until such time as the believers see the Faith touch more and more of the hearts of receptive souls responding to its Divine Call.

4. As the numbers fixed last year for settlers in goal areas were minimum figures, each National Assembly should carefully assess the needs of the territories assigned to its jurisdiction. If more pioneers are needed for any of these territories, a full report should at once be sent to the House of Justice, including recommendations as to numbers required and preferred nationalities of the prospective pioneers.

5. The practical aspects of these pioneering projects are of vital importance. The financial responsibilities assigned under last year's goals do not end because the pioneer has arrived at his post. These responsibilities continue until the objectives are permanently and securely attained. In any case where the National Assembly assigned the responsibility is unable to meet its obligation, application to fill the ascertained need should at once be made to the House of Justice for assistance from the Deputization Fund.

36.3 We assure you of our prayers at the Holy Shrines that the friends in every land may rise above their local and personal problems, realize the needs of the

Cause of God at this juncture of its inexorable onward development, and offer on the altar of sacrifice their measure of service and assistance with complete self-abnegation and wholehearted devotion to His infinitely precious Cause.

With loving Bahá'í greetings,
THE UNIVERSAL HOUSE OF JUSTICE

37
Message to Youth—Three Fields of Service
10 JUNE 1966

To the Bahá'í Youth in every Land

Dear Bahá'í Friends,

37.1 In country after country the achievements of Bahá'í youth are increasingly advancing the work of the Nine Year Plan and arousing the admiration of their fellow believers. From the very beginning of the Bahá'í Era, youth have played a vital part in the promulgation of God's Revelation. The Báb Himself was but twenty-five years old when He declared His Mission, while many of the Letters of the Living were even younger. The Master, as a very young man, was called upon to shoulder heavy responsibilities in the service of His Father in Iraq and Turkey, and His brother, the Purest Branch, yielded up his life to God in the Most Great Prison at the age of twenty-two that the servants of God might "be quickened, and all that dwell on earth be united."[1] Shoghi Effendi was a student at Oxford when called to the throne of his guardianship,[2] and many of the Knights of Bahá'u'lláh, who won imperishable fame during the Ten Year Crusade, were young people. Let it, therefore, never be imagined that youth must await their years of maturity before they can render invaluable services to the Cause of God.

A Time of Decision

37.2 For any person, whether Bahá'í or not, his youthful years are those in which he will make many decisions which will set the course of his life. In these years he is most likely to choose his life's work, complete his education, begin to earn his own living, marry and start to raise his own family. Most important of all, it is during this period that the mind is most questing and that the spiritual values that will guide the person's future behavior are adopted. These factors present Bahá'í youth with their greatest opportunities, their greatest challenges, and their greatest tests—opportunities to truly apprehend

---

37-1. GPB, p. 188. See also MA, p. 34.
37-2. At the time of 'Abdu'l-Bahá's passing in 1921, Shoghi Effendi was twenty-four years old.

the Teachings of their Faith and to give them to their contemporaries, challenges to overcome the pressures of the world and to provide leadership for their and succeeding generations, and tests enabling them to exemplify in their lives the high moral standards set forth in the Bahá'í Writings. Indeed the Guardian wrote of the Bahá'í youth that it is they "who can contribute so decisively to the virility, the purity, and the driving force of the life of the Bahá'í community, and upon whom must depend the future orientation of its destiny, and the complete unfoldment of the potentialities with which God has endowed it."[3]

An Opportunity Unique in Human History

37.3 Those who now are in their teens and twenties are faced with a special challenge and can seize an opportunity that is unique in human history. During the Ten Year Crusade—the ninth part of that majestic process described so vividly by our beloved Guardian—the Community of the Most Great name spread with the speed of lightning over the major territories and islands of the globe, increased manifoldly its manpower and resources, saw the beginning of the entry of the peoples by troops into the Cause of God, and completed the structure of the Administrative Order of Bahá'u'lláh.[4] Now, firmly established in the world, the Cause, in the opening years of the tenth part of that same process, is perceptibly emerging from the obscurity that has, for the most part, shrouded it since its inception, and is arising to challenge the outworn concepts of a corrupt society and proclaim the solution for the agonizing problems of a disordered humanity. During the lifetime of those who are now young the condition of the world, and the place of the Bahá'í Cause in it, will change immeasurably, for we are entering a highly critical phase in this era of transition.

Three Fields of Service

37.4 Three great fields of service lie open before young Bahá'ís, in which they will simultaneously be remaking the character of human society and preparing themselves for the work that they can undertake later in their lives.

37.5 First, the foundation of all their other accomplishments is their study of the teachings, the spiritualization of their lives and the forming of their characters in accordance with the standards of Bahá'u'lláh. As the moral standards of the people around us collapse and decay, whether of the centuries-old civilizations of the East, the more recent cultures of Christendom and Islam; or of the rapidly changing tribal societies of the world, the Bahá'ís must increasingly stand out as pillars of righteousness and forbearance. The life of a Bahá'í

37-3. ADJ, p. 22.
37-4. For Shoghi Effendi's description of the vast and "majestic" ten-part process that began at the dawn of the Adamic Cycle and will continue into the Golden Age of the Faith, see the entry on Ten Part Process in the glossary.

will be characterized by truthfulness and decency; he will walk uprightly among his fellowmen, dependent upon none save God, yet linked by bonds of love and brotherhood with all mankind; he will be entirely detached from the loose standards, the decadent theories, the frenetic experimentation, the desperation of present-day society, will look upon his neighbors with a bright and friendly face and be a beacon light and a haven for all those who would emulate his strength of character and assurance of soul.

37.6 The second field of service, which is linked intimately with the first, is teaching the Faith, particularly to their fellow-youth, among whom are some of the most open and seeking minds in the world. Not yet having acquired all the responsibilities of a family or a long-established home and job, youth can the more easily choose where they will live and study or work. In the world at large young people travel hither and thither seeking amusement, education and experiences. Bahá'í youth, bearing the incomparable treasure of the Word of God for this Day, can harness this mobility into service for mankind and can choose their places of residence, their areas of travel and their types of work with the goal in mind of how they can best serve the Faith.

37.7 The third field of service is the preparation by youth for their later years. It is the obligation of a Bahá'í to educate his children; likewise it is the duty of the children to acquire knowledge of the arts and sciences and to learn a trade or a profession whereby they, in turn, can earn their living and support their families. This, for a Bahá'í youth, is in itself a service to God, a service, moreover, which can be combined with teaching the Faith and often with pioneering. The Bahá'í community will need men and women of many skills and qualifications; for, as it grows in size the sphere of its activities in the life of society will increase and diversify. Let Bahá'í youth, therefore, consider the best ways in which they can use and develop their native abilities for the service of mankind and the Cause of God, whether this be as farmers, teachers, doctors, artisans, musicians or any one of the multitude of livelihoods that are open to them.

37.8 When studying at school or university Bahá'í youth will often find themselves in the unusual and slightly embarrassing position of having a more profound insight into a subject than their instructors. The Teachings of Bahá'u'lláh throw light on so many aspects of human life and knowledge that a Bahá'í must learn, earlier than most, to weigh the information that is given to him rather than to accept it blindly. A Bahá'í has the advantage of the divine Revelation for this Age, which shines like a searchlight on so many problems that baffle modern thinkers; he must therefore develop the ability to learn everything from those around him, showing proper humility before his teachers, but always relating what he hears to the Bahá'í teachings, for they will enable him to sort out the gold from the dross of human error.

Bahá'í Consultation—Tracing New Paths of Human Corporate Action

37.9 Paralleling the growth of his inner life through prayer, meditation, service and study of the teachings, Bahá'í youth have the opportunity to learn in practice the very functioning of the Order of Bahá'u'lláh. Through taking part in conferences and summer schools as well as Nineteen Day Feasts, and in service on committees, they can develop the wonderful skill of Bahá'í consultation, thus tracing new paths of human corporate action. Consultation is no easy skill to learn, requiring as it does the subjugation of all egotism and unruly passions, the cultivation of frankness and freedom of thought as well as courtesy, openness of mind and wholehearted acquiescence in a majority decision. In this field Bahá'í youth may demonstrate the efficiency, the vigor, the access of unity which arise from true consultation and, by contrast, demonstrate the futility of partisanship, lobbying, debate, secret diplomacy and unilateral action which characterize modern affairs. Youth also take part in the life of the Bahá'í community as a whole and promote a society in which all generations—elderly, middle-aged, youth, children—are fully integrated and make up an organic whole. By refusing to carry over the antagonisms and mistrust between the generations which perplex and bedevil modern society they will again demonstrate the healing and life-giving nature of their religion.

37.10 The Nine Year Plan has just entered its third year. The youth have already played a vital part in winning its goals. We now call upon them, with great love and highest hopes and the assurance of our fervent prayers, to consider, individually and in consultation, wherever they live and whatever their circumstances, those steps which they should take now to deepen themselves in their knowledge of the divine message, to develop their characters after the pattern of the Master, to acquire those skills, trades and professions in which they can best serve God and man, to intensify their service to the Cause of Bahá'u'lláh and to radiate its message to the seekers among their contemporaries.

THE UNIVERSAL HOUSE OF JUSTICE

38
Formation of Three More National Spiritual Assemblies at Riḍván 1967
1 September 1966

To all National Spiritual Assemblies

38.1 JOYFULLY ANNOUNCE FORMATION AT RIḌVÁN 1967 ADDITIONAL NEW NATIONAL ASSEMBLIES BELIZE SEAT BELIZE, LAOS SEAT VIENTIANE, SIKKIM SEAT GANGTOK, CALLING UPON NATIONAL ASSEMBLIES GUATEMALA, THAILAND, INDIA RESPECTIVELY CALL FIRST CONVENTIONS ELECTION NATIONAL ASSEMBLIES. SIKKIM ASSEMBLY SUPPLEMENTARY ACHIEVEMENT NINE YEAR PLAN. CHANGED SITUATION CAMBODIA REQUIRES POSTPONEMENT FORMATION NATIONAL ASSEMBLY THAT COUNTRY. ADDITION ABOVE NATIONAL ASSEMBLIES RAISES TOTAL THROUGHOUT THE WORLD TO EIGHTY-ONE WHOSE MEMBERS WILL PARTICIPATE SECOND INTERNATIONAL CONVENTION. OFFERING PRAYERS OF GRATITUDE BAHÁ'U'LLÁH SUPPLICATING DIVINE CONFIRMATIONS EXPANSION CONSOLIDATION THESE TERRITORIES ASSURING SOLID FOUNDATION FUTURE PILLARS UNIVERSAL HOUSE OF JUSTICE.[1]

The Universal House of Justice

39
Incorporation of the National Spiritual Assembly of Italy
22 December 1966

To all National Spiritual Assemblies

39.1 JOYFULLY ANNOUNCE INCORPORATION ITALIAN NATIONAL ASSEMBLY SIGNIFICANT MILESTONE PROCESS RECOGNITION FAITH HEART CHRISTENDOM.

The Universal House of Justice

38-1. The pillars referred to are the National Spiritual Assemblies.

40

Development of the Bahá'í World Center and the Needs of the Bahá'í International Fund
7 March 1967

To all National Spiritual Assemblies

Beloved Friends,

The time has now come in the progress of the Nine Year Plan when the Bahá'í world must devote a greater effort towards the development of the Faith at its World Center. 40.1

Nearly all the accessible unsettled territories of the Plan have now been settled, bases have been established throughout the world for the future expansion of the Faith; a program of progressive consolidation is being pursued hand-in-hand with continued expansion; plans for the construction of the Panama Temple are well advanced; the Ḥaẓíratu'l-Quds, Temple sites and endowments called for in the Plan are being steadily acquired; by the end of the next Riḍván period 81 out of the 108 National Spiritual Assemblies called for by 1973 will have been established; and the opening of the period of the proclamation of the Faith is fast approaching. 40.2

Since the Universal House of Justice came into being in 1963, its primary concern at the World Center of the Faith has been with the basic, minimum essentials of undertaking repairs to the Holy Places; establishing its administrative offices; reorganizing the accommodation of pilgrims; gathering its staff, developing a suitable housing program for the Hands of the Cause and their families, the members of the House of Justice and their families, and all other believers serving at the World Center; formulating plans for the expansion of the Gardens and taking the first steps in their initiation; collating the Sacred Texts and the letters of Shoghi Effendi and indexing them; and fostering relations with the Government of the State of Israel and with the United Nations. 40.3

The increased burden which these essential steps have imposed upon the International Fund we have endeavored to keep at a minimum so that, in the early stages of the Plan, the maximum resources could be utilized in the teaching work throughout the world. 40.4

However, we must now embark upon certain major undertakings vital to the future progress of the Cause. Extensive beautification of the sacred endowments surrounding the Holy Shrines in Bahjí and Haifa, as well as at the site of the future Mashriqu'l-Adhkár on Mount Carmel must be undertaken, both for its own sake and for the protection of these lands which are situated within the boundaries of rapidly expanding cities; the work of classifying and codifying the Holy Texts must be urgently prosecuted; the arrangements for pil- 40.5

grimage may have to be greatly expanded to provide for the ever-increasing number of applications from East and West; the Intercontinental Conferences and the International Convention must be held and paid for; and the auxiliary institutions of the Universal House of Justice must begin to unfold so that the ever-growing and increasingly complex work of the World Center of the Faith may continue to be efficiently discharged. Moreover, the vital assistance given by the International Fund to the work of the Hands of the Cause and National Spiritual Assemblies must be maintained.

40.6 The minimum budget requirements of the International Fund have nearly doubled since 1963, and if in addition we are to be enabled to undertake these developments, a much greater flow of funds will be needed than is now available.

40.7 We call upon every National Spiritual Assembly to consider now the amount that it can allocate as a contribution to the International Fund in its budget for the coming year. In some cases this may mean that contributions made hitherto will be doubled, trebled, or even more greatly increased. Please write as soon as your decision has been made, and not later than 21st April, telling us the estimated amount of your allocation.

40.8 This is a vitally important matter, and we shall pray in the Holy Shrines that the friends throughout the world will respond wholeheartedly to this call.

With loving Bahá'í greetings,
UNIVERSAL HOUSE OF JUSTICE

41
Commemoration of the Revelation of the Súriy-i-Mulúk
17 MARCH 1967

To all National Spiritual Assemblies

Dear Bahá'í Friends,

41.1 In the 1965 Riḍván Message from the Universal House of Justice we announced that in September 1967, on the Feast of Mashíyyat, a few appointed representatives of the Bahá'í world would visit the site of the House of Bahá'u'lláh in Adrianople where the Súriy-i-Mulúk was revealed. We have decided that the six Hands of the Cause of God who will represent the Universal House of Justice at the Intercontinental Conferences in October are the ones to make this visit. Immediately following this historic act they will proceed to their respective Conferences.[1]

41-1. The Hands of the Cause of God Amatu'l-Bahá Rúḥíyyih Khánum, Ugo Giachery, Ṭarázu'lláh Samandarí, 'Alí-Akbar Furútan, Paul Haney, and Abu'l-Qásim Faizi visited Adrianople

For the protection of the Faith it is essential that no one, except the Hands, travel to Turkey on this occasion. 41.2

National Spiritual Assemblies are requested to make this announcement at the National Convention this year, and to repeat it in their newsletters and at other times and places as may be appropriate. 41.3

With loving Bahá'í greetings,
THE UNIVERSAL HOUSE OF JUSTICE

42
Riḍván Message 1967
RIḌVÁN 1967

The Bahá'ís of the World

Dearly loved Friends,

At the conclusion of the third year of the Nine Year Plan we acknowledge with thankful hearts the evidences of Divine favor with which Bahá'u'lláh unfailingly sustains and confirms the dedicated efforts of His servants everywhere, and we unhesitatingly affirm our confidence that the community of the Most Great Name can and will, by its determination and sacrificial efforts, achieve complete victory. 42.1

Worldwide Achievements

Last year the call was raised for the formation, in 1967, of eleven new National Spiritual Assemblies. All will be elected during the Riḍván period. We welcome with great joy the National Spiritual Assemblies of the Bahá'ís of Algeria and Tunisia with its seat in Algiers; Cameroon Republic with its seat in Victoria; Swaziland, Lesotho and Mozambique with its seat in Mbabane; Zambia with its seat in Lusaka; Belize with its seat in Belize; the Leeward, Windward and Virgin Islands with its seat in Charlotte Amalie; Eastern and Southern Arabia with its seat in Baḥrayn; Laos with its seat in Vientiane; Sikkim with its seat in Gangtok; Taiwan with its seat in Taipei; the Gilbert and Ellice Islands with its seat in Tarawa. The World Center of the Faith will be represented at each National Convention by a Hand of the Cause of God, who will present a message from the Universal House of Justice welcoming the new national community and assigning it its share of the goals of the Nine Year Plan. 42.2

---

on 27 September 1967 (the Feast of Will) to mark Bahá'u'lláh's revelation of His Tablet to the Kings (Súriy-i-Mulúk), referred to by Shoghi Effendi as "the most momentous Tablet revealed by Bahá'u'lláh" (GPB, p. 171). See PB, pp. 7–12, 47–54, and 102–03, for passages from the Súriy-i-Mulúk that have been translated into English.

42.3 At this Riḍván eighty-one of the 108 National Spiritual Assemblies, and more than six thousand of the 13,737 Local Spiritual Assemblies called for by 1973 will have been established; of a required 54,102 localities where Bahá'ís reside, 28,217 are reported; fifteen of the sixty-five National Incorporations called for have been achieved; seventeen of fifty-two National Ḥaẓíratu'l-Quds, seven of sixty-two Temple sites, thirteen of fifty-four National Endowments, fourteen of thirty-two Teaching Institutes, have been acquired; of the 973 Local Incorporations called for in the Plan, 123 have been completed; Local Ḥaẓíratu'l-Quds acquired are, twenty-four in India, seventeen in Kenya, nine in Uganda, two in South Africa, two in Turkey and a number in Congo (Kinshasa), while land for eight others has been acquired in Kenya, for four in Cameroon, for two in Pakistan and for one in Mauritius; in eight countries Local Endowments supplementary to those called for in the Plan have been acquired.

42.4 Iceland, Korea, Liberia, Luxembourg and Rhodesia now recognize the Bahá'í Marriage Certificate; the Dominican Republic, Guyana, Hawaii, Iceland, Italy, Kenya and Luxembourg recognize Bahá'í Holy Days. A Summer School has been established in Liberia, and one, beyond the requirements of the Plan, in Canada, while land for others has been acquired in Argentina, Ethiopia and Samoa. Twenty-five new languages have been added to the list of those in which Bahá'í literature is available, bringing the total number to 397. The number of territories now opened to the Faith has reached 311, including the recently settled virgin areas of Chiloé Archipelago, Bonaire, Phoenix Islands and St. Martin, and two territories in addition to those called for in the Plan, namely Melville Island in Australasia and Montserrat in the Windward Islands.

42.5 After protracted frustration the National Spiritual Assembly of Persia has finally gained possession of the historic fortress of Chihríq, that bleak and lonely citadel which was the last earthly residence of the blessed Báb, and from which He was led forth to His martyrdom in Tabríz.[1] Realization of the long-sought recognition of the Faith in Italy is a wonderful victory, resulting not only in the incorporation of the National Spiritual Assembly, but also of all Local Spiritual Assemblies in Italy and the ability to establish that National Spiritual Assembly's Publishing Trust. In Iceland the Faith has been recognized as one of the island's religions. This provides not only for incorporation of the Local Spiritual Assembly of Reykjavik but authorizes the chairman of that Assembly to perform Bahá'í marriages and Bahá'í burials, exempts the Faith from certain taxes, permits the observance of Bahá'í Holy Days and paves the

---

42-1. Chihríq, designated the "Grievous Mountain" by the Báb, is the remote fortress to which He was transferred on 10 April 1848. Tabríz is the town in northwestern Iran where the Báb was martyred 9 July 1850.

way for incorporation of the National Spiritual Assembly of that country when it will be formed. The full number of Local Spiritual Assemblies, Groups and Localities called for in the Plan has been established in fifty-three territories and islands under the direction of twenty-six National Spiritual Assemblies; five territories have formed the required number of Local Spiritual Assemblies and seven have reached the specified number of Localities.

42.6 Since the call was raised a year ago, international traveling-teaching, ranging over the five continents and affecting nearly all national communities, has been undertaken. Seventy-eight projects have been completed in Europe, forty-three in America, twenty-seven in Asia, twenty-five in Australasia which, with those in Africa brings the total number to about two hundred. It is greatly hoped that this stimulating activity, so dear to the beloved Master's heart, will be constantly expanded.

42.7 Sustaining all these visible achievements is a constant activity throughout the world of teaching and administration—a perpetual movement, like the ceaseless surge of the sea, within the Bahá'í community, which is the real cause of its growth. National and Local Spiritual Assemblies facing difficult problems, devising new plans, shouldering responsibility for a community growing in numbers and consciousness, Committees striving to accomplish objectives, Bahá'í Youth in eager and dedicated activity, individual Bahá'ís and families making efforts for the Cause, to give the Message, or hold a fireside, these constant services attract the confirmation of Bahá'u'lláh, and the more they are supported by prayers and intense dedication and the more extensive they become, the more they release into the world a spiritual charge which no force on earth can resist, and which must eventually bring about the complete triumph of the Cause. It is this organic vitality of the Faith, so readily felt at the World Center, whose exhilaration we wish every believer to share.

Work at the World Center

42.8 At the World Center of the Faith codification of the Kitáb-i-Aqdas and collation of other important Texts has continued. Work on the highly important task of formulating the Constitution of the Universal House of Justice is well advanced. Development and extension of the gardens surrounding the sacred Shrines in both Haifa and Bahjí is continuing. Publication of *The Bahá'í World,* Volume 13 has been undertaken; this book covers nine years, from 1954 to 1963, almost the entire period of the Ten Year Crusade, and includes a comprehensive article on the beloved Guardian by Amatu'l-Bahá Rúḥíyyih Khánum. A planned development of relationships with the United Nations is being actively pursued. An important supplementary achievement is the establishment of an International Bahá'í Audio-Visual Center whose function is to provide teaching and deepening aids to all National Spiritual Assemblies, as well as to store and index audiovisual records.

Service of the Hands of the Cause of God

42.9 Throughout the year the services of the beloved Hands of the Cause have shone with an unfailing light. Their constant encouragement of National Spiritual Assemblies and of believers everywhere to pursue the goals of the Plan and to obtain a deeper understanding of the true meaning of Bahá'u'lláh's Revelation is contributing in no small measure to the progress of that Plan and must exercise a lasting effect on the development of the Bahá'í community. These few gallant and dedicated believers, whose place in history is forever assured by virtue of their appointment to their high office, are indeed a precious legacy left to us by our beloved Guardian, and as the years go by there is increasingly added to the honor and respect which is their due by reason of their exalted rank, the love and admiration of the friends evoked by their constant services.

42.10 In response to special needs two changes have been made in the disposition of the Hands during the year, Hand of the Cause John Robarts returning to the Western Hemisphere with a special assignment to his native Canada, and Hand of the Cause William Sears returning to Africa. In addition we are delighted to announce that Hand of the Cause Ṭarázu'lláh Samandarí, whose eyes were blessed by beholding Bahá'u'lláh, will represent the Universal House of Justice at the Intercontinental Conference in Chicago, replacing the late Hand of the Cause Leroy Ioas.

The Panama Temple

42.11 In the international sphere the great project of raising the Panama Temple has begun with choice of a design submitted by Mr. Peter Tillotson, an English architect. Mr. Robert McLaughlin, sometime member of the National Spiritual Assembly of the United States and Dean Emeritus of the School of Architecture of Princeton University, who served as a member of the Technical Advisory Board for the construction of the interior of the Mother Temple of the West in Wilmette, has been appointed Architectural Consultant to the Universal House of Justice for the building of the Temple. He and Mr. Tillotson have visited the site together, and are working in close cooperation. Pictures and drawings of the new Temple will be published shortly, and the friends will be kept informed of the progress of construction of this House of Worship "situated between the two great oceans," a location which 'Abdu'l-Bahá indicated would become very important in the future and whence the Teachings, once established, "will unite the East and the West, the North and the South."[2]

42-2. TDP 14.8. For the message announcing the laying of the foundation stone of the Panama Temple, see letter no. 50; for the message sent on the occasion of its dedication, see letter no. 108.

1964–1973 • THE NINE YEAR PLAN

Pioneer Goals

The brilliant pioneering feat of the second year of the Plan is beginning to reveal its beneficent effects, but pioneers are still urgently needed and will continue to be needed in all parts of the world for consolidation and development of the Faith in the newly won territories as well as for those resettled during the opening years of the Plan. The immediate requirement is for 209 pioneers to settle in eighty-seven territories named on the attached list, and the call is now raised for the speedy achievement of this task.[3] Service in this highly meritorious field is open to every believer and all those who are moved to respond to this particular call are asked to consult the list of territories and to make their offers to their own National Spiritual Assembly. Full details of the requirements in each territory have been sent to the National Spiritual Assemblies concerned and to the Pioneer Committees. 42.12

The Homefronts—Solid Bases for Expansion

The constant need for pioneers no less than the approaching worldwide proclamation render it imperative to pay special attention, in every continent, to the homefronts, for they are the sources of manpower and of administrative experience, the solid bases from which all expansion begins, both at home and abroad. The largest increases in numbers of Local Spiritual Assemblies, of Groups, and of believers, are called for on the homefronts, and these tasks must be vigorously pursued. Some National Spiritual Assemblies have phased these important goals, assigning a specified number for achievement each year, thus ensuring a planned and flexible approach to the total requirements. Such a systematic and determined prosecution of the homefront goals is highly recommended. 42.13

The Fund—Our Honor and Challenge

The pressing and ever-growing needs of the Bahá'í Fund are called to the attention of all believers. There are great projects already under way or lying ahead which require very large amounts of money for their realization. The Panama Temple—the first only of the two called for in the Nine Year Plan—the beautification and development of the World Center itself, involving a necessary and inevitable increase in facilities to serve the growing needs of the Faith; support of the vital teaching program in many parts of the world; establishment and development of new National Spiritual Assemblies—all these urgently require the support of the friends everywhere through sustained and sacrificial contributions. As inflation spreads around the world, the consequent increase in the cost of living is balanced, at least in the more affluent countries, by a corresponding increase in personal incomes. The expenses of the Bahá'í Fund are inevitably and seriously affected by this inflationary condi- 42.14

42-3. The list is too lengthy to include.

tion which can only be relieved by contributions, both of larger amounts and from a larger number of contributors. The House of Justice believes that the financial needs of the Cause should be met by universal participation in giving and urges National and Local Spiritual Assemblies to pursue this goal with vigor and imagination, recalling to the friends the plea of the beloved Guardian to every believer "unhesitatingly, to place, each according to his circumstances, his share on the altar of Bahá'í sacrifice."4 The fact that only we, the Bahá'ís, can contribute financially to the Cause is both our honor and our challenge.

Centenary of Bahá'u'lláh's Summons to the Kings

42.15 As we approach the third phase of the Nine Year Plan there opens before us a prospect of enthralling opportunities such as to thrill the heart of every ardent follower of Bahá'u'lláh. For more than a century we have toiled to teach the Cause; heroic sacrifices, dedicated services, prodigious efforts have been made in order to establish the outposts of the Faith in the chief countries, territories and islands of the earth and to raise the framework of the Administrative Order around the planet. But the Faith of Bahá'u'lláh remains, as yet, unknown to the generality of men. Now at last, at long last, the worldwide community of the Most Great Name is called upon to launch, on a global scale and to every stratum of human society, an enduring and intensive proclamation of the healing message that the Promised One has come and that the unity and well-being of the human race is the purpose of His Revelation. This long-to-be-sustained campaign, commencing next October in commemoration of the centenary of the sounding of the "opening notes" of Bahá'u'lláh's own proclamation,5 and gathering momentum throughout the remainder of the Nine Year Plan, may well become the spearhead of other plans to be launched continually until humanity has recognized and gratefully acclaimed its Redeemer and its Lord.

42.16 A hundred years ago Bahá'u'lláh Himself addressed the kings, rulers, religious leaders and peoples of the world. The Universal House of Justice feels it its bounden duty to bring that Message to the attention of the world's leaders today. It is therefore presenting to them, in the form of a book, the essence of Bahá'u'lláh's announcement. Entitled *The Proclamation of Bahá'u'lláh,* a special edition will be presented to Heads of State during the opening of the proclamation period and a general edition will be available to the friends in English, French, German, Italian and Spanish.

42.17 The Hands of the Cause of God, Amatu'l-Bahá Rúḥíyyih Khánum, Ugo Giachery, Ṭarázu'lláh Samandarí, 'Alí-Akbar Furútan, Paul Haney, Abu'l-Qásim Faizi, who will represent the Universal House of Justice at the Inter-

---

42-4. CF, p. 131.
42-5. GPB, p. 212.

continental Conferences in October to be held in Panama, Sydney, Chicago, Kampala, Frankfurt and New Delhi respectively, will gather at the World Center in September, a few days before the Feast of Mashíyyat. The members of the House of Justice will join these Hands in supplication at the Shrine of Bahá'u'lláh in Bahjí and will meet with them for consultation in the Mansion. From that Holy Spot these Hands of the Cause will make a special pilgrimage on behalf of the entire Bahá'í world to Adrianople where the Súriy-i-Mulúk was revealed.[6] One hundred years after the historic event which it is their purpose to commemorate, they will, on September 27th gather in the House of Bahá'u'lláh for prayer and meditation, while the members of the Universal House of Justice will, in the Most Holy Shrine at Bahjí, share in the same commemoration and pray for the success of the Conferences and of the Proclamation program. The entire Bahá'í world will, between the Conferences and Riḍván 1968, commemorate the centenary of the opening of that wonderful period in human history when the clouds of Divine bounty showered in lavish profusion their treasures upon men and the portals of the Kingdom were thrown open, disclosing to all who had eyes to see, a new heaven and a new earth, and the new Jerusalem coming down from God.

Immediately after the Feast of Mashíyyat the Hands of the Cause will travel from Adrianople to their Conferences, each bearing the precious trust of a photograph of the Blessed Beauty, which it will be the privilege of those attending the Conferences to view. These distinguished Hands will, on their own behalf, each address the Conference which they attend, and will bear a message to each Conference from the Universal House of Justice whom they represent. 42.18

These six Conferences, convened to commemorate the opening of Bahá'u'lláh's own Proclamation and to inaugurate a period of proclamation of His message by the entire company of His followers, will doubtless demonstrate yet again the spirit of joy which pervades such gatherings of the friends and will reinforce them in their determination to seize whatever means and opportunities they may find to raise the Divine call. Honored by the presence of Hands of the Cause, these Conferences, focal points of the love and prayers 42.19

42-6. The occasion marked the centenary of Bahá'u'lláh's revelation of His Súriy-i-Mulúk (Tablet to the Kings), which Shoghi Effendi describes as "the most momentous Tablet revealed by Bahá'u'lláh . . . in which He, for the first time, directs His words collectively to the entire company of the monarchs of East and West, and in which the Sulṭán of Turkey, and his ministers, the kings of Christendom, the French and Persian Ambassadors accredited to the Sublime Porte, the Muslim ecclesiastical leaders in Constantinople, its wise men and inhabitants, the people of Persia and the philosophers of the world are separately addressed. . . ." (GPB, pp. 171–72). For a brief account of the proclamation of Bahá'u'lláh to the kings and rulers of the world, see BW 14:195–204. For a full and detailed account and for a depiction of the fate of the recipients in the wake of their neglect of Bahá'u'lláh's summons, see Shoghi Effendi, *Promised Day Is Come*.

of the friends everywhere, magnets to attract the spiritual powers which alone can confirm their work, will, it is confidently hoped, be potent sources of unity, spiritual enthusiasm and realistic planning. National Spiritual Assemblies are called upon to ensure that they are represented at the Conference held in their continent so that they may share their plans for proclamation with other National Spiritual Assemblies as well as discuss with them the remaining goals of the Nine Year Plan.

42.20 To all those friends in so many countries, suffering in varying degrees from restrictions and oppression which will either prevent altogether, or greatly inhibit their public commemoration and subsequent proclamation programs, we send a special message of love and assurance. To them we convey the love and admiration of their fellow believers, who, in gratitude for their greater freedom, are determined to blaze abroad such a proclamation of the Divine Message as may well pave the way for the eventual emancipation of the entire body of the Faith.

Proclamation, Expansion, and Consolidation—Interrelated Processes

42.21 Worldwide proclamation, the unknown sea on which we must soon sail, will add another dimension to our work, a dimension which will, as it develops, complement and reinforce the twin processes of expansion and consolidation. This pattern of teaching, emerging so soon after the completion of the framework of the Administrative Order, may well be the means of advancing the vital work of consolidation and of rendering more effective the teaching wisdom which has been gained in a hundred years, and more particularly since the beloved Guardian called us to systematic and planned activity. Therefore, in those countries where we are free to publicize our religion, this activity must become part of our regular work, included in budgets, assigned to National and Local Committees for study and implementation and above all for coordination with the programs operating to achieve the goals of the Nine Year Plan. Every effort of proclamation must be sustained by teaching, particularly locally where public announcements should be related to such efforts. This coordination is essential, for nothing will be more disheartening than for thousands to hear of the Faith and have nowhere to turn for further information.

The Nature of Deepening

42.22 The beloved Guardian wrote, "To strive to obtain a more adequate understanding of the significance of Bahá'u'lláh's stupendous Revelation must, it is my unalterable conviction, remain the first obligation and the object of the constant endeavor of each one of its loyal adherents," a statement which places the obligation of deepening in the Cause firmly on every believer.[7] It is there-

---

42-7. WOB, p. 100.

fore upon the nature of deepening, rather than upon the desirability of pursuing it, that we wish to comment.

42.23 A detailed and exact knowledge of the present structure of Bahá'í Administration, or of the By-laws of National and Local Spiritual Assemblies, or of the many and varied applications of Bahá'í law under the diverse conditions prevailing around the world, while valuable in itself, cannot be regarded as the sort of knowledge primarily intended by deepening. Rather is suggested a clearer apprehension of the purpose of God for man, and particularly of His immediate purpose as revealed and directed by Bahá'u'lláh, a purpose as far removed from current concepts of human well-being and happiness as is possible. We should constantly be on our guard lest the glitter and tinsel of an affluent society should lead us to think that such superficial adjustments to the modern world as are envisioned by humanitarian movements or are publicly proclaimed as the policy of enlightened statesmanship—such as an extension to all members of the human race of the benefits of a high standard of living, of education, medical care, technical knowledge—will of themselves fulfill the glorious mission of Bahá'u'lláh. Far otherwise. These are the things which shall be added unto us once we seek the Kingdom of God, and are not themselves the objectives for which the Báb gave His life, Bahá'u'lláh endured such sufferings as none before Him had ever endured, the Master and after Him the Guardian bore their trials and afflictions with such superhuman fortitude. Far deeper and more fundamental was their vision, penetrating to the very purpose of human life. We cannot do better, in this respect, than call to the attention of the friends certain themes pursued by Shoghi Effendi in his trenchant statement "The Goal of a New World Order." "The principle of the Oneness of Mankind" he writes, "implies an organic change in the structure of present-day society, a change such as the world has not yet experienced." Referring to the "epoch-making changes that constitute the greatest landmarks in the history of human civilization," he states that ". . . they cannot but appear, when viewed in their proper perspective, except as subsidiary adjustments preluding that transformation of unparalleled majesty and scope which humanity is in this age bound to undergo."[8] In a later document he refers to the civilization to be established by Bahá'u'lláh as one "with a fullness of life such as the world has never seen nor can as yet conceive."[9]

42.24 Dearly loved Friends, this is the theme we must pursue in our efforts to deepen in the Cause. What is Bahá'u'lláh's purpose for the human race? For what ends did He submit to the appalling cruelties and indignities heaped upon Him? What does He mean by "a new race of men"? What are the profound changes which He will bring about? The answers are to be found in

---

42-8. WOB, pp. 42–43, 45, 46.
42-9. PDIC, ¶302.

the Sacred Writings of our Faith and in their interpretation by 'Abdu'l-Bahá and our beloved Guardian. Let the friends immerse themselves in this ocean, let them organize regular study classes for its constant consideration, and as reinforcement to their effort, let them remember conscientiously the requirements of daily prayer and reading of the Word of God enjoined upon all Bahá'ís by Bahá'u'lláh.

42.25 Such dedicated striving on the part of all the friends to deepen in the Cause becomes imperative with the approach of the proclamation program. As this becomes effective more and more attention will be directed to the claims of Bahá'u'lláh and opposition must be expected. "HOW GREAT, HOW VERY GREAT IS THE CAUSE!" wrote the Master; "HOW VERY FIERCE THE ONSLAUGHT OF ALL THE PEOPLES AND KINDREDS OF THE EARTH! ERELONG SHALL THE CLAMOR OF THE MULTITUDE THROUGHOUT AFRICA, THROUGHOUT AMERICA, THE CRY OF THE EUROPEAN AND OF THE TURK, THE GROANING OF INDIA AND CHINA BE HEARD FROM FAR AND NEAR. ONE AND ALL THEY SHALL ARISE WITH ALL THEIR POWER TO RESIST HIS CAUSE. THEN SHALL THE KNIGHTS OF THE LORD, ASSISTED BY HIS GRACE FROM ON HIGH, STRENGTHENED BY FAITH, AIDED BY THE POWER OF UNDERSTANDING AND REINFORCED BY THE LEGIONS OF THE COVENANT, ARISE AND MAKE MANIFEST THE TRUTH OF THE VERSE: 'BEHOLD THE CONFUSION THAT HATH BEFALLEN THE TRIBES OF THE DEFEATED!'"[10]

42.26 Mindful of the countless expressions of Divine love found in our Scriptures and aware of the extraordinary nature of the crisis facing humanity, we call the friends to a new realization of the very great things which are expected from us in this Day. We recall that the Blessed Beauty, Bahá'u'lláh, as well as His "Best-Beloved"[11] before Him and 'Abdu'l-Bahá after Him bore Their sufferings in this world in order that mankind might be freed from material fetters and "attain unto true liberty," "might prosper and flourish," "attain unto abiding joy, and be filled with gladness,"[12] and we pray that the endeavors of the friends may be the means by which this glory and felicity will speedily come to pass.

THE UNIVERSAL HOUSE OF JUSTICE

42-10. WOB, p. 17.
42-11. The Báb.
42-12. GWB, pp. 99, 100, 99.

43
Release of a Compilation on Teaching the Masses
11 May 1967

To all National Spiritual Assemblies

Dear Bahá'í Friends,

We have recently made a compilation from the writings of the Guardian and from letters written on his behalf regarding the importance and nature of the teaching work among the masses. Attached is a copy of this compilation, which we sincerely recommend to you for your earnest study.[1] The extracts have been arranged chronologically. 43.1

It is our hope and belief that this compilation will guide and assist you in better appreciating the manner of the presentation of the teachings of the Faith; the attitude that must govern those responsible for enrolling new believers; the need to educate the newly enrolled Bahá'ís, to deepen them in the teachings and to wean them gradually away from their old allegiances; the necessity of keeping a proper balance between expansion and consolidation; the significance of the participation of the native believers of each country in the teaching work and in the administration of the affairs of the community; the formulation of budgets within the financial capabilities of the community; the importance of fostering the spirit of self-sacrifice in the hearts of the friends; the worthy goal for each national community to become self-supporting; the preferability of individuality of expression to absolute uniformity, within the framework of the Administrative Order; and the lasting value of dedication and devotion when engaged in the teaching work. 43.2

We are confident that the implementation of these principles as set forth in the writings of the Guardian will aid you in your teaching work. 43.3

With loving Bahá'í greetings,
THE UNIVERSAL HOUSE OF JUSTICE

43-1. See CC 2:61–71.

44

Sesquicentennial of the Birth of Bahá'u'lláh

25 June 1967

To all National Spiritual Assemblies

Dear Bahá'í Friends,

44.1 November 12, 1967, will mark the 150th anniversary of Bahá'u'lláh's birth. We call the entire Bahá'í world to joyful celebration, befitting an event so momentous to the fortunes of humanity.

44.2 The Universal House of Justice feels that the coincidence of this great occasion with the opening of the proclamation period provides a splendid opportunity for bringing to public attention both the spiritual and social import of the Cause. Not only its message, but the historical fact of a new Revelation, with all its implications of a new and worldwide civilization, should be made clear.

44.3 Let the friends not hesitate to welcome to their observances, even to those of a devotional character, the non-Bahá'í public, many of whom may well be attracted by the prayers and expressions of gratitude of the believers, no less than by the exalted tone of passages from Bahá'í writings.

With loving Bahá'í greetings,
THE UNIVERSAL HOUSE OF JUSTICE

45

The Nature and Purpose of Proclamation

2 July 1967

To all National Spiritual Assemblies

Dear Bahá'í Friends,

45.1 In just over three months the period of the worldwide proclamation of the Faith will be opened at the six Intercontinental Conferences called to celebrate the centenary of the revelation of the Súriy-i-Mulúk.[1] Those conferences will provide an opportunity for representatives of the National Spiritual Assemblies to exchange ideas and coordinate plans for the proclamation which will continue throughout the remaining four and a half years of the Plan.

45.2 The stimulating effect of this interchange of ideas will produce greatly increased momentum throughout the world, but inasmuch as many projects

---

45-1. The Súriy-i-Mulúk is Bahá'u'lláh's Tablet to the Kings. For further information about its significance, see messages no. 24, 41, and 42. See also PB, pp. 7–12, 47–54, and 102–03 for the passages of the Súriy-i-Mulúk that have been translated into English.

must be worked out before that date, we feel a few additional comments on the nature and purpose of proclamation will be helpful now.

Proclamation comprises a number of activities, of which publicity is only one. The Universal House of Justice itself will be conveying the Message of Bahá'u'lláh to the heads of all states, but, in addition to this, one of the most important duties of each National Spiritual Assembly is to acquaint leaders of thought and prominent men and women in its country with the fundamental aims, the history and the present status and achievements of the Cause. Such an activity must be carried out with the utmost wisdom, discretion and dignity. Publicity connected with such approaches must be weighed very carefully, as it may be unwise or discourteous. This is, of course, a long-range program, for such things cannot be rushed, but it must be given constant attention. 45.3

Another aspect of proclamation is a series of teaching programs designed to reach every stratum of human society—programs that should be pursued diligently and wisely, using every available resource. 45.4

Publicity itself should be well-conceived, dignified and reverent. A flamboyant approach which may succeed in drawing much initial attention to the Cause, may ultimately prove to have produced a revulsion which would require great effort to overcome. The standard of dignity and reverence set by the beloved Guardian should always be upheld, particularly in musical and dramatic items; and photographs of the Master should not be used indiscriminately. This does not mean that activities of the youth, for example, should be stultified; one can be exuberant without being irreverent or undermining the dignity of the Cause. 45.5

Every land has its own conditions, thus the kind of proclamation activity to be followed in each country should be decided by its National Spiritual Assembly. National Spiritual Assemblies need not follow or copy programs initiated in other countries. 45.6

In all proclamation activities, follow-up is of supreme importance. Proclamation, expansion and consolidation are mutually helpful activities which must be carefully interrelated. In some places it is desirable to open a teaching campaign with publicity—in others it is wiser to establish first a solid local community before publicizing the Faith or encouraging contacts with prominent people. Here, again, wisdom is needed. 45.7

We have been elated by the enthusiasm with which the Bahá'í community is preparing for the challenging months and years ahead, and we eagerly await those days but a few short months away which will open a period of such promise for the diffusion of God's Word. 45.8

With loving Bahá'í greetings,
THE UNIVERSAL HOUSE OF JUSTICE

46
Message to the Six Intercontinental Conferences
OCTOBER 1967

To the Six Intercontinental Conferences[1]

Dearly loved Friends,

46.1 On this, the hundredth anniversary of the sounding in Adrianople of the opening notes of Bahá'u'lláh's proclamation to the rulers, leaders and peoples of the world, we recall with profound emotion the circumstances surrounding the Faith of God at that time. In a land termed by Him the "Land of Mystery," the Bearer of God's Revelation had arisen to carry that Faith a stage further in its divinely ordained destiny.[2]

46.2 Internally, the infant Cause of God was convulsed by a crisis from whose shadows emerged the majestic figure of Bahá'u'lláh, the visible Center and Head of a newly established Faith.[3] The first pilgrimages were made to His Residence, a further stage in the transfer of the remains of the Báb was achieved, and above all the first intimations were given of the future station of 'Abdu'l-Bahá as the Center of the Covenant and of the revelation of the new laws for the New Day. Externally, the full significance of the new Reve-

---

46-1. For an account of the six conferences, see BW 14:223–28.

46-2. Bahá'u'lláh was exiled to Adrianople in December 1863, some eight months after proclaiming His mission to His followers, friends, and companions in the Garden of Riḍván outside of Baghdad, Iraq. Shoghi Effendi explains that toward the latter part of His stay in Adrianople "a period of prodigious activity ensued. . . ." "Such are the outpourings . . . from the clouds of Divine Bounty," Bahá'u'lláh Himself wrote, "that within the space of an hour the equivalent of a thousand verses hath been revealed." "I swear by God! In those days the equivalent of all that hath been sent down aforetime unto the Prophets hath been revealed" (GPB, pp. 170–71). Among the Tablets revealed during this period is the Súriy-i-Mulúk (Tablet to the Kings), "the most momentous Tablet revealed by Bahá'u'lláh in which He, for the first time, directs his words collectively to the entire company of the monarchs of East and West. . . ." (GPB, p. 171). For further information on the Súriy-i-Mulúk, see messages no. 24, 41, and 42.

46-3. The crisis was caused by Mírzá Yaḥyá, Bahá'u'lláh's half-brother, whose covetousness and jealousy of Bahá'u'lláh prompted him to openly challenge Bahá'u'lláh's authority and to issue his own claim to be the Promised One. Mírzá Yaḥyá went so far as to attempt to murder Bahá'u'lláh by serving Him tea in a cup smeared with poison while Bahá'u'lláh was a guest in his home. The poison induced an illness that lasted over a month and was accompanied by severe pains and a high fever and left Bahá'u'lláh with a shaking hand for the rest of His life. This incident and other actions of Mírzá Yaḥyá's led to his separation from Bahá'u'lláh's family and companions. "This supreme crisis," Shoghi Effendi wrote, was designated by Bahá'u'lláh as the "Days of Stress," "during which 'the most grievous veil' was torn asunder, and the 'most great separation' was irrevocably effected. It immensely gratified and emboldened its [the Bahá'í Faith's] external enemies, both civil and ecclesiastical, played into their hands, and evoked their unconcealed derision. It perplexed and confused the friends and supporters of Bahá'u'lláh, and seriously damaged the prestige of the Faith in the eyes of its western admirers. . . . It brought incalculable sorrow to Bahá'u'lláh, visibly aged Him, and inflicted, through its repercussions, the heaviest blow ever sustained by Him in His lifetime" (GPB, pp. 163–64).

lation was proclaimed by no one less than its Divine Bearer, His followers began openly to identify themselves with the Most Great Name, the independent character of the Faith became established and its fearless exponents took up their pens in defense of its fair name.⁴

46.3 Now, a hundred years later, the friends gathered in the six Intercontinental Conferences to commemorate the events of the past, privileged to gaze upon the portrait of their Beloved, must consider the urgent needs of the Cause today. As the Bahá'í world enters the third phase of the Nine Year Plan we are called upon to proclaim once again that Divine Message to the leaders and masses of the world, to aid the Faith of God to emerge from obscurity into the arena of public attention, to demonstrate through steadfast adherence to its laws the independent character of its mission and to brace ourselves in preparation for the attacks that are bound to be directed against its victorious onward march. Upon our efforts depends in very large measure the fate of humanity. The hundred years' respite having ended, the struggle between the forces of darkness—man's lower nature—and the rising sun of the Divine teachings which draw him on to his true station, intensifies day by day.

46.4 The Centenary campaign has been opened by the Universal House of Justice presenting to 140 Heads of State a compilation of Bahá'u'lláh's Own proclamation. The friends must now take the Message to the rest of humanity. The time is ripe and the opportunities illimitable. We are not alone nor helpless. Sustained by our love for each other and given power through the Administrative Order—so laboriously erected by our beloved Guardian—the Army of Light can achieve such victories as will astonish posterity.

46.5 We pray at the Holy Shrines that these Intercontinental Conferences will be centers of spiritual illumination inspiring the friends to redouble their efforts in further expanding and consolidating the Faith of God, to arise to

---

46-4. GPB, pp. 176–77. The first pilgrimages were made to the House of Amru'lláh (the House of God's Command) and foreshadowed later pilgrimages made by Bahá'ís from the East and the West to 'Akká. The remains of the Báb were moved secretly at Bahá'u'lláh's instruction by two Bahá'ís from Tehran from the Shrine of the Imám-Zádih Ma'ṣúm to some other place of safety, an act that proved providential when the shrine later underwent reconstruction. Concerning the station of 'Abdu'l-Bahá, Bahá'u'lláh revealed the Tablet of the Branch in which 'Abdu'l-Bahá is extolled as the "Branch of Holiness," the "Limb of the Law of God," and the "Trust of God," "sent down in the form of a human temple" (quoted in WOB, pp. 134–35). The new laws Bahá'u'lláh revealed were those of pilgrimage and fasting, later set forth in the Kitáb-i-Aqdas.

With respect to external developments, during the days in Adrianople Bahá'u'lláh revealed Tablets to the kings and rulers of the world; the terms "Bábí" and "the people of the Bayán" gave way to "Bahá'í" and "the people of Bahá"; the greeting "Alláh-u-Abhá" (God is Most Glorious) replaced "Alláh-u-Akbar" (God is Most Great); and certain disciples of Bahá'u'lláh arose to defend the Faith by refuting "in numerous and detailed apologies" the arguments of its opponents and "to expose their odious deeds." See GPB, pp. 176–77, and Hasan Balyuzi, *Bahá'u'lláh: The King of Glory*, p. 250.

fill the remaining pioneer goals, to undertake traveling teaching projects, and to offer generously of their substance to the various funds, particularly to the vital project of erecting the Panama Temple, the foundation stone of which is being laid by Amatu'l-Bahá Rúḥíyyih Khánum during the course of these Conferences.

46.6 As humanity enters the dark heart of this age of transition our course is clear—the achievement of the assigned goals and the proclamation of Bahá'u'lláh's healing Message. It is our ardent hope that from these Conferences valiant souls may arise with noble resolve and in loving service to ensure the successful and early accomplishment of the sacred tasks that lie ahead.

With loving Bahá'í greetings,
THE UNIVERSAL HOUSE OF JUSTICE

47
Inauguration of the Third Phase of the Nine Year Plan
15 OCTOBER 1967

To all National Spiritual Assemblies

Dear Bahá'í Friends,

47.1 The following cable has just been sent to the United States National Spiritual Assembly for publication in *Bahá'í News*. Please share it with all friends in your jurisdiction.

47.2 HEARTS FILLED PROFOUND GRATITUDE REJOICE ANNOUNCE INAUGURATION THIRD PHASE NINE YEAR PLAN THROUGH SUCCESSFUL CONSUMMATION SIX INTERCONTINENTAL CONFERENCES ATTENDED BY 9,200 BELIEVERS INCLUDING NEARLY ALL HANDS CAUSE LARGE NUMBER BOARD MEMBERS REPRESENTATIVES ALMOST ALL NATIONAL ASSEMBLIES BAHÁ'Í WORLD OVER 140 TERRITORIES AND HOST OF ASIAN AFRICAN AMERINDIAN TRIBES. INESTIMABLE PRIVILEGE CONFERRED PARTICIPANTS THROUGH VIEWING PORTRAIT ABHÁ BEAUTY.[1] SPIRIT HOLY LAND AND ADRIANOPLE CONVEYED SIX DISTINGUISHED REPRESENTATIVES HOUSE JUSTICE. FIRST PRESENTATIONS BEHALF HOUSE JUSTICE PROCLAMATION BOOK HEADS OF STATE MADE BEFORE AND DURING CONFERENCE. FRUITFUL DELIBERATIONS HELD PROCLAMATION EXECUTION REMAINING GOALS PLAN. SOLIDARITY BAHÁ'Í WORLD FURTHER EVINCED THROUGH INGENIOUS SCHEME TELEPHONIC EXCHANGE GREETINGS ALL SIX CONFERENCES. SPIRITUAL POTENCIES THIS NEW PHASE REINFORCED THROUGH FORMAL LAYING BY AMATU'L-BAHÁ OF CORNERSTONE MOTHER TEMPLE LATIN AMERICA. OVER 230 OFFERS MADE AT CONFERENCES JOIN RANKS VALIANT PIONEERS CAUSE. RAISE

47-1. Bahá'u'lláh.

SUPPLIANT HANDS BAHÁ'U'LLÁH ENDOW FRIENDS EVERY LAND FRESH MEASURE CELESTIAL STRENGTH ENABLE THEM PURSUE WITH INCREASED VISION UNABATED RESOLVE GLORIOUS GOALS AHEAD UNTIL THIS NEW PERIOD PROCLAMATION YIELDS ITS SHARE IN DIVINELY PROPELLED PROCESS ESTABLISHMENT KINGDOM GOD HEARTS MEN.

With loving Bahá'í greetings,
THE UNIVERSAL HOUSE OF JUSTICE

48
Announcement of the Resignation of Dr. Luṭfu'lláh Ḥakím from the Universal House of Justice
15 OCTOBER 1967

To all National Spiritual Assemblies

Dear Bahá'í Friends,

48.1 After a lifetime of devoted and self-sacrificing service to the Cause of God Dr. Luṭfu'lláh Ḥakím has asked the Universal House of Justice to accept his resignation from that Institution because his health and advancing age make it increasingly difficult for him to participate as effectively as he would wish in its work.

48.2 The Universal House of Justice has regretfully accepted Dr. Ḥakím's resignation, but in view of the imminence of the next election, has asked him to continue to serve as a member until that time, and Dr. Ḥakím has kindly consented to do so.

48.3 Having served the Master Himself in the Holy Land, as well as accompanying Him during His historic visits in England and Scotland, and been intimately associated with Shoghi Effendi in his youth, Dr. Ḥakím was called again to the World Center by the beloved Guardian in 1950 for important service at the World Center and was later appointed to the first International Bahá'í Council, of which he was the Eastern Assistant Secretary.[1] He continued to serve on that body and then on the Universal House of Justice with undiminished devotion but with increasing difficulty during the subsequent sixteen years, earning the love and admiration of his co-workers.

48.4 Please share this announcement with the friends in your area.

With loving Bahá'í greetings,
THE UNIVERSAL HOUSE OF JUSTICE

48-1. For an explanation of the International Bahá'í Council, see the glossary.

49
Assumption by the Universal House of Justice of Representation of the Bahá'í International Community at the United Nations
17 October 1967

To all National Spiritual Assemblies

49.1 MORROW SIX INTERCONTINENTAL CONFERENCES INAUGURATING PROCLAMATION PERIOD ANNOUNCE BAHÁ'Í WORLD SIGNIFICANT STEP DEVELOPMENT RELATIONS UNITED NATIONS THROUGH ASSUMPTION BY UNIVERSAL HOUSE OF JUSTICE FUNCTION REPRESENTATION BAHÁ'Í INTERNATIONAL COMMUNITY CAPACITY NONGOVERNMENTAL ORGANIZATION AT UNITED NATIONS. TAKE THIS OCCASION EXPRESS NATIONAL ASSEMBLY UNITED STATES AND MILDRED MOTTAHEDEH GRATEFUL LOVING APPRECIATION MANY YEARS DEVOTED TIRELESS SUCCESSFUL SERVICES AS REPRESENTATIVE AND OBSERVER RESPECTIVELY.[1]

UNIVERSAL HOUSE OF JUSTICE

50
Laying of the Foundation Stone of the Panama Temple
23 October 1967

To National Spiritual Assemblies

Dear Bahá'í Friends,

50.1 We are happy to share with you the joyous news that the Foundation Stone of the Mother Temple of Latin America has been laid by Amatu'l-Bahá Rúhíyyih Khánum on behalf of the Universal House of Justice.

49-1. In 1947 the National Spiritual Assembly of the Bahá'ís of the United States and Canada was accredited to the United Nations as a national nongovernmental organization qualified to be represented at United Nations Conferences through a designated observer. One year later the eight existing National Spiritual Assemblies were recognized collectively as an international nongovernmental organization under the title "The Bahá'í International Community." Each National Spiritual Assembly designated the National Spiritual Assembly of the United States as its representative at the United Nations. Mrs. Mildred Mottahedeh, who also served as a member of the International Bahá'í Council from 1961 to 1963, was the observer for the Bahá'í International Community for nearly twenty years. The Bahá'í International Community, which now includes at least five million believers, 165 National Spiritual Assemblies, and approximately twenty thousand Local Spiritual Assemblies, maintains offices in New York and Geneva that are responsible for relations between the Bahá'í International Community and the United Nations. For more information about the development of the relationship between the Bahá'í International Community and the United Nations during the years 1963–73, see BW 15:364–73.

This inaugurates the next phase in the planning and construction of this important edifice which will culminate in the commencement of the erection of the Temple itself in January, 1969. During the current phase the final working plans and specifications will be drafted, bids will be obtained and contracts for construction will be placed. All this will call for funds in ever-increasing amounts.

We therefore call upon all National Spiritual Assemblies to:

1. Inform the National Spiritual Assembly of Panama of the amount of your budget allocation for the Panama temple during this year and when they may expect contributions to be received.
2. Transmit as soon as possible all available funds, either accumulated or earmarked, directly to the National Spiritual Assembly of Panama.
3. Encourage all Local Spiritual Assemblies and individuals to contribute to this project.

With loving Bahá'í greetings,
THE UNIVERSAL HOUSE OF JUSTICE

51
Selection of Traveling Teachers
26 OCTOBER 1967

To National Spiritual Assemblies Engaged in Mass Teaching
Dear Bahá'í Friends,

We have been watching with keen interest the development of the teaching work among the masses, and from time to time have offered suggestions to guide and assist the efforts of National Spiritual Assemblies engaged in this highly meritorious activity.

Many National Spiritual Assemblies in carrying out their plans for expansion and consolidation have found it necessary to select a number of believers for service as traveling teachers. While we appreciate the valuable services these traveling teachers have already rendered we are nevertheless deeply conscious of the problems facing your National Assemblies in your desire to carry out your teaching programs with as much dispatch as possible. The purpose of this letter is to draw your attention to the fact that these problems could well be minimized if the selection of such teachers were done with great care and discretion.

It must be realized that people who are mostly illiterate cannot have the benefit of reading for themselves the written word and of deriving directly from it the spiritual sustenance they need for the enrichment of their Bahá'í lives. They become dependent, therefore, to a large extent on their contacts with visiting teachers. The spiritual caliber or moral quality of these teachers assumes, therefore, great importance. The National Spiritual Assembly or the Teaching Committees responsible for the selection of these teachers should bear in mind that their choice must depend, not only on the knowledge or grasp of the teachings on the part of the teachers, but primarily upon their pure spirit and their true love for the Cause, and their capacity to convey that spirit and love to others.

51.4 We are enclosing some extracts from the writings which will no doubt assist you in your deliberations on this vital subject.[1] What wonderful results will soon be witnessed in the areas under your jurisdiction if you devise ways and means to ensure, as far as circumstances permit, that the traveling teachers you are encouraging to circulate among the friends will all be of the standard called for in these quotations—pure and sanctified souls, with nothing but true devotion and self-sacrifice motivating them in their services to God's Holy Cause. We also suggest that the study of quotations such as these should form part of the courses offered at your Teaching Institutes for the deepening of the friends.

51.5 We wish to assure you once again of our fervent prayers at the Holy Shrines for the solution of your problems and the removal of all obstacles from the path you are so valiantly pursuing.

With loving Bahá'í greetings,
THE UNIVERSAL HOUSE OF JUSTICE

52
Extracts on Teaching the Masses
31 OCTOBER 1967

To all National Spiritual Assemblies

Dear Bahá'í Friends,

52.1 We have recently sent to those National Spiritual Assemblies which are engaged in mass teaching the enclosed extracts from the Writings of Bahá'u'lláh and 'Abdu'l-Bahá and from the letters of Shoghi Effendi. We feel that they will also be of great assistance to all other National Spiritual Assemblies.

52.2 The paramount goal of the teaching work at the present time is to carry the message of Bahá'u'lláh to every stratum of human society and every walk of life. An eager response to the teachings will often be found in the most unexpected quarters, and any such response should be quickly followed up,

---

51-1. The same extracts were also included in the Universal House of Justice's letter dated 31 October 1967 (message no. 52).

for success in a fertile area awakens a response in those who were at first uninterested.

52.3 The same presentation of the teachings will not appeal to everybody; the method of expression and the approach must be varied in accordance with the outlook and interests of the hearer. An approach which is designed to appeal to everybody will usually result in attracting the middle section, leaving both extremes untouched. No effort must be spared to ensure that the healing Word of God reaches the rich and the poor, the learned and the illiterate, the old and the young, the devout and the atheist, the dweller in the remote hills and islands, the inhabitant of the teeming cities, the suburban businessman, the laborer in the slums, the nomadic tribesman, the farmer, the university student; all must be brought consciously within the teaching plans of the Bahá'í Community.

52.4 Whereas plans must be carefully made, and every useful means adopted in the furtherance of this work, your Assemblies must never let such plans eclipse the shining truth expounded in the enclosed quotations: that it is the purity of heart, detachment, uprightness, devotion and love of the teacher that attracts the divine confirmations and enables him, however ignorant he be in this world's learning, to win the hearts of his fellowmen to the Cause of God.

With loving greetings,
THE UNIVERSAL HOUSE OF JUSTICE

Teaching the Masses
Annex

52.5 Whoso ariseth, in this Day, to aid Our Cause, and summoneth to his assistance the hosts of a praiseworthy character and upright conduct, the influence flowing from such an action will, most certainly, be diffused throughout the whole world.

(Bahá'u'lláh, *Gleanings from the Writings of Bahá'u'lláh*, Rev. ed. [Wilmette: Bahá'í Publishing Trust, 1983], sec. 131, p. 287)

52.6 Whoso ariseth to teach Our Cause must needs detach himself from all earthly things, and regard, at all times, the triumph of Our Faith as his supreme objective.... And when he determineth to leave his home, for the sake of the Cause of his Lord, let him put his whole trust in God, as the best provision for his journey, and array himself with the robe of virtue....

52.7 If he be kindled with the fire of His love, if he forgoeth all created things, the words he uttereth shall set on fire them that hear him....

(Bahá'u'lláh, *Gleanings from the Writings of Bahá'u'lláh*, sec. 157, pp. 334–35)

52.8 I swear by Him Who is the Most Great Ocean! Within the very breath of such souls as are pure and sanctified far-reaching potentialities are hidden.

So great are these potentialities that they exercise their influence upon all created things.
(Bahá'u'lláh, cited in Shoghi Effendi, *The Advent of Divine Justice* [Wilmette: Bahá'í Publishing Trust, 1984], p. 23)

52.9 He is the true servant of God, who, in this day, were he to pass through cities of silver and gold, would not deign to look upon them, and whose heart would remain pure and undefiled from whatever things can be seen in this world, be they its goods or its treasures. I swear by the Sun of Truth! The breath of such a man is endowed with potency, and his words with attraction.
(Bahá'u'lláh, cited in Shoghi Effendi, *The Advent of Divine Justice*, p. 23)

52.10 The most vital duty, in this day, is to purify your characters, to correct your manners, and improve your conduct. The beloved of the Merciful must show forth such character and conduct among His creatures, that the fragrance of their holiness may be shed upon the whole world, and may quicken the dead, inasmuch as the purpose of the Manifestation of God and the dawning of the limitless lights of the Invisible is to educate the souls of men, and refine the character of every living man . . .
('Abdu'l-Bahá, cited in Shoghi Effendi, *The Advent of Divine Justice*, p. 26)

52.11 Whensoever ye behold a person whose entire attention is directed toward the Cause of God; whose only aim is this, to make the Word of God to take effect; who, day and night, with pure intent, is rendering service to the Cause; from whose behavior not the slightest trace of egotism or private motives is discerned—who, rather, wandereth distracted in the wilderness of the love of God, and drinketh only from the cup of the knowledge of God, and is utterly engrossed in spreading the sweet savours of God, and is enamored of the holy verses of the Kingdom of God—know ye for a certainty that this individual will be supported and reinforced by heaven; that like unto the morning star, he will forever gleam brightly out of the skies of eternal grace. But if he show the slightest taint of selfish desires and self love, his efforts will lead to nothing and he will be destroyed and left hopeless at the last.
('Abdu'l-Bahá, *Selections from the Writings of 'Abdu'l-Bahá*, [Rev. ed.] [Haifa: Bahá'í World Center, 1982], pp. 71–72)

52.12 The aim is this: The intention of the teacher must be pure, his heart independent, his spirit attracted, his thought at peace, his resolution firm, his magnanimity exalted and in the love of God a shining torch. Should he become as such, his sanctified breath will even affect the rock; otherwise there will be no result whatsoever. As long as a soul is not perfected, how can he efface the

defects of others. Unless he is detached from aught else save God, how can he teach severance to others!
> ('Abdu'l-Bahá, *Tablets of the Divine Plan: Revealed by 'Abdu'l-Bahá to the North American Bahá'ís*, Rev. ed. [Wilmette: Bahá'í Publishing Trust, 1980], p. 51)

One thing and only one thing will unfailingly and alone secure the undoubted triumph of this sacred Cause, namely the extent to which our own inner life and private character mirror forth in their manifold aspects the splendor of those eternal principles proclaimed by Bahá'u'lláh. 52.13
> (Shoghi Effendi, from a letter dated 24 September 1924 to the Bahá'ís of America, published in *Bahá'í Administration: Selected Messages, 1922–1932* [Wilmette: Bahá'í Publishing Trust, 1980], p. 66)

. . . having attained sufficiently that individual regeneration—the essential requisite of teaching—let us arise to teach His Cause with righteousness, conviction, understanding and vigor. Let this be the paramount and most urgent duty of every Bahá'í. . . . 52.14
> (Shoghi Effendi, from a letter dated 24 November 1924 to the National Spiritual Assembly of the United States and Canada, published in *Bahá'í Administration*, p. 67)

The first and most important qualification of a Bahá'í teacher is, indeed, unqualified loyalty and attachment to the Cause. Knowledge is, of course, essential, but compared to devotion it is secondary in importance. 52.15

What the Cause now requires is not so much a group of highly cultured and intellectual people who can adequately present its Teachings, but a number of devoted, sincere and loyal supporters who, in utter disregard of their own weaknesses and limitations, and with hearts afire with the love of God, forsake their all for the sake of spreading and establishing His Faith. . . . 52.16
> (From a letter dated 14 November 1935 written on behalf of Shoghi Effendi to the National Teaching Committee of the United States and Canada, published in *Bahá'í News*, No. 102 [August 1936], p. 2)

They must remember the glorious history of the Cause, which . . . was established by dedicated souls who, for the most part, were neither rich, famous, nor well educated, but whose devotion, zeal and self-sacrifice overcame every obstacle and won miraculous victories for the Faith of God. . . . 52.17
> (From a letter dated 29 June 1941 written on behalf of Shoghi Effendi to the National Spiritual Assembly of India and Burma, published in *Dawn of a New Day* [New Delhi: Bahá'í Publishing Trust, n.d. (1970)], p. 89)

. . . what raised aloft the banner of Bahá'u'lláh was the love, sacrifice, and devotion of His humble followers and the change that His teachings wrought in their hearts and lives. 52.18
> (From a letter dated 20 June 1942 written on behalf of Shoghi Effendi to the National Spiritual Assembly of the British Isles, published in *Unfolding Destiny*, [London: Bahá'í Publishing Trust, 1981], p. 152)

52.19 It is the quality of devotion and self-sacrifice that brings rewards in the service of this Faith rather than means, ability or financial backing.
(From a letter dated 11 May 1948 written on behalf of Shoghi Effendi to the National Spiritual Assembly of Australia and New Zealand)

52.20 One wise and dedicated soul can so often give life to an inactive community, bring in new people and inspire to greater sacrifice. He hopes that whatever else you are able to do during the coming months, you will be able to keep in circulation a few really good Bahá'í teachers.
(From a letter dated 30 June 1952 written on behalf of Shoghi Effendi to the National Spiritual Assembly of Central America)

53
Plans for Commemorating the Sesquicentennial of the Birth of Bahá'u'lláh at the First Oceanic Conference, Palermo, Sicily
12 November 1967

To all National Spiritual Assemblies

53.1 OCCASION HUNDRED FIFTIETH ANNIVERSARY BIRTH BLESSED BEAUTY WE CONTEMPLATE WITH HEARTS OVERFLOWING GRATITUDE INESTIMABLE BOUNTIES CONFERRED BY GOD THROUGH HIS SUPREME MANIFESTATION ENSURING FULFILLMENT GLORIOUS LONG PROMISED KINGDOM NOW EVOLVING WOMB TRAVAILING AGE DESTINED CONFER PEACE UNDREAMT FELICITY MANKIND.[1] ANNOUNCE CONVOCATION TWENTY-THIRD TO TWENTY-FIFTH AUGUST 1968 FIRST OCEANIC CONFERENCE BAHÁ'Í WORLD PALERMO SICILY HEART SEA TRAVERSED GOD'S MANIFESTATION CENTURY AGO PROCEEDING INCARCERATION MOST GREAT PRISON.[2] TWOFOLD PURPOSE CONFERENCE CONSIDER MOMENTOUS FULFILLMENT AGE-OLD PROPHECIES TRIUMPH GOD'S MESSENGER OVER EVERY GRIEVOUS CALAMITY AND CONSULT PLANS PROPAGATION CAUSE ISLANDS LANDS BORDERING MEDITERRANEAN SEA. PARTICIPANTS INVITED HOLY LAND IMMEDIATELY FOLLOWING CONFERENCE ATTEND COMMEMORATION ARRIVAL LORD HOSTS THESE SACRED SHORES RECONSECRATE THEMSELVES THRESHOLD HIS SHRINE PROSECUTION GLORIOUS TASKS AHEAD.

THE UNIVERSAL HOUSE OF JUSTICE

53-1. The Blessed Beauty, Bahá'u'lláh, was born on 12 November 1817.

53-2. On 12 August 1868 Bahá'u'lláh left Adrianople and journeyed four days to Gallipoli, a city in Turkey on the north side of the Dardenelles at the mouth of the Sea of Marmara. After a few days in Gallipoli, Bahá'u'lláh sailed in an Austrian steamer for Alexandria, where He was transferred to another ship. He arrived in 'Akká on 31 August 1868. For a fuller account of the journey, see GPB, pp. 178–82, and H. M. Balyuzi, *Bahá'u'lláh: The King of Glory*, pp. 255–79.

54
Safeguarding the Letters of Shoghi Effendi
DECEMBER 1967

To all National Spiritual Assemblies

Dear Bahá'í Friends,

As the friends are already aware, one of the goals of the Nine Year Plan is continued collation and classification of the Bahá'í Sacred Scriptures, as well as the writings of Shoghi Effendi.[1]

We have already pointed out to the friends on several occasions that the application of Bahá'í laws, the elucidation and extension of basic administrative principles, and the all-important function of legislating on matters not explicitly recorded in our teachings are dependent upon a careful study by the Universal House of Justice of the revealed and pertinent words of Bahá'u'lláh and 'Abdu'l-Bahá, as well as the illuminating interpretations and directions of Shoghi Effendi.

Through the labors of the beloved Guardian himself, and the collaboration of the National Spiritual Assembly of Persia, great strides have already been taken to collate the Writings of Bahá'u'lláh and 'Abdu'l-Bahá. Even up to the present time, a special National Committee in Persia is assiduously and regularly engaged in classifying the Holy Texts of the Founder of our Faith, and the Center of the Covenant in fulfillment of the goal of the Nine Year Plan referred to above. The writings of the Guardian, however, apart from his general communications already published, which consist to a large extent of letters written by him or on his behalf to various National Spiritual Assemblies, Local Spiritual Assemblies, groups of believers and individual Bahá'ís, have only been partly collated. Although many National Spiritual Assemblies and individual believers have already sent the originals or copies of their letters from the Guardian to the Holy Land, and quite a number of these have been published at different times in *Bahá'í News* and other news bulletins, it is our confirmed opinion that there are many such letters that have still not been shared with us which may contain a new application of a principle, an already enunciated guideline rephrased to apply in a new situation, a further elucidation of a Bahá'í law, a fresh light thrown on a many-sided issue or some wise counsel in a personal problem.

Your National Spiritual Assembly is requested, therefore, to check carefully first in its own archives or files of correspondence with the Guardian, for any letters addressed to its body or to its subsidiary institutions, written by Shoghi Effendi or on his behalf, and not yet forwarded to the Holy Land. At the end

---

54-1. For additional letters on the topic, see messages no. 161 and 409.

of this letter we indicate the number and description of Shoghi Effendi's letters falling into this category and pertaining to your area, which we have in our archival files, so that you need to take action in sending us only the text of letters not included in this listing.

54.5 You are also requested to appeal to the friends under your jurisdiction, calling on those who were privileged to have received letters from the Guardian but have not as yet sent their texts to the Holy Land to take immediate steps to do so. In assisting the friends and your National Spiritual Assembly in carrying out this important project we offer the following points for your consideration:

54.5a 1. Recipients of letters from the Guardian have the inherent right of deciding to keep the letters themselves, or to have them preserved for the future in their families. To assist the Universal House of Justice, however, in its efforts to study and compile the letters of the Guardian, the friends are urged to provide, for dispatch to the Holy Land, photostatic copies of their communications from the Guardian if they wish to keep the originals themselves.

54.5b 2. If they are not in a position to provide such copies, they should kindly allow National Spiritual Assemblies to undertake this project on our behalf.

54.5c 3. If facilities for obtaining clear photostatic or Xerox copies are not available in any area, National Spiritual Assemblies are requested to make carefully typed or handwritten copies of such letters, with a certification on behalf of the National Spiritual Assembly concerned that the copies are true and exact.

54.5d 4. Should any believer possess letters so personal and confidential that he does not wish to disclose their contents to any institution other than the Universal House of Justice, he is invited to send either the originals or copies of such letters, marked confidential, directly to the Universal House of Justice, by registered mail, with any instructions he wishes to be followed.

54.6 We hope these guidelines will help you in promoting a project directly linked with the vital functions of the Universal House of Justice. You are free to quote from this letter as you wish, in any appeal you address to the believers.

With loving Bahá'í greetings,
THE UNIVERSAL HOUSE OF JUSTICE

55
Relationship of Baháʼís to Politics
8 December 1967

To an individual Believer[1]

Dear Baháʼí Friend,

... we will gladly attempt to clarify some of the points which bewilder you 55.1 in the relationship of Baháʼís to politics. This is a matter of very great importance, particularly in these days when the world situation is so confused; an unwise act or statement by a Baháʼí in one country could result in a grave setback for the Faith there or elsewhere—and even loss of the lives of fellow-believers.

Viewing the World's Problems in the Light of God's Purpose for Man

The whole conduct of a Baháʼí in relation to the problems, sufferings and 55.2 bewilderment of his fellowmen should be viewed in the light of God's purpose for mankind in this age and the processes He has set in motion for its achievement.

When Baháʼuʼlláh proclaimed His Message to the world in the nineteenth 55.3 century He made it abundantly clear that the first step essential for the peace and progress of mankind was its unification. As He says, "The well-being of mankind, its peace and security are unattainable unless and until its unity is firmly established." (*The World Order of Baháʼuʼlláh*, p. 203) To this day, however, you will find most people take the opposite point of view: they look upon unity as an ultimate almost unattainable goal and concentrate first on remedying all the other ills of mankind. If they did but know it, these other ills are but various symptoms and side effects of the basic disease—disunity.

Baháʼuʼlláh has, furthermore, stated that the revivification of mankind and 55.4 the curing of all its ills can be achieved only through the instrumentality of His Faith. "The vitality of men's belief in God is dying out in every land; nothing short of His wholesome medicine can ever restore it. The corrosion of ungodliness is eating into the vitals of human society; what else but the Elixir of His potent Revelation can cleanse and revive it?" (*Gleanings*, XCIX) "That which the Lord hath ordained as the sovereign remedy and mightiest instrument for the healing of all the world is the union of all its peoples in one universal Cause, one common Faith. This can in no wise be achieved

---

55-1. The Universal House of Justice permitted publication of portions of this letter to an individual who asked about the relationship of Baháʼís to the social and political forces presently operating in the world, as it felt the explanation had general application in many parts of the world.

except through the power of a skilled, an all-powerful and inspired Physician. This, verily, is the truth, and all else naught but error." (*Gleanings*, CXX) In similar vein, the beloved Guardian wrote:

55.4a Humanity, whether viewed in the light of man's individual conduct or in the existing relationships between organized communities and nations, has, alas, strayed too far and suffered too great a decline to be redeemed through the unaided efforts of the best among its recognized rulers and statesmen—however disinterested their motives, however concerted their action, however unsparing in their zeal and devotion to its cause. No scheme which the calculations of the highest statesmanship may yet devise, no doctrine which the most distinguished exponents of economic theory may hope to advance, no principle which the most ardent of moralists may strive to inculcate, can provide, in the last resort, adequate foundations upon which the future of a distracted world can be built. No appeal for mutual tolerance which the worldly-wise might raise, however compelling and insistent, can calm its passions or help restore its vigor. Nor would any general scheme of mere organized international cooperation, in whatever sphere of human activity, however ingenious in conception or extensive in scope, succeed in removing the root cause of the evil that has so rudely upset the equilibrium of present-day society. Not even, I venture to assert, would the very act of devising the machinery required for the political and economic unification of the world—a principle that has been increasingly advocated in recent times—provide in itself the antidote against the poison that is steadily undermining the vigor of organized peoples and nations. What else, might we not confidently affirm, but the unreserved acceptance of the Divine Program enunciated, with such simplicity and force as far back as sixty years ago, by Bahá'u'lláh, embodying in its essentials God's divinely appointed scheme for the unification of mankind in this age, coupled with an indomitable conviction in the unfailing efficacy of each and all of its provisions, is eventually capable of withstanding the forces of internal disintegration which, if unchecked, must needs continue to eat into the vitals of a despairing society.
(*The World Order of Bahá'u'lláh*, pp. 33–34)

Two Processes at Work

55.5 We are told by Shoghi Effendi that two great processes are at work in the world: the great Plan of God, tumultuous in its progress, working through mankind as a whole, tearing down barriers to world unity and forging humankind into a unified body in the fires of suffering and experience. This process will produce in God's due time, the Lesser Peace, the political unification of the world.² Mankind at that time can be likened to a body that is unified but without life. The second process, the task of breathing life into this unified body—of creating true unity and spirituality culminating in the Most Great Peace—is that of the Bahá'ís, who are laboring consciously, with detailed instructions and continuing divine guidance, to erect the fabric of the Kingdom of God on earth, into which they call their fellowmen, thus conferring upon them eternal life.³

55.6 The working out of God's Major Plan proceeds mysteriously in ways directed by Him alone, but the Minor Plan that He has given us to execute, as our part in His grand design for the redemption of mankind, is clearly delineated.⁴ It is to this work that we must devote all our energies, for there is no one else to do it. So vital is this function of the Bahá'ís that Bahá'u'lláh has written: "O friends! Be not careless of the virtues with which ye have been endowed, neither be neglectful of your high destiny. Suffer not your labors to be wasted through the vain imaginations which certain hearts have devised. Ye are the stars of the heaven of understanding, the breeze that stirreth at the break of day, the soft-flowing waters upon which must depend the very life of all men, the letters inscribed upon His sacred scroll. With the utmost unity, and in a spirit of perfect fellowship, exert yourselves, that ye may be enabled to achieve that which beseemeth this Day of God." (*Gleanings*, XCVI)

55.7 Because love for our fellowmen and anguish at their plight are essential parts of a true Bahá'í's life, we are continually drawn to do what we can to help them. It is vitally important that we do so whenever the occasion presents itself, for our actions must say the same thing as our words—but this compassion for our fellows must not be allowed to divert our energies into channels which are ultimately doomed to failure, causing us to neglect the most important and fundamental work of all. There are hundreds of thousands of well-wishers of mankind who devote their lives to works of relief and charity, but a pitiful few to do the work which God Himself most wants done: the spiritual awakening and regeneration of mankind.

Our Task—Building Up the Bahá'í System

55.8 It is often through our misguided feeling that we can somehow aid our fellows better by some activity outside the Faith, that Bahá'ís are led to indulge

---

55-2. For an explanation of the Lesser Peace, see the glossary.
55-3. For an explanation of the Most Great Peace, see the glossary.
55-4. The Major Plan is God's plan for humanity that Bahá'ís believe He Himself operates, which is tumultuous in its progress, which works through humanity as a whole, and which forges mankind into a unified body through the fires of suffering and tribulation. Its ultimate object is the Kingdom of God on earth. The Minor Plan is that part of God's plan which the Bahá'ís are called upon to carry out. It is clear and orderly and operates in the world through the plans, instructions, and guidance given by 'Abdu'l-Bahá, Shoghi Effendi, and now by the Universal House of Justice.

in politics. This is a dangerous delusion. As Shoghi Effendi's secretary wrote on his behalf: "What we Bahá'ís must face is the fact that society is rapidly disintegrating—so rapidly that moral issues which were clear a half century ago are now hopelessly confused, and what is more, thoroughly mixed up with battling political interests. That is why the Bahá'ís must turn all their forces into the channel of building up the Bahá'í Cause and its administration. They can neither change nor help the world in any other way at present. If they become involved in the issues the governments of the world are struggling over, they will be lost. But if they build up the Bahá'í pattern they can offer it as a remedy when all else has failed." (*Bahá'í News*, No. 241, p. 14) "We must build up our Bahá'í system, and leave the faulty systems of the world to go their own way. We cannot change them through becoming involved in them; on the contrary they will destroy us." (*Bahá'í News*, No. 215, p.1)

55.9 Other instructions from the Guardian, covering the same theme in more detail, can be found on pages 24 and 29 to 32 of *Principles of Bahá'í Administration* (1963 edition); you are no doubt already familiar with these.[5]

55.10 The key to a true understanding of these principles seems to be in these words of Bahá'u'lláh: "O people of God! Do not busy yourselves in your own concerns; let your thoughts be fixed upon that which will rehabilitate the fortunes of mankind and sanctify the hearts and souls of men. This can best be achieved through pure and holy deeds, through a virtuous life and a goodly behavior. Valiant acts will ensure the triumph of this Cause, and a saintly character will reinforce its power. Cleave unto righteousness, O people of Bahá! This, verily, is the commandment which this wronged One hath given unto you, and the first choice of His unrestrained Will for every one of you." (*Gleanings*, XLIII).

With loving Bahá'í greetings,
THE UNIVERSAL HOUSE OF JUSTICE

55-5. In the 1982 edition see pp. 24, 29–33.

56
Election of the Universal House of Justice— Riḍván 1968
22 April 1968

To all National Spiritual Assemblies

ANNOUNCE BAHÁ'Í WORLD NEWLY ELECTED MEMBERS UNIVERSAL HOUSE OF JUS- 56.1
TICE AMOZ GIBSON 'ALÍ NAKHJAVÁNÍ HUSHMAND FATHEAZAM IAN SEMPLE CHARLES
WOLCOTT DAVID HOFMAN H. BORRAH KAVELIN HUGH CHANCE DAVID RUHE.

THE UNIVERSAL HOUSE OF JUSTICE

57
Message to National Conventions—1968
9 May 1968

To all National Bahá'í Conventions

WITH JOYFUL MEMORY OF DEDICATED SPIRIT MATURE DELIBERATIONS SECOND 57.1
INTERNATIONAL CONVENTION HAIL GOLDEN OPPORTUNITY NATIONAL CONVENTIONS AS CRUCIAL MIDWAY POINT NINE YEAR PLAN APPROACHES GALVANIZE
BELIEVERS DIRECT ALL EFFORTS ACHIEVEMENT EVERY REMAINING GOAL AND SIMULTANEOUSLY EXTEND ACCELERATE UNIVERSAL PROCLAMATION DIVINE MESSAGE.
WITH UTMOST LOVE CALL UPON ALL BAHÁ'ÍS FOR SACRIFICIAL OUTPOURING ENERGIES RESOURCES ADVANCEMENT REDEEMING ORDER BAHÁ'U'LLÁH SOLE REFUGE
MISDIRECTED HEEDLESS MILLIONS. WORLD CENTER FAITH SCENE PROLONGED
PRAYERFUL CONSULTATION WITH ASSEMBLED HANDS CAUSE GOALS PLAN
INCLUDING FUNDAMENTAL OBJECTIVE DEVELOPMENT INSTITUTION HANDS VIEW
EXTENSION FUTURE GOD-GIVEN DUTIES PROTECTION PROPAGATION. SUPPLICATING
CONTINUALLY HOLY SHRINES LORD HOSTS BOUNTIFULLY REWARD DEDICATED
ARDENT LOVERS COMPLETE GLORIOUS VICTORY.

THE UNIVERSAL HOUSE OF JUSTICE

58
Announcement of Decision to Establish Eleven Continental Boards of Counselors
21 June 1968

To all National Spiritual Assemblies

58.1 REJOICE ANNOUNCE MOMENTOUS DECISION ESTABLISH ELEVEN CONTINENTAL BOARDS COUNSELORS PROTECTION PROPAGATION FAITH THREE EACH FOR AFRICA AMERICAS ASIA ONE EACH FOR AUSTRALASIA EUROPE. ADOPTION THIS SIGNIFICANT STEP FOLLOWING CONSULTATION WITH HANDS CAUSE GOD ENSURES EXTENSION FUTURE APPOINTED FUNCTIONS THEIR INSTITUTION. CONTINENTAL BOARDS ENTRUSTED IN CLOSE COLLABORATION HANDS CAUSE WITH RESPONSIBILITY DIRECTION AUXILIARY BOARDS AND CONSULTATION NATIONAL SPIRITUAL ASSEMBLIES. HANDS CAUSE GOD WILL HENCEFORTH INCREASE INTERCONTINENTAL SERVICES ASSUMING WORLDWIDE ROLE PROTECTION PROPAGATION FAITH. MEMBERS AUXILIARY BOARDS WILL REPORT BE RESPONSIBLE TO CONTINENTAL BOARDS COUNSELORS. HANDS CAUSE RESIDING HOLY LAND IN ADDITION SERVING LIAISON BETWEEN UNIVERSAL HOUSE JUSTICE AND CONTINENTAL BOARDS COUNSELORS WILL ASSIST FUTURE ESTABLISHMENT INTERNATIONAL TEACHING CENTER HOLY LAND FORESHADOWED WRITINGS BELOVED GUARDIAN. DETAILS NEW DEVELOPMENTS BEING CONVEYED BY LETTER. FERVENTLY SUPPLICATING HOLY THRESHOLD DIVINE CONFIRMATIONS FURTHER STEP IRRESISTIBLE UNFOLDMENT MIGHTY ADMINISTRATIVE ORDER BAHÁ'U'LLÁH.

UNIVERSAL HOUSE OF JUSTICE

59
Establishment of Continental Boards of Counselors
24 June 1968

To the Bahá'ís of the World

Dear Bahá'í Friends,

59.1 The majestic unfoldment of Bahá'u'lláh's world-redeeming administrative system has been marked by the successive establishment of the various institutions and agencies which constitute the framework of that divinely created Order. Thus, more than a quarter of a century after the emergence of the first National Spiritual Assemblies of the Bahá'í world the Institution of the Hands of the Cause of God was formally established, with the appointment by the beloved Guardian, in conformity with the provisions of 'Abdu'l-Bahá's Will and Testament, of the first contingent of these high-ranking officers of the

Faith. Following the passing of the Guardian of the Cause of God, it fell to the House of Justice to devise a way, within the Administrative Order, of developing "the Institution of the Hands of the Cause with a view to extension into the future of its appointed functions of protection and propagation," and this was made a goal of the Nine Year Plan.[1] Much thought and study has been given to the question over the past four years, and the texts have been collected and reviewed. During the last two months, this goal, as announced in our cable to the National Conventions, has been the object of prolonged and prayerful consultation between the Universal House of Justice and the Hands of the Cause of God. All this made evident the framework within which this goal was to be achieved, namely:

> The Universal House of Justice sees no way in which additional Hands of the Cause of God can be appointed. 59.1a

> The absence of the Guardian of the Faith brought about an entirely new relationship between the Universal House of Justice and the Hands of the Cause and called for the progressive unfoldment by the Universal House of Justice of the manner in which the Hands of the Cause would carry out their divinely conferred functions of protection and propagation. 59.1b

> Whatever new development or institution is initiated should come into operation as soon as possible in order to reinforce and supplement the work of the Hands of the Cause while at the same time taking full advantage of the opportunity of having the Hands themselves assist in launching and guiding the new procedures. 59.1c

> Any such institution must grow and operate in harmony with the principles governing the functioning of the Institution of the Hands of the Cause of God. 59.1d

In the light of these considerations the Universal House of Justice decided, as announced in its recent cable, to establish Continental Boards of Counselors for the protection and propagation of the Faith. Their duties will include directing the Auxiliary Boards in their respective areas, consulting and collaborating with National Spiritual Assemblies, and keeping the Hands of the Cause and the Universal House of Justice informed concerning the conditions of the Cause in their areas. 59.2

Initially eleven Boards of Counselors have been appointed, one for each of the following areas: Northwestern Africa, Central and East Africa, Southern Africa, North America, Central America, South America, Western Asia, Southeastern Asia, Northeastern Asia, Australasia and Europe. 59.3

59-1. See message no. 14.

59.4 The members of these Boards of Counselors will serve for a term, or terms, the length of which will be determined and announced at a later date, and while serving in this capacity, will not be eligible for membership on national or local administrative bodies. One member of each Continental Board of Counselors has been designated as Trustee of the Continental Fund for its area.

59.5 The Auxiliary Boards for Protection and Propagation will henceforth report to the Continental Boards of Counselors who will appoint or replace members of the Auxiliary Boards as circumstances may require. Such appointments and replacements as may be necessary in the initial stages will take place after consultation with the Hand or Hands previously assigned to the continent or zone.

59.6 The Hands of the Cause of God have the prerogative and obligation to consult with the Continental Boards of Counselors and National Spiritual Assemblies on any subject which, in their view, affects the interests of the Cause. The Hands residing in the Holy Land will act as liaison between the Universal House of Justice and the Continental Boards of Counselors, and will also assist the Universal House of Justice in setting up, at a propitious time, an international teaching center in the Holy Land, as anticipated in the Guardian's writings.

59.7 The Hands of the Cause of God are one of the most precious assets the Bahá'í world possesses. Released from administration of the Auxiliary Boards, they will be able to concentrate their energies on the more primary responsibilities of general protection and propagation, "PRESERVATION [of the] SPIRITUAL HEALTH [of the] BAHÁ'Í COMMUNITIES" and the "VITALITY [of the] FAITH" of the Bahá'ís throughout the world.[2] The House of Justice will call upon them to undertake special missions on its behalf, to represent it on both Bahá'í and other occasions and to keep it informed of the welfare of the Cause. While the Hands of the Cause will, naturally, have special concern for the affairs of the Cause in the areas in which they reside, they will operate increasingly on an intercontinental level, a factor which will lend tremendous impetus to the diffusion throughout the Bahá'í world of the spiritual inspiration channeled through them—the Chief Stewards of Bahá'u'lláh's embryonic World Commonwealth.

59.8 With joyful hearts we proclaim this further unfoldment of the Administrative Order of Bahá'u'lláh and join our prayers to those of the friends throughout the East and the West that Bahá'u'lláh may continue to shower His confirmations upon the efforts of His servants in the safeguarding and promotion of His Faith.

With loving Bahá'í greetings,
THE UNIVERSAL HOUSE OF JUSTICE

59-2. MBW, p. 123.

60
First Appointments to Continental Boards of Counselors
24 JUNE 1968

To all National Spiritual Assemblies

Dear Bahá'í Friends,

We list below the names of those who have been appointed to the first Continental Boards of Counselors for the Protection and Propagation of the Faith:

Northwestern Africa
Ḥusayn Ardikání (Trustee, Continental Fund), Muḥammad Kebdani, William Maxwell.

Central and East Africa
Oloro Epyeru, Kolonario Oule, Isobel Sabri, Mihdí Samandarí, 'Azíz Yazdí (Trustee, Continental Fund).

Southern Africa
Seewoosumbur-Jeehoba Appa, Shidan Fat'he-Aazam (Trustee, Continental Fund), Bahíyyih Ford.

North America
Lloyd Gardner, Florence Mayberry, Edna True (Trustee, Continental Fund).

Central America
Carmen de Burafato, Artemus Lamb, Alfred Osborne (Trustee, Continental Fund).

South America
Athos Costas, Hooper Dunbar (Trustee, Continental Fund), Donald Witzel.

Western Asia
Masíḥ Farhangí, Mas'úd Khamsí, Hádí Raḥmání (Trustee, Continental Fund), Manúchihr Salmánpúr, Sankaran-Nair Vasudevan.

Southeast Asia
Yan Kee Leong, Khudáraḥm Paymán (Trustee, Continental Fund), Chellie Sundram.

Northeast Asia
Rúḥu'lláh Mumtází (Trustee, Continental Fund), Vicente Samaniego.

Australasia

Suhayl 'Alá'í, Howard Harwood, Thelma Perks (Trustee, Continental Fund).

Europe

Erik Blumenthal, Dorothy Ferraby (Trustee, Continental Fund), Louis Hénuzet.

60.2 Please share this list with the friends.

With loving Bahá'í greetings,
THE UNIVERSAL HOUSE OF JUSTICE

61
Message to First National Youth Conference in the United States
26 JUNE 1968

To the National Spiritual Assembly of the Bahá'ís of the United States

61.1 WARMLY ACKNOWLEDGE CONFIDENT JOYOUS MESSAGE FROM BAHÁ'Í YOUTH GATHERED PRECINCTS MOTHER TEMPLE WEST. MOVED THEIR DETERMINATION SEIZE OPPORTUNITIES SERVE BELOVED FAITH CALL ON THEM BOLDLY CHALLENGE INVITE CONFUSED CONTEMPORARIES ENTRAPPED MORASS MATERIALISM TO EXAMINE PARTAKE LIFEGIVING POWER CAUSE JOIN ARMY BAHÁ'U'LLÁH CONFRONT NEGATIVE FORCES OF A SOCIETY SADLY LACKING SPIRITUAL VALUES. ASSURE YOUTH ARDENT PRAYERS HOLY SHRINES GUIDANCE CONFIRMATION THEIR COURAGEOUS EFFORTS.

UNIVERSAL HOUSE OF JUSTICE

1964–1973 · THE NINE YEAR PLAN

62
Passing of the Hand of the Cause of God Hermann Grossmann
9 JULY 1968

To all National Spiritual Assemblies

DEEPLY REGRET ANNOUNCE PASSING HAND CAUSE HERMANN GROSSMANN.[1]   62.1
GREATLY ADMIRED BELOVED GUARDIAN HIS GRIEVOUS LOSS DEPRIVES COMPANY HANDS CAUSE OUTSTANDING COLLABORATOR AND BAHÁ'Í WORLD COMMUNITY STAUNCH DEFENDER PROMOTER FAITH. HIS COURAGEOUS LOYALTY DURING CHALLENGING YEARS TESTS PERSECUTIONS GERMANY OUTSTANDING SERVICES SOUTH AMERICA IMMORTALIZED ANNALS FAITH. INVITE ALL NATIONAL SPIRITUAL ASSEMBLIES HOLD MEMORIAL GATHERINGS BEFITTING HIS EXALTED RANK EXEMPLARY SERVICES. REQUEST THOSE RESPONSIBLE MOTHER TEMPLES ARRANGE SERVICES AUDITORIUM.

UNIVERSAL HOUSE OF JUSTICE

63
Message to the First Oceanic Conference, Palermo, Sicily
AUGUST 1968

To the Hands of the Cause of God and the Bahá'í Friends assembled in Palermo, Sicily, at the First Bahá'í Oceanic Conference

Dearly loved Friends,

The event which we commemorate at this first Bahá'í Oceanic Conference   63.1
is unique. Neither the migration of Abraham from Ur of the Chaldees to the region of Aleppo, nor the journey of Moses towards the Promised Land, nor the flight into Egypt of Mary and Joseph with the infant Jesus, nor yet the Hegira of Muḥammad can compare with the voyage made by God's Supreme Manifestation one hundred years ago from Gallipoli to the Most Great Prison.[1] Bahá'u'lláh's voyage was forced upon Him by the two despots who were His chief

---

62-1. For an account of the life and services of Hermann Grossmann, see BW 15:416–21.

63-1. Shoghi Effendi explains that during the stay in Gallipoli, which lasted three nights, "no one knew what Bahá'u'lláh's destination would be. Some believed that He and His brothers would be banished to one place, and the remainder dispersed, and sent into exile. Others thought that His companions would be sent back to Persia, while still others expected their immediate extermination. . . ."

"So grievous were the dangers and trials confronting Bahá'u'lláh at the hour of His departure from Gallipoli that He warned His companions that 'this journey will be unlike any of the previous journeys,' and that whoever did not feel himself 'man enough to face the future' had

adversaries in a determined attempt to extirpate once and for all His Cause, and the decree of His fourth banishment came when the tide of His prophetic utterance was in full flood.[2] The proclamation of His Message to mankind had begun; the sun of His majesty had reached its zenith and, as attested by the devotion of His followers, the respect of the population and the esteem of officials and the representatives of foreign powers, His ascendancy had become manifest. At such a time He was confronted with the decree of final exile to a remote, obscure and pestilential outpost of the decrepit Turkish empire.

63.2   Bahá'u'lláh knew, better than His royal persecutors, the magnitude of the crisis, with all its potentiality for disaster, which confronted Him. Consigned to a prison cell, debarred from access to those to whom His Message must be addressed, cut off from His followers save for the handful who were to accompany Him, and deprived even of association with them, it was apparent that by all earthly standards the ship of His Cause must founder, His mission wither and die.

63.3   But it was the Lord of Hosts with Whom they were dealing. Knowing the sufferings which faced Him His one thought was to instill confidence and fortitude into His followers, to whom He immediately dispatched sublime Tablets asserting the power of His Cause to overcome all opposition. "Should they attempt to conceal its light on the continent," is one of His powerful utterances on this theme, "it will assuredly rear its head in the midmost heart of the ocean, and, raising its voice, proclaim 'I am the lifegiver of the world!'"[3] All the afflictions which men could heap upon Him were thrown back from the rock of His adamantine will like spray from the ocean. His patient submission to the affronts of men, His fortitude, His divine genius transformed the somber notes of disaster into the diapason of triumph. At the nadir of His worldly fortunes He raised His standard of victory above the Prison City and poured forth upon mankind the healing balm of His laws and ordinances revealed in His Most Holy Book.[4] "Until our time," comments 'Abdu'l-Bahá, "no such thing has ever occurred."[5]

---

best 'depart to whatever place he pleaseth, and be preserved from tests, for hereafter he will find himself unable to leave'—a warning which His companions unanimously chose to disregard" (GPB, pp. 181–82). For a fuller account of Bahá'u'lláh's journey from Adrianople to the Most Great Prison in 'Akká, see GPB, pp. 178–82, and H. M. Balyuzi, *Bahá'u'lláh: The King of Glory*, pp. 255–68.

63-2. Bahá'u'lláh's chief adversaries were Sulṭán 'Abdu'l-'Azíz of Turkey and Náṣiri'd-Dín Sh̲áh of Persia. Bahá'u'lláh's banishments were to Baghdad, Iraq, January 1853–April 1863; to Constantinople (now Istanbul), Turkey, August–December 1863; to Adrianople (now Edirne), Turkey, December 1863–August 1868; and to 'Akká, Palestine, August 1868. The "full flood" of Bahá'u'lláh's utterances refers to, among other things, His letters to the kings and rulers of the world.

63-3. GPB, p. 253.

63-4. Bahá'u'lláh revealed the Kitáb-i-Aqdas, the chief repository of His laws and the Mother Book of His Revelation, in 'Akká circa 1873.

63-5. GPB, p. 196.

1964–1973 · THE NINE YEAR PLAN

Our Part—Building Bahá'u'lláh's Order

Contemplating this awe-inspiring, supernal episode, we may obtain a clearer understanding of our own times, a more confident view of their outcome and a deeper apprehension of the part we are called upon to play. That the violent disruption which has seized the entire planet is beyond the ability of men to assuage, unaided by God's revelation, is a truth repeatedly and forcibly set forth in our Writings. The old order cannot be repaired; it is being rolled up before our eyes. The moral decay and disorder convulsing human society must run their course; we can neither arrest nor divert them. 63.4

Our task is to build the Order of Bahá'u'lláh. Undeflected by the desperate expedients of those who seek to subdue the storm convulsing human life by political, economic, social or educational programs, let us, with single-minded devotion and concentrating all our efforts on our objective, raise His Divine System and sheltered within its impregnable stronghold, safe from the darts of doubtfulness, demonstrate the Bahá'í way of life. Wherever a Bahá'í community exists, whether large or small, let it be distinguished for its abiding sense of security and faith, its high standard of rectitude, its complete freedom from all forms of prejudice, the spirit of love among its members and for the closely knit fabric of its social life. The acute distinction between this and present-day society will inevitably arouse the interest of the more enlightened, and as the world's gloom deepens the light of Bahá'í life will shine brighter and brighter until its brilliance must eventually attract the disillusioned masses and cause them to enter the haven of the Covenant of Bahá'u'lláh, Who alone can bring them peace and justice and an ordered life. 63.5

The Mediterranean—Past History and Spiritual Potential

The great sea, on one of whose chief islands you are now gathered, within whose hinterland and islands have flourished the Jewish, the Christian and Islamic civilizations is a befitting scene for the first Oceanic Bahá'í Conference. Two millenniums ago, in this arena, the disciples of Christ performed such deeds of heroism and self-sacrifice as are remembered to this day and are forever enshrined in the annals of His Cause. A thousand years later the lands, bordering the southern and western shores of this sea witnessed the glory of Islam's Golden Age.[6] 63.6

In the day of the Promised One this same sea achieved eternal fame through its association with the Heroic and Formative Ages of His Cause. It bore upon its bosom the King of kings Himself; the Center of His Covenant crossed and recrossed it in the course of His epoch-making journeys to the West, during which He left the indelible imprint of His presence upon European and African lands; the Sign of God on earth frequently journeyed upon it.[7] It 63.7

63-6. The classical age of Islamic civilization, the eighth through thirteenth centuries.

63-7. Bahá'u'lláh sailed upon the Mediterranean Sea in 1868 during His journey from Gallipoli to 'Akká. 'Abdu'l-Bahá, the Center of Bahá'u'lláh's Covenant, accompanied Bahá'u'lláh on

enshrines within its depths the mortal remains of the Hand of the Cause of God Dorothy Baker and around its shores lies the dust of apostles, martyrs and pioneers. Forty-six Knights of Bahá'u'lláh are identified with seven of its islands and five of its territories. Through such and many other episodes, Mediterranean lands—ancient home of civilizations—have been endowed with spiritual potentiality to dissolve the encrustations of those once glorious but now moribund social orders and to radiate once again the light of Divine guidance.

63.8 Through dedicated, heroic and sacrificial deeds during the course of the beloved Guardian's ministry, the Faith of Bahá'u'lláh was established in this area. Eight pillars of the Universal House of Justice were raised, the first of an even larger number to be established now and during the course of future plans, to include, as envisioned by Shoghi Effendi, National Spiritual Assemblies in major islands of that historic sea.

The Need for a Dramatic Upsurge in Effective Teaching

63.9 The timing of such exciting developments is dependent upon the outcome of the Nine Year Plan. At this midway point of that Plan, although great strides have been made, more than half the goals are still to be won. The greatest deficiencies are in the opening of new centers where Bahá'ís reside and the formation of Local Spiritual Assemblies, which inevitably affects the ability to establish National Spiritual Assemblies. A dramatic upsurge of teaching—effective teaching—is necessary to make up the leeway; pioneers are needed, teachers must travel, funds must be provided. It is our hope that there will be engendered at this Conference, through your enthusiasm, prayers and spirit of devotion, a great spiritual dynamic to reinforce that grand momentum which, mounting steadily during the next four years, must carry the community of the Most Great Name to overwhelming victory in 1973.

63.10 Dear friends, within a few short days the observance of the Centenary of Bahá'u'lláh's arrival in the Holy Land will take place. The hearts and minds of the entire Bahá'í world will be focused on the Most Holy Shrine, where those privileged to attend this commemoration will circumambulate that Holy Spot and raise their prayers to the Lord of the Age.[8] Let them remember their fellow-believers at home and supplicate from the depths of their souls for such bounties and favors to descend upon the friends of God everywhere

---

that journey and later sailed upon the Mediterranean in the course of His travels to Egypt, Europe, and North America, 1910–13. Shoghi Effendi, the Sign of God on earth, traversed the Mediterranean in his travels to England to study at Oxford University and in the course of later visits to Europe.

63-8. The Centenary of Bahá'u'lláh's arrival in the Holy Land was observed at the Bahá'í World Center 26–31 August 1968. About 1,800 Bahá'ís who attended the First Oceanic Conference in Palermo, Sicily, 23–25 August 1968, went to Israel to participate in the commemoration. For an account of the commemoration, see BW 15:81–86.

as to cause them to rise as one man to demonstrate their love for Him Who suffered for them, by such deeds of sacrifice and devotion as shall outshine the deeds of the past and sweep away every obstacle from the onward march of the Cause of God.

With loving Bahá'í greetings,
THE UNIVERSAL HOUSE OF JUSTICE

64
Passing of Dr. Luṭfu'lláh Ḥakím, Member of the Universal House of Justice
12 AUGUST 1968

To all National Spiritual Assemblies

Dear Bahá'í Friends,

We share with you the following cable which we have just sent to the National Spiritual Assembly of Persia: 64.1

> GRIEVE ANNOUNCE PASSING LUṬFU'LLÁH ḤAKÍM DEDICATED SERVANT CAUSE GOD.[1] SPECIAL MISSIONS ENTRUSTED HIM FULL CONFIDENCE REPOSED IN HIM BY MASTER AND GUARDIAN HIS CLOSE ASSOCIATION WITH EARLY DISTINGUISHED BELIEVERS EAST WEST INCLUDING HIS COLLABORATION ESSLEMONT HIS SERVICES PERSIA BRITISH ISLES HOLY LAND HIS MEMBERSHIP APPOINTED AND ELECTED INTERNATIONAL BAHÁ'Í COUNCIL HIS ELECTION UNIVERSAL HOUSE JUSTICE WILL ALWAYS BE REMEMBERED IMMORTAL ANNALS FAITH BAHÁ'U'LLÁH. INFORM BELIEVERS HOLD BEFITTING MEMORIAL MEETINGS ALL CENTERS. CONVEY ALL MEMBERS HIS FAMILY EXPRESSIONS LOVING SYMPATHY ASSURANCE PRAYERS PROGRESS HIS RADIANT SOUL ABHÁ KINGDOM.

In view of Dr. Ḥakím's long and devoted record of services to the Faith other National Spiritual Assemblies are requested to hold memorial gatherings. Special commemorative services should also be held in the four Mother Temples of the Bahá'í World.[2] 64.2

With loving Bahá'í greetings,
THE UNIVERSAL HOUSE OF JUSTICE

64-1. For an account of the life and services of Luṭfu'lláh Ḥakím, see BW 15:430–34. See also message no. 48.

64-2. Wilmette, near Chicago, U.S.A.; Ingleside, near Sydney, Australia; Kampala, Uganda; Langenhain, near Frankfurt am Main, Germany.

65
Passing of the Hand of the Cause of God Ṭarázu'lláh Samandarí
4 September 1968

To all National Spiritual Assemblies

65.1 WITH SORROWFUL HEARTS ANNOUNCE PASSING HAND CAUSE GOD SHIELD HIS FAITH DEARLY LOVED ṬARÁZU'LLÁH SAMANDARÍ NINETY-THIRD YEAR HIS LIFE ON MORROW COMMEMORATION CENTENARY BAHÁ'U'LLÁH'S ARRIVAL HOLY LAND.[1] FAITHFUL TO LAST BREATH INSTRUCTIONS HIS LORD HIS MASTER HIS GUARDIAN HE CONTINUED SELFLESS DEVOTED SERVICE UNABATED UNTIL FALLING ILL DURING RECENT TEACHING MISSION. UNMINDFUL ILLNESS HE PROCEEDED HOLY LAND PARTICIPATE CENTENARY. EVER REMEMBERED HEARTS BELIEVERS EAST WEST TO WHOSE LANDS HE TRAVELED BEARING MESSAGE HIS LORD WHOSE COMMUNITIES HE FAITHFULLY SERVED THIS PRECIOUS REMNANT HEROIC AGE WHO ATTAINED PRESENCE BLESSED BEAUTY YEAR HIS ASCENSION NOW LAID REST FOOT MOUNTAIN GOD AMIDST THRONG BELIEVERS ASSEMBLED VICINITY VERY SPOT BAHÁ'U'LLÁH FIRST TROD THESE SACRED SHORES. REQUEST ALL NATIONAL ASSEMBLIES HOLD MEMORIAL SERVICES INCLUDING FOUR MOTHER TEMPLES BAHÁ'Í WORLD BEFITTING LONG LIFE DEDICATED EXEMPLARY SERVICE LORD HOSTS BY ONE ASSURED CENTER COVENANT LOVING WELCOME PRESENCE BAHÁ'U'LLÁH ABHÁ KINGDOM.[2] EXTEND LOVING SYMPATHY ASSURANCE PRAYERS MEMBERS DISTINGUISHED FAMILY.

UNIVERSAL HOUSE OF JUSTICE

66
Results of the Palermo Conference
8 September 1968

To the Bahá'ís of the World

Dear Bahá'í Friends,

66.1 The glorious Conference in Palermo concluded with a burst of eager enthusiasm of determined and dedicated believers who have pledged to do their part in winning the remaining goals of the Nine Year Plan. More than 125 offered to pioneer and more than 100 volunteered to do travel teaching. In addition, there was a generous outpouring of material resources to finance

---

65-1. For an account of the life and services of Ṭarázu'lláh Samandarí, see BW 15:410–16.

65-2. Memorial services were held in the Houses of Worship in Wilmette, near Chicago, U.S.A.; Ingleside, near Sydney, Australia; Kampala, Uganda; and Langenhain, near Frankfurt am Main, Germany.

teaching projects. Had the entire Bahá'í world been able to participate in the Mediterranean Conference we have no doubt that all the goals would be quickly won.

With this in mind we wish to impress upon the friends who could not attend the Conference, and who will surely—through reports and personal contact with those who did—sense the enthusiasm generated there, that all believers have the privilege to share in the pioneering work, in the travel teaching program and in contributing to the Fund. 66.2

We announced at the Conference that the International Deputization Fund, so far used to aid pioneering and travel teaching on an international level, will henceforth be available to assist such projects on the national level in those areas where support is vitally important to the winning of the goals of the Nine Year Plan. We are concerned that, although we are now approaching the midway point of the Plan we must yet form an additional 6,997 Local Spiritual Assemblies (76% of the goal), and take the Faith to over 22,800 new localities (59% of the goal). Obviously, hundreds of pioneers and traveling teachers will be required, many of whom will serve in their own countries. 66.3

Those who cannot pioneer or do travel teaching will want to participate by contributing to the International Deputization Fund. Let them remember Bahá'u'lláh's injunction: "Center your energies in the propagation of the Faith of God. Whoso is worthy of so high a calling, let him arise and promote it. Whoso is unable, it is his duty to appoint him who will, in his stead, proclaim this Revelation . . ."[1] Let the Bahá'ís of the world join in the true spirit of universal participation and win all the victories while there is yet time. Let each assume his full measure of responsibility that all may share the laurels of accomplishment at the end of the Plan. 66.4

Our fervent prayer is that this one-hundredth anniversary of the final banishment of Bahá'u'lláh will mark a significant turning-point in the fortunes of the Nine Year Plan.[2] 66.5

With loving Bahá'í greetings,
THE UNIVERSAL HOUSE OF JUSTICE

66-1. GWB, pp. 196–97.
66-2. See the August 1968 message to the First Oceanic Conference—Palermo, Sicily, (no. 63) for an explanation of the final banishment of Bahá'u'lláh and its significance.

67
Letter to Youth—Pioneering and Education
9 October 1968

To the Bahá'í Youth in Every Land

Dear Bahá'í Friends,

67.1 In the two years since we last addressed the youth of the Bahá'í world many remarkable advances have been made in the fortunes of the Faith. Not the least of these is the enrollment under the banner of Bahá'u'lláh of a growing army of young men and women eager to serve His Cause. The zeal, the enthusiasm, the steadfastness and the devotion of the youth in every land has brought great joy and assurance to our hearts.

67.2 During the last days of August and the first days of September, when nearly two thousand believers from all over the world gathered in the Holy Land to commemorate the Centenary of Bahá'u'lláh's arrival on these sacred shores, we had an opportunity to observe at first hand those qualities of good character, selfless service and determined effort exemplified in the youth who served as volunteer helpers, and we wish to express our gratitude for their loving assistance and for their example.

67.3 Many of them offered to pioneer, but one perplexing question recurred: Shall I continue my education, or should I pioneer, now? Undoubtedly this same question is in the mind of every young Bahá'í wishing to dedicate his life to the advancement of the Faith. There is no stock answer which applies to all situations; the beloved Guardian gave different answers to different individuals on this question. Obviously circumstances vary with each individual case. Each individual must decide how he can best serve the Cause. In making this decision, it will be helpful to weigh the following factors:

67.3a
- Upon becoming a Bahá'í one's whole life is, or should become devoted to the progress of the Cause of God, and every talent or faculty he possesses is ultimately committed to this overriding life objective. Within this framework he must consider, among other things, whether by continuing his education now he can be a more effective pioneer later, or alternatively whether the urgent need for pioneers, while possibilities for teaching are still open, outweighs an anticipated increase in effectiveness. This is not an easy decision, since oftentimes the spirit which prompts the pioneering offer is more important than one's academic attainments.

67.3b
- One's liability for military service may be a factor in timing the offer of pioneer service.

- One may have outstanding obligations to others, including those who may be dependent on him for support. 67.3c

- It may be possible to combine a pioneer project with a continuing educational program. Consideration may also be given to the possibility that a pioneering experience, even though it interrupts the formal educational program, may prove beneficial in the long run in that studies would later be resumed with a more mature outlook. 67.3d

- The urgency of a particular goal which one is especially qualified to fill and for which there are no other offers. 67.3e

- The fact that the need for pioneers will undoubtedly be with us for many generations to come, and that therefore there will be many calls in future for pioneering service. 67.3f

- The principle of consultation also applies. One may have the obligation to consult others, such as one's parents, one's Local and National Assemblies, and the pioneering committees. 67.3g

- Finally, bearing in mind the principle of sacrificial service and the unfailing promises Bahá'u'lláh ordained for those who arise to serve His Cause, one should pray and meditate on what his course of action will be. Indeed, it often happens that the answer will be found in no other way. 67.3h

We assure the youth that we are mindful of the many important decisions they must make as they tread the path of service to Bahá'u'lláh. We will offer our ardent supplications at the Holy Threshold that all will be divinely guided and that they will attract the blessings of the All-Merciful. 67.4

Deepest Bahá'í love,
THE UNIVERSAL HOUSE OF JUSTICE

68
Riḍván Message 1969
RIḌVÁN 126 [1969 A.D.]

To the Bahá'ís of the World

Dearly loved Friends,

68.1 The continued progress of the Cause of God stands in vivid contrast to the chronic unrest afflicting human society, a contrast which the events of the past year, both within and without the Faith, have only served to intensify. Amidst the disintegration of the old order the Cause of God has pursued its majestic course, extending the range of its activities and influence and accomplishing a further development of its administrative system.

A Year of Remarkable Activity

68.2 Opening with the convening, in the Holy Land, of the Second International Convention for the election of the Universal House of Justice, the year has witnessed a remarkable activity in the Cause. The most significant and far-reaching development was undoubtedly the appointment of the eleven Continental Boards of Counselors, which fulfilled the goal of the Nine Year Plan calling for the development of the Institution of the Hands of the Cause of God with a view to the extension into the future of its appointed functions of protection and propagation. This step, taken after full consultation with the Hands of the Cause, has, at one and the same time, strongly reinforced the activities of that Institution and made it possible for the Hands themselves to extend the range of their individual services beyond the continental sphere, thereby making universally available to the friends the love, the wisdom and the spirit of dedication animating the Guardian's appointees. We wish to pay tribute at this time to the exemplary manner in which the Counselors, under the guidance of the Hands, have embarked upon their high duties.

The Palermo Conference

68.3 In August, the first Oceanic Bahá'í Conference, held in Palermo, commemorated Bahá'u'lláh's voyage on the Mediterranean Sea on His way to the Most Great Prison. Attendants at this Conference came immediately afterwards to the Qiblih of their Faith to pay homage at the Shrine of its Founder[1] and to commemorate with deep awareness of its spiritual import the long prophesied arrival of the Lord of Hosts on the shores of the Holy Land. This gathering of more than two thousand believers presented an inexpressibly poignant contrast to the actual arrival of Bahá'u'lláh one hundred years before,

---

68-1. The Qiblih (point of adoration) is the place toward which the faithful turn in prayer. The Qiblih for Bahá'ís is the Shrine of Bahá'u'lláh at Bahjí, outside of 'Akká.

1964–1973 • THE NINE YEAR PLAN

rejected by the rulers of this earth and derided by the local populace. Such is the conquering power of His Message, such is the undefeatable might of the King of Kings.

68.4 That same message is now being proclaimed by His followers from end to end of the world. Already one hundred and twenty-two Heads of State have been presented with the special edition of *The Proclamation of Bahá'u'lláh*, and copies have been received by thousands more officials and leaders.[2]

68.5 Taking full advantage of the designation of 1968 as Human Rights Year by the United Nations, Bahá'í communities throughout the world have not only strengthened the ties between the Bahá'í International Community and the United Nations, but have at the same time proclaimed the Faith and its healing message. In country after country the Cause has been featured for the first time in modern mass communications media. The volume of this call to the peoples of the world is increasing day by day and must so continue, penetrating every stratum of society, until the conclusion of the Plan and beyond.

Eight Oceanic and Continental Conferences

68.6 As a stimulus and aid to this vital work as well as to the promotion of all the goals of the Plan, we announce the holding between August 1970 and September 1971 of a series of eight Oceanic and Continental Conferences, as follows: La Paz, Bolivia, and Rose Hill, Mauritius, in August 1970; Monrovia, Liberia, and Djakarta, Indonesia, in January 1971; Suva, Fiji, and Kingston, Jamaica, in May 1971; Sapporo, Japan, and Reykjavik, Iceland, in September 1971.

Areas of Progress in the Nine Year Plan

68.7 A review of the progress of the Nine Year Plan discloses that great strides have been made in the acquisition of Ḥaẓíratu'l-Quds, Temple sites and Teaching Institutes, in translation of Bahá'í literature into more languages and in the incorporation of Local and National Spiritual Assemblies. The site of the Panama Temple has been prepared for construction which will begin as soon as final plans and specifications and the placing of the contract have been approved.

Formation of Twelve National Spiritual Assemblies

68.8 As a result of the accelerated pace of expansion and consolidation which has been initiated, and which, if fostered and fed, will become a full tide of victorious achievement, we joyfully announce the formation of twelve more National Spiritual Assemblies, two during Riḍván 1969: the National Spiritual

---

68-2. The Bahá'í World Center published Bahá'u'lláh's messages to the kings, rulers, religious leaders, and peoples of the world under the title *The Proclamation of Bahá'u'lláh;* a special edition was presented to heads of state. See the Riḍván 1967 message (no. 42) for more details. For an account of the delivery of *The Proclamation of Bahá'u'lláh* to heads of state and to humanity in general, see BW 14:204–20.

Assembly of the Bahá'ís of Burundi and Rwanda with its seat in Bujumbura and the National Spiritual Assembly of the Bahá'ís of Papua and New Guinea with its seat in Lae, and ten during Riḍván 1970: six in Africa, the National Spiritual Assemblies of the Bahá'ís of the Congo Republic (Kinshasa); Ghana; Dahomey, Togo and Niger; Malawi; Botswana; and Gambia, Senegal, Portuguese Guinea and the Cape Verde Islands; one in the Americas, the National Spiritual Assembly of the Bahá'ís of the Guianas; one in Asia, the National Spiritual Assembly of the Bahá'ís of the Near East; and two in Australasia, the National Spiritual Assemblies of the Bahá'ís of Tonga and the Cook Islands; and Samoa. Thus at Riḍván 1970 the number of National Spiritual Assemblies will be raised to ninety-three.

Opening the Door of Pilgrimage

68.9   In harmony with the worldwide growth of the Cause the World Center of the Faith is also developing rapidly. The pilgrims, the beloved Guardian has said, are the lifeblood of this World Center and it has long been our cherished hope and desire to be able to grant the bounty of pilgrimage to the Holy Land to all who can avail themselves of it. It is therefore with great joy that we now find it possible to open the door of pilgrimage to a much greater number of believers. Beginning in October of this year the size of each group of friends to be invited will be quadrupled and the number of groups each year will be increased so that nearly six times the present number of pilgrims will have the opportunity each year to pray in the Shrines of the Central Figures of their Faith, to visit the places hallowed by the footsteps, sufferings and triumphs of Bahá'u'lláh and 'Abdu'l-Bahá, and to meditate in the tranquillity of these sacred precincts, beautified with so much loving care by our beloved Guardian.

68.10   This increased flow of pilgrims will greatly augment the spiritual development of the Bahá'í World Community which now, after five years of strenuous labor and bearing the laurels of outstanding victories, is entering the fourth phase of the Nine Year Plan.

The Great Need for More Believers, Localities, Assemblies

68.11   The great, the most pressing need, at this stage of the Plan, is a rapid increase in the number of believers, and a major advance in the opening of the additional localities as well as in the formation of the well-grounded Local Spiritual Assemblies called for in the Plan. This worldwide activity, the hallmark of the fourth phase of the Plan, answering the tremendous opportunities offered by the present condition of mankind, will be strongly reinforced by the continuance of proclamation, is the essential foundation for the erection of the remaining National Spiritual Assemblies, and will increasingly witness to the benefits of international traveling teaching and inter-Assembly cooperation. Above all, it requires a sacrificial outpouring by the friends of contributions in support of the Funds of the Faith, and the raising up of a mighty host of pioneers.

During the second year of the Plan the Bahá'í world achieved its greatest feat of organized pioneering when a total of five hundred and five believers arose to settle in the unopened and weakly held territories of the earth.³ This magnificent achievement must now be surpassed. The call is raised for seven hundred and thirty-three believers to leave their homes and settle in territories of the globe in dire need of pioneer support or as yet unopened to the Faith. These devoted believers, who should arise without delay, are needed to settle, during the fourth phase of the Plan, in 184 specified territories of the globe: 48 in Africa, 40 in the Americas, 40 in Asia, 18 in Australasia and 38 in Europe. Although primary responsibility has been assigned to those national Bahá'í communities most able to provide pioneers, all should ponder in their hearts whether they too cannot respond to this call, either by going themselves or by deputizing, in response to Bahá'u'lláh's injunction, those who can go in their stead. Detailed information is being sent to National Spiritual Assemblies to ensure that this vital mobilization of Bahá'í warriors is accomplished as quickly as possible.

Our Commitment to Complete Victory

Beloved Friends, the Nine Year Plan is well advanced, our work is blessed by the never-ceasing confirmations of Bahá'u'lláh, and the entire Bahá'í World Community is committed to complete victory. That happy consummation, now faintly discernible on the far horizon, will be reached through hard work, realistic planning, sacrificial deeds, intensification of the teaching work and, above all, through constant endeavor on the part of every single Bahá'í to conform his inner life to that glorious ideal set for mankind by Bahá'u'lláh and exemplified by 'Abdu'l-Bahá. In contemplating the Master's divine example we may well reflect that His life and deeds were not acted to a pattern of expediency, but were the inevitable and spontaneous expression of His inner self. We, likewise, shall act according to His example only as our inward spirits, growing and maturing through the disciplines of prayer and practice of the Teachings, become the wellsprings of all our attitudes and actions. This will promote the accomplishment of God's purpose; this will ensure the triumph of His Faith and enable us to build up the present motion of the Cause into a grand momentum whose force will carry the community of the Most Great Name to glorious victory in 1973 and onwards to the as yet unapprehended vistas of the Most Great Peace.

THE UNIVERSAL HOUSE OF JUSTICE

68-3. April 1965–April 1966.

69
Guidance on Self-Defense
26 May 1969

The National Spiritual Assembly of the Bahá'ís of Canada

Dear Bahá'í Friends,

69.1 We have reviewed your letter of April 11th, asking about the teachings of the Faith on self-defense and any guidance on individual conduct in the face of increasing civil disorder in North American cities.

69.2 From the texts you already have available it is clear that Bahá'u'lláh has stated that it is preferable to be killed in the path of God's good pleasure than to kill, and that organized religious attack against Bahá'ís should never turn into any kind of warfare, as this is strictly prohibited in our Writings.

69.3 A hitherto untranslated Tablet from 'Abdu'l-Bahá, however, points out that in the case of attack by robbers and highwaymen, a Bahá'í should not surrender himself, but should try, as far as circumstances permit, to defend himself, and later on lodge a complaint with the government authorities. In a letter written on behalf of the Guardian, he also indicates that in an emergency when there is no legal force at hand to appeal to, a Bahá'í is justified in defending his life. In another letter the Guardian has further pointed out that the assault of an irresponsible assailant upon a Bahá'í should be resisted by the Bahá'í, who would be justified, under such circumstances, in protecting his life.

69.4 The House of Justice does not wish at the present time to go beyond the guidelines given in the above-mentioned statements. The question is basically a matter of conscience, and in each case the Bahá'í involved must use his judgment in determining when to stop in self-defense lest his action deteriorate into retaliation.

69.5 Of course the above principles apply also in cases when a Bahá'í finds himself involved in situations of civil disorder. We have, however, advised the National Spiritual Assembly of the United States that under the present circumstances in that country it is preferable that Bahá'ís do not buy nor own arms for their protection or the protection of their families.

With loving Bahá'í greetings,
THE UNIVERSAL HOUSE OF JUSTICE

70

New Appointments to Continental Boards of Counselors

10 JULY 1969

To National Spiritual Assemblies

Dear Bahá'í Friends,

With great joy we announce that we have decided to increase the total number of members of the Continental Boards of Counselors for the Protection and Propagation of the Faith to thirty-eight by adding John McHenry III to the Continental Board of Counselors in North East Asia and Mas'úd Khamsí to the Continental Board of Counselors in South America, raising the number of Counselors on each Board to three and four, respectively. 70.1

We also rejoice to announce the appointment of Mrs. Shirin Boman to the Continental Board of Counselors of Western Asia to fill a vacancy on that Board. 70.2

The devoted efforts of all eleven Continental Boards of Counselors during the first year of their service to the Faith of Bahá'u'lláh have been most exemplary and praiseworthy. We are deeply grateful for the loyalty, steadfastness and devotion which have characterized the activities of all members in reinforcing the vitally important work of the Hands of the Cause of God. 70.3

Please share these glad tidings with the friends. 70.4

With loving Bahá'í greetings,
THE UNIVERSAL HOUSE OF JUSTICE

71

Formation of an Additional National Spiritual Assembly during Riḍván 1970

11 AUGUST 1969

To all National Spiritual Assemblies

Dear Bahá'í Friends,

In the brief space of time following the announcement of the formation of six new National Spiritual Assemblies in Africa next Riḍván, the succession of victories, resulting from the prodigious efforts exerted by the devoted friends, impels us to announce that a seventh National Spiritual Assembly will be formed in Africa at Riḍván, 1970. The new National Spiritual Assembly including Congo (Brazzaville), Chad, Central African Republic and Gabon, 71.1

will have its seat in Bangui. This will leave Uganda with its own separate National Spiritual Assembly.

72.2 Please share this joyous news with the believers. We know the friends throughout the world join us in our supplications for the continued, uninterrupted prosecution and speedy fulfillment of the goals, terminating in the ultimate triumph of the Cause of Bahá'u'lláh.

With loving Bahá'í greetings,
THE UNIVERSAL HOUSE OF JUSTICE

72
Work of Continental Boards of Counselors and Their Auxiliary Board Members
1 OCTOBER 1969

To the Continental Boards of Counselors and National Spiritual Assemblies

Dear Bahá'í Friends,

72.1 A number of questions have been raised concerning the work of the Counselors and Auxiliary Board members, and it has been suggested that Auxiliary Board members be permitted to work regularly with National Spiritual Assemblies and national committees. We have carefully considered again the various factors involved and have decided that we must uphold the principle that such direct consultations should be exceptional rather than the rule.

**Spiritual Assemblies and Auxiliary Board Members:
A Clarification of Roles**

72.2 It is the responsibility of Spiritual Assemblies, assisted by their committees, to organize and direct the teaching work, and in doing so they must, naturally, also do all they can to stimulate and inspire the friends. It is, however, inevitable that the Assemblies and committees, being burdened with the administration of the teaching work as well as with all other aspects of Bahá'í community life, will be unable to spend as much time as they would wish on stimulating the believers.

72.3 Authority and direction flow from the Assemblies, whereas the power to accomplish the tasks resides primarily in the entire body of the believers. It is the principal task of the Auxiliary Boards to assist in arousing and releasing this power. This is a vital activity, and if they are to be able to perform it adequately they must avoid becoming involved in the work of administration. For example, when Auxiliary Board members arouse believers to pioneer, any believer who expresses his desire to do so should be referred to the appropriate committee which will then organize the project. Counselors and Auxiliary

Board members should not, themselves, organize pioneering or travel teaching projects. Thus it is seen that the Auxiliary Boards should work closely with the grass roots of the community: the individual believers, groups and Local Spiritual Assemblies, advising, stimulating and assisting them. The Counselors are responsible for stimulating, counseling and assisting National Spiritual Assemblies, and also work with individuals, groups and Local Assemblies.

It is always possible, of course, for Counselors to depute an Auxiliary Board member to meet with a National Spiritual Assembly for a particular purpose, but this should not become a regular practice. Similarly, if the National Spiritual Assembly agrees, it may be advisable for an Auxiliary Board member to meet occasionally with a national committee to clarify the situation in the area and share information and ideas thoroughly. But this also should not become regular. Were it to do so there would be grave danger of inhibiting the proper working of these two institutions, vitiating and undermining the collaboration that must essentially exist between the Continental Boards of Counselors and National Spiritual Assemblies. It would diffuse the energies and time of the Auxiliary Board members through their becoming involved in the administration of teaching. It could lead to the Auxiliary Board member's gradually taking over the direction of the national committee, usurping the function of the National Assembly, or to his becoming merely a traveling teacher sent hither and thither at the direction of the committee or National Assembly. 72.4

Sharing Information, Reports, and Recommendations

It is, of course vital that information be shared fully and promptly, as has been explained in the compilation on the work of Auxiliary Board members that was circulated on March 25, 1969. The ways of ensuring this should be worked out by the Counselors and National Spiritual Assemblies and methods may vary from area to area. 72.5

Reports and recommendations for action, however, are quite different. Auxiliary Board members should send theirs to the Counselors and not to National Assemblies or national committees directly. It is possible that the Counselors may reject or modify the recommendation; or, if they accept it and pass it on to the National Spiritual Assembly, the National Assembly may decide to refuse it. For an Auxiliary Board member to make recommendations directly to a national committee would lose the benefit of knowledge and experience in a wider field than that of which the Auxiliary Board member is aware, and would short-circuit and undermine the authority of both the Counselors and the National Assembly. 72.6

Advice—The Province of Counselors and Auxiliary Board

Similarly, although an Auxiliary Board member can and should receive information from the National Assemblies and national committees, his primary source of information about the community should be his own direct 72.7

contacts with Local Spiritual Assemblies, groups and individual believers. In this way the Counselors as well as the National Spiritual Assemblies have the benefit of two independent sources of information about the community: through the Auxiliary Board members on the one hand, and through the national committees on the other.

72.8 Assemblies sometimes misunderstand what is meant by the statement that Counselors and Auxiliary Board members are concerned with the teaching work and not with administration. It is taken to mean that they may not give advice on administrative matters. This is quite wrong. One of the things that Counselors and Auxiliary Board members should watch and report on is the proper working of administrative institutions. The statement that they do not have anything to do with administration means, simply, that they do not administer. They do not direct or organize the teaching work nor do they adjudicate in matters of personal conflict or personal problems. All these activities fall within the sphere of responsibility of the Spiritual Assemblies. But if an Auxiliary Board member finds a Local Spiritual Assembly functioning incorrectly he should call its attention to the appropriate Texts; likewise if, in his work with the community, an Auxiliary Board member finds that the teaching work is being held up by inefficiency of national committees, he should report this in detail to the Counselors who will then decide whether to refer it to the National Spiritual Assembly concerned. Similarly, if the Counselors find that a National Spiritual Assembly is not functioning properly, they should not hesitate to consult with the National Spiritual Assembly about this in a frank and loving way.

72.9 It is the Spiritual Assemblies who plan and direct the work, but these plans should be well known to the Counselors and Auxiliary Board members, because one of the ways in which they can assist the Assemblies is by urging the believers continually to support the plans of the Assemblies. If a National Spiritual Assembly has adopted one goal as preeminent in a year, the Auxiliary Board members should bear this in mind in all their contacts with the believers and should direct their attention to the plans of the National Assembly, and stimulate them to enthusiastically support them.

72.10 The Counselors in each continental zone have wide latitude in the carrying out of their work. Likewise they should give to each Auxiliary Board member considerable freedom of action within his own allocated area. Although the Counselors should regularly direct the work of the Auxiliary Board members, the latter should realize that they need not wait for direction; the nature of their work is such that they should be continually engaged in it according to their own best judgment, even if they are given no specific tasks to perform. Above all the Auxiliary Board members should build up a warm and loving relationship between themselves and the believers in their area so that the Local Spiritual Assemblies will spontaneously turn to them for advice and assistance.

We assure you all of our fervent prayers in the Holy Shrines for the blessings of Bahá'u'lláh upon the strenuous and highly meritorious services that you are performing with such devotion in His path. 72.11

THE UNIVERSAL HOUSE OF JUSTICE

73

Appeal to Increase Teaching Efforts amidst Catastrophic Events of the Day
16 NOVEMBER 1969

To the Bahá'ís of the World

Dear Friends,

In the worsening world situation, fraught with pain of war, violence and the sudden uprooting of long-established institutions, can be seen the fulfillment of the prophecies of Bahá'u'lláh and the oft-repeated warnings of the Master and the beloved Guardian about the inevitable fate of a lamentably defective social system, an unenlightened leadership and a rebellious and unbelieving humanity.[1] Governments and peoples of both the developed and developing nations, and other human institutions, secular and religious, finding themselves helpless to reverse the trend of the catastrophic events of the day, stand bewildered and overpowered by the magnitude and complexity of the problems facing them. At this fateful hour in human history many, unfortunately, seem content to stand aside and wring their hands in despair or else join in the babel of shouting and protestation which loudly objects, but offers no solution to the woes and afflictions plaguing our age. 73.1

Nevertheless a greater and greater number of thoughtful and fair-minded men and women are recognizing in the clamor of contention, grief and 73.2

73-1. The years 1968 and 1969 were racked with war, violence, terrorism, and civil unrest around the world. Wars raged in Vietnam and Nigeria; Soviet and Chinese troops skirmished in a continuing border dispute; Soviet troops entered Czechoslovakia to quell a movement toward liberalization; and El Salvador invaded Honduras. Coups d'état toppled governments in Iraq, Syria, Sierra Leone, Dahomey, the Congo, Mali, the Sudan, Libya, the Netherlands Antilles, Peru, Panama, and Bolivia; states of emergency were declared in Spain, Malaysia, and Chile; violent unrest occurred in West Germany, Spain, Bombay, Pakistan, Argentina, Kenya, and the United States; and student protests erupted in Paris, Mexico City, Czechoslovakia, Argentina, and the United States. United States and Israeli airliners were hijacked, and two Israeli airliners were attacked by terrorists. Moreover, a number of leaders were assassinated, including Somalian president Abdirascid Ali Scermarche; U.S. civil rights leader Martin Luther King, Jr.; U.S. presidential candidate Robert F. Kennedy; U.S. ambassador to Guatemala John Gordon Mein; and Mozambique Liberation Front leader Eduardo Chivambo Mondlane.

destruction, now reaching such horrendous proportions, the evidences of Divine chastisement, and turning their faces towards God are becoming increasingly receptive to His Word. Doubtless the present circumstances, though tragic and awful in their immediate consequences, are serving to sharpen the focus on the indispensability of the Teachings of Bahá'u'lláh to the needs of the present age, and will provide many opportunities to reach countless waiting souls, hungry and thirsty for Divine guidance.

73.3 It is these opportunities which we must seize before it is too late. What is needed now is the awakening of all believers to the immediacy of the challenge so that each may assume his share of the responsibility for taking the Teachings to all humanity. Universal participation, a salient objective of the Nine Year Plan, must be pressed toward attainment in every continent, country and island of the globe. Every Bahá'í, however humble or inarticulate, must become intent on fulfilling his role as a bearer of the Divine Message. Indeed, how can a true believer remain silent while around us men cry out in anguish for truth, love and unity to descend upon this world?

73.4 We all know how often the Master and the beloved Guardian called upon the friends to consciously strive to be more loving, more united, more dedicated and prayerful than ever before in order to overcome the atmosphere of present-day society which is unloving, disunited, careless of right and wrong and heedless of God. ". . . when we see the increasing darkness in the world today," the Guardian's secretary wrote on his behalf, "we can fully realize that unless the Message of Bahá'u'lláh reaches into the hearts of men and transforms them, there can be no peace and no spiritual progress in the future."[2]

The Necessity of Individual Teaching Goals

73.5 The Nine Year Plan is the current stage in the achievement of that sublime objective. It is now imperative for every Bahá'í to set for himself individual teaching goals. The admonition of 'Abdu'l-Bahá to lead at least one new soul to the Faith each year and the exhortation of Shoghi Effendi to hold a Bahá'í fireside in one's home every Bahá'í month are examples of individual goals. Many have capacities to do even more, but this alone will assure final and complete victory for the Plan.

73.6 We call upon the friends to join with us in prayer during the Feast of Sulṭán that we will all become so imbued with zeal, courage and enthusiasm that from this day to the end of the Nine Year Plan nothing will be able to stay the victorious onward march of the followers of the Most Great Name.[3] May our efforts be worthy of the blessings and confirmations of Bahá'u'lláh.

With loving Bahá'í greetings,
THE UNIVERSAL HOUSE OF JUSTICE

73-2. From a letter that was later published in the compilation "Living the Life."
73-3. The Feast of Sulṭán is the Feast of Sovereignty, 19 January.

74
Acquisition of Property Adjacent to Bahjí
18 November 1969

To National Spiritual Assemblies

Dear Bahá'í Friends,

Enclosed please find our letter of 16 November 1969 addressed to the Bahá'ís of the World.¹ Please share this letter with all believers in your jurisdiction as soon as possible.

After several years of protracted negotiations with agencies of the Israel Government both in Jerusalem and Haifa, an important property adjacent to Bahjí and embracing the Master's teahouse has been acquired. On 17 November we cabled the National Spiritual Assembly of the United States as follows:

> WITH GRATEFUL HEARTS ANNOUNCE SUCCESSFUL CONCLUSION FORMAL NEGOTIATIONS INITIATED NEARLY TWO DECADES AGO BY BELOVED GUARDIAN WITH AUTHORITIES STATE ISRAEL RESULTING OWNERSHIP VITALLY NEEDED PROPERTY SURROUNDING 'ABDU'L-BAHÁ'S TEAHOUSE IMMEDIATE NEIGHBORHOOD MOST HOLY TOMB FOUNDER FAITH. ACQUISITION MUCH DESIRED LAND EXTENDING GARDENS BAHJÍ FACILITATED THROUGH EXCHANGE PROPERTY DEDICATED SOME THIRTY-SIX YEARS AGO TO HOLY TOMB BAHÁ'U'LLÁH BY DEVOTED SERVANT CAUSE ḤÁJÍ 'ALÍ YAZDÍ.
>
> UNIVERSAL HOUSE OF JUSTICE

The successful conclusion of these negotiations initiated during the lifetime of the beloved Guardian was made possible through the acceptance by the Government, as even exchange, of an endowment property given to the Faith in 1933 by the late Ḥájí 'Alí Yazdí. The significance of the specific piece of land donated by this venerable soul becomes apparent when reading the following quotation from the IN MEMORIAM article about him in *The Bahá'í World*, Volume 9 [624–25]:

> He will forever be remembered, amongst other things, as the establisher of Bahá'í endowments in the vicinity of 'Akká through his gift of a tract of land dedicated to Bahá'u'lláh's Holy Tomb in Bahjí. . . .

It is a glowing tribute to the memory of this devoted servant of the Blessed Beauty that his gift should play such an important part in securing this valuable additional safeguard for the Most Holy Tomb.

Please also convey the news of this victory to the friends.

With loving Bahá'í greetings,
THE UNIVERSAL HOUSE OF JUSTICE

74-1. See message no. 73.

75
Comments on the Guardianship and the Universal House of Justice
7 December 1969

To an individual Bahá'í

Dear Bahá'í Friend,

75.1 Your recent letter, in which you share with us the questions that have occurred to some of the youth in studying "The Dispensation of Bahá'u'lláh," has been carefully considered, and we feel that we should comment both on the particular passage you mention and on a related passage in the same work, because both bear on the relationship between the Guardianship and the Universal House of Justice.

75.2 The first passage concerns the Guardian's duty to insist upon a reconsideration by his fellow members in the Universal House of Justice of any enactment which he believes conflicts with the meaning and departs from the spirit of the Sacred Writings. The second passage concerns the infallibility of the Universal House of Justice without the Guardian, namely Shoghi Effendi's statement that "Without such an institution [the Guardianship] . . . the necessary guidance to define the sphere of the legislative action of its elected representatives would be totally withdrawn."[1]

75.3 Some of the youth, you indicate, were puzzled as to how to reconcile the former of these two passages with such statements as that in the Will of 'Abdu'l-Bahá which affirms that the Universal House of Justice is "freed from all error."[2]

Seeking the Writings' Unity of Meaning

75.4 Just as the Will and Testament of 'Abdu'l-Bahá does not in any way contradict the Kitáb-i-Aqdas but, in the Guardian's words, "confirms, supplements, and correlates the provisions of the Aqdas," so the writings of the Guardian contradict neither the revealed Word nor the interpretations of the Master.[3] In attempting to understand the Writings, therefore, one must first realize that there is and can be no real contradiction in them, and in the light of this we can confidently seek the unity of meaning which they contain.

75.5 The Guardian and the Universal House of Justice have certain duties and functions in common; each also operates within a separate and distinct sphere. As Shoghi Effendi explained, ". . . it is made indubitably clear and evident that the Guardian of the Faith has been made the Interpreter of the Word

---

75-1. WOB, p. 148.
75-2. WT, p. 14.
75-3. WOB, p. 19.

and that the Universal House of Justice has been invested with the function of legislating on matters not expressly revealed in the teachings. The interpretation of the Guardian, functioning within his own sphere, is as authoritative and binding as the enactments of the International House of Justice, whose exclusive right and prerogative is to pronounce upon and deliver the final judgment on such laws and ordinances as Bahá'u'lláh has not expressly revealed." He goes on to affirm, "Neither can, nor will ever, infringe upon the sacred and prescribed domain of the other. Neither will seek to curtail the specific and undoubted authority with which both have been divinely invested." It is impossible to conceive that two centers of authority, which the Master has stated "are both under the care and protection of the Abhá Beauty, under the shelter and unerring guidance of His Holiness, the Exalted One," could conflict with one another, because both are vehicles of the same Divine Guidance.[4]

75.6 The Universal House of Justice, beyond its function as the enactor of legislation, has been invested with the more general functions of protecting and administering the Cause, solving obscure questions and deciding upon matters that have caused difference. Nowhere is it stated that the infallibility of the Universal House of Justice is by virtue of the Guardian's membership or presence on that body. Indeed, 'Abdu'l-Bahá in His Will and Shoghi Effendi in his "Dispensation of Bahá'u'lláh" have both explicitly stated that the elected members of the Universal House of Justice in consultation are recipients of unfailing Divine Guidance. Furthermore the Guardian himself in *The World Order of Bahá'u'lláh* asserted that "It must be also clearly understood by every believer that the institution of Guardianship does not under any circumstances abrogate, or even in the slightest degree detract from, the powers granted to the Universal House of Justice by Bahá'u'lláh in the Kitábu'l-Aqdas, and repeatedly and solemnly confirmed by 'Abdu'l-Bahá in His Will. It does not constitute in any manner a contradiction to the Will and Writings of Bahá'u'lláh, nor does it nullify any of His revealed instructions."[5]

75.7 While the specific responsibility of the Guardian is the interpretation of the Word, he is also invested with all the powers and prerogatives necessary to discharge his function as Guardian of the Cause, its Head and supreme protector. He is, furthermore, made the irremovable head and member for life of the supreme legislative body of the Faith. It is as the head of the Universal House of Justice, and as a member of that body, that the Guardian takes part in the process of legislation. If the following passage, which gave rise to your query, is considered as referring to this last relationship, you will see that there is no contradiction between it and the other texts: "Though the

---

75-4. WOB, pp. 149–50; WT, p. 11.
75-5. WOB, p. 8.

Guardian of the Faith has been made the permanent head of so august a body he can never, even temporarily, assume the right of exclusive legislation. He cannot override the decision of the majority of his fellow members, but is bound to insist upon a reconsideration by them of any enactment he conscientiously believes to conflict with the meaning and to depart from the spirit of Bahá'u'lláh's revealed utterances."[6]

75.8 Although the Guardian, in relation to his fellow members within the Universal House of Justice, cannot override the decision of the majority, it is inconceivable that the other members would ignore any objection he raised in the course of consultation or pass legislation contrary to what he expressed as being in harmony with the spirit of the Cause. It is, after all, the final act of judgment delivered by the Universal House of Justice that is vouchsafed infallibility, not any views expressed in the course of the process of enactment.

75.9 It can be seen, therefore, that there is no conflict between the Master's statements concerning the unfailing divine guidance conferred upon the Universal House of Justice and the above passage from "The Dispensation of Bahá'u'lláh."

The Process of Legislation

75.10 It may help the friends to understand this relationship if they are aware of some of the processes that the Universal House of Justice follows when legislating. First, of course, it observes the greatest care in studying the Sacred Texts and the interpretations of the Guardian as well as considering the views of all the members. After long consultation the process of drafting a pronouncement is put into effect. During this process the whole matter may well be reconsidered. As a result of such reconsideration the final judgment may be significantly different from the conclusion earlier favored, or possibly it may be decided not to legislate at all on that subject at that time. One can understand how great would be the attention paid to the views of the Guardian during the above process were he alive.

The Universal House of Justice in the Absence of the Guardian

75.11 In considering the second passage we must once more hold fast to the principle that the teachings do not contradict themselves.

75.12 Future Guardians are clearly envisaged and referred to in the Writings, but there is nowhere any promise or guarantee that the line of Guardians would endure forever; on the contrary there are clear indications that the line could be broken. Yet, in spite of this, there is a repeated insistence in the Writings on the indestructibility of the Covenant and the immutability of God's Purpose for this Day.

75-6. WOB, p. 150.

One of the most striking passages which envisage the possibility of such a 75.13
break in the line of Guardians is in the Kitáb-i-Aqdas itself:

> The endowments dedicated to charity revert to God, the Revealer of Signs. No one has the right to lay hold on them without leave from the Dawning-Place of Revelation.[7] After Him the decision rests with the Aghsán [Branches],[8] and after them with the House of Justice—should it be established in the world by then—so that they may use these endowments for the benefit of the Sites exalted in this Cause, and for that which they have been commanded by God, the Almighty, the All-Powerful. Otherwise the endowments should be referred to the people of Bahá, who speak not without His leave and who pass no judgment but in accordance with that which God has ordained in this Tablet, they who are the champions of victory betwixt heaven and earth, so that they may spend them on that which has been decreed in the Holy Book by God, the Mighty, the Bountiful.[9]

The passing of Shoghi Effendi in 1957 precipitated the very situation pro- 75.14
vided for in this passage, in that the line of Aghsán ended before the House of Justice had been elected. Although, as is seen, the ending of the line of Aghsán at some stage was provided for, we must never underestimate the grievous loss that the Faith has suffered. God's purpose for mankind remains unchanged, however, and the mighty Covenant of Bahá'u'lláh remains impregnable. Has not Bahá'u'lláh stated categorically, "The Hand of Omnipotence hath established His Revelation upon an unassailable, an enduring foundation."[10] While 'Abdu'l-Bahá confirms: "Verily, God effecteth that which He pleaseth; naught can annul His Covenant; naught can obstruct His favor nor oppose His Cause!" "Everything is subject to corruption; but the Covenant of thy Lord shall continue to pervade all regions." "The tests of every dispensation are in direct proportion to the greatness of the Cause, and as heretofore such a manifest Covenant, written by the Supreme Pen, hath not been entered upon, the tests are proportionately severe. . . . These agitations of the violators are no more than the foam of the ocean, . . . This foam of the ocean shall not endure and shall soon disperse and vanish, while the ocean of the Covenant shall eternally surge and roar."[11] And Shoghi Effendi has clearly stated: "The bedrock on which this Administrative Order is founded is God's immutable Purpose for mankind in this day." ". . . this priceless gem of Divine

---

75-7. "The Dawning-Place of Revelation" is a reference to the Manifestation of God; here, a specific reference to Bahá'u'lláh.
75-8. Aghsán (Branches) denotes the sons and male descendants of Bahá'u'lláh.
75-9. See KA ¶42.
75-10. WOB, p. 109.
75-11. TABA 2:598; *Star of the West*, vol. 4, no. 10, p. 170; SWAB, pp. 210-11.

Revelation, now still in its embryonic state, shall evolve within the shell of His law, and shall forge ahead, undivided and unimpaired, till it embraces the whole of mankind."[12]

Two Authoritative Centers

75.15 In the Bahá'í Faith there are two authoritative centers appointed to which the believers must turn, for in reality the Interpreter of the Word is an extension of that center which is the Word itself. The Book is the record of the utterance of Bahá'u'lláh, while the divinely inspired Interpreter is the living Mouth of that Book—it is he and he alone who can authoritatively state what the Book means. Thus one center is the Book with its Interpreter, and the other is the Universal House of Justice guided by God to decide on whatever is not explicitly revealed in the Book. This pattern of centers and their relationships is apparent at every stage in the unfoldment of the Cause. In the Kitáb-i-Aqdas Bahá'u'lláh tells the believers to refer after His passing to the Book, and to "Him Whom God hath purposed, Who hath branched from this Ancient Root."[13] In the Kitáb-i-'Ahdí (the Book of Bahá'u'lláh's Covenant), He makes it clear that this reference is to 'Abdu'l-Bahá.[14] In the Aqdas Bahá'u'lláh also ordains the institution of the Universal House of Justice, and confers upon it the powers necessary for it to discharge its ordained functions. The Master in His Will and Testament explicitly institutes the Guardianship, which Shoghi Effendi states was clearly anticipated in the verses of the Kitáb-i-Aqdas, reaffirms and elucidates the authority of the Universal House of Justice, and refers the believers once again to the Book: "Unto the Most Holy Book everyone must turn, and all that is not expressly recorded therein must be referred to the Universal House of Justice," and at the very end of the Will He says: "All must seek guidance and turn unto the Center of the Cause and the House of Justice. And he that turneth unto whatsoever else is indeed in grievous error."[15]

75.16 As the sphere of jurisdiction of the Universal House of Justice in matters of legislation extends to whatever is not explicitly revealed in the Sacred Text, it is clear that the Book itself is the highest authority and delimits the sphere of action of the House of Justice. Likewise, the Interpreter of the Book must also have the authority to define the sphere of the legislative action of the elected representatives of the Cause. The writings of the Guardian and the advice given by him over the thirty-six years of his Guardianship show the way in which he exercised this function in relation to the Universal House of Justice as well as to National and Local Spiritual Assemblies.

75-12. WOB, p. 156, 23.
75-13. KA ¶ 121.
75-14. TB, pp. 217–23.
75-15. WT, pp. 19, 26.

The fact that the Guardian has the authority to define the sphere of the 75.17
legislative action of the Universal House of Justice does not carry with it the
corollary that without such guidance the Universal House of Justice might
stray beyond the limits of its proper authority; such a deduction would conflict with all the other texts referring to its infallibility, and specifically with
the Guardian's own clear assertion that the Universal House of Justice never
can or will infringe on the sacred and prescribed domain of the Guardianship. It should be remembered, however, that although National and Local
Spiritual Assemblies can receive divine guidance if they consult in the manner
and spirit described by 'Abdu'l-Bahá, they do not share in the explicit guarantees of infallibility conferred upon the Universal House of Justice. Any careful student of the Cause can see with what care the Guardian, after the passing of 'Abdu'l-Bahá, guided these elected representatives of the believers in the
painstaking erection of the Administrative Order and in the formulation of
Local and National Bahá'í Constitutions.

We hope that these elucidations will assist the friends in understanding 75.18
these relationships more clearly, but we must all remember that we stand too
close to the beginnings of the System ordained by Bahá'u'lláh to be able fully
to understand its potentialities or the interrelationships of its component parts.
As Shoghi Effendi's secretary wrote on his behalf to an individual believer on
25 March 1930, "The contents of the Will of the Master are far too much for
the present generation to comprehend. It needs at least a century of actual
working before the treasures of wisdom hidden in it can be revealed. . . ."

With loving Bahá'í greetings,
THE UNIVERSAL HOUSE OF JUSTICE

76

Release of a Compilation on Bahá'í Funds
1 JANUARY 1970

National Spiritual Assemblies of the Bahá'ís of the World

Dear Bahá'í Friends,

76.1 In order to assist the friends everywhere in the proper appreciation of the importance and meaning of contributing to Bahá'í Funds, and to remind them as well as all Assemblies of the underlying principles that must govern the offering and administration of these funds, we have made a compilation of extracts from the Guardian's letters on this subject which we are now sharing with you.[1]

76.2 You may use these extracts in any manner you deem advisable, at conferences, in summer schools, in deepening classes, and in your newsletters and circular letters.

With loving Bahá'í greetings,
THE UNIVERSAL HOUSE OF JUSTICE

77

Noninterference in Political Affairs
8 FEBRUARY 1970

To National Spiritual Assemblies in Africa

Dear Bahá'í Friends,

77.1 For long centuries the African Continent, or rather that great part of it which lies south of the Sahara, remained relatively isolated from the rest of the world, untroubled and scarcely touched by the surging conflicts of the nations to the north and east. Now, rapidly emerging into the main stream of international interest, the African peoples, who were compared by Bahá'u'lláh to the black pupil of the eye through which "the light of the spirit shineth forth,"[1] are being swept by the heady enthusiasms of new-found independence, torn by the conflicting forces of divergent political interests, their vision obscured by the haze of materialism and the dust of nationalistic passions and age-old tribal rivalries.

77.2 In the midst of the storm and stress of the battles of selfish interests being waged about them, stand the followers of the Most Great Name, their sight attracted to the rising Sun of God's Holy Cause, their hearts welded together

---

76-1. See CC 1:529–50.
77-1. ADJ, p. 37.

in a bond of true unity with all the children of men, and their voices raised in a universal song of praise to the Glory of God and the oneness of mankind, calling on their fellowmen to forget and forgo their differences and join them in obedience and service to God's Holy Command in this Day.

The Army of the Cause, advancing at the bidding of the Lord, to conquer the hearts of men, can never be defeated, but its rate of advance can be slowed down by acts of unwisdom and ignorance on the part of its supporters. We are writing you this letter to help in clarifying some of the issues that have, in the past, blurred the vision of some of the believers, and caused them to commit errors of judgment which have retarded the progress of the Faith in their countries.

The Principle of Noninterference in Political Affairs

One of these issues, and by far the most important, is a lack of appreciation of the implications of the Bahá'í principle of noninterference in political affairs. We find that 'Abdu'l-Bahá and Shoghi Effendi have given us clear and convincing reasons why we must uphold this principle. These reasons are summarized below for the study and deepening of the friends. It is our hope that these observations will not only help the friends to intelligently and radiantly follow the holy teachings on this matter, but will help them to explain the Bahá'í attitude to those who may question its wisdom and usefulness:

> The Faith of God is the sole source of salvation for mankind today. The true cause of the ills of humanity is its disunity. No matter how perfect may be the machinery devised by the leaders of men for the political unity of the world, it will still not provide the antidote to the poison sapping the vigor of present-day society. These ills can be cured only through the instrumentality of God's Faith. There are many well-wishers of mankind who devote their efforts to relief work and charity and to the material well-being of man, but only Bahá'ís can do the work which God most wants done. When we devote ourselves to the work of the Faith we are doing a work which is the greatest aid and only refuge for a needy and divided world.

> The Bahá'í Community is a worldwide organization seeking to establish true and universal peace on earth. If a Bahá'í works for one political party to overcome another it is a negation of the very spirit of the Faith. Membership in any political party, therefore, necessarily entails repudiation of some or all of the principles of peace and unity proclaimed by Bahá'u'lláh. As 'Abdu'l-Bahá stated: "Our party is God's party—we don't belong to any party."[2]

77-2. Quoted in Shoghi Effendi, letter dated 15 July 1955 to the National Spiritual Assembly of the Bahá'ís of Central America.

77.4c   If a Bahá'í were to insist on his right to support a certain political party, he could not deny the same degree of freedom to other believers. This would mean that within the ranks of the Faith, whose primary mission is to unite all men as one great family under God, there would be Bahá'ís opposed to each other. Where, then, would be the example of unity and harmony which the world is seeking?

77.4d   If the institutions of the Faith, God forbid, became involved in politics, the Bahá'ís would find themselves arousing antagonism instead of love. If they took one stand in one country, they would be bound to change the views of the people in another country about the aims and purposes of the Faith. By becoming involved in political disputes, the Bahá'ís instead of changing the world or helping it, would themselves be lost and destroyed. The world situation is so confused and moral issues which were once clear have become so mixed up with selfish and battling factions, that the best way Bahá'ís can serve the highest interests of their country and the cause of true salvation for the world, is to sacrifice their political pursuits and affiliations and wholeheartedly and fully support the divine system of Bahá'u'lláh.

77.4e   The Faith is not opposed to the true interests of any nation, nor is it against any party or faction. It holds aloof from all controversies and transcends them all, while enjoining upon its followers loyalty to government and a sane patriotism. This love for their country the Bahá'ís show by serving its well-being in their daily activity, or by working in the administrative channels of the government instead of through party politics or in diplomatic or political posts. The Bahá'ís may, indeed are encouraged to, mix with all strata of society, with the highest authorities and with leading personalities as well as with the mass of the people, and should bring the knowledge of the Faith to them; but in so doing they should strictly avoid becoming identified, or identifying the Faith, with political pursuits and party programs.

77.5   So vital is this principle of noninterference in political matters, which must govern the acts and words of Bahá'ís in every land, that Shoghi Effendi has written that "Neither the charges which the uninformed and the malicious may be led to bring against them, nor the allurements of honors and rewards" would ever induce the true believers to deviate from this path, and that their words and conduct must proclaim that the followers of Bahá'u'lláh "are actuated by no selfish ambition, that they neither thirst for power, nor mind any wave of unpopularity, of distrust or criticism, which a strict adherence to their standards might provoke."[3]

77-3. WOB, pp. 66–67.

"Difficult and delicate though be our task," he continues, "the sustaining power of Bahá'u'lláh and of His Divine guidance will assuredly assist us if we follow steadfastly in His way, and strive to uphold the integrity of His laws. The light of His redeeming grace, which no earthly power can obscure, will if we persevere, illuminate our path, as we steer our course amid the snares and pitfalls of a troubled age, and will enable us to discharge our duties in a manner that would redound to the glory and the honor of His blessed Name."⁴

The Problem of Tribalism

The second issue which causes difficulties for the African friends in these days is the matter of tribalism. As Bahá'ís they are convinced that mankind is one and must be viewed as one entity, yet, as members of their respective tribes, they find themselves expected by their non-Bahá'í brothers to give their first loyalty to, and even aggressively pursue the interests of their tribe. They live, moreover, in an atmosphere which is only too often one of mistrust, fear and even hatred against the members of other tribes.

The Bahá'í attitude in such a situation is clearly set forth in the Writings. As Bahá'ís we are attached to our tribes and clans, just as we are to our families and, on a larger scale, to our nations, but we do not allow this attachment to conflict with our wider loyalty to humanity. The followers of the Faith, the Guardian has clearly stated, "will not hesitate to subordinate every particular interest, be it personal, regional or national, to the overriding interests of the generality of mankind, knowing full well that in a world of interdependent peoples and nations the advantage of the part is best to be reached by the advantage of the whole, and that no lasting result can be achieved by any of the component parts if the general interests of the entity itself are neglected."⁵

In further elucidating this theme he has written: "Let there be no misgivings as to the animating purpose of the worldwide Law of Bahá'u'lláh. . . . It does not ignore nor does it attempt to suppress the diversity of ethnical origins, of climate, of history, of language and tradition, of thought and habit, that differentiate the peoples and nations of the world. It calls for a wider loyalty, for a larger aspiration than any that has animated the human race. It insists upon the subordination of national impulses and interests to the imperative claims of a unified world. It repudiates excessive centralization on one hand, and disclaims all attempts at uniformity on the other. Its watchword is unity in diversity . . ."⁶

77-4. WOB, p. 67.
77-5. PDIC, ¶ v.
77-6. WOB, pp. 41–42.

The Example of a Unified Community

77.10 In these days when tribal tensions are increasing in Africa the friends should be vigilant lest any trace of prejudice or hatred, God forbid, may enter their midst. On the contrary, they should endeavor to bring into the Faith an ever larger representation of the various tribes in each country, and through complete lack of prejudice as well as through the love that Bahá'ís have for each other and for their non-Bahá'í neighbors, demonstrate to their countrymen what the Word of God can do. They will thus provide, for the scrutiny of the leaders and rulers of their countries, a shining example of a unified community, working together in full concord and harmony, demonstrating a hope that is attainable, and a pattern worthy to be emulated.

77.11 To discriminate against any tribes because they are in a minority is a violation of the spirit that animates the Faith of Bahá'u'lláh. As followers of God's Holy Faith it is our obligation to protect the just interests of any minority element within the Bahá'í community. In fact in the administration of our Bahá'í affairs, representatives of minority groups are not only enabled to enjoy equal rights and privileges, but they are even favored and accorded priority. Bahá'ís should be careful never to deviate from this noble standard, even if the course of events or public opinion should bring pressure to bear upon them.

77.12 The principles in the Writings are clear, but usually it is when these principles are applied that questions arise. In all cases where the correct course of action is not clear believers should consult their National Spiritual Assembly who will exercise their judgment in advising the friends on the best course to follow.

77.13 It is the hope and prayer of the Universal House of Justice that National Spiritual Assemblies in Africa will, in full collaboration with the Continental Boards of Counselors and Auxiliary Boards in their areas, act as loving shepherds to the divine flock in that mighty Continent, protect the friends from the evil influences surrounding them, guide them in the true and right path, and assist them to attain a continuously deeper understanding, a firmer conviction and a more consuming love for the Cause they are so devotedly seeking to promote and serve.

With loving Bahá'í greetings,
THE UNIVERSAL HOUSE OF JUSTICE

1964–1973 • THE NINE YEAR PLAN

78
Attainment of Consultative Status with the United Nations Economic and Social Council
18 FEBRUARY 1970

To all National Spiritual Assemblies

JOYFULLY ANNOUNCE BAHÁ'Í WORLD ATTAINMENT CONSULTATIVE STATUS UNITED NATIONS ECONOMIC AND SOCIAL COUNCIL THEREBY FULFILLING LONG CHERISHED HOPE BELOVED GUARDIAN AND WORLD CENTER GOAL NINE YEAR PLAN.[1] SUSTAINED PERSISTENT EFFORTS MORE THAN TWENTY YEARS ACCREDITED REPRESENTATIVES BAHÁ'Í INTERNATIONAL COMMUNITY UNITED NATIONS DEVOTED SUPPORT BAHÁ'Í COMMUNITIES THROUGHOUT WORLD FINALLY REWARDED. SIGNIFICANT ACHIEVEMENT ADDS PRESTIGE INFLUENCE RECOGNITION EVER ADVANCING FAITH BAHÁ'U'LLÁH. OFFERING PRAYERS GRATITUDE HOLY SHRINES.

UNIVERSAL HOUSE OF JUSTICE

79
The Spirit of Bahá'í Consultation
6 MARCH 1970

The National Spiritual Assembly of the Bahá'ís of Canada

Dear Bahá'í Friends,

We have your letter of 14 January 1970 asking questions about the decision-making process of Spiritual Assemblies.

It is important to realize that the spirit of Bahá'í consultation is very different from that current in the decision-making processes of non-Bahá'í bodies.

The ideal of Bahá'í consultation is to arrive at a unanimous decision. When this is not possible a vote must be taken. In the words of the beloved Guardian: ". . . when they are called upon to arrive at a certain decision, they should, after dispassionate, anxious, and cordial consultation, turn to God in prayer, and with earnestness and conviction and courage record their vote and abide by the voice of the majority, which we are told by our Master to be the voice of truth, never to be challenged, and always to be wholeheartedly enforced."[1]

78-1. In a meeting on 12 February 1970 the Committee on Non-Governmental Organizations, the functional committee of the Economic and Social Council (ECOSOC) responsible for its relationship with nongovernmental organizations, had unanimously recommended that the Bahá'í International Community's application for consultative status be approved. ECOSOC formally accepted that recommendation on 27 May 1970.

79-1. BA, p. 64.

79.4 As soon as a decision is reached it becomes the decision of the whole Assembly, not merely of those members who happened to be among the majority.

79.5 When it is proposed to put a matter to the vote, a member of the Assembly may feel that there are additional facts or views which must be sought before he can make up his mind and intelligently vote on the proposition. He should express this feeling to the Assembly, and it is for the Assembly to decide whether or not further consultation is needed before voting.

79.6 Whenever it is decided to vote on a proposition all that is required is to ascertain how many of the members are in favor of it; if this is a majority of those present, the motion is carried; if it is a minority, the motion is defeated. Thus the whole question of "abstaining" does not arise in Bahá'í voting. A member who does not vote in favor of a proposition is, in effect, voting against it, even if at that moment he himself feels that he has been unable to make up his mind on the matter.

With loving Bahá'í greetings,
THE UNIVERSAL HOUSE OF JUSTICE

80

Commemoration of the Centenary of the Martyrdom of Mírzá Mihdí, the Purest Branch
25 MARCH 1970

To all National Spiritual Assemblies

Dear Bahá'í Friends,

80.1 In commemoration of the centenary of the martyrdom of the Purest Branch, which falls on June 23, 1970, we call upon the Bahá'ís of the world to unite in prayer for "the regeneration of the world and the unification of its peoples."[1]

80.2 During those days one hundred years ago Bahá'u'lláh was enduring His imprisonment in the Barracks of 'Akká. Upon the tribulations which weighed Him down was heaped the fatal accident which befell His young son, His companion and amanuensis, Mírzá Mihdí, the Purest Branch, whose dying supplication to his Father was to accept his life "as a ransom for those of His loved ones who yearned for, but were unable to attain, His presence."[2] In a Tablet revealed in that grievous hour Bahá'u'lláh sorrows that "This is the day whereon he that was created of the light of Bahá has suffered martyrdom, at

---

80-1. GPB, p. 348. For an account of the commemoration of the passing of the Purest Branch, see BW 15:163.
80-2. MA, p. 31.

a time when he lay imprisoned at the hands of his enemies." Yet He makes clear that the youth's passing has a far profounder meaning than His acceptance of the simple request, declaring that "Thou art, verily, the trust of God and His treasure in this land. Erelong will God reveal through thee that which He hath desired." In a prayer revealed for His son He proclaims the purpose underlying the tragedy: "I have, O my Lord, offered up that which Thou hast given Me, that Thy servants may be quickened, and all that dwell on earth be united."³ Thus upon a youth of consummate devotion who demonstrated such beauty of spirit and total dedication was conferred a unique station in the Cause of God.

80.3 In your recalling the bereavement of Bahá'u'lláh upon the loss of His loved son, and honoring a highly significant event in the Faith, we leave it to the discretion of the Assemblies whether they choose to hold special gatherings of prayer. In the Holy Land at the World Center on Mount Carmel there will be an observance at the grave of Mírzá Mihdí, at which time his pure example and sacrifice for all mankind will be remembered through the words of his glorious Father.

With loving Bahá'í greetings,
THE UNIVERSAL HOUSE OF JUSTICE

80.4 For your background and appreciation of the nature of the martyrdom of the Purest Branch, we refer you to *God Passes By*, pp. 188–89, and to *Bahá'í Holy Places at the World Center*, pp. 60, and 70–74, from which the several quotations above are derived.

81
Riḍván Message 1970
RIḌVÁN 1970

To all National Spiritual Assemblies

81.1 BAHÁ'Í WORLD COMMUNITY ENTERING SEVENTH YEAR NINE YEAR PLAN HAS AMPLY DEMONSTRATED ABILITY SCALE HEIGHTS DEVOTION SACRIFICE WIN ASTONISHING VICTORIES WORLD-REDEEMING WORLD-HEALING WORLD-UNITING FAITH. AT THIS RIḌVÁN EXTEND LOVING WELCOME ELEVEN NEW NATIONAL SPIRITUAL ASSEMBLIES NOW FORMING SEVEN IN AFRICA ONE IN AMERICAS ONE IN ASIA TWO IN AUSTRALASIA RAISING TO NINETY-FOUR NUMBER SUPPORTING PILLARS UNIVERSAL HOUSE JUSTICE. MOVED PAY LOVING TRIBUTE HANDS CAUSE GOD THEIR BRILLIANT SERVICES BLAZING TEACHING TRAILS SURFACE PLANET UPLIFTING ADVISING ASSEMBLIES FRIENDS ALL CONTINENTS. IN VIEW EFFECTIVE REINFORCEMENT THIS

80-3. MA, pp. 33, 34.

NOBLE WORK BY ABLE DEDICATED CONTINENTAL BOARDS COUNSELORS THEIR AUXILIARY BOARDS TOGETHER WITH GROWING NEED AND EXPANSION WORLD COMMUNITY ANNOUNCE AUGMENTATION VITAL INSTITUTION THROUGH APPOINTMENT THREE ADDITIONAL COUNSELORS IRAJ AYMAN WESTERN ASIA ANNELIESE BOPP BETTY REED EUROPE AND AUTHORIZATION APPOINTMENT FORTY-FIVE ADDITIONAL AUXILIARY BOARD MEMBERS NINE AFRICA SIXTEEN ASIA TWO AUSTRALASIA EIGHTEEN WESTERN HEMISPHERE. CALLING FORMATION FOUR NATIONAL SPIRITUAL ASSEMBLIES RIḌVÁN 1971 LESOTHO SEAT MASERU IVORY COAST MALI AND UPPER VOLTA SEAT ABIDJAN TRINIDAD AND TOBAGO SEAT PORT OF SPAIN SOLOMON ISLANDS SEAT HONIARA. NINE YEAR PLAN ALREADY MARKED GREAT ACHIEVEMENTS PIONEERING PROCLAMATION RECOGNITION FAITH UPSURGE YOUTH ACQUISITION PROPERTIES COMMENCEMENT CONSTRUCTION PANAMA TEMPLE DEVELOPMENTS WORLD CENTER. URGENT IMMEDIATE VITAL NEED CONCENTRATE ATTENTION INCREASE NUMBER LOCALITIES LOCAL SPIRITUAL ASSEMBLIES BELIEVERS FILL REMAINING PIONEER POSTS. LAST RIḌVÁN CALL RAISED SEVEN HUNDRED AND THIRTY-THREE PIONEERS MINIMUM REQUIREMENT. FOUR HUNDRED AND SEVENTY-NINE SPECIFIC POSTS STILL UNFILLED. TOTAL VICTORY REQUIRES MORE PIONEERS MORE FUNDS MORE NEW BELIEVERS. HANDS CAUSE COUNSELORS BOARD MEMBERS NATIONAL LOCAL SPIRITUAL ASSEMBLIES EVERY SINGLE FOLLOWER BAHÁ'U'LLÁH SUMMONED UTMOST EFFORT REMAINING YEARS NINE YEAR PLAN. ACHIEVEMENT THIS STEP MASTER'S DIVINE PLAN WILL ENDOW COMMUNITY CAPACITY ADMINISTRATIVE AGENCIES UNDERTAKE NEXT STAGE IMPLEMENTATION SUPREME PURPOSE BAHÁ'U'LLÁH'S REVELATION UNIFICATION MANKIND ESTABLISHMENT LONG PROMISED KINGDOM GOD THIS EARTH. ASSURE ARDENT LOVING PRAYERS HOLY SHRINES.

THE UNIVERSAL HOUSE OF JUSTICE

82
Message to Bolivia and Mauritius Conferences—August 1970
AUGUST 1970

To the Continental Conference in La Paz, Bolivia,
and the Oceanic Conference in Rose-Hill, Mauritius

Beloved Friends,

82.1 Our hearts turn with eager expectancy to the twin Conferences now in session in the southern hemisphere. Their convocation so shortly after the worldwide commemoration of the Centenary of the Martyrdom of the Purest Branch, calls to mind that the promotion and establishment of the Faith of

God have always been through sacrifice and dedicated service.[1] Indeed, these very Conferences testify to the creative power, the fruitfulness, the invocation of Divine confirmations which result from sacrificial service to the Cause of God. Although both Bolivia and Mauritius are mentioned specifically in the Tablets of the Divine Plan, the Cause, even thirty-five years ago, was virtually unknown in those areas; today we witness the holding of these historic Conferences.[2]

82.2 Little wonder that South America, whose rulers and presidents were addressed by Bahá'u'lláh in His Kitáb-i-Aqdas, of whose indigenous believers the Master, in those Tablets already referred to, wrote "should they be educated and guided, there can be no doubt that they will become so illumined as to enlighten the whole world,"[3] should have exerted a magnetic attraction upon a number of ardent souls in the northern continent, eager to serve in so promising a field.[4] A band of heroic pioneers, bearing the Message of Bahá'u'lláh, gradually penetrated its wide territories, its jungles and mountains. They were followed by others under systematic crusades of two Seven Year Plans and the beloved Guardian's Ten Year Plan and together they became the spiritual conquerors of that continent.[5] The Latin American communities which arose as a result of their pioneer efforts were described by the beloved Guardian as "associates in the execution" of 'Abdu'l-Bahá's Divine Plan.[6] May Maxwell, one of the great heroines of the Faith, attained her longed-for crown of martyrdom in Buenos Aires; Panama became the site of the sixth Mashriqu'l-Adhkár of the Bahá'í world, and La Paz, Bolivia, is now the scene of this Continental Conference.

82.3 The Indian Ocean, whose furthermost waves lap the shores of the Cradle of our Faith, upon whose waters the Divine Báb traveled in the course of His pilgrimage to Mecca, the heart of Islam, where He openly announced His Mission; whose mighty subcontinent from which it derives its name was the home and assigned province of the ninth Letter of the Living; whose major islands were severally mentioned by 'Abdu'l-Bahá in the seventh of His Tablets of the Divine Plan, lay, for most of a century, fallow to the Word of God, a

---

82-1. The Purest Branch is Mírzá Mihdí, youngest son of Bahá'u'lláh. See message no. 80 regarding the 23 June 1970 commemoration of the centenary of his martyrdom.
82-2. See TDP 6.11, 7.11, 14.7.
82-3. TDP 6.8.
82-4. TDP 6.8.
82-5. One of the goals of the first Seven Year Plan (1937–44) was to establish a center in each republic in Latin America and the Caribbean; an objective of the second Seven Year Plan (1946–53) was the consolidation and expansion of the Faith throughout the Americas. The two plans were pursued by the Bahá'ís of the United States and Canada under Shoghi Effendi's direction. At the beginning of the Ten Year Crusade (1953–63) regional National Spiritual Assemblies had formed in both Central America and South America.
82-6. MBW, p. 146.

challenge to the promotion of His Faith.[7] This challenge was answered by half a hundred Knights of Bahá'u'lláh, who, in response to the beloved Guardian's call left their homes and wholeheartedly gave themselves to the establishment of the Cause in those parts. They implanted the banner of Bahá'u'lláh upon its atolls, its great islands and bordering territories. Now, in the midmost heart of that huge expanse of sea, Mauritius, an island whose name was enshrined in Bahá'í history during the Heroic Age of our Faith as the source, two years before 'Abdu'l-Bahá's arrival in America, of a contribution to the purchase of the site of the Mother Temple of the West, has been chosen as the venue of this oceanic Conference.

82.4 Not only have the institutions of the Faith been established in this ocean and this continent, but the spirit of the New Day, brilliant even at this early dawn with the light of Bahá'u'lláh's gifts to man, is apparent in the diversity of the attendants, in the brotherhood of erstwhile strangers—even enemies—and above all in the noble purposes for which you have gathered.

82.5 Your aim is the redemption of mankind from its godlessness, its ignorance, its confusion and conflict. You will succeed, as those before you succeeded, by sacrifice to the Cause of God. The deeds and services required of you now, will shine in the future, even as those of your spiritual predecessors shine today and will forever shine in the annals of the Cause.

82.6 We share with you the spiritual delight of these occasions and assure you of our constant and ardent prayers that your deliberations upon the objectives of the Cause in your areas and the spiritual fellowship which you will enjoy will result in immediate and determined plans to complete the tasks assigned to you ere the rapidly approaching end of the Nine Year Plan. This Plan is the current stage of the Master's Divine Plan and its success must precede those greater triumphs when, as the result of your labors, the divine outpourings will raise up a vast concourse of radiant and devoted servants of Bahá'u'lláh who will establish His Kingdom in this world.

With loving Bahá'í greetings,
THE UNIVERSAL HOUSE OF JUSTICE

82-7. TDP 7.10.

83
Call for Pioneers
2 August 1970

To all National Spiritual Assemblies

Dear Bahá'í Friends,

All National Spiritual Assemblies have been aware of the urgency of our Riḍván 1969 message to the Bahá'ís of the world when the call was raised for pioneers to settle in territories in need of pioneer support or as yet unopened to the Faith, and have been cognizant of the emphasis which was placed on the need for the believers to arise quickly to ensure the success of the Nine Year Plan in the pioneer field.

Since that call was raised, no less than 330 of the pioneer posts in the 184 specified territories in the globe have been filled, and in a few of those territories additional pioneers have arrived to supplement the ranks of Bahá'u'lláh's followers at those posts.

As you will note from the attached list[1] showing the current status of pioneer goals, some 417 pioneers must yet arise and settle in the posts previously assigned. After a recent review of pioneer needs we find it is necessary to call upon the valiant, constantly swelling community of believers throughout all continents to fill yet another 204 pioneer posts where manpower is desperately needed, in some territories in order to win the minimum number of Assemblies or localities called for in the Plan, and in others where vast new mass teaching areas have been opened to the Faith, thus necessitating additional reinforcements who must arrive soon if the precious gains are to be retained. These 204 new pioneer goals have been assigned to specific National Spiritual Assemblies.

Despite the magnitude of this undertaking and the grave challenge which your communities face in ensuring the homefront goals, we are compelled to point out that each of those goals assigned must be considered as a minimum requirement. Pioneers unable to go to the goals assigned by their own National Spiritual Assembly should be encouraged to fill goals assigned to other National Assemblies. Of course a self-supporting believer is free to settle as a pioneer in any country he chooses.

We call upon the friends to act promptly and decisively in this vital international undertaking in which the followers of Bahá'u'lláh are, in all continents of the globe, summoned to participate. The time is short and the effort required is truly formidable.

83-1. The list is too lengthy to include in this volume.

83.6 We shall offer ardent prayers at the Holy Shrines, supplicating that the waves of pioneers required to complete this urgent task of the present hour shall arise and quickly rush forth into the arena of service.

With loving Bahá'í greetings,
THE UNIVERSAL HOUSE OF JUSTICE

84
Release of a Compilation on the Local Spiritual Assembly
11 AUGUST 1970

To all National Spiritual Assemblies

Dear Bahá'í Friends,

84.1 As the Bahá'í Administrative Order rapidly expands throughout the world it behooves everyone associated with it to familiarize himself with its principles, to understand its import and to put its precepts into practice. Only as individual members of Local Spiritual Assemblies deepen themselves in the fundamental verities of the Faith and in the proper application of the principles governing the operation of the Assembly will this institution grow and develop toward its full potential.

84.2 It is because the principles of Bahá'í Administration are new to so many who are now being called upon to serve as members of Local Spiritual Assemblies that we felt the need to make available in brief form some of the texts and instructions which apply.[1] No attempt has been made to put together a complete compilation of all texts from the writings of Bahá'u'lláh, 'Abdu'l-Bahá and Shoghi Effendi, but it is hoped that the enclosed extracts will suffice as an introduction to a more profound study of the subject, and lead to a more efficient functioning of Local Spiritual Assemblies everywhere.

84.3 We call upon you to consider ways and means of sharing this material with the friends, and especially members of Local Spiritual Assemblies, as quickly as possible. In many instances, of course, the material will need to be translated; in other instances many of the quotations included in the compilation may already be available to the friends in their own language.

With loving Bahá'í greetings,
THE UNIVERSAL HOUSE OF JUSTICE

84-1. See CC 2:39–60.

85
Formation of Seven National Spiritual Assemblies during Riḍván 1971
12 August 1970

To all National Spiritual Assemblies

Dear Bahá'í Friends,

The following cable has just been sent to Hands of the Cause Rúḥíyyih Khánum and William Sears representing the Universal House of Justice at Conferences in Bolivia and Mauritius:

85.1

> PLEASE ANNOUNCE TO PARTICIPANTS CONFERENCE JOYOUS NEWS DECISION CALL THREE ADDITIONAL NATIONAL CONVENTIONS NEXT RIḌVÁN NAMELY SUDAN CHAD AND CONGO BRAZZAVILLE GABON BRINGING TO SEVEN NEW NATIONAL SPIRITUAL ASSEMBLIES BEING FORMED AT CLOSE OF SEVENTH YEAR NINE YEAR PLAN. FERVENTLY PRAYING HOLY SHRINES BEHALF NATIONAL COMMUNITIES BAHÁ'Í WORLD REACHING ONE HUNDRED ONE BY NEXT RIḌVÁN SUPPLICATING REINFORCEMENT TIES UNITING THEM GREATER CONSECRATION CHALLENGING TASKS STILL AHEAD WIDER PARTICIPATION ALL RANKS FAITHFUL. COMMUNICATING TEXT CABLE ALL NATIONAL SPIRITUAL ASSEMBLIES.

Please share this news with the friends.

85.2

With loving Bahá'í greetings,
THE UNIVERSAL HOUSE OF JUSTICE

86
Message to Pioneers
29 November 1970

To all Pioneers[1]

Dearly loved Friends,

The spirit of self-sacrifice and devotion that has animated so large a number of the followers of Bahá'u'lláh to leave their homes, move to posts far and near, to foreign lands and on the homefronts, to hoist the banner of the Faith and promote the divine teachings in well-nigh every populated area of the globe, uplifts our hearts and evokes our profound pride and admiration. We are now entering the most challenging and crucial closing period of the Nine Year Plan, that will culminate in the joyous celebration of the hundredth

86.1

86-1. This letter was sent to all National Spiritual Assemblies on 29 November 1970 for distribution to "all pioneers both on homefronts and overseas."

anniversary of the revelation of Bahá'u'lláh's Most Holy Book, the Kitáb-i-Aqdas. What greater gift can we lay at the feet of our Beloved, at that historic moment, than the proclamation of Victory in His Name!

86.2 Our deep appreciation of the vital role which the pioneers play in the onward march of the Army of Light towards victory arouses in us the desire to comfort their hearts, upraise their spirits, and strengthen their loins by calling to their minds the stirring appeal which flowed from the Pen of 'Abdu'l-Bahá:

86.2a > O that I could travel, even though on foot and in the utmost poverty, to these regions, and, raising the call of "Yá Bahá'u'l-Abhá" in cities, villages, mountains, deserts and oceans, promote the Divine teachings! This, alas, I cannot do. How intensely I deplore it! Please God, ye may achieve it.[2]

and the following words of guidance from our beloved Guardian:

86.2b > Theirs, at this present hour, unpropitious and unpromising though the immediate prospects may appear, is the duty to plod on, confident and unsparing in their daily efforts, undimmed in their vision, alert and conscious of the sublimity of their calling and of the future glory of their Mission, undistracted by the petty pursuits and temptations of the environment in which they live, exerting their utmost, and playing, each independently, as well as through their concerted efforts, their part in hastening the advent of the day when their dearly beloved Faith will, at long last, have revealed the full measure of its potentialities, and soared, as destined by Providence, to new heights of power, of eminence and glory.[3]

> It is hard for the friends to appreciate, when they are isolated in one of these goal territories, and see that they are making no progress in teaching others, are living in inhospitable climes for the most part, and are lonesome for Bahá'í companionship and activity, that they represent a force for good, that they are like a lighthouse of Bahá'u'lláh shining at a strategic point and casting its beam out into the darkness. This is why he [Shoghi Effendi] so consistently urges these pioneers not to abandon their posts.[4]

However gigantic the task may be, no matter how insuperable the obstacles standing in the way of its accomplishment may appear, and however restricted the means, capacity, and numbers of those called upon

---

86-2. TDP 7.8.
86-3. Letter dated 12 August 1957 to Italy and Switzerland.
86-4. MC, p. 68.

to ensure its fulfillment, it surely cannot, by virtue of the divine potency with which it is charged, but be successfully achieved in due time. God's redemptive grace, flowing through the small yet infinitely resourceful band of His faithful servants will, as in the days past, gradually permeate the world, and infuse into the consciousness of peoples and nations alike the realization that nothing short of the divine panacea He Himself has prescribed can cure the ills now so sadly afflicting the whole of mankind. What a higher privilege therefore than to be the instrument, the channel for the transmission of such divine grace. Let us then take courage, and faithfully pursue our mission, and rest ever assured that the promised day of victory, foretold by Bahá'u'lláh as marking the golden age of His Cause, will dawn upon us and upon a world as yet unconscious of the divine potency of His Message.[5]

And finally from the Pen of Glory Itself: 86.2c

> They that have forsaken their country for the purpose of teaching Our Cause—these shall the Faithful Spirit strengthen through its power. A company of Our chosen angels shall go forth with them, as bidden by Him Who is the Almighty, the All-Wise. How great the blessedness that awaiteth him that hath attained the honor of serving the Almighty! By My life! No act, however great, can compare with it, except such deeds as have been ordained by God, the All-Powerful, the Most Mighty. Such a service is, indeed, the prince of all goodly deeds, and the ornament of every goodly act. Thus hath it been ordained by Him Who is the Sovereign Ruler, the Ancient of Days.[6]

To each and every one of you we send our love and assurance of our prayers on your behalf in the Holy Shrines. 86.3

With loving Bahá'í greetings,
THE UNIVERSAL HOUSE OF JUSTICE

86-5. Letter dated 21 October 1939 to an individual.
86-6. GWB, p. 334.

87
Grave Crisis in Bahá'í International Fund
29 December 1970

To the Followers of Bahá'u'lláh in every land

Dear Bahá'í Friends,

87.1 We have reached a critical point in the progress of the Nine Year Plan. In many lands multitudes are thirsty and eager to embrace the Message of Bahá'u'lláh. In others, materially advanced but spiritually backward, a great effort is needed to awaken the people to the light of this New Day. The recently established National Spiritual Assemblies in many lands are occupied in acquiring the Ḥaẓíratu'l-Quds, Temple Sites, National Endowments and Teaching Institutes essential for the proper development of the Administrative Order and the deepening of the Bahá'í knowledge of their believers, while in the heart of the Western Hemisphere, the Mashriqu'l-Adhkár of Panama requires several hundred thousand dollars for its completion. To accomplish these many essential tasks the resources of the Cause are being stretched to their uttermost.

87.2 At this crucial moment, when the activities of the believers and the expenditure of funds should be increased to seize the opportunities which lie before us, the Bahá'í International Fund finds itself plunged into a grave crisis by a steep reduction in contributions. Undoubtedly worldwide economic difficulties are one of the causes of this, but we are confident that the believers throughout the world will respond to this challenge and will make every sacrifice to ensure that the work of the Cause of God goes forward unimpeded.

Demands on the International Fund

87.3 Since 1963 when there were 56 National Spiritual Assemblies, to the present time when there are 94 (soon to be 101), the work of the Cause has expanded so rapidly, both in the teaching field and at the World Center, that the Universal House of Justice has had to increase more than fourfold the annual international budget of the Cause. This year fifty-eight percent of the International Fund is being expended outside the Holy Land on projects such as assistance to National Spiritual Assemblies (56 of which receive a large part, if not all, of their budgets from the World Center), contributions to the work of the Hands of the Cause and the Continental Boards of Counselors, defense of the Cause in lands where it is facing persecution, and our expanded activities at the United Nations.

87.4 In order to meet the present situation the Universal House of Justice must drastically reduce the expenditure of the Bahá'í International Fund until the flow of contributions is restored. While the work on the International Archives Building necessary to protect the precious Tablets and relics from the high humidity and increasingly polluted atmosphere of Haifa city has been completed, the projects of further developing the Gardens in Bahjí and of starting upon an extension of the Terraces below the Shrine of the Báb, as well as additional developments to the office facilities of the World Center, must now be postponed. In addition we are reluctantly compelled to reduce by ten percent the next two quarterly remittances of assistance to National Spiritual Assemblies, and we call upon these Assemblies now to reduce their own expenditure to take account of this.

These, however, can but be temporary measures designed to minimize the present emergency. The real answer lies, not in restricting the activities of the friends at this time when mankind stands in such dire need of the Message of Bahá'u'lláh, but in the universal participation of every believer in the work of the Cause.

Backbone of the Fund: Universal Participation

The poor believers vastly outnumber the wealthy ones, and this majority will grow rapidly as mass teaching spreads. Thus, although the work in mass teaching areas will continue to be assisted by the contributions of the friends in prosperous lands, and these believers must for the immediate future continue to be the main support of the International Fund, it becomes ever more urgent for the friends in mass teaching areas to finance their own activities to an ever greater degree. The backbone of the Fund must be the regular contributions of every believer. Even though such contributions may be small because of the poverty of the donors, large numbers of small sums combine into a mighty river that can carry along the work of the Cause. Moreover the unity of the friends in sacrifice draws upon them the confirmations of the Blessed Beauty.

The universal participation of the believers in every aspect of the Faith—in contributing to the Fund, in teaching, deepening, living the Bahá'í life, administering the affairs of the community, and, above all, in the life of prayer and devotion to God—will endow the Bahá'í community with such strength that it can overcome the forces of spiritual disintegration which are engulfing the non-Bahá'í world, and can become an ocean of oneness that will cover the face of the planet.

We ask every one of you to ponder these matters deeply, and to join us in fervent prayer that this momentary crisis will prove to have been a providential test that will spur the community of the Greatest Name to new heights of dedication and triumphant achievement.

With loving Bahá'í greetings,
THE UNIVERSAL HOUSE OF JUSTICE

88
Message to the Monrovia Conference—January 1971
JANUARY 1971

To the Friends of God assembled
in the Conference in Monrovia, Liberia

Dearly loved Friends,

88.1 The emergence on the African Continent of a widely spread, numerous, diversified and united Bahá'í community, so swiftly after the initiation of organized teaching plans there, is of the utmost significance and a signal evidence of the bounties which God has destined for its peoples in this day.

88.2 The great victories in Africa, which brought such joy to the Guardian's heart in the last years of his life, resulted from the self-sacrificing devotion of a handful of pioneers, gradually assisted by the first few native believers, all laboring under the loving shadow of the Hand of the Cause Músá Banání. From their efforts there has been raised up an increasing army of African teachers, administrators, pioneers and valiant promoters of the Divine Cause, whose main task is to bring to all Africa the bounties conferred by the Word of God, bounties of enlightenment, zeal, devotion and eventually the true civilization of Bahá'u'lláh's World Order.

88.3 Many of the gravest ills now afflicting the human race appear in acute form on the African Continent. Racial, tribal and religious prejudice, disunity of nations, the scourge of political factionalism, poverty and lack of education are obvious examples. Bahá'ís have a great part to play—greater than they may realize—in the healing of these sicknesses and the abatement of their worst effects. By their radiant unity, by their "bright and shining"[1] faces, their self-discipline in zealously following all the requirements of Bahá'í law, their abstention from politics, their constant study and proclamation of the Great Message, they will hasten the advent of that glorious day when all mankind will know its true brotherhood and will bask in the sunshine of God's love and blessing.

88.4 That the African believers are fully capable of taking their full share in building the Kingdom of God on earth, their natural abilities and present deeds have fully demonstrated. An African Hand of the Cause of God, even now in the course of a brilliant, triumphal teaching tour of the planet, African Counselors, Board members, national and local administrators and an ever-increasing army of believers testify to the vigor and immense capacity of this highly blessed continent to serve its Lord in the great day of His appearance.[2]

88-1. See ABL, p. 28, or PT, p. 61.
88-2. The African Hand of the Cause of God referred to is Enoch Olinga. For an account of his life and services, see BW 18:618–35. For the message about his death, see no. 237.

That the African believers, so beloved by the Guardian of the Faith, will rise to the challenge facing them and earn the gratitude and goodwill of all mankind by their deeds of dedication and self-sacrifice is the longing of our hearts.

88.5 May this Conference become a sun from which will stream forth to all parts of the vast continent rays of spiritual energy and inspiration, galvanizing the friends to action in the fields of teaching and pioneering in such manner that they will rapidly achieve all the tasks assigned to them under the Nine Year Plan.

88.6 Our thoughts and prayers are with you.

THE UNIVERSAL HOUSE OF JUSTICE

89
Message to the Oceanic Conference of the South China Seas, Singapore—January 1971
JANUARY 1971

To the Friends of God assembled in the
Oceanic Conference in Singapore, Malaysia

Dearly loved Friends,

89.1 The wonderful progress made by the Bahá'í communities of South East Asia towards achievement of the tasks assigned to them under the Nine Year Plan fills our hearts with thankfulness to God and arouses our keenest admiration for the capacities and dedicated services of the friends in all those vast and varied territories. Indeed, so bountiful have been the divine confirmations rewarding their efforts that we are confident of their ability to far exceed the stated objectives and to initiate the opening phase of the next stage of their development, a massive increase in the establishment of the Cause of God among the teeming millions of the islands and ocean-bordering countries of so huge an area of the earth.

89.2 South East Asia, whose gifted and industrious peoples have embraced four of the world's major religions, have produced in all ages civilizations and cultures representative of the highest accomplishments of the human race, now experiencing with the rest of the world the disruptive, revolutionizing, "vibrating influence of this most great, this new World Order . . . the like of which mortal eyes have never witnessed," lies open and receptive to the Word of God, ready once more to nourish in its fertile soil that potent seed and to bring forth, in its own characteristic manner and as an integral part of the world civilization, the institutions, the fabric, the brilliant edifice of Bahá'u'lláh's World Order.¹

89.3 We now summon the believers of this highly promising area, flushed with the tide of approaching victory, to launch a three-pronged campaign, the main feature of which is to achieve an immediate expansion of the Faith, exceeding the aims of the Nine Year Plan. In addition you are called upon to raise a corps of traveling teachers, whose main objective will be to visit all the communities and groups in the area for the purpose of deepening and consolidating their Bahá'í life, thus preserving the victories won and reinforcing the base for future development. Simultaneously a number of Chinese-speaking believers must arise who, as pioneers and traveling teachers in all the countries of South East Asia, will attract large numbers of the talented Chinese race to embrace and serve the Faith of Bahá'u'lláh.

89.4 Recognizing your current achievements and fully confident in your determination and ability to continue to attract the divine confirmations of Bahá'u'lláh, we are happy to announce as a supplementary goal of the Nine Year Plan, the establishment, at Riḍván, 1972, of the National Spiritual Assembly of the Bahá'ís of Singapore, an additional supporting pillar of the Universal House of Justice and a new bastion of the Faith in so vital a cross-roads of human activity.

89.5 We pray that your deliberations will engender a new wave of enthusiasm, cement ever more firmly the bonds of love between the many and various national communities of your area and result in practical plans for the implementation of the above tasks.

89.6 We send you all our most loving greetings and look forward eagerly to the report of your conference.

THE UNIVERSAL HOUSE OF JUSTICE

90

Passing of the Hand of the Cause of God Agnes Alexander

4 JANUARY 1971

To all National Spiritual Assemblies

90.1 PROFOUNDLY GRIEVE PASSING ILLUMINED SOUL HAND CAUSE AGNES ALEXANDER LONG-STANDING PILLAR CAUSE FAR EAST FIRST BRING FAITH HAWAIIAN ISLANDS.¹ HER LONG DEDICATED EXEMPLARY LIFE SERVICE DEVOTION CAUSE GOD ANTICIPATED BY CENTER COVENANT SELECTING HER SHARE MAY MAXWELL IMPERISH-

---

89-1. GWB, p. 136.
90-1. For an account of the life and services of Agnes Alexander, see BW 15:423–30.

ABLE HONOR MENTION TABLETS DIVINE PLAN. HER UNRESTRAINED UNCEASING PURSUIT TEACHING OBEDIENCE COMMAND BAHÁ'U'LLÁH EXHORTATIONS MASTER GUIDANCE BELOVED GUARDIAN SHINING EXAMPLE ALL FOLLOWERS FAITH. HER PASSING SEVERS ONE MORE LINK HEROIC AGE. ASSURE FAMILY FRIENDS ARDENT PRAYERS HOLIEST SHRINE PROGRESS RADIANT SOUL REQUEST ALL NATIONAL SPIRITUAL ASSEMBLIES HOLD MEMORIAL MEETINGS AND THOSE RESPONSIBLE HOLD SERVICES MOTHER TEMPLES.

UNIVERSAL HOUSE OF JUSTICE

91
Participation of the Hands of the Cause of God in First National Conventions
1 FEBRUARY 1971

To all National Spiritual Assemblies

HAPPY ANNOUNCE FOLLOWING HANDS CAUSE WILL REPRESENT UNIVERSAL HOUSE OF JUSTICE FIRST NATIONAL CONVENTIONS COMING RIḌVÁN AMATU'L-BAHÁ RÚḤÍYYIH KHÁNUM IVORY COAST UPPER VOLTA MALI ZIKRULLAH KHADEM TRINIDAD TOBAGO ADELBERT MÜHLSCHLEGEL LESOTHO 'ALI-MUḤAMMAD VARQÁ CONGO BRAZZAVILLE GABON ENOCH OLINGA BOTH SUDAN CHAD COLLIS FEATHERSTONE BOTH SOLOMON ISLANDS SOUTHWEST PACIFIC OCEAN. CONFIDENT PRESENCE PARTICIPATION THESE STANDARD-BEARERS NINE YEAR PLAN HISTORIC FIRST CONVENTIONS WILL ATTRACT DIVINE BLESSINGS ASSIST NEW NATIONAL COMMUNITIES BEFITTINGLY ASSUME SACRED RESPONSIBILITIES.

91.1

UNIVERSAL HOUSE OF JUSTICE

92
Formation of Nine Additional National Spiritual Assemblies during Riḍván 1971
11 FEBRUARY 1971

To all National Spiritual Assemblies

REJOICE ANNOUNCE ALL FRIENDS FORMATION DURING RIḌVÁN 1972 NINE ADDITIONAL NATIONAL SPIRITUAL ASSEMBLIES RAISING TOTAL NUMBER PILLARS UNIVERSAL HOUSE JUSTICE TO ONE HUNDRED AND TEN. THREE IN AFRICA MALAGASY REPUBLIC RÉUNION SEYCHELLES THREE IN ASIA EAST PAKISTAN NEPAL SINGAPORE ONE IN AUSTRALASIA NORTHWEST PACIFIC OCEAN COMPRISING GUAM CAROLINES MARIANAS MARSHALLS TWO IN EUROPE ICELAND AND REPUBLIC IRELAND.

92.1

FOUR OF THESE SEYCHELLES EAST PAKISTAN SINGAPORE NORTHWEST PACIFIC CONSTITUTE SUPPLEMENTARY ACHIEVEMENTS NINE YEAR PLAN. URGE PIONEERS SCHEDULED ALL THESE AREAS SETTLE POSTS WITHOUT DELAY. CALL UPON RESPECTIVE COMMUNITIES BRACE THEMSELVES EXERT SUPREME EFFORT FAST FLEETING WEEKS BEFORE COMING RIḌVÁN ESTABLISH AS MANY ASSEMBLIES AS POSSIBLE THEREBY BROADENING STRENGTHENING FOUNDATIONS PROJECTED NATIONAL INSTITUTIONS. FERVENTLY PRAYING HOLY SHRINES FOLLOWERS MOST GREAT NAME MAY SEIZE UNIQUE OPPORTUNITIES PRESENT HOUR AND SPARE NO EFFORT UNTIL GOALS PLAN ARE FULLY ACCOMPLISHED THEREBY ATTRACTING TO THEMSELVES AND THEIR COMMUNITIES INESTIMABLE BLESSINGS ANCIENT BEAUTY.

UNIVERSAL HOUSE OF JUSTICE

93
Warning against the Misuse of Recordings of 'Abdu'l-Bahá's Voice
23 FEBRUARY 1971

To all National Spiritual Assemblies

Dear Bahá'í Friends,

93.1 The advent and liberal supply of tape and cassette recorders in the markets of the world have opened new doors and placed in almost every land at the disposal of the friends new methods for the dissemination of Bahá'í material. It is the hope of the Universal House of Justice that the recording of Bahá'í talks, and other audio features, and their wide use among Bahá'ís and non-Bahá'ís alike, will prove to be a powerful new instrument in the teaching and deepening work everywhere. There is one area, however, where great care must be exercised, and this is in the use of the record of 'Abdu'l-Bahá's voice.

93.2 The Guardian, when referring to this record, requested the friends "to exercise restraint and caution." "In my view," he added, "it should be used only on special occasions and be listened to with the utmost reverence. The dignity of the Cause, I am sure, would suffer from too wide and indiscriminate use of one of the most precious relics of our departed Master."[1]

93.3 We request you to share the contents of this letter, in any manner you deem advisable, with the friends residing under your jurisdiction. We are con-

---

93-1. BA, p. 55.

fident that all the friends will strictly observe the Guardian's exhortation and will not overstep the bounds of courtesy and moderation in the use of a precious relic so lovingly left to us by the Center of God's Covenant.

> With loving Bahá'í greetings,
> THE UNIVERSAL HOUSE OF JUSTICE

94
Principles of Bahá'í Publishing
28 MARCH 1971

To the National Spiritual Assemblies of the Bahá'í World

Dear Bahá'í Friends,

Recognizing the need for a great increase in the provision of Bahá'í literature in all languages, we have reviewed the whole process of Bahá'í publishing including such matters as reviewing, standards of production, sales and distribution, relationships between National Spiritual Assemblies, the international needs of the teaching work and the position of Bahá'í authors. We are both to stimulate the supply of new works and to liberate the channels of publication and distribution. We wish to encourage Bahá'í authors as well as to promote production of the basic texts of the Faith. 94.1

We therefore ask you to study the attached memorandum yourselves, pass it on to your Publishing Trusts and/or other agencies concerned, and make it available generally to the friends in whatever way you may find practicable. 94.2

> With loving Bahá'í greetings,
> THE UNIVERSAL HOUSE OF JUSTICE

Memorandum on Bahá'í Publishing—Riḍván 1971

The following principles and observations are called to the attention of National Spiritual Assemblies and all those concerned with the production of Bahá'í publications: 94.3

Reviewing

Obligatory

At this early stage of the Cause all works by Bahá'ís which deal with the Faith, whether in the form of books, pamphlets, translations, poems, songs, radio and television scripts, films, recordings, etc. must be approved before submission for publication, whether to a Bahá'í or non-Bahá'í publisher. In 94.3a the case of material for purely local consumption the competent authority is the Local Spiritual Assembly, otherwise the National Spiritual Assembly (through its Reviewing Committee) is the approving authority.

A Temporary Measure

94.3b That this measure is both obligatory and temporary is borne out by the following statements of the Guardian:

> They must supervise in these days when the Cause is still in its infancy all Bahá'í publications and translations, and provide in general for a dignified and accurate presentation of all Bahá'í literature and its distribution to the general public.
> (*Principles of Bahá'í Administration*, pp. 38–39)

> ... the administration of the Cause ... should guard against such rigidity as would clog and fetter the liberating forces released by His Revelation. ... ... the present restrictions imposed on the publication of Bahá'í literature will be definitely abolished; ...
> (*The World Order of Bahá'u'lláh*, p. 9)

Purpose of Review

94.3c The purpose of review is to protect the Faith from misrepresentation and to ensure dignity and accuracy in its presentation. In general the function of a Reviewing Committee is to say whether the work submitted gives an acceptable presentation of the Cause or not. Reviewers may win the gratitude and good will of authors by calling attention to such things as occasional grammatical or spelling errors, but approval should not be refused on such grounds; all such details are editorial matters for agreement between author and publisher.

Translations

94.3d As regards English, the beloved Guardian's translations are obviously the most authentic and should be used. If, for some particular reason, a Bahá'í author when quoting a passage of the Sacred Text which has been rendered into English by the Guardian, wishes to use a translation other than that made by the beloved Guardian, his request may be referred to the Universal House of Justice. Passages from the Sacred Text not translated by Shoghi Effendi, but already in English and published with approval, may be used. If an author wishes to make his own translation of a passage not already translated by Shoghi Effendi, the new translation may be submitted to the Universal House of Justice for approval.

With the exception of certain oriental languages such as Turkish, Arabic 94.3e
and Urdu, which are related to the original Persian or Arabic, new translations of the Sacred Text into languages other than English must be made from the Guardian's English translation where it exists. When there is no translation into English by Shoghi Effendi of a particular passage, the National Spiritual Assembly concerned should seek the advice of the Universal House of Justice. When translations already exist, which are not made from the Guardian's English text, but have been published and approved, they may be used.

Reviewing Committees

It is recommended that Reviewing Committees be small, composed of two 94.3f
or three believers with adequate education and knowledge of the Cause. It is essential that works submitted be dealt with promptly. The standards to be upheld by reviewers are the following: (a) conformity with the Teachings, (b) accuracy, (c) dignity in presentation. The Spiritual Assembly, on the basis of its Reviewing Committee's report, gives or withholds approval of the work.

Approval of Works Already Reviewed Elsewhere

While a National Spiritual Assembly intending to publish Bahá'í literature 94.3g
is encouraged to accept the review of another National Spiritual Assembly, it is not required to do so and has the right to review any work prior to authorizing its publication or republication by its own Publishing Trust or publisher in its area of jurisdiction. This does not apply to works by Hands of the Cause, which are reviewed in the Holy Land.

A National Spiritual Assembly which receives for approval a manuscript 94.3h
from outside its area of jurisdiction should inquire whether it has already been submitted for review elsewhere, and in the case of its having been refused approval, the reasons for such refusal.

Bahá'í Publishers

Bahá'í publishers may not publish any work about the Faith until it has 94.3i
been approved by the National Spiritual Assembly of the country where it is to be published.

Approval of a work imposes no obligation upon any Bahá'í publisher to 94.3j
publish it.

Whatever "house styles" Publishing Trusts and other Bahá'í publishers may 94.3k
adopt, transliteration of oriental terms into languages using the Roman alphabet must at present be according to the system chosen by the Guardian and described in volumes of *The Bahá'í World*.

Cables

94.31 Cables in English should be printed exactly as received, without interpolation.[1]

Editing

94.3m Bahá'í publishers, when accepting a work for publication, will make their own arrangements with the author on all such matters as accuracy of quotations, documentation, grammar and spelling, dates and even the rewriting of passages which the publisher may consider need improving, or he may ask the author to write additional material or to delete part of the original manuscript. Although such matters are entirely between the author and publisher, any addition, deletion or changes which affect the meaning must be submitted for review with the relative context.

Approval Notice

94.3n Although no Bahá'í work may be published without approval, it is not mandatory to print an approval notice in any publication.

Bahá'í Authors

94.3o Bahá'í authors should welcome review of their works, and can greatly assist promptness in review by supplying a sufficient number of copies of the manuscript for each member of the Reviewing Committee to have one.

94.3p Bahá'í authors may submit their works for review to any National Spiritual Assembly, and may send their works, once approved, to any publisher they like, Bahá'í or non-Bahá'í, at home or abroad. It should be remembered, however, that the approval should be given by the National Spiritual Assembly of the country where the work is to be first published. And in the case of a non-Bahá'í publisher the author should insist on use of the system of transliteration at present used by the Faith for languages employing the Roman alphabet.

94.3q It is hoped that Bahá'í authors will provide a constant stream of new works. Introductory books, commentaries, dissertations on various aspects of the Revelation, text books, histories, reviews, audiovisual material are all needed to stimulate study of the Faith and to promote the vital teaching work.

94-1. The Universal House of Justice, in messages dated 16 July 1974 and 18 March 1981, has explained what editing is permissible when publishing its cables: (1) the word "STOP," when used to indicate the end of a sentence, may be replaced with a period; (2) transliteration of Persian and Arabic words may be added; (3) apostrophes may be added to possessive case nouns; (4) spelling mistakes made in transmission may be corrected; (5) capitalization may follow house style; (6) when both a cable and its letter of transmission are in hand, the date of the cable should be used, and only the text of the cable itself should be printed, unless the letter contains other pertinent information.

Sale and Distribution of Bahá'í Literature

1. Bahá'í publications reviewed and published in one country may be sold or offered for sale anywhere in the world. This includes the right of the publisher or the author to promote the sale of the publication in any legitimate manner including the right to advise the Bahá'ís in any country of its contents, price and availability. It does not include the right to insist that National Assemblies, their Publishing Trusts or Publishing Committees stock, promote or advertise the publication or offer it for sale. If any National Spiritual Assembly feels that a book would be damaging to the Faith in its country, it may represent this fact to the publisher and author and ask them not to promote it in that particular country.

   It is hoped that there will be great cooperation among those publishing Bahá'í literature, and Publishing Trusts are encouraged to supply to believers, the book trade and libraries, all Bahá'í publications from any country.

2. Believers should not be prevented from purchasing Bahá'í books reviewed and published in other countries.

3. National Spiritual Assemblies are not obliged to furnish mailing lists of believers to publishers, but publishers may compile their own mailing lists and use them for the announcement and promotion of sale of their Bahá'í books and literature.

4. Five copies of every new book and every new edition (not reprints) should be sent to the World Center.[2]

95

Call for Deepening on the Significance of the Formative Age

15 April 1971

To all National Spiritual Assemblies

Dear Bahá'í Friends,

Our Riḍván message this year[1] called attention to the impending fiftieth anniversary of the passing of 'Abdu'l-Bahá, "an event which signalized at once the end of the Heroic Age of our Faith, the opening of the Formative Age and the birth of the Administrative Order, the nucleus and pattern of the World Order of Bahá'u'lláh."

94-2. For current instructions, consult the Bahá'í World Center Library.
95-1. See the following message (no. 96).

95.2 We now call upon all National Spiritual Assemblies to formulate and implement plans designed to educate the friends everywhere in their understanding of the significance of the Formative Age of our Faith. As an aid to this program we attach extracts from the writings of the beloved Guardian on this general theme, and we suggest that these and similar excerpts from the writings be studied and expounded at the forthcoming Summer Schools, at special sessions of Teaching Institutes, at conferences of the friends, and indeed on any occasions which you may deem suitable.

95.3 We leave it to you to use the wonderful material bequeathed to us by the beloved Guardian on this theme in whatever manner you deem best for your own communities. The study and understanding of this subject will immensely strengthen the faith of the believers as well as their ability to present the message to a waiting world.

With loving Bahá'í greetings,
THE UNIVERSAL HOUSE OF JUSTICE

Extracts from the Writings of Shoghi Effendi on the Significance of the Formative Age of Our Faith
APRIL 1971

95.4 The passing of 'Abdu'l-Bahá, so sudden in the circumstances which caused it, so dramatic in its consequences, could neither impede the operation of such a dynamic force nor obscure its purpose. Those fervid appeals, embodied in the Will and Testament of a departed Master, could not but confirm its aim, define its character and reinforce the promise of its ultimate success.

95.5 Out of the pangs of anguish which His bereaved followers have suffered, amid the heat and dust which the attacks launched by a sleepless enemy had precipitated, the Administration of Bahá'u'lláh's invincible Faith was born. The potent energies released through the ascension of the Center of His Covenant crystallized into this supreme, this infallible Organ for the accomplishment of a Divine Purpose. The Will and Testament of 'Abdu'l-Bahá unveiled its character, reaffirmed its basis, supplemented its principles, asserted its indispensability, and enumerated its chief institutions. . . .

"America and the Most Great Peace"—21 April 1933, *The World Order of Bahá'u'lláh: Selected Letters*, rev. ed. (Wilmette: Bahá'í Publishing Trust, 1982), p. 89.

95.6 With 'Abdu'l-Bahá's ascension, and more particularly with the passing of His well-beloved and illustrious sister the Most Exalted Leaf[2]—the last survivor of a glorious and heroic age—there draws to a close the first and most

---

95-2. Bahíyyih Khánum, also known as the Greatest Holy Leaf, the daughter of Bahá'u'lláh and the sister of 'Abdu'l-Bahá.

moving chapter of Bahá'í history, marking the conclusion of the Primitive, the Apostolic Age of the Faith of Bahá'u'lláh.³ It was 'Abdu'l-Bahá Who, through the provisions of His weighty Will and Testament, has forged the vital link which must for ever connect the age that has just expired with the one we now live in—the Transitional and Formative period of the Faith—a stage that must in the fullness of time reach its blossom and yield its fruit in the exploits and triumphs that are to herald the Golden Age of the Revelation of Bahá'u'lláh.

95.7 Dearly beloved friends! The onrushing forces so miraculously released through the agency of two independent and swiftly successive Manifestations are now under our very eyes and through the care of the chosen stewards of a far-flung Faith being gradually mustered and disciplined. They are slowly crystallizing into institutions that will come to be regarded as the hallmark and glory of the age we are called upon to establish and by our deeds immortalize. For upon our present-day efforts, and above all upon the extent to which we strive to remodel our lives after the pattern of sublime heroism associated with those gone before us, must depend the efficacy of the instruments we now fashion—instruments that must erect the structure of that blissful Commonwealth which must signalize the Golden Age of our Faith.⁴

"The Dispensation of Bahá'u'lláh"—8 February 1934, *The World Order of Bahá'u'lláh*, p. 98.

95.8 'Abdu'l-Bahá, Who incarnates an institution for which we can find no parallel whatsoever in any of the world's recognized religious systems, may be said to have closed the Age to which He Himself belonged and opened the one in which we are now laboring. His Will and Testament should thus be regarded as the perpetual, the indissoluble link which the mind of Him Who is the Mystery of God has conceived in order to insure the continuity of the three ages⁵ that constitute the component parts of the Bahá'í Dispensation. The period in which the seed of the Faith had been slowly germinating is thus intertwined both with the one which must witness its efflorescence and the subsequent age in which that seed will have finally yielded its golden fruit.

95.9 The creative energies released by the Law of Bahá'u'lláh, permeating and evolving within the mind of 'Abdu'l-Bahá, have, by their very impact and close interaction, given birth to an Instrument which may be viewed as the Charter of the New World Order which is at once the glory and the promise of

---

95-3. The Apostolic Age of the Bahá'í Faith, also referred to as the Heroic Age, began in 1844 and concluded in 1921 with the passing of 'Abdu'l-Bahá. For more details, see the entry on Ages in the glossary.

95-4. The "blissful Commonwealth" is the future worldwide community of Bahá'í nations, states, and localities that will give birth to a Bahá'í civilization.

95-5. The three Ages of the Bahá'í Dispensation are the Heroic, Formative, and Golden Ages. For more information, see the entries for Ages and Dispensation in the glossary.

this most great Dispensation. The Will may thus be acclaimed as the inevitable offspring resulting from that mystic intercourse between Him Who communicated the generating influence of His divine Purpose and the One Who was its vehicle and chosen recipient. Being the Child of the Covenant—the Heir of both the Originator and the Interpreter of the Law of God—the Will and Testament of 'Abdu'l-Bahá can no more be divorced from Him Who supplied the original and motivating impulse than from the One Who ultimately conceived it. Bahá'u'lláh's inscrutable purpose, we must ever bear in mind, has been so thoroughly infused into the conduct of 'Abdu'l-Bahá, and their motives have been so closely wedded together, that the mere attempt to dissociate the teachings of the former from any system which the ideal Exemplar of those same teachings has established would amount to a repudiation of one of the most sacred and basic truths of the Faith.

95.10 The Administrative Order, which ever since 'Abdu'l-Bahá's ascension has evolved and is taking shape under our very eyes in no fewer than forty countries of the world, may be considered as the framework of the Will itself, the inviolable stronghold wherein this newborn child is being nurtured and developed. This Administrative Order, as it expands and consolidates itself, will no doubt manifest the potentialities and reveal the full implications of this momentous Document—this most remarkable expression of the Will of One of the most remarkable Figures of the Dispensation of Bahá'u'lláh. It will, as its component parts, its organic institutions, begin to function with efficiency and vigor, assert its claim and demonstrate its capacity to be regarded not only as the nucleus but the very pattern of the New World Order destined to embrace in the fullness of time the whole of mankind.

"The Dispensation of Bahá'u'lláh"—8 February 1934, *The World Order of Bahá'u'lláh*, pp. 143–44.

95.11 Dearly beloved friends: Though the Revelation of Bahá'u'lláh has been delivered, the World Order which such a Revelation must needs beget is as yet unborn. Though the Heroic Age of His Faith is passed, the creative energies which that Age has released have not as yet crystallized into that world society which, in the fullness of time, is to mirror forth the brightness of His glory. Though the framework of His Administrative Order has been erected, and the Formative Period of the Bahá'í Era has begun, yet the promised Kingdom into which the seed of His institutions must ripen remains as yet uninaugurated. . . .

95.12 "The heights," Bahá'u'lláh Himself testifies, "which, through the most gracious favor of God, mortal man can attain in this Day are as yet unrevealed to his sight. The world of being hath never had, nor doth it yet possess, the capacity for such a revelation. The day, however, is approaching when the potentialities of so great a favor will, by virtue of His behest, be manifested unto men."

For the revelation of so great a favor a period of intense turmoil and widespread suffering would seem to be indispensable. Resplendent as has been the Age that has witnessed the inception of the Mission with which Bahá'u'lláh has been entrusted, the interval which must elapse ere that Age yields its choicest fruit must, it is becoming increasingly apparent, be overshadowed by such moral and social gloom as can alone prepare an unrepentant humanity for the prize she is destined to inherit. . . . 95.13

As we view the world around us, we are compelled to observe the manifold evidences of that universal fermentation which, in every continent of the globe and in every department of human life, be it religious, social, economic or political, is purging and reshaping humanity in anticipation of the Day when the wholeness of the human race will have been recognized and its unity established. A twofold process, however, can be distinguished, each tending, in its own way and with an accelerated momentum, to bring to a climax the forces that are transforming the face of our planet. The first is essentially an integrating process, while the second is fundamentally disruptive. The former, as it steadily evolves, unfolds a System which may well serve as a pattern for that world polity towards which a strangely disordered world is continually advancing; while the latter, as its disintegrating influence deepens, tends to tear down, with increasing violence, the antiquated barriers that seek to block humanity's progress towards its destined goal. The constructive process stands associated with the nascent Faith of Bahá'u'lláh, and is the harbinger of the New World Order that Faith must erelong establish. The destructive forces that characterize the other should be identified with a civilization that has refused to answer to the expectation of a new age, and is consequently falling into chaos and decline. 95.14

A titanic, a spiritual struggle, unparalleled in its magnitude yet unspeakably glorious in its ultimate consequences, is being waged as a result of these opposing tendencies, in this age of transition through which the organized community of the followers of Bahá'u'lláh and mankind as a whole are passing. . . . 95.15

It is not my purpose to call to mind, much less to attempt a detailed analysis of, the spiritual struggles that have ensued, or to note the victories that have redounded to the glory of the Faith of Bahá'u'lláh since the day of its foundation. My chief concern is not with the happenings that have distinguished the First, the Apostolic Age of the Bahá'í Dispensation, but rather with the outstanding events that are transpiring in, and the tendencies which characterize, the formative period of its development, this Age of Transition, whose tribulations are the precursors of that Era of blissful felicity which is to incarnate God's ultimate purpose for all mankind. 95.16

"The Unfoldment of World Civilization"—11 March 1936, *The World Order of Bahá'u'lláh,* pp. 168–71.

95.17 The moment had now arrived for that undying, that world-vitalizing Spirit that was born in S͟híráz, that had been rekindled in Ṭihrán, that had been fanned into flame in Bag͟hdád and Adrianople, that had been carried to the West, and was now illuminating the fringes of five continents, to incarnate itself in institutions designed to canalize its outspreading energies and stimulate its growth. The Age that had witnessed the birth and rise of the Faith had now closed. The Heroic, the Apostolic Age of the Dispensation of Bahá'u'lláh, that primitive period in which its Founders had lived, in which its life had been generated, in which its greatest heroes had struggled and quaffed the cup of martyrdom, and its pristine foundations been established—a period whose splendors no victories in this or any future age, however brilliant, can rival—had now terminated with the passing of One Whose mission may be regarded as the link binding the Age in which the seed of the newborn Message had been incubating and those which are destined to witness its efflorescence and ultimate fruition.

95.18 The Formative Period, the Iron Age, of that Dispensation was now beginning, the Age in which the institutions, local, national and international, of the Faith of Bahá'u'lláh were to take shape, develop and become fully consolidated, in anticipation of the third, the last, the Golden Age destined to witness the emergence of a world-embracing Order enshrining the ultimate fruit of God's latest Revelation to mankind, a fruit whose maturity must signalize the establishment of a world civilization and the formal inauguration of the Kingdom of the Father upon earth as promised by Jesus Christ Himself. . . .

95.19 The last twenty-three years of the first Bahá'í century may thus be regarded as the initial stage of the Formative Period of the Faith, an Age of Transition to be identified with the rise and establishment of the Administrative Order, upon which the institutions of the future Bahá'í World Commonwealth must needs be ultimately erected in the Golden Age that must witness the consummation of the Bahá'í Dispensation. The Charter which called into being, outlined the features and set in motion the processes of, this Administrative Order is none other than the Will and Testament of 'Abdu'l-Bahá, His greatest legacy to posterity, the brightest emanation of His mind and the mightiest instrument forged to insure the continuity of the three ages which constitute the component parts of His Father's Dispensation. . . .

95.20 The Administrative Order which this historic Document has established, it should be noted, is, by virtue of its origin and character, unique in the annals of the world's religious systems. . . .

95.21 The Document establishing that Order, the Charter of a future world civilization, which may be regarded in some of its features as supplementary to no less weighty a Book than the Kitáb-i-Aqdas; . . .

God Passes By, rev. ed. (Wilmette: Bahá'í Publishing Trust, 1987), pp. 324–28.

The first seventy-seven years of the preceding century, constituting the 95.22
Apostolic and Heroic Age of our Faith, fell into three distinct epochs, of nine,
of thirty-nine and of twenty-nine years' duration, associated respectively with
the Bábí Dispensation and the ministries of Bahá'u'lláh and of 'Abdu'l-Bahá.[6]
This Primitive Age of the Bahá'í Era, unapproached in spiritual fecundity by
any period associated with the mission of the Founder of any previous Dispensation, was impregnated, from its inception to its termination, with the
creative energies generated through the advent of two independent Manifestations and the establishment of a Covenant unique in the spiritual annals of
mankind.

The last twenty-three years[7] of that same century coincided with the first 95.23
epoch of the second, the Iron and Formative, Age of the Dispensation of
Bahá'u'lláh—the first of a series of epochs which must precede the inception
of the last and Golden Age of that Dispensation—a Dispensation which, as
the Author of the Faith has Himself categorically asserted, must extend over
a period of no less than one thousand years, and which will constitute the
first stage in a series of Dispensations, to be established by future Manifestations, all deriving their inspiration from the Author of the Bahá'í Revelation,
and destined to last, in their aggregate, no less than five thousand centuries. . . .

During this Formative Age of the Faith, and in the course of present and 95.24
succeeding epochs, the last and crowning stage in the erection of the framework of the Administrative Order of the Faith of Bahá'u'lláh—the election of
the Universal House of Justice—will have been completed, the Kitáb-i-Aqdas,
the Mother-Book of His Revelation, will have been codified and its laws promulgated, the Lesser Peace will have been established, the unity of mankind
will have been achieved and its maturity attained, the Plan conceived by
'Abdu'l-Bahá will have been executed, the emancipation of the Faith from the
fetters of religious orthodoxy will have been effected, and its independent religious status will have been universally recognized, whilst in the course of the
Golden Age, destined to consummate the Dispensation itself, the banner of
the Most Great Peace, promised by its Author, will have been unfurled, the
World Bahá'í Commonwealth will have emerged in the plenitude of its power
and splendor, and the birth and efflorescence of a world civilization, the child
of that Peace, will have conferred its inestimable blessings upon all mankind.

"The Challenging Requirements of the Present Hour"—5 June 1947, *Citadel
of Faith: Messages to America 1947–1957* (Wilmette: Bahá'í Publishing Trust,
1980), pp. 4–6.

95-6. The ministry of the Báb lasted from 1844 to 1853; the ministry of Bahá'u'lláh, 1853 to
1892; and the ministry of 'Abdu'l-Bahá, 1892 to 1921.

95-7. 1921–44.

96
Riḍván Message 1971
RIḌVÁN 1971

To the Bahá'ís of the World

Dearly loved Friends,

96.1 On November 28th 1971 the Bahá'í World will commemorate the fiftieth anniversary of the Passing of 'Abdu'l-Bahá, the Center of the Covenant, the Ensign of the Oneness of Mankind, the Mystery of God, an event which signalized at once the end of the Heroic Age of our Faith, the opening of the Formative Age and the birth of the Administrative Order, the nucleus and pattern of the World Order of Bahá'u'lláh.¹ As we contemplate the fruits of the Master's Ministry harvested during the first fifty years of the Formative Age, a period dominated by the dynamic and beloved figure of Shoghi Effendi, whose life was dedicated to the systematic implementation of the provisions of the Will and Testament of 'Abdu'l-Bahá and of the Tablets of the Divine Plan—the two charters provided by the Master for the administration and the teaching of the Cause of God—we may well experience a sense of awe at the prospect of the next fifty years. That first half-century of the Formative Age has seen the Bahá'í Community grow from a few hundred centers in 35 countries in 1921, to over 46,000 centers in 135 independent states and 182 significant territories and islands at the present day, has been marked by the raising throughout the world of the framework of the Administrative Order, which in its turn has brought recognition of the Faith by many governments and civil authorities and accreditation in consultative status to the Economic and Social Council of the United Nations, and has witnessed the spread to many parts of the world of that "entry by troops" promised by the Master and so long and so eagerly anticipated by the friends.²

96.2 A new horizon, bright with intimations of thrilling developments in the unfolding life of the Cause of God, is now discernible. The approach to it is complete victory in the Nine Year Plan. For we should never forget that the beloved Guardian's Ten Year Crusade, the current Nine Year Plan, other plans to follow throughout successive epochs of the Formative Age of the Faith, are all phases in the implementation of the Divine Plan of 'Abdu'l-Bahá, set out in fourteen of His Tablets to North America.

96-1. The Heroic Age (1844–1921) spanned the ministries of the Báb, Bahá'u'lláh, and 'Abdu'l-Bahá. For a fuller explanation of the Heroic and the Formative Ages, see the entry on Ages in the glossary. For information on the Administrative Order and the World Order of Bahá'u'lláh, see the glossary.

96-2. CF, p. 117.

Review of Significant Achievements

96.3 The Nine Year Plan is well advanced, and this Riḍván will witness the establishment of seven more National Spiritual Assemblies, five in Africa, one in South America and one in the Pacific, bringing the total number of these exalted bodies to 101. Next Riḍván the nine already announced will be formed, together with 4 more, one each in Afghanistan, Arabia, the Windward Islands and Puerto Rico, bringing the total to 114, six more than called for in the Nine Year Plan. The members of all National Spiritual Assemblies which will be elected at Riḍván 1972 will take part in the election of the Universal House of Justice at Riḍván 1973, when an international convention will be held at the World Center.

96.4 The Mother Temple of Latin America, the Mashriqu'l-Adhkár of Panama, is scheduled to be completed by December 1971 and its dedication will take place at the following Riḍván.[3]

96.5 The wonderful spirit released at the four Oceanic and Intercontinental Conferences, together with the practical benefits which accrued to the Cause from them, reinforce our high hopes that the four Conferences to be held this year will be resounding successes and result in more pioneers, more traveling teachers, greater proclamation of the Message and a raising of the spirits and devotion of the friends.

96.6 Our appeal to the friends in December 1970 for support of the Bahá'í International Fund,[4] which had reached a serious condition due to various unforeseen circumstances, has had a magnificent response from many quarters of the worldwide Bahá'í Community, and we are heartened to believe that this manifestation of devotion and sacrifice, as it continues and becomes more widespread, will resolve the condition that had threatened to adversely affect the attainment of cherished goals of the Nine Year Plan.

Services of the Hands of the Cause of God

96.7 The travels and other services of the Hands of the Cause of God continually evoke our thankfulness and delight, even wonder and astonishment. Their deeds are such as to eclipse the acts of the apostles of old and to confer eternal splendor on this period of the Formative Age. On behalf of all the friends everywhere, we offer them our reverent love and gratitude. It is fitting to record here the passing, after seventy years' exemplary service to the Faith, of the Hand of the Cause Agnes Alexander, whose early services in Hawaii were said by the Master to be greater than if she had founded an empire.[5]

96-3. The Bahá'í House of Worship in Panama was dedicated 29–30 April 1972. See message no. 108.
96-4. See message no. 87.
96-5. See message no. 90 on Agnes Alexander's passing.

96.8 Restrictive measures, directed against the Faith, and varying in severity from outright oppression to imposition of disabilities make virtually impossible the achievement of the goals of the Nine Year Plan in a number of countries, particularly in the Middle East, in North West Africa, along the fringes of East Africa and certain areas in South East Asia. It is hoped that those Bahá'í communities which enjoy freedom to teach their Faith will so far surpass their own goals as to amply compensate for the disabilities suffered by their less fortunate brothers. The army of traveling teachers must be reinforced and the friends, particularly Bahá'í youth, are called to seriously consider how much time they can offer to the Faith during the remaining two years of the Nine Year Plan. Teaching visits of brief or long duration, deputization of others, the undertaking of such tasks as would free other friends for teaching work, are all means of building up, in unison, that final surge which will carry the Plan to victory.

Two Objectives—Forming Assemblies and Opening Localities

96.9 Two major objectives of the Plan are the formation of new Local Spiritual Assemblies and the opening of new localities. 14,966 Local Spiritual Assemblies are called for; 10,360 are now in existence. 54,503 localities must claim a Bahá'í resident; 46,334 do so now. The goal is in sight, the time short. However, the growth reflected in the above statistics has not taken place at all levels and in all areas. For while a number of national communities have already achieved, or even surpassed the goals assigned to them, many face extreme difficulties in attaining theirs. With mutual help and an increase in the momentum already generated there is no doubt that the community of the Most Great Name is capable of sweeping on to total victory, thereby gaining a view of those enthralling vistas at present beyond the horizon.

Immediate and Future Tasks

96.10 The twin processes so clearly described by the beloved Guardian in his essay "The Unfoldment of World Civilization"—the steady progress and consolidation of the Cause of God on the one hand and the progressive disintegration of a moribund world on the other—will undoubtedly impose upon us new tasks, the obligation of devising new approaches to teaching, of demonstrating more clearly to a disillusioned world the Bahá'í way of life and making more effective the administrative institutions of the Faith.[6] The authority and influence of National and Local Spiritual Assemblies will have to be strengthened in order to deal with larger Bahá'í communities; the international character of the Cause will need to be developed, while the international teaching agency at the World Center, already referred to in previous general letters, will be established.

96-6. WOB, pp. 161–206.

However fascinating such considerations, which are likely to be forced upon our attention in the near future, may be, they must not deflect our energies and will from the immediate task—the goals of the Nine Year Plan. Their achievement is the best preparation for the future and the means of developing new powers and capacities in the Bahá'í Community. We are confident that the Army of Light, growing in strength and unity will, by 1973, the centenary year of the revelation of the Kitáb-i-Aqdas, have scaled the heights of yet another peak in the path leading ultimately to the broad uplands of the Most Great Peace. 96.11

With loving Bahá'í greetings,
THE UNIVERSAL HOUSE OF JUSTICE

97
Message to the Caribbean Conference—May 1971
MAY 1971

To the Friends of God gathered in the Caribbean Conference

Warmest greetings!

How propitious that on its mountaintop between the two greatest oceans and the two American continents the Mother Temple of Latin America is rising now in Panama, a land blessed by 'Abdu'l-Bahá's prophecy that "in the future it will gain most great importance."[1] How splendid that the vision projected in the Divine Plan for the Americas has sprung into such vibrant life in this Caribbean basin, in country after country upon its verdant shores, in island after island across its expanse, all named by the Master in His Tablets. What shall we not witness erelong in these places so charged with destiny through the Master's utterances! 97.1

The Nine Year Plan, the current stage in the unfoldment of the Divine Plan of 'Abdu'l-Bahá, is approaching its triumphant end. This Conference is an occasion to sum up what has been won, to determine to achieve the remaining goals for expansion in these blessed lands, and to consolidate the old and new communities of the Most Great Name. Indeed, the winning of our grand Bahá'í objectives began just yesterday when, in the early years of the Formative Age, a few travelers crossed the Caribbean. Yet it was not until the successive Plans of the beloved Guardian, culminating in the Ten Year Crusade, when 27 Knights of Bahá'u'lláh settled throughout this vast area, that the Cause took firm root. By 1963 the countries and islands of the Caribbean claimed less than 400 localities and only 147 Local Spiritual Assemblies. Now 97.2

97-1. TDP 6.9.

Bahá'ís are to be found in over 2500 localities, more than 500 Local Assemblies and 16 National Spiritual Assemblies have been formed, and there have been hundreds of concrete achievements which have brought about our recognition as an independent Faith.

97.3 The Americas have been a melting pot and a meeting place for the races of men, and the need is acute for the fulfillment of God's promises of the realization of the oneness of mankind. Particularly do the Master and the Guardian point to the Afro-Americans and the Amerindians, two great ethnic groups whose spiritual powers will be released through their response to the Creative Word. But our Teachings must touch all, must include all peoples. And, in this hour of your tireless activity, what special rewards shall come to those who will arise, summoned by 'Abdu'l-Bahá's Words: "Now is the time for you to divest yourselves of the garment of attachment to this world that perisheth, to be wholly severed from the physical world, become heavenly angels, and travel to these countries."²

97.4 The time is short, the needs many. No effort can be foregone, no opportunity wasted. Praised be God that you have gathered in this Conference to consult upon the vital requirements of this highly significant moment. Our prayers ascend at the Holy Threshold that every session of this historic meeting will attract Divine Blessings, and that each soul, armed with the love of God and imbued with His purpose for a struggling mankind, will arise to activate, beyond all present hopes, the vast spiritual potentialities of the Americas.

97.5 To each of you we send our deepest love.

THE UNIVERSAL HOUSE OF JUSTICE

98
Message to the South Pacific Oceanic Conference—May 1971
MAY 1971

To the Friends of God assembled
in the Conference of the South Pacific Ocean

Dearly loved Friends,

98.1 We send our warmest greetings and deepest love on the occasion of the first Conference in the heart of the Pacific Ocean. Praise be to God that you have gathered to consult on the vital needs of the hour!

98.2 Recalling the promise of Bahá'u'lláh "Should they attempt to conceal His light on the continent, He will assuredly rear His head in the midmost heart

---

97-2. TDP 6.13.

of the ocean and, raising His voice, proclaim: 'I am the lifegiver of the world!'"¹ we now witness its fulfillment in the vast area of the Pacific Ocean, in island after island mentioned by the Master in the Tablets of the Divine Plan.² How great is the potential for the Faith in localities blessed by these references!

98.3 At the inception of the Formative Age,³ the Cause was little known here. Agnes Alexander had brought the Teachings to the Hawaiian Islands. Father and Mother Dunn had only recently arrived in Australia. Later the name of Martha Root was to be emblazoned across the Pacific.⁴ Still later, at the beginning of the Ten Year Crusade, a vanguard of twenty-one Knights of Bahá'u'lláh raised His call as they settled in the islands of this great Ocean. The names of these valiant souls, together with the names of the army of pioneers and teachers who followed, will be forever enshrined in the annals of the Faith.

98.4 Their mighty endeavors brought about the enrollment of thousands of the peoples of Polynesia, Micronesia and Melanesia under the banner of the Most Great Name, the opening in Australasia of more than 800 centers and the establishment of ten pillars of the Universal House of Justice.⁵ We can but marvel at such triumphs attained despite great difficulties imposed by the vast expanse of ocean separating the island communities, especially when it is recalled that in many of these islands even the Christian Gospel was unknown as late as the 1830s.

98-1. WOB, p. 108.
98-2. TDP 6.13.
98-3. The second Age of the Bahá'í Dispensation, also called the Age of Transition or the Iron Age, the Formative Age began in 1921. For details, see the entry on Ages in the glossary.
98-4. Agnes Alexander became a Bahá'í in Rome in 1900 and on 26 December 1901 returned to her home in the Hawaiian Islands. After receiving a Tablet from 'Abdu'l-Bahá encouraging her to travel to Japan to teach the Faith, she went there in 1914 and remained for twenty-three years. John Henry Hyde Dunn and his wife, Clara Dunn, also arose spontaneously after reading 'Abdu'l-Bahá's statement in *Tablets of the Divine Plan* (7.8), "O that I could travel, even though on foot and in the utmost poverty, to these regions, and, raising the call of 'Yá Bahá'u'l-Abhá' in cities, villages, mountains, deserts and oceans, promote the Divine teachings! This, alas, I cannot do. How intensely I deplore it! Please God, ye may achieve it." Mr. Dunn resigned his position, and the Dunns left their home in San Francisco for Australia, arriving on 18 April 1920. Mr. Dunn was sixty-two; Mrs. Dunn was fifty. They remained in Australia until their deaths in 1941 and 1960, respectively. Martha Root arose in 1919 in response to 'Abdu'l-Bahá's call and embarked upon the first of many journeys to which she dedicated herself throughout the remaining twenty years of her life. All four were elevated to the rank of Hand of the Cause of God, Miss Root being called by Shoghi Effendi the "foremost Hand which 'Abdu'l-Bahá's will has raised up first Bahá'í century." For accounts of their lives and services, see: Agnes Alexander—BW 15:423–30; Hyde Dunn—BW 9:593–97; Clara Dunn—BW 13:859–62; and Martha Root—BW 13:643–48, and Mabel Garis, *Martha Root*. See also O. Z. Whitehead, "Father and Mother Dunn," in *Some Bahá'ís to Remember*, pp. 153–75. For the Universal House of Justice's message about Miss Alexander's passing, see no. 90.
98-5. The "ten pillars" referred to are the National Spiritual Assemblies of Australia, Fiji Islands, Gilbert and Ellice Islands, Hawaiian Islands, New Zealand, North West Pacific Ocean, Samoa, Solomon Islands, South West Pacific Ocean, and Tonga and the Cook Islands.

98.5 How great is the responsibility to continue spreading the Word of God throughout the Pacific. It was in the Tablets of the Divine Plan that 'Abdu'l-Bahá called for teachers "speaking their languages, severed, holy, sanctified and filled with the love of God," to "turn their faces to and travel through the three great Island groups of the Pacific Ocean—Polynesia, Micronesia and Melanesia . . . With hearts overflowing with the love of God, with tongues commemorating the mention of God" to "deliver the Glad Tidings of the manifestation of the Lord of Hosts to all the people."[6]

98.6 The Nine Year Plan, the current phase of the unfoldment of the Divine Plan, is now approaching its final stages. It is incumbent on the friends to assess what has been accomplished and to anticipate and plan for such rapid acceleration of the teaching and consolidation work as is necessary to win all goals by 1973. Time is short; the needs critical. No effort must be spared; no opportunity overlooked.

98.7 Our prayers ascend at the Holy Threshold that every session of this historic meeting will attract Divine blessings, and that the friends will go forth, armed with the love of God and enthusiasm born of the Spirit, fully prepared to scale the heights of victory!

With loving Bahá'í greetings,
The Universal House of Justice

99
Commemoration of the Fiftieth Anniversary of the Passing of 'Abdu'l-Bahá
12 July 1971

To all National Spiritual Assemblies

Dear Bahá'í Friends,

99.1 We have noted with deep satisfaction that some National Spiritual Assemblies have already initiated plans to befittingly commemorate the Fiftieth Anniversary of the passing of 'Abdu'l-Bahá and the inception of the Formative Age of the Bahá'í Dispensation.[1]

99.2 We feel it would be highly fitting for the three days, November 26 to 28, during which the Day of the Covenant and the Anniversary of the Ascension

---

98-6. TDP 7.5.

99-1. For an account of the commemoration in the Holy Land and throughout the world of the fiftieth anniversary of 'Abdu'l-Bahá's passing, see BW 15:125–28.

For an abridged version of the account by Shoghi Effendi and Lady Blomfield of the passing of 'Abdu'l-Bahá, see BW 15:113–24.

of 'Abdu'l-Bahá occur, to be set aside this year by all National Spiritual Assemblies for specially arranged gatherings and conferences, convened either nationally or locally or both, on the three following main themes: The Bahá'í Covenant, The Formative Age, and The Life of 'Abdu'l-Bahá.

We hope that these gatherings will serve to intensify the consecration of the workers in the Divine Vineyard in every land, and provide them with the opportunity, especially in the watches of the night of that Ascension, when they will be commemorating the passing hour of our Beloved Master, to renew their pledge to Bahá'u'lláh and to rededicate themselves to the accomplishment of the as yet unfulfilled goals of the Nine Year Plan. 99.3

The Hands of the Cause in the Holy Land, the members of the Universal House of Justice, and all resident and visiting believers at the World Center will, on that memory-laden night, visit the Shrine of that Mystery of God on behalf of the entire Community of the Blessed Beauty and will supplicate for the stalwart champions of the Faith laboring in the forefront of so many fields of service and winning fresh triumphs in His Name, for the self-sacrificing believers without whose support and sustained assistance most of these victories could not be achieved, and for those who will be inspired to join the ranks of the active and dedicated promoters of His glorious Cause at this crucial stage in the development of the Plan, that we may all meet our obligations and discharge our sacred trust, thus making it possible in the latter months of the Plan for our entire resources to be devoted to an even greater expansion of the Faith in its onward march towards the spiritual conquest of the planet. 99.4

With loving Bahá'í greetings,
THE UNIVERSAL HOUSE OF JUSTICE

100
Message to the North Pacific Oceanic Conference—September 1971
SEPTEMBER 1971

To the Friends of God assembled
in the Conference of the North Pacific Ocean

Dearly loved Friends,

On the eve of the Fiftieth Anniversary of the opening of the Formative Age of our Faith we call to mind the high hopes often expressed by the beloved Master for the spread of the Cause in this region, His mention in the Tablets of the Divine Plan of many of the territories represented in this Conference, 100.1

and the faithful and devoted services of that maidservant of Bahá'u'lláh, the Hand of the Cause Agnes Alexander, who brought the Teachings to these shores in the early years of this century.[1]

100.2 In these days we are witnessing an unprecedented acceleration of the teaching work in almost every part of the globe. In the North Pacific Ocean area great strides have been made in the advancement of the Cause since that historic Asia Regional Teaching Conference in Nikko just sixteen years ago. The next two years witnessed the formation of the National Spiritual Assembly of Alaska and of the Regional National Spiritual Assembly of North East Asia. To the Convention in Tokyo at Riḍván 1957 the Guardian addressed these prophetic words:

100.2a This auspicious event, which posterity will regard as the culmination of a process initiated, half a century ago, in the capital city of Japan, ... marks the opening of the second chapter in the history of the evolution of His Faith in the North Pacific area. Such a consummation cannot fail to lend a tremendous impetus to its onward march in the entire Pacific Ocean ...

100.3 Since that time National Spiritual Assemblies have also been firmly established in Korea and Taiwan.

100.4 Hokkaido, the site of this Conference, first heard of the Teachings less than fifteen years ago, and the first aboriginal peoples of this land accepted Bahá'u'lláh just over a decade ago. Now you are the witnesses to the beginnings of a rapid increase in the number of believers. Peoples in other islands and lands of the North Pacific, including the Ryukyus, Guam, the Trust Territories, the western shores of Canada and Alaska and the Aleutians are also enrolling under the banner of the Most Great Name, and next Riḍván yet another pillar of the Universal House of Justice is to be raised in Micronesia. We are heartened at the prospect that from the indigenous peoples of this vast oceanic area—the Ainu, the Japanese, the Chinese, the Koreans, the Okinawans, the Micronesians, the American Indians, the Eskimos, and the Aleuts—vast numbers will soon enter the Faith.

100.5 The final hours of the Nine Year Plan are fast fleeting. Praise be to God that you have gathered to consult on ways and means of assuring complete victory so that from these outposts the Teachings may spread to those nearby lands where teeming millions have not as yet heard of the advent of this Most Great Dispensation.

100-1. Agnes Alexander traveled to Japan at 'Abdu'l-Bahá's encouragement and remained there for twenty-three years (1914–37). For an account of her life and services, see BW 15:423–30. For the Universal House of Justice's message upon her passing, see no. 90; see also the May 1971 message to the South Pacific Oceanic Conference, (no. 98).

The sweet perfume of victory is in the air, and we must hasten to achieve it while there is yet time. Vital goals, particularly on the homefronts of Taiwan and Japan, remain to be won, and everywhere the roots of the faith of the believers must sink deeper and deeper into the firm earth of the Teachings lest tempests and trials as yet unforeseen shake or uproot the tender plants so lovingly raised in the islands of this great Ocean and the lands surrounding it. 100.6

As you and the friends in the sister Conference in Reykjavik bring this series of eight Oceanic and Continental Conferences to a triumphant close, our prayers for the success of your deliberations ascend at the Holy Threshold. May God grant you the resources, the strength, and the determination to attain your highest hopes, and enable you to open a new and glorious chapter in the evolution of His Faith in the North Pacific area. 100.7

With loving Bahá'í greetings,
THE UNIVERSAL HOUSE OF JUSTICE

101
Message to the North Atlantic Conference—September 1971
SEPTEMBER 1971

To the Friends assembled
in the North Atlantic Conference in Reykjavik

Dearly loved Friends,

To each and every one of you in this historic Conference we send our most cordial and loving greetings. The famous island in which you are now gathered, so strategically placed between the two great continents flanking the vast oceanic area which surrounds it, to which the Teachings of Christ were brought a millennium ago, and which, in this Dispensation, was mentioned by the Center of the Covenant in His Tablets of the Divine Plan,[1] first heard the Name of Bahá'u'lláh in 1924 when the Hand of the Cause Amelia Collins stopped briefly in Reykjavik and made the acquaintance of Hólmfríður Árnadóttir who subsequently became the first Bahá'í of Iceland. Eleven years later the beloved Martha Root spent a month in this land which she loved so well. On that occasion, with the help of Hólmfríður, the Cause of Bahá'u'lláh was widely proclaimed in the press, on the radio and from the lecture platform. 101.1

101-1. TDP 7.15.

101.2   The great Ocean extending from the equator to the Pole and from Europe to North America, which has been both the barrier and the link between the Old and the New Worlds, has played a highly significant part in the later history of mankind. Long before Columbus arrived in the West Indies the Vikings, forebears of Icelanders of today, were plying its northern waters. In later centuries wave upon wave of Europeans sailed from east to west, engaging in one of the most significant migrations in human history. In the twentieth century 'Abdu'l-Bahá Himself sailed across it and back, a voyage unique in the religious history of mankind and creating a remarkable parallel with the Light of the Cause itself, beaming from the East across the great Ocean to the heart of the North American Continent, being reflected back again, firing new beacon lights in Europe and in later years diffusing its radiance throughout the world. The great Republic[2] whose eastern shore forms part of the boundary of this Ocean has become the Cradle of the Administrative Order and at this present time the banner of the Most Great Name is being raised in island after island of this Ocean, two of which—Iceland and Ireland—will raise, next Riḍván, new pillars of the Universal House of Justice.

101.3   The Faith of God is flourishing in the lands around the North Atlantic; a new wind is blowing, promoting an upsurge of proclamation and teaching. In Europe the youth are afire with enthusiasm and vigor. In Canada and the United States a ground swell of unknown proportions is carrying Bahá'í communities to heights of unprecedented achievement.

101.4   You are gathered in this Conference to consult on ways and means of winning, in the few fleeting months ahead, the remaining goals of the Nine Year Plan. In Europe particularly there is much to be done, but we have full faith that the friends, galvanized by their love for Bahá'u'lláh and fortified by His promises of Divine assistance, will, with the enthusiasm which they already display, commit their resources to the tasks ahead and will surely attain the victory.

101.5   The beloved Master prayed that holy souls would arise from the Northern Territories of the West and become signs of God's guidance and standards of the Supreme Concourse. In one of the Tablets of the Divine Plan He refers to an inhospitable island of that area saying:

101.5a   Should the fire of the love of God be kindled in Greenland, all the ice of that country will be melted, and its cold weather become temperate—that is, if the hearts be touched with the heat of the love of God, that territory will become a divine rose garden and a heavenly paradise, and the souls, even as fruitful trees, will acquire the utmost freshness and beauty. Effort, the utmost effort is required. . . .[3]

101-2. The United States of America.
101-3. TDP 5.2.

As the friends gathered in Reykjavik and Sapporo bring this worldwide series of Oceanic and Continental Conferences to a triumphant close our thoughts are with you and our prayers on your behalf rise from the Sacred Threshold. May untold blessings and confirmations be showered upon you as you go forth to labor for the advancement of the Cause of God and may your brows be crowned with victory.

THE UNIVERSAL HOUSE OF JUSTICE

102
Passing of the Hand of the Cause of God Músá Banání
5 SEPTEMBER 1971

To the Hands of the Cause of God,
the Continental Boards of Counselors,
the National Spiritual Assemblies

Dear Bahá'í Friends,

The Hand of the Cause Músá Banání passed to the Abhá Kingdom at noon on Saturday, September 4th. The following cable was sent by the Universal House of Justice:

> PROFOUNDLY MOURN PASSING DEARLY LOVED HAND CAUSE MÚSÁ BANÁNÍ RECALL WITH DEEP AFFECTION HIS SELFLESS UNASSUMING PROLONGED SERVICES CRADLE FAITH HIS EXEMPLARY PIONEERING UGANDA CULMINATING HIS APPOINTMENT AS HAND CAUSE AFRICA AND PRAISE BELOVED GUARDIAN AS SPIRITUAL CONQUEROR THAT CONTINENT.[1] INTERMENT HIS REMAINS AFRICAN SOIL UNDER SHADOW MOTHER TEMPLE ENHANCES SPIRITUAL LUSTER THAT BLESSED SPOT. FERVENTLY PRAYING SHRINES PROGRESS HIS NOBLE SOUL. MAY AFRICA NOW ROBBED STAUNCH VENERABLE PROMOTER DEFENDER FAITH FOLLOW HIS EXAMPLE CHEER HIS HEART ABHÁ KINGDOM. CONVEY FAMILY MOST TENDER SYMPATHIES ADVISE HOLD MEMORIAL MEETINGS ALL COMMUNITIES BAHÁ'Í WORLD BEFITTING GATHERINGS MOTHER TEMPLES.

With loving Bahá'í greetings,
THE UNIVERSAL HOUSE OF JUSTICE

102-1. For an account of the life and services of Músá Banání, see BW 15:421–23.

103
Spiritual Character of Bahá'í Elections
11 November 1971

To all National Spiritual Assemblies

Dear Bahá'í Friends,

103.1 A compilation has been made from the writings of the beloved Guardian about the spiritual character of Bahá'í elections.[1] We are sharing these extracts with you to bring to the attention of the friends under your jurisdiction in any manner you deem advisable.

With loving Bahá'í greetings,
THE UNIVERSAL HOUSE OF JUSTICE

104
Formation of Thirteen New National Spiritual Assemblies during Riḍván 1972
7 December 1971

To all National Spiritual Assemblies

104.1 HAPPY ANNOUNCE DECISION ADD RWANDA LIST NATIONAL SPIRITUAL ASSEMBLIES TO BE FORMED NEXT RIḌVÁN. REPRESENTATIVES HOUSE JUSTICE FIRST CONVENTIONS AS FOLLOWS HANDS CAUSE AMATU'L-BAHÁ RÚḤÍYYIH KHÁNUM WINDWARD ISLANDS UGO GIACHERY PUERTO RICO 'ALÍ-AKBAR FURÚTAN NEPAL SHU'Á'U'LLÁH 'ALÁ'Í EAST PAKISTAN ADELBERT MÜHLSCHLEGEL RWANDA SEYCHELLES JALÁL KHÁZEH SINGAPORE ENOCH OLINGA ICELAND WILLIAM SEARS IRELAND COLLIS FEATHERSTONE NORTHWEST PACIFIC RAḤMATU'LLÁH MUHÁJIR MALAGASY REPUBLIC RÉUNION COUNSELOR HÁDÍ RAḤMÁNÍ AFGHANISTAN. PRAYING SHRINES OUTSTANDING SUCCESS THESE HISTORIC GATHERINGS ELECTING MEMBERS NEW PILLARS HOUSE JUSTICE FOCAL CENTERS SPIRITUAL INVIGORATION ILLUMINATION NEWLY EMERGING NATIONAL COMMUNITIES.

UNIVERSAL HOUSE OF JUSTICE

103-1. See CC 1:315–18 for a revised version provided by the Universal House of Justice in 1989.

105
World Center Developments—Erection of Obelisk and Extension of Gardens
13 December 1971

To all National Spiritual Assemblies

JOYOUSLY ANNOUNCE FURTHER DEVELOPMENTS WORLD CENTER. AFTER MANY YEARS DIFFICULT NEGOTIATIONS ERECTION OBELISK MARKING SITE FUTURE MA<u>SH</u>RIQU'L-A<u>DH</u>KÁR MOUNT CARMEL COMPLETED THUS FULFILLING PROJECT INITIATED BELOVED GUARDIAN EARLY YEARS CRUSADE. GARDENS BAHJÍ HAIFA EXTENDED BY DEVELOPMENT QUADRANT SOUTHEAST MANSION BAHÁ'U'LLÁH AND ESTABLISHMENT FORMAL GARDEN SOUTHWEST CORNER PROPERTY SURROUNDING SHRINE BÁB. 105.1

UNIVERSAL HOUSE OF JUSTICE

106
Surpassing the Nine Year Plan Goal for the Number of Bahá'í Localities
14 February 1972

To all National Spiritual Assemblies

OVERJOYED ANNOUNCE FRIENDS EVERY LAND NUMBER LOCALITIES NOW 56645 EXCEEDING BY OVER 2500 ORIGINAL GOAL NINE YEAR PLAN. OFFERING PRAYERS THANKSGIVING SACRED THRESHOLD FOR DIVINE BOUNTIES SURROUNDING SACRIFICIAL EFFORTS LOVE-INTOXICATED SUPPORTERS HIS BLESSED NAME. URGE BELIEVERS THOSE AREAS WHOSE TEACHING GOALS ARE STILL OUTSTANDING EXERT UTMOST EFFORT COURSE SWIFTLY PASSING REMAINING MONTHS PLAN WIN THEIR GOALS ENABLING THEM JOIN RANKS THEIR VICTORIOUS BRETHREN WHO ARE URGED CONTINUE THEIR VIGOROUS BRILLIANT EXPLOITS IN SERVICE GOD'S INFINITELY GLORIOUS CAUSE. 106.1

UNIVERSAL HOUSE OF JUSTICE

107
Release of a Compilation on Music and Singing
1 March 1972

To all National Spiritual Assemblies

Dear Bahá'í Friends,

107.1 In these days when music and singing are playing such an important and effective part in the teaching work, we feel it appropriate and timely to share with you the enclosed compilation of extracts from the writings of our Faith on the subject.[1]

107.2 We leave it to your discretion to decide in which manner and to what extent to share the attached extracts with the friends and communities under your jurisdiction.

With loving Bahá'í greetings,
THE UNIVERSAL HOUSE OF JUSTICE

108
Dedication of the Mother Temple of Latin America, Panama City, Panama
19 March 1972

To the Beloved of God gathered in
the Conference called on the occasion of the
Dedication of the Mother Temple of Latin America

Dear Bahá'í Friends,

108.1 With praise and gratitude to God the whole Bahá'í world acclaims the dedication of the Mother Temple of Latin America, an edifice which glorifies the Cause of Bahá'u'lláh at that point where, the beloved Master asserted, "the Occident and the Orient find each other united through the Panama Canal," where "The teachings, once established . . . , will unite the East and the West, the North and the South."[1]

108.2 This historic project, in a hemisphere of infinite spiritual potentiality, fulfills one of the most important goals of the Nine Year Plan, and brings untold joy to the hearts of the friends in every land. Privileged are they who share in the raising of this glorious Silent Teacher with deeds of loving generosity and sacrifice. A crown to the labors of all those who have striven to establish the Faith of Bahá'u'lláh in Latin America, this Mashriqu'l-Adhkár, the rally-

---

107-1. See CC 2:73–82.
108-1. TDP 14.8.

ing point for the Bahá'ís of those lands, whether they are of the blessed Indian peoples or represent the other races whose diversity enriches the nations of that hemisphere, will be a fountainhead of spiritual confirmations, and this mighty achievement will endow the Bahá'í Community with new and greater capacities, enabling the friends in Latin America, and particularly in this privileged land of Panama, to win victories that will eclipse all their past achievements.

108.3 The threefold task to which your attention is now directed comprises the proclamation, expansion and consolidation of the Faith. We urge you to concentrate your deliberations not only on the exchange of ideas for the prosecution of this task, but on ways and means for fostering collaboration among the Bahá'í Communities of Central and South America so that the most fruitful harvest may be gathered in all three aspects of the teaching work and enable you to achieve your remaining goals of the Nine Year Plan.

108.4 Our loving, ardent prayers will be offered at the Sacred Threshold, that the Almighty may inspire your discussions in this historic Conference and crown all your efforts with victory.[2]

THE UNIVERSAL HOUSE OF JUSTICE

109
Release of a Compilation on Bahá'í Schools and Institutes
10 APRIL 1972

To all National Spiritual Assemblies

Dear Bahá'í Friends,

109.1 In view of the increasing importance of Bahá'í summer schools, a number of excerpts from the writings of the beloved Guardian have been compiled which set forth objectives to be attained by and principles underlying the operation of this institution.[1] Many of these objectives and principles apply to Teaching Institutes, and a section of quotations from the Universal House of Justice on these Institutes is included.

109.2 We share these extracts with the friends in the hope that they will be assisted and guided in the development of these vital institutions of the Faith.

With loving Bahá'í greetings,
THE UNIVERSAL HOUSE OF JUSTICE

108-2. The International Teaching Conference held in conjunction with the dedication of the House of Worship was held in Panama City, 28 April–2 May 1972.
109-1. See CC 1:25–44.

110
Riḍván Message 1972
RIḌVÁN 1972

To the Bahá'ís of the World

Dearly loved Friends,

110.1 The opening of the final year of the Nine Year Plan sees the Bahá'í world community poised for overwhelming victory. With grateful hearts we acknowledge the continuing confirmations which have attended its efforts and the Divine bounties which have never ceased to rain down upon this blessed, this ever-developing embryonic world order.

Review of Significant Achievements

110.2 The Mashriqu'l-Adhkár of Panama, the Mother Temple of Latin America, will be dedicated this Riḍván. Three beloved Hands of the Cause, Amatu'l-Bahá Rúḥíyyih Khánum representing the Universal House of Justice, Ugo Giachery and Zikrullah Khadem will attend this historic ceremony. The imaginative and inspiring concept of the architect, Peter Tillotson, has been wonderfully realized and we extend to the National Spiritual Assembly of Panama on behalf of the entire Bahá'í world, loving congratulations on their achievement.

110.3 Although the dissolution of the National Spiritual Assembly of Iraq has, unhappily, resulted from the persecution of the Faith in that land, the thirteen new National Spiritual Assemblies which will come into being this Riḍván will bring the total number of these pillars of the Universal House of Justice to 113.

110.4 The goals requiring acquisition of properties and establishment of Teaching Institutes are well in hand and, in those countries where legal circumstances permit, incorporation of Assemblies and recognition of Bahá'í marriage and Holy Days are making good progress.

Teaching Goals

110.5 It is the teaching goals which must engage our attention and effort. Although more than 260 territories have achieved their assigned goals of localities where Bahá'ís reside, and in some cases have exceeded them, enabling the Bahá'í world community to rejoice in having outstripped on a world scale the total number of localities envisaged in the Plan, there are still some 60 territories where this goal is yet to be won and where its attainment must be given absolute priority between now and Riḍván 1973. It is expected that a large number of new Local Spiritual Assemblies will be established at Riḍván and immediately after the position of this goal is ascertained a detailed listing of all territories throughout the world which have not yet won their goals for localities and Local Spiritual Assemblies will be sent to every National Spiritual Assembly for urgent release to the friends.

International Cooperation

110.6 It is hoped that during this last year of the Plan the principle of collaboration between National Spiritual Assemblies will be extended far beyond the special tasks set in the Nine Year Plan. Those communities which have already attained their goals or are in clear sight of them should consider the world picture as disclosed by the listing mentioned above and do everything they can, without jeopardizing their own success, to assist their fellow communities with pioneers and traveling teachers, or in any other way possible. Such a process will greatly consolidate the unity and brotherhood of the Bahá'í world community.

Pioneering Goals

110.7 In the meantime we call on all believers everywhere to prayerfully consider their personal circumstances, and to arise while there is yet time, to fill the international pioneer goals of the Plan. There are 267 pioneer needs still to be answered—75 in Africa, 57 in the Americas, 40 in Asia, 30 in Australasia and 65 in Europe.

New Requirements of the Ever-Growing World Order

110.8 The extraordinary advances made since that Riḍván of 1964 when the Nine Year Plan was begun, continuing the organized and purposeful process of teaching on a world scale instituted by our beloved Guardian when he launched the Ten Year Crusade, force upon our attention new requirements of this ever-growing world order both for its own organic life and in relation to the disintegrating world society in which it is set. The divergence between the ways of the world and of the Cause of God becomes ever wider. And yet the two must come together. The Bahá'í community must demonstrate in ever-increasing measure its ability to redeem the disorderliness, the lack of cohesion, the permissiveness, the godlessness of modern society; the laws, the religious obligations, the observances of Bahá'í life, Bahá'í moral principles and standards of dignity, decency and reverence, must become deeply implanted in Bahá'í consciousness and increasingly inform and characterize this community. Such a process will require a great development in the maturity and effectiveness of Local Spiritual Assemblies. The purposes and standards of the Cause must be more and more understood and courageously upheld. The influence of the Continental Boards of Counselors and the work of their Auxiliary Boards must develop and spread through the entire fabric of the Bahá'í community. A vast systematic program for the production of Bahá'í literature must be promoted.

Our Task—Achieving Every Attainable Goal of the Plan

110.9 Our immediate and inescapable task, however, is to ensure that every attainable goal of the Nine Year Plan is achieved. This must be done at all costs. No sacrifice, no deferment of cherished plans must be refused in order to discharge this "most important" of the many "important" duties facing us. Who can doubt that one last supreme effort will be crowned with success? Even now the national community to bear the laurels of first achieving every task assigned to it, Fiji, leads the procession of rejoicing and victorious communities within the Army of Light. We may well emulate Bahá'í youth whose recent surge forward into the van of proclamation and teaching is one of the most encouraging and significant trends in the Faith, and who storm the gates of heaven for support in their enterprises by long-sustained, precedent and continuing prayer. We are all able to call upon Bahá'u'lláh for His Divine, all-powerful aid, and He will surely help us. For He is the Hearer of prayers, the Answerer.

THE UNIVERSAL HOUSE OF JUSTICE

111
Elucidation of the Nature of the Continental Boards of Counselors
24 APRIL 1972

To the Continental Boards of Counselors
and National Spiritual Assemblies

Beloved Friends,

111.1 Recently we have received queries from several sources about the nature of the Institution of the Continental Boards of Counselors and its relationship to the Institution of the Hands of the Cause, and we feel it is timely for us to give further elucidation.

111.2 As with so many aspects of the Administrative Order, understanding of this subject will develop and clarify with the passage of time as that Order grows organically in response to the power and guidance of Almighty God and in accordance with the needs of a rapidly developing worldwide community. However, certain aspects are already so clear as to require a proper understanding by the friends.

The Rulers and the Learned

111.3 In the Kitáb-i-'Ahdí (the Book of His Covenant) Bahá'u'lláh wrote "Blessed are the rulers and the learned among the people of Bahá,"[1] and referring to this very passage the beloved Guardian wrote on 4 November 1931:

III-1. TB, p. 221.

In this holy cycle the "learned" are, on the one hand, the Hands of 111.3a
the Cause of God, and, on the other, the teachers and diffusers of His
teachings who do not rank as Hands, but who have attained an eminent position in the teaching work. As to the "rulers" they refer to the
members of the Local, National and International Houses of Justice.
The duties of each of these souls will be determined in the future.
(Translated from the Persian)

The Learned

The Hands of the Cause of God, the Counselors and the members of the 111.4
Auxiliary Boards fall within the definition of the "learned" given by the
beloved Guardian. Thus they are all intimately interrelated and it is not incorrect to refer to the three ranks collectively as one institution.

However, each is also a separate institution in itself. The Institution of the 111.5
Hands of the Cause of God was brought into existence in the time of Bahá'u'lláh and when the Administrative Order was proclaimed and formally established by 'Abdu'l-Bahá in His Will, it became an auxiliary institution of the
Guardianship. The Auxiliary Boards, in their turn, were brought into being
by Shoghi Effendi as an auxiliary institution of the Hands of the Cause.

When, following the passing of Shoghi Effendi, the Universal House of 111.6
Justice decided that it could not legislate to make possible the appointment
of further Hands of the Cause, it became necessary for it to create a new institution, appointed by itself, to extend into the future the functions of protection and propagation vested in the Hands of the Cause and, with that in view,
so to develop the Institution of the Hands that it could nurture the new institution and function in close collaboration with it as long as possible. It was
also vital so to arrange matters as to make the most effective use of the unique
services of the Hands themselves.

The first step in this development was taken in November 1964 when the 111.7
Universal House of Justice formally related the Institution of the Hands to
itself by stating that "Responsibility for decisions on matters of general policy
affecting the institution of the Hands of the Cause, which was formerly exercised by the beloved Guardian, now devolves upon the Universal House of
Justice as the supreme and central institution of the Faith to which all must
turn." At that time the number of members of the Auxiliary Boards was
increased from 72 to 135, and the Hands of the Cause in each continent were
called upon to appoint one or more members of their Auxiliary Boards to act
in an executive capacity on behalf of and in the name of each Hand, thereby
assisting him in carrying out his work.

In June 1968 the Institution of the Continental Boards of Counselors was 111.8
brought into being, fulfilling the goal of extending the aforementioned functions of the Hands into the future, and this momentous decision was accompanied by the next step in the development of the Institution of the Hands

of the Cause: the continental Hands were to serve henceforth on a worldwide basis and operate individually in direct relationship to the Universal House of Justice; the Hands ceased to be responsible for the direction of the Auxiliary Boards, which became an auxiliary institution of the Continental Boards of Counselors; the Hands of the Cause residing in the Holy Land were given the task of acting as liaison between the Universal House of Justice and the Boards of Counselors; and the working interrelationships between the Hands and the Boards of Counselors were established. Reference was also made to the future establishment by the Universal House of Justice, with the assistance of the Hands residing in the Holy Land, of an international teaching center in the Holy Land.[2]

111.9 In July 1969 and at Riḍván 1970 further increases in the numbers of Counselors and Auxiliary Board members were made.

111.10 Other developments in the Institution of the Hands of the Cause and the Institution of the Continental Boards of Counselors will no doubt take place in future as the international teaching center comes into being and as the work of the Counselors expands.

Distinctions between the Rulers and the Learned

111.11 We have noted that the Hands, the Counselors and the Auxiliary Boards are sometimes referred to by the friends as the "appointive arm" of the Administrative Order in contradistinction to the Universal House of Justice and the National and Local Assemblies which constitute the "elective arm." While there is truth in this description as it applies to the method used in the creation of these institutions, the friends should understand that it is not only the fact of appointment that particularly distinguishes the institutions of the Hands, Counselors and Auxiliary Boards. There are, for instance, many more believers appointed to committees in the "elective arm" than are serving in the so-called "appointive arm." A more striking distinction is that whereas the "rulers" in the Cause function as corporate bodies, the "learned" operate primarily as individuals.

Exclusion of Past Evils and Retention of Beneficial Elements

111.12 In a letter written on 14 March 1927 to the Spiritual Assembly of the Bahá'ís of Istanbul, the Guardian's Secretary explained, on his behalf, the principle in the Cause of action by majority vote. He pointed out how, in the past, it was certain individuals who "accounted themselves as superior in knowledge and elevated in position" who caused division, and that it was those "who pretended to be the most distinguished of all" who "always proved themselves to be the source of contention." "But praise be to God," he continued, "that the Pen of Glory has done away with the unyielding and dictatorial views of the

---

111-2. For messages about the establishment and duties of the International Teaching Center, see nos. 131, 132, and 361.

learned and the wise, dismissed the assertions of individuals as an authoritative criterion, even though they were recognized as the most accomplished and learned among men and ordained that all matters be referred to authorized centers and specified Assemblies. Even so, no Assembly has been invested with the absolute authority to deal with such general matters as affect the interests of nations. Nay rather, He has brought all the assemblies together under the shadow of one House of Justice, one divinely appointed Center, so that there would be only one Center and all the rest integrated into a single body, revolving around one expressly designated Pivot, thus making them all proof against schism and division." (Translated from the Persian.)

111.13 Having permanently excluded the evils admittedly inherent in the institutions of the "learned" in past dispensations, Bahá'u'lláh has nevertheless embodied in His Administrative Order the beneficent elements which exist in such institutions, elements which are of fundamental value for the progress of the Cause, as can be gauged from even a cursory reading of the Guardian's message of 4 June 1957.

111.14 The existence of institutions of such exalted rank, comprising individuals who play such a vital role, who yet have no legislative, administrative or judicial authority, and are entirely devoid of priestly functions or the right to make authoritative interpretations, is a feature of Bahá'í administration unparalleled in the religions of the past. The newness and uniqueness of this concept make it difficult to grasp; only as the Bahá'í Community grows and the believers are increasingly able to contemplate its administrative structure uninfluenced by concepts from past ages, will the vital interdependence of the "rulers" and "learned" in the Faith be properly understood, and the inestimable value of their interaction be fully recognized.

With loving Bahá'í greetings,
THE UNIVERSAL HOUSE OF JUSTICE

112
Message to Bahá'í Unity Conference, Ganado, Arizona
18 May 1972

Bahá'í Unity Conference
Ganado, Arizona

Beloved Friends,

112.1 Praise be to the Almighty that you have gathered in that beautiful spot in a spirit of love and harmony for the purpose of strengthening the bonds of unity between yourselves and among all men.

112.2 The All-Wise Creator of earth and heaven has from the beginning which has no beginning sent to His peoples Divine Messengers to guide them to the Straight Path. These Wise Ones have come to establish the unity of the Kingdom in human hearts. This great evolutionary process of building the organic unity of the human race has entered a new stage with this mighty message of Bahá'u'lláh. His voice is the voice of the Great Spirit. His love for humankind is the force of the New Age.

112.3 He who sends the rain, who causes the sun and the stars to shine, the rivers to flow, the winds to blow and the earth to give forth her bounties has in this Great Day sent to all mankind Bahá'u'lláh. It is this Great One who has opened the door of divine knowledge to every soul. It is His teachings that will establish world unity and bring about universal peace.

112.4 The people of the world are the tools in His hand. They must strive to understand His message and to walk in the path of His divine guidance. Every human being is responsible in this day to seek the truth for himself and thereafter to live according to that wise counsel. The old ones have all longed for this sweet message. Praise God that you have found it.

112.5 Now awakened to new wisdom, now guided to the straight path, now illumined with this mighty message, strive you day and night to guide and assist the thirsty ones in all lands to the ever-flowing fountain, the wandering ones to this fortress of certainty, the ignorant ones to this source of knowledge and the seekers to that One for whom their hearts long.

112.6 May your consultation reach so high a level of endeavor and purpose that the Great One will open before your faces the doors of the paradise of wisdom and love and cause the light of the Abhá Beauty to shine in your midst.

With loving Bahá'í greetings,
THE UNIVERSAL HOUSE OF JUSTICE

113
Establishment of Local Spiritual Assemblies during the Final Year of the Nine Year Plan
28 MAY 1972

To all National Spiritual Assemblies

Dear Bahá'í Friends,

113.1 In order to stimulate the teaching work in every land and encourage the friends during this last year of the Nine Year Plan we have decided that as soon as the number of adult believers in any locality reaches or exceeds nine they are permitted to form their Local Spiritual Assembly immediately, rather than wait until 21 April 1973.

We hope moreover that, especially in the areas where the people are entering the Cause in troops, the implementation of this decision will increase the number of those communities which will, without the need for outside assistance, reelect their Assemblies on the first day of Riḍván in 1973 and in succeeding years.

It is our prayer at the Sacred Threshold that during the months ahead the steadily mounting number of these divine institutions will tremendously reinforce the labors of the valiant servants of the Blessed Beauty in every clime.

With loving Bahá'í greetings,
THE UNIVERSAL HOUSE OF JUSTICE

114
Release of a Compilation on the National Spiritual Assembly
4 JUNE 1972

To all National Spiritual Assemblies

Dear Bahá'í Friends,

We have just made a compilation of extracts from letters written by Shoghi Effendi or by his secretaries on his behalf about the institution of the National Spiritual Assembly.[1] We have not attempted a complete compilation of all the available texts on the subject, but it is hoped that the enclosed material will be of assistance to members of National Spiritual Assemblies and the friends generally in their appreciation of this vital institution of the Faith.

We leave it to your discretion to determine in what manner you may wish to share the material with the friends under your jurisdiction.

With loving Bahá'í greetings,
THE UNIVERSAL HOUSE OF JUSTICE

114-1. See CC 2:83–136.

115

Announcement of the Decision to Build the Seat of the Universal House of Justice

7 June 1972

To all National Spiritual Assemblies

115.1 JOYFULLY INFORM BAHÁ'Í WORLD RANGE AND ACCELERATION GROWTH CAUSE BAHÁ'U'LLÁH LOCAL NATIONAL LEVELS AND RESULTANT EXPANSION ACTIVITIES WORLD CENTER IMPEL US NOW ANNOUNCE ERE COMPLETION NINE YEAR PLAN DECISION INITIATE PROCEDURE SELECT ARCHITECT DESIGN BUILDING FOR SEAT UNIVERSAL HOUSE JUSTICE ENVISAGED BELOVED GUARDIAN ON FAR FLUNG ARC HEART MOUNT CARMEL CENTERING SPOT CONSECRATED RESTING PLACES SISTER BROTHER MOTHER BELOVED MASTER.[1] CONSTRUCTION THIS CENTER LEGISLATION GOD'S WORLD-REDEEMING ORDER WILL CONSTITUTE FIRST MAJOR STEP DEVELOPMENT AREA SURROUNDING HOLY SHRINE SINCE COMPLETION INTERNATIONAL ARCHIVES BUILDING. MOVED PAY TRIBUTE EXPRESS HEARTFELT GRATITUDE OUTSTANDING SERVICES ROBERT MCLAUGHLIN IN PREPARATION FOR THIS HISTORIC UNDERTAKING. FERVENTLY PRAYING PROJECT NOW INITIATED MAY DURING YEARS IMMEDIATELY AHEAD PROGRESS UNINTERRUPTEDLY SPEEDILY ATTAIN MAJESTIC CONSUMMATION.[2]

Universal House of Justice

116

Plans for Commemorating the Centenary of the Revelation of the Kitáb-i-Aqdas

30 June 1972

To all National Spiritual Assemblies

Dear Bahá'í Friends,

116.1 At Riḍván 1973 we shall, God willing, witness the successful conclusion of the nine-year-long, world-encircling enterprise which inaugurated the second epoch of 'Abdu'l-Bahá's Divine Plan.[1] The victorious culmination of this initial stage of the tenth part of the majestic process set in motion over six thousand years ago will prepare the way for the next stage in the series of crusades

---

115-1. Bahíyyih Khánum, the Greatest Holy Leaf, is 'Abdu'l-Bahá's sister; Mírzá Mihdí, the Purest Branch, His brother; and Ásíyih Khánum, the Most Exalted Leaf, His mother.

115-2. For an explanation of the significance of the Seat of the Universal House of Justice, see message no. 164. For information about steps in the process of its erection and occupation, see messages no. 136, 140, 165, 186, and 354.

116-1. The second epoch of the Divine Plan began in 1963 at the commencement of the Nine Year Plan. See the entries on Divine Plan and Epochs in the glossary.

destined to achieve the penetration of the Light of God's Faith into the remaining territories of the planet and the erection of the entire machinery of Bahá'u'lláh's Administrative Order throughout the world.²

On the Twelfth Day of Riḍván, the members of the National Spiritual Assemblies gathered in the Holy Land for the Third International Convention will participate in the commemoration of the One Hundredth Anniversary of the Revelation of the Kitáb-i-Aqdas.³ To enable all believers throughout the world to share in the observance of this highly significant centenary, national and local celebrations for Bahá'ís only should be held on that day. 116.2

We call upon each community to undertake, during the period between the First Day of the Most Great Festival and the commemoration of the Declaration of the Blessed Báb,⁴ a widespread proclamation campaign which would include publicity on the completion of the Nine Year Plan and the holding of the third International Convention at the Bahá'í World Center in the Holy Land. 116.3

With loving Bahá'í greetings,
THE UNIVERSAL HOUSE OF JUSTICE

117
Exhortation to Blot Out Every Last Trace of Prejudice
13 JULY 1972

To all National Spiritual Assemblies

Dear Bahá'í Friends,

The blessings of the Ancient Beauty are being showered upon the followers of the Greatest Name. Our efforts to serve Him and humanity are being crowned with victories throughout the world. As we give thanks for these splendid achievements, as the Cause of God spreads in every land, as our insti- 117.1

116-2. For Shoghi Effendi's description (see MBW, pp. 153–55) of that "vast," "majestic" ten-part process that began "at the dawn of the Adamic cycle" and will continue into the Golden Age of the Faith, see the entry on Ten Part Process in the glossary.

116-3. The Kitáb-i-Aqdas (the Most Holy Book) was referred to by Shoghi Effendi as "the Mother Book" of Bahá'u'lláh's Dispensation and "the Charter of His New World Order." The publication of a copiously annotated English translation was a goal of the Six Year Plan (see message no. 456). In its Riḍván 1992 letter the Universal House of Justice announced the forthcoming publication of the annotated English translation of the Kitáb-i-Aqdas during the Holy Year 1992–93. For background on the translation and publication of the Kitáb-i-Aqdas, see message no. 27. For the announcement of the completion of *A Synopsis and Codification of the Kitáb-i-Aqdas*, see message no. 125.

116-4. 21 April (the First Day of Riḍván)–23 May (the Declaration of the Báb). The Third International Convention was held 26 April–2 May 1973, on the sixth through twelfth days of the Riḍván Festival.

tutions become more perfected, as the number of believers increases over the face of the planet, our individual lives must increasingly mirror forth each day the teachings of Bahá'u'lláh and we must so live our lives that all will see in us a different people. The acts we perform, the attitudes we manifest, the very words we speak should be an attraction, a magnet, drawing the sincere to the Divine Teachings.

117.2   Bahá'u'lláh tells us that prejudice in its various forms destroys the edifice of humanity. We are adjured by the Divine Messenger to eliminate all forms of prejudice from our lives. Our outer lives must show forth our beliefs. The world must see that, regardless of each passing whim or current fashion of the generality of mankind, the Bahá'í lives his life according to the tenets of his Faith. We must not allow the fear of rejection by our friends and neighbors to deter us from our goal: to live the Bahá'í life. Let us strive to blot out from our lives every last trace of prejudice—racial, religious, political, economic, national, tribal, class, cultural, and that which is based on differences of education or age. We shall be distinguished from our non-Bahá'í associates if our lives are adorned with this principle.

117.3   If we allow prejudice of any kind to manifest itself in us, we shall be guilty before God of causing a setback to the progress and real growth of the Faith of Bahá'u'lláh. It is incumbent upon every believer to endeavor with a fierce determination to eliminate this defect from his thoughts and acts. It is the duty of the institutions of the Faith to inculcate this principle in the hearts of the friends through every means at their disposal including summer schools, conferences, institutes and study classes.

117.4   The fundamental purpose of the Faith of Bahá'u'lláh is the realization of the organic unity of the entire human race. Bearing this glorious destiny in mind, and with entire reliance on the promises of the Blessed Beauty, we should follow His exhortation:

117.4a   We love to see you at all times consorting in amity and concord within the paradise of My good-pleasure, and to inhale from your acts the fragrance of friendliness and unity, of loving-kindness and fellowship....[1]

With loving Bahá'í greetings,
THE UNIVERSAL HOUSE OF JUSTICE

117-1. GWB, p. 315.

118
Embryonic Nature of Local Spiritual Assemblies
30 July 1972

The National Spiritual Assembly of the Bahá'ís of Bolivia

Dear Bahá'í Friends,

118.1 In reply to your letter of July 4th asking guidance as to what is a functioning Local Spiritual Assembly, we offer you the following comments:

118.2 Local Spiritual Assemblies are at the present newly born institutions, struggling for the most part to establish themselves both in the Bahá'í community and in the world. They are as yet only embryos of the majestic institutions ordained by Bahá'u'lláh in His Writings. This is also true of National Spiritual Assemblies. In the following passage written by the Secretary of the Guardian on his behalf this point is elucidated:

> 118.2a The Bahá'í Administration is only the first shaping of what in future will come to be the social life and laws of community living. As yet the believers are only just beginning to grasp and practice it properly. So we must have patience if at times it seems a little self-conscious and rigid in its workings. It is because we are learning something very difficult but very wonderful—how to live together as a community of Bahá'ís, according to the glorious teachings.
>
> (From letter dated 14 October 1941 to two believers)

118.3 What we find expounded in the writings of our Faith is the lofty station Local Spiritual Assemblies must attain in their gradual and at times painful development. In encouraging these Assemblies to attain this aim, there is no harm in the National Spiritual Assembly mentioning certain minimum requirements from time to time, provided it is clear that nonattainment of such standards, which by their very nature must be continuously revised with changing conditions, do not justify the withdrawal of recognition from any weak Assemblies. It would not be profitable therefore for the Universal House of Justice to lay down universal minimum standards for properly functioning Local Spiritual Assemblies, as these must necessarily differ from country to country, and even from district to district within the same country, in the process of the evolution of these Assemblies into Houses of Justice, as envisaged by Bahá'u'lláh.

Salient Objectives of the Local Spiritual Assembly

118.4 Among the more salient objectives to be attained by the Local Spiritual Assembly in its process of development to full maturity are to act as a loving shepherd to the Bahá'í flock, promote unity and concord among the friends, direct the teaching work, protect the Cause of God, arrange for Feasts, anniversaries and regular meetings of the community, familiarize the Bahá'ís with its plans, invite the community to offer its recommendations, promote the welfare of youth and children, and participate, as circumstances permit, in humanitarian activities. In its relationship to the individual believer, the Assembly should continuously invite and encourage him to study the Faith, to deliver its glorious message, to live in accordance with its teachings, to contribute freely and regularly to the Fund, to participate in community activities, and to seek refuge in the Assembly for advice and help, when needed.

118.5 In its own meetings it must endeavor to develop skill in the difficult but highly rewarding art of Bahá'í consultation, a process which will require great self-discipline on the part of all members and complete reliance on the power of Bahá'u'lláh. It should hold regular meetings and ensure that all its members are currently informed of the activities of the Assembly, that its secretary carries out his duties, and its treasurer holds and disburses the funds of the Faith to its satisfaction, keeping proper accounts and issuing receipts for all contributions. Many Assemblies find that some of their activities such as teaching, observance of Feasts and anniversaries, solution of personal problems, and other duties are best dealt with by committees appointed by the Assembly and responsible to it.

The Bahá'í Quality of Leadership

118.6 In all cases submitted for its consideration the Assembly must uphold the standard of justice in delivering its verdict, and in all its dealings with the community and the outside world it must strive to evince the qualities of leadership. The following quotation from a letter of the Guardian summarizes in simple terms the immediate goal every Assembly should set for itself in its efforts to pursue the exalted standard of perfection inculcated in our writings:

118.6a The first quality for leadership, both among individuals and Assemblies, is the capacity to use the energy and competence that exists in the rank and file of its followers. Otherwise the more competent members of the group will go at a tangent and try to find elsewhere a field of work where they could use their energy.

118.6b Shoghi Effendi hopes that the Assemblies will do their utmost in planning such teaching activities and that every single soul will be kept busy.
(From letter dated 30 August 1930 written on behalf of the Guardian to the National Spiritual Assembly of the United States and Canada)

118.7 In the compilation of texts we sent to all National Spiritual Assemblies in August 1970, and in the By-Laws of a Local Spiritual Assembly, you will find all the objectives Local Spiritual Assemblies must aim at achieving in their process of growth and development.¹ We recommend that you restudy these documents carefully and discuss this highly important problem with the Counselors of your zone, who will be only too glad to help you encourage the development of Local Spiritual Assemblies in your country.

With loving Bahá'í greetings,
THE UNIVERSAL HOUSE OF JUSTICE

119
Passing of Ishráq-Khávarí, Preeminent Bahá'í Scholar
6 AUGUST 1972

To the National Spiritual Assembly of the Bahá'ís of Iran

GRIEVED LOSS PREEMINENT SCHOLAR VALUED PROMOTER FAITH ISHRÁQ-KHÁVARÍ.¹ HIS PRECIOUS INDEFATIGABLE SERVICES OVER SEVERAL DECADES WON HIM APPRECIATION BELOVED GUARDIAN. HIS SCHOLARLY CONTRIBUTIONS IMMORTALIZED THROUGH NUMEROUS USEFUL COMPILATIONS TREATISES BEARING ELOQUENT TRIBUTE HIS DEVOTION DEDICATION CAUSE GOD. URGE HOLD APPROPRIATE MEMORIAL GATHERINGS ASSURE RELATIVES FRIENDS FERVENT PRAYERS HOLY SHRINES.

119.1

UNIVERSAL HOUSE OF JUSTICE

120
Release of a Compilation on Newsletters
24 AUGUST 1972

To all National Spiritual Assemblies

Dear Bahá'í Friends,

It is clear from the writings of our beloved Guardian that the initiation, regular publication and distribution of a Bahá'í newsletter by each National Spiritual Assembly is one of its vital functions and a means of promoting understanding and unity among the friends, of stimulating their interest and deepening their knowledge of the teachings, and of coordinating the activities of the Faith.

120.1

118-1. The compilation accompanied a letter dated 11 August 1970 (no. 84). For the compilation, see CC 2:39–60. The by-laws of a local spiritual assembly can be found in *The Bahá'í World*, beginning with volume 3.

119-1. For an account of the life and services of Ishráq-Khávarí, see BW 15:518–20.

120.2 In order to assist National Spiritual Assemblies to properly assess the importance of this activity and take effective steps to ensure that such a newsletter is issued and widely distributed among the friends, we attach a compilation of extracts on this subject from the letters written by the Guardian or on his behalf.[1] We urge you to study these extracts carefully and consider, as called for, ways and means of improving the standard, ensuring the regular release and facilitating the circulation of your news bulletin. We realize, of course, that in some countries it is necessary to issue the bulletin in more than one language.

120.3 A section of the attached extracts also deals with local newsletters. In some countries it may be feasible and desirable to encourage the issue of such bulletins by responsible Local Spiritual Assemblies.

With loving Bahá'í greetings,
THE UNIVERSAL HOUSE OF JUSTICE

121
Martyrdom of Three Iranian Bahá'í Students in the Philippine Islands
19 SEPTEMBER 1972

To the Bahá'ís of the World

121.1 With feelings of deep sorrow we relate to the Bahá'í world the distressing circumstances surrounding the murder of three Iranian Bahá'í students, pioneers to the Philippine Islands.

121.2 Parvíz Ṣádiqí, Farámarz Vujdání and Parvíz Furúghí were among a number of Iranian Bahá'í youth who answered the call for pioneers. With eleven others they registered at the Universities in Mindanao with the intention of completing their studies and proclaiming the Faith of Bahá'u'lláh. These three had conceived the plan of making teaching trips to a rural area inhabited by Muslims. When on July 31st the authorities of Mindanao State University were notified that they had left the campus the previous day and had not yet returned, search parties were immediately formed and the assistance of the police and local authorities obtained. After inquiries and search, led entirely by President Tamano of Mindanao State University, the bodies of the three young men were found in a shallow grave. They had been shot, grievously mutilated and two had been decapitated. The bodies were removed and given Bahá'í burial in a beautiful plot donated for the purpose.

120-1. The compilation is too lengthy to include in this volume.

Immediately upon receipt of the tragic news, Vicente Samaniego, Counselor in Northeast Asia, in close cooperation with the National Spiritual Assembly of the Philippines, acted vigorously on behalf of the Baháʼís and was given the utmost cooperation and sympathy by the authorities, police, military and civil. A convocation was called, attended by more than 900 students, faculty members and University officials. Prayers were said in English, Arabic and Persian. The President of the University gave a talk in which he said that the murdered Iranian students are not ordinary students, for with them is the Message of Baháʼuʼlláh which is the way to unity. The Council of the Student Body asked that their new Social Hall be renamed Iranian Student Memorial Hall. Three thousand people marched in the funeral procession and six hundred went to the burial site to attend the interment.

A dignified burial was conducted by the Baháʼís in the presence of University authorities and friends.

The relatives and friends of these three young men, who gave their lives in the service of the Blessed Beauty, are assured of the loving sympathy and prayers of their fellow believers. The sacrifice made by these youth adds a crown of glory to the wonderful services now being performed by Baháʼí youth throughout the world. Baháʼuʼlláh Himself testifies:

> They that have forsaken their country in the path of God and subsequently ascended unto His presence, such souls shall be blessed by the Concourse on High and their names recorded by the Pen of Glory among such as have laid down their lives as martyrs in the path of God, the Help in peril, the Self-Subsistent.¹

THE UNIVERSAL HOUSE OF JUSTICE

122
Release of a Compilation on Baháʼí Life
24 NOVEMBER 1972

To all National Spiritual Assemblies

Dear Baháʼí Friends,

We are asking the National Spiritual Assembly of the United Kingdom to send you by surface mail one copy of their compilation *The Pattern of Baháʼí Life*, and attached we are sending you an addendum consisting of quotations from the letters and writings of the beloved Guardian.¹

121-1. From an unpublished Tablet.

122-1. See CC 2:1–27. *The Pattern of Baháʼí Life* is a compilation from the writings of Baháʼuʼlláh, ʻAbduʼl-Bahá, and Shoghi Effendi that was published by the Baháʼí Publishing Trust of the British Isles in 1948.

122.2 These extracts are shared with you with the thought that you may be able to use them in assisting the believers to attain a fuller understanding of what it means to be a Bahá'í and to guide and aid them to pattern their personal lives in accordance with the Teachings.

122.3 You are free to use this material in whatever way is appropriate and best suited to the needs of your community. For example, you may wish to purchase copies of *The Pattern of Bahá'í Life* and use them with printed or mimeographed copies of the attached compilation, or you may prefer to issue a compilation of your own. The important thing is that the moral and spiritual admonitions contained in our writings be widely disseminated, properly understood, and the friends encouraged to follow them. Shoghi Effendi has pointed out:

122.3a Humanity, through suffering and turmoil, is swiftly moving on towards its destiny; if we be loiterers, if we fail to play our part surely others will be called upon to take up our task as ministers to the crying needs of this afflicted world.

122.3b Not by the force of our numbers, not by the mere exposition of a set of new and noble principles, not by an organized campaign of teaching—no matter how worldwide and elaborate in its character—not even by the staunchness of our faith or the exaltation of our enthusiasm, can we ultimately hope to vindicate in the eyes of a critical and skeptical age the supreme claim of the Abhá Revelation. One thing and only one thing will unfailingly and alone secure the undoubted triumph of this sacred Cause, namely the extent to which our own inner life and private character mirror forth in their manifold aspects the splendor of those eternal principles proclaimed by Bahá'u'lláh.[2]

With loving Bahá'í greetings,
THE UNIVERSAL HOUSE OF JUSTICE

122-2. BA, p. 66.

123
Adoption of the Constitution of the Universal House of Justice
26 NOVEMBER 1972

To all National Spiritual Assemblies

Dear Bahá'í Friends,

The following cable has just been sent to the United States *Bahá'í News* for publication.

WITH GRATEFUL JOYOUS HEARTS ANNOUNCE ENTIRE BAHÁ'Í WORLD ADOPTION PROFOUNDLY SIGNIFICANT STEP IN UNFOLDMENT MISSION SUPREME ORGAN BAHÁ'Í WORLD COMMONWEALTH THROUGH FORMULATION CONSTITUTION UNIVERSAL HOUSE JUSTICE. AFTER OFFERING HUMBLE PRAYERS GRATITUDE ON DAY COVENANT AT THREE SACRED THRESHOLDS BAHJÍ HAIFA MEMBERS GATHERED COUNCIL CHAMBER PRECINCTS HOUSE BLESSED MASTER APPENDED THEIR SIGNATURES FIXED SEAL ON INSTRUMENT ENVISAGED WRITINGS BELOVED GUARDIAN HAILED BY HIM AS MOST GREAT LAW FAITH BAHÁ'U'LLÁH.[1] FULLY ASSURED MEASURE JUST TAKEN WILL FURTHER REINFORCE TIES BINDING WORLD CENTER TO NATIONAL LOCAL COMMUNITIES THROUGHOUT WORLD RELEASE FRESH ENERGIES INCREASE ENTHUSIASM CONFIDENCE VALIANT WORKERS HIS DIVINE VINEYARD LABORING ASSIDUOUSLY BRING MANKIND UNDER SHELTER HIS ALL-GLORIOUS COVENANT.

Please share this joyous news with the friends. It is anticipated that the Constitution will be published at Riḍván.[2]

With loving Bahá'í greetings,
THE UNIVERSAL HOUSE OF JUSTICE

123-1. On the Day of the Covenant, 26 November, prayers were offered at the Shrine of Bahá'u'lláh at Bahjí, outside 'Akká, and at the Shrine of the Báb on Mount Carmel in Haifa, where 'Abdu'l-Bahá is also buried.

123-2. See *The Constitution of the Universal House of Justice*, published by the Bahá'í World Center (1972).

124
Activities for the Year Preceding the Global Plan to be Launched Riḍván 1974
14 January 1973

To all National Spiritual Assemblies

Dear Bahá'í Friends,

124.1 As the Bahá'í world approaches the triumphant conclusion of the Nine Year Plan it gives us the utmost gratification to see that a few National Spiritual Assemblies have already formulated plans for activity during the coming Bahá'í year.

124.2 The next global plan will be launched at Riḍván 1974 and you will therefore have twelve months to prepare for it. We call upon you all to take the greatest possible advantage of that year to:

124.2a Strengthen the foundations of your achievements through developing and enriching Bahá'í community life, fostering youth activity and through all means suited to your circumstances; and

124.2b Continue expansion of the Faith, trying new openings and possibilities not fully explored when you were under the pressure of other priorities.

124.3 Obviously conditions differ in the various areas under the jurisdiction of the National Spiritual Assemblies, and the goals which each Assembly adopts must be suited to its particular circumstances and possibilities, but, as the beloved Guardian once pointed out, "The broader the basis" of such a campaign, and "the deeper its roots, the finer the flower into which it shall eventually blossom."[1]

124.4 We ask you to make your plans now and to send us your report of them to reach us as soon as possible and not later than 1st April 1973 so that we may present a consolidated summary to the International Convention. We feel that such a summary will be an inspiration and a source of new ideas to the delegates when they are consulting upon the challenges that lie before the Bahá'í community in the years ahead and which must be faced during the next global plan. Moreover, the achievements of the coming year, added to the great victories of the Nine Year Plan, will enable the worldwide Bahá'í community to enter with even greater assurance upon the next stage of its ever-unfolding destiny.

124-1. MA, p. 29.

We pray at the Holy Shrines that the blessings of Bahá'u'lláh may guide 124.5
and assist you with a fresh measure of His divine grace in the few months
separating us from the glorious festivities of next Riḍván.

With loving Bahá'í greetings,
UNIVERSAL HOUSE OF JUSTICE

125
Announcement of the Completion of the Synopsis and Codification of the Kitáb-i-Aqdas
19 JANUARY 1973

To all National Spiritual Assemblies

JOYFULLY ANNOUNCE COMPLETION SYNOPSIS CODIFICATION KITÁB-I-AQDAS FOR 125.1
PUBLICATION RIḌVÁN SYNCHRONIZING CELEBRATION HUNDREDTH ANNIVERSARY
REVELATION MOST HOLY BOOK FULFILLING WORLD CENTER GOAL NINE YEAR
PLAN.[1] CONFIDENT RELEASE THIS PUBLICATION ENVISAGED BY BELOVED GUARDIAN
AND WHOSE MAIN FEATURES HE OUTLINED WILL CONSTITUTE ANOTHER SIGNIF-
ICANT STEP PATH LEADING BAHÁ'Í COMMUNITY FULL MATURITY ESTABLISHMENT
WORLD ORDER BAHÁ'U'LLÁH.

THE UNIVERSAL HOUSE OF JUSTICE

126
Obeying the Law of God in One's Personal Life
6 FEBRUARY 1973

To all National Spiritual Assemblies

Dear Bahá'í Friends,

The following is an excerpt from a letter written recently in response to 126.1
questions from an individual believer. As it is of general interest we are sending it to you so that you may share it with the friends within your jurisdiction in whatever manner you judge wise and necessary.

Just as there are laws governing our physical lives, requiring that we 126.2
must supply our bodies with certain foods, maintain them within a certain range of temperatures, and so forth, if we wish to avoid physical
disabilities, so also there are laws governing our spiritual lives. These

---

125-1. For the letter calling for worldwide observances of the centenary of the revelation of
the Kitáb-i-Aqdas, see message no. 116. A copiously annotated English translation of the Kitáb-
i-Aqdas and related texts was published in 1992.

laws are revealed to mankind in each age by the Manifestation of God, and obedience to them is of vital importance if each human being, and mankind in general, is to develop properly and harmoniously. Moreover, these various aspects are interdependent. If an individual violates the spiritual laws for his own development he will cause injury not only to himself but to the society in which he lives. Similarly, the condition of society has a direct effect on the individuals who must live within it.

126.3 As you point out, it is particularly difficult to follow the laws of Bahá'u'lláh in present-day society whose accepted practice is so at variance with the standards of the Faith. However, there are certain laws that are so fundamental to the healthy functioning of human society that they must be upheld whatever the circumstances. Realizing the degree of human frailty, Bahá'u'lláh has provided that other laws are to be applied only gradually, but these too, once they are applied, must be followed, or else society will not be reformed but will sink into an ever worsening condition. It is the challenging task of the Bahá'ís to obey the law of God in their own lives, and gradually to win the rest of mankind to its acceptance.

The Effect of Obedience to the Laws of Bahá'u'lláh on One's Life

126.4 In considering the effect of obedience to the laws on individual lives, one must remember that the purpose of this life is to prepare the soul for the next. Here one must learn to control and direct one's animal impulses, not to be a slave to them. Life in this world is a succession of tests and achievements, of falling short and of making new spiritual advances. Sometimes the course may seem very hard, but one can witness, again and again, that the soul who steadfastly obeys the law of Bahá'u'lláh, however hard it may seem, grows spiritually, while the one who compromises with the law for the sake of his own apparent happiness is seen to have been following a chimera: he does not attain the happiness he sought, he retards his spiritual advance and often brings new problems upon himself.

126.5 To give one very obvious example: the Bahá'í law requiring consent of parents to marriage. All too often nowadays such consent is withheld by non-Bahá'í parents for reasons of bigotry or racial prejudice; yet we have seen again and again the profound effect on those very parents of the firmness of the children in the Bahá'í law, to the extent that not only is the consent ultimately given in many cases, but the character of the parents can be affected and their relationship with their child greatly strengthened.

126.6 Thus, by upholding Bahá'í law in the face of all difficulties we not only strengthen our own characters but influence those around us.

The Bahá'í Teachings on Sexual Intercourse

The Bahá'í teaching on sexual intercourse is very clear. It is permissible only between a man and the woman who is his wife. In this connection we share with you extracts from four letters written on behalf of the Guardian which throw light on various aspects of the matter. One of them contains the paragraph that you quote in your letter.

126.7

> With reference to the question you have asked concerning the Bahá'í attitude towards the problem of sex and its relation to marriage:
>
> The Bahá'í Teachings on this matter, which is of such vital concern and about which there is such a wide divergency of views, are very clear and emphatic. Briefly stated the Bahá'í conception of sex is based on the belief that chastity should be strictly practiced by both sexes, not only because it is in itself highly commendable ethically, but also due to its being the only way to a happy and successful marital life. Sex relationships of any form, outside marriage, are not permissible therefore, and whoso violates this rule will not only be responsible to God, but will incur the necessary punishment from society.
>
> The Bahá'í Faith recognizes the value of the sex impulse, but condemns its illegitimate and improper expression such as free love, companionate marriage and others, all of which it considers positively harmful to man and to the society in which he lives. The proper use of the sex instinct is the natural right of every individual, and it is precisely for this very purpose that the institution of marriage has been established. The Bahá'ís do not believe in the suppression of the sex impulse but in its regulation and control.
>
> (From a letter dated 5 September 1938, to an individual believer)

126.7a

> The question you raise as to the place in one's life that a deep bond of love with someone we meet other than our husband or wife can have is easily defined in view of the teachings. Chastity implies both before and after marriage an unsullied, chaste sex life. Before marriage absolutely chaste, after marriage absolutely faithful to one's chosen companion. Faithful in all sexual acts, faithful in word and in deed.
>
> The world today is submerged, amongst other things, in an overexaggeration of the importance of physical love, and a dearth of spiritual values. In as far as possible the believers should try to realize this and rise above the level of their fellowmen who are, typical of all decadent periods in history, placing so much overemphasis on the purely physical side of mating. Outside of their normal, legitimate married life they should seek to establish bonds of comradeship

126.7b

and love which are eternal and founded on the spiritual life of man, not on his physical life. This is one of the many fields in which it is incumbent on the Bahá'ís to set the example and lead the way to a true human standard of life, when the soul of man is exalted and his body but the tool for his enlightened spirit. Needless to say this does not preclude the living of a perfectly normal sex life in its legitimate channel of marriage.

(From a letter dated 28 September 1941, to an individual believer)

126.7c Concerning your question whether there are any legitimate forms of expression of the sex instinct outside of marriage: according to the Bahá'í Teachings no sexual act can be considered lawful unless performed between lawfully married persons. Outside of marital life there can be no lawful or healthy use of the sex impulse. The Bahá'í youth should, on the one hand, be taught the lesson of self-control which, when exercised, undoubtedly has a salutary effect on the development of character and of personality in general, and on the other should be advised, nay even encouraged, to contract marriage while still young and in full possession of their physical vigor. Economic factors, no doubt, are often a serious hindrance to early marriage, but in most cases are only an excuse, and as such should not be overstressed.

(From a letter dated 13 December 1940, to an individual believer)

126.7d As regards your question whether it would be advisable and useful for you to marry again: he feels unable to give you any definite answer on that point, as this is essentially a private affair about which you, and the friends around you or your Local Assembly, are in a much better position to judge. Of course, under normal circumstances, every person should consider it his moral duty to marry. And this is what Bahá'u'lláh has encouraged the believers to do. But marriage is by no means an obligation. In the last resort it is for the individual to decide whether he wishes to lead a family life or live in a state of celibacy.

(From a letter dated 3 May 1936, to an individual believer)

126.8 You express surprise at the Guardian's reference to "the necessary punishment from society." In the Kitáb-i-Aqdas Bahá'u'lláh prohibits sexual immorality and in the Annex to that Book states that the various degrees of sexual offenses and the punishments for them are to be decided by the Universal House of Justice. In this connection it should be realized that there is a distinction drawn in the Faith between the attitudes which should characterize individuals in their relationship to other people, namely, loving forgiveness, forbearance, and concern with one's own sins, not the sins of others, and those attitudes which should be shown by the Spiritual Assemblies, whose duty is to administer the law of God with justice.

A number of sexual problems, such as homosexuality and transsexuality can well have medical aspects, and in such cases recourse should certainly be had to the best medical assistance. But it is clear from the teaching of Bahá'u'lláh that homosexuality is not a condition to which a person should be reconciled, but is a distortion of his or her nature which should be controlled and overcome. This may require a hard struggle, but so also can be the struggle of a heterosexual person to control his or her desires. The exercise of self-control in this, as in so very many other aspects of life, has a beneficial effect on the progress of the soul. It should, moreover, be borne in mind that although to be married is highly desirable, and Bahá'u'lláh has strongly recommended it, it is not the central purpose of life. If a person has to wait a considerable period before finding a spouse, or if ultimately, he or she must remain single, it does not mean that he or she is thereby unable to fulfill his or her life's purpose.

One's Attitude toward the Laws of Bahá'u'lláh

In all this we have been speaking about the attitude that Bahá'ís should have towards the law of Bahá'u'lláh. You, however, as a doctor working mainly as a counselor in family and sexual problems, will mostly be concerned with advising non-Bahá'ís, who do not accept, and see no reason to follow, the laws of Bahá'u'lláh. You are already a qualified practitioner in your field, and no doubt you give advice on the basis of what you have learned from study and experience—a whole fabric of concepts about the human mind, its growth, development and proper functioning, which you have learned and evolved, without reference to the teachings of Bahá'u'lláh. Now, as a Bahá'í, you know that what Bahá'u'lláh teaches about the purpose of human life, the nature of the human being and the proper conduct of human lives, is divinely revealed and therefore true. However, it will inevitably take time for you not only to study the Bahá'í teachings so that you clearly understand them, but also to work out how they modify your professional concepts. This is, of course, not an unusual predicament for a scientist. How often in the course of research is a factor discovered which requires a revolution in thinking over a wide field of human endeavor. You must be guided in each case by your own professional knowledge and judgment as illuminated by your growing knowledge of the Bahá'í teachings; undoubtedly you will find that your own understanding of the human problems dealt with in your work will change and develop and you will see new and improved ways of helping the people who come to you.

Psychology is still a very young and inexact science, and as the years go by Bahá'í psychologists, who know from the teachings of Bahá'u'lláh the true pattern of human life, will be able to make great strides in the development of this science, and will help profoundly in the alleviation of human suffering.

With loving Bahá'í greetings,
THE UNIVERSAL HOUSE OF JUSTICE

127
Purchase of Mazra'ih Mansion
15 MARCH 1973

To all National Spiritual Assemblies

127.1 OCCASION NAW-RÚZ 130[1] JOYOUSLY ANNOUNCE BAHÁ'Í WORLD ACQUISITION BY PURCHASE MANSION MAZRA'IH RESULT SEVERAL YEARS PATIENT PERSISTENT DETERMINED NEGOTIATIONS THEREBY ADDING TO BAHÁ'Í ENDOWMENTS HOLY LAND FIRST RESIDENCE BAHÁ'U'LLÁH AFTER NINE YEARS SPENT WALLED PRISON CITY 'AKKÁ. CONTROL THIS HOLY SITE REACQUIRED BY BELOVED GUARDIAN AFTER LAPSE MORE THAN FIFTY YEARS WHEN HE SECURED LEASE MANSION 1950 EXTENDED TO PRESENT TIME. PURCHASE INCLUDES LAND AREA APPROXIMATING TWENTY-FOUR THOUSAND SQUARE METERS HIGHLY SUITABLE EXTENSION GARDENS CULTIVATION. OFFERING PRAYER THANKSGIVING SACRED THRESHOLD THIS GREATLY CHERISHED BOUNTY.

UNIVERSAL HOUSE OF JUSTICE

128
Riḍván Message 1973
RIḌVÁN 1973

To the Bahá'ís of the World

Dearly loved Friends,

128.1 We announce with joyful and thankful hearts the completion in overwhelming victory of the world-encircling Nine Year Plan. The Army of Light has won its second global campaign; it has surpassed the goals set for expansion and has achieved a truly impressive degree of universal participation, the twin objectives of the Plan. With gratitude and love we testify to the unceasing confirmations which Bahá'u'lláh has showered upon His servants, enabling

---

127-1. 21 March 1973.

each and every one of us to offer Him some part of the labor, the devotion, the sacrifice, the supplication which He has so bountifully rewarded. At this Centenary of the Revelation of the Most Holy Book, the Community of the Most Great Name lays its tribute of victory at His feet, acknowledging that it is He Who has bestowed it.[1]

Victories of the Nine Year Plan

The Cause of God at the end of the Nine Year Plan is immensely more widespread, more firmly founded, and its own international relations more closely knit than in 1964 when the Plan was launched. Ninety-five new territories have been opened to the Faith; the 69 National Spiritual Assemblies which shouldered the world community's task have become 113, 5 more than called for. These embryonic secondary Houses of Justice are supported by more than 17,000 Local Spiritual Assemblies, 3,000 in excess of the goal and 12,000 more than at the beginning of the Plan. Bahá'ís reside in 69,500 localities, 15,000 more than called for, and 54,000 more than in 1964. Bahá'í literature has been translated into 225 more languages bringing the total number to 571; 63 Temple sites, 56 National Ḥaẓíratu'l-Quds, and 62 National Endowments have been acquired bringing the total numbers of these properties to 98, 112 and 104 respectively; 50 Teaching Institutes and Summer and Winter Schools are playing their part in Bahá'í education and 15 Publishing Trusts produce Bahá'í literature in major languages of the world. The Mother Temple of Latin America has been built and dedicated. Among those goals whose achievement is dependent on favorable circumstances outside our control are the incorporation of Assemblies and recognition of Bahá'í Holy Days. It is gratifying to record that 90 National Spiritual Assemblies and 1,556 Local Spiritual Assemblies—181 more than the total number called for—are incorporated, while Bahá'í Holy Days are recognized in 64 countries and Bahá'í certification of marriage in 40.

128.2

This great expansion of the Faith required an army of international pioneers. Two major calls were raised, for 461 and 733, which together with others for particular posts made an overall total of 1,344. The Community of the Most Great Name responded with 3,553 who actually left their homes, 2,265 of whom are still at their posts.

128.3

Developments at the World Center

At the World Center of the Faith the collation and classification of the Bahá'í Sacred Scriptures and of the writings of Shoghi Effendi have been car-

128.4

128-1. The Most Holy Book is the Kitáb-i-Aqdas, the chief repository of Bahá'u'lláh's laws and the Mother Book of His revelation. It was revealed in 1873. For messages pertaining to its translation and publication, see no. 27; to the commemoration of its revelation, see no. 116; to the announcement of the completion of *A Synopsis and Codification of the Kitáb-i-Aqdas*, see no. 125.

ried forward in ever increasing volume, a task supported and enriched by the labors of a special committee appointed by the Persian National Spiritual Assembly. The material at the World Center, includes some 2,600 original Tablets by Bahá'u'lláh, 6,000 by 'Abdu'l-Bahá and 2,300 letters of Shoghi Effendi. There are in addition some 18,000 authenticated copies of other such Tablets and letters. All these have been studied, important passages from them excerpted and classified, and the subject matter indexed under 400 general headings.

128.5 A Synopsis and Codification of the Laws and Ordinances of the Kitáb-i-Aqdas—completing the considerable progress made by the beloved Guardian in this task—is being published on the Centenary of the Revelation of the Most Holy Book, which, as already announced, is to be celebrated both in the Holy Land and throughout the Bahá'í world during this Riḍván.

128.6 The Constitution of the Universal House of Justice, hailed by Shoghi Effendi as the Most Great Law of the Faith of Bahá'u'lláh, has been formulated and published.

128.7 The gardens in Bahjí and on Mount Carmel have been significantly extended and plans have been approved for the befitting development and beautification of the entire area of Bahá'í property surrounding the Holy Shrines in Bahjí and Haifa.

Worldwide Proclamation

128.8 The worldwide proclamation of the Faith, an intensive and long-to-be-sustained process initiated during the third phase of the Plan, opened in October 1967 with the commemoration of the Centenary of Bahá'u'lláh's Proclamation to the kings and rulers which had centered around His revelation of the Súriy-i-Mulúk in Adrianople.[2] This historic event was commemorated at six Intercontinental Conferences held simultaneously around the planet. A further nine Oceanic and Continental Conferences held during the Plan gave great impetus to this proclamation program.[3] The fifteen Conferences were attended by nearly 17,000 believers and attracted great publicity by press and radio and were made the occasion of acquainting dignitaries and notabilities with the Divine Message. The presentation, on behalf of the Universal House of Justice, to 142 Heads of State, of a specially produced book containing the translation into English of the Tablets and passages of Scripture in which

---

128-2. For information about the significance of the Súriy-i-Mulúk, see the Riḍván messages of 1965 and 1967 (nos. 24 and 42).

128-3. For the message to the six intercontinental conferences, see message no. 46; for the messages to other conferences, see First Oceanic Conference—Palermo, Sicily, August 1968 (no. 63); Bolivia and Mauritius—August 1970 (no. 82); Monrovia, Liberia—January 1971 (no. 88); South China Seas, Singapore—January 1971 (no. 89); Caribbean Sea, Kingston, Jamaica—May 1971 (no. 97); South Pacific Ocean, Suva, Fiji—May 1971 (no. 98); North Pacific Oceanic Conference, Sapporo, Japan—September 1971 (no. 100); and North Atlantic Ocean, Reykjavik, Iceland—September 1971 (no. 101).

Bahá'u'lláh, some hundred years before, had issued His mighty Proclamation to mankind, initiated this campaign, which will continue long beyond the end of the Nine Year Plan.⁴

The outstanding development in the relationship of the Bahá'í International Community to the United Nations was the accreditation of that Community as a nongovernmental organization with consultative status to the Economic and Social Council of the United Nations. The Bahá'í International Community now has a permanent representative at the United Nations and maintains an office in New York.⁵

Tribute to the Hands of the Cause of God

The loved and revered Hands of the Cause have rendered sacrificial and distinguished service throughout the Nine Year Plan. They have, in all parts of the world, inspired the friends, assisted National Spiritual Assemblies, promoted the teaching work and played a vital part in the success of the Plan. The lagging fortunes of more than one national community have been revolutionized by a visit of a Hand of the Cause; swift and energetic action, inspired by the Hand, has been followed by astonishing results, completely reversing that community's prospects. They have added distinguished works to the literature of the Faith.

The goal of the Plan to develop "The institution of the Hands of the Cause of God, in consultation with the body of the Hands of the Cause, with a view to the extension into the future of its appointed functions of protection and propagation," was accomplished in stages, leading to the establishment of eleven Continental Boards of Counselors, whose members were appointed by the Universal House of Justice and who assumed responsibility for the Auxiliary Boards for protection and propagation. The beloved Hands no longer remained individually identified with any particular continent—except insofar as their residence was concerned—but extended their sphere of action to the whole planet. The Continental Boards of Counselors, advised and guided by the Hands of the Cause of God and working in close collaboration with them, have already, in their brief period of office, performed outstanding and distinguished services.⁶

Three Portentous Developments

Three highly portentous developments have taken place during the Nine Year Plan, namely, the advance of youth to the forefront of the teaching work, a great increase in the financial resources of the Faith, and an astonishing proliferation of inter-National Assembly assistance projects.

128-4. The book was published by the Bahá'í World Center under the title *The Proclamation of Bahá'u'lláh to the kings and leaders of the world* (1967).

128-5. For the message announcing the attainment of this goal, see no. 78.

128-6. For messages about the role and function of the Continental Boards of Counselors, see nos. 60, 72, 111, 132, 206, and 267.

The Advance of Youth

128.13 The first, the heartwarming upsurge of Bahá'í youth, has changed the face of the teaching work; impenetrable barriers have been broken or overpassed by eager teams of young Bahá'ís, dedicated and prayerful, presenting the Divine Message in ways acceptable to their own generation from which it has spread and is spreading throughout the social structure. The entire Bahá'í world has been thrilled by this development. Having rejected the values and standards of the old world, Bahá'í youth are eager to learn and adapt themselves to the standards of Bahá'u'lláh and so to offer the Divine Program to fill the gap left by the abandonment of the old order.

An Increase in Financial Resources

128.14 The vast increase in the financial resources of the Faith called for under the Plan has evoked a heartwarming response from the entire Bahá'í community. Not only the Bahá'í International Fund but the local, national and continental Funds of the Faith have been sacrificially supported. This practical proof of the love which the friends bear for the Faith has enabled all the work to go forward—the support of pioneers and traveling teachers, the raising of Mashriqu'l-Adhkárs and acquisition of Bahá'í properties, the purchase of Holy Places in the Cradle of the Faith and at the World Center, the development of educational institutions and all the multifarious activities of a vigorous, onward-marching, constructive world community. It is of interest that sixty percent of the international funds of the Faith is used to assist the work of National Spiritual Assemblies, to promote the teaching work and to defend the Cause against attacks in many parts of the world. Without such help from the Bahá'í world community many National Assemblies would be paralyzed in their efforts of expansion and deepening. The administration of Ḥuqúqu'lláh has been strengthened in preparation for its extension to other parts of the world.[7] An International Deputization Fund was established at the World Center to assist pioneers and traveling teachers who were ready to serve but unable to provide their own expenses, and this Fund was later extended to the support of projects on national homefronts. Contribution to the Fund is a service which will never cease to be open to all believers; the growth of the Faith and the rise of its Administrative Order require an ever-increasing outpouring of our substance, commensurate in however small a measure with the bounty and liberality of the outpouring confirmations of Bahá'u'lláh.

International Collaboration

128.15 When the Plan was launched 219 assistance projects were specified whereby national communities would render financial, pioneering or teaching aid to others, generally remote from them geographically. The intention was to

---

128-7. For information on Ḥuqúqu'lláh, see the glossary.

strengthen the bonds of unity between distant parts of the Bahá'í world with different social, cultural and historical backgrounds. At the end of the Plan more than 600 such projects had been carried out.[8] Intercommunity cooperation has been further developed in the field of publishing Bahá'í literature, notably in Spanish and French and the languages of Africa. A vast field of fruitful endeavor lies open in this respect.

In some countries due to lack of freedom, to actual repression in others, to legal and physical obstacles in yet others, certain particular goals—mainly those requiring incorporation or recognition—could not be won. Foreseeing this, the Universal House of Justice called upon national communities in lands where there is freedom to practice and promote the Faith, to exceed their own goals and thus ensure that the overall goals would be won. It has proved still impossible to begin work on the erection of the Mashriqu'l-Adhkár in Tehran, but contracts have been signed for the preparation of detailed drawings, geological surveys are being made, and everything made ready for immediate action whenever the situation in Persia becomes propitious.

Additional Developments at the World Center

During the period of the Nine Year Plan a number of important and interesting events, not directly associated with it, have taken place. First and foremost was the commemoration, in the precincts of the Qiblih of the Bahá'í world, of the centenary of the arrival at the prison city of 'Akká, as foretold in former Scriptures, of the Promised One of all ages.[9]

The Mansion of Mazra'ih, often referred to by the beloved Guardian as one of the "twin mansions" in which the Blessed Beauty resided after nine years within the walled prison city of 'Akká, and dear to the hearts of the believers by reason of its associations with their Lord, has at last been purchased together with 24,000 square meters of land extending into the plain on its eastward side.[10]

The raising of the obelisk, marking the site of the future Mashriqu'l-Adhkár on Mount Carmel, completes a project initiated by the beloved Guardian.[11]

The decision has been made and announced to the Bahá'í world, and the initial steps have been taken for the erection on Mount Carmel, at a site on the Arc as purposed by Shoghi Effendi, of the building which shall serve as the Seat of the Universal House of Justice.[12]

128-8. For a list of the projects, see the Universal House of Justice, *1964–1973 Analysis of the Nine Year International Teaching Plan*, (1964 [sic]), pp. 18–24.
128-9. The commemoration took place 26–31 August 1968.
128-10. The other Mansion is that at Bahjí, outside of 'Akká.
128-11. For the message announcing the erection of the obelisk, see no. 105.
128-12. For an explanation of the significance of the Seat of the Universal House of Justice, see the letter dated 5 June 1975 (no. 164); for messages about steps in the process of its erection and occupation, see nos. 136, 140, 186, 165, and 354.

Raising God's Kingdom on Earth

128.21 The progress of the Cause of God gathers increasing momentum and we may with confidence look forward to the day when this Community, in God's good time, shall have traversed the stages predicated for it by its Guardian, and shall have raised on this tormented planet the fair mansions of God's Own Kingdom wherein humanity may find surcease from its self-induced confusion and chaos and ruin, and the hatreds and violence of this time shall be transmuted into an abiding sense of world brotherhood and peace. All this shall be accomplished within the Covenant of the everlasting Father, the Covenant of Bahá'u'lláh.

THE UNIVERSAL HOUSE OF JUSTICE

129
Election of the Universal House of Justice—Riḍván 1973
3 MAY 1973

To all National Spiritual Assemblies

129.1 NEWLY ELECTED MEMBERS UNIVERSAL HOUSE OF JUSTICE 'ALÍ NAKHJAVÁNÍ HUSHMAND FATHEAZAM AMOZ GIBSON IAN SEMPLE DAVID HOFMAN CHARLES WOLCOTT BORRAH KAVELIN DAVID RUHE HUGH CHANCE.

UNIVERSAL HOUSE OF JUSTICE

1973–1974

A Year of Preparation

130
Announcement of the Acceptance of the Faith by His Highness Malietoa Tanumafili II of Western Samoa
7 May 1973

To the Bahá'ís of the World

Dear Bahá'í Friends,

130.1 It is now possible to share with you all the news of an event which crowns the victories with which Bahá'u'lláh has blessed His followers during the Nine Year Plan, an event of which the true significance will be fully understood only in the course of centuries to come: a reigning monarch has accepted the Message of Bahá'u'lláh.

130.2 Among those to whom *The Proclamation of Bahá'u'lláh* was presented in 1967 was His Highness Malietoa Tanumafili II, the Head of State of the independent nation of Western Samoa in the heart of the Pacific Ocean.¹ His Highness, who had already heard of the Faith, showed immediately that the sacred Words had touched his heart, and the Universal House of Justice thereupon asked the Hand of the Cause Dr. Ugo Giachery, who had presented the book to him, to return to Western Samoa for further audiences with His Highness. Following this visit the Malietoa conveyed his acceptance of the Faith of Bahá'u'lláh to the Universal House of Justice and became the first reigning sovereign to enter beneath the shade of this Cause.

130.3 His Highness decided, with the full agreement of the Universal House of Justice, that it was not propitious to make his declaration public at that time. He has been visited from time to time by Hands of the Cause and other believers, and continual touch with His Highness has been maintained by the House of Justice through Mr. Suhayl 'Alá'í, a member of the Continental Board of Counselors for Australasia. Gradually the Malietoa has let it be known to those around him that he has accepted Bahá'u'lláh. Now he has judged the time ripe to share this wondrous news with his fellow-believers in all parts of the world, by addressing to the International Bahá'í Convention the gracious and inspiring message of which a copy is enclosed with this letter. . . .²

With loving Bahá'í greetings,

THE UNIVERSAL HOUSE OF JUSTICE

130-1. In its Riḍván 1967 message (no. 42) the Universal House of Justice announced that it would present to Heads of State around the world a collection of Tablets Bahá'u'lláh addressed to the kings and rulers of the world a century before. His Highness Malietoa Tanumafili II was one of the recipients.

130-2. See BW 15:183 for a photograph of the Malietoa's letter.

131
Establishment of the International Teaching Center
5 June 1973

To all National Spiritual Assemblies

131.1 ANNOUNCE ESTABLISHMENT HOLY LAND LONG ANTICIPATED INTERNATIONAL TEACHING CENTER DESTINED EVOLVE INTO ONE THOSE WORLD-SHAKING WORLD-EMBRACING WORLD-DIRECTING ADMINISTRATIVE INSTITUTIONS ORDAINED BY BAHÁ'U'LLÁH ANTICIPATED BY 'ABDU'L-BAHÁ ELUCIDATED BY SHOGHI EFFENDI. MEMBERSHIP THIS NASCENT INSTITUTION COMPRISES ALL HANDS CAUSE GOD AND INITIALLY THREE COUNSELORS WHO WITH HANDS PRESENT HOLY LAND WILL CONSTITUTE NUCLEUS ITS VITAL OPERATIONS. CALLING UPON HOOPER DUNBAR FLORENCE MAYBERRY 'AZÍZ YAZDÍ PROCEED HOLY LAND ASSUME THIS HIGHLY MERITORIOUS SERVICE. OFFERING PRAYERS HEARTFELT GRATITUDE SACRED THRESHOLD THIS FURTHER EVIDENCE ORGANIC EVOLUTION ADMINISTRATIVE ORDER BAHÁ'U'LLÁH.

With loving Bahá'í greetings,
THE UNIVERSAL HOUSE OF JUSTICE

132
Elucidation of the Duties of the International Teaching Center and Expansion of the Continental Boards of Counselors
8 June 1973

To the Bahá'ís of the World

Dear Bahá'í Friends,

132.1 The centennial year of the revelation of the Kitáb-i-Aqdas has already witnessed events of such capital significance in the annals of the Bahá'í Dispensation as to cause us to contemplate with awe the rapidity with which Divine Providence is advancing the Cause of the Most Great Name. The time is indeed propitious for the establishment of the International Teaching Center, a development which, at one and the same time, brings to fruition the work of the Hands of the Cause residing in the Holy Land and provides for its extension into the future, links the institution of the Boards of Counselors even more intimately with that of the Hands of the Cause of God, and powerfully reinforces the discharge of the rapidly growing responsibilities of the Universal House of Justice.

132.2 This International Teaching Center now established will, in due course, operate from that building designated by the Guardian as the Seat for the

Hands of the Cause, which must be raised on the Arc on Mount Carmel in close proximity to the Seat of the Universal House of Justice.¹

The duties now assigned to this nascent institution are:

- To coordinate, stimulate and direct the activities of the Continental Boards of Counselors and to act as liaison between them and the Universal House of Justice.

- To be fully informed of the situation of the Cause in all parts of the world and to be able, from the background of this knowledge, to make reports and recommendations to the Universal House of Justice and give advice to the Continental Boards of Counselors.

- To be alert to possibilities, both within and without the Bahá'í community, for the extension of the teaching work into receptive or needy areas, and to draw the attention of the Universal House of Justice and the Continental Boards of Counselors to such possibilities, making recommendations for action.

- To determine and anticipate needs for literature, pioneers and traveling teachers and to work out teaching plans, both regional and global, for the approval of the Universal House of Justice.

All the Hands of the Cause of God will be members of the International Teaching Center. Each Hand will be kept regularly informed of the activities of the Center through reports or copies of its minutes, and will be able, wherever he may be residing or traveling, to convey suggestions, recommendations and information to the Center and, whenever he is in the Holy Land, to take part in the consultations and other activities of the Center.

In addition, we now appoint Mr. Hooper Dunbar, Mrs. Florence Mayberry and Mr. 'Azíz Yazdí to membership of the International Teaching Center, with the rank of Counselor. These believers, who have been serving with distinction on the Continental Boards of Counselors in South America, North America and Central and East Africa respectively, will henceforth reside in Haifa and will, together with the Hands present in the Holy Land, constitute the nucleus of the operations of the Center.

Authority for the expulsion and reinstatement of Covenant-breakers remains with the Hands of the Cause of God. All such matters will be investigated locally by the relative Continental Board of Counselors in consultation with any Hand or Hands who may be in the area. The Continental Board of Counselors and the Hands concerned will then make their reports to the International Teaching Center where they will be considered. The decision

---

132-1. For a discussion of the Arc, see the glossary.

whether or not to expel or reinstate will be made by the Hands of the Cause residing in the Holy Land who will, as at present, submit their decision to the Universal House of Justice for approval.

132.7 The following changes to the zones of the Continental Boards of Counselors are now made:

132.7a — The number of zones has been raised to twelve by the removal of India, Tibet, Nepal, Sikkim, Bhutan, Bangladesh, Sri Lanka and the Laccadive, Maldive, Andaman and Nicobar Islands from Central Asia.

132.7b — The Philippines, Hong Kong and Macau are transferred from Northeastern Asia to Southeastern Asia.

132.7c — The Caroline Islands and all other Pacific islands lying north of the equator and between longitudes 140° east and 140° west, with the exception of the Gilbert Islands, will be transferred from the zone of Australasia to the zone of Northeastern Asia. Islands under the jurisdiction of the National Spiritual Assembly of Alaska remain in the zone of North America.

132.8 The number of Counselors is now raised to fifty-seven by the appointment of Mr. Friday Ekpe and Mr. Zekrullah Kazemi in Northwestern Africa, Mr. Hushang Ahdieh and Mr. Peter Vuyiya in Central and East Africa, Dr. Sarah Pereira and Mrs. Velma Sherrill in North America, Mr. Rowland Estall and Mr. Paul Lucas in Central America, Mrs. Leonora Armstrong, Mr. Peter McLaren and Mr. Raúl Pavón in South America, Mr. Dipchand Khianra and Mrs. Zena Sorabjee in South Central Asia, Mr. Firaydún Mítháqíyán in Southeastern Asia, Mr. Richard Benson and Miss Elena Marsella in Northeastern Asia and Miss Violet Hoehnke in Australasia. Dr. William Maxwell who has been rendering distinguished service as a member of the Continental Board of Counselors in Northwestern Africa has been obliged to return to the United States.

132.9 Mrs. Zena Sorabjee is appointed Trustee of the new Continental Fund of South Central Asia, while Mr. Hushang Ahdieh and Mr. Mas'úd Khamsí are appointed the new Trustees of the Continental Funds of Central and East Africa and South America respectively.

132.10 Beyond these significant developments at the World Center of the Faith and on the continental level, it is becoming increasingly necessary in many parts of the world for the Auxiliary Boards to be reinforced. The nature of the work differs from zone to zone and the Universal House of Justice is now consulting the Boards of Counselors on this matter before making an announcement.

132.11 The decisions now announced are the outcome of deliberation extending over a number of years, reinforced by consultations with the Hands of the

Cause of God, and especially with the Hands residing in the Holy Land who were requested in 1968 to assist the Universal House of Justice in the establishment of the International Teaching Center, a task that now increases in magnitude as that Center begins its work.

It is our fervent prayer that the Blessed Beauty will abundantly confirm this latest unfoldment of His divinely-purposed Administrative Order.

THE UNIVERSAL HOUSE OF JUSTICE

Zones of the Continental Boards of Counselors As Revised by the Universal House of Justice in May 1973

1. **Northwestern Africa**
   All the continent of Africa west of the eastern frontiers of Tunisia, Algeria, Niger and Nigeria plus the Cape Verde Islands.

2. **Central and East Africa**
   All the continent of Africa east of the western frontiers of Libya, Chad and the United Cameroon Republic and north of the southern frontiers of Zaïre and Tanzania plus the islands of Fernando Póo, Príncipe, São Tomé and Annobón in the Atlantic Ocean and Zanzibar, Pemba and Mafia Islands in the Indian Ocean.

3. **Southern Africa**
   All the continent of Africa south of the northern frontiers of Angola, Zambia, Malawi and Mozambique plus the Island of Madagascar and all islands in the Atlantic and Indian Oceans between longitudes 20° west and 80° east and south of the equator with the exception of the Islands of Annobón, Zanzibar, Pemba, and Mafia which are assigned to the zone of Central and East Africa.

4. **North America**
   All the continent of America north of the southern frontier of the United States plus all offshore islands in the Pacific and Arctic Oceans including the Aleutian chain and all islands under the jurisdiction of the National Spiritual Assembly of Alaska, also Greenland and all offshore islands politically belonging to Greenland, all islands in the Atlantic Ocean west of longitude 40° west and between latitude 60° north and the Tropic of Cancer plus those Bahama Islands lying south of the Tropic of Cancer.

5. **Central America**
   All the continent of America south of the northern frontier of Mexico and north of the southern frontier of Panama plus the offshore islands

in the Pacific Ocean belonging politically to countries of this zone plus Clipperton Island, all islands in the Gulf of Mexico and the Caribbean Sea south of the Tropic of Cancer except the Bahama Islands which are allocated to the zone of North America and islands belonging politically to Colombia and Venezuela, the islands of Curaçao, Bonaire, Aruba, Trinidad and Tobago which are all allocated to the zone of South America.

6. South America

All the continent of South America, the Galápagos Islands, Curaçao, Bonaire, Aruba, Trinidad and Tobago, all islands in the Caribbean and North Atlantic Oceans belonging politically to countries of this zone plus all islands in the Pacific and Atlantic Oceans south of the equator and between longitude 120° west and longitude 20° west.

7. Western Asia

All the continent of Asia west of the eastern boundaries of Pakistan, Sinkiang, the Mongolian Republic, the Oblasts of Chita and Irkutsk and the Kray of Krasnoyarsk and east of the western boundaries of Lebanon, Syria, Jordan and Saudi Arabia, plus those parts of Turkey and Kazakhstan which lie in Europe and including the Transcaucasian S.S.R.s of Georgia, Armenia and Azerbaijan, as well as all islands in the Persian Gulf and islands in the Arabian Sea belonging politically to countries of this zone.

8. South Central Asia

India, Tibet, Nepal, Sikkim, Bhutan, Bangladesh, Sri Lanka and the Laccadive, Maldive, Andaman and Nicobar Islands.

9. Southeastern Asia

China south of the northern boundaries of Yünnan, Szechwan, Hupeh, Anhwei and Kiangsu, as well as Burma, Thailand, Laos, Vietnam, Cambodia, Malaysia, Brunei, Indonesia, the Philippines, Hong Kong and Macau. Excluding Portuguese Timor.

10. Northeastern Asia

All the Soviet Union east of the western boundary of the Yakutsk A.S.S.R. and the Oblast of Amur, China east of Sinkiang and north of the southern boundaries of Tsinghai, Kansu, Shensi, Honan and Shantung; Korea, Japan, Taiwan and all islands belonging politically to those nations plus all islands in the Pacific Ocean north of the equator and between the longitudes of 140° east and 140° west with the exception of the Gilbert Islands and those islands under the jurisdiction of the National Spiritual Assembly of Alaska, but including those Caroline Islands lying west of longitude 140° east.

11. Australasia
Australia and New Zealand plus all islands in the Indian and Pacific Oceans lying south of the equator and between the longitudes of 80° east and 120° west including Portuguese Timor and the Gilbert and Ellice Islands but excepting Indonesia.

12. Europe
The entire continent of Europe less those portions of Kazakhstan and Turkey which lie in Europe, plus Iceland and all islands in the Atlantic Ocean north of latitude 60° north which belong politically to nations of the European continent plus all islands in the Atlantic Ocean east of longitude 40° west and between latitude 60° north and the Tropic of Cancer plus all islands of the Mediterranean Sea including Cyprus but excluding islands belonging politically to nations of the African and Asiatic continents.

Membership of the Continental Boards of Counselors, According to the New Boundaries of May 1973
(New appointments are indicated with an asterisk.)

Northwestern Africa
    Ḥusayn Ardikání (Trustee of Continental Fund)
  * Friday Ekpe
  * Zekrullah Kazemi
    Muḥammad Kebdani

Central and East Africa
  * Hushang Ahdieh (Trustee of Continental Fund)
    Oloro Epyeru
    Kolonario Oule
    Isobel Sabri
    Mihdí Samandarí
  * Peter Vuyiya

Southern Africa
    Seewoosumbur-Jeehoba Appa
    Shidan Fat'he-Aazam (Trustee of Continental Fund)
    Bahíyyih Winckler

North America
    Lloyd Gardner
  * Sarah Pereira
  * Velma Sherrill
    Edna True (Trustee of Continental Fund)

Central America
  Carmen de Burafato
  * Rowland Estall
  Artemus Lamb
  * Paul Lucas
  Alfred Osborne (Trustee of Continental Fund)

South America
  * Leonora Armstrong
  Athos Costas
  Mas'úd Khamsí (Trustee of Continental Fund)
  * Peter McLaren
  * Raúl Pavón
  Donald Witzel

Western Asia
  Iraj Ayman
  Masíḥ Farhangí
  Hádí Raḥmání (Trustee of Continental Fund)
  Manúchihr Salmánpúr

South Central Asia
  Shirin Boman
  * Dipchand Khianra
  * Zena Sorabjee (Trustee of Continental Fund)
  Sankaran-Nair Vasudevan

Southeastern Asia
  * Firaydún Mítháqíyán
  Khudáraḥm Paymán (Trustee of Continental Fund)
  Vicente Samaniego
  Chellie Sundram
  Yan Kee Leong

Northeastern Asia
  * Richard Benson
  John McHenry III
  * Elena Marsella
  Rúḥu'lláh Mumtází (Trustee of Continental Fund)

Australasia
  Suhayl 'Alá'í
  * Violet Hoehnke
  Howard Harwood
  Thelma Perks (Trustee of Continental Fund)

Europe
    Erik Blumenthal
    Anneliese Bopp
    Dorothy Ferraby
    Louis Hénuzet (Trustee of Continental Fund)
    Betty Reed

133
Youth and Bahá'í Standards of Behavior
9 JULY 1973

To a Local Spiritual Assembly

Dear Bahá'í Friends,

133.1 We have received your letter of 19 June 1973 and can sympathize with the problems that Bahá'í youth face when trying to live up to the Bahá'í standards of behavior. It is, perhaps, natural that in the bewildering amoral environment in which Bahá'í youth are growing up they feel the need for specific instructions on which intimacies are permissible and which are not. However, we feel it would be most unwise for any Bahá'í institution to issue detailed instructions about this.

133.2 The Bahá'í youth should study the teachings on chastity and, with these in mind, should avoid any behavior which would arouse passions which would tempt them to violate them. In deciding what acts are permissible to them in the light of these considerations the youth must use their own judgment, following the guidance of their consciences and the advice of their parents.

133.3 If Bahá'í youth combine such personal purity with an attitude of uncensorious forbearance towards others they will find that those who may have criticized or even mocked them will come, in time, to respect them. They will, moreover, be laying a firm foundation for future married happiness.

                        With loving Bahá'í greetings,
                        THE UNIVERSAL HOUSE OF JUSTICE

134

Formation of the National Spiritual Assembly of Equatorial Guinea
18 July 1973

To all National Spiritual Assemblies

134.1 DELIGHTED ANNOUNCE SUCCESSFUL ELECTION NATIONAL SPIRITUAL ASSEMBLY EQUATORIAL GUINEA COUNTRY OPENED FAITH GUARDIAN'S TEN YEAR CRUSADE. NEW ASSEMBLY FORMED NOW RESPONSE REQUIREMENT GOVERNMENT RECOGNITION FAITH. PRAYERS OFFERED HOLY SHRINES THANKSGIVING AND GUIDANCE ASSISTANCE FRIENDS NEWLY INDEPENDENT COMMUNITY.

THE UNIVERSAL HOUSE OF JUSTICE

135

Passing of the Hand of the Cause of God John Ferraby
9 September 1973

To the Hands of the Cause of God,
the Continental Boards of Counselors,
the National Spiritual Assemblies

Dear Bahá'í Friends,

135.1 The Hand of the Cause John Ferraby passed away suddenly on September 5th. The following cable was sent by the Universal House of Justice:

135.1a REGRET SUDDEN PASSING HAND CAUSE JOHN FERRABY.[1] RECALL LONG SERVICES FAITH BRITISH ISLES CROWNED ELEVATION RANK HAND CAUSE VALUABLE CONTRIBUTION BAHÁ'Í LITERATURE THROUGH HIS BOOK ALL THINGS MADE NEW. REQUESTING BEFITTING GATHERINGS MA<u>SH</u>RIQU'L-A<u>DH</u>KÁRS MEMORIAL MEETINGS ALL COMMUNITIES BAHÁ'Í WORLD.

With loving Bahá'í greetings,
THE UNIVERSAL HOUSE OF JUSTICE

135-1. For an account of the life and services of John Ferraby, see BW 16:511–12.

1973–1974 • A YEAR OF PREPARATION

136
Appointment of the Architect of the Seat of the Universal House of Justice
17 September 1973

To all National Spiritual Assemblies

DELIGHTED ANNOUNCE APPOINTMENT ḤUSAYN AMÁNAT BRILLIANT YOUNG BAHÁ'Í ARCHITECT CRADLE FAITH AS ARCHITECT OF BUILDING FOR UNIVERSAL HOUSE OF JUSTICE. 136.1

THE UNIVERSAL HOUSE OF JUSTICE

137
Expansion and Reinforcement of the Auxiliary Boards
7 October 1973

To the Bahá'ís of the World

Dear Bahá'í Friends,

In order to meet the growing needs of an ever-expanding Bahá'í World Community we have taken two decisions designed to reinforce and extend the services of the Auxiliary Boards. 137.1

First, the number of Auxiliary Board members throughout the world is to be raised to two hundred and seventy, of whom eighty-one will serve on the Auxiliary Boards for the Protection of the Faith and one hundred and eighty-nine will serve on the Auxiliary Boards for the Propagation of the Faith. In all there will be fifty-four Auxiliary Board members in Africa, eighty-one in the Western Hemisphere, eighty-one in Asia, eighteen in Australasia and thirty-six in Europe. 137.2

Secondly, we have decided to take a further step in the development of the institution by giving to each Continental Board of Counselors the discretion to authorize individual Auxiliary Board members to appoint assistants. Such authorization does not have to be given to all the Auxiliary Board members in a zone nor does the number assigned have to be the same for all Board members; indeed certain Boards of Counselors may decide that the present circumstances in their zones do not require them to take advantage of this possibility. Such matters are left entirely to the discretion of each Continental Board of Counselors. 137.3

The exact nature of the duties and the duration of the appointment of the assistants is also left to each Continental Board to decide for itself. Their aims should be to activate and encourage Local Spiritual Assemblies, to call the attention of Local Spiritual Assembly members to the importance of holding 137.4

regular meetings, to encourage local communities to meet for the Nineteen Day Feasts and Holy Days, to help deepen their fellow-believers' understanding of the Teachings, and generally to assist the Auxiliary Board members in the discharge of their duties. Appointments may be made for a limited period, such as a year or two, with the possibility of reappointment. Believers can serve at the same time both as assistants to Auxiliary Board members and on administrative institutions.

137.5 It is our prayer at the Sacred Threshold that this new development in the institution of the Auxiliary Boards will lead to an unprecedented strengthening of the Local Spiritual Assemblies throughout the world.

THE UNIVERSAL HOUSE OF JUSTICE

137.6
Numbers of Auxiliary Board Members by Zones
October 1973

	Auxiliary Board for Protection	Auxiliary Board for Propagation
Africa		
Northwestern	3	9
Central and East	13	19
Southern	2	8
	18	36
Western Hemisphere		
North America	9	18
Central America	9	9
South America	9	27
	27	54
Asia		
Western	9	18
South Central	3	15
Southeastern	3	15
Northeastern	3	15
	18	63
Australasia	9	9
Europe	9	27
TOTAL	**81**	**189**

138
Request for Actions in Preparation for Launching Five Year Plan at Riḍván 1974
21 NOVEMBER 1973

To all National Spiritual Assemblies

Dear Bahá'í Friends,

Five months separate us from Riḍván 1974 when the next global plan will be launched. For a period of five years the attention, resources and energies of the Bahá'í World Community will be directed to achieving the aims of this plan.

By Naw-Rúz 1974 you will have been notified of the overall goals of the plan and the specific tasks assigned to each of your national communities. Each one of you is therefore urged to arrange for a meeting, at Naw-Rúz or soon after, to which you will invite the Board of Counselors in your zone to be represented and at which the plan can be considered and thorough consultation held on the manner in which each one of your communities will launch it.

You are asked to give careful consideration as soon as possible to the advisability of holding one or more conferences in conjunction with your Convention or soon after. You may wish to consult the Counselors on this matter. We believe that such conferences would greatly assist in acquainting the friends with the nature and aims of the plan and in enlisting their enthusiasm and resolution to achieve it. All details as to the number of such conferences, their timing, their agendas are left entirely to your discretion, but we recommend that in planning them you attach great importance to the participation of youth so that they may feel wholly identified with the tasks assigned and give their immediate and maximum support to their accomplishment.

Now is the time to begin directing the thoughts and plans of the friends to the next great demand which will be made upon them, and we assure you of our prayers at the Sacred Threshold that you may be guided and strengthened to take such decisions and make such plans as will enable your communities to anticipate with eagerness and receive with joy the new tasks to be offered them, tasks whose wholehearted and united accomplishment will raise the Community of the Most Great Name to a position where it may have far greater effect upon men's minds and prepare it for further thrilling and awe-inspiring achievements in the pursuit of its ultimate goal of the redemption of mankind.

With loving Bahá'í greetings,
THE UNIVERSAL HOUSE OF JUSTICE

139

Extension of Gardens at Bahjí
4 December 1973

To all National Spiritual Assemblies

139.1 REJOICE ANNOUNCE FRIENDS BEAUTIFICATION DURING CONFLICT AGITATING MIDDLE EAST FOURTH QUADRANT AREA SURROUNDING MOST HOLY SHRINE EMBRACING OLIVE GROVE SOUTHWEST PILGRIM HOUSE BAHJÍ. BLESSED SHRINE AND MANSION NOW COMPLETELY ENCIRCLED BEAUTIFUL GARDENS INSPIRED BY PATTERN ḤARAM-I-AQDAS CREATED BY BELOVED GUARDIAN.¹ PRAYING SHRINES SUPPORTERS MOST GREAT NAME EVERY LAND MAY REDOUBLE EFFORTS PROMOTE INTERESTS PRECIOUS FAITH IN ANTICIPATION FIVE YEAR GLOBAL PLAN SOON TO BE LAUNCHED.

Universal House of Justice

140

Acceptance of Design for the Seat of the Universal House of Justice
7 February 1974

To all National Spiritual Assemblies

140.1 JOYFULLY ANNOUNCE ACCEPTANCE EXQUISITE DESIGN CONCEIVED BY ḤUSAYN AMÁNAT FOR BUILDING TO SERVE AS PERMANENT SEAT UNIVERSAL HOUSE OF JUSTICE MOUNT CARMEL. DECISION MADE TO PROCEED NEGOTIATE CONTRACTS CONSTRUCTION THIS NOBLE EDIFICE SECOND THOSE BUILDINGS DESTINED ARISE AROUND ARC CONSTITUTE ADMINISTRATIVE CENTER BAHÁ'Í WORLD.¹

The Universal House of Justice

139-1. Ḥaram-i-Aqdas (the Most Holy Court) is a designation Shoghi Effendi gave to the northwestern quadrant of the garden surrounding the Shrine of Bahá'u'lláh.

140-1. For a discussion of the Arc, see the glossary.

1974–1979

The Five Year Plan

141
Launching of the Five Year Plan—Naw-Rúz 1974
NAW-RÚZ 1974

To the Bahá'ís of the World

Dearly loved Friends,

A span of eighteen years separates us from the centenary of Bahá'u'lláh's Ascension and the unveiling of His Almighty Covenant.[1] The fortunes of humanity in that period no man can foretell. We can, however, confidently predict that the Cause of God, impelled by the mighty forces of life within it, must go on from strength to strength, increasing in size and developing greater and greater powers for the accomplishment of God's purpose on earth.

The abundant evidences of Divine confirmation which have rewarded the strenuous and dedicated efforts of the Bahá'í community during the past decade are apparent throughout the earth and give incontrovertible assurance of its capacity to win the good pleasure of Bahá'u'lláh and answer every call made upon it in His service.

The Five Year Plan to which this community is now summoned is the opening campaign of these critical years. It is the third global plan embarked upon by the Army of Light in its implementation of 'Abdu'l-Bahá's Divine Plan, that world-encompassing program disclosed in His perspicuous Tablets and described by the Guardian of the Cause of God as the Charter for the propagation of the Faith throughout the world.[2] It was the Guardian himself, the beloved "sign of God,"[3] who, through his exposition and interpretation of the Revelation, through his discipline and education of the Bahá'í community and through a series of national plans assigned to the various units of that community, forged the Administrative Order of the Faith and made it an instrument for the carrying out of this great Charter, and he himself designed and launched the first global plan, the unique, brilliant and spiritually glorious Ten Year Crusade. The victories of that crusade implanted the banner of Bahá'u'lláh throughout the planet and the following Nine Year Plan reinforced and extended the bastions of the Faith and raised the number of National Spiritual Assemblies—the supporting pillars of the Universal House of Justice—to one hundred and thirteen, a number increased to one hundred and fifteen by the formation at this Riḍván of the National Spiritual Assemblies of Hong Kong and South East Arabia.

141-1. The centenary of both Bahá'u'lláh's Ascension and the unveiling of His Covenant were celebrated in 1992 with a World Congress in New York City, the City of the Covenant, so named by 'Abdu'l-Bahá on 19 June 1912 when He declared His station as the Center of the Covenant.
141-2. See *Tablets of the Divine Plan* (1993).
141-3. WT, p. 11.

Major Objectives of the Plan

141.4 This Five Year Plan has three major objectives: preservation and consolidation of the victories won; a vast and widespread expansion of the Bahá'í community; development of the distinctive character of Bahá'í life particularly in the local communities. The achievement of these overall aims requires the accomplishment of particular tasks at the World Center of the Faith, and by national and local communities.

World Center Goals

141.5 At the World Center work will continue on the collation and classification of the Sacred Texts; authorized translations of three compilations of Scripture will be made and published, namely, Tablets of Bahá'u'lláh revealed after the Kitáb-i-Aqdas, prayers and extracts from the Writings of the Báb, greatly augmenting the fragments of His Utterance now available in the West, and of the Master's works comprising a wide selection from the vast range of subjects illumined by His Divine wisdom; construction will begin on the building on Mount Carmel to serve as the seat of the Universal House of Justice and it is hoped to complete it during the Five Year Plan; further extension and beautification of the gardens and lands surrounding the Holy Places will take place; strengthening of the relationship between the Bahá'í International Community and the United Nations will continue; and efforts will be constantly made to protect the Faith from persecution and to free it from the restraints imposed by religious orthodoxy.[4]

International Conferences

141.6 In the international sphere the erection of two Mashriqu'l-Adhkárs—one in India and one in Samoa—will be initiated; eight International Teaching Conferences will be held during the middle part of the Five Year Plan; two for the Arctic, one in Anchorage and one in Helsinki during July 1976, one in Paris in August 1976, one in Nairobi in October 1976, one in Hong Kong in November 1976, one in Auckland and one in Bahia, Brazil in January 1977 and one in Mérida, Mexico in February 1977.

National Goals

141.7 Sixteen new National Spiritual Assemblies will be formed, namely the National Spiritual Assemblies of the Bahamas, Burundi, Cyprus, the French Antilles, Greece, Jordan, Mali, Mauritania, the New Hebrides, Niger, Sénégal, Sierra Leone, Somalia, Surinam and French Guiana, Togo, and Upper Volta; their national Ḥaẓíratu'l-Quds, Temple sites and endowments must be acquired; the dissemination of news and messages, so vital to the knowledge,

---

141-4. *Tablets of Bahá'u'lláh* was published in 1978; *Selections from the Writings of the Báb*, in 1976; and *Selections from the Writings of 'Abdu'l-Bahá*, in 1978. The Seat of the Universal House of Justice was completed in 1983.

encouragement and unity of the Bahá'í community, must be made efficient and rapid, and in anticipation of a vast expansion in the number of believers, of Local Spiritual Assemblies and of localities where Bahá'ís reside a coordinated program of translating and publishing Bahá'í literature with the eventual aim of providing the Sacred Text and the teachings of the Faith to all mankind is to be developed—a program which will include the founding of six Bahá'í Publishing Trusts and the continued subvention of Bahá'í literature, 409 inter-Assembly assistance projects are scheduled and, at the outset of the Plan, 557 pioneers are called for.

Financial Self-Sufficiency

One of the distinguishing features of the Cause of God is its principle of nonacceptance of financial contributions for its own purposes from non-Bahá'ís; support of the Bahá'í Fund is a bounty reserved by Bahá'u'lláh to His declared followers. This bounty imposes full responsibility for financial support of the Faith on the believers alone, every one of whom is called upon to do his utmost to ensure that the constant and liberal outpouring of means is maintained and increased to meet the growing needs of the Cause. Many Bahá'í communities are at present dependent on outside help, and for them the aim must be to become self-supporting, confident that the Generous Lord will, as their efforts increase, eventually enable them to offer for the progress of His Faith material wealth as well as their devotion, their energy and love. 141.8

Proclamation

The proclamation of the Faith, following established plans and aiming to use on an increasing scale the facilities of mass communication must be vigorously pursued. It should be remembered that the purpose of proclamation is to make known to all mankind the fact and general aim of the new Revelation, while teaching programs should be planned to confirm individuals from every stratum of society. 141.9

Youth

The vast reservoir of spiritual energy, zeal and idealism resident in Bahá'í youth, which so effectively contributed to the success of the Nine Year Plan, must be directed and lavishly spent for the proclamation, teaching, and consolidation of the Cause. Spiritual Assemblies are urged to provide consultation and the offer of guidance to Bahá'í youth who seek to plan their lives in such a way as to be of utmost service to the Cause of God. 141.10

Education of Children

The education of children in the teachings of the Faith must be regarded as an essential obligation of every Bahá'í parent, every local and national community and it must become a firmly established Bahá'í activity during the 141.11

course of the Plan. It should include moral instruction by word and example and active participation by children in Bahá'í community life.

Distinctive Bahá'í Characteristics

141.12   This Five Year Plan must witness the development in the worldwide Bahá'í community of distinctive Bahá'í characteristics implanted in it by Bahá'u'lláh Himself. Unity of mankind is the pivotal principle of His Revelation; Bahá'í communities must therefore become renowned for their demonstration of this unity. In a world becoming daily more divided by factionalism and group interests, the Bahá'í community must be distinguished by the concord and harmony of its relationships. The coming of age of the human race must be foreshadowed by the mature, responsible understanding of human problems and the wise administration of their affairs by these same Bahá'í communities. The practice and development of such Bahá'í characteristics are the responsibility alike of individual Bahá'ís and administrative institutions, although the greatest opportunity to foster their growth rests with the Local Spiritual Assemblies.

Development of Local Spiritual Assemblies

141.13   The divinely ordained institution of the Local Spiritual Assembly operates at the first levels of human society and is the basic administrative unit of Bahá'u'lláh's World Order. It is concerned with individuals and families whom it must constantly encourage to unite in a distinctive Bahá'í society, vitalized and guarded by the laws, ordinances and principles of Bahá'u'lláh's Revelation. It protects the Cause of God; it acts as the loving shepherd of the Bahá'í flock.

141.14   Strengthening and development of Local Spiritual Assemblies is a vital objective of the Five Year Plan. Success in this one goal will greatly enrich the quality of Bahá'í life, will heighten the capacity of the Faith to deal with entry by troops which is even now taking place and, above all, will demonstrate the solidarity and ever-growing distinctiveness of the Bahá'í community, thereby attracting more and more thoughtful souls to the Faith and offering a refuge to the leaderless and hapless millions of the spiritually bankrupt, moribund present order.

141.15   "These Spiritual Assemblies," wrote 'Abdu'l-Bahá, "are aided by the Spirit of God. Their defender is 'Abdu'l-Bahá. Over them He spreadeth His Wings. What bounty is there greater than this?" Likewise, "These Spiritual Assemblies are shining lamps and heavenly gardens, from which the fragrances of holiness are diffused over all regions, and the lights of knowledge are shed abroad over all created things. From them the spirit of life streameth in every direction. They, indeed, are the potent sources of the progress of man, at all times and under all conditions."[5]

141-5. Quoted in GPB, p. 332.

During the Five Year Plan Local Spiritual Assemblies which are being formed for the first time are to be formed whenever there are nine or more adult believers in the relevant area; thereafter they must be elected or declared at Riḍván.[6] National Spiritual Assemblies are called upon to assign, and encourage the Local Spiritual Assemblies to adopt, goals within the overall framework of the Five Year Plan, to consult with them and to assist them to make great efforts to gradually assume their proper function and responsibilities in the World Order of Bahá'u'lláh. The friends are called upon to give their wholehearted support and cooperation to the Local Spiritual Assembly, first by voting for the membership and then by energetically pursuing its plans and programs, by turning to it in time of trouble or difficulty, by praying for its success and taking delight in its rise to influence and honor. This great prize, this gift of God within each community must be cherished, nurtured, loved, assisted, obeyed and prayed for. 141.16

Such a firmly founded, busy and happy community life as is envisioned when Local Spiritual Assemblies are truly effective, will provide a firm home foundation from which the friends may derive courage and strength and loving support in bearing the Divine Message to their fellowmen and conforming their lives to its benevolent rule. 141.17

The Hands of the Cause of God and the International Teaching Center

The deeds and programs, all these multifarious worldwide activities to which you are summoned have but one aim—the establishment of God's Kingdom on earth. At every stage of this process and at all levels of Bahá'í responsibility, whether individual, local or national, you will be encouraged, advised and assisted by the divinely ordained institution of the Hands of the Cause of God, an institution powerfully reinforced by the successful establishment of the International Teaching Center. Through the emergence of this Center the seal has been set on the accomplishment of the goal, announced nearly ten years ago, of ensuring the extension into the future of the specific functions of protection and propagation conferred upon the Hands of the Cause in the Sacred Text. Through the work of the International Teaching Center, which supervises and coordinates the work of the Boards of Counselors around the world, the love, the guidance, the assistance of the Hands, through the Boards of Counselors, their Auxiliary Board members and their assistants, permeates the entire structure of Bahá'í society. 141.18

The Chief Stewards of Bahá'u'lláh's embryonic world commonwealth have indeed assured to that growing community, the care for its welfare, for the development of its character, for its spiritual encouragement which are among the duties of their high office. 141.19

141-6. For further guidance on the formation of Local Spiritual Assemblies, see messages no. 189, 199, and 219.

Our Opportunities

141.20 As the old order gives way to the new, the changes which must take place in human affairs are such as to stagger the imagination. This is the opportunity for the hosts of the Lord. Undismayed and undeterred by the wreckage of "long-cherished ideals and time-honored institutions," now being "swept away and relegated to the limbo of obsolescent and forgotten doctrines," the world community of Bahá'ís must surge forward eagerly, and with ever-increasing energy, to build those new, God-given institutions from which will be diffused the light of the holy principles and teachings sent down by God in this day for the salvation of all mankind.[7]

THE UNIVERSAL HOUSE OF JUSTICE

142
Elucidation of Five Year Plan Goals
NAW-RÚZ 1974

To all National Spiritual Assemblies

Dear Bahá'í Friends,

142.1 To supplement the message which is being addressed to each of your Communities giving its specific goals under the Five Year Plan, we now share with you a number of elucidations. Certain of the paragraphs which follow may apply to goals which have not been allotted to your community, but it will no doubt be of interest to you to read them in relation to the worldwide scope of the Plan.

Opening Localities

142.2 When choosing localities to be opened to the Faith and when deciding which localities should have Local Spiritual Assemblies, you should bear in mind the need to have the Bahá'í community represented broadly across the area under your jurisdiction. It is likely that some areas will show themselves particularly receptive and numerous Bahá'í communities will speedily arise there, but while fostering such growth you should not neglect those areas in which the Faith is as yet unrepresented.

The Development of Local Spiritual Assemblies

142.3 The institution of the Local Spiritual Assembly is of primary importance in the firm establishment of the Faith, and we hope that you will give particular attention to ensuring that as many as possible, and in increasing num-

---

141-7. WOB, p. 42.

bers, are, in the words of the beloved Guardian, "broad-based, securely grounded" and "efficiently functioning."[1]

The time has come, we believe, when increasing numbers of Local Spiritual Assemblies should assume responsibility for helping the teaching work of groups, isolated believers, and other Spiritual Assemblies in their neighborhood. Such extension teaching goals should be assigned by the National Spiritual Assembly or one of its teaching committees, or can be spontaneously adopted by Local Spiritual Assemblies, and should be carried out within the framework of the overall teaching plans of the country. It should also be made clear that by being given such goals a Spiritual Assembly is not being given any jurisdiction over believers outside its area, still less over other Local Spiritual Assemblies, but is being called upon to collaborate with them in their work. 142.4

The Recognition of Bahá'í Marriage and Holy Days

The Five Year Plan does not include specific goals for the recognition of Bahá'í marriage certificates or of Bahá'í Holy Days because, in most countries where these goals are not already won, achievement depends upon circumstances beyond our control. Nevertheless, National Spiritual Assemblies should bear in mind the need to increase recognition of the Faith and should be alert to possibilities of winning these goals where they are as yet unattained. 142.5

National Incorporation Goals

There are a number of national incorporation goals of the Nine Year Plan towards the attainment of which considerable progress has already been made. These have not been included as goals of the Five Year Plan although they are still pending, but of course they should be pursued to completion. 142.6

Property Acquisitions

If acquisition of a National Ḥaẓíratu'l-Quds is a responsibility assigned to you under the Five Year Plan, you should treat it as an urgent matter in view of the worldwide condition of inflation and rising property costs. Such a building, which must be suitable to serve as the seat of the National Spiritual Assembly, should be purchased as economically as possible. Preferably it should be a freehold detached building, although if such is not obtainable, a semi-detached house or an apartment may be considered, or even a property on a long-term lease. 142.7

A site for a future Mashriqu'l-Adhkár can be as small as 8,000 square meters in area if a larger property would be too expensive. It should, if possible, be situated within the city designated or, if this is not feasible, within 25 kilometers from the city. 142.8

142-1. CF, p. 22.

142.9 A national endowment should be regarded as an investment in real estate owned by the National Spiritual Assembly. It may be anywhere in the country and can be a small, inexpensive piece of land donated by one of the friends, or else acquired out of the resources of the National Fund.

142.10 Where we have given a goal to acquire a Ḥaẓíratu'l-Quds which is to serve the entire community in a certain country, it is to be a local Ḥaẓíratu'l-Quds at the present time but should be of a size and quality to serve as an administrative center and focal point for the whole community. We envisage that some of such Ḥaẓíratu'l-Quds may, at a later date, be converted into National Ḥaẓíratu'l-Quds, and this fact should be borne in mind when acquiring them.

142.11 In the goal for local Ḥaẓíratu'l-Quds given to some communities we state that a certain number should be large enough to accommodate activities of a number of communities in the surrounding district. While not being at all in the same category as the Ḥaẓíratu'l-Quds described in the last paragraph above, these particular buildings are intended to be rather more substantial structures than the average local Ḥaẓíratu'l-Quds, and should be located in areas which form easily accessible, central gathering places for districts in which large numbers of Bahá'ís are living. In addition to serving as a local Ḥaẓíratu'l-Quds for its own town or village, such a building can be used for district gatherings, for the holding of teaching institutes, conferences, deepening classes, etc., for the larger area, and could possibly accommodate the office of the district teaching committee.

142.12 In general we intend that the local Ḥaẓíratu'l-Quds called for in the Plan should be very simple structures to serve as focal points and meeting places for the local communities. It is hoped that land for them can be provided by local believers and that they can be built, for the most part, by the local friends. In certain instances the National Spiritual Assembly may feel justified in giving a small amount of assistance from the National Fund.

142.13 The acquisition of local endowments, which is given as a specific goal to some national communities, is intended to assist in the consolidation of local communities and to foster the spirit of unity and collaboration among the believers. A local endowment can be quite a small piece of land; it can be purchased by the Local Spiritual Assembly or is more usually the gift of one or more of the believers. If the Local Spiritual Assembly is incorporated, the endowment should be registered in its name, but if it is not, the endowment can be held by one or more of the believers on behalf of the community. For example, if one of the believers gives a small piece of land he can continue to hold it in his name, but it will be known that he does so on behalf of the Local Spiritual Assembly and that the land will in time be transferred legally to the Assembly when that is possible. In some countries land is owned by the state or the tribe and only the use of the land can be assigned; in such places the goal can be considered achieved if the Local Spiritual Assembly can obtain the use of a plot of land in its own name. In some countries, even if the land can be purchased, government regulations require that within a specific time a building must be erected on land held by religious institutions. This problem can be met in several ways: it may be possible for the Spiritual Assembly to obtain the use of, or acquire, a plot of land for agricultural purposes, thus avoiding the need to erect a building; or if the most practical course is to erect on the land a Baháʼí institution such as a local Ḥaẓíratuʼl-Quds, the Assembly could, in its own records, demarcate a portion of the land to be the endowment, distinct from the portion on which the Ḥaẓíratuʼl-Quds stands.

Dawn Prayers

One of the characteristics of Baháʼí society will be the gathering of the believers each day during the hours between dawn and two hours after sunrise to listen to the reading and chanting of the Holy Word. In many communities at the present time, especially in rural ones, such gatherings would fit naturally into the pattern of the friends' daily life, and where this is the case it would do much to foster the unity of the local community and deepen the friends' knowledge of the Teachings if such gatherings could be organized by the Local Spiritual Assembly on a regular basis. Attendance at these gatherings is not to be obligatory, but we hope that the friends will more and more be drawn to take part in them. This is a goal which can be attained gradually. 142.14

National Teaching Conferences

The holding of regular national teaching conferences has proved to be a valuable stimulus to the work in a number of countries, as well as a means for forging more strongly the bonds of unity among the believers. Beyond this, many national communities are presented with a special opportunity to hold a highly effective teaching conference at the time of the eight Intercontinental Conferences which are being called at the midway point of the Plan. Believers traveling to and from these Intercontinental Conferences are likely to be eager to assist the work in the countries through which they pass. Therefore, if you hold a national conference shortly after the Intercontinental Conference which is nearest to you, it may well be attended by believers from other lands who will bring with them the spirit of that Conference, and, by augmenting the numbers attending your national conference will greatly assist its effectiveness as a means of proclaiming the Faith and enthusing those believers who will have been unable to attend the Intercontinental Conferences. 142.15

Youth—Specific Periods of Service

Baháʼí youth should be encouraged to think of their studies and of their training for a trade or profession as part of their service to the Cause of God 142.16

and in the context of a lifetime that will be devoted to advancing the interests of the Faith. At the same time, during their years of study, youth are often able to offer specific periods of weeks or months, or even of a year or more, during which they can devote themselves to travel teaching or to serving the Bahá'í community in other ways, such as conducting children's classes in remote villages. They should be encouraged to offer such service, which will in itself be admirable experience for the future, and the National Assembly should instruct an appropriate committee to receive such offers and to organize their implementation so as to derive the greatest possible advantage from them.

External Affairs Work

142.17 A very important activity which has been pursued effectively in all too few countries, is the undertaking by the National Spiritual Assembly of a sustained, planned effort to foster cordial relations with prominent people and responsible government officials and to familiarize them personally with the basic tenets and the teachings of the Faith. Such an activity must be carried out with wisdom and discretion, and requires the constant attention of a responsible committee as well as periodic review by the National Spiritual Assembly itself. Where successful it can effectively forestall opposition to the Faith and smooth the way for many essential aspects of the development of the Bahá'í community.

Pioneer Goals

142.18 Enclosed with this letter you will receive a list of pioneer assistance initially called for at the opening of the Plan.[2] Any National Spiritual Assembly which has pioneers abroad from previous plans is still responsible for helping them to remain at their posts, or for replacing them, if the services they have been rendering are still needed. However, if you have any still unfilled pioneer goals from the Nine Year Plan or from the current year, you may consider them canceled, because such unfilled goals have been taken into consideration in assigning the goals of the Five Year Plan. Best results can be obtained when pioneer projects are arranged in consultation between the sending and receiving National Spiritual Assemblies or their appropriate committees.

With loving Bahá'í greetings,
THE UNIVERSAL HOUSE OF JUSTICE

142-2. The list is too lengthy to include in this volume.

143
Call for Architects for Houses of Worship in India and Western Samoa
1 April 1974

To all National Spiritual Assemblies

Dear Bahá'í Friends,

143.1 The Universal House of Justice will soon be considering the selection of architects for the Mashriqu'l-Adhkárs to be erected in India and Samoa.

143.2 Those wishing to be considered as architects for either of these Temples are invited to submit statements of their qualifications. Such submissions may include examples of work previously designed and/or executed and, if desired, any thoughts or concepts of proposed designs for the Temples may be expressed in whatever way the applicant chooses.

143.3 The design of each Temple will be developed by the architect selected in relation to the climate, environment and culture of the area where it is to be built.

143.4 The initiation of construction of these Temples is a goal of the current Five Year Plan, and consequently those interested should forward their submissions at an early date to the Universal House of Justice, Bahá'í World Center, P.O. Box 155, Haifa 31-000, Israel.

143.5 Please convey the above message to the friends assembled at your Convention and thereafter to the community at large in whatever way you see fit.

With loving Bahá'í greetings,
The Universal House of Justice

144
Passing of Covenant-Breaker Charles Mason Remey
5 April 1974

To all National Spiritual Assemblies

144.1 CHARLES MASON REMEY WHOSE ARROGANT ATTEMPT USURP GUARDIANSHIP AFTER PASSING SHOGHI EFFENDI LED TO HIS EXPULSION FROM RANKS FAITHFUL HAS DIED IN FLORENCE ITALY IN HUNDREDTH YEAR OF HIS LIFE BURIED WITHOUT RELIGIOUS RITES ABANDONED BY ERSTWHILE FOLLOWERS. HISTORY THIS PITIABLE DEFECTION BY ONE WHO HAD RECEIVED GREAT HONORS FROM BOTH MASTER AND GUARDIAN CONSTITUTES YET ANOTHER EXAMPLE FUTILITY ALL ATTEMPTS UNDERMINE IMPREGNABLE COVENANT CAUSE BAHÁ'U'LLÁH.

The Universal House of Justice

145

Laws of the Kitáb-i-Aqdas Concerning Men and Women

28 April 1974

To an individual Believer

Dear Bahá'í Friend,

145.1 The various questions you set forth in your letter of 18 February were noted, and we offer you the following comments.

145.2 The Laws of the Kitáb-i-Aqdas, and indeed all the Teachings of the Faith, form a coherent whole; therefore in order to understand their implications they must be considered in their own context. For example, in the case of intestacy, as you have noted, the eldest son receives preferential treatment in certain respects but, as 'Abdu'l-Bahá has explained in one of His Tablets, he should take into consideration the needs of the other heirs.

145.3 Furthermore it should be remembered that, as Shoghi Effendi has explained (see *The World Order of Bahá'u'lláh,* page 148), Bahá'u'lláh has deliberately left gaps in the body of His legislative ordinances, to be filled in due course by the Universal House of Justice.

145.4 You should, therefore, when studying the *Synopsis and Codification of the Laws and Ordinances of the Kitáb-i-Aqdas,* bear these factors in mind, and always remember Bahá'u'lláh's exhortation to "Weigh not the Book of God with such standards and sciences as are current amongst you, for the Book itself is the unerring balance established amongst men. In this most perfect balance whatsoever the peoples and kindreds of the earth possess must be weighed, while the measure of its weight should be tested according to its own standard, did ye but know it."[1]

145.5 The equality of men and women, as 'Abdu'l-Bahá has often explained, is a fundamental principle of Bahá'u'lláh; therefore the Laws of the Aqdas should be studied in the light of this. Equality between men and women does not, indeed physiologically it cannot, mean identity of function. In some things women excel men, in others men are better than women, while in very many things the difference in sex is of no effect at all. The differences are most apparent in family life. The capacity for motherhood has many far-reaching effects. For example, because of this, daughters receive preference in education over sons. Again, for physiological reasons, women are granted exemptions from fasting that are not applied to men.

145.6 It is apparent from the Guardian's writings that where Bahá'u'lláh has expressed a law as between a man and a woman it applies, mutatis mutan-

---

145-1. SC, p. 22; see also KA ¶99.

dis,² between a woman and a man unless the context should make this impossible. For example, the text of the Kitáb-i-Aqdas forbids a man to marry his father's wife (i.e., his stepmother), and the Guardian has indicated that likewise a woman is forbidden to marry her stepfather. In the case you cite, however, that of a wife who is found by her husband not to have been a virgin, the dissolution of the marriage can be demanded only "If the marriage has been conditioned on virginity";³ presumably, therefore, if the wife wishes to exercise such a right in respect to the husband, she would have to include a condition as to his virginity in the marriage contract, and this would seem to be one of those matters on which the Universal House of Justice will have to legislate in due course.

Although the Universal House of Justice has to apply and supplement the laws of the Aqdas, it has no right at all to change any law that Bahá'u'lláh has specifically revealed. As clearly stated by the Guardian, the provisions of the Kitáb-i-Aqdas "remain inviolate" during the entire Dispensation. . . .⁴ 145.7

With loving Bahá'í greetings,
THE UNIVERSAL HOUSE OF JUSTICE

146
Memorandum on Establishing and Operating a Bahá'í Publishing Trust
13 MAY 1974

The National Spiritual Assembly of the Bahá'ís of the United States

Dear Bahá'í Friends,

The goal of the Five Year Plan to establish six new Publishing Trusts is by now known to you; these new publishing agencies are to be established in Australia, the Fiji Islands, Japan, Korea, the Philippines and Malaysia. 146.1

We have just sent to these six National Spiritual Assemblies the attached Memorandum on Establishing and Operating a Bahá'í Publishing Trust, together with our Memorandum of March 28, 1971. We now enclose both these memoranda solely for your information. It is possible that some of the six National Spiritual Assemblies charged with this goal may apply to any one of you for information about the structure and operation of your own publishing agency and we feel sure you will answer any questions they may ask. 146.2

With loving Bahá'í greetings,
THE UNIVERSAL HOUSE OF JUSTICE

145-2. A Latin term meaning "with due alteration of details."
145-3. KA Q47, pp. 151–52.
145-4. GPB, p. 213.

Memorandum on Establishing and Operating a Bahá'í Publishing Trust
May 1974

146.3 1. The name "Bahá'í Publishing Trust" does not require the establishment of a Trust in the legal sense, and, in fact, more than one Bahá'í publishing agency is not called a Trust.

146.4 2. By whatever name it is called the objective is to establish a publishing agency, under the complete control and direction of the National Spiritual Assembly.

146.5 3. The difference between a Bahá'í Publishing Trust and any other Committee of the National Spiritual Assembly lies chiefly in the fact that the publishing agency does not operate on a budget from the National Spiritual Assembly but is established as a business with its own capital (whose sources are listed at 6 below), trading in the publishing and sale of Bahá'í literature and allied items, and the results of this trading remain within its own financial structure. It is a business, owned by the National Spiritual Assembly, to carry out its publishing requirements.

146.6 4. While it may first be set up as a Committee the aim should be to form some association, legally established, by which the National Spiritual Assembly may act as a publisher. This may be achieved either through the National Spiritual Assembly's own incorporation or by the establishment of a separate legal entity with the National Spiritual Assembly having full control. But in any case legal advice must be sought.

146.7 5. The Company or Trust must be a non-profit-making organization, that is to say all proceeds from its transactions must be used for such things as paying salaries and other operational expenses, royalties and interests on loans and augmenting its own capital. It is not operated for individual profit.

Capitalization

146.8 6. Since the agency is to be operated solely for Bahá'í purposes, capital funds may not be received from non-Bahá'ís, although of course the aim is to sell books to the largest possible public. Capital may be obtained from:

   a) Grants from the National Spiritual Assembly
   b) Gifts from individual Bahá'ís or from Spiritual Assemblies
   c) Profit from trading

d) Loans from Bahá'ís or Bahá'í institutions. Such loans may be interest free or interest bearing but for every loan there must be a written contract setting out the terms of the loan, its duration, condition of repayment and all details.

e) Taking over any publishing assets (stock, outstanding accounts, etc.) which your National Spiritual Assembly or one of your Committees may at present have.

Production

7. Publishing is not the same as printing or manufacturing books. The publisher engages manufacturing firms to produce his books according to his—the publisher's—design and specifications. The actual production and distribution of books need not be confined to the country in which the Publishing Trust operates. The printing and binding may be done anywhere it is deemed most feasible economically, and from the point of view of control, quality, economy and financial arrangements.

Publishing Program

8. Bahá'í literature comprises in general the Sacred Text (works of Bahá'u'lláh, the Báb, 'Abdu'l-Bahá); the Guardian's writings; letters and publications of the Universal House of Justice; introductory and explanatory works; historical works; teaching pamphlets and other teaching literature. The purpose of establishing Bahá'í publishing agencies throughout the world is to make a wide range of such material available to everybody.

The specific program you must devise will therefore take into consideration the following factors:

a) What are the prevailing languages in your area of jurisdiction.

b) Will other National Spiritual Assemblies be interested in your publications.

c) What are your immediate needs for teaching and study of the Faith.

d) What literature useful to you already exists.

e) Reviewing.

Under

a) You will need to consider a program of translation and we refer you to our Memorandum of 28 March 1971.[1]

146-1. See message no. 94.

    b) If the answer is yes, you will need to consult any such National Spiritual Assembly with a view to establishing priorities and enlisting their help in translating.

    c) Together with b) and d) should enable you to establish a publishing program by priority of need.

    d) If you can, with reasonable ease, obtain needed literature from other Bahá'í Publishing Trusts you should obviously do so and use your own resources for publishing items not available elsewhere. Your own publishing agency should buy such material at wholesale prices and re-sell it to Local Spiritual Assemblies and individuals.

    e) Everything published must be approved; see our Memorandum of 28 March 1971.

Financial Program

146.12   9. An appeal may be made to all the believers under your jurisdiction, as well as to any National Spiritual Assemblies under 8 b) above, to support the new publishing agency. In addition to such an appeal a general invitation to the friends may be issued to take up loans, see 6 d) above.

146.13   Proper accounts must be kept and a Profit and Loss Account and a Balance Sheet drawn up and audited every year.

146.14   *Pricing of publications.* Two objectives have to be balanced against each other, namely, to make Bahá'í literature available at as low a price as possible and to build up a sound business. Retail prices will have to cover

    a) production costs

    b) operating expenses (see 5)

    c) discounts allowed

    d) a small profit to repay loans and build up capital.

146.15   *Postage on books.* Cost of postage or freight may either be charged directly to the customer or included in the selling price.

146.16   If the National Spiritual Assembly wishes to sell a book at less than the commercial retail price, it should subsidize its Publishing Trust so that the Trust itself will incur no loss.

Management

146.17   10. The Publishing Trust should be managed by a Committee, appointed by your National Spiritual Assembly and directly responsible to you. Ideally it should have in its membership one believer capable of acting as general manager and conducting the business of the Trust on behalf of the Committee. At the outset it may not be possible to make this a full-time position or to offer a salary to the manager, but this point should be borne in mind as the business of the Trust increases and its volume of sales justifies such an expense. Perhaps it may be possible to find some competent believer who, for the present, would make the management of the Publishing Trust his or her Bahá'í service.

The above are not hard and fast rules but guidelines for consideration. The important thing is to tackle at once the problem of supplying literature to support the all-important work of teaching and study of the Cause. The Sacred Text, the Guardian's writings, expository and historical works are all essential to the propagation and promotion of the Faith. 146.18

147
Laws of the Kitáb-i-Aqdas Not Binding in the West
9 June 1974

The National Spiritual Assembly of the Bahá'ís of Iceland

Dear Bahá'í Friends,

Thank you for your letter of 4 March 1974 enclosing the inquiry from the Bahá'í Group of Ísafjörður. It has become apparent from a number of questions we have received that many believers are not clear which are those laws already binding upon the Bahá'ís in the West. We therefore feel it is timely to clarify the situation, and the simplest way is to state those laws listed in the *Synopsis and Codification of the Kitáb-i-Aqdas* which are *not* at present binding upon the friends in the western world. For ease of reference we give the numbers of the sections listed. 147.1

IV.A.(4)(c) The law regarding the exemption from obligatory prayer granted to women in their courses. 147.2

IV.A.(10) The law concerning ablutions, with the exception of the ablutions required for the Medium Obligatory Prayer which are described in Section CLXXXII of *Prayers and Meditations* and are required for the recitation of that prayer. 147.3

IV.A.(12) The law concerning actions to be taken in place of an Obligatory Prayer missed on account of insecure conditions. 147.4

IV.B.(5)(a) The definition of travelers for the purpose of exemption from fasting. Instead of these definitions the believers in the West should observe the following guidance given by the beloved Guardian's 147.5

secretary on his behalf: "travelers are exempt from fasting, but if they want to fast while they are traveling, they are free to do so. You are exempt the whole period of your travel, not just the hours you are in a train or car, etc. . . ."[1]

147.6 IV.B.(5)(f) The law regarding the exemption from fasting granted to women in their courses.

147.7 IV.C.(1)(i) The laws governing betrothal.

147.8 IV.C.(1)(j) The law concerning the payment of a dowry by the groom to the bride on marriage.

147.9 IV.C.(1)(l) and (m) The laws concerning the traveling of a husband away from his wife.

147.10 IV.C.(1)(n) and (o) The laws relating to the virginity of the wife.

147.11 IV.C.(2)(b) That part of the divorce law relating to fines payable to the House of Justice.

147.12 IV.C.(3) The law of inheritance. This is normally covered by civil laws of intestacy at the present time.

147.13 IV.D.(1)(a) The law of pilgrimage.

147.14 IV.D.(1)(b) The law of Ḥuqúqu'lláh is not yet applied to the western friends.[2]

147.15 IV.D.(1)(d) The law of the Mashriqu'l-Adhkár is gradually being put into effect.

147.16 IV.D.(1)(f) The Bahá'í Festivals are being celebrated by the western friends on their anniversaries in the Gregorian calendar until such time as the Universal House of Justice deems it desirable to pass supplementary legislation necessary for the full implementation of the Badí' calendar.

147.17 IV.D.(1)(j) The age of maturity applies only to Bahá'í religious duties as yet.[3] On other matters it is subject to the civil law of each country. The age of administrative maturity in the Bahá'í community has, for the time being, been fixed at 21.

147-1. LG, p. 234.
147-2. See message dated 6 August 1984 on the introduction of Ḥuqúqu'lláh to the West (no. 404), and message dated 4 July 1985 introducing a compilation on Ḥuqúqu'lláh (no. 430). For the compilation, see CC 1:489–527.
147-3. For more information on the responsibilities of youth at the age of maturity, see message no. 426.

IV.D.(1)(k) For the burial of the dead the only requirements now binding in the West are to bury the body (not to cremate it), not to carry it more than a distance of one hour's journey from the place of death, and to say the Prayer for the Dead if the deceased is a believer over the age of 15.   147.18

IV.D.(1)(p) The law of tithes.   147.19

IV.D.(1)(q) The law concerning the repetition of the Greatest Name 95 times a day.   147.20

IV.D.(1)(r) The law concerning the hunting of animals.   147.21

IV.D.(1)(t), (u), (v) and (w) The laws relating to the finding of lost property, the disposition of treasure trove, the disposal of objects held in trust and compensation for manslaughter are all designed for a future state of society. These matters are usually covered by the civil law of each country.   147.22

IV.D.(1)(y) (xiv),(xv),(xvi) and (xvii) Arson, adultery, murder and theft are all forbidden to Bahá'ís, but the punishments prescribed for them in the Kitáb-i-Aqdas are designed for a future state of society. Such matters are usually covered by the civil laws of each country.   147.23

IV.D.(1)(y) (xxv), (xxx), (xxxi) and (xxxii) The laws prohibiting the use of the type of pools which used to be found in Persian baths, the plunging of one's hand in food, the shaving of one's head and the growth of men's hair below the lobe of the ear.   147.24

All the exhortations, listed in section IV.D.(3), are applicable universally at the present time insofar as it is possible for the friends to implement them; for example, the exhortation to teach one's children to chant the Holy Verses in the Mashriqu'l-Adhkár can be literally carried out only on a limited scale at the present time, but the friends should, nevertheless, teach their children the Holy Writings as far as possible.   147.25

With loving Bahá'í greetings,
THE UNIVERSAL HOUSE OF JUSTICE

148
Revision of the Functions of Continental Pioneer Committees
22 July 1974

To all National Spiritual Assemblies

Dear Bahá'í Friends,

148.1 In view of the ever-increasing number of pioneers and traveling teachers now arising from various countries to serve the Cause of God in widely scattered lands throughout all continents the Universal House of Justice has considered ways of deriving maximum benefit from the services of these devoted believers, coordinating their efforts and anticipating the needs of the future.

148.2 The Continental Boards of Counselors will soon be approaching you about the need for pioneers and traveling teachers for the period ending Riḍván 1976.

148.3 The functions of the Continental Pioneer Committees have been reviewed and developed in a way that will enable them to operate in closer collaboration with the Continental Boards of Counselors and the National Spiritual Assemblies of their areas. A copy of the statement outlining the functions of the Continental Pioneer Committees as now revised is attached for your information. As you will note, the members of these Committees will henceforth be appointed by the Universal House of Justice. Nothing in the functions now assigned to the Continental Pioneer Committees in any way detracts from the primary responsibility of National Spiritual Assemblies to foster and promote pioneering and traveling teaching.

148.4 It is our hope and prayer that as the Five Year Plan unfolds evidences of closer ties of cooperation among the various institutions of the Faith will be increasingly witnessed in every land.

With loving Bahá'í greetings,
THE UNIVERSAL HOUSE OF JUSTICE

149

The Lesser Peace and "the Calamity"
29 July 1974

The National Spiritual Assembly of the Bahá'ís of the United States

Dear Bahá'í Friends,

149.1 We have received your letter of 19 June 1974 describing the preoccupation of some American believers with the date of the Lesser Peace, and with their feeling that "the calamity," as a prelude to that peace, is imminent.

149.2 It is true that 'Abdu'l-Bahá made statements linking the establishment of the unity of nations to the twentieth century. For example: "The fifth candle is the unity of nations—a unity which, in this century, will be securely established, causing all the peoples of the world to regard themselves as citizens of one common fatherland."[1] And, in *The Promised Day Is Come*, following a similar statement quoted from *Some Answered Questions*, Shoghi Effendi makes this comment: "This is the stage which the world is now approaching, the stage of world unity, which, as 'Abdu'l-Bahá assures us, will, in this century, be securely established."[2]

149.3 There is also this statement from a letter written in 1946 to an individual believer on behalf of the beloved Guardian by his secretary:

> All we know is that the Lesser and the Most Great Peace *will* come—their *exact* dates we do not know. The same is true as regards the possibility of a future war; we cannot state dogmatically it will or will not take place—all we know is that mankind must suffer and be punished sufficiently to make it turn to God.

149.4 It is apparent that the disintegration of the old order is accelerating, but the friends should not permit this inevitable process to deter them from giving their undivided attention to the tasks lying immediately before them. Let them take heart from the reassuring words of Shoghi Effendi contained in the closing paragraphs of his momentous message of June 5, 1947, and concentrate on the challenging tasks of this hour.[3]

With loving Bahá'í greetings,
THE UNIVERSAL HOUSE OF JUSTICE

149-1. SWAB, p. 32.
149-2. PDIC ¶298; see also SAQ, p. 65.
149-3. CF, pp. 37–38.

150
Passing of Laura Dreyfus-Barney, Compiler of *Some Answered Questions*
22 August 1974

To the National Spiritual Assembly of the Bahá'ís of France

150.1 ASCENSION DISTINGUISHED MAIDSERVANT LAURA DREYFUS-BARNEY FURTHER DEPLETES SMALL BAND PROMOTERS FAITH HEROIC AGE.[1] MEMBER FIRST HISTORIC GROUP PARIS TAUGHT BY MAY MAXWELL SHE ACHIEVED IMMORTAL FAME THROUGH COMPILATION SOME ANSWERED QUESTIONS UNIQUE ENTIRE FIELD RELIGIOUS HISTORY. OFFERING ARDENT PRAYERS SACRED THRESHOLD PROGRESS HER SOUL ABHÁ KINGDOM URGE ALL COMMUNITIES FRANCE HOLD MEMORIAL GATHERINGS GRATITUDE OUTSTANDING ACHIEVEMENT.

UNIVERSAL HOUSE OF JUSTICE

151
Comments on the Bahá'í Attitude toward Material Suffering
19 November 1974

The National Spiritual Assembly of the Bahá'ís of Italy

Dear Bahá'í Friends,

151.1 In your letter of 11 September you say that the questions of how to help the Third World or the poor who are suffering under calamities are much discussed in your community and you wish to know whether to create a special fund for such needs, to ask for special contributions from time to time, or whether there are other ways in which you could help.

151.2 It is understandable that Bahá'ís who witness the miserable conditions under which so many human beings have to live, or who hear of a sudden disaster that has struck a certain area of the world, are moved to do something practical to ameliorate those conditions and to help their suffering fellow-mortals.

150-1. Laura Clifford Dreyfus-Barney was born in the United States in 1879 into a family of scholars and artists. She learned about the Bahá'í Faith from May Bolles Maxwell in Paris, circa 1900, during the Heroic Age of the Faith (1844–1921). *Some Answered Questions,* first published in London in 1908 and issued five times since by the Bahá'í Publishing Trust of the United States, consists of 'Abdu'l-Bahá's responses to questions put to Him at table by Miss Barney between 1904 and 1906. In 1911 she married the distinguished Hippolyte Dreyfus, the first French Bahá'í. She died in Paris on 18 August 1974. For an account of her life and services, see BW 16:535–38.

There are many ways in which help can be rendered. Every Bahá'í has the 151.3
duty to acquire a trade or profession through which he will earn that wherewith he can support himself and his family; in the choice of such work he can seek those activities which are of benefit to his fellowmen and not merely those which promote his personal interests, still less those whose effects are actually harmful.

There are also the situations in which an individual Bahá'í or a Spiritual 151.4
Assembly is confronted with an urgent need which neither justice nor compassion could allow to go unheeded and unhelped. How many are the stories told of 'Abdu'l-Bahá in such situations, when He would even take off a garment He was wearing and give it to a shivering man in rags.

But in our concern for such immediate obvious calls upon our succor we 151.5
must not allow ourselves to forget the continuing, appalling burden of suffering under which millions of human beings are always groaning—a burden which they have borne for century upon century and which it is the mission of Bahá'u'lláh to lift at last. The principal cause of this suffering, which one can witness wherever one turns, is the corruption of human morals and the prevalence of prejudice, suspicion, hatred, untrustworthiness, selfishness and tyranny among men. It is not merely material well-being that people need. What they desperately need is to know how to live their lives—they need to know who they are, to what purpose they exist, and how they should act towards one another; and, once they know the answers to these questions they need to be helped to gradually apply these answers to everyday behavior. It is to the solution of this basic problem of mankind that the greater part of all our energy and resources should be directed. There are mighty agencies in this world, governments, foundations, institutions of many kinds with tremendous financial resources which are working to improve the material lot of human beings. Anything we Bahá'ís could add to such resources in the way of special funds or contributions would be a negligible drop in the ocean. However, alone among men we have the divinely given remedy for the real ills of mankind; no one else is doing or can do this most important work, and if we divert our energy and our funds into fields in which others are already doing more than we can hope to do, we shall be delaying the diffusion of the Divine Message which is the most important task of all.

Because of such an attitude, and also because of our refusal to become 151.6
involved in politics, Bahá'ís are often accused of holding aloof from the "real problems" of their fellowmen. But when we hear this accusation let us not forget that those who make it are usually idealistic materialists to whom material good is the only "real" good, whereas we know that the working of the material world is merely a reflection of spiritual conditions and until the spiritual conditions can be changed there can be no lasting change for the better in material affairs.

151.7 We should also remember that most people have no clear concept of the sort of world they wish to build, nor how to go about building it. Even those who are concerned to improve conditions are therefore reduced to combating every apparent evil that takes their attention. Willingness to fight against evils, whether in the form of conditions or embodied in evil men, has thus become for most people the touchstone by which they judge a person's moral worth. Bahá'ís, on the other hand, know the goal they are working towards and know what they must do, step by step, to attain it. Their whole energy is directed towards the building of the good, a good which has such a positive strength that in the face of it the multitude of evils—which are in essence negative—will fade away and be no more. To enter into the quixotic tournament of demolishing one by one the evils in the world is, to a Bahá'í, a vain waste of time and effort. His whole life is directed towards proclaiming the Message of Bahá'u'lláh, reviving the spiritual life of his fellowmen, uniting them in a divinely created World Order, and then, as the Order grows in strength and influence, he will see the power of that Message transforming the whole human society and progressively solving the problems and removing the injustices which have so long bedeviled the world.

With loving Bahá'í greetings,
The Universal House of Justice

152
Release of a Compilation on Opposition
26 November 1974

To all National Spiritual Assemblies

Dear Bahá'í Friends,

152.1 Five months before he passed away, the beloved Guardian in his cable to the Bahá'í world, dated 4 June 1957, drew our attention to the fact that from both without and within the Faith evidences of "INCREASING HOSTILITY" and "PERSISTENT MACHINATIONS" were apparent, and that they foreshadowed the "DIRE CONTEST" predicted by 'Abdu'l-Bahá, which was destined to "RANGE [the] ARMY [of] LIGHT [against the] FORCES [of] DARKNESS, BOTH SECULAR [and] RELIGIOUS."[1]

152.2 The marvelous victories won in the name of Bahá'u'lláh, since those words were written; and the triumphs increasingly being achieved by His dedicated and ardent lovers in every land, will no doubt serve to rouse the internal and external enemies of the Faith to fresh attempts to attack the Faith and dampen

---

152-1. MBW, p. 123.

the enthusiasm of its supporters, as evidenced by the book attacking Shoghi Effendi recently published in Germany by Hermann Zimmer, a Covenant-breaker, and the new book misrepresenting the Faith written by William Miller, a longtime enemy of the Faith who used to be a missionary in Persia.

We felt, therefore, that we could contribute to your devoted and incessant efforts to protect our precious Cause by placing in your hands a compilation from the Writings of Bahá'u'lláh, of 'Abdu'l-Bahá and of Shoghi Effendi, clearly outlining the principle that the progressive unfoldment and onward march of the Faith of God are bound to raise up adversaries, indubitably foreshadowing the worldwide opposition which is to come, and unequivocally giving the assurance of ultimate victory. This compilation is far from complete and exhaustive, but provides a basis for the study of this all-important subject.[2]

We leave it to your discretion, in consultation with a Hand or Hands of the Cause who may be available, as well as with the Counselors, to decide in what manner, and how much of this material should be shared with the friends. In some areas it may be best for National Spiritual Assemblies to publish these extracts in Bahá'í newsletters gradually, in others the circulation or even publication of the entire compilation, with other pertinent texts, if called for, may be desirable; in yet other areas it may be enough to draw the attention of the friends to this important subject, through courses and lectures based on these texts and given in conferences and summer schools.

We feel strongly that, whatever method is chosen to inform the friends, the time has come for them to clearly grasp the inevitability of the critical contests which lie ahead, give you their full support in repelling with confidence and determination "the darts" which will be leveled against them by "their present enemies, as well as those whom Providence will, through His mysterious dispensations, raise up, from within or from without," and aid and enable the Faith of God to scale loftier heights, win more signal triumphs, and traverse more vital stages in its predestined course to complete victory and worldwide ascendancy.[3]

With loving Bahá'í greetings,
THE UNIVERSAL HOUSE OF JUSTICE

152-2. See CC 2:137–50. For further information on the subject of opposition to the Faith, see the compilation prepared by the Research Department of the Universal House of Justice in 1987 on crisis and victory in CC 1:131–85.

152-3. MBW, p. 39.

153

Formation of Five New National Spiritual Assemblies during Riḍván 1975 and Readjustment of the Zones of African Continental Boards of Counselors

6 January 1975

To all National Spiritual Assemblies

Dear Bahá'í Friends,

153.1 We are glad to announce that preparations are being made for next Riḍván by the friends in several countries in West Africa and one in the Near East to form, in accordance with the provisions of the Five Year Plan, their new National Spiritual Assemblies. In Western Africa, the National Spiritual Assembly of Dahomey, Togo and Niger will divide into three separate national communities for each of the three countries which presently compose the region, with their seats in Cotonou, Lomé and Niamey respectively, while the National Spiritual Assemblies of West Africa and of Upper West Africa will each split into two units, the former into Liberia and Guinea, with its seat in Monrovia, and Sierra Leone, with its seat in Freetown, and the latter into the Gambia, with its seat in Banjul, and a new National Spiritual Assembly with the name of Upper West Africa comprising Sénégal, Mauritania, Guinea-Bissau and the Cape Verde Islands, with its seat in Dakar. In the Near East the National Spiritual Assembly of Jordan will be formed, with its seat in 'Ammán. These developments on the national level will result in a net increase next Riḍván of five National Spiritual Assemblies, but in view of the inability of the friends in Indonesia to maintain national administrative activities, the total number of National Spiritual Assemblies will thus be raised throughout the world to 119.

153.2 Of the five new National Spiritual Assemblies, four will have their seats in Western Africa. Three more National Spiritual Assemblies are scheduled to be formed in this area in the course of the Plan. The mighty potentialities for growth and expansion in the western regions of Africa are such as to justify a corresponding development of the institution of the Continental Boards of Counselors in that vast and promising area. The decision has been taken, therefore, after consultation with the International Teaching Center, to break the present zone of Northwestern Africa into two separate zones of Northern and Western Africa, to each of which will be transferred parts of the Central and East African zone. The zone of Northern Africa will comprise Egypt, Libya, Tunisia, Algeria, Morocco and Spanish Sahara. The zone of Western Africa will consist of Mauritania, Sénégal, the Gambia, Guinea-Bissau, the Cape Verde Islands, Guinea, Mali, Sierra Leone, Liberia, Ivory Coast, Upper Volta, Niger, Ghana, Togo, Dahomey, Nigeria, Chad, Cameroon, Equatorial Guinea, Gabon, and São Tomé and Príncipe.

Because of the creation of a new Board for Northern Africa, the Counselors in this and the one for Western Africa must be regrouped, new appointments made to the Northern Board, and the number of Auxiliary Board members increased. We decided, therefore, that the Board for Northern Africa will consist of Mr. Muḥammad Kebdani, already serving as a Counselor, Mr. Muḥammad Muṣṭafá, and Mr. 'Imád Ṣábirán. The Board for Western Africa will consist of Mr. Ḥusayn Ardikání (Trustee), Mr. Friday Ekpe, Mr. Zekrullah Kazemi, and Dr. Mihdí Samandarí (transferred from the Central and East African Board).

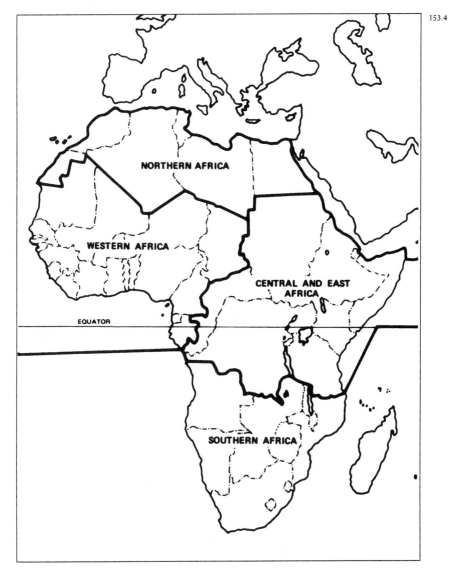

153.5 We are also increasing the number of Auxiliary Board members in Africa, adding 9 members to the Board for Protection, and 9 to that for Propagation, bringing the totals for that continent to 27 and 45 respectively, allocated according to the following schedule:

	Auxiliary Board members for Protection	Auxiliary Board members for Propagation
Central and East Africa	13	19
Southern Africa	4	10
Northern Africa	5	5
Western Africa	5	11
	27	45

153.6 We pray at the Holy Shrines that these decisions, which reflect the growth of our beloved Faith in Africa, will pave the way for speedier progress, wider expansion and greater consolidation, as the friends of that mighty continent forge ahead in their efforts to promote and protect the precious Cause of Bahá'u'lláh.

With loving Bahá'í greetings,
The Universal House of Justice

154
Acquisition of the House of 'Abdu'lláh Páshá in 'Akká
9 January 1975

To all National Spiritual Assemblies

154.1 JOYOUSLY ANNOUNCE SUCCESSFUL CONCLUSION LENGTHY DELICATE NEGOTIATIONS RESULTING ACQUISITION BY PURCHASE HOLY HOUSE CENTER COVENANT 'ABDU'L-BAHÁ BIRTHPLACE BELOVED GUARDIAN SHOGHI EFFENDI. HISTORIC PROPERTY ADJACENT BARRACKS MOST GREAT PRISON COMPRISES LAND AREA APPROXIMATING SEVEN THOUSAND SQUARE METERS INCLUDES OTHER STRUCTURES WITHIN COMPLEX ASSURING PERMANENT PROTECTION HOUSE VISITED BY MANY PILGRIMS TURN CENTURY SCENE HISTORIC VISIT FIRST GROUP WESTERN PILGRIMS.[1] PLANS BEING PREPARED RESTORATION HOLY HOUSE BEAUTIFICATION GROUNDS AS ADDITIONAL PLACE PILGRIMAGE WORLD CENTER WHEN CIRCUMSTANCES FUNDS PERMIT. OFFER HUMBLE THANKSGIVING BAHÁ'U'LLÁH THIS GREAT BLESSING.

Universal House of Justice

154-1. For information on the first group of Western pilgrims, see the entry on Pilgrimage in the glossary. For an account of the significance of the House of 'Abdu'lláh Páshá, see message no. 157.

155
Call for Pioneers
13 January 1975

To the Bahá'ís of the World

Dear Bahá'í Friends,

The striking progress made during the first eight months of the Five Year Plan and the urgent needs of the work as disclosed in a survey made by the International Teaching Center impel us to raise anew the call for pioneers made at Riḍván, increasing the number from 557 to 933. The details of the allocations are now being sent to your National Spiritual Assemblies for immediate action.

The eager response of the friends to the initial call has already resulted in 279 pioneers settled or in process of becoming so. The remainder are urged to arise as quickly as possible before the confusion and chaos which are engulfing the old order disrupt transportation and communications and cause doors which are now open to be closed in our faces. It is our ardent hope that most, if not all, of the 933 posts will be filled by the midway point of the Five Year Plan, which coincides with the Anniversary of the Birth of the Báb, on 20th October 1976.[1]

We renew our plea to individual believers, as well as to National and Local Spiritual Assemblies, to give generous support to the International Deputization Fund, which will not only be an essential factor in the speedy settlement of this urgently needed army of pioneers, but will also stimulate and assist the flow of traveling teachers, whose labors will provide strong reinforcement to the work of the followers of Bahá'u'lláh in all parts of the world.

Our prayers for your guidance and confirmation are offered at the Sacred Threshold. May Bahá'u'lláh inspire those who arise and guide their feet in the path of His service.

With loving Bahá'í greetings,
THE UNIVERSAL HOUSE OF JUSTICE

155-1. See cable of 21 October 1976 (message no. 179) reporting that the majority of the pioneer goals had indeed been achieved by the midpoint of the Five Year Plan.

156

Acquisition of Land Adjacent to the Guardian's Resting-Place in London
4 February 1975

To all National Spiritual Assemblies

156.1 ANNOUNCE PURCHASE STRIP LAND GREAT NORTHERN LONDON CEMETERY FACING BELOVED GUARDIAN'S RESTING PLACE ENSURING PROTECTION SACRED PLOT. PRAYERS GRATITUDE OFFERED DIVINE THRESHOLD.

UNIVERSAL HOUSE OF JUSTICE

157

The Significance of the House of 'Abdu'lláh Páshá
4 March 1975

To all National Spiritual Assemblies

Dear Bahá'í Friends,

157.1 Immediately after sending the cable announcing the joyful news of the acquisition of this property [the house of 'Abdu'lláh Páshá], the Universal House of Justice had the enclosed article prepared at the World Center, and it is sent for you to disseminate as you see fit.[1]

With loving Bahá'í greetings,
DEPARTMENT OF THE SECRETARIAT

The House of 'Abdu'lláh Páshá

157.2 Some of the most poignant, dramatic and historically significant events of the Heroic Age of our Faith are associated with this house, which derives its name from the Governor of 'Akká who built it and used it as his official residence during his term of office, from 1820 to 1832. It stands just inside the northwestern corner of the sea wall of 'Akká, in the close neighborhood of the citadel where Bahá'u'lláh was confined. The main building is L-shaped, facing south and east on its outer prospects. The structure, though chiefly on two stories, is irregular and on the inside angle has balconies, uncovered stairways, a bathhouse and a well. The entire property comprises large courtyards and is bounded on the west, or seaward, side by a wall, which turns due east at its southern angle and continues towards the heart of 'Akká, forming after

---

157-1. For the announcement of the acquisition of the House of 'Abdu'lláh Páshá, see message dated 9 January 1975 (no. 154).

a few yards, the wall of a narrow street; at the eastern terminus of this wall, and within the property, is an imposing house which was occupied by that Governor of 'Akká whose incumbency coincided with 'Abdu'l-Bahá's residence in the main building, and whose northern windows permitted him to maintain a constant surveillance of 'Abdu'l-Bahá's activities. Beyond this house is a small mosque. The eastern boundary of the property is a row of houses giving directly, on its western aspect, to the courtyard and offering many additional vantage points for observing the Master. A similar row of houses extends from the northeastern corner along the northern boundary until they terminate at the longitudinal wing of the main building which, at this point, projects northwards into several conjoined buildings, making a large irregular outcrop on the northern boundary. The western end of the northern boundary is a short stretch of wall completing the enclosure at the northwestern corner of the west wall. Large stables, coach houses and storerooms line the southern boundary.

In this house, fifty lunar years after the Báb's martyrdom, in January, 1899, the casket containing His sacred and precious remains was received by 'Abdu'l-Bahá, Who successfully concealed it until it was possible to inter it, with all honors, in its permanent resting-place in the bosom of Carmel.[2] In this house 'Abdu'l-Bahá was confined during the period of His renewed incarceration.[3] Shoghi Effendi, in *God Passes By*, testifies to the conditions of His life at that time:

157.3

> Even His numerous friends and admirers refrained, during the most turbulent days of this period, from calling upon Him, for fear of being implicated and of incurring the suspicion of the authorities. On certain days and nights, when the outlook was at its darkest, the house in which He was living, and which had for many years been a focus of activity, was completely deserted. Spies, secretly and openly, kept watch around it, observing His every movement and restricting the freedom of His family.[4]

157.3a

Yet during these troublous times, and from this house, He directed the construction of the Báb's sepulcher on Mount Carmel, erected under its shadow His own house in Haifa and later the Pilgrim House,[5] issued instructions for the restoration of the Báb's holy House in Shíráz and for the erection of the

157.3b

157-2. 'Abdu'l-Bahá interred the remains of the Báb on Mount Carmel on 21 March 1909. For an account of this event, see GPB, p. 276.

157-3. In August 1901 the restrictions on 'Abdu'l-Bahá that had gradually been relaxed were reimposed so that He was incarcerated in 'Akká until September 1908.

157-4. GPB, p. 267.

157-5. 'Abdu'l-Bahá's house is located at 7 Haparsim Street in Haifa. The Western Pilgrim House, later the Seat of the Universal House of Justice and later still the Seat of the International Teaching Center, is located across the street at 10 Haparsim Street.

first Ma<u>sh</u>riqu'l-A<u>dh</u>kár of the world in the city of 'I<u>sh</u>qábád.⁶ Again the Guardian is our reference for the Master's ceaseless activity at that time:

157.3c Eyewitnesses have testified that, during that agitated and perilous period of His life, they had known Him to pen, with His own hand, no less than ninety Tablets in a single day, and to pass many a night, from dusk to dawn, alone in His bedchamber engaged in a correspondence which the pressure of His manifold responsibilities had prevented Him from attending to in the daytime.⁷

157.4 It was in this house that His celebrated table talks were given and compiled, to be published later under the title *Some Answered Questions*.⁸ In this house and in the darkest hours of a period which the beloved Guardian describes as "the most dramatic period of His ministry," "in the heyday of His life and in the full tide of His power" He penned the first part of His Will and Testament, which delineates the features and lays the foundations of the Administrative Order to arise after His passing.⁹ In this house He revealed the highly significant Tablet addressed to the Báb's cousin and chief builder of the 'I<u>sh</u>qábád Temple, a Tablet whose import can be appreciated and grasped only as future events unfold before our eyes, and in which, as testified by Shoghi Effendi, 'Abdu'l-Bahá "in stirring terms proclaimed the immeasurable greatness of the Revelation of Bahá'u'lláh, sounded the warnings foreshadowing the turmoil which its enemies, both far and near, would let loose upon the world, and prophesied, in moving language, the ascendancy which the torch-bearers of the Covenant would ultimately achieve over them."¹⁰

157.5 During the twelve years of His residence in this house, 'Abdu'l-Bahá demonstrated the true nobility of His divine nature; overcame hatred with love; pursued without rest, against ever-mounting opposition, the direction of His Father's Cause; maintained in the face of fanaticism, jealousy and bitterness His unceasing care of the poor and sick; and overcame, with unruffled equanimity, the severest crisis of His life. The Guardian's words testify to these things:

157.5a At His table, in those days, whenever there was a lull in the storm raging about Him, there would gather pilgrims, friends and inquirers from most of the aforementioned countries [Persia, the United States, Can-

---

157-6. For the announcement of the demolition of the House of Worship in 'I<u>sh</u>qábád, see letter dated 25 August 1963 (message no. 4).
157-7. GPB, p. 267.
157-8. 'Abdu'l-Bahá, *Some Answered Questions,* collected and trans. Laura Clifford Barney (Wilmette, Ill.: Bahá'í Publishing Trust, 1984). The book was first published in London in 1908 by Keegan, Paul, Trench, Trübner & Co.
157-9. GPB, pp. 267–68.
157-10. GPB, p. 268.

ada, France, England, Germany, Egypt, Iraq, Russia, India, Burma, Japan, and the Pacific Islands], representative of the Christian, the Muslim, the Jewish, the Zoroastrian, the Hindu and Buddhist Faiths. To the needy thronging His doors and filling the courtyard of His house every Friday morning, in spite of the perils that environed Him, He would distribute alms with His own hands, with a regularity and generosity that won Him the title of "Father of the Poor." Nothing in those tempestuous days could shake His confidence, nothing would be allowed to interfere with His ministrations to the destitute, the orphan, the sick, and the downtrodden, nothing could prevent Him from calling in person upon those who were either incapacitated, or ashamed to solicit His aid. . . .

So imperturbable was 'Abdu'l-Bahá's equanimity that, while rumors were being bruited about that He might be cast into the sea, or exiled to Fízán in Tripolitania, or hanged on the gallows, He, to the amazement of His friends and the amusement of His enemies, was to be seen planting trees and vines in the garden of His house, whose fruits when the storm had blown over, He would bid His faithful gardener, Ismá'íl Áqá, pluck and present to those same friends and enemies on the occasion of their visits to Him.[11]

In this house was born the child ordained to hold the destiny of the Faith in his hands for thirty-six years and to become its "beloved Guardian," the child named "Shoghi" by his Grandfather, who grew up under His loving and solicitous care and became the recipient of His Tablets.

When Bahá'u'lláh ascended, in 1892, the Mansion at Bahjí remained in the occupancy of the arch-breaker of the Covenant, the Master's half-brother Muḥammad-'Alí, and members of that branch of Bahá'u'lláh's family. 'Abdu'l-Bahá and the members of His family, including His illustrious sister the Greatest Holy Leaf, remained in the House of 'Abbúd, which continued to be 'Abdu'l-Bahá's official residence.[12] In the course of the fifth year after Bahá'u'lláh's passing, the marriage of 'Abdu'l-Bahá's two eldest daughters took place, and it quickly became apparent that the portion of the House of 'Abbúd available for occupation was woefully inadequate to the enlarged family. With characteristic vigor 'Abdu'l-Bahá took action and in the months preceding the birth of Shoghi Effendi arranged to rent the main building, and subsequently the

---

157-11. GPB, p. 269.
157-12. The building now known as the House of 'Abbúd comprises two houses: the House of 'Udí-Khammár, in which the Holy Family was confined initially, and the House of 'Abbúd itself, which they were later able to rent and to join to the former. The House of 'Abdu'lláh Páshá is some distance away on the same street, which follows the wall of the city of 'Akká next to the sea.

subsidiary wings, of 'Abdu'lláh Páshá's house, and He established it as His official residence. Thus it came about that, in 1897, Shoghi Effendi was born in the same house (in an upper room of the wing facing south) that witnessed events of such vital importance to the Faith and the future of mankind.

157.8 The Guardian's childhood and upbringing in that house are referred to by Amatu'l-Bahá Rúḥíyyih Khánum in *The Priceless Pearl:*

157.8a It may sound disrespectful to say the Guardian was a mischievous child, but he himself told me he was the acknowledged ringleader of all the other children. Bubbling with high spirits, enthusiasm and daring, full of laughter and wit, the small boy led the way in many pranks; whenever something was afoot, behind it would be found Shoghi Effendi! This boundless energy was often a source of anxiety as he would rush madly up and down the long flight of high steps to the upper story of the house, to the consternation of the pilgrims below, waiting to meet the Master. His exuberance was irrepressible and was in the child the same force that was to make the man such an untiring and unflinching commander-in-chief of the forces of Bahá'u'lláh, leading them to victory after victory, indeed, to the spiritual conquest of the entire globe. We have a very reliable witness to this characteristic of the Guardian, 'Abdu'l-Bahá Himself, Who wrote on a used envelope a short sentence to please His little grandson: "Shoghi Effendi is a wise man—but he runs about very much!"

157.8b In those days of Shoghi Effendi's childhood it was the custom to rise about dawn and spend the first hour of the day in the Master's room, where prayers were said and the family all had breakfast with Him. The children sat on the floor, their legs folded under them, their arms folded across their breasts, in great respect; when asked they would chant for 'Abdu'l-Bahá; there was no shouting or unseemly conduct. Breakfast consisted of tea, brewed on the bubbling Russian brass samovar and served in little crystal glasses, very hot and very sweet, pure wheat bread and goats' milk cheese. . . .[13]

157.9 It was to this house that that historic first group of pilgrims from the West came to see the Master in the winter of 1898–99, and in which many more from both East and West sought His presence.[14] Some of them have left memorable descriptions of their experiences with 'Abdu'l-Bahá and His household in that home. Ella Goodall Cooper, one of the very earliest American believers, records the following:

157-13. PP, pp. 7–8.
157-14. For information on the first group of Western pilgrims, see the entry on Pilgrimage in the glossary.

> One day . . . I had joined the ladies of the Family in the room of the Greatest Holy Leaf for early morning tea, the beloved Master was sitting in His favourite corner of the divan where, through the window on His right, He could look over the ramparts and see the blue Mediterranean beyond. He was busy writing Tablets, and the quiet peace of the room was broken only by the bubble of the samovar, where one of the young maidservants, sitting on the floor before it, was brewing the tea.[15]

Thornton Chase, the first American believer, records in his memoir, *In Galilee:*

> We did not know we had reached our destination until we saw a Persian gentleman, and then another and another, step out at the entrance and smile at us. We alighted and they conducted us through the arched, red brick entrance to an open court, across it to a long flight of stone steps, broken and ancient, leading to the highest story and into a small walled court open to the sky, where was the upper chamber assigned to us, which adjoined the room of 'Abdu'l-Bahá. The buildings are all of stone, whitewashed and plastered, and it bears the aspect of a prison.
>
> Our windows looked out over the garden and tent of 'Abdu'l-Bahá on the sea side of the house. That garden is bounded on one side by the house of the Governor, which overlooks it, and on another by the inner wall of fortification. A few feet beyond that is the outer wall upon the sea, and between these two are the guns and soldiers constantly on guard. A sentry house stands at one corner of the wall and garden, from which the sentry can see the grounds and the tent where 'Abdu'l-Bahá meets transient visitors and the officials who often call on him. Thus all his acts outside of the house itself are visible to the Governor from his windows and to the men on guard. Perhaps that is one reason why the officials so often become his friends. No one, with humanity, justice, or mercy in his heart, could watch 'Abdu'l-Bahá long without admiring and loving him for the beautiful qualities constantly displayed.[16]

Mary Hanford Ford published an account of her pilgrimage to this house in *Star of the West,* vol. XXIV:

> The little room in which I stayed and in which the significant conversations with 'Abdu'l-Bahá took place, was of the simplest description. The floor was covered with matting, the narrow iron bed and the iron

---

157-15. Quoted in PP, p. 5.
157-16. See Thornton Chase, "In Galilee," in Thornton Chase and Arthur S. Agnew, *In Galilee and In Wonderland,* (Los Angeles: Kalimát Press, 1985), pp. 22–24.

wash stand with larger and smaller holes for bowl and pitcher were of that vermin proof description with which I had become familiar. Everything was scrupulously clean, and there was an abundant supply of sparkling water for bathing and drinking. A wide window looked over the huge town wall upon the blue Mediterranean and before this stretched a divan upon which 'Abdu'l-Bahá sat when He came to see me.¹⁷

157.12 The palpable victory which 'Abdu'l-Bahá had wrested from the persecution, intrigue, hatred, vilification even, directed against Him during His twelve years in the House of 'Abdu'lláh Páshá, was signally apparent when, upon His release from incarceration in 1908, He moved to His new residence in Haifa. At that time the future Guardian was a boy of eleven, but his appointment, although a carefully guarded secret, had already been made by 'Abdu'l-Bahá in the part of His Will and Testament revealed in that house.¹⁸

157.13 As we contemplate the extraordinary focusing of powerful forces and events upon this house, we eagerly anticipate the day when it will be restored and made ready for pilgrims, who may inhale from its atmosphere, its grounds and sacred walls, the fragrances of a glorious past.¹⁹

158
A Plan for International Collaboration in Traveling Teaching
25 March 1975

To all National Spiritual Assemblies

Dear Bahá'í Friends,

158.1 As we approach the threshold of the second year of the Five Year Plan, it is evident that the need for traveling teachers as indicated in the message launching that Plan is acquiring greater urgency and importance.

158.2 During the past year steps have been taken to revise the functions, broaden the base and strengthen the work of the Continental Pioneer Committees and to bring them into much closer collaboration with the Continental Boards of

---

157-17. Mary Hanford Ford, "An Interview With 'Abdu'l-Bahá," *Star of the West* 24, no. 4 (July 1933): 105.

157-18. See WT, Part I, pp. 3–15.

157-19. The House of 'Abdu'lláh Páshá was restored under the direction of the Hand of the Cause of God Amatu'l-Bahá Rúḥíyyih Khánum and opened to pilgrims in 1983. For the announcement of the appointment of the architect to help restore the House, see message dated 14 October 1977 (no. 198).

Counselors. Already, with their assistance an army of pioneers has moved and is moving towards its objectives, and a general readiness has been evinced by the friends, particularly the youth, to serve as itinerant teachers.

158.3 The strenuous efforts being made to fill the pioneer goals by the midway point of the Plan must now be paralleled by well-considered and determined efforts to swell to a mighty river the stream of those friends who will travel to foreign lands to reinforce the efforts of those who are laboring so valiantly to expand and consolidate the widely scattered Bahá'í communities and to proclaim the Message of Bahá'u'lláh to every stratum of society.

158.4 At our request the International Teaching Center has evolved a plan, which we have warmly approved, comprising specific goals of international collaboration in the field of traveling teaching. This plan is now being sent to the Continental Boards of Counselors who will, in turn, present it to the National Spiritual Assemblies, whose task it will be to implement it. In consultation with the Counselors each National Spiritual Assembly is to work out specific proposals which it should then present to the other National Assemblies with whom it is to collaborate, so that, as soon as possible, actual projects can be worked out and set in motion, thus inaugurating a process which should rapidly gather momentum and be prosecuted with undiminished vigor in the years ahead.

158.5 The Continental Pioneer Committees should be kept closely informed of all projects so that they may know how best to reinforce the flow with those many volunteers who will undoubtedly arise outside the framework of the specific projects now to be conceived. It is our hope that, as far as possible, travel teaching projects will be self-supporting or can be assisted by the National Funds involved, but where necessary, the International Deputization Fund is available to assist. Whenever assistance from the Deputization Fund is required, the request should be made to the Continental Pioneer Committee, giving details of the project. If the sum required is small the Committee may be able to help immediately; otherwise it will pass the request, together with its recommendation, to the Universal House of Justice for consideration.

158.6 We sincerely hope that in the forefront of the volunteers, the Bahá'í youth will arise for the sake of God and, through their driving force, their ability to endure inhospitable and arduous conditions, and their contentment with the bare necessities of life, they will offer an inspiring example to the peoples and communities they set out to serve, will exert an abiding influence on their personal lives, and will promote with distinction the vital interests of God's Cause at this crucial stage in the fortunes of the Plan.

158.7 We shall offer our ardent prayers at the Holy Shrines for the confirmation of the efforts of all those who will heroically respond to this call.

With loving Bahá'í greetings,
THE UNIVERSAL HOUSE OF JUSTICE

159
Riḍván Message 1975
4 April 1975

To all National Spiritual Assemblies

159.1 OCCASION MOST GREAT FESTIVAL WE CONTEMPLATE WITH THANKFUL HEARTS ACHIEVEMENTS FIRST YEAR FIVE YEAR PLAN ELECTION THIS RIḌVÁN FIVE NEW NATIONAL SPIRITUAL ASSEMBLIES FOUR IN AFRICA ONE IN ASIA. DESPITE WORSENING PLIGHT MORIBUND CIVILIZATION EVIDENCES GATHERING CLOUDS WIDESPREAD OPPOSITION DIVINE MESSAGE BELIEVERS THROUGHOUT WORLD FORGING AHEAD ACCOMPLISHMENT GOALS. THREE HUNDRED EIGHTY-SIX PIONEERS ALREADY SETTLED ONE HUNDRED FIFTY-TWO PREPARING PROCEED POSTS. NEW WORLDWIDE TRAVEL TEACHING PROGRAM DESIGNED BY INTERNATIONAL TEACHING CENTER NOW BEING LAUNCHED BY NATIONAL SPIRITUAL ASSEMBLIES IN CONSULTATION COUNSELORS. BELOVED HANDS CAUSE ADVANCING VANGUARD ARMY LIGHT LENDING CONSTANT LOVING GUIDANCE ENCOURAGEMENT PROTECTION FRIENDS LABORING DIVINE VINEYARD. WORLD CENTER FAITH RICHLY BLESSED THROUGH ACQUISITION HOLY HOUSE MASTER BIRTHPLACE SHOGHI EFFENDI WITHIN WALLS 'AKKÁ[1] WILL EARLY WITNESS ON CONSECRATED SOIL SLOPES MOUNT CARMEL INITIATION EXCAVATION FOUNDATIONS PERMANENT SEAT UNIVERSAL HOUSE JUSTICE AND IN ITALY SIGNATURE CONTRACT MARBLE REQUIRED MAJESTIC EDIFICE.[2] AT THIS CRITICAL JUNCTURE HUMAN HISTORY THREE MAJOR OBJECTIVES PLAN AND ITS SPECIFIC GOALS PRESENT DISTINCT INSISTENT CHALLENGE TO EACH INDIVIDUAL BAHÁ'Í ADULT YOUTH CHILD TO EACH BAHÁ'Í FAMILY TO EACH LOCAL COMMUNITY AND ABOVE ALL TO EACH LOCAL SPIRITUAL ASSEMBLY WHOSE DEVELOPMENT IS VITAL SUCCESS FIVE YEAR PLAN AND PROGRESSIVE UNFOLDMENT DIVINELY ORDAINED BAHÁ'Í SOCIETY. MAY REMAINING THREE HUNDRED NINETY-FIVE PIONEERS SPEEDILY ARISE AND ARMY VOLUNTEERS RESPOND NEWLY LAUNCHED TRAVEL TEACHING PROGRAM. NATIONAL LOCAL ASSEMBLIES INDIVIDUAL BELIEVERS URGED CONTRIBUTE UNSTINTINGLY TIME EFFORT OUTPOURING MATERIAL RESOURCES SUPPORT EVERY PHASE IMPLEMENTATION PLAN COMING YEAR. APPEAL BELIEVERS EVERY LAND JOIN US PRAYERS SUPPLICATION BLESSED BEAUTY GUIDE SUSTAIN PROTECT HIS DEVOTED FOLLOWERS IN THEIR DEDICATED EFFORTS PURIFY THEIR SOULS RAISE HIS BANNER SERVE HIS CAUSE.

UNIVERSAL HOUSE OF JUSTICE

159-1. See message dated 9 January 1975 about the acquisition of the house of 'Abdu'lláh Páshá (no. 154). For the announcement of the completion of the restoration of the upper floor of the house and its opening to visitors, see message dated Riḍván 140 B.E. (no. 358). For the announcement of the completion of restoration and opening to visitors of the southern wing of the house, see message dated Riḍván 1986 (no. 456).

159-2. For further information on the significance of the Seat of the Universal House of Justice, see message dated 5 June 1975 (no. 164); on the announcement of the decision to build, see message dated 7 June 1972 (no. 115); on the appointment of the architect, see message dated 17 September 1973 (no. 136); on the acceptance of the design, see message dated 7 February 1974 (no. 140); on the excavation of the site, see message dated 17 June 1975 (no. 165); and on the occupation of the Seat, see message dated 2 February 1983 (no. 354).

160

Fiftieth Anniversary of the National Spiritual Assembly of the United States

24 April 1975

To the National Spiritual Assembly of the Bahá'ís of the United States

PORTENTOUS OCCASION FIFTIETH ANNIVERSARY FORMAL ESTABLISHMENT NATIONAL SPIRITUAL ASSEMBLY BAHÁ'ÍS UNITED STATES RECALL WITH PRIDE ADMIRATION PREEMINENT STATION CONFERRED AMERICAN BAHÁ'Í COMMUNITY CENTER COVENANT PREPONDERATING ROLE ALREADY ASSUMED WORLDWIDE PROMOTION FAITH GLORIOUS ACHIEVEMENTS UNMATCHED RANGE MAGNITUDE ENTIRE BAHÁ'Í WORLD PAST HALF CENTURY. CALL UPON THIS HIGHLY BLESSED DISTINGUISHED COMMUNITY CONTEMPLATE ITS UNDOUBTED DUTIES PRIVILEGES PURGE ITS SOUL ALL WORLDLY ENTANGLEMENTS REDEDICATE ITS ENERGIES RESOURCES IMMEDIATE TASKS ARISE SINGLE-MINDEDLY PERFORM HISTORIC MISSION RENDER SUCH SERVICES ATTAIN SUCH SACRIFICIAL HEIGHTS WORTHY BRILLIANT FOREBEARS HEROIC AGE FAITH. 160.1

UNIVERSAL HOUSE OF JUSTICE

161

Safeguarding the Letters of Shoghi Effendi

14 May 1975

To all National Spiritual Assemblies

Dear Bahá'í Friends,

In December 1967 the Universal House of Justice wrote to all National Spiritual Assemblies expressing the need of the World Center for letters written by the Guardian to them, or to their subsidiary institutions, as well as to the friends under their jurisdiction.[1] 161.1

The response to this request was encouraging, but it is obvious that there are many letters which have not yet been received. 161.2

The Universal House of Justice requests you, therefore, to check again in your archives or files of correspondence with the Guardian for any further letters which have not yet been forwarded to the World Center and to appeal to the believers under your jurisdiction, calling upon those who were privileged to have received letters from the Guardian to send the text of such letters to the World Center. 161.3

161-1. For the December 1967 letter, see message no. 54. See also the 26 August 1984 letter (no. 409), in which the need is reiterated.

161.4 To assist your National Spiritual Assembly and the friends to carry out this urgent project the following points from the letter of December 1967 from the Universal House of Justice are here repeated for your consideration.

1. Recipients of letters from the Guardian have the inherent right of deciding to keep the letters themselves, or to have them preserved for the future in their families. To assist the Universal House of Justice, however, in its efforts to study and compile the letters of the Guardian, the friends are urged to provide, for dispatch to the Holy Land, photostatic copies of their communications from the Guardian if they wish to keep the originals themselves.

2. If they are not in a position to provide such copies, they should kindly allow National Spiritual Assemblies to undertake this project on our behalf.

3. Should any believer possess letters so personal and confidential that he does not wish to disclose their contents to any institution other than the Universal House of Justice, he is invited to send either the originals or copies of such letters, marked confidential, directly to the Universal House of Justice, by registered mail, with any instructions he wishes to be followed.

161.5 Will you please give this matter your early attention. The Universal House of Justice thanks you warmly for your assistance.

With loving Bahá'í greetings,
DEPARTMENT OF THE SECRETARIAT

162
Comments on the Progress of the Five Year Plan
25 MAY 1975

To all National Spiritual Assemblies

Dear Bahá'í Friends,

162.1 A fifth of the span allotted to the Five Year Plan has run its course and we have passed a major milestone in the destinies of that Plan. It is appropriate for every National Spiritual Assembly to pause in order to appraise its position, and that of the community which it represents and serves, and to determine its progress in relation to the goals with which it stands identified.

162.2 To help each National Spiritual Assembly in this appraisal we send you the following statement which, under various headings, outlines the impressions we have gathered and comments we are prompted to make on the prosecution of certain goals of the Plan. Although some of the items may not be directly applicable to you, you may find them of interest. Each National Spiritual Assembly should determine, in the light of the goals assigned to it, to what extent each of our observations is applicable to its work.

Teaching—Expansion and Consolidation

Teaching the Faith embraces many diverse activities, all of which are vital to success, and each of which reinforces the other. Time and again the beloved Guardian emphasized that expansion and consolidation are twin and inseparable aspects of teaching that must proceed simultaneously, yet one still hears believers discussing the virtues of one as against the other. The purpose of teaching is not complete when a person declares that he has accepted Bahá'u'lláh as the Manifestation of God for this age; the purpose of teaching is to attract human beings to the divine Message and so imbue them with its spirit that they will dedicate themselves to its service, and this world will become another world and its people another people. Viewed in this light a declaration of faith is merely a milestone along the way—albeit a very important one. Teaching may also be likened to kindling a fire, the fire of faith, in the hearts of men. If a fire burns only so long as the match is held to it, it cannot truly be said to have been kindled; to be kindled it must continue to burn of its own accord. Thereafter more fuel can be added and the flame can be fanned, but even if left alone for a period, a truly kindled fire will not be extinguished by the first breath of wind.

The aim, therefore, of all Bahá'í institutions and Bahá'í teachers is to advance continually to new areas and strata of society, with such thoroughness that, as the spark of faith kindles the hearts of the hearers, the teaching of the believers continues until, and even after, they shoulder their responsibilities as Bahá'ís and participate in both the teaching and administrative work of the Faith.

There are now many areas in the world where thousands of people have accepted the Faith so quickly that it has been beyond the capacity of the existing Bahá'í communities to consolidate adequately these advances. The people in these areas must be progressively deepened in their understanding of the Faith, in accordance with well-laid plans, so that their communities may, as soon as possible, become sources of great strength to the work of the Faith and begin to manifest the pattern of Bahá'í life.

Reaching Remote Areas—an Immediate Challenge

At the same time there is a challenge of great urgency facing the worldwide Bahá'í community. When launching the Ten Year Crusade, Shoghi Effendi urged the believers to "carry the torch of the Faith to regions so remote, so backward, so inhospitable that neither the light of Christianity or Islam has, after the revolution of centuries, as yet penetrated."[1] A number of

---

162-1. CF, p. 114.

such regions still exist in places like New Guinea, the heart of Africa and the Amazon Basin in South America. As the influence of civilization spreads, the age-old ways of life of the inhabitants of these regions will inevitably perish, and they will rapidly be infected with the materialistic ideas of a decadent civilization. It is our pressing duty to carry the Message of Bahá'u'lláh to such people while they are still purehearted and receptive, and through it to prepare them for the changed world which will come upon them.

Teaching Tribal Peoples and Minorities

162.7 In addition to the tribes in these remote regions of the world, there are tribes and minorities who still live in their traditional ways in the midst of other cultures. All too often such peoples are despised and ignored by the nations among whom they dwell, but we should seek them out, teach them the Cause of God, and enrich through their membership the Bahá'í communities of the lands in which they live. So important is this goal that each National Spiritual Assembly should study the requirements for teaching each of the different tribes and groups within its area, appoint a committee for this purpose—even a special committee for each tribe or minority where this is feasible and desirable—and launch a series of well-conceived, far-reaching campaigns to bring about the enrollment of these peoples within the Cause of Bahá'u'lláh, and the establishment among them of the Bahá'í Administrative Order.

Pioneering

162.8 Great challenges and opportunities for teaching often occur far from large well-established Bahá'í communities; this is especially true in respect of many of the tribal peoples. Pioneering and travel teaching are therefore of the greatest importance for the accomplishment of teaching plans. It is not always difficult to see what the ideal solution for any particular teaching problem may be; however, ideal solutions are seldom available, and the Assemblies which achieve the most outstanding results are those which have developed the skill of using to their best advantage whatever means they have at their command and whatever assistance can be given to them. Pioneers, for example, all have different capacities, different skills, different problems and different responsibilities. A National Assembly may see that its most urgent need is for a financially independent married couple who can live in a remote village area to conduct regular classes for the believers there; but what it actually receives are two single middle-aged ladies who need to work to support themselves and can only get jobs in one of the large towns. Instead of despairing, a resourceful Assembly will immediately see whether the presence of either or both of these ladies in such a town would enable one or more native believers to pioneer to the village area. Even if this does not work out, it will nevertheless do all it can to assist the two pioneers to settle down and will make the utmost use of whatever services they can render, services which may well, in the long run, be of inestimable benefit to that national community.

162.9 There are several ways of pioneering, and all are entirely valid and are of great help to the teaching work. There is, first of all, the pioneer who goes to a particular country, devotes the remainder of his life to the service of the Faith in that land and finally lays his bones to rest in its soil. Secondly, there is the pioneer who goes to a post, serves valiantly there until the native Bahá'í community is strongly established, and then moves on to new fields of service. Thirdly, there are those, for example youth between the completion of their schooling and the starting of their chosen profession, who go pioneering for a specific limited period.

162.10 Ideally, of course, a pioneer should be, or become as soon as possible, financially independent of the Fund in his chosen post, not only to husband the financial resources of the Faith but because it is a Bahá'í principle that everyone should work and support himself and his family whenever possible, and there is no such profession as pioneer or teacher in the Bahá'í Faith as there are professional missionaries and clergymen in other religions. Nevertheless it must be recognized that in some posts where pioneers are desperately needed there is no possibility for them to get work. Either there is no work available in the area or else the pioneer is refused a work permit because he is a foreigner. In such cases it is essential for the Assemblies to provide financial assistance to support the pioneer for as long as is necessary.

162.11 There are a number of methods of financing pioneers in areas where work is unobtainable. Believers can be found who have independent means and are willing to pioneer to the area and live on whatever income they have, however slender. There are those who, in accordance with Bahá'u'lláh's injunction, have been deputized by friends who are unable to go themselves. Believers may be found who are willing to go to such an area for a specific period supported by the meager budget that the Fund can afford, with the clear understanding that at the end of that period they will return from the pioneer post and become self-supporting again; in such a way an area can be serviced with a succession of pioneers. Then there are those believers who are willing to serve in a remote and inhospitable area, but whose age or situation makes it clear from the outset that they will not be able to become self-supporting again; when the need is great and cannot be met in any other way, an Assembly would be fully justified in supporting them, but it should realize from the outset the extent of the responsibility it is incurring for an indefinite period into the future.

162.12 Naturally these ways of financing pioneering are not mutually exclusive. A person, for example, can be partially self-supporting and assisted to only a limited degree; or a pioneer may go to an area with the intention of finding work but is unable to do so and the Assembly repeatedly extends the period of financial support until the time comes when he is no longer able to become self-supporting anywhere. In such a case the Assembly needs to watch the process very carefully so that, on the one hand, it does not incur a permanent responsibility it had not intended, and on the other, does not commit the injustice of terminating the financial support extended to a pioneer at a time when he has become unemployable, and is unable to obtain any other means of support.

Traveling Teaching

162.13   While pioneers provide a very valuable long-term reinforcement of a community and are often the only feasible means for opening new areas—and here we are speaking not only of pioneers from foreign lands but of homefront pioneers as well, the use of whom must be greatly developed in most countries—a second vital reinforcement of the work is provided by traveling teachers. As mentioned in the message sent to all believers at Riḍván, a new international travel teaching program is now being launched. National Assemblies and their committees, therefore, need to develop a threefold integrated program for travel teaching. Firstly, there should be within each national community regular circuits of local traveling teachers, that is to say of believers who are members of that national community, whether native or pioneers, who are able and willing to devote time to this activity. Secondly, and integrated within these circuits, provision should be made for planned visits of traveling teachers from abroad. Thirdly, each National Assembly should establish an agency and a procedure for taking advantage of the unheralded arrival of visitors from abroad, or of sudden offers from believers on the homefront, who would be able to give valuable help in the fields of travel teaching or proclamation if properly organized. Such an agency would, of course, be responsible for evaluating the capacity of those who offer services because while an unexpected offer can often provide a very valuable teaching opportunity, it is also true to say that some Bahá'í communities have been exhausted and their work hindered by the arrival of a succession of traveling Bahá'ís who were not really suited, for lack of a language or for other reasons, to assist with teaching in the area concerned. Friends who travel spontaneously in this way can do valuable teaching themselves but should not expect the assistance of local administrative institutions if they have not arranged the trip in advance.

Correspondence Courses

162.14   Only a few National Spiritual Assemblies have been given the specific goal of developing and conducting correspondence courses; however, those National Assemblies who have the goal of training selected believers to assist in consolidating local communities would find it worthwhile to consider how the use of correspondence courses could help in the fulfillment of this goal. For example, once the selection of trainees has been made, the first stage in their training could well be a correspondence deepening course which would ascertain the degree of interest and capacity of each trainee and also prepare him to attend a series of lectures or classes which would follow as a second stage. The entire training process could consist of several stages interspersed in this way. This combination of two methods has the advantage of helping the Assembly to ascertain at the outset which trainees have the capacity and desire to continue with the course, thus leading to a better selection and helping to ensure that the costs of holding classes and bringing trainees to them are incurred only in respect of those whose interest and capacity have been established.

Economy can be exercised by holding the deepening classes in smaller gatherings by grouping several neighboring local communities together and sending one or more teachers to the area. This might prove more economical than inviting the selected trainees to, say, the capital, and having to accommodate and feed them during the period of the course.

162.15

Teaching Conferences

Teaching Conferences can have a great value for the advance of the Faith. Their aim is to strengthen the bonds of unity and fellowship among the friends, to increase their involvement in the teaching work and their interest in its progress, and to serve as magnets to attract divine confirmations. They are also rallying points for the believers, evidences of the vitality of their love for Bahá'u'lláh, and potent instruments for generating enthusiasm and spiritual drive for advancing the interests of the Faith.

162.16

Certain National Spiritual Assemblies, which are not among the majority who are already doing so, have been assigned the goal of holding at least one National Teaching Conference during each year. The purpose of this is to provide a national event of major importance in addition to the annual National Convention to stimulate the interest and reorientate the efforts of the friends, focusing their attention upon the current urgent needs of the Plan. These National Teaching Conferences should, therefore, be held some months away from Riḍván, or they will lose a great part of the intended effect.

162.17

As the eight International Conferences will soon be upon us, it is important for National Assemblies to decide as soon as possible, in consultation with the Counselors, whether it would be feasible and helpful to hold a national conference soon after, or possibly immediately before, the International Conference nearest to their area. The sooner this study is made and decisions taken and announced, the greater will be the participation of the friends, locally and from abroad.

162.18

Newsletters

Although during the past year a marked improvement has been noticed in certain countries in the standard and regularity of the Bahá'í newsletters, the development of this organ of Bahá'í communication still needs great atten-

162.19

tion in most national communities. A special committee should be appointed, on which members of the National Spiritual Assembly could well serve, with the task of making the national newsletter a powerful instrument of direct and regular contact with the friends, which will disseminate news among them, stimulate and maintain their interest in the growth of the Faith in the world and throughout the area of national jurisdiction, share with them the National Spiritual Assembly's plans, hopes and aspirations, convey to them its comments on Bahá'í developments of special significance, and cause the believers to anticipate the future with feelings of excitement and confidence. The doors of communication between the friends, the Local Spiritual Assemblies and the National Spiritual Assembly should always be open. The one means which will contribute most to the promotion of this open-door policy is the regular issue of an interesting and heartwarming newsletter. In certain countries, we are glad to see, there are in addition to the national newsletter, news bulletins issued on regional or district levels. The importance of these secondary organs of Bahá'í communication acquires added weight in areas where differences of language make the issue of bulletins in a local language of each area highly desirable, if not essential.

Literature

162.20   When each National Spiritual Assembly carefully compares the demands of the waiting public and the needs of the believers for Bahá'í literature with the current supply, it will realize how urgent is the need for it to multiply its efforts to ensure that a comprehensive range of our literature is made constantly available. The basic literature of the Faith must be translated into languages that are most suitable and in demand for the spread and development of the Faith in accordance with the goals of the Plan. In each national area the agencies for obtaining and disseminating Bahá'í literature should be greatly strengthened so that they will efficiently ensure an uninterrupted supply of the literature which is available from the various Publishing Trusts and organize its distribution throughout the area, through Local Assemblies and groups, by sale at conferences and summer schools, and directly to individuals. At the same time these agencies should ensure that the monies received from the sale of literature are kept separate from other funds of the Faith and are used for the replenishment of stocks of books and the widening of the range of literature available. National Assemblies must also give consideration to the need to cover the cost of certain literature out of the National Fund, so that it can be supplied free or sold at a price within the reach of those who urgently require it.

Radio and Television

162.21   A compilation has recently been made from the letters written on behalf of the Guardian and a copy is attached for your information. This brief compilation shows the importance that Shoghi Effendi attached to the use of radio as a means of teaching and proclaiming the Faith in countries where such activity is possible.

The Universal House of Justice has initiated a pilot project in Ecuador for the purchase and operation of a Bahá'í radio station, and at the present time this is the only one for which sufficient funds are available.[2] However, the actual owning of a radio station is not the only way of making use of this medium. National Spiritual Assemblies responsible for countries where Bahá'í radio programs would raise no objection from the civil authorities, should regard it as their bounden duty to explore, if they have not already done so, whatever options are open to them to utilize radio to sow the seeds of the Faith as widely as they can and to broadcast its divine teachings, as well as to assist in the consolidation of the local Bahá'í communities.

Where the use of television broadcasts is open to Bahá'í communities they should also take the utmost advantage of this opportunity.

Contact with the Authorities

The events of the past year have demonstrated clearly that the enemies of the Faith are intensifying their attacks on the precious Cause of God. The Five Year Plan calls for a planned and sustained effort, under the close supervision of each National Spiritual Assembly, to foster cordial relations with responsible government officials and prominent people. In every country where the doors of contact with those in authority are open to the friends, the National Spiritual Assembly should, as indicated in our letter of Naw-Rúz 131, appoint a special committee to be given the task of finding effective ways of informing the authorities about the Faith, of dispelling any misgivings and of removing any misapprehensions which may be deceitfully created by those who are striving to extinguish the fire of God's Faith.[3] We cannot overemphasize the necessity of this activity and the need to use utmost tact and wisdom in pursuing it, for, not only will it facilitate the further proclamation and recognition of the Faith, but, as opposition to and misconceptions about the aims and purposes of the Bahá'ís increase, when a moment of crisis arrives the institutions of the Faith may know where to turn, whose advice and assistance to seek and how to minimize the effects of opposition.

Closely linked with the above undertaking, in countries where the Faith is not yet recognized, is the need to apply for such recognition if the laws of the country permit and if the Universal House of Justice has approved that

---

162-2. For announcements of the inauguration of Bahá'í radio stations, see messages dated 15 December 1977, Naw-Rúz 1978, and 28 August 1978 (messages no. 201, 205, and 213); 13 December 1982 (no. 348); 2 April 1984 (no. 391); and 31 January 1986 (no. 450). For the message to the Hemispheric Bahá'í Radio-Television Conference, see telex dated 15 December 1977 (no. 201).

162-3. See "Elucidation of Five Year Plan Goals," Naw-Rúz 1974 (message no. 142).

an approach be made to the authorities on the subject. In other countries where some measure of recognition, such as the incorporation of Assemblies, has been obtained, National Spiritual Assemblies should be alert to the possibilities which are open to them to widen the scope and broaden the base of the recognition obtained for Bahá'í institutions, the Bahá'í marriage certificate and Bahá'í Holy Days. These measures will not only secure for the Faith a higher degree of legal protection, but will enhance its stature in the eyes of the authorities and the general public.

Wisdom in the Use of Bahá'í Funds

162.26 The Five Year Plan emphasizes the obligation of the friends, in view of the growing needs of the Faith to ensure that a generous outpouring of contributions is offered in support of Bahá'í Funds, and encourages Bahá'í communities at present dependent on outside help to aim at becoming self-supporting. While all National Spiritual Assemblies have the obligation to administer Bahá'í funds wisely and judiciously, those National Spiritual Assemblies which depend to a large extent on budgetary assistance from the World Center have an even greater responsibility, so to speak, to carefully supervise expenditures. The more rigorous the exercise of economy on the part of National Spiritual Assemblies, the sooner will the body of the friends be encouraged to feel financial responsibility toward the progress of the Faith in their areas, to place greater reliance upon the wise administration of the National Spiritual Assembly, and to offer their resources, however modest they may be, for the furthering of its plans and activities.

162.27 National Spiritual Assemblies must uphold economy not only because the funds at their disposal are limited but, as experience has repeatedly shown, because lack of proper control and supervision in the expenditure of these funds is both an unfair temptation to the untrustworthy and a test to the body of the believers, causing them to become disenchanted with Bahá'í administration and weakening their resolve to fulfill their sacred obligation of contributing to the Fund.

162.28 In the attitudes seen at the National Office, in the appropriations made to committees and other agencies of the National Assembly, in any budgetary assistance given to pioneers and traveling teachers, in the holding of conferences and deepening courses, and in all aspects of the work of the Cause for which the National Assembly is responsible, supervision, careful planning and lack of extravagance should be observed and be seen to be upheld.

Local Spiritual Assemblies

162.29 It is becoming increasingly understood by the friends why the Five Year Plan places such great emphasis upon the firmness of the foundation and the efficiency of the operation of the Local Spiritual Assemblies. This is very heartening, for upon the degree to which the members of these Assemblies grasp the true significance of the divine institution on which they serve, arise selflessly to fulfill their prescribed and sacred duties, and persevere in their endeavors, depends to a large extent the healthy growth of the worldwide community of the Most Great Name, the force of its outward thrust, and the strength of its supporting roots.

162.30 We long to see every Local Spiritual Assembly either spontaneously adopt its own goals or warmly welcome those it has been or will be given by its National Spiritual Assembly, swell the number of the adherents who compose its local community and, guided by the general policy outlined by its National Spiritual Assembly, proclaim the Faith more effectively, energetically pursue its extension teaching and consolidation goals, arrange the observances of the Holy Days, regularly hold its Nineteen Day Feasts and its sessions for deepening, initiate and maintain community projects, and encourage the participation of every member of its community in giving to the Fund and undertaking teaching activities and administrative services, so as to make each locality a stronghold of the Faith and a torchbearer of the Covenant.

162.31 We are confident that the institution of the Boards of Counselors will lend its vital support and, through the Counselors' own contacts with the friends, through their Auxiliary Boards and their assistants, will nourish the roots of each local community, enrich and cultivate the soil of knowledge of the teachings and irrigate it with the living waters of love for Bahá'u'lláh. Thus will the saplings grow into mighty trees, and the trees bear their golden fruit.

Women

162.32 'Abdu'l-Bahá has pointed out that "Among the miracles which distinguish this sacred dispensation is this, that women have evinced a greater boldness than men when enlisted in the ranks of the Faith." Shoghi Effendi has further stated that this "boldness" must, in the course of time, "be more convincingly demonstrated, and win for the beloved Cause victories more stirring than any it has as yet achieved."[4] Although obviously the entire Bahá'í world is committed to encouraging and stimulating the vital role of women in the Bahá'í community as well as in society at large, the Five Year Plan calls specifically on eighty National Spiritual Assemblies to organize Bahá'í activities for women. In the course of the current year which has been designated "International Women's Year" as a worldwide activity of the United Nations, the Bahá'ís, particularly in these eighty national communities, should initiate and implement programs which will stimulate and promote the full and equal participation of women in all aspects of Bahá'í community life, so that through their accomplishments the friends will demonstrate the distinction of the Cause of God in this field of human endeavor.

162-4. ADJ, p. 69.

Youth

162.33　It is our hope that in the international travel teaching program now being launched the youth will assume a major role by devoting time during their vacations, and particularly during the long vacation at the end of the academic year, to the promotion of the teaching work in all its aspects, not only within their own national communities but farther afield. Some youth may have financial resources of their own, others may be able and willing to work and save the funds necessary for such projects, still others may have the financial backing of their parents, relatives or friends. In other cases the Bahá'í funds may be able to supplement whatever resources the prospective traveling teacher may be able to supply.

162.34　The endurance of youth under arduous conditions, their vitality and vigor, and their ability to adapt themselves to local situations, to meet new challenges, and to impart their warmth and enthusiasm to those they visit, combined with the standard of conduct upheld by Bahá'í youth, make them potent instruments for the execution of the contemplated projects. Indeed, through these distinctive qualities they can become the spearhead of any enterprise and the driving force of any undertaking in which they participate, whether local or national. Our expectant eyes are fixed on Bahá'í youth!

Children

162.35　How often have well-organized Bahá'í children's classes given parents, even those who are not Bahá'ís, the incentive to learn more and study more deeply the Teachings of the Faith! How often have the children, through their songs and recitation of prayers during Feasts and at other gatherings of the friends, added luster and inspiration to the program and created a true sense of belonging to the community in the hearts of those present! How many are the children who have grown into active and enkindled youth, and later into wholly dedicated adults, energetically supporting the work of the Cause and advancing its vital interests!

162.36　Certain National Spiritual Assemblies have been given the specific goal of organizing children's activities, and many of these Assemblies have been assigned assistance in the form of at least one helper who will have received some training in the education of Bahá'í children. The National Assemblies to receive such helpers, however, should not await their arrival before initiating activities. Through the services of a committee chosen from among those interested in this area of service, simple lessons could be improvised, suitable extracts from the Writings and Prayers chosen for the children to study and memorize, and local talent called upon to carry out this vital activity which will assuredly exert a far-reaching influence on the well-being and strength of each community.

Dawn Prayers

162.37 We have been watching with profound interest the manner in which the goal of encouraging the friends to meet for dawn prayers is being carried out. In some rural areas this has become already an established practice of the friends and indeed a source of blessing and benefit to them as they pursue their activities during the day, as well as increasing the consciousness of community solidarity. In other areas, the friends have found that, because of the distances involved, better results are obtained by meeting for prayer in smaller groups. In yet other areas, as a first step, plans have been made to meet for dawn prayers once a week.

162.38 May the Blessed Beauty sustain you bountifully as you prepare yourselves to discharge the commitments and surmount the challenges of the year which has just begun.

With loving Bahá'í greetings,
THE UNIVERSAL HOUSE OF JUSTICE

Use of Radio and Television in Teaching
(Extracts from letters written on behalf of Shoghi Effendi)

162.39 In regard to your wish of broadcasting the Message, Shoghi Effendi would advise you to consult with the Spiritual Assembly as to whether such an action meets their approval, and if so to ask their assistance and help for finding the best means through which to carry out your plan. The idea of a wireless station is rather ambitious and requires much financial expenditure. If, however, you find it feasible and within your financial capacity you should not hesitate to do so, inasmuch as this will enable you to spread the Cause in a much easier and more efficient manner.

(13 August 1933 to an individual believer)

162.40 Your suggestion regarding the installation of a radio station in the Temple is truly splendid. But it remains to be seen whether the National Spiritual Assembly finds it financially feasible to undertake such a project, which is, beyond doubt, a very costly enterprise. Whatever the expenditure involved in this project, there is no reason why the believers should not start now considering seriously the possibility of such a plan, which, when carried out and perfected, can lend an unprecedented impetus to the expansion of the teaching work throughout America.

162.41 It is for the National Spiritual Assembly, however, to take the final decision in this matter, and to determine whether the national fund of the Cause is at present sufficiently strong to permit them to install a radio station in the Temple.

162.42 The Guardian feels, nevertheless, confident that this plan will receive the careful consideration of the National Spiritual Assembly members, and hopes that, if feasible, they will take some definite action in this matter.
(31 January 1937 to an individual believer)

162.43 He read with interest the various suggestions you made to the National Spiritual Assembly, and feels they are fundamentally sound, especially the wider use of the radio. Unfortunately at the present time anything that would make a fresh demand on the financial resources of the Cause in America—such as a Bahá'í-owned broadcasting station—is out of the question, as the friends are finding it difficult to meet the great needs of the teaching and Temple Funds. However the idea should, he feels, be kept in mind for future realization.
(14 October 1942 to an individual believer)

162.44 In connection with the radio work . . . he would suggest that the main consideration is to bring to the attention of the public the fact that the Faith exists, and its teachings. Every kind of broadcast, whether of passages from the Writings, or on topical subjects, or lectures, should be used. The people need to hear the word "Bahá'í" so that they can, if receptive, respond and seek the Cause out. The primary duty of the friends everywhere in the world is to let the people know such a Revelation is in existence; their next duty is to teach it.
(24 July 1943 to an individual believer)

162.45 He feels it would be excellent if the Cause could be introduced more to the people through the medium of radio, as it reaches the masses, especially those who do not take an interest in lectures or attend any type of meeting.
(7 March 1945 to an individual believer)

162.46 The matter of obtaining free time on the radio is one which the Radio Committee and the National Spiritual Assembly must decide upon: but the principle is that every effort should be made to present the teachings over the air as often as possible as long as the manner in which it is done is compatible with the dignity of our beloved Faith.
(15 August 1945 to an individual believer)

162.47 He was sorry to learn through your cable that the project for a Bahá'í radio station can not be carried out at present; he considers that such a station would be a very great asset to the Cause, not only as a teaching medium and a wonderful form of publicity, but also as an enhancement of its prestige. He feels your Assembly should not drop the matter, but go on investigating ways to make such a project materialize as soon as possible.
(20 March 1946 to the National Spiritual Assembly of the United States and Canada)

He hopes that a Bahá'í radio station will prove feasible during the coming years, as he considers it of great importance. 162.48
(4 May 1946 to the Radio Committee
of the National Spiritual Assembly of the United States and Canada)

The Bahá'ís should not always be the last to take up new and obviously excellent methods, but rather the first, as this agrees with the dynamic nature of the Faith which is not only progressive, but holds within itself the seeds of an entirely new culture and civilization. 162.49
(5 May 1946 to an individual believer)

The Guardian approves in principle of a radio station, and sees no objection to its being in the Temple; but he considers the cost you quote too much of a burden at the present time for the Fund to bear, in view of the multiple expenses of the new Seven Year Plan.[5] If there is any way it can be done for a price you feel the Fund could pay, and which would be more reasonable, he approves of your doing it. In any case the National Spiritual Assembly should strongly press for recognition as a Religious Body, and claim full rights to be represented on the air on an equal footing with other established Churches. 162.50
(20 July 1946 to the National Spiritual Assembly
of the United States and Canada)

He approves of your desire to teach the principles of the Faith through radio. But he urges you to do all you can to always, however small the reference you are able to make to it may be, clearly identify or associate what you are giving out with Bahá'u'lláh. The time is too short now for us Bahá'ís to be able to first educate humanity and then tell it that the Source is this new World Faith. For their own spiritual protection people must hear of the name Bahá'í—then, if they turn blindly away, they cannot excuse themselves by saying they never even knew it existed! For dark days seem still ahead of the world, and outside of this Divine Refuge the people will not, we firmly believe, find inner conviction, peace and security. So they have a right to at least hear of the Cause as such! 162.51
(24 April 1949 to an individual believer)

162-5. The second Seven Year Plan, 1946–53.

163
Representation of the Universal House of Justice by Hands of the Cause of God at International Conferences
27 May 1975

To all National Spiritual Assemblies
Dear Bahá'í Friends,

International Teaching Conferences
1976–1977

163.1 We joyfully announce that the following Hands of the Cause of God have been named as our representatives to the International Conferences:

Amatu'l-Bahá Rúḥíyyih <u>Kh</u>ánum	Paris, France	3–6 August 1976
Ugo Giachery	Helsinki, Finland	6–8 July 1976
'Alí-Akbar Furútan	Hong Kong	5–8 November 1976
Paul Haney	Mérida, Mexico	4–6 February 1977
Enoch Olinga	Bahia, Brazil	28–30 January 1977
William Sears	Nairobi, Kenya	15–17 October 1976
Collis Featherstone	Anchorage, Alaska	23–25 July 1976
Abu'l-Qásim Faizi	Auckland, New Zealand	19–22 January 1977

With loving Bahá'í greetings,
THE UNIVERSAL HOUSE OF JUSTICE

1974–1979 • THE FIVE YEAR PLAN

164
Significance of the Seat of the Universal House of Justice
5 JUNE 1975

To the Followers of Bahá'u'lláh throughout the World

Dear Bahá'í Friends,

As the Five Year Plan gathers momentum in all parts of the world, with the followers of the Blessed Perfection firmly embarked on the course that will lead to victory, the time has come for us to contemplate, in preparation for its imminent initiation, the project which will rank as the greatest single undertaking of that Plan, the construction of a befitting seat for the Universal House of Justice in the heart of God's Holy Mountain.[1]

Nearly thirty-six years ago, after overcoming a multitude of difficulties, the beloved Guardian succeeded in transferring to Mount Carmel the sacred remains of the Purest Branch and Navváb, interring them in the immediate neighborhood of the resting-place of the Greatest Holy Leaf, and alluded, in these words, to the "capital institutional significance" that these events constituted in the unfoldment of the World Center of the Faith:[2]

> For it must be clearly understood, nor can it be sufficiently emphasized, that the conjunction of the resting-place of the Greatest Holy Leaf with those of her brother and mother incalculably reinforces the spiritual potencies of that consecrated Spot which, under the wings of the Báb's overshadowing Sepulcher, and in the vicinity of the future Mashriqu'l-Adhkár[3] which will be reared on its flank, is destined to

---

164-1. In a message dated 31 August 1987 to the Bahá'ís of the world, the Universal House of Justice outlined plans for completing the "world-shaking, world-embracing, world-directing administrative institutions" that Shoghi Effendi (MA, p. 32) envisioned on Mount Carmel, God's Holy Mountain. The buildings to be constructed include the International Bahá'í Library and the seats of the International Teaching Center and the Center for the Study of the Texts. Additional projects include constructing an extension to the International Archives Building to accommodate the ever-growing World Center archives and constructing eighteen monumental terraces from the foot of Mount Carmel to its crest, nine leading to the terrace on which the Shrine of the Báb stands and nine rising above it.

164-2. The Purest Branch is Mírzá Mihdí, Bahá'u'lláh's youngest son; Navváb is Ásíyih Khánum, titled the Most Exalted Leaf, wife of Bahá'u'lláh and mother of 'Abdu'l-Bahá, Mírzá Mihdí, and Bahíyyih Khánum; the Greatest Holy Leaf is Bahíyyih Khánum, daughter of Bahá'u'lláh. See MA, p. 31.

164-3. A Mashriqu'l-Adhkár is to be raised on a site Shoghi Effendi described as "the head of the Mountain of God, in close proximity to the Spot hallowed by the footsteps of Bahá'u'lláh, near the time-honored Cave of Elijah, and associated with the revelation of the Tablet of Carmel" (MBW, p. 63). The Universal House of Justice announced the erection of an obelisk marking the site of the future House of Worship in its cable of 13 December 1971 (message no. 105).

evolve into the focal center of those world-shaking, world-embracing, world-directing administrative institutions, ordained by Bahá'u'lláh and anticipated by 'Abdu'l-Bahá, and which are to function in consonance with the principles that govern the twin institutions of the Guardianship and the Universal House of Justice. Then, and then only, will this momentous prophecy which illuminates the concluding passages of the Tablet of Carmel be fulfilled: "Erelong will God sail His Ark upon thee (Carmel), and will manifest the people of Bahá who have been mentioned in the Book of Names."[4]

164.2b   To attempt to visualize, even in its barest outline, the glory that must envelop these institutions, to essay even a tentative and partial description of their character or the manner of their operation, or to trace however inadequately the course of events leading to their rise and eventual establishment is far beyond my own capacity and power. Suffice it to say that at this troubled stage in world history the association of these three incomparably precious souls who, next to the three Central Figures of our Faith, tower in rank above the vast multitude of the heroes, Letters, martyrs, hands, teachers and administrators of the Cause of Bahá'u'lláh, in such a potentially powerful spiritual and administrative Center, is in itself an event which will release forces that are bound to hasten the emergence in a land which, geographically, spiritually and administratively, constitutes the heart of the entire planet, of some of the brightest gems of that World Order now shaping in the womb of this travailing age.[5]

164.3   The first of the majestic edifices constituting this mighty Center, was the building for the International Archives of the Faith which was completed in the summer of 1957 as one of the last major achievements of Shoghi Effendi's Guardianship and which set the style for the remaining structures which, as described by him, were to be raised in the course of time in the form of a far-flung arc on the slope of Mount Carmel. In the eighteen years since that achievement, the community of the Most Great Name has grown rapidly in size and influence: from twenty-six National Spiritual Assemblies to one hundred and nineteen, from some one thousand to seventeen thousand Local Spiritual Assemblies, and from four thousand five hundred localities to over seventy thousand, accompanied by a corresponding increase in the volume of the work carried on at the World Center of the Faith and in the complexity of

---

164-4. For the Tablet of Carmel, which Shoghi Effendi called "the Charter of the World Spiritual and Administrative Centers of the Faith" (MBW, p. 63), see GWB, pp. 14–17, or TB, pp. 3–5. The Ark in this context is a reference to the World Administrative Center of the Faith on Mount Carmel.

164-5. MA, pp. 32–33.

its institutions. It is now both necessary and possible to initiate construction of a building that will not only serve the practical needs of a steadily consolidating administrative center but will, for centuries to come, stand as a visible expression of the majesty of the divinely ordained institutions of the Administrative Order of Bahá'u'lláh.

164.4 Faced, like the Archives Building, with stone from Italy, and surrounded by a stately colonnade of sixty Corinthian columns,[6] the seat for the Universal House of Justice will contain, in addition to the council chamber of the House of Justice, a library, a concourse for the reception of pilgrims and dignitaries, storage vaults with air purification for the preservation of original Tablets and other precious documents, accommodation for the secretariat and the many ancillary services that will be required. Conceived in a style of enduring beauty and majesty, and faced with stone that will weather the centuries, the building in its interior arrangements will be very simple and capable of adaptation in the generations ahead to whatever technological advances will be made by the rapid growth of human knowledge.

164.5 The erection of this building which, comprising five and a half stories, far surpasses in size and complexity any building at present in existence at the World Center presents a major challenge to the Bahá'í community, whose resources are already all too meager in relation to the great tasks that lie before it. But the spirit of sacrifice has been the hallmark of the followers of Bahá'u'lláh of every race and clime and as they unite to raise this second of the great edifices of the Administrative Center of their Faith they will rejoice at having the inestimable privilege of taking part in a "vast and irresistible process" which Shoghi Effendi stated is "unexampled in the spiritual history of mankind," a process "which will synchronize with two no less significant developments—the establishment of the Lesser Peace and the evolution of Bahá'í national and local institutions—the one outside and the other within the Bahá'í world—will attain its final consummation, in the Golden Age of the Faith, through the raising of the standard of the Most Great Peace, and the emergence, in the plenitude of its power and glory, of the focal Center of the agencies constituting the World Order of Bahá'u'lláh."

THE UNIVERSAL HOUSE OF JUSTICE

164-6. The building, as finally completed, has fifty-eight columns.

165
Excavation of the Site of the Seat of the Universal House of Justice
17 JUNE 1975

To all National Spiritual Assemblies

165.1 JOYFULLY ANNOUNCE COMMENCEMENT EXCAVATION SITE UNIVERSAL HOUSE JUSTICE BUILDING ARC MOUNT CARMEL. CONTRACT ENTAILS REMOVAL FORTY THOUSAND CUBIC METERS ROCK AND EARTH AT COST APPROXIMATELY TWO HUNDRED THOUSAND DOLLARS. INVITE ALL BELIEVERS CONTRIBUTE UNSTINTINGLY BUILDING FUND ENSURE UNINTERRUPTED PROGRESS HISTORIC UNDERTAKING.

UNIVERSAL HOUSE OF JUSTICE

166
Laws of the Kitáb-i-Aqdas Concerning Men and Women; Membership on the Universal House of Justice
24 JULY 1975

To an individual Bahá'í

Dear Bahá'í Friend,

166.1 Your letter of 16 March 1975 has been received and we have studied the various questions arising from your study of the *Synopsis and Codification of the Kitáb-i-Aqdas.* . . .

166.2 Concerning your questions about the equality of men and women, this, as 'Abdu'l-Bahá has often explained, is a fundamental principle of Bahá'u'lláh; therefore the Laws of the Aqdas should be studied in the light of it. Equality between men and women does not, indeed physiologically it cannot, mean identity of functions. In some things women excel men, for others men are better fitted than women, while in very many things the difference of sex is of no effect at all. The differences of function are most apparent in family life. The capacity for motherhood has many far-reaching implications which are recognized in Bahá'í Law. For example, when it is not possible to educate all one's children, daughters receive preference over sons, as mothers are the first educators of the next generation. Again, for physiological reasons, women are granted certain exemptions from fasting that are not applicable to men.

166.3 You mention the provision in the Kitáb-i-Aqdas regarding inheritance, in which the eldest son receives preferential treatment. As you no doubt know, the duty of making a will is enjoined upon all Bahá'ís, and in such a will a believer is free to bequeath his or her property in whatever way he or she wishes (see note 25 on page 60 of the *Synopsis and Codification*). Every system of law, however, needs to make provision for the disposal of a person's property if he or she dies without having made a will, and it is in cases of intestacy that the specific provisions stated in the Kitáb-i-Aqdas are applied. These provisions give expression to the law of primogeniture, which, as 'Abdu'l-Bahá has stated, has invariably been upheld by the Law of God. In a Tablet to a follower of the Faith in Persia He wrote: "In all the Divine Dispensations the eldest son hath been given extraordinary distinctions. Even the station of prophethood hath been his birthright."[1] With the distinctions given to the eldest son, however, go concomitant duties. For example, with respect to the law of inheritance 'Abdu'l-Bahá has explained in one of His Tablets that the eldest son has the responsibility to take into consideration the needs of the other heirs. Similar considerations no doubt apply to the provisions that, in intestacy, limit the shares due to half-brothers and half-sisters of the deceased on his or her mother's side; they will, of course, be due to receive inheritance from their own father's estate.

166.4 Your statement that "Gifts to a wife are included in the man's property to be given away after his death" is incorrect. It is clear from the passage in the Kitáb-i-Aqdas that certain things that a husband buys for his wife are intended to be for the general household and certain are intended to be the wife's personal property. These latter, that is to say the wife's used clothing and gifts which have been made to her, are *not* included in the husband's property.

166.5 The husband's duty to send his wife home if differences arise between them while traveling is a part of the law of divorce, and relates to the husband's obligation to support his wife during the year of waiting. The *Synopsis and Codification of the Kitáb-i-Aqdas* reads as follows (at section (g) on pages 42 and 43):

> 166.5a Should differences arise between husband and wife while traveling, he is required to send her home, or entrust her to a dependable person, who will escort her there, paying her journey and her full year's expenses.

166.6 You have also asked for an explanation of why, in view of the Bahá'í principle of equality of men and women, women are not allowed to serve on the Universal House of Justice. We share with you the following passages about this subject, taken from letters written on behalf of the beloved Guardian to a National Spiritual Assembly and to an individual believer.

166.6a As regards the membership of the International House of Justice, 'Abdu'l-Bahá states in a Tablet that it is confined to men, and that the wisdom of it will be revealed as manifest as the sun in the future. In

---

166-1. Quoted in WOB, p. 148.

any case the believers should know that, as 'Abdu'l-Bahá Himself has explicitly stated that sexes are equal except in some cases, the exclusion of women from the International House of Justice should not be surprising. From the fact that there is no equality of functions between the sexes one should not, however, infer that either sex is inherently superior or inferior to the other, or that they are unequal in their rights.— 14 December 1940[2]

166.6b Regarding your question, the Master said the wisdom of having no women on the International House of Justice would become manifest in the future. We have no other indication than this.—17 September 1952

166.7 We must always remember Bahá'u'lláh's exhortation, which is quoted on page 22 of the *Synopsis and Codification:* "Weigh not the Book of God with such standards and sciences as are current amongst you, for the Book itself is the unerring balance established amongst men. In this most perfect balance whatsoever the peoples and kindreds of the earth possess must be weighed, while the measure of its weight should be tested according to its own standard, did ye but know it."

166.8 It is hoped that the foregoing will be helpful to your own understanding of the matters about which you have asked.

With loving Bahá'í greetings,
THE UNIVERSAL HOUSE OF JUSTICE

167
Release of a Compilation of Prayers and Tablets for Children and Youth
25 SEPTEMBER 1975

To all National Spiritual Assemblies

Dear Bahá'í Friends,

167.1 We are very happy to send you the enclosed selection of prayers revealed by Bahá'u'lláh and 'Abdu'l-Bahá for children, together with some Tablets of the Master intended for children and youth, translated into English.[1]

167.2 You may use this translation as you wish, adding from it to the prayer books you have already published, using it as a basis of prayer books or other

---

166-2. See DND, p. 86.
167-1. The compilation was published under the title *Let Thy breeze refresh them . . .* by the Bahá'í Publishing Trust of the United Kingdom (1976) and under the title *Bahá'í Prayers and Tablets for the Young* by the Bahá'í Publishing Trust of the United States (1978).

literature published specially for children, or sharing these precious words with the friends in any other manner you deem wise and useful. You are, of course, at liberty to translate these prayers and Tablets into other languages, and we hope that this will be done.

167.3 The raising of children in the Faith of God and the spiritualization of their lives from their earliest years is of prime importance in the life of the Bahá'í community, and the firm establishment of activities to promote these aims is one of the vital goals of the Five Year Plan.

With loving Bahá'í greetings,
THE UNIVERSAL HOUSE OF JUSTICE

168
Release of a Compilation on Bahá'í Meetings and the Nineteen Day Feast
30 NOVEMBER 1975

To all National Spiritual Assemblies

Dear Bahá'í Friends,

168.1 The Research Department has recently prepared two compilations from the Writings of Bahá'u'lláh and 'Abdu'l-Bahá and the letters of Shoghi Effendi on the subject of "Bahá'í Meetings" and "The Nineteen Day Feast," and copies are sent to you herewith.[1]

168.2 The Universal House of Justice leaves it to your discretion to decide in which manner these texts may be shared with the friends under your jurisdiction.

With loving Bahá'í greetings,
DEPARTMENT OF THE SECRETARIAT

168-1. The compilation was published under the title *Bahá'í Meetings/The Nineteen Day Feast* by the Bahá'í Publishing Trust of the United States (1976). See also CC 1:419–58 for another compilation from the Universal House of Justice on the Nineteen Day Feast.

169
Naw-Rúz Message 1976
18 March 1976

To all National Spiritual Assemblies

169.1 ANNOUNCE DELEGATES ASSEMBLED NATIONAL CONVENTIONS GLAD TIDINGS COMPLETION EXCAVATION MOUNT CARMEL PREPARATORY RAISING MAJESTIC CENTER LEGISLATION GOD'S FAITH THAT SACRED SPOT, SIGNATURE ITALY FIVE AND HALF MILLION DOLLAR CONTRACT FOR SUPPLYING OVER TWO THOUSAND FIVE HUNDRED CUBIC METERS PENTELIKON MARBLE FROM GREECE AND FASHIONING THEREFROM THE COLUMNS FACINGS ORNAMENTATION BEFITTING MONUMENTAL BUILDING. DEEPLY MOVED ENTHUSIASTIC RESPONSE BELIEVERS ALL PARTS WORLD THIS CHALLENGING GLORIOUS TASK.[1] DEVELOPMENTS WORLD CENTER PARALLELED FURTHER UNFOLDMENT ADMINISTRATIVE STRUCTURE CONTINENTAL NATIONAL LEVELS THROUGH RAISING NUMBER CONTINENTAL COUNSELORS TO SIXTY-ONE BY APPOINTMENT THELMA KHELGHATI WESTERN AFRICA, WILLIAM MASEHLA SOUTHERN AFRICA, BURHÁNI'D-DÍN AFSHÍN SOUTH CENTRAL ASIA, HIDEYA SUZUKI NORTHEASTERN ASIA, OWEN BATTRICK AUSTRALASIA AND ADIB TAHERZADEH EUROPE, AUTHORIZATION BOARDS COUNSELORS APPOINT NINETY MORE MEMBERS AUXILIARY BOARDS, AND CALL FOR ELECTION AT RIḌVÁN 1977 OF SEVEN NEW NATIONAL SPIRITUAL ASSEMBLIES: TWO IN AFRICA, MALI WITH ITS SEAT IN BAMAKO AND UPPER VOLTA WITH ITS SEAT IN OUAGADOUGOU, TWO IN THE AMERICAS, THE FRENCH ANTILLES WITH ITS SEAT IN POINT-À-PITRE AND SURINAM AND FRENCH GUIANA WITH ITS SEAT IN PARAMARIBO, ONE IN EUROPE, GREECE WITH ITS SEAT IN ATHENS, AND TWO IN THE PACIFIC, THE NEW HEBRIDES WITH ITS SEAT IN PORT VILA AND THE MARSHALL ISLANDS WITH ITS SEAT IN MAJURO, THE LATTER BEING SUPPLEMENTARY ACHIEVEMENT OF PLAN. NUMBER NATIONAL SPIRITUAL ASSEMBLIES THUS RAISED ONE HUNDRED TWENTY-FOUR FOLLOWING DISSOLUTION ACCOUNT LOCAL RESTRICTIONS NATIONAL ASSEMBLIES EQUATORIAL GUINEA NEPAL. OF NINE HUNDRED FIFTY-THREE PIONEERS CALLED FOR SPECIFIC POSTS FOUR HUNDRED NINETY-TWO ALREADY SETTLED. ALSO FOUR HUNDRED SEVENTY-SEVEN OTHER PIONEERS PROCEEDED GOAL COUNTRIES. GREAT OUTFLOW INTERNATIONAL TRAVELING TEACHERS RECORDED. MOVED PAY TRIBUTE INDEFATIGABLE SERVICES HANDS CAUSE GOD PAST YEAR IN PROMOTING ABOVE SUCCESSES AND IN FIELDS TEACHING PROTECTION PRESERVATION PROCLAMATION AND LITERATURE FAITH AS WELL AS SIGNAL SERVICES INTERNATIONAL TEACHING CENTER CONSTITUTING GREAT ACCESSION STRENGTH WORLD CENTER RELIEF BURDENS RESTING UNIVERSAL HOUSE JUSTICE. MIDDLE YEAR FIVE YEAR PLAN NOW OPENING WILL WITNESS GATHERING FOLLOWERS BAHÁ'U'LLÁH EIGHT INTERNA-

---

169-1. For the announcement of the decision to build the Seat of the Universal House of Justice, see message no. 115; for the explanation of its significance, see message no. 164; for the announcement of its occupation, see message no. 354.

TIONAL TEACHING CONFERENCES DESIGNED GENERATE TREMENDOUS IMPETUS PROGRESS PLAN ACCOMPLISHMENT WHOSE GOALS NOW LAGGING SERIOUSLY BEHIND. MOST PRESSING NEED FAITH THIS CRITICAL JUNCTURE ITS MISSION REDEEM MANKIND IS FOR EVERY BELIEVER ALL ASSEMBLIES NATIONAL LOCAL CONCENTRATE ATTAINMENT GOALS PLACED BEFORE BAHÁ'Í WORLD, PROMOTE PROCESS ENTRY BY TROOPS, ACHIEVE VAST INCREASE SIZE COMMUNITY, INCREASE NUMBER STEADFAST SELF-SACRIFICING BELIEVERS DEDICATED CONFORM EVERY ASPECT THEIR LIVES HIGH STANDARDS SET SACRED TEXTS. THE FIELD IS VAST THE TIME SHORT THE LABORERS LAMENTABLY FEW BUT ON THE EFFORTS WE FOLLOWERS OF THE BLESSED BEAUTY NOW EXERT, ON THE DEGREE TO WHICH WE SUCCESSFULLY AND SPEEDILY PROCLAIM AND TEACH HIS MESSAGE TO OUR FELLOW HUMAN BEINGS, DEPENDS IN GREAT MEASURE THE COURSE OF HUMAN HISTORY IN THE DECADES IMMEDIATELY AHEAD.

THE UNIVERSAL HOUSE OF JUSTICE

170
Appointments to Continental Boards of Counselors and Auxiliary Boards
24 MARCH 1976

To all National Spiritual Assemblies

Dear Bahá'í Friends,

170.1 As you will have seen in the Convention message, the Universal House of Justice has appointed six new Counselors and has authorized the appointment of ninety more Auxiliary Board members.

170.2 On the instruction of the House of Justice we now enclose for your information a complete list of the members of the Continental Boards of Counselors following the above appointments, and a list of the Auxiliary Boards showing the increases.

With loving Bahá'í greetings,
DEPARTMENT OF THE SECRETARIAT

170.3 Membership of the Continental Boards of Counselors March 1976

AFRICA

Northern Africa	Muḥammad Kebdani, Muḥammad Muṣṭafá, 'Imád Ṣábirán
Western Africa	Ḥusayn Ardikání, Friday Ekpe, Zekrullah Kazemi, Thelma Khelghati, Mihdí Samandarí
Central & East Africa	Hushang Ahdieh, Oloro Epyeru, Kolonario Oule, Isobel Sabri, Peter Vuyiya
Southern Africa	Seewoosumbur-Jeehoba Appa, Shidan Fat'he-Aazam, William Masehla, Bahíyyih Winckler

WESTERN HEMISPHERE

North America	Lloyd Gardner, Sarah Pereira, Velma Sherrill, Edna True
Central America	Carmen de Burafato, Rowland Estall, Artemus Lamb, Paul Lucas, Alfred Osborne
South America	Leonora Armstrong, Athos Costas, Mas'úd Khamsí, Peter McLaren, Raúl Pavón, Donald Witzel

ASIA

Western Asia	Íraj Ayman, Masíḥ Farhangí, Hádí Raḥmání, Manúchihr Salmánpúr
South Central Asia	Burhání'd-Dín Afshín, Shirin Boman, Salisa Kermani, Dipchand Khianra, Zena Sorabjee
Northeastern Asia	Richard Benson, Elena Marsella, Rúḥu'lláh Mumtází, Hideya Suzuki
Southeastern Asia	Yan Kee Leong, Firaydún Mítháqíyán, Khudáraḥm Paymán, Vicente Samaniego, Chellie Sundram

AUSTRALASIA

Suhayl 'Alá'í, Owen Battrick, Howard Harwood, Violet Hoehnke, Thelma Perks

EUROPE

Erik Blumenthal, Anneliese Bopp, Dorothy Ferraby, Louis Hénuzet, Betty Reed, Adib Taherzadeh

1974–1979 • THE FIVE YEAR PLAN

Auxiliary Boards—Riḍván 1976

170.4

		Former Number	Present Increase	New Total
Africa				
Protection:	Northern	5	–	5
	Western	5	6	11
	Central and East	13	–	13
	Southern	4	3	7
		27		36
Propagation:	Northern	5	–	5
	Western	11	3	14
	Central and East	19	–	19
	Southern	10	6	16
		45		54
Western Hemisphere				
Protection:	North America	9	–	9
	Central America	9	2	11
	South America	9	7	16
		27		36
Propagation:	North America	18	–	18
	Central America	9	7	16
	South America	27	11	38
		54		72
Asia				
Protection:	Western	9	–	9
	South Central	3	3	6
	Northeastern	3	–	3
	Southeastern	3	6	9
		18		27
Propagation:	Western	18	–	18
	South Central	15	15	30
	Northeastern	15	–	15
	Southeastern	15	12	27
		63		90
Australasia				
Protection:		9	–	9
Propagation:		9	9	18
Europe				
Protection:		9	–	9
Propagation:		27	–	27
TOTAL		**288**	**90**	**378**
Total Protection		90	27	117
Total Propagation		198	63	261

171
Message to the International Teaching Conference, Helsinki, Finland—July 1976
July 1976

To the Followers of Bahá'u'lláh gathered at
the International Teaching Conference in Helsinki
Dearly loved Friends,

171.1 With eager hearts we hail the convocation of this first of the twin Arctic Conferences inaugurating the series of eight International Bahá'í Conferences to be held during the middle part of the Five Year Plan. The northern regions of the world were alluded to by Bahá'u'lláh in the Kitáb-i-Aqdas, the Mother Book of this Revelation. Their names were recorded in the Tablets of the Divine Plan by the pen of 'Abdu'l-Bahá, Who, in one of His other Tablets, supplicated God to "raise up sanctified, pure and spiritual souls in the countries of the West and the territories of the North, and make them signs of His guidance, ensigns of the Concourse on High and angels of the Abhá Kingdom."[1] These lands received the constant attention of Shoghi Effendi, who repeatedly urged the friends to carry the Faith to their uttermost inhabited areas, and who joyfully announced every advance of the Bahá'ís that established a center closer to the North Pole.

171.2 Already touched by the morning light of God's Cause by the nineteen-twenties, the lands of the North were blessed by visits from the indomitable Martha Root, whose love warmed and encouraged the hearts of the handful of believers then laboring in a few scattered centers in Scandinavia and illumined the soul of Hólmfríður Árnadóttir, Iceland's first Bahá'í.[2] Bursting into blossom under the impact of the rays of the second Seven Year Plan, these communities received a major impetus from the Ten Year Crusade, of which the European campaign was launched at the never-to-be-forgotten conference in Stockholm in 1953, and which established centers as far north as Thule in Greenland and Sassen in the islands of Spitzbergen. Yet another stage of growth was reached with the Nine Year Plan and the convocation of the North Atlantic Conference in Reykjavik, which marked the opening

---

171-1. From an unpublished Tablet.
171-2. In a cable dated 3 October 1939 to the National Spiritual Assembly of the United States, Shoghi Effendi designated Martha Root a Hand of the Cause of God, describing her as the "foremost Hand which 'Abdu'l-Bahá's Will has raised up first Bahá'í century." See BW 13:643–48 and Mabel Garis, *Martha Root*. For an account of Hólmfríður Árnadóttir's life and services, see BW 13:943.

of a new phase in the collaboration between the northern communities on both sides of that ocean.³

171.3 Only thirty-eight years have passed since Väinö Rissanen, the first Bahá'í in Finland, accepted with radiant heart the life-giving message brought to him by Josephine Kruka, the "Mother of Finland," in July 1938,⁴ and now the city of Helsinki, the seat of the National Spiritual Assembly of the Bahá'ís of Finland, is the scene of an International Bahá'í Conference whose deliberations are focused on the diffusion of the light of God's Faith throughout the entire arctic and subarctic regions of the world.

171.4 The followers of the Blessed Perfection gathered in Helsinki must direct their attention to the urgent tasks of the second half of the Five Year Plan: to the reopening of Spitzbergen; the winning of the 34 Local Spiritual Assemblies still to be formed in Greenland, Iceland, Norway, Sweden, Denmark and Finland; the acceleration of the translation and publication of Bahá'í literature; the forging of still closer links of collaboration with the Bahá'í communities of Alaska, Canada and in the continent of Europe; the enlistment under the banner of Bahá'u'lláh of increasing numbers of the Eskimo, Lapp and Gypsy peoples; and the pursuit of the vital and challenging objectives of the Plan beyond the frontiers of their homelands.

171.5 It is our fervent prayer at the Threshold of Bahá'u'lláh that this Conference will produce an upsurge of Bahá'í activity throughout the northern lands and in the islands of the North Sea and the Baltic that will outshine every achievement made in those promising regions, and be an inspiration to your fellow-believers in every country of the world.

THE UNIVERSAL HOUSE OF JUSTICE

171-3. The second Seven Year Plan included the years 1946–53; the Ten Year Crusade, 1953–63; and the Nine Year Plan, 1964–73. The North Atlantic Conference was held in September 1971.
171-4. For an account of Josephine Kruka's life and services, see BW 15:493–96.

172

Message to the International Teaching Conference, Anchorage, Alaska—July 1976

JULY 1976

To the Friends assembled at the
International Teaching Conference in Anchorage

Dearly loved Friends,

172.1 Sixty years ago 'Abdu'l-Bahá summoned the valiant North American believers to open the remote and inhospitable regions which form the climatic frontiers of the Arctic and the sub-Arctic. The full implementation of His wish had to be postponed for some years, until, under the leadership of His beloved grandson, well-grounded administrative bases were established from which Bahá'í crusaders could set out in conquest of these prized and virgin lands.

172.2 As early as 1915, however, we see a first wave of itinerant teachers and short-time settlers directing their steps towards Alaska in an attempt to open it to the light of Bahá'u'lláh. This was followed by a second wave of determined pioneers and spiritual conquerors who, ever since the first Seven Year Plan, demonstrated their exemplary enthusiasm and caused "the breezes . . . of the love of God" to "perfume the nostrils of the inhabitants" of that "vast country."[1] In Canada, in response to the Master's call, a succession of homefront pioneers settled and opened the length and breadth of their land, so rich in promise "whether from a material or a spiritual standpoint," and whose destiny is to "become the object of the glance of Providence."[2] To the fringes of Greenland North American and European pioneers brought the light of God's Faith, and provided the means for the fire of His love to be kindled in that land, in anticipation of the day when it will become "a divine rose garden and a heavenly paradise." Iceland, specifically mentioned by 'Abdu'l-Bahá in the Tablets of the Divine Plan, was opened and consolidated, and has, through the warm response of its inhabitants to the Call of God, undoubtedly become one of the shining beacons of the "lights of the Most Great Guidance" in the North.[3]

172.3 As a result of these movements and organized activities, the call of the Kingdom reached "the ears of the Eskimos," and the divine spark was struck in their lands.[4] Praise be to God, today there are many who justly belong to the rank of heroes from among that noble race, and whose hearts are burning with His love. Upon the zeal and endurance of these enkindled believers

---

172-1. TDP 6.15. The first Seven Year Plan included the years 1937–44.
172-2. TDP 13.6, 13.2.
172-3. TDP 5.2, 6.5.
172-4. TDP 5.2.

will depend the early fulfillment of the Master's glowing promises. The teaching work among the Indians of the northern lands of the Western Hemisphere has likewise borne rich fruit, as tribe upon tribe has been enlisted under the banner of Bahá'u'lláh. Whether in Alaska's southeastern islands and rugged mountains, or in Canada's huge Indian reserves from the west to the east, many Amerindian believers have arisen to serve the Cause, and through their joint efforts, their sacrificial endeavors and distinctive talents they bid fair to accelerate the dawn of the day when they will be so "illumined as to enlighten the whole world."[5]

172.4 Many are the goals which now challenge the peoples of the North under the Five Year Plan: encouraging and educating the children and stimulating and guiding the youth; a wider participation of women in Bahá'í services; a greater assumption by the indigenous inhabitants of these regions of responsibilities in the leadership and administration of the community; a bolder proclamation of the Faith by radio and television; and a more far-flung and intensified campaign of teaching, audaciously conceived by National Spiritual Assemblies and their agencies and vigorously executed by Local Spiritual Assemblies and individual believers, aiming at a vast increase in the number of adherents to the Faith from every segment of society, a multiplication of Bahá'í administrative institutions, and a richer and more diverse range of publications in all media. What will set the seal on the success of the Plan and pave the way for the long-awaited and divinely promised glories of the future, is a mightier effort by every supporter of the Most Great Name in those climes to increasingly deepen themselves in the teachings, to pour forth their substance in the path of His love, to resolve to conform their personal lives to the high standards set in His teachings, and to undertake more daring tasks however great the sacrifice, and more extensive travels however arduous the voyage. In this wise will they draw nearer to the Spirit of Bahá'u'lláh and become true and radiant signs of His Most Great Guidance. These are the tasks! This is the work!

172.5 We pray at the Sacred Threshold that the Bahá'ís of the North may in the not-too-distant future transform the Arctic into that spiritual rose garden and heavenly paradise longed and yearned for by 'Abdu'l-Bahá, and that its peoples may be bountifully blessed and lovingly guided in their selfless services to promote the Faith of Bahá'u'lláh.

THE UNIVERSAL HOUSE OF JUSTICE

172-5. TDP 6.8.

173

Comments on the Subject of Politics
7 July 1976

To an individual Baháʾí

Dear Baháʾí Friend,

173.1 The Universal House of Justice received your letter of May 15 conveying your thoughts on the need for Baháʾís to become involved as may be necessary in political affairs and to participate in activities aimed at the eradication of injustice. The sincerity which prompted you to write such a letter and to candidly express your sentiments deeply touched the Universal House of Justice. We have been asked to convey its comments to you.

The Divine Physician's Remedy

173.2 You ask if silence on the part of Baháʾís will not allow chaos and human humiliation to be a permanent feature on earth, and state that shunning of politics by the Baháʾís can but weaken the freedom fighters of the world. When viewing the conditions of our society we see a world beset by ills and groaning under the burden of suffering. This suffering, Baháʾuʾlláh has Himself testified, is because the "body" of the world, "though created whole and perfect, has been afflicted, through divers causes, with grave ills and maladies," and "its sickness waxed more severe, as it fell under the treatment of unskilled physicians who have spurred on the steed of their worldly desires and have erred grievously." Baháʾuʾlláh's statement in this passage concludes with the assertion that the "sovereign remedy" lies in turning and submitting to the "skilled," the "all-powerful," and "inspired Physician. This, verily, is the truth, and all else naught but error."[1]

173.3 This Divine Physician has assured us in His writings that God is All-Seeing and All-Knowing and has willed to establish in this Day and among men His everlasting Kingdom. "The whole earth," Baháʾuʾlláh has stated, "is now in a state of pregnancy. The day is approaching when it will have yielded its noblest fruits, when from it will have sprung forth the loftiest trees, the most enchanting blossoms, the most heavenly blessings."[2] In order to achieve this purpose God sent us the spirit and message of the New Day through two successive Manifestations,[3] both of Whom the generality of mankind have rejected, and have, alas, preferred to continue in their own blindness and perversity. Commenting on such a world spectacle, Baháʾuʾlláh wrote: "Soon will the present-day order be rolled up, and a new one spread out in its stead." "After a time,"

173-1. WOB, pp. 39–40; GWB, pp. 254–55.
173-2. WOB, p. 169.
173-3. Baháʾuʾlláh and the Báb.

He further wrote, "all the governments on earth will change. Oppression will envelop the world. And following a universal convulsion, the sun of justice will rise from the horizon of the unseen realm."[4]

When we turn to His other writings to learn more of His warning that this "present-day order" is to be "rolled up," we read statements and predictions such as these: "The time for the destruction of the world and its people hath arrived." "The hour is approaching when the most great convulsion will have appeared." "The promised day is come, the day when tormenting trials will have surged above your heads, and beneath your feet, saying: 'Taste ye what your hands have wrought!'" "Soon shall the blasts of His chastisement beat upon you and the dust of hell enshroud you." "And when the appointed hour is come, there shall suddenly appear that which shall cause the limbs of mankind to quake." "The day is approaching when its [civilization's] flame will devour the cities, when the Tongue of Grandeur will proclaim: 'The Kingdom is God's, the Almighty, the All-Praised!'" "The day is approaching when the wrathful anger of the Almighty will have taken hold of them. He, verily, is the Omnipotent, the All-Subduing, the Most Powerful. He shall cleanse the earth from the defilement of their corruption, and shall give it for an heritage unto such of His servants as are nigh unto Him."[5]

The Hand of Providence at Work

From the above it becomes clear that the Hand of Providence is at work, and is engaged in fulfilling God's purpose for mankind in this Age. "God's purpose," Shoghi Effendi assures us, "is none other than to usher in, in ways He alone can bring about, and the full significance of which He alone can fathom, the Great, the Golden Age of a long-divided, a long-afflicted humanity. Its present state, indeed even its immediate future, is dark, distressingly dark. Its distant future, however, is radiant, gloriously radiant—so radiant that no eye can visualize it."[6]

Let us consider the First World War, which Shoghi Effendi has described in his writings as "the first stage in a titanic convulsion long predicted by Bahá'u'lláh."[7] Although it ended outwardly in a Treaty of Peace, 'Abdu'l-Bahá remarked: Peace, Peace, the lips of potentates and peoples unceasingly proclaim, whereas the fire of unquenched hatreds still smolders in their hearts. And then in 1920, He wrote: "The ills from which the world now suffers will multiply; the gloom which envelops it will deepen." And again: "another war, fiercer than the last, will assuredly break out."[8] After this Second World War broke out in 1939, Shoghi Effendi called it a "tempest, unprecedented in its

---

173-4. Quoted in PDIC ¶287.
173-5. Quoted in PDIC ¶3.
173-6. PDIC ¶286.
173-7. GPB, p. 305.
173-8. Quoted in WOB, pp. 29–30, 46.

violence," and the "great and mighty wind of God invading the remotest and fairest regions of the earth." After the termination of this War and the creation of the United Nations, the Guardian wrote in 1948, anticipating "still more violent convulsions" and referred to the "wings of yet another conflict" destined to "darken the international horizon."⁹ And finally in his last Riḍván Message of April 1957, he left for posterity the following analysis of world conditions in the light of the prophecies and predictions recorded in the writings of the Faith:

173.6a Indeed, as we gaze in retrospect beyond the immediate past, and survey, in however cursory a manner, the vicissitudes afflicting an increasingly tormented society, and recall the strains and stresses to which the fabric of a dying Order has been increasingly subjected, we cannot but marvel at the sharp contrast presented, on the one hand, by the accumulated evidences of the orderly unfoldment, and the uninterrupted multiplication of the agencies, of an Administrative Order designed to be the harbinger of a world civilization, and, on the other, by the ominous manifestations of acute political conflict, of social unrest, of racial animosity, of class antagonism, of immorality and of irreligion, proclaiming, in no uncertain terms, the corruption and obsolescence of the institutions of a bankrupt Order.

173.6b Against the background of these afflictive disturbances—the turmoil and tribulations of a travailing age—we may well ponder the portentous prophecies uttered well-nigh four score years ago, by the Author of our Faith, as well as the dire predictions made by Him Who is the unerring Interpreter of His teachings, all foreshadowing a universal commotion, of a scope and intensity unparalleled in the annals of mankind.

173.6c The violent derangement of the world's equilibrium; the trembling that will seize the limbs of mankind; the radical transformation of human society; the rolling up of the present-day Order; the fundamental changes affecting the structure of government; the weakening of the pillars of religion; the rise of dictatorships; the spread of tyranny; the fall of monarchies; the decline of ecclesiastical institutions; the increase of anarchy and chaos; the extension and consolidation of the Movement of the Left; the fanning into flame of the smoldering fire of racial strife; the development of infernal engines of war; the burning of cities; the contamination of the atmosphere of the earth—these stand out as the signs and portents that must either herald or accompany the retributive calamity which, as decreed by Him Who is the Judge and Redeemer of mankind, must, sooner or later, afflict a society which, for the most part, and for over a century, has turned a deaf

---

173-9. PDIC ¶2; CF, p. 58.

ear to the Voice of God's Messenger in this day—a calamity which must purge the human race of the dross of its age-long corruptions, and weld its component parts into a firmly knit world-embracing Fellowship . . ."[10]

Thus we see how the Divine Physician is both the "Judge" of mankind and its "Redeemer."[11]

Noninterference in Political Affairs

This same Physician, addressing His followers, "the beloved of the one true God," wrote: "Forbear ye from concerning yourselves with the affairs of this world and all that pertaineth unto it, or from meddling with the activities of those who are its outward leaders. The one true God, exalted be His glory, hath bestowed the government of the earth upon the kings. To none is given the right to act in any manner that would run counter to the considered views of them who are in authority."[12]

In another Tablet He laid on His followers the obligation to "behave towards the government of the country in which they reside with loyalty, honesty and truthfulness."[13] 'Abdu'l-Bahá reaffirmed the same principles. When in America He explained: "The essence of the Bahá'í spirit is that, in order to establish a better social order and economic condition, there must be allegiance to the laws and principles of government." And in a Tablet He referred to the "irrefutable command that the Blessed Perfection hath given" in His Tablets, namely, "that the believers must obey the kings with the utmost sincerity and fidelity, and He hath forbidden them [the believers] to interfere at all with political problems. He hath even prohibited the believers from discussing political affairs."[14]

And finally in His last Will and Testament He wrote: "We must obey and be the well-wishers of the government of the land . . ."[15]

The Role of Bahá'ís in Today's World

We have also been asked to share with you at this juncture the following two extracts from letters written by the Universal House of Justice, and it is hoped these will help you in appreciating the significant and vital role Bahá'ís can and must play in the world today:

> We are told by Shoghi Effendi that two great processes are at work in the world: the great Plan of God, tumultuous in its progress, work-

---

173-10. MBW, pp. 102–03.
173-11. PDIC ¶6.
173-12. GWB, p. 241.
173-13. GPB, p. 219.
173-14. PUP, p. 238; TABA 3:498.
173-15. WT, p. 8.

ing through mankind as a whole, tearing down barriers to world unity and forging humankind into a unified body in the fires of suffering and experience. This process will produce in God's due time, the Lesser Peace, the political unification of the world. Mankind at that time can be likened to a body that is unified but without life. The second process, the task of breathing life into this unified body—of creating true unity and spirituality culminating in the Most Great Peace—is that of the Bahá'ís, who are laboring consciously, with detailed instructions and continuing divine guidance, to erect the fabric of the Kingdom of God on earth, into which they call their fellowmen, thus conferring upon them eternal life.

173.11b The working out of God's Major Plan proceeds mysteriously in ways directed by Him alone, but the Minor Plan that He has given us to execute, as our part in His grand design for the redemption of mankind, is clearly delineated. It is to this work that we must devote all our energies, for there is no one else to do it. . . .[16]

173.11c . . . Bahá'ís are often accused of holding aloof from the "real problems" of their fellowmen. But when we hear this accusation let us not forget that those who make it are usually idealistic materialists to whom material good is the only "real" good, whereas we know that the working of the material world is merely a reflection of spiritual conditions and until the spiritual conditions can be changed there can be no lasting change for the better in material affairs.

173.11d We should also remember that most people have no clear concept of the sort of world they wish to build, nor how to go about building it. Even those who are concerned to improve conditions are therefore reduced to combating every apparent evil that takes their attention. Willingness to fight against evils, whether in the form of conditions or embodied in evil men, has thus become for most people the touchstone by which they judge a person's moral worth. Bahá'ís, on the other hand, know the goal they are working towards and know what they must do, step by step, to attain it. Their whole energy is directed towards the building of the good, a good which has such a positive strength that in the face of it the multitude of evils—which are in essence negative—will fade away and be no more. To enter into the quixotic tournament of demolishing one by one the evils in the world is, to a Bahá'í, a vain waste of time and effort. His whole life is directed towards proclaiming the Message of Bahá'u'lláh, reviving the spiritual life of his fellowmen, uniting them in a divinely-created World

---

173-16. Letter dated 8 December 1967 to an individual. See message no. 55.

Order, and then, as the Order grows in strength and influence, he will see the power of that Message transforming the whole human society and progressively solving the problems and removing the injustices which have so long bedeviled the world.[17]

173.12 You have asked whether it is possible to have a World Federation when not all countries have attained their independence. The answer is in the negative. Both 'Abdu'l-Bahá and Shoghi Effendi likened the emergence of the American Republic and the unification of the "diversified and loosely related elements" of its "divided" community into one national entity, to the unity of the world and the incorporation of its federated units into "one coherent system."[18] Just as the American Constitution does not allow one state to be more autonomous than another, so must the nations of the world enjoy equal status in any form of World Federation. Indeed one of the "candles" of unity anticipated by 'Abdu'l-Bahá is "unity in freedom."[19]

173.13 Yet another question is whether it is morally right to remain silent when equality is being abused. The beloved Guardian has given us the following guidelines in letters written on his behalf. "Much as the friends must guard against in any way ever seeming to identify themselves or the Cause with any political party, they must also guard against the other extreme of never taking part with other progressive groups, in conferences or committees designed to promote some activity in entire accord with our teachings—such as, for instance, better race relations."[20]

173.14 The Universal House of Justice hopes that you and your Bahá'í coworkers in that land will steep yourselves in the teachings of Bahá'u'lláh, endeavor to follow in your personal lives the noble standards set by Him, attract the multitudes to the radiance of His glorious Faith, and enable them to warm their hearts and ignite their souls with the flames of that undying Fire which "blazeth and rageth in the world of creation."[21]

173.15 We have been asked by the Universal House of Justice to assure you of its prayers on your behalf in the Holy Shrines.

With loving Bahá'í greetings,
DEPARTMENT OF THE SECRETARIAT

173-17. Letter dated 19 November 1974 to the National Spiritual Assembly of the Bahá'ís of Italy. See message no. 151.
173-18. WOB, p. 165.
173-19. WOB, p. 165.
173-20. Shoghi Effendi, letter dated 21 November 1948 to an individual.
173-21. Bahá'u'lláh, in BP, p. 53, or PM, p. 76.

174

Message to the International Teaching Conference, Paris, France—August 1976
AUGUST 1976

To the Friends assembled at the International Teaching Conference in Paris

Dearly loved Friends,

174.1 The brilliance of Paris in the story of European civilization bids fair to be renewed with even greater splendor during the Day of the Lord of Hosts and the establishment of God's Kingdom on earth.¹ The annals of Paris in this Day have already acquired eternal luster from events of mightier import, of greater universal significance and more sacred character than any which its past history has witnessed. The visits of 'Abdu'l-Bahá, the Center of God's inviolable Covenant, alone outshine in historical importance anything in the long history of France, and are immortalized in the greatly loved collection of His discourses given in that capital city.² Beyond this, we recall with awe and pride that it was at 'Abdu'l-Bahá's instruction that the illustrious May Maxwell succeeded in establishing in Paris the first Bahá'í center on the European Continent, a continent described by Shoghi Effendi as "the cradle of a civilization to some of whose beneficent features the Pen of Bahá'u'lláh has paid significant tribute; on whose soil both the Greek and Roman civilizations were born and flourished; which has contributed so richly to the unfoldment of American civilization; the fountainhead of American culture; the mother of Christendom, and the scene of the greatest exploits of the followers of Jesus Christ," and experiencing "the first stirrings of that spiritual revolution" which must culminate in the permanent establishment throughout its diversified lands of the divinely ordained Order of Bahá'u'lláh.³ This first center was rapidly reinforced by the conversion of the first English believer and of the first Frenchman to accept the Faith—the distinguished Hippolyte Dreyfus, whose "preeminent role" it was to kindle "the torch which is destined to shed eternal illumination upon his native land and its people"—and by Laura Barney, whose "imperishable service" was to transmit to posterity *Some Answered Questions*.⁴ The steadfastness and devotion of the Paris Bahá'í Community during the dark and somber days of the Second World War earned great praise from the beloved Guardian of the Faith, while the

---

174-1. "The Day of the Lord of Hosts" is a reference to the Dispensation of Bahá'u'lláh.
174-2. See 'Abdu'l-Bahá, *Paris Talks*, (1969). 'Abdu'l-Bahá visited Paris three times. The first visit lasted nine weeks, 3 October through 2 December 1911; the second was 21 January through 30 March 1913; the third was early May through 12 June 1913.
174-3. CF, pp. 26–27.
174-4. UD, p. 84; GPB, p. 260.

recent signs of widespread and effective teaching work throughout France lend wings to the hope that this veteran, sorely tested and steadfast community is about to gather the harvest of those potent seeds sown and nourished so lovingly by 'Abdu'l-Bahá.

174.2 It is highly propitious that this city, thrice blessed by the presence of the Master, should be the scene of the first International Bahá'í Conference in France and one of the eight International Conferences to be held around the world during the Five Year Plan. You are gathered in this historic spot to deliberate on the fortunes of that Plan, to derive inspiration from the deeds performed there in the Heroic Age of our Faith and from your association together, to rededicate yourselves to the service of Bahá'u'lláh and to determine, each and every one, how best you can promote the victory of the Five Year Plan. We call to your attention 'Abdu'l-Bahá's words:

> 174.2a The call of Yá Bahá'u'l-Abhá can be heard far and wide. It is my hope that this soul-stirring melody of the Abhá Kingdom may also be raised high in Paris, for Paris is tumultuous in all things. I pray the Almighty that the music and singing of the beloved of God will be so loud that the vibrations thereof may cause the limbs of Paris to quake. I await very joyful tidings from the friends in Paris. Unquestionably the divine melody will in the future be raised in that city, but I long that this may happen in these days of the Covenant, and that you will be the enchanting songsters and the sweet-singing nightingales of that land.[5]

174.3 Our hopes are high and we pray at the Sacred Threshold that from this Conference will surge throughout Europe a wave of such sacrificial teaching as will impel large numbers of its diverse and highly talented peoples to embrace the Faith of God and dedicate themselves to the redemption of mankind under the glorious banner of the Prince of Peace.

THE UNIVERSAL HOUSE OF JUSTICE

174-5. From an unpublished Tablet.

175
Release of a Compilation on Bahá'í Education
31 August 1976

To all National Spiritual Assemblies

Dear Bahá'í Friends,

175.1 Among the goals of the Five Year Plan given to certain National Spiritual Assemblies is that of the promotion of the Bahá'í education of children. But this subject is of profound interest to Bahá'ís in every land, not only in those where, as the Plan indicates, specific goals must be attained.

175.2 At our request the Research Department at the World Center has prepared this compilation of Bahá'í teachings on the subject of Bahá'í education for the friends everywhere to study and to begin to apply as their circumstances permit.[1] In the main part of the compilation all extracts are from Tablets or other authenticated writings. In addition there is a supplement of extracts from talks of 'Abdu'l-Bahá recorded in *Star of the West* and *The Promulgation of Universal Peace*. It is hoped that in future it will be possible to have such well-known addresses verified against authenticated transcripts of the talks in the original language, but until that time the friends may use them in their present form for their reference and study. You are free to share this compilation, in whole or in part, with the believers under your jurisdiction in whatever way you judge best.

175.3 A letter written on behalf of the Guardian to an individual believer on 7 June 1939 explained:

175.3a ... the Teachings of Bahá'u'lláh and 'Abdu'l-Bahá do not present a definite and detailed educational system, but simply offer certain basic principles and set forth a number of teaching ideals that should guide future Bahá'í educationalists in their efforts to formulate an adequate teaching curriculum which would be in full harmony with the spirit of the Bahá'í Teachings, and would thus meet the requirements and needs of the modern age.

175.3b These basic principles are available in the sacred writings of the Cause, and should be carefully studied, and gradually incorporated in various college and university programs. But the task of formulating a system of education which would be officially recognized by the Cause, and enforced as such throughout the Bahá'í world, is one which the present-day generation of believers cannot obviously undertake, and which has to be gradually accomplished by Bahá'í scholars and educationalists of the future.

175-1. See CC 1:245–313.

In reading the compilation the friends should bear the above explanation 175.4
in mind and not interpret as a universal instruction what may, in fact, be
simply an example of the sort of course that could advantageously be followed.
For example, the quotations from the Tablets of 'Abdu'l-Bahá and the letters
of Shoghi Effendi written in the early years of his Guardianship were addressed
to the Bahá'ís in Persia when that community, although oppressed, was already
large in numbers, and was living in a society where schooling was rudimentary and scarce. At the present time in most countries, compulsory education
and state school systems are widespread and meet the general need for material education, so the resources of the Faith in that field have to be concentrated on the spiritual and moral education of our children and on providing
primary and tutorial schools in mass-teaching areas where illiteracy is still the
rule. Also, the various extracts must be read in the context of the whole.

The proper education of children is of vital importance to the progress of 175.5
mankind, and the heart and essential foundation of all education is spiritual
and moral training. When we teach our fellowmen the truths and way of life
of the Bahá'í Faith we have to struggle against barriers of indifference, materialism, superstition and a multitude of erroneous preconceptions; but in our
newborn children we are presented with pure souls, untarnished by the world.
As they grow they will face countless tests and difficulties. From their earliest
moments we have the duty to train them, both spiritually and materially, in
the way that God has shown, and thus, as they come to adulthood, they can
become champions of His Cause and spiritual and moral giants among
mankind, equipped to meet all tests, and will be, indeed, "stars of the heaven
of understanding," "soft-flowing waters upon which must depend the very life
of all men."[2]

It is our prayer at the Sacred Threshold that the provision of the gems of 175.6
divine guidance contained in this compilation will be a stimulus and a source
of inspiration for Bahá'í parents, teachers and Spiritual Assemblies throughout the world, and a blessing for the rising generations.

With loving Bahá'í greetings,
THE UNIVERSAL HOUSE OF JUSTICE

175-2. GWB, p. 196.

176

Message to the International Teaching Conference, Nairobi, Kenya—September 1976
26 SEPTEMBER 1976

To the Friends gathered at the International Conference in Nairobi

Beloved Friends,

176.1 The flames of enthusiasm which ignited the hearts of the followers and lovers of the Most Great Name in Helsinki, in Anchorage and in Paris are now being kindled in a city which occupies a central and envied position at the very crossroads of the vast African mainland and are destined to illumine its horizons. This Conference marking the imminent approach of the midway point of the Five Year Plan which coincides with the anniversary of the birth of the Blessed Báb,[1] will no doubt go down in Bahá'í history as a further landmark in the irresistible march of events which have characterized the impact of the Faith of God upon that continent.

176.2 We recall that in addition to Quddús the only other companion of the Báb on His pilgrimage to Mecca was an Ethiopian, and that he and his wife were intimately associated with Him and His household in Shiraz.[2] During the Ministry of Bahá'u'lláh a few of His stalwart disciples reached the northeastern shores of Africa, and under His direct guidance, announced the glad tidings of the New Day to the people of the Nile, thus opening to the Faith two countries of the African mainland. Soon afterwards, His blessed person approached those shores in the course of His exile to the Holy Land. Still later He voiced His significant utterance in which He compared the colored people to "the black pupil of the eye," through which "the light of the spirit shineth forth."[3] Just over six years after His ascension, the first member of the black race to embrace His Cause in the West, who was destined to become a disciple of 'Abdu'l-Bahá, a herald of the Kingdom, and the door through which numberless members of his race were to enter that Kingdom, came on pilgrimage to the Holy Land with the first group of Western friends who arrived in 'Akká to visit the Center of the Covenant.[4] This was followed by a steady extension of the teaching work among the black people of North America, and the opening to the Faith, by the end of the Heroic Age, of two more

---

176-1. 20 October.

176-2. The Ethiopian was Ḥájí Mubárak; Quddús was the eighteenth Letter of the Living, the title given to the first eighteen people to recognize the Báb's prophetic station. Quddús was distinguished for his serenity and sagacity and was elevated by Bahá'u'lláh to a rank second only to that of the Báb.

176-3. Quoted in ADJ, p. 37.

176-4. Robert Turner was among the first group of Western pilgrims to arrive in 'Akká on 10 December 1898, some six years after Bahá'u'lláh's ascension on 29 May 1892.

countries in Africa, under the watchful care of the Master, Whose three visits to Egypt have blessed the soil of that Continent.⁵ Prior to the conclusion of the first Bahá'í century the number of countries opened to the Faith had been raised to seven, and the teaching work among the black race in North America had entered a new phase of development through the continuous guidance flowing from the pen of Shoghi Effendi, who himself traversed the African continent twice from south to north, and who, in the course of his ministry, elevated two members of the black race to the rank of Hand of the Cause, appointed three more believers residing in Africa to that high office, and there raised up four National Spiritual Assemblies.⁶

176.3 At the beginning of the Ten Year Crusade the number of countries opened to the Faith had reached twenty-four, including those opened under the aegis of the Two Year African Campaign coordinated by the British National Spiritual Assembly. The Ten Year Plan opened the rest of Africa to the light of God's Faith, and today we see with joy and pride in that vast continent and its neighboring islands the establishment of four Boards of Counselors, thirty-four National Spiritual Assemblies—firm pillars of God's Administrative Order—and over 2800 Local Spiritual Assemblies, nuclei of a growing Bahá'í society.

176.4 Africa, a privileged continent with a past rich in cherished associations, has reached its present stage of growth through countless feats of heroism and dedication. Before us unfolds the vision of the future. "Africa," the beloved Guardian assures us in one of the letters written on his behalf, "is truly awakening and finding herself, and she undoubtedly has a great message to give, and a great contribution to make to the advancement of world civilization. To the degree to which her peoples accept Bahá'u'lláh will they be blessed, strengthened and protected."⁷

176.5 The realization of this glorious destiny requires that the immediate tasks be worthily discharged, and the pressing challenges and urgent requirements of the Five Year Plan be wholeheartedly and effectively met and satisfied. As the forces of darkness in that part of the world wax fiercer, and the problems facing its peoples and tribes become more critical, the believers in that con-

---

176-5. 'Abdu'l-Bahá visited Egypt circa August/September 1910 through 11 August 1911, circa December 1911 through 25 March 1912, and 17 June 1913 through 2 December 1913.

176-6. By 1944, the seven African countries opened to the Faith were Abyssinia (Ethiopia), Belgian Congo, Egypt, South Africa, Southern Rhodesia, Sudan, and Tunisia. Shoghi Effendi traveled in Africa circa September–October 1929 and in 1940. The two black Hands of the Cause of God were Louis Gregory of the United States, appointed posthumously in August 1951, and Enoch Olinga of Uganda, appointed in October 1957. The additional three Hands appointed while living in Africa were Músá Banání (29 February 1952, Kampala, Uganda), and William Sears and John Robarts (October 1957, South Africa). The four National Spiritual Assemblies Shoghi Effendi raised up are those of Central and East Africa, North East Africa, North West Africa, and South and West Africa.

176-7. UD, p. 330.

tinent must evince greater cohesion, scale loftier heights of heroism and self-sacrifice and demonstrate higher standards of concerted effort and harmonious development.

176.6 During the brief thirty months separating us from the end of the Plan, Africa must once again distinguish itself among its sister continents through a vast increase in the number of its believers, its Local Spiritual Assemblies and its localities opened to the Faith, and by accelerating the process of entry by troops throughout its length and breadth. The deepening of the faith, of the understanding and of the spiritual life of its individual believers must gather greater momentum; the foundations of its existing Local Spiritual Assemblies must be more speedily consolidated; the number of local Ḥaẓíratu'l-Quds and of local endowments called for in the Plan must be soon acquired; the Baháʼí activities of women and of youth must be systematically stimulated; the Baháʼí education of the children of the believers must continuously be encouraged; the basis of the recognition that the institutions of the Faith have succeeded in obtaining from the authorities must steadily be broadened; mass communication facilities must be used far more frequently to teach and proclaim the Faith; and the publication and dissemination of the essential literature of the Faith must be given much greater importance. Above all it is imperative that in ever greater measure each individual believer should realize the vital need to subordinate his personal advantages to the overall welfare of the Cause, to awaken and reinforce his sense of responsibility before God to promote and protect its vital interests at all costs, and to renew his total consecration and dedication to His glorious Faith, so that, himself enkindled with the flames of its holy fire, he may, in concert with his fellow-believers, ignite the light of faith and certitude in the hearts of his family, his tribe, his countrymen and all the peoples of that mighty continent, in preparation for the day when Africa's major contribution to world civilization will become fully consummated.

176.7 We fervently pray at the Holy Shrines that these hopes and aspirations may soon come true, and that the "pure-hearted" and "spiritually receptive" people of Africa may draw ever nearer to the spirit of Baháʼu'lláh, and may become shining examples of self-abnegation, of courage and of love to the supporters of the Most Great Name in every land.[8]

THE UNIVERSAL HOUSE OF JUSTICE

176-8. MBW, p. 135.

177

Visit of His Highness Malietoa Tanumafili II to the Resting-Place of Shoghi Effendi
5 OCTOBER 1976

To all National Spiritual Assemblies

WITH FEELINGS IMMENSE ELATION ANNOUNCE BAHÁ'Í COMMUNITIES ALL CONTINENTS JOYFUL NEWS FIRST VISIT BY BAHÁ'Í REIGNING MONARCH TO RESTING PLACE BELOVED SHOGHI EFFENDI WELL NIGH NINETEEN YEARS FOLLOWING HIS PASSING.[1] HIS HIGHNESS MALIETOA TANUMAFILI II OF WESTERN SAMOA COURSE HIS RECENT VISIT LONDON ATTAINED THIS INESTIMABLE BOUNTY. HIS HIGHNESS ACCOMPANIED BY SMALL DELEGATION FRIENDS HEADED BY HAND CAUSE GIACHERY INCLUDING HANDS CAUSE KHADEM VARQÁ AND FIVE MEMBERS NATIONAL ASSEMBLY UNITED KINGDOM PROCEEDED CEMETERY PARTICIPATED DEEPLY MOVING HIGHLY DIGNIFIED VISITATION RESTING PLACE SIGN GOD ON EARTH WHOSE LIFELONG HEROIC LABORS ACHIEVED WORLDWIDE SPREAD GLORIOUS CAUSE BAHÁ'U'LLÁH.[2] FOLLOWING PRAYERS DEVOTIONS AND EXPRESSIONS APPRECIATION GRATITUDE BY HIS HIGHNESS HE MET WITH RADIANT SPIRIT LARGE GATHERING BELIEVERS ASSEMBLED PRECINCTS GRAVE TO WELCOME HIM AND TO WHOM HE CONVEYED GREETINGS HIS FELLOW SAMOAN BAHÁ'ÍS AND HIS HOPE GROWTH FAITH FROM STRENGTH TO STRENGTH ALL OVER WORLD. IN HOLY LAND SYNCHRONIZING WITH THESE EVENTS HANDS CAUSE OTHER MEMBERS INTERNATIONAL TEACHING CENTER JOINED MEMBERS HOUSE JUSTICE SPECIAL VISIT BAHJÍ WHERE FERVENT PRAYERS WERE OFFERED SACRED THRESHOLD MOST HOLY SHRINE IN THANKSGIVING HIGHLY SIGNIFICANT DEVELOPMENT WHICH REPRESENTS PRELUDE TO FULFILLMENT LONG CHERISHED DESIRE 'ABDU'L-BAHÁ AND SHOGHI EFFENDI WITNESS PILGRIM KINGS PAY THEIR HUMBLE TRIBUTE AT HOLY SHRINES FOUNDERS OUR FAITH IN SPIRITUAL HEART PLANET. FERVENTLY PRAYING THIS JOYOUS NEWS WILL STRENGTHEN RESOLVE SUPPORTERS MOST GREAT NAME EVERY LAND REDOUBLE THEIR EFFORTS TO FORGE AHEAD HOWEVER ARDUOUS THE TASKS HOWEVER SEEMINGLY INSURMOUNTABLE THE OBSTACLES UNTIL EVERY GOAL WORLDWIDE PLAN IS SPEEDILY AND FULLY CONSUMMATED.

UNIVERSAL HOUSE OF JUSTICE

177-1. For the announcement of His Highness's acceptance of the Bahá'í Faith, see message no. 130.

177-2. Later information confirms that six members of the National Spiritual Assembly of the United Kingdom were present. The visit took place on 12 September 1976.

MESSAGES FROM THE UNIVERSAL HOUSE OF JUSTICE

178
Outstanding Achievements in Khurásán, Iran[1]
15 October 1976

To the Friends gathered at the
International Teaching Conference in Nairobi, Kenya

178.1 REJOICE SHARE WITH FRIENDS AT NAIROBI CONFERENCE JOYOUS TIDINGS RECEIVED FROM CRADLE FAITH FRIENDS PROVINCE KHURÁSÁN WHERE MIGHTY EVENTS TOOK PLACE EARLY YEARS HEROIC AGE HAVE WON UNIQUE DISTINCTION. EVERY LOCAL SPIRITUAL ASSEMBLY EVEN EVERY GROUP THAT PROVINCE HAS FORMULATED ITS LOCAL GOALS AND EVERY INDIVIDUAL BELIEVER HAS ADOPTED PERSONAL TEACHING GOALS. OBJECTIVE ENVISAGED FIVE YEAR PLAN BRILLIANTLY OUTSTRIPPED. FERVENTLY PRAYING SHRINES SPIRITUAL ASSEMBLIES GROUPS INDIVIDUAL BELIEVERS EVERY LAND WILL BE INSPIRED FOLLOW THIS EXAMPLE.

UNIVERSAL HOUSE OF JUSTICE

179
Achievement of the Majority of Pioneer Goals at Midpoint of Five Year Plan
21 October 1976

To all National Spiritual Assemblies

179.1 JOYFULLY ANNOUNCE REALIZATION HOPE EXPRESSED IN JANUARY 1975 THAT BY MIDWAY POINT FIVE YEAR PLAN MOST PIONEER POSTS BE SETTLED.[1] OUT OF THESE 962 POSTS 612 ALREADY FILLED THROUGH SETTLEMENT 1374 PIONEERS. 350 POSTS STILL REMAIN UNFILLED BUT 227 VOLUNTEERS HAVE ARISEN AND ARE BEING PROCESSED FOR SETTLEMENT THESE GOALS. CALL WITH URGENT INSISTENCE ALL NATIONAL ASSEMBLIES WITH UNFILLED PIONEER GOALS EXPEDITE FULFILLMENT ASSIGNMENTS COUNTRIES STILL NEEDING URGENT SUPPORT. PRAYING HOLY SHRINES BOUNTIFUL BLESSINGS ALL WHO HAVE ARISEN PIONEER FIELDS VARIOUS LANDS AND WHO WILL ARISE DURING FAST FLEETING MONTHS BEFORE RIḐVÁN 1977 WIN OUTSTANDING PIONEER GOALS FIVE YEAR PLAN.

UNIVERSAL HOUSE OF JUSTICE

178-1. This message was cabled to Hand of the Cause of God William Sears, representative of the Universal House of Justice at the International Teaching Conference in Nairobi, Kenya.
179-1. See message no. 155.

180
Additional Appointments to Continental Boards of Counselors
31 OCTOBER 1976

To all National Spiritual Assemblies

Dear Bahá'í Friends,

The Universal House of Justice announces with pleasure that it has appointed three new Counselors as follows:

Dr. Peter Khan—to the Continental Board of Counselors in Australasia

Dr. Hidáyatu'lláh Ahmadíyyih—to the Continental Board of Counselors in Central America

Mr. Angus Cowan—to the Continental Board of Counselors in North America.

With loving Bahá'í greetings,
DEPARTMENT OF THE SECRETARIAT

181
Message to the International Teaching Conference, Hong Kong—November 1976
NOVEMBER 1976

To the Friends assembled at the
International Teaching Conference in Hong Kong

Dearly loved Friends,

With grateful and joyous hearts we extend our warmest greetings and express our highest hopes to the followers of Bahá'u'lláh gathered, at this critical point of the Five Year Plan, in this Conference on Asia's eastern shore. This great continent has nourished mighty civilizations; above its horizons the suns of major Revelations of God have risen; on its soil many of the heroes of this New Age have shed their blood and offered their lives in token of their love for Him Who is the Beloved of the World and the Desire of the Nations.

What an imperishable glory has been bestowed upon the people of Asia, the first to be illumined by the rays of God's Faith, the first recipients of His Call and the first promoters of His Cause. Their spiritual capacity is extolled and the great role they are destined to play in the strengthening of the foundation of the New World Order of Bahá'u'lláh unhesitatingly affirmed by 'Abdu'l-Bahá in His Tablets of the Divine Plan.[1]

181-1. See TDP 7.5, 7.10, 8.21.

181.3 This great continent contains within its boundaries the heart of the Faith and its Cradle, the lands wherein its Founders toiled and suffered, and not only the great majority of the human race but the great majority of the followers of Bahá'u'lláh. The potentiality of such a situation cannot be underestimated, nor must the great force latent within so large a proportion of the Army of Light be neglected. They must be mobilized to accelerate the expansion of the beloved Cause, to consolidate its victories, to enhance its prestige and to augment its influence. We appeal to every participant in this historic Conference to become conscious of these tremendous but hidden potentialities which, if properly tapped and directed, can hasten the process of the spiritualization of the nations of Asia, the influence of which will extend far beyond the confines of that continent—even to the entire planet.

181.4 We eagerly await news that from this Conference will surge throughout Asia a wave of vigorous activity devoted to the execution on the individual, local and national levels, of systematic plans designed to attract the great masses of that continent to the life-giving, all-embracing Message of Bahá'u'lláh. Such plans must provide opportunities for those thousands who in recent years have swelled the rank and file, as well as for those veterans who have, for so many years, striven in the path of service to the Cause. It is through active participation of more and more new believers in both teaching and administration that the consolidation of the Bahá'í community can best be achieved.

181.5 In a continent so richly endowed, so greatly blessed, new generations of Bahá'ís must continually be raised up, trained from childhood in the school of the love of God, and nurtured under the shadow of His Cause. Bahá'ís in every country must be constantly urged and, wherever necessary, assisted to pass on to their children as their most cherished legacy, the precious Faith they themselves have embraced. Those new generations of Bahá'ís will have a vital role in consolidating the Cause of God on a firm foundation.

181.6 The establishment of Local Assemblies as the basic administrative unit of the World Order of Bahá'u'lláh and as rallying points for the Bahá'ís of every community should, in accordance with the objectives of the Five Year Plan, be multiplied no matter at what cost of effort and endurance. The process of invigorating the Local Assemblies requires, during the Five Year Plan, the consultation of all the institutions of the Cause.

181.7 Publication of literature in many languages as part of a determined campaign to win thousands upon thousands of diverse peoples in all spheres of life must be vigorously pursued. Participation of all believers in supporting the Bahá'í Funds, the lifeblood of the Cause, must be given adequate attention, and the blessings which reward the act of voluntary giving for the promotion of the Faith, no matter how small the amount may be, must be lovingly and wisely explained.

May this Conference become a landmark in the process of attracting vast numbers of the great Chinese race scattered throughout the world. May it be a prelude to the unprecedented expansion of the Faith in all the countries of Asia. May it become a source of strength to the supporters of the Most Great Name so that despite the rising tide of trials and upheavals afflicting the world, and whatever forces of opposition may be mounted against the Cause of God itself, the believers will not waver or become diverted from their course, but be ever more confirmed in their determination to raise the edifice of the Faith of God as the last bastion of hope to a lost and wayward humanity.

THE UNIVERSAL HOUSE OF JUSTICE

182

Grappling with the Challenge of Consolidation
2 DECEMBER 1976

The National Spiritual Assembly of the Bahá'ís of the United States

Dear Bahá'í Friends,

Your letter of 12 October 1976 proposing the establishment of an International Human Development Center has caused us concern for a number of reasons.¹ It is true that in our letter of 23 December 1975 we stated: "We feel that it is for your Assembly to decide what methods should be employed to bring about the desired result," but your present proposals are a great enlargement and development of the concepts outlined in your letter of 5 December 1975 and involve the establishment of an institution with international ramifications requiring financial assistance from beyond the confines of your own community.

We are acutely aware of the varied problems of community and character development that the American Bahá'í community faces. They are problems that in varying ways and to different degrees face every Bahá'í community in the world. In many countries they are further complicated by grinding poverty, widespread illiteracy, religious persecution or compulsory political indoctrination.

182-1. After Davison Bahá'í School in Davison, Michigan, closed in 1974, the National Spiritual Assembly of the Bahá'ís of the United States made plans to use the property for an International Human Development Center that was to include, among other things, a demonstration school and formal study in education at the postgraduate level. In response to guidance from the Universal House of Justice, a more modest facility, the Louhelen Bahá'í School, was opened in 1982. It was designed to serve as a Bahá'í conference center and as a site for summer and winter school programs.

182.3 As you quite correctly observe, Bahá'í Administration should make use of whatever expertise or appropriate instruments are available, whether Bahá'í or non-Bahá'í, for the attainment of its objectives. But this is not the same as establishing a quasi-Bahá'í institution under Bahá'í auspices based on one particular theory. It is far too early in the development of the Faith and of the social sciences for the Administrative Order thus to promote one particular system or theory of education. A similar situation exists in the field of psychology. As you are well aware, many people come into the Faith needing psychiatric treatment, and it is often very difficult for them to find a psychiatrist who will not urge them to some course of behavior which is contrary to the teachings of the Faith. There are a number of Bahá'í psychologists and psychiatrists who are endeavoring to develop their skills in the light of the Revelation of Bahá'u'lláh, and use can certainly be made of their services where available—but it would be premature to consider establishing a Bahá'í School of Psychology. . . .

The Gradual Unfoldment of Bahá'í Education

182.4 The grave problems faced by Bahá'í parents and children, when the children must attend schools that are strongly influenced by the degradation of present-day society, are fully appreciated. However, the only ways to completely overcome these dangers would seem to be either to effect a reform of the entire non-Bahá'í educational system or to provide a worldwide network of Bahá'í schools. Both ways are very long-term projects beyond the capacity of the Bahá'í community at this time. Already, of course, Bahá'í communities are establishing primary or tutorial schools in many parts of the world, but these are small and few in number and are located where there are such conditions as general illiteracy among the believers or where no other schools are available to them. Undoubtedly, in time, this process will gain momentum, and Bahá'í schools of ever higher quality and scope will be established in country after country, as has already occurred in India, but necessarily, this must now be a gradual process related, among other things, to the resources of the community, the number of Bahá'í children needing education, and the availability of other suitable schools. Perhaps in certain parts of the United States there are sufficiently large concentrations of Bahá'í children to make the running of a private Bahá'í school feasible—such a proposal has, indeed, been made by a number of individual believers in Alaska, principally teachers, but we stressed in that instance that, if implemented, it should be conducted as a private venture and that the people concerned should give very careful consideration to all the factors involved before initiating it; furthermore we pointed out to them their opportunities for improving the schools in which they themselves worked.

182.5 Failing a nationwide system of Bahá'í schools, the establishment of which is clearly out of the question at this stage of the growth of the Cause in the

United States, Bahá'í parents will continue to be faced with the problems caused by the exposure of their children to irreligious and immoral attitudes, behavior, and even instruction, from their fellow pupils and their teachers. This is a great challenge to Bahá'í parents, to the Bahá'í children themselves, and to the Spiritual Assemblies. It was to assist in meeting such challenges that we recently issued the compilation of Bahá'í prayers for children and that on Bahá'í education.[2] Your Assembly is correct in its view that a major effort will have to be exerted to raise the number and quality of Bahá'í children's classes, and to assist Bahá'í parents to bring up their children as firm Bahá'ís able to withstand the moral and spiritual poisons and temptations of the society around them. . . .

The Spiritual Nature of Personal Transformation

182.6 In addition to the specific problems of child education, you instance the difficulties of local communities which are faced with the task of reorienting and integrating into the Cause new believers who enter with all sorts of immoral and even criminal tendencies from their former life. This is indeed difficult, but this is the very stuff of the work of the Cause. The Bahá'í Faith not only provides teachings in accordance with which the behavior of human beings can be reformed, but also makes available a spiritual power which reinforces the devoted efforts of every believer, whether veteran or neophyte. Arising to serve the Cause has, itself, a transforming effect upon believers, as the beloved Guardian wrote with respect to service upon Spiritual Assemblies: "If we but turn our gaze to the high qualifications of the members of Bahá'í Assemblies, as enumerated in 'Abdu'l-Bahá's Tablets, we are filled with feelings of unworthiness and dismay, and would feel truly disheartened but for the comforting thought that if we arise to play nobly our part every deficiency in our lives will be more than compensated by the all-conquering spirit of His grace and power."[3] Thus, what is most imperative for the promotion of the spiritual life of local Bahá'í communities is the stimulation of the believers to increase their devotion to Bahá'u'lláh, their absolute reliance upon Him and upon His love, and their determination to apply His teachings in every aspect of their lives. This stimulation can be conveyed from heart to heart and mind to mind by devoted Bahá'ís without the need of formal training. . . .

The Auxiliary Board: A Potent Aid to Consultation

182.7 As we pointed out previously, you have already initiated excellent programs; we continually receive evidence of the enthusiasm with which they have been received by local communities in the United States. You should persevere with these programs, expanding and supplementing them as nec-

---

182-2. *Bahá'í Prayers and Tablets for the Young* and *Bahá'í Education.*
182-3. BA, p. 88.

essary with others that you may judge desirable for the work of the Cause in the conditions of each of the widely diverse areas of your vast national territory. One of the most potent aids to the consolidation of local communities and Assemblies and the deepening of the faith of the believers, is the services of the Auxiliary Board members and their assistants. Here is an institution of the Faith, reaching into every locality, composed of firm believers who know the area they have to serve and are familiar with its problems and potentialities—an institution expressly designed to encourage and reinforce the work of the Spiritual Assemblies, to enthuse the believers, to stimulate them to study the Teachings and apply them in their lives—a body of Bahá'ís whose efforts and services will complement and support the work being done by your committees and by the Local Assemblies themselves in every sphere of Bahá'í endeavor. . . .

With loving Bahá'í greetings,
THE UNIVERSAL HOUSE OF JUSTICE

183
Call for a Vigorous Traveling Teaching Program
19 DECEMBER 1976

To all National Spiritual Assemblies

Dear Bahá'í Friends,

183.1 Now that we have passed the midway point of the Five Year Plan and through the grace of the Blessed Beauty the processes which it has set in motion in every continent are gaining momentum, we observe almost daily with awe and admiration how abundant are the blessings which surround the efforts of the devoted believers who are delivering His divine Message with dedication, enthusiasm and perseverance, and how infinitely vast and challenging are the possibilities for the future.

183.2 One of the activities which must be given greater attention, both nationally and internationally, is travel teaching to localities which are in need of consolidation and stimulation. While teaching projects of short duration, in terms of days and one or two weeks, are useful, the time has come for long-term teaching projects of three to six months and possibly more, as envisaged in the Five Year Plan, to be more vigorously prosecuted in every continent, so that more lasting results may accrue from these teaching trips and the rich harvest anticipated at the outset of the Plan may become a reality, imparting its inestimable benefits to the devoted laborers in His divine Vineyard.

183.3 It is hoped that those who will arise will be mostly self-supporting or supported by private deputization, and that the funds of the Faith, nationally for projects on the homefront, and internationally for projects in foreign fields, will be requested only when the individual concerned is regarded as well qualified and there is no other source of funds available to him.

183.4 We hope that National Assemblies whose assignments include international travel teaching projects will make a special appeal to the friends under their jurisdiction calling on them to meet this new challenge. If this urgent call, which is directed to all believers in every land, is answered with promptness and enthusiasm by those whose personal circumstances permit, great advances will be made towards fulfilling the hitherto sorely neglected teaching goals of the Plan. While the youth can through their creative resourcefulness and energetic labors effectively support the work ahead, believers from the older age groups can be a valuable asset in the achievement of the goals through their experience, knowledge and wisdom.

183.5 Our prayers are offered frequently at the Holy Shrines for the infinite blessings of the Abhá Beauty to reinforce every step you take in strengthening and broadening the foundations of the Faith in the areas entrusted to your care.

With loving Bahá'í greetings,
THE UNIVERSAL HOUSE OF JUSTICE

184

Message to the International Teaching Conference, Auckland, New Zealand—January 1977

JANUARY 1977

To the Friends assembled at the
International Teaching Conference in Auckland

Dearly loved Friends,

184.1 With hearts full of love and admiration for the followers of the Most Great Name in Australasia we send our warmest greetings to all assembled in this historic gathering in the heart of the Antipodes.

184.2 How great is your place in Bahá'í history! How bright are the prospects for the future of the Cause so lovingly nurtured for more than half a century by hundreds of stalwart steadfast believers, spiritual heirs of Hyde and Clara Dunn, who in direct response to the Tablets of the Divine Plan forsook their home and went to pioneer in Australia, and whose names, Shoghi Effendi wrote, were "graven in letters of gold" upon his heart.[1] In March 1951, when

---

184-1. Mr. and Mrs. Dunn left their home in San Francisco and sailed to Sydney, arriving on 18 April 1920. They remained in Australia until their deaths in 1941 and 1960, respectively. Shoghi Effendi appointed Mrs. Dunn a Hand of the Cause of God on 29 February 1952; Mr. Dunn was named a Hand of the Cause of God posthumously. For an account of the life and services of Mr. Dunn, see BW 9:593–97; for Mrs. Dunn, see BW 13:859–62.

in the entire Pacific area there was but one National Spiritual Assembly, the beloved Guardian predicted that "The prizes destined for the heroic warriors, battling for the Cause of Bahá'u'lláh throughout the Southern Hemisphere, and particularly Australasia, are glorious beyond compare. The assistance to be vouchsafed to them from on high in their struggle for its establishment, its recognition and triumph is ready to be poured forth in astonishing abundance."[2]

184.3 Now, twenty-five years later, the achievements are truly astounding. Beginning with the establishment of the National Spiritual Assembly of the Bahá'ís of New Zealand at Riḍván 1957, the number of National Assemblies has increased elevenfold; the Mashriqu'l-Adhkár of the Antipodes has been erected near Sydney; His Highness Malietoa Tanumafili II of Western Samoa has become the first reigning monarch to embrace the Cause of Bahá'u'lláh;[3] the number of Local Spiritual Assemblies now stands at over 360; and the number of localities where Bahá'ís reside in this vast oceanic area covering well-nigh one-eighth of the earth's surface is more than 1,800. These accomplishments doubtless have been a source of great joy to the immortal soul of Shoghi Effendi, whose esteem and affection for the followers of Bahá'u'lláh laboring for His Cause in the Antipodes was frequently expressed in glowing terms in his letters to the Assemblies and friends in Australasia.

184.4 Dear friends, we have now passed the midpoint of the Five Year Plan. You are met in the beautiful city of Auckland to take stock and to make plans for attaining the victories which will surely be yours.

184.5 The National Spiritual Assemblies of the New Hebrides and of the Marshall Islands are to be raised up next Riḍván; plans for the soon-to-be-erected Mashriqu'l-Adhkár of Samoa are in process; but although the goal of establishing Bahá'í centers totaling 2,188 is within easy reach, the Local Assembly goals assigned to each national community, totaling 613, need prompt and decisive attention. The divine assistance spoken of by the beloved Guardian in 1951 has ever been available, and is still "ready to be poured forth in astonishing abundance."[4] It is within your power during the coming year to win all assigned teaching goals, leaving the final year of the Plan for consolidation and the winning of supplementary victories.

184.6 This will be achieved, not by resting on laurels, but by manifesting those qualities of faith, judgment, vision, loyalty, courage and self-sacrifice which

---

184-2. LFG, p. 94.

184-3. For the announcement of His Highness's acceptance of the Bahá'í Faith, see message dated 7 May 1973 (no. 130); for his visit to the resting-place of Shoghi Effendi, see message dated 5 October 1976 (no. 177).

184-4. LFG, p. 96. For reports on the appointment of the architect for the House of Worship in Samoa, see message dated 15 May 1978 (no. 210); on the excavation of the construction site, see message dated 6 December 1979 (no. 242); for the dedication, see message dated August 1984 (no. 403).

earned the Guardian's praise so frequently in past decades. Let the valiant Australasian Bahá'í communities vie once more with their sister communities throughout the world for the palm of victory and maintain their position in the vanguard of the Army of Light.

Pioneers, traveling teachers and a fresh outpouring of funds are essential ingredients to the onward march of the Cause throughout Australia, New Zealand and the islands of the South Pacific. Let those who can offer their valued services to the teaching work arise without delay; let those who cannot travel or pioneer deputize those who can go in their stead.

We cherish the highest hopes for the success of your endeavors and it is our constant prayer that Bahá'u'lláh will shower His richest blessings and confirmations upon you.

THE UNIVERSAL HOUSE OF JUSTICE

185

Message to the International Teaching Conference, Bahia, Brazil—January 1977

JANUARY 1977

To the Followers of Bahá'u'lláh gathered
at the International Teaching Conference in Bahia, Brazil

Dearly loved Friends,

With joyous hearts we hail the convocation of this first of the twin Latin American Conferences closing the series of eight International Bahá'í Conferences held during the midway period of the Five Year Plan.

The ringing call of Bahá'u'lláh in His Most Holy Book to the Rulers of America and the Presidents of the Republics therein was followed after an interval of more than four decades by the revelation of 'Abdu'l-Bahá's Tablets of the Divine Plan in which the beloved Master stressed the importance of the Republics of the South American Continent.[1]

The first believer to respond to 'Abdu'l-Bahá's divine call was that star-servant of the Cause of Bahá'u'lláh, valiant, indomitable Martha Root, who in 1919 visited many important cities in South America.[2] Two years later Leonora Holsapple Armstrong, mother of the Bahá'ís of Brazil, settled in Bahia.[3] The

---

185-1. See PB, p. 63; see also CF, pp. 18–19, and TDP 6.11, 14.7.
185-2. Upon her passing in 1939 Martha Root was appointed a Hand of the Cause of God by Shoghi Effendi. See Mabel Garis, *Martha Root*, pp. 87–112.
185-3. Leonora Armstrong, a member of the Continental Board of Counselors from 1973 through 1980, died in Bahia, Brazil, on 17 October 1980. For an account of her life and service to the Faith, see BW 18:733–38.

teaching work in the continent progressed steadily to the point where, in 1937, the beloved Guardian launched his first Seven Year Plan paving the way for the raising in subsequent Plans of the institution of Bahá'u'lláh's Administrative Order in every one of its republics and in its islands. It was in the course of that first Seven Year Plan that 'Abdu'l-Bahá's beloved handmaid, May Maxwell, in 1940 won a martyr's crown when she laid down her life in Argentina, thereby adding further luster to the spiritual history of South America.4

185.4 How truly fitting, then, that this auspicious Conference have its venue in the city of Bahia, singled out for special mention by 'Abdu'l-Bahá in His Tablets of the Divine Plan.5 And how timely it is at this crucial point in the Five Year Plan that the friends gathered in Bahia from many lands prayerfully consider, and arise to prosecute expeditiously all measures aimed at achieving glorious victory in all goals of the Plan.

185.5 Noteworthy progress has been achieved in many fields of service throughout the South American Bahá'í community, particularly in attracting to the Cause large numbers of its indigenous peoples. But myriads of pure-hearted souls have not yet heard the clarion call of Bahá'u'lláh and hungrily await the spiritual nourishment that only His followers can give them.

185.6 Steps must be taken to attract members of every stratum of society to the divine circle of the Faith through effective proclamation and teaching. Greater utilization of radio broadcasts is necessary, not only to reach all levels of society but also to deepen the believers themselves. The valuable and dynamic services of Bahá'í youth must be multiplied in the fields of pioneering and travel teaching. A far wider dissemination of Bahá'í literature must be accompanied by a continuous program of translation of the Sacred Text into the major indigenous languages of the continent.

185.7 The continental goals for South America of 8,670 centers and 2,293 Local Spiritual Assemblies must be won, and may even be surpassed, for every country must achieve the goals assigned to it. This calls for the dedicated effort of every National Spiritual Assembly, every Local Spiritual Assembly and indeed every believer. All the divinely ordained instruments of the Administrative Order of Bahá'u'lláh must now unite in executing a symphony of victory in all the unfinished goals of the Five Year Plan, winning thereby the good pleasure of the Blessed Beauty.

185.8 Beloved friends, go forward with complete assurance that a continent so rich in spiritual promise, so diverse in its peoples and races, so fertile for the

---

185-4. In 1940 at the age of seventy, after a life of dedicated service, May Maxwell arose to pioneer and died shortly after her arrival in Buenos Aires. For an account of her life and service to the Faith, see BW 18:631–42.

185-5. See TDP 6.11, 14.7.

planting of the seeds of Bahá'u'lláh's Faith will yield a brilliant harvest for all who labor in that Divine Vineyard.

It is our fervent loving prayer at the Sacred Threshold that Bahá'u'lláh's 185.9 bountiful confirmations and richest blessings may be showered upon you.

THE UNIVERSAL HOUSE OF JUSTICE

186

Signature of Contract for Erection of the Seat of the Universal House of Justice
10 JANUARY 1977

To all National Spiritual Assemblies

JOYFULLY ANNOUNCE SIGNATURE AGREEMENT GENERAL CONTRACTOR FOR EREC- 186.1
TION SEAT UNIVERSAL HOUSE OF JUSTICE AT COST JUST OVER TWO MILLION
DOLLARS.[1] FIRST SHIPMENT FINISHED MARBLE DELIVERED SITE. LOVING GREETINGS.

UNIVERSAL HOUSE OF JUSTICE

187

Message to the International Teaching Conference, Mérida, Mexico—February 1977
FEBRUARY 1977

To the Followers of Bahá'u'lláh gathered at the
International Teaching Conference in Mérida, Mexico

Dearly loved Friends,

With joyous hearts and eager anticipation we send warmest greetings to 187.1 you the participants in the last of the eight great International Teaching Conferences marking the halfway point of the Five Year Plan.

The convening of this Conference in the Republic of Mexico, in the cap- 187.2 ital city of a state that was once an important part of a great Indian empire, provides a unique opportunity to initiate what may well become the widespread reawakening of a people whose ancestors more than 1,200 years ago developed one of the most brilliant pre-Columbian civilizations known to modern man. These present-day descendants, many of whom have already embraced the Faith of Bahá'u'lláh and who consider the Yucatán Peninsula

---

186-1. This general contract related to the initial phase of construction and totaled $2,250,000.

and the seacoast lowlands and rugged spine of mountains joining North and South America to be their homeland, are among the very people mentioned by 'Abdu'l-Bahá in His Tablets of the Divine Plan as having a great destiny once they have accepted His Father's Cause.¹ Here, too, and throughout Middle America, are those whose forefathers came from the Iberian Peninsula, Africa, and the Far East linking the Old with the New World.

187.3 Conscious of 'Abdu'l-Bahá's impassioned plea to promulgate the oneness of mankind to a spiritually impoverished humanity, a handful of itinerant Bahá'í teachers set forth four decades ago, traversed the land bridge connecting the two continents of the Western Hemisphere and carried the healing Message of Bahá'u'lláh to the Spanish-American Republics. Their dedicated efforts were rewarded when, in 1938, the first Local Spiritual Assembly in Latin America was formed in Mexico City. This initial triumph at the inception of the first of the teaching plans formulated by Shoghi Effendi spearheaded other victories leading to the formation of two, then of four Regional Spiritual Assemblies and ultimately to the establishment of National Spiritual Assemblies in each of the republics of Latin America and in the islands of the Caribbean.

187.4 Praiseworthy indeed were these achievements but the Bahá'í communities of Central America and the West Indies must not be content to rest on these laurels. The beloved Guardian during the last months of his precious life continually urged the friends of Latin America to pursue what he described as "the paramount task," the teaching work.² How much more does that injunction apply today! In less than thirty months, approximately 900 groups and isolated centers and over 400 Local Spiritual Assemblies must be added to those already existing in the mainland and island nations of Middle America!

187.5 To accomplish this challenging task, intensive effort to attract new believers, be they black, brown, red or white, from all strata of society, must be exerted. Hand in hand with this endeavor, particularly in local communities, goes the development of the distinctive character of Bahá'í life. Prompt attention must also be given to the acquisition of local Ḥaẓíratu'l-Quds and endowments; and the translation and publication of Bahá'í literature, especially in indigenous languages, must be accelerated.

187.6 Dear friends, if at the close of the Five Year Plan we are to witness the ensigns of victory lifted high, the wholehearted support of the followers of Bahá'u'lláh must be enlisted now and their energies systematically channeled into areas most in need. We cherish the hope that at this final Conference the friends will arise with enthusiasm and determination not only to win the remaining goals of the Plan but to carry out Shoghi Effendi's injunction to win the allegiance of members of the various tribes of American Indians to

---

187-1. See TDP 6.8.
187-2. From an unpublished letter.

the Cause, thereby hastening the period prophesied by the Master when the Indian peoples of America would become a source of spiritual illumination to the world.

187.7 Our hearts, our hopes and our prayers will be with you during all the days of your deliberations. May Bahá'u'lláh inspire each and every one of you.

THE UNIVERSAL HOUSE OF JUSTICE

188
Release of a Compilation on the Individual and Teaching
3 MARCH 1977

To all National Spiritual Assemblies

Dear Bahá'í Friends,

188.1 The cornerstone of the foundation of all Bahá'í activity is teaching the Cause. As 'Abdu'l-Bahá has categorically proclaimed in His Will and Testament, "the guidance of the nations and peoples of the world" is "the most important of all things," and "Of all the gifts of God the greatest is the gift of Teaching."[1]

188.2 The friends likewise are in varying degrees aware of the repeated exhortations found in the writings of our Faith that divine confirmations are dependent upon the active pursuit of the teaching work. In the words of the beloved Master, "confirmations from the unseen world are encompassing all those who deliver the divine Message." He further states, "Should the work of teaching lapse, these confirmations would be entirely cut off, since it is impossible for the loved ones of God to receive assistance unless they teach."[2]

188.3 While the friends are generally conscious of the vital importance of teaching, yet, because of their frailties, many for the most part lack confidence, and feel they do not know what course of action to follow, or how to bring their efforts to a conclusion. Since guidance on such fundamental issues comes from the writings of the Faith, we asked the Research Department to prepare a compilation of texts on the subject. This is now ready and a copy is enclosed.[3]

188.4 A study of the compilation will provide the friends with stimulating information on general guidelines to be followed by them when engaged in the teaching work. While many will be inspired, after reading the compilation, to cast aside their fears and misgivings and their sense of inadequacy, and will

---

188-1. WT, pp. 10, 25.
188-2. A translation from TABA 2:390 has been replaced by one from SWAB, sec. 209, pp. 264–65.
188-3. See CC 2:293–326.

arise to speak forth announcing the glad-tidings of the Kingdom to their fellowmen, many more will still be in need of loving education and more detailed guidance on the part of the institutions of the Faith, and patient and wise prodding before they are aroused to action. And since the primary purpose for which Local Spiritual Assemblies are established is to promote the teaching work, it is clear that every National Spiritual Assembly must give careful consideration to ways and means to encourage each Local Assembly under its jurisdiction to fulfill its principal obligation. For instance, Local Assemblies could be urged to organize special meetings when texts, such as those included in the compilation, would be studied. Furthermore, it is important that Local Assemblies share with the local friends stories of successes achieved by some of them, descriptions of effective presentations found useful by them, examples of various ways that a Bahá'í subject could be introduced to inquirers, or illustrations of methods which would enable the believer to relate the needs of society to our teachings. Such information and suggestions could be offered to the friends at Nineteen Day Feasts, through a local newsletter, or by any other means open to each Local Assembly. In all these contacts with the believers, each Local Spiritual Assembly should impress upon the friends the unique and irreplaceable role the individual plays in the prosecution of any Bahá'í undertaking. Quotations from the writings on this point, such as the following passage from one of the letters of Shoghi Effendi, should be repeatedly presented and explained to the friends:

188.4a   He [the individual believer] it is who constitutes the warp and woof on which the quality and pattern of the whole fabric must depend. He it is who acts as one of the countless links in the mighty chain that now girdles the globe. He it is who serves as one of the multitude of bricks which support the structure and insure the stability of the administrative edifice now being raised in every part of the world. Without his support, at once wholehearted, continuous and generous, every measure adopted, and every plan formulated, by the body which acts as the national representative of the community to which he belongs is foredoomed to failure. The World Center of the Faith itself is paralyzed if such a support on the part of the rank and file of the community is denied it. The Author of the Divine Plan Himself[4] is impeded in His purpose if the proper instruments for the execution of His design are lacking. The sustaining strength of Bahá'u'lláh Himself, the Founder of the Faith, will be withheld from every and each individual who fails in the long run to arise and play his part.[5]

188-4. 'Abdu'l-Bahá.
188-5. CF, pp. 130–31.

When the friends realize that the hosts of the Kingdom are waiting to rush forth and assist them, that others from their own ranks have arisen and have been successful, that everyone can find some effective method of teaching according to his own particular capacities and talents, they will then no doubt arise with greater confidence to take the first step, and this, we know, will be aided and guided from on high, for the very act of striving to respond to God's call will bring in its wake countless divine blessings. 188.5

It is the hope and prayer of the Universal House of Justice that each National Spiritual Assembly will do its utmost to constantly encourage the friends to participate in what Shoghi Effendi calls "the most essential, the most urgent of all our obligations," and what must be "the dominating passion of our life," and follow the example of the Apostles of Christ who, as testified by 'Abdu'l-Bahá, "forgot themselves and all earthly things, forsook all their cares and belongings, purged themselves of self and passion . . . till at last they made the world another world, illumined the surface of the earth and even to their last hour proved self-sacrificing in the pathway of that beloved One of God. . . . Let them that are men of action follow in their footsteps!"[6] 188.6

With loving Bahá'í greetings,
THE UNIVERSAL HOUSE OF JUSTICE

189
Consolidation of Local Spiritual Assemblies; Formation at Any Time during Riḍván
6 MARCH 1977

To National Spiritual Assemblies

Dear Bahá'í Friends,

The establishment and strong growth of Local Spiritual Assemblies is one of the most fundamental requirements for the spread of the Message of Bahá'u'lláh, the development of Bahá'í community life and the emergence of a transformed society. This theme has been made one of the central goals of the Five Year Plan, and National Spiritual Assemblies, aided by their committees, have been making strenuous efforts to establish new Local Spiritual Assemblies and to consolidate those which have lapsed or are in need of strengthening. 189.1

Already a number of specific steps have been taken by the Universal House of Justice to assist National Spiritual Assemblies towards the attainment of these objectives. The most far-reaching of these steps is the authority given to members of the Auxiliary Boards to appoint assistants whose primary aim is 189.2

188-6. BA, p. 42, 68; WT, pp. 10–11.

to stimulate and assist the believers to bring into being and to consolidate Local Spiritual Assemblies in all localities where nine or more Bahá'ís reside, and to advise and assist these Assemblies in the performance of their God-given duties. The effects of the appointment of assistants by Auxiliary Board members are beginning to appear and will undoubtedly bear more and more fruit as the months pass.

189.3 Nevertheless, it is of concern to the Universal House of Justice that, in spite of the efforts of the National Spiritual Assemblies and the Auxiliary Boards and of all the believers who are laboring in the field, there are many areas in which there are communities of nine or more believers who are left, for year after year, without the blessing of the divine institution of a Local Spiritual Assembly. This is a phenomenon of the present stage of the spread of the Faith where there has been a rapid acceptance of the Message of Bahá'u'lláh by people who, because of factors such as illiteracy, unfamiliarity with the concepts of Bahá'í administration, or an attitude to the calendar and the passage of time that is different from that of city-dwellers, fail to reelect their Spiritual Assemblies on the First Day of Riḍván. National Assemblies are striving to send pioneers and traveling teachers to deepen such believers in their understanding of the teachings and administrative principles of the Faith, but often the localities in which they dwell are remote from the other friends or difficult to reach, and there are, in any case, too few well-grounded believers who can be sent on such projects.

189.4 Not wishing such communities to be deprived of the bounty and experience of having Local Spiritual Assemblies, we have decided that, in such cases, when the local friends fail to elect their Spiritual Assembly on the First Day of Riḍván, they should do so on any subsequent day of the Riḍván Festival.[1] This is not a general permission to all Local Spiritual Assemblies; it is intended only for those which are affected by factors such as those mentioned above, and it is for your Assembly to decide the areas or Assemblies in your country to which it will apply. The aim is still to so consolidate all communities that they will elect their Assemblies regularly on the First Day of Riḍván.

189.5 It is hoped that this decision will do much to help you consolidate the communities throughout your area of jurisdiction. It does not change the permission given at the beginning of the Five Year Plan for new Assemblies, being formed for the first time, to be formed at any time during the year.[2]

189-1. Referring to the provision for forming Local Spiritual Assemblies in certain localities anytime during the Riḍván Festival, the Universal House of Justice, in a letter dated 19 February 1979 to all National Spiritual Assemblies (no. 219), explained that, "until further notice, it is permissible to follow this procedure every Riḍván where it is called for." This ruling remained in effect until Riḍván 1997.

189-2. See message no. 141.

We shall supplicate the Blessed Perfection to reinforce with the Hosts of 189.6
Heaven every effort you make towards this vital objective.

With loving Bahá'í greetings,
THE UNIVERSAL HOUSE OF JUSTICE

190
Riḍván Message 1977
24 MARCH 1977

To all National Spiritual Assemblies

REJOICE OUTSTANDING EVENTS AND ACHIEVEMENTS MARKING MID YEAR FIVE 190.1
YEAR PLAN: PUBLICATION IN ENGLISH SELECTION WRITINGS BLESSED BÁB OPENING
TO EYES WESTERN FOLLOWERS FAITH A PRICELESS TREASURY HIS IMMORTAL UTTERANCES, A BOUNTY WHICH CANNOT FAIL DRAW HEARTS EVER NEARER YOUTHFUL
MARTYR-PROPHET;[1] SUCCESSFUL HOLDING EIGHT INTERNATIONAL CONFERENCES,
BLAZONING NAME CAUSE GOD BEFORE A RECEPTIVE PUBLIC, KNITTING MORE
CLOSELY TIES LINKING FRIENDS ALL LANDS, RESULTING UPSURGE INTENSE ACTIVITY
TEACHING PIONEERING, CONFERRING AWARENESS URGENT CHALLENGE PRESENTED
BY GOALS PLAN THIS CRITICAL PERIOD;[2] PROFOUNDLY SIGNIFICANT VISIT TO
RESTING-PLACE BELOVED GUARDIAN BY FIRST REIGNING MONARCH TO ACCEPT
FAITH BAHÁ'U'LLÁH;[3] DEPARTURE FORMER TENANTS HOUSE 'ABDU'LLÁH PÁSHÁ
ENABLING FAITH OBTAIN POSSESSION RECENTLY PURCHASED HOLY PLACE INITIATE
PROCESS RESTORATION PREPARATION EVENTUAL OPENING TO VISITS BY PILGRIMS;[4]
ARRIVAL HAIFA FIRST FOUR CONSIGNMENTS MARBLE AND INITIATION ACTUAL
CONSTRUCTION SEAT UNIVERSAL HOUSE JUSTICE MOUNT CARMEL; APPOINTMENT
THREE ADDITIONAL COUNSELORS NORTH CENTRAL AMERICA AUSTRALASIA; ATTAINMENT GOAL SETTLEMENT MIDWAY POINT PLAN MAJORITY PIONEERS CALLED FOR
DURING FIRST PHASE ACCOMPANIED GREAT OUTFLOW INTERNATIONAL TRAVELING
TEACHERS; DRAMATIC RESURGENCE TEACHING WORK CRADLE FAITH BY INDIVIDUAL
BELIEVERS UNDER LOCAL PLANS; FINALLY, ELECTION THIS RIḌVÁN SIX NEW PILLARS
UNIVERSAL HOUSE JUSTICE, NATIONAL SPIRITUAL ASSEMBLIES OF UPPER VOLTA IN
AFRICA, OF THE FRENCH ANTILLES IN THE CARIBBEAN, OF SURINAM AND FRENCH
GUIANA IN SOUTH AMERICA, OF THE MARSHALL ISLANDS AND OF THE NEW

190-1. See *Selections from the Writings of the Báb* (1976).
190-2. The eight conferences were held between July 1976 and February 1977 in Helsinki, Finland; Anchorage, Alaska; Paris, France; Nairobi, Kenya; Hong Kong; Auckland, New Zealand; Bahia, Brazil; and Mérida, Mexico.
190-3. For the announcement of the visit of His Highness Malietoa Tanumafili II of Western Samoa to Shoghi Effendi's resting-place, see message dated 5 October 1976 (no. 177).
190-4. See message dated 9 January 1975 (no. 154) regarding the acquisition of the House of 'Abdu'lláh Páshá. For an account of the significance of the House of 'Abdu'lláh Páshá, see message no. 157.

HEBRIDES IN PACIFIC OCEAN, AND OF GREECE IN EUROPE, RAISING TOTAL NUMBER NATIONAL SPIRITUAL ASSEMBLIES TO 123 TO TAKE PART IN FOURTH ELECTION UNIVERSAL HOUSE OF JUSTICE IN HOLY LAND DURING RIḌVÁN 1978.

190.2 NATIONAL CONVENTIONS IN 1978 WILL BE ON WEEKEND PRECEDING OR FOLLOWING 23 MAY FEAST DECLARATION BÁB. CALL FOR FORMATION AT THAT TIME SIX MORE NATIONAL ASSEMBLIES: BURUNDI AND MAURITANIA IN AFRICA, THE BAHAMAS IN AMERICA, OMAN AND QATAR IN ASIA, AND THE MARIANA ISLANDS IN THE PACIFIC.

190.3 PRESENT RATE GROWTH COMMUNITY PROSPECT ACCELERATION PROCESS ENTRY BY TROOPS ITS SPREAD NEW AREAS IMPEL US STRENGTHEN STILL FURTHER THE AUXILIARY BOARDS WHOSE SERVICES SO VITAL SOUND DEVELOPMENT COMMUNITY. ANNOUNCE AUTHORIZATION INCREASE MEMBERSHIP BOARDS BY 297 RAISING TOTAL TO 675 OF WHICH 279 ARE AUXILIARY BOARD MEMBERS FOR PROTECTION AND 396 FOR PROPAGATION OF FAITH.

190.4 IN EARLY DAYS OF JUNE 1877 BAHÁ'U'LLÁH LEFT CITY 'AKKÁ AND TOOK UP RESIDENCE IN MAZRA'IH. TO MARK CENTENARY THIS TERMINATION CONFINEMENT ANCIENT BEAUTY WITHIN WALLS PRISON CITY WE CALL UPON HIS FOLLOWERS ALL LANDS DEVOTE NINETEEN DAY FEAST OF NÚR[5] COMMEMORATION HISTORIC EVENT, REDEDICATING THEMSELVES URGENT TASKS BEFORE THEM, SO THAT PENT-UP ENERGIES HIS PRECIOUS FAITH MAY BE RELEASED TO REACH EVER GREATER NUMBER SEEKING SOULS IN EVER WIDER CIRCLE THEIR FELLOWMEN.

190.5 GREATEST CHALLENGE FACING FOLLOWERS BAHÁ'U'LLÁH LAST TWO YEARS PLAN IS IN FIELDS EXPANSION CONSOLIDATION. TREMENDOUS UPSURGE NEEDED IN SERVICES INDIVIDUAL BELIEVERS ON WHOSE DEEDS ULTIMATELY ALL PROGRESS DEPENDS. MOMENTUM GENERATED BY INTERNATIONAL CONFERENCES MUST BE ACCELERATED WITHOUT DELAY AND SPIRIT RELEASED MUST PERMEATE ALL COMMUNITIES. GREAT INCREASE MUST TAKE PLACE IN ENTHUSIASTIC TEACHING CARRIED OUT WITH CONFIDENCE, IMAGINATION AND PERSEVERANCE BY YOUNG AND OLD, RICH AND POOR, LEARNED AND ILLITERATE, WHETHER AT HOME OR TRAVELING. PARTICULARLY CALL UPON BAHÁ'Í WOMEN, WHOSE CAPACITIES IN MANY LANDS STILL LARGELY UNUSED, AND WHOSE POTENTIAL FOR SERVICE CAUSE SO GREAT, TO ARISE AND DEMONSTRATE IMPORTANCE PART THEY ARE TO PLAY IN ALL FIELDS SERVICE FAITH.

190.6 BOUNTIES IN ABUNDANCE WAITING DESCEND FROM SUPREME CONCOURSE. THAT THE FRIENDS OF GOD WILL NOW SURGE AHEAD WITH RESOLUTE RADIANT SPIRITS IN EVERY CONTINENT ISLANDS OF THE SEAS, TO BRING MESSAGE OF BAHÁ'U'LLÁH TO WAITING SOULS WIN THEIR ALLEGIANCE HIS CAUSE, ENSURING OVERWHELMING VICTORY PLAN TO WHICH THEY ARE NOW COMMITTED, IS OUR HIGH HOPE AND ARDENT PRAYER AT SACRED THRESHOLD.

THE UNIVERSAL HOUSE OF JUSTICE

190-5. The Feast of Light, 5 June 1977.

1974–1979 • THE FIVE YEAR PLAN

Auxiliary Boards—Riḍván 1977

190.7

		Former Number	Present Increase	New Total
Africa				
Protection:	Northern	5	–	5
	Western	11	13	24
	Central and East	13	23	36
	Southern	7	9	16
		36		81
Propagation:	Northern	5	–	5
	Western	14	19	33
	Central and East	19	17	36
	Southern	16	9	25
		54		99
Western Hemisphere				
Protection:	North America	9	9	18
	Central America	11	7	18
	South America	16	2	18
		36		54
Propagation:	North America	18	9	27
	Central America	16	11	27
	South America	38	25	63
		72		117
Asia				
Protection:	Western	9	9	18
	South Central	6	30	36
	Northeastern	3	15	18
	Southeastern	9	18	27
		27		99
Propagation:	Western	18	18	36
	South Central	30	6	36
	Northeastern	15	3	18
	Southeastern	27	–	27
		90		117
Australasia				
Protection:		9	9	18
Propagation:		18	9	27
Europe				
Protection:		9	18	27
Propagation:		27	9	36
TOTAL		**378**	**297**	**675**
Total Protection		117	162	279
Total Propagation		261	135	396

191
Call for Pioneers
26 May 1977

To all National Spiritual Assemblies

Dear Bahá'í Friends,

191.1 We are gratified that so many pioneers have settled in the territories scheduled under the two pioneer calls of the Five Year Plan.

191.2 We have just completed a review of the current status of the Plan in consultation with the International Teaching Center, and it became obvious that a new pioneer call is needed. A new list of pioneer assignments is enclosed, from which you will see that 462 pioneers must arise to settle in specified territories. These pioneers should proceed to their posts as soon as possible to lend their support to the vital local teaching work, in time to participate in winning the remaining goals and in consolidating the work prior to the conclusion of the Plan.

191.3 In view of the difficulties that nationals of certain countries are experiencing in settling in the goal countries assigned to their national community, some adjustments, and in certain cases deletions, have been made in the few unfilled goals. Therefore this pioneer call includes all outstanding goals. Any National Assembly finding that its prior unfilled assignments have not been carried forward should realize that alternative solutions have been found. However, if at the time National Assemblies receive this list some of their prospective pioneers are well advanced in their preparations to proceed to their posts, nothing should be done to prevent their going forward with their plans. We hope that the pioneers already in the field will remain at their posts and that the National Spiritual Assemblies will encourage them, and if necessary provide the means required to enable them to continue their worthy labors in their respective fields of service.

191.4 It is our fervent hope and prayer that this new outflow of pioneers will provide a stimulus to the teaching work throughout the world. We realize the heavy commitments that many National Spiritual Assemblies called upon to provide pioneer manpower have in respect to their homefront goals, but in view of the worldwide needs of the Plan, we trust that every effort will be made to respond to this call, and not only pave the way for a triumphal conclusion of the Plan in 1979, but reinforce the efforts now being exerted to lay firm foundations for the spread and development of the precious Faith of God in every land.

With loving Bahá'í greetings,
THE UNIVERSAL HOUSE OF JUSTICE

192
Commemoration of the Centenary of the Termination of Bahá'u'lláh's Confinement in 'Akká
JUNE 1977

IN EARLY DAYS OF JUNE 1877 BAHÁ'U'LLÁH LEFT CITY 'AKKÁ AND TOOK UP RESIDENCE IN MAZRA'IH. TO MARK CENTENARY THIS TERMINATION CONFINEMENT ANCIENT BEAUTY WITHIN WALLS PRISON CITY WE CALL UPON HIS FOLLOWERS ALL LANDS DEVOTE NINETEEN DAY FEAST OF NÚR[1] COMMEMORATION HISTORIC EVENT, REDEDICATING THEMSELVES URGENT TASKS BEFORE THEM, SO THAT PENT-UP ENERGIES HIS PRECIOUS FAITH MAY BE RELEASED TO REACH EVER GREATER NUMBER SEEKING SOULS IN EVER WIDER CIRCLE THEIR FELLOWMEN.

The Universal House of Justice, Riḍván 1977

In the early afternoon of Saturday, 11 June 1977, the pilgrims and the friends serving at the World Center made their way to Mazra'ih to visit the Mansion which 'Abdu'l-Bahá had rented a century ago for the use of Bahá'u'lláh, His first residence after leaving the prison city of 'Akká.[2] While there, each was privileged to visit and offer prayers in the very room occupied by the Blessed Beauty; later they repaired to the gardens at Bahjí.

Meanwhile the Hands of the Cause present in the Holy Land, the members of the Universal House of Justice and the Counselor members of the International Teaching Center were paying their respects to the memory of Bahá'u'lláh at the House of 'Abbúd in 'Akká and at the Garden of Riḍván. The rooms Bahá'u'lláh had occupied were visited and prayers of thanksgiving for His release from confinement were offered in these Holy Places. They then journeyed north to Mazra'ih for prayers at that Holy Spot, and afterwards joined the other friends in the Ḥaram-i-Aqdas at Bahjí for the formal program of the Commemoration of this great event in the history of the Heroic Age of the Cause.[3]

Prayers were recited; 'Abdu'l-Bahá's account of the end of Bahá'u'lláh's confinement and Shoghi Effendi's narrative from *God Passes By* were read; and finally, just as the sun was casting its last light over the Mediterranean, the friends made their way in the utmost reverence to the Most Holy Shrine for the chanting of the Tablet of Visitation.

192-1. The Feast of Light, 5 June 1977.
192-2. For an account of the commemoration at the Bahá'í World Center of the centenary of the termination of Bahá'u'lláh's confinement in 'Akká, see BW 17:64.
192-3. Ḥaram-i-Aqdas (the Most Holy Court) is a designation Shoghi Effendi gave to the northwestern quadrant of the garden surrounding the Shrine of Bahá'u'lláh in Bahjí.

192.5 The remembrance of God and His praise, and the glory of God and His splendor, rest upon Thee, O Thou Who art His Beauty! I bear witness that the eye of creation hath never gazed upon one wronged like Thee. Thou wast immersed all the days of Thy life beneath an ocean of tribulations. At one time Thou wast in chains and fetters; at another Thou wast threatened by the sword of Thine enemies. Yet, despite all this, Thou didst enjoin upon all men to observe what had been prescribed unto Thee by Him Who is the All-Knowing, the All-Wise.

Bahá'u'lláh[4]

193
Murder of an Iranian Homefront Pioneer
14 June 1977

The National Spiritual Assembly of the Bahá'ís of the United States

Dear Bahá'í Friends,

193.1 The Universal House of Justice has asked us to send you the enclosed article for the *Bahá'í News* concerning a recent martyrdom in Persia. A photograph of Mr. Rúḥu'lláh Taymúrí is also enclosed.

With loving Bahá'í greetings,
Department of the Secretariat

193.2 The National Spiritual Assembly of the Bahá'ís of Iran has reported that a 37 year old man, Mr. Rúḥu'lláh Taymúrí-Muqaddam, a steadfast pioneer for some twenty years in Fáḍilábád in the Province of Gurgán, has been martyred at the hands of misguided people of that village. Some members of a fanatical element of the area had been plotting against the Faith and had collected money to finance a large-scale attack on the Bahá'ís. In starting the campaign they went to the house of Taymúrí and attacked him and his sister. The blows of a knife and other weapons caused the death of Mr. Taymúrí and serious injury to his younger sister, Miss Parvín Taymúrí.

193.3 Fortunately the authorities in Iran have arrested the assassins and they now await trial. At present the area is calm and the Government has the situation under control.

193.4 The National Spiritual Assembly reports that it is confident that this persecution in the cradle of the Faith will attract divine confirmations and will bring new victories to the steadfast and devoted believers in the land of Bahá'u'lláh.

192-4. BP, p. 232.

194

Collaboration between Auxiliary Board Members and National and Regional Teaching Committees
6 July 1977

To all National Spiritual Assemblies

Dear Bahá'í Friends,

It has become apparent that in some areas the progress of the teaching work requires closer collaboration between Auxiliary Board members and National or Regional Teaching Committees than heretofore. Following consultation with the International Teaching Center on the matter, we have concluded that the possibilities provided by the present policy are adequate and that where a lack of collaboration has been felt it has arisen from an insufficiently full and frequent exchange of information between the institutions.

While the members of the Auxiliary Boards and their assistants should never attempt to direct the work of committees or become involved in the administrative work associated with the committees' functions, it is absolutely vital that they be kept fully informed of the committees' activities and plans and their hopes for the work in the area. Only then can the members of the Auxiliary Boards be confident that the services to which they are exhorting the believers and the projects in which they are encouraging them are in harmony with the overall plans and objectives of the National Spiritual Assembly and its committees.

The existing policy and the reasons for it were conveyed to the Continental Boards of Counselors and all National Spiritual Assemblies in our letter of 1 October 1969, a copy of which is attached.[1] It should be noted that under this policy it is permissible and highly desirable to have a direct and regular exchange of information between the committees and the Auxiliary Board members. Moreover, at the outset of the work of the year or at times during the year when new plans are being evolved, it is often helpful to arrange for consultations to be held between the Auxiliary Board members and the National or Regional Teaching Committees before such plans are finalized.

We are confident that a greater awareness of the importance of close collaboration between the two arms of the Administrative Order and of the ways available to achieve this will lead to a much-needed intensification of the teaching work in every land.

With loving Bahá'í greetings,
THE UNIVERSAL HOUSE OF JUSTICE

194-1. See message no. 69.

195
Reconstruction of Society
21 August 1977

To an individual Bahá'í

Dear Bahá'í Friend,

195.1 The Universal House of Justice has studied your long letter of 19 May 1977. With many of your observations it thoroughly agrees; others, it believes are founded on erroneous information, on an inaccurate assessment of the current status of the Bahá'í community, or on misconceptions about the objectives towards which it is working. The House of Justice does not have the time which would be required to formulate a detailed reply to all the various points in your letter. It reaffirms, however, the decisions conveyed to your National Spiritual Assembly in its letter of 2 December 1976, and has instructed us to add the following comments.

195.2 Mankind's response to the Message of Bahá'u'lláh has been dangerously, one might say disastrously, slow. From the earliest days it has been brought to the notice of leaders and scholars, but few of these, very few, have rallied to its support. The most profound and most widespread response has been from the middle classes and indeed from the poor, the unlettered, the deprived and the suffering. But, as the Guardian's secretary wrote on his behalf on 20 June 1942,

195.2a That is perhaps what is most glorious about our present activities all over the world, that we, a band not large in numbers, not possessing financial backing or the prestige of great names, should, in the name of our beloved Faith, be forging ahead at such a pace, and demonstrating to future and present generations that it is the God-given qualities of our religion that are raising it up and not the transient support of worldly fame and power. All that will come later, when it has been made clear beyond the shadow of a doubt that what raised aloft the banner of Bahá'u'lláh was the love, sacrifice, and devotion of His humble followers and the change that His teachings wrought in their hearts and lives.

195.3 Already the situation is changing, and larger numbers of believers are occupying positions of eminence and distinction in the world, but, in comparison with the overwhelming majority of the Bahá'ís, they are still a small handful. The process of changing the hearts and lives is also a gradual one, but while we should strive to hasten it, we should not let the problems dismay us. On 5 July 1947 the Guardian's secretary wrote on his behalf to an individual believer:

The primary reason for anyone becoming a Bahá'í must of course be because he has come to believe the doctrines, the teachings and the Order of Bahá'u'lláh are the correct thing for this stage in the world's evolution. The Bahá'ís themselves as a body have one great advantage: they are sincerely convinced Bahá'u'lláh is right; they have a plan, and they are trying to follow it. But to pretend they are perfect, that the Bahá'ís of the future will not be a hundred times more mature, better balanced, more exemplary in their conduct, would be foolish. 195.3a

The Universal House of Justice is aware of the magnitude of the problems that the Bahá'í communities face, but as the response to the Message of Bahá'u'lláh increases and as the Bahá'í community throughout the world shows its ability to overcome these problems, the attention of men and women of every stratum of society will increasingly be drawn to the Faith. The most urgent need now—so late is the hour—is for the Bahá'ís to spread the Message, while they are still able to do so, to the largest possible number of their fellow human beings, simultaneously expanding and consolidating the Bahá'í community as quickly as they can with the resources at their disposal. As mankind passes through the darkest phase of its history, the Bahá'í community will have to face not only entry by troops, which it is now experiencing, but, before too long, mass conversion. 195.4

The first step in the reconstruction of human society is for individuals to accept Bahá'u'lláh as the Manifestation of God for this age and to begin to strive, as well as they can, to follow His Teachings in their individual and in their communal lives. Conversion is but the first step, yet it is the essential one. Without it no amount of expertise or scientifically based knowledge will have a lasting effect, because the fundamental motivating and sustaining power will be lacking. 195.5

As the Bahá'í community grows it will acquire experts in numerous fields— both by Bahá'ís becoming experts and by experts becoming Bahá'ís. As these experts bring their knowledge and skill to the service of the community and, even more, as they transform their various disciplines by bringing to bear upon them the light of the Divine Teachings, problem after problem now disrupting society will be answered. In such developments they should strive to make the utmost use of non-Bahá'í resources and should collaborate fully with non-Bahá'ís who are working in the same fields. Such collaboration will, in the long run, be of far more benefit than any attempt now to treat such scientific endeavors as specifically Bahá'í projects operating under Bahá'í institutions and financed by investment of Bahá'í funds. 195.6

Paralleling this process, Bahá'í institutional life will also be developing, and as it does so the Assemblies will draw increasingly upon scientific and expert 195.7

knowledge—whether of Bahá'ís or of non-Bahá'ís—to assist in solving the problems of their communities.

195.8 In time great Bahá'í institutions of learning, great international and national projects for the betterment of human life will be inaugurated and flourish.[1]

195.9 The Bahá'í work for the reconstruction of human society can thus be seen to comprise three streams: the most fundamental is the spreading of the Word of God, the winning of the allegiance of ever-greater numbers of men and women to the Cause of Bahá'u'lláh and the establishment of the Bahá'í Administrative Order; concurrent with this is the contribution to human advancement and to the progress of the Bahá'í community made by individual Bahá'ís in the pursuit of their daily work; and then there are the projects and institutions for human advancement launched and operated by Bahá'í Spiritual Assemblies as their resources grow and the range of their activities expands. It is for the Universal House of Justice to direct the energies of the believers in these various channels and to make known what activities are timely and have priority. It considers that the establishment of an International Human Development Center now as a Bahá'í-affiliated institution would be untimely and ill-advised.

195.10 The House of Justice assures you of its prayers for the confirmation of your endeavors on behalf of the Faith and in your professional work.

With loving Bahá'í greetings,
DEPARTMENT OF THE SECRETARIAT

195-1. For the call to the Bahá'í world to initiate activities intended to foster social and economic development, see message dated 20 October 1983 (no. 379).

1974–1979 • THE FIVE YEAR PLAN

196

Ceremony for Laying the Foundation Stone of the Mother Temple of the Indian Subcontinent

10 OCTOBER 1977

To the National Spiritual Assembly of the Bahá'ís of India

OVERJOYED REMOVAL OBSTACLES USE TEMPLE SITE.[1] WELCOME PRESENCE AMATU'L-BAHÁ RÚḤÍYYIH KHÁNUM IN YOUR MIDST OCCASION WOMEN'S CONFERENCE ENABLING YOU HOLD BEFITTING CEREMONY MARKING INITIATION PROJECT CONSTRUCTION MOTHER TEMPLE INDIAN SUBCONTINENT.[2] CALLING ON AMATU'L-BAHÁ REPRESENT HOUSE JUSTICE MOMENTOUS OCCASION LAY FOUNDATION STONE HISTORIC EDIFICE. FERVENTLY PRAYING NOBLE INSTITUTION SOON TO BE REARED YOUR SOIL WILL ATTRACT ADDED DIVINE BLESSINGS UPON COMMUNITY WHOSE TEACHING SUCCESS STANDS UNEQUALED ENTIRE BAHÁ'Í WORLD.

196.1

UNIVERSAL HOUSE OF JUSTICE

197

Message to Women's Conference, New Delhi

13 OCTOBER 1977

To the Bahá'í Women's Conference in New Delhi, India

WITH UTMOST JOY HAIL BAHÁ'Í WOMEN'S CONFERENCE GRACED PRESENCE AMATU'L-BAHÁ RÚḤÍYYIH KHÁNUM AS ANOTHER MAJOR STEP IN ASIA LEADING TO FULL RECOGNITION NOBLE STATION FULFILLMENT THEIR GREAT POTENTIALITIES THEIR SIGNIFICANT ROLE PROMOTION CAUSE IMMENSE RESPONSIBILITIES THEY ARE CALLED UPON TO DISCHARGE IN COOPERATION WITH THEIR BRETHREN IN BUILDING NEW CIVILIZATION SHELTERING ALL MANKIND. ARDENTLY PRAYING SACRED THRESHOLD THIS GATHERING MAY BECOME LANDMARK PROGRESS FAITH ASIA STRENGTHEN WING HUMANITY SO TENDERLY ENCOURAGED BY BLESSED BEAUTY ENABLE COMMUNITIES CONTINENT SOAR HEIGHTS GLORIOUS VICTORIES.

197.1

UNIVERSAL HOUSE OF JUSTICE

196-1. During the Ten Year World Crusade, twenty-two acres of land in New Delhi were acquired as a site for a House of Worship. In the process of demarcating the boundaries, it was discovered that the land was part of an ancient village named Bahapur, meaning "the Abode of Bahá." In Hindustani, a dialect of Hindi, *Bahá* means "a channel" or "a water course." By a happy coincidence, *Bahá* is Arabic for "light," "splendor," or "glory" and is a form of Bahá'u'lláh's name. Later the government requisitioned the property for use as a greenbelt area. After several years of constant negotiation by the National Spiritual Assembly of India, and upon viewing the beautiful design for the Temple, the government agreed to release the entire twenty-two acres.

196-2. See message of 13 October 1977 (no. 177).

198

Appointment of Architects for Mother Temple of the Indian Subcontinent and the Restoration of the House of 'Abdu'lláh Páshá

14 October 1977

To all National Spiritual Assemblies

198.1 HAVE MUCH JOY IN ANNOUNCING APPOINTMENT TWO DISTINGUISHED BAHÁ'Í ARCHITECTS FROM CRADLE FAITH: FARÍBURZ ṢAHBÁ AS ARCHITECT MASHRIQU'L-ADHKÁR INDIA, AND RIḌVÁNU'LLÁH ASHRAF AS ARCHITECT FOR RESTORATION SACRED RESIDENCE BELOVED MASTER KNOWN AS HOUSE OF 'ABDU'LLÁH PÁSHÁ.[1] SUPPLICATING DIVINE CONFIRMATIONS THESE TWO HISTORIC ENTERPRISES.

Universal House of Justice

199

Formation of New and Lapsed Assemblies throughout the Year

20 November 1977

To all National Spiritual Assemblies

Dear Bahá'í Friends,

199.1 The Universal House of Justice has decided that during the last year of the Five Year Plan, i.e., from April 21, 1978 until April 20, 1979 inclusive, Local Spiritual Assemblies being established for the first time, as well as lapsed Assemblies which achieve adequate strength to regain their Assembly status, may be formed at any time during that year.[1] This means that Local Assem-

---

198-1. For the announcement of the acquisition of the House of 'Abdu'lláh Páshá, see message dated 9 January 1975 (no. 154); for an account of its significance, see message dated 4 March 1975 (no. 157).

199-1. It had long been the practice to permit, during the final year of a teaching plan, the formation or re-formation of any Local Spiritual Assembly at *any* time during that year. This practice is confirmed by this letter for the Five Year Plan.

With respect to Spiritual Assemblies that are being formed for the first time, the Universal House of Justice gave a more general permission in its letter dated Naw-Rúz 1974 launching the Five Year Plan (no. 141), saying that such Local Spiritual Assemblies "are to be formed whenever there are nine or more adult believers in the relevant area; thereafter they must be elected or declared at Riḍván." In a message dated 19 February 1979 (no. 219), the Universal House of Justice informed all National Spiritual Assemblies that this provision would continue to apply after the completion of the Five Year Plan. This ruling remained in effect until Riḍván 1997.

Special permission for the formation of certain Local Spiritual Assemblies at any time during the twelve days of Riḍván was given in a message dated 6 March 1977 (no. 189) and extended in the message dated 19 February 1979 (no. 219).

blies formed at Riḍván 1979 will not be counted towards the fulfillment of the goals of the Five Year Plan.

It is the hope of the Universal House of Justice that this information will enable you to plan your teaching activities intelligently and realistically over the period of time separating us from the end of the Plan, and to intensify your efforts in order to achieve maximum results.

The Universal House of Justice will offer prayers at the Holy Shrines that the process of forming firmly grounded Local Spiritual Assemblies, which is one of the vital goals of the Five Year Plan, will be pursued with outstanding success through the dedicated efforts of the friends in every land.

With loving Bahá'í greetings,
DEPARTMENT OF THE SECRETARIAT

200

Appointment of New Counselor in Southeastern Asia
25 NOVEMBER 1977

To all National Spiritual Assemblies

Dear Bahá'í Friends,

The Universal House of Justice announces with pleasure that it has appointed a new Counselor in the zone of Southeastern Asia, Mr. Inparaju Chinniah of Malaysia. Mr. Chinniah replaces Mr. Firaydún Mítháqíyán, who ceased to be a Counselor upon leaving the zone and pioneering to Korea, where he is continuing his devoted services to the Cause of God.

With loving Bahá'í greetings,
DEPARTMENT OF THE SECRETARIAT

201

Announcement of Initiation of Broadcasting by First Bahá'í Radio Station
15 DECEMBER 1977

To all National Spiritual Assemblies

REJOICE ANNOUNCE INITIATION FULL TIME BROADCASTING FIRST RADIO STATION BAHÁ'Í WORLD DECEMBER 12 IN OTAVALO, ECUADOR. HAIL VISION LABORS ASSEMBLY COMMUNITY ECUADOR IN ACHIEVING THIS MILESTONE BAHÁ'Í PROCLAMATION TEACHING DEEPENING. OFFERING PRAYERS SACRED THRESHOLD BAHÁ'Í RADIO ECUADOR WILL FULFILL ITS PROMISE AS LANDMARK CAUSE AND SERVICE PROGRESS PEOPLES LATIN AMERICA.

UNIVERSAL HOUSE OF JUSTICE

202

Message of Greetings to Hemispheric Bahá'í Radio-Television Conference

15 December 1977

To the Hemispheric Radio-Television Conference

202.1 EXTEND WARM GREETINGS ATTENDANTS HEMISPHERIC RADIO TELEVISION CONFERENCE. GREAT OPPORTUNITIES AFFORDED THOSE BAHÁ'ÍS TRAINED USE POWERS RADIO TELEVISION REACH HEARTS MINDS NUMBERLESS PERSONS AWAITING COMING KINGDOM OF GOD ON EARTH. MAY VOICE CAUSE BE RAISED BEFORE MILLIONS PROCLAIMING MESSAGE BAHÁ'U'LLÁH THROUGHOUT HEMISPHERE. SHARING OF TALENTS RESOURCES ZEAL WILL ASSUREDLY BRING GREAT CONFIRMATIONS. ASSURE LOVING PRAYERS HOLY SHRINES SUCCESS YOUR IMPORTANT DELIBERATIONS.

UNIVERSAL HOUSE OF JUSTICE

203

Comments on First International Bahá'í Women's Conference in South America

3 January 1978

To the First International Bahá'í Women's Conference in South America

203.1 DELIGHTED GREAT SUCCESS WIDESPREAD ATTENDANCE EXCELLENT PUBLICITY PROCLAMATION FIRST INTERNATIONAL BAHÁ'Í WOMEN'S CONFERENCE SOUTH AMERICA WILL OFFER PRAYERS SACRED THRESHOLD DETERMINATION WIN GOALS PLAN WILL BE CONFIRMED.

UNIVERSAL HOUSE OF JUSTICE

204
Release of a Compilation on Bahá'í Consultation
1 FEBRUARY 1978

To all National Spiritual Assemblies

Dear Bahá'í Friends,

Recently the Research Department made a compilation on *Bahá'í Consultation*, and we have been asked by the Universal House of Justice to send you herewith a copy.[1] You may share the contents of this compilation with the friends in whatever manner you consider advisable.

With loving Bahá'í greetings,
DEPARTMENT OF THE SECRETARIAT

205
Inauguration of Radio Bahá'í Ecuador
NAW-RÚZ 1978

To the National Spiritual Assembly of the Bahá'ís of Ecuador

WITH JOYOUS HEARTS WE HAIL THE OFFICIAL INAUGURATION IN ECUADOR OF THE FIRST RADIO STATION IN THE BAHÁ'Í WORLD, AUSPICIOUS MAJOR STEP FULFILLMENT GOAL OF THE FIVE YEAR PLAN OF INCREASING THE USE OF MASS COMMUNICATION IN PROCLAIMING AND TEACHING THE FAITH OF BAHÁ'U'LLÁH.[1]

MAY THIS HAPPY EVENT BE THE FORERUNNER AND THE INSPIRATION FOR THE EARLY ESTABLISHMENT OF RADIO STATIONS IN OTHER PARTS OF THE WORLD AS A NOTABLE SERVICE NOT ONLY TO THE FOLLOWERS OF THE BAHÁ'Í FAITH BUT TO ALL PEOPLES.

WE WILL SUPPLICATE BAHÁ'U'LLÁH IN THE HOLY SHRINES THAT HIS BLESSINGS AND CONFIRMATIONS WILL DESCEND UPON ALL WHO LABOR FOR THE SUCCESS OF THIS WORTHY UNDERTAKING.

THE UNIVERSAL HOUSE OF JUSTICE

204-1. See CC 1:93–110.

205-1. The initial broadcast took place on 12 December 1977. This message was prepared for a formal Naw-Rúz inauguration of the station, but the ceremony was delayed until 28 August 1978 (see message no. 213).

206
Elucidation of Bahá'í Teachings on Ranks and Stations
27 March 1978

To all National Spiritual Assemblies

Dear Bahá'í Friends,

206.1 One of the believers wrote recently to the Universal House of Justice requesting an elucidation of a statement made by it in one of its letters about the relationship between the Boards of Counselors and National Spiritual Assemblies. The House of Justice instructed us to send the following reply, which is now being shared with all National Assemblies as it will undoubtedly be of interest to the believers in general.

206.2 The statement that the Boards of Counselors outrank the National Institutions of the Faith has a number of implications. A Board of Counselors has the particular responsibility of caring for the protection and propagation of the Faith throughout a continental zone which contains a number of national Bahá'í communities. In performing these tasks it neither directs nor instructs the Spiritual Assemblies or individual believers, but it has the necessary rank to enable it to ensure that it is kept properly informed and that the Spiritual Assemblies give due consideration to its advice and recommendations. However, the essence of the relationships between Bahá'í institutions is loving consultation and a common desire to serve the Cause of God rather than a matter of rank or station.

206.3 It is clear from the Writings of Bahá'u'lláh, as well as from those of 'Abdu'l-Bahá and the interpretations of the Guardian, that the proper functioning of human society requires the preservation of ranks and classes within its membership. The friends should recognize this without envy or jealousy, and those who occupy ranks should never exploit their position or regard themselves as being superior to others. About this Bahá'u'lláh has written:

> 206.3a And amongst the realms of unity is the unity of rank and station. It redoundeth to the exaltation of the Cause, glorifying it among all peoples. Ever since the seeking of preference and distinction came into play, the world hath been laid waste. It hath become desolate. Those who have quaffed from the ocean of divine utterance and fixed their gaze upon the Realm of Glory should regard themselves as being on the same level as the others and in the same station. Were this matter to be definitely established and conclusively demonstrated through the power and might of God, the world would become as the Abhá Paradise.

> 206.3b Indeed, man is noble, inasmuch as each one is a repository of the sign of God. Nevertheless, to regard oneself as superior in knowledge, learning or virtue, or to exalt oneself or seek preference, is a grievous

transgression. Great is the blessedness of those who are adorned with the ornament of this unity and have been graciously confirmed by God.[1]

In similar vein, Shoghi Effendi gave this warning to those who are elected to serve on National Spiritual Assemblies:

> They should never be led to suppose that they are the central ornaments of the body of the Cause, intrinsically superior to others in capacity or merit, and sole promoters of its teachings and principles. They should approach their task with extreme humility, and endeavor by their open-mindedness, their high sense of justice and duty, their candor, their modesty, their entire devotion to the welfare and interests of the friends, the Cause, and humanity, to win not only the confidence and the genuine support and respect of those whom they should serve, but also their esteem and real affection. . . .[2]

Courtesy, reverence, dignity, respect for the rank and achievements of others are virtues which contribute to the harmony and well-being of every community, but pride and self-aggrandizement are among the most deadly of sins.

The House of Justice hopes that all the friends will remember that the ultimate aim in life of every soul should be to attain spiritual excellence—to win the good pleasure of God. The true spiritual station of any soul is known only to God. It is quite a different thing from the ranks and stations that men and women occupy in the various sectors of society. Whoever has his eyes fixed on the goal of attaining the good pleasure of God will accept with joy and radiant acquiescence whatever work or station is assigned to him in the Cause of God, and will rejoice to serve Him under all conditions.

There are many passages on this theme in the Holy Writings, and the Universal House of Justice hopes that these remarks will help the friends to turn to them and understand their purport.

With loving Bahá'í greetings,
DEPARTMENT OF THE SECRETARIAT

206-1. From an unpublished Tablet.
206-2. BA, p. 64.

207
Message to the International Convention—Riḍván 1978
RIḌVÁN 1978

To the International Bahá'í Convention

Dearly Loved Friends,

207.1 The Universal House of Justice takes great pleasure in addressing the members of National Spiritual Assemblies gathered in the Holy Land, in the presence of Hands of the Cause of God and Counselors from all continents, at this fourth International Convention, pausing with you to review the course and needs of the Five Year Plan as we cross the threshold of its final year.

Review of the Progress of the Five Year Plan

207.2 The opening of the Plan witnessed the eager response of the friends, careful study made by the national institutions of the Faith of its implications and requirements, the establishment of machinery and the setting up of projects to achieve its goals, and the often arduous struggle to fulfill the first of its three major objectives—the safeguarding and consolidation of all prizes won in earlier campaigns. This phase extended in many countries over a period of several months, and in others continued as far as the midway point of the Plan.

207.3 The middle year of the Plan saw the holding of the International Conferences and those many regional and national conferences which were held concurrently and diffused far and wide the inspiration flowing from these eight major assemblages of the believers. These gatherings motivated a great acceleration of the work and helped the believers throughout the world to arrive at a new realization of the responsibility entrusted to the followers of the Most Great Name for the spiritual regeneration of their fellowmen.

207.4 We are now in the last stage of the Plan, and this Convention provides us with a welcome and auspicious hour in which to assess our progress and to direct our thoughts to the complete achievement of the goals.

207.5 Of the 130 National Spiritual Assemblies which will be operating during the last year of the Plan, 50 have either achieved or nearly achieved their teaching goals. Of the remaining 80 National Spiritual Assemblies, some 40 are confidently forging ahead and are assured of victory if the present tempo in their teaching work is maintained. Nine National Assemblies are restricted by conditions which make the fulfillment of their homefront goals dependent upon circumstances beyond their control. The remaining 30 national communities are, alas, seriously lagging behind, and only strenuous and sacrificial effort will enable them to win their goals.

Review of the Second Objective: Vast and Widespread Expansion

207.6 The second of the three major objectives of the Plan—a vast and widespread expansion of the Bahá'í community—has seen great but geographically uneven progress. There are now more than 19,000 Local Spiritual Assemblies and the number of localities where Bahá'ís reside is over 83,000. This expansion has been accompanied by an intensification of proclamation efforts and by increased use of mass media such as radio and television.

207.7 There have been notable advances in the process of gaining wider recognition for the Cause of God and in fostering cordial relations with civil authorities, a matter of vital importance in these days when there is a growth of opposition to the Faith from those who, misconstruing its true nature and aims, take alarm at its progress.

Review of the Third Objective: Developing the Distinctive Character of Bahá'í Life

207.8 Some of the most significant achievements of the Plan have been towards its third major objective—the development of the distinctive character of Bahá'í life—and in the consolidation and strengthening of the structure of the Bahá'í community. The beloved Hands of the Cause of God, who have been in the forefront of so many aspects of the work of the Faith, have rendered far-reaching services in this field.

207.9 The Local Spiritual Assemblies, focal centers for the teaching of the Faith and the consolidation of the community, are growing in experience, maturity and wisdom, are proving to be potent instruments for nurturing the Bahá'í life and are, in increasing numbers, carrying out plans for the establishment of the Faith in areas outside their own range of jurisdiction, under the overall guidance of their National Spiritual Assemblies, and with the encouragement and help of the Auxiliary Boards and their assistants. The work of developing Local Spiritual Assemblies is a task without end in the foreseeable future. As the Bahá'í community, which is still very thinly spread around the world, moves continually and with increasing rapidity into new areas, new Assemblies will come into being and will need patient help and training in their sacred duties.

207.10 The devotion and self-sacrifice of the friends, which have drawn to them the confirmations of Bahá'u'lláh, have resulted in the very great advances made so far. Evidences of this striving are apparent in the growing number of national communities which, under the wise stewardship and challenging leadership of their National Spiritual Assemblies, are becoming financially self-supporting; in the fact that ever more individual believers are adopting for themselves specific goals and plans of service for the advancement of the Faith; in the settlement of more than 2,000 pioneers during the course of the Plan; in the upsurge of travel teaching individually and in teams; in a greater awareness of the power of prayer; and in many other ways. Three vital aspects of Bahá'í community life which have seen marked progress during the past four years are the development of the services of women and of youth, and the Bahá'í education of children. The youth have long been in the forefront of the teaching work, and now our hearts rejoice to see the women, in so many lands where previously their capacities were largely left unused, devoting their capable services to the life of the Bahá'í community. The education of Bahá'í children is also receiving much attention, which bodes well for the future generations of Bahá'ís.

The Continental Boards of Counselors

207.11 Experience has shown that active and loving collaboration between the Continental Boards of Counselors and National Spiritual Assemblies has been a particularly invigorating and strengthening factor in the progress of the Cause in all aspects of the work. Reflecting the growth of the community, the number of Continental Counselors has been raised to 64 during the Plan, and the number of Auxiliary Boards to 675. Under the authorization given to them, the members of the Auxiliary Boards have till now appointed 3,358 assistants, who are already playing a significant role in the formation and consolidation of Local Spiritual Assemblies and the fostering of the Bahá'í way of life in local communities. Coordinating and directing the work of these Continental Boards from the Holy Land, the International Teaching Center is now well established in the conduct of its responsibilities, foreshadowing the mighty role that it is destined to play in the functioning of the Administrative Order of Bahá'u'lláh.

Seizing Opportunities to Steer the Course of History

207.12 The Faith is passing through a time of tremendous opportunity and development, as well as of increasing opposition and of growing complexity in the problems confronting it. These opportunities must be seized and these problems overcome, for so crucial are these times that the future course of human history is daily in the balance. During this year the Universal House of Justice will be consulting on the nature, duration and goals of the next stage in the implementation of the Divine Plan. The firm base of the achievement of the Five Year Plan goals, both those of quality and those of quantity, is therefore the burning necessity of the months now before us. Let us go forward in a spirit of optimism, with confidence, determination, courage and unity. The greater the love and unity among the friends, the more speedily will the work advance.

207.13 May the Almighty bless the endeavors of His servants and inspire their hearts to arise in His Cause with that degree of radiant faith and self-sacrifice which will draw to their aid the conquering hosts of the Supreme Concourse.

THE UNIVERSAL HOUSE OF JUSTICE

208

Message to the National Conventions—Riḍván 1978
RIḌVÁN 1978

To the Friends gathered at National Baháʼí Conventions
Beloved Friends,

We joyfully hail the formation of seven more National Spiritual Assemblies, those of Burundi, Mauritania, the Bahamas, Oman, Qatar, the Mariana Islands and Cyprus; two in Africa, one in the Americas, two in Asia, one in the Pacific and one in Europe, raising to one hundred and thirty the number of pillars of the Universal House of Justice. 208.1

Your National Spiritual Assemblies will be sharing with you the message addressed to the International Baháʼí Convention and the news of the progress of the Five Year Plan that was released on that occasion.[1] As you will see, many national communities have already completed, or virtually completed, their Five Year Plan goals. These communities must now ensure that the pace of expansion and consolidation which brought them victory is maintained so that they will advance strongly into the next plan. They can also, by pioneering and travel teaching, rally to the assistance of their sister communities which still have months of intensive work before them in order to win their goals. It is to these latter communities that we now address our call to redoubled, united and sacrificial effort. We are fervently supplicating at the Sacred Threshold that the followers of the Blessed Beauty will arise with enthusiasm, confidence and consecration to ensure that every goal is attained. 208.2

With loving Baháʼí greetings,
THE UNIVERSAL HOUSE OF JUSTICE

209

Election of the Universal House of Justice—Riḍván 1978
1 MAY 1978

To all National Spiritual Assemblies
Dear Baháʼí Friends,

NEWLY ELECTED MEMBERS UNIVERSAL HOUSE OF JUSTICE ʻALÍ NAKHJAVÁNÍ HUSHMAND FATHEAZAM AMOZ GIBSON IAN SEMPLE DAVID RUHE CHARLES WOLCOTT DAVID HOFMAN HUGH CHANCE BORRAH KAVELIN. 209.1

With loving Baháʼí greetings,
THE UNIVERSAL HOUSE OF JUSTICE

208-1. See message no. 207.

210
Appointment of Architect for House of Worship in Western Samoa
15 May 1978

To all National Spiritual Assemblies

210.1 ANNOUNCE APPOINTMENT ḤUSAYN AMÁNAT ARCHITECT FOR MAS͟HRIQU'L-AD͟HKÁR SAMOA.

UNIVERSAL HOUSE OF JUSTICE

211
Love for God and Bahá'u'lláh
25 May 1978

To an individual Bahá'í

Dear Bahá'í Friend,

211.1 The Universal House of Justice has received your letter of March 21, 1978 in which you express concern that you do not yet feel in your heart the degree of love for God and for Bahá'u'lláh that you wish to have and which you witness in others, and is touched by the depth of your longing and effort. We have been requested to convey the following.

211.2 One source of true joy and happiness which you would do well to concentrate upon is that you have been able to recognize and accept God's supreme Manifestation in the Day of His appearance. There is no greater bounty than this and the souls of all Bahá'ís should be filled with gratitude for this supreme gift.

211.3 The House of Justice encourages you to continue your reading of the Words of Bahá'u'lláh and the Master and adds that spiritual growth has been likened to organic growth. Everything living must change. Growth and change can be imperceptible or dramatic and rapid. It is stated in a letter dated 6 October 1954 written on behalf of the beloved Guardian to an individual believer:

211.3a When a person becomes a Bahá'í, actually what takes place is that the seed of the spirit starts to grow in the human soul. This seed must be watered by the outpourings of the Holy Spirit. These gifts of the spirit are received through prayer, meditation, study of the Holy Utterances and service to the Cause of God.

We have been directed to assure you of the prayers of the House of Justice for your spiritual advancement and that you may be so strengthened in your faith that you will be enabled to devotedly serve the Cause of Bahá'u'lláh.

With loving Bahá'í greetings,
THE DEPARTMENT OF THE SECRETARIAT

212
Inclusion of Mother Temple of the West in National Register of Historic Places
22 JUNE 1978

The National Spiritual Assembly of the Bahá'ís of the United States

DELIGHTED ACTION BY FEDERAL AUTHORITIES TO INCLUDE MOTHER TEMPLE OF THE WEST IN NATIONAL REGISTER OF HISTORIC PLACES. FRUITION YOUR EFFORTS ON 134TH ANNIVERSARY OF DECLARATION HIS MISSION BY BLESSED BÁB OBTAIN THIS SIGNIFICANT RECOGNITION DESERVES WARM COMMENDATIONS AND IS AN OUTSTANDING ACHIEVEMENT.

UNIVERSAL HOUSE OF JUSTICE

213
Inauguration of Radio Bahá'í Ecuador
28 AUGUST 1978

The National Spiritual Assembly of the Bahá'ís of Ecuador

DELIGHTED YOUR CABLE MAY FORMAL INAUGURATION STATION RADIO BAHÁ'Í BRING WIDESPREAD REALIZATION NOBLE AIMS PURPOSES FAITH SERVE PEOPLES LOCAL AREA BRING ADVANCEMENT ECUADOR.[1] DELIGHTED PROGRESS TOWARD GOALS PLAN AIDED BY HAPPY EVENTS OTAVALO PRAYING SUCCESSES CONFIRMATIONS BEYOND EXPECTATION.

Loving greetings,
UNIVERSAL HOUSE OF JUSTICE

213-1. See also messages dated 15 December 1977 (no. 201) and Naw-Rúz 1978 (no. 205).

214
Directing the Course of One's Life
11 October 1978

To an individual Bahá'í

Dear Bahá'í Friend,

214.1 The Universal House of Justice has received your moving appeal for guidance in your letter of 5 September 1978, and has instructed us to convey to you the following advice.

214.2 Each individual is unique and has a unique path to tread in his lifetime. In espousing the Bahá'í Faith you have defined the direction of that path, for your recognition of God's Manifestation for this Day and your devotion to His Message provide the spiritual and ethical basis for all aspects of your life of service to mankind, while the continuing guidance that He has provided for the community of His followers enables you to know the directions in which the most effort is required at the present time.

214.3 While, during the early years of the development of the Faith, Bahá'u'lláh, 'Abdu'l-Bahá and Shoghi Effendi sometimes gave specific instructions to individual believers on how they should serve the Cause, the Universal House of Justice seldom does this. It is, indeed, the precious privilege of the individual human being to direct the course of his own life. Through exercising this privilege while striving always to conform his conduct to the divine Teachings and devote his talents in the best possible way to the service of the Cause and mankind, a soul deepens his understanding of God and His will.

214.4 This does not mean that you are left to make your decisions without guidance. This you will find from several sources. Firstly, in general, you will find it in the Writings. Secondly, and more specifically, in the teaching plans issued by the Universal House of Justice. Thirdly, in the plans and projects of your own National Spiritual Assembly. All these, it would seem from your letter, you have been striving to follow. Fourthly, with regard to your own personal goals and actions, is the guidance you can receive through consultation—with your wife, with friends of your choice whose opinions you value, with your Local Spiritual Assembly, with such committees of your National Assembly as are concerned with the fields of activity towards which your inclinations lie. Fifthly, there is prayer and meditation.

214.5 You mention that the answers to your prayers never seem to have come through clearly. Mrs. Ruth Moffett has published her recollection of five steps of prayer for guidance that she was told by the beloved Guardian. When asked about these notes, Shoghi Effendi replied, in letters written by his secretary on his behalf, that the notes should be regarded as "personal suggestions," that he considered them to be "quite sound," but that the friends need not adopt them "strictly and universally."¹ The House of Justice feels that they may be helpful to you and, indeed, you may already be familiar with them. They are as follows:

> ... use these five steps if we have a problem of any kind for which we desire a solution, or wish help.
> Pray and meditate about it. Use the prayers of the Manifestations, as they have the greatest power. Learn to remain in the silence of contemplation for a few moments. During this deepest communion take the next step.
> Arrive at a decision and hold to this. This decision is usually born in a flash at the close or during the contemplation. It may seem almost impossible of accomplishment, but if it seems to be an answer to prayer or a way of solving the problem, then immediately take the next step.
> Have determination to carry the decision through. Many fail here. The decision, budding into determination, is blighted and instead becomes a wish or a vague longing. When determination is born, immediately take the next step.
> Have faith and confidence, that the Power of the Holy Spirit will flow through you, the right way will appear, the door will open, the right message, the right principle or the right book will be given to you. Have confidence, and the right thing will come to meet your need. Then as you rise from prayer take immediately the fifth step.
> Act as though it had all been answered. Then act with tireless, ceaseless energy. And, as you act, you yourself will become a magnet which will attract more power to your being, until you become an unobstructed channel for the Divine Power to flow through you.²

214.6 Also the Guardian's secretary wrote to an individual believer on his behalf: "The Master said guidance was when the doors opened after we tried. We can pray, ask to do God's will only, try hard, and then if we find our plan is not working out, assume it is not the right one, at least for the moment."³

214.7 The Universal House of Justice deeply appreciates your candor and spirit of devotion, and assures you of its prayers in the Holy Shrines on your behalf.

With loving Bahá'í greetings,
DEPARTMENT OF THE SECRETARIAT

214-1. Letter dated 30 June 1938 to Wilfrid Barton, in BN, no. 134 (March 1940): 2; letter dated 29 October 1952 to an individual, quoted in Moffett, *Du'á*, p. 27 n.
214-2. Moffett, *Du'á*, pp. 27–28.
214-3. Letter dated 29 October 1952 to an individual, quoted in Moffett, *Du'á*, p. 27 n.

215
The Grave Peril Facing Bahá'ís and Holy Places in Iran
15 December 1978

To National Spiritual Assemblies

215.1 FRIENDS IRAN AND MOST HOLY PLACES IN SHIRAZ AND TEHRAN IN GRAVE PERIL. BAHÁ'ÍS HAVE BEEN THREATENED OVER SEVERAL WEEKS MOST PARTS IRAN WITH IMMINENT DANGERS. THIS THREAT IS NOW MATERIALIZING IN FORMS OF LOOTING, BURNING BAHÁ'Í HOUSES AND FURTHER THREATS OF ASSASSINATION. IN NAYRÍZ 25 BAHÁ'Í HOMES BURNED, IN SHIRAZ 60 HOMES LOOTED. SIMILAR ATTACKS REPORTED IN OTHER PROVINCES. IN SARVISTÁN BAHÁ'ÍS TAKEN TO MOSQUES AND FORCIBLY REQUIRED TO RECANT THEIR FAITH. . . . URGE FRIENDS JOIN US PRAYERS PROTECTION FRIENDS HOLY PLACES CRADLE FAITH.

UNIVERSAL HOUSE OF JUSTICE

216
Announcement of Decision to Launch a Seven Year Plan
26 December 1978

To all National Spiritual Assemblies

216.1 ANNOUNCE WITH UTMOST JOY DECISION TO LAUNCH DURING FORTHCOMING RIḌVÁN FESTIVITIES A SEVEN YEAR GLOBAL PLAN CONSTITUTING NEXT STAGE MASTER'S STEADILY UNFOLDING DIVINE PLAN. CONFIDENT MOMENTOUS DECISION TAKEN IN MIDST SEVERE CRISIS SHAKING CRADLE FAITH AND WHILE EFFORTS BAHÁ'Í WORLD COMMUNITY ARE STRENUOUSLY BENT UPON FULFILLMENT GOALS FIVE YEAR PLAN WILL RELEASE OUTPOURING SPIRITUAL ENERGY ACCELERATE DESTINED PROGRESS BAHÁ'Í WORLD COMMUNITY NOW GRADUALLY APPEARING IN SHARPER RELIEF BEFORE EYES OF A BEWILDERED HUMANITY FLOUNDERING IN DEPTHS OF CONFLICT AND MORAL DEGRADATION.

216.2 DETAILS PLANS NATIONAL COMMUNITIES FOR INITIAL TWO YEAR PHASE NEW PLAN NOW BEING EVOLVED IN CONSULTATION WITH INTERNATIONAL TEACHING CENTER WILL SHORTLY BE ANNOUNCED TO EACH NATIONAL ASSEMBLY. THIS INITIAL PHASE WILL CALL FOR GREATER PROCLAMATION, CONTINUED CONSOLIDATION AND WIDER EXPANSION. NATIONAL ASSEMBLIES ARE THEREFORE URGED ENSURE THAT TEACHING ACTIVITIES ARE PURSUED WITH CONTINUING VIGOR INTO OPENING YEARS NEW PLAN, THAT PIONEERS ARE ENCOURAGED TO REMAIN AT THEIR POSTS, THAT PROCESS DEVELOPMENT COMMUNITY LIFE IS UNINTERRUPTEDLY SUSTAINED, AND THAT MOMENTUM NOW IMPELLING BAHÁ'Í COMMUNITY FORWARD IS MAINTAINED.

URGE ALL NATIONAL ASSEMBLIES SHARE THIS MESSAGE IMMEDIATELY WITH FRIENDS UNDER THEIR JURISDICTION INVITING THEM MAKE SPECIAL EFFORT ATTEND NATIONAL CONVENTIONS NEXT RIḌVÁN CELEBRATE VICTORIOUS CONCLUSION FIVE YEAR PLAN SIMULTANEOUSLY INAUGURATE SEVEN YEAR PLAN. 216.3

OWING IMPORTANCE NEXT CONVENTION REQUEST NATIONAL SPIRITUAL ASSEMBLIES CONSIDER EXTENDING WHEREVER PRACTICABLE ITS DURATION BY ONE OR TWO DAYS. WE ARE CALLING ON COUNSELORS IN ADDITION TO THEIR OWN PARTICIPATION TO ENCOURAGE AUXILIARY BOARD MEMBERS TO ATTEND THESE PORTENTOUS CONVENTIONS. 216.4

AS THE TURMOIL OF AN AGITATED WORLD SURGES ABOUT THEM THE SUPPORTERS OF BAHÁ'U'LLÁH'S MAJESTICALLY RISING FAITH MUST, AS THE BELOVED GUARDIAN SO CLEARLY INDICATED, SCALE NOBLER HEIGHTS OF HEROISM, SERENELY CONFIDENT THAT THE HOUR OF THEIR MIGHTIEST EXERTIONS MUST COINCIDE WITH THE LOWEST EBB OF MANKIND'S FAST DECLINING FORTUNES. 216.5

FERVENTLY SUPPLICATING BAHÁ'U'LLÁH BOUNTIFULLY BLESS STRENUOUS EFFORTS HIS DEVOTED SERVANTS EVERY LAND WIN GOAL FIVE YEAR PLAN ENSURE FIRM FOUNDATION NEXT STAGE WORLDWIDE DEVELOPMENT GOD'S HOLY CAUSE. 216.6

THE UNIVERSAL HOUSE OF JUSTICE

217
Bahá'í Scholarship
3 JANUARY 1979

To the Participants in the Bahá'í Studies Seminar
held in Cambridge on 30 September and 1 October 1978

Dear Bahá'í Friends,

The Universal House of Justice has read with great interest the report of your seminar. It regards Bahá'í scholarship as of great potential importance for the development and consolidation of the Bahá'í community as it emerges from obscurity. It noted that there are a number of problems with which you have been grappling, and while it feels that it should, in general, leave the working out of solutions to Bahá'í scholars themselves, the House of Justice has the impression that it would be helpful to provide you, at this relatively early stage of the development of Bahá'í scholarship, with a few thoughts on matters raised during your seminar. Reports of your seminar were therefore referred to the Research Department, and the Universal House of Justice commends to your study the enclosed memorandum which that Department has prepared.[1] 217.1

217-1. The memorandum was revised for general application and published at the request of the Universal House of Justice in BN, no. 579 (June 1979): 2–3. The memorandum was also published in BW 17:195–96 under the title "The Challenge and Promise of Bahá'í Scholarship."

217.2 The House of Justice also urges you not to feel constrained in any way in consulting it about problems, whether theoretical or practical, that you meet in your work. It has noted, for example, the difficulties presented by the current temporary requirement for the review of publications, and in this connection it asks us to inform you that it has already established the policy that doctoral theses do not have to be reviewed unless there is a proposal to publish them in larger quantities than is required by the examining body.

217.3 You are still in the early stages of a very challenging and promising development in the life of the Bahá'í community, and the Universal House of Justice is eager to foster and assist your work in whatever ways it can. We are to assure you of its prayers in the Sacred Shrines on behalf of you all and of the progress of Bahá'í scholarship.

With loving Bahá'í greetings,
DEPARTMENT OF THE SECRETARIAT

The Bahá'í Studies Seminar on Ethics and Methodology Held in Cambridge on 30 September and 1 October 1978
Comments by the Research Department at the World Center

217.4 This seminar seems to have provided a very valuable forum for the discussion of a number of aspects of Bahá'í scholarship, and the airing of certain problems which have been worrying some of the friends in relationship to their work and to their fellow believers. We believe that many of the problems arise from an attempt by some Bahá'í scholars to make use of methodologies devised by non-Bahá'ís without thinking through the implications of such a course and without working out a methodology which would be in consonance with the spirit of the Faith. The seminar itself may well prove to be an initial step in such a working out. The following remarks are intended merely to draw attention to certain aspects which we believe can help to advance this process.

The Harmony of Science and Religion

217.5 It has become customary in the West to think of science and religion as occupying two distinct—and even opposed—areas of human thought and activity. This dichotomy can be characterized in the pairs of antitheses: faith and reason; value and fact. It is a dichotomy which is foreign to Bahá'í thought and should, we feel, be regarded with suspicion by Bahá'í scholars in every field. The principle of the harmony of science and religion means not only that religious teachings should be studied with the light of reason and evidence as well as of faith and inspiration, but also that everything in this creation, all aspects of human life and knowledge, should be studied in the light of revelation as well as in that of purely rational investigation. In other words, a Bahá'í scholar, when studying a subject, should not lock out of his mind any aspect of truth that is known to him.

It has, for example, become commonplace to regard religion as the product of human striving after truth, as the outcome of certain climates of thought and conditions of society. This has been taken, by many non-Bahá'í thinkers, to the extreme of denying altogether the reality or even the possibility of a specific revelation of the Will of God to mankind through a human Mouthpiece. A Bahá'í who has studied the Teachings of Bahá'u'lláh, who has accepted His claim to be the Manifestation of God for this Age, and who has seen His Teachings at work in his daily life, knows as the result of rational investigation, confirmed by actual experience, that true religion, far from being the product solely of human striving after truth, is the fruit of the creative Word of God which, with divine power, transforms human thought and action. 217.6

The Distinction between Divine Revelation and What People Know and Do about It

A Bahá'í, through this faith in, this "conscious knowledge"[2] of, the reality of divine Revelation, can distinguish, for instance, between Christianity, which is the divine message given by Jesus of Nazareth, and the development of Christendom, which is the history of what men did with that message in subsequent centuries, a distinction which has become blurred if not entirely obscured in current Christian theology. A Bahá'í scholar conscious of this distinction will not make the mistake of regarding the sayings and beliefs of certain Bahá'ís at any one time as being the Bahá'í Faith. The Bahá'í Faith is the Revelation of Bahá'u'lláh: His Own Words as interpreted by 'Abdu'l-Bahá and the Guardian. It is a revelation of such staggering magnitude that no Bahá'í at this early stage in Bahá'í history can rightly claim to have more than a partial and imperfect understanding of it. Thus, Bahá'í historians would see the overcoming of early misconceptions held by the Bahá'í community, or by parts of the Bahá'í community, not as "developments of the Bahá'í Faith"—as a non-Bahá'í historian might well regard them—but as growth of that community's understanding of the Bahá'í Revelation.[3] 217.7

A Unity of Faith and Reason

It has been suggested that the words of Bahá'u'lláh that a true seeker should "so cleanse his heart that no remnant of either love or hate may linger therein, lest that love blindly incline him to error, or that hate repel him away from 217.8

217-2. See TABA 3:549.
217-3. In a message dated 27 May 1966 (no. 35) the Universal House of Justice explains the clear distinction the Bahá'í Faith makes between authoritative interpretation and the interpretation or understanding of individuals. The Bahá'í Faith has two sources of authoritative interpretation: 'Abdu'l-Bahá, Whose authority is derived from His appointment in the Kitáb-i-Aqdas and the Kitáb-i-'Ahd (Book of the Covenant) as the Center of Bahá'u'lláh's Covenant, and the Guardian, whose authority is derived from 'Abdu'l-Bahá's Will and Testament.

the truth," support the viewpoint of methodological agnosticism. But we believe that on deeper reflection it will be recognized that love and hate are emotional attachments or repulsions that can irrationally influence the seeker; they are not aspects of the truth itself. Moreover, the whole passage concerns taking "the step of search in the path leading to the knowledge of the Ancient of Days" and is summarized by Bahá'u'lláh in the words: "Our purpose in revealing these convincing and weighty utterances is to impress upon the seeker that he should regard all else beside God as transient, and count all things save Him, Who is the Object of all adoration, as utter nothingness." It is in this context that He says, near the beginning of the passage, that the seeker must, "before all else, cleanse and purify his heart . . . from the obscuring dust of all acquired knowledge, and the allusions of the embodiments of satanic fancy."[4] It is similar, we think, to Bahá'u'lláh's injunction to look upon the Manifestation with His Own eyes.[5] In scientific investigation when searching after the facts of any matter a Bahá'í must, of course, be entirely openminded, but in his interpretation of the facts and his evaluation of evidence we do not see by what logic he can ignore the truth of the Bahá'í Revelation which he has already accepted; to do so would, we feel, be both hypocritical and unscholarly.

217.9 Undoubtedly the fact that Bahá'í scholars of the history and teachings of the Faith believe in the Faith that they are studying will be a grave flaw in the eyes of many non-Bahá'í academics, whose own dogmatic materialism passes without comment because it is fashionable; but this difficulty is one that Bahá'í scholars share with their fellow believers in many fields of human endeavor.

217.10 If Bahá'í scholars will try to avoid this snare of allowing a divorce between their faith and their reason, we are sure that they will also avoid many of the occasions for tension arising between themselves and their fellow believers.

The Spiritual Qualities of Bahá'í Scholars

217.11 The sundering of science and religion is but one example of the tendency of the human mind (which is necessarily limited in its capacity) to concentrate on one virtue, one aspect of truth, one goal, to the exclusion of others. This leads, in extreme cases, to fanaticism and the utter distortion of truth, and in all cases to some degree of imbalance and inaccuracy. A scholar who is imbued with an understanding of the broad teachings of the Faith will always remember that being a scholar does not exempt him from the primal duties and purposes for which all human beings are created. All men, not scholars alone, are exhorted to seek out and uphold the truth, no matter how uncomfortable it may be. But they are also exhorted to be wise in their utter-

---

217-4. KI, pp. 192, 195, 192.
217-5. See GWB, pp. 90–91, 272.

ance, to be tolerant of the views of others, to be courteous in their behavior and speech, not to sow the seeds of doubt in faithful hearts, to look at the good rather than at the bad, to avoid conflict and contention, to be reverent, to be faithful to the Covenant of God, to promote His Faith and safeguard its honor, and to educate their fellowmen, giving milk to babes and meat to those who are stronger.

Scholarship has a high station in the Bahá'í teachings, and Bahá'í scholars have a great responsibility. We believe that they would do well to concentrate upon the ascertainment of truth—of a fuller understanding of the subject of their scholarship, whatever its field—not upon exposing and attacking the errors of others, whether they be of non-Bahá'í or of their fellow believers. Inevitably the demonstration of truth exposes the falsity of error, but the emphasis and motive are important. We refer to these words of Bahá'u'lláh:

> Consort with all men, O people of Bahá, in a spirit of friendliness and fellowship. If ye be aware of a certain truth, if ye possess a jewel, of which others are deprived, share it with them in a language of utmost kindliness and goodwill. If it be accepted, if it fulfill its purpose, your object is attained. If any one should refuse it, leave him unto himself, and beseech God to guide him. Beware lest ye deal unkindly with him. A kindly tongue is the lodestone of the hearts of men. It is the bread of the spirit, it clotheth the words with meaning, it is the fountain of the light of wisdom and understanding. . . .
> (*Gleanings from the Writings of Bahá'u'lláh* CXXXII)

and again:

> Should any one among you be incapable of grasping a certain truth, or be striving to comprehend it, show forth, when conversing with him, a spirit of extreme kindliness and goodwill. Help him to see and recognize the truth, without esteeming yourself to be, in the least, superior to him, or to be possessed of greater endowments.
> (*Gleanings from the Writings of Bahá'u'lláh* V)

In our view there are two particular dangers to which Bahá'í scholars are exposed, and which they share with those believers who rise to eminent positions in the administration of the Cause. One danger is faced by only a few: those whose work requires them to read the writings of Covenant-breakers. They have to remember that they are by no means immune to the spiritual poison that such works distill, and that they must approach this aspect of their work with great caution, alert to the danger that it presents. The second danger, which may well be as insidious, is that of spiritual pride and arrogance. Bahá'í scholars, especially those who are scholars in the teachings and history of the Faith itself, would be well advised to remember that scholars have often been most wrong when they have been most certain that they were right. The virtues of moderation, humility and humor in regard to one's own work and ideas are a potent protection against this danger.

217.14 We feel that by following such avenues of approach as those described in this memorandum Bahá'í scholars will find that many of the "fears, doubts and anxieties" which were aired at the seminar will be dispelled.

218

Persecution of the Bahá'ís of Iran
12 January 1979

To all National Spiritual Assemblies

Dear Bahá'í Friends,

218.1 From reports in the news media you have no doubt learned of the disturbances in Iran. The followers of the Faith of Bahá'u'lláh have in the land of its birth once again been subjected to severe persecution and active repression.

218.2 The National Spiritual Assembly compiled during the month of October 1978 a list of 93 cases dealing with personal injuries inflicted upon individual believers and of damages to houses, shops, crops and livestock, as well as to local Ḥaẓíratu'l-Quds. During the month of December organized mobs attacked Bahá'ís and their properties in Shiraz and its environs. As a result of these attacks over 300 homes were either burned or destroyed, and some 200 looted. In these events 15 believers were beaten or wounded, and two were killed.[1] Fortunately the intention of the attackers to destroy the Holy House of the Báb was not carried out,[2] but the spirit of aggressive animosity towards the Bahá'ís spread to several centers throughout the province of Fárs, including the town of Marvdasht, where 31 Bahá'í homes were looted and the imposing structure of the local Ḥaẓíratu'l-Quds reared by that community was razed to the ground.

218.3 Following these events, the wave of persecution spread to the north of the country. In several towns and villages of Ádhirbáyján, and particularly in Míyán-Duáb, the onslaught was severe. In the latter town the first target was the local Ḥaẓíratu'l-Quds, which was totally destroyed, and this was followed by the burning or looting of 80 homes and the brutal murder of two believers, a father and his son, whose bodies were then dragged through the streets, cut in pieces, and consigned to the flames.[3]

218-1. The two Bahá'ís killed were Mr. Hatan Ruzbihi and Mr. Jan-Ali Ruzbihi, both from Shiraz.

218-2. For information about attacks on the House of the Báb and about its eventual destruction, see messages dated 10 May 1979 (no. 225), 9 September 1979 (no. 235), and 26 May 1981 (no. 282).

218-3. The father and son were Mr. Parvíz Afnání and Mr. Khusraw Afnání.

The organized and violent assaults on Bahá'í lives and properties have 218.4 emboldened and incited hooligans all over the country, and the oppressed Bahá'ís are constantly under threat of mass aggression and assault.

These acts of hostility against the Bahá'ís have so far cost four lives, mil- 218.5 lions of dollars in loss of property, and the displacement of some 700 individuals who have become homeless. The spirit of the Bahá'ís, however, is very high, and acts of heroism and magnanimity have been reported, which historians will record for posterity.

The National Spiritual Assembly of Iran has instituted a special fund for 218.6 the relief of the needy and suffering from among the believers in that country. The House of Justice has already contributed a sum of $135,000.00 to this fund, and it calls upon all friends in every land to offer of their substance, at this hour of need, to help their tormented brethren in the Cradle of the Faith. All contributions should preferably be sent to the Universal House of Justice, which will ensure that the contributions are transmitted safely to the National Spiritual Assembly of Iran.

The House of Justice further calls on the friends the world over to join it 218.7 in fervent prayers for the protection of the Faith and the Holy Places and for the relief and deliverance of the beloved and steadfast friends in Iran. It particularly invites the friends to pray daily during the period of the Fast,[4] supplicating Bahá'u'lláh that the distressing plight of the Persian Community may be mitigated and that their sorrows and deprivations may be transmuted into comfort and joy through His grace and bounty.

With loving Bahá'í greetings,
THE UNIVERSAL HOUSE OF JUSTICE

219
Extension of Permission to Form Local Spiritual Assemblies under Certain Conditions during Riḍván
19 FEBRUARY 1979

To National Spiritual Assemblies

Dear Bahá'í Friends,

We have been asked by the Universal House of Justice to draw your atten- 219.1 tion to its letter sent to you and other selected National Spiritual Assemblies of the world on March 6, 1977.[1] In that letter, a copy of which is enclosed for your ease of reference, permission was given to hold the election of Local

---

218-4. 2–20 March.
219-1. See message no. 189.

Spiritual Assemblies during the twelve days of Riḍván if certain conditions existed in some localities. This is to inform you that, until further notice, it is permissible to follow this procedure every Riḍván where it is called for.

219.2 Furthermore, in view of the conditions of the Faith at this stage in its development, the House of Justice has decided that the provision set forth under the Five Year Plan permitting the establishment of Local Spiritual Assemblies which are being formed for the first time, whenever there are nine or more adult believers residing in a locality, will continue to apply after the completion of the Five Year Plan.[2]

With loving Baháʼí greetings,
DEPARTMENT OF THE SECRETARIAT

220
Refutation of False Accusations against Iranian Baháʼís
26 FEBRUARY 1979

To all National Spiritual Assemblies

Dear Baháʼí Friends,

220.1 Recent events in Iran have focused the attention of the world's news media on that country, and the Baháʼís and the Baháʼí Faith have been mentioned frequently. Our enemies have spread many misleading statements and calumnies through the media.

220.2 Already in the United States the prompt reaction of Local Spiritual Assemblies and the National Spiritual Assembly to an attack on the Faith made during a national television program has resulted in the greatest publicity for the Cause for many years.[1]

220.3 The friends are urged through their National and Local Assemblies, and individually, and without in any way criticizing or confronting editors and program directors, to offer articles, letters to editors, statements to radio and television producers and to take occasional advertisements. All these should be linked to any mention of the Faith, particularly though not necessarily a

---

219-2. The Universal House of Justice announced in a letter dated 26 December 1995 to the Conference of the Continental Boards of Counselors that, beginning at Riḍván 1997, the practice of electing all Local Spiritual Assemblies on the First Day of Riḍván would be reinstituted.

220-1. On 8 February 1979 Mansur Farhang of Sacramento State University appeared on the public television network's nationally syndicated "The MacNeil/Lehrer NewsHour" and put forth the usual false accusations against the Iranian Baháʼí community to justify the oppressive measures instigated against it. Local Spiritual Assemblies and individuals wrote letters defending the Iranian Baháʼís to the program's producers. As a result, the Secretary of the National Spiritual Assembly of the Baháʼís of the United States, Glenford E. Mitchell, was invited to respond to the allegations on a later edition of the show.

misleading one, and should be solely concerned with repudiating falsehoods and giving the truth about the Faith, not indulging in argument or complaint.

220.4 The Universal House of Justice is aware of a pattern in the false statements being circulated about the Faith and sends the following information to enable the friends everywhere to take action whenever the opportunity arises.

220.5 The allegations against us are mainly that in Iran the Bahá'ís have been political supporters of the previous regime, that one of the Prime Ministers and some Ministers have been Bahá'ís, and that even the head of SAVAK (the Iranian Secret Police) and other of its high-ranking officers have been members of the Bahá'í Faith. The Bahá'ís are also accused of being against Islam and of supporting causes which are hostile to Muslim nations.

220.6 It is obvious that these allegations are entirely unfounded. The established nonpolitical nature of the Faith, as well as the principle that whoever among the friends participates in partisan politics or becomes a member of a political party is expelled from the Bahá'í community, support this. These false accusations by the enemies of the Cause are being deliberately spread for two main reasons: one, to discredit their political opponents who have been or are in power by associating them with the Faith; and two, to incite further hatred of the fanatical sections of the population against the Bahá'ís.

220.7 During the previous regime, when a one-party system, Rastakhiz, was in force in Iran and the people were induced and often compelled to become members of it, the Bahá'ís of Iran were perhaps the only community who, on the grounds of their religious beliefs, firmly refused to join this party. They declared that although, as an act of faith, they are loyal to the government of the country in which they reside, they cannot accept membership in any political party. Threats of the consequences of such refusal did not deter the Bahá'ís from standing firm in their conviction.

220.8 Regarding the false allegation that Mr. Abbas Hoveida, the ex-Prime Minister of Iran, was a Bahá'í, the facts are that his grandfather was a Bahá'í at the time of Bahá'u'lláh and his father was also a member of the Faith for some time. However, because the latter accepted a political assignment in the foreign ministry of Iran, he was expelled from the Bahá'í community. Mr. Hoveida himself never became a Bahá'í, and asserted that he was a Muslim. In fact, during his term of office, he created many difficulties for the Bahá'í community in order to counter the accusations of his alleged affiliation with the Faith. It was during his regime that many Bahá'ís were dismissed from their administrative posts in the government because of their Faith, and an anti-Bahá'í bias was fostered in respect of employment.

220.9 There was another Minister, Mansour Rouhani, whose father was a Bahá'í and mother a Muslim, but he was not, nor had ever been a Bahá'í. Further, some years ago, a Bahá'í accepted a cabinet post just for a brief time, and he was promptly expelled from the Bahá'í community.

220.10 It should be categorically denied and refuted that General Nasiri, the late head of SAVAK, as well as his assistants, were ever Bahá'ís, and it can be stated that their organization was responsible for the dismissal of many of the friends from government offices in Iran.

220.11 It is true that Bahá'ís must show loyalty to their respective governments, and it is also true that a number of the friends in Iran, while demonstrating this principle, as well as rectitude of conduct and trustworthiness, became known for these qualities and obtained high-ranking, nonpolitical, financial, and administrative positions in the government. However, loyalty and obedience to the government has never meant that the Bahá'ís agreed with or promoted political principles and policies.

220.12 Another principle of the Faith may be cited, namely, that Bahá'ís are forbidden to deny their Faith, even if their very lives are at stake. It is an historical fact that thousands of martyrs, given the choice to deny their faith so that their lives would be spared, refused to do so and proclaimed their faith openly, suffering the consequences. Therefore the public should know that whoever denies that he is a Bahá'í cannot be a member of the Bahá'í community.

220.13 In Iran the officially recognized religious minorities are the adherents of the Jewish, Christian, and Zoroastrian Faiths, although the Bahá'ís outnumber them all. The enemies of the Cause in Iran consider the Bahá'ís as heretical, a "sect," "cult," or similar group. This is because the Muslims, unlike Bahá'ís who believe in progressive and continuous divine revelation, believe that no prophet will appear after Muḥammad. Therefore, whenever Bahá'ís are referred to as a sect or group, the friends should try to remove this misunderstanding and proclaim the independent nature of the Faith to the non-Bahá'í public.

220.14 The Bahá'ís are also accused of being against Islam, whereas it is easy to explain to the public that we believe that all religions of the past, including Islam, are divine in origin and are revered and respected by the followers of Bahá'u'lláh. Indeed, the Author of this Revelation Himself states this fact time and time again in His Writings.

220.15 One of the excuses given by Muslims for hostility to the Faith is the location of our world administrative center in Israel; in the conflict between some Islamic nations and Israel, the Bahá'ís have been accused of being Zionists. It should be made clear that Bahá'ís, who believe in the oneness of humanity and who do not show enmity to any nation, people or creed, cannot take sides in any political controversy. As promoters of genuine love and proclaimers of the unity of mankind, taking sides in such disputes would be diametrically opposed to their religious beliefs. It can be explained, whenever necessary, that Bahá'u'lláh was sent, in 1868, as a Prisoner to the Holy Land by the Ottoman Emperor. For the remainder of His life He was a Prisoner and

Exile, and He subsequently passed away near 'Akká in 1892. The holiest Shrines of the Bahá'í Faith, around which its world administrative center has been established, are situated in the Holy Land because of events which occurred more than half a century before the establishment of Israel and other countries in this part of the world as independent nations. Holy Shrines of the Muslims, Christians and Jews are also located in the Holy Land. Therefore, it is simple enmity to attack the Bahá'í Faith on the basis of the geographical location of its World Center.

220.16 The Universal House of Justice has requested us to bring these facts to your attention so that you may use them whenever necessary in refuting falsehoods and in answering questions and writing articles and letters to the press.

With loving Bahá'í greetings,
DEPARTMENT OF THE SECRETARIAT

1979–1986

The Seven Year Plan

221

Launching of the Seven Year Plan—Naw-Rúz 1979
Naw-Rúz 1979

To the Bahá'ís of the World

Dearly loved Friends,

The decline of religious and moral restraints has unleashed a fury of chaos and confusion that already bears the signs of universal anarchy. Engulfed in this maelstrom, the Bahá'í world community, pursuing with indefeasible unity and spiritual force its redemptive mission, inevitably suffers the disruption of economic, social and civil life which afflicts its fellowmen throughout the planet. It must also bear particular tribulations. The violent disturbances in Persia, coinciding with the gathering in of the bountiful harvest of the Five Year Plan, have brought new and cruel hardships to our long-suffering brethren in the Cradle of our Faith and confronted the Bahá'í world community with critical challenges to its life and work. As the Bahá'í world stood poised on the brink of victory, eagerly anticipating the next stage in the unfoldment of the Master's Divine Plan, Bahá'u'lláh's heroic compatriots, the custodians of the Holy Places of our Faith in the land of its birth, were yet again called upon to endure the passions of brutal mobs, the looting and burning of their homes, the destruction of their means of livelihood, and physical violence and threats of death to force them to recant their faith.[1] They, like their immortal forebears, the Dawn-Breakers, are standing steadfast in face of this new persecution and the ever-present threat of organized extermination. 221.1

Remembering that during the Five Year Plan the Persian friends far surpassed any other national community in their outpouring of pioneers and funds, we, in all those parts of the world where we are still free to promote the Cause of God, have the responsibility to make good their temporary inability to serve. Therefore, with uplifted hearts and radiant faith, we must arise with redoubled energy to pursue our mighty task, confident that the Lord of Hosts will continue to reward our efforts with the same bountiful grace He vouchsafed to us in the Five Year Plan. 221.2

Teaching Victories in the Five Year Plan

The teaching victories in that Plan have been truly prodigious; the points of light, those localities where the Promised One is recognized, have increased from sixty-nine thousand five hundred to over ninety-six thousand; the number of Local Spiritual Assemblies has grown from seventeen thousand to 221.3

221-1. The next stage in the unfoldment of 'Abdu'l-Bahá's Divine Plan was the Seven Year Plan, 1979–86, introduced in this letter. Its inauguration marked the conclusion of the Five Year Plan, 1974–79.

over twenty-five thousand; eighteen new National Spiritual Assemblies have been formed. The final report will disclose in all their manifold aspects the magnitude of the victories won.

221.4 In the world at large the Bahá'í community is now firmly established. The Institution of the Hands of the Cause of God, the Chief Stewards of Bahá'u'lláh's embryonic World Commonwealth, is bearing a precious fruit in the development of the International Teaching Center as a mighty institution of the World Center of the Faith; an institution blessed by the membership of all the Hands of the Cause; an institution whose beneficent influence is diffused to all parts of the Bahá'í community through the Continental Boards of Counselors, the members of the Auxiliary Boards and their assistants.

221.5 Advised, stimulated and supported by this vital arm of the Administrative Order, 125 National Spiritual Assemblies are rapidly acquiring experience and growing in wisdom as they administer the complex affairs of their respective communities as organic parts of one worldwide fellowship. More and more Local Spiritual Assemblies are becoming strong focal centers of local Bahá'í communities and firm pillars of the National Spiritual Assembly in each land. Even in those countries where the Bahá'í Administration cannot operate or has had to be disbanded, countries to which have now been added Afghanistan, the Congo Republic, Niger, Uganda and Vietnam, the believers, while obedient to their governments, nevertheless staunchly keep alive the flame of faith.

Spiritual Development of the Bahá'í Community

221.6 Beyond the expansion of the community, vital as it is, the Five Year Plan witnessed great progress in the spiritual development of the friends, the growing maturity and wisdom of Local and National Assemblies, and in the degree to which Bahá'í communities embody the distinguishing characteristics of Bahá'í life and attract, by their unity, their steadfastness, their radiance and good reputation, the interest and eventual wholehearted support of their fellow citizens. This is the magnet which will attract the masses to the Cause of God, and the leaven that will transform human society.

Obstacles and Opportunities

221.7 The conditions of the world present the followers of Bahá'u'lláh with both obstacles and opportunities. In an increasing number of countries we are witnessing the fulfillment of the warnings that the writings of our Faith contain. "Peoples, nations, adherents of divers faiths," the beloved Guardian wrote, "will jointly and successively arise to shatter its unity, to sap its force, and to degrade its holy name. They will assail not only the spirit which it inculcates, but the administration which is the channel, the instrument, the embodiment of that spirit. For as the authority with which Bahá'u'lláh has invested the future Bahá'í Commonwealth becomes more and more apparent, the fiercer shall be the challenge which from every quarter will be thrown at the verities it enshrines."² In different countries, in varying degrees, the followers of Baháʼuʼlláh at this very hour are undergoing such attacks, and are facing imprisonment and even martyrdom rather than deny the Truth for whose sake the Báb and Baháʼuʼlláh drained the cup of sacrifice.

221.8 In other lands, such as those in Western Europe, the faithful believers have to struggle to convey the message in the face of widespread indifference, materialistic self-satisfaction, cynicism and moral degradation. These friends, however, still have freedom to teach the Faith in their homelands, and in spite of the discouraging meagerness of outward results they continue to proclaim the Message of Baháʼuʼlláh to their fellow-citizens, to raise high the reputation of the Cause in the public eye, to acquaint leaders of thought and those in authority with its true tenets, and to spare no effort to seek out those receptive souls in every town and village who will respond to the divine summons and devote their lives to its service.

221.9 In many lands, however, there is an eager receptivity for the teachings of the Faith. The challenge for the Baháʼís is to provide these thousands of seeking souls, as swiftly as possible, with the spiritual food that they crave, to enlist them under the banner of Baháʼuʼlláh, to nurture them in the way of life He has revealed, and to guide them to elect Local Spiritual Assemblies which, as they begin to function strongly, will unite the friends in firmly consolidated Baháʼí communities and become beacons of guidance and havens of refuge to mankind.

221.10 Faced by such a combination of danger and opportunity, the Baháʼís, confident in the ultimate triumph of God's purpose for mankind, raise their eyes to the goals of a new Seven Year Plan.

World Center Goals

221.11 In the Holy Land the strengthening of the World Center and the augmentation of its worldwide influence must continue:

221.11a
- The Seat of the Universal House of Justice will be completed and designs will be adopted for the remaining three buildings of the World Administrative Center of the Faith.³

221-2. WOB, p. 18.
221-3. The Seat of the Universal House of Justice was completed and occupied on 2 February 1983. In a letter dated 31 August 1987 to the Baháʼís of the world, the Universal House of Justice outlined plans for the completion of the "world-shaking, world-embracing, world-directing administrative institutions" Shoghi Effendi envisioned on Mount Carmel (see letter dated 21 December 1939, in MA, p. 32). The buildings yet to be constructed are the International Baháʼí Library, the seat of the International Teaching Center, and the Center for the Study of the Texts. Additional projects include extending the present International Archives Building and constructing eighteen monumental terraces from the foot of Mount Carmel to its crest, nine leading to the terrace on which the Shrine of the Báb stands, and nine rising above it.

- The Institution of the International Teaching Center will be developed and its functions expanded. This will require an increase in its membership and the assumption by it and by the Continental Boards of Counselors of wider functions in the stimulation on an international scale of the propagation and consolidation of the Faith, and in the promotion of the spiritual, intellectual and community aspects of Bahá'í life.[4]
- The House of 'Abdu'lláh Páshá in 'Akká will be opened to pilgrimage.[5]
- Work will be continued on the collation and classification of the Sacred Texts and a series of compilations gleaned and translated from the writings of the Faith will be sent out to the Bahá'í world to help in deepening the friends in their understanding of the fundamentals of the Faith, enriching their spiritual lives, and reinforcing their efforts to teach the Cause.[6]
- The ties binding the Bahá'í International Community to the United Nations will be further developed.
- Continued efforts will be made to protect the Faith from opposition and to emancipate it from the fetters of persecution.

International Goals

Each National Spiritual Assembly has been given goals for these first two years of the Plan, designed to continue the process of expansion, to consolidate the victories won, and to attain, where circumstances permit, any goals that may have had to remain unaccomplished at the end of the Five Year Plan. During these first two years we shall be examining, with the Continental

---

221-4. For information on the establishment of the International Teaching Center, see message dated 5 June 1973 (no. 131); for the elucidation of its duties, see message dated 8 June 1973 (no. 132); for information about its further evolution, see message dated 19 May 1983 (no. 361).

221-5. The House in 'Akká that served as the residence of 'Abdu'l-Bahá and His family from 1896 until their move to Haifa between 1907 and 1910. The House of 'Abdu'lláh Páshá was restored under the direction of Hand of the Cause of God Amatu'l-Bahá Rúḥíyyih Khánum and was opened to pilgrims in April 1983. For the message announcing its acquisition, see cable dated 9 January 1975 (no. 154). For an account of its significance, see message dated 4 March 1975 (no. 157).

221-6. During the Seven Year Plan the Universal House of Justice sent to the Bahá'í world compilations on the following subjects: inspiring the heart (24 October 1979); divorce (18 January 1980); the importance of prayer, meditation, and the devotional attitude (31 March 1980); attendance at National Spiritual Assembly meetings (26 October 1980); the assistance of God (24 August 1981); prayers and passages from the holy writings (16 September 1981); excellence in all things (23 November 1981); family life (18 February 1982); the importance of deepening our knowledge and understanding of the Faith (13 January 1983); the Law of Ḥuqúqu'lláh (4 July 1985); peace (9 August 1985); women (1 January 1986); and the epochs of the Formative Age (5 February 1986). Translations included the Long Healing Prayer and the Fire Tablet (13 August 1980) and a prayer of 'Abdu'l-Bahá for martyred Bahá'ís and their relatives (11 January 1983). All of the compilations except those on inspiring the heart, attendance at National Spiritual Assembly meetings, and prayers and passages from the holy writings, can be found in *The Compilation of Compilations*.

1979–1986 • THE SEVEN YEAR PLAN

Boards of Counselors and National Spiritual Assemblies, the conditions and possibilities in each country, and shall be considering in detail the capacities and needs of each of the rapidly differentiating national Bahá'í communities before formulating the further goals towards which each community is to work following the opening phase of the Plan.

Throughout the world the Seven Year Plan must witness the attainment of the following objectives: 221.13

- The Mas̲h̲riqu'l-Ad̲h̲kár of Samoa is to be completed and progress will be made in the construction of the Mas̲h̲riqu'l-Ad̲h̲kár in India.[7] 221.13a
- Nineteen new National Spiritual Assemblies are to be brought into being: eight in Africa, those of Angola, Bophuthatswana, the Cape Verde Islands, Gabon, Mali, Mozambique, Namibia and Transkei; eight in the Americas, those of Bermuda, Dominica, French Guiana, Grenada, the Leeward Islands, Martinique, St. Lucia and St. Vincent; and three in the Pacific, those of the Cook Islands, Tuvalu and the West Caroline Islands. Those National Spiritual Assemblies which have had to be dissolved will, circumstances permitting, be reestablished. 221.13b
- The Message of Bahá'u'lláh must be taken to territories and islands which are as yet unopened to His Faith. 221.13c
- The teaching work, both that organized by institutions of the Faith and that which is the fruit of individual initiative, must be actively carried forward so that there will be growing numbers of believers, leading more countries to the stage of entry by troops and ultimately to mass conversion. 221.13d
- This teaching work must include prompt, thorough and continuing consolidation so that all victories will be safeguarded, the number of Local Spiritual Assemblies will be increased and the foundations of the Cause reinforced. 221.13e
- The interchange of pioneers and traveling teachers, which contributes so importantly to the unity of the Bahá'í world and to a true understanding of the oneness of mankind, must continue, especially between neighboring lands. At the same time, each national Bahá'í community must aspire to a rapid achievement of self-sufficiency in carrying out its vital activities, thus acquiring the capacity to continue to function and grow even if outside help is cut off. 221.13f
- Especially in finance is the attainment of independence by national Bahá'í communities urgent. Already the persecutions in Iran have deprived the believers in that country of the bounty of contributing to 221.13g

---

221-7. For messages on the excavation of the site of the House of Worship in Samoa and on its dedication, see messages dated 6 December 1979 (no. 242) and August 1984 (no. 403), respectively. For the message on the dedication of the House of Worship in India, see letter dated 24 February 1986 (no. 452).

405

the international funds of the Faith, of which they have been a major source.⁸ Economic disruption in other countries threatens further diminution of financial resources. We therefore appeal to the friends everywhere to exercise the utmost economy in the use of funds and to make those sacrifices in their personal lives which will enable them to contribute their share, according to their means, to the local, national, continental and international funds of the Faith.

221.13h • For the prompt achievement of all the goals and the healthy growth of Bahá'í community life National Spiritual Assemblies must pay particular attention to the efficient functioning, in the true spirit of the Faith, of their national committees and other auxiliary institutions, and, in consultation with the Continental Boards of Counselors, must conceive and implement programs that will guide and reinforce the efforts of the friends in the path of service.

221.13i • National Spiritual Assemblies must promote wise and dignified approaches to people prominent in all areas of human endeavor, acquainting them with the nature of the Bahá'í community and the basic tenets of the Faith, and winning their esteem and friendship.

221.13j • At the heart of all activities, the spiritual, intellectual and community life of the believers must be developed and fostered, requiring: the prosecution with increased vigor of the development of Local Spiritual Assemblies so that they may exercise their beneficial influence and guidance on the life of Bahá'í communities; the nurturing of a deeper understanding of Bahá'í family life; the Bahá'í education of children, including the holding of regular Bahá'í classes and, where necessary, the establishment of tutorial schools for the provision of elementary education; the encouragement of Bahá'í youth in study and service; and the encouragement of Bahá'í women to exercise to the full their privileges and responsibilities in the work of the community—may they befittingly bear witness to the memory of the Greatest Holy Leaf, the immortal heroine of the Bahá'í Dispensation, as we approach the fiftieth anniversary of her passing.⁹

A Time of Testing: A Time for Clinging to the Covenant

221.14 As lawlessness spreads in the world, as governments rise and fall, as rival groups and feuding peoples struggle, each for its own advantage, the plight

---

221-8. The Bahá'ís of Iran were unable at the time to contribute to the Faith's international funds because of their inability to send funds to Israel. Also, many had been dismissed from their jobs, had had their savings and pensions seized, and had lost their homes and possessions. Thus the community's financial resources were virtually exhausted.

221-9. For messages on the fiftieth anniversary of the passing of the Greatest Holy Leaf, see messages no. 275, 313, 321, 329, 334, 337, 338, 340, 341.

of the oppressed and the deprived wrings the heart of every true Bahá'í, tempting him to cry out in protest or to arise in wrath at the perpetrators of injustice. For this is a time of testing which calls to mind Bahá'u'lláh's words, "O concourse of the heedless! I swear by God! The promised day is come, the day when tormenting trials will have surged above your heads, and beneath your feet, saying: 'Taste ye what your hands have wrought!'"[10]

221.15 Now is the time when every follower of Bahá'u'lláh must cling fast to the Covenant of God, resist every temptation to become embroiled in the conflicts of the world, and remember that he is the holder of a precious trust, the Message of God which, alone, can banish injustice from the world and cure the ills afflicting the body and spirit of man. We are the bearers of the Word of God in this day and, however dark the immediate horizons, we must go forward rejoicing in the knowledge that the work we are privileged to perform is God's work and will bring to birth a world whose splendor will outshine our brightest visions and surpass our highest hopes.

THE UNIVERSAL HOUSE OF JUSTICE

222
Elucidation of Seven Year Plan Goals
Naw-Rúz 1979

To National Spiritual Assemblies

Beloved Friends,

222.1 In the message of the Universal House of Justice to the Bahá'ís of the world and in its letters to individual communities setting the goals of the first phase of the Seven Year Plan are a number of references which it wishes to amplify for your guidance. Not all will apply to every national Bahá'í community, but you will all undoubtedly find interest in reading even those which do not immediately apply to your specific situation. The points we have been asked to set forth are as follows.

Local Spiritual Assemblies

222.2 In August 1970 the House of Justice sent to all National Spiritual Assemblies a compilation of the words of Bahá'u'lláh, 'Abdu'l-Bahá and Shoghi Effendi on the Local Spiritual Assembly. To supplement this fundamental and most important guidance we now enclose a compilation of extracts from the letters of the Universal House of Justice written between 1966 and 1975, covering the importance of Local Spiritual Assemblies, their development, the

---

221-10. Quoted in ADJ, p. 81.

supporting role of the Auxiliary Board members and their assistants, and suggested goals for Local Assemblies.¹

222.3 In selecting goal towns for the formation of Local Spiritual Assemblies a National Assembly should ensure that there will be a wide distribution of Local Assemblies throughout the country.

222.4 National Assemblies should consider calling upon every Local Assembly to meet at least once every Bahá'í month, and to appoint a local teaching committee wherever it is desirable to do so and has not already been done.

Pioneers and Traveling Teachers

222.5 The need for the services of pioneers and traveling teachers remains very great. In the goals for the initial two-year phase of the Plan few specific assignments for the sending of pioneers and traveling teachers have been made. In recent years a steady stream has begun to flow, and the Universal House of Justice calls upon the followers of Bahá'u'lláh in the stronger national communities to arise to join this stream. Enclosed are two lists showing those countries which are particularly in need of pioneers and traveling teachers at the present time.² You should publish these as soon as possible. They are also being supplied to the Continental Pioneer Committees, and those friends who arise will be able to decide upon their area of service in consultation with their National Assembly and the appropriate committees. The international funds of the Faith are now very limited, and this adds to the need for pioneers and traveling teachers to be self-supporting.

Youth Teaching

222.6 Experience has shown that youth can render valuable service in many activities of the community, and particularly in taking the message to the members of their own generation. Those in schools and universities have many opportunities to teach their fellow students and faculty members, and many can be particularly effective by attending a school or university in a pioneer goal. During vacations youth can often render outstanding services as traveling teachers. Traveling in teams has been very useful.

Border Teaching

222.7 It is very important that there be collaborative teaching between national Bahá'í communities in border areas, both by travel teaching across the border and in the organization of joint teaching campaigns on both sides of it. Each National Spiritual Assembly should study this possibility and, if it finds such projects profitable, should seek the collaboration of its sister National Assemblies and request the advice and assistance of the Continental Board of Counselors.

222-1. See CC 2:29–30 and 2:39–60. (For the announcement of the first compilation's release, see message dated 11 August 1970, no. 84.)
222-2. The lists are too lengthy to include.

Teaching Conferences

222.8 These conferences, whether national or regional, in addition to providing good opportunities for fanning the enthusiasm of the friends and fostering their unity, have been effectively used by many National Assemblies as working conferences where reports are given of the status of the goals of the Plan and of the urgent needs and priorities; and, where necessary, calls are raised for pioneers, traveling teachers and funds.

Summer and Winter Schools

222.9 The Guardian once described the institution of the Summer School in a letter written on his behalf, as "a vital and inseparable part of any teaching campaign."[3] In April 1972 the House of Justice issued a compilation on the importance of Bahá'í Summer Schools, and it commends this to every National Spiritual Assembly for study.[4] In only a few countries has it been possible or timely to acquire properties to house Summer and Winter Schools; in most they are still held in rented premises, and the House of Justice stresses the importance of holding them at as low a cost as possible in a place that is easily accessible to the friends, so that large numbers of believers and inquirers can attend. It is hoped that this activity will become at least an annual feature of the Bahá'í community life in every land.

The Bahá'í Education of Children

222.10 It is important to hold regular Bahá'í children's classes to give the children a thorough grounding in knowledge of the teachings and history of the Faith, to imbue them with its spirit, to establish loving ties between them and to provide them with that firm foundation in the Faith which will enable them to grow up as staunch and enlightened servants of Bahá'u'lláh. Non-Bahá'í parents will often welcome the opportunity of having their children take part in such classes, and this, in addition to the benefit it confers upon the children, may well be a means of attracting their parents to the Faith.

Tutorial Schools

222.11 This is a term, originally adopted in the Bahá'í community of India, to describe the simple type of school, organized and conducted under the auspices of the Bahá'í administrative institutions, wherein one teacher is employed to conduct classes in reading and writing and elementary subjects for the Bahá'í and non-Bahá'í children in a village. In addition to the academic subjects he also conducts Bahá'í classes for the children and, in his spare time, makes a valuable contribution to the teaching and consolidation work in his own and neighboring Bahá'í communities. The school may be held in the open air, in one of the houses of the Bahá'ís, in the local Ḥaẓíratu'l-Quds, or in a simple building constructed for the purpose, as conditions allow.

222-3. UD, p. 110.
222-4. See CC 1:25–44. For the message announcing the compilation's release, see no. 109.

222.12 The teacher's salary as well as the other costs of the school are provided out of fees paid by the parents, supplemented, if necessary and possible, by allocations from the local or national funds.

222.13 In the Tablet of the World Bahá'u'lláh states that "Everyone, whether man or woman, should hand over to a trusted person a portion of what he or she earneth through trade, agriculture or other occupation, for the training and education of children, to be spent for this purpose with the knowledge of the Trustees of the House of Justice."[5] In many countries this duty is fulfilled through the taxes that the government levies for the support of the state educational system, but there are other lands where no such facilities are provided and the Local Spiritual Assemblies may well begin to fulfill this aspect of their duties by encouraging the local friends to contribute to a special education fund which can be used for the support of tutorial schools or to assist the children of indigent believers to obtain schooling.

Publications

222.14 Every National Spiritual Assembly should have a well conceived plan for the provision and dissemination of a balanced supply of Bahá'í literature for the believers and for the teaching work. In translation and publication, priority should be given to the Sacred Texts[6] and the writings of Shoghi Effendi, for without access to the life-giving waters of the Holy Word, how are the believers to deepen in their understanding of the Teachings and convey them accurately to others?

Recordings

222.15 In addition to the publication of Texts and teaching materials for the friends, it would be helpful in areas where the degree of literacy is not high, to find ways to teach the friends Bahá'í songs, poems, stories and brief quotations from the Writings as well as prayers. This can be done through the use of cassette tapes or radio broadcasts.

222.16 The goal given to certain national communities to make recordings of the Holy Texts is not intended to imply the large-scale production of cassette tapes but rather the development of locally based programs for the recording on cassette tapes of passages in the indigenous languages. Such tapes can then be carried by traveling teachers to outlying areas, used in the teaching work, or left behind if there are tape-recorders locally available.

Communications

222.17 Keeping the friends informed of the news of the Faith is so important that every National Assembly is urged to devote attention to the prompt and regular dissemination of its national newsletter, supplemented, where necessary and

---

222-5. TB, p. 90.
222-6. The writings of Bahá'u'lláh, the Báb, and 'Abdu'l-Bahá.

feasible, by regional and local news organs.⁷ Some National Assemblies have also found that cassette recordings can be useful for communicating with friends in outlying areas, and radio programs can, of course, fulfill a similar purpose.

Correspondence Courses

Such courses have proved their usefulness both for teaching the Faith and deepening the knowledge of the believers, and their production has been given as a goal to some national communities. If any National Assembly assigned this goal is not certain how to proceed, it may consult with the Continental Board of Counselors or write to the Universal House of Justice which will put it in touch with those National Assemblies most likely to be able to help.

Properties

Many properties have already been acquired in the course of previous plans. It is important that these properties be properly maintained in good repair. National Spiritual Assemblies should set aside sums annually in their budgets for the maintenance of national properties so that when a repair becomes necessary the funds will be available without creating a sudden crisis for the national fund. As far as possible, local Ḥaẓíratu'l-Quds and other local properties should be kept up by the local friends themselves.

It is also important to make full use of the properties of the Faith for the purposes for which they were acquired. Well maintained and regularly used properties will not only be a means of fostering Bahá'í community life, but will add to the prestige and dignity of the Faith in the eyes of the non-Bahá'í public.

A number of properties called for in the Five Year Plan, such as district and local Ḥaẓíratu'l-Quds and local endowments, have not yet been acquired, usually as a result of local circumstances beyond the control of the friends. These goals should continue to be diligently pursued so that they will be attained as soon as conditions permit. If there are insuperable difficulties which make such a property unobtainable in the foreseeable future, a full report should be sent to the Universal House of Justice.

For goals requiring the acquisition of additional local Ḥaẓíratu'l-Quds during the initial phase of the Seven Year Plan, no budget has been provided for assistance from the International Fund.

The Universal House of Justice is eagerly anticipating an upsurge of activity in the years ahead, and assures you all of its fervent prayers in the Holy Shrines for the rapid progress of all aspects of the new Plan.

With loving Bahá'í greetings,
DEPARTMENT OF THE SECRETARIAT

222-7. For the announcement of a compilation on newsletters, see message dated 24 August 1972 (no. 120).

223

Member of the Universal House of Justice to Address Significant Meetings in North America and Europe
8 MAY 1979

To the National Spiritual Assembly of the Bahá'ís of the United States

223.1 IN THE WAKE OF JOYOUS WORLDWIDE CELEBRATIONS VICTORIES FIVE YEAR PLAN OUR HEARTS TURN TO OUR BELEAGUERED BRETHREN CRADLE FAITH, TO DEEPENING CRISIS INTERNATIONAL FUND AND ITS SPECIAL IMPACT ON CHALLENGES FACING BAHÁ'Í WORLD IN OPENING TWO-YEAR PHASE SEVEN YEAR PLAN. OUR PRAYERFUL CONSIDERATION MEASURES DESIGNED MITIGATE GRAVE PROBLEM RESULTED IN DECISION ARRANGE A SERIES OF SIGNIFICANT WELL ATTENDED MEETINGS WITH FRIENDS IN SEVERAL KEY CITIES NORTH AMERICA AND EUROPE TO BE PLANNED BY RESPECTIVE NATIONAL ASSEMBLIES AND ADDRESSED BY MEMBER UNIVERSAL HOUSE OF JUSTICE.[1] THIS MISSION ASSIGNED TO MR. BORRAH KAVELIN. TIME IS OF THE ESSENCE AND WE DEEM MONTHS OF JULY AND AUGUST MOST AUSPICIOUS CARRY OUT PROGRAM, PREFERABLY COVERING UNITED STATES AND CANADA IN JULY AND EUROPEAN COUNTRIES IN AUGUST. AMONG CITIES IN UNITED STATES, WE SUGGEST NEW YORK METROPOLITAN AREA, HOUSTON, WILMETTE, SAN DIEGO, LOS ANGELES, SAN FRANCISCO. WE WOULD LEAVE TO YOUR CONVENIENCE THE TIME AND LOCATION OF MEETINGS. YOU MAY ALSO CONSIDER HAVING OUR REPRESENTATIVE JOIN YOUR JUNE MEETING FOR CONSULTATION AND PROCEED FROM THERE TO INITIATE PROGRAM IN STAGES, THENCE TO CANADA. THERE IS FLEXIBILITY IN ARRANGEMENTS TO MEET CONDITIONS IN BOTH COUNTRIES. WE DEEM IT HIGHLY FITTING AND WORTHY THAT THIS PROGRAM BE LAUNCHED IN THE MUCH-LOVED COMMUNITY CALLED BY THE BELOVED MASTER APOSTLES OF BAHÁ'U'LLÁH AND NAMED BY SHOGHI EFFENDI PRINCIPAL BUILDERS AND DEFENDERS OF A MIGHTY ORDER. IN VIEW OF THE URGENCY IN MAKING APPROPRIATE ARRANGEMENTS ON BOTH CONTINENTS, WE WOULD APPRECIATE AN EARLY RESPONSE FROM YOU.

UNIVERSAL HOUSE OF JUSTICE

223-1. A similar telex was sent to the National Spiritual Assembly of the Bahá'ís of Canada, and a general letter about the visit was sent to the National Spiritual Assemblies of Austria, Belgium, Denmark, France, Germany, the Republic of Ireland, Italy, Spain, and the United Kingdom. A telex dated 18 June 1979 from Mr. Borrah Kavelin, member of the Universal House of Justice, was also sent in response to a telex from the National Spiritual Assembly of Alaska to inform them that he was prepared to meet with the Bahá'ís there on the evening of July 23. For a report of the results of the meetings, see message dated 16 September 1979 (no. 236).

224
Living a Chaste and Holy Life
8 May 1979

To an individual Bahá'í

Dear Bahá'í Friend,

The Universal House of Justice has received your letter of 12 September 1978 and is impressed with the eager desire you show to train your behavior in accordance with the standards of the Faith. It has asked us to send you the following comments in answer to your questions. 224.1

On page 25 of *The Advent of Divine Justice*[1] the beloved Guardian is describing the requirements not only of chastity, but of "a chaste and holy life"—both the adjectives are important.[2] One of the signs of a decadent society, a sign which is very evident in the world today, is an almost frenetic devotion to pleasure and diversion, an insatiable thirst for amusement, a fanatical devotion to games and sport, a reluctance to treat any matter seriously, and a scornful, derisory attitude towards virtue and solid worth. Abandonment of "a frivolous conduct" does not imply that a Bahá'í must be sour-faced or perpetually solemn. Humor, happiness, joy are characteristics of a true Bahá'í life. Frivolity palls and eventually leads to boredom and emptiness, but true happiness and joy and humor that are parts of a balanced life that includes serious thought, compassion and humble servitude to God are characteristics that enrich life and add to its radiance. 224.2

Shoghi Effendi's choice of words was always significant, and each one is important in understanding his guidance. In this particular passage, he does not forbid "trivial" pleasures, but he does warn against "excessive attachment" to them and indicates that they can often be "misdirected." One is reminded of 'Abdu'l-Bahá's caution that we should not let a pastime become a waste of time. 224.3

Concerning the positive aspects of chastity, the Universal House of Justice states that the Bahá'í Faith recognizes the value of the sex impulse and holds that the institution of marriage has been established as the channel of its rightful expression. Bahá'ís do not believe that the sex impulse should be suppressed but that it should be regulated and controlled. 224.4

Chastity in no way implies withdrawal from human relationships. It liberates people from the tyranny of the ubiquity of sex. A person who is in control of his sexual impulses is enabled to have profound and enduring friendships with many people, both men and women, without ever sullying that unique and priceless bond that should unite man and wife. 224.5

224-1. See p. 30 in the 1984 or 1990 U.S. Bahá'í Publishing Trust editions.
224-2. Shoghi Effendi's discussion of "a chaste and holy life" appears on pp. 29–33 in the 1984 or 1990 editions.

224.6 A believer cannot fulfill his true mission in life as a follower of the Blessed Perfection merely by living according to a set of rigid regulations, as you will recognize. It is neither possible nor desirable for the House of Justice to lay down a set of rules covering every situation. Rather is it the task of the individual believer to determine, according to his own prayerful understanding of the Writings, precisely what his course of conduct should be in relation to situations which he encounters in his daily life. He must continually study the sacred Writings and the instructions of the beloved Guardian, striving always to attain a new and better understanding of their import to him, and orient his life towards service to Bahá'u'lláh, praying fervently for divine guidance, wisdom and strength to do what is pleasing to God.

224.7 The House of Justice hopes that these comments will be of help to you, and assures you of its prayers in the Holy Shrines that you will be guided and confirmed in your efforts to put into practice the divine teachings.

With loving Bahá'í greetings,
DEPARTMENT OF THE SECRETARIAT

225
Seizure of the House of the Báb
10 MAY 1979

To National Spiritual Assemblies

225.1 NEWS JUST RECEIVED BLESSED HOUSE SHIRAZ[1] AND FOUR ADJACENT HOUSES SEIZED BY ARMED MEN LAST WEEK OCCUPIED BY THEM FEW DAYS, BAHÁ'Í CUSTODIANS THEN SENT AWAY AND DOORS LOCKED AND SEALED. . . .

UNIVERSAL HOUSE OF JUSTICE

225-1. The House of the Báb was ordained by Bahá'u'lláh in the Kitáb-i-Aqdas as a place of pilgrimage. In this House the Báb held His celebrated interview with Mullá Ḥusayn (the first to seek out the Báb and independently recognize His station) on the evening of 22 May 1844.

226
Call for Prayers for the Bahá'ís of Iran
23 MAY 1979

To all National Spiritual Assemblies

Dear Bahá'í Friends,

Since you were last informed of the recent events in the Cradle of the Faith, news has been received of further persecutions directed against the beloved and steadfast friends. 226.1

Historic sites of the Faith, Ḥaẓíratu'l-Quds of local communities, and other Bahá'í properties in several provinces of Iran have been seized and occupied by forces of the new revolutionary government.[1] Among the more sacred of these properties are the Most Holy House of the Báb in Shíráz, ordained by Bahá'u'lláh to be a place of pilgrimage for His followers and regarded by them as the most hallowed Spot in that land, the Síyáh-Chál and two ancestral Houses of Bahá'u'lláh in Tehran and Tákur. Ominous signs of the intensification of active repression of the Faith are becoming evident, and the defenseless and vulnerable Persian Bahá'í community faces mounting perils to its Holy Places, institutions, properties and even the lives of its members. 226.2

We call on all believers in every land to offer special prayers for the protection of the Faith and the believers in Iran on the forthcoming anniversary of the Martyrdom of the Báb, on 9 July 1979, supplicating that through God's loving grace this fresh wave of persecution may not seriously harm the interests of His Cause, and that He may remove the obstacles from the path of the friends, and provide the means for the protection of the Holy Places and institutions of His Faith. 226.3

As soon as the Secretariat of your Assembly receives this letter, immediate steps should be taken to inform all the friends under your jurisdiction of the contents of the letter so that as many believers as possible in every land may participate in this day of prayer on behalf of their beleaguered brethren in Iran. 226.4

With loving Bahá'í greetings,
THE UNIVERSAL HOUSE OF JUSTICE

226-1. On 1 February 1979 Ayatollah Khomeini returned to Iran from exile in France and was greeted by three million supporters. On 16 January 1979 Shah Muhammed Reza Pahlavi left Iran, never to return. On 31 March–1 April the public, in a referendum, approved the installation of the Islamic Republic.

227
The Increasing Dangers Facing Iranian Bahá'ís
15 June 1979

To all National Spiritual Assemblies

Dear Bahá'í Friends,

227.1 Further to the letter of the House of Justice to you of May 23rd, the situation in Persia continues to be a cause for deep concern and the friends and Holy Places in the country are in serious danger. We have been directed by the House of Justice to inform you of the following developments in the Cradle of our Faith.

227.2 1. An order has been issued by the authorities requiring the Umaná Company to cease functioning under its Bahá'í manager and to operate henceforth under a new non-Bahá'í management.[1] This company holds on behalf of the Bahá'í community all the properties of the Faith, including the Holy Places. This step is ominous in its implications as it forebodes total confiscation of all our properties, including Bahá'í cemeteries. A similar step has been taken in respect to the Bahá'í hospital in Tehran, known as the Mítháqíyyih Hospital.

227.3 2. As a result of the recent disturbances, local revolutionary committees in Iran have instigated, in rural areas, the looting of the homes of several hundred Bahá'í families and the deprivation of their means of livelihood. Although a partial restitution of these properties has taken place, adequate compensation for the losses sustained by the Bahá'ís has yet to be made.

227.4 3. Efforts are made to silence the religious conscience of the Bahá'ís, as they are threatened with dismissal from their jobs and loss of their retirement allowances if they refuse to recant their Faith.

227.5 4. Shirkat-i-Nawnahálán, a commercial company of sixty years' standing, in which over 15,000 Bahá'ís have shares and investments, is occupied, its assets frozen, and its staff prevented from work, and denied their salaries. This action contradicts public proclamations of the new regime as well as accepted international standards.

227.6 5. The proposed drafts of the new constitution as published in the press recognize three religious minorities[2] but omit mention of the Bahá'ís, in spite of the fact that they are the largest religious minority in the country.

227-1. For information on the Umaná Company, see the glossary.
227-2. Zoroastrianism, Judaism, and Christianity.

6. The true aims and principles of the Faith are being maliciously misrepresented by a group of fanatical Shi'ah fundamentalists, established over twenty years ago, and one of whose chief aims has been and is to harass the Bahá'í community in Iran. This group is presently spreading false allegations against the Bahá'ís, unjustly accusing them of being enemies of Islam, agents of Zionism and political tools of the previous regime. Such allegations have aroused the passions of uninformed mobs, and created misunderstandings with the authorities. As the Bahá'ís are not a recognized entity in Iran, they have no opportunity to deny or disprove these false accusations. . . .

One of the Persian friends has also written to the House of Justice a eulogy of the spirit of his fellow-believers at this moment of deep agitation and turmoil in the Cradle of our Faith. A copy of extracts from his letter is enclosed.

With loving Bahá'í greetings,
DEPARTMENT OF THE SECRETARIAT

Extracts from a Report of One of the Friends from Persia

The enemies of the Faith, filled with hatred and cruelty, have once again attacked the wronged and homeless believers and the Bahá'í properties. They are truly the return of their bloodthirsty predecessors. The friends have encountered such persecution and have manifested such courage and steadfastness that in every detail they have become the return of the martyrs and the heroes of the Cause of God.[3] The events of history have become alive once again and are reoccurring. No day passes without the shedding of tears of blood and the anguish of hearts. The news of sad events, like a weighty hammer continually descends upon the Bahá'ís. No pen is able to describe the degree of afflictions and difficulties inflicted upon these wronged believers.

About 2,000 men, women, children and youth have sought refuge in the mountains and deserts and live in tents. They have spent many cold and rainy days in the caves of the mountains. Many are injured with broken arms and legs. The small children have lost their ability to talk, having been frightened so much because of the incidents, and the milk of the nursing mothers has dried up. These believers, without having any means of livelihood, pass their days with utmost difficulty and are banished from place to place.

227-3. The "return of the martyrs and the heroes of the Cause of God" is a reference to the Iranian Bábís and Bahá'ís of the Heroic Age of the Faith (1844–1921) who suffered extreme persecution at the hands of their enemies. During the late 1840s and 1850s, some twenty thousand Bábís were killed.

227.11 When the believers, hungry and grief-stricken, had gathered together in the wilderness, the enemies sent them chilaw-kabáb (kabáb with rice) to win their hearts. But those beloved ones did not accept the food and returned it. It is easy to say or write these words, but the bearing of these afflictions is only possible through the power of God. Those few who have denied their faith have escaped to Isfahan, crying and lamenting that they were threatened with the raping of the women of their households. They sit and cry for hours saying they did not know what else to do.

227.12 This is only a glimpse into one incident. Every day, from every corner, there is another cry of grief. The Ḥaẓíratu'l-Quds of Ábádih, where the heads of the early martyrs of the Faith have been buried, has been leveled to the dust. Many other Ḥaẓíratu'l-Quds have been destroyed. The number of Bahá'ís in prisons for one reason or another has increased to 20. Many have been discharged from their jobs. Many have lost their retirement allowances. The Ministry of Education has officially sent a circular that those Bahá'ís who do not deny their faith should be immediately discharged.

227.13 Facing these difficulties, in the midst of the darkness of this oppression and tyranny, are the illumined faces of the National Spiritual Assembly members: the sources of hope. Truly they are angels of God; no, more exalted. Every minute of their lives deserves the reward of a martyr, and each one of them, the reward of a thousand martyrs. They are the personification of steadfastness, courage, and sacrifice, with nothing but the service of the Cause in their hearts and souls. Whenever I looked upon the faces of these illuminated and beloved ones in the meetings of the Assembly, my tears would uncontrollably pour from my eyes.

227.14 There are many such examples amongst the Auxiliary Board members, members of the Assemblies and the youth. Truly, the new creations of God are beyond our imaginations. Unless one witnesses such events in person, the extent of the sacrifice and steadfastness that these friends have manifested cannot be comprehended.

227.15 Whenever I witnessed what befell these believers, the words of God would find meaning in front of my eyes. I had looked up the meaning of these words in the dictionaries, but I did not know that in addition to their obvious meaning they describe, or better even, create new realities. Now that the tempest of trials and afflictions has encircled the community of the beloved ones, the believers who have remained behind and who steadfastly and firmly are bearing the burden of this storm, only can sing the eternal epic of the second century of the Faith. Truly, all of them are the children and descendants of those who watered the tree of the Faith with their pure blood. This tree is still bearing fruit, is still growing! What a glory! What a glory!

227.16 There is so much to say and tell, but the mental anguish is so severe, the conditions are so dark and confused, and the outpouring of the difficulties so abundant that my tongue is not able to utter a word and my mind is bewildered. I can only cry. May my soul be sacrificed for the faithful followers of Bahá'u'lláh who have created the greatest epic in the history of the second century of this new Day. One of the guards who had gone to the house of one of the friends had told her that he could not believe the forbearance and patience of the Bahá'ís and had asked how we could ever do it!

227.17 These are events to remember. Whenever the bloodthirsty enemies or others have returned a part of the looted belongings of the Bahá'ís, they have refused to accept them, crying that they will not take back what they have given in the path of God. One of the friends . . . lost absolutely everything, and refusing the help of the Assembly on the grounds that there were many more needy than he, started to work as a laborer to make a living for his family. He could not bring himself to accept any help whatever from non-Bahá'ís, or to tell of his situation to friends. A friend related that when he saw him, he was so touched that his knees could no more bear the weight of his body.

227.18 And yet another story. A believer who had incurred a loss of Rls. 170,000,000,[4] wrote on the questionnaire form of the Assembly that he did not need any help! When everything is gone with the wind, only faith remains.

227.19 At present thousands of friends in Iran have lost everything, or have lost their jobs and are meeting their expenses by the sale of their belongings. And then there are those who are fleeing from one place to another and in grave danger. This is only the beginning of the journey of love, and its end is not known.

227.20 For five months the National Spiritual Assembly has been meeting at least three or four times a week, for about six to eight continuous hours each time, and devotes 90 per cent of its time to discussion of urgent matters relating to the situation.

227.21 The staff of the Nawnahálán Company have not received their salaries for three months now and about 40 families are affected by this situation.[5] Many other families who had given whatever they had to the Nawnahálán Company and were dependent on the interest received to pay for their expenses, are left without any income. All petitions and complaints have remained unanswered. Whatever was lost is lost and nothing has been recovered.

227.22 God willing, I will write a book instead of a letter and present it to you so that perhaps a drop of this ocean of difficulties may be recorded. The request of this servant and every one of the believers is to express our servitude and beseech the House of Justice for its prayers in the Holy Shrines. From whomever I asked whether they had any special request to be conveyed to you, I was told to beg for your prayers that God may give them the power and worthiness to accept and bear the difficulties.

227-4. Equivalent to approximately $2,414,000.00 (U.S.) in 1979.
227-5. For more information about the Nawnahálán Company, see the glossary.

228
Advice on Use of Pioneers and Traveling Teachers
28 June 1979

To all National Spiritual Assemblies

Dear Bahá'í Friends,

228.1 One of the objectives of the Seven Year Plan is the continued settlement of pioneers in needed areas and the movement of traveling teachers. We have been asked by the Universal House of Justice to share with you its advice on these vital tasks.

228.2 The House of Justice feels that, while ultimate decisions regarding the selection of pioneers and traveling teachers and the manner in which their services are to be utilized remain, of course, in the hands of National Spiritual Assemblies, a closer degree of communication with Boards of Counselors in these matters should be maintained. For example, the number of pioneers and traveling teachers to foreign lands which a National Spiritual Assembly can supply during the current year and every subsequent year of the Plan; the degree of dependence of a national community on outside workers and, if needed, how many and from which countries; and when necessary the evaluation of the services of certain pioneers, are among issues that every National Spiritual Assembly can usefully discuss with the Counselors in its zone. The House of Justice is sure that such consultations would most certainly be conducive to excellent results.

228.3 On a different level, a National Spiritual Assembly may need assistance from Continental Pioneer Committees in the movement of pioneers to their posts or in the coordination of visits by traveling teachers. In order to obtain the best results from the collaboration of these Committees, it is important that information and views be exchanged with them speedily and efficiently. The House of Justice feels that it is highly desirable for each National Spiritual Assembly to make arrangements for Continental Pioneer Committees to deal directly with agencies of National Spiritual Assemblies responsible for pioneers and traveling teachers. When a decision is taken in this regard, the name and address of the correspondent or correspondents should be immediately conveyed to the Continental Pioneer Committees concerned.

228.4 The Universal House of Justice assures you of its loving prayers as you exert yourselves to fulfill the goals and tasks ahead.

With loving Bahá'í greetings,
Department of the Secretariat

229
The Setting of Counselors' Terms of Service
29 JUNE 1979

To all National Spiritual Assemblies

MOMENT PROPITIOUS ANNOUNCE DURATION TERMS SERVICE MEMBERS CONTINENTAL BOARDS COUNSELORS AS ANTICIPATED IN ANNOUNCEMENT ESTABLISHMENT THAT INSTITUTION AND IN CONSTITUTION UNIVERSAL HOUSE OF JUSTICE. DECISION NOW TAKEN THAT TERMS WILL BE OF FIVE YEARS STARTING DAY COVENANT 26 NOVEMBER 1980. SUPPLICATING ANCIENT BEAUTY DIVINE BLESSINGS DEVELOPMENT THIS ESSENTIAL INSTITUTION BAHÁ'Í ADMINISTRATIVE ORDER. 229.1

UNIVERSAL HOUSE OF JUSTICE

230
New Member of International Teaching Center
4 JULY 1979

To all National Spiritual Assemblies

JOYFULLY ANNOUNCE APPOINTMENT COUNSELOR ANNELIESE BOPP TO MEMBERSHIP INTERNATIONAL TEACHING CENTER. 230.1

UNIVERSAL HOUSE OF JUSTICE

231
Conversion of Iranian National Bahá'í Headquarters into an Islamic University
17 JULY 1979

To all National Spiritual Assemblies

SITUATION FRIENDS IRAN STEADILY DETERIORATING. REUTERS[1] HAS SENT WORLD PRESS NEWS RELEASE TWELVE JULY QUOTING OFFICIAL PARIS NEWS AGENCY THAT NATIONAL BAHÁ'Í HEADQUARTERS BEING CONVERTED INTO ISLAMIC UNIVERSITY.... 231.1

UNIVERSAL HOUSE OF JUSTICE

231-1. An international news service.

232
Call for Pioneers
28 July 1979

To National Spiritual Assemblies

Dear Bahá'í Friends,

232.1 In order to expedite the dispatch of pioneers called for under the first phase of the Seven Year Plan to the territories named by the Universal House of Justice as having high priority, the House of Justice has decided to make specific assignments to selected National Spiritual Assemblies. A list of these assignments is attached.[1]

232.2 It is the hope of the House of Justice that the arrival of these pioneers will be of valuable assistance to you in the promotion of the teaching work in the areas under your jurisdiction. However, as you are aware, the process of preparing and sending pioneers is often time-consuming, and you should not wait for the arrival of these pioneers before making your teaching plans. The success of the efforts of your Assembly is not dependent solely upon the presence and assistance of pioneers, but must in the long term be firmly rooted in the devoted efforts of the local believers themselves.

232.3 You may wish to contact directly the National Assemblies scheduled to provide pioneers for your territory in order to expedite achievement of these important goals.

232.4 You are assured of the loving prayers of the Universal House of Justice at the Holy Threshold for the success of all your efforts in the service of the Cause, and also for all those who arise to fulfill the goals.

With loving Bahá'í greetings,
DEPARTMENT OF THE SECRETARIAT

233
Release of Bahá'í Prisoners in Baghdad
20 August 1979

To all National Spiritual Assemblies

Dear Bahá'í Friends,

233.1 The following message has just been telexed by the Universal House of Justice to the Bahá'í International Community office in New York:

232-1. The list is too lengthy to include.

HIGHLY GRATIFIED NEWS JUST RECEIVED ALL BAHÁ'Í PRISONERS BAGHDAD RELEASED AS PART GENERAL AMNESTY. WE OFFER THANKSGIVING BAHÁ'U'LLÁH HIS BOUNTIFUL GRACE ENABLING HIS STEADFAST FRIENDS DEMONSTRATE THEIR UNWAVERING LOYALTY HIS PRECIOUS FAITH.

233.2 The Universal House of Justice is sure that this news will delight the hearts of the friends.

With loving Bahá'í greetings,
Department of the Secretariat

234
Effects of Bahá'í Actions on the Situation in Iran
20 August 1979

To National Spiritual Assemblies

Dear Bahá'í Friends,

234.1 Since writing to you on 15 June on behalf of the Universal House of Justice about events and persecutions in Iran, several developments have transpired, and we have been asked to share the information set forth below with you.

234.2 The cumulative effect of the cables sent by Local Spiritual Assemblies throughout the world, as well as the issuance and quiet circulation of open letters directed to the public in Iran by the National Spiritual Assembly of that country, refuting the false allegations made against the Bahá'í Community, seem to have produced, among others, the following results.

234.3 1. The Iranian press has stopped its daily vituperation against the Bahá'ís.

234.4 2. The official government explanation for the seizure of properties belonging to the Bahá'í Community, including the Holy Places, is that these have been seized in order to protect them.

234.5 3. Individual government officials, contacted by representatives of the National Spiritual Assembly of Iran, have expressed personal sympathy with the Bahá'ís; however, they seem incapable of effectively helping the friends in their present plight. . . .

With loving Bahá'í greetings,
Department of the Secretariat

235
Demolition of the House of the Báb
9 September 1979

To all National Spiritual Assemblies

235.1 PERSECUTION OF THE BAHÁ'ÍS, THE LARGEST RELIGIOUS MINORITY IN IRAN, HAS TAKEN A NEW TURN. EARLY YESTERDAY MORNING A CROWD OF OVER 100 PEOPLE, INCLUDING THE HEAD OF THE GOVERNMENT DEPARTMENT FOR RELIGIOUS ENDOWMENTS IN SHIRAZ, AND ACCOMPANIED BY 25 REVOLUTIONARY GUARDSMEN AND 10 OTHER ARMED MEN, ATTACKED THE MOST HOLY HOUSE OF THE BÁB WHICH WAS ORDAINED BY BAHÁ'U'LLÁH, THE FOUNDER OF THE BAHÁ'Í FAITH, TO BE A PLACE OF PILGRIMAGE FOR HIS FOLLOWERS THROUGHOUT THE WORLD AND IS REGARDED BY THEM AS THE MOST HALLOWED SPOT IN IRAN. THIS CROWD, WHICH HAD THE KEY TO THE HOUSE, SMASHED AND DISMANTLED DOORS AND WINDOWS, DESTROYED ORNAMENTAL PLASTERWORK, BREACHED THE WALLS AND HACKED TO PIECES A TREE IN THE COURTYARD. THIS MORNING THE WORK OF DEMOLITION IS BEING CONTINUED BY A GROUP OF WORKMEN AND IT IS CLEAR THAT THE PURPOSE IS TO RAZE TO THE GROUND THE HOUSE OF THE BÁB AND TWO ADJACENT HOUSES WHICH ALSO BELONG TO THE BAHÁ'Í COMMUNITY.

235.2 A WAVE OF ANGUISHED INDIGNATION IS SWEEPING THE BAHÁ'Í COMMUNITY THROUGHOUT THE WORLD. WHEN ALL BAHÁ'Í HOLY PLACES IN IRAN WERE SEIZED BY THE AUTHORITIES IN RECENT MONTHS, THE PROTESTATIONS OF THE BAHÁ'ÍS WERE MET WITH BLAND ASSURANCES, CONFIRMED IN WRITING, THAT THE TAKEOVER WAS FOR THE PROTECTION OF THESE SACRED PROPERTIES.

235.3 BAHÁ'ÍS IN EAST AND WEST ARE REGISTERING VEHEMENT PROTESTS WITH THE IRANIAN AUTHORITIES.[1]

UNIVERSAL HOUSE OF JUSTICE

235-1. The Universal House of Justice sent this press release to the National Spiritual Assembly of the Bahá'ís of the United States to be shared with the media.

236
Results of Significant Meetings Addressed by Member of the Universal House of Justice
16 SEPTEMBER 1979

To the National Spiritual Assemblies of the Bahá'ís of Alaska, Austria, Belgium, Canada, Denmark, France, Germany, Republic of Ireland, Italy, Spain, United Kingdom, United States

Dear Bahá'í Friends,

236.1 We have been directed by the Universal House of Justice to convey its loving appreciation for the devoted and wholehearted cooperation each of you, as a host National Assembly, extended to its official representative, Mr. Borrah Kavelin, in the series of meetings arranged in North America and in Europe extending over a ten-week period that ended on 31st August.[1]

236.2 In all, 42 meetings were held in 33 cities. It is estimated that a total of more than twelve thousand friends were in attendance at the series of meetings, and that all national Bahá'í communities in Europe were represented in that number. It is hoped that the intended distribution by all host National Assemblies to Local Spiritual Assemblies of copies of the taped message of the Universal House of Justice's representative will reach the widest possible number of friends in all countries within the scope of the assigned mission.

236.3 Already, the magnificent response by National and Local Spiritual Assemblies, and by untold numbers of friends, received by the World Center in the form of commitments and generous and sacrificial contributions, is a noble testament to the wholehearted acceptance of the challenge facing the Bahá'í world in the opening two-year phase of the Seven Year Plan, and to the redemption of the ransom imposed upon the beloved friends in the Cradle of the Faith, even as the clouds of persecution and oppression continue to darken the horizon in that troubled land, as evidenced by the tragic news of the severe damage recently inflicted on the Blessed House of the Báb in Shiraz.

236.4 The Universal House of Justice is serenely confident that each of you, in collaboration with the institution of the Continental Board of Counselors, will maintain, and indeed enhance, the high level of activity called for in successfully overcoming the crisis we face. Ardent and loving prayers at the Divine Threshold will be offered for the blessings and confirmations of Bahá'u'lláh to surround you in your devoted labors.

With loving Bahá'í greetings,
DEPARTMENT OF THE SECRETARIAT

236-1. For the 8 May 1979 message announcing the series of meetings to be addressed by the representative of the Universal House of Justice, see no. 223.

237

Assassination of the Hand of the Cause of God Enoch Olinga

17 September 1979

To all National Spiritual Assemblies

237.1 WITH GRIEF-STRICKEN HEARTS ANNOUNCE TRAGIC NEWS BRUTAL MURDER DEARLY LOVED GREATLY ADMIRED HAND CAUSE GOD ENOCH OLINGA BY UNKNOWN GUNMEN COURTYARD HIS KAMPALA HOME.[1] HIS WIFE ELIZABETH AND THREE OF HIS CHILDREN BADÍʻ, LENNIE AND ṬÁHIRIH HAVE ALSO FALLEN INNOCENT VICTIMS THIS CRUEL ACT. MOTIVE ATTACK NOT YET ASCERTAINED. HIS RADIANT SPIRIT, HIS UNWAVERING FAITH, HIS ALL-EMBRACING LOVE, HIS LEONINE AUDACITY IN THE TEACHING FIELD, HIS TITLES KNIGHT BAHÁ'U'LLÁH, FATHER VICTORIES CONFERRED BELOVED GUARDIAN, ALL COMBINE DISTINGUISH HIM AS PREEMINENT MEMBER HIS RACE IN ANNALS FAITH AFRICAN CONTINENT. URGE FRIENDS EVERYWHERE HOLD MEMORIAL GATHERINGS BEFITTING TRIBUTE HIS IMPERISHABLE MEMORY. FERVENTLY PRAYING HOLY SHRINES PROGRESS HIS NOBLE SOUL AND SOULS FOUR MEMBERS HIS PRECIOUS FAMILY.[2]

UNIVERSAL HOUSE OF JUSTICE

238

Omission of Baháʼís as Recognized Religious Minority from Iranian Constitution

20 September 1979

To National Spiritual Assemblies

238.1 FRAMERS NEW CONSTITUTION IRAN APPROVED CLAUSE RECOGNIZING MINORITY RELIGIONS SPECIFYING JEWS CHRISTIANS ZOROASTRIANS BUT OMITTING MENTION BAHÁʼÍS. . . .

UNIVERSAL HOUSE OF JUSTICE

237-1. For an account of the life and services of Enoch Olinga, see BW 18:618–35.

237-2. In separate cables the Universal House of Justice directed that memorial services be held in all Houses of Worship.

239
Refutation of Accusations against Iranian Bahá'ís
17 OCTOBER 1979

To National Spiritual Assemblies

Dear Bahá'í Friends,

The Universal House of Justice has asked us to send you the attached material. . . . 239.1

With loving Bahá'í greetings,
DEPARTMENT OF THE SECRETARIAT

IRANIAN GOVERNMENT AGENCIES OUTSIDE IRAN HAVE APPARENTLY ADOPTED A UNIFORM STAND IN THEIR REPLIES TO APPEALS BEING MADE ON BEHALF OF THE IRANIAN BAHÁ'ÍS. THEY SAY THAT BAHÁ'ÍS IN IRAN, UNLIKE BAHÁ'ÍS ELSEWHERE, HAVE BEEN INVOLVED IN POLITICS, CONSPIRED WITH AND WERE FAVORED AND SUPPORTED BY THE PREVIOUS REGIME, AND WERE PROMINENT MEMBERS OF SAVAK.[1] IN SUPPORT OF THESE STATEMENTS THESE OFFICIALS NAME AS BAHÁ'ÍS: FORMER PRIME MINISTER ABBAS AMIR HOVEIDA, A FORMER MINISTER OF AGRICULTURE MANSOUR ROUHANI, A SAVAK SENIOR OFFICER PARVIZ SABETI, AND THE PHYSICIAN TO THE SHAH DR. AYADI. 239.2

THE FACTS ARE THAT HOVEIDA'S GRANDFATHER WAS A BAHÁ'Í, HIS FATHER WAS EXPELLED FROM THE BAHÁ'Í COMMUNITY BECAUSE HE BECAME INVOLVED IN POLITICAL ACTIVITY, AND HOVEIDA HIMSELF WAS NEVER A BAHÁ'Í. ROUHANI'S FATHER WAS A BAHÁ'Í, HIS MOTHER A DEVOUT MUSLIM, BUT ROUHANI WAS NEVER A BAHÁ'Í. SABETI'S PARENTS WERE BAHÁ'ÍS, AND THEY REGISTERED HIM IN THE COMMUNITY AS A BAHÁ'Í CHILD. HOWEVER, WHEN HE CAME OF AGE HE DID NOT HIMSELF REGISTER AS A BAHÁ'Í AND NEVER BECAME A MEMBER OF THE COMMUNITY. UNLIKE CHILDREN OF OTHER RELIGIONS, BAHÁ'Í CHILDREN DO NOT AUTOMATICALLY INHERIT THE FAITH OF THEIR PARENTS. WHEN THEY COME OF AGE THEY MUST OF THEIR OWN VOLITION EXPRESS THEIR BELIEF IN BAHÁ'U'LLÁH AND HIS TEACHINGS. DR. AYADI, HOWEVER, IS A BAHÁ'Í. HE HELD TWO POSITIONS: ONE AS DIRECTOR OF THE ARMY MEDICAL SERVICE, THE OTHER AS PRIVATE PHYSICIAN OF THE COURT. NEITHER OF THESE TWO POSITIONS WAS REGARDED BY THE BAHÁ'Í COMMUNITY AS POLITICAL IN NATURE. 239.3

ALTHOUGH SOME BAHÁ'ÍS WITH UNIQUE QUALIFICATIONS WERE PLACED IN POSITIONS OF TRUST BECAUSE OF THEIR ABILITY AND INTEGRITY, IT IS NOT TRUE TO SAY THAT BAHÁ'ÍS WERE FAVORED BY THE PREVIOUS REGIME. ON THE CONTRARY, THEY WERE DENIED CIVIL RIGHTS, SUCH AS PERMISSION TO REGISTER THEIR BAHÁ'Í MARRIAGES, PRIVILEGE TO HOLD BAHÁ'Í RELIGIOUS ENDOWMENTS IN NAME 239.4

239-1. SAVAK is a Persian acronym meaning in English "National Security and Information Organization." SAVAK functioned as the Iranian secret police during the Shah's reign.

OF BAHÁ'Í COMMUNITY, AND FREEDOM TO PUBLISH BAHÁ'Í LITERATURE OR ESTABLISH BAHÁ'Í SCHOOLS (INDEED DURING THE REIGN OF MUHAMMAD REZA SHAH'S FATHER, OVER THIRTY BAHÁ'Í SCHOOLS THROUGHOUT THE COUNTRY WERE PERMANENTLY CLOSED). MANY OF RANK AND FILE OF BAHÁ'ÍS WERE DENIED JOBS AND SOMETIMES EVEN THEIR RIGHTS TO PENSIONS BECAUSE OF THEIR REFUSAL TO DENY THEIR FAITH.

239.5 AS TO THE ALLEGED ROLE OF BAHÁ'ÍS IN SAVAK, THIS IS LIKEWISE UNTRUE. FOR EXAMPLE, IN JANUARY 1979, THROUGH THE MACHINATIONS OF SAVAK, AN ORDER WAS GIVEN TO SYSTEMATICALLY LOOT AND BURN OR OTHERWISE DESTROY HUNDREDS OF HOMES OF BAHÁ'ÍS. THIS IS A FACT ATTESTED TO BY MUSLIM CLERICS BELONGING TO THE PRESENT REGIME[2] WHO, DURING THAT PERIOD OF TERROR AND VIOLENCE AGAINST THE BAHÁ'ÍS, WERE AMONG THE FIRST TO TRY TO DISSUADE THE MOBS FROM PARTICIPATING IN THE SAVAK PLAN, SINCE THE CLERGY KNEW THAT THE AIM OF THIS PLAN WAS TO GIVE EXCUSE TO SAVAK TO DISCREDIT AND SUPPRESS THEM.

239.6 SUMMARIZING THE FOREGOING—IT IS FEARED THAT THE PRESENT REGIME, AS INDICATED BY THE SIMILARITY OF THE STATEMENTS BEING GIVEN OUT BY IRANIAN DIPLOMATIC AGENCIES, IS ATTEMPTING TO JUSTIFY ACTIONS BEING TAKEN AGAINST THE BAHÁ'ÍS BY ASSERTING THAT THE BAHÁ'Í FAITH IS NOT A RELIGION BUT A POLITICAL PARTY, AND THAT THE BAHÁ'Í COMMUNITY SUPPORTED THE PREVIOUS REGIME AND THEREBY BECAME POWERFUL AND WEALTHY. THE TRUTH OF THE MATTER CAN BE FOUND IN THE BAHÁ'Í PRINCIPLE AND PRACTICE OF COMPLETE ABSTENTION FROM PARTICIPATION IN PARTISAN POLITICS WHICH WAS DEMONSTRATED IN IRAN IN 1975 WHEN BAHÁ'ÍS EVEN IN THE FACE OF THREATS REFUSED TO BECOME MEMBERS OF THE RASTAKHIZ PARTY PROMOTED BY THE PREVIOUS REGIME.[3] IN ONE CASE WHEN A BAHÁ'Í ACCEPTED A CABINET POST UNDER DURESS HE WAS DEPRIVED OF MEMBERSHIP IN THE BAHÁ'Í COMMUNITY. AS TO THE ALLEGATION THAT THE BAHÁ'Í COMMUNITY REAPED FINANCIAL REWARD BECAUSE OF ACTIVE INVOLVEMENT WITH THE PREVIOUS REGIME THE FACT IS THAT THE VAST MAJORITY OF IRANIAN BAHÁ'ÍS ARE OF THE POORER CLASSES LIVING IN VILLAGES. FEW ARE WEALTHY, AND AMONG THEM A NUMBER WERE BUSINESSMEN WHO PROVIDED FACILITIES FOR EMPLOYMENT OF THOUSANDS OF WORKERS. THE FEW WHO RIGHTLY OR WRONGLY ARE BEING ACCUSED OF CORRUPTION AND OTHER OFFENSES SHOULD NOT BE REGARDED AS REPRESENTATIVE OF THE BAHÁ'Í COMMUNITY AS A WHOLE. IT IS AN INJUSTICE TO HOLD ANY RELIGIOUS COMMUNITY RESPONSIBLE FOR THE ILL-DOINGS OF ANY ONE OF ITS MEMBERS WHO FAILS TO REFLECT THE PRINCIPLES PROMULGATED BY THAT RELIGION.

239.7 AS THE NEW CONSTITUTION MAKES NO REFERENCE TO THE BAHÁ'ÍS, WAYS AND MEANS SHOULD BE SOUGHT TO EXTEND TO THE BAHÁ'Í COMMUNITY PROTEC-

---

239-2. The Islamic Republic, which was led at that time by Ayatollah Khomeini.

239-3. In 1975 the Shah founded the Rastakhiz (Resurrection) Party, the only government-authorized political party. In the face of government pressure to join, Bahá'ís refused, in adherence to the Bahá'í principle of noninvolvement in political activities.

TION OF ITS INTERESTS, AND TO ENSURE FOR ITS INDIVIDUAL MEMBERS BASIC CIVIL RIGHTS THUS AVOIDING FRICTION AND FRUSTRATION IN SUCH OFT-RECURRING PERSONAL PROBLEMS RELATED TO REGISTRATION OF MARRIAGES AND BIRTHS, EMPLOYMENT, TRAVELING DOCUMENTS, ETC.

UNLESS THESE DISABILITIES CURRENTLY AFFLICTING IRAN'S LARGEST RELIGIOUS MINORITY ARE REMEDIED, FANATICAL ELEMENTS WILL BE GIVEN FREE REIN TO REPEATEDLY RESORT TO MOB VIOLENCE AGAINST THE BAHÁ'ÍS, EMBARRASSING THE GOVERNMENT AND PREVENTING HOPED-FOR PEACE AND TRANQUILLITY IN THAT COUNTRY. . . . 239.8

ONE OF THE OFT-REPEATED ACCUSATIONS AGAINST BAHÁ'ÍS IS THAT THEY ARE ENEMIES OF ISLAM. THIS CHARGE ASSUMES NOW NEW PROPORTIONS AS MANY RIGHTS AND LIBERTIES IN NEW CONSTITUTION APPLY ONLY IF INDIVIDUALS AND COMMUNITIES CONCERNED ARE NOT REGARDED AS ANTI-ISLAMIC. HENCE OFFICIAL BRANDING BAHÁ'Í FAITH AS ANTI-ISLAMIC MAY BE CONVENIENT DEVICE TO DENY BAHÁ'ÍS ESSENTIAL HUMAN RIGHTS. FURTHERMORE, SOMETIMES DISTINCTION IS MADE BETWEEN BAHÁ'Í FAITH AND OTHER RELIGIONS BY STATING THAT OUR FAITH APPEARED AFTER ISLAM AND THEREFORE IS NOT CONSIDERED BY MUSLIMS AS A RELIGION ENTITLED TO RIGHTS OF OTHER RELIGIONS. SUCH THEOLOGICAL DIFFERENCES SHOULD NOT BE CAUSE DENIAL CIVIL RIGHTS. SAME SITUATION APPLIES TO MUSLIMS WHO RESIDE IN CHRISTIAN COUNTRIES, AND ENJOY FULL RELIGIOUS AND CIVIL RIGHTS. 239.9

REGARDING PROPERTIES HELD BY UMANÁ COMPANY:[4] THESE PROPERTIES CONSIST PRIMARILY OF BAHÁ'Í HOLY AND HISTORICAL SITES HELD IN TRUST BY IRANIAN BAHÁ'ÍS ON BEHALF THEIR CORELIGIONISTS THROUGHOUT WORLD AS WELL AS PROPERTIES OF PURELY RELIGIOUS SIGNIFICANCE SUCH AS TEMPLE LAND, COMMUNITY CENTERS AND CEMETERIES AND MANY OF THESE HAVE BEEN IN BAHÁ'Í POSSESSION FOR OVER A CENTURY. 239.10

REGARDING NAWNAHÁLÁN:[5] THIS COMPANY WAS FOUNDED PRIOR TO RULE PAHLAVI DYNASTY. OVERWHELMING MAJORITY OF FIFTEEN THOUSAND BAHÁ'ÍS WHO HAVE SHARES AND INVESTMENTS IN COMPANY ARE NOT OF WEALTHY CLASS, AND DEPEND FOR THEIR LIVELIHOOD ON INCOME THEY WERE DERIVING FROM THEIR ASSETS IN THE COMPANY. . . . 239.11

239-4. For information about the Umaná Company, see the glossary.
239-5. For information about Nawnahálán, see the glossary.

240
Release of a Compilation on Inspiring the Heart
24 OCTOBER 1979

To all National Spiritual Assemblies

Dear Bahá'í Friends,

240.1 The provision and dissemination of a balanced supply of Bahá'í literature is one of the aims set forth in the Seven Year Plan. The Universal House of Justice has been considering this aspect of the Plan, and has asked us to convey its comments to you.

240.2 The House of Justice hopes that every National Spiritual Assembly will provide the believers under its jurisdiction with publications of the Words of Bahá'u'lláh, the Báb, and 'Abdu'l-Bahá. It is explicit in the Holy Text that the followers of the Most Great Name should recite the verses daily. How is this possible for the thousands of Bahá'ís who do not have access to these Holy Words in a language which they can understand? Furthermore, the way is open for the consolidation and maturing of the Bahá'í community when the hearts of its members can be exposed to the Divine Teachings in their pure form.

240.3 With these principles in mind, the House of Justice asked a committee at the World Center to prepare a compilation from previously published texts covering a broad range of subjects dealt with by the Central Figures of the Faith, including material which can be easily comprehended, inspire the heart, strengthen the spirit of faith, and enrich the spiritual understanding of the reader.

240.4 Such a compilation has now been prepared, and it is being sent to you by airmail under separate cover.¹ It is not meant to supersede any material you may have already compiled. It is a sample of what can be done in this vital area of Bahá'í activity. You should, therefore, feel free to use or translate as much or as little of this material as you wish, to add selections from the Writings which you feel are particularly applicable to the friends in your area, and to publish and distribute your own compilations as quickly and inexpensively as possible.

240.5 It is the hope of the Universal House of Justice that the workers in the Divine Vineyard in every land will always and increasingly have recourse to the Writings revealed by the Central Figures of our Faith, will appreciate the potency of the Holy Word, and will allow its ennobling and spiritualizing influence to stimulate and direct their personal lives and guide them in devoted services to the Cause of God.

With loving Bahá'í greetings,
DEPARTMENT OF THE SECRETARIAT

240-1. The compilation was published under the title *Inspiring the Heart* by the Bahá'í Publishing Trust of the United Kingdom (n.d.).

241
Demolition of the House of the Báb and Adjacent Bahá'í Properties
19 November 1979

To National Spiritual Assemblies

241.1 WITH SORROWFUL HEARTS SHARE NEWS FRESH ATTACK BÁB'S HOUSE HAS RESULTED DEMOLITION REMNANTS BLESSED HOUSE AND DESTRUCTION TWO OTHER ADJACENT BAHÁ'Í PROPERTIES. WALLS SEPARATING COURTYARDS THESE HOUSES AND ALL WALLS BORDERING STREET ALSO DISMANTLED. DEMOLITION ADDITIONAL BAHÁ'Í HOUSES IMMEDIATE VICINITY GRAVELY THREATENED. . . .

UNIVERSAL HOUSE OF JUSTICE

242
Excavation of Temple Site in Western Samoa
6 December 1979

To all National Spiritual Assemblies

242.1 JOYFULLY ANNOUNCE COMMENCEMENT EXCAVATION TEMPLE SITE SAMOA PRESENCE HIS HIGHNESS MALIETOA TANUMAFILI II[1] SPECIAL CEREMONY DECEMBER FIRST. THIS HISTORIC EVENT COINCIDES COMPLETION EXCAVATION AND LETTING TENDERS FOR CONSTRUCTION MOTHER TEMPLE INDIA. CONFIDENT THESE VICTORIES WILL RAISE SPIRITS BRING JOY HEARTS FRIENDS DISTRESSED RECENT EVENTS CRADLE FAITH. OFFERING ARDENT PRAYERS THANKSGIVING HOLY THRESHOLD.

UNIVERSAL HOUSE OF JUSTICE

243
Passing of the Hand of the Cause of God Raḥmatu'lláh Muhájir
30 December 1979

To all Hands of the Cause of God and National Spiritual Assemblies

243.1 PROFOUNDLY LAMENT UNTIMELY PASSING IN QUITO ECUADOR BELOVED HAND CAUSE RAḤMATU'LLÁH MUHÁJIR FOLLOWING HEART ATTACK COURSE HIS LATEST

242-1. The first reigning monarch to become a Bahá'í. For the announcement of his acceptance of the Faith, see message dated 7 May 1973 (no. 130); for the announcement of his visit to the resting-place of Shoghi Effendi, see message dated 5 October 1976 (no. 177).

SOUTH AMERICAN TOUR.¹ UNSTINTED UNRESTRAINED OUTPOURING OF PHYSICAL SPIRITUAL ENERGIES BY ONE WHO OFFERED HIS ALL PATH SERVICE HAS NOW CEASED. POSTERITY WILL RECORD HIS DEVOTED SERVICES YOUTHFUL YEARS CRADLE FAITH HIS SUBSEQUENT UNIQUE EXPLOITS PIONEERING FIELD SOUTHEAST ASIA WHERE HE WON ACCOLADE KNIGHTHOOD BAHÁ'U'LLÁH HIS CEASELESS EFFORTS OVER TWO DECADES SINCE HIS APPOINTMENT HAND CAUSE STIMULATING IN MANY LANDS EAST WEST PROCESS ENTRY BY TROOPS. FRIENDS ALL CONTINENTS WHO MOURN THIS TRAGIC LOSS NOW SUDDENLY DEPRIVED COLLABORATION ONE WHO ENDEARED HIMSELF TO THEM THROUGH HIS GENTLENESS HIS LUMINOUS PERSONALITY HIS EXEMPLARY UNFLAGGING ZEAL HIS CREATIVE ENTHUSIASTIC APPROACH TO FULFILLMENT ASSIGNED GOALS. URGE FRIENDS EVERYWHERE HOLD MEMORIAL GATHERINGS BEFITTING HIS HIGH STATION UNIQUE ACHIEVEMENTS. MAY HIS RADIANT SOUL ABHÁ KINGDOM REAP RICH HARVEST HIS DEDICATED SELF-SACRIFICING SERVICES CAUSE GOD. . . .

UNIVERSAL HOUSE OF JUSTICE

244

Release of a Compilation on Divorce
18 January 1980

To all National Spiritual Assemblies

Dear Bahá'í Friends,

244.1 The Universal House of Justice has noted with increasing concern that the undisciplined attitude of present-day society towards divorce is reflected in some parts of the Bahá'í World Community. Our Teachings on this subject are clear and in direct contrast to the loose and casual attitude of the "permissive society," and it is vital that the Bahá'í Community practice these Teachings.

244.2 In order to help the believers appreciate the need to preserve the sacred marital bond, the Research Department has, at the instruction of the House of Justice, prepared a compilation of texts on the reprehensibility of divorce in the light of our Teachings.¹ This compilation is now ready and is being shared with all National Spiritual Assemblies, who are left free to use the texts in any manner they deem advisable.

With loving Bahá'í greetings,
DEPARTMENT OF THE SECRETARIAT

243-1. For an account of the life and services of Raḥmatu'lláh Muhájir, see BW 18:651–69.
244-1. See CC 1:235–44.

245
Passing of Inparaju Chinniah, Continental Counselor
7 February 1980

To the National Spiritual Assembly of the Bahá'ís of Malaysia

DEEPLY GRIEVED UNTIMELY PASSING DEVOTED COWORKER INPARAJU CHINNIAH.[1] HIS OUTSTANDING UNTIRING SERVICES INSTITUTIONS FAITH BOTH MALAYSIA AND SOUTHEAST ASIA SHED LUSTER ANNALS CAUSE GOD ENTIRE REGION. PRAYING HOLY THRESHOLD PROGRESS SOUL ABHÁ KINGDOM MAY BELOVED FRIENDS MALAYSIA INCREASE FERVOR SERVITUDE BAHÁ'U'LLÁH FOLLOW EXAMPLE DEPARTED FRIEND COMPENSATE HIS LOSS THEIR MIDST. ASSURE FAMILY FRIENDS SYMPATHY. ADVISE HOLD BEFITTING MEMORIAL MEETINGS.

Universal House of Justice

246
Message to Iranian Bahá'ís throughout the World
10 February 1980

To the dear Iranian believers
resident in other countries throughout the world[1]

In these tumultuous days when the lovers of the Best Beloved are remote from their homeland, associated with their fellow-believers in other lands, and participating in the services of the loyal supporters throughout the world, we felt it necessary to convey our thoughts to those distinguished friends, with absolute sincerity and affection, and invite them to that which we believe can guarantee their tranquillity and happiness, as well as their eternal salvation and redemption, so that with firm steps and sure hearts they may, God willing, withstand the onslaughts which have and will afflict all the countries of the globe. Thus they may fix their gaze on the dawn of the fulfillment of the soul-vitalizing promises of God and remain certain that behind these dark clouds the Sun of the Will of God is shining resplendent from its height of glory and might. Before long these dark clouds of contention, negligence,

---

245-1. For an account of the life and services of Inparaju Chinniah, see BW 18:711–13.

246-1. This is an English translation, prepared by the National Spiritual Assembly of the Bahá'ís of the United Kingdom, of a letter originally written in Persian. In a letter dated 29 July 1980 in which it forwarded the English translation to all National Spiritual Assemblies, the Universal House of Justice wrote: "The message includes several quotations from the Writings of Bahá'u'lláh, 'Abdu'l-Bahá and Shoghi Effendi hitherto untranslated into English. The English texts of these passages, as they appear in the attached translated message, have been checked and approved at the World Center, and may be regarded by the friends as authorized texts."

fanaticism, and rebellion shall disperse, the day of victory shall dawn above the horizon, and a new age shall illumine the world. It should not be surmised that the events which have taken place in all corners of the globe, including the sacred land of Iran, have occurred as isolated incidents without any aim and purpose. According to the words of our beloved Guardian, "The invisible hand is at work and the convulsions taking place on earth are a prelude to the proclamation of the Cause of God."[2] This is but one of the mysterious forces of this supreme Revelation which is causing the limbs of mankind to quake and those who are drunk with pride and negligence to be thunderstruck and shaken. To the truth of this testifies the sacred verse: "The world's equilibrium hath been upset through the vibrating influence of this most great, this new World Order,"[3] and the repeated warnings of the Pen of the Most High, such as:

246.1a The world is encircled with calamities. Even if at times some good may be evident, it is inevitable that a great calamity followeth—and yet no one on earth hath perceived its origin.[4]

246.1b The world is in travail, and its agitation waxeth day by day. Its face is turned towards waywardness and unbelief. Such shall be its plight, that to disclose it now would not be meet and seemly. Its perversity will long continue. And when the appointed hour is come, there shall suddenly appear that which shall cause the limbs of mankind to quake. Then and only then will the Divine Standard be unfurled, and the Nightingale of Paradise warble its melody.[5]

The Calamities Ahead

246.2 Similarly, the Pen of the Center of the Covenant has repeatedly prophesied the intolerable calamities which must beset this wayward humanity ere it heeds the life-giving Teachings of Bahá'u'lláh.

246.3 Chaos and confusion are daily increasing in the world. They will attain such intensity as to render the frame of mankind unable to bear them. Then will men be awakened and become aware that religion is the impregnable stronghold and the manifest light of the world, and its laws, exhortations and teachings, the source of life on earth.[6]

246.4 Every discerning eye clearly sees that the early stages of this chaos have daily manifestations affecting the structure of human society; its destructive

---

246-2. From an unpublished letter.
246-3. GWB, p. 136.
246-4. From an unpublished Tablet.
246-5. Quoted in WOB, p. 33.
246-6. From an unpublished Tablet.

forces are uprooting time-honored institutions which were a haven and refuge for the inhabitants of the earth in bygone days and centuries, and around which revolved all human affairs. The same destructive forces are also deranging the political, economic, scientific, literary, and moral equilibrium of the world and are destroying the fairest fruits of the present civilization. Political machinations of those in authority have placed the seal of obsolescence upon the root principles of the world's order. Greed and passion, deceit, hypocrisy, tyranny, and pride are dominating features afflicting human relations. Discoveries and inventions, which are the fruit of scientific and technological advancements, have become the means and tools of mass extermination and destruction and are in the hands of the ungodly. Even music, art, and literature, which are to represent and inspire the noblest sentiments and highest aspirations and should be a source of comfort and tranquillity for troubled souls, have strayed from the straight path and are now the mirrors of the soiled hearts of this confused, unprincipled, and disordered age. Perversions such as these shall result in the ordeals which have been prophesied by the Blessed Beauty in the following words: ". . . the earth will be tormented by a fresh calamity every day and unprecedented commotions will break out." "The day is approaching when its [civilization's] flame will devour the cities."[7]

The Vision

246.5 In such an afflicted time, when mankind is bewildered and the wisest of men are perplexed as to the remedy, the people of Bahá, who have confidence in His unfailing grace and divine guidance, are assured that each of these tormenting trials has a cause, a purpose, and a definite result, and all are essential instruments for the establishment of the immutable Will of God on earth. In other words, on the one hand humanity is struck by the scourge of His chastisement which will inevitably bring together the scattered and vanquished tribes of the earth; and on the other, the weak few whom He has nurtured under the protection of His loving guidance are, in this Formative Age and period of transition, continuing to build amidst these tumultuous waves an impregnable stronghold which will be the sole remaining refuge for those lost multitudes. Therefore, the dear friends of God who have such a broad and clear vision before them are not perturbed by such events, nor are they panic-stricken by such thundering sounds, nor will they face such convulsions with fear and trepidation, nor will they be deterred, even for a moment, from fulfilling their sacred responsibilities.

The Responsibility of Providing an Example

246.6 One of their sacred responsibilities is to exemplify in their lives those attributes which are acceptable at His Sacred Threshold. Others must inhale from

---

246-7. TB, p. 166; Bahá'u'lláh, quoted in PDIC ¶3.

them the holy fragrances of the homeland of Bahá'u'lláh, the land which is the birthplace of self-sacrificing martyrs and devoted lovers of the Omnipotent Lord. They must not forget that Bahá'ís throughout the world expect much from the Iranian believers. They should hearken to the life-giving clarion call which their Peerless Beloved has given to the friends in Iran:

246.6a   The wish of 'Abdu'l-Bahá, that which attracts His good pleasure, and, indeed, His binding command, is that Bahá'ís, in all matters, even in small daily transactions and dealings with others, should act in accordance with the divine Teachings. He has commanded us not to be content with lowliness, humility and meekness, but rather to become manifestations of selflessness and utter nothingness. Of old, all have been exhorted to loyalty and fidelity, compassion and love; in this supreme Dispensation, the people of Bahá are called upon to sacrifice their very lives. Notice the extent to which the friends have been required in the Sacred Epistles and Tablets, as well as in our Beloved's Testament, to be righteous, well-wishing, forbearing, sanctified, pure, detached from all else save God, severed from the trappings of this world and adorned with the mantle of a goodly character and godly attributes.[8]

246.6b   First and foremost, one should use every possible means to purge one's heart and motives, otherwise, engaging in any form of enterprise would be futile. It is also essential to abstain from hypocrisy and blind imitation, inasmuch as their foul odor is soon detected by every man of understanding and wisdom. Moreover, the friends must observe the specific times for the remembrance of God, meditation, devotion and prayer, as it is highly unlikely, nay impossible, for any enterprise to prosper and develop when deprived of divine bestowals and confirmation. One can hardly imagine what a great influence genuine love, truthfulness and purity of motives exert on the souls of men. But these traits cannot be acquired by any believer unless he makes a daily effort to gain them. Let not the stranger, the envious and the enemy have cause to attribute the sublimity of the Faith in the past and in its early days to the appearance of outstanding and sanctified souls and the perseverance of martyrs whose absence today implies the necessary decline, weakening, scattering and annihilation of the Faith of Bahá'u'lláh.[9]

246-8. "Our Beloved's Testament" refers to 'Abdu'l-Bahá's Will and Testament.

246-9. Shoghi Effendi, letter dated 19 December 1923 to the Bahá'ís of the East. The letter was published in its entirety in Persian in *Tawqí'át-i-Mubárakih* ("Blessed Letters"), vol. 1 (Tehran: Bahá'í Publishing Trust, [1972]): 158–79. Parts of the letter translated into English appear in the compilations *Living the Life* (see CC 2: no. 1267), *Excerpts from the Writings of the Guardian on the Bahá'í Life*, comp. the Universal House of Justice (Toronto, Ont.: National Spiritual Assembly of the Bahá'ís of Canada, 1973), and *The Gift of Teaching* (London: Bahá'í Publishing Trust, 1977), albeit in slightly variant translations.

We beseech God to aid and assist them daily to center their attentions on these divine admonitions and to tread the path of faithfulness so as to secure abiding happiness. 246.6c

The Responsibility of Serving God

Another of the sacred responsibilities of the believers is their spiritual commitment to serve God's Sacred Threshold at all times and under all conditions so that they may dedicate the few, fleeting days of their lives—particularly in this age of transition—to the Cause of God, unmindful of the vicissitudes of fortune, trusting in Providence, and relieved of worries and anxieties. Witness what joyful tidings the Pen of the Most High has given to such blessed souls: 246.7

> Whatsoever occurreth in the world of being is light for His loved ones and fire for the people of sedition and strife. Even if all the losses of the world were to be sustained by one of the friends of God, he would still profit thereby, whereas true loss would be borne by such as are wayward, ignorant and contemptuous. Although the author[10] of the following saying had intended it otherwise, yet We find it pertinent to the operation of God's immutable Will: "Even or odd, thou shalt win the wager." The friends of God shall win and profit under all conditions, and shall attain true wealth. In fire they remain cold, and from water they emerge dry. Their affairs are at variance with the affairs of men. Gain is their lot, whatever the deal. To this testifieth every wise one with a discerning eye, and every fair-minded one with a hearing ear.[11] 246.7a

The Responsibility of Moderation

Yet another sacred duty is that of clinging to the cord of moderation in all things, lest they who are to be the essence of detachment and moderation be deluded by the trappings of this nether world or set their hearts on its adornments and waste their lives. If they are wealthy, they should make these bestowals a means of drawing nigh unto God's Threshold, rather than being so attached to them that they forget the admonitions of the Pen of the Most High. The Voice of Truth has said, "Having attained the stage of fulfillment and reached his maturity, man standeth in need of wealth, and such wealth as he acquireth through crafts or professions is commendable and praiseworthy in the estimation of men of wisdom."[12] If wealth and prosperity become the means of service at God's Threshold, it is highly meritorious; otherwise it would be better to avoid them. Turn to the Book of the Covenant, the Hidden 246.8

246-10. Sa'dí, Muṣliḥu'd-Dín of Shiraz (d. 691 A.H./1291 A.D.), famed author of the "Gulistán" and other poetical works.
246-11. See CC 1:153–54.
246-12. TB, p. 35.

Words, and other Tablets, lest the cord of your salvation become a rope of woe which will lead to your own destruction. How numerous are those negligent souls, particularly from among your own compatriots, who have been deprived of the blessings of faith and true understanding. Witness how, no sooner had they attained their newly amassed wealth and status, than they became so bewitched by them as to forget the virtues and true perfections of man's station. They clung to their empty and fruitless lifestyle. They had naught else but their homes, their commercial success, and their ornamental trappings of which to be proud. Behold their ultimate fate. Many a triumphal arch was reduced to a ruin, many an imperial palace was converted into a barn. Many a day of deceit turned into a night of despair. Vast treasures changed hands and, at the end of their lives, they were left only with tears of loss and regret. ". . . all that perisheth and changeth is not, and hath never been, worthy of attention, except to a recognized measure."[13] Therefore the people of Bahá must not fall prey to the corruption of the ruthless, but rather cling to contentment and moderation. They must make their homes havens for the believers, folds for their gatherings and centers for the promulgation of His Cause and the diffusion of His love, so that people of all strata, whether high or low, may feel at home and be able to consort in an atmosphere of love and fellowship.

The Responsibility of Resettling

246.9   Another sacred responsibility of those dear Iranian friends now living abroad is to consult with the Assemblies and Bahá'í Institutions so that their settlement in needy areas may help the establishment and consolidation of the Faith. They must serve on the pioneer front wherever they reside. They must not allow themselves to be drawn to and congregate in areas where their relatives or friends reside, unaware of the pioneering needs of the Faith. Praised be God that, through the blessings of the Greatest Name, the believers have been imbued with a love and unity which transcends the ties of kinship and friendship and overcomes the barriers of language and culture. Therefore there is no need for the Iranian friends to congregate in one place. Often such a congregation creates problems. For example, should the number of Iranians exceed the number of native believers in a community, they would inadvertently bring about such difficulties as might hamper the progress of the Cause of God, and the world-conquering religion of the Abhá Beauty might appear to others as a religion which is limited and peculiar to Iranians. This could but lead to a waste of time and the disenchantment of both Bahá'ís and non-Bahá'ís. Under such circumstances the dear Iranian friends would neither enjoy their stay in that place nor would they be able to serve the Faith in a befit-

---

246-13. TB, p. 219.

ting manner. It is our ardent hope that, wherever possible, the Iranian friends may settle in those towns or villages which are pioneering goals, so that through their stay the foundation of the Cause may be strengthened. They must encourage each other to pioneer and disperse in accordance with the teaching plans wherever they reside, and sacrifice the happiness and joy which they may otherwise obtain from companionship with each other for the sake of the vital interests of the Cause.

The Responsibility of Avoiding Political Involvement

246.10 Another of the sacred duties incumbent upon the believers is that of avoiding participation in political discussions and intrigues which have become popular nowadays. What have the people of Bahá to do with political contention and controversy? With absolute certainty they must prove to the world that Bahá'ís, by virtue of their beliefs, are loyal citizens of whatever country they reside in and are far removed from the machinations of conspirators and the perpetrators of destruction and chaos. Their ideal is the happiness of all the peoples of the world and sincere and wholehearted service to them. In administrative positions they are obedient to their governments and carry out their duties with the utmost honesty and trustworthiness. They regard no faction as superior to another and prefer no individual above another. They oppose no one, for the Divine Pen has prohibited sedition and corruption and enjoined peace and harmony upon us. For more than a century Bahá'ís have proven by their deeds that they regard servitude and service to their fellowman as being more worthy than the privileges of power which can be gained from politics. In administering their own affairs, they rely on God rather than on the influence of those in power and authority. Particularly in these days when the enemies of the Faith have afflicted the Cause in the sacred land of Iran with the darts of calumny and slander on every side, the dear Iranian friends should be vigilant, both in their contact with other Iranians abroad and with people in general, and behave in such a way as to leave no doubt as to the independence and nonalignment of the Bahá'ís and their good will to all people, whether in Iran or elsewhere. They must not give a new excuse to cause trouble to those mischief-makers who have always sought to further their own unworthy ends by making the Bahá'í community a target for their malicious accusations.

The Strengths of the Iranian Bahá'ís

246.11 O beloved of God, and compatriots of the Abhá Beauty! Your relationship to the Blessed Perfection merits befitting gratitude. Having appreciated the true value of so inestimable a bounty, your forefathers regarded the offering of their lives in the path of their kind Beloved as easy to make. They were put in chains, became captives of the sword, lost their homes and belongings, yet no sound was heard from their lacerated throats but the cry of "Yá Bahá'u'l-Abhá" and "Yá 'Alíyyu'l-A'lá."[14] The vibration of the sound of that same soul-burning cry gradually noised abroad the call of this world-illuminating Great Announcement and the ringing notes of that call resounded in all regions of the world, and, now too, the beloved friends in Iran, who are the devoted dwellers in the courtyard of the Beloved, stand firm in the same Covenant and Testament. Behold the courage, firmness, detachment, unity, cooperation, zeal and enthusiasm with which these loyal lovers of the Beloved daily face their tests and prove and demonstrate to the world, with radiant and shining faces, their purity, their heritage, their quality, and their virtue. With the utmost meekness, truthfulness, wisdom, and courage they meet the challenges presented to them, the challenge of defying the enemies, dispelling misunderstandings which are a result of the proliferation of calumnies and false accusations. They have met their fate with acquiescence, have bowed their heads in the valley of submission and resignation, and have borne every tribulation with radiance, for they know with absolute certainty that the fulfillment of divine prophecies will coincide with dire events and the bearing of innumerable afflictions. The beloved Guardian says: "If in the days to come, adversities of various kinds should encircle that land and national upheavals should further aggravate the present calamities, and intensify the repeated afflictions," the dear friends in that country should not feel "sorrowful and grieved" and must not be deflected "from their straight path and chosen highway."[15] He then continues to address the dear friends in these words:

246.11a The liberation of this meek and innocent band of His followers from the fetters of its bondage and the talons of the people of tyranny and enmity must needs be preceded by the clamor and agitation of the masses. The realization of glory, of tranquillity, and of true security for the people of Bahá will necessitate opposition, aggression and commotion on the part of the people of malevolence and iniquity. Therefore, should the buffeting waves of the sea of tribulation intensify and the storms of trials and tribulations assail that meek congregation from all six sides,[16] know of a certainty and without a moment's hesitation that the time for their deliverance has drawn nigh, that the age-old promise of their assured glory will soon be fulfilled, and that at long last the

---

246-14. *Yá Bahá'u'l-Abhá* (O Thou the Glory of the Most Glorious!) is a form of the Greatest Name and is used as an invocation to Bahá'u'lláh; *Yá 'Alíyyu'l-A'lá* (O Thou the Exalted of the Most Exalted!) is an invocation addressed to the Báb.

246-15. From an unpublished letter.

246-16. "All six sides" means all sides, or every side, as the prayer of the Báb for protection (BP, p. 135) indicates: "in front of us and behind us, above our heads, on our right, on our left, below our feet and every other side to which we are exposed."

means are provided for the persecuted people of Bahá in that land to attain salvation and supreme triumph. A firm step and an unshakable resolve are essential so that the remaining stages may come to pass and the cherished ideals of the people of Bahá may be realized on the loftiest summits, and be made manifest in astounding brilliance. "Such is God's method, and no change shalt thou find in His method."[17]

246.11b That is why those royal falcons who soar in the firmament of God's love have arisen with such joy, tranquillity, and dignity that their serenity has become a magnet for the attraction of the confirmations of the Concourse on High and has brought such a resounding success to them as has astonished and startled the people of Bahá throughout the world. Others have been inspired by the example of those treasured brethren to renew their pledge to their All-Glorious Beloved and to serve His Sacred Threshold with high endeavor. Thus they endeavor, as far as possible, to make good the temporary disability of the believers in Iran. Inspired by the courage, constancy, sincerity, and devotion of those enamored friends in the path of their Beloved, they are increasing their services and renewing and strengthening their resolve so that they may arise in the arena of the love of God as it beseems true lovers. That is why in these days the followers of the Greatest Name in different parts of the world have undertaken to win new victories in remembrance and on behalf of their dear friends in Iran. They have made new plans and their efforts have been confirmed with resounding success, which they attribute to the influences of the high endeavors and the constancy of the friends in the Cradle of the Faith. What then will you do, dear friends who come from that sacred land of Iran? You are the birds of that rose garden. You should sing such a song that the hearts of others will rejoice with gladness. You are the candles of that Divine Sanctuary. You should shed such a light that it will illumine the eyes of the intimates of God's mysteries. Our eager hearts in these days are expectant to see the rays of loyalty and integrity from amidst these dark and threatening clouds, so that your blessed names, like those of your self-sacrificing compatriots, may be recorded in gold upon the Tablet of Honor. This is dependent upon your own high endeavor.

THE UNIVERSAL HOUSE OF JUSTICE

246-17. From an unpublished letter.

247
Passing of the Hand of the Cause of God Hasan M. Balyuzi
12 February 1980

To all National Spiritual Assemblies

247.1 HOLD SPECIAL MEMORIAL GATHERING IN MOTHER TEMPLE. WITH BROKEN HEARTS ANNOUNCE PASSING DEARLY LOVED HAND CAUSE HASAN BALYUZI.[1] ENTIRE BAHÁ'Í WORLD ROBBED ONE OF ITS MOST POWERFUL DEFENDERS MOST RESOURCEFUL HISTORIANS. HIS ILLUSTRIOUS LINEAGE HIS DEVOTED LABORS DIVINE VINEYARD HIS OUTSTANDING LITERARY WORKS COMBINE IN IMMORTALIZING HIS HONORED NAME IN ANNALS BELOVED FAITH. CALL ON FRIENDS EVERYWHERE HOLD MEMORIAL GATHERINGS. PRAYING SHRINES HIS EXEMPLARY ACHIEVEMENTS HIS STEADFASTNESS PATIENCE HUMILITY HIS OUTSTANDING SCHOLARLY PURSUITS WILL INSPIRE MANY DEVOTED WORKERS AMONG RISING GENERATIONS FOLLOW HIS GLORIOUS FOOTSTEPS.

UNIVERSAL HOUSE OF JUSTICE

248
Naw-Rúz Message 1980
Naw-Rúz 1980

To the Bahá'ís of the World

Dearly loved Friends,

248.1 The successful launching of the Seven Year Plan and the advances made in the first year of its opening phase mitigate, in some degree, the disasters and calamities which, in the past year, have assailed the struggling Faith of God. The newest wave of persecution unleashed against us in the Cradle of our Faith has been compounded by Divine decree afflicting the entire Bahá'í world community. In the full tide of their brilliant services to the Faith of God, and within the short span of twenty weeks three Chief Stewards of Bahá'u'lláh's embryonic World Order, the Hands of the Cause of God Enoch Olinga, Rahmatu'lláh Muhájir and Hasan Balyuzi were summoned to the Abhá Kingdom, leaving the rest of us bereft and shocked by the enormity of our loss and the tragic brutality of the circumstances attending the murder of beloved Enoch Olinga and members of his family.

247-1. For an account of the life and services of Hasan Balyuzi, see BW 18:635–51.

The Dialectic of Disaster and Triumph

248.2 In Iran, the confusion which has seized the whole country opened the way for the fierce and inveterate enemies of the Faith, unrestrained by any effective authority, to indulge their fanatical hatred. The Holy House of the Báb has been demolished and proposals have been made to erase its very site. The Síyáh-Chál and Bahá'u'lláh's Home in Tehran have been seized, together with all other Holy Places and properties. One member of the National Spiritual Assembly and two of the Local Spiritual Assembly of Tehran have been kidnapped and the whereabouts of two of them is still unknown, while the third is still in prison. Also, a Counselor and some friends who are associated with the National Office or are members of the Local Spiritual Assembly of Tehran have been imprisoned. Bahá'ís have been heavily pressed to recant their faith and in one case a believer, who refused to do so, followed the glorious path of the martyrs and was executed. Beyond all this a campaign of vilification and false charges has been conducted against the friends in an effort to make them the scapegoat of unrestrained mobs.

248.3 And yet, as ever in the Cause of God, the beneficent operation of the dialectic of disaster and triumph is clearly apparent. The unwavering faith of the dearly loved, severely tested, ever-steadfast Persian Bahá'í community, guided by the heroic stand and example of its National Spiritual Assembly, supported and inspired by the Counselors and their Auxiliary Board members, has effected a spiritual revitalization of the beloved friends. They have united as one man to present a front of refulgent spirituality and assurance and appear, as one observer reports, like a dazzling community of eager, uplifted, radiant new believers.

248.4 Nor is the influence of their response to the sufferings engulfing them confined to their homeland. From farthest east to farthest west, from pole to pole, wherever the Standard of Bahá'u'lláh has been implanted, the friends have felt the impulse of sacrifice and risen to assume that enormous share of the work of the Faith in the fields of teaching, pioneering and financial contribution which the Persian friends, for the time being, are no longer able to shoulder.

248.5 The wonderful love aroused in Bahá'í hearts everywhere by the sudden, untimely passing of the beloved Hands of the Cause has moved the believers to dedicate themselves anew with increased ardor and self-sacrifice to the promotion of the work to which all the Hands of the Cause of God have dedicated their lives.

248.6 The worldwide response of the friends to these tragedies is the more heartening in view of the clear warnings voiced by 'Abdu'l-Bahá and the beloved Guardian of the fierce and widespread opposition which the increasing growth of the Cause of God will arouse. There is no doubt of this. Shoghi Effendi called attention to "the extent and character of the forces that are destined to contest with God's holy Faith," and supported his argument with "these prophetic and ominous words" from 'Abdu'l-Bahá: "HOW GREAT, HOW VERY GREAT IS THE CAUSE! HOW VERY FIERCE THE ONSLAUGHT OF ALL THE PEOPLES AND KINDREDS OF THE EARTH! ERELONG SHALL THE CLAMOR OF THE MULTITUDE THROUGHOUT AFRICA, THROUGHOUT AMERICA, THE CRY OF THE EUROPEAN AND OF THE TURK, THE GROANING OF INDIA AND CHINA, BE HEARD FROM FAR AND NEAR. ONE AND ALL THEY SHALL ARISE WITH ALL THEIR POWER TO RESIST HIS CAUSE. THEN SHALL THE KNIGHTS OF THE LORD, ASSISTED BY HIS GRACE FROM ON HIGH, STRENGTHENED BY FAITH, AIDED BY THE POWER OF UNDERSTANDING, AND REINFORCED BY THE LEGIONS OF THE COVENANT, ARISE AND MAKE MANIFEST THE TRUTH OF THE VERSE: 'BEHOLD THE CONFUSION THAT HATH BEFALLEN THE TRIBES OF THE DEFEATED!' "[1]

248.7 The beloved Guardian expatiated at length upon this theme and its inevitable outcome: "Stupendous as is the struggle which His words foreshadow, they also testify to the complete victory which the upholders of the Greatest Name are destined eventually to achieve."[2]

248.8 Now, therefore, it is our sacred duty to make the utmost use of our freedom, wherever it exists, to promote the Cause of God while we may. The surest way to do this and to win the good-pleasure of Bahá'u'lláh is to pursue, with dedication and unrelenting vigor, the goals of whatever Plan is in force, for Bahá'u'lláh has stated: "To assist Me is to teach My Cause."[3]

Accomplishments during the First Year of the Seven Year Plan

248.9 A good start has been made with the Seven Year Plan. At the World Center of the Faith the uninterrupted progress in raising the Seat of the House of Justice, repairing and refurbishing the House of 'Abdu'lláh Páshá,[4] further extension of the gardens surrounding the Ḥaram-i-Aqdas at Bahjí, and the initiation of a general reorganization of the work of the World Center to accommodate its ever-growing needs and make use of the most up-to-date technological developments, have taken place.

248.10 In the international sphere the enthusiasm with which the friends everywhere greeted the launching of the Seven Year Plan and girded themselves to achieve the goals of the first two-year phase, their generous and sacrificial outpouring of funds, the confident and sustained efforts exerted to carry forward the two sacred enterprises initiated in the Indian subcontinent and at the heart of the vast Pacific Ocean, the constant activity of the Bahá'í International Community in fostering its relations with the United Nations,[5] the great in-

---

248-1. WOB, p. 17; 'Abdu'l-Bahá, quoted in WOB, p. 17.
248-2. WOB, pp. 17–18.
248-3. TB, p. 196.
248-4. See messages dated 9 January 1975 (no. 154), 4 March 1975 (no. 157), and 14 October 1977 (no. 198) for further information on the House of 'Abdu'lláh Páshá.
248-5. For announcements concerning the relationship between the Bahá'í International Community and the United Nations, see messages dated 17 October 1967 (no. 49), 18 February 1970 (no. 78), 25 November 1985 (no. 442), and 17 December 1985 (no. 444).

crease in the number of children's Bahá'í classes and the innumerable victories won in the teaching field, recorded by the establishment of the worldwide community of the Most Great Name in over 106,000 localities, all testify to the unassailable, and indeed ever-increasing vigor of the Cause of God.

248.11 The number of pioneers and traveling teachers who have entered the field during the first year of the Seven Year Plan, and the increase in the number of national communities which have sent them out are highly encouraging. This stream of pioneers and traveling teachers must be increased and more widely diffused, and we fervently hope that, at the very least, all those pioneers filling the assigned goals of the first phase of the Seven Year Plan will be at their posts by Riḍván 1981.[6]

248.12 In the field of proclamation unprecedented publicity has been accorded the Cause of God, chiefly as a result of the persecutions in Iran. In addition significant gains have been made in the Bahá'í radio operation in South America, where short wave transmission has greatly extended the range of Radio Bahá'í in Otavalo, Ecuador, and where a new station is being established in Puno, in Peru, on the shores of Lake Titicaca. Both these achievements offer immeasurable new opportunities for the teaching, proclamation and consolidation of the Cause in that area.

248.13 In 88 languages of the world the supply of Bahá'í literature has been enriched, while three new languages have been added to bring to 660 the number of those in which Bahá'í material is available.

Formation of Six National Assemblies at Riḍván 1981

248.14 The National Spiritual Assembly of Transkei with its seat in Umtata will be formed at Riḍván 1980. At Riḍván 1981 six new National Spiritual Assemblies will be formed; two in Africa, Namibia with its seat in Windhoek, and Bophuthatswana with its seat in Mmabatho; three in the Americas, the Leeward Islands with its seat in St. John's, Antigua, the Windward Islands with its seat in Kingstown, St. Vincent, and Bermuda with its seat in Hamilton; one in Australasia, Tuvalu with its seat in Funafuti. With great joy we announce the reformation of the National Spiritual Assembly of Uganda, to take place at Riḍván 1981.

248.15 In the course of the coming year, the Universal House of Justice, in consultation with the International Teaching Center, will review the accomplishments of the initial phase and will then announce to all National Spiritual Assemblies the goals towards which they should strive in the next stage of the Seven Year Plan.

Approaching People of Prominence

248.16 During this final year of the initial phase National Spiritual Assemblies are urged to continue their wise and dignified approaches to people prominent in

---

248-6. 21 April–2 May.

all areas of human endeavor in order to acquaint them with the nature and spirit of the Faith and to win their esteem and friendship. At the same time vigorous campaigns must be continually mounted to proclaim more and more directly and to as large audiences as possible the existence and basic principles of the Faith of God. Now is the time, as all human endeavors to repair the old order only result in deeper and deeper confusion, to proclaim constantly and openly the claims of the Faith and the redemptive power of Bahá'u'lláh.

An Acceleration of Momentum

248.17 The marvelous momentum generated at the beginning of the Plan and now propelling the Bahá'í world community forward to the achievement of the immediate objectives of the initial phase must be maintained and indeed accelerated, so that firm foundations in the spiritual life of the community may be laid and its forces gathered for the winning of the specific tasks with which it will be challenged in the major part of the Plan.

248.18 Our hearts go out in love and admiration to the friends in Iran and in gratitude to the believers throughout the world for their spontaneous defense of their persecuted brethren and their shouldering of the load which must, at all costs, be borne.

With loving Bahá'í greetings,
THE UNIVERSAL HOUSE OF JUSTICE

249

Release of a Compilation on the Importance of Prayer, Meditation, and the Devotional Attitude
31 MARCH 1980

To all National Spiritual Assemblies

Dear Bahá'í Friends,

249.1 The Research Department has just prepared a compilation of *The Importance of Prayer, Meditation, and the Devotional Attitude,* and we have been asked to send you a copy.[1]

249.2 The Universal House of Justice leaves it to your discretion to determine the manner in which you share the contents of the compilation with the friends under your jurisdiction.

With loving Bahá'í greetings,
DEPARTMENT OF THE SECRETARIAT

249-1. See CC 2:225–43.

1979–1986 · THE SEVEN YEAR PLAN

250
Contract for Construction of the Indian Temple
1 May 1980

To the National Spiritual Assembly of the Bahá'ís of India

GRATEFUL SENTIMENTS EXPRESSED PARTICIPANTS CONVENTION HELD AUSPICIOUS 250.1 OCCASIONS SIGNING CONTRACT CONSTRUCTION BEAUTIFUL TEMPLE HEART SUBCONTINENT INAUGURATION CENTENARY CELEBRATIONS DESIGNED PROMOTE LIFE-GIVING MESSAGE BAHÁ'U'LLÁH GREATER NUMBER RECEPTIVE SOULS ALL STRATA SOCIETY LENGTH BREADTH WIDTH INDIA.[1] ARDENTLY PRAYING DIVINE THRESHOLD SUPPLICATING BLESSINGS ASSISTANCE OUTSTANDING SERVICES BAHÁ'Í COMMUNITY INDIA ALREADY DISTINGUISHED BY MAGNIFICENT ACHIEVEMENTS WHICH HAVE BRIGHTENED HOPES ATTRACTED ADMIRATION ENTIRE BAHÁ'Í WORLD.

UNIVERSAL HOUSE OF JUSTICE

251
Comments on the Kitáb-i-Aqdas
7 May 1980

The National Spiritual Assembly of the Bahá'ís of Germany

Dear Bahá'í Friends,

The Universal House of Justice has received your letter of 14 April 1980 251.1 enclosing the copy of that part of page 15 of the issue of the *Hannoversche Allgemeine Zeitung* for 14 March 1980, on which the letter from Francesco Ficicchia was published.

The House of Justice feels that the best course is to ignore this disgraceful 251.2 attack on the Faith, rather than to make any rebuttal. Your Assembly should, however, be prepared to answer any inquiries that you may receive as a result of this article, whether from Bahá'ís or non-Bahá'ís. To assist you in this, we enclose extracts from a letter written on behalf of the Universal House of Justice, which touch on matters raised by Mr. Ficicchia. You are also, no doubt, aware that a Russian translation of the Kitáb-i-Aqdas by A. G. Tumansky was published in 1899. A translation of the complete text into English, made by two former Presbyterian missionaries in Iran, was published by The Royal Asiatic Society in 1961; it is accompanied by highly prejudiced and misleading footnotes and introduction, and the inadequacy of the translation itself is

---

250-1. The centenary referred to is that of the establishment of a nucleus of the Faith in India in 1880, although Jamál Effendi was sent there by Bahá'u'lláh as early as 1872–73 and arrived in New Delhi in 1876 after touring the country.

447

immediately apparent to anyone who compares passages with those that Shoghi Effendi translated. The existence of these two published translations, however, demonstrates the falsity of Ficicchia's statement that no complete translation exists. A far clearer understanding of the contents of the Most Holy Book, however, than given by either translation, is presented in the *Synopsis and Codification of the Kitáb-i-Aqdas* which gives not only the laws of the Aqdas themselves, but also includes the elucidations given by Bahá'u'lláh Himself in the "Questions and Answers," and contains many explanatory annotations provided by the House of Justice.

With loving Bahá'í greetings,
DEPARTMENT OF THE SECRETARIAT

27 MAY 1980

Extracts from a reply written on behalf of the Universal House of Justice to questions about the Kitáb-i-Aqdas

251.3 The institution of the Covenant has a direct bearing on the implementation of the laws of the Kitáb-i-Aqdas. This Book is the repository of the basic laws for the Dispensation to be implemented gradually in accordance with the guidance given by God through those infallible Institutions which lie at the heart of the Covenant. Indeed, one of those Institutions, the Universal House of Justice, has been given by Bahá'u'lláh the task not only of applying the laws but of supplementing them and of making laws on all matters not explicitly covered in the Sacred Text. An English translation of the Kitáb-i-Aqdas was made by Dr. Earl E. Elder and Dr. William McE. Miller, two men who were Presbyterian missionaries in Persia and have long been strongly antagonistic to the Faith. A great many of the statements that they make about its history are based on the assertions of Covenant-breakers or opponents of the Faith—rather like a history of Christianity based primarily on statements by enemies of Jesus Christ. Dr. Miller, for example, places great reliance on a document called the "Nuqṭatu'l-Káf," which is, in fact, spurious, as is fully demonstrated by the Hand of the Cause Hasan Balyuzi in his book *Edward Granville Browne and the Bahá'í Faith*.[1]

251-1. The Nuqṭatu'l-Káf—Arabic for "The Point of Káf" (the letter K)—is a short chronicle of events of the Bábí Faith originally written by Ḥájí Mírzá Jání, a merchant from Káshán who was martyred in 1852. Mr. Balyuzi explains that it was later tampered with and was denounced as a forgery by Mírzá Abu'l-Faḍl, the preeminent Bahá'í scholar of the East during the Faith's Heroic Age. The Nuqṭatu'l-Káf presents a distorted history of the Bábí Faith and its doctrines. See *Edward Granville Browne*, pp. 62–88.

The reasons for the delay in the translation of the Kitáb-i-Aqdas are given in the introduction to the *Synopsis and Codification*.[2] The Kitáb-i-Aqdas itself is the kernel of a vast structure of Bahá'í law that will have to come into being in the years and centuries ahead as the unity of mankind is established and develops. Thus to properly understand the contents of that Book one should also read many other Tablets of Bahá'u'lláh relating to them, as well as the interpretations of 'Abdu'l-Bahá and the Guardian, and realize that great areas of detail have been left by Bahá'u'lláh for the Universal House of Justice to fill in and to vary in accordance with the needs of a developing society. For example: 251.4

1. The law of divorce in the Aqdas seems to apply only to a husband divorcing his wife, and not vice versa. 'Abdu'l-Bahá and the Guardian have made it quite clear that the principle enunciated by Bahá'u'lláh in the Kitáb-i-Aqdas applies equally to men and women, and the law has always been implemented in this way. Such elucidations are one of the specific functions intended by Bahá'u'lláh for the authoritative Interpreter. 251.4a

2. The Kitáb-i-Aqdas appears to allow bigamy. This is explained in Note 17 on page 59 of the *Synopsis and Codification:* "The text of the Kitáb-i-Aqdas upholds monogamy, but as it appears also to permit bigamy, the Guardian was asked for a clarification, and in reply his secretary wrote on his behalf: 'Regarding Bahá'í marriage: in the light of the Master's Tablet interpreting the provision in the Aqdas on the subject of the plurality of wives, it becomes evident that monogamy alone is permissible, since, as 'Abdu'l-Bahá states, bigamy is conditioned upon justice and as justice is impossible, it follows that bigamy is not permissible, and monogamy alone should be practiced.'" 251.4b

This is an authoritative interpretation, and as an interpretation states what is intended by the original text, it is correct to say that the Kitáb-i-Aqdas prohibits plurality of wives. This method of establishing monogamy as the law of the Faith is one example of the process referred to in the introduction to the *Synopsis and Codification* whereby there is a progressive disclosure of the full meaning of the laws of the Faith as the Dispensation unfolds. 251.4c

3. The punishments prescribed for theft, murder and arson are given only in barest outline. It is explained in Note 42 on page 64 of the *Synopsis and Codification* that these punishments are intended for a future condition of 251.4d

---

251-2. *Synopsis and Codification*, a book published by the Universal House of Justice in 1973 that contains those passages of the Kitáb-i-Aqdas that had been translated into English by Shoghi Effendi and a codification of the contents of the Kitáb-i-Aqdas and the "Questions and Answers" with explanatory notes. An annotated English translation of the Kitáb-i-Aqdas was published by the Universal House of Justice at Naw-Rúz 1993.

society and will have to be supplemented and applied by the Universal House of Justice. The punishment for theft, for example, says that for the third offense a mark must be placed on the thief's forehead (nothing is said about branding), so that people will be warned of his proclivities. All details of how the mark is to be applied, how long it must be worn, on what conditions it may be removed, as well as the seriousness of various degrees of theft have been left by Bahá'u'lláh for the Universal House of Justice to decide when the law has to be applied. Similarly, merely the fundamental principles of the punishments for murder and arson are given in the Kitáb-i-Aqdas. Willful murder is to be punished either by capital punishment or life imprisonment. Such matters as degrees of offense and whether any extenuating circumstances are to be taken into account, and which of the two prescribed punishments is to be the norm are left to the Universal House of Justice to decide in light of prevailing conditions when the law is in operation. Arson, as you yourself can see from the newspapers, is becoming an increasingly frequent offense—scarcely a day passes without some building being burned or blown up, often causing agonizing death to innocent people. Bahá'u'lláh prescribes that a person who burns a house intentionally is to be burned or imprisoned for life, but again, the application of these punishments, the method of carrying them out and the fixing of degrees of offense are left to the Universal House of Justice. Obviously there is a tremendous difference in the degree of the offense of a person who burns down an empty warehouse from that of one who sets fire to a school full of children.

251.4e From the above examples it should be clear why a translation of the Kitáb-i-Aqdas made without proper comprehensive footnotes referring to other Tablets of Bahá'u'lláh which elucidate His laws as well as to interpretations made by 'Abdu'l-Bahá and the Guardian, can give a very misleading impression—quite apart from the problem of achieving a beauty of style in the English which can approach that of the original, an aspect in which the Elder-Miller translation falls woefully short.

251.5 Although there is no explicit reference to the Guardianship in the Kitáb-i-Aqdas, the *Synopsis and Codification* lists "Anticipation of the Institution of the Guardianship." On page 214 of *God Passes By*, when summarizing the contents of the Aqdas, Shoghi Effendi states that in it Bahá'u'lláh "anticipates by implication the institution of Guardianship," and again, on page 147 of *The World Order of Bahá'u'lláh* the Guardian refers to "the verses of the Kitáb-i-Aqdas the implications of which clearly anticipate the institution of the Guardianship." One such implication is in the matter of Ḥuqúqu'lláh (The Right of God),[3]

251-3. For information on Ḥuqúqu'lláh, see the glossary. See also the letter dated 6 August 1984 (no. 404) introducing Ḥuqúqu'lláh to the West.

which is ordained in the Kitáb-i-Aqdas without provision being made for who is to receive it; in His Will and Testament 'Abdu'l-Bahá fills this gap by stating "It is to be offered through the Guardian of the Cause of God . . ."[4] Other implications of this institution can be seen in the terms in which 'Abdu'l-Bahá is appointed as the Successor of Bahá'u'lláh and the Interpreter of His Teachings. The faithful are enjoined to turn their faces towards the one whom "God hath purposed" and who "hath branched from this Ancient Root" and are bidden to refer whatsoever they do not understand in the Bahá'í writings to him who "hath branched from this mighty Stock."[5] Yet another can be seen in the provision of the Aqdas concerning the disposition of international endowments—a passage which not only refers this matter to the Aghsán (male descendants of Bahá'u'lláh) but also provides for what should happen should the line of Aghsán end before the coming into being of the Universal House of Justice. Thus the "Anticipation of the Institution of the Guardianship" is correctly included in the *Synopsis and Codification of the Kitáb-i-Aqdas*.

252

Our Attitude and Actions toward the Impending Catastrophe
13 May 1980

To an individual Bahá'í

Dear Bahá'í Friend,

252.1 Your letter of January 13 to the Universal House of Justice has been received and we are instructed to convey its response.

252.2 As a concert pianist you are uniquely endowed for service to God and mankind, for the Master states that "the musician's art is among those arts worthy of the highest praise, and it moveth the hearts of all who grieve."[1] Further, the pursuit of excellence in your art both fulfills Bahá'í admonitions and is worship manifested in your profession.

252.3 Your concern about the future in these troubled times is understandable. There is every reason to expect that the world will experience travail and testing as never before, but we do not know what form these upheavals will take, when exactly they will come, how severe they will be, nor how long they will last. In a letter dated 30 September 1950 written on behalf of the beloved Guardian to an individual believer, it is stated:

251-4. WT, p. 15.
251-5. TB, p. 221; SC, p. 27.
252-1. SWAB, p. 112.

252.3a He does not feel that fear—for ourselves or for others—solves any problems, or enables us to better meet it if it ever does arise. We do not know what the future holds exactly, or how soon we may all pass through another ordeal worse than the last one.

252.4 The important aspect for the Bahá'ís is that their attitude and actions in response to the pending catastrophe be correct. We all know that the Cause of Bahá'u'lláh is the world's only salvation, and that our duty is to actively teach receptive souls, and to do our utmost to help in the consolidation of the institutions of the Faith. Only in this way can we contribute our share of servitude at His Threshold, and we should then leave the rest to Him. . . .

252.5 The House of Justice will offer prayers at the Holy Shrines that you may have abundant opportunities to uplift the souls of men through your music and through your teaching of the Cause of the Blessed Beauty.

> With loving Bahá'í greetings,
> DEPARTMENT OF THE SECRETARIAT

253
Ominous Increase in Pressures on Iranian Bahá'ís
30 JUNE 1980

To all National Spiritual Assemblies

253.1 . . . PRESSURES ON BELEAGUERED FRIENDS OMINOUSLY INCREASING. IN ISSUE *LE MONDE* NO. 11009 TUESDAY 24 JUNE IN ARTICLE ABOUT IRAN BEGINNING FRONT PAGE ENDING SIXTH PAGE FOLLOWING STATEMENTS MADE ". . . LE VÉNÉRABLE AYATOLLAH (SADOUGHI) A INVITÉ FOULE DES FIDÈLES À 'CHASSER LES BAHAÏS QUE VOUS CONNAISSEZ DE TOUTES LES ADMINISTRATIONS ET DE LES LIVRER AU PARQUET RÉVOLUTIONNAIRE.' LA PEUR S'EST EMPARÉE DE LA COMMUNAUTÉ BAHAÏ (ENVIRON CINQUANTE MILLE PERSONNES) ET POUR CAUSE. LES PROPOS DE L'AYATOLLAH SADOUGHI ONT ÉTÉ REPRODUITS DANS LE QUOTIDIEN *INGUILAB ISLAMI* (ORGANE DU PRÉSIDENT BANI-SADR) LEUR DONNANT AINSI UNE DANGEREUSE PUBLICITÉ. UNE VAGUE D'ARRESTATIONS DE BAHAÏS A COMMENCÉ À DÉFERLER DANS DEUX VILLES: À YAZD ET À CHIRAZ.

253.2 "DIMANCHE 22 JUIN, *ETELAAT* PUBLIE UN COMMUNIQUÉ DE L'ASSOCIATION ISLAMIQUE DES FONCTIONNAIRES DE L'ETAT EXIGEANT ENTRE AUTRES LE RENVOI DE TOUS LES EMPLOYÉS DES CONFESSIONS BAHAÏES EN LEUR INTERDISANT DÉSORMAIS D'EXERCER TOUTES ACTIVITÉS RÉMUNÉRÉES."[1]

253-1. The English translation of the quotation from *Le Monde* is as follows: ". . . the venerable Ayatollah [Sadoughi] invited the mass of the faithful to 'hunt out the Bahá'ís throughout the public services and deliver them to the revolutionary prosecution department.' Fear has gripped the Bahá'í community (about fifty thousand persons), and with good cause. The remarks

SUCH PRONOUNCEMENTS AND COMMUNIQUES CAN BUT PROVOKE EASILY 253.3
EXCITABLE PASSIONS LEAD TRAGIC CONSEQUENCES FATE INDIVIDUAL BAHÁ'ÍS
INFLICT UPON MEMBERS OPPRESSED COMMUNITY UNTOLD SUFFERINGS PARTICU-
LARLY ON OCCASION HOLY DAY CURRENT MONTH SHA'BÁN AND MONTH RAMAḌÁN
ENDING MID-AUGUST. . . .[2]

UNIVERSAL HOUSE OF JUSTICE

254
Execution of Two Members of the Tabríz Spiritual Assembly
16 July 1980

To the Bahá'ís of the World

WITH HEARTS LADEN WITH SORROW YET FILLED WITH PRIDE AND ADMIRA- 254.1
TION ANNOUNCE MARTYRDOM IN CRADLE FAITH OF YADU'LLÁH ÁSTÁNÍ AND
FARÁMARZ SAMANDARÍ RESPECTIVELY CHAIRMAN MEMBER LOCAL ASSEMBLY TABRÍZ.
THESE TWO HEROIC SOULS, TOGETHER WITH BAHÁR VUJDÁNÍ OF MAHÁBÁD,
GHULÁM-ḤUSAYN A'ẒAMÍ OF SANGSAR AND MÍR-ASADU'LLÁH MUKHTÁRÍ OF BÍRJAND,
WHO DURING PAST SEVERAL MONTHS ALSO SUFFERED CRUEL MARTYRDOM, HAVE
JOINED RANKS THEIR NOBLE FOREBEARS HEROIC AGE. BAHÁ'Í WORLD WITNESSING
WITH AWE AND WONDER VALIANT BELIEVERS CRADLE FAITH, WORTHY HEIRS OF
DAWN-BREAKERS, ONCE AGAIN BEARING SUCCESSIVE WAVES OF REPRESSION AND
HARASSMENT, ARRESTS AND KILLINGS, AND CONFISCATION OF PERSONAL
PROPERTIES.

MAY FOLLOWERS MOST GREAT NAME, INSPIRED BY PRODIGIOUS ACTS HEROISM 254.2
THEIR LONG-SUFFERING BRETHREN IRAN, ARISE, WITH RENEWED CONSECRATION
AND VIGOR, WIN FRESH VICTORIES, REWARD SACRIFICES, COMPENSATE DISABILI-
TIES, CHEER LONGING HEARTS STAUNCH STEADFAST BELIEVERS HOMELAND
BAHÁ'U'LLÁH.

UNIVERSAL HOUSE OF JUSTICE

---

of the Ayatollah Sadoughi have been reproduced in the daily paper *Inquilab Islami* (organ of President Bani-Sadr), thus giving them dangerous publicity. A wave of arrests of Bahá'ís has broken out in two cities: Yazd and Shiraz.

"Sunday, 22 June, *Etelaat* publishes a communiqué of the Islamic Association of Civil Servants requiring, among other things, the dismissal of all employees who are of the Bahá'í religion, prohibiting them henceforth from practicing any remunerated activity."

253-2. The fifteenth day of the month of Sha'bán is believed to be the birthday of the Twelfth and Hidden Imám, who, according to the teachings of Shí'ah Islam, was the last in the series of successors to the Prophet Muḥammad. There are two holy days during Ramaḍán, the Muslim month of fasting—the nineteenth, which commemorates the day the Imám Ḥusayn was wounded on the battlefield of Karbilá, and the twenty-first, which commemorates his death.

255
Vigorous Steps Taken in Defense of the Bahá'ís of Iran
24 July 1980

To all National Spiritual Assemblies

255.1 VIGOROUS STEPS BEING TAKEN ENERGETICALLY BY FRIENDS ALL CONTINENTS FOLLOWING NEWS MARTYRDOM TWO FRIENDS TABRÍZ HIGHLY GRATIFYING. ALREADY INFLUENTIAL PAPERS HAVE PUBLISHED EXCEPTIONALLY FAVORABLE ARTICLES IN SUPPORT PERSECUTED FRIENDS. JUST RECEIVED NEWS CANADIAN PARLIAMENT VOTED UNANIMOUSLY DEPLORING PERSECUTION IRANIAN BAHÁ'ÍS CALLING ON CANADIAN GOVERNMENT REFER UNITED NATIONS HUMAN RIGHTS COMMISSION. CANADIAN GOVERNMENT HAS FORWARDED RESOLUTION DIRECTLY TO SECRETARY-GENERAL UNITED NATIONS. IN IRAN WAVE PERSECUTION CONTINUES UNABATED. NOT SATISFIED WITH CONFISCATION COMMUNITY PROPERTIES INCLUDING HOLY PLACES STEPS NOW TAKEN FREEZE FINANCIAL ASSETS SCORES BAHÁ'ÍS BANNING THEM FROM ALL BUSINESS DEALINGS. IN YAZD AND SHIRAZ MORE MEMBERS LOCAL ASSEMBLIES ARRESTED IMPRISONED. IN MAN<u>SH</u>ÁD ONE OF OUTLYING VILLAGES YAZD FIVE BAHÁ'ÍS THREE MEN TWO WOMEN CRUELLY MOBBED RESULTING THEIR HOSPITALIZATION YAZD. BAHÁ'ÍS RESIDENT VILLAGE IN CONSTANT DANGER. . . .

UNIVERSAL HOUSE OF JUSTICE

256
Passing of the Hand of the Cause of God Adelbert Mühlschlegel
29 July 1980

To all National Spiritual Assemblies

256.1 WITH SORROWFUL HEARTS ANNOUNCE PASSING BELOVED HAND CAUSE ADELBERT MÜHLSCHLEGEL.[1] GRIEVOUS LOSS SUSTAINED ENTIRE BAHÁ'Í WORLD PARTICULARLY FELT EUROPE MAIN ARENA HIS DISTINGUISHED SERVICES CAUSE GOD. SERVING FOR MANY YEARS NATIONAL SPIRITUAL ASSEMBLY GERMANY HE BECAME AFTER ELEVATION RANK HAND CAUSE ONE OF CHAMPION BUILDERS EMERGING EUROPEAN BAHÁ'Í COMMUNITY CONSTANTLY TRAVELING ENCOURAGING RAISING SPIRITS FRIENDS RESIDING WHEREVER SERVICES MOST NEEDED FINALLY PIONEERING GREECE AND SURRENDERING HIS SOUL PIONEER POST. HIS CONSTANT WILLINGNESS SERVE HIS ABILITY ENDEAR HIMSELF BELIEVERS AND OTHERS ALIKE BY HIS LOVING GENTLENESS SERENE HUMILITY RADIANT CHEERFULNESS HIS NEVER-

---

256-1. For an account of the life and services of Adelbert Mühlschlegel, see BW 18:611–13.

CEASING PURSUIT KNOWLEDGE AND TOTAL DEDICATION BLESSED BEAUTY PROVIDE WONDERFUL EXAMPLE BAHÁ'Í LIFE. ADVISE FRIENDS COMMEMORATE HIS PASSING AND REQUEST BEFITTING MEMORIAL SERVICES ALL MOTHER TEMPLES. . . .

UNIVERSAL HOUSE OF JUSTICE

257
Commencement of Construction of Indian Temple
30 JULY 1980

To the National Spiritual Assembly of the Bahá'ís of India

REJOICE NEWS COMMENCEMENT CONSTRUCTION MASHRIQU'L-ADHKÁR HEART INDIAN SUBCONTINENT. THIS ACHIEVEMENT IS IN SHARP CONTRAST SAD HAPPENINGS CRADLE FAITH AND WILL BE SOURCE DELIGHT BELIEVERS ALL LANDS. SUPPLICATING SACRED THRESHOLD THAT TEMPLE OF LIGHT WILL SOON BECOME INSTRUMENT RADIATE DIVINE GUIDANCE ILLUMINE HEARTS OF THE MULTITUDES. MAY BAHÁ'ÍS OF INDIA ARISE TO MEET RESPONSIBILITIES RENDER UTMOST ASSISTANCE RAISING THIS BEAUTIFUL EDIFICE IN THE NAME OF BAHÁ'U'LLÁH.

257.1

UNIVERSAL HOUSE OF JUSTICE

258
Translation of the Long Healing Prayer and the Fire Tablet
13 AUGUST 1980

To all National Spiritual Assemblies

Dear Bahá'í Friends,

The Universal House of Justice recently commissioned the translation into English of two of the important works of Bahá'u'lláh, namely, the Long Healing Prayer and His Tablet entitled "Qad-Iḥtaraqa'l-Mukhliṣún," known to many in the West as the "Fire Tablet."

258.1

These translations have now been checked and approved, and a copy of each is enclosed.[1] The Universal House of Justice leaves it entirely to your discretion to determine how these two works may be shared with the friends under your jurisdiction. There is no objection to their inclusion in prayer books published by National Spiritual Assemblies.

258.2

With loving Bahá'í greetings,
DEPARTMENT OF THE SECRETARIAT

258-1. Both prayers are published in BP, pp. 91–99, 214–20.

259
Arrest of National Spiritual Assembly of Iran and Two Auxiliary Board Members
24 August 1980

To selected National Spiritual Assemblies

259.1 WITH HEARTS BURNING INDIGNATION DISTRESS ANNOUNCE ALL MEMBERS NATIONAL ASSEMBLY IRAN TOGETHER WITH TWO AUXILIARY BOARD MEMBERS FORCIBLY AND PEREMPTORILY TAKEN FROM MEETING IN PRIVATE HOME ON EVENING 21 AUGUST TO UNKNOWN DESTINATION. VIEWING RISING TIDE PERSECUTIONS STEADFAST HEROIC LEADERS HARASSED COMMUNITY HAD ALREADY ANTICIPATED POSSIBILITY SUCH AN UNWARRANTED ACTION AND HAD MADE ARRANGEMENTS FOR NINE ALTERNATIVE MEMBERS ASSUME RESPONSIBILITIES NATIONAL ASSEMBLY IN CASE THEY WERE UNABLE FUNCTION. NINE MEMBERS APPOINTED BODY NOW AT HELM DETERMINED DISCHARGE SACRED DUTIES FOLLOW IN FOOTSTEPS THEIR COURAGEOUS PREDECESSORS, REMINISCENT SPIRIT EVINCED UNDAUNTED HEROES FAITH DURING MEMORABLE EPISODE BÁRFURÚSH.[1] NEW NATIONAL ASSEMBLY TOGETHER WITH WIVES CLOSE RELATIVES ARRESTED FRIENDS HAVE APPEALED AUTHORITIES FOR RELEASE LAW-ABIDING CITIZENS. THROUGH THEIR CONSTANCY AND VALOR PERSIAN FRIENDS HAVE ONCE AGAIN SET AN EXAMPLE OF SERVITUDE AT SACRED THRESHOLD WORTHY EMULATION ALL FRIENDS EVERY LAND PARTICULARLY THOSE PRIVILEGED SERVE ACTIVELY ON BAHÁ'Í INSTITUTIONS. CALL ON FRIENDS EVERYWHERE JOIN US IN FERVENT PRAYERS THAT SELF-SACRIFICING PERSIAN BRETHREN MAY AT LAST YIELD GOLDEN FRUIT THEIR RELEASE FROM YOKE TRIBULATION OPPRESSION SO NOBLY BORNE IN PATH HIS LOVE MORE THAN ONE CENTURY. . . .

UNIVERSAL HOUSE OF JUSTICE

259-1. In 1848 Mullá Ḥusayn set out with a company of Bábís for the province of Mázindarán to rescue Quddús, a Letter of the Living who was being held in the town of Sárí. Nabíl reports in *The Dawn-Breakers* (pp. 337–38) that, after a skirmish with the citizens of Bárfurúsh that ended with the villagers' begging for peace, Mullá Ḥusayn and his companions proceeded to the caravansary. Detecting signs of continued hostility, Mullá Ḥusayn had the gates of the inn closed. As evening approached, he asked whether any of his companions was willing to arise and renounce his life for the sake of the Faith by ascending to the roof of the caravansary to sound the call to prayer. A youth gladly responded. No sooner had he uttered the opening words of a prayer than a bullet struck and killed him. Another youth arose, and, in the same spirit, proceeded with the prayer. He, too, was struck down by an enemy bullet. A third youth attempted to complete the unfinished prayer; but he, too, suffered the same fate as he neared the end of the prayer.

260
Another False Accusation against the Iranian Bahá'ís
25 August 1980

To selected National Spiritual Assemblies

SUBSEQUENT ARREST ENTIRE MEMBERSHIP NATIONAL ASSEMBLY IRAN AND TWO AUXILIARY BOARD MEMBERS OFFICIAL ANNOUNCEMENT JUST RELEASED THROUGH MASS MEDIA IRAN ACCUSES LEADERS BAHÁ'Í COMMUNITY INVOLVEMENT IN RECENTLY ATTEMPTED COUP. THIS FALSE ACCUSATION ANOTHER EVIDENCE DELIBERATE SCHEME AUTHORITIES ELIMINATE BAHÁ'Í COMMUNITY AND ITS INSTITUTIONS IN LAND ITS ORIGIN. . . . 260.1

ASSURE YOU PRAYERS SUCCESS EFFORTS ALLEVIATE SUFFERINGS OPPRESSED BRETHREN IRAN. 260.2

Universal House of Justice

261
Martyrdom of Seven Bahá'ís in Yazd
9 September 1980

To National Spiritual Assemblies

ANNOUNCE WITH DEEP SORROW HEARTRENDING NEWS MARTYRDOM ON MORNING MONDAY SEPTEMBER EIGHT IN YAZD SEVEN DEVOTED STEADFAST BELIEVERS TWO OF WHOM WERE AUXILIARY BOARD MEMBERS AND FIVE PROMINENT BAHÁ'ÍS YAZD INCLUDING ONE SEVENTY-EIGHT YEAR OLD BELIEVER. LOCAL REVOLUTIONARY GUARDS WHO PERPETRATED THIS DASTARDLY ACT CHARGED INNOCENT FRIENDS ON LOCAL RADIO WITH FALSE IMPUTATIONS SUCH AS SPYING AND SUBVERSIVE ACTIVITIES. THIS EPISODE WILL GO DOWN IN HISTORY AS ANOTHER EVIDENCE PATIENCE LONG-SUFFERING INNOCENT PERSIAN BRETHREN IN FACE OF UNABATED HATRED VIRULENT MALICE INVETERATE ENEMIES FAITH. . . . 261.1

NAMES MARTYRS AS FOLLOWS: 261.2

    'AZÍZU'LLÁH DHABÍḤÍYÁN
    FIRAYDÚN FARÍDÁNÍ
    NÚRU'LLÁH AKHTAR-KHÁVARÍ
    JALÁL MUSTAQÍM
    MAHMÚD ḤASANZÁDIH
    'ALÍ MUṬAHHARÍ
    'ABDU'L-VAHHÁB KÁẒIMÍ-MANSHÁDÍ. . . .

Universal House of Justice

262
Defense of the Bahá'ís of Iran in *Le Monde*
11 September 1980

To all National Spiritual Assemblies

Dear Bahá'í Friends,

262.1 The Universal House of Justice has asked us to convey to you the following. . . .

262.2 An official statement published in Iran in the columns of newspapers of nationwide circulation charges that those executed [the seven martyrs of Yazd]¹ were guilty of spying and subversion and denounces the Local Spiritual Assembly of Yazd as an instrument of espionage. These accusations are nothing but downright falsehood. . . .

262.3 Recently the world press has carried several excellent articles written by fair-minded and well-known non-Bahá'í writers in support of our beloved Faith. *Le Monde* of Paris in its 29 August issue published a remarkable article earnestly defending the Bahá'í community in Iran. The Universal House of Justice has asked us to send you a copy of the article in French, with a translation of its text into English, for your information and use in any manner you deem advisable.

262.4 The situation of the friends in Iran is perilous. The total number of believers imprisoned in the country has exceeded sixty. The friends have no refuge save their trust in Bahá'u'lláh. We must steadfastly continue to cling to the Hem of His Garment, fervently pray for the alleviation of the sufferings of our oppressed brethren, resolutely arise to win victories in His name, and be confident that the promised deliverance of the Faith of God from the shackles of blind orthodoxy will, in the fullness of time, be fully realized with the aid of His invisible hosts and as decreed by His invincible Will.

With loving Bahá'í greetings,
DEPARTMENT OF THE SECRETARIAT

The Bahá'ís: An Accursed Community²

262.5 The Iranian press is still silent concerning the arrest, on 21st August last, of the nine members of the supreme governing body of the Bahá'í community, "the National Spiritual Assembly" (*Le Monde* of 28th August). No one knows who proceeded to arrest them, except that it was a group of armed men "claiming authority," who took refuge in pregnant silence. Nothing is known, either, of the place of their confinement. Neither President Bani-Sadr,

---

262-1. See the preceding cable dated 9 September 1980 (message no. 261).
262-2. Translation of an article in *Le Monde*, 29 August 1980, p. 6.

nor the Attorney-General, Ayatollah Ghodoussi, could, or would, reply to the anguished appeals of the families and co-religionists of the nine persons.

Accused of "plotting against the security of the State," they are in danger of execution. In accordance with the procedure of the "Islamic revolutionary tribunals," the arraignment and the trial will take place in camera without the presence of an advocate; the verdict could be pronounced and executed before even the Attorney General is informed. 262.6

Thousands of "counterrevolutionaries" are no doubt in the same situation. But the situation of the Bahá'ís is, in general, far more grave inasmuch as the repression falls not only on individuals who could have been guilty of subversive activities, but on an entire group of people who lack the benefit of any legal protection. In the eyes of the Constitution, indeed, the Bahá'ís . . . do not exist. Christians, Jews, Zoroastrians, although far less numerous than the followers of this humanist religion, are mentioned in the fundamental law and benefit from all the rights accorded to other citizens of Muslim faith, including that of being represented in Parliament. Thus some three hundred thousand Bahá'ís have been reduced to the condition of pariahs. 262.7

The anarchy produced by the revolution—which was in principle directed against imperial autocracy and foreign domination—has favored the unleashing of the fanaticism, not of the population, but of ultraconservative religious organizations like the brotherhood of the Tablighat-i-Islamí (an Islamic propaganda group), who already under the monarchy were conducting persecutions with the active collaboration of SAVAK. Since the installation of the Republic a score of Bahá'í Holy Places have been destroyed or confiscated by the local authorities; centers of worship in eighty towns and villages have been destroyed or burned down; some forty cemeteries have been profaned and in most cases confiscated; museums, shops, or places of business belonging to Bahá'ís have been attacked or looted in more than three hundred and eighty urban or rural settlements, according to a report compiled two months ago by the National Spiritual Assembly of the community. 262.8

The high authorities of the Republic would find it difficult to exonerate themselves from all responsibility for these collective crimes. Not one of them, as far as is known, has explicitly condemned them. Worse, the Revolutionary Council, on which, moreover, some "liberals" are represented, decreed last winter (see the daily *Etelaat* of 20th December) that the army must expel from its ranks all persons who do not belong to one of the four "official" religions of the State (Muslim, Christian, Jewish, and Zoroastrian). Following this, in addition to the officers and soldiers, civil servants, employees of nationalized industries, university professors and teachers have been dismissed from one day to the next without any indemnity or pension. Again by order of the government, Bahá'í places of worship in several cities, notably in Tehran and Mashhad, have been converted into "free Islamic Universities." 262.9

262.10 The Bahá'ís are denounced as "heretics," as "renegades" from Islam, because the Founders of their religion, in the last century, were for the most part, Persian Muslims, because their Prophet, the Báb (1819–1850) had the misfortune to be born after Muḥammad, who is considered to be the last Messenger of God. However, the millions of Bahá'ís scattered today in some eighty-eight thousand localities across the five continents are neither of Persian origin nor are they Muslim converts. Beyond this, the Bahá'í Faith recognizes and respects Islam.

262.11 However, the persecutors have "refined" their accusations by "politicizing" them. The Bahá'ís, according to them, are "agents of Israel" because the World Center of the community is situated in Haifa. But the seat was established in that city in 1868, almost a century before the foundation of the Jewish State.

262.12 The Bahá'ís, it is said again, were "supporters of the regime and of SAVAK!" Yet they refused—in face of demands of the Shah—to join the single Rastakhiz party, in spite of the pogroms organized against them by SAVAK, notably in 1955 and in 1963.[3] In the last months of the monarchy the agents of the secret police had caused some three hundred houses belonging to Bahá'ís to be burnt. In accordance with their teachings, the Bahá'ís are required to be obedient to the established authority, whatever it may be, and refuse—under pain of being expelled from the community—to involve themselves in any activity or to accept any office of a political nature. This shows how absurd appears the accusation of "plotting against the security of the State" formulated against the nine elected members of the supreme body of the community in Iran.

262.13 They knew that they were in danger. They could, like so many others, have chosen exile. As one of them said to us recently, they had nevertheless decided "to remain at the side of our fellow believers in distress." Their fidelity can cost them their lives. But who, in Iran or abroad, will have the courage or the wish to intervene in favor of a community accursed among all others?

—Eric Rouleau

262-3. The Rastakhiz (Resurrection) Party, formed in 1975 by the Shah, was the only government-authorized political party. In the face of government pressure on all citizens to join the party, Bahá'ís refused because of the Bahá'í principle of political noninvolvement. In 1955 a wave of persecution engulfed the Iranian Bahá'í community. A Muslim cleric, Shaykh Muḥammad Taqí Falsafí preached sermons at his mosque in Tehran and over the radio during the month of Ramaḍán (fasting). As a result of the anti-Bahá'í fervor he aroused, private homes, businesses, and Bahá'í institutions were plundered, cemeteries were desecrated and corpses disinterred, and mobs conducted an orgy of murder, rape, and pillage. The House of the Báb in Shiraz was partly demolished, and the Bahá'í National Center in Tehran was taken over by members of the police, army, and clergy. Its imposing dome was ripped to pieces. Bahá'ís were again singled out for attack in 1963 during antigovernment demonstrations.

1979–1986 • THE SEVEN YEAR PLAN

263
Worldwide Response to Iranian Persecutions
23 SEPTEMBER 1980

To all National Spiritual Assemblies

MEMBERS VENERABLE COMMUNITY OF BAHÁ'U'LLÁH'S FOLLOWERS IN HIS NATIVE 263.1 LAND ARE FACING IN UTMOST NOBILITY HEROISM AND IN TRADITION THEIR ILLUSTRIOUS FOREBEARS ORDEALS SEVERITY OF WHICH IS REMINISCENT OF FEROCITY PERSECUTIONS HEROIC AGE OUR FAITH. INNOCENT BLOOD THEY ARE SHEDDING ALTAR SACRIFICE, INTENSE SUFFERINGS THEY ARE SERENELY BEARING IN LOVE HIS PATH ARE RELEASING SPIRITUAL ENERGIES WHICH ARE ACCELERATING MOMENTUM PROGRESS FAITH OPENING UP NEW HORIZONS AS IT IRRESISTIBLY FORGES AHEAD TOWARDS ITS ULTIMATE DESTINY.

UNITED NATIONS SUBCOMMISSION ON PREVENTION DISCRIMINATION PROTEC- 263.2 TION MINORITIES MEETING IN GENEVA FOLLOWING PRESENTATION MADE BY REPRESENTATIVE BAHÁ'Í INTERNATIONAL COMMUNITY DISCUSSED APPROVED ON 10 SEPTEMBER RESOLUTION EXPRESSING PROFOUND CONCERN OVER SAFETY BAHÁ'ÍS IRAN AND REQUESTING UNITED NATIONS SECRETARY-GENERAL CONVEY THIS CONCERN TO AUTHORITIES IRAN AND CALL ON THEM PROTECT FUNDAMENTAL RIGHTS AND FREEDOMS BAHÁ'Í COMMUNITY.

EUROPEAN PARLIAMENT MEETING IN STRASBOURG AND COMPRISING OVER 400 263.3 MEMBERS UNANIMOUSLY ADOPTED ON SEPTEMBER 19 HISTORIC RESOLUTION WHICH DENOUNCES SYSTEMATIC CAMPAIGN PERSECUTION IRANIAN BAHÁ'ÍS AND VIOLATION THEIR ELEMENTARY HUMAN RIGHTS, CALLS UPON GOVERNMENT IRAN GRANT BAHÁ'Í COMMUNITY LEGAL RECOGNITION AND PROTECTION, AND UPON FOREIGN MINISTERS EUROPEAN COMMUNITIES MAKE URGENT PRESENTATIONS TO IRANIAN AUTHORITIES CEASE PERSECUTION BAHÁ'ÍS ALLOWING THEM PRACTICE THEIR RELIGION FREELY AND ENJOY FUNDAMENTAL HUMAN RIGHTS, AND REQUESTS MEMBER STATES IMPOSE EMBARGO ON ALL SALES OF SUBSIDIZED SURPLUS AGRICULTURAL PRODUCTS TO IRAN UNTIL FULL HUMAN RIGHTS ARE RESTORED TO IRANIAN CITIZENS. THIS REMARKABLE DOCUMENT ENDS WITH STATEMENT THAT EVIDENCE SUGGESTS ARRESTS EXECUTIONS OF BAHÁ'ÍS ARE CONTINUING AND INCREASING ACCORDING TO PREARRANGED PLAN AND ONLY SPEEDY RESPONSE BY EUROPEAN PARLIAMENT CAN STOP THESE DEVELOPMENTS.

REPORTS RECEIVED FROM ALL CONTINENTS ELOQUENTLY BEAR TESTIMONY TO 263.4 SENSE OF UNITY FELT BY FRIENDS EVERYWHERE IN FACE DEPRIVATIONS THEIR PERSIAN BRETHREN, GENEROUS OFFERINGS OF TIME, EFFORT, AND RESOURCES IN MEMORY MARTYRS AND ON BEHALF OF IMPRISONED FRIENDS, AMONG WHOM ARE A COUNSELOR, AUXILIARY BOARD MEMBERS, ENTIRE MEMBERSHIP NATIONAL SPIRITUAL ASSEMBLY AS WELL AS MEMBERS SEVERAL LOCAL ASSEMBLIES. THESE AFFLICTIONS HAVE ENABLED FRIENDS IN MOST COUNTRIES WORLD OVER PROCLAIM FAITH TO HIGHEST OFFICIALS THEIR COUNTRY AND NOISE ABROAD ITS TEACHINGS THROUGH MASS MEDIA. COUNTLESS PLEDGES OF LOYALTY AND OF RENEWED DEDI-

CATION HAVE BEEN MADE TO SHOW IN THE DAYS AHEAD GREATER SOLIDARITY AUDACITY ACHIEVE MORE STIRRING VICTORIES FOR GOD'S HOLY FAITH. ALREADY DURING PAST SEVERAL WEEKS IN ONE COUNTRY LATIN AMERICA NUMBER BELIEVERS HAS MORE THAN DOUBLED RESULT EFFORTS INSPIRED BLOOD INNOCENT MARTYRS. LET MEN OF VALOR IN OTHER COUNTRIES FOLLOW IN THEIR FOOTSTEPS.

263.5 CALL ON FRIENDS EVERYWHERE JOIN US IN PRAYERS THANKSGIVING FOR THESE VICTORIES WORTHY RESPONSES OF BAHÁ'Í COMMUNITY TO BITTER PERSECUTIONS BEING METED OUT ITS FOLLOWERS IN NATIVE LAND BAHÁ'U'LLÁH. URGE ALL FRIENDS CONTINUE THEIR SUPPLICATIONS TO HIM THAT UNYIELDING INTRANSIGENT ATTITUDE OF IRANIAN AUTHORITIES TOWARD BAHÁ'ÍS MAY BE TRANSMUTED INTO CONFIDENCE IN AND GOODWILL TOWARDS PEACE-LOVING, LAW-ABIDING MEMBERS PRESENTLY PROSCRIBED COMMUNITY.

Universal House of Justice

264
Property Acquisitions in the Holy Land
24 September 1980

To the Bahá'ís of the World

264.1 WITH HEARTS BRIMMING WITH GRATITUDE FOR BOUNTIFUL CONFIRMATIONS BLESSED BEAUTY ANNOUNCE FOLLOWING RECENT ACCOMPLISHMENTS HOLY LAND:

264.2 IN BAHJÍ JUST ACQUIRED STRIP OF LAND 13,150 SQUARE METERS IN AREA BORDERING DRIVEWAY FROM WESTERN GATE BAHÁ'Í PROPERTY. AREA ACQUIRED ADJACENT TO AND SOUTH OF PRESENT BOUNDARY OLIVE GROVE ENABLES SOUTHWEST QUADRANT GARDENS SURROUNDING MOST HOLY SHRINE BE COMPLETED. THIS ACQUISITION WAS MADE POSSIBLE BY EXCHANGE AGAINST PROPERTY ON MAIN ROAD TO NAZARETH DONATED WORLD CENTER BY DESCENDANTS LATE ḤUSAYN BÁQIR KÁSHÁNÍ.

264.3 ALSO PURCHASED NEARLY 50,000 SQUARE METERS AGRICULTURAL LAND ADJACENT TO AND NORTH OF MAZRA'IH PROPERTY AS PROTECTION TO MANSION IN RAPIDLY DEVELOPING AREA.

264.4 SHIPMENT OVER 1,000 PIECES COMPRISING 120 CUBIC METERS KATRINA CEDAR WOOD FROM TURKEY JUST CLEARED FROM HAIFA PORT FOR USE RESTORATION HOUSE 'ABDU'LLÁH PÁSHÁ IN 'AKKÁ.

Universal House of Justice

265

Passing of Leonora Stirling Armstrong, Spiritual Mother of South America
20 OCTOBER 1980

The National Spiritual Assembly of the Bahá'ís of Brazil

HEARTS SADDENED PASSING DISTINGUISHED COUNSELOR LEONORA STIRLING ARMSTRONG, HERALD OF THE KINGDOM, BELOVED HANDMAIDEN 'ABDU'L-BAHÁ, SPIRITUAL MOTHER SOUTH AMERICA.[1] HER SIXTY YEARS VALIANT DEVOTED SERVICES CAUSE BRAZIL SHEDS LUSTER ANNALS FAITH THAT PROMISING LAND. REQUESTING MEMORIAL SERVICES MA<u>SH</u>RIQU'L-A<u>DH</u>KÁRS WILMETTE PANAMA URGE ALL COMMUNITIES BRAZIL LIKEWISE HOLD SERVICES. OFFERING ARDENT SUPPLICATIONS MOST HOLY SHRINE PROGRESS HER RADIANT SPIRIT ABHÁ KINGDOM. 265.1

UNIVERSAL HOUSE OF JUSTICE

266

Attendance of the Members of a Spiritual Assembly at Its Meetings
26 OCTOBER 1980

To all National Spiritual Assemblies

Dear Bahá'í Friends,

... The attached compilation on the importance of participation in meetings of a National Spiritual Assembly by its entire membership has been prepared to assist members of these divinely ordained bodies to understand the nature and importance of the responsibilities they must assume when they are elected by their fellow believers.[1] The House of Justice requests that every member of your Assembly be provided with a copy of this compilation, and that it be the subject of prayerful consultation at an early meeting of your Assembly. 266.1

Many of the passages included in the compilation apply equally to Local Spiritual Assemblies. 266.2

With loving Bahá'í greetings,
DEPARTMENT OF THE SECRETARIAT

265-1. For an account of the life and services of Leonora Stirling Armstrong, see BW 18:733–38.
266-1. The compilation is too lengthy to include in this volume.

267

Further Development of the Continental Boards of Counselors
3 NOVEMBER 1980

To the Bahá'ís of the World

Dearly loved Friends,

267.1 One of the greatest sources of consolation for the Universal House of Justice amid the tribulations of the past twelve years, has been the establishment and growth of the Continental Boards of Counselors, and the assistance that this institution has been rendering, in ever-increasing measure, to the sound development of the worldwide Bahá'í community. We cannot pay too high a tribute to the indefatigable labors of the devoted souls who have been called upon to shoulder this onerous responsibility, and who have followed with such fidelity the path of self-sacrificing service that has been blazed for them by the beloved Hands of the Cause of God.

267.2 In June 1979 we were moved to announce that the duration of the terms of office of Continental Counselors would be five years, to start on the Day of the Covenant of this year.[1] As this date approaches, we have decided that the time is ripe for a further step in the development of the institution itself that will, at one and the same time, accord greater discretion and freedom of action to the Continental Boards of Counselors in the carrying out of their duties, and widen the scope of each Board to embrace an entire continent. In accordance with this decision, the zones of the Continental Boards of Counselors will, from the Day of the Covenant of the year 137[2] (26 November 1980), be as follows:

1. Africa, comprising the areas of the four present zones of that continent.
2. The Americas, comprising the present zones of North, Central and South America.
3. Asia, comprising the present zones of Western, South Central and Southeastern Asia, together with the present zone of Northeastern Asia without the Hawaiian Islands and Micronesia.
4. Australasia, comprising the present zone of Australasia plus the Hawaiian Islands and Micronesia.
5. Europe.

267.3 Those who are now appointed as Counselors to serve on these Continental Boards for the next five years are:

267-1. See message no. 229. The five-year terms were to begin on 26 November 1980.
267-2. Of the Bahá'í Era.

Africa: Dr. Hushang Ahdieh (Trustee of the Continental Fund), Mr. Ḥusayn Ardíkání, Mr. Friday Ekpe, Mr. Oloro Epyeru, Mr. Shidan Fat'he-Aazam, Mr. Zekrullah Kazemi, Mr. Muḥammad Kebdani, Mrs. Thelma Khelghati, Mr. William Masehla, Mr. Muḥammad Muṣṭafá, Mr. Kolonario Oule, Mrs. Isobel Sabri, Dr. Mihdí Samandarí, Mr. Peter Vuyiya, Mrs. Bahíyyih Winckler. 267.3a

The Americas: Dr. Hidáyatu'lláh Aḥmadíyyih, Dr. Farzam Arbáb, Mrs. Carmen de Burafato, Mr. Athos Costas, Mr. Angus Cowan, Mr. Lloyd Gardner (Trustee of the Continental Fund), Mr. Mas'úd K͟hamsí, Mrs. Lauretta King, Mr. Artemus Lamb, Mr. Peter McLaren, Mr. Raúl Pavón, Dr. Sarah Pereira, Mrs. Ruth Pringle, Mr. Fred Schechter, Mrs. Velma Sherrill, Mr. Donald Witzel. 267.3b

Asia: Mr. Burháni'd-Dín Afs͟hín, Mrs. Shirin Boman, Dr. Masíḥ Farhangí, Dr. John Fozdar, Mr. Zabíḥu'lláh Gulmuḥammadí, Mr. Aydin Güney, Mr. Dipchand Khianra, Mr. Rúḥu'lláh Mumtází, Mr. S. Nagaratnam, Mr. K͟hudárahm Paymán (Trustee of the Continental Fund), Mr. Manúc͟hihr Salmánpúr, Mr. Vicente Samaniego, Mrs. Zena Sorabjee, Dr. Chellie Sundram, Mr. Hideya Suzuki, Mr. Yan Kee Leong. 267.3c

Australasia: Mr. Suhayl 'Alá'í, Mr. Ben Ayala, Mr. Owen Battrick (Trustee of the Continental Fund), Mr. Richard Benson, Mrs. Tinai Hancock, Dr. Peter Khan, Mr. Lisiate Maka. 267.3d

Europe: Mr. Erik Blumenthal, Mrs. Dorothy Ferraby, Dr. Agnes Ghaznavi, Mr. Hartmut Grossmann, Mr. Louis Hénuzet (Trustee of the Continental Fund), Mrs. Ursula Mühlschlegel, Dr. Leo Niederreiter, Mrs. Betty Reed, Mr. Adib Taherzadeh. 267.3e

A number of friends who have rendered highly valued services as Counselors are not being reappointed for the coming term, and we wish to express here our profound gratitude for the devoted labors they have rendered and are rendering in the path of the Cause. These dearly loved believers are: 267.4

Mr. Seewoosumbur-Jeehoba Appa, Dr. Iraj Ayman, Mr. Rowland Estall, Mr. Howard Harwood, Miss Violet Hoehnke, Mrs. Salisa Kermani, Mr. Paul Lucas, Miss Elena Marsella, Mr. Alfred Osborne, Miss Thelma Perks, Mr. Hádí Raḥmání, Mr. 'Imád Sábirán, Miss Edna True. 267.4a

Henceforth the Board of Counselors in each continent will have wider discretion to decide such matters as whether to divide its area into zones, and what the boundaries of such zones should be, the number and location of the Board's offices, and the manner in which the members of the Auxiliary Boards will report to and operate under the Counselors. The principles and policies 267.5

governing the operation of the Continental Boards of Counselors, however, and their relationships with the National and Local Spiritual Assemblies and the individual believers will remain unchanged. As the Bahá'í world experiences the manifold interactions of these two vital and complementary arms of the Administrative Order of Bahá'u'lláh, the unique benefits of this divinely ordained System become ever more apparent. The harmonious interaction and the proper discharge of the duties of these institutions representing the rulers and the learned among the people of Bahá is the essential basis at this time for the protection of the Cause of Bahá'u'lláh and the fulfillment of its God-given mandate.[3]

267.6 Events of the most profound significance are taking place in the world. The river of human history is flowing at a bewildering speed. Age-old institutions are collapsing. Traditional ways are being forgotten, and newly born ideologies which were fondly expected to take their place, are withering and decaying before the eyes of their disillusioned adherents. Amidst this decay and disruption, assailed from every side by the turmoil of the age, the Order of Bahá'u'lláh, unshakably founded on the Word of God, protected by the shield of the Covenant and assisted by the hosts of the Concourse on High, is rising in every part of the world.

267.7 Every institution of this divinely created Order is one more refuge for a distraught populace; every soul illumined by the light of the sacred Message is one more link in the oneness of mankind, one more servant ministering to the needs of an ailing world. Even should the Bahá'í communities, in the years immediately ahead, be cut off from the World Center or from one another—as some already have been—the Bahá'ís will neither halt nor hesitate; they will continue to pursue their objectives, guided by their Spiritual Assemblies and led by the Counselors, the members of the Auxiliary Boards and their assistants. It is our prayer at the Sacred Threshold that the new and challenging development now taking place in the evolution of the institution of the Counselors will release great energies for the advancement of the Cause of God in every land.

THE UNIVERSAL HOUSE OF JUSTICE

267-3. For an elucidation of the roles of the rulers and the learned, see message no. 111.

268

Martyrdom of Riḍá Fírúzí in Tabríz; Uncertainty of Whereabouts of National Assembly Members

14 NOVEMBER 1980

To all National Spiritual Assemblies

SORROWFUL NEWS JUST RECEIVED ANOTHER BELIEVER RIḌA FÍRÚZÍ MARTYRED TABRÍZ 9 NOVEMBER. NEWS ITEM ISSUED PARS NEWS AGENCY AND REUTERS[1] AND PUBLISHED IN TEHRAN PAPERS CHARGES HIM WITH BEING PROMINENT BAHÁ'Í AND INFIDEL AND INCLUDES USUAL LIST TOTALLY UNFOUNDED ACCUSATIONS. WHEREABOUTS NINE MEMBERS NATIONAL ASSEMBLY AND FEW OTHERS STILL UNCERTAIN. STRONG FEARS HARASSED COMMUNITY ONCE AGAIN FACING AFTER BRIEF LULL FRESH WAVE OPPRESSION LEADING TO SHAM TRIALS MERCILESS ATROCITIES. . . . 268.1

UNIVERSAL HOUSE OF JUSTICE

269

Passing of the Hand of the Cause of God Abu'l-Qasim Faizi

20 NOVEMBER 1980

To all National Spiritual Assemblies

HEARTS FILLED WITH SORROW PASSING INDEFATIGABLE SELF-SACRIFICING DEARLY LOVED HAND CAUSE GOD ABU'L-QASIM FAIZI.[1] ENTIRE BAHÁ'Í WORLD MOURNS HIS LOSS. HIS EARLY OUTSTANDING ACHIEVEMENTS IN CRADLE FAITH THROUGH EDUCATION CHILDREN YOUTH STIMULATION FRIENDS PROMOTION TEACHING WORK PROMPTED BELOVED GUARDIAN DESCRIBE HIM AS LUMINOUS DISTINGUISHED ACTIVE YOUTH. HIS SUBSEQUENT PIONEERING WORK IN LANDS BORDERING IRAN WON HIM APPELLATION SPIRITUAL CONQUEROR THOSE LANDS. FOLLOWING HIS APPOINTMENT HAND CAUSE HE PLAYED INVALUABLE PART WORK HANDS HOLY LAND TRAVELED WIDELY PENNED HIS LITERARY WORKS CONTINUED HIS EXTENSIVE INSPIRING CORRESPONDENCE WITH HIGH AND LOW YOUNG AND OLD UNTIL AFTER LONG ILLNESS HIS SOUL WAS RELEASED AND WINGED ITS FLIGHT ABHÁ KINGDOM. CALL ON FRIENDS EVERYWHERE HOLD BEFITTING MEMORIAL GATHERINGS HIS HONOR, INCLUDING SPECIAL COMMEMORATIVE MEETINGS HIS NAME 269.1

268-1. An international news service.
269-1. For an account of the life and services of Abu'l-Qasim Faizi, see BW 18:659–65.

IN HOUSES WORSHIP ALL CONTINENTS. MAY HIS SHINING EXAMPLE CONSECRATION CONTINUE INSPIRE HIS ADMIRERS EVERY LAND. PRAYING HOLY SHRINES HIS NOBLE RADIANT SOUL MAY BE IMMERSED IN OCEAN DIVINE MERCY CONTINUE ITS UNINTERRUPTED PROGRESS IN INFINITE WORLDS BEYOND.

UNIVERSAL HOUSE OF JUSTICE

270

Immolation of Bahá'í Couple in Núk Village
8 December 1980

To all National Spiritual Assemblies

Dear Bahá'í Friends,

270.1 On 1 December 1980 the Universal House of Justice cabled certain National Spiritual Assemblies as follows:

> WITH FEELINGS HORROR INDIGNATION INFORM YOU MARTYRDOM MUHAMMAD-HUSAYN MA'ṢÚMÍ AND HIS WIFE[1] BRUTALLY PUT TO DEATH 23RD NOVEMBER NÚK VILLAGE NEAR BÍRJAND IN KHURÁSÁN. FIFTEEN MASKED MEN ATTACKED COUPLE THEIR HOME MIDDLE NIGHT FIRST POURED KEROSENE ON HUSBAND AND SET HIM ON FIRE THEN FORCED HIM RUN FEW METERS FINALLY HEAPED WOOD UPON HIM BURNING HIM TO DEATH. HIS WIFE SUBJECT SIMILAR TREATMENT DIED TWO DAYS LATER HOSPITAL. SO FAR THIS ATROCITY NOT REPORTED IN NEWS MEDIA. . . .

270.2 . . . The Universal House of Justice calls on the friends in every land to continue to remember the oppressed friends of Iran in their prayers, supplicating Bahá'u'lláh to grant His divine aid and assist that persecuted community to face the trials that God has ordained for them, in accordance with His inscrutable wisdom.

With loving Bahá'í greetings,
DEPARTMENT OF THE SECRETARIAT

270-1. Shikkar Nisá.

271
Message to Counselors' Convocations on Five Continents
22 DECEMBER 1980

To gatherings of Counselors in all countries

HAIL CONVOCATION IN ALL CONTINENTS FIVE GATHERINGS CONTINENTAL BOARDS 271.1
COUNSELORS ON MORROW THEIR APPOINTMENT FOR NEW TERM. GRATEFUL DIVINE BLESSINGS ENABLING THESE GATHERINGS BE GRACED BY PRESENCE HANDS CAUSE GOD WHO IN ACCORDANCE 'ABDU'L-BAHÁ'S WILL TESTAMENT WERE APPOINTED BY BELOVED GUARDIAN AND DESCRIBED BY HIM AS CHIEF STEWARDS BAHÁ'U'LLÁH'S EMBRYONIC WORLD COMMONWEALTH.[1]

INVESTED WITH DUAL FUNCTION DISTINGUISHING MEMBERS THAT AUGUST 271.2
INSTITUTION AND FOLLOWING IN THEIR FOOTSTEPS EACH NEWLY APPOINTED BOARD COUNSELORS IS CALLED UPON AMIDST TURMOIL PRESENT HOUR ASSUME WITH RENEWED VIGOR AND DETERMINATION UNDER GUIDANCE INTERNATIONAL TEACHING CENTER ITS VITAL DUTIES GUARD OVER SECURITY ENSURE PROPAGATION GOD'S FAITH IN CONSULTATION NATIONAL ASSEMBLIES WITHIN CONFINES ITS CONTINENT IN STRICT CONFORMITY PRINCIPLES OUTLINED BY GUARDIAN IN HIS CABLE TO BAHÁ'Í WORLD DATED 4 JUNE 1957.[2] AS HE CLEARLY ENVISAGED THEN AND AS IS BECOMING INCREASINGLY APPARENT NOW SECURITY CAUSE PRESERVATION SPIRITUAL HEALTH COMMUNITY VITALITY FAITH INDIVIDUALS PROPER FUNCTIONING BAHÁ'Í INSTITUTIONS FRUITION WORLDWIDE ENTERPRISES FULFILLMENT ULTIMATE DESTINY FAITH ALL DEPENDENT UPON BEFITTING DISCHARGE WEIGHTY RESPONSIBILITIES MEMBERS TWIN ARMS ADMINISTRATION.[3]

IF AT ANY TIME FOR ANY REASON COMMUNICATION WITH WORLD CENTER IS 271.3
CUT OFF IN FUTURE, COUNSELORS EACH CONTINENT SHOULD COLLECTIVELY INDIVIDUALLY ASSIST NATIONAL ASSEMBLIES ENSURE CONTINUATION NORMAL ADMINISTRATION FAITH BY THESE ASSEMBLIES WITHOUT INTERRUPTION UNTIL COMMUNICATIONS CAN BE RESTORED. FURTHERMORE IF IT SHOULD PROVE UNFEASIBLE AT END OF ANY FIVE YEAR TERM OF OFFICE FOR HOUSE JUSTICE TO REVIEW AND RENEW MEMBERSHIP CONTINENTAL BOARDS COUNSELORS, THESE BOARDS SHOULD CONTINUE IN OFFICE EVEN IF ONE OR MORE THEIR MEMBERS IS UNABLE FUNCTION, FAITHFULLY DISCHARGING THEIR RESPONSIBILITIES, UNTIL PROPITIOUS

271-1. See WT, pp. 12–13, and MBW, p. 127.
271-2. See MBW, pp. 122–23.
271-3. These "twin arms" are described in the Constitution of the Universal House of Justice as follows: "This Administrative Order consists, on the one hand, of a series of elected councils, universal, secondary and local, in which are vested legislative, executive and judicial powers over the Bahá'í community and, on the other, of eminent and devoted believers appointed for the specific purposes of protecting and propagating the Faith of Bahá'u'lláh under the guidance of the Head of that Faith." For an explanation of the roles of the two arms, see message no. 111.

CONDITIONS PREVAIL FOR HOUSE JUSTICE CONSIDER APPOINTMENT SUCCESSORS. FERVENTLY PRAYING HOLY SHRINES GOD'S PRECIOUS FAITH MAY BE PROTECTED FROM ONSLAUGHT ITS ENEMIES WITHIN ITS ANTAGONISTS BOTH RELIGIOUS AND SECULAR WITHOUT, THAT MEMBERS RECONSTITUTED INSTITUTION COUNSELORS MAY BE RECIPIENTS FRESH MEASURE HEAVENLY CONFIRMATIONS AND THAT FIRST MEETINGS NOW CONVENED MAY LAY FIRM FOUNDATIONS FOR EFFICIENT OPERATION THIS ESSENTIAL ORGAN STEADILY UNFOLDING ADMINISTRATIVE ORDER FAITH BAHÁ'U'LLÁH. REQUEST COUNSELORS SHARE COPY THIS MESSAGE WITH ALL NATIONAL SPIRITUAL ASSEMBLIES THEIR CONTINENT.

UNIVERSAL HOUSE OF JUSTICE

272
Relationship between Husband and Wife
28 December 1980

The National Spiritual Assembly of the Bahá'ís of New Zealand

Dear Bahá'í Friends,

272.1 The Universal House of Justice has received your letter of 16 October 1980 enclosing a letter from the Spiritual Assembly of . . . posing questions which have arisen as a result of reading the book *When We Grow Up* by Bahíyyih Nakhjavání, and it has instructed us to convey the following.

272.2 The House of Justice suggests that all statements in the Holy Writings concerning specific areas of the relationship between men and women should be considered in the light of the general principle of equality between the sexes that has been authoritatively and repeatedly enunciated in the Sacred Texts. In one of His Tablets 'Abdu'l-Bahá asserts: "In this divine age the bounties of God have encompassed the world of women. Equality of men and women, except in some negligible instances, has been fully and categorically announced. Distinctions have been utterly removed."[1] That men and women differ from one another in certain characteristics and functions is an inescapable fact of nature; the important thing is that 'Abdu'l-Bahá regards such inequalities as remain between the sexes as being "negligible."

272.3 The relationship between husband and wife must be viewed in the context of the Bahá'í ideal of family life. Bahá'u'lláh came to bring unity to the world, and a fundamental unity is that of the family. Therefore, one must believe that the Faith is intended to strengthen the family, not weaken it, and one of the keys to the strengthening of unity is loving consultation. The atmosphere within a Bahá'í family as within the community as a whole should express

---

272-1. From an unpublished Tablet.

"the keynote of the Cause of God" which, the beloved Guardian has stated, "is not dictatorial authority, but humble fellowship, not arbitrary power, but the spirit of frank and loving consultation."[2]

A family, however, is a very special kind of "community." The Research Department has not come across any statements which specifically name the father as responsible for the "security, progress and unity of the family" as is stated in Bahíyyih Nakhjavání's book, but it can be inferred from a number of the responsibilities placed upon him, that the father can be regarded as the "head" of the family.[3] The members of a family all have duties and responsibilities towards one another and to the family as a whole, and these duties and responsibilities vary from member to member because of their natural relationships. The parents have the inescapable duty to educate their children—but not vice versa; the children have the duty to obey their parents—the parents do not obey the children; the mother—not the father—bears the children, nurses them in babyhood, and is thus their first educator; hence daughters have a prior right to education over sons and, as the Guardian's secretary has written on his behalf, "The task of bringing up a Bahá'í child, as emphasized time and again in Bahá'í Writings, is the chief responsibility of the mother, whose unique privilege is indeed to create in her home such conditions as would be most conducive to both his material and spiritual welfare and advancement. The training which a child first receives through his mother constitutes the strongest foundation for his future development . . ."[4] A corollary of this responsibility of the mother is her right to be supported by her husband—a husband has no explicit right to be supported by his wife. This principle of the husband's responsibility to provide for and protect the family can be seen applied also in the law of intestacy which provides that the family's dwelling place passes, on the father's death, not to his widow, but to his eldest son; the son at the same time has the responsibility to care for his mother.

It is in this context of mutual and complementary duties, and responsibilities that one should read the Tablet in which 'Abdu'l-Bahá gives the following exhortation:

> O Handmaids of the All-Sufficing God!
>
> Exert yourselves, that haply ye may be enabled to acquire such virtues as shall honor and distinguish you amongst all women. Of a surety, there is no greater pride and glory for a woman than to be a handmaid in God's Court of Grandeur; and the qualities that shall merit her this station are an alert and wakeful heart; a firm conviction of the unity of

---

272-2. BA, p. 63.
272-3. For a further discussion of the concept of the husband as the head of the family, see message no. 402.
272-4. See CC 1:303–04.

God, the Peerless; a heartfelt love for all His maidservants; spotless purity and chastity; obedience to and consideration for her husband; attention to the education and nurturing of her children; composure, calmness, dignity and self-possession; diligence in praising God, and worshiping Him both night and day; constancy and firmness in His holy Covenant; and the utmost ardor, enthusiasm, and attachment to His Cause. . . .[5]

272.5c This exhortation to the utmost degree of spirituality and self-abnegation should not be read as a legal definition giving the husband absolute authority over his wife, for, in a letter written to an individual believer on 22 July 1943, the beloved Guardian's secretary wrote on his behalf:

272.5d The Guardian, in his remarks . . . about parents' and children's, wives' and husbands' relations in America, meant that there is a tendency in that country for children to be too independent of the wishes of their parents and lacking in the respect due to them. Also wives, in some cases, have a tendency to exert an unjust degree of domination over their husbands, which, of course, is not right, any more than that the husband should unjustly dominate his wife.

272.5e In any group, however loving the consultation, there are nevertheless points on which, from time to time, agreement cannot be reached. In a Spiritual Assembly this dilemma is resolved by a majority vote. There can, however, be no majority where only two parties are involved, as in the case of a husband and wife. There are, therefore, times when a wife should defer to her husband, and times when a husband should defer to his wife, but neither should ever unjustly dominate the other. In short, the relationship between husband and wife should be as held forth in the prayer revealed by 'Abdu'l-Bahá which is often read at Bahá'í weddings: "Verily, they are married in obedience to Thy command. Cause them to become the signs of harmony and unity until the end of time."[6]

272.6 These are all relationships within the family, but there is a much wider sphere of relationships between men and women than in the home, and this too we should consider in the context of Bahá'í society, not in that of past or present social norms. For example, although the mother is the first educator of the child, and the most important formative influence in his development, the father also has the responsibility of educating his children, and this responsibility is so weighty that Bahá'u'lláh has stated that a father who fails to exercise it forfeits his rights of fatherhood. Similarly, although the primary responsibility for supporting the family financially is placed upon the husband, this

---

272-5. The quotation in the original letter has been replaced by this revised translation.
272-6. BP, p. 107.

does not by any means imply that the place of woman is confined to the home. On the contrary, 'Abdu'l-Bahá has stated:

> In the Dispensation of Bahá'u'lláh, women are advancing side by side with men. There is no area or instance where they will lag behind: they have equal rights with men, and will enter, in the future, into all branches of the administration of society. Such will be their elevation that, in every area of endeavor, they will occupy the highest levels in the human world. . . .[7]

and again:

> So it will come to pass that when women participate fully and equally in the affairs of the world, when they enter confidently and capably the great arena of laws and politics, war will cease; . . .
> (*The Promulgation of Universal Peace*, p. 135)

In the Tablet of the World, Bahá'u'lláh Himself has envisaged that women as well as men would be breadwinners in stating:

> Everyone, whether man or woman, should hand over to a trusted person a portion of what he or she earneth through trade, agriculture or other occupation, for the training and education of children, to be spent for this purpose with the knowledge of the Trustees of the House of Justice.
> (*Tablets of Bahá'u'lláh Revealed after the Kitáb-i-Aqdas*, p. 90)

A very important element in the attainment of such equality is Bahá'u'lláh's provision that boys and girls must follow essentially the same curriculum in schools.

It is hoped that the above explanations and comments will help the Local Spiritual Assembly of . . . to resolve the questions set forth in its letter.

With loving Bahá'í greetings,
DEPARTMENT OF THE SECRETARIAT

272-7. The quotation in the original letter, which was taken from *Paris Talks*, p. 182, has been replaced by this revised translation.

273

Assassination of Dr. Manúchihr Ḥakím, Chairman of Iranian National Spiritual Assembly

13 January 1981

To National Spiritual Assemblies

273.1 PROFOUNDLY SHOCKED TRAGIC NEWS JUST RECEIVED OF ASSASSINATION IN HIS CLINIC OF PROFESSOR MANÚCHIHR ḤAKÍM DISTINGUISHED PHYSICIAN EMINENT PROFESSOR UNIVERSITY TEHRAN AND FOR SUCCESSIVE DECADES INTERMITTENTLY MEMBER CHAIRMAN NATIONAL SPIRITUAL ASSEMBLY IRAN. SPIRIT COURAGE DEDICATION WITH WHICH THIS MOST RECENT MARTYR DISCHARGED SACRED RESPONSIBILITIES ON SUPREME NATIONAL BAHÁ'Í INSTITUTION CRADLE FAITH WORTHY EMULATION BAHÁ'Í ADMINISTRATORS AND FRIENDS EVERY LAND. DEEPLY DISTURBED THIS LATEST OMINOUS SIGN GATHERING CLOUDS FRESH CALAMITIES HARD-PRESSED FRIENDS IRAN. . . .

UNIVERSAL HOUSE OF JUSTICE

274

Events Related to the Execution of Dr. Manúchihr Ḥakím

30 January 1981

To National Spiritual Assemblies

Dear Bahá'í Friends,

274.1 We have been asked by the Universal House of Justice to share with you the following comments about events related to the martyrdom of Professor Manúchihr Ḥakím, and the general situation as it affects the safety of our brethren in Iran.

274.2 The National Spiritual Assembly of Iran believes that ominous signs foreshadow an intensification of the persecution of the Bahá'ís, particularly since the release of the American hostages, and it fears that they will again become the target of their traditional adversaries who are presently in power.[1]

274.3 In reply to complaints from the Bahá'ís concerning the recent assassination of Professor Ḥakím, the authorities have insisted that the murderer is unknown to them, and they categorically deny any involvement in the case.

274-1. On 4 November 1979, seven months after the Islamic Republic was established, a mob surged over the wall of the United States Embassy compound in Tehran. Diplomatic and military personnel were held hostage until their release on 18 January 1981.

It is very significant to note, however, that a few days after the assassination, the government officially produced documentation authorizing the confiscation of Professor Ḥakím's house, and officials went to that house, took an inventory of its furnishings, and sealed it on Wednesday, 14 January. A lawyer who was protecting the interests of the Ḥakím family objected to this action, saying that the house belonged to and was in the name of Mrs. Germaine Ḥakím. On the next day about twenty Revolutionary guards entered the house and removed everything, including the car in the garage, leaving the premises entirely bare!

274.4 Many non-Bahá'ís in Iran have expressed sympathy to the Bahá'ís for the cruel death of Professor Ḥakím, who was renowned for his gentleness and his services to the community.

274.5 Mrs. Ḥakím has sent a number of cables of protest to Iran, but these have not thus far produced any favorable results. She is in Paris and has been advised to remain there since it is highly unsafe for her to return to Iran.

274.6 The National Spiritual Assembly is apprehensive about an official bill which is presently being drafted in the office of the Prime Minister, Mr. Rajá'í. This bill, if passed, will prevent the employment of Bahá'ís by any government or quasi-government institution. A circular letter has already been issued to all branches of the government, instructing them to dismiss all Bahá'ís. When the Bahá'ís do not fill out the blank marked "religion" on employment forms, they are asked again to complete the form so that there will be reason for their dismissal, namely being Bahá'ís. It should be recalled that the dismissal of Bahá'ís took place previously on a large scale in the field of education, and at that time the Minister of Education was the present Prime Minister, Mr. Rajá'í. A considerable number of dismissals also occurred in the Army, despite protests made by a cadre of officers in the Ministry of Defense.

274.7 The National Assembly also reports that a number of local mullahs are continuously instigating the populace against the Bahá'ís in their respective areas. . . .

274.8 The Universal House of Justice will continue to keep you apprised from time to time of the latest developments concerning our beleaguered brethren in the Cradle of the Faith.

With loving Bahá'í greetings,
DEPARTMENT OF THE SECRETARIAT

275
Second Phase of the Seven Year Plan
MARCH 1981

To the Bahá'ís of the World

Dearly loved Friends,

275.1 The successes of the initial phase of the Seven Year Plan are heartening evidence of the Divine care with which the growth of the Cause of God is so lovingly invigorated and sheltered. This still infant Cause, harassed and buffeted over these two years by relentless enemies, experiencing in swift succession a number of sharply contrasting crises and victories, surrounded by the increasing turmoil of a disintegrating world, has raised its banner, reinforced its foundations, and extended the range of its administrative institutions.

275.2 The resurgence of bitter and barbaric persecution of the Faith in the land of its birth, the passing to the Abhá Kingdom of five Hands of the Cause of God, the darkening of the horizons of the world as the somber shadows of universal convulsions and chaos extinguish the lights of justice and order, are among the factors which have chiefly affected the conditions and fortunes of the worldwide army of God.[1]

The Bahá'ís in Iran

275.3 The Bahá'í community in the Cradle of the Faith, having witnessed the destruction of its holiest Shrine,[2] the sequestration of its Holy Places, confiscation of its endowments and even personal properties, the martyrdom of many of its adherents, the imprisonment and holding without trial or news of the members of its National Spiritual Assembly and other leading figures of its community, the deprivation of the means of livelihood, vilification and slander of its cherished tenets, has stood staunch as the Dawn-Breakers of old[3] and emerged spiritually united and steadfast, the pride and inspiration of the entire Bahá'í world. In all continents of the globe, their example and hapless plight has led the friends to proclaim the Name of Bahá'u'lláh as never before, personally, locally, and through all the media of mass communication. The Bahá'í world community, acting through its representatives at the United

---

275-1. The resurging persecution of the Bahá'ís of Iran, the country in which the Bahá'í Faith originated, began in 1978 and came into full force following the fall of the Shah on 17 January 1979 and the declaration of the Islamic Republic on 1 April 1979. For an overview of the situation, see messages no. 215, 227, 234, and 262. The five Hands of the Cause of God who died were Enoch Olinga (see message no. 237), Raḥmatu'lláh Muhájir (see message no. 243), Hasan M. Balyuzi (see message no. 247), Adelbert Mühlschlegel (see message no. 256), and Abu'l-Qasim Faizi (see message no. 269).

275-2. The House of the Báb in Shiraz.

275-3. The Bábís and Bahá'ís of the early years of the Heroic Age who lived during the ministries of the Báb and Bahá'u'lláh.

Nations and through its National Spiritual Assemblies, has brought to the attention of governments and world leaders in many spheres the tenets and character of the Faith of God. The world's parliaments, its federal councils, its humanitarian agencies have considered the Bahá'í Cause and in many instances have extended their support and expressed their sympathy.

In the midst of this time- and energy-consuming activity on behalf of our beloved Persian brethren, the community of the Most Great Name, far from lessening its pursuit of the objectives of the initial phase of the Seven Year Plan, has promoted them with increasing vigor. Added to the burning desire of the friends everywhere to show their love for their brethren in Persia by teaching the Cause with redoubled fervor, has been the further inspiration to teach derived from the loss of the beloved Hands of the Cause, an inspiration which has been fostered by the travels of those dear Hands still able to extend this loving service to the believers. 275.4

Developments within the Faith during the First Phase

The broadening, during this opening phase of the Seven Year Plan, of the foundations of the Boards of Counselors and the consolidation of the thirteen zonal Boards to five continental ones have greatly reinforced this vital institution of the Faith. It has been further developed by the setting of a specified term of office for Continental Counselors, as was envisaged in the original appointments.[4] 275.5

Progress on the Seat of the Universal House of Justice and on the Temples of India and Samoa has continued. Six new National Spiritual Assemblies will be formed during this Riḍván: two in Africa, that of South West Africa/Namibia with its seat in Windhoek and that of Bophuthatswana with its seat in Mmabatho; three in the Americas, Bermuda with its seat in Hamilton, the Leeward Islands with its seat in St. John's, Antigua, and the Windward Islands with its seat in Kingstown, St. Vincent; one in the Pacific, namely that of Tuvalu with its seat in Funafuti; and the National Spiritual Assembly of Uganda will be reconstituted. To those to be formed during the remainder of the Seven Year Plan, the following have been added: two in Africa, Equatorial Guinea with its seat in Malabo, Somalia with its seat in Mogadishu, and one in Asia, that of the Andaman and Nicobar Islands with its seat in Port Blair. 275.6

Increases in the total number of Local Spiritual Assemblies and localities have been registered during the opening phase, and Bahá'í communities in all parts of the world have demonstrated greater unity and maturity in their collective activities. 275.7

275-4. In a message dated 29 June 1979 (no. 229), the Universal House of Justice fixed the term at five years.

The Second Phase of the Plan—1981–84

275.8 The second phase of the Seven Year Plan, now opening, will last for three years and will be followed by the final phase of two years, ending at Riḍván 1986. The twenty-fifth anniversary of the passing of our beloved Guardian will occur during the second year of the second phase of the Plan and that same year will also witness the fiftieth anniversary of the passing of the Greatest Holy Leaf.[5] The House of Justice plans to issue a compilation of letters to her and of statements about her by Bahá'u'lláh, 'Abdu'l-Bahá, and the beloved Guardian, and of her own letters.[6]

Worldwide Goals for the Second Phase

275.9 All National Spiritual Assemblies have been sent the goals assigned to their communities for the second phase, for the prosecution of which the Bahá'í world community now stands poised and ready. Among the major developments envisioned during this phase are:

275.9a Occupation by the Universal House of Justice of its permanent Seat on the slopes of Mount Carmel above the Arc;[7]

275.9b Completion of the Temple in Samoa and continued progress on the work of the Temple in India;[8]

275.9c Further development of the functions of the International Teaching Center and the Boards of Counselors, with special reference to the promotion of the spiritual, intellectual, and social life of the Bahá'í community;[9]

275.9d The holding, during the first nine months of 1982, of five international conferences, in Lagos, Nigeria; Montreal, Canada; Quito, Ecuador; Dublin, Ireland; and Manila, the Philippines, this last one taking place at the midpoint of an axis, referred to by the beloved Guardian, whose poles are Japan and Australia;

275.9e Preparation of architect's plans for the first dependency of the European Mashriqu'l-Adhkár,[10] namely, a Home for the Aged, and an increase in the number of national and local Ḥaẓíratu'l-Quds; the latter, which will be particularly in rural areas, are to be acquired or built through the efforts of the local friends;

275-5. Shoghi Effendi died on 4 November 1957. The Greatest Holy Leaf died at 1:00 A.M. on 15 July 1932.

275-6. The book was published under the title *Bahíyyih Khánum* by the Bahá'í World Center in Haifa (1982).

275-7. For a discussion of the Arc, see the glossary. The Seat of the Universal House of Justice was occupied on 2 February 1983 (see message no. 354).

275-8. The Bahá'í House of Worship in Samoa was dedicated on 1 September 1984 (see message no. 403); the Bahá'í House of Worship in India was dedicated 23–27 December 1986 (see message no. 452).

275-9. For the Universal House of Justice's messages announcing, establishing, elucidating, and expanding the duties of the International Teaching Center, see messages dated 21 June 1968 (no. 58), 5 June 1973 (no. 131), 8 June 1973 (no. 132), and 19 May 1983 (no. 361).

275-10. The Bahá'í House of Worship at Langenhain, near Frankfurt am Main, Germany.

Acquisition of six new Temple sites, five in Africa and one in Australasia; and of five new national endowments, four in Africa and one in the Americas; 275.9f

Formation of two Publishing Trusts, one in the Ivory Coast and one in Nigeria; 275.9g

A great increase in the production of Baháʼí literature in an increasing number of languages, the ultimate aim being to enable every believer to have some portion of the Sacred Text available in his native tongue; 275.9h

Completion of three more radio stations in South America;[11] 275.9i

Great attention to the development and consolidation of Local Spiritual Assemblies throughout the world; 275.9j

Development of Baháʼí community life with special attention to the Baháʼí education of children and the spiritual enrichment of communities; 275.9k

The settlement of 279 pioneers in 80 countries during the first year of the second phase. 275.9l

The Need for Increased Contributions to Baháʼí Funds

Liberal and increased contributions to the various Funds of the Faith will be essential if the above-mentioned tasks are to be successfully pursued. Furthermore, the now observable emergence from obscurity of our beloved Faith will impose the necessity of new undertakings involving large calls on the Funds. The growing awareness of the friends throughout the world in the past few years that the Funds of the Faith are indeed the lifeblood of its activities is a heartening augury for the future. We are confident that this awareness will increase, that more National Spiritual Assemblies will make great strides towards financial independence, that national budgets will be met, and the Baháʼí International Fund will receive an ever-increasing outpouring of contributions enabling that Fund to keep pace with the ever-increasing international needs of the Faith. 275.10

A Spiritual Service

Beloved friends, the world moves deeper into the heart of darkness as its old order is rolled up. Pursing our objectives with confidence, optimism, and an unshakable resolve, we must never forget that our service is a spiritual one. Mankind is dying for lack of true religion and this is what we have to offer to humanity. It is the love of God, manifest in the appearance of Baháʼuʼlláh, which will feed the hungry souls of the world and eventually lead the peoples out of the present morass into the orderly, uplifting, and soul-inspiring task of establishing God's Kingdom on earth. 275.11

THE UNIVERSAL HOUSE OF JUSTICE

275-11. Baháʼí radio stations were established in Peru (see message dated 29 November 1981, no. 304), Bolivia (see message dated 2 April 1984, no. 391), and Panama (see message dated 31 January 1986, no. 450). During the second phase of the Seven Year Plan a radio station was also established in the United States (see message no. 348).

276
Return of the Hand of the Cause of God William Sears to Africa
4 March 1981

To all National Spiritual Assemblies

276.1 AFRICA REJOICES RETURN HAND CAUSE WILLIAM SEARS FULFILLMENT PLEDGE MADE RECEIPT TRAGIC NEWS ASSASSINATION HAND CAUSE ENOCH OLINGA.[1] CONTINENT WHICH EARNED SUCH HIGH PRAISE BELOVED GUARDIAN AGAIN GRACED PRESENCE HAND. NOBLE ACTION ENRICHES RECORD HIS HISTORIC SERVICES. OFFERING ARDENT PRAYERS SACRED THRESHOLD GREAT NEW VICTORIES AFRICA.

Universal House of Justice

277
Execution of Two Bahá'ís in Shiraz
17 March 1981

To all National Spiritual Assemblies

277.1 SORROWFUL NEWS JUST RECEIVED TWO BELIEVERS FROM ÁBÁDIH IMPRISONED SHIRAZ WERE EXECUTED MONDAY NIGHT 16 MARCH BY GOVERNMENT FIRING SQUAD SHIRAZ CHARGE LIST INCLUDES USUAL FALSE IMPUTATIONS LEVELED AGAINST DEFENSELESS BELIEVERS.[1] NAMES THESE TWO VALIANT FRIENDS MIDHÍ ANVARÍ AND HIDÁYATU'LLÁH DIHQÁNÍ ADDED LIST OTHER SELFLESS MARTYRS WHO WITH THEIR BLOOD HAVE TESTIFIED TRUTH GOD'S GLORIOUS REVELATION. OMINOUS SIGNS VISIBLE FURTHER AGGRAVATION PERILOUS SITUATION FACING BELEAGUERED PERSIAN COMMUNITY. . . .

Universal House of Justice

276-1. For the announcement of Enoch Olinga's assassination, see message dated 17 September 1979 (no. 237).

277-1. Two days later, 19 March 1981, the Universal House of Justice cabled fifty-four National Spiritual Assemblies, saying that the "verdict Shiraz court clearly indicates sentence based membership Bahá'í Faith engagement Bahá'í activities."

278

Attendance of Bahá'ís from Rural Areas at International Conferences
30 March 1981

To the National Spiritual Assemblies in
Africa, Asia, Australasia, South and Central America

Dear Bahá'í Friends,

The Universal House of Justice wishes to stress the importance of encouraging as many believers as possible living in rural areas to attend the International Conferences, and to this end it is requesting the National Spiritual Assemblies of Nigeria, the Philippines and Ecuador to investigate the possibilities of providing inexpensive accommodation so that more will be able to attend. 278.1

Cost of travel is another important item, and it is the hope of the Universal House of Justice that National Spiritual Assemblies can find ways to assist a number of indigenous believers to travel to the Conferences, as attendance could be considered as a part of the deepening program. National Assemblies should also start now to encourage believers to make plans to attend the Conference nearest them and to save money for their fares. The Universal House of Justice requests you to give this matter your urgent attention, and report to it the steps you are taking to implement the contents of this letter. 278.2

With loving Bahá'í greetings,
DEPARTMENT OF THE SECRETARIAT

279

Deployment of Pioneers and Traveling Teachers
16 April 1981

To all Continental Pioneer Committees

Dear Bahá'í Friends,

As we enter the second phase of the Seven Year Plan, the Universal House of Justice has given consideration to the vital role that Continental Pioneer Committees can and should increasingly play in helping to fulfill the pioneer goals that have been set, and in coordinating international traveling teaching projects. We have been asked to send you the following comments. 279.1

There is no doubt that closer collaboration of the Continental Pioneer Committees with the Continental Boards of Counselors and with the National 279.2

Spiritual Assemblies, both those supplying pioneers and travel teachers and those receiving them, will increase the number of pioneers effectively and expeditiously settled at their posts, and will improve the results of the labors of international traveling teachers. This collaboration should always be uppermost in the minds of each Continental Pioneer Committee, particularly its secretary, so that efforts are increasingly made to widen the scope of the relationships and to strengthen the ties which bind the Continental Pioneer Committees to the institutions which they are called upon to serve.

Pioneers

279.3 In the message of the Universal House of Justice to the Bahá'ís of the world, intended to be read at the National Conventions, a call is raised for a total of 279 pioneers to settle in 80 countries. A list of these countries is attached, showing the number to be settled in each.[1] The number of pioneers on this list is additional to those called for at the outset of the Plan. No assignment of specific quotas has been made to National Spiritual Assemblies, although it is generally expected that the National Spiritual Assemblies originally made responsible for sending pioneers to these countries will respond favorably to this new call and spontaneously feel the spiritual responsibility to fill the supplementary pioneer needs. It will be the duty of Pioneer Committees to keep a close tally of pioneers settling in the countries named by the House of Justice, and to ensure, to the extent possible, that no goals remain unfilled.

Traveling Teachers

279.4 Experience has shown that traveling teachers from abroad can be of tremendous assistance to the teaching work in the fields of proclamation, expansion, and consolidation. It is important that the best use possible be made of these friends who are sacrificing their time and resources to serve the Faith in foreign fields.

279.5 Care must also be taken that traveling teachers do not prove to be a burden on the receiving community, and a cause of problems. The two most frequently occurring problems caused by traveling teachers can, the House of Justice feels, be greatly reduced by prior advice and information provided by your Committees.

279.6 The first is the arrival in a country, in rapid succession, of foreign traveling teachers who do not speak the language. Sometimes the net result is that the time of all the best local teachers (who may well be better teachers than the visitor) is occupied by translating for the traveler, and the community, instead of being helped and stimulated by visitors, is exhausted and becomes reluctant to accept future help. This can be avoided by proper advance planning and by explaining to both the Spiritual Assemblies and to the traveling

---

279-1. See message no. 275. The list referred to is too lengthy to include in this volume.

teachers themselves, that the Assemblies should not feel obliged to provide assistance for a visitor who arrives without prior agreement, and that a teacher who arrives unannounced may well have to concentrate on doing his teaching work unaided and without burdening the local friends.

The second problem occurs most frequently in countries such as those in Africa, where there is entry by troops. In such countries it is comparatively easy to bring large numbers of new believers into the Faith, and this is such a thrilling experience that visiting teachers often tend to prefer to do this rather than help with the consolidation work. Yet it is in consolidation that traveling teachers from abroad can often be most useful to the community. The House of Justice believes that this problem can be eased by your Committees' impressing upon traveling teachers that they must adhere strictly to the guidance given to them by the teaching committees and Spiritual Assemblies on the spot, and subordinate their own wishes to the need to render their services in the fields where they are most urgently required. It should be pointed out that, especially if they are assigned to expansion work, they must remember that consolidation is an essential and inseparable element of teaching, and if they go to a remote area and enroll believers whom no one is going to be able to visit again in the near future, they may well be doing a disservice to those people and to the Faith. To give people this glorious Message and then leave them in the lurch, produces disappointment and disillusionment, so that, when it does become possible to carry out properly planned teaching in that area, the teachers may well find the people resistant to the Message. The first teacher who was careless of consolidation, instead of planting and nourishing the seeds of faith has, in fact, "inoculated" the people against the Divine Message and made subsequent teaching very much harder. 279.7

While the caveats given above should be carefully considered by the Continental Pioneer Committees, nothing should be done to dampen the zeal of the friends to arise in order to carry out the injunction of Bahá'u'lláh to move from place to place. Their desire to offer themselves as travel teachers should be encouraged by the Continental Pioneer Committees to the extent that this lies within their power, and when the friends have volunteered, they should be lovingly guided so that the maximum results are obtained from their visits. 279.8

International Conferences 1982

The Universal House of Justice has called International Conferences during the first nine months in 1982 in the following locations: Montreal, Canada; Quito, Ecuador; Dublin, Ireland; Lagos, Nigeria; and Manila, Philippines. As soon as specific dates have been set, you will be notified. It is anticipated that many friends attending these Conferences will be able to undertake travel teaching assignments in connection with their travel to and from Conferences. 279.9

The House of Justice has asked us to alert you to the opportunities, and to request you to devise ways and means of encouraging such offers and of tak- 279.10

ing advantage of them, preferably by advance planning and routing of volunteers in consultation with the National Spiritual Assemblies of the countries they will visit. You should plan to have a representative of your Committee at the Conference nearest you to assist in the processing of pioneer offers and also in the routing of late traveling teacher offers. If it is not possible for a member of your Committee to attend, you may delegate another believer to represent you who, in such case, must be thoroughly briefed by you.

With loving Bahá'í greetings,
DEPARTMENT OF THE SECRETARIAT

280
Comments on Aspects of the Seven Year Plan
17 APRIL 1981

To all National Spiritual Assemblies

Dear Bahá'í Friends,

280.1 The message of the Universal House of Justice to the Bahá'ís of the world, dated March 1981, must have reached you by now, and, likewise, the specific message addressed to the Bahá'ís under your jurisdiction, outlining the goals of the second phase of the Seven Year Plan. The Universal House of Justice feels it is important that we share with you now, on its behalf, the following comments on certain aspects of the Plan. Each National Spiritual Assembly should be able to determine what portion of these comments is applicable to its work in the light of the goals it has been assigned.

The Local Spiritual Assembly

280.2 As you note from statements of guidelines and goals, a great deal of emphasis has been placed on the activities of local communities. It is obvious that through the consolidation of the foundations of the Administrative Order on the local level, the national institutions of the Faith will receive support and strength in the conduct of their activities. In turn, the National Spiritual Assembly and its agencies should not only oversee the activities of the local communities, but it has the duty and privilege to coordinate the efforts and to stimulate and give direction to the spirit of enterprise and initiative of the individual friends. When a proper and balanced relationship is maintained between these two levels of Bahá'í activity, and a healthy interaction takes place between them, a foundation is laid for the community to become "spiritually welded into a unit at once dynamic and coherent."[1]

280-1. MBW, p. 140.

The broad outlines of duties and functions of Local Spiritual Assemblies 280.3
are set forth clearly in the instructions of Bahá'u'lláh, 'Abdu'l-Bahá and Shoghi
Effendi, and these instructions have already been sent to National Assemblies
in the form of compilations.[2] Similar statements have also been made by the
Universal House of Justice, and these, too, have been shared with the friends.
If any National Assembly does not have these compilations at hand, it should
write to the World Center at once so that copies may be sent.

Consolidation

Consolidation is as vital a part of the teaching work as expansion. It is that 280.4
aspect of teaching which assists the believers to deepen their knowledge and
understanding of the Teachings, and fans the flame of their devotion to Bahá'u'lláh and His Cause, so that they will, of their own volition, continue the
process of their spiritual development, promote the teaching work, and
strengthen the functioning of their administrative institutions. Proper consolidation is essential to the preservation of the spiritual health of the community, to the protection of its interests, to the upholding of its good name, and
ultimately to the continuation of the work of expansion itself.

If a National Spiritual Assembly finds that its National Teaching Com- 280.5
mittee cannot devote sufficient attention to the work of consolidation, it
should not hesitate to appoint, in addition, special committees whose tasks
would be the conduct of the various activities which are essential for consolidation. Activities falling within this category include the organization of circuits of traveling teachers skilled in consolidation work; the holding of summer
and winter schools, weekend institutes and conferences; the initiation and
operation of tutorial schools; the dissemination of Bahá'í literature and the
encouragement of its study by the friends; and the organization of special
courses and institutes for Local Spiritual Assembly members.

In the courses for Local Assembly members special attention should be paid 280.6
to the significance of the Assembly and the importance of attending its meetings; the functions and duties of the Assembly's officers, especially those of
the secretary, upon the proper discharge of whose responsibilities the efficient
functioning of the Assembly largely depends; the importance of making the
Word of God easily accessible to the friends and of holding regular deepening classes where the Teachings can be studied and discussed; the vital necessity of prayer, and the value of holding gatherings for dawn prayers where and
when feasible; the proper holding of Nineteen Day Feasts and the observance
of Bahá'í Holy Days and anniversaries; the need to pay particular attention
to the education of children; and the value of organizing social gatherings,
such as picnics, encouraging the friends to associate together and with their
non-Bahá'í friends in love and fragrance.

280-2. For the compilations referred to, see CC 2:29–38, 2:39–60.

280.7 Consolidation activities promote the individual spiritual development of the friends, help to unite and strengthen Bahá'í community life, establish new social patterns for the friends, and stimulate the teaching work.

Bahá'í Literature

280.8 The question of making the Sacred Texts available to the friends is so important that the House of Justice commissioned a special committee a year or so ago to prepare three compilations from the Writings of Bahá'u'lláh, the Báb, and 'Abdu'l-Bahá. These compilations have already been sent to all National Spiritual Assemblies for publication in vernacular languages.[3]

280.9 Recently, the House of Justice instructed that a compilation in the form of a small booklet be prepared for use by the friends, especially in areas where literature is not easily available in print. A copy of this compilation, which consists of basic prayers and passages from the Writings, will soon be sent to you. It would be highly desirable for every believer to have easy access to at least a compilation of this type in a language he can understand, and it is sincerely hoped that by the reading of the Sacred Texts and the exposure of the believer's soul to their influence, his spiritual growth will be stimulated. He will thereby not only increase his own spiritual joy and understanding, but also contribute to the consolidation of the entire community.[4]

The Bahá'í Family

280.10 Another aspect of Bahá'í life emphasized in the provisions of the Seven Year Plan is the development of the Bahá'í family life. If the believer is the only one of his family who has embraced the Faith, it is his duty to endeavor to lead as many other family members as possible to the light of divine guidance. As soon as a Bahá'í family unit emerges, the members should feel responsible for making the collective life of the family a spiritual reality, animated by divine love and inspired by the ennobling principles of the Faith. To achieve this purpose, the reading of the Sacred Writings and prayers should ideally become a daily family activity. As far as the teaching work is concerned, just as individuals are called upon to adopt teaching goals, the family itself could adopt its own goals. In this way the friends could make of their families strong healthy units, bright candles for the diffusion of the light of the Kingdom, and powerful centers to attract the heavenly confirmations.

Recognition of the Faith

280.11 In the goals assigned to National Spiritual Assemblies, no specific reference has been made to goals for the recognition of the Bahá'í marriage certificate

---

280-3. The three compilations were published in one volume under the title *Inspiring the Heart: Selections from the Writings of the Báb, Bahá'u'lláh and 'Abdu'l-Bahá* by the Bahá'í Publishing Trust of the United Kingdom (n.d.).

280-4. The compilation was published under the title *Words of God* by the Bahá'í World Center (1981).

and of Bahá'í Holy Days, for the incorporation of Local Spiritual Assemblies, or for the obtaining of exemptions from state and municipal taxes on Bahá'í properties. Conditions in every country differ, and it is the duty of every National Spiritual Assembly to consider carefully the means whereby it can continually increase the degree of recognition officially accorded to Bahá'í institutions. It is important that any accomplishments in this vital area be reported to the World Center when National Spiritual Assemblies submit their semi-annual statistical reports.

280.12 The process of obtaining added recognition for the Faith can be stimulated if National Assemblies give adequate attention in their work to the ever-present need to continually foster cordial relations with government officials. Many of these officials have but a scanty and sometimes faulty knowledge of the Faith, and there is no doubt that when they become familiar with our aims and tenets and are assured of their beneficent effect on their society, they will be well-disposed to accord the Faith and its institutions at least such rights and privileges as are given to other religious organizations in the country.

Mass Media

280.13 The importance of mass media to Bahá'í proclamation has risen sharply in past years, and acutely so since the crisis in Iran. Mass media, and particularly radio, have proven to be potent instruments for the deepening of the friends and the promotion of the teaching work in mass teaching areas. Thus in the second phase of the Plan, 92 National Spiritual Assemblies have been given goals calling for increased use of the mass media. Countries not included in the list are those where the use of mass media for Bahá'í purposes is not permitted, or where the friends are already so actively engaged in such projects that they did not need the inclusion of such a goal in their assignments.

280.14 It is important for National Spiritual Assemblies generally to be aware that the use of mass media is becoming international in scope and that there is a need for National Spiritual Assemblies to share materials, methods, and experiences, and even personnel, in order to achieve best results.

280.15 In order to facilitate this exchange of information about the promotion of the Bahá'í work through radio and television, as well as about the availability of audiovisual materials, the Universal House of Justice, as it has been announced already, has established the International Bahá'í Audio-Visual Center (IBAVC) now in Toronto, Canada. This Center can be called upon by National Spiritual Assemblies for program materials and for advice on personnel training.

280.16 Every National Spiritual Assembly engaged in the teaching work in rural areas, where means of communication are scarce, is strongly advised to consider the possibility of applying for radio time on the local radio station, provided the cost is not too high. It may be possible to obtain such time in certain areas free of charge, especially when such a privilege is given to other groups or religions, or when the radio station is anxious to fill its time with worthwhile and helpful programs.

280.17 The House of Justice feels that in countries where the doors of publicity are open to the friends, every effort should be made to make the most of the attention being drawn to the Faith by the present situation, and exploit fully the potential for enlisting large numbers under the banner of the Cause.

Newsletters

280.18 The dissemination of Bahá'í news, local, national, and international, should be pursued with added vigor. Bahá'í newsletters should be issued regularly to the friends, however great the sacrifice, for news of Bahá'í activities in other communities has always been a source of encouragement and has given the friends a sense of belonging to a vital, growing, and united worldwide Bahá'í family.

280.19 In some areas, it may be found that more than one newsletter is necessary in order to reach all the friends with news about the work of the Faith. According to reports received at the World Center, a number of National Assemblies, in addition to issuing a national news bulletin, have made provisions for their Teaching Committees to issue regional or district newsletters in languages understood by the friends. Indeed, many a Local Spiritual Assembly has its own newsletter to ensure that information about developments of the Cause can reach every believer.

Marriages, Births, Burials

280.20 Local Spiritual Assemblies, which are embryonic Local Houses of Justice, should develop as rallying centers of the community. They must concern themselves not only with teaching the Faith, with the development of the Bahá'í way of life and with the proper organization of the Bahá'í activities of their communities, but also with those crucial events which profoundly affect the life of all human beings: birth, marriage, and death. When a Bahá'í couple has a child it is a matter of joy to the whole local community as well as to the couple, and each Local Spiritual Assembly should be encouraged to keep a register of such births, issuing a birth certificate to the parents. Such a practice will foster the consolidation of the community and of the Assembly itself. Even if only one of the parents is a Bahá'í, the Assembly could register the birth of the child, and upon application of the Bahá'í parent, issue the certificate.

280.21 The carrying out of the Bahá'í marriage laws, as given to the friends throughout the world, is a vital obligation of every believer who wishes to marry, and it is an important duty of every Local Spiritual Assembly to ensure that these laws are known to, and obeyed by, the believers within their jurisdiction, whether or not the Bahá'í marriage ceremony is recognized by civil law. Each Assembly, therefore, must conscientiously carry out its responsibilities in connection with the holding of Bahá'í marriage ceremonies, the recording of Bahá'í marriages in a register kept for this purpose, and the issuing of Bahá'í marriage certificates.

280.22 The burial of the dead is an occasion of great solemnity and importance, and while the conduct of the funeral service and the arrangements for the interment may be left to the relatives of the deceased, the Local Spiritual Assembly has the responsibility for educating the believers in the essential requirements of the Bahá'í law of burial as at present applied, and in courteously and tactfully drawing these requirements to the attention of the relatives if there is any indication that they may fail to observe them. These requirements are: that the body not be cremated; that it not be transported more than an hour's journey from the place of death to the place of burial; that the Prayer for the Dead be recited if the deceased is a Bahá'í of fifteen years of age or more; and that the funeral be carried out in a simple and dignified manner that would be a credit to the community.

280.23 In some parts of the world, if Local Spiritual Assemblies fail to carry out these sacred duties, some believers might gradually drift away from the Faith and even pay dues to churches or other religious organizations to ensure that, when they require to register the birth of a child, to solemnize a marriage or to have a funeral service, there will be a religious institution ready to perform the necessary services. Conversely, when Local Assemblies have arisen to carry out these responsibilities, the believers have acquired a sense of security and solidarity, and have become confident that in such matters they can rely upon the agencies of the World Order of Bahá'u'lláh.

Education of Children

280.24 The House of Justice has noted with deep gratification the increased number of Local Spiritual Assemblies which are organizing Bahá'í classes for children. In order to make these classes effective, it is important to have a graduated system of lesson plans suited to different age groups. Usually, such material is prepared by each National Spiritual Assembly in the manner suited to its conditions. However, to assist National Assemblies in benefiting from the fruits of the labors of Bahá'ís in other countries, we have been asked to inform you that the National Spiritual Assemblies of Colombia, India, Malaysia, and the United States have reported the availability of literature prepared by them for this purpose. You should feel free to correspond with these National Spiritual Assemblies.

280.25 Regarding tutorial schools, some National Assemblies engaged in this activity have reported excellent results, which have helped both in the expansion work and in the consolidation of the Faith. A report on this type of activity has recently been received at the World Center, and a digest of the report is attached for the study of those National Assemblies who have been assigned this goal.

Pioneers

280.26 Certain National Spiritual Assemblies have been assigned the goal of raising self-supporting homefront pioneers. This activity has great potential for the spread and consolidation of the Faith, and it is the hope of the House of Justice that this type of service will be encouraged in all national communities, including those which have not been given this goal.

280.27 In addition to homefront pioneers, there is a need for 279 pioneers to settle in 80 countries and islands of the world. A list of these pioneer needs is attached, and it is hoped that the required number of dedicated souls will arise to fill the posts that are in need of pioneer support.[5] It is suggested that National Assemblies keep in close touch with the Continental Pioneer Committees, who will be in a position to keep National Assemblies informed of progress towards these goals.

Continental Pioneer Committees

280.28 Continental Pioneer Committees working in close collaboration with National Spiritual Assemblies are assuming greater importance as the work of the Faith unfolds on every continent. The House of Justice is writing to all Continental Pioneer Committees, outlining their added responsibilities in relation both to the newly formed Continental Boards of Counselors and the National Spiritual Assemblies. It is the hope of the House of Justice that the services of these important Continental Committees will in the future be made available to the friends with ever-greater effectiveness.

280.29 The Universal House of Justice has asked us to assure you of its prayers for the blessings of Bahá'u'lláh to confirm your efforts as you face the next three years with optimism and confidence, and respond to the challenging opportunities ahead with determination and vigor.

With loving Bahá'í greetings,
DEPARTMENT OF THE SECRETARIAT

Bahá'í Tutorial Schools: An Example
(Summarized from a report received concerning the functioning of such schools in a particular country)

Special Conditions

280.30 The Bahá'í tutorial schools in this country are called "Bahá'í Educational Centers" to prevent confusion with the schools supported by the State and managed by the Department of National Education. They were established in part in response to the goal assigned during the last Plan, to deepen at least one person in each Bahá'í locality to assist in consolidating the community

---

280-5. The list is too lengthy to include in this volume.

and the Local Spiritual Assemblies. This need was felt particularly because in many of the villages, Bahá'í activities take place only during visits from pioneers or other visitors from the large cities.

Definition

"The Bahá'í Educational Center is a place where Bahá'í and non-Bahá'í children and adults receive, under the direction of their Local Spiritual Assemblies, first a spiritual education, then a basic literacy education, and finally a technical education such as training in crafts. This is also a center [for] numerous other social activities for youth and women. The purpose is to obtain quasi-universal participation of the whole village." 280.31

Relationship to Government

The Centers "are allowed to function on the provincial level by the [Government] Department of Social Affairs, which was seeking every possible means to provide literacy education for the masses. The authorities of that Department are so happy with the development of these Educational Centers which cost them nothing that they gave permission to award certificates of participation in these courses, signed, on the one hand, by themselves and, on the other hand, by the Local Spiritual Assembly concerned, which acts as the principal. For their part, the chiefs of villages also stated that when a Local Spiritual Assembly wants an Educational Center, they will themselves erect huts in which classes may be held and encourage all the children to attend day classes and adults evening classes." The chiefs also witness the signing of the contract between the volunteer teachers (see below) and the Local Spiritual Assembly. 280.32

Teachers

The teachers are Bahá'í volunteers, "often youth almost illiterate themselves who, after some time of unemployment in large cities, return disheartened to their native village but they become important when they are engaged in Educational Centers." These teachers sign an official contract with the Local Assembly before the chief of the civil community, who represents the tribe, stating that they are volunteers and will not demand any salary later on. "The teacher is recompensed by the Local Spiritual Assembly, which either provides labor to raise crops for him, provides him with food through pupils' contributions, or gives him financial assistance according to each particular case." 280.33

In addition, local villagers may offer to teach practical subjects in which they are knowledgeable (e.g., weaving of baskets or mats, canoe building, fishing, pottery, "cure by medicinal plants," etc.). (See below, "Program.") 280.34

Books and Materials

A committee assists the Local Assemblies in the preparation of programs and teaching materials. Books and materials are now in preparation, and are 280.35

being adapted to the "low instructional level of the teachers. This material will permit them to deepen themselves so that they may maintain a higher level than their pupils." Basic books for the teacher's use and such materials as chalk are bought by the Local Spiritual Assembly; materials for each pupil are bought by the parents at a wholesale price offered by the committee.

Registration and Fees

280.36 Fees are set by each Local Spiritual Assembly. At the time of registration, which may take place at 3-month, 6-month or yearly intervals according to the decision of the Local Assembly, a registration card is issued to each pupil.

Program

280.37 The program includes three parts: a "spiritual part which includes the teachings and laws, history of the Faith, and Bahá'í administration, all adapted to the level of each class;" a program which "includes basic literacy and general knowledge. This level will develop into post-literacy and permanent education in order to attain more advanced levels of evening classes, high school and even university." The third part "is the apprenticeship in crafts or other trades useful in the development of the village. Any villager can offer to teach crafts or trades in which he is knowledgeable: for instance, weaving of baskets and mats . . . fishing, cure by medicinal plants, pottery, embroidery, etc. . . ."

280.38 The schedule of courses is determined by the Local Spiritual Assembly, which takes into account harvest and fishing seasons, Nineteen Day Feasts and Bahá'í Holy Days, national holidays, and the like.

280.39 The duration of the program is not yet determined, but the first literacy phase is expected to last approximately two years, to be followed by "post-literacy and permanent education, and even to university."

Results to date

280.40 Effects already seen in various Local Assembly areas include the following:
1. The Assemblies are obliged to meet, and are thus strengthened.
2. Educational activities for children are established.
3. Youth activities, "such as playgrounds, choirs, lectures, homefront pioneers, team teaching and others," are organized.
4. Women's activities, "such as embroidery, sewing, hygiene, study of the importance of children's education, etc.," are organized.
5. "A more universal participation of the whole village in the spiritual activities, such as prayers, Nineteen Day Feast, contribution to the fund, teaching, etc."
6. "Increase and expansion" in the number of believers.
7. "Great prestige in the eyes of the government for the Bahá'í education."

Proposals for Future Developments
1. Recording of prayers, Holy Writings, and songs on tapes, to be used as deepening material;
2. Recording on a local basis by the believers, with duplication of tapes for use by other Educational Centers;
3. Radio broadcasts addressed directly to Bahá'í Educational Centers, either through the establishment of a radio station or through rental of time on existing stations;
4. Creation of an audiovisual center able to adapt material to the needs of the development of the Educational Centers;
5. A regional bulletin dealing with the needs of the "post-literacy" students;
6. Publication of a book on the Messengers of God in the major languages of the country, intended for the Educational Center but which could then also be sold to "lay schools" for countrywide distribution;
7. Construction of durable buildings for the Educational Centers;
8. A mobile institute for the training of teachers.

281
Execution of Three Bahá'ís in Shiraz
4 May 1981

To all National Spiritual Assemblies

WITH SADDENED HEARTS ANNOUNCE MARTYRDOM THREE MORE HEROES CRADLE FAITH WHO SURRENDERED LIVES AS THEIR LAST OFFERING SACRED THRESHOLD. YADU'LLÁH VAḤDAT, IḤSÁNU'LLÁH MIHDÍZÁDIH, SATTÁR K͟HUSHK͟HÚ, ALL THREE MEMBERS BAHÁ'Í INSTITUTIONS, WERE EXECUTED IN SHIRAZ BY FIRING SQUAD ON NIGHT APRIL 29 BY ORDER ISLAMIC REVOLUTIONARY COURT SHIRAZ AND WITH APPROVAL HIGH COURT JUSTICE TEHRAN. VERDICT SHIRAZ COURT PUBLISHED ALL MAJOR NEWSPAPERS IRAN SPECIFICALLY MENTIONS USUAL FALSE CHARGES GIVING DISTORTED IMAGE ACTIVITIES BAHÁ'ÍS MISREPRESENTING THEM AS ZIONISTS IMPERIALIST AGENTS PROMOTERS GODLESSNESS ENEMIES ISLAM. . . .

UNDAUNTED BY THIS FRESH OUTBREAK ANIMOSITY AGAINST THEM STAUNCH FRIENDS CRADLE FAITH PREPARING THEMSELVES WITH JOY UNFOLDMENT THEIR GLORIOUS DESTINY, CONFIDENT PRECIOUS LIVES THEY ARE OFFERING ALTAR SACRIFICE WILL GALVANIZE BELIEVERS EVERY LAND INSPIRE THEM REDOUBLE THEIR DEVOTED EXERTIONS IN SERVITUDE BELOVED FAITH.

SUMMON FRIENDS EVERY COUNTRY HOLD SPECIAL PRAYERS DURING NIGHT AND DAY OF ASCENSION BAHÁ'U'LLÁH MAY 29 BESEECHING DIVINE PROTECTION BELEAGUERED FRIENDS IRAN.

UNIVERSAL HOUSE OF JUSTICE

282
Imminent Obliteration of the Site of the House of the Báb
26 May 1981

To all National Spiritual Assemblies

282.1 PROFOUNDLY DISTRESSED OMINOUS NEWS IMMINENT OBLITERATION SITE BLESSED HOUSE BÁB BY AUTHORITIES SHIRAZ IN IMPLEMENTATION PLANS DRAWN UP SEVERAL MONTHS AND BUILD ROAD AND PUBLIC SQUARE.[1] OCCUPANTS ADJACENT HOUSES MOST OF WHICH HAD BEEN ACQUIRED BY BAHÁ'Í COMMUNITY AS PROTECTION HOLY HOUSE NOW ORDERED VACATE HOUSES PRELIMINARY COMMENCEMENT PROJECT THIS WEEK. RECALL WHEN BÁB'S HOUSE WAS CONFISCATED GOVERNMENT ALLEGED STEP TAKEN AS PROTECTION HOLY PLACE, WHEN HOUSE DESTROYED GOVERNMENT STATEMENT ATTRIBUTED ACT TO UNRULY MOB, BUT THIS DEVELOPMENT NOW TOTALLY BELIES SUCH ALLEGATIONS SHOWS OFFICIAL DELIBERATE SYSTEMATIC DESIGN ERADICATION BAHÁ'Í HOLY PLACES AFTER THEIR CONFISCATION, IN ADDITION TO HARASSMENT PRESSURE ON INDIVIDUAL BELIEVERS RECANT THEIR FAITH ON PAIN LOSING JOBS PENSIONS DEPRIVATION CIVIL RIGHTS IMPRISONMENT EXECUTION ASSASSINATION. . . .

UNIVERSAL HOUSE OF JUSTICE

283
Establishment of European Branch Office of Bahá'í International Community
26 May 1981

To the European National Spiritual Assemblies

Dear Bahá'í Friends,

283.1 Further to the letter sent to you on behalf of the Universal House of Justice on 28 December 1980 seeking your views on the proposal to establish in Europe a branch office of the Bahá'í International Community, we are now instructed to tell you that the House of Justice has decided to establish the office now, with its address in Geneva. Mr. Giovanni Ballerio has been appointed full-time representative of the Bahá'í International Community in Europe, and will be moving, with his family, to Geneva or its vicinity. When the address is known and the office is operating an announcement will be made to the Bahá'í world.[1]

With loving Bahá'í greetings,
DEPARTMENT OF THE SECRETARIAT

282-1. For more about the destruction of the House of the Báb, see messages dated 9 September 1979 (no. 235) and 19 November 1979 (no. 241).

283-1. On August 10, 1981, a letter was sent to the United Nations Offices notifying them of Mr. Ballerio's appointment and office address.

284

Bahá'í International Fund—Recent Victories and Immediate Challenges

8 June 1981

To the Bahá'ís of the World

Dear Bahá'í Friends,

At the outset of the Seven Year Plan, faced with tremendous tasks to be accomplished by the Bahá'í world, and confronted by the seemingly crippling financial losses that resulted from the savage onslaught of inveterate enemies upon the valiant believers in the Cradle of the Faith, the Universal House of Justice turned with fervent hope to the believers in the rest of the world, calling upon them to arise and champion the cause of their persecuted brethren in the international arena and, through self-sacrifice and the exercise of wise stewardship of the funds of the Faith, to enable its work to go forward unhindered by the sudden inability of the believers in Iran to continue their major role in providing the lifeblood of the Cause. In both fields, these past two years have witnessed astonishing victories. 284.1

The manner in which the case of the persecuted Faith of Bahá'u'lláh has been blazoned in the media, conveying its message to millions of souls who had scarcely if at all heard of it before; and the degree to which world authorities have risen to plead its case and call for its vindication, have both been witnessed with eager and uplifted hearts by Bahá'ís in all lands. Now the Universal House of Justice has instructed us to inform you that in supporting the Bahá'í International Fund the self-sacrificing followers of the Blessed Beauty have won similar victories. 284.2

The manifold acts of devotion and service that have been so distinctive a mark of progress in the opening phase of the Seven Year Plan have not only laid a firm foundation for the development of the institutions of the Faith worldwide, but have also been manifested in an outpouring of financial substance that has made possible the setting of goals for the second phase of the Plan that will enhance the prestige of our beloved Faith and hasten the day of its complete emergence from obscurity. The financial needs of the first phase of the Plan have been fully met. The teaching work has continued with unabated zeal. The Seat of the Universal House of Justice is now nearing completion and the funds required for that tremendous task are in hand. The Mashriqu'l-Adhkár of the Indian subcontinent is rising and the work on that for Samoa will shortly begin. 284.3

Now the Universal House of Justice turns with loving confidence to those beloved ones of God who have responded with such zeal to fulfill the commitments that it has been guided to undertake, and has asked us to lay before you the needs of the year which has just begun. 284.4

284.5 Although the task of raising the Seat of the Universal House of Justice on God's Holy Mountain is now well-nigh behind us, the work on the two Mashriqu'l-Adhkárs has acquired additional urgency. The unstable condition of the world and rapidly rising prices make it essential to complete these two enterprises at the earliest possible date. Thus, major expenditures which it had been hoped could be spread over a number of years must be met within the next twelve months. The public attention drawn to the Faith by the Iranian situation and the many valuable friendships and contacts that have been made with those in authority demand, if the ground now gained is not to be rapidly lost, expansion and intensification of the activities of the representatives of the Bahá'í International Community with the United Nations and its specialized agencies, as well as with other international bodies such as the Parliament of Europe. The Universal House of Justice has therefore taken the decision to establish an office of the Bahá'í International Community at Geneva, with a full-time representative. It has also become necessary to allocate large sums to the provision of a wider range of Bahá'í literature in many languages and to develop the Bahá'í use of radio.

284.6 The Universal House of Justice has estimated that to meet all these urgent goals as well as carrying on the current work of the Faith during the year 1981/82, an increase of 50% over the amount of contributions for the year just past will be required. It therefore asks every believer and every community to consider prayerfully the degree to which they can take part in this mighty effort, and to strain every sinew to ensure that the tasks placed by an omniscient and all-wise Ordainer on the shoulders of His privileged lovers will be worthily and speedily performed.

284.7 The House of Justice assures you all of its loving, fervent prayers in the Holy Shrines for your guidance and assistance.

With loving Bahá'í greetings,
DEPARTMENT OF THE SECRETARIAT

1979–1986 • THE SEVEN YEAR PLAN

285
Execution of Seven Members of the Spiritual Assembly of Hamadan
15 June 1981

To all National Spiritual Assemblies

WITH STRICKEN HEARTS SHARE NEWS SEVEN MEMBERS LOCAL SPIRITUAL ASSEMBLY HAMADAN MARTYRED AFTER BEING TORTURED. EXECUTIONS CARRIED OUT DAWN JUNE 14 WITH APPROVAL SUPREME JUDICIAL COUNCIL ON USUAL TRUMPED-UP CHARGES. THIS FRESH BLOW ANOTHER STEP IN PERSECUTION SCHEME TRADITIONAL ADVERSARIES UPROOT FAITH IN LAND ITS BIRTH. . . . 285.1

NAMES OF SEVEN MARTYRS ARE AS FOLLOWS: ḤUSAYN MUTLAQ, MUḤAMMAD-BÁQIR (SUHAYL) ḤABÍBÍ, MUḤAMMAD (SUHRÁB) ḤABÍBÍ, DR. NÁṢIR VAFÁ'Í, DR. FÍRÚZ NA'ÍMÍ, ḤUSAYN KHÁNDIL, AND ṬARÁẒU'LLÁH KHUZAYN. . . . 285.2

UNIVERSAL HOUSE OF JUSTICE

286
Execution of Three Bahá'ís in Tehran
23 June 1981

To all National Spiritual Assemblies

TRAGIC NEWS JUST RECEIVED THREE MORE PROMINENT PERSIAN BAHÁ'ÍS, BUZURG 'ALAVÍYÁN, HÁSHIM FARNÚSH AND FARHANG MAVADDAT EXECUTED TEHRAN YESTERDAY. CHARGES COMPLETELY MISREPRESENT BAHÁ'Í SERVICES AS POLITICALLY MOTIVATED ACTIVITIES. . . . OFFERING PRAYERS HOLY SHRINES BLESSED BEAUTY MAY CONFIRM ALL EFFORTS ALLEVIATE SUFFERINGS OPPRESSED INNOCENT COMMUNITY. 286.1

UNIVERSAL HOUSE OF JUSTICE

287
Execution of Four Bahá'ís in Iran
24 June 1981

To all National Spiritual Assemblies

WITH HEARTS BURNING WITH ANGUISH SHARE NEWS SOULS ANOTHER FOUR DISTINGUISHED BELIEVERS NOW GATHERED ABHÁ KINGDOM ON BEING MARTYRED YESTERDAY BY FIRING SQUAD: DR. MASÍḤ FARHANGÍ MEMBER BOARD COUNSELORS ASIA, BADÍ'U'LLÁH FARÍD, YADU'LLÁH PÚSTCHÍ, VARQÁ TIBYÁNÍYÁN. . . . 287.1

UNIVERSAL HOUSE OF JUSTICE

288
Announcement of Representatives to the International Conferences
20 July 1981

To all National Spiritual Assemblies

Dear Bahá'í Friends,

288.1 It is with great pleasure we announce that the following Hands of the Cause of God have been named as our representatives to the International Conferences:

Amatu'l-Bahá Rúḥíyyih Khánum	Montreal, Canada	2–5 September 1982
Ugo Giachery	Manila, Philippines	7–9 May 1982
Paul Haney	Quito, Ecuador	6–8 August 1982
William Sears	Lagos, Nigeria	19–22 August 1982
Collis Featherstone	Dublin, Ireland	25–27 June 1982

With loving Bahá'í greetings,
THE UNIVERSAL HOUSE OF JUSTICE

289
Tests of Bahá'í Community Life— A Balanced Perspective
22 July 1981

To an individual Bahá'í

Dear Bahá'í Friend,

289.1 The Universal House of Justice has received your letter of 6 March 1981 and has instructed us to send you the following comments on the issues you have raised.

289.2 The House of Justice feels that your questions are very perceptive and that, in many instances, you have, yourself, provided the answers. As 'Abdu'l-Bahá so often points out, the Manifestation of God is a Divine Educator. He attracts the hearts of men, pours out His spirit upon those who respond to Him, instructs them in the right way of life, uses them to carry forward the development of human society, and disciplines them by His law. We Bahá'ís, we who have answered His call, bear the responsibility of carrying forward His work among mankind, and in spite of our innumerable failings His plan is irresistibly progressing. The great tragedy of mankind at this time is the failure of the vast majority of human beings to heed the Divine Call, and this is in large part occasioned by the failure of most of those who have believed to live up to the high standard that Bahá'u'lláh has set. This is the condition in which we must work in our service to mankind, turning a sin-covering eye to the faults of others, and striving in our own inmost selves to purify our lives in accordance with the divine Teachings.

Differentiating the Roles of Spiritual Assemblies and Individuals

289.3 The Day of God is a Day of Joy, but also a Day of Judgment. Every man is guided both by the Love of God and by the Fear of God. In their relationships with one another individual believers should be loving and forgiving, overlooking one another's faults for the sake of God, but the Spiritual Assemblies are the upholders of the law of God. They are embryonic Houses of Justice. The education of a child requires both love and discipline; so also does the education of believers and the education of a community. One of the failings of Bahá'ís, however, is to confuse these two roles, individuals behaving like little Spiritual Assemblies, and Spiritual Assemblies forgetting that they must exercise justice.

Deepening—Developing a Spiritual Attitude

289.4 Great love and patience are needed towards new believers, especially those who have come from very troubled backgrounds, but ultimately they too have to learn the responsibilities they have taken upon themselves by accepting Bahá'u'lláh and must uphold the principles that Bahá'u'lláh has revealed. If they do not do so, how can the condition of mankind be improved? Some people accept the Faith, not as a response to the divine Summons to God's service, but as a way to find love and happiness and companionship and understanding for themselves. At the beginning this is only natural, for people are sorely in need of such spiritual strengths, but if such people do not soon progress to the point where they are more concerned about what they can do for God and His Cause than what it can do for them, they will surely become disillusioned and drift away. Arousing in the hearts of the friends the enthusiasm and spirit of selfless service that will carry them over this transition is one of the most fundamental aspects of deepening and consolidation. Deepening is far more a matter of developing a spiritual attitude, devotion and selflessness than it is of acquiring information, although this, of course, is also important.

289.5 In a letter to an individual Bahá'í, dated 5 April 1956, the beloved Guardian's secretary wrote on his behalf:

289.5a He was very sorry to hear that you have had so many tests in your Bahá'í life. There is no doubt that many of them are due to our own nature. In other words, if we are very sensitive, or if we are in some way brought up in a different environment from the Bahá'ís amongst whom we live, we naturally see things differently and may feel them more acutely; and the other side of it is that the imperfections of our fellow-Bahá'ís can be a great trial to us.

289.5b   We must always remember that in the cesspool of materialism, which is what modern civilization has to a certain extent become, Bahá'ís—that is some of them—are still to a certain extent affected by the society from which they have sprung. In other words, they have recognized the Manifestation of God, but they have not been believers long enough, or perhaps tried hard enough, to become "a new creation."

289.5c   All we can do in such cases is to do our duty; and the Guardian feels very strongly that your duty is towards Bahá'u'lláh and the Faith you love so dearly; and certainly is not to take the weaker course and sever yourself from the Bahá'í Community.

289.5d   He feels that, if you close your eyes to the failings of others, and fix your love and prayers upon Bahá'u'lláh, you will have the strength to weather this storm, and will be much better for it in the end, spiritually. Although you suffer, you will gain a maturity that will enable you to be of greater help to both your fellow-Bahá'ís and your children.

Spiritual Assembly Intervention—Loving and Patient

289.6   The ideal of human life is described again and again and in multitudes of ways in the Writings. These aspects of the Teachings are discussed in Teaching Institutes and Summer Schools and elaborated in many books. Then, in general, it is left to the individual believer, as a responsibility between himself and God, to follow these Teachings. It is not the business either of the believers or of the Spiritual Assemblies to pry into the lives of individual friends to ascertain the degree to which they are living up to the standards of the Cause. Only if misbehavior becomes blatant and flagrant does it become a matter for action, and then it is a matter for action by the Assembly and not by individuals. Even then the Assembly must be loving and patient, and exhort the believer to follow the Path of the Cause, but, if he persists in openly and flagrantly flouting Bahá'í law, the Assembly has no alternative to ultimately depriving him of his voting rights.

The Relationship between Supporting an Assembly and Its Maturation

289.7   Applying these principles requires mature understanding and judgment, and great love for one's fellowmen. It is a weighty responsibility which rests upon the shoulders of the members of Spiritual Assemblies. Undoubtedly errors are made and will continue to be made, but the more the friends are united and wholeheartedly support their Assemblies, the sooner will these mature in their decisions and actions, outgrow their mistakes, and become strong magnets for the Faith.

Briefly, then, one can say that the Bahá'ís, while in the process of improving their own lives, are engaged in attracting their fellowmen to the Love of God, educating them through the Teachings of God, introducing them to the vivifying discipline of the Law of God, and enlisting them as fellow-warriors in the Army of God. The difficulties that you describe are the result of the problem of properly balancing these many aspects of following the Bahá'í Cause and of training new believers from the point of acceptance of the Message to being champions of the Faith. 289.8

With loving Bahá'í greetings,
DEPARTMENT OF THE SECRETARIAT

290
Execution of Two Bahá'ís in Káshmar
27 JULY 1981

To all National Spiritual Assemblies

GRIEVED ANNOUNCE TWO MORE ACTIVE DEDICATED SUPPORTERS FAITH BAHÁ'U'- 290.1
LLÁH IRAN KAMÁLI'D-DÍN BAKHTÁVAR AND NI'MATU'LLÁH KÁTIBPÚR-SHAHÍDÍ MARTYRED BY FIRING SQUAD IN KÁSHMAR, KHURÁSÁN PROVINCE CHARGED WITH TOTALLY FALSE ACCUSATIONS INVOLVEMENT POLITICAL ACTIVITIES. IRANIAN BAHÁ'ÍS SPECIALLY IN RURAL AREAS FACED WITH FRESH WAVE ARRESTS AND CONFISCATION PERSONAL PROPERTIES. AROUND ISFAHAN ALSO KHURÁSÁN KÁSHÁN IN NUMBER TOWNS VILLAGES SEVERAL HUNDRED BAHÁ'ÍS FORCED FLEE THEIR HOMES TAKING REFUGE IN NEARBY LARGER TOWNS. SUCH RUTHLESS ATTACKS REMAIN UNCHECKED. ATTEMPTS TERRORIZE BAHÁ'ÍS INTENSIFIED. MORE INNOCENT LIVES AT STAKE. . . .

UNIVERSAL HOUSE OF JUSTICE

291
Message to First National Bahá'í Women's Conference, United Kingdom
27 JULY 1981

To the First Women's Conference, United Kingdom

KINDLY CONVEY PARTICIPANTS FIRST NATIONAL WOMEN'S CONFERENCE LOVING 291.1
CONGRATULATIONS SUCCESS ASSURANCE PRAYERS BOUNTIFUL CONFIRMATIONS ALL EFFORTS BAHÁ'Í WOMEN MAKE NOBLE CONTRIBUTION PROMOTION SPIRITUAL LIFE COMMUNITY.[1]

UNIVERSAL HOUSE OF JUSTICE

291-1. The conference, attended by 230 individuals, was held on 25 July 1981 in Manchester, England, at the Manchester Institute of Science and Religion.

292

Execution of Auxiliary Board Member and Seven Assembly Members in Tabríz; Abduction of Two Teenage Girls

30 July 1981

To all National Spiritual Assemblies

292.1 WITH HEAVY-LADEN HEARTS ANNOUNCE NEWS JUST RECEIVED NINE HEROIC FRIENDS EXECUTED TABRÍZ BY FIRING SQUAD ONE OF WHOM AUXILIARY BOARD MEMBER AND SEVEN MEMBERS LOCAL SPIRITUAL ASSEMBLY:

MASRÚR DAKHÍLÍ
ḤUSAYN ASADU'LLÁHZÁDIH
ALLÁH-VIRDÍ MÍTHÁQÍ
MANÚCHIHR KHÁDÍ'Í
'ABDU'L-'ALÍ ASADYÁRÍ
ISMÁ'ÍL ZIHTÁB
PARVÍZ FÍRÚZÍ
MIHDÍ BÁHIRÍ
ḤABÍBU'LLÁH TAḤQÍQÍ.

292.2 TWO TEENAGE GIRL STUDENTS. . . . ABDUCTED FROM SCHOOL BY TEACHERS IN RELIGIOUS INSTRUCTION. PARENTS UNABLE DETERMINE FATE DISAPPEARED CHILDREN. TEACHERS CLAIM GIRLS CONVERTED ISLAM REFUSE MEET BAHÁ'Í PARENTS. LOCAL AUTHORITIES UNCOOPERATIVE.

292.3 GRIEVED AT RUTHLESSNESS RAPIDITY WITH WHICH PRECIOUS LIVES DISTINGUISHED VIRTUOUS MEMBERS COMMUNITY BEING SNUFFED OUT, THEIR HONOR VIOLATED, THEIR HOMES POSSESSIONS PLUNDERED. WE PRAY BAHÁ'U'LLÁH BEHALF ENTIRE BAHÁ'Í WORLD ATTAIN GREATER CAPACITY SERENELY BEAR WEIGHT ORDEALS, WITNESS EARLY DELIVERANCE HIS PERSECUTED LOVERS CRADLE FAITH FROM SHACKLES REPRESSION BIGOTRY, AS PROMISED HIS SACRED WRITINGS.

UNIVERSAL HOUSE OF JUSTICE

293
A New Phase of Persecution in Iran
11 August 1981

To all National Spiritual Assemblies

PERSECUTION BAHÁ'ÍS IRAN GAINING MOMENTUM ENTERING NEW PHASE: IN YAZD A FEW DAYS AGO GOVERNMENT FROZE ALL ASSETS 117 BELIEVERS. ON 8 AUGUST ANNOUNCEMENT ON LOCAL RADIO SUMMONED HEADS 150 PROMINENT BAHÁ'Í FAMILIES TO REPORT WITHIN ONE WEEK TO REVOLUTIONARY AUTHORITIES. IN ABSENTIA DECREES TO BE ISSUED RESPECT ANY NAMED BELIEVER WHO FAILS PRESENT HIMSELF BY 15 AUGUST. AMONG NAMES ARE FEW WHO PASSED AWAY, CONFIRMING DETERMINATION AUTHORITIES PERSECUTE BAHÁ'ÍS PURELY FOR THEIR BELIEF, NOT BECAUSE OF ANY ALLEGED CRIME. ONE OF THOSE NAMED WAS ARRESTED AS HE WAS PROCEEDING FOR NECESSARY TEMPORARY JOURNEY OUTSIDE YAZD. ACTIONS TAKEN FORESHADOW PLAN AUTHORITIES FORCE BAHÁ'ÍS RECANT THEIR FAITH PURELY FOR THEIR FAITH ON PAIN CONFISCATION ALL THEIR PROPERTIES, OTHER DIRE CONSEQUENCES. . . . 293.1

FURTHER REPORT JUST RECEIVED INDICATES IN MANSHÁD VILLAGE NEAR YAZD GOVERNMENT OFFICIAL FROM YAZD ACCOMPANIED REVOLUTIONARY GUARDS HAS PEREMPTORILY SEIZED FURNITURE CROPS LIVESTOCK LOCAL BELIEVERS. 293.2

FOLLOWING DETAILS ADDITIONAL PERSECUTION OTHER PROVINCES NOW IN HAND: 293.3

IN MASJID SULAYMÁN AUTHORITIES HAVE INSTRUCTED BANKS SUBMIT LIST ALL CHECKING AND DEPOSIT ACCOUNTS BAHÁ'ÍS.

IN NAYSHÁBÚR WHERE TWO BELIEVERS WERE RECENTLY MARTYRED MOB HAD DESTROYED WALL BAHÁ'Í CEMETERY. AUTHORITIES NOW CLAIM TWO MILLION RIALS FROM LOCAL COMMUNITY TO RESTORE WALL.[1]

IN HIMMATÁBÁD NEAR ABÁDIH WIVES OF BELIEVERS WHO HAD FLED FROM THEIR HOMES HAVE BEEN GIVEN NOTICE CALL THEIR HUSBANDS. WIVES THREATENED GRAVE REPERCUSSIONS IF HUSBANDS FAIL PRESENT THEMSELVES. . . .

UNIVERSAL HOUSE OF JUSTICE

293-1. Two million rials was equivalent to about $28,400.00 in 1981.

294
Release of a Compilation on the Assistance of God
24 August 1981

To all National Spiritual Assemblies

Dear Bahá'í Friends,

294.1 The Universal House of Justice commissioned the Research Department to compile texts which convey the assurance that the assistance of God will surround and confirm the efforts of the friends when they arise to deliver His divine message and teach His Cause. This compilation is now ready, and a copy is attached.[1] The Universal House of Justice leaves it to your discretion to determine how best the contents of this compilation or other texts on this subject should be shared with the friends.

294.2 It is the conviction of the House of Justice that the powers of heaven and earth will, as repeatedly asserted in the attached extracts, mysteriously and unfailingly assist all those who will arise with love, dedication, and trust in their hearts to teach the Cause, to promote the Word of God, to deliver its healing message to receptive souls, and to serve its vital interests.

With loving Bahá'í greetings,
DEPARTMENT OF THE SECRETARIAT

295
Release of a Compilation of Prayers and Passages from the Bahá'í Writings
16 September 1981

To all National Spiritual Assemblies

Dear Bahá'í Friends,

295.1 You will recall that in its letters outlining goals of successive Teaching Plans for the expansion and consolidation of the Faith, the Universal House of Justice has repeatedly stressed the importance of the friends' having access to the Sacred Text. In our letter of April 17, 1981, written on behalf of the House of Justice to all National Spiritual Assemblies, it was mentioned that steps were being taken to prepare a compilation which would consist of prayers and passages from the Holy Writings of our Faith.[1]

294-1. See CC 2:201–24.
295-1. See message no. 280.

This material is now ready and is being sent to you in the form of a book- 295.2
let.² The contents are suitable for use by individual believers, as well as for
occasions when the friends gather for meetings. Your National Assembly may
have already published a compilation of this order and size for the use of the
friends. If not, and if your Assembly is among those specifically assigned the
goal to make basic Bahá'í literature available, and if you decide that the material included in the enclosed booklet is suitable for your purposes, the Universal House of Justice is prepared to send you, upon request, a proof copy
of the English text in flat sheets ready for offset printing, as this will facilitate the printing of the booklet in English, if such is your need.

The material in the booklet, it is hoped, will soon be ready also in French 295.3
and in Spanish, as the House of Justice is asking the respective Publishing
Trusts in Belgium and Argentina to undertake its translation and publication.
However, if the language or languages in which you should publish your Bahá'í
literature is other than the three mentioned above, you should consider the
material, such as the passages included in the enclosed compilation, for translation and dissemination among the friends, thus enabling them, as envisaged
in the Plan, to enrich their spiritual lives through exposure to the Sacred Word.

With loving Bahá'í greetings,
DEPARTMENT OF THE SECRETARIAT

296
Execution of Six Bahá'ís in Tehran and Dáryún
18 SEPTEMBER 1981

To all National Spiritual Assemblies

RUTHLESS PERSECUTION DEFENSELESS COMMUNITY BAHÁ'ÍS IRAN FURTHER INTEN- 296.1
SIFIED THROUGH RECENT EXECUTION BY FIRING SQUADS OF SIX STAUNCH
MARTYRS, ONE IN TEHRAN ḤABÍBU'LLÁH 'AZÍZÍ, FIVE IN DÁRYÚN NEAR ISFAHAN,
BAHMAN 'ÁṬIFÍ, 'IZZAT 'ÁṬIFÍ, 'AṬÁ'U'LLÁH RAWḤÁNÍ, AḤMAD RIḌVÁNÍ, AND
GUSHTÁSB THÁBIT-RÁSIKH. LAST FIVE WERE IMPRISONED FOLLOWING LARGE-SCALE
SIMULTANEOUS ATTACKS ON BAHÁ'ÍS IN THEIR HOMES AND ARREST OF SEVERAL
OF THEM IN VILLAGES NEAR ISFAHAN. FAMILIES MARTYRS WERE NOT INFORMED
OF EXECUTIONS WHILE RELATIVES THOSE EXECUTED IN DÁRYÚN ALSO NOT
PERMITTED CONDUCT BAHÁ'Í FUNERALS THEIR LOVED ONES, AND LAST THREE
NAMED WERE BURIED UNCEREMONIOUSLY IN MUSLIM CEMETERY. NO ANNOUNCEMENT BY AUTHORITIES WAS MADE ABOUT EXECUTION LAST FIVE. . . .

UNIVERSAL HOUSE OF JUSTICE

295-2. The compilation was published under the title *Words of God* by the Bahá'í World
Center (1981).

297

The Worsening Situation in Yazd
28 September 1981

To all National Spiritual Assemblies

297.1 SITUATION PERSECUTION FRIENDS YAZD ENTERING NEW PERILOUS PHASE: FURTHER INFORMATION CONVEYED TO YOU OUR CABLE 11 AUGUST ABOUT RADIO SUMMONS ISSUED PROMINENT YAZD BAHÁ'ÍS THESE FRIENDS ARE AGAIN BEING CALLED PRESENT THEMSELVES BEFORE REVOLUTIONARY COURTS. 14 OF THEM HAVE SO FAR REPORTED. WHILE FATE THESE IN BALANCE TEN OTHER BELIEVERS YAZD WHO HAD BEEN PREVIOUSLY DETAINED HAVE JUST BEEN TRANSFERRED TO PRISONS DIRECTLY ASSOCIATED REVOLUTIONARY COURTS. FOUR OF THESE ARE WOMEN. ONE OF THEM IS 75 YEARS OLD, THE OTHER THREE IN THEIR LATE FIFTIES. TRANSFER NEW PRISON OMINOUS SIGN THAT LOCAL REVOLUTIONARY COURT INTENDS ISSUE ITS FINAL VERDICT. FRIENDS IN CAPITAL HAVE REPEATEDLY IN PAST APPEALED WITHOUT SUCCESS FOR VINDICATION RIGHTS BAHÁ'ÍS TO CENTRAL AUTHORITIES TEHRAN. IN RECENT WEEKS HIGH-RANKING AUTHORITIES EVEN REFUSE GIVE AUDIENCE TO BAHÁ'ÍS AS SOON AS THEY LEARN SUBJECT MATTER. OFFICIALS IN LOWER RANKS SO FAR CONTACTED HAVE IN RECENT MONTHS WARNED BELIEVERS OF GRAVE PERSECUTIONS IN STORE FOR BAHÁ'ÍS. ONE GOVERNMENT REPRESENTATIVE HAS UNOFFICIALLY INFORMED BAHÁ'ÍS THAT DEVELOPMENTS IN YAZD MAY WELL FORM PATTERN FOR TREATMENT BAHÁ'ÍS ELSEWHERE IN COUNTRY, THAT AUTHORITIES ARE FULLY AWARE ALL MEASURES TAKEN BY BAHÁ'ÍS IN INTERNATIONAL CIRCLES, AND CONSIDER THESE AS ANTI-GOVERNMENT ACTIONS, THAT GOVERNMENT IRAN HAS NO REGARD FOR INTERNATIONAL OPINION, THAT BAHÁ'ÍS ARE ENEMIES ISLAM, AND BAHÁ'Í TEACHINGS SUCH AS EQUALITY SEXES AND REMOVAL VEIL CONSTITUTE VIOLATION ISLAMIC PRINCIPLES, AND THAT GOVERNMENT CAPABLE ORGANIZE MOBS TO PUBLICLY DEMAND EXPULSION, EXECUTION, EXTERMINATION BAHÁ'ÍS. THESE THREATS SHOULD NOT BE TAKEN LIGHTLY. . . .

UNIVERSAL HOUSE OF JUSTICE

298

Further Deterioration of the Situation in Iran
16 October 1981

To all National Spiritual Assemblies

SITUATION FRIENDS IRAN STEADILY DETERIORATING. ALTHOUGH REVOLUTIONARY COURT YAZD ORDERED RELEASE FOUR OF THOSE DETAINED PRESSURES IN DIFFERENT FORMS ARE MOUNTING IN YAZD AND IN OTHER PARTS COUNTRY. BAHÁ'Í WORKERS BEING INCREASINGLY EXPELLED FROM FACTORIES, BAHÁ'Í SHOPKEEPERS IN ONE PROVINCE ORDERED CLOSE DOWN FOLLOWING WITHDRAWAL BUSINESS LICENSES, SCHOOL AUTHORITIES SCRUTINIZING RELIGION STUDENTS PRIOR TO REGISTRATION RESULTING IN REFUSAL REGISTER BAHÁ'Í STUDENTS IN ELEMENTARY AND SECONDARY SCHOOLS. DECREE ANNOUNCED BY MINISTRY OF EDUCATION PUBLISHED IN *KAYHÁN* DAILY NUMBER 11397 DATED 9 MIHR 1360 (SEPTEMBER 30, 1981) LISTS CRIMES WHICH BAR PROFESSORS AND STUDENTS FROM BEING EMPLOYED OR REGISTERED AT UNIVERSITIES. AMONG CRIMES LISTED IS WHAT TEXT OF DECREE DESCRIBES AS MEMBERSHIP IN SECT WHICH IS RECOGNIZED BY MUSLIMS AS MISLED AND HERETICAL SECT. THIS IS OBVIOUS REFERENCE TO MEMBERSHIP IN BAHÁ'Í COMMUNITY. BAHÁ'Í CHILDREN AND YOUTH EXEMPLIFYING HIGH SPIRIT HEROISM STEADFASTNESS PREFERRING DEPRIVATION FROM SCHOOL AND UNIVERSITY EDUCATION TO RECANTATION FAITH. . . .

298.1

UNIVERSAL HOUSE OF JUSTICE

299

Voice of America Report on Campaign of Arrests against Iranian Bahá'ís
22 October 1981

To all National Spiritual Assemblies

A REPORT FROM TEHRAN SAYS IRAN'S CENTRAL REVOLUTIONARY COMMITTEE IS PLANNING A NEW CAMPAIGN TO ROUND UP MEMBERS OF THE BAHÁ'Í FAITH. THE BAHÁ'ÍS REPORTEDLY WILL BE ARRESTED ON GROUNDS THAT THEIR MARRIAGES ARE ILLEGAL AND THEIR CHILDREN ILLEGITIMATE SINCE IRAN DOES NOT RECOGNIZE THE BAHÁ'Í FAITH. THE COMMITTEE SAYS SOME 96 BAHÁ'ÍS HAVE BEEN EXECUTED SO FAR AND ANOTHER 200 ARE UNDER ARREST. SOME 10,000 BAHÁ'Í FAMILIES ARE BELIEVED TO HAVE FLED IRAN SINCE THE CAMPAIGN AGAINST THEM BEGAN. THE COMMITTEE HAS ANOTHER 20,000 BAHÁ'Í NAMES ON ITS ARREST LIST COMPILED FROM CAPTURED BAHÁ'Í OFFICE LISTS AND DOCUMENTS FROM THE SHAH'S OLD SECRET POLICE, SAVAK. MANY BAHÁ'ÍS HAVE CHANGED THEIR NAMES OR GONE INTO HIDING TO ESCAPE CAPTURE.

299.1

UNIVERSAL HOUSE OF JUSTICE

300
Further Acts of Persecution in Iran
23 October 1981

To all National Spiritual Assemblies

300.1 FOLLOWING NEWS FURTHER ACTS PERSECUTION INFLICTED BRETHREN IRAN JUST RECEIVED:

— IN URÚMÍYYIH, ÁDHIRBÁYJÁN TWENTY-SIX BAHÁ'ÍS SUMMARILY ARRESTED, FIVE OF WHOM ARE WOMEN.

— IN BÍRJAND, KHURÁSÁN HOUSES ALL LOCAL BELIEVERS RANSACKED FORCING FRIENDS DESERT THEIR HOMES AND FLEE FOR SAFETY.

— IN TEHRAN HOME PROMINENT BELIEVER OCCUPIED ALL HER FURNITURE CONFISCATED. LISTING OF ITEMS ALLEGEDLY FOUND IN HOUSE FALSELY INCLUDES ARMAMENTS DRUGS.

— FRIENDS IRAN LIVING IN ATMOSPHERE CONSTANT FEAR THEIR PROPERTIES AND LIVES. . . .

UNIVERSAL HOUSE OF JUSTICE

301
Grave Developments in Iran
5 November 1981

To all National Spiritual Assemblies

301.1 FOLLOWING GRAVE DEVELOPMENTS REPORTED FROM IRAN:

1. REFERENCE OUR CABLE 26 MAY 1981 WORK ON BUILDING ROAD DESIGNED RUN THROUGH SITE BÁB'S HOLY HOUSE SHIRAZ BEING ACTIVELY RESUMED.[1] RECENT REPORT INDICATES WORK STEADILY PROGRESSING APPROACHING SACRED PRECINCTS.

2. OFFICE NATIONAL ASSEMBLY BROKEN INTO PAPERS FILES REMOVED.

3. SIX MEMBERS LOCAL ASSEMBLY TEHRAN SUMMARILY ARRESTED WHILE IN SESSION. COUPLE IN WHOSE HOME MEETING WAS HELD HAVE ALSO BEEN DETAINED.

4. GOVERNMENT IRAN HAS RECENTLY INSTRUCTED ITS CONSULAR REPRESENTATIVES EVERYWHERE COMPILE LIST BAHÁ'ÍS RESIDING AREAS THEIR RESPONSIBILITY AND HENCEFORTH REFRAIN FROM EXTENDING PASSPORTS IRANIAN BAHÁ'ÍS. . . .

UNIVERSAL HOUSE OF JUSTICE

301-1. The message of 26 May 1981 (no. 282) reported the obliteration of the House of the Báb. For more about its seizure and demolition, see messages dated 10 May 1979 (no. 225), 9 September 1979 (no. 235), and 19 November 1979 (no. 241).

302

Call for Pioneers
13 NOVEMBER 1981

To all National Spiritual Assemblies

Dear Bahá'í Friends,

The Universal House of Justice has asked us to convey to you the following with respect to the settlement of pioneers since the opening of the second phase of the Seven Year Plan. 302.1

Out of a total of 285 pioneers (originally 279 plus 6 added later) assigned for the first year of the second phase, 137 posts have been filled. While many additional pioneers have settled around the world, a net balance of 148 pioneers is still needed to fill the posts originally envisioned. 302.2

In consultation with the International Teaching Center the goals of the Plan in terms of pioneer needs have been reviewed, and a number of new goals have been established and added to the net outstanding goals. The consolidated list of the pioneer needs has now been prepared and is attached.[1] As you see it calls for 216 pioneers to arise, God willing, by the end of this first year of the second phase of the Plan. The list also shows the National Spiritual Assemblies responsible for providing the manpower in each case and, if necessary, arranging finance for the projects. 302.3

The Universal House of Justice is praying at the Holy Shrines that the friends everywhere may respond to this call, and that the Continental Pioneer Committees throughout the world will offer their services in a manner to facilitate the early fulfillment of these goals. 302.4

With loving Bahá'í greetings,
DEPARTMENT OF THE SECRETARIAT

303

Release of a Compilation on Excellence in All Things
23 NOVEMBER 1981

To all National Spiritual Assemblies

Dear Bahá'í Friends,

Attached is a compilation prepared by the Research Department which is aimed at presenting texts and extracts which encourage the friends to attain distinction and excellence in all their undertakings. The compilation is enti- 303.1

302-1. The list is too lengthy to include in this volume.

tled *Excellence in All Things*.¹ The Universal House of Justice hopes that the contents of this compilation will guide the friends everywhere in the conduct of their individual lives so that they may follow Bahá'u'lláh's exhortation to distinguish themselves from others through deeds, and that their "light can be shed upon the whole earth." "Happy is the man," He assures us, "that heedeth My counsel . . ."²

303.2 The Universal House of Justice leaves it to your discretion to determine how to share the contents of this compilation with the friends and communities for whom you are responsible.

With loving Bahá'í greetings,
DEPARTMENT OF THE SECRETARIAT

304

Inauguration of Radio Bahá'í Peru

29 NOVEMBER 1981

To the National Spiritual Assembly of the Bahá'ís of Peru

304.1 WARMEST CONGRATULATIONS AUSPICIOUS INAUGURATION RADIO BAHÁ'Í DEL LAGO TITICACA DAY COVENANT AND SIMULTANEOUS OPENING MUHÁJIR TEACHING INSTITUTE FIRST MUSIC FESTIVAL PRAYING HOLY THRESHOLD INCREASING CONFIRMATIONS YOUR ASSEMBLY COOPERATING ASSEMBLIES ECUADOR BOLIVIA RADIO COMMISSION ALL STAFF LOVING GREETINGS THIS IMPORTANT STEP GROWTH FAITH ALTIPLANO. KINDLY ASSURE MICHAEL STOKES OUR PRAYERS HIS BEHALF.¹

UNIVERSAL HOUSE OF JUSTICE

305

Demolition of the House of Bahá'u'lláh in Tákur; Seizure of Cemetery in Tehran

10 DECEMBER 1981

To all National Spiritual Assemblies

305.1 DISTRESSED REPORT RECENT DEVELOPMENTS IRAN EVIDENCE FURTHER PERSECUTIONS ATROCITIES AGAINST DEFENSELESS BELIEVERS CRADLE FAITH:

1. HOUSE OF BAHÁ'U'LLÁH IN TÁKUR, PREVIOUSLY CONFISCATED, HAS NOW BEEN TOTALLY DEMOLISHED AND THIS BAHÁ'Í HOLY PLACE INCLUDING LAND AND GARDENS OFFERED FOR SALE TO PUBLIC BY AUTHORITIES.

303-1. See CC 1:367–84.
303-2. GWB, p. 305.
304-1. Mr. Stokes assisted the National Spiritual Assembly of Peru as a project planner for the administration of the station.

2. BAHÁ'Í CEMETERY TEHRAN SEIZED ON SATURDAY 5 DECEMBER BY ORDER REVOLUTIONARY COURT, 5 CARETAKERS AND 8 TEMPORARY WORKERS ARRESTED, AND CEMETERY CLOSED. BAHÁ'ÍS FEARFUL DESECRATION GRAVES. TENS OF THOUSANDS BAHÁ'ÍS TEHRAN NOW WITHOUT BURIAL GROUNDS. THIS RECENT DEVELOPMENT INDECENT ACT STILL ANOTHER EXAMPLE EVIL DESIGNS ELIMINATE BAHÁ'Í COMMUNITY THAT COUNTRY.
3. OTHER PERSECUTIONS SUCH AS SUMMARY ARRESTS, HARASSMENT, CONFISCATION PROPERTIES, CONTINUE. . . .

Universal House of Justice

306
Arrest of Eight Members of the Iranian National Spiritual Assembly
14 December 1981

To all National Spiritual Assemblies

306.1 HAVE JUST RECEIVED DISTRESSING REPORT EIGHT MEMBERS NATIONAL ASSEMBLY IRAN WHILE IN SESSION ARRESTED TOGETHER WITH TWO OTHER BELIEVERS IN HOME WHERE MEETING WAS HELD. TWO WOMEN AMONG TEN DETAINEES. AUTHORITY RESPONSIBLE FOR ARRESTS AND LOCATION PRISON STILL UNIDENTIFIED. . . .

Universal House of Justice

307
Execution of Eight Members of the Iranian National Spiritual Assembly
29 December 1981

To National Spiritual Assemblies

307.1 WITH HEAVY HEARTS INFORM FRIENDS THROUGHOUT WORLD EIGHT MEMBERS NATIONAL ASSEMBLY IRAN ARRESTED 13 DECEMBER WERE EXECUTED 27 DECEMBER. THEY ARE:

MR. KÁMRÁN ṢAMÍMÍ
MRS. ZHÍNÚS MAḤMÚDÍ
MR. MAḤMÚD MAJDHÚB
MR. JALÁL 'AZÍZÍ
MR. MIHDÍ AMÍN AMÍN
MR. SÍRÚS RAWSHANÍ
MR. 'IZZATU'LLÁH FURÚHÍ
MR. QUDRATU'LLÁH RAWḤÁNÍ

307.2 FAMILIES NOT NOTIFIED OF ARRESTS, TRIAL, EXECUTIONS. BODIES BURIED UNCEREMONIOUSLY IN BARREN FIELD RESERVED BY GOVERNMENT FOR INFIDELS. INFORMATION DISCOVERED FORTUITOUSLY. GOVERNMENT AUTHORITIES TOTALLY SILENT, UNCOOPERATIVE.

307.3 THIS HEINOUS ACT CAUSES US FEAR THAT MEMBERS PREVIOUS NATIONAL ASSEMBLY AND TWO AUXILIARY BOARD MEMBERS WHO DISAPPEARED AUGUST 1980, AS WELL AS TWO OTHERS WHOSE WHEREABOUTS UNKNOWN OVER TWO YEARS, HAVE SUFFERED SAME FATE. NAMES THESE HEROIC DEDICATED SERVANTS BLESSED BEAUTY ARE:

AUXILIARY BOARD MEMBERS
DR. YÚSIF 'ABBÁSÍYÁN
DR. HISHMATU'LLÁH RAWHÁNÍ

NATIONAL ASSEMBLY MEMBERS
DR. 'ALÍMURÁD DÁVÚDÍ
MR. 'ABDU'L-HUSAYN TASLÍMÍ
MR. HÚSHANG MAHMÚDÍ
MR. IBRÁHÍM RAHMÁNÍ
DR. HUSAYN NAJÍ
MR. MANÚCHIHR QÁ'IM MAQÁMÍ
MR. 'ATÁ'U'LLÁH MUQARRABÍ
MR. YÚSIF QADÍMÍ
MRS. BAHÍYYIH NÁDIRÍ
DR. KÁMBÍZ SADIQZÁDIH

MEMBER LOCAL ASSEMBLY TEHRAN
MR. RÚHÍ RAWSHANÍ

PROMINENT TEACHER
MR. MUHAMMAD MUVAHHID

307.4 EXEMPLARY CHARACTER THESE SELFLESS GLORIOUS SOULS SOURCE INSPIRATION TO BAHÁ'ÍS OF WORLD. WHILE NOT ABLE WIN CROWN MARTYRDOM LIKE PERSIAN BRETHREN, VALIANT DETACHED FRIENDS EVERY LAND UNDOUBTEDLY ARE ENDEAVORING EVINCE SAME SPIRIT FOLLOW SAME PATH CONSECRATION DEDICATION GOD'S HOLY FAITH. WE ARE CONFIDENT THAT SANCTIFIED BLOOD OF THESE DESCENDANTS DAWN-BREAKERS WILL SERVE STRENGTHEN BODY CAUSE GOD THROUGHOUT GLOBE, PRODUCE UNPRECEDENTED VICTORIES TO COMPENSATE LOSSES SUSTAINED CRADLE FAITH. . . .

UNIVERSAL HOUSE OF JUSTICE

308
Proselytizing, Development, and the Covenant
3 JANUARY 1982

To an individual Bahá'í

Dear Bahá'í Friend,

The Universal House of Justice has received your letter and has asked us to assure you that you should feel no diffidence in raising the sort of questions that you have expressed. It seems clear from your letter that you have been greatly attracted to the Message of Bahá'u'lláh and have accepted His Faith before, as you say, becoming "fully committed," and are, therefore, now having to face and resolve problems that many believers overcome before they declare their faith. The House of Justice urges you not to let it worry you. All through life Bahá'ís are faced with tests of many kinds, and problems and doubts, but it is through facing and overcoming them that we grow spiritually. 308.1

On the particular issues that you raise, the House of Justice has instructed us to send you the following comments. 308.2

Teaching vs. Proselytizing

It is true that Bahá'u'lláh lays on every Bahá'í the duty to teach His Faith. At the same time, however, we are forbidden to proselytize, so it is important for all believers to understand the difference between teaching and proselytizing. It is a significant difference and, in some countries where teaching a religion is permitted, but proselytizing is forbidden, the distinction is made in the law of the land. Proselytizing implies bringing undue pressure to bear upon someone to change his Faith. It is also usually understood to imply the making of threats or the offering of material benefits as an inducement to conversion. In some countries mission schools or hospitals, for all the good they do, are regarded with suspicion and even aversion by the local authorities because they are considered to be material inducements to conversion and hence instruments of proselytization. 308.3

Bahá'u'lláh, in *The Hidden Words*, says, "O Son of Dust! The wise are they that speak not unless they obtain a hearing, even as the cup-bearer, who proffereth not his cup till he findeth a seeker, and the lover who crieth not out from the depths of his heart until he gazeth upon the beauty of his beloved. . . .", and on page 55 of *The Advent of Divine Justice*, a letter which is primarily directed towards exhorting the friends to fulfill their responsibilities in teaching the Faith, Shoghi Effendi writes: "Care, however, should, at all times, be exercised, lest in their eagerness to further the international interests of the Faith they rustrate their purpose, and turn away, through any act that might be misconstrued as an attempt to proselytize and bring undue pres- 308.4

sure upon them, those whom they wish to win over to their Cause."¹ Some Bahá'ís sometimes overstep the proper bounds, but this does not alter the clear principle.

308.5 The responsibility of the Bahá'ís to teach the Faith is very great. The contraction of the world and the onward rush of events require us to seize every chance open to us to touch the hearts and minds of our fellowmen. The Message of Bahá'u'lláh is God's guidance for mankind to overcome the difficulties of this age of transition and move forward into the next stage of its evolution, and human beings have the right to hear it. Those who accept it incur the duty of passing it on to their fellowman. The slowness of the response of the world has caused and is causing great suffering; hence the historical pressure upon Bahá'ís to exert every effort to teach the Faith for the sake of their fellowmen. They should teach with enthusiasm, conviction, wisdom and courtesy, but without pressing their hearer, bearing in mind the words of Bahá'u'lláh: "Beware lest ye contend with any one, nay, strive to make him aware of the truth with kindly manner and most convincing exhortation. If your hearer respond, he will have responded to his own behoof, and if not, turn ye away from him, and set your faces towards God's sacred Court, the seat of resplendent holiness." (*Gleanings* CXXVIII)

Considerations in the Application of Bahá'í Social Teachings

308.6 The application and development of the social aspects of the Teachings is dependent on the stage of growth of the Bahá'í community in each area, and on worldwide priorities. We are living in an age of transition, and as 'Abdu'l-Bahá explained, we must, in order to succeed in our aims, sacrifice the important for the most important. The House of Justice, for example, had to turn down the request of certain believers to establish Bahá'í schools in a Western country which already had a functioning state educational system; those Bahá'í funds which are available for educational projects must be spent on the establishment and running of schools in areas where there are large Bahá'í communities of poor people, with no adequate system of education available to them. In its answer, the House of Justice pointed out that if these friends, on their own initiative, wished to establish their own school, run on Bahá'í lines, and financially self-supporting, they were entirely free to do so. This highlights an aspect of the matter which is often overlooked. The social services of Bahá'ís are not restricted to what they do as a community. Every Bahá'í has a duty to work and earn his living, and in choosing a career a Bahá'í should consider not only its earning capacity but also the benefit of the work to his fellowmen. All over the world Bahá'ís are rendering outstanding services in this way.

308-1. ADJ, p. 66.

When a Bahá'í community is very small, there is little that it can do to implement the social teachings of the Faith (beyond their impact on the behavior of individual believers), because such a community with the resources in funds and manpower at its disposal is but a drop in the ocean in comparison with the many large agencies, governmental and private, which are engaged in social improvement. When the Bahá'í community grows sufficiently large, however, its activities can and must proliferate and diversify. This development is already taking place in many parts of the world. In India, for example, the New Era School in Panchgani, which has been developing remarkably for a number of years, is closely associated with a rural development project in the villages close by that is having dramatically favorable results in the life of the villagers. In the province of Madhya Pradesh, where there are hundreds of thousands of Bahá'ís, the Rabbani School in Gwalior is educating children from the villages of the area in the Teachings of the Faith, in academic subjects and in agriculture, so that when they return to their home villages, these pupils not only promote the Faith but will influence their growth and development in every way. In Ecuador, as you no doubt know, the size of the Bahá'í community, scattered over inaccessible terrain in the high Andes, made it both necessary and possible some years ago to establish a Bahá'í radio station.² "Radio Bahá'í," as it is known, broadcasts not only about the Faith, but has programs concerning health, agriculture, literacy and so on. It has now become so well established and highly regarded that it has been able to apply for and receive a Canadian Government grant through C.I.D.A.³ to finance the development of certain social service activities. Thus it can be seen that once the Bahá'í community attains a certain stature it is able to work in fruitful collaboration with non-Bahá'í agencies in its social activities. 308.7

A further aspect of this kind of work is the collaboration between the Bahá'í International Community and the United Nations. Having consultative status with both ECOSOC and UNICEF, and long association with the Department of Public Information, the Bahá'í International Community is able to take part in conferences and consultations on many aspects of human development, both from the point of view of the Bahá'í Teachings and with the background of its extensive experience in meeting the problems of developing countries, such as illiteracy, the status of women, tribalism, racial prejudice, and so on. 308.8

As you can see, all these developments relate directly to the teaching work inasmuch as the Bahá'í communities must reach a certain size before they can begin to implement many of them. How, for example, can a Bahá'í community demonstrate effectively the abolition of prejudices which divide the inhab- 308.9

308-2. See messages dated 15 December 1977 (no. 201), Naw-Rúz 1978 (no. 205), and 28 August 1978 (no. 213).
308-3. Canadian International Development Agency.

itants of a country until it has a cross-section of those inhabitants within its ranks? A seed is the vital origin of a tree and of a tremendous importance for that reason, but it cannot produce fruit until it has grown into a tree and flowered and fruited. So a Bahá'í community of nine believers is a vital step, since it can bring into being for that locality the divine institution of the Local Spiritual Assembly, but it is still only a seed, and needs to grow in size and in the diversity of its members before it can produce really convincing fruit for its fellow citizens.

308.10 One could say, however, that the Bahá'í communities could assist in social development from a very early stage in their development by supporting the activities of other groups who are, at this point, more numerous and powerful. To some extent this is true, provided that such involvement does not divert the efforts of the friends from the more fundamentally important teaching work or involve them in the disputes of non-Bahá'í rival groups.

Humanity's Most Urgent Need

308.11 The teaching work is of primary importance for this reason: the most urgent need of human beings is to recognize the Manifestation of God and thereby to learn how to collaborate constructively. All over the world tremendous efforts are being made to improve the lot of mankind—or of parts of mankind, but most of these efforts are frustrated by the conflicts of aims, by corruption of the morals of those involved, by mistrust, or by fear. There is no lack of material resources in the world if they are properly used. The problem is the education of human beings in the ultimate and most important purpose of life and in how to weld the differences of opinion and outlook into a united constructive effort. Bahá'ís believe that God has revealed the purpose of life, has shown us how to attain it, has provided the ways in which we can work together and, beyond that, has given mankind the assurance both of continuing divine guidance and of divine assistance. As people learn and follow these teachings their efforts will produce durable results. In the absence of these teachings, a lifetime of effort only too often ends in disillusionment and the collapse of all that has been built.

308.12 It is not easy for people to learn the Bahá'í way, to overcome their inherited prejudices or to resist their personal temptations. This way takes time, is subject to checks and backsliding, but one can see, looking at the past 138 years, that there is an overall advance that is astonishing in the light of the obstacles to be overcome, and is accelerating with every passing decade.

Obstacle to Progress: Getting Sucked Into Prevailing Attitudes

308.13 One of the great obstacles to progress is the tendency of Bahá'ís to be sucked into the general attitudes and disputes that surround them, to be influenced, for example, as you yourself pointed out, by the prevailing attitude to marriage so that the divorce rate becomes a problem within the Bahá'í community itself which should be an example to the rest of society in such matters. Involvement in politics and controversial questions is another aspect of the same phenomenon. In one of His Tablets Bahá'u'lláh warns the Bahá'ís: "Dispute not with any one concerning the things of this world and its affairs, for God hath abandoned them to such as have set their affection upon them. Out of the whole world He hath chosen for Himself the hearts of men—hearts which the hosts of revelation and of utterance can subdue." (*Gleanings* CXXVIII) As you realize, this cannot mean that Bahá'ís must not be controversial since, in many societies, being a Bahá'í is itself a controversial matter. The central importance of this principle of avoidance of politics and controversial matters is that Bahá'ís should not allow themselves to be drawn into the disputes of the many conflicting elements of the society around them. The aim of the Bahá'ís is to reconcile, to heal divisions, to bring about tolerance and mutual respect among men, and this aim is undermined if we allow ourselves to be swept along by the ephemeral passions of others. This does not mean that Bahá'ís cannot collaborate with any non-Bahá'í movement; it does mean that good judgment is required to distinguish those activities and associations which are beneficial and constructive from those which are divisive.

The Uniqueness of the Bahá'í Covenant

308.14 The House of Justice hopes that these explanations will help you to understand some of the aspects of the Faith that have been troubling you. The crux of the matter, as you realize, is the acceptance of spiritual authority and what this implies. You express the fear that the authority conferred upon 'Abdu'l-Bahá, the Guardian and the Universal House of Justice could lead to a progressive reduction in the "available scope for personal interpretation," and that "the actual writings of the Manifestation will have less and less import," and you instance what has happened in previous Dispensations. The House of Justice suggests that, in thinking about this, you contemplate the way the Covenant of Bahá'u'lláh has actually worked, and you will be able to see how very different its processes are from those of, say, the development of the law in Rabbinical Judaism or the functioning of the Papacy in Christianity. The practice in the past in these two religions, and also to a great extent in Islam, has been to assume that the Revelation given by the Founder was the final, perfect revelation of God's Will to mankind, and all subsequent elucidation and legislation has been interpretative in the sense that it aimed at applying this basic Revelation to the new problems and situations that have arisen. The Bahá'í premises are quite different. Although the Revelation of Bahá'u'lláh is accepted as the Word of God and His Law as the Law of God, it is understood from the outset that Revelation is progressive, and that the Law, although the Will of God for this Age, will undoubtedly be changed by the next Manifestation of God. Secondly, only the written text of the Revelation is regarded as authoritative. There is no Oral Law as in Judaism, no Tradition of the Church as in Christianity, no Hadíth as in Islam. Thirdly, a clear distinction is drawn between interpretation and legislation. Authoritative interpretation is the exclusive prerogative of 'Abdu'l-Bahá and the Guardian, while infallible legislation is the function of the Universal House of Justice.

308.15   If you study the Writings of 'Abdu'l-Bahá and of the Guardian, you will see how tremendously they differ from the interpretations of the Rabbis and the Church. They are not a progressive fossilization of the Revelation, they are for the most part expositions which throw a clear light upon passages which may have been considered obscure, they point up the intimate interrelationship between various teachings, they expound the implications of scriptural allusions, and they educate the Bahá'ís in the tremendous significances of the Words of Bahá'u'lláh. Rather than in any way supplanting the Words of the Manifestation, they lead us back to them time and again.

Authoritative vs. Individual Interpretation

308.16   There is also an important distinction made in the Faith between authoritative interpretation, as described above, and the interpretation which every believer is fully entitled to voice. Believers are free, indeed are encouraged, to study the Writings for themselves and to express their understanding of them. Such personal interpretations can be most illuminating, but all Bahá'ís, including the one expressing the view, however learned he may be, should realize that it is only a personal view and can never be upheld as a standard for others to accept, nor should disputes ever be permitted to arise over differences in such opinions.

Interpretation and Legislation

308.17   The legislation enacted by the Universal House of Justice is different from interpretation. Authoritative interpretation, as uttered by 'Abdu'l-Bahá and the Guardian, is a divinely guided statement of what the Word of God means. The divinely inspired legislation of the Universal House of Justice does not attempt to say what the revealed Word means—it states what must be done in cases where the revealed Text or its authoritative interpretation is not explicit. It is, therefore, on quite a different level from the Sacred Text, and the Universal House of Justice is empowered to abrogate or amend its own legislation whenever it judges the conditions make this desirable. Moreover, the attitude to legislation is different in the Bahá'í Faith. The human tendency in past Dispensations has been to want every question answered and to arrive at a binding decision affecting every small detail of belief or practice. The tendency in the Bahá'í Dispensation, from the time of Bahá'u'lláh Himself, has been to clarify the governing principles, to make binding pronouncements on details which are considered essential, but to leave a wide area to the conscience of the individual. The same tendency appears also in administrative matters. The Guardian used to state that the working of National Spiritual

Assemblies should be uniform in essentials but that diversity in secondary matters was not only permissible but desirable. For this reason a number of points are not expressed in the National Baháʼí Constitution (the Declaration of Trust and By-Laws of National Assemblies); these are left to each National Spiritual Assembly to decide for itself.[4]

The Covenant of Baháʼuʼlláh

308.18 The Covenant is the "axis of the oneness of the world of humanity" because it preserves the unity and integrity of the Faith itself and protects it from being disrupted by individuals who are convinced that only their understanding of the Teachings is the right one—a fate that has overcome all past Revelations. The Covenant is, moreover, embedded in the Writings of Baháʼuʼlláh Himself. Thus, as you clearly see, to accept Baháʼuʼlláh is to accept His Covenant; to reject His Covenant is to reject Him.

308.19 The House of Justice asks us to assure you of its loving prayers at the Sacred Threshold for your guidance in your efforts to arrive at a greater understanding of this wonderful Revelation.

With loving Baháʼí greetings,
DEPARTMENT OF THE SECRETARIAT

309
Iranian Supreme Court's Denial of Execution of Iranian National Spiritual Assembly Members
5 JANUARY 1982

To all National Spiritual Assemblies

309.1 PRESIDENT SUPREME COURT AYATOLLAH ARDEBILI IS REPORTED BY NEWS SERVICES TO HAVE DENIED EXECUTION EIGHT MEMBERS NATIONAL ASSEMBLY IRAN.... WE CONVINCED EXECUTIONS WERE PLANNED TO BE KEPT SECRET BUT WERE DISCOVERED FORTUITOUSLY. BAHÁʼÍ INTERNATIONAL COMMUNITY REQUESTING SECRETARY GENERAL UNITED NATIONS INVESTIGATE....

UNIVERSAL HOUSE OF JUSTICE

308-4. The Declaration of Trust and By-Laws of the National Spiritual Assembly are published in each volume of *The Baháʼí World*, beginning with volume 3.

MESSAGES FROM THE UNIVERSAL HOUSE OF JUSTICE

310
Appeal by Bahá'í International Community to Iranian Leaders
7 January 1982

To all National Spiritual Assemblies

310.1 BAHÁ'Í INTERNATIONAL COMMUNITY CABLING FOLLOWING TEXT TO AYATOLLAH KHOMEINI, PRIME MINISTER MIR ḤUSAYN MUSAVI, AND PRESIDENT SUPREME COURT AYATOLLAH MUSAVI ARDEBILI:

310.2 "NEWS RECENT SECRET EXECUTION EIGHT MEMBERS NATIONAL ASSEMBLY BAHÁ'ÍS IRAN, AND SEVEN OTHERS, SIX OF WHOM WERE MEMBERS LOCAL ASSEMBLY TEHRAN, HAS SHOCKED BAHÁ'ÍS ENTIRE WORLD. TOTAL NUMBER BAHÁ'ÍS WHOSE MARTYRDOMS OFFICIALLY ACKNOWLEDGED HAS NOW REACHED NINETY-SEVEN. FOURTEEN OTHERS, KNOWN TO HAVE DISAPPEARED, FEARED TO HAVE SUFFERED SAME FATE. HUNDREDS IMPRISONED THROUGHOUT COUNTRY ON CHARGES WITHOUT SUBSTANTIATING EVIDENCE. WE CATEGORICALLY DENY TRUTH THESE ACCUSATIONS.

310.3 AS BAHÁ'ÍS IRAN ARE RIGIDLY DENIED OPPORTUNITY PUBLICLY PRESENT THEIR CASE, DEFEND THEIR RIGHTS, PROVE THEIR INNOCENCE, AND ALL DOORS APPEAL THEIR CASE CLOSED BEFORE THEM, WE THEREFORE APPEAL TO YOU ON THEIR BEHALF IN NAME OF BAHÁ'Í COMMUNITIES IN 164 INDEPENDENT COUNTRIES OF THE WORLD:

1. TO ISSUE IMMEDIATE INSTRUCTIONS STOP SUMMARY ARRESTS EXECUTIONS,
2. TO REQUIRE THOSE RESPONSIBLE PRODUCE PUBLISH DOCUMENTS WHICH HAVE FORMED BASIS CONVICTION BAHÁ'ÍS AS SO-CALLED SPIES,
3. TO EXTEND TO BAHÁ'ÍS AS LAW-ABIDING CITIZENS, AND AS A COMMUNITY, INALIENABLE RIGHT TO PUBLICLY DEFEND THEMSELVES DISPROVE MALICIOUS ACCUSATIONS FALSE CHARGES.

310.4 WE LAY BEFORE YOU FATE THESE LOYAL CITIZENS ABOUT WHOSE INNOCENCE, TRUE LOVE FOR IRAN AND REVERENCE FOR SPIRIT ISLAM YOU SHOULD HAVE NO DOUBT. WE PRAY THE ALMIGHTY MAY GUIDE YOU THIS ELEVENTH HOUR TO DISCHARGE SACRED INESCAPABLE RESPONSIBILITIES BEFORE GOD AND MAN.

BAHÁ'Í INTERNATIONAL COMMUNITY"

UNIVERSAL HOUSE OF JUSTICE

311
Secret Execution of Six Members of Tehran Assembly and Hostess
7 January 1982

To all National Spiritual Assemblies

INFORMATION JUST RECEIVED SIX MEMBERS LOCAL SPIRITUAL ASSEMBLY TEHRAN TOGETHER WITH WOMAN BELIEVER IN WHOSE HOME ARRESTS WERE MADE ON SECOND NOVEMBER WERE SECRETLY EXECUTED ON FOURTH JANUARY. INFORMATION OBTAINED FORTUITOUSLY BY RELATIVES FRIENDS MARTYRS. NAMES THESE VALIANT SOULS ARE:

MR. KÚRUSH ṬALÁ'Í
MR. KHUSRAW MUHANDISÍ
MR. ISKANDAR 'AZÍZÍ
MR. FATḤU'LLÁH FIRDAWSÍ
MR. 'AṬÁ'U'LLÁH YÁVARÍ
MRS. SHÍVÁ MAḤMÚDÍ ASADU'LLÁHZÁDIH
HOSTESS: MRS. SHÍDRUKH AMÍR-KÍYÁ BAQÁ . . .

UNIVERSAL HOUSE OF JUSTICE

312
Using *A Cry from the Heart* to Explain the Tragic Events in Iran
21 January 1982

To all National Spiritual Assemblies

Dear Bahá'í Friends,

The Hand of the Cause William Sears, deeply moved by the savage onslaught on the dearly loved and heroic Bahá'í community in Iran, has written a dramatic account of the current persecution.[1]

Mr. Sears's story acquires deep poignancy from the account of his travels in Iran when he came to know and love the Persian Bahá'ís, both individually and collectively, among whom are a number of the martyrs. In a white heat of indignation, he came to the World Center to write the first draft of his book, using the detailed information available here to present—as he is so well able—the successive tragedies which, even in these days of universal terrorism, are beginning to shock the world.

312-1. The book, *A Cry from the Heart*, was published by George Ronald (1982).

312.3 The House of Justice believes that Mr. Sears's book will not only be warmly welcomed by the friends, but can also be a potent instrument for interesting large numbers of people in the tragic events now taking place in Persia, winning their sympathy for the harassed Bahá'ís and leading them to inquire further about a Faith which can inspire such fortitude in ordinary people. He does not miss the opportunities offered, in refuting the trumped-up charges made against the Bahá'ís, for presenting basic Bahá'í teachings and conveying a persuasive sense of why Bahá'u'lláh is worth dying for. Furthermore, widespread knowledge among the public of the persecutions in Iran can be an added deterrent to their continuation, reinforcing the many efforts being made by governments and other agencies as a result of the constant appeals of the Bahá'í International Community and National Spiritual Assemblies.

312.4 The House of Justice requests you to give the greatest possible support to distribution of this book, calling it to the attention of the friends and urging them to use it as a topical introduction for their contacts, to try to persuade their local newspapers and radio stations to review it and their local bookshops to display it. We have asked that it be published at as low a price as possible so that it can be used by all the friends. It is now at press and you will shortly hear from the publisher.

With loving Bahá'í greetings,
DEPARTMENT OF THE SECRETARIAT

313
Plans for Commemorating the Fiftieth Anniversary of the Ascension of Bahíyyih Khánum
24 JANUARY 1982

To all National Spiritual Assemblies

Dear Bahá'í Friends,

313.1 The fiftieth anniversary of the ascension of Bahíyyih Khánum, eldest daughter of Bahá'u'lláh and designated by Him the Greatest Holy Leaf, will occur on July 15th of this year. We summon the entire Bahá'í world to a befitting commemoration of the life of the greatest woman in the Bahá'í Dispensation.

313.2 National Spiritual Assemblies are requested to arrange national commemorative services, and to ensure that all local communities hold befitting meetings. These services should be held on the date of the anniversary or on the weekend immediately following it, and in those countries where Mashriqu'l-Adhkárs are in existence, they should be held in the Temple.

In order to provide for your arrangement of the services, we have requested 313.3
our Research Department to compile a bibliography of references to the Greatest Holy Leaf in Bahá'í literature in English, and this is enclosed for your use.[1]
In addition, the book announced in our message of March 1981 is now at press, and will be available by the time of the commemoration.[2]

With loving Bahá'í greetings,
THE UNIVERSAL HOUSE OF JUSTICE

Some References to the Greatest Holy Leaf Found in Published Works

Bahá'u'lláh and 'Abdu'l-Bahá, *Tablets Revealed in Honor of the Greatest Holy Leaf* (New York: National Spiritual Assembly of the Bahá'ís of the United States and Canada, 1933). 313.4

Shoghi Effendi: 313.5
The Advent of Divine Justice (Wilmette: Bahá'í Publishing Trust, 1990), p. 37.
Bahá'í Administration (Wilmette: Bahá'í Publishing Trust, 1974), pp. 25, 57, 70, 93, 187–196.
The Dawn-Breakers (Wilmette: Bahá'í Publishing Trust, 1974), dedication.
God Passes By (Wilmette: Bahá'í Publishing Trust, 1974), pp. 108, 347, 350, 392.
Guidance for Today and Tomorrow (London: Bahá'í Publishing Trust, 1973), pp. 58–71.
Messages to America: Selected Letters and Cablegrams Addressed to the Bahá'ís of North America 1932–1946 (Wilmette: Bahá'í Publishing Committee, 1947), pp. 1, 31, 37.
Messages to the Bahá'í World (Wilmette: Bahá'í Publishing Trust, 1971), p. 74.
The World Order of Bahá'u'lláh (Wilmette: Bahá'í Publishing Trust, 1991), pp. 67–68, 81–82, 93–94, 98.

313-1. In a letter dated 25 February 1982 the Universal House of Justice sent to all National Spiritual Assemblies a revised bibliography of references in Bahá'í literature to the Greatest Holy Leaf. The revised bibliography published here contains references to Persian and Arabic texts in addition to the original list of references in English. The latest editions are cited for the reader's convenience.

313-2. The book, *Bahíyyih Khánum: The Greatest Holy Leaf*, was published by the Bahá'í World Center in Haifa (1982). For an account of the commemoration at the Bahá'í World Center of the fiftieth anniversary of Bahíyyih Khánum's passing, see BW 18:53–54.

313.6 Others:

Balyuzi, H.M., *'Abdu'l-Bahá, the Centre of the Covenant of Bahá'u'lláh* (London: George Ronald, 1971), pp. 12, 54–55, 74, 332, 401, 416, 454–455, 463–464, 482.

Balyuzi, H.M., *Edward Granville Browne and the Bahá'í Faith* (London: George Ronald, 1970), pp. 119–220.

Blomfield, Lady Sarah, *The Chosen Highway* (Wilmette: Bahá'í Publishing Trust, 1967), pp. 37–69, 73.

Gail, Marzieh, <u>Kh</u>ánum: *the Greatest Holy Leaf as Remembered by Marzieh Gail* (London: George Ronald, 1982).

Maxwell, May, *An Early Pilgrimage* (London: George Ronald, 1969), pp. 18–19.

Muhájir, Írán Furútan, comp., *The Mystery of God* (London: Bahá'í Publishing Trust, 1979), pp. 278–304.

Rabbaní, Rúḥíyyih, *The Priceless Pearl* (London: Bahá'í Publishing Trust, 1969), pp. 6–7, 10–11, 13–15, 21–22, 39, 44, 46–51, 57–58, 63, 90, 102–03, 112, 115, 129–30, 139–40, 144–48, 151–52, 168, 199, 218, 236, 259, 261–62, 266–67, 273, 279–80, 430, 438.

Universal House of Justice, *Bahá'í Holy Places at the World Centre* (Haifa: Bahá'í World Centre, 1968), pp. 62–70.

313.7 *The Bahá'í World, an International Record*

vol. 2, 1926–1928, p. 83, 132

vol. 3, 1928–1930, p. 64

vol. 5, 1932–1934, pp. 22–23, 114–15, 169–88

vol. 8, 1938–1940, p. 5, 8, 206, 255–56, 262, 266

vol. 9, 1940–1944, p. 329

vol. 10, 1944–1946, p. 536

vol. 11, 1946–1950, p. 474, 492

vol. 16, 1973–1976, p. 54, 66, 73

313.8 *Bahá'í News*, published by the National Spiritual Assembly of the Bahá'ís of the United States

no. 18, June 1927, p. 5

no. 36, December 1929, p. 1

no. 52, May 1931, pp. 1–2

no. 62, May 1932, p. 2

no. 65, August 1932, pp. 1–2

no. 66, September 1932, p. 1

no. 72, March 1933, p. 3

no. 121, December 1938, p. 3

no. 124, April 1939, p. 1

no. 128, August 1939, p. 4

no. 133, February 1940, p. 1
no. 135, April 1940, insert

Star of the West (Chicago: Bahá'í News Service)
vol. 10, no. 17, pp. 312–14
vol. 12, no. 10, pp. 163–67; no. 11, pp. 186–88; no. 13, pp. 211–14; no. 15, p. 245; no. 19, pp. 302–03
vol. 13, no. 4, pp. 68–69, 82–83, 88; no. 8, pp. 207–10, 219–20; no. 11, p. 314
vol. 17, no. 8, pp. 256–60
vol. 18, no. 9, pp. 278–82
vol. 20, no. 1, p. 18; no. 4, p. 104
vol. 23, no. 5, p. 134; no. 7, pp. 202–04; no. 12, pp. 374–77
vol. 24, no. 1, pp. 18–20; no. 3, pp. 90–93
vol. 25, no. 4, pp. 118–22

313.9

(Many of these references are accounts of early pilgrimages, and give only a brief mention of the Greatest Holy Leaf)

314
Golden Opportunity Offered by Reign of Relentless Terror; Actions Taken to Defend the Bahá'ís of Iran
26 January 1982

The Bahá'ís of the World

Dearly loved Friends,

With indignation and anguish the Bahá'ís of the world, over the past three years, have received continuously tragic news of the sufferings and martyrdoms of their brethren in Iran, where a reign of relentless terror is now encompassing that long-abused and downtrodden community. The inhuman cruelties heaped on the followers of the Most Great Name—worthy descendants of their forebears, the Dawn-Breakers—in that land where the heroes and martyrs of the Faith have shed such luster on their generations, are increasing daily. We have seen how the House of the Báb in Shíráz and Bahá'u'lláh's ancestral home in Tákur have been demolished,[1] all Bahá'í endowments, including our Holy Places, have been seized, and the main financial assets of the community sequestered. We have seen with what callousness Bahá'í chil-

314.1

314-1. For more about the destruction of the House of the Báb, see messages dated 9 September 1979 (no. 235), 19 November 1979 (no. 241), and 26 May 1981 (no. 282). For the message about the destruction of the House of Bahá'u'lláh in Tákur, see telex dated 10 December 1981 (no. 305).

dren have been refused admission to schools, Bahá'í employees dismissed from government positions, and the essential human rights of the sorely tried Bahá'ís violated, their means of livelihood undermined or destroyed, their homes plundered, their properties confiscated, their very lives snuffed out.

An Acceleration of Persecution

314.2 Contemplating the history of the persecution of the Bahá'ís of Iran, we note an alarming acceleration in the degree of blatancy with which the traditional enemies of the Faith pursue their single purpose of extirpating the Faith in the land of its birth. In the past, with the exception of a few specific instances, the persecution of the members of the Bahá'í community by those traditionally inimical to the Faith, was random and sporadic, resulting from the incitement of easily aroused mobs to attack the lives and properties of the Bahá'ís.

314.3 Now the enemies of God's precious Cause who, as they themselves attest, have in the past twenty-five years organized themselves to counteract the influence of the Faith, to vilify and misrepresent its purpose and teachings, to inflame religious passions leading to the harassment and intimidation of the believers, to sow seeds of doubt among the friends and sympathizers, have infiltrated the ranks of officialdom, where, from this more advantageous position, they continue to instigate the persecution of the Bahá'ís. The incidence of violation of the rights of the Bahá'ís is thus becoming more frequent as is well evidenced by reports published in the press of Iran in recent months. Examples abound. For instance, formerly when Bahá'ís were arrested they were given an opportunity to defend themselves in some form of judicial proceedings held for the sake of appearances. On one occasion part of the proceedings which resulted in the execution of seven believers in Yazd, as late as September 1980, was televised. But recently the court proceedings, if any, have been held in camera,[2] and reports have even been received of the torture of Bahá'ís before their execution. No longer are the relatives of imprisoned Bahá'ís permitted to visit them, as they were until recently; no longer are the condemned permitted to solace their families with letters of farewell or the making of wills before their execution; and, more tragically, disturbed by the large number of Bahá'ís and sympathetic people of other religions who attended the funeral services of the slain Bahá'ís, the authorities have now seized the Bahá'í cemetery in Tehran and do not permit burial there. Indeed, the families of those most recently martyred were not even notified of the secret execution of their loved ones, whose bodies, unceremoniously deposited in graves for "infidels," were only fortuitously discovered.

314.4 Although the oppressors maintain that they are killing the Bahá'ís because they are guilty of serving as political agents and spies, it has been ascertained

---

314-2. In private, secretly.

that in almost every instance of execution, the accused Bahá'í was offered recantation as a means of release.

Oppression—The Cause of Stability and Firmness

The inveterate enemies of the Faith imagine that their persecutions will disrupt the foundations of the Faith and tarnish its glory. Alas! Alas for their ignorance and folly! These acts of oppression, far from weakening the resolve of the friends, have always served to inflame their zeal and galvanize their beings. In the words of 'Abdu'l-Bahá, ". . . they thought that violence and interference would cause extinction and silence and lead to suppression and oblivion, whereas interference in matters of conscience causes stability and firmness and attracts the attention of men's sight and souls, which fact has received experimental proof many times and often."[3]

Every drop of blood shed by the valiant martyrs, every sigh heaved by the silent victims of oppression, every supplication for divine assistance offered by the faithful, has released, and will continue mysteriously to release, forces over which no antagonist of the Faith has any control, and which, as marshaled by an All-Watchful Providence, have served to noise abroad the name and fame of the Faith to the masses of humanity in all continents, millions of whom had previously been totally ignorant of the existence of the Faith or had but a superficial, and oft-times erroneous, understanding of its teachings and history.

Worldwide Publicity

The current persecution has resulted in bringing the name and character of our beloved Faith to the attention of the world as never before in its history. As a direct result of the protests sent by the worldwide community of the Most Great Name to the rulers in Iran, of the representations made to the media when those protests were ignored, of direct approach by Bahá'í institutions at national and international levels to governments, communities of nations, international agencies and the United Nations itself, the Faith of Bahá'u'lláh has not only been given sympathetic attention in the world's councils, but also its merits and violated rights have been discussed and resolutions of protest sent to the Iranian authorities by sovereign governments, singly and in unison. The world's leading newspapers, followed by the local press, have presented sympathetic accounts of the Faith to millions of readers, while television and radio stations are increasingly making the persecutions in Iran the subject of their programs. Commercial publishing houses are beginning to commission books about the Faith.

But in spite of this great wave of publicity now bringing the name of the Faith to the attention of large masses of mankind, and in spite of the many

---

314-3. TN, p. 6.

representations made to the authorities in Iran, the persecution of the Bahá'ís there continues. The world stands helpless before the imperviousness of that country to outside opinion or criticism. In face of this tragic impasse we can only redouble our efforts to teach the Cause, taking advantage of the increasing interest in the character and principles of our beloved Faith created by the sufferings of the Persian community.

A Blessing in Disguise

314.9   Indeed, this new wave of persecution sweeping the Cradle of the Faith may well be seen as a blessing in disguise, a "providence" whose "calamity" is, as always, borne heroically by the beloved Persian community. It may be regarded as the latest move in God's Major Plan, another trumpet blast to awaken the heedless from their slumber and a golden opportunity offered to the Bahá'ís to demonstrate once again their unity and fellowship before the eyes of a declining and skeptical world, to proclaim with full force the Message of Bahá'u'lláh to high and low alike, to establish the reverence of our Faith for Islam and its Prophet, to assert the principles of noninterference in political activities and obedience to government which stand at the very core of our Faith, and to provide comfort and solace to the breasts of the serene sufferers and steadfast heroes in the forefront of a persecuted community. Our motto in these days of world-encircling gloom should be the Words of God addressed to the Blessed Beauty Himself: "When the swords flash, go forward! When the shafts fly, press onward!"[4]

314.10   Future historians will have to assess the impact of this crisis on the onward march of a triumphant Faith. A detailed list of the steps that have already been taken by the Bahá'ís of the world during the past three years is attached for the study of the friends.

314.11   Our fervent prayers are offered most ardently at the Holy Shrines for the blessings of Bahá'u'lláh to surround His lovers and loved ones in every land, and to assist and confirm them as they face with certitude and confidence the challenges of the future.

With loving Bahá'í greetings,
THE UNIVERSAL HOUSE OF JUSTICE

314-4. Bahá'u'lláh, in BP, pp. 219–20.

1979–1986 • THE SEVEN YEAR PLAN

Summary of Actions Taken by the Bahá'í International Community, National and Local Bahá'í Institutions, Governments, Non-Bahá'í Organizations and Prominent People in Connection with the Persecution of the Bahá'ís of Iran

The Bahá'í International Community

- Issued official statements to the press;
- kept the Secretary-General and appropriate offices of the United Nations apprised of developments as they occurred;
- cabled the Ayatollah Khomeini, the President and Prime Minister of Iran, and the President of the Iranian Supreme Court, urging their intervention and refuting accusations made against the Faith;
- prepared materials and made statements in connection with the adoption of resolutions by the United Nations General Assembly, the United Nations Commission on Human Rights Subcommission on Prevention of Discrimination and Protection of Minorities, the European Parliament, and the Parliamentary Assembly of the Council of Europe;
- made statements at the United Nations Commission on Human Rights on the Question of Enforced or Involuntary Disappearances;
- contacted the Iranian representative to the United Nations in New York in order to repudiate falsehoods made about the Faith and provide him with the true facts;
- and prepared the "White Paper" and "Update" and arranged for their translation in three languages, the "Chronological Summary of Individual Acts of Persecution in Iran," and other documents for submission to high-ranking officials, government and United Nations offices, and worldwide distribution to National Spiritual Assemblies.

314.12

Bahá'í Institutions

- National Assemblies throughout the world cabled the Ayatollah Khomeini on four occasions, the Prime Minister and Head of the Iranian Supreme Court three times each, and the Secretary of the Revolutionary Council of Iran once.
- 118 National Spiritual Assemblies cabled the Secretary-General of the United Nations, as did thousands of Local Assemblies, Bahá'í groups and isolated centers. It was estimated that some 10–15,000 cables reached him, protesting the execution of seven Bahá'ís in Hamadan.
- Over 10,000 Local Spiritual Assemblies cabled the Ayatollah Khomeini, urging his intervention regarding the expropriation of Bahá'í properties in Iran.
- Most National Assemblies contacted by letter or delegation, or sent cables to their respective Iranian Embassy or Consulate, on five occa-

314.13

sions and kept their government officials continually informed of developments.
- Selected National Assemblies cabled the Ayatollah Khomeini, the President and the Prime Minister of Iran, and the Secretary of the Revolutionary Council on several occasions. They also cabled or contacted their respective Iranian Embassy or Consulate at least eleven times; approached humanitarian, business or professional organizations; and were in constant touch with government offices and the media.
- A large number of National Assemblies pursued a well organized campaign of approaching the mass media, providing them with accurate information about the Faith and refuting false accusations made by enemies of the Cause. As a result, an unprecedented volume of publicity occurred in leading newspapers and periodicals throughout the world, as well as in newspapers having modest circulations. Well-known journalists wrote articles, some of which were distributed through international news agencies. Interviews were held with families of the martyrs, individual Bahá'ís wrote letters to editors of newspapers, and many radio and television programs were aired, including "Iran's Secret Pogrom" on W5 TV in Canada, and "Day One" and "John Craven's Newsround" on BBC1 TV.
- Many National Assemblies contacted immigration authorities and appropriate government offices in efforts to assist the displaced Iranian Bahá'ís in extending their visas and obtaining work permits and travel documents. They established special committees to work specifically to assist the Iranian friends, and they set up Persian Relief Funds on a national scale to aid deserving cases. The National Assemblies of Australia and Canada worked out with their respective immigration offices procedures whereby the process of immigration by Iranian Bahá'ís would be facilitated.

314.14
- Bahá'í communities the world over have assisted Iranian students abroad, who have been faced with the termination of their education because they are unable to receive funds from their families in Iran whose assets in Nawnahálán Company were frozen,[5] or did not receive funds because the Iranian government prevented the transfer of money from Iran to Bahá'í students abroad. In some areas, Iranian embassies have refused to extend the visas of Bahá'í students. Certain universities and colleges have allowed the Bahá'í students to continue their studies, and in some instances their tuition fees have even been waived.

314-5. For further information on the Nawnahálán Company, see the glossary.

1979–1986 • THE SEVEN YEAR PLAN

Resolutions Adopted on Behalf of the Bahá'ís in Iran 314.15
- Canadian Parliament (2)
- House of Representatives, Australia
- Senate, Australia
- German Federal Parliament
- A meeting held in a committee room of the House of Commons, United Kingdom
- United Nations General Assembly, Third Committee on the Elimination of All Forms of Religious Intolerance
- United Nations Subcommission on the Prevention of Discrimination and Protection of Minorities, Commission on Human Rights (2)
- European Parliament (2)
- Parliamentary Assembly of the Council of Europe (2)
- House of Representatives of the State of Alaska, U.S.A.
- House of Representatives, State of Illinois, U.S.A.
- International Association for Religious Freedom

Statements and Letters from Governments, World Leaders and Others
To name just a few 314.16
- Prime Minister's Office of the United Kingdom
- President Mitterrand of France
- Offices of the King and Minister for Foreign Affairs of Belgium
- President and Minister of Cultural Affairs of Luxembourg
- All three parliamentary parties in Luxembourg
- Foreign Minister Hans-Dietrich Genscher of Germany
- Prime Minister Indira Gandhi of India
- 148 out of 150 Members of Parliament in the Netherlands
- Swiss Parliamentarians
- Western Samoan Government
- Minister of Foreign Affairs, Australia
- Governor of the Hawaiian Islands, U.S.A.
- Governor of the Commonwealth of the Northern Mariana Islands

Some Non-Bahá'í Individuals and Organizations that Issued Statements, Letters, Cables, or Press Releases
- Human Rights Commission of the Federation of Protestant Churches in Switzerland 314.17
- Amnesty International
- Trinidad and Tobago Bureau on Human Rights
- Former Chief Justice, India
- Commission on Social Action of Reform Judaism
- Pacific Conference of Churches
- 13 Heads of Colleges in Oxford, U.K.

- The Master, Balliol College, Oxford, England
- Iran Committee for Democratic Action and Human Rights (based in the United States)
- Action by Christians for the Abolition of Torture (based in France)
- A large number of Senators and Congressmen of the United States

315
Call for a Day of Prayer for the Bahá'ís of Iran; Inviting Sympathetic Organizations and Individuals
29 JANUARY 1982

To National Spiritual Assemblies

Dear Bahá'í Friends,

315.1 At this time when our thoughts are so often directed toward our beleaguered brethren in the Cradle of the Faith, where no relenting of the persecutions is in sight, the Universal House of Justice is moved to call on National Spiritual Assemblies to schedule a day of prayer in order to extol the memory of the glorious martyrs, and to invoke the power of God to emancipate the Iranian believers from the fetters of oppression and hardship. Each National Assembly should select a suitable day and request all Bahá'í communities under its jurisdiction to hold appropriate gatherings on that specific day.

315.2 There is no doubt that the persecutions in Iran have attracted the interest of the public in a number of countries to the Cause of God and many have even expressed deep and genuine sympathy for the plight of the Persian friends. In such countries, if the National Assembly considers it desirable, it will be entirely permissible to invite to the prayer meetings, in the spirit and manner befitting the occasion, members of sympathetic organizations or selected individual non-Bahá'ís. These meetings could confirm and deepen the love and respect that many already cherish in their hearts for the Cause of God.

315.3 The following suggestions may be helpful to you:

315.3a If invitations are to be extended to non-Bahá'í friends, consideration should be given to those in authority both on national and local levels, leaders of thought, prominent people of all walks of life, and those sympathetic to the Faith. A general invitation to the public could also be issued in some localities, if appropriate and advisable.

315.3b Where Bahá'í centers are unavailable or inadequate for such meetings, public or community halls could be hired. If all efforts fail to obtain such secular facilities, it would also be permissible, as a last resort, to hold the gatherings in public halls owned by, or places of worship of, other religious communities, if such places are spontaneously offered or easily available, and provided it is absolutely clear that the meetings are to be conducted by, and under the auspices of, the Bahá'ís.

In each country where a Mashriqu'l-Adhkár is located the national meeting should be held by the National Assembly concerned in the Mother Temple. 315.4

It is important that the gatherings be carefully planned well in advance and the meetings be held with utmost dignity. 315.5

In countries where it would not be propitious to invite non-Bahá'ís to the proposed prayer meetings, such gatherings should as usual be confined to Bahá'ís. 315.6

The House of Justice is confident that the power of prayer emanating from these meetings will attract divine blessings, will cause the souls of those who have joined the hosts on high to rejoice, and will assist the endeavors of the believers everywhere in expediting the triumph of the Cause of God. 315.7

With loving Bahá'í greetings,
DEPARTMENT OF THE SECRETARIAT

316
Release of a Compilation on Family Life
18 FEBRUARY 1982

To all National Spiritual Assemblies

Dear Bahá'í Friends,

Many of the National Spiritual Assemblies have been given a specific goal, during the second phase of the Seven Year Plan, to organize programs for the development of family life and to nurture in the friends a deeper understanding of the nature of an institution which is at the very base of Bahá'í society. In order to help the friends everywhere to strengthen and enrich their family ties, and enable the family unit to reflect the glory of the Words of God, the Universal House of Justice asked its Research Department to prepare a compilation of suitable texts on this subject. 316.1

The compilation is now ready and a copy is attached.[1] 316.2

The Universal House of Justice leaves it to your discretion to determine how best the contents of this compilation may be shared with the friends under your jurisdiction. 316.3

With loving Bahá'í greetings,
DEPARTMENT OF THE SECRETARIAT

316-1. See CC 1:385–416.

317

Mounting Cruelties and the Heightened Steadfastness of the Iranian Bahá'ís

9 March 1982

To all National Spiritual Assemblies

317.1 ACCOUNTS HEROISM BELIEVERS CRADLE FAITH FILL OUR HEARTS WITH FEELINGS OF AWE, GRATITUDE, ADMIRATION. MOUNTING CRUELTIES OPPRESSORS MATCHED BY HEIGHTENED ENDURANCE STEADFASTNESS STAUNCH SUPPORTERS GREATEST NAME.

317.2 SINCE LAST REPORT ON 11 JANUARY 1982 CROWN MARTYRDOM HAS ADORNED TWO MORE LOVING SOULS, ḤUSAYN VAḤDAT-I-ḤAQQ OF TEHRAN AND IBRÁHÍM KHAYRKHÁH OF BÁBULSAR.[1] THE FORMER, A HIGHLY QUALIFIED ELECTRONICS ENGINEER, WAS EXECUTED ON 28 FEBRUARY, EVE OBSERVANCE DECLARATION BÁB ACCORDING LUNAR CALENDAR, AND MR. KHAYRKHÁH, ACTIVE BELIEVER CASPIAN AREA, TWO DAYS EARLIER. BOTH EXECUTED BURIED UNCEREMONIOUSLY WITHOUT RELATIVES FRIENDS BEING INFORMED.

317.3 CONFISCATION OF HOMES INNOCENT BAHÁ'ÍS WITHOUT PROVOCATION IS CONTINUING. LOOTING AND AUCTIONING OF FURNISHINGS OF BAHÁ'Í HOME IN ARDIKÁN NEAR YAZD WAS PRELUDE TO SERIES SIMILAR RAIDS ON HOMES OTHER BAHÁ'ÍS THAT TOWN. IN SHIRAZ 17 MORE HOMES EITHER CONFISCATED OR IN PROCESS CONFISCATION. 35 ADDITIONAL BANK ACCOUNTS OF BAHÁ'ÍS IN SHIRAZ NOW FROZEN. SCORES OF BAHÁ'ÍS HAVE LOST THEIR JOBS OR BEEN DEPRIVED THEIR BUSINESS AND TRADE LICENSES. ON ONE OCCASION A HIGH-RANKING AUTHORITY DECREED, IN REPLY TO QUESTION FROM INSURANCE COMPANY, THAT A BAHÁ'Í WIDOW HAD NO RIGHT COLLECT HALF HER HUSBAND'S PENSION DUE HER NOR RETAIN CUSTODY HER CHILDREN. HISTORIC BAHÁ'Í SITES PROGRESSIVELY BEING DEMOLISHED INCLUDING HOUSE BÁBÍYYIH IN MASHHAD.[2]

317.4 IN FACE SUCH OPPRESSIVE MEASURES, THOUSANDS BAHÁ'ÍS IRAN UNMINDFUL OF POSSIBLE DIRE CONSEQUENCES, HAVE COURAGEOUSLY APPEALED BY LETTER OR CABLE TO VARIOUS HIGH OFFICIALS AT NATIONAL AND LOCAL LEVELS COMPLAINING ABOUT BARBARIC ACTS GROSS INJUSTICE, HAVE REVEALED THEIR NAMES AND ADDRESSES, AND HAVE EXPRESSED HOPE THAT FEAR GOD WILL ULTIMATELY AWAKEN BLOODTHIRSTY AND HATE-FILLED INDIVIDUALS TO DISGRACEFUL ABUSE THEIR POWERS AND INDUCE THEM CEASE BEHAVIOR ABHORRENT ALL CIVILIZED PEOPLE.

317.5 BAHÁ'ÍS IRAN ARE GRATEFUL THEIR BRETHREN THROUGHOUT WORLD BECAUSE THEY HAVE NOT ONLY SUCCESSFULLY RAISED THEIR VOICES IN NATIONAL AND INTERNATIONAL FORUMS BUT ALSO HAVE PLEDGED REDOUBLE THEIR EFFORTS SERVE

317-1. In a letter dated 11 January 1982 the Universal House of Justice sent to all National Spiritual Assemblies the text of a 5 January 1982 cable and two 7 January 1982 cables (see messages no. 309, 310, and 311).

317-2. For information on the Bábíyyih, see the glossary.

BAHÁ'U'LLÁH IN NAME COWORKERS CRADLE FAITH, IN ORDER TO COUNTERACT EVIL MACHINATIONS ENEMIES CAUSE DESIGNED ERADICATE FAITH IN LAND ITS BIRTH.

UNIVERSAL HOUSE OF JUSTICE

318
Relocation of Manila Conference
11 MARCH 1982

To National Spiritual Assemblies

ON RECOMMENDATION NATIONAL SPIRITUAL ASSEMBLY PHILIPPINES FOLLOWING ITS CONSULTATIONS AUTHORITIES VIEW PRESENT DISTURBED CONDITIONS HAVE DECIDED TRANSFER MANILA CONFERENCE ELSEWHERE. NEW DATE VENUE WILL BE ANNOUNCED SHORTLY TO BAHÁ'Í WORLD.

318.1

UNIVERSAL HOUSE OF JUSTICE

319
Transfer of International Conference to Canberra, Australia
29 MARCH 1982

To all National Spiritual Assemblies

Dear Bahá'í Friends,

As you learned from an earlier communication, it has been necessary to transfer the International Conference planned for Manila, Philippines to another location.¹ The Universal House of Justice requested the National Spiritual Assembly of Australia to act as host since it is located at the southern pole of the spiritual axis referred to by the beloved Guardian as extending from "the Antipodes to the northern islands of the Pacific Ocean."² It is pleased to announce that the National Assembly has made arrangements to hold the Conference in Canberra from the 2nd through the 5th of September 1982. The Hand of the Cause Dr. Ugo Giachery will attend as representative of the House of Justice.

319.1

Kindly circulate this information to the friends as quickly as possible. It is hoped that all those who had been planning to attend the Manila Conference will find it possible to support the one now scheduled for Canberra.

319.2

With loving Bahá'í greetings,
DEPARTMENT OF THE SECRETARIAT

319-1. See message dated 11 March 1982 (no. 318).
319-2. LFG, p. 138.

320
Representation of the Universal House of Justice at Lagos Conference by Hand of the Cause of God John Robarts
29 MARCH 1982

To all National Spiritual Assemblies

Dear Bahá'í Friends,

320.1 The Universal House of Justice has asked us to inform you that it has become clear that in the present circumstances the representation of the World Center at the Lagos Conference by anyone currently residing in South Africa may give rise to unfavorable publicity for the Faith.

320.2 It has been decided, therefore, by the Universal House of Justice to invite the Hand of the Cause of God John Robarts to represent it at the Lagos Conference.

320.3 Four special conferences are being called for the benefit of the friends in Southern Africa, as these friends, because of the prevailing situation, are for the most part unable to attend the Lagos Conference. At these Conferences the Hand of the Cause William Sears will represent the Universal House of Justice.

With loving Bahá'í greetings,
DEPARTMENT OF THE SECRETARIAT

321
Riḍván Message 1982
RIḌVÁN 1982

To the Bahá'ís of the World

Dearly loved Friends,

321.1 Triumphs of inestimable portent for the unfoldment of the Cause of God, many of them resulting directly from the steadfast heroism of the beloved Persians in face of the savage persecutions meted out to them, have characterized the year just ending. The effect of these developments is to offer such golden opportunities for teaching and further proclamation as can only lead, if vigorously and enthusiastically seized, to large-scale conversion and an increasing prestige.

Notable Achievements

321.2 Heartwarming progress in the construction of the Indian and Western Samoan Mashriqu'l-Adhkárs, the opening of the second Bahá'í radio station of Latin America in Peru,[1] the establishment of the European office of the Bahá'í International Community in Geneva, steady advances in the second phase of the Seven Year Plan, encouraging expansion of the systematized Bahá'í education of children, sacrifice and generous outpouring of funds from a growing number of friends, all testify to the abundant confirmations with which Bahá'u'lláh rewards the dedicated efforts of His loved ones throughout the world. The worldwide attention accorded the Faith in the media, which has opened wide the doors of mass proclamation of the divine Message, and the sympathetic discussion of it in the highest councils of mankind with the resulting actions taken by sovereign governments and international authorities,[2] are unprecedented in Bahá'í history.

A Year Rich in Bahá'í Occasions

All this, dear friends, augurs well for the coming year which is rich in Bahá'í occasions. The fiftieth anniversary of the passing of the Greatest Holy Leaf will be commemorated at the five International Conferences and by the publication of a book, compiled at the World Center, comprising texts about her and some hundred of her own letters;[3] the move to the permanent Seat of the Universal House of Justice will take place; in November the twenty-fifth anniversary of the passing of our beloved Guardian will coincide with the midway point of the Seven Year Plan[4] and the year will terminate with the fifth International Convention when members of National Spiritual Assemblies throughout the world will come to Haifa to elect the Universal House of Justice.[5]

Tribute to the Hands of the Cause of God and the Continental Boards of Counselors

The distinguished and invaluable activities of the beloved Hands of the Cause are a source of pride and joy to the entire Bahá'í world. The assumption of wider responsibilities by each Continental Board of Counselors is proving an unqualified success and we express our warm thanks and admiration to the International Teaching Center and all the Counselors for the great contribution they are making, in increasing measure, to the stability and development of the embryonic World Order of Bahá'u'lláh.

321-1. For the message sent on the occasion of the inauguration of Radio Bahá'í Peru, see the telex dated 29 November 1981 (no. 304).

321-2. See letter dated 26 January 1982 (no. 314) and its addendum for a description of actions taken to defend the Iranian Bahá'ís.

321-3. The Greatest Holy Leaf passed away at 1:00 A.M. on 15 July 1932. The book referred to, *Bahíyyih Khánum*, was published by the Bahá'í World Center (1982).

321-4. Shoghi Effendi passed away on 4 November 1957; the midpoint of the Seven Year Plan was November 1982.

321-5. Riḍván 1983.

Youth

As to Bahá'í youth, legatees of the heroic early believers and now standing on their shoulders, we call upon them to redouble their efforts, in this day of widespread interest in the Cause of God, to enthuse their contemporaries with the divine Message and thus prepare themselves for the day when they will be veteran believers able to assume whatever tasks may be laid upon them. We offer them this passage from the Pen of Bahá'u'lláh:

> Blessed is he who in the prime of his youth and the heyday of his life will arise to serve the Cause of the Lord of the beginning and of the end, and adorn his heart with His love. The manifestation of such a grace is greater than the creation of the heavens and of the earth. Blessed are the steadfast and well is it with those who are firm.[6]

Our Response: A Mighty Upsurge in Effective Teaching

The rising sun of Bahá'u'lláh's Revelation is having its visible effect upon the world and upon the Bahá'í community itself. Opportunities, long dreamed of for teaching, attended by showering confirmations, now challenge in ever-increasing numbers, every individual believer, every Local and National Spiritual Assembly. The potent seeds sown by 'Abdu'l-Bahá are beginning to germinate within the divinely ordained Order expounded and firmly laid by the beloved Guardian. Humanity is beaten almost to its knees, bewildered and shepherdless, hungry for the bread of life. This is our day of service; we have that heavenly food to offer. The peoples are disillusioned with deficient political theories, social systems and orders; they crave, knowingly or unknowingly, the love of God and reunion with Him. Our response to this growing challenge must be a mighty upsurge of effective teaching, imparting the divine fire which Bahá'u'lláh has kindled in our hearts until a conflagration arising from millions of souls on fire with His love shall at last testify that the Day for which the Chief Luminaries of our Faith so ardently prayed has at last dawned.

THE UNIVERSAL HOUSE OF JUSTICE

321-6. From an unpublished Tablet.

322

Further Developments of the Situation in Iran
3 May 1982

To all National Spiritual Assemblies

Dear Bahá'í Friends,

We have been instructed by the Universal House of Justice to announce the sorrowful news that since our last report of 11 March 1982, two more steadfast and devoted supporters of the Greatest Name in Iran, Mr. Iḥsánu'lláh Khayyámí of Urúmíyyih and Mr. 'Azízu'lláh Gulshaní of Mashhad, have joined the ranks of the martyrs of His Cause.[1] After a few months of imprisonment and intense but unsuccessful pressure to recant his Faith, Mr. Khayyámí, a staunch believer of humble means, was executed on 12 April 1982. Mr. Gulshaní, a prominent Bahá'í of Khurásán province who was imprisoned for many months, was executed on 29 April. The following developments are also shared with you.

- The friends in the village of Saysán, in one of the outlying districts of Tabríz, are under extreme pressure by a mullah of the neighboring town of Bustánábád, and have been given a month to decide whether they will convert to Islam or face grave consequences. The members of the Bahá'í community in this village, which consists of several generations of Bahá'ís, have remained steadfast and approached the religious authorities in Qum who have recommended to the Imám-Jum'ih of Tabríz that he intervene and try to relieve the Bahá'ís from such harassment.
- All the Bahá'ís of Ḥiṣár in Khurásán have been forced to leave their homes and had to flee to Mashhad.
- Although a few believers have been released from prison during the recent period of amnesty, many have been arrested and imprisoned, including nine active Bahá'ís in Qazvín and its surrounding areas, seven in Shiraz, five in Tehran, and seven in Zanján.

The Universal House of Justice urges the friends throughout the world to continue their prayers for the Bahá'ís in Iran and to intensify their services in the path of the Blessed Beauty to compensate for the plight of our oppressed brethren in the Cradle of the Faith. . . .

With loving Bahá'í greetings,
DEPARTMENT OF THE SECRETARIAT

322-1. For the last report, see the cable dated 9 March 1982 (no. 317), which was transmitted in a letter dated 11 March 1982.

323

Six Martyrdoms, More Arrests Resulting from Fresh Outburst of Persecutions

10 May 1982

To all National Spiritual Assemblies

323.1 AFTER BRIEF LULL AND RELEASE FEW BAHÁ'Í PRISONERS, FRESH OUTBURST PERSECUTIONS AGAINST MEMBERS DEFENSELESS COMMUNITY IN CRADLE FAITH CLEARLY EVIDENT.

323.2 HEARTS GRIEVED ANNOUNCE THAT SINCE EARLY APRIL SIX STAUNCH BELIEVERS HAVE OFFERED THEIR LIVES ALTAR SACRIFICE. ALL MERCILESSLY EXECUTED SOLELY DUE THEIR ADHERENCE BELOVED FAITH, BUT OUTWARDLY ON BASIS TRUMPED-UP ACCUSATIONS.

323.3 THEIR NAMES ARE:

1. MR. 'ASKAR MUḤAMMADÍ OF RAḤÍMKHÁN VILLAGE IN BÚKÁN, KURDISTÁN, WHO WAS ASSASSINATED AT HIS HOME IN EARLY APRIL BY REVOLUTIONARY GUARDS AFTER INTRODUCING HIMSELF AS BAHÁ'Í WHILE THEY WERE SEARCHING HOUSES THAT AREA FOR WEAPONS.
2. MR. IḤSÁNU'LLÁH KHAYYÁMÍ OF URÚMÍYYIH, WHO WAS EXECUTED BY FIRING SQUAD ON 12 APRIL AFTER INTENSE BUT UNSUCCESSFUL PRESSURE TO RECANT HIS FAITH.
3. MR. 'AZÍZU'LLÁH GULSHANÍ OF MASHHAD, KHURÁSÁN, WHO WAS HANGED ON 29 APRIL. COURT VERDICT CLEARLY BASED HIS AFFILIATION FAITH AND BAHÁ'Í ACTIVITIES. NEWSPAPER *KAYHÁN* REPORTS VERDICT COURT REFERS TO HIM AS A HERETIC, WHICH IS PUNISHABLE BY DEATH. THIS IMPORTANT DOCUMENT PROVES BAHÁ'ÍS ARE BEING KILLED BECAUSE OF THEIR RELIGION.
4. MR. BADÍ'U'LLÁH ḤAQQ-PAYKAR OF KARAJ NEAR TEHRAN, AN ACTIVE MEMBER OF THE LOCAL ASSEMBLY, WHO WAS EXECUTED BY FIRING SQUAD ON 8 MAY.
5. AND 6. MR. MAḤMÚD FARÚHAR AND HIS WIFE ISHRÁQÍYYIH, HIGHLY QUALIFIED AND EDUCATED ACTIVE BELIEVERS, AND MEMBERS OF THE LOCAL ASSEMBLY, WHO WERE ALSO EXECUTED IN KARAJ ON 8 MAY.

323.4 FURTHER, MANY PROMINENT BAHÁ'ÍS, INCLUDING 23 MEMBERS LOCAL SPIRITUAL ASSEMBLIES IN VARIOUS PARTS COUNTRY ARRESTED IN RECENT WEEKS. . . .

UNIVERSAL HOUSE OF JUSTICE

1979–1986 • THE SEVEN YEAR PLAN

324
Execution of Two Members of the Spiritual Assembly of Urúmíyyih
11 MAY 1982

To all National Spiritual Assemblies

FURTHER RECENT MESSAGE[1] ON RECRUDESCENCE PERSECUTIONS IRAN, NEWS JUST RECEIVED TWO OTHER STAUNCH BELIEVERS MARTYRED BY FIRING SQUAD IN IRAN. THEY ARE: MR. ÁGÁHU'LLÁH TÍZFAHM AND MISS JALÁLÍYYIH MUSHTÁ'IL-USKÚ'Í. BOTH MEMBERS LOCAL SPIRITUAL ASSEMBLY URÚMÍYYIH. . . .

UNIVERSAL HOUSE OF JUSTICE

325
Appointment of New Counselor in Asia
13 MAY 1982

To all National Spiritual Assemblies

Dear Bahá'í Friends,

The Universal House of Justice announces with pleasure the appointment of Dr. Ṣábir Áfáqí to the Continental Board of Counselors in Asia to replace Dr. Masíḥ Farhangí, who was martyred last year.

With loving Bahá'í greetings,
DEPARTMENT OF THE SECRETARIAT

326
Passing of Amoz Gibson, Member of the Universal House of Justice
15 MAY 1982

To National Spiritual Assemblies

WITH SORROWFUL HEARTS LAMENT LOSS OUR DEARLY LOVED BROTHER AMOZ GIBSON WHO PASSED AWAY AFTER PROLONGED HEROIC STRUGGLE FATAL ILLNESS.[1] EXEMPLARY SELF-SACRIFICING PROMOTER FAITH ACHIEVED BRILLIANT UNBLEMISHED RECORD CONSTANT SERVICE FOUNDED ON ROCK-LIKE STAUNCHNESS AND

324-1. 10 May 1982.
326-1. For an account of Amoz Gibson's life and services, see BW 18:665–69.

DEEP INSATIABLE LOVE FOR TEACHING WORK PARTICULARLY AMONG INDIAN AND BLACK MINORITIES WESTERN HEMISPHERE AND INDIGENOUS PEOPLES AFRICA. HIS NOTABLE WORK ADMINISTRATIVE FIELDS NORTH AMERICA CROWNED FINAL NINETEEN YEARS INCALCULABLE CONTRIBUTION DEVELOPMENT WORLD CENTER WORLD-EMBRACING FAITH. PRAYING SHRINES BOUNTIFUL REWARD HIS NOBLE SOUL THROUGHOUT PROGRESS ABHÁ KINGDOM. EXPRESS LOVING SYMPATHY VALIANT BELOVED WIDOW PARTNER HIS SERVICES AND BEREAVED CHILDREN ADVISE HOLD BEFITTING MEMORIAL GATHERINGS EVERYWHERE BAHÁ'Í WORLD AND COMMEMORATIVE SERVICES ALL MA<u>SH</u>RIQU'L-A<u>DH</u>KÁRS.

Universal House of Justice

327
Call for Ballots to Elect One Member of the Universal House of Justice
18 May 1982

To National Spiritual Assemblies

327.1 CALL UPON MEMBERS YOUR ASSEMBLY CAST THEIR BALLOTS FOR ELECTION ONE MEMBER UNIVERSAL HOUSE OF JUSTICE REPLACE AMOZ GIBSON. EXHORT ALL ELECTORS PONDER SACRED RESPONSIBILITY NOW RESTING THEIR SHOULDERS UPHOLD SCRUPULOUSLY PRAYERFULLY SANCTIFIED SPIRIT BAHÁ'Í ELECTIONS. EACH MEMBER SHOULD RECORD VOTE FOR ONE MALE ADULT BELIEVER ON PIECE PLAIN PAPER ENCLOSE IT IN A SEALED UNMARKED ENVELOPE AND SEAL THAT IN ANOTHER ENVELOPE ON WHICH ELECTOR SHOULD WRITE OWN NAME AND NATIONAL SPIRITUAL ASSEMBLY NAME AND IMMEDIATELY SEND TO NATIONAL ASSEMBLY OFFICE. AS SOON AS ALL NINE BALLOTS RECEIVED NATIONAL OFFICE SHOULD AIRMAIL THEM IN ONE PACKAGE WITH LIST BALLOTS ENCLOSED TO UNIVERSAL HOUSE OF JUSTICE NOTIFYING WORLD CENTER OF DESPATCH AT ONCE BY CABLE. IF NOT ALL NINE RECEIVED AT NATIONAL OFFICE BY TWENTY-SECOND JUNE BALLOTS RECEIVED MUST BE SENT THAT DAY. LATE ARRIVALS SHOULD BE FORWARDED AS SOON AS RECEIVED CABLING WORLD CENTER IN EACH CASE NUMBER BALLOTS DESPATCHED. RESULT WILL BE ANNOUNCED FIFTEENTH JULY. SUPPLICATING BAHÁ'U'LLÁH GUIDE PROTECT YOU IN EXERCISE THIS DIVINELY CONFERRED RESPONSIBILITY.

Universal House of Justice

328

Execution of Two Members of the Kháníábád Spiritual Assembly

26 MAY 1982

To all National Spiritual Assemblies

WITH UTMOST SADNESS ANNOUNCE EXECUTION ON 16 MAY TWO MORE VALIANT SOULS CRADLE FAITH SA'DU'LLÁH BÁBÁZÁDIH AND NAṢRU'LLÁH AMÍNÍ, MEMBERS SPIRITUAL ASSEMBLY KHÁNÍÁBÁD VILLAGE NEAR TEHRAN. DEATH SENTENCE HANDED DOWN OVER TWO MONTHS AGO, HOWEVER EXECUTION POSTPONED WHILE INTENSE PRESSURE BROUGHT TO BEAR UPON TWO CONDEMNED BAHÁ'ÍS RECANT FAITH OBTAIN FREEDOM. THESE TWO STEADFAST FRIENDS WHO PREFERRED MARTYRDOM TO DENIAL FAITH WERE BURIED UNCEREMONIOUSLY WITHOUT FAMILY FRIENDS BEING INFORMED. AFTER SEVEN DAYS RELATIVES FORTUITOUSLY DISCOVERED EXECUTION THEIR LOVED ONES. 328.1

ALTHOUGH FEW BAHÁ'Í PRISONERS RELEASED, ARRESTS CONTINUE THROUGHOUT COUNTRY.... 328.2

UNIVERSAL HOUSE OF JUSTICE

329

Message to the International Conference in Dublin, Ireland—2 June 1982

2 JUNE 1982

To the Friends gathered at the International Conference in Dublin

Dearly loved Friends,

"The world is in travail, and its agitation waxeth day by day.... Such shall be its plight, that to disclose it now would not be meet and seemly."[1] The shattering blows dealt to the old, divisive system of the planet and the constantly accelerating decline in civilized life since that dire warning was uttered by Bahá'u'lláh a hundred years ago, have brought mankind to its present appalling condition. Consideration of how the Bahá'ís of Europe, confronted by this situation, can meet their responsibilities, spiritually and actively, is the main purpose of this Conference. 329.1

The holding of this Conference in Dublin calls to mind the historic and heroic services of Ireland in spreading the divine religion throughout pagan Europe. Europe's response was to develop, through many vicissitudes, the most widespread and effective civilization known. That civilization, together with 329.2

329-1. GWB, p. 118.

all other systems in the world, is now being rolled up, and Europe's plight in proportion to her former preeminence, is desperate indeed. By the same token her opportunity is correspondingly great. The challenges to her resilience, to her deep-seated spiritual vitality, nourished over the centuries by the Teachings of Christ—now, alas, neglected and even contemned—can and must call forth a more magnificent response than was ever made by the divided and contending peoples of olden times. Yours is the task to arouse that response. The power of Bahá'u'lláh is with you and this Day, as attested by the Báb, is "immensely exalted . . . above the days of the Apostles of old."[2]

329.3 In this great Day Europe is blessed as never before in its history, for the Manifestation of God, the Lord of Hosts, spent five years of His exiles within its borders, sending forth from His "remote prison" the first of those challenging, world-shaking addresses to the kings and rulers, six of whom were European potentates.[3] There is no authenticated record of a Manifestation of God ever before setting foot in Europe.

329.4 You are engaged on a Seven Year Plan and have made devoted and sacrificial efforts to attain its objectives. But its ultimate purpose, as that of all other plans, namely the attracting of the masses of mankind to the all-embracing shelter of the Cause of God, still evades us. Particularly in Europe. We have not, as yet, found the secret of setting aglow the hearts of great numbers of Europeans with the divine fire. This must now be your constant preoccupation, the subject of your deliberations at this Conference, the purpose of your lives, to which you will attain "only if you arise to trample beneath your feet every earthly desire."[4] We call upon every Bahá'í in Europe to ponder this vital matter in his inmost soul, to consider what each may do to attract greater power to his efforts, to radiate more brilliantly and irresistibly the joyous, regenerating power of the Cause, so that the Bahá'í community in every country of Europe may stand out as a beacon light repelling the dark shadows of godlessness and moral degradation now threatening to obliterate the last remnants of a dying order. We call upon the Continental Board of Counselors to consult following this Conference with every National Spiritual Assembly in Europe, and together, launch such a campaign of spiritualization of the Bahá'í community, allied with intensified personal teaching, as has never been witnessed in your continent. The goals of the Seven Year Plan can all be accomplished as the result of such a program and the European Bahá'í com-

---

329-2. Quoted in DB, p. 93.

329-3. Bahá'u'lláh was banished to Adrianople (now Edirne, Turkey), which He called the "remote prison" (see BP, p. 211, and Balyuzi, *Bahá'u'lláh: The King of Glory*, p. 217). The European potentates whom Bahá'u'lláh addressed were Queen Victoria of the United Kingdom, Napoleon III of France, Czar Alexander II of Russia, Kaiser Wilhelm I of Germany, Emperor Francis Joseph of Austria, and Pope Pius IX. The addresses are published in *Proclamation of Bahá'u'lláh*. For an account of the revelation of these Tablets, see letter dated Riḍván 1965 (no. 24).

329-4. The Báb, quoted in DB, p. 93.

munity may achieve through it the spiritual force and character to demonstrate to a stricken and declining civilization the peace and joy and order of the long-awaited, Christ-promised Kingdom of God on earth.

329.5 May the loving spirit and saintly life of the Greatest Holy Leaf, the fiftieth anniversary of whose ascension is commemorated in this Conference, imbue your thoughts and aspirations and resolves with that dedicated, self-sacrificing, utter devotion to Bahá'u'lláh and His Cause which she so greatly exemplified.[5]

With loving Bahá'í greetings,
THE UNIVERSAL HOUSE OF JUSTICE

330
Comments on the Station of 'Abdu'l-Bahá and the Language Used by the Central Figures
3 JUNE 1982

To individual believers

Dear Bahá'í Friends,

330.1 The Universal House of Justice has asked us to acknowledge your letter of April 28 and to make the following comments concerning your three questions.

1. 330.2 It was the express wish of Bahá'u'lláh that after Him the friends should "turn" to 'Abdu'l-Bahá. Bahá'u'lláh also said in His Book of Laws that anything that was not clear in His Writings should be "referred" to His Most Mighty Branch springing from the Ancient Root. (See *The World Order of Bahá'u'lláh*, pages 134–35.) In one of the Tablets of 'Abdu'l-Bahá published in *Selections from the Writings of 'Abdu'l-Bahá* (page 214) He quotes the passages mentioned above and interprets them to mean that "whatever He ['Abdu'l-Bahá] saith is the very truth." 'Abdu'l-Bahá further says, referring to those who do not accept Him as the Interpreter of the Word of God, "Whoso deviates from my interpretation is a victim of his own fancy" (*The World Order of Bahá'u'lláh*, page 138). Moreover, in the *Star of the West*, Volume 12, page 227, 'Abdu'l-Bahá interprets the verses from the "Tablet of the Branch" to mean "whatsoever His ['Abdu'l-Bahá's] pen records, that is correct."

2. 330.3 There is nothing in the Writings that would lead us to the conclusion that what Shoghi Effendi says about himself concerning statements on

---

329-5. The Greatest Holy Leaf (Bahíyyih Khánum), the daughter of Bahá'u'lláh and sister of 'Abdu'l-Bahá, was born in Iran in 1846 and died in Haifa at 1:00 A.M. on 15 July 1932.

subjects not directly related to the Faith also applies to 'Abdu'l-Bahá.¹ Instead we have assertions which indicate that 'Abdu'l-Bahá's position in the Faith is one for which we find "no parallel" in past Dispensations.² For example, Bahá'u'lláh, in addition to His reference to the Center of His Covenant as the "Mystery of God," states that 'Abdu'l-Bahá should be regarded as God's "exalted Handiwork" and "a Word which God hath adorned with the ornament of His Own Self, and made it sovereign over the earth and all that there is therein."³ And from Shoghi Effendi we have the incontrovertible statement that the Guardian of the Faith while "overshadowed" by the "protection" of Bahá'u'lláh and of the Báb, "remains essentially human," whereas in respect of 'Abdu'l-Bahá Shoghi Effendi categorically states that "in the person of 'Abdu'l-Bahá the incompatible characteristics of a human nature and superhuman knowledge and perfection have been blended and are completely harmonized."⁴

330.4 3. With reference to your question about the "ether," the various definitions of this word as given in the Oxford English Dictionary all refer to a physical reality, for instance, "an element," "a substance," "a medium," all of which imply a physical and objective reality and, as you say, this was the concept posited by nineteenth century scientists to explain the propagation of light waves. It would have been understood in this sense by the audiences whom 'Abdu'l-Bahá was addressing. However, in Chapter XVI of *Some Answered Questions,* 'Abdu'l-Bahá devotes a whole chapter to explaining the difference between things which are "perceptible to the senses" which He calls "objective or sensible," and realities of the "intellect" which have "no outward form and no place," and are "not perceptible to the senses." He gives examples of both "kinds" of "human knowledge." The first kind is obvious and does not need elaboration. To illustrate the second kind the examples He gives are: love, grief, happiness, the power of the intellect, the human spirit and "ethereal matter." (In the original Persian the word "ethereal" is the same as "etheric.") He states clearly that "Even ethereal matter, the forces of which are said in physics to be heat, light, electricity and magnet-

---

330-1. The following extract from a letter dated 17 October 1944 on behalf of Shoghi Effendi to an individual believer is a clarification of the Guardian's infallibility: "The infallibility of the Guardian is confined to matters which are related strictly to the Cause and interpretation of the teachings; he is not an infallible authority on other subjects, such as economics, science, etc. When he feels that a certain thing is essential for the protection of the Cause, even if it is something that affects a person personally, he must be obeyed, but when he gives *advice,* such as that he gave you in a previous letter about your future, it is not binding; you are free to follow it or not as you please."
330-2. WOB, p. 143.
330-3. Quoted in WOB, pp. 134, 135.
330-4. WOB, pp. 151, 134.

ism, is an intellectual reality, and is not sensible." In other words, the "ether" is a concept arrived at intellectually to explain certain phenomena. In due course, when scientists failed to confirm the physical existence of the "ether" by delicate experiments, they constructed other intellectual concepts to explain the same phenomena.

330.5 In considering the whole field of divinely conferred "infallibility" one must be careful to avoid the literal understanding and petty-mindeness that has so often characterized discussions of this matter in the Christian world. The Manifestation of God (and, to a lesser degree, 'Abdu'l-Bahá and Shoghi Effendi,) has to convey tremendous concepts covering the whole field of human life and activity to people whose present knowledge and degree of understanding are far below His. He must use the limited medium of human language against the limited and often erroneous background of His audience's traditional knowledge and current understanding to raise them to a wholly new level of awareness and behavior. It is a human tendency, against which the Manifestation warns us, to measure His statements against the inaccurate standard of the acquired knowledge of mankind. We tend to take them and place them within one or other of the existing categories of human philosophy or science while, in reality, they transcend these and will, if properly understood, open new and vast horizons to our understanding.

330.6 Some sayings of the Manifestation are clear and obvious. Among these are laws of behavior. Others are elucidations which lead men from their present level of understanding to a new one. Others are pregnant allusions, the significance of which only becomes apparent as the knowledge and understanding of the reader grow. And all are integral parts of one great Revelation intended to raise mankind to a new level of its evolution.

330.7 It may well be that we shall find some statement is couched in terms familiar to the audience to which it was first addressed, but is strange now to us. For example, in answer to a question about Bahá'u'lláh's reference to the "fourth heaven" in the *Kitáb-i-Íqán*,[5] the Guardian's secretary wrote on his behalf:

330.7a As to the ascent of Christ to the fourth heaven, as revealed in the glorious "Book of Íqán," he [the Guardian] stated that the "fourth heaven" is a term used and a belief held by the early astronomers. The followers of the Shí'ih sect likewise held this belief. As the *Kitáb-i-Íqán* was revealed for the guidance of that sect, this term was used in conformity with the concepts of its followers.

(Translated from the Arabic)[6]

330-5. KI, p. 89.
330-6. From an unpublished letter.

330.7b In studying such statements, however, we must have the humility to appreciate the limitations of our own knowledge and outlook, and strive always to understand the purpose of Bahá'u'lláh in making them, trying to look upon Him with His own eyes, as it were.

330.8 It is hoped that the above explanations will prove useful to you in your study of the subjects in which you have expressed interest.

With loving Bahá'í greetings,
DEPARTMENT OF THE SECRETARIAT

331
Martyrdom of Four Bahá'ís in Qazvín and Surrounding Areas
12 July 1982

To National Spiritual Assemblies

331.1 WITH GREAT SORROW INDIGNATION ANNOUNCE FURTHER MARTYRDOMS FOUR MORE VALIANT DEVOTED SERVANTS BAHÁ'U'LLÁH, ALL PROMINENT BAHÁ'ÍS IN QAZVÍN AND SURROUNDING AREAS.[1]

MUḤAMMAD MANṢÚRÍ
JADÍDU'LLÁH ASHRAF
MUḤAMMAD 'ABBÁSÍ
MANÚCHIHR FARZÁNIH-MU'AYYAD

331.2 PRESSURES AGAINST BAHÁ'ÍS IRAN MOUNTING, THEIR SCOPE WIDENING, ENGULFING BAHÁ'ÍS ALL WALKS OF LIFE....

UNIVERSAL HOUSE OF JUSTICE

332
Election of Glenford Mitchell to the Universal House of Justice
15 July 1982

To National Spiritual Assemblies

332.1 WARMLY WELCOME NEWLY ELECTED MEMBER HOUSE JUSTICE GLENFORD MITCHELL.

UNIVERSAL HOUSE OF JUSTICE

331-1. The executions took place on Friday, 9 July 1982.

333
Guidelines Concerning the Registration of Children and Youth
19 JULY 1982

The National Spiritual Assembly of the Bahá'ís of the United Kingdom
Dear Bahá'í Friends,

333.1 The Universal House of Justice has received your letter of 3 June 1982 mentioning some of the misunderstandings which still exist concerning the registration of Bahá'í children and youth, and it has instructed us to send you the following clarifications.

333.2 The beloved Guardian's secretary wrote on his behalf to your National Spiritual Assembly on 17 June 1954:

333.2a Although the children of Bahá'í parents are considered to be Bahá'ís, there is no objection at the present time, for purposes of keeping a correct census, and also ascertaining whether the young people are, sincerely, believers, and willing to do their share in service to the Faith, to asking them to make a declaration of their intention, at the age of fifteen or so. Originally the Guardian understands this was adopted in America to enable young Bahá'í men to make certain arrangements in connection with their application for noncombatant status, upon their attaining the age of military service. There is really nothing about it in the Teachings or in the Administration. Your Assembly is free to do as it pleases in this matter.

333.2b He has also written that in deciding who is to be regarded as a believer, Assemblies must refrain as far as possible from drawing rigidly the line of demarcation.

333.3 At the present time, with so many more young people entering the Faith, and with such a great increase in mobility from country to country, the House of Justice feels that it is necessary to establish certain guidelines which will apply to all National Spiritual Assemblies in this delicate issue, while matters of procedure are left to each National Assembly to decide according to the conditions in its own area of jurisdiction. In the current phase of the Seven Year Plan, for example, the House of Justice has called on Assemblies to keep registers of the birth of Bahá'í children, and great stress is being laid on the Bahá'í education of children and on the activities of Bahá'í youth.

333.4 In letters replying to questions on the registration of children and youth the Universal House of Justice has attempted to avoid laying down rulings that are universally applicable. However, for the assistance of National Spiritual Assemblies it is now providing the following summary of guidelines and elucidations that have been given. We are to emphasize that no hard and fast lines should be drawn, and procedural matters must never be allowed to eclipse the spiritual reality of belief, which is an intensely personal relationship between the soul and its Creator.

Bahá'í Children

333.5 Unlike the children of some other religions, Bahá'í children do not automatically inherit the Faith of their parents. However, the parents are responsible for the upbringing and spiritual welfare of their children, and Spiritual Assemblies have the duty to assist parents, if necessary, in fulfilling these obligations, so that the children will be reared in the light of the Revelation of Bahá'u'lláh and from their earliest years will learn to love God and His Manifestations and to walk in the way of God's Law. It is natural, therefore, to regard the children of Bahá'ís as Bahá'ís unless there is a reason to conclude the contrary. It is quite wrong to think of Bahá'í children as existing in some sort of spiritual limbo until the age of fifteen at which point they can "become" Bahá'ís. In the light of this one can conclude the following:

333.6 Children born to a Bahá'í couple are regarded as Bahá'ís from the beginning of their lives, and their births should be registered by the Spiritual Assembly.

333.7 The birth of a child to a couple, one of whom is a Bahá'í, should also be registered unless the non-Bahá'í parent objects.

333.8 A Spiritual Assembly may accept the declaration of faith of a child of non-Bahá'í parents, and register him as a Bahá'í child, provided the parents give their consent.

333.9 In the cases of children whose parents become Bahá'ís, much depends upon the ages and reactions of the children concerned. They will require great love and understanding, and each case must be judged on its own merits. This applies to an added degree, of course, if only one of the parents has accepted the Faith, in which case the attitude of the other parent is an important factor; the aim of the Bahá'ís should be to foster family unity. The important thing is that the children, whether registered as Bahá'ís or not, should be made to feel welcome at Bahá'í children's classes and other community gatherings.

333.10 It is within a Spiritual Assembly's discretion to request Bahá'í children to undertake work of which they are capable in service to the Faith, such as service on suitable committees.

Bahá'í Youth

333.11 Fifteen is the age at which a child attains spiritual maturity, and thus it is at the age of fifteen that a Bahá'í child assumes the responsibility for obeying such laws as those of fasting and prayer, and for affirming of his own volition his faith in Bahá'u'lláh.

333.12 At the present time the Universal House of Justice prefers to leave it to each National Spiritual Assembly to decide what method is to be followed in ascertaining the attitude of Bahá'í children when they reach fifteen, provided that it is clear that a Bahá'í child is not becoming a Bahá'í at that age, but is simply affirming his faith on his own behalf. One Spiritual Assembly, for example, sends a very kind letter to each Bahá'í child in its community on the occasion of his fifteenth birthday (unless, of course, it has reason to doubt that the child in question is a Bahá'í), explaining the meaning of attaining the age of maturity, and extending the good wishes of the Assembly for his future services to the Cause. This does not require an active response from every child but does provide each with an opportunity to make his position clear if desired. In whatever procedure it adopts a National Spiritual Assembly must wisely steer a course between seeming to doubt the faith of a child who has been brought up as a devout Bahá'í on the one hand, and seeming to compel a child to be a member of the Bahá'í community against his will, on the other.

333.13 If the Assembly ascertains from a youth that he does not, in fact, accept the Faith, even if he has been brought up in a Bahá'í family, it should not register him as a Bahá'í youth, and such a youth, since he is now mature and responsible for his own actions, would be in the same situation as any other non-Bahá'í youth who is in close contact with the Bahá'í community. He should be treated with warmth and friendship.

333.14 It may happen that a Bahá'í child, on reaching the age of fifteen is not entirely sure in his own mind. This can well happen if one of the parents is not a Bahá'í or if the parents have accepted the Faith not long before. In such a case the Assembly should not assume automatically that he is not a Bahá'í. If the youth wishes to attend Feasts and is content to continue to be regarded as a Bahá'í as he was when a child, this should be permitted, but in the process of deepening his understanding of the Faith his parents and the Assembly should explain to him that it is his responsibility to soon make his position clear.

333.15 Declarations of faith from non-Bahá'í youth between the ages of 15 and 21, whose parents are not Bahá'ís, may be accepted without the consent of their parents unless this is contrary to the civil law. However, the importance of respect for one's parents must not be forgotten, and such youth may need to be counseled to give heed to their parents' wishes as far as the degree of their activity on behalf of the Faith is concerned, and even, if the parents are very antagonistic, to be completely inactive for a time.

Administrative Maturity

333.16 It is stated in the Constitution of a National Spiritual Assembly that upon attaining the age of 21 years a Bahá'í is eligible to vote and to hold elective office.[1] Some National Spiritual Assemblies in the past, and indeed a few at

---

333-1. See *Declaration of Trust and By-Laws of the National Spiritual Assembly of the Bahá'ís of the United States,* p. 14. For a model Declaration of Trust and By-Laws upon which the constitution of every National Spiritual Assembly is based, see *Bahá'í World,* Volumes 14, 15, 16, 17, or 18, under "The Institution of the National Spiritual Assembly," within the part titled "The World Order of Bahá'u'lláh."

the present time, have required a specific action on the part of a Bahá'í youth to claim his voting rights on attaining the age of 21. The House of Justice does not wish to make a general ruling on this point, since in some circumstances this may be a necessary protection to the Faith. It does, however, discourage National Assemblies from making such a requirement unless they deem it essential.

With loving Bahá'í greetings,
DEPARTMENT OF THE SECRETARIAT

334
Commemoration at the World Center of the Fiftieth Anniversary of the Passing of the Greatest Holy Leaf and Inauguration of the Seat of the Universal House of Justice
26 JULY 1982

To all National Spiritual Assemblies

Dear Bahá'í Friends,

334.1 We have been asked by the Universal House of Justice to share with you news of the observances at the World Center which marked the Fiftieth Anniversary of the passing of the Greatest Holy Leaf.

334.2 Before gathering around her monument for prayers, at the hour coincident with that of her passing, the friends at the World Center visited the Holy Shrines at midnight for recitation of the Tablets of Visitation. This observance was followed by an all-day seminar, on Saturday, July 17, held in the main Hall on the ground floor of the Seat of the Universal House of Justice.[1]

334.3 It was a source of deep satisfaction to the Universal House of Justice that the first gathering of the friends in this beautiful setting should have been such a meeting, dedicated to the memory of the Greatest Holy Leaf, whose resting-place is embosomed in the heart of that consecrated spot enfolded by the "far-flung arc" on which now already stand two of those structures constituting the World Administrative Center of the Faith.[2]

334-1. The monument to the Greatest Holy Leaf (Bahíyyih Khánum) is located in the Monument Gardens near the Shrine of the Báb. For a full account of the commemoration held at midnight on 14 July 1982, see BW 18:53–54; for the Tablets of Visitation said at the Shrines of the Báb and 'Abdu'l-Bahá, see BP, pp. 230–35; for excerpts from the address given by the member of the Universal House of Justice Mr. 'Alí Nakhjavání at the all-day seminar, see BW 18:59–66. For the address on the life and services of the Greatest Holy Leaf given by Dr. Bahíyyih Nakhjavání at the Montreal Conference in September 1982, see BW 18:68–73.

334-2. For a discussion of the Arc, see the glossary.

This highly befitting inauguration of the use of the building, held in the presence of the three Hands of the Cause residing in the Holy Land,[3] is a prelude to the fast-approaching day when the Universal House of Justice and its several Departments will occupy this majestic structure.

With loving Bahá'í greetings,
DEPARTMENT OF THE SECRETARIAT

335
Execution of 'Abbas-'Alí Ṣádiqípúr in Shiraz
2 AUGUST 1982

To all National Spiritual Assemblies

SADDENED REPORT YET ANOTHER STALWART SUPPORTER GREATEST NAME IRAN 'ABBAS-'ALÍ ṢÁDIQÍPÚR EXECUTED 15 JULY SHIRAZ. CHARGES PUBLISHED LEADING NEWSPAPER IRAN WERE BASED HIS CONNECTION FAITH, INCLUDING BEING ACTIVE BAHÁ'Í. SUCH CHARGES SIMILAR THOSE RESULTING EXECUTION OTHER PROMINENT BAHÁ'ÍS CONFIRM EVIL INTENTIONS AGAINST FRIENDS CRADLE FAITH. . . .

UNIVERSAL HOUSE OF JUSTICE

336
Elucidation of the Principle of Confidentiality
2 AUGUST 1982

To a National Spiritual Assembly

Dear Bahá'í Friends,

The Universal House of Justice has received your letter of 17 June 1982 and has instructed us to send you the following reply.

The House of Justice is very sorry to learn that the problems which you instance have become a threat to the unity of your Assembly, and it hopes and prays that this difficulty will quickly be overcome. It feels that there are a number of distinct but related principles which are involved in the situations you describe, and that the issues will become clearer if they are considered separately.

Every institution in the Faith has certain matters which it considers should be kept confidential, and any member who is privy to such confidential information is obliged to preserve the confidentiality within the institution where

---

334-3. Amatu'l-Bahá Rúḥíyyih Khánum, 'Alí-Akbar Furútan, and Paul Haney.

he learned it. Such matters, however, are but a small portion of the business of any Bahá'í institution. Most subjects dealt with are of common interest and can be discussed openly with anyone. Where no confidentiality is involved the institutions must strive to avoid the stifling atmosphere of secrecy; on the other hand, every believer must know that he can confide a personal problem to an institution of the Faith, with the assurance that knowledge of the matter will remain confidential.

336.4 Members of Assemblies, whether they are assistants or not, are obviously in a position to receive confidential information as individuals from several sources. It is an important principle of the Faith that one must not promise what one is not going to fulfill. Therefore, if a Bahá'í accepts confidential information either by virtue of his profession (e.g., as a doctor, a lawyer, etc.), or by permitting another person to confide in him, he is in duty bound to preserve that confidentiality.

336.5 In the relationship between assistants and the National Spiritual Assembly no problems should arise, because the functions are entirely separate. An assistant is appointed by an Auxiliary Board member to help him in a specified area of the territory and he functions as an assistant only in relation to that area. Assistants, like Auxiliary Board members, function individually, not as a consultative body. Assistants who are members of a National Assembly or a national committee do not function as assistants in relation to that body, and they have the same duty to observe the confidentiality of its consultations, and of matters considered by the Assembly to be confidential, as does any other member. An assistant can, of course, be a member of a Local Spiritual Assembly, but his task here as an assistant is to help the Spiritual Assembly to function harmoniously and efficiently in the discharge of its duties and this will hardly succeed if he gives the Assembly the feeling that he is reporting privately everything it does to the Auxiliary Board member. He should, on the contrary, do all he can to foster an atmosphere of warm and loving collaboration between the Local Assembly and the Board member.

336.6 In answer to your fourth question the House of Justice instructs us to say that an element of judgment is required in deciding what are and what are not "administrative" matters. Immoral actions of believers, for example, generally become subjects for administrative action only when they are blatant or flagrant, and reflect on the good name of the Faith. If a believer turns to an assistant or Auxiliary Board member for advice on a personal matter it is for the assistant or Auxiliary Board member to decide whether he should advise the believer to turn to his Spiritual Assembly, whether he should himself give advice and, in either case, whether he should report the matter to the Counselors, or to the Local Assembly, which, of course, would depend upon the degree of confidentiality he had undertaken to observe. Likewise, it is for the Counselor to decide whether it is a matter of which the National Assembly should be informed. All this is, of course, within the general context that, apart from matters which ought to remain confidential, the more freely information is shared between the institutions of the Faith the better.

336.7 National Assembly members themselves must exercise such discretion, and it should be clear to the believers that they are not justified in assuming that because a matter is known to individual members of the Assembly it is therefore before the Assembly itself. If a believer wishes to bring a matter to the Assembly's attention he should do so explicitly and officially. If a member of the Assembly knows of a personal problem, and if he has not undertaken to keep it confidential, he may bring it to the Assembly's attention if he feels it would be in the interests of the Faith for him to do so, but he is not obliged to.

336.8 The House of Justice does not wish to elaborate these comments beyond the above, and believes that your Assembly will be able to answer the questions that you pose in the light of these principles. The House of Justice will pray in the Holy Shrines for speedy strengthening of the unity of your Assembly and for the growth of closer collaboration between your body and the Board of Counselors.

With loving Bahá'í greetings,
DEPARTMENT OF THE SECRETARIAT

337

Message to the International Conference in Quito, Ecuador—6–8 August 1982

6 AUGUST 1982

To the Followers of Bahá'u'lláh gathered at
the International Conference in Quito, Ecuador

Beloved Friends,

337.1 We hail with joyous hearts and eager anticipation the soldiers of Bahá'u'lláh's army of light gathered together in Quito, the capital city of the Republic of Ecuador, to do honor and homage to the blessed memory of Bahíyyih Khánum, the Greatest Holy Leaf, the most outstanding heroine of the Bahá'í Dispensation, the fiftieth anniversary of whose ascension was so recently commemorated throughout the world.[1]

337.2 Conscious of the beloved Master's plea to promulgate the oneness of mankind to a spiritually impoverished humanity, inspired by the memory of the Hand of the Cause Dr. Raḥmatu'lláh Muhájir whose mortal remains are

---

337-1. The Greatest Holy Leaf, the daughter of Bahá'u'lláh, was born in Iran in 1846 and died in Haifa at 1:00 A.M. on 15 July 1932.

interred in the soil of Quito, and deriving spiritual stimulus from the Mother Temple for Latin America,² the friends are reminded of the galvanizing words of our beloved Guardian addressed to "the eager, the warm-hearted, the spiritually minded and staunch members of these Latin American Bahá'í communities": "Let them ponder the honor which the Author of the Revelation Himself has chosen to confer upon their countries, the obligations which that honor automatically brings in its wake, the opportunities it offers, the power it releases for the removal of all obstacles, however formidable, which may be encountered in their path, and the promise of guidance it implies . . ."³

337.3 Praiseworthy indeed are the achievements thus far made by the communities of South and Central America and the islands of the Caribbean in the first half of the Seven Year Plan. Full advantage should be taken of the current high tide of proclamation engendered by the crisis in Iran to attract to the Cause of Bahá'u'lláh earnest and seeking souls from every stratum of society, thereby enriching the spiritual and material diversity of our communities. Great effort should be made to utilize more fully the valuable possibilities of radio and television as a means of reaching the vast multitudes whose hearts and minds offer fertile soil for the planting of the seeds of the Faith. All elements of the Bahá'í community, particularly the women and youth, should arise as one soul to shoulder the responsibilities laid upon them. All outstanding goals of the Seven Year Plan should be pursued with enthusiasm and assurance of their accomplishment.

337.4 All National Spiritual Assemblies during the remaining fast-fleeting years of this radiant century, in collaboration with the Institutions of the Faith standing ready and eager to assist them, must greatly reinforce the foundations of maturing National and Local Spiritual Assemblies to enable them to cope successfully with the multifarious and challenging problems that will confront them.

337.5 At a moment in Bahá'í history when the persecuted, beleaguered friends in the Cradle of the Faith heroically continue to face the trials ordained for them in the Major Plan of God, meeting martyrdom, as need be, with joyous acceptance, it behooves the friends throughout the Bahá'í world to endeavor by their own greatly increased acts of self-abnegation to make fruitful the spiritual energies released by the sacrifices of their stricken brethren.

337.6 May you all immerse yourselves in the spirit of the saintly life of the Greatest Holy Leaf, whose self-sacrificing devotion to her beloved Father's Cause is a worthy example for every believer to emulate.

THE UNIVERSAL HOUSE OF JUSTICE

337-2. Dr. Muhájir died in Quito in October 1979. The Mother Temple of Latin America is outside of Panama City, Panama.
337-3. CF, p. 19.

338

Message to the International Conference in Lagos, Nigeria—19–22 August 1982
19 August 1982

To the Friends gathered at the Bahá'í International Conference at Lagos
Dearly loved Friends,

With hearts overflowing with love for the people of Africa, so richly endowed with the gifts of the spirit, so abundantly and repeatedly blessed since the dawn of this Revelation, and so gloriously promising in the unfoldment of their hidden potentialities, we welcome the friends gathered at this Conference held in one of the most important capitals of their emergent continent. 338.1

As we review the annals of our Faith we see that since the days of the Blessed Beauty and up to the early 1950s, the activities of the friends in Africa had produced the formation of one National Spiritual Assembly with its seat in Cairo, Egypt, the opening of 12 countries to the light of the Faith, and some 50 localities established throughout its vast lands. It was at such a time that the beloved Guardian ushered in the first African Teaching Plan, to be followed during the remaining years of his ministry and in subsequent years after his passing, by a series of challenging and bravely executed plans designed to implant the banner of the Faith throughout the length and breadth of that continent and its neighboring islands. Today, after the lapse of a little over three decades, we stand in awe as we view with admiration one of the most valiant contingents of the Army of Light, guided by its own Board of Counselors, led and administered by 37 National Spiritual Assemblies and 4990 Local Spiritual Assemblies, privileged to serve an eager and radiant community of believers drawn from 1152 African tribes residing in 29,000 localities. 338.2

How wonderful that it has been possible to convene this Conference on African soil with such a large number of African friends in attendance, in loving memory of the most distinguished heroine of the Bahá'í Dispensation, the eldest daughter of the King of Glory, who lived a long life of sacrificial service to the Cause of her Beloved Father. Her meekness, her unassuming nature, the purity of her soul, the sensitivity of her heart, the calmness of her demeanor, her patience and long-suffering in trials, and above all, her unshakable faith, her tenderness and love, and the spirit of self-renunciation which she evinced throughout her blessed life, are outstanding characteristics that we can well emulate, particularly in Africa, where these heavenly qualities play such an important part in attracting the souls and winning the hearts to the Cause of Bahá'u'lláh. 338.3

We rejoice in the knowledge that some communities have already initiated in her name teaching and consolidation campaigns of far-reaching magnitude; 338.4

that many Bahá'í women, inspired by her example, are accepting an ever-greater share of responsibility in running the affairs of the community; and that numerous newsletters are reflecting eulogies of the station she occupied, the sufferings she endured, and the heroism she demonstrated in her love for the glorious Cause of her Lord.

338.5 The fortunes of the Seven Year Plan in Africa are in the balance. As we draw near to the midway point in the unfoldment of the processes it has set in motion, we call upon its valiant promoters on the African mainland and its surrounding islands, to take stock of their position, to reappraise their progress, and to concentrate their resources on whatever portions of the goals are as yet unachieved. Chief among its objectives are a widespread recruitment of many more supporters of the Most Great Name, the deepening of the individual believers, for the fulfillment of all goals ultimately depends upon them, and a notable increase in the number of newly formed as well as firmly rooted Local Spiritual Assemblies, to serve as bases for the manifold activities of the community, including the Bahá'í education of children, a greater participation of women and youth in Bahá'í activities, and the formulation of ways and means to enrich the spiritual lives of the "noble" and "purehearted" believers of a "FAST-AWAKENING CONTINENT."[1]

338.6 May the participants in this Conference carry to the mass of their devoted fellow believers, whose personal circumstances have made it impossible for them to attend, the spirit of joy and optimism which we hope will be generated at this gathering and the flames of enthusiasm which we pray will be enkindled in their hearts.

338.7 May the memory of the Greatest Holy Leaf, who through her life of heroic self-sacrifice has left to us "a legacy that time can never dim," inspire the friends in every country of the continent to rededicate themselves to the Cause of God, not to allow any opportunity for mentioning the Faith to slip by unutilized, and not to permit one day of their lives to pass without a noble effort to draw nearer to the good pleasure of the Blessed Beauty.[2]

338.8 Our fervent prayers surround you as you proceed with your deliberations.

THE UNIVERSAL HOUSE OF JUSTICE

338-1. Shoghi Effendi, letter dated 11 May 1954 to the Uganda Teaching Committee; Shoghi Effendi, letter dated 5 January 1953 to the National Spiritual Assembly of the Bahá'ís of the United States; UD, p. 290; MBW, p. 93.

338-2. BA, p. 187. For more information on the Greatest Holy Leaf, see the entry on Bahíyyih Khánum in the glossary.

339
Execution of 'Alí Na'ímíyán in Urúmíyyih
30 August 1982

To all National Spiritual Assemblies

WITH SORROWFUL HEARTS ANNOUNCE EXECUTION ON 11 AUGUST IN URÚMÍYYIH 339.1
ACTIVE BAHÁ'Í 'ALÍ NA'ÍMÍYÁN AFTER BEING IMPRISONED ONE YEAR.

PRESSURES INTENSIFYING AGAINST BELIEVERS IRAN. . . . 339.2

UNIVERSAL HOUSE OF JUSTICE

340
Message to the International Conference in Montreal, Canada—3–5 September 1982
2 September 1982

To the Friends gathered at
the Bahá'í International Conference in Montreal

Dearly loved Friends,

Seventy years ago 'Abdu'l-Bahá visited Montreal,[1] hallowing it forever. The visit of the beloved Master to America, the laying by Him of the cornerstone of the first Mashriqu'l-Adhkár of the West and the revelation by Him five years later of the Tablets of the Divine Plan, which invest its chief executors and their allies with spiritual primacy, constitute successive stages in the gradual disclosure of a mission whose seeds can be found in the Báb's address to the people of the West, urging them to aid God's Holy Cause.[2] This mission was given specific direction through Bahá'u'lláh's summons to the rulers of America, calling on them to heal the injuries of the oppressed and, with the "rod of the commandments" of their Lord, to bring their corrective influence to bear upon the injustices perpetrated by the tyrannical and the ungodly. 'Abdu'l-Bahá revealed in clearer details than those given by either the Báb or Bahá'u'lláh the nature and scope of that glorious mission. In His eternal 340.1

340-1. 30 August–8 September 1912.
340-2. 'Abdu'l-Bahá visited North America from 11 April through 5 December 1912. On 1 May 1912 He laid the cornerstone of the House of Worship in Wilmette, Illinois. He revealed the Tablets of the Divine Plan between 26 March 1916 and 8 March 1917 after returning to the Holy Land. Shoghi Effendi writes that the mandate 'Abdu'l-Bahá gave to the Bahá'ís of North America in the Tablets of the Divine Plan "may be said to have in a sense originated with the mandate issued by the Báb in his 'Qayyúmu'l-Asmá,' [Commentary on the Súrih of Joseph] one of His earliest and greatest works, . . . directed specifically to the 'peoples of the West' to 'issue forth' from their 'cities' and aid His Cause" (MA, p. 90).

Tablets unveiling America's spiritual destiny the Master wrote, "The continent of America is, in the eyes of the one true God, the land wherein the splendors of His light shall be revealed, where the mysteries of His Faith shall be unveiled, where the righteous will abide and the free assemble. Therefore, every section thereof is blessed . . ." and, referring to Canada, He asserted that its future "is very great, and the events connected with it infinitely glorious." Even more specifically, He expressed the "hope that in the future Montreal may become so stirred, that the melody of the Kingdom may travel to all parts of the world from that Dominion and the breaths of the Holy Spirit may spread from that center to the East and the West of America."[3]

340.2 After the passing of 'Abdu'l-Bahá and under the guidance of the Guardian the Bahá'ís of the world witnessed with awe and admiration the North American community arising as one man to champion the Administrative Order taking shape on their own soil, to embark upon the first collective teaching plan in the annals of the Faith, to lead the entire Bahá'í world in intercontinental teaching campaigns, to demonstrate with devotion their exemplary firmness in the Covenant, to extend their support and protection and relief to the oppressed followers of Bahá'u'lláh throughout the East and particularly in His native land, and to send forth valiant pioneers and traveling teachers to every continent of the globe. These marvelous and noble exertions, calling for the expenditure of resources almost beyond their means, paved the way for the achievement of glorious victories which synchronized with a series of world convulsions, signs of universal commotion and travail, and with repeated crises within the Faith.[4] And in this day, while the blood of the martyrs of Persia is once again watering the roots of the Cause of God and when the international outlook is impenetrably and ominously dark, the Bahá'ís of North America are in the van of the embattled legions of the Cause.

340.3 Less than a score of years remain until the end of this century which the Master called "the century of light," and He clearly foresaw that ere its ter-

---

340-3. KA ¶88; TDP 9.3, 13.2, 13.4. For a fuller account of Bahá'u'lláh's summons to the kings and rulers of the world, see messages dated Riḍván 1965 (no. 24), Riḍván 1967 (no. 42), and October 1967 (no. 46).

340-4. Foremost among the victories won by the North American Bahá'ís during the first and second Seven Year Plans (1937–44 and 1946–53) was the establishment of the Administrative Order in Central and South America and Europe. The North American Bahá'ís have also played a major role in establishing the Administrative Order in Africa, Asia, and Australasia. The series of world convulsions with which the Plans synchronized included World War II, the outbreak of which threatened the security of the World Center itself. The German Bahá'ís faced grave danger throughout the Nazi period; their activities as a community were banned, and some of their members were killed. Instances of crises within the Faith were those instigated by the relatives of Shoghi Effendi who rebelled against his authority as Guardian of the Cause of God and were ultimately expelled from the Faith as Covenant-breakers. Also troublesome during the 1930s and 1940s were the activities of a former secretary to 'Abdu'l-Bahá, Ahmad Sohrab, who also violated the provisions of Bahá'u'lláh's Covenant and was expelled from the Faith by Shoghi Effendi.

mination an advanced stage would have been reached in the striving towards the political, racial, and religious unity of the peoples of the world, unfolding new horizons in scientific accomplishments, universal undertakings and world solidarity.[5] The calls of the Master and the Guardian plainly summon the Bahá'ís of the Americas to prodigies of proclamation, of teaching and of service. The American melting pot of peoples needs the unifying power of the new Faith of God to achieve its fusion. The representative character of the Bahá'í community should therefore be reinforced through the attraction, conversion and support of an ever-growing number of new believers from the diverse elements constituting the population of that vast mainland and particularly from among Indians and Eskimos about whose future the Master wrote in such glowing terms.[6] In the glorious freedom which enables you to proclaim, to teach and confirm, to educate and deepen yourselves and others in the verities of the Faith, you have precious opportunities of service denied to many of your fellow believers elsewhere. If your blessed communities are to lead the world spiritually, as the Master envisaged,[7] then the Faith must strike deeper roots in your hearts, the spirit of its teachings must be exemplified in ever greater measure in your lives, and God's Holy Cause must be taught and proclaimed with ever greater intensity. In His immortal Tablets addressed to the Bahá'ís of North America 'Abdu'l-Bahá assures each one of you that "whosoever arises in this day to diffuse the divine fragrances, the cohorts of the Kingdom of God shall confirm him . . ."[8]

340.4 You are met in this Conference to review the progress of the Seven Year Plan, to be confirmed, galvanized and sent into action. It is not enough for the North American believers to stand at the forefront of the Bahá'í world; the scope of their exertions must be steadily widened. In the words of 'Abdu'l-Bahá, "The range of your future achievements still remains undisclosed. I fervently hope that in the near future the whole earth may be stirred and shaken by the results of your achievements." "Exert yourselves; your mission is unspeakably glorious. Should success crown your enterprise, America will assuredly evolve into a centre from which waves of spiritual power will emanate . . ."[9] The valiant countries of North America should in the second half of the Seven Year Plan ensure that an ever-swelling number of pioneers and traveling teachers will arise and travel to and settle in countries which need their support, however inhospitable the local conditions may be, ceaselessly endeavoring to contribute to the expansion of the teaching work and the strengthening of the foundations of the communities they are called upon

---

340-5. SWAB, p. 32.
340-6. See TDP 6.8.
340-7. See WOB, pp. 75–76.
340-8. TDP 7.7.
340-9. TDP 7.4, 11.11.

to assist. They should, moreover, continue their defence of the downtrodden, open their doors to their Bahá'í brethren who are seeking refuge in their lands, provide technological expertise to communities which need it, and supply an uninterrupted flow of resources to support the ever-increasing international projects of the Faith.

340.5 In their respective homefronts the Bahá'ís of North America should intensify the drive to attract the masses to God's Holy Cause, to provide the means for their integration into the work of the Faith, and should become standard-bearers of an embryonic Bahá'í society which is destined to gradually emerge under the influence of the integrating and civilizing forces emanating from the Source of God's Revelation. Such noble objectives cannot be fully achieved unless and until local communities become those collective centers of unity ordained in our Writings, and every individual earnestly strives to support the structure and ensure the stability of the Administrative Edifice of the Faith.

340.6 How fitting that this Conference, and the one held for Bahá'í children on a scale unprecedented in North America, should commemorate the fiftieth anniversary of the passing of Bahíyyih Khánum, the Greatest Holy Leaf, whose love for the North American believers and whose admiration for their heroism were so deep and so sustained and whose natural fondness for children was so characteristic of Bahá'u'lláh.[10] May each of you emulate her unswerving devotion and loyalty to the Covenant of God and her perseverance in the path of His love. We shall mark the first day of your Conference, together with the one being held concurrently in Canberra, with prayers at the Holy Shrines that all may "be assisted in . . . service and, like unto brilliant stars, shine in these regions with the light of . . . guidance."[11]

With loving Bahá'í greetings,
THE UNIVERSAL HOUSE OF JUSTICE

340-10. For the address given by Dr. Bahíyyih Nakhjavání at the Montreal Conference on the life and services of the Greatest Holy Leaf, see BW 18:68–73.
340-11. TDP 9.13.

341

Message to the International Conference in Canberra, Australia—2–5 September 1982
2 September 1982

To the Friends gathered in the
Asian-Australasian Bahá'í Conference in Canberra

Dearly loved Friends,

These are momentous times. The institutions of the old world order are 341.1
crumbling and in disarray. Materialism, greed, corruption and conflict are
infecting the social order with a grave malaise from which it is helpless to
extricate itself. With every passing day it becomes more and more evident that
no time must be lost in applying the remedy prescribed by Bahá'u'lláh, and
it is to this task that Bahá'ís everywhere must bend their energies and commit
their resources.

New conditions now present themselves, making it easier to accomplish 341.2
our purpose. Galvanized by the fires of fierce opposition and nurtured by the
blood of the martyrs, the forces of the Cause of Bahá'u'lláh are, at long last,
emerging from obscurity. Never before in history has the Faith been the subject of such universal attention and comment. Eminent statesmen, parliamentarians, journalists, writers, educators, commentators, clergymen and
other leaders of thought have raised their voices and set their pens to expressions of horror and revulsion at the persecutions of our brethren in Iran on
the one hand, and to paeans of praise and admiration of the noble principles
which motivate the followers of the Most Great Name on the other.

The five international conferences of the Seven Year Plan were called to 341.3
commemorate the fiftieth anniversary of the passing of the Greatest Holy
Leaf,[1] to discuss anew the present condition of the Faith in a turbulent world
society, to examine the great opportunities for its future growth and development, and to focus attention on the unfulfilled goals of the Plan. We are certain that the contemplation of the gathered friends on the sterling qualities
which distinguished the heroic life of the Greatest Holy Leaf will help them
to persevere in their noble endeavors.

This particular Conference is unique in many ways. The geographical area 341.4
of concern spans over half the globe, including within its purview all the vast
continent of Asia as well as the water hemisphere which comprises all of Australasia. Within the continent of Asia is the "cradle of the principal religions
of mankind . . . above whose horizons, in modern times, the suns of two

---

341-1. For information on the Greatest Holy Leaf, see the entry on Bahíyyih Khánum in the glossary.

independent Revelations . . . have successively arisen . . . on whose Western extremity the Qiblih of the Bahá'í world has been definitely established . . ."[2] The first Mashriqu'l-Adhkár of the Bahá'í world was erected on this continent under the direction of 'Abdu'l-Bahá, and now another is arising on the Indian subcontinent in the midst of the world's largest Bahá'í community. In Australasia the Mother Temple of the Antipodes, dedicated to the Glory of God just two decades ago, looks out across the vast Pacific Ocean in whose "midmost heart" still another Mashriqu'l-Adhkár is being built on the mountain slope above Apia in the country of the first reigning monarch to embrace the Faith of Bahá'u'lláh.[3]

341.5　　The population of Asia and Australasia is well over half the world population. The area includes Asiatic U.S.S.R. and mainland China, accounting for more than one thousand million souls who are, for the most part, untouched by the Revelation of Bahá'u'lláh. Obviously present conditions in these areas call for the exercise of the utmost wisdom and circumspection. Yet this vast segment of humanity cannot be ignored.

341.6　　Canberra, where you are now meeting, is at the southern pole of the spiritual axis referred to in the beloved Guardian's last message to the Bahá'ís of Australia as "extending from the Antipodes to the northern islands of the Pacific Ocean."[4] Referring to the National Spiritual Assemblies at the northern and southern poles of that axis, Shoghi Effendi went on to say:

341.6a　　A responsibility, at once weighty and inescapable, must rest on the communities which occupy so privileged a position in so vast and turbulent an area of the globe. However great the distance that separates them; however much they differ in race, language, custom, and religion; however active the political forces which tend to keep them apart and foster racial and political antagonisms, the close and continued association of these communities in their common, their peculiar and paramount task of raising up and of consolidating the embryonic World Order of Bahá'u'lláh in those regions of the globe, is a matter of vital and urgent importance, which should receive on the part of the elected representatives of their communities, a most earnest and prayerful consideration.[5]

341-2. MBW, p. 168.
341-3. Construction of the first Bahá'í House of Worship in 'Ishqábád, Russia, began in 1902 and was completed circa 1921. The House of Worship in India is near New Delhi; it was dedicated 23–27 December 1986. The House of Worship in Australasia, outside of Sydney in Ingleside, was dedicated on 16 September 1961. The extract from the writings of Bahá'u'lláh about the "midmost heart" of the ocean is quoted by Shoghi Effendi in WOB, p. 108. The House of Worship in Apia is in Western Samoa, whose head of state, His Highness Malietoa Tanumafili II, embraced the Bahá'í Faith in 1968.
341-4. LFG, p. 138.
341-5. LFG, pp. 138–39.

These guidelines, penned a quarter of a century ago, are as valid today as 341.7
when they were written, and can be taken to heart by all Bahá'í communities on either side of the axis.

We are approaching the midway point of the Seven Year Plan. As we review 341.8
our accomplishments with respect to the goals of that Plan, it is essential that
we fortify ourselves for the tasks ahead, and that we rededicate ourselves to
that Cause for which our beloved martyrs rendered their last full measure of
devotion. We can do no less!

We shall be with you in spirit during your important deliberations. Our 341.9
prayers ascend at the Holy Threshold for the success of your Conference and
the International Conference being held concurrently in Montreal. We shall
ardently supplicate that the blessings and confirmations of Bahá'u'lláh will
descend upon you and surround you wherever you go in service to His Faith.

With loving Bahá'í greetings,
THE UNIVERSAL HOUSE OF JUSTICE

342

Comments on the First Danish Bahá'í Women's Conference

20 SEPTEMBER 1982

The National Spiritual Assembly of the Bahá'ís of Denmark

Dear Bahá'í Friends,

The Universal House of Justice was very pleased to receive your report of 342.1
the first Danish Women's Seminar, submitted with your letter of 31 August
1982, and also a photograph of those attending.

Judging by the reactions of the participants, this seems to have been a very 342.2
valuable experience for which there was a definite need. From the earliest days
of the Faith women have played a prominent role, and especially in the West
where in the beginning their influence predominated. It is the hope of the
House of Justice that the Bahá'í women of Denmark will exert a growing
influence on the community, increasing its spirituality, forging stronger ties
of unity, consolidating its administrative institutions and increasing the rate
of its expansion throughout Danish society.

The House of Justice assures your Assembly of its prayers at the Sacred 342.3
Threshold for the confirmation of your noble endeavors and for the blessings
of Bahá'u'lláh to reinforce the efforts of every member of the Danish Bahá'í
community.

With loving Bahá'í greetings,
DEPARTMENT OF THE SECRETARIAT

343

Guidelines for Behavior of Children at Feast
14 October 1982

The National Spiritual Assembly of the Bahá'ís of Canada

Dear Bahá'í Friends,

343.1 The Universal House of Justice has received your letter of 14 September 1982 concerning the role of Local Spiritual Assemblies in guiding parents and children in standards of behavior for children at community gatherings, such as Nineteen Day Feasts and Bahá'í Holy Day observances.

343.2 Further to the letter we wrote on its behalf on 28 June 1977, the House of Justice has instructed us to say that children should be trained to understand the spiritual significance of the gatherings of the followers of the Blessed Beauty, and to appreciate the honor and bounty of being able to take part in them, whatever their outward form may be. It is realized that some Bahá'í observances are lengthy and it is difficult for very small children to remain quiet for so long. In such cases one or other of the parents may have to miss part of the meeting in order to care for the child. The Spiritual Assembly can also perhaps help the parents by providing for a children's observance, suited to their capacities, in a separate room during part of the community's observance. Attendance at the whole of the adult celebration thus becomes a sign of growing maturity and a distinction to be earned by good behavior.

343.3 In any case, the House of Justice points out that parents are responsible for their children and should make them behave when they attend Bahá'í meetings. If children persist in creating a disturbance they should be taken out of the meeting. This is not merely necessary to ensure the properly dignified conduct of Bahá'í meetings but is an aspect of the training of children in courtesy, consideration for others, reverence, and obedience to their parents.

With loving Bahá'í greetings,
DEPARTMENT OF THE SECRETARIAT

344

Intensification of Oppression of Iranian Bahá'ís
18 November 1982

To all National Spiritual Assemblies

344.1 CRUEL SYSTEMATIC OPPRESSION LONG-SUFFERING IRANIAN FRIENDS HAS REACHED NEW LEVEL INTENSITY. DWINDLING SOURCES THEIR LIVELIHOOD FURTHER SEVERELY CURTAILED. DISMISSAL FROM JOBS, CANCELLATION TRADE LICENSES, CONFISCATION PRIVATE PROPERTIES UNABATED. HOMELESS THOUSANDS THROWN

ON MERCY RELATIVES FRIENDS. DOORS SCHOOLS CLOSED TO INCREASING NUMBER CHILDREN. FRESH BLOOD WANTONLY SPILLED AFTER LULL EXECUTIONS.

344.2 WITH HEAVY HEARTS ANNOUNCE ḤABÍBU'LLÁH AWJÍ, ENTHUSIASTIC ACTIVE BELIEVER, HANGED SHIRAZ 16 NOVEMBER. RECENT EVIDENCE CONFIRMS MARTYRDOM SOME TIME AGO OF TWO STEADFAST UPHOLDERS CAUSE: YADU'LLÁH SIPIHRARFA' EXECUTED BY FIRING SQUAD TEHRAN, MANÚCHIHR VAFÁ'Í MURDERED IN HIS HOME TEHRAN BY UNKNOWN ASSAILANT WHO ATTACHED NOTE TO BODY GIVING AS REASON FOR DASTARDLY DEED INNOCENT VICTIM'S BAHÁ'Í BELIEF. . . .

Universal House of Justice

345
Execution of Dr. Ḍíyá'u'lláh Aḥrárí in Shiraz
23 November 1982

To all National Spiritual Assemblies

345.1 HEARTS SORELY GRIEVED EXECUTION DOCTOR ḌÍYÁ'U'LLÁH AḤRÁRÍ BY FIRING SQUAD SHIRAZ 21 NOVEMBER. TRAGEDY OCCURRING SO QUICKLY UPON HEELS RECENTLY ANNOUNCED HANGING ḤABÍBU'LLÁH AWJÍ AROUSES DEEP CONCERN FATE THREE OTHER FAITHFUL SOULS TOGETHER WITH WHOM THESE TWO WERE CONDEMNED TO DEATH BY RELIGIOUS COURT SHIRAZ LAST SEPTEMBER. JUDGE AT TRIAL OFFERED ALL FIVE 30 MINUTES TO RECANT AND BE FREE OR FACE DEATH SENTENCE. THEY INSTANTLY REAFFIRMED THEIR FAITH. FRESH MARTYRDOMS HIGHLIGHT INTENSITY PERSECUTIONS SHIRAZ WHERE WITHIN MONTHS 40 FRIENDS WERE ROUNDED UP AND IMPRISONED ADDING TO STILL OTHERS PREVIOUSLY ARRESTED. . . .

Universal House of Justice

346
Passing of the Hand of the Cause of God Paul Haney
5 December 1982

To National Spiritual Assemblies

346.1 WITH STRICKEN HEARTS ANNOUNCE SUDDEN IRREPARABLE LOSS THROUGH AUTOMOBILE ACCIDENT 3 DECEMBER HIGHLY DISTINGUISHED GREATLY PRIZED HAND CAUSE GOD STAUNCH DEFENDER COVENANT PAUL HANEY.[1]

346.2 THIS DISTINGUISHED SERVANT BAHÁ'U'LLÁH WAS BLESSED CHILDHOOD THROUGH ATTAINMENT PRESENCE 'ABDU'L-BAHÁ. HIS NATURAL GENTLENESS, GENUINE HUMILITY, UNAFFECTED UNBOUNDED LOVE, HIS UPRIGHTNESS, INTEGRITY,

---

346-1. For an account of the life and services of Paul Haney, see BW 18:613–18.

HIS SINGLE-MINDED DEVOTION CAUSE SINCE YOUTHFUL YEARS, HIS UNFAILING RELIABILITY, METICULOUS ATTENTION DETAIL, CHARACTERIZED HIS HISTORIC SERVICES BOTH NATIONAL AND INTERNATIONAL LEVELS. SPANNING MORE THAN HALF CENTURY HIS TIRELESS LABORS INCLUDED LONGTIME MEMBERSHIP AMERICAN NATIONAL ASSEMBLY. SINCE 1954 HE CONSECRATED HIS ENERGIES AS MEMBER UNIQUE COMPANY CHIEF STEWARDS FAITH AND LATER AS MEMBER BODY HANDS CAUSE RESIDING HOLY LAND AT ONE OF MOST CRITICAL PERIODS BAHÁ'Í HISTORY. LAST DECADE HIS EARTHLY LIFE WAS FULLY DEDICATED DEVELOPMENT NEWLY FORMED INTERNATIONAL TEACHING CENTER. GENERATIONS YET UNBORN WILL GLORY IN HIS IMPERISHABLE ACHIEVEMENTS AND BE INSPIRED BY HIS UNIQUE FORTITUDE.

346.3 ARDENTLY SUPPLICATING HOLY THRESHOLD PROGRESS HIS NOBLE SOUL ABHÁ KINGDOM. ADVISE HOLD THROUGHOUT BAHÁ'Í WORLD INCLUDING ALL MAS͟HRIQU'L-AD͟HKÁRS MEMORIAL GATHERINGS BEFITTING HIS HIGH RANK AND HIS MERITORIOUS SERVICES.

UNIVERSAL HOUSE OF JUSTICE

347
Imprisonment of More Than Eighty Bahá'ís in Shiraz
6 December 1982

To all National Spiritual Assemblies

347.1 DISTRESSED FURTHER ARRESTS SHIRAZ INCREASING NUMBER IMPRISONED FRIENDS TO MORE THAN 80 LAST WEEK. . . .

UNIVERSAL HOUSE OF JUSTICE

348
Establishment of First Bahá'í Radio Station in North America
13 December 1982

To the National Spiritual Assembly of the Bahá'ís of United States

348.1 HEARTY CONGRATULATIONS ACHIEVEMENT LONG AWAITED STEP ESTABLISHMENT FIRST BAHÁ'Í RADIO STATION NORTH AMERICA ENABLING YOU PIONEER MODEL PROGRAM CONSOLIDATION UPLIFTMENT BAHÁ'Í COMMUNITY SOUTH CAROLINA.[1] PRAYING YOUR WISE MOBILIZATION BOUNTEOUS RESOURCES FOR MEDIA DEVELOPMENT IN UNITED STATES BEHALF BELOVED FAITH PROVIDING FURTHER IMPETUS TEACHING SOUTHERN STATES.

LOVING GREETINGS,
UNIVERSAL HOUSE OF JUSTICE

348-1. The radio station, the call-letters of which are WLGI, is located at the Louis Gregory Bahá'í Institute in Hemingway, South Carolina.

349

Torture of Bahá'ís Imprisoned in Shiraz
28 December 1982

To National Spiritual Assemblies

Dear Bahá'í Friends,

The Universal House of Justice has received distressing news that the friends who are imprisoned in Shiraz are being tortured. It is not known how many Bahá'ís are being subjected to this cruelty or what form of punishment is involved. Reliable reports indicate, however, that the reason for the torture is to force them to recant their faith and induce them to give the names of active members of the Bahá'í community. . . .

With loving Bahá'í greetings,
Department of the Secretariat

349.1

350

Call for Another Day of Prayer
29 December 1982

To all National Spiritual Assemblies

Dear Bahá'í Friends,

As you know, the persecution of our friends in Iran continues unabated, and our oppressed brethren are bearing this ordeal with resignation and valor. The Universal House of Justice feels that the time has come to once again call for a day of prayer, and it asks that you inform the members of your community to offer special prayers on the Feast of Jamál (Beauty) on Thursday, 28 April 1983.

350.1

The House of Justice is confident that the friends will also want to pray for the believers in Iran during the Fast, particularly at dawn, and it would be useful for National Spiritual Assemblies to remind the friends of the privilege of joining universally in prayers during this period about which Bahá'u'lláh has written, "Thou hast endowed every hour of these days with a special virtue, inscrutable to all except Thee, . . ."[1]

350.2

In addition, your Assembly should set aside time during your National Convention for such prayers, as will be done at the International Convention.

350.3

It is hoped that these manifold supplications, reverberating from every corner of the globe, will attract in ever greater measure the blessings of the

350.4

350-1. PM, p. 143.

Almighty upon His lovers in the Cradle of His Faith, and will reinvigorate our determination to accelerate the tempo of our activities, in order to compensate for and indeed more than offset the setbacks being suffered by our beleaguered co-workers in Iran.

With loving Bahá'í greetings,
DEPARTMENT OF THE SECRETARIAT

351
Execution of Hidáyat Síyávu<u>sh</u>í in Shiraz
3 JANUARY 1983

To all National Spiritual Assemblies

351.1 GREATLY DISMAYED LAMENTABLE NEWS EXECUTION HIDÁYAT SÍYÁVU<u>SH</u>Í HANGED SHIRAZ JANUARY 1, THIRD OF FIVE BAHÁ'ÍS SENTENCED TO DEATH 23 SEPTEMBER 1982. . . .

UNIVERSAL HOUSE OF JUSTICE

352
Prayer of 'Abdu'l-Bahá for Martyred Friends and Relatives
11 JANUARY 1983

To all National Spiritual Assemblies

Dear Bahá'í Friends,

352.1 We have been directed by the Universal House of Justice to send you the enclosed copy of a prayer revealed by 'Abdu'l-Bahá in which mention is made of persecutions, the martyred friends and their relatives.[1] This is for general use, and it is felt that it would be specially suitable for inclusion in the program of prayers being arranged by National Spiritual Assemblies all over the world for the Feast of Jamál (Beauty) on Thursday, 28 April 1983, as was requested in our letter of 29 December 1982.[2]

352.2 A copy of the text in the original Arabic is also enclosed so that it may be shared with those friends who may be able to use it.

With loving Bahá'í greetings,
DEPARTMENT OF THE SECRETARIAT

352-1. The prayer is published in English in BP, pp. 265–67.
352-2. See message no. 250.

353

Release of a Compilation on the Importance of Deepening

13 January 1983

To all National Spiritual Assemblies
Dear Bahá'í Friends,

353.1 At the request of the Universal House of Justice, and with the aim of fulfilling the goal assigned to all National Assemblies to pursue a program of spiritual enrichment designed to inspire and enlighten the friends regarding the Bahá'í way of life, the Research Department has prepared a compilation from the Writings of Bahá'u'lláh, 'Abdu'l-Bahá and Shoghi Effendi on the importance of deepening our knowledge and understanding of the Faith.[1]

353.2 It is hoped that the loving exhortations and guidelines given in these passages will kindle in the hearts of the believers a spark of longing to delve more deeply into the rich mine of the Writings of the Faith. Such efforts will undoubtedly be abundantly rewarded as the friends attain increased spirituality and understanding, aim at acquiring greater perfection, draw nearer to attracting the good pleasure of their Lord, and seek to become more effective promoters of the Cause of God.

353.3 A copy of this compilation is attached and the Universal House of Justice leaves it to your discretion to determine how best its contents may be shared with the friends under your jurisdiction.

With loving Bahá'í greetings,
DEPARTMENT OF THE SECRETARIAT

354

Occupation of the Seat of the Universal House of Justice

2 February 1983

To all National Spiritual Assemblies
TO THE FOLLOWERS OF BAHÁ'U'LLÁH IN EVERY LAND

354.1 WE BOW OUR HEADS IN INFINITE GRATITUDE TO THE BLESSED BEAUTY FOR HIS ALL-EMBRACING CONFIRMATIONS ENABLING HOUSE JUSTICE OCCUPY ITS NEWLY CONSTRUCTED PERMANENT SEAT. THIS AUSPICIOUS EVENT SIGNALIZES ANOTHER

353-1. See CC 1:187–234.

PHASE IN PROCESS FULFILLMENT SAILING GOD'S ARK ON MOUNTAIN OF THE LORD AS ANTICIPATED IN TABLET CARMEL, WONDROUS CHARTER WORLD SPIRITUAL AND ADMINISTRATIVE CENTERS FAITH BAHÁ'U'LLÁH.[1]

354.2 THIS HIGH POINT HISTORY STRUGGLING FAITH NOW EMERGING FROM OBSCURITY, THIS CRUCIAL HOUR SUFFUSED WITH UNTOLD POTENCIES GENERATED BY SOUL-STIRRING SACRIFICES BELOVED BRETHREN IRAN, MARKED BY VISIT MEMBERS HOUSE JUSTICE TWIN HOLY SHRINES AND FIRST GATHERING COUNCIL CHAMBER TOGETHER WITH HANDS CAUSE AMATU'L-BAHÁ RÚḤÍYYIH KHÁNUM 'ALÍ-AKBAR FURÚTAN AND COUNSELOR MEMBERS INTERNATIONAL TEACHING CENTER TO OFFER PRAYERS HUMBLE THANKSGIVING. LET ALL REJOICE. LET PRAISES ANCIENT BEAUTY RESOUND. MAY UNRELENTING EFFORTS FRIENDS EVERYWHERE HASTEN ADVENT THAT DAY WHEN WONDROUS POTENTIALITIES ENSHRINED IN TABLET CARMEL WILL BE FULLY REVEALED AND WHEN FROM GOD'S HOLY MOUNTAIN, AS ENVISAGED BELOVED GUARDIAN, WILL STREAM FORTH RIVERS OF LAWS AND ORDINANCES WITH ALL-CONQUERING POWER AND MAJESTY.

UNIVERSAL HOUSE OF JUSTICE

355
Execution of Two Bahá'ís in Shiraz
14 MARCH 1983

To all National Spiritual Assemblies

355.1 DISTRESSED ANNOUNCE EXECUTION BY HANGING ON 12 MARCH TWO INNOCENT FRIENDS SHIRAZ YADU'LLÁH MAḤMÚDNIZHÁD AND RAḤMATU'LLÁH VAFÁ'Í. THIS HEINOUS CRIME PERPETRATED ON MORROW PASSAGE RESOLUTION UNITED NATIONS HUMAN RIGHTS COMMISSION EXPRESSING CONCERN VIOLATION FUNDAMENTAL FREEDOMS AND REQUESTING SECRETARY GENERAL CONTINUE EFFORTS SAFEGUARD RIGHTS BAHÁ'ÍS IRAN. . . .

355.2 SUPPLICATING DIVINE THRESHOLD DELIVERANCE BRETHREN IRAN BRAVELY FACING INTENSE CRUELTIES.

UNIVERSAL HOUSE OF JUSTICE

354-1. See GWB, pp. 14–17; TB, pp. 1–5.

356

Execution of Mrs. Ṭúbá Zá'irpúr in Shiraz

15 March 1983

To all National Spiritual Assemblies

FURTHER OUR TELEX 14 MARCH JUST LEARNED DISTRESSING NEWS MRS. ṬÚBÁ ZÁ'IRPÚR WAS EXECUTED BY HANGING TOGETHER TWO FRIENDS MENTIONED OUR PREVIOUS MESSAGE. BODIES OF THREE WERE BURIED IN BAHÁ'Í CEMETERY BY PRISON GUARDS WITHOUT KNOWLEDGE, PRESENCE RELATIVES. EXECUTIONS WERE NOT ANNOUNCED EVEN TO FAMILIES. . . . 356.1

Universal House of Justice

357

Message to Youth Conferences in Costa Rica and Honduras

17 March 1983

The Bahá'í Youth Conferences in Costa Rica and Honduras

WARMLY WELCOME OCCASION SIMULTANEOUS CONFERENCES COSTA RICA AND HONDURAS TO GREET VIBRANT BAHÁ'Í YOUTH CENTRAL AMERICA. YOUR ENTHUSIASTIC EXERTIONS IN SERVICE CAUSE BAHÁ'U'LLÁH AS SHOWN BY SUBSTANTIAL INCREASE YOUR NUMBERS BRING GLADNESS TO OUR HEARTS AND INSPIRE EXHILARATING THOUGHT THAT BRIGHT PROSPECTS SUCCESS LIE IMMEDIATELY BEFORE YOU. 357.1

YOU MEET AT HIGHLY CRITICAL MOMENT HISTORY WHEN TURMOIL ASSOCIATED WITH THIS ERA OF TRANSITION INTENSIFIES. WITHIN CAUSE ITSELF CAN BE SEEN ON ONE HAND UNPRECEDENTED CAMPAIGN PERSECUTION LONG-SUFFERING IRANIAN BRETHREN AND ON OTHER HAND RESOUNDING TRIUMPHS SEVEN YEAR PLAN INDUCED BY THEIR SACRIFICES AND SYMBOLIZED BY OCCUPANCY PERMANENT SEAT UNIVERSAL HOUSE OF JUSTICE. MANKIND RAPIDLY APPROACHES RECKONING WITH BAHÁ'U'LLÁH'S INJUNCTION THAT IT BE UNITED. FROM FAR AND NEAR ANGUISHED MULTITUDES CRY FOR PEACE BUT BEING LARGELY IGNORANT HIS LIFE-REDEEMING MESSAGE THEY FEEL NO HOPE. SITUATION THUS PRESENTS BAHÁ'Í YOUTH WITH GREAT OPPORTUNITIES INESCAPABLE CHALLENGE TO RESCUE THEIR PEERS FROM SLOUGH DESPONDENCY POINTING THEM TOWARDS HOPE-RESTORING BANNER MOST GREAT NAME. HOW FITTING THEN THAT YOU SHOULD CONSIDER AT THESE CONFERENCES BEST MEANS EQUIP YOURSELVES SPIRITUALLY TO FULFILL TEACHING MISSION PARTICULARLY SUITED TO YOUR CAPACITIES FOR SERVICE, YOUR ABOUNDING ZEAL AND ENERGY. 357.2

ARDENTLY SUPPLICATING AT HOLY THRESHOLD ON YOUR BEHALF THAT IN ADDITION TO PRAYING, ABSORBING HOLY PRINCIPLES AND TEACHING FAITH, YOU WILL 357.3

BE SO IMBUED BY BELOVED MASTER'S EXAMPLE SERVICE TO HUMANITY AS TO BE ABLE THROUGH YOUR INDIVIDUAL AND COLLECTIVE DEEDS TO DEMONSTRATE CIVILIZING POWER OUR SACRED CAUSE AND CONVEY VISION ITS SPIRITUAL AND SOCIALLY CONSTRUCTIVE BENEFITS TO YOUR COMPATRIOTS OF ALL AGES.

UNIVERSAL HOUSE OF JUSTICE

358
Riḍván Message 1983
RIḌVÁN 140 [1983 A.D.]

To the Baháʼís of the World

Dearly loved friends,

358.1 The observable acceleration, during the past decade, of the two processes described by our beloved Guardian, the disintegration of the old order and the progress and consolidation of the new World Order of Baháʼuʼlláh, may well come to be regarded by future historians as one of the most remarkable features of this period. The recent increase in this very acceleration is even more remarkable. Both within and without the Cause of God, powerful forces are operating to bring to a climax the twin tendencies of this portentous century. Among the many evidences which reveal this process may be cited, on the one hand, the continual increase of lawlessness, terrorism, economic confusion, immorality and the growing danger from the proliferation of weapons of destruction, and on the other, the worldwide, divinely propelled expansion, consolidation and rapid emergence into the limelight of world affairs of the Cause itself, a process crowned by the wonderful efflorescence of Mount Carmel, the mountain of God, whose Divine springtime is now so magnificently burgeoning.

The Dialectic of Triumph and Disaster

358.2 During the past five years, the historical dialectic of triumph and disaster has operated simultaneously within the Cause of God. The Army of Light has sustained the loss of six Hands of the Cause and waves of bitter persecution which have again engulfed the long-suffering community in Iran, and have resulted in the razing of the House of the Báb, the demolition of Baháʼuʼlláh's ancestral home in Tákur, and the martyrdom of scores of valiant souls.[1] Yet these disasters have called forth fresh energies in the hearts of the friends, have fed the deep roots of the Cause and given rise to a great harvest of signal victories. Chief among these are the successful conclusion of

---

358-1. The deceased Hands of the Cause of God include Hasan M. Balyuzi (see cable dated 12 February 1980, no. 247), Abu'l-Qásim Faizí (see cable dated 20 November 1980, no. 269), Paul Haney (see cable dated 5 December 1982, no. 346), Raḥmatuʼlláh Muhájir (see cable dated

the Five Year Plan;² the launching of the Seven Year Plan,³ now in the final year of its second phase and unprecedented proclamation of the Faith to Heads of States, parliaments and parliamentarians, government ministers and officials, leaders of thought and people prominent in the professions, resulting in a change of attitude on the part of the mass media, which now increasingly approach us for information about the Cause.

358.3 To these movements must be added the worldwide observances commemorating the fiftieth anniversary of the passing of the Greatest Holy Leaf; the completion of the restoration of the upper floor of the House of 'Abdu'lláh Páshá, and its opening, at this very time, to its first visitors; the occupation by the Universal House of Justice of its permanent Seat, in further fulfillment of the great prophecy in the Tablet of Carmel; steady progress on the construction of the first Mashriqu'l-Adhkár of the Pacific Islands in Samoa and the Mother Temple of the Indian Subcontinent in New Delhi.⁴

358.4 Among the outstanding features of the teaching and consolidation work are the continuing effective results of the participation of more than sixteen thousand believers from all parts of the world in the five International Conferences; intensive teaching campaigns carried out with the active support of all levels of the community and drawing upon the enthusiasm and capacity of Bahá'í youth; the establishment of a second radio station in South America; the re-formation of the National Spiritual Assemblies of Uganda and Nepal, and the establishment of nine new National Spiritual Assemblies, two of which will be elected during the month of May this year, bringing the total of these secondary Houses of Justice to 135.

358.5 Above and beyond all these is the unity in action achieved by the Bahá'í world community in its efforts to enlist public support for the dearly loved, greatly admired, cruelly beleaguered Iranian believers, a unity further manifested in an outpouring of funds to replace their former liberal contributions, and an upsurge of personal dedication rarely seen on so universal a scale and holding the highest promise for the future.

---

30 December 1979, no. 243), Adelbert Mühlschlegel (see cable dated 29 July 1980, no. 256), and Enoch Olinga (see cable dated 17 September 1979, no. 237).

For messages on the destruction of the House of the Báb, see letters dated 9 September 1979, 19 November 1979, and 26 May 1981 (nos. 235, 241, and 282).

For the message on the demolition of the House of Bahá'u'lláh in Tákur, see telex dated 10 December 1981 (no. 205).

358-2. The Five Year Plan began in 1974 and ended in 1979.
358-3. The Seven Year Plan began in 1979 and ended in 1986.
358-4. For messages about the fiftieth anniversary of the passing of Bahíyyih Khánum, the Greatest Holy Leaf, see letters dated March 1981 (no. 275), 24 January 1982 (no. 313), Riḍván 1982 (no. 321), and 26 July 1982 (no. 334). For the message announcing the acquisition of the House of 'Abdu'lláh Páshá, see cable dated 9 January 1975 (no. 154); for an account of the significance of the House of 'Abdu'lláh Páshá, see letter dated 4 March 1975 (no. 157). For the Tablet of Carmel, see GWB, pp. 14–17, or TB, pp. 1–5. See also Shoghi Effendi's description of the Tablet of Carmel in GPB, p. 194.

Social and Economic Development

358.6 The growing maturity of a worldwide religious community which all these processes indicate is further evidenced in the reaching out, by a number of national communities, to the social and economic life of their countries, exemplified by the founding of tutorial schools, the inception of radio stations, the pursuit of rural development programs and the operation of medical and agricultural schemes. To these early beginnings must be added the undoubted skills acquired, as a result of the Iranian crisis, in dealing with international organizations, national governments and the mass media—the very elements of society with which it must increasingly collaborate toward the realization of peace on earth.

A Wider Horizon Beckons

358.7 A wider horizon is opening before us, illumined by a growing and universal manifestation of the inherent potentialities of the Cause for ordering human affairs. In this light can be discerned not only our immediate tasks but, more dimly, new pursuits and undertakings upon which we must shortly become engaged. At present we must complete the objectives of the Seven Year Plan, paying great attention to those inner spiritual developments which will be manifested in greater unity among the friends and in National and Local Spiritual Assemblies functioning "harmoniously, vigorously and efficiently" as the Guardian desired.[5]

358.8 We have no doubt that the Bahá'í world community will accomplish all these tasks and go forward to new achievements. The powers released by Bahá'u'lláh match the needs of the times. We may therefore be utterly confident that the new throb of energy now vibrating throughout the Cause will empower it to meet the oncoming challenges of assisting, as maturity and resources allow, the development of the social and economic life of peoples, of collaborating with the forces leading towards the establishment of order in the world, of influencing the exploitation and constructive uses of modern technology, and in all these ways enhancing the prestige and progress of the Faith and uplifting the conditions of the generality of mankind.

A Time for Rejoicing

358.9 It is a time for rejoicing. The Sun of Bahá'u'lláh is mounting the heavens, bringing into ever clearer light the contrast between the gloom, the despair, the frustrations and bewilderment of the world, and the radiance, confidence, joy and certitude of His lovers. Lift up your hearts. The Day of God is here.

With loving Bahá'í greetings,
THE UNIVERSAL HOUSE OF JUSTICE

358-5. BA, p. 41.

1979–1986 • THE SEVEN YEAR PLAN

359
Election of the Universal House of Justice Riḍván 1983
1 MAY 1983

To all National Spiritual Assemblies

NEWLY ELECTED MEMBERS UNIVERSAL HOUSE OF JUSTICE 'ALÍ NAKHJAVÁNÍ HUSH- 359.1
MAND FATHEAZAM IAN SEMPLE DAVID RUHE GLENFORD MITCHELL DAVID HOFMAN
BORRAH KAVELIN CHARLES WOLCOTT HUGH CHANCE.

UNIVERSAL HOUSE OF JUSTICE

360
Execution of Two Bahá'ís in Isfahan
12 MAY 1983

To all National Spiritual Assemblies

WITH SADDENED HEARTS ANNOUNCE EXECUTION ON 1 MAY TWO MORE ACTIVE 360.1
SUPPORTERS GREATEST NAME CRADLE FAITH, MR. SUHAYL ṢAFÁ'Í AND MR. JALÁL
ḤAKÍMÁN, IMPRISONED SINCE OCTOBER 1982 IN ISFAHAN. THEIR FAMILIES, FRIENDS
LEARNED AFTER MORE THAN A WEEK OF THEIR EXECUTION IN TEHRAN. . . .

UNIVERSAL HOUSE OF JUSTICE

361
Further Evolution of the International Teaching Center
19 MAY 1983

To the Followers of Bahá'u'lláh throughout the World

Dearly loved Friends,

For ten years the International Teaching Center has rendered invaluable 361.1
services at the World Center of the Faith, and it is with great joy that we now
announce a number of major steps in the evolution of this vital institution of
the Administrative Order of Bahá'u'lláh.

Since the tragic death of Mr. Paul Haney there have been only two Hands 361.2
of the Cause residing in the Holy Land. We have therefore decided to call
upon Dr. 'Alí Muḥammad Varqá and Mr. Collis Featherstone to participate
in the discharge of the special duties of the Hands of the Cause residing in
the Holy Land when the occasion requires, as for example, in dealing with

matters of Covenant-breaking. They will be able to perform these functions either by correspondence or by periodic sojourns at the World Center.

Appointment of New Counselor Members

361.3 We have decided to raise the number of resident members of the International Teaching Center to nine. For reasons of health Mrs. Florence Mayberry is leaving the World Center, bringing to an end her highly valued services on this institution. Four new Counselor members have therefore been appointed: Dr. Magdalene Carney, Mr. Mas'úd Khamsí, Dr. Peter Khan and Mrs. Isobel Sabri, whom we now call upon to transfer their residences to the Holy Land, where they will join the Hands of the Cause Amatu'l-Bahá Rúḥíyyih Khánum and 'Alí-Akbar Furútan and Counselors Anneliese Bopp, Hooper Dunbar and 'Azíz Yazdí.

Institution of a Five-Year Term

361.4 We have further decided, as foreshadowed in previous announcements, to institute a five-year term for the Counselor members of the International Teaching Center. Each term will start on 23 May immediately following the International Bahá'í Convention, and the current term will end on 23 May 1988. Should circumstances prevent the Universal House of Justice from making new appointments at the end of any five-year term, the Counselors will remain in office until such time as new appointments can be made.

An Increase of Responsibility

361.5 With the rapid growth of the Faith, its emergence from obscurity, and the diversification of the activities that the believers in many lands must undertake in such fields as education, rural development, radio and public relations—matters which must increasingly occupy the attention of the Universal House of Justice—we have decided that the time is ripe to devolve increased responsibility upon the International Teaching Center in the fields of protection and propagation of the Faith. The duties of the International Teaching Center, including those announced previously and those now being assigned to it, are as follows:

- To assume full responsibility for coordinating, stimulating and directing the Continental Boards of Counselors, acting also as liaison between them and the Universal House of Justice.
- To be fully informed of the situation of the Cause in all parts of the world and, from this knowledge, to make reports and recommendations to the Universal House of Justice and give advice to the Continental Boards of Counselors.
- To watch over the security and ensure the protection of the Faith of God.

- To be alert to possibilities for the extension of the teaching work and the development of economic and social life both within and without the Bahá'í community, and to draw the attention of the Universal House of Justice and the Continental Boards of Counselors to such possibilities, making recommendations for action.
- To determine and anticipate needs for literature, pioneers and traveling teachers and to work out teaching plans, both regional and global, for the approval of the Universal House of Justice.
- To direct the work of the Continental Pioneer Committees.
- To administer the expenditure of the International Deputization Fund.
- To administer an annual budget that will be provided from the Bahá'í International Fund, allocating therefrom to the Continental Boards of Counselors monies for special teaching projects and literature subvention, and, when necessary, contributions to the Continental Funds.

The transfer of functions and responsibilities in implementation of the above decisions will be made gradually as the new members are able to settle in the Holy Land. National Spiritual Assemblies and Continental Pioneer Committees will be notified, as necessary, of any changes in procedure that will be required; in the meantime they should continue to operate as before.

New Offices

In the near future the International Teaching Center will be moving into its new offices near the House of the Master, in the building which served for several decades as the Western Pilgrim House, later as the seat of the International Bahá'í Council and, for the past twenty years as that of the Universal House of Justice.[1] Now, most befittingly, it will serve as the office of the International Teaching Center until the permanent building for that mighty institution can be raised on Mount Carmel in close proximity to the Universal House of Justice.

It is our ardent prayer that the decisions now taken will be blessed by Bahá'u'lláh and will enable the World Center of the Faith to coordinate and direct with ever greater effectiveness the self-sacrificing and assiduous labors of the friends of God in every part of the world during the challenging years which lie before us.

With loving Bahá'í greetings,
THE UNIVERSAL HOUSE OF JUSTICE

361-1. The Western Pilgrim House is located in Haifa at 10 Haparsim Street, across the street from the House of 'Abdu'l-Bahá (7 Haparsim Street).

362

Statement by the President of the United States about the Bahá'ís of Iran
24 May 1983

To the National Spiritual Assembly of the Bahá'ís of the United States

362.1 OUTSTANDING STATEMENT PRESIDENT REAGAN BEHALF BAHÁ'ÍS IRAN FILLS HEARTS BELIEVERS WORLD OVER WITH JOY AND GRATITUDE. THIS ACHIEVEMENT, INDICATIVE SPIRITUAL POWER RELEASED BY PURE BLOOD MARTYRS AND SUSTAINED EFFORTS AMERICAN BAHÁ'ÍS THROUGH THEIR ELECTED REPRESENTATIVES, FURTHER STEP IMPLEMENTATION ROLE AMERICAN NATION DESTINED PLAY PROTECTION OPPRESSED PEOPLES. CONFIDENT BAHÁ'Í COMMUNITY UNITED STATES WILL SEIZE UNIQUE OPPORTUNITY PROCLAIM FAITH WIDELY AND WISELY ENTIRE NATION UNDER GUIDANCE NATIONAL ASSEMBLY. EXTEND HEARTFELT CONGRATULATIONS ASSURE LOVING PRAYERS FURTHER VICTORIES.

UNIVERSAL HOUSE OF JUSTICE

363

Execution of Six Bahá'ís in Shiraz
18 June 1983

To all National Spiritual Assemblies

363.1 WITH GREAT SORROW IMPART NEWS EXECUTION BY HANGING LATE HOURS 16 JUNE IN SHIRAZ ANOTHER SIX VALIANT SERVANTS CAUSE:

DR. BAHRÁM AFNÁN, PROMINENT PHYSICIAN, 48 YEARS OLD
MR. BAHRÁM YALDÁ'Í, STUDENT, 23 YEARS OLD
MR. JAMSHÍD SÍYÁVUSHÍ, MERCHANT, 39 YEARS OLD
MR. 'INÁYATU'LLÁH ISHRÁQÍ, RETIRED OFFICER OIL COMPANY, 60 YEARS OLD
MR. KÚRUSH HOQBÍN, ELECTRICAL TECHNICIAN, 27 YEARS OLD
MR. 'ABDU'L-HUSAYN ÁZÁDÍ, EMPLOYEE HEALTH MINISTRY, 60 YEARS OLD

363.2 GRAVELY CONCERNED LIVES OTHER PRISONERS THREATENED BE SUBJECTED SIMILAR FATE IF REFUSE RECANT FAITH AND EMBRACE ISLAM. THIS RUTHLESS TREATMENT BY FANATICS NOW TAKING REINS JUSTICE THEIR HANDS IN DEFIANCE WORLD PUBLIC OPINION, DEMANDS SPECIAL CONSIDERATION BY GOVERNMENTS AND PEOPLE OF PROMINENCE TO EXERT UTMOST EFFORTS PREVENT CONTINUATION SUCH ACTS WHICH VIOLATE PRINCIPLES JUSTICE AND HUMAN RIGHTS. . . .

UNIVERSAL HOUSE OF JUSTICE

1979–1986 • THE SEVEN YEAR PLAN

364
Execution by Hanging of Ten Bahá'í Women in Shiraz
19 JUNE 1983

To all National Spiritual Assemblies

FOLLOWING OUTRAGEOUS EXECUTION SIX BAHÁ'ÍS IN SHIRAZ ON 16 JUNE, FURTHER HIDEOUS CRIME HAS BEEN PERPETRATED BY AUTHORITIES THAT CITY BY HANGING TEN INNOCENT WOMEN NIGHT OF 18 JUNE. THEY ARE: 364.1

    MRS. NUṢRAT YALDÁ'Í, 54 YEARS OLD, MOTHER OF BAHRÁM, HANGED 16 JUNE
    MRS. 'IZZAT JÁNAMÍ ISHRÁQÍ, 50 YEARS OLD, WIFE OF 'INÁYATU'LLÁH, HANGED 16 JUNE
    MISS RU'YÁ ISHRÁQÍ, IN EARLY 20S, DAUGHTER OF ABOVE
    MRS. ṬÁHIRIH SÍYÁVUSHÍ, 32 YEARS OLD, WIFE OF JAMSHÍD, HANGED 16 JUNE
    MISS MUNÁ MAḤMÚDNIZHÁD, 18 YEARS OLD, DAUGHTER OF YADU'LLÁH, EXECUTED 12 MARCH
    MISS ZARRÍN MUQÍMÍ-ABYÁNIH, UNDER 25 YEARS OLD
    MISS SHAHÍN (SHÍRÍN) DÁLVAND, EARLY 20S
    MISS AKHTAR THÁBIT, 19 YEARS OLD
    MISS SÍMÍN ṢÁBIRÍ, IN EARLY 20S
    MISS MAHSHÍD NÍRÚMAND, 18 YEARS OLD

THE EXECUTION OF THESE GUILTLESS WOMEN IN THE NAME OF RELIGION MUST SHOCK CONSCIENCE HUMANITY. THEY WERE ARRESTED FOR ACTIVITIES IN BAHÁ'Í COMMUNITY INCLUDING EDUCATION OF YOUTH. 364.2

FOLLOWING LONG INTERROGATION IN PRISON THEY WERE WARNED THEY WOULD BE SUBJECTED TO FOUR SESSIONS PRESSURING THEM RECANT THEIR FAITH ACCEPT ISLAM AND IF BY FOURTH TIME THEY HAD NOT SIGNED PREPARED STATEMENT RECANTING FAITH THEY WOULD BE KILLED. ALL PREFERRED DIE RATHER THAN DENY THEIR FAITH. 364.3

FEW HOURS PRIOR EXECUTION WOMEN MET WITH FAMILIES, NONE OF WHOM KNEW IMPENDING EXECUTION. NEWS THIS DASTARDLY CRIME NOT PUBLICLY ANNOUNCED OR FORMALLY GIVEN TO FAMILIES. AUTHORITIES REFUSED ALLOW FAMILIES RECEIVE BODIES FOR BURIAL OR EVEN TO SEE THEM. 364.4

IT SHOULD BE RECALLED THAT BETWEEN OCTOBER AND NOVEMBER 1982 OVER 80 BAHÁ'ÍS WERE ARRESTED IN SHIRAZ. AUTHORITIES LATER REVEALED THAT 22 PERSONS AMONG THE 80 WERE CONDEMNED TO DEATH IF WOULD NOT RECANT. NAMES OF THESE 22 HOWEVER WERE NEVER REVEALED, INTENSIFYING PSYCHOLOGICAL STRESS AMONG BAHÁ'Í PRISONERS. 364.5

IN DEFIANCE APPEALS WORLD LEADERS AND WORLD PUBLIC OPINION, 21 OF THESE BAHÁ'ÍS HAVE THUS FAR BEEN EXECUTED, CASTING SHADOW ON FATE REMAINING BELIEVERS LANGUISHING IN PRISON. . . . 364.6

                    UNIVERSAL HOUSE OF JUSTICE

365
Message to Youth—
Summons to Vindicate Martyred Youth
23 June 1983

To Bahá'í Youth Throughout the World:

365.1 RECENT MARTYRDOMS COURAGEOUS STEADFAST YOUTH IN SHIRAZ, SCENE INAUGURATION MISSION MARTYR-PROPHET, REMINISCENT ACTS VALOR YOUTHFUL IMMORTALS HEROIC AGE. CONFIDENT BAHÁ'Í YOUTH THIS GENERATION WILL NOT ALLOW THIS FRESH BLOOD SHED ON VERY SOIL WHERE FIRST WAVE PERSECUTION FAITH TOOK PLACE REMAIN UNVINDICATED OR THIS SUBLIME SACRIFICE UNAVAILING. AT THIS HOUR OF AFFLICTION AND GRIEF, AND AS WE APPROACH ANNIVERSARY MARTYRDOM BLESSED BÁB CALL ON BAHÁ'Í YOUTH TO REDEDICATE THEMSELVES TO URGENT NEEDS CAUSE BAHÁ'U'LLÁH. LET THEM RECALL BLESSINGS HE PROMISED THOSE WHO IN PRIME OF YOUTH WILL ARISE TO ADORN THEIR HEARTS WITH HIS LOVE AND REMAIN STEADFAST AND FIRM. LET THEM CALL TO MIND EXPECTATIONS MASTER FOR EACH TO BE A FEARLESS LION, A MUSK-LADEN BREEZE WAFTING OVER MEADS VIRTUE. LET THEM MEDITATE OVER UNIQUE QUALITIES YOUTH SO GRAPHICALLY MENTIONED IN WRITINGS GUARDIAN WHO PRAISED THEIR ENTERPRISING AND ADVENTUROUS SPIRIT, THEIR VIGOR, THEIR ALERTNESS, OPTIMISM AND EAGERNESS, AND THEIR DIVINELY APPOINTED, HOLY AND ENTHRALLING TASKS. WE FERVENTLY PRAY AT SACRED THRESHOLD THAT ARMY OF SPIRITUALLY AWAKENED AND DETERMINED YOUTH MAY IMMEDIATELY ARISE RESPONSE NEEDS PRESENT HOUR DEVOTE IN EVER GREATER MEASURE THEIR VALUED ENERGIES TO PROMOTE, BOTH ON HOMEFRONTS AND IN FOREIGN FIELDS, CAUSE THEIR ALL-WATCHFUL AND EXPECTANT LORD. MAY THEY MANIFEST SAME SPIRIT SO RECENTLY EVINCED THEIR MARTYR BRETHREN CRADLE FAITH, SCALE SUCH HEIGHTS OF ENDEAVOR AS TO BECOME PRIDE THEIR PEERS CONSOLATION HEARTS PERSIAN BELIEVERS, AND DEMONSTRATE THAT THE FLAME HIS OMNIPOTENT HAND HAS KINDLED BURNS EVER BRIGHT AND THAT ITS LIFE-IMPARTING WARMTH AND RADIANCE SHALL SOON ENVELOP PERMEATE WHOLE EARTH.

UNIVERSAL HOUSE OF JUSTICE

366
Appointment of Continental Counselors
27 June 1983

To all National Spiritual Assemblies

Dear Bahá'í Friends,

366.1 The Universal House of Justice announces with pleasure the appointment of the following Continental Counselors:

In Africa:	Mr. Gila Michael Bahta
	Mr. Kassimi Fofana
In the Americas:	Mr. Shapoor Monadjem
In Australasia:	Mrs. Joy Stevenson

With loving Bahá'í greetings,
DEPARTMENT OF THE SECRETARIAT

367
Execution of Suhayl Hú<u>sh</u>mand in Shiraz
30 June 1983

To all National Spiritual Assemblies

367.1 SWIFT ON HEELS EXECUTION YOUNG BAHÁ'ÍS SHIRAZ ANNOUNCE DISTRESSING NEWS YET ANOTHER YOUNG SERVANT BAHÁ'U'LLÁH THAT CITY, 24 YEAR OLD SUHAYL HÚ<u>SH</u>MAND, HANGED 28 JUNE, BRINGING TO 142 TOTAL NUMBER BAHÁ'ÍS KILLED SINCE BEGINNING ISLAMIC REVOLUTION, NOT INCLUDING 14 WHO DISAPPEARED. FATE OTHER PRISONERS SHIRAZ IN BALANCE.

UNIVERSAL HOUSE OF JUSTICE

368
Attacks against Bahá'í Villagers in Mázindarán; Kidnapping of Two Bahá'ís in Tehran
4 July 1983

To all National Spiritual Assemblies

368.1 ATROCITIES MOUNTING AGAINST BRETHREN CRADLE FAITH, NOW DIRECTED TOWARDS DEFENSELESS VILLAGERS NEAR SARI IN MÁZINDARÁN. IN VILLAGE OF ÍVÁL OVER 130 BAHÁ'ÍS INCLUDING WOMEN AND CHILDREN WERE HELD CAPTIVE FOR THREE DAYS IN WALLED-IN OPEN FIELD WITHOUT FOOD AND WATER. WHEN PRESSURES TO RECANT FAITH ACCEPT ISLAM FAILED THEY WERE ALLOWED TO RETURN TO THEIR HOMES. HOWEVER SAME NIGHT, JULY 1ST, THEY WERE ATTACKED BY VILLAGERS AND FORCED HIDE IN NEARBY FOREST.

368.2 FURTHER DISTRESSING NEWS TWO PROMINENT BAHÁ'ÍS TEHRAN, JAHÁNGÍR HIDÁYATÍ AND AḤMAD BAS͟HÍRÍ, KIDNAPPED.[1] APPEALS TO AUTHORITIES SO FAR UNHEEDED, ANY KNOWLEDGE THEIR ABDUCTION OR WHEREABOUTS DENIED. . . .

UNIVERSAL HOUSE OF JUSTICE

369
Message to the European Youth Conference in Innsbruck
4 July 1983

To the European Youth Conference in Innsbruck

Dear Bahá'í Friends,

369.1 With high hopes we greet the representatives of the Bahá'í youth of Europe gathered in conference in Innsbruck. This generation of Bahá'í youth enjoys a unique distinction. You will live your lives in a period when the forces of history are moving to a climax, when mankind will see the establishment of the Lesser Peace, and during which the Cause of God will play an increasingly prominent role in the reconstruction of human society. It is you who will be called upon in the years to come to stand at the helm of the Cause in face of conditions and developments which can, as yet, scarcely be imagined.

369.2 European Bahá'í youth in particular face tremendous and challenging tasks in the immediate future. Can one doubt that the manner in which the governments of the European nations have rallied to the defense of the persecuted Bahá'ís in Iran will draw down blessings from on high upon this con-

---

368-1. Jahángír Hidáyatí was killed in Tehran on 15 May 1984. Aḥmad Bas͟hírí was killed in Tehran on 1 November 1984.

tinent? And who among the people of Europe are more likely to be kindled by the challenge and hope of the Message of Bahá'u'lláh than the youth? Now is an opportunity to awaken the interest, set afire the hearts and enlist the active support of young people of every nation, class and creed in that continent. The key to success in this endeavor is, firstly, to deepen your understanding of the Teachings of the Cause so that you will be able to apply them to the problems of individuals and society, and explain them to your peers in ways that they will understand and welcome; secondly, to strive to model your behavior in every way after the high standards of honesty, trustworthiness, courage, loyalty, forbearance, purity and spirituality set forth in the Teachings; and, above all, to live in continual awareness of the presence and all-conquering power of Bahá'u'lláh, which will enable you to overcome every temptation and surmount every obstacle.

A vibrant band of Bahá'í youth on the European continent, committed to the promotion of the Cause of Bahá'u'lláh and the upholding of His laws and principles, determined to work in harmony and unity with their fellow believers of all ages and classes, can revolutionize the progress of the Cause. With a rapid increase in the size of the Bahá'í communities in Europe, the believers of that continent, the cradle of Western civilization, will be the better able to serve as a fountainhead of pioneers, traveling teachers and financial assistance to the Bahá'í communities of the Third World. 369.3

When deciding what course of training to follow, youth can consider acquiring those skills and professions that will be of benefit in education, rural development, agriculture, economics, technology, health, radio and in many other areas of endeavor that are so urgently needed in the developing countries of the world. You can also devote time in the midst of your studies, or other activities, to travel teaching or service projects in the Third World. 369.4

A particular challenge to the Bahá'í youth of Europe is the vast eastern half of the continent that is as yet scarcely touched by the light of the Faith of Bahá'u'lláh. It is not easy to settle in those lands, but with ingenuity, determination and reliance upon the confirmations of Bahá'u'lláh it is certainly possible both to settle and to persevere in service in goals which demand a spirit of self-sacrifice, detachment and purity of heart worthy of those who would emulate the shining example set by the martyrs in Iran, so many of whom are youth, who have given their lives rather than breathe one word that would be a betrayal of the trust of God placed upon them. 369.5

With love and utmost longing we call upon you to immerse yourselves in the divine Teachings, champion the Cause of God and His law, and arise for the quickening of mankind. 369.6

THE UNIVERSAL HOUSE OF JUSTICE

MESSAGES FROM THE UNIVERSAL HOUSE OF JUSTICE

370
Continuing Harassments and Arrests in Iran
26 July 1983

To all National Spiritual Assemblies

370.1 PRESSURES MOUNTING AGAINST BAHÁ'ÍS IRAN, PARTICULARLY IN TEHRAN WHERE MEMBERS OF BAHÁ'Í INSTITUTIONS ARE BEING HUNTED, THEIR HOMES RAIDED. IF NOT FOUND, THEIR FAMILIES ARE BEING HARASSED.

370.2 TWENTY-TWO BELIEVERS INCLUDING 11 WOMEN WERE ARRESTED BETWEEN 11 AND 20 JULY. . . .

UNIVERSAL HOUSE OF JUSTICE

371
Formation of Nine New National Spiritual Assemblies at Riḍván
31 July 1983

To all National Spiritual Assemblies

371.1 OVERJOYED ANNOUNCE FORMATION AT RIḌVÁN 1984 FOLLOWING NINE NEW NATIONAL SPIRITUAL ASSEMBLIES:

THREE IN AFRICA:	CAPE VERDE ISLANDS, EQUATORIAL GUINEA AND GABON
THREE IN AMERICAS:	FRENCH GUIANA, GRENADA AND MARTINIQUE
TWO IN ASIA:	ANDAMAN AND NICOBAR ISLANDS, AND YEMEN (SAN'A)
ONE IN EUROPE:	CANARY ISLANDS

371.2 LAST TWO NATIONAL ASSEMBLIES NAMED ABOVE ARE SUPPLEMENTARY ACHIEVEMENTS SEVEN YEAR PLAN. PRAYING SHRINES BOUNTIFUL BLESSINGS MAY SURROUND SUPPORT FRIENDS EVERYWHERE IN THEIR ENDEAVORS PROMOTE GOD'S HOLY FAITH.

UNIVERSAL HOUSE OF JUSTICE

372
Intensification of Waves of Arrests in Iran
2 August 1983

To all National Spiritual Assemblies

PERSECUTIONS IRANIAN FRIENDS REMAIN UNABATED. WAVES ARRESTS PROMINENT BAHÁ'ÍS RECENTLY INTENSIFIED. . . . 372.1

UNIVERSAL HOUSE OF JUSTICE

373
Arrest of More Bahá'ís in Iran
16 August 1983

To all National Spiritual Assemblies

ARRESTS IRANIAN BRETHREN CONTINUING. . . . 24 INCLUDING 10 WOMEN IMPRISONED BETWEEN 3 AND 5 AUGUST. 373.1

UNIVERSAL HOUSE OF JUSTICE

374
Passing of Sylvia Ioas, Member of the International Bahá'í Council
25 August 1983

To all National Spiritual Assemblies

SADDENED PASSING DEVOTED MAIDSERVANT BAHÁ'U'LLÁH SYLVIA IOAS.[1] HER LONG YEARS SERVICE DIVINE THRESHOLD CONSTANT SUPPORT CLOSE COLLABORATION HER DISTINGUISHED HUSBAND CROWNED BY HER APPOINTMENT BY BELOVED GUARDIAN AS MEMBER INTERNATIONAL BAHÁ'Í COUNCIL AND HER SUBSEQUENT ELECTION SAME HISTORIC INSTITUTION AND AS ITS VICE-PRESIDENT. HER GRACIOUS MANNER, CHEERFUL DISPOSITION, HOSPITABLE SPIRIT REMAIN AS INDELIBLE IMPRESSIONS HER FRUITFUL LIFE. FERVENTLY PRAYING HOLY SHRINES HER RADIANT SOUL MAY BE RICHLY REWARDED ABHÁ KINGDOM. URGE NATIONAL ASSEMBLIES HOLD BEFITTING MEMORIAL SERVICES. 374.1

UNIVERSAL HOUSE OF JUSTICE

374-1. For an account of the life and services of Sylvia Ioas, see BW 19:611.

375

Requisites for Spiritual Growth
1 September 1983

The National Spiritual Assembly of the Bahá'ís of Norway

Dear Bahá'í Friends,

375.1 On several occasions there has been correspondence between your Assembly and the Universal House of Justice on meditation and kindred subjects. The House of Justice is aware that such matters have been a cause of differences of opinion among the Norwegian Bahá'ís. It has now come to the attention of the House of Justice that there was a session of group meditation of a particular kind at your summer school under the aegis of the National Teaching Committee. We have, therefore, been instructed to send you the following comments which, it is hoped, will help to resolve this long-standing problem.

Spiritualization of the Bahá'í Community

375.2 In its message to the Dublin Conference the Universal House of Justice called upon the Continental Board of Counselors and the National Spiritual Assemblies of Europe to launch together "such a campaign of spiritualization of the Bahá'í community, allied with intensified personal teaching, as has never been witnessed in your continent."[1] It realizes that the session at your Summer School referred to above may well have been intended as an aspect of this campaign, and it feels that it would be helpful to explain more fully what it intended by "spiritualization of the Bahá'í community."

375.3 Europe has suffered so appallingly in past centuries from persecutions and conflicts inspired by religious differences and fanaticism that there has been a revulsion against religion. Many Europeans have become skeptical, scornful of religious practices, and reluctant either to discuss religious subjects or to give credence to the power of faith. This turning away from religion has been powerfully reinforced by the growth of materialism, and has produced a combination of physical well-being and spiritual aridity that is having catastrophic results, socially and psychologically, on the population.

375.4 This intellectual and emotional atmosphere creates problems for the Bahá'í community in two ways. Its effect upon a large proportion of the non-Bahá'í population makes it difficult for Bahá'ís to convey the Message to others. Its effect upon the Bahá'ís is more subtle, but no less harmful; if not consciously combated it can lead the believers to neglect those spiritual exercises which are the very fountainhead of their spiritual strength and the nourishment of their souls.

375-1. See letter dated 2 June 1982 (no. 329).

Essential Requisites for Spiritual Growth

Bahá'u'lláh has stated quite clearly in His Writings the essential requisites for our spiritual growth, and these are stressed again and again by 'Abdu'l-Bahá in His talks and Tablets. One can summarize them briefly in this way:

1. The recital each day of one of the Obligatory Prayers with pure-hearted devotion.
2. The regular reading of the Sacred Scriptures, specifically at least each morning and evening, with reverence, attention and thought.
3. Prayerful meditation on the Teachings, so that we may understand them more deeply, fulfill them more faithfully, and convey them more accurately to others.
4. Striving every day to bring our behavior more into accordance with the high standards that are set forth in the Teachings.
5. Teaching the Cause of God.
6. Selfless service in the work of the Cause and in the carrying on of our trade or profession.

These points, expressed in other words, have already been conveyed to the friends in Europe by the Counselors, but the House of Justice wishes to stress them, because they represent the path towards the attainment of true spirituality that has been laid down by the Manifestation of God for this age.

Prayer and Meditation—Personal Exercises

It is striking how private and personal the most fundamental spiritual exercises of prayer and meditation are in the Faith. Bahá'ís do, of course, have meetings for devotions, as in the Mashriqu'l-Adhkár or at Nineteen Day Feasts, but the daily obligatory prayers are ordained to be said in the privacy of one's chamber, and meditation on the Teachings is, likewise, a private individual activity, not a form of group therapy. In His talks 'Abdu'l-Bahá describes prayer as "conversation with God," and concerning meditation He says that "while you meditate you are speaking with your own spirit. In that state of mind you put certain questions to your spirit and the spirit answers: the light breaks forth and the reality is revealed."[2]

There are, of course, other things that one can do to increase one's spirituality. For example, Bahá'u'lláh has specified no procedures to be followed in meditation, and individual believers are free to do as they wish in this area, provided that they remain in harmony with the Teachings, but such activities are purely personal and should under no circumstances be confused with those actions which Bahá'u'lláh Himself considered to be of fundamental importance for our spiritual growth. Some believers may find that it is beneficial to

---

375-2. SBS, p. 250; PT, p. 174.

them to follow a particular method of meditation, and they may certainly do so, but such methods should not be taught at Bahá'í Summer Schools or be carried out during a session of the School because, while they may appeal to some people, they may repel others. They have nothing to do with the Faith and should be kept quite separate so that inquirers will not be confused.

375.9 It would seem that there are in Norway many believers who draw particular benefit from meditation. The House of Justice suggests that for their private meditations they may wish to use the repetition of the Greatest Name, Alláh-u-Abhá, ninety-five times a day which, although not yet applied in the West, is among the Laws, Ordinances and Exhortations of the Kitáb-i-Aqdas. (See p. 46 of the *Synopsis and Codification of the Kitáb-i-Aqdas*.)

375.10 The House of Justice is confident that if the believers throughout Europe will conscientiously strive to increase their spirituality in the six ways outlined above, and become aware in their inmost beings that in all their services they are but vehicles for the confirming power of God, they will attract the hearts of their fellow citizens and penetrate the miasma of materialism that veils the sight of so many of their countrymen. Effort, activity, unity and constant reliance on the power of Bahá'u'lláh will assuredly overcome all obstacles.

With loving Bahá'í greetings,
DEPARTMENT OF THE SECRETARIAT

376
Preparations for International Youth Year
7 SEPTEMBER 1983

To all National Spiritual Assemblies

Dear Bahá'í Friends,

376.1 As you are probably aware, 1985 has been designated by the United Nations as International Youth Year, and the Bahá'í International Community is taking preliminary steps to set up a program for the involvement of the worldwide Bahá'í community in this event, in close cooperation with the United Nations and its agencies, such as UNICEF.[1] The purpose of this letter is to request your National Assembly to determine how the Bahá'í youth in your country can, to the extent practicable, participate in general youth activities which are being proposed by the United Nations for International Youth Year, as well as in

---

376-1. United Nations International Children's Emergency Fund, established by the United Nations' General Assembly in 1946 to provide relief for children in devastated areas. It also seeks to feed destitute children and to prevent childhood diseases. In 1953 it was made a permanent UN organization. The Bahá'í International Community has consultative status with UNICEF.

specific Bahá'í projects of service to the local community or to the country as a whole.

Your National Spiritual Assembly should be alert as to whether your government is making any plans to observe International Youth Year, and if so, you should consider the steps you could take in setting up or participating in projects associated with this event. Such activities should be planned, of course, in consultation with the Bahá'í International Community, which is coordinating and monitoring such efforts. Participation of this kind by the Bahá'í youth would not only add to the prestige of the Faith and demonstrate its humanitarian character, but would also act as a rallying point for the youth, strengthening them in their role in the Faith and training them to organize themselves for future services to this blessed Cause. 376.2

The Universal House of Justice will be anxious to learn of your plans to participate in this important event. 376.3

With loving Bahá'í greetings,
DEPARTMENT OF THE SECRETARIAT

377
Banning of Bahá'í Administration in Iran
13 September 1983

To all National Spiritual Assemblies

SORELY TRIED COMMUNITY GREATEST NAME IRAN HAS IN RECENT DAYS SUSTAINED YET ANOTHER CRUEL BLOW OPENING NEW CHAPTER ITS TURBULENT HISTORY. ON 29 AUGUST IN UNPRECEDENTED MOVE REVOLUTIONARY GOVERNMENT THROUGH STATEMENT ISSUED ATTORNEY-GENERAL ANNOUNCED BAN BAHÁ'Í ADMINISTRATION, RECITING USUAL FALSE ACCUSATIONS STATING EXISTENCE ADMINISTRATION OFFICIALLY CONSIDERED TO BE AGAINST LAWS CONSTITUTION COUNTRY. HOWEVER STATEMENT SAID BAHÁ'ÍS MAY PRACTICE BELIEFS AS PRIVATE INDIVIDUALS PROVIDED THEY DO NOT TEACH OR INVITE OTHERS TO JOIN FAITH, THEY DO NOT FORM ASSEMBLIES OR HAVE ANYTHING TO DO WITH ADMINISTRATION. SERVING IN BAHÁ'Í ADMINISTRATION NOW SPECIFIED AS CRIMINAL ACT. THIS LATEST ONSLAUGHT DEFENSELESS COMMUNITY CLEARLY DEMONSTRATES IMPLACABILITY FANATICAL ELEMENTS IN THEIR DRIVE SUPPRESS LIGHT GOD'S INFANT FAITH IN LAND ITS FIRST GLEAMING. 377.1

IN CONFORMITY PRINCIPLE LOYALTY OBEDIENCE GOVERNMENT NATIONAL SPIRITUAL ASSEMBLY IRAN IMMEDIATELY TOOK ACTION DISSOLVE BAHÁ'Í ADMINISTRATION THROUGHOUT COUNTRY THUS UPHOLDING INTEGRITY COMMUNITY DESPITE HEAVY YOKE CRUELTIES BORNE BY ITS MEMBERS FOR SEVERAL GENERATIONS. CONFIDENT THAT STEADFAST TRIED AND DEVOTED FRIENDS THAT LONG-AGITATED LAND WILL FACE NEW SITUATION WITH RADIANT FORTITUDE. AT SAME 377.2

TIME BAHÁ'ÍS ELSEWHERE ENJOYING FREEDOM PRACTICE FAITH ACUTELY CHALLENGED TO VINDICATE BY THEIR RECONSECRATION TO IMMEDIATE SACRED TASKS UNABATED SUFFERING THEIR GRIEVOUSLY WRONGED IRANIAN BRETHREN. INDEED ALL NATIONAL ASSEMBLIES URGED TAKE STEPS STRENGTHEN FOUNDATION BAHÁ'Í INSTITUTIONS THEIR COUNTRIES AS TRIBUTE SACRIFICES COURAGEOUSLY ACCEPTED MEMBERS COMMUNITY BAHÁ'U'LLÁH'S NATIVE LAND.

UNIVERSAL HOUSE OF JUSTICE

378
Open Letter to Iranian Authorities from the National Spiritual Assembly of Iran
19 OCTOBER 1983

To all National Spiritual Assemblies

Dear Bahá'í Friends,

378.1 You have already been informed of the disbanding of the National Spiritual Assembly and other Bahá'í institutions in Iran as a result of the new policy of the Iranian Government. The National Spiritual Assembly, before disbanding itself, decided to issue an open letter to the authorities and the public of their country explaining the action of their Government and pointing out the unjust attitudes and accusations of the Government against the Bahá'í administration in that land.[1] Over two thousand of these letters were dispatched to important officials and other prominent people in Iran.

378.2 A copy of this letter has been received by the Universal House of Justice, and we have been instructed to enclose a translation for your information and possible use. Publication of the letter is permissible in whatever manner your National Spiritual Assembly deems fit.

With loving Bahá'í greetings,
DEPARTMENT OF THE SECRETARIAT

378-1. This is not the first open letter written in defense of the Bahá'í community and circulated by the National Spiritual Assembly of Iran. In a letter dated 20 August 1979 (no. 234) the Universal House of Justice referred to "the issuance and quiet circulation of open letters directed to the public in Iran by the National Spiritual Assembly of that country, refuting the false allegations made against the Bahá'í community."

An Open Letter from the National Spiritual Assembly of the Bahá'ís of Iran about the Banning of the Bahá'í Administration
(12 S͟HÁHRÍVAR 1362)
[3 SEPTEMBER 1983]

378.3 Recently the esteemed Public Prosecutor of the Islamic Revolution of the country, in an interview that was published in the newspapers, declared that the continued functioning of the Bahá'í religious and spiritual administration is banned and that membership in it is considered to be a crime. This declaration has been made after certain unjustified accusations have been leveled against the Bahá'í community of Iran and after a number of its members—ostensibly for imaginary and fabricated crimes but in reality merely for the sake of their beliefs—have been either executed, or arrested and imprisoned. The majority of those who have been imprisoned have not yet been brought to trial.

378.4 The Bahá'í community finds the conduct of the authorities and the judges bewildering and lamentable—as indeed would any fair-minded observer who is unblinded by malice. The authorities are the refuge of the people; the judges in pursuit of their work of examining and ascertaining the truth and facts in legal cases devote years of their lives to studying the law and, when uncertain of a legal point spend hours poring over copious tomes in order to cross a "t" and dot an "i." Yet these very people consider themselves to be justified in brazenly bringing false accusations against a band of innocent people, without fear of the Day of Judgment, without even believing the calumnies they utter against their victims, and having exerted not the slightest effort to investigate to any degree the validity of the charges they are making. "Methinks they are not believers in the Day of Judgment." (Ḥáfiẓ)

378.5 The honorable Prosecutor has again introduced the baseless and fictitious story that Bahá'ís engage in espionage, but without producing so much as one document in support of the accusation, without presenting proof in any form, and without any explanation as to what is the mission in this country of this extraordinary number of "spies": what sort of information do they obtain and from what sources? Whither do they relay it, and for what purpose? What kind of "spy" is an eighty-five-year-old man from Yazd who has never set foot outside his village? Why do these alleged spies not hide themselves, conceal their religious beliefs and exert every effort to penetrate, by every stratagem, the Government's information centers and offices? Why has no Bahá'í "spy" been arrested anywhere else in the world? How could students, housewives, innocent young girls, and old men and women, like those blameless Bahá'ís who have recently been delivered to the gallows in Iran, or who have become targets for the darts of prejudice and enmity, be "spies"? How could the Bahá'í farmers of the villages of Afús, Chígán, Qal'ih Malik (near Isfahan), and those of the village of Núk in Bírjand, be "spies"? What Secret Intelligence documents have been found in their possession? What espionage equipment has come to hand? What "spying" activities were engaged in by the primary schoolchildren who have been expelled from their schools?

378.6 And how strange! The Public Prosecutor perhaps does not know, or does not care to know, that spying is an element of politics, while noninterference in politics is an established principle of the Bahá'í Faith. On the contrary, Bahá'ís love their country and never permit themselves to be traitors. 'Abdu'l-Bahá, the successor of the Founder of the Bahá'í Cause, says: "Any abasement is bearable except betraying one's own country, and any sin is forgivable other than dishonoring the government and inflicting harm upon the nation."[2]

378.7 All the other accusations made against the Bahá'ís by the honorable Public Prosecutor of the Revolution are similarly groundless. He brands the Bahá'í community with accusations of subversion and corruption. For example, on the basis of a manifestly forged interview, the falsity of which has been dealt with in a detailed statement, he accuses the Bahá'í community of hoarding, an act which its members would consider highly reprehensible. The Prosecutor alleges that the Bahá'í administration sanctioned the insensible act of hoarding, yet he subtly overlooks the fact that with the proceeds that might be realized from the sale of unusable automobile spare parts whose total value is some 70 million túmáns—the value of the stock of any medium-size store for spare parts—it would be impossible to overthrow a powerful government whose daily expenditures amount to hundreds of millions of túmáns. If the Public Prosecutor chooses to label the Bahá'í administration as a network of espionage, let him at least consider it intelligent enough not to plan the overthrow of such a strong regime by hoarding a few spare parts! Yes, such allegations of corruption and subversion are similar to those hurled against us at the time of the Episcopalian case in Isfahan when this oppressed community was accused of collaboration with foreign agents as a result of which seven innocent Bahá'ís of Yazd were executed.[3] Following this the falsity of the

---

378-2. From an unpublished Tablet.

378-3. The executions took place on 8 September 1980. The Ayatollah Khomeini, in his lectures on "Islamic Government" (1973), had for his own purposes falsely linked the Christian missions in Iran with the Bahá'ís as corrupters of Muslim youth. In August 1980 pressure against the Christian community in general began to mount. The 27 August 1980 issue (No. 11079) (5 Sháhrívar 1359 A.H.) of the Tehran-based newspaper *Kayhán* carried a story released through the Pars agency about the discovery in an Episcopalian church in Iran of a document purporting to be a receipt for U.S. $500,000,000 signed by a clergyman who was said to be spying for the C.I.A. The sum was, the report said, to be split among various agencies including "the head of the Bahá'ís and antirevolutionary groups" and was described as a preliminary step in anticipated terrorist activities including bombing the residence of the Ayatollah Khomeini. Soon after the hanging of seven Bahá'ís, and it may be concluded because groups other than Bahá'ís were affected, the Public Prosecutor announced that the document was a forgery.

charges was made known and the Public Prosecutor announced the episode to be the outcome of a forgery.

Bahá'ís are accused of collecting contributions and transferring sums of money to foreign countries. How strange! If Muslims, in accordance with their sacred and respected spiritual beliefs, send millions of túmáns to Karbilá, Najaf and Jerusalem, or to other Muslim Holy Places outside Iran to be spent on the maintenance and upkeep of the Islamic Sacred Shrines, it is considered very praiseworthy; but if a Bahá'í—even during the time in which the transfer of foreign currency was allowed—sends a negligible amount for his international community to be used for the repair and maintenance of the Holy Places of his Faith, it is considered that he has committed an unforgivable sin and it is counted as proof that he has done so in order to strengthen other countries. 378.8

Accusations of this nature are many but all are easy to investigate. If just and impartial people and God-fearing judges will only do so, the falsity of these spurious accusations will be revealed in case after case. The Bahá'í community emphatically requests that such accusations be investigated openly in the presence of juries composed of judges and international observers so that, once and for all, the accusations may be discredited and their repetition prevented. 378.9

The basic principles and beliefs of the Bahá'ís have been repeatedly proclaimed and set forth in writing during the past five years. Apparently these communications, either by deliberate design or by mischance, have not received any attention, otherwise accusations such as those described above would not have been repeated by one of the highest and most responsible authorities. This in itself is a proof that the numerous communications referred to were not accorded the attention of the leaders; therefore, we mention them again. 378.10

The Bahá'í Faith confesses the unity of God and the justice of the divine Essence. It recognizes that Almighty God is an exalted, unknowable and concealed entity, sanctified from ascent and descent, from egress and regress, and from assuming a physical body. The Bahá'í Faith, which professes the existence of the invisible God, the One, the Single, the Eternal, the Peerless, bows before the loftiness of His Threshold, believes in all divine Manifestations, considers all the Prophets from Adam to the Seal of the Prophets[4] as true divine Messengers Who are the Manifestations of Truth in the world of creation, accepts Their Books as having come from God, believes in the continuation of the divine outpourings, emphatically believes in reward and punishment and, uniquely among existing revealed religions outside Islam, accepts the Prophet Muḥammad as a true Prophet and the Qur'án as the Word of God. 378.11

378-4. Muḥammad.

378.12   The Bahá'í Faith embodies independent principles and laws. It has its own Holy Book. It prescribes pilgrimage and worship. A Bahá'í performs obligatory prayers and observes a fast. He gives, according to his beliefs, tithes and contributions. He is required to be of upright conduct, to manifest a praiseworthy character, to love all mankind, to be of service to the world of humanity and to sacrifice his own interests for the good and well-being of his fellow kind. He is forbidden to commit unbecoming deeds. 'Abdu'l-Bahá says: "A Bahá'í is known by the attributes manifested by him, not by his name; he is recognized by his character, not by his person."[5]

378.13   Shoghi Rabbaní,[6] the Guardian of the Bahá'í Faith, says: ". . . a person who is not adorned with the ornaments of virtue, sanctity, and morality, is not a true Bahá'í, even though he may call himself one and be known as such."[7]

378.14   He also says: ". . . the friends have been required . . . to be righteous, well-wishing, forbearing, sanctified, pure, detached from all else save God, severed from the trappings of this world and adorned with the mantle of a goodly character and godly attributes."[8]

378.15   The teachings and laws of the Bahá'í religion testify to this truth. Fortunately, the books and writings which have been plundered in abundance from the homes of Bahá'ís and are available to the authorities, bear witness to the truth of these assertions. Bahá'ís, in keeping with their spiritual beliefs, stay clear of politics; they do not support or reject any party, group or nation; they do not champion or attack any ideology or any specific political philosophy; they shrink from and abhor political agitations. The Guardian of the Bahá'í Cause says, "The people of Bahá, under the jurisdiction of whatsoever state or government they may be residing, should conduct themselves with honesty and sincerity, trustworthiness and rectitude. . . . They are neither thirsty for prominence, nor acquisitive of power; they are neither adepts at dissimulation and hypocrisy, nor are they seekers after wealth and influence; they neither crave for the pomp and circumstance of high office, nor do they lust after the glory of titles and ranks. They are averse to affectation and ostentation, and shrink from the use of coercive force; they have closed their eyes to all but God, and set their hearts on the firm and incontrovertible promises of their Lord . . . Oblivious to themselves, they have occupied their energies in working towards the good of society . . . While vigilantly refusing to accept political posts, they should wholeheartedly welcome the chance to assume admin-

---

378-5. From an unpublished Tablet.
378-6. Shoghi Effendi; Rabbaní is his surname. "Effendi" is a Turkish honorific comparable to the English "Mr." or "Sir."
378-7. From an unpublished letter.
378-8. From a letter dated 19 December 1923 to the Bahá'ís of the East (translated from the Persian).

istrative positions; for the primary purpose of the people of Bahá is to advance the interests and promote the welfare of the nation . . . Such is the method of the Bahá'ís; such is the conduct of all spiritually illuminated souls; and aught else is manifest error."⁹

378.16 Also, Bahá'ís, in accordance with their exalted teachings, are duty bound to be obedient to their government. Elucidating this subject, Shoghi Rabbaní says: "The people of Bahá are required to obey their respective governments, and to demonstrate their truthfulness and good will towards the authorities. . . . Bahá'ís, in every land and without any exception, should . . . be obedient and bow to the clear instructions and the declared decrees issued by the authorities. They must faithfully carry out such directives."¹⁰

378.17 Bahá'í administration has no aim except the good of all nations and it does not take any steps that are against the public good. Contrary to the conception the word "administration" may create in the mind because of the similarity in name, it does not resemble the current organizations of political parties; it does not interfere in political affairs; and it is the safeguard against the involvement of Bahá'ís in subversive political activities. Its high ideals are "to improve the characters of men; to extend the scope of knowledge; to abolish ignorance and prejudice; to strengthen the foundations of true religion in all hearts; to encourage self-reliance and discourage false imitation; . . . to uphold truthfulness, audacity, frankness, and courage; to promote craftsmanship and agriculture; . . . to educate, on a compulsory basis, children of both sexes; to insist on integrity in business transactions; to lay stress on the observance of honesty and piety; . . . to acquire mastery and skill in the modern sciences and arts; to promote the interests of the public; . . . to obey outwardly and inwardly and with true loyalty the regulations enacted by state and government; . . . to honor, to extol and to follow the example of those who have distinguished themselves in science and learning." And again, ". . . to help the needy from every creed or sect, and to collaborate with the people of the country in all welfare services."¹¹

378.18 In brief, whatever the clergy in other religions undertake individually and by virtue of their appointment to their positions, the Bahá'í administration performs collectively and through an elective process.

378.19 The statements made by the esteemed Public Prosecutor of the Revolution do not seem to have legal basis, because in order to circumscribe individuals and deprive them of the rights which have not been denied them by the Constitution, it is necessary to enact special legislation, provided that legislation is not contradictory to the Constitution. It was hoped that the recent years

---

378-9. The quotation in the original letter has been replaced by this revised translation.
378-10. From an unpublished letter.
378-11. From unpublished Tablets.

would have witnessed, on the one hand, the administration of divine justice—a principle promoted by the true religion of Islam and prescribed by all monotheistic religions—and, on the other, and coupled with an impartial investigation of the truths of the Bahá'í Faith, the abolition or at least mitigation of discriminations, restrictions and pressures suffered by Bahá'ís over the past 135 years. Alas, on the contrary, because of long-standing misunderstandings and prejudices, the difficulties increased immensely and the portals of calamity were thrown wide open in the faces of the long-suffering and sorely oppressed Bahá'ís of Iran who were, to an even greater degree, deprived of their birthrights through the systematic machinations of Government officials who are supposed to be the refuge of the public, and of some impostors in the garb of divines, who engaged in official or unofficial spreading of mischievous and harmful accusations and calumnies, and issued, in the name of religious and judicial authorities, unlawful decrees and verdicts.

378.20 Many are the pure and innocent lives that have been snuffed out; many the distinguished heads that have adorned the hangman's noose; and many the precious breasts that became the targets of firing squads. Vast amounts of money and great quantities of personal property have been plundered or confiscated. Many technical experts and learned people have been tortured and condemned to long-term imprisonment and are still languishing in dark dungeons, deprived of the opportunity of placing their expertise at the service of the Government and the nation. Numerous are the self-sacrificing employees of the Government who spent their lives in faithful service but who were dismissed from work and afflicted with poverty and need because of hatred and prejudice. Even the owners of private firms and institutions were prevented from engaging Bahá'ís. Many privately owned Bahá'í establishments have been confiscated. Many tradesmen have been denied the right to continue working by cancellation of their business licenses. Bahá'í youth have been denied access to education in many schools and in all universities and institutions of higher education. Bahá'í university students abroad are deprived of receiving money for their education, and others who wish to pursue their studies outside Iran have been denied exit permits. Bahá'ís, including the very sick whose only hope for cure was to receive medical treatment in specialized medical centers in foreign lands, have been prevented from leaving the country. Bahá'í cemeteries have been confiscated and bodies rudely disinterred. Numerous have been the days when a body has remained unburied while the bereaved family pleaded to have a permit issued and a burial place assigned so that the body might be decently buried. As of today, thousands of Bahá'ís have been divested of their homes and forced to live as exiles. Many have been driven from their villages and dwelling places and are living as wanderers and stranded refugees in other parts of Iran with no other haven and refuge but the Court of the All-Merciful God and the loving-kindness of their friends and relatives.

It is a pity that the mass media, newspapers and magazines, either do not 378.21
want or are not allowed to publish any news about the Bahá'í community of
Iran or to elaborate upon what is happening. If they were free to do so and
were unbiased in reporting the daily news, volumes would have been compiled describing the inhumane cruelty to and oppression of the innocent. For
example, if they were allowed to do so, they would have written that in Shiraz,
seven courageous men and ten valiant women—seven of whom were girls in
the prime of their lives—audaciously rejected the suggestion of the religious
judge that they recant their Faith or, at least, dissemble their belief, and preferred death to the concealment of their Faith. The women, after hours of
waiting with dried lips, shrouded themselves in their chádurs,[12] kissed the
noose of their gallows, and with intense love offered up their souls for the
One Who proffereth life. The observers of this cruel scene might well ask forgiveness for the murderers at Karbilá, since they, despite their countless atrocities, did not put women to the sword nor harass the sick and infirm. Alas,
tongues are prevented from making utterance and pens are broken and the
hidden cause of these brutalities is not made manifest to teach the world a
lesson. The Public Prosecutor alleges that they were spies. Gracious God!
Where in history can one point to a spy who readily surrendered his life in
order to prove the truth of his belief?

Unfortunately it is beyond the scope of this letter to recount the atrocities 378.22
inflicted upon the guiltless Bahá'ís of Iran or to answer, one by one, the accusations leveled against them. But let us ask all just and fair-minded people
only one question: If, according to the much publicized statements of the
Public Prosecutor, Bahá'ís are not arrested and executed because of their belief,
and are not even imprisoned on that account, how is it that, when a group
of them are arrested and each is charged with the same "crime" of "spying,"
if one of them recants his belief, he is immediately freed, a photograph of
him and a description of his defection are victoriously featured in the newspapers, and respect and glory are heaped upon him? What kind of spying,
subversion, illegal accumulation of goods, aggression or conspiracy or other
"crime" can it be that is capable of being blotted out upon the recantation of
one's beliefs? Is this not a clear proof of the absurdity of the accusations?

In spite of all this, the Bahá'í community of Iran, whose principles have 378.23
been described earlier in this statement, announces the suspension of the
Bahá'í organizations throughout Iran in order to establish its good intentions
and in conformity with its basic tenets concerning complete obedience to the
instructions of the Government. Henceforth, until the time when, God willing, the misunderstandings are eliminated and the realities are at last made

---

378-12. The cloth with which Muslim women in the Middle East cover themselves when they
go out in public and which Bahá'í women in Iran are now forced to wear.

manifest to the authorities, the National Assembly and all Local Spiritual Assemblies and their Committees are disbanded, and no one may any longer be designated a member of the Bahá'í Administration.

378.24   The Bahá'í community of Iran hopes that this step will be considered a sign of its complete obedience to the Government in power. It further hopes that the authorities—including the esteemed Public Prosecutor of the Islamic Revolution who says that there is no opposition to and no enmity towards individual Bahá'ís, who has acknowledged the existence of a large Bahá'í community and has, in his interview, guaranteed its members the right to live and be free in their acts of worship—will reciprocate by proving their good intentions and the truth of their assurances by issuing orders that pledge, henceforth:

1. To bring to an end the persecutions, arrests, torture and imprisonment of Bahá'ís for imaginary crimes and on baseless pretexts, because God knows—and so do the authorities—that the only "crime" of which these innocent ones are guilty is that of their beliefs, and not the unsubstantiated accusations brought against them;
2. To guarantee the safety of their lives, their personal property and belongings, and their honor;
3. To accord them freedom to choose their residence and occupation and the right of association based on the provisions of the Constitution of the Islamic Republic;
4. To restore all the rights which have been taken away from them in accordance with the groundless assertions of the Public Prosecutor of the country;
5. To restore to Bahá'í employees the rights denied them by returning them to their jobs and by paying them their due wages;
6. To release from prison all innocent prisoners;
7. To lift the restrictions imposed on the properties of those Bahá'ís who, in their own country, have been deprived of their belongings;
8. To permit Bahá'í students who wish to continue their studies abroad to benefit from the same facilities that are provided to others;
9. To permit those Bahá'í youth who have been prevented from continuing their studies in the country to resume their education;
10. To permit those Bahá'í students stranded abroad who have been deprived of foreign exchange facilities to receive their allowances as other Iranian students do;
11. To restore Bahá'í cemeteries and to permit Bahá'ís to bury their dead in accordance with Bahá'í burial ceremonies;
12. To guarantee the freedom of Bahá'ís to perform their religious rites; to conduct funerals and burials including the recitation of the Prayer for the Dead; to solemnize Bahá'í marriages and divorces, and to carry out

all acts of worship and laws and ordinances affecting personal status; because although Bahá'ís are entirely obedient and subordinate to the Government in the administration of the affairs which are in the jurisdiction of Bahá'í organizations, in matters of conscience and belief, and in accordance with their spiritual principles, they prefer martyrdom to recantation or the abandoning of the divine ordinances prescribed by their Faith;

13. To desist henceforth from arresting and imprisoning anyone because of his previous membership in Bahá'í organizations.

Finally, although the order issued by the Public Prosecutor of the Islamic Revolution was unjust and unfair, we have accepted it. We beseech God to remove the dross of prejudice from the hearts of the authorities so that, aided and enlightened by His confirmations, they will be inspired to recognize the true nature of the affairs of the Bahá'í community and come to the unalterable conviction that the infliction of atrocities and cruelties upon a pious band of wronged ones, and the shedding of their pure blood, will stain the good name and injure the prestige of any nation or government, for what will, in truth, endure are the records of good deeds, and of acts of justice and fairness, and the names of the doers of good. These will history preserve in its bosom for posterity. 378.25

Respectfully,
(SIGNED) THE NATIONAL SPIRITUAL
ASSEMBLY OF THE BAHÁ'ÍS OF IRAN

379
Social and Economic Development— New Field of Bahá'í Service
20 OCTOBER 1983

To the Bahá'ís of the World

Dear Bahá'í Friends,

The soul-stirring events in Bahá'u'lláh's native land and the concomitant advance into the theater of world affairs of the agencies of His Administrative Order have combined to bring into focus new possibilities in the evolution of the Bahá'í world community. Our Riḍván message this year captured these implications in its reference to the opening before us of a wider horizon in whose light can dimly be discerned new pursuits and undertakings upon which we must soon embark. These portend our greater involvement in the development of the social and economic life of peoples. 379.1

Dynamic Coherence between the Spiritual and Practical Requirements of Life

379.2　From the beginning of His stupendous mission, Bahá'u'lláh urged upon the attention of nations the necessity of ordering human affairs in such a way as to bring into being a world unified in all the essential aspects of its life. In unnumbered verses and tablets He repeatedly and variously declared the "progress of the world" and the "development of nations" as being among the ordinances of God for this day.[1] The oneness of mankind, which is at once the operating principle and ultimate goal of His Revelation, implies the achievement of a dynamic coherence between the spiritual and practical requirements of life on earth. The indispensability of this coherence is unmistakably illustrated in His ordination of the Mashriqu'l-Adhkár, the spiritual center of every Bahá'í community round which must flourish dependencies dedicated to the social, humanitarian, educational and scientific advancement of mankind. Thus, we can readily appreciate that although it has hitherto been impracticable for Bahá'í institutions generally to emphasize development activities, the concept of social and economic development is enshrined in the sacred Teachings of our Faith. The beloved Master, through His illuminating words and deeds, set the example for the application of this concept to the reconstruction of society. Witness, for instance, what social and economic progress the Iranian believers attained under His loving guidance and, subsequently, with the unfailing encouragement of the Guardian of the Cause.

Self-Sufficiency, Self-Reliance, and the Preservation of Human Honor

379.3　Now, after all the years of constant teaching activity, the Community of the Greatest Name has grown to the stage at which the processes of this development must be incorporated into its regular pursuits; particularly is action compelled by the expansion of the Faith in Third World countries where the vast majority of its adherents reside. The steps to be taken must necessarily begin in the Bahá'í Community itself, with the friends endeavoring, through their application of spiritual principles, their rectitude of conduct and the practice of the art of consultation, to uplift themselves and thus become self-sufficient and self-reliant. Moreover, these exertions will conduce to the preservation of human honor so desired by Bahá'u'lláh. In the process and as a consequence, the friends will undoubtedly extend the benefits of their efforts to society as a whole, until all mankind achieves the progress intended by the Lord of the Age.

An Office of Social and Economic Development at the World Center

379.4　It is indeed propitious that systematic attention be given to this vital sphere of Bahá'í endeavor. We are happy, therefore, to announce the establishment

---

379-1. TB, pp. 129–30.

at the World Center of the Office of Social and Economic Development, which is to assist the Universal House of Justice to promote and coordinate the activities of the friends throughout the world in this new field.

The International Teaching Center and, through it, the Continental Boards of Counselors are poised for the special responsibilities which devolve upon them to be alert to possibilities for extending the development of social and economic life both within and outside the Bahá'í Community, and to advise and encourage the Assemblies and friends in their strivings. 379.5

Working from the Grass Roots

We call now upon National Spiritual Assemblies to consider the implications of this emerging trend for their respective communities, and to take well-conceived measures to involve the thought and actions of Local Spiritual Assemblies and individuals in the devising and implementing of plans within the constraints of existing circumstances and available resources. Progress in the development field will largely depend on natural stirrings at the grass roots, and it should receive its driving force from those sources rather than from an imposition of plans and programs from the top. The major task of National Assemblies, therefore, is to increase the local communities' awareness of needs and possibilities, and to guide and coordinate the efforts resulting from such awareness. Already in many areas the friends are witnessing the confirmations of their initiatives in such pursuits as the founding of tutorial and other schools, the promotion of literacy, the launching of rural development programs, the inception of educational radio stations, and the operation of agricultural and medical projects. As they enlarge the scope of their endeavors other modes of development will undoubtedly emerge. 379.6

This challenge evokes the resourcefulness, flexibility and cohesiveness of the many communities composing the Bahá'í world. Different communities will, of course, perceive different approaches and different solutions to similar needs. Some can offer assistance abroad, while, at the outset, others must of necessity receive assistance; but all, irrespective of circumstances or resources, are endowed with the capacity to respond in some measure; all can share; all can participate in the joint enterprise of applying more systematically the principles of the Faith to upraising the quality of human life. The key to success is unity in spirit and in action. 379.7

Development Activities—A Reinforcement of the Teaching Work

We go forward confident that the wholehearted involvement of the friends in these activities will ensure a deeper consolidation of the community at all levels. Our engagement in the technical aspects of development should, however, not be allowed to supplant the essentials of teaching, which remains the primary duty of every follower of Bahá'u'lláh. Rather should our increased activities in the development field be viewed as a reinforcement of the teach- 379.8

ing work, as a greater manifestation of faith in action. For, if expansion of the teaching work does not continue, there can be no hope of success for this enlarged dimension of the consolidation process.

Call to Action

379.9 Ultimately, the call to action is addressed to the individual friends, whether they be adult or youth, veteran or newly enrolled. Let them step forth to take their places in the arena of service where their talents and skills, their specialized training, their material resources, their offers of time and energy and, above all, their dedication to Bahá'í principles, can be put to work in improving the lot of man.

379.10 May all derive enduring inspiration from the following statement written in 1933 by the hand of our beloved Guardian:

379.10a The problems which confront the believers at the present time, whether social, spiritual, economic or administrative will be gradually solved as the number and the resources of the friends multiply and their capacity for service and for the application of Bahá'í principles develops. They should be patient, confident and active in utilizing every possible opportunity that presents itself within the limits now necessarily imposed upon them. May the Almighty aid them to fulfill their highest hopes . . .[2]

With loving Bahá'í greetings,
THE UNIVERSAL HOUSE OF JUSTICE

380
Passing of Continental Counselor Raúl Pavón
23 OCTOBER 1983

To all National Spiritual Assemblies

380.1 DEEPLY DEPLORE LOSS ZEALOUS SERVANT CAUSE BAHÁ'U'LLÁH RAÚL PAVÓN DISTINGUISHED PROMOTER FAITH AND INDEFATIGABLE WORKER IN TEACHING INDIGENOUS PEOPLES LATIN AMERICA.[1] HIS OUTSTANDING SERVICES AS MEMBER BOARD COUNSELORS AMERICAS AND UNIQUE ENDEAVORS ESTABLISHMENT FIRST BAHÁ'Í RADIO STATION WARMLY REMEMBERED. PRAYING HOLY SHRINES PROGRESS HIS RADIANT SOUL WORLDS GOD AND SOLACE LOVING COMFORT HIS BEREAVED FAMILY. FEEL CONFIDANT HIS DEARLY CHERISHED PARENTS REJOICE ABHÁ KINGDOM RANGE HIS DEDICATED SERVICES. URGE ALL COMMUNITIES AMERICAS HOLD MEMORIAL GATHERINGS BEFITTING HIS HIGHLY VALUED CONTRIBUTIONS BELOVED FAITH.

UNIVERSAL HOUSE OF JUSTICE

379-2. Letter dated 11 March 1933 to an individual.
380-1. For an account of the life and services of Raúl Pavón, see BW 19:616.

381
Continued Pressure against the Bahá'ís of Iran
7 November 1983

To all National Spiritual Assemblies

Dear Bahá'í Friends,

381.1 Recent events in Iran following the banning of the Bahá'í administration indicate, unfortunately, the continuation of intense pressures against the defenseless Bahá'í community in that country. Many of these pressures are being exerted by the authorities in the hope that the Bahá'ís will recant their faith and trade their love of Bahá'u'lláh for the comfort and security which the authorities offer to them in exchange.

381.2 With great sadness the Universal House of Justice announces that many friends in prison are being tortured when they refuse to submit to the will of the authorities to deny their love for Bahá'u'lláh. In addition, while it is true that no Bahá'ís have been executed since the statement calling for the disbanding of the administration of the Faith was made by the Attorney-General on 29 August, it has been reported that on 19 September a Bahá'í farmer in the town of Khuy, Mr. Akbar Ḥaqíqí, died as a result of a beating by a mob instigated by the clergy. Moreover, at least 140 Bahá'ís have been arrested in all parts of the country following the Attorney-General's statement, 50 of whom were detained on 30 October in the Caspian Sea area. Although a number of the friends have been released, the total number of Bahá'ís still imprisoned in Iran, according to our records, stands at over 450.

381.3 Three believers who were advanced in age have died in prison and thus have joined the ranks of those who have laid down their lives in service to the Cause. They are:

> Mr. Ḥusayn Nayyirí-Iṣfahání, 64 years old—imprisoned in Isfahan and died just as he was going into court for his trial on 29 November 1982.
>
> Mr. Aḥmad-'Alí Thábit-Sarvistání, 67 years old—died in prison in Shiraz on 30 June 1983.
>
> Mr. Muḥammad Ishráqí, 81 years old, an Auxiliary Board member—died in prison in Tehran on 31 August 1983.

381.4 Word has also recently been received that in the city of Dizfúl, a Bahá'í woman, Mrs. Írán Raḥímpúr (Khurmá'í), was executed on 12 May 1982 after giving birth to her child. The baby was taken away by the Muslims and his fate is unknown.

381.5 One of the most obvious examples of persecution and proof of the evil intention of the Iranian authorities to uproot the Faith in that land is the destruction and desecration of Bahá'í cemeteries. Recently there was an official advertisement in the newspapers in Iran indicating that the tombstones in the Bahá'í cemetery in Tehran were being put up for sale. Since all markers on the graves are apparently being eliminated, it is possible that no trace of the Bahá'í cemetery will remain in future. . . .

With loving Bahá'í greetings,
DEPARTMENT OF THE SECRETARIAT

382
Appointment of Continental Counselors
6 DECEMBER 1983

To all National Spiritual Assemblies

Dear Bahá'í Friends,

382.1 The Universal House of Justice announces with pleasure the appointment of the following Continental Counselors:

In Africa:	Mr. Mabuku Wingi
In the Americas:	Mrs. Isabel P. de Calderón

With loving Bahá'í greetings,
DEPARTMENT OF THE SECRETARIAT

383
Response of Individuals to Persecution of Iranian Bahá'ís
7 DECEMBER 1983

The National Spiritual Assembly of the Bahá'ís of the United States

Dear Bahá'í Friends,

383.1 In recent months the Universal House of Justice has received comments from individual American believers expressing their distress over the continuing persecution of their beloved co-workers in Iran, and proposing such ways of registering their protest as public demonstrations and the wearing of armbands and ribbons. Consideration of these comments has prompted the instruction of the House of Justice that we write you thus.

The Need for a Proper Perspective

383.2 It is indeed difficult, given the heartbreaking disabilities imposed upon the Iranian Bahá'í Community and the seeming impotence of the American Community directly to effect a positive change, for the friends to be at ease.[1] But that the situation in Iran, grave as it is, should lead to feelings of depression and alienation on the part of the American believers, as has been reported, or that it should be allowed to hamper their success in teaching on the home front, suggests the need for a proper perspective. You will sense in the comments and appeals of the beloved Guardian addressed to the American Community during 1955–56 a striking resemblance between the reactions and attitudes of the friends towards the crisis then and now. A rereading of his letter of 20 August 1955 describing the situation then and the opportunities it created for the proclamation of the Faith, and his cables of 5 January, 2 February and 22 June 1956 (*Citadel of Faith,* pages 133–42) urging action is most instructive. The American Community has displayed in the past a tendency towards periodic immobility, a condition the Guardian commented upon from time to time and that was the main concern of his last letter to the America dated 21 September 1957 (*Citadel of Faith,* pages 151–58). While the House of Justice does not now have the impression that the American believers as a whole are depressed, it feels it might be helpful to all concerned to make the following comments on the basis of that assumption, as conveyed in the correspondence received.

383.3 The American Bahá'í Community has for many years been in the forefront of defending the weak and oppressed. Its distinction in this respect won the repeated praise of the beloved Guardian, as, for example, *The Advent of Divine Justice,* his celebrated letter of 25 December 1938, affirms. Against an enumeration of afflictions that at that time threatened to force the majority of the existing Bahá'í communities into the shadows of retreat, he described the Bahá'í Community in North America as the "one chief remaining citadel, the mighty arm which still raises aloft the standard of an unconquerable Faith."[2] The many instances during the subsequent years in which that Community came to the aid of other defenseless communities are a testimony to the effectiveness of the American responses in times of need and trouble. But the situation in the world and in the Bahá'í community has changed. Consequently, the accustomed reactions to American interventions have also changed. While this change does not nullify the preeminent role destined for America in the eventual efflorescence of Bahá'u'lláh's system in the world, it does require the American believers to obtain a deeper understanding of their situation in relation to the changed circumstances.

383-1. Relations between the United States and Iran at the time were antagonistic. On 4 November 1980 Iranians stormed the American Embassy and held United States diplomatic and military personnel hostage until 18 January 1981. The United States government responded by freezing millions of dollars of Iranian assets held by United States banks and companies. The enmity between the two governments made it all the more difficult for the American Bahá'í community to initiate effective measures to ease the suffering of the Bahá'ís of Iran.

383-2. ADJ, p. 6.

A Combination of Pragmatism and Spirituality

383.4 The friends' response to the Iranian situation should neither be solely pragmatic nor solely spiritual, but a combination of both; moreover, it must not only meet the test of the immediate crisis, it must also match the challenge of the historic moment. In both these respects a fair appraisal of the results thus far should steel their resolve, not induce depression. Even though when viewed from the truculent reactions of the powers in Iran, the petitions and protests of the American Community in particular, the Bahá'í communities in general, appear to have been ineffective in stemming the persecutions of our brethren, there is strong reason to believe that had the Bahá'ís around the world not taken these actions, the plight of the Iranian friends would be far more tragic. And while there is a vast array of direct action that could be, or could have been, contemplated, not every proposed action is fitting or timely. Actions perceived to be appropriate within the framework of American society can be counterproductive when viewed in the broader framework of a world community. The Bahá'í world, in its complex diversity, has been guided to act according to the assessment by the House of Justice of the immediate circumstances and the resource at hand, and in consideration of the opinions and judgments of the National Spiritual Assembly of Iran, which until its recent disbanding, functioned with heroic verve in defending the rights of those under its charge. Since the Bahá'í Community must operate as a cohesive entity, it is not conducive to its success for any single unit of that world-embracing organism to go off at a tangent in its reactions to the situation, as well-intentioned as that unit might be.

A Review of Significant Achievements

383.5 The House of Justice suggests that the American friends look at the great contrast between the relatively low profile the Faith kept before the crisis and the obvious prominence it has achieved since then; that they consider the remarkable impression that the national Bahá'í communities throughout the world have made upon their national governments and international organizations; that they contemplate the extraordinary breakthroughs for the Faith in the media of the world compared to the media's reticence toward the Cause a few years ago; and, finally, that they think of the implications of all these new advantages for the teaching work and determine dispassionately whether, in fact, the opportunities for the progress of the Faith are greater or lesser than before.

Opposition—Creating Opportunities for Progress

383.6 There is no gainsaying the insight set forth in the Teachings that opposition to the Faith creates opportunities for its progress. By their radiant submission to this insight, the Iranian Bahá'ís have surely demonstrated their conviction as to its potency. Motivated by their understanding of it, a number of

Bahá'í communities in different parts of the world have engaged in activities that are producing concrete results in the expansion and consolidation of the Faith. For the American believers to give true expression to the anguish they feel over the persecution of their brethren, they too must capture the value of this insight and act upon it. The vision, the admonitions, the encouragement which *The Advent of Divine Justice* conveyed so many years ago, and which the Guardian elaborated in so many of his subsequent messages, are as appropriate, perhaps even more so, to the current situation as when they were first given. The responses he attempted long ago to evoke are no less desirable and necessary now; a deeper spiritual awareness, a deeper spiritual maturity are called for if the American community must pursue successfully its destined course. It should not be forgotten that the beloved Master promised, as the Guardian recalled in his early letters to America, to send intellectual, rather than physical, tests to the friends there in order to purify and better prepare them for their divinely appointed destiny. Nor should the prerequisites of success the Guardian outlined for the American Bahá'í Community be overlooked. It is in the framework of these prerequisites that the success or failure of the American Community must ultimately be measured, not from the standpoint of any frustration in effecting a desired change in the Iranian situation, which, despite its admitted distress for the community, has done more to proclaim the Cause abroad than any other experience in this century.

Toward A Response Worthy of the Valorous Victims

In a society whose people are as protestant and demonstrative as the Americans, such public displays and symbolic gestures as street demonstrations and the wearing of armbands and ribbons may have a certain appeal to public attention and may even prove to be effective in the proper context and under the proper circumstances. But the evocative power of such activities is difficult to sustain over a long period. Particularly is this so in your country where the public demonstrations of a myriad groups constantly compete for attention. The House of Justice feels that while these ephemeral activities might relieve the immediate anxieties of some of the friends, they would have no measurable effect on the course of events in Iran. Our human resources are so limited, they must be devoted to the most effective means of responding to the situation. Of infinitely more value are actions that reflect the spiritual profundity underlying these persecutions and that match the dignity, radiance and optimism for the Faith of the valorous victims, who, as their published testimonies show, are quite clear about the reasons for their suffering and dying. Moreover, in the long view, it would not serve the best interests of the Faith for its members, at the very time of their emergence from obscurity, to impress themselves upon the consciousness of the public as a community identified with such symbols as armbands and ribbons.

383.8 An important point to bear in mind is that our activities in defense of the Iranian believers must be supported by those toward the accomplishment of our stated goals. Preoccupation with the Iranian crisis, at the expense of neglecting the Seven Year Plan, would divert the Bahá'í world community from achieving the very success necessary to the strengthening of the Faith and the confounding of its enemies. Such a neglect would be unworthy of the sufferings of the Iranian friends. Indeed, the ideal being constantly pursued is to defend them vigorously, while, at the same time, exploiting the opportunities created by their sacrifices to promote the Cause of God. The unprecedented publicity, the unremitting appeals to governments and international bodies, the increased contacts with leaders of thought and, above all, the redoubling of teaching activities and the deeper consolidation of the Bahá'í Community, as called for by the House of Justice, are ultimately the best means of defending and securing the relief of the beleaguered Iranian Community; besides, by these means will the world community of Bahá'ís be better prepared to meet the inevitable opposition yet to come elsewhere.

383.9 It is the fervent prayer of the House of Justice that the American friends will be sustained by the same unconquerable spirit that fortifies the resolute faith of their brothers and sisters in Iran, that they will be refreshed from the same fountain of hope inspired by Bahá'u'lláh's incontrovertible assurance to the loved ones of God in that land, whom He exhorts "to patience, to acquiescence, and to tranquillity," and addresses in these consoling words:

> Whatever hath befallen you, hath been for the sake of God. This is the truth, and in this there is no doubt. You should, therefore, leave all your affairs in His Hands, place your trust in Him, and rely upon Him. He will assuredly not forsake you. In this, likewise, there is no doubt. No father will surrender his sons to devouring beasts; no shepherd will leave his flock to ravening wolves. He will most certainly do his utmost to protect his own.
>
> If, however, for a few days, in compliance with God's all-encompassing wisdom, outward affairs should run their course contrary to one's cherished desire, this is of no consequence and should not matter. Our intent is that all the friends should fix their gaze on the Supreme Horizon, and cling to that which hath been revealed in the Tablets. . . .[3]

With loving Bahá'í greetings,
DEPARTMENT OF THE SECRETARIAT

383-3. Excerpts from "Fire and Light," in BW 18:10.

384

Service in Voluntary Nonsectarian Organizations
13 December 1983

To the National Spiritual Assemblies of the Bahá'ís of Alaska, Australia, Austria, Belgium, Canada, Denmark, Finland, France, Germany, the Hawaiian Islands, Iceland, Ireland, Italy, Luxembourg, Netherlands, New Zealand, Norway, Portugal, Spain, Sweden, Switzerland, United Kingdom and the United States

Dear Bahá'í Friends,

384.1 The Universal House of Justice has been consulting upon aspects of youth service in pioneering throughout the Bahá'í world, and has requested that we convey its views on service in other lands undertaken by Bahá'í youth with voluntary nonsectarian organizations.

384.2 In the past, the policy adopted by some National Assemblies was to discourage young Bahá'ís from enrolling to serve in activities sponsored by non-Bahá'í voluntary organizations, as the Assemblies were under the impression that these young people would not be able to engage in direct teaching, nor participate, for the most part, in Bahá'í activities while serving abroad in such programs. Perhaps in some instances the Bahá'ís involved were not sure how to function as members of the Bahá'í community in order to give each aspect of their lives its proper due.

384.3 In the light of experience, however, it is now clear that we should have no misgivings in encouraging young Bahá'ís to enroll in such voluntary service organization programs as the United Nations Volunteers, United States Peace Corps, Canadian University Services Overseas (CUSO) and similar Canadian agencies, the British Volunteer Program (BVP) of the United Kingdom, and other voluntary service organizations. Other countries such as Germany, the Netherlands, and the Scandinavian lands are understood to have similar service organizations which are compatible with Bahá'í development goals as now tentatively envisaged.

384.4 Some of the advantages of such service to the Faith are worth mentioning. Volunteers will receive thorough orientation and sometimes will be taught basic skills which will enable them to help the Bahá'í community in projects undertaken in developing countries. Wherever they serve, these volunteers should be able to participate in Bahá'í activities and contribute to the consolidation of the Bahá'í community. The freedom to teach is to a large extent dependent upon the local interpretation of the group leader, but even if volunteers do not engage in direct teaching, being known as Bahá'ís and showing the Bahá'í spirit and attitude towards work and service should attract favorable attention and may, in many instances, be instrumental in attracting individuals to the Faith of Bahá'u'lláh. And finally, the period of overseas service often produces a taste for such service, and volunteers may well offer to directly promote the pioneer work either in the same country or in another developing country.

384.5 It is well known that a considerable number of Bahá'ís have already gone abroad to serve with these agencies and that others have espoused the Faith while serving in foreign lands with voluntary service organizations. . . .

384.6 National Spiritual Assemblies which hold orientation courses for pioneers may benefit from including the subject of rural development in their program, and, as in the past, from inviting people who have served in voluntary service organizations to participate in the planning of orientation programs and in having them share their experiences as volunteer workers in developing countries.

384.7 The House of Justice expresses the hope that the information contained in this letter will dispel the misunderstandings that have in the past surrounded the question of participation of Bahá'í youth in projects sponsored by non-Bahá'í voluntary organizations.

With loving Bahá'í greetings,
DEPARTMENT OF THE SECRETARIAT

385
Acute Needs of the International Funds
2 JANUARY 1984

To the Followers of Bahá'u'lláh in every Land

Dearly loved Friends,

385.1 The gathering of the representatives of the Bahá'í world at the International Convention last Riḍván was held in an atmosphere charged with awareness of the sacrifices being made by our fellow believers in Iran and with eager anticipation of the new prospects opening before the Cause as a result of changing conditions in the world, the widespread publicity that the Faith has received in all continents, and the growing maturity of its administrative institutions.

385.2 During the succeeding eight months we have been developing the agencies and formulating the plans to enable the Faith to seize the unprecedented opportunities now before it, but we are confronted with a shortage of funds which, if not remedied, could frustrate these plans. For the last two years there has been a decline in the amount of contributions to the international funds of the Faith, and we note that many national funds also are facing the danger of deficits.

Four Areas Requiring Immediate Action

Beyond carrying on the general work of the Cause there are four areas where immediate action is required.   385.3

The first is the completion of the Mashriqu'l-Adhkárs in India and Samoa.[1] Any delay in this work can but make it more expensive and also seriously injure the reputation of the Faith in these two vital areas.   385.4

The second is the development of the World Center, the focal point of the entire Administrative Order of the Faith where, in the words of Shoghi Effendi, "the dust of its Founders reposes, where the processes disclosing its purposes, energizing its life and shaping its destiny all originate."[2]   385.5

The third is in the prosecution of programs of social and economic development. Bahá'í communities in many lands have attained a size and complexity that both require and make possible the implementation of a range of activities for their social and economic development which will not only be of immense value for the consolidation of these communities and the development of their Bahá'í life, but will also benefit the wider communities within which they are embedded and will demonstrate the beneficial effects of the Bahá'í Message to the critical gaze of the world. Funds for the initiation and carrying out of these projects will be dispensed very gradually and with great care in order not to undermine the natural growth and sense of responsibility of these communities, but the field is so vast, the opportunities so far-reaching, that the need will stretch the resources of the Cause to the uttermost.   385.6

The fourth area is in the development and coordination of worldwide efforts to present to a far more extensive audience than ever before the divine remedy for the problems besetting society and its individual members, to establish the universality of the Faith and the implications of its teachings in the eyes of statesmen, and to ensure that the leaders of thought become thoroughly aware of the Bahá'í Revelation and the profundity of its message.   385.7

The work on the Temples is already well advanced and must not be stopped; the development of the agencies of the World Center, located in one of the principal trouble-spots of the world, cannot be indefinitely held back; the time for the expansion of social and economic development as an aspect of the work of the Cause has arrived and it cannot be neglected without grave consequences to the life of Bahá'í communities; the unprecedented opportunity for proclamation of the Faith has been given to us as a direct result of the persecutions inflicted on the believers in the Cradle of the Faith. If we are to be worthy of the sacrifices of these valiant friends, and if we are not to betray the trust that Bahá'u'lláh has placed upon us for the redemption of   385.8

385-1. The House of Worship in New Delhi, India, was dedicated 23–27 December 1986; the House of Worship in Apia, Western Samoa, was dedicated on 1 September 1984.
385-2. GPB, p. 355.

mankind in this hour of its acute need, we must not fail to seize the opportunities now before us.

385.9 This fourfold challenge faces us at the very time when the world is in the midst of an economic crisis and is overshadowed with threats of war and other disasters. These conditions, far from daunting the followers of Bahá'u'lláh, can only drive home to us the urgency for our response.

Call for Renewed Consecration

385.10 We therefore call upon every true-hearted Bahá'í to consecrate his life anew to the service of God and the betterment of the lot of mankind, so that manpower will not be lacking in the fields of pioneering, teaching and administrative service. Most urgently, may every believer give sacrificially of his substance, each in accordance with his means, to the funds of the Cause, local, national, continental and international, so that the material resources—the lifeblood of all activities—will be adequate to the tremendous work that we have to perform in the months and years immediately ahead. It requires a concentration of effort, a unity of purpose and a degree of self-sacrifice to match the heroic exertions of the victors of past plans in the progress of the Cause.

With loving Bahá'í greetings,
UNIVERSAL HOUSE OF JUSTICE

386
Message to Youth—International Youth Year
3 JANUARY 1984

To the Bahá'í Youth of the World

Dear Bahá'í Friends,

386.1 The designation of 1985 by the United Nations as International Youth Year opens new vistas for the activities in which the young members of our community are engaged. The hope of the United Nations in thus focusing on youth is to encourage their conscious participation in the affairs of the world through their involvement in international development and such other undertakings and relationships as may aid the realization of their aspirations for a world without war.

386.2 These expectations reinforce the immediate, vast opportunities begging our attention. To visualize, however imperfectly, the challenges that engage us now, we have only to reflect, in the light of our sacred Writings, upon the confluence of favorable circumstances brought about by the accelerated unfolding of the Divine Plan over nearly five decades, by the untold potencies of the spiritual drama being played out in Iran, and by the creative energy stimulated by awareness of the approaching end of the twentieth century.¹ Undoubtedly, it is within your power to contribute significantly to shaping the societies of the coming century; youth can move the world.

How apt, indeed how exciting, that so portentous an occasion should be presented to you, the young, eager followers of the Blessed Beauty, to enlarge the scope of your endeavors in precisely that arena of action in which you strive so conscientiously to distinguish yourselves! For in the theme proposed by the United Nations—"Participation, Development, Peace"—can be perceived an affirmation that the goals pursued by you, as Bahá'ís, are at heart the very objects of the frenetic searchings of your despairing contemporaries.

You are already engaged in the thrust of the Seven Year Plan, which provides the framework for any further course of action you may now be moved by this new opportunity to adopt. International Youth Year will fall within the Plan's next phase; thus the activities you will undertake, and for which you will wish to prepare even now, cannot but enhance your contributions to the vitality of that Plan, while at the same time aiding the proceedings for the Youth Year. Let there be no delay, then, in the vigor of your response.

Providing a Living Example of Virtue

A highlight of this period of the Seven Year Plan has been the phenomenal proclamation accorded the Faith in the wake of the unabating persecutions in Iran; a new interest in its Teachings has been aroused on a wide scale. Simultaneously, more and more people from all strata of society frantically seek their true identity, which is to say, although they would not so plainly admit it, the spiritual meaning of their lives; prominent among these seekers are the young. Not only does this knowledge open fruitful avenues for Bahá'í

386-1. The unfolding of 'Abdu'l-Bahá's Divine Plan, the purpose of which is to spread the Bahá'í Faith throughout the world, began in 1937. This was under Shoghi Effendi's direction with the launching of the first Seven Year Plan (1937–44) pursued by the Bahá'ís of the United States and Canada. The unfoldment continued with the second Seven Year Plan (1946–53) in North America and plans of varying durations in other countries, which were followed by the Ten Year World Crusade (1953–63). The Universal House of Justice, following the pattern Shoghi Effendi initiated, launched the Nine Year Plan (1964–73), the Five Year Plan (1974–79), the Seven Year Plan (1979–86), the Six Year Plan (1986–92), and the Three Year Plan (1993–96). The execution of these plans resulted in an enormous increase in the size of the Bahá'í community, in the number of its institutions, and in its ability to reflect the distinctive characteristics of Bahá'í life.

The persecution of the Iranian Bahá'ís and the destruction of Bahá'í holy places have released spiritual potencies that have further accelerated the unfolding of the Divine Plan. Moreover, the Universal House of Justice, in its Riḍván 1987 letter, identified the persecution of the Iranian Bahá'ís as "the mainspring" of the worldwide attention being focused on the Bahá'í Faith.

The approaching end of the twentieth century is significant, for 'Abdu'l-Bahá predicted that the unity of nations will be achieved in the twentieth century. For a discussion of the imminence of the Lesser Peace, see letter dated 29 July 1974 (no. 149). For Shoghi Effendi's explanation of the synchronization of the Lesser Peace, the development of Bahá'í institutions, and the completion of the buildings on the Arc at the Bahá'í World Center, see message dated 5 June 1975 on the significance of the Seat of the Universal House of Justice (no. 164).

initiative, it also indicates to young Bahá'ís a particular responsibility so to teach the Cause and live the life as to give vivid expression to those virtues that would fulfill the spiritual yearning of their peers.

386.6　For the sake of preserving such virtues much innocent blood has been shed in the past, and much, even today, is being sacrificed in Iran by young and old alike. Consider, for example, the instances in Shiraz last summer of the six young women, their ages ranging from 18 to 25 years, whose lives were snuffed out by the hangman's noose. All faced attempted inducements to recant their faith; all refused to deny their Beloved. Look also at the accounts of the astounding fortitude shown over and over again by children and youth who were subjected to the interrogations and abuses of teachers and mullahs and were expelled from school for upholding their beliefs. It, moreover, bears noting that under the restrictions so cruelly imposed on their community, the youth, rendered signal services, placing their energies at the disposal of Bahá'í institutions throughout the country. No splendor of speech could give more fitting testimony to their spiritual commitment and fidelity than these pure acts of selflessness and devotion. In virtually no other place on earth is so great a price for faith required of the Bahá'ís. Nor could there be found more willing, more radiant bearers of the cup of sacrifice than the valiant Bahá'í youth of Iran. Might it, then, not be reasonably expected that you, the youth and young adults living at such an extraordinary time, witnessing such stirring examples of the valor of your Iranian fellows, and exercising such freedom of movement, would sally forth, "unrestrained as the wind," into the field of Bahá'í action?[2]

Reinvigorating Individual Efforts

386.7　May you all persevere in your individual efforts to teach the Faith, but with added zest, to study the Writings, but with greater earnestness. May you pursue your education and training for future service to mankind, offering as much of your free time as possible to activities on behalf of the Cause. May those of you already bent on your life's work and who may have already founded families, strive toward becoming the living embodiments of Bahá'í ideals, both in the spiritual nurturing of your families and in your active involvement in the efforts on the home front or abroad in the pioneering field. May all respond to the current demands upon the Faith by displaying a fresh measure of dedication to the tasks at hand.

386.8　Further to these aspirations is the need for a mighty mobilization of teaching activities reflecting regularity in the patterns of service rendered by young Bahá'ís. The native urge of youth to move from place to place, combined with their abounding zeal, indicates that you can become more deliberately and numerously involved in these activities as travelling teachers. One pattern of

---

386-2. GWB, p. 339.

this mobilization could be short-term projects, carried out at home or in other lands, dedicated to both teaching the Faith and improving the living conditions of people. Another could be that, while still young and unburdened by family responsibilities, you give attention to the idea of volunteering a set period, say, one or two years, to some Bahá'í service, on the home front or abroad, in the teaching or development field. It would accrue to the strength and stability of the community if such patterns could be followed by succeeding generations of youth. Regardless of the modes of service, however, youth must be understood to be fully engaged, at all times, in all climes and under all conditions. In your varied pursuits you may rest assured of the loving support and guidance of the Bahá'í institutions operating at every level.

386.9 Our ardent prayers, our unshakable confidence in your ability to succeed, our imperishable love surround you in all you endeavor to do in the path of service to the Blessed Perfection.

The Universal House of Justice

387
Death of Three More Iranian Bahá'ís and Arrest of 250 More
17 January 1984

To all National Spiritual Assemblies

387.1 PRESSURES AGAINST BELEAGUERED BRETHREN CRADLE FAITH UNCEASING. THREE MORE STALWART SUPPORTERS GREATEST NAME JOINED RANKS MARTYRS, AS FOLLOWS:

- MR. 'ABDU'L-MAJÍD MUTAHHAR IMPRISONED ISFAHAN 4 SEPTEMBER 1983 DIED SHORTLY AFTER CONFINEMENT.
- MR. RAHMATU'LLÁH HAKÍMÁN MYSTERIOUSLY PASSED AWAY IN PRISON IN KIRMÁN EARLY JANUARY 1984 FEW DAYS AFTER BEING IMPRISONED
- ON 19 NOVEMBER IN VILLAGE OF MUHAMMADÍYYIH NEAR ISFAHAN, MR. BAHMAN DIHQÁNÍ, WELL KNOWN AND RESPECTED BAHÁ'Í, DIED AS RESULT OF MOB ATTACK. SINCE BURIAL NOT ALLOWED HIS VILLAGE, BAHÁ'ÍS CARRIED BODY TO NAJAFÁBÁD AND BURIED HIM THERE.
- FULL DETAILS CIRCUMSTANCES ALL THREE DEATHS UNKNOWN.

387.2 SINCE LAST REPORT 7 NOVEMBER OVER 250 BAHÁ'ÍS INCLUDING INFANTS AND CHILDREN HAVE BEEN ARRESTED IN ALL PARTS COUNTRY. NEARLY 70 OF THESE WERE DETAINED BETWEEN 31 DECEMBER 1983 AND 3 JANUARY 1984.

387.3 OTHER DESPICABLE ACTS AGAINST BAHÁ'ÍS INCLUDE THE FOLLOWING:

REVOLUTIONARY GUARDS PERMITTED ENTER BAHÁ'Í HOMES WITHOUT WARRANTS, CONFISCATE VALUABLE PERSONAL POSSESSIONS. COMPLAINTS TO AUTHORITIES UNAVAILING.

SOME IMPRISONED BAHÁ'ÍS WHO WERE EXPELLED FROM THEIR JOBS HAVE BEEN PROMISED RELEASE IF THEY REPAY ALL SALARIES PAID TO THEM FROM BEGINNING THEIR EMPLOYMENT, SOMETIMES UP TO 30 YEARS. . . .

Universal House of Justice

388
Passing of Ethel Revell, Member of the International Bahá'í Council
9 February 1984

To all National Spiritual Assemblies

388.1 DEEPLY GRIEVED ANNOUNCE PASSING ETHEL REVELL SAINTLY STEADFAST SELF-SACRIFICING PROMOTER CAUSE GOD.[1] BLESSED BY ASSOCIATION 'ABDU'L-BAHÁ COURSE HIS VISIT AMERICA AND RECEIPT TABLETS FROM HIM. HER TIRELESS LABORS STERLING QUALITIES EARNED ADMIRATION SHOGHI EFFENDI WHO APPOINTED HER INTERNATIONAL BAHÁ'Í COUNCIL AS ITS WESTERN ASSISTANT SECRETARY. THIS CROWN HER SERVICES CONTINUED MEMBERSHIP ELECTED COUNCIL SUBSEQUENT SERVICES MANY CAPACITIES WORLD CENTER INCLUDING SECRETARY HANDS HOLY LAND. URGE NATIONAL ASSEMBLIES HOLD BEFITTING MEMORIAL GATHERINGS HER HONOR IN ALL MASHRIQU'L-ADHKÁRS OTHER CENTERS. PRAYING HOLY SHRINES PROGRESS HER RADIANT SOUL ABHÁ KINGDOM.

Universal House of Justice

389
Message to National Bahá'í Women's Conference, Kenya
26 March 1984

To the National Bahá'í Women's Conference, Nairobi, Kenya

389.1 OCCASION NATIONAL WOMEN'S CONFERENCE WE GREET PARTICIPANTS EXPRESSING HOPE DELIBERATIONS WILL GIVE DIRECTION AND STIMULATION TO EFFORTS BAHÁ'Í WOMEN THROUGHOUT COUNTRY TO EXEMPLIFY BAHÁ'Í WAY OF LIFE, TO STEP FORWARD AS SUSTAINING AND DRIVING FORCE COMMUNITY AND TO PROVIDE LEADERSHIP THEIR PEERS. PRAYING HOLY SHRINES BOUNTIFUL BLESSING, FULFILLMENT HIGHEST ASPIRATIONS.

Universal House of Justice

388-1. For an account of the life and services of Ethel Revell, see BW 19:626–33.

390

Persecution of Bahá'ís in Morocco

29 March 1984

To National Spiritual Assemblies

FOR YOUR INFORMATION AND GUIDANCE FOLLOWING DEVELOPMENTS HAVE OCCURRED RE MOROCCAN SITUATION:

1. COURT OF APPEAL CASABLANCA POSTPONED HEARING UNTIL 12 APRIL.
2. INTERNATIONAL FEDERATION FOR HUMAN RIGHTS, A NONGOVERNMENTAL ORGANIZATION BASED IN PARIS WHICH HAD SENT ITS REPRESENTATIVE TO MOROCCO FOR FIRST HAND COVERAGE DEVELOPMENTS AFFECTING BAHÁ'Í CASES, HAS JUST ISSUED PRESS RELEASE IN PARIS OPENLY REPORTING SENTENCES, DECLARING THESE INCOMPATIBLE WITH STANDARD TOLERANCE PRACTICED IN MOROCCO AS WELL AS INCORPORATED ITS CONSTITUTION, AND WARNING THAT IF JUSTICE IS NOT DONE, FEDERATION IS PREPARED CALL ON OTHER INTERNATIONAL AGENCIES TAKE ACTION BEHALF BAHÁ'ÍS.
3. IF YOU ARE APPROACHED BY MEDIA YOU SHOULD STATE THAT ACTION TAKEN BY FEDERATION ARISES FROM FINDINGS ITS OWN REPRESENTATIVE; HOWEVER BAHÁ'ÍS THEMSELVES HAVE SO FAR REFRAINED FROM PUBLICITY AS THEY HOPE HEARINGS COURTS OF APPEAL WILL REDRESS WRONGS AND UPHOLD JUSTICE.

Universal House of Justice

391

Inaugural Broadcast of Radio Bahá'í Bolivia

2 April 1984

To the National Spiritual Assembly of the Bahá'ís of Bolivia

HAPPY NEWS INITIAL BROADCAST RADIO BAHÁ'Í CARACOLLO MARKED BY FAVORABLE LOCAL RECOGNITION SIGNIFICANCE EVENT ENTIRE REGION. ASSURE PRAYERS AMPLE REWARDS YOUR LONG EFFORTS. ARDUOUS DAYS DEVELOPMENT FULL POTENTIALS POWERFUL INSTRUMENT TEACHING DEEPENING UPLIFTMENT PEOPLES ALTIPLANO NOW BEFORE YOU. MAY GUIDANCE BLESSED BEAUTY BE YOURS WIN LAURELS IN SERVICE BELOVED FAITH.

Universal House of Justice

MESSAGES FROM THE UNIVERSAL HOUSE OF JUSTICE

392

Insidious Turn of Persecution of Iranian Bahá'ís

10 APRIL 1984

To all National Spiritual Assemblies

392.1 PERSECUTION FRIENDS CRADLE FAITH PERSISTS, TAKING EVEN MORE INSIDIOUS TURN. IN MARCH AT LEAST THREE PRISONERS DIED UNDER MYSTERIOUS CIRCUMSTANCES, TWO IN TEHRAN AND ONE IN BÁFT, KIRMÁN. BODY OF MUḤSIN RAḌAVÍ, 55 YEARS OLD, SHOWED EVIDENCE OF HANGING. OTHER TWO, GHULÁM-ḤUSAYN ḤASANZÁDIH-SHÁKIRÍ, 80, AND NUṢRATU'LLÁH ḌÍYÁ'Í, 61, WERE BURIED WITHOUT FRIENDS RELATIVES BEING INFORMED.

392.2 ARRESTS CONTINUE WITH LEAST AMOUNT PUBLICITY. SINCE LAST REPORT 17 JANUARY ALTHOUGH SOME BAHÁ'Í PRISONERS RELEASED, 111 HAVE BEEN ARRESTED, MOST OF WHOM WERE MEMBERS OF SPIRITUAL ASSEMBLIES BEFORE THEIR DISSOLUTION LAST YEAR. NUMBER BAHÁ'ÍS KNOWN TO BE IN PRISONS IN IRAN TOTALS 704.

392.3 ACCURATE INFORMATION IN HAND DESCRIBES TORTURES PERPETRATED AGAINST SOME VERY PROMINENT BAHÁ'ÍS. FOR EXAMPLE ONE BAHÁ'Í SEVERELY TORTURED UNDER EXTREME CIRCUMSTANCES FORCED CONFESS TO FALSE CHARGES. OTHERS SIMILARLY TORTURED RESISTED PRESSURES EXERTED ON THEM TO MAKE FALSE PUBLIC CONFESSIONS FOR BENEFIT RADIO TELEVISION. . . .

UNIVERSAL HOUSE OF JUSTICE

393

Execution of Three by Firing Squad

13 APRIL 1984

To all National Spiritual Assemblies

393.1 FURTHER OUR MESSAGE 10/11 APRIL JUST RECEIVED DISTRESSING NEWS EXECUTION BY FIRING SQUAD IN EVIN PRISON TEHRAN FOLLOWING BELIEVERS:

393.2 MR. KÁMRÁN LUṬFÍ, 32 YEAR OLD UNIVERSITY PROFESSOR, IMPRISONED 5 MAY 1983

393.3 MR. RAḤÍM RAḤÍMÍYÁN, 52 YEAR OLD BUSINESSMAN, IMPRISONED 5 MAY 1983

393.4 MR. YADU'LLÁH ṢÁBIRÍYÁN, 60 YEAR OLD PRINTING PRESS MANAGER, IMPRISONED 9 FEBRUARY 1982

393.5 SINCE EXECUTIONS NOT ANNOUNCED AND BODIES UNCEREMONIOUSLY BURIED WITHOUT FAMILIES RELATIVES BEING INFORMED, EXACT DATE EXECUTIONS UNKNOWN. FEARFUL OTHER BAHÁ'Í PRISONERS RECEIVED SAME FATE.

393.6 NOW ESTABLISHED FACT THAT GHULÁM-ḤUSAYN ḤASANZÁDIH-SHÁKIRÍ WAS ALSO EXECUTED BY FIRING SQUAD INSTEAD OF DYING MYSTERIOUSLY IN PRISON AS REPORTED. . . .

UNIVERSAL HOUSE OF JUSTICE

394
Riḍván Message 1984
RIḌVÁN 1984

To the Bahá'ís of the World

Dearly loved Friends,

The emergence from obscurity, which has been so marked a feature of the Cause of God during the first five years of the Seven Year Plan, has been attended by changes, both external and internal, affecting the Bahá'í world community. Externally, there are signs of a crystallization of a public image of the Cause—largely uninformed, however friendly—while internally growing maturity and confidence are indicated by increased administrative ability, a desire for Bahá'í communities to render service to the larger body of mankind and a deepening understanding of the relevance of the divine Message to modern problems. Both these aspects of change must be taken into consideration as we enter the third and final phase of the Seven Year Plan. 394.1

The Bahá'ís of Iran

The year just closing has been overshadowed by the continued persecution of the friends in Iran. They have been forced to disband their administrative structure, they have been harassed, dispossessed, dismissed from employment, made homeless and their children are refused education. Some six hundred men, women and children are now in prison, some denied any contact with their friends and relatives, some subjected to torture and all under pressure to recant their faith. Their heroic and exemplary steadfastness has been the mainspring in bringing the Cause out of obscurity, and it is the consolation of their hearts that their suffering results in unprecedented advances in teaching and proclaiming the divine Message to a world so desperately in need of its healing power. For this they embrace the final service of martyrdom. Our obligation is crystal clear. We cannot fail them now. Sacrificial action in teaching and promoting the Cause of God must follow every new instance of publicity arising from their persecution. Let this be our message to them of love and spiritual union. 394.2

Developments in the International Sphere

In the international sphere, the beloved Hands of the Cause, ever growing in our love and admiration, have, whenever their health has permitted, continued to uplift and encourage the friends and to promote the unity and onward march of the army of life. The International Teaching Center, operating from its world seat, has provided loving and wise leadership and direction to the Boards of Counselors.[1] Its sphere of service has been immensely 394.3

394-1. For messages concerning the establishment and duties of the International Teaching Center, see messages dated 5 June 1973 (no. 131), 8 June 1973 (no. 132), and 19 May 1983 (no. 361).

extended by the assignment of new responsibilities and by raising the number of its Counselor members to seven. The dedicated services of the Counselors in all the continents, ably supported by the Auxiliary Board members, have been invaluable in fostering the spiritual health and integrity of the worldwide community. To develop further this vital organ of the Administrative Order, it has been decided to establish a term of five years' service for those appointed to the Auxiliary Boards, commencing November 26, 1986. The work of the Bahá'í International Community in relationship with the United Nations has brought increasing appreciation of our social attitudes and principles, and in some instances—notably the sessions on human rights—the Bahá'í participation has been spectacular, again resulting from the heroism of the Persian friends. The Geneva office has been consolidated and additional staff engaged to deal with its expanding activities. In spite of severe problems the construction of the Indian and Samoan Houses of Worship has progressed satisfactorily, and the latter will be dedicated and opened to public worship between August 30th and September 3rd 1984, when the Universal House of Justice will be represented by the Hand of the Cause Amatu'l-Bahá Rúḥíyyih Khánum. Immediately following the International Convention last Riḍván, two new National Spiritual Assemblies were formed—in St. Lucia and Dominica. Two new radio stations will make their inaugural broadcasts this year, namely Radio Bahá'í of Bolivia, at Caracollo, and WLGI, the Bahá'í radio station at the Louis Gregory Institute, in the United States.[2] Bahá'í membership in eleven countries, all in the Third World and nine of them island communities, have reached or surpassed one per cent of the total population.

394.4   During the final months of the second phase of the Seven Year Plan a generous response has been made by believers and institutions alike to an appeal which set out the increasing needs of the International Fund. We are confident that sustained and regular contributions during the final phase of the Plan will enable its aims and objectives to be fully accomplished.

Entrance of the Cause onto the World Scene

394.5   The entrance of the Cause onto the world scene is apparent from a number of public statements in which we have been characterized as "model citizens," "gentle," "law-abiding," "not guilty of any political offense or crime"— all excellent but utterly inadequate insofar as the reality of the Faith and its aims and purposes are concerned. Nevertheless people are willing to hear about the Faith, and the opportunity must be seized. Persistently greater and greater efforts must be made to acquaint the leaders of the world, in all departments of life, with

---

394-2. The Louis Gregory Bahá'í Institute is a permanent Bahá'í teaching institute near Hemingway, South Carolina, that operates under the aegis of the National Spiritual Assembly of the Bahá'ís of the United States.

the true nature of Bahá'u'lláh's revelation as the sole hope for the pacification and unification of the world. Simultaneous with such a program must be unabated, vigorous pursuit of the teaching work, so that we may be seen to be a growing community, while universal observance by the friends of the Bahá'í laws of personal living will assert the fullness of, and arouse a desire to share in, the Bahá'í way of life. By all these means the public image of the Faith will become, gradually but constantly, nearer to its true character.

Social and Economic Development

The upsurge of zeal throughout the Bahá'í world for exploration of the new dimension of social and economic development is both heartwarming and uplifting to all our hopes. This energy within the community, carefully and wisely directed, will undoubtedly bring about a new era of consolidation and expansion, which in turn will attract further widespread attention, so that both aspects of change in the Bahá'í world community will be interactive and mutually propelling.

Consolidation of the Community

A prime element in the careful and wise direction needed is the achievement of victory in the Seven Year Plan, paying great attention to the development and strengthening of Local Assemblies. Great efforts must be made to encourage them to discharge their primary duties of meeting regularly, holding the Nineteen Day Feasts and observing Holy Days, organizing children's classes, encouraging the practice of family prayers, undertaking extension teaching projects, administering the Bahá'í Fund and constantly encouraging and leading their communities in all Bahá'í activities. The equality of men and women is not, at the present time, universally applied. In those areas where traditional inequality still hampers its progress we must take the lead in practicing this Bahá'í principle. Bahá'í women and girls must be encouraged to take part in the social, spiritual and administrative activities of their communities. Bahá'í youth, now rendering exemplary and devoted service in the forefront of the army of life, must be encouraged, even while equipping themselves for future service, to devise and execute their own teaching plans among their contemporaries.

New National Spiritual Assemblies

Now, as we enter the final, two-year phase of the Seven Year Plan, we rejoice in the addition of nine new National Spiritual Assemblies; three in Africa, three in the Americas, two in Asia, one in Europe, bringing the total number to 143. Five more are to be established in Riḍván 1985. They are Ciskei, Mali and Mozambique in Africa and the Cook Islands and the West Caroline Islands in Australasia. Thus the Plan will end with a minimum of 148 National Spiritual Assemblies. By that time plans must be approved for the completion of the Arc around the Monument Gardens on Mount Carmel, including the siting and designs of the three remaining buildings to be constructed around that Arc.

An Increasing Relationship to the Non-Bahá'í World

394.9 There can be no doubt that the progress of the Cause from this time onward will be characterized by an ever increasing relationship to the agencies, activities, institutions and leading individuals of the non-Bahá'í world. We shall acquire greater stature at the United Nations, become better known in the deliberations of governments, a familiar figure to the media, a subject of interest to academics, and inevitably the envy of failing establishments. Our preparation for and response to this situation must be a continual deepening of our faith, an unwavering adherence to its principles of abstention from partisan politics and freedom from prejudices, and above all an increasing understanding of its fundamental verities and relevance to the modern world.

Goals for the Final Phase of the Plan

394.10 Accompanying this Riḍván message are a call for 298 pioneers to settle in 79 national communities, and specific messages addressed to each of the present 143 national communities. They are the fruit of intensive study and consultation by the Universal House of Justice and the International Teaching Center, and set out the goals to be won and the objectives to be pursued by each national community so that Riḍván 1986 may witness the completion in glorious victory of this highly significant Plan. It will have run its course through a period of unprecedented world confusion, bearing witness to the vitality, the irresistible advance and socially creative power of the Cause of God, standing out in sharp contrast to the accelerating decline in the fortunes of the generality of mankind.

394.11 Beloved Friends, the bounties and protection with which the Blessed Beauty is nurturing and sheltering the infant organism of His new world order through this violent period of transition and trial, give ample assurance of victories to come if we but follow the path of His guidance. He rewards our humble efforts with effusions of grace which bring not only advancement to the Cause but assurance and happiness to our hearts, so that we may indeed look upon our neighbors with bright and shining faces, confident that from our services now will eventuate that blissful future which our descendants will inherit, glorifying Bahá'u'lláh, the Prince of Peace, the Redeemer of Mankind.

With loving Bahá'í greetings,
THE UNIVERSAL HOUSE OF JUSTICE

395
Use of Torture to Extract False Confessions from Iranian Bahá'ís
10 May 1984

To all National Spiritual Assemblies

Dear Bahá'í Friends,

On Thursday, 10 May 1984, the Universal House of Justice instructed the Bahá'í International Community in New York to issue the following press release and to keep the appropriate United Nations agencies informed of the continuing persecution of the Bahá'ís in Iran.

395.1

> THE PERSECUTION OF THE BAHÁ'ÍS OF IRAN HAS TAKEN AN EXTREMELY SEVERE AND CRITICAL TURN.

395.2

> OVER THE LAST FIVE YEARS ABDUCTIONS, EXECUTIONS, THE IMPRISONMENT OF HUNDREDS, DISMISSAL FROM JOBS, EXPULSION OF BAHÁ'Í CHILDREN FROM SCHOOLS, DESTRUCTION OF HOMES HAVE BEEN THE COMMON LOT OF THIS OPPRESSED COMMUNITY. NOW, HOWEVER, THE AUTHORITIES IN IRAN SEEM TO HAVE DECIDED TO SUBJECT PROMINENT BAHÁ'ÍS TO BARBARIC METHODS OF TORTURE TO EXTRACT FROM THEM CONFESSIONS TO THE FALSE CHARGES LEVELED AGAINST THEM.

395.3

> TORTURE HAS BEEN USED ON BAHÁ'Í PRISONERS BEFORE, BUT HITHERTO THE PURPOSE HAS BEEN TO FORCE THEM TO RECANT THEIR FAITH OR REVEAL INFORMATION ABOUT THEIR FELLOW BELIEVERS. ALL BUT A FEW OF THE BAHÁ'ÍS REMAINED FIRM AND PREFERRED IMPRISONMENT AND DEATH TO RECANTATION OF THEIR FAITH. RECENTLY AN ADDED OBJECTIVE SEEMS TO INSPIRE THE GOVERNMENT'S TORTURE-MONGERS.

395.4

> ALTHOUGH IN ALL THESE YEARS, SINCE THE BEGINNING OF THE REVOLUTION, THE GOVERNMENT HAS BEEN IN POSSESSION, THROUGH CONFISCATION, OF ALL THE RECORDS AND FILES OF THE ENTIRE BAHÁ'Í ADMINISTRATIVE SYSTEM IN IRAN, IT HAS FAILED TO PRODUCE A SINGLE SHRED OF EVIDENCE TO SUPPORT ITS REITERATED ACCUSATION THAT THE BAHÁ'Í COMMUNITY IN IRAN IS A NETWORK OF FOREIGN SPIES AND AGENTS. IT WOULD SEEM THAT NOW IT HAS DETERMINED TO MANUFACTURE BY MEANS OF FALSE CONFESSIONS EXTRACTED UNDER TORTURE, EVIDENCE TO ENABLE IT TO PUBLICLY JUSTIFY ITS INHUMAN PERSECUTION OF THE BAHÁ'Í COMMUNITY OF IRAN. WE KNOW OF THREE BAHÁ'ÍS WHOSE WILLS HAVE BEEN BROKEN UNDER THIS BARBARIC TORTURE AND WHO HAVE SUBMITTED TO THE DEMANDS OF THE AUTHORITIES, MAKING STATEMENTS ADMITTING TO A SERIES OF CRIMES. DOUBTLESS THE AUTHORITIES HOPE TO PUBLISH SIMILAR "CONFESSIONS" IN THE DAYS TO COME, MADE BY OTHER HAPLESS VICTIMS OF THIS FLAGITIOUS TREATMENT.

395.5

395.6 WE KNOW THE NAMES OF MANY BAHÁ'ÍS WHO ARE NOW BEING TORTURED, AND ALTHOUGH, FOR OBVIOUS REASONS, THESE CANNOT BE PUBLICIZED, THEY HAVE BEEN GIVEN TO SOME INTERNATIONAL ORGANIZATIONS AND GOVERNMENTS.

395.7 In addition to the above press release, supplementary information was telexed to the Bahá'í International Community and selected National Spiritual Assemblies:

395.7a FOLLOWING IS SOME DESCRIPTION TORTURES TO ENABLE YOU ANSWER QUESTIONS GIVE FURTHER BACKGROUND AS NECESSARY. THIS INFORMATION OBTAINED FROM RELIABLE SOURCES INCLUDING EYEWITNESSES.

395.7b REVOLUTIONARY GUARDS, USUALLY LATE AT NIGHT, WILL SUDDENLY CONVERGE UPON HOME BAHÁ'Í FAMILY. AFTER ENGAGING IN VARIETY VERBALLY AND PHYSICALLY ABUSIVE ACTIONS INVOLVING THREATS AND RANSACKING HOUSEHOLD IN GUISE OF CONDUCTING SEARCH, THEY SEIZE ALL BAHÁ'Í BOOKS AND PAPERS AND FAMILY PHOTOGRAPHS AND TAKE AS PRISONER WHOMEVER THEY ARE INTERESTED IN APPREHENDING.

395.7c INTERROGATIONS OFTEN TAKE PLACE WHILE VICTIM SITS FACING WALL OR IS BLINDFOLDED SO THAT HE OR SHE MAY NOT IDENTIFY INTERROGATORS. USUALLY SUCH VICTIMS ARE CURSED, REVILED, BEATEN AND SUBJECTED TO EXTREME PSYCHOLOGICAL PRESSURES DURING COURSE THESE INTERROGATIONS.

395.7d THERE ARE INSTANCES OF PRISONERS BEING PLACED IN TOTAL SOLITARY CONFINEMENT IN ISOLATED CELL 1.70 METERS BY 2 METERS AND HELD THERE FOR WEEKS OR EVEN MONTHS WITHOUT BEING ABLE TO SPEAK TO ANYONE WHATEVER, NOT EVEN PRISON GUARD.

395.7e IN OTHER INSTANCES BAHÁ'Í PRISONER, MAN OR WOMAN, WILL BE TIED FIRMLY TO SPECIALLY DESIGNED TABLE AND WHIPPED MERCILESSLY ON ALL PARTS BODY, BUT PARTICULARLY ON SOLES FEET. PERIODICALLY PRISONER WILL BE CHECKED AS TO WHETHER HIS WILL HAS BEEN BROKEN AND HE IS WILLING EMBRACE PURPOSES PRISON AUTHORITIES. AS LONG AS HE RESISTS, THIS PHYSICAL ASSAULT IS REPEATED MORNING AND EVENING, RESULTING IN SWELLING PRISONER'S LEGS ON WHICH HE IS COMPELLED TO WALK. PRISONER EXPERIENCES GREAT THIRST IN SUCH CIRCUMSTANCES AND IS FORCED IN THIS CONDITION TO WALK UNAIDED TO SLAKE THIRST WITH GLASS WATER ALLURINGLY DISPLAYED SOME DISTANCE FROM HIM. INTERNAL BODILY INJURY ALSO RESULTS FROM SEVERE BEATINGS—INVARIABLY PRISONER PASSES BLOOD IN URINE. TO KEEP PRISONER ALIVE SOMETIMES DOCTOR WILL BE SUMMONED TO APPLY SUPERFICIAL TREATMENT.

395.7f IN CASES WHEN BOTH WIFE AND HUSBAND ARE PRISONERS, ONE WILL BE SHOWN DAMAGED BODY OF THE OTHER DURING COURSE OF INTERROGATION. CRIES AND MOANS OF TORTURED PRISONERS REVERBERATE THROUGH PRISON AND ARE HEARD BY INMATES WHOSE TURN WILL SOON COME.

UNIVERSAL HOUSE OF JUSTICE

1979–1986 • THE SEVEN YEAR PLAN

Kindly share this information with the believers in your community, and use it as a basis for answering inquiries which you may receive from news media or government officials. 395.8

With loving Bahá'í greetings,
DEPARTMENT OF THE SECRETARIAT

396
Summary of the Situation in Iran
21 May 1984

To all National Spiritual Assemblies

FURTHER TO OUR MESSAGE OF 10 MAY REGARDING TORTURE BEING INFLICTED BAHÁ'ÍS IRAN WE SEND FOLLOWING SUMMARY OVERALL SITUATION THAT COUNTRY FROM MOST RECENT INFORMATION PROVIDED BY RELIABLE SOURCES. 396.1

1. SINCE THE BEGINNING OF THE ISLAMIC REVOLUTION MORE THAN 300 RESIDENCES OF BAHÁ'ÍS HAVE BEEN PLUNDERED OR SET ON FIRE. 396.2

2. SOME 170 BAHÁ'ÍS, MOST OF THEM PROMINENT MEMBERS OF THE BAHÁ'Í COMMUNITY, HAVE BEEN KILLED BY A VARIETY OF METHODS, BUT PRINCIPALLY THROUGH EXECUTION BY FIRING SQUADS AND BY HANGING. 396.3

3. IN URBAN AREAS PROPERTIES BELONGING TO SEVERAL HUNDRED FAMILIES HAVE BEEN SEIZED, WHILE IN RURAL AREAS MANY ORCHARDS HAVE BEEN DESTROYED AND FARMS AND ARABLE LANDS CONFISCATED. PETITIONS TO THE AUTHORITIES FOR REDRESS OF GRIEVANCES HAVE BEEN IGNORED. 396.4

4. THE MINISTRY OF WORKS AND SOCIAL AFFAIRS FORMALLY INSTRUCTED INDUSTRIAL AND COMMERCIAL INSTITUTIONS NOT TO PAY THEIR BAHÁ'Í STAFF. 396.5

5. MORE THAN 10,000 BAHÁ'ÍS EMPLOYED IN GOVERNMENT OFFICES OR IN THE PRIVATE SECTOR HAVE BEEN SUMMARILY DISCHARGED, THEIR RIGHTS TO PENSIONS AND OTHER EMPLOYMENT BENEFITS REVOKED. DEMANDS WERE MADE OF A NUMBER OF THEM TO REFUND THE SALARIES THEY HAD RECEIVED FOR THE DURATION OF THEIR EMPLOYMENT. 396.6

6. BAHÁ'Í STUDENTS HAVE BEEN DISMISSED FROM ALL UNIVERSITIES AND OTHER INSTITUTIONS OF HIGHER LEARNING. 396.7

7. IN MOST CITIES AND PROVINCES, BAHÁ'Í CHILDREN HAVE BEEN DENIED ENTRY TO SCHOOLS AND THEREFORE HAVE NO ACCESS TO BASIC EDUCATION. 396.8

8. SOME 700 BAHÁ'ÍS, INCLUDING MEN, WOMEN AND CHILDREN, ARE BEING HELD IN VARIOUS PRISONS THROUGHOUT IRAN. 396.9

9. FOR MORE THAN NINE MONTHS VISITS TO 40 BAHÁ'Í PRISONERS HAVE BEEN STRICTLY PROHIBITED BY THE AUTHORITIES. THEIR FATE IS THEREFORE UNKNOWN. 396.10

10. IN EVIN AND GAWHAR-DA<u>SH</u>T PRISONS A NUMBER OF BAHÁ'Í PRISONERS ARE UNDERGOING RELENTLESS TORTURE IN AN EFFORT ON THE PART OF THE AUTHORITIES TO FORCE THEM TO ADMIT TO FALSE CHARGES OF ENGAGING IN ESPIONAGE AND ACTING AGAINST THE ISLAMIC REPUBLIC OF IRAN. FOR A PERIOD 396.11

OF MONTHS THEY HAVE BEEN SUBJECTED TO FLOGGINGS OF ALL PARTS OF THE BODY, PARTICULARLY THE LEGS AND FEET. SOMETIMES UP TO 400 STROKES BY WIRE CABLES HAVE BEEN ADMINISTERED TO ONE PRISONER, THEN HE OR SHE HAS BEEN MADE TO WALK. FINDING THIS IMPOSSIBLE, THE UNFORTUNATE PRISONER HAS BEEN FORCED TO CRAWL ON HANDS AND KNEES BACK TO A DARK CELL. IN MASHHAD AND YAZD BAHÁ'Í PRISONERS ARE REGULARLY WHIPPED ON THE HEAD AND FACE WITH THICK PLASTIC TUBES, SIMILAR PROCEDURES ARE USED TO A LESSER DEGREE IN OTHER PRISONS. A NUMBER OF THESE VICTIMS OF TORTURE HAVE LOST THEIR SIGHT AND HEARING, OTHERS THEIR MENTAL COMPETENCE. THE BODIES OF FOUR PRISONERS SUBJECTED TO SUCH TREATMENT WERE SEEN BEFORE BEING BURIED IN UNKNOWN GRAVES. IT IS THEREFORE FEARED THAT OTHER PRISONERS WHOSE BODIES HAVE BEEN SIMILARLY BURIED WITHOUT THEIR FAMILIES BEING NOTIFIED SUFFERED THE SAME FATE.

396.12 11. BAHÁ'ÍS ARE UNSAFE IN THEIR OWN HOMES, WHICH ARE ENTERED AT WILL, DAY OR NIGHT, BY REVOLUTIONARY GUARDS, WHO HARASS THE INHABITANTS BY INSULTING, THREATENING AND BEATING THEM. WHEN THE REVOLUTIONARY GUARDS INVADE A HOME WITH THE INTENTION OF ARRESTING A PARTICULAR BAHÁ'Í, IF THAT PERSON HAPPENS TO BE ABSENT, THEIR PRACTICE IS TO SEIZE AS HOSTAGES OTHER MEMBERS OF THE HOUSEHOLD, EVEN CHILDREN, AND TO RANSACK THE PLACE, CONFISCATING WHATEVER THEY PLEASE.

396.13 12. WHENEVER THE HEAD OR SOME OTHER IMPORTANT MEMBER OF THE FAMILY HAS BEEN KILLED, AND OFTEN WHEN SUCH A PERSON HAS BEEN IMPRISONED, THOSE REMAINING BEHIND HAVE BEEN FORCED FROM THEIR HOMES AND NOT PERMITTED TO TAKE ANY BELONGINGS, EVEN IN THE DEAD OF WINTER. THE VICTIMS OF SUCH TREATMENT HAVE NO RECOURSE TO JUSTICE SINCE THEIR PETITIONS TO THE AUTHORITIES ARE IGNORED. BAHÁ'Í FAMILIES IN ISFAHAN, MASHHAD, TEHRAN, URÚMÍYYIH AND YAZD IN PARTICULAR ARE AFFECTED BY THESE CONDITIONS. . . .

396.14 FOLLOWING INFORMATION JUST RECEIVED: EXECUTION FOUR MORE COURAGEOUS BRETHREN IRAN

TABRÍZ, 5 MAY — MR. JALÁL PAYRAVÍ, IMPRISONED 22 OCTOBER 1981
MR. MAQSÚD 'ALÍZÁDIH, IMPRISONED 27 JANUARY 1982
TEHRAN, 15 MAY — MR. 'ALÍ-MUHAMMAD ZAMÁNÍ, 45 YEARS OLD
MR. JAHÁNGÍR HIDÁYATÍ, 61 YEAR OLD CONSTRUCTION ENGINEER, MEMBER NOW DISSOLVED NATIONAL ASSEMBLY KIDNAPPED JUNE 1983

396.15 BOTH TORTURED DURING IMPRISONMENT

396.16 IN ADDITION, MR. ASADU'LLÁH KÁMIL-MUQADDAM DIED IN PRISON 2 MAY, CIRCUMSTANCES UNKNOWN.

396.17 FRIENDS IRAN FEEL SITUATION DANGEROUS GREAT NUMBER OTHER BAHÁ'Í PRISONERS. . . .

UNIVERSAL HOUSE OF JUSTICE

397

The Development of the Soul and the Reconstruction of Society
12 JUNE 1984

To an individual Bahá'í

Dear Bahá'í Friend,

Your letter of 18 October 1983 was received by the Universal House of Justice and it was very glad to note that many of the problems seemed to have been resolved. It regrets the delay in replying to you, but, as you will understand, the pressure of work has been very heavy and, as the additional clarification that you requested did not seem to be a matter of urgency, a reply has been postponed until now. On the points you raised we have now been instructed to convey the following comments. 397.1

The Purpose of Religion—The Development of the Soul

As you quite correctly appreciate, the fundamental purpose of all religion is the spiritual development of the souls of human beings. This is expressed in the Short Obligatory Prayer, and also there is the following very clear statement written on behalf of the beloved Guardian to an individual believer on 8 December 1935: 397.2

> How to attain spirituality is, indeed, a question to which every young man and woman must sooner or later try to find a satisfactory answer. It is precisely because no such satisfactory reply has been given or found, that modern youth finds itself bewildered, and is being consequently carried away by the materialistic forces that are so powerfully undermining the foundation of man's moral and spiritual life. 397.2a

> Indeed, the chief reason for the evils now rampant in society is a lack of spirituality. The materialistic civilization of our age has so much absorbed the energy and interest of mankind, that people in general no longer feel the necessity of raising themselves above the forces and conditions of their daily material existence. There is not sufficient demand for things that we should call spiritual to differentiate them from the needs and requirements of our physical existence. The universal crisis affecting mankind is, therefore, essentially spiritual in its causes. The spirit of the age, taken on the whole, is irreligious. Man's outlook upon life is too crude and materialistic to enable him to elevate himself into the higher realms of the spirit. 397.2b

> It is this condition, so sadly morbid, into which society has that fallen, religion seeks to improve and transform. For the core of religious faith is that mystic feeling that unites man with God. This state of spiritual communion can be brought about and maintained by means of 397.2c

meditation and prayer. And this is the reason why Bahá'u'lláh has so much stressed the importance of worship. It is not sufficient for a believer to merely accept and observe the teachings. He should, in addition, cultivate the sense of spirituality which he can acquire chiefly by the means of prayer. The Bahá'í Faith, like all other Divine religions, is thus fundamentally mystic in character. Its chief goal is the development of the individual and society, through the acquisition of spiritual virtues and powers. It is the soul of man that has first to be fed. And this spiritual nourishment prayer can best provide. Laws and institutions, as viewed by Bahá'u'lláh, can become really effective only when our inner spiritual life has been perfected and transformed. Otherwise religion will degenerate into a mere organization, and become a dead thing.

397.2d   The believers, particularly the young ones, should therefore realize fully the necessity of praying. For prayer is absolutely indispensable to their inner spiritual development, and this, [as] already stated, is the very foundation and purpose of the religion of God.

(Published in U.S. *Bahá'í News*, No. 102, August 1936, p. 3)

The Central Purpose of Bahá'u'lláh's Dispensation

397.3   In addition to this fundamental purpose underlying all Revelation, there is a particular central purpose for each Dispensation. The one for this Dispensation is the establishment of the oneness of the world of humanity, and it is a Bahá'í teaching that the spiritual development of the soul requires not merely prayer and meditation, but also active service to one's fellowmen in accordance with the laws and principles of the Revelation of God. The reconstruction of human society and the spiritual advancement of individual souls go hand in hand.

Attaining Unity in Spirit and Action

397.4   People are in many different conditions, come from many different backgrounds, and face many different problems in attaining unity in spirit and practice. Our guiding light is the Message of Bahá'u'lláh. The Administrative Order is the strong framework within which we work and the channel for the outflowing of the spirit into the world.

397.5   All Bahá'ís are fallible human beings, each one has his own insights, enthusiasms and degree of wisdom and understanding. Differences of viewpoint could cause the community to fragment into a thousand pieces, if it were not cemented together by the strong bond of the Covenant, and if the friends were not willing to subordinate their own ideas to the considered decisions that issue from the divinely ordained process of consultation and, at the same time, exercise the utmost forbearance towards their fellow believers, their individual characteristics and their shortcomings. One of the most potent statements of the Guardian, which illuminates many issues, is:

Let us also bear in mind that the keynote of the Cause of God is 397.5a
not dictatorial authority, but humble fellowship, not arbitrary power,
but the spirit of frank and loving consultation. *Nothing short of the spirit
of a true Bahá'í can hope to reconcile the principles of mercy and justice,
of freedom and submission, of the sanctity of the right of the individual and
of self-surrender, of vigilance, discretion and prudence on the one hand and
fellowship, candor, and courage on the other.*[1]

Since we are all imperfect and have to learn the perfect standard which 397.6
Bahá'u'lláh has unveiled, there are often things in the Teachings themselves
which individual believers find difficult, and which they have to strive to
learn and understand. All the believers are growing and this is a gradual
process. Each one, as you say, must develop wisdom, and with this must
realize the fundamental importance of the unity of the community and the
bond of love and affection among the believers, for the sake of which he
will sacrifice many things. . . .

With loving Bahá'í greetings,
DEPARTMENT OF THE SECRETARIAT

398
Message to Bahá'í Women's Conference in Cotonou, Benin
28 JUNE 1984

To the Bahá'í Women's Conference in Cotonou, Benin

AS FINAL YEAR UNITED NATIONS DECADE WOMEN APPROACHES BEFITTING YOU 398.1
GATHER THIS HISTORIC CONFERENCE REVIEW ACHIEVEMENTS PLAN ENLARGE
SCOPE ACTIVITIES FULFILL NOBLE ROLE WOMANKIND ENVISAGED BAHÁ'U'LLÁH.[1]
PRAYING HOLY SHRINES DIVINE CONFIRMATIONS NOBLE ASPIRATIONS FRUITFUL
DELIBERATIONS.

UNIVERSAL HOUSE OF JUSTICE

397-1. BA, pp. 63–64. Italics added.
398-1. The conference was held 6–9 July 1984.

399
Execution of Nuṣratu'lláh Vaḥdat in Ma_sh_had
3 July 1984

To all National Spiritual Assemblies

399.1 FURTHER DISTRESSING NEWS RECEIVED FROM IRAN OF EXECUTION BY HANGING OF NUṢRATU'LLÁH VAḤDAT IN MA_SH_HAD JUNE 17. AN ADDITIONAL 51 BELIEVERS NOW HELD IN PRISONS IRAN, MAKING TOTAL OF 751, SOME OF WHOM ARE SUBJECTED TO CRUEL TORTURE. A NUMBER OF FRIENDS ARE NOW IN PARTICULAR DANGER. . . .

Universal House of Justice

400
Execution of Iḥsánu'lláh Ka_th_írí in Tehran
5 July 1984

To all National Spiritual Assemblies

400.1 FURTHER OUR TELEX DATED JULY 3 WE HAVE JUST RECEIVED DISTRESSING NEWS OF EXECUTION IN TEHRAN ON 27 JUNE OF IḤSÁNU'LLÁH KA_TH_ÍRÍ, 40 YEARS OLD, AFTER 11 MONTHS IMPRISONMENT. HIS BODY UNCEREMONIOUSLY BURIED BY AUTHORITIES WITHOUT INFORMING FAMILY. . . .

Universal House of Justice

401
Results of Bahá'í Women's Conference in Cotonou, Benin
20 July 1984

To the National Spiritual Assembly of the Bahá'ís of Benin

401.1 HEARTS UPLIFTED NEWS FIRST INTERREGIONAL WOMEN'S CONFERENCE. OFFERING PRAYERS THANKSGIVING FOR BOUNTIFUL CONFIRMATIONS. CONFIDENT SPIRIT GENERATED WILL BENEFIT BAHÁ'Í WOMEN BENIN, INDEED ENTIRE REGION. PRAYING ADDED BLESSINGS BRILLIANT VICTORIES.

Universal House of Justice

402
Relationship between Husband and Wife— Further Comments
25 JULY 1984

To an individual Bahá'í

Dear Bahá'í Friend,

The Universal House of Justice received your letter which raised questions concerning the status of men and women especially within the family setting. We are requested by the House of Justice to convey to you the following points. 402.1

The Universal House of Justice invites you once again to carefully read the letter written on its behalf and addressed to the National Spiritual Assembly of New Zealand, dated 28 December 1980.[1] This letter is published in the compilation *Bahá'í Marriage and Family Life,* pp. 57–61. 402.2

Authoritative interpretation of the Writings was the exclusive domain of 'Abdu'l-Bahá and Shoghi Effendi. When the House of Justice stated that the "father can be regarded as the 'head' of the family," it was giving expression to its own inference as you indicate. This inference, as the letter to New Zealand points out, is based on the clear and primary responsibility of the husband to provide for the financial support of the wife and family, and on the provisions of the law of intestacy, which assigns special functions and rights to the eldest son. 402.3

The description of the husband as "head" of the family does not confer superiority upon the husband nor does it give him special rights to undermine the rights of the other members of his family. 'Abdu'l-Bahá says: 402.4

> The integrity of the family bond must be constantly considered, and the rights of the individual members must not be transgressed. The rights of the son, the father, the mother—none of them must be transgressed, none of them must be arbitrary. . . .[2] 402.4a

The relationship between family members represents a complex of mutual and complementary duties and responsibilities that are implemented within the framework of the Bahá'í ideal of family life and are conducive to its unity. The concept of a Bahá'í family is based on the principle that the man is charged with the responsibility of supporting the entire family financially, and the woman is the chief and primary educator of the children. This does not mean that these functions are inflexibly fixed and cannot be changed and 402.5

402-1. See message no. 272.
402-2. PUP, p. 168.

adjusted to suit particular family situations. Furthermore, while primary responsibility is assigned, it is anticipated that fathers would play a significant role in the education of the children and women would be breadwinners. (See pages 59–60 of *Bahá'í Marriage and Family Life*)[3]

402.6 The principle of the equality between women and men, like the other teachings of the Faith, can be effectively and universally established among the friends when it is pursued in conjunction with all the other aspects of Bahá'í life. Change is an evolutionary process requiring patience with one's self and others, loving education and the passage of time as the believers deepen their knowledge of the principles of the Faith, gradually discard long-held traditional attitudes and progressively conform their lives to the unifying teachings of the Cause.

402.7 There are a number of Tablets on marriage. You are referred, for example, to *Selections from the Writings of 'Abdu'l-Bahá*, pp. 117–22, and *Bahá'í Prayers* (United States, 1982 Edition), pp. 104–08.

402.8 In relation to the particular "Marriage Tablet" to which you refer, we provide, for your information, an extract from a letter written on behalf of the Universal House of Justice to a National Spiritual Assembly on 4 April 1976, which deals with the use of this "Tablet":

402.8a The so-called "Marriage Tablet" . . . is not a Tablet at all but is an unauthenticated record by Ahmad Sohrab of a talk by 'Abdu'l-Bahá. The friends may use it, but it should be made clear that this is not Scripture . . .

402.9 *The Promulgation of Universal Peace* (1982 Edition) provides much useful source material on the subject of the equality of men and women. For talks of 'Abdu'l-Bahá citing historical examples, see particularly:

– Talk to the Federation of Women's Clubs, pp. 74–77
– Talk to a woman's suffrage meeting, pp. 133–37
– Talk at Franklin Square House, pp. 280–84

402.10 We trust that the foregoing information will resolve your concerns and help to reinforce your sense of certitude.

With loving Bahá'í greetings,
DEPARTMENT OF THE SECRETARIAT

402-3. See message dated 28 December 1980 (no. 272).

403

Dedication of the Mother Temple of the Pacific Islands, Apia, Western Samoa
AUGUST 1984

To the Friends gathered in Apia, Western Samoa on the occasion of the Dedication of the Mother Temple of the Pacific Islands[1]

Dearly loved Friends,

In these historic days we are witnessing a major triumph in the development of the Faith of Bahá'u'lláh. Not only is the raising up of this House of Worship a further significant fulfillment of the Blessed Beauty's promise,[2] it also presages a brilliant future in the Pacific for His Faith, whose quickening light is casting its rays on the peoples of this vast ocean. In his message to the Bahá'í world in April 1957 the beloved Guardian, referring to the Pacific, wrote, ". . . Bahá'í exploits bid fair to outshine the feats achieved in any other ocean, and indeed in every continent of the globe. . ."[3]

In reviewing the religious history of the Pacific during the Bahá'í Dispensation we recall that it was only a short time prior to the Declaration of the Báb that the Teachings of Christ spread throughout these islands;[4] that the Teachings of Bahá'u'lláh were first proclaimed there when the Hand of the Cause Agnes Alexander arrived in the Hawaiian Islands in December 1901;[5] that at the beginning of the World Crusade in 1953 only a handful of islands had had any contact with the Faith; and that at Riḍván 1959 when the first regional National Spiritual Assembly of the South Pacific was established in Suva there were in its area but twelve Local Spiritual Assemblies in nine island groups. Witness now what has happened in the quarter century since 1959.

– The first reigning monarch in the world to embrace the Faith of Bahá'u'lláh is the Head of State of Western Samoa whose official residence is near the Mashriqu'l-Adhkár.[6]

403-1. The Mother Temple of the Pacific Islands was dedicated on 1 September 1984.

403-2. The promise referred to is the following statement from Bahá'u'lláh: "Should they attempt to conceal its [His Revelation's] light on the continent, it will assuredly rear its head in the midmost heart of the ocean, and, raising its voice, proclaim: 'I am the life-giver of the world!'" (WOB, p. 79)

403-3. MBW, p. 111.

403-4. The Báb declared His mission on 23 May 1844. According to information obtained from the public library of Wilmette, Illinois, the first Christian mission in Samoa was established in 1830 by the London Missionary Society on the island of Savai'i.

403-5. For information on Miss Alexander's life, see footnote 98-4.

403-6. The Head of State of Western Samoa referred to is Malietoa Tanumafili II. For announcements of his acceptance of the Bahá'í Faith, his visit to the resting-place of Shoghi Effendi, and his participation in the dedication of the Samoan House of Worship, see messages dated 7 May 1973, and 5 October 1976 (nos. 130 and 177).

403.2b — There are now a total of thirteen National Spiritual Assemblies in the homelands of the Polynesians, the Melanesians and the Micronesians.

> The Caroline Islands
> Fiji
> The Hawaiian Islands
> Kiribati
> The Mariana Islands
> The Marshall Islands
> New Caledonia and the Loyalty Islands
> Papua New Guinea
> Samoa
> The Solomon Islands
> Tonga
> Tuvalu
> Vanuatu

403.2c — Two additional National Spiritual Assemblies are to be formed at Riḍván 1985: those of The Cook Islands and the Western Caroline Islands.

403.2d — Nearly five hundred Local Spiritual Assemblies are established.

403.2e — Bahá'ís reside in nearly 2,000 localities.

403.2f — There are more than 31,000 believers in the Pacific.

403.2g — Bahá'ís constitute more than one percent of the total population in eight national communities in the Pacific.

403.2h — The seventh Mashriqu'l-Adhkár of the world stands in all its glory on the mountainside at Tiapapata.[7] This spot has now become the spiritual heart of the Bahá'í communities in the Pacific basin.

403.3 The new Temple of light is the outward expression of the illumination of hearts and minds by the Revelation of Bahá'u'lláh, that process initiated in these islands at the turn of the century and now reaching a mighty crescendo. The effulgent light of His Teachings has aroused stirrings in the friends, impelling them to apply these heavenly principles in their personal lives and in the daily activities of their communities. Just as the spiritual illumination of the Bahá'ís is symbolized in the Temple, so the development of local community life in all its aspects, whether administrative or social, which flow from the fountainhead of faith in God's Revelation, will reach a fuller expression in the form of various dependencies around this edifice—dependencies which the Guardian described as "institutions of social service as shall afford relief

---

403-7. The six other Mashriqu'l-Adhkárs that have been constructed are found in Wilmette, near Chicago, U.S.A.; Kampala, Uganda; Ingleside, near Sydney, Australia; Langenhain, near Frankfurt am Main, Germany; Panama City, Panama; and New Delhi, India.

to the suffering, sustenance to the poor, shelter to the wayfarer, solace to the bereaved, and education to the ignorant."⁸

403.4 The dedication of this noble edifice will undoubtedly mark the beginning of a new phase in the growth of the Faith in this hemisphere. The portents of success are many and varied. The friendliness and spiritual perceptiveness of the people of the Pacific, the freedom enjoyed by the friends in traveling and sharing their faith with others, the absence hitherto of organized and concerted opposition to the Faith, the relatively low level of infection by the forces of materialism so rapidly invading every part of the world, and the genuine simplicity of the mode of life of the peoples of these islands—all combine to make of the Pacific a region rich with promise for the further dramatic spread of the Faith and for the efflorescence of its community life.

403.5 Let the joyous spirit generated at this conference, reinforced by your common endeavors and crowned by your participation in the dedicatory ceremony, be the generating impulse for a mighty surge of teaching activity which will yield a vast increase in the number of followers of Bahá'u'lláh throughout the far-flung islands of the Pacific. As the Seven Year Plan hastens towards its conclusion, you have a priceless opportunity to take advantage of the favorable conditions in your region to win victories in the Plan which will astonish the world and will fulfill the great promises enshrined in our Writings. Our prayers ascend at the Holy Threshold for the success of your selfless endeavors in the path of His Cause.

With loving Bahá'í greetings,
THE UNIVERSAL HOUSE OF JUSTICE

404
Introduction of the Law of Ḥuqúqu'lláh to the West
6 AUGUST 1984

The National Spiritual Assembly of the Bahá'ís of the United States
Dear Bahá'í Friends,

404.1 We are requested by the Universal House of Justice to thank you for your letter of 10 July 1984 recounting the electrifying events which took place at the closing session of your last National Convention. The House of Justice was deeply moved to learn of the enthusiastic response that the delegates and others present made to the inspiring talk given by the Hand of the Cause of God Zikrullah Khadem, and by their desire, expressed on the scroll which

---

403-8. BA, p. 184.

you have forwarded to the World Center, for the Law of Ḥuqúqu'lláh to be applied to all believers in the United States.¹

404.2 In response to this petition the Universal House of Justice has decided that, although it is not yet timely to apply this mighty law in the West, it will send to the believers in the United States and other Western countries a translation into English of a compilation of texts so that they will be able to familiarize themselves with this subject. This translation is now being prepared and when it is complete an appropriate announcement will be made which will constitute the first step in the process of applying this law of God to the Western Bahá'í communities.²

404.3 Kindly share this letter with the friends in your area of jurisdiction.

With loving Bahá'í greetings,
DEPARTMENT OF THE SECRETARIAT

405
Understanding the Laws of the Faith
DATE: 12 AUGUST 1984

To: An individual Bahá'í
From: The Universal House of Justice

405.1 ... We are very glad to see that the Assembly is taking steps to educate the friends in the laws of the Faith and to emphasize the importance of obedience to them. ...

405.2 1. It would seem to be important to make clear to the friends that the "laws" of the Faith must be regarded in various lights. There are laws, ordinances, exhortations and principles, all of which are sometimes loosely referred to as "laws." All are very important for the life of the community and the spiritual life of the individual, but they are applied differently. Some affect the society and the social relationships, and the Spiritual Assemblies are responsible for their enforcement. If a believer breaks such a law, he is subject to the imposition of sanctions. Others,

---

404-1. At the Riḍván 1984 National Convention the Hand of the Cause of God Zikrullah Khadem spoke of the spiritual bounties that accrue when one obeys the law of Ḥuqúqu'lláh (Right of God). Delegates to the convention were moved to recommend that the National Spiritual Assembly ask the Universal House of Justice to make the law of Ḥuqúqu'lláh binding upon the Bahá'ís of the United States. The delegates, together with hundreds of visitors attending the convention, expressed the ardor of their request by signing their names to a scroll. For information on Ḥuqúqu'lláh, see the glossary.

404-2. The compilation was sent to selected National Spiritual Assemblies with the 4 July 1985 letter (no. 430). For a copy of the compilation, see CC 1:489–527.

although of very great importance, are not sanctionable, because their observance is a matter of conscience between the individual and God; among these fall the laws of prayer and fasting and the law of Ḥuqúqu'lláh. Then there are those high ethical standards to which Bahá'u'lláh calls His followers, such as trustworthiness, abstention from backbiting, and so on; generally speaking, obedience to these is a matter for individual conscience, and the Assemblies should not pry into people's lives to see whether or not they are following them; nevertheless, if a believer's conduct falls so far below the standard set by Bahá'u'lláh that it becomes a flagrant disgrace and brings the name of the Faith into disrepute, the Assembly would have to intervene, to encourage the believer to correct his ways, to warn him of the consequences of continued misconduct, and possibly, if he does not respond, to deprive him of his administrative rights.

In other words, the friends should realize the importance of following all the teachings and not assume that merely because an offense is not punishable it is therefore less grave. Assemblies, on the other hand, should distinguish clearly between those laws which it is their duty to enforce, those which should be left strictly to the conscience of the individual, and those in which it may have to intervene if the misbehavior is blatant and injurious to the good name of the Faith.... 405.3

406
Execution of Manúchihr Rúḥí
22 August 1984

To all National Spiritual Assemblies

MANÚCHIHR RÚḤÍ, WELL-RESPECTED 52 YEAR OLD PHARMACIST, EXECUTED BY FIRING SQUAD ON 16 AUGUST, IN BUJNÚRD, KHURÁSÁN PROVINCE, IRAN, AFTER SPENDING ELEVEN MONTHS IN PRISON. ALL HIS PROPERTIES CONFISCATED. LIVES TWENTY-FIVE OTHER BAHÁ'ÍS IMPRISONED IN DIFFERENT PARTS COUNTRY AT STAKE AS DEATH SENTENCES ALREADY PRONOUNCED AGAINST THEM.... 406.1

UNIVERSAL HOUSE OF JUSTICE

407
Roles of Parents within the Bahá'í Family
23 August 1984

To an individual Bahá'í

Dear Bahá'í Friends,

407.1 The Universal House of Justice has received your letter of 29 July 1984 and has instructed us to send you the following reply.

407.2 The seeker to whom you refer seems to have misconstrued the Bahá'í teachings about the responsibility of the parents for the education of their children. The father certainly has a very important role to play. In the Kitáb-i-Aqdas itself, Bahá'u'lláh revealed:

407.2a > Unto every father hath been enjoined the instruction of his son and daughter in the art of reading and writing and in all that hath been laid down in the Holy Tablet. . . . He that bringeth up his son or the son of another, it is as though he hath brought up a son of Mine; upon him rest My Glory, My loving-kindness, My Mercy, that have compassed the world.[1]

407.3 The great importance attached to the mother's role derives from the fact that she is the *first* educator of the child. Her attitude, her prayers, even what she eats and her physical condition have a great influence on the child when it is still in the womb. When the child is born, it is she who has been endowed by God with the milk which is the first food designed for it, and it is intended that, if possible, she should be with the baby to train and nurture it in its earliest days and months. This does not mean that the father does not also love, pray for, and care for his baby, but as he has the primary responsibility of providing for the family, his time to be with his child is usually limited, while the mother is usually closely associated with the baby during this intensely formative time when it is growing and developing faster than it ever will again during the whole of its life. As the child grows older and more independent, the relative nature of its relationship with its mother and father modifies and the father can play a greater role.

407.4 It may be helpful to stress to your seeker that the Bahá'í principle of the equality of men and women is clearly stated in the teachings, and the fact that there is diversity of function between them in certain areas does not negate this principle.

With loving Bahá'í greetings,
DEPARTMENT OF THE SECRETARIAT

407-1. KA ¶48.

408

Message to Youth Conference, London, Ontario
23 August 1984

To the Bahá'í Youth Conference in London, Ontario

WE HAIL WITH JOY AND HOPE THE ENNOBLING PURPOSES OF YOUR CONFERENCE IN LONDON, ONTARIO. YOU ARE GATHERED AT A MOMENT WHICH RESOUNDS WITH THE SIGNIFICANCES AND CHALLENGES POSED BY THE WORLD-SHAKING EVENTS ENVELOPING THE COMMUNITY OF THE GREATEST NAME IN BAHÁ'U'LLÁH'S NATIVE LAND. THE OUTPOURING GRACE PROVIDENTIALLY VOUCHSAFED THE ONWARD MARCH OF OUR HOLY CAUSE AS A CONSEQUENCE OF THESE EVENTS IS CLEARLY EVIDENT. 408.1

OUR HEARTS LEAP AT THE INNUMERABLE IMMEDIATE OPPORTUNITIES FOR THE FURTHER UNFOLDMENT OF THE ORDER OF BAHÁ'U'LLÁH TO WHICH UNDOUBTEDLY, YOU CAN AND WILL APPLY YOUR ABUNDANT TALENTS, YOUR ZEST FOR ACTION AND, ABOVE ALL, THE ENTHUSIASM OF YOUR DEVOTION. SURELY, YOU WILL SEE THAT THE HEROIC DEEDS OF SACRIFICE ON THE PART OF YOUR IRANIAN BRETHREN ARE MATCHED WITH CORRESPONDING EFFORTS ON YOUR PART IN THE VAST FIELDS OF TEACHING AND SERVICE LYING OPEN BEFORE YOU. 408.2

THE EXHORTATIONS ESPECIALLY ADDRESSED TO YOUTH BY OUR BELOVED MASTER AND THE GALVANIZING INFLUENCE OF THE GUARDIAN'S GUIDANCE WILL ECHO EVEN MORE LOUDLY IN YOUR HEARTS NOW. INDEED, WE WILL PRAY ARDENTLY AT THE HOLY SHRINES THAT YOU MAY REALIZE IN YOUR LIVES THE IDEALS THEY SO PERSISTENTLY UPHELD, THAT YOU MAY THUS "ACQUIRE BOTH INNER AND OUTER PERFECTIONS" AS YOU INCREASE YOUR STUDY OF THE HEAVENLY WRITINGS, STRIVE TOWARDS EXCELLENCE IN THE SCIENCES AND ARTS AND BECOME KNOWN FOR YOUR INDEPENDENCE OF SPIRIT, YOUR KNOWLEDGE AND YOUR SELF-CONTROL.[1] MAY YOU, AS 'ABDU'L-BAHÁ WISHED, BE "FIRST AMONG THE PURE, THE FREE AND THE WISE."[2] 408.3

UNIVERSAL HOUSE OF JUSTICE

408-1. See CC 1:247.
408-2. SWAB, p. 150.

409
Safeguarding the Letters of Shoghi Effendi
26 August 1984

To all National Spiritual Assemblies

Dear Bahá'í Friends,

409.1 In December 1967 the Universal House of Justice wrote to all National Spiritual Assemblies requesting them to send to the World Center originals or photocopies of letters from the beloved Guardian, or written on his behalf, and addressed to Bahá'í institutions or individual believers. The response to this request and to one issued in May 1975 was encouraging, but it is now clear that the originals and photocopies held at the World Center represent only a portion of the letters the Guardian is known to have written.[1]

409.2 The House of Justice is eager to pursue, as swiftly as possible, the task of tracing such vitally important documents, and has therefore given the Archives Office the urgent duty of collecting information which will enable it to work towards the completion of the collection held at the World Center.

409.3 In order to avoid giving institutions and believers a great deal of unnecessary labor, the Archives Office will be directing letters with specific guidelines to certain National Assemblies and individuals who are known to have received large numbers of letters from the Guardian, sending a list of the letters of which originals or satisfactory photocopies are already held at the World Center. The recipient will then be able to check whether he has any which have not been supplied to the World Center.

409.4 National Assemblies which have in their national archives only a small number of letters written by the Guardian, or on his behalf, could assist greatly by sending to the Archives Office at the World Center immediately a list of such letters, specifying in each case the name of the addressee and the date.

409.5 Although the majority of National Spiritual Assemblies have come into being since the passing of Shoghi Effendi, they may well have within their jurisdiction Local Spiritual Assemblies and individual believers who corresponded with the Guardian, or the children and grandchildren of such believers, who may have the letters in their possession. Therefore, all National Spiritual Assemblies are asked to publish a statement in their newsletters, requesting all Local Spiritual Assemblies and individuals who have in their possession original letters from the Guardian, or written on his behalf, to notify the World Center of that fact either directly or through their National Assembly. They should be asked to state the number of letters that they have and, if these are only a few, to list the date and the exact name of the addressee of each one.

409-1. See letter dated 14 May 1975 (no. 161).

In this announcement the National Assembly should stress the point made in earlier appeals: that any such letter is the property of the person to whom the beloved Guardian wrote, and that that person is perfectly entitled to keep the original or to pass it down in his family. He does not have to give it up to any Bahá'í Archives. In such a case, however, the Archives needs to have a good photocopy, and arrangements will be made for this to be done. Alternatively, if the individual wishes to give the original document into the safekeeping of the Archives, but would like to have a photocopy, this can be sent to him. It should also be mentioned that whenever the contents of a letter are seen to be confidential that confidentiality is respected. 409.6

Since many believers may have already sent such letters to their National Archives, National Assemblies are requested to send to the World Center a list of the names of those persons and other institutions, such as committees, for whom they hold originals or photocopies. This list will be checked here and the Archives Office will write in due course to inquire further as appears necessary. 409.7

In closing, the House of Justice asks us to stress that it is not asking for any of the Guardian's letters to be sent now. It merely wishes to collect information so that the Archives Office can systematically work to trace those letters which are lacking from the World Center collection. 409.8

With loving Bahá'í greetings,
DEPARTMENT OF THE SECRETARIAT

410
Call for Pioneers
2 OCTOBER 1984

To all National Spiritual Assemblies

Dear Bahá'í Friends,

The Universal House of Justice has asked us to convey the following with respect to settlement of pioneers since the opening of the third phase of the Seven Year Plan. 410.1

Of a total number of 298 pioneers called for in the last phase of the Plan, 23 posts have so far been filled. Although many additional pioneers have also settled around the world, 275 pioneers are still needed to fill the posts originally assigned. 410.2

Furthermore, since the announcement last Riḍván of these pioneering goals, the Universal House of Justice has received reports of the need for additional pioneers to reinforce the Bahá'í work in many countries. 410.3

410.4 In consultation with the International Teaching Center, outstanding goals have been reviewed, pioneer needs have been considered, and a supplementary pioneer call has been prepared and is attached.[1] These additional assignments, which call for 88 pioneers to be sent by 18 National Spiritual Assemblies to 29 national communities, raise the total number of pioneer goals for this last phase of the Plan to 386. Those pioneers now preparing to leave for their posts should, of course, proceed with their plans.

410.5 As you are aware, less than two years remain in the current Plan. The countries assigned to supply pioneers are urged to send the quota of pioneers to the territories named as quickly as circumstances permit. It is the hope of the Universal House of Justice that these goals may be filled for the most part by the end of the first year of the last phase, thereby helping to usher in the triumphant conclusion of the Seven Year Plan.

410.6 We have also been asked to emphasize that the pioneer goals, as announced at Riḍván and in this supplementary call, represent the minimum needs of some younger national communities. Settlement of pioneers above and beyond the goals assigned would be highly meritorious.

410.7 The Continental Pioneer Committees stand ready to do all in their power to collaborate with you in the fulfillment of these important goals, and you should feel free to call upon them for any assistance you may find necessary.

410.8 You may be assured that the Universal House of Justice will supplicate the Almighty that you may succeed in discharging your sacred responsibilities.

With loving Bahá'í greetings,
DEPARTMENT OF THE SECRETARIAT

411
Martyrdom of Iranian National Spiritual Assembly Member Shápúr Markazí; Death of Amínu'lláh Qurbánpúr in Prison
11 October 1984

To all National Spiritual Assemblies

411.1 WITH HEAVY HEARTS ANNOUNCE MARTYRDOM SHÁPÚR MARKAZÍ OUTSTANDING SERVANT FAITH IN IRAN MEMBER PREVIOUS NATIONAL SPIRITUAL ASSEMBLY AND AUXILIARY BOARD MEMBER. HE SUFFERED CRUEL TORTURE PAST FEW MONTHS. PURPOSE THESE TORTURES WAS FORCE HIM TO ADMIT FALSE CHARGES IMPLICATING BAHÁ'Í INSTITUTIONS AS NETWORK ESPIONAGE AND HIMSELF AS SPY. HIS GROWING RESISTANCE INCREASED INTENSITY TORTURES WHICH MAY HAVE CAUSED HIS DEATH 23 SEPTEMBER. HE WAS BURIED 25 SEPTEMBER WITHOUT KNOWLEDGE RELATIVES FRIENDS.

410-1. The list is too lengthy to be included in this volume.

REGRET ANNOUNCE ALSO PASSING AWAY IN PRISON OF AMÍNU'LLÁH QURBÁNPÚR 60 YEAR OLD MASON ON 25 AUGUST 1984. CAUSE DEATH UNKNOWN. HOWEVER BLOOD-STAINED CLOTHES RETURNED TO FAMILY TOGETHER WITH HIS RECENTLY WASHED SHOES MAKE CIRCUMSTANCES HIS DEATH SUSPICIOUS. HIS BODY WAS ALSO BURIED BY AUTHORITIES WITHOUT KNOWLEDGE RELATIVES. . . .

UNIVERSAL HOUSE OF JUSTICE

412
The Universal House of Justice's Power of Elucidation
25 OCTOBER 1984

To an individual Bahá'í

Dear Bahá'í Friend,

The Universal House of Justice has received your letter dated 4 September 1984 in which you seek further clarification about the qualitative difference between the Guardian's prerogative of interpretation and the power of elucidation of the Universal House of Justice, and raise questions about other aspects of the Teachings. We are directed to convey the following comments.

As you are aware, the Universal House of Justice has written three major messages which explain, among other things, the duties and functions shared by the Guardian and the Universal House of Justice, and those functions that are unique to each specific Institution. These messages are published in *Wellspring of Guidance*, pp. 44–56, and pp. 81–91, and in *Messages of the Universal House of Justice: 1968–1973*, pp. 37–44.[1] In relation to their specific functions, Shoghi Effendi explained that "it is made indubitably clear and evident that the Guardian of the Faith has been made the Interpreter of the Word and that the Universal House of Justice has been invested with the function of legislating in matters not expressly revealed in the teachings."[2]

The use of the term "elucidation" by the Universal House of Justice and the process by which it is implemented are based on passages in the Will and Testament of 'Abdu'l-Bahá and statements in the writings of the Guardian. For example, in the Will and Testament, 'Abdu'l-Bahá states:

> It is incumbent upon these members (of the Universal House of Justice) to . . . deliberate upon all problems which have caused difference, questions that are obscure and matters that are not expressly recorded in the Book . . . and bear upon daily transactions, . . . (p. 20)

412-1. The three messages are included in this volume; see 9 March 1965 (no. 23), 27 May 1966 (no. 35), and 7 December 1969 (no. 75).
412-2. WOB, pp. 149–50.

412.3b Further, in response to a question raised by the American National Spiritual Assembly about the Universal Court of Arbitration, the Guardian in a letter dated 9 April 1923, defined such explanation as being in the domain of the Universal House of Justice and anticipated its function of elucidation:

412.3c ... regarding the nature and scope of the Universal Court of Arbitration, this and other similar matters will have to be explained and elucidated by the Universal House of Justice, to which, according to the Master's explicit Instructions, all important fundamental questions must be referred.... (*Bahá'í Administration,* p. 47)

412.4 In a letter dated 9 March 1965, the Universal House of Justice stresses the "profound difference" that exists between the "interpretations of the Guardian and the elucidations of the House of Justice in exercise of its function to 'deliberate upon all problems which have caused difference, questions that are obscure, and matters that are not expressly recorded in the Book.'" (*Wellspring of Guidance,* p. 52) Among these is the outlining of such steps as are necessary to establish the World Order of Bahá'u'lláh on this earth. The elucidations of the Universal House of Justice stem from its legislative function, while the interpretations of the Guardian represent the true intent inherent in the Sacred Texts. The major distinction between the two functions is that legislation with its resultant outcome of elucidation is susceptible of amendment by the House of Justice itself, whereas the Guardian's interpretation is a statement of truth which cannot be varied.

412.5 Shoghi Effendi has given categorical assurances that neither the Guardian nor the Universal House of Justice "can, nor will ever, infringe upon the sacred and prescribed domain of the other."[3] Therefore, the friends can be sure that the Universal House of Justice will not engage in interpreting the Holy Writings....

With loving Bahá'í greetings,
DEPARTMENT OF THE SECRETARIAT

412-3. WOB, pp. 149–50.

413
Execution of Member of Disbanded Iranian National Spiritual Assembly and of Bahá'ís from Karaj and Ma<u>sh</u>had
8 NOVEMBER 1984

To all National Spiritual Assemblies

WITH SORROWFUL HEARTS WE ANNOUNCE EXECUTION OF THREE MORE BAHÁ'ÍS IN IRAN, MR. A<u>H</u>MAD BA<u>SH</u>ÍRÍ, MEMBER NATIONAL ASSEMBLY DISBANDED 29 AUGUST 1983, MR. YÚNIS NAWRÚZÍ-ÍRÁNZÁD, MEMBER LOCAL ASSEMBLY KARAJ. 413.1

THESE TWO EXECUTED BY HANGING AND THE THIRD, MR. FÍRÚZ PURDIL, AN ENGINEER FROM MA<u>SH</u>HAD DETAILS OF WHOSE EXECUTION NOT YET KNOWN. IT IS CERTAIN THAT MR. BA<u>SH</u>ÍRÍ IN HIS FIFTEEN MONTHS' IMPRISONMENT SUFFERED CRUEL INHUMAN TORTURES DESIGNED TO OBTAIN FALSE DECLARATION FROM HIM IMPLICATING DISBANDED BAHÁ'Í ADMINISTRATION IN IRAN AS ESPIONAGE NETWORK. HIS ENDURANCE, STEADFASTNESS, LIKE THAT OF HIS HEROIC FELLOW BELIEVERS, THWARTED INFAMOUS DESIGNS. MR. BA<u>SH</u>ÍRÍ AND MR. NAWRÚZÍ-ÍRÁNZÁD ALONG WITH MR. <u>SH</u>ÁPÚR MARKAZÍ PREVIOUSLY REPORTED WERE INCLUDED IN LIST BAHÁ'ÍS ALREADY CONDEMNED TO DEATH. THIS CAUSES GRAVE CONCERN FATE REMAINING VALIANT SOULS LANGUISHING IN PRISON. . . . 413.2

UNIVERSAL HOUSE OF JUSTICE

414
Passing of the Hand of the Cause of God <u>Sh</u>u'á'u'lláh 'Alá'í
18 NOVEMBER 1984

To all National Spiritual Assemblies

GRIEVED ANNOUNCE PASSING HAND CAUSE <u>SH</u>U'Á'U'LLÁH 'ALÁ'Í 16 NOVEMBER THUS ENDING MORE THAN 70 YEARS UNINTERRUPTED DEDICATED SERVICES THRESHOLD BAHÁ'U'LLÁH.[1] HE WAS TOWER STRENGTH CRADLE FAITH WHERE HE SERVED EMINENTLY, DEVOTEDLY IN ITS EMERGING ADMINISTRATIVE INSTITUTIONS SINCE THEIR INCEPTION. HIS MEMBERSHIP MANY DECADES NATIONAL ASSEMBLY, FREQUENTLY AS CHAIRMAN, BEARS WITNESS TRUST BAHÁ'ÍS IRAN PLACED HIS NOBLE PERSON. HIS EXEMPLARY COURAGE REPRESENTING INTERESTS FAITH HIGH 414.1

414-1. For an account of the life and services of <u>Sh</u>u'á'u'lláh 'Alá'í, see BW 19:593.

PLACES, HIS INTEGRITY PERFORMING OFFICIAL DUTIES ENHANCED PRESTIGE BELOVED FAITH HE SO DILIGENTLY SINCERELY CHAMPIONED ENTIRE LIFE. HIS MANIFOLD ACHIEVEMENTS CROWNED HONOR APPOINTMENT HAND CAUSE 29 FEBRUARY 1952. THIS ENABLED HIM EXTEND SERVICES FAITH INTERNATIONAL ARENA. SUPPLICATING SACRED THRESHOLD PROGRESS RADIANT SOUL ABHÁ KINGDOM. ADVISE HOLD MEMORIAL GATHERINGS BAHÁ'Í WORLD INCLUDING ALL MA<u>SH</u>RIQU'L-A<u>DH</u>KÁRS.

UNIVERSAL HOUSE OF JUSTICE

415
Invitation to Participate in the Formulation of the Next Plan
25 November 1984

To all National Spiritual Assemblies

Dear Bahá'í Friends,

415.1 At the Fifth International Convention held at Riḍván 1983, the delegates consulted on the challenging needs and opportunities of the Faith during the remaining years of this century. Since then, some National Spiritual Assemblies have continued this consultation within the framework of their respective communities. As the Seven Year Plan draws to a close, the Universal House of Justice, with the aid of the International Teaching Center, is beginning to consider the next stage in the development of the Bahá'í world community, and the results of your consultation on the subject carried over from the Convention will be valuable to the formulation of the objectives of the next Plan. The House of Justice therefore invites your participation in the process of formulating a new Plan, and we are asked to request each of your Assemblies to send it your comments on the following points in view of the prevailing situation in the area under your jurisdiction.

415.2 1. **Current Status and Specific Needs of the Bahá'í Community.** What distinctive strengths and weaknesses do you see in your community? Are there any unusual features of the condition of the Faith in your community which must be taken into account in the formulation of the next Plan?

415.3 2. **Condition of the Nation.** Do the social, political and economic trends indicate any unusual opportunities or difficulties for the development of the Bahá'í community in the immediate years ahead?

415.4 The House of Justice would appreciate concise but adequate comments, which it would like to receive at your earliest convenience, but no later than

31 March 1985. Kindly send a copy of your comments to the Continental Board of Counselors.

415.5 The House of Justice anticipates with gratitude your cooperation in this matter.

With loving Bahá'í greetings,
DEPARTMENT OF THE SECRETARIAT

416

Execution of Six Bahá'ís in Tehran
17 DECEMBER 1984

To National Spiritual Assemblies

416.1 REGRETFULLY ANNOUNCE SAD NEWS EXECUTION OF SIX BAHÁ'Í FRIENDS IN TEHRAN: DR. RÚḤÚ'LLÁH TA'LÍM FROM KIRMÁNSHÁH, MR. FÍRÚZ AT̲H̲ARÍ, MR. 'INÁYATU'LLÁH ḤAQÍQÍ, MR. JAMS̲H̲ÍD PÚR-USTÁDKÁR, MR. JAMÁL KÁS̲H̲ÁNÍ, MR. G̲H̲ULÁM ḤUSAYN FARHAND. THE LAST FIVE PREVIOUSLY NOTIFIED TO YOU AS AMONG THOSE CONDEMNED TO DEATH. FATE OF REMAINING 19 IN BALANCE. ALL EXECUTIONS TOOK PLACE ON ONE DAY, 9 DECEMBER. DETAILS ARE UNKNOWN AS NEITHER RELATIVES FRIENDS WERE INFORMED. FACT OF EXECUTIONS DISCOVERED 8 DAYS LATER. . . .

UNIVERSAL HOUSE OF JUSTICE

417

Execution of Dr. Farhád Aṣdaqí after Four Months of Torture; Death of Two Bahá'ís in Tabríz after More Than Two Years' Imprisonment
18 DECEMBER 1984

To all National Spiritual Assemblies

417.1 FURTHER OUR TELEX 17 DECEMBER WE SORROWFULLY ANNOUNCE NEWS EXECUTION BY HANGING OF DR. FARHÁD AṢDAQÍ ON 19 NOVEMBER AFTER FOUR MONTHS IMPRISONMENT AND TORTURE. MR. ḌÍYÁ'U'LLÁH MANÍ'I-USKÚ'Í AND MR. 'ALÍ-RIḌÁ NÍYÁKÁN DIED IN TABRÍZ PRISON 13TH AND 11TH NOVEMBER RESPECTIVELY AFTER TWO AND A HALF YEARS IMPRISONMENT. CAUSE DEATHS NOT ESTABLISHED.

UNIVERSAL HOUSE OF JUSTICE

418

Adoption of Another Oppressive Measure by the Iranian Government

3 JANUARY 1985

To all National Spiritual Assemblies

418.1 DISTRESSED INFORM YOU THAT RECENTLY IRANIAN GOVERNMENT ANNOUNCED ANOTHER OPPRESSIVE MEASURE AGAINST BAHÁ'Í PRISONERS. IN ORDER TO BE RELEASED PRISONER MUST SIGN FOLLOWING UNDERTAKING: "I THE UNDERSIGNED [DETAILS OF PERSONAL STATUS INCLUDING RELIGION] UNDERTAKE NOT TO HAVE IN MY POSSESSION ANY BOOK, PAMPHLET, DOCUMENT, SYMBOL OR PICTURE OF THIS MISGUIDED, ZIONIST, ESPIONAGE GROUP OF BAHÁ'ÍS. IF ANY OF THE ABOVE-MENTIONED ARTICLES BELONGING TO THIS HATED UNDERGROUND MOVEMENT IS FOUND ON MY PERSON OR IN MY HOME, THIS WILL BE TANTAMOUNT TO MY BEING OF THOSE 'WHO WAR AGAINST GOD' AND THE ATTORNEY-GENERAL WOULD BE FREE TO DECIDE AGAINST ME IN THE MANNER HE DEEMS FIT." THE TERMINOLOGY "WHO WAR AGAINST GOD" IS A TERM USED BY PRESENT REGIME TO SIGNIFY A CRIME DESERVING SENTENCE OF DEATH. BAHÁ'Í PRISONERS HAVE REFUSED SIGN SUCH AN INFAMOUS DOCUMENT FALSELY IMPUGNING THEIR FAITH. MOREOVER SIGNING SUCH A DOCUMENT WOULD LEAVE BAHÁ'ÍS OPEN TO HAVING SUCH ITEMS PLANTED ON THEIR PERSON OR IN THEIR HOMES. WAVE OF RECENT ARRESTS OF BAHÁ'ÍS SUGGESTS THAT AUTHORITIES PLAN EXERT PRESSURE ALL BAHÁ'ÍS TO SIGN SUCH AN IMPOSSIBLE UNDERTAKING. OBVIOUSLY FAITHFUL FRIENDS WILL REFUSE SUCCUMB SUCH CONTEMPTIBLE PLAN. . . .

UNIVERSAL HOUSE OF JUSTICE

419

Renewed Call for Support of International Funds

3 JANUARY 1985

To the Followers of Bahá'u'lláh in every Land

Dearly loved Friends,

419.1 Twelve months have passed since we addressed to the devoted followers of the Blessed Beauty throughout the world a message in which we outlined the major challenges which face the Cause of God and the thrilling opportunities which are presenting themselves for us to use in His Service.[1]

419.2 There was an immediate and heartwarming response in offers of service, in plans of action put promptly into effect, and in contributions to the Fund.

419-1. See message no. 385.

The activities of the friends are still increasing, and evidences of rich har- 419.3
vests are appearing. In India alone, over 150,000 new believers have joined the
Bahá'í community; in Samoa the Mashriqu'l-Adhkár has been dedicated
amidst unprecedented recognition of the Cause; in Canada, at the conference
held in London, Ontario, an upsurge of activity among the Bahá'í youth has
started a movement which has caught the imagination of the friends far and
wide.[2] In relation to the Fund, however, the rate of contributions during the
second six months of the year has slowed seriously, and we feel it is timely to
draw to your attention that our letter of 2 January 1984 was not an appeal for
a one-time herculean effort, but was intended to inform the whole world community of the present great challenges and opportunities which are not only
immediate but require also a long-range, sustained increase in the efforts and
self-sacrifices of the friends, both in service and in contributing from their
financial resources to the advancement of the Faith.

The challenges which we enumerated then are by no means met, nor the 419.4
opportunities wholly seized. The Indian Mashriqu'l-Adhkár, a building of great
size and beauty, is still unfinished; plans for the further development of the
World Center, for the design and erection of the three remaining buildings
on the Arc must be laid in full confidence that the funds for their completion will be made available;[3] projects for social and economic development,
for the establishment of Bahá'í schools and radio stations, for agricultural
advancement, and a wide range of other urgently needed activities are multi-
plying; worldwide attention to the Faith has increased during the past year
with even greater rapidity than before, demanding new measures to coordinate public information services and contacts with governments and leaders
of thought; and last, but near to the hearts of all, is the need of funds to
assist in the relief of those hard-pressed believers who have been forced to
leave Iran, often penniless and in great distress, seeking to build a new life in
other parts of the world.

Last April we were deeply touched by receiving a petition from the dele- 419.5
gates gathered at the National Convention of the Bahá'ís of the United States,
requesting that the Law of Ḥuqúqu'lláh be made binding on all the believers
in that country.[4] Although it is not yet the time to take this far-reaching step,

---

419-2. For the dedication of the House of Worship in Western Samoa, see message dated August 1984 (no. 403). For the message to the youth conference in London, Ontario, see telex dated 23 August 1984 (no. 408).

419-3. The Indian Temple was dedicated on 23–27 December 1986. The three buildings yet to be constructed on the Arc on Mount Carmel are the International Bahá'í Library and the seats for the International Teaching Center and the Center for the Study of the Texts.

419-4. For information on the law of Ḥuqúqu'lláh, see the glossary. For information on the introduction of Ḥuqúqu'lláh to the West, see message dated 6 August 1984 (no. 404). In its Riḍván message of 1991, the Universal House of Justice noted the "exceptional confluence" of a number of imminent achievements (the publication of the Kitáb-i-Aqdas, the progress of the

we were moved to decide that, as a preliminary measure, the texts relating to the Law of Ḥuqúqu'lláh will be translated into English for general information against the time when this law will be applied more widely.⁵

419.6 However, important as is the Law of Ḥuqúqu'lláh, the devoted followers of Bahá'u'lláh have, even without it, every opportunity to contribute regularly and sacrificially to the work of the Cause. It is to a greater realization of the privilege and responsibility of supporting the multiple activities of our beloved Faith that we call you all at this critical time in world history, and remind you that to support the Bahá'í funds is an integral part of the Bahá'í way of life. The need is not only now, but throughout the years to come, until our exertions, reinforced by confirmations from on high, will have overcome the great perils now facing mankind and have made this world another world—a world whose splendor and grace will surpass our highest hopes and greatest dreams.

With loving Bahá'í greetings,
THE UNIVERSAL HOUSE OF JUSTICE

420

International Year of Peace
23 JANUARY 1985

To all National Spiritual Assemblies

Dear Bahá'í Friends,

420.1 1986 has been named the International Year of Peace by the United Nations. Considering the dangers surrounding mankind and the remedial prospects of the Lesser Peace to which Bahá'u'lláh has summoned the nations, we embrace this God-sent opportunity to proclaim ever more widely and convincingly the vitalizing principles upon which, as our Teachings emphatically assert, a lasting peace must be founded.¹ The nature and variety of the proclamation activities which the Bahá'í community will undertake, during 1986 and beyond, will be outlined in detail later. We wish now to indicate some of the ideas we are contemplating, so that you may sense what to expect and how to prepare for your own participation.

---

building projects on Mount Carmel, the conclusion of the Six Year Plan, and the inception of the Holy Year 1992) and announced that, as of Riḍván 1992, the law of Ḥuqúqu'lláh would become universally applicable to "all who profess their belief in the Supreme Manifestation of God."

419-5. The compilation was sent to selected National Spiritual Assemblies with the 4 July 1985 message (no. 430). For a copy of the compilation, see CC 1:489–527.

420-1. For a discussion of the Lesser Peace, see the glossary.

1979–1986 · THE SEVEN YEAR PLAN

In addition to projects to be initiated at the World Center, these ideas include: 420.2

- Calling upon local and national Baháʼí communities to sponsor a wide range of activities which will engage the attention of people from all walks of life to various topics relevant to peace, such as: the role of women, the elimination of racism, the eradication of prejudice, the promotion of education, the extension of social and economic development, the adoption of a world auxiliary language, the establishment of world government; 421.2a
- Mounting a publicity campaign which will make use of such themes as "world peace through world religion," "world peace through world education," "world peace through world language," "world peace through world law"—a campaign which could lead to discussion of these subjects in small or large gatherings, at local or national levels, and perhaps in collaboration with organizations promoting such ideas; 421.2b
- Urging the publishing within and without the Baháʼí community of a wide assortment of literature, posters and other graphic materials on peace; 421.2c
- Requesting Baháʼí magazines—children, youth, adult—whether intended for internal or external circulation, to carry special features on peace; 421.2d
- Inviting Baháʼí radio stations to devote particular attention to this theme; 421.2e
- Asking the Associations for Baháʼí Studies to conduct programs on peace; 421.2f
- Encouraging Baháʼí artists and musicians to contribute, and consider inviting their non-Baháʼí colleagues to contribute, to the effectiveness of such activities by giving expression through the various arts to important themes relating to world peace. 421.2g

In effect, we envision a proclamation campaign which will not only involve large public events and the use of the mass media, but will also engage people at the grassroots and at all other levels of society in a broad range of profoundly effective activities through which they will interact with the Baháʼí community in a sustained, worldwide effort to attend to the fundamental issues of peace, aided by the unique insights provided by the Teachings of Baháʼuʼlláh. 420.3

As you contemplate what possibilities these and similar ideas suggest for your own plans, we advise you to take preparatory steps to hold within your jurisdiction, during 1986, local and national peace conferences to which public officials and other prominent persons should be invited. In those places where 420.4

national conferences may not be possible, local conferences should certainly be held.

420.5 In some regions, neighboring National Assemblies may find it convenient to pool their resources and hold regional conferences instead of national ones. These need not be very large, but should be effective enough to make a good impression on the public as well as on the national Bahá'í communities involved. It is left to the initiative of the National Assemblies, in consultation with Continental Counselors, to hold such conferences.

420.6 Simultaneously as you make initial arrangements for the conferences, you will also want to find out what plans are being made by the governments and organizations in your respective countries, so that you will know beforehand how to coordinate your own programs with the programs of others in ways most conducive to the proclamation of the Faith and the mutual benefit of all concerned.

420.7 We would welcome any thoughts and suggestions you may have on the activities to be undertaken by you during the International Year of Peace.

420.8 Your planning efforts for 1986 must not, of course, interrupt the work of the Seven Year Plan. Indeed, the activities associated with the economic and social development of the Bahá'í community, the observance during 1985 of International Youth Year, and the anticipated activities for the peace campaign to begin a year hence are mutually reinforcing and go far to enhance the teaching opportunities necessary to the successful completion of the Plan. We have every confidence that your continuing exertions to meet the new challenges resulting from the emergence of the Faith from obscurity will be richly rewarded by the Blessed Beauty; and we shall renew our supplications at the Holy Threshold that your brightest expectations may be surpassed by resounding triumph.

With loving Bahá'í greetings,
THE UNIVERSAL HOUSE OF JUSTICE

421

Execution of Rúḥu'lláh Ḥaṣúrí in Yazd; Confirmation of Death in prison of Rustam Varjávandí

29 January 1985

To all National Spiritual Assemblies

JUST RECEIVED NEWS EXECUTION YAZD, RÚḤU'LLÁH ḤAṢÚRÍ, 35 YEARS OLD. HE WAS SHOT IN PRISON. NEWS DEATH IN PRISON TEHRAN RUSTAM VARJÁVANDÍ 15 SEPTEMBER CONFIRMED. THIS BRINGS TO 178 NUMBER THOSE WHO HAVE GIVEN THEIR LIVES FOR FAITH. SINCE 3 SEPTEMBER 16 BAHÁ'ÍS HAVE BEEN KILLED OR HAVE DIED IN PRISON, 101 IMPRISONED AND 79 RELEASED, SOME OF WHOM HAD NOT PREVIOUSLY BEEN REPORTED IMPRISONED. TOTAL NUMBER NOW IN PRISON, AS FAR AS CAN BE ASCERTAINED, IS 707. GOVERNMENT EMPLOYEES WHO WERE DISMISSED OR THEIR PENSIONS STOPPED BECAUSE OF THEIR FAITH ARE NOW BEING SUMMONED BY ATTORNEY-GENERAL AND COMPELLED PAY BACK ALL THE SALARIES RECEIVED FOR MANY YEARS, EVEN DECADES, FAILING WHICH THEY WILL BE IMPRISONED. CAUSE MANY RECENT IMPRISONMENTS IS INABILITY TO PAY BACK SALARIES BAHÁ'ÍS HAD RECEIVED WHILE LAWFULLY EMPLOYED BY GOVERNMENT. . . .

UNIVERSAL HOUSE OF JUSTICE

422

Elucidation of the Lesser Peace and the Supreme Tribunal

31 January 1985

To an individual Bahá'í

Dear Bahá'í Friend,

The Universal House of Justice has received your letter of 13 December 1984 inquiring about the Lesser Peace and the Supreme Tribunal referred to in the writings of the Faith. We are asked to convey the following comments.

Bahá'u'lláh's principal mission in appearing at this time in human history is the realization of the oneness of mankind and the establishment of peace among the nations; therefore, all the forces which are focused on accomplishing these ends are influenced by His Revelation. We know, however, that peace will come in stages. First, there will come the Lesser Peace, when the unity of nations will be achieved, then gradually the Most Great Peace—the spiritual as well as social and political unity of mankind, when the Bahá'í World Commonwealth, operating in strict accordance with the laws and ordinances of the Most Holy Book of the Bahá'í Revelation,[1] will have been established through the efforts of the Bahá'ís.

422.3 As to the Lesser Peace, Shoghi Effendi has explained that this will initially be a political unity arrived at by decision of the governments of various nations; it will not be established by direct action of the Bahá'í community. This does not mean, however, that the Bahá'ís are standing aside and waiting for the Lesser Peace to come before they do something about the peace of mankind. Indeed, by promoting the principles of the Faith, which are indispensable to the maintenance of peace, and by fashioning the instruments of the Bahá'í Administrative Order, which we are told by the beloved Guardian is the pattern for future society, the Bahá'ís are constantly engaged in laying the groundwork for a permanent peace, the Most Great Peace being their ultimate goal.

422.4 The Lesser Peace itself will pass through stages; at the initial stage the governments will act entirely on their own without the conscious involvement of the Faith; later on, in God's good time, the Faith will have a direct influence on it in ways indicated by Shoghi Effendi in his "The Goal of a New World Order."[2] In connection with the steps that will lead to this latter stage, the Universal House of Justice will certainly determine what has to be done, in accordance with the guidance in the Writings, such as the passage you quoted from *Tablets of Bahá'u'lláh*, page 89. In the meantime, the Bahá'ís will undoubtedly continue to do all in their power to promote the establishment of peace.

422.5 The Universal House of Justice is greatly pleased with the initiative you have taken with others to start an organization through which youth can contribute their considerable energies and creative abilities towards fostering world peace. It fully appreciates the wisdom of your approach in not affiliating your organization directly with the Faith and finds your leaflet describing the aims of Youth for World Peace most impressive. Since 1986 has been designated the International Year of Peace by the United Nations, your efforts are timely and will blend with the activities to be undertaken by the Bahá'í community during that year and beyond.

422.6 In your worthy endeavors to pursue your goal to create a peace movement of youth around the world which would draw non-Bahá'í youth within the influence of Bahá'í concepts, you should seek the advice of your National Spiritual Assembly, which is in a position to give you necessary advice.

422.7 Regarding your question about the Supreme Tribunal, enclosed are a few extracts from statements by 'Abdu'l-Bahá and letters written on behalf of Shoghi Effendi on this subject; also included is a statement on the Lesser Peace.

422-1. The Kitáb-i-Aqdas, Bahá'u'lláh's Book of Laws.
422-2. See WOB, pp. 29–48.

We are asked by the House of Justice to assure you of its ardent prayers in 422.8
the Holy Shrines that your noble efforts may contribute significantly to the
upraising of the banner of peace in the world.

> With loving Bahá'í greetings,
> DEPARTMENT OF THE SECRETARIAT

From the Writings of 'Abdu'l-Bahá

So long as these prejudices survive, there will be continuous and fearsome wars. 422.9

To remedy this condition there must be universal peace. To bring this 422.10
about, a Supreme Tribunal must be established, representative of all governments and peoples; questions both national and international must be referred thereto, and all must carry out the decrees of this Tribunal. Should any government or people disobey, let the whole world arise against that government or people.

(*Selections from the Writings of 'Abdu'l-Bahá,* [Rev. ed.] [Haifa: Bahá'í World Center, 1982], p. 249)

For example, the question of universal peace, about which Bahá'u'lláh says 422.11
that the Supreme Tribunal must be established: although the League of Nations has been brought into existence, yet it is incapable of establishing universal peace.[3] But the Supreme Tribunal which Bahá'u'lláh has described will fulfill this sacred task with the utmost might and power. And His plan is this: that the national assemblies of each country and nation—that is to say parliaments—should elect two or three persons who are the choicest of that nation, and are well informed concerning international laws and the relations between governments and aware of the essential needs of the world of humanity in this day. The number of these representatives should be in proportion to the number of inhabitants of that country. The election of these souls who are chosen by the national assembly, that is, the parliament, must be confirmed by the upper house, the congress and the cabinet and also by the president or monarch so these persons may be the elected ones of all the nation and the government. The Supreme Tribunal will be composed of these

---

422-3. The League of Nations was an organization for international cooperation established at the initiative of the Allied powers in 1919 at the end of World War I. It was based on a covenant calling for joint action by League members against an aggressor, arbitration of international disputes, reduction of armaments, and open diplomacy. The League was weakened by the United States' failure to join as a result of the Congress's rejection of membership and by the dissatisfaction of Japan, Italy, and Germany with the terms of the Versailles Treaty following World War I. Moreover, the League had no power to enforce its covenant apart from the ability of member states to take united and determined action against a belligerent state. Thus its failure to prevent Japanese expansion in Manchuria and China, Italy's conquest of Ethiopia, or Hitler's repudiation of the Versailles Treaty thoroughly discredited the League and rendered it powerless to avert World War II.

people and all mankind will thus have a share therein, for every one of these delegates is fully representative of his nation. When the Supreme Tribunal gives a ruling on any international question, either unanimously or by majority rule, there will no longer be any pretext for the plaintiff or ground of objection for the defendant. In case any of the governments or nations, in the execution of the irrefutable decision of the Supreme Tribunal, be negligent or dilatory, the rest of the nations will rise up against it, because all the governments and nations of the world are the supporters of this Supreme Tribunal. Consider what a firm foundation this is! But by a limited and restricted League the purpose will not be realized as it ought and should. This is the truth about the situation, which has been stated. . . .

(Tablet to the Executive Committee of the Central Organization for a Durable Peace, *Selections from the Writings of 'Abdu'l-Bahá*, pp. 306–7)

422.12 A Supreme Tribunal shall be established by the peoples and Governments of every nation, composed of members elected from each country and Government. The members of this Great Council shall assemble in unity. All disputes of an international character shall be submitted to this Court, its work being to arrange by arbitration everything which otherwise would be a cause of war. The mission of this Tribunal would be to prevent war.

(*Paris Talks: Addresses Given by 'Abdu'l-Bahá in Paris in 1911–1912*, 11th ed. [London: Bahá'í Publishing Trust, 1979], p. 155)

From letters written on behalf of the Guardian to individual believers

The Supreme Tribunal

422.13 The Universal Court of Arbitration and the International Tribunal are the same. When the Bahá'í State will be established they will be merged in the Universal House of Justice.

(17 June 1933)

422.14 As regards the International Executive referred to by the Guardian in his "Goal of a New World Order," it should be noted that this statement refers by no means to the Bahá'í Commonwealth of the future, but simply to that world government which will herald the advent and lead to the final establishment of the World Order of Bahá'u'lláh. The formation of this International Executive, which corresponds to the executive head or board in present-day national governments, is but a step leading to the Bahá'í world government of the future, and hence should not be identified with either the institution of the Guardianship or that of the International House of Justice.

(17 March 1934)
(refers to pp. 40–41 of *The World Order of Bahá'u'lláh*)

1 and 2 The Supreme Tribunal is an aspect of a World Super-state; the exact nature of its relationship to that State we cannot at present foresee.

(19 November 1945)

The above statement was in reply to the following questions:

1. Is the Supreme Tribunal the world court or world tribunal referred to in "The Unfoldment of World Civilization," p. 43, and "Goal of a New World Order," p. 20? [See *The World Order of Bahá'u'lláh*, pp. 203 and 41.] Is it part of the world Super-State just as our Supreme Court is part of the federal government at Washington?
2. Will the Supreme Tribunal (a world court) exist apart from the world government?

The Lesser Peace

With reference to the question you have asked concerning the time and means through which the Lesser and Most Great Peace, referred to by Bahá'u'lláh, will be established, following the coming World War:[4] Your view that the Lesser Peace will come about through the political efforts of the states and nations of the world, and independently of any direct Bahá'í plan or effort, and the Most Great Peace be established through the instrumentality of the believers, and by the direct operation of the laws and principles revealed by Bahá'u'lláh and the functioning of the Universal House of Justice as the supreme organ of the Bahá'í superstate—your view on this subject is quite correct and in full accord with the pronouncements of the Guardian as embodied in "The Unfoldment of World Civilization."[5]

(14 March 1939)

422-4. World War II.
422-5. See WOB, pp. 161–206.

423
Passing of Lloyd Gardner, Continental Counselor
7 March 1985

To all National Spiritual Assemblies

423.1 DEEPLY GRIEVED UNTIMELY PASSING ESTEEMED LLOYD GARDNER STALWART DEFENDER INTERESTS FAITH AMERICAS.[1] HIS DISTINGUISHED LONGTIME SERVICE NATIONAL ASSEMBLY CANADA AND MEMBERSHIP BOARD COUNSELORS AMERICAS SINCE INCEPTION MARKED BY INDEFATIGABLE LABORS ALL ASPECTS TEACHING WORK AND COMMUNITY LIFE, NOTABLY YOUTH ACTIVITIES. HIS STERLING CHARACTER, HIGH INTEGRITY, WARM-HEARTED NATURE, TOTAL DEDICATION FAITH WORTHY EMULATION. FERVENTLY PRAYING HOLY SHRINES PROGRESS HIS LUMINOUS SPIRIT ABHÁ KINGDOM AND SOLACE HIS SORROWING FAMILY. ADVISING ALL COMMUNITIES AMERICAS OFFER PRAYERS IN HIS NAME. REQUESTING NATIONAL ASSEMBLY UNITED STATES HOLD MEMORIAL GATHERING TEMPLE WILMETTE.

UNIVERSAL HOUSE OF JUSTICE

424
Execution of Rúhu'lláh Bahrámsháhí in Yazd and Nuṣratu'lláh Subḥání in Tehran
14 March 1985

To all National Spiritual Assemblies

424.1 RECENTLY SAD NEWS OF FURTHER EXECUTIONS OF TWO BAHÁ'ÍS IN IRAN RECEIVED. MR. RÚḤU'LLÁH BAHRÁMSHÁHÍ, AGE 50, EXECUTED YAZD 25 FEBRUARY, MR. NUṢRATU'LLÁH ṢUBḤÁNÍ IN TEHRAN 5 MARCH. . . .

UNIVERSAL HOUSE OF JUSTICE

423-1. For an account of the life and services of Lloyd Gardner, see BW 19:663–65.

425
Human Suffering and the Reconstruction of Society
14 March 1985

To an individual Bahá'í

Dear Bahá'í Friend,

Your letter of 16 January 1985 in which you share the anguish of your heart and express deep concern for the fate of the suffering masses of mankind has been received by the Universal House of Justice. We are instructed to convey this reply to you. 425.1

The world is clearly beset by ills and is groaning under the burden of appalling suffering. The trials of the innocent are indeed heartrending and constitute a mystery that the mind of man cannot fathom. Even the Prophets of God Themselves have borne Their share of grievous afflictions in every age. Yet in spite of the evidence of all this suffering, God's Manifestations, Whose lives and wisdom show Them to have been far above human beings in understanding, unitedly bear testimony to the justice, love and mercy of God. 425.2

To understand the condition of the world it is necessary to step back, so to speak, to gain a clearer view of the panorama of God's great redemptive Major Plan, which is shaping the destiny of mankind according to the operation of the divine Will.[1] It should not be surmised that the calamitous events transpiring in all corners of the globe are random and lack purpose, though individually they may be difficult to comprehend. According to the words of our beloved Guardian: "The invisible hand is at work and the convulsions taking place on earth are a prelude to the proclamation of the Cause of God."[2] We can confidently anticipate therefore, the arrival of the "new life-giving spring" once the destructive icy blasts of winter's tempests have run their course.[3] 425.3

As Bahá'ís, we know that the "sovereign remedy" for each and every one of these ills lies in turning and submitting to the "skilled," the "all-powerful" and "inspired Physician." Bahá'u'lláh has assured us in His writings that God has not forsaken us. He is the All-Seeing and All-Knowing, the "prayer-hearing, prayer-answering God" to those who turn to Him in supplication, and He intervenes actively in human history by sending His Manifestations, Sources of knowledge and spiritual truth to "liberate the children of men from 425.4

425-1. God's Major Plan is the plan for humanity that He Himself directs, that is tumultuous in its progress, that works through humanity as a whole and welds humankind into a unified body through the fires of suffering and tribulation. Its first goal is the Lesser Peace, the political unification of the world (see glossary). Its ultimate object is the Kingdom of God on earth.

425-2. From an unpublished letter.

425-3. SAQ, p. 74.

the darkness of ignorance" and to "ensure the peace and tranquillity of mankind."4 In this Age, God has determined to establish His everlasting Kingdom among men, and so, to this end, He sent us the spirit and message of the New Day through two successive Manifestations, Who alas, were rejected by the generality of people.

425.5 When we contemplate the fate of mankind, it is important to reflect on the very complex arena in which man plays out the drama of his existence. There are a number of elements involved. For example, man is a spiritual being located within the material creation; hence he is subjected to opposing forces, and has to live in accordance with values which refer to two worlds, the material world with all its imperfections and the spiritual world with its perfections. Tension derives from the fact that "In man there are two natures; his spiritual or higher nature and his material or lower nature. In one he approaches God, in the other he lives for the world alone."5 Man's actions then have both a material and spiritual consequence. While the material effect of his actions is usually clearly perceptible, their spiritual effect can only be determined by reference to spiritual principles revealed by the Manifestation of God. Suffering and trials, sent by God to test and perfect His creatures, are another integral part of life. They contain the potential for man's progress or retrogression, depending on the individual's response. As 'Abdu'l-Bahá explains:

425.5a The souls who bear the tests of God become the manifestations of great bounties; for the divine trials cause some souls to become entirely lifeless, while they cause the holy souls to ascend to the highest degree of love and solidity. . . .6

425.5b In addition to the factors associated with man's station and nature, the Writings indicate that man's soul "is independent of all infirmities of body or mind," and not only continues to exist "after departing from this mortal world," but progresses "through the bounty and grace of the Lord."7 Therefore, an evaluation of man's material existence and achievements cannot ignore the potential spiritual development stimulated by the individual's desire to manifest the attributes of God and his response to the exigencies of his life, nor can it exclude the possibility of the operations of God's mercy in terms of compensation for earthly suffering, in the next life.

425.6 God in His bounty has endowed every created thing, however humble, "with the capacity to exercise a particular influence, and been made to possess a distinct virtue."8 And, reminiscent of the parable of the talents (Matthew

---

425-4. GWB, p. 255; BP, p. 163; GWB, pp. 79–80.
425-5. PT, p. 60.
425-6. TABA 2:324.
425-7. GWB, p. 154; SAQ, p. 240.
425-8. GWB, p. 189.

25:14–30), Bahá'u'lláh, in the *Gleanings* (page 149), draws our attention to the need to make efforts to develop and demonstrate in action our God-given potential:

> All that which ye potentially possess can, however, be manifested only as a result of your own volition. Your own acts testify to this truth. . . .   425.6a

Is it not an evidence of the justice of God that each of us, whether materially comfortable or struggling for physical survival, is assessed in terms of the efforts we have made to seize whatever opportunities existed in our lives, to develop and use our allotted talent, be it large or small? "Each shall receive his share from thy Lord," is Bahá'u'lláh's assurance.[9] Thus, if we bestir ourselves, we will all have access to the rewards of this life and the next.   425.6b

In the same passage from the *Gleanings*, Bahá'u'lláh also raises the possibility that possessing free will, human beings may well commit evil and "wittingly" break "His law." By the exercise of his free will, man either affirms his spiritual purpose in life or chooses to perpetuate evil by living below his highest station. The question is asked: "Is such a behavior to be attributed to God, or to their proper selves?" And concludes: 425.7

> Every good thing is of God, and every evil thing is from yourselves. . . .[10]   425.7a

The amelioration of the conditions of the world requires the reconstruction of human society and efforts to improve the material well-being of humanity. The Bahá'í approach to this task is evolutionary and multifaceted, involving not only the spiritual transformation of individuals but the establishment of an administrative system based on the application of justice, a system which is at once the "nucleus" and the "pattern" of the future World Order, together with the implementation of programs of social and economic development that derive their impetus from the grass roots of the community.[11] Such an integrated approach will inevitably create a new world, a world where human dignity is restored and the burden of inequity is lifted from the shoulders of humanity. Then will the generations look back with heartfelt appreciation, for the sacrifices made by Bahá'ís and non-Bahá'ís alike, during this most turbulent period in human history.   425.8

With regard to your concern that certain remote geographical regions have historically been deprived of Divine Revelation, the following extracts indicate that there are many other Prophets Who have appeared in the world, but Whose names are not mentioned in the Scriptures with which we are familiar, for example: 425.9

425-9. GWB, p. 170.
425-10. GWB, p. 149.
425-11. WOB, p. 144.

425.9a And to every people have we sent an apostle. . .
(*Qur'án*, XVI, 38)

425.9b God hath raised up Prophets and revealed Books as numerous as the creatures of the world, and will continue to do so to everlasting.
(*Selections from the Writings of the Báb*, p. 125)

425.9c Know thou that the absence of any reference to them [Prophets before Adam] is no proof that they did not actually exist. That no records concerning them are now available, should be attributed to their extreme remoteness, as well as to the vast changes which the earth hath undergone since their time.
(*Gleanings*, p. 172)

425.9d While Asia has clearly been blessed as the birthplace of many Manifestations of God, we have found nothing in the Writings to suggest exactly where in the world these Messengers of God in the remote past, may have arisen. In a letter written on behalf of the Guardian, we have the promise that ". . . there always have been Manifestations of God, but we do not have any record of Their names."[12]

425.10 We are instructed to assure you that the Universal House of Justice will offer prayers at the Holy Shrines that your faith may be deepened and your perplexities resolved, and we share with you these solacing words of Bahá'u'lláh:

425.10a O My servants! Sorrow not if, in these days and on this earthly plane, things contrary to your wishes have been ordained and manifested by God, for days of blissful joy, of heavenly delight, are assuredly in store for you. Worlds, holy and spiritually glorious, will be unveiled to your eyes. You are destined by Him, in this world and hereafter, to partake of their benefits, to share in their joys, and to obtain a portion of the sustaining grace. To each and every one of them you will, no doubt, attain.
(*Gleanings*, p. 329)

With loving Bahá'í greetings,
DEPARTMENT OF THE SECRETARIAT

425-12. LG, p. 503.

426
Responsibilities of Youth at the Age of Maturity
11 April 1985

To an individual Bahá'í

Dear Bahá'í Friend,

The Universal House of Justice received your very thoughtful letter and instructs us to convey the following answer to you. 426.1

While some opportunities for service in the Administrative Order are clearly reserved for those who are over twenty-one years of age, the importance of attaining spiritual maturity at the age of fifteen is that it marks that point in life at which the believer takes firmly into his own hands the responsibility for his spiritual destiny. At age fifteen, the individual has the privilege of affirming, in his own name, his faith in Bahá'u'lláh. For while the children of Bahá'í parents are considered to be Bahá'ís, they do not automatically inherit the Faith of their parents. Therefore, when they come of age, they must, of their own volition, express their belief. 426.2

Having reached the age of fifteen, Bahá'í youth are personally responsible for certain spiritual activities such as observing the obligation of daily prayer, keeping the Fast, and they are invited to participate in Bahá'í youth activities. The significance of the age of maturity, however, goes far beyond the fulfillment of responsibilities. The following extract from a Tablet of 'Abdu'l-Bahá links the attainment of maturity with the deepening of one's understanding and comprehension of the realities of life, and the enhancement of one's very capacity for understanding: 426.3

> Know thou that before maturity man liveth from day to day and comprehendeth only such matters as are superficial and outwardly obvious. However, when he cometh of age he understandeth the realities of things and the inner truths. Indeed, in his comprehension, his feelings, his deductions and his discoveries, every day of his life after maturity is equal to a year before it. 426.3a

The signing of a card is simply the means by which the individual indicates his desire to be registered as a Bahá'í youth, as a member of the Bahá'í community, and it enables the National Spiritual Assembly to keep an accurate record of the membership of the community.... 426.4

The Universal House of Justice assures you that it will offer prayers at the Holy Shrines that you will strive to become an active and enlightened servant of the Cause of God. 426.5

With loving Bahá'í greetings,
DEPARTMENT OF THE SECRETARIAT

427
Riḍván Message 1985
RIḌVÁN 1985

To the Bahá'ís of the World

Dearly loved Friends,

427.1 As we enter the final year of the Seven Year Plan, confidence of victory and a growing sense of the opening of a new stage in the onward march of the Faith must arouse in every Bahá'í heart feelings of gratitude and eager expectation. Victory in the Plan is now within sight, and at its completion the summation of its achievements may well astonish us all. But the great, the historic feature of this period is the emergence of the Faith from obscurity, promoted by the steadfast heroism of the renowned, the indefatigable, dearly loved Bahá'í community of Bahá'u'lláh's and the Báb's native land.[1]

427.2 This dramatic change in the status of the Faith of God, occurring at so chaotic a moment in the world's history when statesmen and leaders and governors of human institutions are witnessing, with increasing despair, the bankruptcy and utter ineffectiveness of their best efforts to stay the tide of disruption, forces upon us, the Bahá'ís, the obligation to consider anew and ponder deeply the beloved Guardian's statement that "The principle of the Oneness of Mankind—the pivot round which all the teachings of Bahá'u'lláh revolve— . . . implies an organic change in the structure of present-day society, a change such as the world has not yet experienced."[2]

427.3 Intimations in the non-Bahá'í world of a rapidly growing realization that mankind is indeed entering a new stage in its evolution present us with unprecedented opportunities to show that the Bahá'í world community is not only "the nucleus but the very pattern" of that world society which it is the purpose of Bahá'u'lláh to establish and towards which a harassed humanity, albeit largely unconsciously, is striving.[3]

More Involvement in the Life of Society

427.4 The time has come for the Bahá'í community to become more involved in the life of the society around it, without in the least supporting any of the world's moribund and divisive concepts, or slackening its direct teaching efforts, but rather, by association, exerting its influence towards unity, demonstrating its ability to settle differences by consultation rather than by confrontation, violence or schism, and declaring its faith in the divine purpose of human existence.

427-1. Iran.
427-2. WOB, pp. 42–43.
427-3. WOB, p. 144.

427.5 Bahá'í Youth are taking advantage of the United Nations' designation of 1985 as the Year of Youth to launch their own campaign of active cooperation with other youth groups, sharing with them Bahá'í ideals and a vision of what they intend to make of the world. The Bahá'í community will be strongly represented at the culminating event of the United Nations' Decade of Women in this same year. 1986 has been named the Year of Peace, and the Faith will be far from silent or obscure on that issue. Even now the House of Justice is making plans for the presentation of the Bahá'í concepts on peace to the governments and leaders of the world and, through the Bahá'í world community, to its national and local authorities and to all sections of the variegated world society.[4] But it is in the local Bahá'í communities that the most widespread presentation of the Faith can take place. It is here that the real pattern of Bahá'í life can be seen. It is here that the power of Bahá'u'lláh to organize human affairs on a basis of spiritual unity can be most apparent. Every Local Spiritual Assembly which unitedly strives to grow in maturity and efficiency and encourages its community to fulfill its destiny as a foundation stone of Bahá'u'lláh's World Order can add to a growing ground swell of interest in and eventual recognition of the Cause of God as the sole hope for mankind.

427.6 Such considerations as these are now occupying the earnest attention of the Universal House of Justice. Their specific implementation will form a large part of the next Plan which will follow immediately on the completion of the present one and will be of six years' duration. By winning the Seven Year Plan, by consolidating our local communities, and above all by strengthening and deepening our understanding of the purpose of Bahá'u'lláh's Revelation we shall be preparing ourselves to play our part in bringing about that transformation of human life on this planet which must take place ere it becomes fit to receive the bounties and blessings of God's own Kingdom.

With loving Bahá'í greetings,
THE UNIVERSAL HOUSE OF JUSTICE

427-4. See message dated 4 July 1985 (no. 429) announcing the completion of the peace statement and the October 1985 statement of the Universal House of Justice addressed to the peoples of the world (no. 438).

428

Message to Youth—Scaling the Heights of Excellence
8 May 1985

To the Bahá'í Youth of the World

Dear Bahá'í Friends,

428.1 We extend our loving greetings and best wishes to all who will meet in youth conferences yet to be held during International Youth Year. So eager and resourceful have been the responses of the Bahá'í youth in many countries to the challenges of this special year that we are moved to expressions of delight and high hope.

428.2 We applaud those youth who, in respect of this period, have already engaged in some activity within their national and local communities or in collaboration with their peers in other countries, and call upon them to persevere in their unyielding efforts to acquire spiritual qualities and useful qualifications. For if they do so, the influence of their high-minded motivations will exert itself upon world developments conducive to a productive, progressive and peaceful future.

428.3 May the youth activities begun this year be a fitting prelude to and an ongoing, significant feature throughout the International Year of Peace, 1986.

428.4 The present requirements of a Faith whose responsibilities rapidly increase in relation to its rise from obscurity impose an inescapable duty on the youth to ensure that their lives reflect to a marked degree the transforming power of the new Revelation they have embraced. Otherwise, by what example are the claims of Bahá'u'lláh to be judged? How is His healing Message to be acknowledged by a skeptical humanity if it produces no noticeable effect upon the young, who are seen to be among the most energetic, the most pliable and promising elements in any society?

428.5 The dark horizon faced by a world which has failed to recognize the Promised One, the Source of its salvation, acutely affects the outlook of the younger generations; their distressing lack of hope and their indulgence in desperate but futile and even dangerous solutions make a direct claim on the remedial attention of Bahá'í youth, who, through their knowledge of that Source and the bright vision with which they have thus been endowed, cannot hesitate to impart to their despairing fellow youth the restorative joy, the constructive hope, the radiant assurances of Bahá'u'lláh's stupendous Revelation.

428.6 The words, the deeds, the attitudes, the lack of prejudice, the nobility of character, the high sense of service to others—in a word, those qualities and actions which distinguish a Bahá'í must unfailingly characterize their inner life and outer behavior, and their interactions with friend or foe.

Rejecting the low sights of mediocrity, let them scale the ascending heights of excellence in all they aspire to do. May they resolve to elevate the very atmosphere in which they move, whether it be in the school rooms or halls of higher learning, in their work, their recreation, their Bahá'í activity or social service. 428.7

Indeed, let them welcome with confidence the challenges awaiting them. Imbued with this excellence and a corresponding humility, with tenacity and a loving servitude, today's youth must move towards the front ranks of the professions, trades, arts and crafts which are necessary to the further progress of humankind—this to ensure that the spirit of the Cause will cast its illumination on all these important areas of human endeavor. Moreover, while aiming at mastering the unifying concepts and swiftly advancing technologies of this era of communications, they can, indeed they must also guarantee the transmittal to the future of those skills which will preserve the marvelous, indispensable achievements of the past. The transformation which is to occur in the functioning of society will certainly depend to a great extent on the effectiveness of the preparations the youth make for the world they will inherit. 428.8

We commend these thoughts to your private contemplation and to the consultations you conduct about your future. 428.9

And we offer the assurance of our prayerful remembrances of you, our trust and confidence. 428.10

THE UNIVERSAL HOUSE OF JUSTICE

429
Announcement of Completion of a Statement on Peace Addressed to the Peoples of the World
4 JULY 1985

To all National Spiritual Assemblies

JOYFULLY ANNOUNCE COMPLETION STATEMENT ON PEACE ADDRESSED TO PEOPLES WORLD.[1] ARRANGEMENTS BEING MADE PRESENTATION WORLD LEADERS BEGINNING 24 OCTOBER. STATEMENT BEING SENT NATIONAL SPIRITUAL ASSEMBLIES SHORTLY. OUR HEARTS FILLED GRATITUDE BAHÁ'U'LLÁH FOR UNPRECEDENTED OPPORTUNITY ENABLING HIS SERVANTS OFFER HIS HEALING MESSAGE AT THIS CRITICAL TIME FORTUNES MANKIND. 429.1

UNIVERSAL HOUSE OF JUSTICE

429-1. See the statement dated October 1985 (message no. 438).

430
Release of a Compilation on the Law of Ḥuqúqu'lláh
4 July 1985

To National Spiritual Assemblies

Dear Bahá'í Friends,

430.1 At the request of the Universal House of Justice, the Research Department has prepared a compilation of texts on the subject of Ḥuqúqu'lláh. A copy of this compilation is attached for your information.[1] You may share its contents with the friends under your jurisdiction as you see fit.

430.2 As you are aware, the law of Ḥuqúqu'lláh is not applicable universally. Study of these extracts by the friends in communities where this law is not yet binding will help them to prepare themselves for the day when the House of Justice will decide to apply the law progressively to all communities in the Bahá'í World.

With loving Bahá'í greetings,
DEPARTMENT OF THE SECRETARIAT

431
Message to International Youth Conference, Columbus, Ohio
11 July 1985

To the International Youth Conference in Columbus, Ohio

431.1 OUR HEARTS UPLIFTED HIGH SPIRIT, LOFTY PURPOSES, MULTIFARIOUS FEATURES, REMARKABLE IMPACT INTERNATIONAL YOUTH CONFERENCE COLUMBUS, OHIO, PARTICULARLY DISTINGUISHED BY IMPRESSIVE EXAMPLES SELFLESS SERVICE ITS PARTICIPANTS. THEIR UPRIGHT BEHAVIOR, NOBLE DEEDS, CREATIVE VIGOR HEIGHTENED PRESTIGE OUR GLORIOUS CAUSE INSPIRED JOY HEARTS BELEAGUERED BRETHREN CRADLE FAITH. OFFERING PRAYERS THANKSGIVING, BESEECHING BLESSED BEAUTY GUIDE, PROTECT, CONFIRM YOUTH IN THEIR EARNEST ENDEAVORS MOVE WORLD TOWARDS UNITY PEACE.

UNIVERSAL HOUSE OF JUSTICE

430-1. See CC 1:489–527.

432

Inauguration of Public Information Office at the World Center
11 JULY 1985

To all National Spiritual Assemblies

GREATLY PLEASED ANNOUNCE INAUGURATION PUBLIC INFORMATION OFFICE WORLD CENTER WITH BRANCH OFFICE SOON TO BE ESTABLISHED NEW YORK. THIS NEW AGENCY BAHÁ'Í INTERNATIONAL COMMUNITY TO EXTEND SCOPE COORDINATE WORLDWIDE ACTIVITIES DESIGNED INCREASE PUBLIC KNOWLEDGE AIMS ACHIEVEMENTS CAUSE BAHÁ'U'LLÁH.[1] PROFOUNDLY GRATEFUL BLESSED BEAUTY FURTHER DEVELOPMENT INSTRUMENTS HIS UNIQUE ORDER. 432.1

UNIVERSAL HOUSE OF JUSTICE

433

Election of Delegates to National Conventions
21 JULY 1985

To all National Spiritual Assemblies

Dear Bahá'í Friends,

As you are aware, some national communities elect their delegates to the National Convention on the basis of areas which have Local Spiritual Assemblies, while in other, larger, national communities delegates are elected on the basis of electoral units in which all adult believers have the vote. 433.1

Electoral Units

In view of the growth of the Faith and the developing life of the Bahá'í communities, the Universal House of Justice has decided that, notwithstanding that in some countries the number of believers and of Local Spiritual Assemblies is still small, the time has come for delegates to National Conventions everywhere to be elected on the basis of electoral units, but with the option of introducing certain differences from the procedures followed to date. These differences are explained below and are designed to make the system 433.2

432-1. The Bahá'í International Community is the name under which the Bahá'í world community conducts its relations on the international level with the United Nations, governmental agencies, the media, and international nongovernmental organizations. For this purpose offices of representation have been established in New York, Geneva, and elsewhere. For messages regarding the role of the Bahá'í International Community in relation to the United Nations, see messages dated 17 October 1967 (no. 49), 18 February 1970 (no. 78), 28 December 1980 (no. 272), and 26 May 1981 (no. 283).

adaptable to the variations in the make-up of the many Bahá'í communities and in the geography of the lands in which they are situated.

433.3 The House of Justice has decided that the number of delegates to each National Convention will remain unchanged for the present. However, if a National Assembly finds that under the new system a change would be advisable, it should feel free to write to the World Center stating the reasons for its view.

433.4 When establishing the electoral unit basis for the election of delegates, a National Spiritual Assembly should divide the territory under its jurisdiction into electoral units, based on the number of adult Bahá'ís in each area, in such a way that each unit will be responsible for electing preferably one delegate only.

Unit Conventions

433.5 In addition to the voting, the opportunity for consultation with the delegates is important. Hitherto this has been achieved by calling a convention in each unit to which all the believers in that electoral unit are invited. The voting for delegates has then taken place at the unit conventions with provision for voting by mail for those who do not attend. In some areas these meetings have been very fruitful and have helped to foster collaboration among the believers in the unit. However, in other areas, no doubt for a number of reasons, attendance at unit conventions has been very low, as has been the voting by mail, and this has meant that the delegates have been elected by a relatively small proportion of the electorate. National Assemblies are free to call unit conventions if they find they are successful, but if they find problems of attendance they may follow the alternative method described below.

Alternatives to Unit Conventions

433.6 Where holding unit conventions has proved ineffective, or does not seem to be a viable procedure, a National Assembly may divide each electoral unit into subunits of a convenient size. A meeting could then be held in each subunit to which all the adult believers residing therein would be invited. This should result in the participation of a large number of the believers. It is important to remember, however, that the delegate to be elected represents the *entire unit* and therefore, although the voting may be carried out in subunits, each voter has all the adult believers resident in the *entire unit* to choose from in voting for the delegate.

433.7 In some countries, it may even be too difficult to expect the believers throughout a subunit to gather together at a certain time, and so it would not be practical to hold subunit meetings. In such places a central point in each subunit could be chosen for the establishment of a polling station to which the friends would come to leave their ballots on the voting day as and when they can do so.

When one considers that there are now national Bahá'í communities varying in size from India to a single small island, some established in highly industrialized thickly populated countries, some in widely scattered archipelagoes, others covering equatorial jungles and still others embracing icebound arctic wastes, one can appreciate that a great deal of discretion must be accorded to each National Spiritual Assembly to establish the most effective means for the election of the delegates to its National Convention within the general principles outlined above. 433.8

Details To Be Worked Out by National Spiritual Assemblies

Each National Spiritual Assembly should study and master the broad outlines of this system. All matters of detail should be decided by the National Assembly which should ensure that the friends are fully informed and thoroughly understand what they are expected to do. The help and advice of the Counselors and their Auxiliary Board members and assistants could be sought in working out these details and in educating the friends. It may also be desirable for the National Assembly to appoint a special national committee to organize the elections and to oversee them through unit or subunit committees or representatives. Such matters of detail could include the following: 433.9

- The number of delegates to be allocated to each unit. Although one for each unit is preferable, this may not be practicable in certain instances, such as in a unit which contains one or more very large local communities. In such cases it may be necessary to make the unit large enough to be the electoral base for two or possibly three delegates. 433.9a
- The number and size of subunits. These could be as many as there are Local Spiritual Assemblies in a unit, the boundaries being so delineated as to include the surrounding isolated believers and Bahá'í groups. It may even be necessary in some remote areas to have subunits in which there are no Local Spiritual Assemblies. 433.9b
- The body to be responsible for organizing a unit convention or subunit meeting or for establishing and supervising a polling station. This could be a centrally located, firmly established Local Spiritual Assembly or a committee. 433.9c
- The day or days on which the elections should take place. Elections could be carried out in different subunits on different days, extended over a reasonable period of time, if this is felt to be desirable. 433.9d
- The manner in which ballots are to be cast, collected, counted, and consolidated with other ballots from the same unit. 433.9e
- Procedures to be followed in consultation, if the procedure chosen allows for consultation. 433.9f
- A method for monitoring the balloting to ensure that proper Bahá'í procedures are followed, that the ballots are safeguarded, and that a Bahá'í voter cannot cast more than one ballot. 433.9g

433.9h   — The procedure for holding a second ballot should there be a tie vote for the delegate.

433.9i   — The means for announcing to the friends in all units the names of their elected delegates.

433.10   It is the hope of the Universal House of Justice that the implementation of these instructions this year and thereafter will promote Bahá'í solidarity, broaden the basis of representation at the National Conventions and that thereby the work of the Faith in each country will be characterized by greater efficiency and enhanced harmony.

433.11   As this further step in the onward march of the Faith of Bahá'u'lláh is taken, you are assured of the prayers of the House of Justice at the Holy Shrines that you will be granted vision and wisdom to carry out your task and be enabled to extend the range of your dedicated services to His Cause.

With loving Bahá'í greetings,
DEPARTMENT OF THE SECRETARIAT

434
Dr. Martin Luther King, Jr., and Bahá'í Peace Activities
5 AUGUST 1985

To National Spiritual Assemblies

Dear Bahá'í Friends,

434.1   In its letter of 23 January 1985 concerning the International Year of Peace, the Universal House of Justice urged Bahá'í communities to reach out to the non-Bahá'í public by finding ways of discussing the important issues of peace with others. One way to make such discussions relevant and effective is for the friends to know and acknowledge and pay just tribute to persons whose lives were dedicated to peaceful means of bettering social conditions.

434.2   One such person was the black American Martin Luther King, Jr., whose promotion of nonviolent means of achieving racial equality in the United States cost him his life.[1] The positive effects of his heroic efforts brought encouragement to downtrodden peoples throughout the world and earned him the Nobel Peace Prize in 1964. Four years later he was assassinated. His aspi-

---

434-1. Dr. Martin Luther King, Jr., was a Baptist minister who led the civil rights movement in the United States from the mid-1950s until his assassination on 4 April 1968 at the age of thirty-nine. In 1963 he led a march on Washington, D.C., to achieve civil rights by nonviolent means, and there delivered his celebrated "I Have a Dream" speech. Toward the end of his life he broadened the scope of his efforts to oppose the war in Vietnam and to include poor people of all races.

rations for a society in which the races can live in harmony are perhaps best expressed in the famous speech he delivered at a gathering of some 250,000 people in the capital of the United States in 1963. A copy is enclosed.[2]

434.3 The House of Justice has asked us to call your attention to Dr. King for these reasons. His widow, Mrs. Coretta Scott King, a non-Bahá'í, has written to the House of Justice that a national public holiday has been officially designated in the United States in honor of Dr. King. She intends to make an appeal that on 20 January 1986, the first observance of this holiday, "nations and liberation movements all over the world cease all violent actions, seek amnesty and reconciliation both within and outside of their national boundaries, and encourage all of their citizens to recommit themselves to work for international peace, universal justice and the elimination of hunger and poverty in the world." The House of Justice feels that Mrs. King has a noble intention to which the friends can lend their moral and spiritual support. Since the date on which action is desired falls within the International Year of Peace, Spiritual Assemblies may consider holding peace conferences on 20 January, or close to that date, and naturally include in the presentations at these conferences references to the life and work of Dr. King. An alternative might be to devote the Bahá'í programs on World Religion Day, 19 January, to peace and on these occasions pay tribute to Dr. King.

434.4 The thought of the House of Justice in suggesting such action is not to promote the holiday for Dr. King, and it does not expect Bahá'í communities everywhere to commemorate his life annually; rather, it wishes to indicate to the friends a legitimate occasion, as illustrated by Mrs. King's plan, when the Bahá'í peace activities can be associated with the worthy activities of others.

434.5 We are to assure you of the continuing prayers of the House of Justice in the Holy Shrines that your energetic efforts to further the cause of peace throughout the earth may be richly confirmed by the Blessed Beauty.

With loving Bahá'í greetings,
DEPARTMENT OF THE SECRETARIAT

434-2. The text of the speech, titled "I Have a Dream," though not included here, is well known and widely published elsewhere.

435
Development of Local and National Bahá'í Funds
7 August 1985

To National Spiritual Assemblies

Dear Bahá'í Friends,

435.1 This letter and the annexed memorandum of comments are addressed primarily to those National Spiritual Assemblies whose communities include large numbers of materially poor people but inasmuch as the principles expressed, as distinct from some of the procedures suggested, are of universal application, they are being sent to all National Assemblies.

435.2 There is a profound aspect to the relationship between a believer and the Fund, which holds true irrespective of his or her economic condition. When a human soul accepts Bahá'u'lláh as the Manifestation of God for this age and enters into the divine Covenant, that soul should progressively bring his or her whole life into harmony with the divine purpose—he becomes a co-worker in the Cause of God and receives the bounty of being permitted to devote his material possessions, no matter how meager, to the work of the Faith.

435.3 Giving to the Fund, therefore, is a spiritual privilege not open to those who have not accepted Bahá'u'lláh, of which no believer should deny himself. It is both a responsibility and a source of bounty. This is an aspect of the Cause which, we feel, is an essential part of the basic teaching and deepening of new believers. The importance of contributing resides in the degree of sacrifice of the giver, the spirit of devotion with which the contribution is made and the unity of the friends in this service; these attract the confirmations of God and enhance the dignity and self-respect of the individuals and the community.

435.4 To reemphasize the spiritual significance of contributing to the Faith by all members of the Bahá'í community, we quote the following extract from a letter of the Guardian to the National Spiritual Assembly of the Bahá'ís of Central and East Africa dated 8 August 1957:

435.4a All, no matter how modest their resources, must participate. Upon the degree of self-sacrifice involved in these individual contributions will directly depend the efficacy and the spiritual influence which these nascent administrative institutions, called into being through the power of Bahá'u'lláh, and by virtue of the Design conceived by the Center of His Covenant, will exert. A sustained and strenuous effort must henceforth be made by the rank and file of the avowed upholders of the Faith . . .

We assure you of our prayers at the Sacred Threshold for your guidance 435.5
and confirmation as you labor to develop this aspect of Bahá'í life in your
communities.

With loving Bahá'í greetings,
THE UNIVERSAL HOUSE OF JUSTICE

Development of the Local and National Funds of the Faith
Some Comments and Suggestions

While the friends have the sacred obligation and privilege to contribute to 435.6
the Fund, each Local and National Assembly also has the inescapable duty of
educating itself and the believers in the spiritual principles related to Bahá'í
contributions, to devise simple methods to facilitate the flow and receipt of
contributions, and to formulate effective procedures to ensure the wise expenditure of the funds of the Faith. The following comments and suggestions
have been compiled at the request of the Universal House of Justice and are
being shared with National Spiritual Assemblies to assist them in these important tasks.

A primary requisite for all who have responsibility for the care of the funds 435.7
of the Faith is trustworthiness. This, as Bahá'u'lláh has stressed, is one of the
most basic and vital of all human virtues, and its exercise has a direct and
profound influence on the willingness of the believers to contribute to the
Fund.

Conditions vary from country to country and, therefore, in educating the 435.8
believers and developing the Fund, each National Spiritual Assembly needs to
tailor its actions to the conditions of its area of jurisdiction.

In many parts of the world gifts of produce and handicrafts may be a large 435.9
potential source of regular donations and could well be encouraged, proper
arrangements being made for their collection and sale and the disposition of
the proceeds.

Pledges can be useful as a means of encouraging contributions and of bringing the financial needs of the Cause to the attention of the friends. This 435.10
method can be particularly helpful in a situation where a Spiritual Assembly
has a major task to perform, such as the building of a Ḥaẓíratu'l-Quds or the
establishment of a tutorial school,[1] and needs to have some idea in advance
of whether the funds for the project will be available. However, it would be
entirely contrary to Bahá'í principles to bring any pressure to bear when calling for pledges or when endeavoring to collect them. Once a pledge has been

---

435-1. A simple school, usually in a rural area where no formal education is available, that
often consists of one teacher and a group of students who learn Bahá'í moral and ethical teachings and basic reading, writing, and computation skills.

given it is permissible to remind the donor, privately, of his expressed intention to contribute and to inquire courteously if it would be possible for him to honor his pledge, but Assemblies must be aware that such pledges are not an obligation in any legal sense; their redemption is entirely a matter of conscience. Lists of those making pledges must not be publicized.

435.11 The beloved Guardian has explained that the general and national interests of the Cause take precedence over local ones; thus contributions to local funds are secondary to those to national funds. However, the stability of the National Assembly rests on the firmness of the Local Spiritual Assemblies, and in the matter of educating the friends in the importance of the Fund, it is often most practical and efficacious to concentrate at first on the development of the local funds and the efficient operation of the Local Spiritual Assemblies. Then, once the friends understand the principle and learn from experience at a local level, they will the more easily understand the importance of the national fund and the work of the National Spiritual Assembly.

435.12 Regarding the local funds, it is suggested that until such time as the friends have developed the habit of contributing regularly and freely, any Local Spiritual Assembly which has a large community might appoint a small committee to assist the local treasurer in the discharge of his responsibilities. Such committees could be appointed after consultation with the Auxiliary Board member or assistant for the area. Great care must be taken in the appointment of the members of the committees; they must be both trustworthy and conscientious and must be imbued with awareness of the importance of maintaining the confidentiality of contributions to the funds. It is envisaged that these Treasury Committees would serve a number of functions:

435.12a — To render general assistance to the treasurer, as needed; for example, members of the committee could assist with issuing receipts or keeping accounts.

435.12b — To arrange for inspirational talks and discussions at Nineteen Day Feasts or at specially called meetings for the education of the friends in the spiritual and practical importance of contributing to the funds.

435.12c — To receive donations of money on behalf of the local treasurer and transmit these to him.

435.12d — To receive gifts of produce and handicrafts. The committee would be responsible for arranging for their sale and for handing over the proceeds to the local treasurer.

435.12e — To receive from the friends written pledges of their hope or intention of making a contribution to the local or national funds, whether in cash or in kind, and to assist in collecting them.

435.13 As to the national fund, in those areas where there are problems as a result of lack of banking facilities, unreliable mail systems and general difficulties of communication, it would be desirable for the National Spiritual Assembly to appoint a national committee to assist the national treasurer in a manner similar to that outlined above for Local Spiritual Assemblies. Further, it may even be necessary to subsidize, from the national fund, one or more trusted individuals, depending on the size of the national community, who would travel to rural areas to meet with the local Treasury Committees, assist them in the execution of their functions, explain the needs of the national fund, collect the donations to the national fund from the local areas and transmit them to the national treasurer.

In considering the above suggestions and their applicability to its national community, each National Assembly should also bear in mind the following points: 435.14

- It may find it valuable to study the methods being used already in those rural areas where notable success has been achieved in bringing about participation in sacrifice and giving. 435.14a
- Voluntary service for the Faith could also be stressed. It has an effect on the Fund by reducing the cost of carrying out the work of the Faith, and should be undertaken with joy by the friends. 435.14b
- It can be useful and helpful for both National and Local Spiritual Assemblies to make plans for financial self-sufficiency, set goals for levels of contributions, and share the news of progress towards such goals. 435.14c
- Assemblies should take the members of their communities into their confidence and regularly inform them of the uses to which the Fund is put and the projects for which money is needed. 435.14d

436
Release of a Compilation on Peace
9 August 1985

To National Spiritual Assemblies

Dear Bahá'í Friends,

The Universal House of Justice has asked us to send you the enclosed copy of a compilation on Peace prepared from the Bahá'í writings and the letters of the House of Justice by the Research Department.[1] 436.1

As indicated in our letter of 2 July 1985,[2] it is intended that the compilation be used to aid study by the friends of the Bahá'í concepts on peace and to facilitate their understanding of the House of Justice's statement on peace 436.2

436-1. See CC 2:151–200.
436-2. The letter contained "additional information and advice" to facilitate preparing for activities in observance of the International Year of Peace.

recently mailed to you. The House of Justice requests you to disseminate the compilation to the friends along with any advice you may have for its use. It should become a regular reference at summer schools and at other gatherings in which Bahá'í adults and youth study the Teachings. Moreover, the compilation should become a ready resource for those preparing summaries and commentaries on the statement addressed by the House of Justice to the peoples of the world.

With loving Bahá'í greetings,
Department of the Secretariat

437
Execution of Two More Bahá'ís; Bahá'í Students Pressured
19 September 1985

To all National Spiritual Assemblies

437.1 AFTER FEW MONTHS' CESSATION OF EXECUTIONS OF BAHÁ'ÍS IN IRAN, GRIEVED ANNOUNCE TWO FURTHER EXECUTIONS. VALIANT SOULS ARE MR. 'ABBÁS AYDIL-KHÁNÍ AND MR. RAHMATU'LLÁH VUJDÁNÍ. FORMER WAS EXECUTED ON 1 AUGUST IN PRISON WITHOUT ANY NOTIFICATION HIS FAMILY. HIS GRAVE WAS ACCIDENTALLY DISCOVERED NEAR TEHRAN. HE HAD BEEN IMPRISONED ON 26 APRIL 1982 IN ZANJÁN WHERE HE REMAINED UNTIL APRIL 1985 WHEN HE WAS TAKEN TO TEHRAN. HE WAS 45 YEARS OLD AND WAS AN AIR-CONDITIONING TECHNICIAN. MANNER HIS EXECUTION STILL UNKNOWN. HIS WIFE IS ALSO IN PRISON IN ZANJÁN.

437.2 MR. RAHMATU'LLÁH VUJDÁNÍ WAS ARRESTED IN JULY 1984 IN BANDAR-'ABBÁS, WHERE HE WAS EXECUTED BY FIRING SQUAD ON 28 AUGUST 1985. HE WAS 57 YEARS OLD. HIS BODY WAS DELIVERED, AND HIS FUNERAL TOOK PLACE IN PRESENCE HIS FAMILY AND FRIENDS. HE WAS A TEACHER BY PROFESSION.

437.3 FROM THE END JANUARY TO SEPTEMBER 1985, 63 BAHÁ'ÍS WERE ARRESTED AND 39 RELEASED. TOTAL NUMBER PRISONERS NOW 741. THIS FIGURE INCLUDES 39 PRISONERS RELEASED DURING PERIOD. BAHÁ'Í STUDENTS OF ALL LEVELS HAVE TO COMPLETE ADMISSION FORMS WHICH INCLUDE SPACE FOR ONLY FOUR OFFICIALLY RECOGNIZED RELIGIONS. BAHÁ'Í STUDENTS WHO STATE THEY ARE BAHÁ'ÍS ARE DENIED SCHOOLING OR IF ADMITTED FACE TREMENDOUS PRESSURE AND HARASSMENT. OTHER FORMS PERSECUTION INNOCENT BAHÁ'ÍS PERSIST. . . .

Universal House of Justice

438

The Promise of World Peace[1]
OCTOBER 1985

To the Peoples of the World

The Great Peace towards which people of goodwill throughout the centuries have inclined their hearts, of which seers and poets for countless generations have expressed their vision, and for which from age to age the sacred scriptures of mankind have constantly held the promise, is now at long last within the reach of the nations. For the first time in history it is possible for everyone to view the entire planet, with all its myriad diversified peoples, in one perspective. World peace is not only possible but inevitable. It is the next stage in the evolution of this planet—in the words of one great thinker, "the planetization of mankind."[2]

Whether peace is to be reached only after unimaginable horrors precipitated by humanity's stubborn clinging to old patterns of behavior, or is to be embraced now by an act of consultative will, is the choice before all who inhabit the earth. At this critical juncture when the intractable problems confronting nations have been fused into one common concern for the whole world, failure to stem the tide of conflict and disorder would be unconscionably irresponsible.

Among the favorable signs are the steadily growing strength of the steps towards world order taken initially near the beginning of this century in the creation of the League of Nations, succeeded by the more broadly based United Nations Organization;[3] the achievement since the Second World War of independence by the majority of all the nations on earth, indicating the

---

438-1. This statement on peace was published in the United States under the title *The Promise of World Peace* (Wilmette, Ill.: Bahá'í Publishing Trust, 1985). It has been published in many editions around the world.

438-2. Pierre Teilhard de Chardin, *The Future of Man*, Ch. 7 (New York: Harper and Row, 1964).

438-3. The League of Nations was an organization for international cooperation established at the initiative of the Allied powers at the Paris Peace Conference of 1919 following the end of World War I. It was based on a covenant calling for joint action by League members against an aggressor, arbitration of international disputes, reduction of armaments, and open diplomacy. The lack of political will of member states to intervene in acts of aggression by one state against another discredited the League and rendered it powerless to prevent World War II. 'Abdu'l-Bahá recognized that the League of Nations was "incapable of establishing universal peace." However, He perceived "the dawn of universal peace" in the fourteen points put forth by United States President Woodrow Wilson in 1918 that included the establishment of a League of Nations (SWAB, pp. 306, 311).

The United Nations was formally inaugurated on 24 October 1945, near the end of World War II, at the initiative of the Allied powers following the United Nations Conference on International Organization that convened in San Francisco on 25 April 1945.

completion of the process of nation building, and the involvement of these fledgling nations with older ones in matters of mutual concern; the consequent vast increase in cooperation among hitherto isolated and antagonistic peoples and groups in international undertakings in the scientific, educational, legal, economic and cultural fields; the rise in recent decades of an unprecedented number of international humanitarian organizations; the spread of women's and youth movements calling for an end to war; and the spontaneous spawning of widening networks of ordinary people seeking understanding through personal communication.

438.4 The scientific and technological advances occurring in this unusually blessed century portend a great surge forward in the social evolution of the planet, and indicate the means by which the practical problems of humanity may be solved. They provide, indeed, the very means for the administration of the complex life of a united world. Yet barriers persist. Doubts, misconceptions, prejudices, suspicions and narrow self-interest beset nations and peoples in their relations one to another.

438.5 It is out of a deep sense of spiritual and moral duty that we are impelled at this opportune moment to invite your attention to the penetrating insights first communicated to the rulers of mankind more than a century ago by Bahá'u'lláh, Founder of the Bahá'í Faith, of which we are the Trustees.

438.6 "The winds of despair," Bahá'u'lláh wrote, "are, alas, blowing from every direction, and the strife that divides and afflicts the human race is daily increasing. The signs of impending convulsions and chaos can now be discerned, inasmuch as the prevailing order appears to be lamentably defective."[4] This prophetic judgment has been amply confirmed by the common experience of humanity. Flaws in the prevailing order are conspicuous in the inability of sovereign states organized as United Nations to exorcise the specter of war, the threatened collapse of the international economic order, the spread of anarchy and terrorism, and the intense suffering which these and other afflictions are causing to increasing millions. Indeed, so much have aggression and conflict come to characterize our social, economic and religious systems, that many have succumbed to the view that such behavior is intrinsic to human nature and therefore ineradicable.

438.7 With the entrenchment of this view, a paralyzing contradiction has developed in human affairs. On the one hand, people of all nations proclaim not only their readiness but their longing for peace and harmony, for an end to the harrowing apprehensions tormenting their daily lives. On the other, uncritical assent is given to the proposition that human beings are incorrigibly selfish and aggressive and thus incapable of erecting a social system at once progressive and peaceful, dynamic and harmonious, a system giving free play to individual creativity and initiative but based on cooperation and reciprocity.

438-4. GWB, p. 216.

As the need for peace becomes more urgent, this fundamental contradiction, which hinders its realization, demands a reassessment of the assumptions upon which the commonly held view of mankind's historical predicament is based. Dispassionately examined, the evidence reveals that such conduct, far from expressing man's true self, represents a distortion of the human spirit. Satisfaction on this point will enable all people to set in motion constructive social forces which, because they are consistent with human nature, will encourage harmony and cooperation instead of war and conflict. 438.8

To choose such a course is not to deny humanity's past but to understand it. The Bahá'í Faith regards the current world confusion and calamitous condition in human affairs as a natural phase in an organic process leading ultimately and irresistibly to the unification of the human race in a single social order whose boundaries are those of the planet. The human race, as a distinct, organic unit, has passed through evolutionary stages analogous to the stages of infancy and childhood in the lives of its individual members, and is now in the culminating period of its turbulent adolescence approaching its long-awaited coming of age. 438.9

A candid acknowledgement that prejudice, war and exploitation have been the expression of immature stages in a vast historical process and that the human race is today experiencing the unavoidable tumult which marks its collective coming of age is not a reason for despair but a prerequisite to undertaking the stupendous enterprise of building a peaceful world. That such an enterprise is possible, that the necessary constructive forces do exist, that unifying social structures can be erected, is the theme we urge you to examine. 438.10

Whatever suffering and turmoil the years immediately ahead may hold, however dark the immediate circumstances, the Bahá'í community believes that humanity can confront this supreme trial with confidence in its ultimate outcome. Far from signaling the end of civilization, the convulsive changes towards which humanity is being ever more rapidly impelled will serve to release the "potentialities inherent in the station of man" and reveal "the full measure of his destiny on earth, the innate excellence of his reality."[5] 438.11

I

The endowments which distinguish the human race from all other forms of life are summed up in what is known as the human spirit; the mind is its essential quality. These endowments have enabled humanity to build civilizations and to prosper materially. But such accomplishments alone have never satisfied the human spirit, whose mysterious nature inclines it towards transcendence, a reaching towards an invisible realm, towards the ultimate reality, that unknowable essence of essences called God. The religions brought to 438.12

438-5. GWB, p. 340.

mankind by a succession of spiritual luminaries have been the primary link between humanity and that ultimate reality, and have galvanized and refined mankind's capacity to achieve spiritual success together with social progress.

438.13 No serious attempt to set human affairs aright, to achieve world peace, can ignore religion. Man's perception and practice of it are largely the stuff of history. An eminent historian described religion as a "faculty of human nature."[6] That the perversion of this faculty has contributed to much of the confusion in society and the conflicts in and between individuals can hardly be denied. But neither can any fair-minded observer discount the preponderating influence exerted by religion on the vital expressions of civilization. Furthermore, its indispensability to social order has repeatedly been demonstrated by its direct effect on laws and morality.

438.14 Writing of religion as a social force, Bahá'u'lláh said: "Religion is the greatest of all means for the establishment of order in the world and for the peaceful contentment of all that dwell therein." Referring to the eclipse or corruption of religion, he wrote: "Should the lamp of religion be obscured, chaos and confusion will ensue, and the lights of fairness, of justice, of tranquillity and peace cease to shine."[7] In an enumeration of such consequences the Bahá'í writings point out that the "perversion of human nature, the degradation of human conduct, the corruption and dissolution of human institutions, reveal themselves, under such circumstances, in their worst and most revolting aspects. Human character is debased, confidence is shaken, the nerves of discipline are relaxed, the voice of human conscience is stilled, the sense of decency and shame is obscured, conceptions of duty, of solidarity, of reciprocity and loyalty are distorted, and the very feeling of peacefulness, of joy and of hope is gradually extinguished."[8]

438.15 If, therefore, humanity has come to a point of paralyzing conflict it must look to itself, to its own negligence, to the siren voices to which it has listened, for the source of the misunderstandings and confusion perpetrated in the name of religion. Those who have held blindly and selfishly to their particular orthodoxies, who have imposed on their votaries erroneous and conflicting interpretations of the pronouncements of the Prophets of God, bear heavy responsibility for this confusion—a confusion compounded by the artificial barriers erected between faith and reason, science and religion. For from a fair-minded examination of the actual utterances of the Founders of the great religions, and of the social milieus in which they were obliged to carry out their missions, there is nothing to support the contentions and prejudices deranging the religious communities of mankind and therefore all human affairs.

438-6. A. Toynbee, in John Cogley, *Religion in a Secular Age* (New York: Praeger, 1968), p. ix.
438-7. Quoted in WOB, pp. 186, 186–87.
438-8. WOB, p. 187.

The teaching that we should treat others as we ourselves would wish to be treated, an ethic variously repeated in all the great religions, lends force to this latter observation in two particular respects: it sums up the moral attitude, the peace-inducing aspect, extending through these religions irrespective of their place or time of origin; it also signifies an aspect of unity which is their essential virtue, a virtue mankind in its disjointed view of history has failed to appreciate. 438.16

Had humanity seen the Educators of its collective childhood in their true character, as agents of one civilizing process, it would no doubt have reaped incalculably greater benefits from the cumulative effects of their successive missions. This, alas, it failed to do. 438.17

The resurgence of fanatical religious fervor occurring in many lands cannot be regarded as more than a dying convulsion. The very nature of the violent and disruptive phenomena associated with it testifies to the spiritual bankruptcy it represents. Indeed, one of the strangest and saddest features of the current outbreak of religious fanaticism is the extent to which, in each case, it is undermining not only the spiritual values which are conducive to the unity of mankind but also those unique moral victories won by the particular religion it purports to serve. 438.18

However vital a force religion has been in the history of mankind, and however dramatic the current resurgence of militant religious fanaticism, religion and religious institutions have, for many decades, been viewed by increasing numbers of people as irrelevant to the major concerns of the modern world. In its place they have turned either to the hedonistic pursuit of material satisfactions or to the following of man-made ideologies designed to rescue society from the evident evils under which it groans. All too many of these ideologies, alas, instead of embracing the concept of the oneness of mankind and promoting the increase of concord among different peoples, have tended to deify the state, to subordinate the rest of mankind to one nation, race or class, to attempt to suppress all discussion and interchange of ideas, or to callously abandon starving millions to the operations of a market system that all too clearly is aggravating the plight of the majority of mankind, while enabling small sections to live in a condition of affluence scarcely dreamed of by our forebears. 438.19

How tragic is the record of the substitute faiths that the worldly-wise of our age have created. In the massive disillusionment of entire populations who have been taught to worship at their altars can be read history's irreversible verdict on their value. The fruits these doctrines have produced, after decades of an increasingly unrestrained exercise of power by those who owe their ascendancy in human affairs to them, are the social and economic ills that blight every region of our world in the closing years of the twentieth century. Underlying all these outward afflictions is the spiritual damage reflected in the apathy 438.20

that has gripped the mass of the peoples of all nations and by the extinction of hope in the hearts of deprived and anguished millions.

438.21 The time has come when those who preach the dogmas of materialism, whether of the east or the west, whether of capitalism or socialism, must give account of the moral stewardship they have presumed to exercise. Where is the "new world" promised by these ideologies? Where is the international peace to whose ideals they proclaim their devotion? Where are the breakthroughs into new realms of cultural achievement produced by the aggrandizement of this race, of that nation or of a particular class? Why is the vast majority of the world's peoples sinking ever deeper into hunger and wretchedness when wealth on a scale undreamed of by the Pharaohs, the Caesars, or even the imperialist powers of the nineteenth century is at the disposal of the present arbiters of human affairs?

438.22 Most particularly, it is in the glorification of material pursuits, at once the progenitor and common feature of all such ideologies, that we find the roots which nourish the falsehood that human beings are incorrigibly selfish and aggressive. It is here that the ground must be cleared for the building of a new world fit for our descendants.

438.23 That materialistic ideals have, in the light of experience, failed to satisfy the needs of mankind calls for an honest acknowledgement that a fresh effort must now be made to find the solutions to the agonizing problems of the planet. The intolerable conditions pervading society bespeak a common failure of all, a circumstance which tends to incite rather than relieve the entrenchment on every side. Clearly, a common remedial effort is urgently required. It is primarily a matter of attitude. Will humanity continue in its waywardness, holding to outworn concepts and unworkable assumptions? Or will its leaders, regardless of ideology, step forth and, with a resolute will, consult together in a united search for appropriate solutions?

438.24 Those who care for the future of the human race may well ponder this advice. "If long-cherished ideals and time-honored institutions, if certain social assumptions and religious formulae have ceased to promote the welfare of the generality of mankind, if they no longer minister to the needs of a continually evolving humanity, let them be swept away and relegated to the limbo of obsolescent and forgotten doctrines. Why should these, in a world subject to the immutable law of change and decay, be exempt from the deterioration that must needs overtake every human institution? For legal standards, political and economic theories are solely designed to safeguard the interests of humanity as a whole, and not humanity to be crucified for the preservation of the integrity of any particular law or doctrine."[9]

438-9. WOB, p. 42.

II

Banning nuclear weapons, prohibiting the use of poison gases, or outlawing germ warfare will not remove the root causes of war. However important such practical measures obviously are as elements of the peace process, they are in themselves too superficial to exert enduring influence. Peoples are ingenious enough to invent yet other forms of warfare, and to use food, raw materials, finance, industrial power, ideology, and terrorism to subvert one another in an endless quest for supremacy and dominion. Nor can the present massive dislocation in the affairs of humanity be resolved through the settlement of specific conflicts or disagreements among nations. A genuine universal framework must be adopted. 438.25

Certainly, there is no lack of recognition by national leaders of the worldwide character of the problem, which is self-evident in the mounting issues that confront them daily. And there are the accumulating studies and solutions proposed by many concerned and enlightened groups as well as by agencies of the United Nations, to remove any possibility of ignorance as to the challenging requirements to be met. There is, however, a paralysis of will; and it is this that must be carefully examined and resolutely dealt with. This paralysis is rooted, as we have stated, in a deep-seated conviction of the inevitable quarrelsomeness of mankind, which has led to the reluctance to entertain the possibility of subordinating national self-interest to the requirements of world order, and in an unwillingness to face courageously the far-reaching implications of establishing a united world authority. It is also traceable to the incapacity of largely ignorant and subjugated masses to articulate their desire for a new order in which they can live in peace, harmony and prosperity with all humanity. 438.26

The tentative steps towards world order, especially since World War II, give hopeful signs. The increasing tendency of groups of nations to formalize relationships which enable them to cooperate in matters of mutual interest suggests that eventually all nations could overcome this paralysis. The Association of South East Asian Nations, the Caribbean Community and Common Market, the Central American Common Market, the Council for Mutual Economic Assistance, the European Communities, the League of Arab States, the Organization of African Unity, the Organization of American States, the South Pacific Forum—all the joint endeavors represented by such organizations prepare the path to world order. 438.27

The increasing attention being focused on some of the most deep-rooted problems of the planet is yet another hopeful sign. Despite the obvious shortcomings of the United Nations, the more than two score declarations and conventions adopted by that organization, even where governments have not been enthusiastic in their commitment, have given ordinary people a sense of 438.28

a new lease on life. The Universal Declaration of Human Rights, the Convention on the Prevention and Punishment of the Crime of Genocide, and the similar measures concerned with eliminating all forms of discrimination based on race, sex or religious belief; upholding the rights of the child; protecting all persons against being subjected to torture; eradicating hunger and malnutrition; using scientific and technological progress in the interest of peace and the benefit of mankind—all such measures, if courageously enforced and expanded, will advance the day when the specter of war will have lost its power to dominate international relations. There is no need to stress the significance of the issues addressed by these declarations and conventions. However, a few such issues, because of their immediate relevance to establishing world peace, deserve additional comment.

438.29 Racism, one of the most baneful and persistent evils, is a major barrier to peace. Its practice perpetrates too outrageous a violation of the dignity of human beings to be countenanced under any pretext. Racism retards the unfoldment of the boundless potentialities of its victims, corrupts its perpetrators, and blights human progress. Recognition of the oneness of mankind, implemented by appropriate legal measures, must be universally upheld if this problem is to be overcome.

438.30 The inordinate disparity between rich and poor, a source of acute suffering, keeps the world in a state of instability, virtually on the brink of war. Few societies have dealt effectively with this situation. The solution calls for the combined application of spiritual, moral and practical approaches. A fresh look at the problem is required, entailing consultation with experts from a wide spectrum of disciplines, devoid of economic and ideological polemics, and involving the people directly affected in the decisions that must urgently be made. It is an issue that is bound up not only with the necessity for eliminating extremes of wealth and poverty but also with those spiritual verities the understanding of which can produce a new universal attitude. Fostering such an attitude is itself a major part of the solution.

438.31 Unbridled nationalism, as distinguished from a sane and legitimate patriotism, must give way to a wider loyalty, to the love of humanity as a whole. Bahá'u'lláh's statement is: "The earth is but one country, and mankind its citizens."[10] The concept of world citizenship is a direct result of the contraction of the world into a single neighborhood through scientific advances and of the indisputable interdependence of nations. Love of all the world's peoples does not exclude love of one's country. The advantage of the part in a world society is best served by promoting the advantage of the whole. Current international activities in various fields which nurture mutual affection and a sense of solidarity among peoples need greatly to be increased.

438-10. GWB, p. 250.

Religious strife, throughout history, has been the cause of innumerable wars and conflicts, a major blight to progress, and is increasingly abhorrent to the people of all faiths and no faith. Followers of all religions must be willing to face the basic questions which this strife raises, and to arrive at clear answers. How are the differences between them to be resolved, both in theory and in practice? The challenge facing the religious leaders of mankind is to contemplate, with hearts filled with the spirit of compassion and a desire for truth, the plight of humanity, and to ask themselves whether they cannot, in humility before their Almighty Creator, submerge their theological differences in a great spirit of mutual forbearance that will enable them to work together for the advancement of human understanding and peace. 438.32

The emancipation of women, the achievement of full equality between the sexes, is one of the most important, though less acknowledged prerequisites of peace. The denial of such equality perpetrates an injustice against one half of the world's population and promotes in men harmful attitudes and habits that are carried from the family to the workplace, to political life, and ultimately to international relations. There are no grounds, moral, practical, or biological, upon which such denial can be justified. Only as women are welcomed into full partnership in all fields of human endeavor will the moral and psychological climate be created in which international peace can emerge. 438.33

The cause of universal education, which has already enlisted in its service an army of dedicated people from every faith and nation, deserves the utmost support that the governments of the world can lend it. For ignorance is indisputably the principal reason for the decline and fall of peoples and the perpetuation of prejudice. No nation can achieve success unless education is accorded all its citizens. Lack of resources limits the ability of many nations to fulfill this necessity, imposing a certain ordering of priorities. The decision-making agencies involved would do well to consider giving first priority to the education of women and girls, since it is through educated mothers that the benefits of knowledge can be most effectively and rapidly diffused throughout society. In keeping with the requirements of the times, consideration should also be given to teaching the concept of world citizenship as part of the standard education of every child. 438.34

A fundamental lack of communication between peoples seriously undermines efforts towards world peace. Adopting an international auxiliary language would go far to resolve this problem and necessitates the most urgent attention. 438.35

Two points bear emphasizing in all these issues. One is that the abolition of war is not simply a matter of signing treaties and protocols; it is a complex task requiring a new level of commitment to resolving issues not customarily associated with the pursuit of peace. Based on political agreements alone, the idea of collective security is a chimera. The other point is that the 438.36

primary challenge in dealing with issues of peace is to raise the context to the level of principle, as distinct from pure pragmatism. For, in essence, peace stems from an inner state supported by a spiritual or moral attitude, and it is chiefly in evoking this attitude that the possibility of enduring solutions can be found.

438.37 There are spiritual principles, or what some call human values, by which solutions can be found for every social problem. Any well-intentioned group can in a general sense devise practical solutions to its problems, but good intentions and practical knowledge are usually not enough. The essential merit of spiritual principle is that it not only presents a perspective which harmonizes with that which is immanent in human nature, it also induces an attitude, a dynamic, a will, an aspiration, which facilitate the discovery and implementation of practical measures. Leaders of governments and all in authority would be well served in their efforts to solve problems if they would first seek to identify the principles involved and then be guided by them.

III

438.38 The primary question to be resolved is how the present world, with its entrenched pattern of conflict, can change to a world in which harmony and cooperation will prevail.

438.39 World order can be founded only on an unshakable consciousness of the oneness of mankind, a spiritual truth which all the human sciences confirm. Anthropology, physiology, psychology, recognize only one human species, albeit infinitely varied in the secondary aspects of life. Recognition of this truth requires abandonment of prejudice—prejudice of every kind—race, class, color, creed, nation, sex, degree of material civilization, everything which enables people to consider themselves superior to others.

438.40 Acceptance of the oneness of mankind is the first fundamental prerequisite for reorganization and administration of the world as one country, the home of humankind. Universal acceptance of this spiritual principle is essential to any successful attempt to establish world peace. It should therefore be universally proclaimed, taught in schools, and constantly asserted in every nation as preparation for the organic change in the structure of society which it implies.

438.41 In the Bahá'í view, recognition of the oneness of mankind "calls for no less than the reconstruction and the demilitarization of the whole civilized world—a world organically unified in all the essential aspects of its life, its political machinery, its spiritual aspiration, its trade and finance, its script and language, and yet infinite in the diversity of the national characteristics of its federated units."[11]

438-11. WOB, p. 43.

1979–1986 • THE SEVEN YEAR PLAN

Elaborating the implications of this pivotal principle, Shoghi Effendi, the Guardian of the Bahá'í Faith, commented in 1931 that: "Far from aiming at the subversion of the existing foundations of society, it seeks to broaden its basis, to remold its institutions in a manner consonant with the needs of an ever-changing world. It can conflict with no legitimate allegiances, nor can it undermine essential loyalties. Its purpose is neither to stifle the flame of a sane and intelligent patriotism in men's hearts, nor to abolish the system of national autonomy so essential if the evils of excessive centralization are to be avoided. It does not ignore, nor does it attempt to suppress, the diversity of ethnical origins, of climate, of history, of language and tradition, of thought and habit, that differentiate the peoples and nations of the world. It calls for a wider loyalty, for a larger aspiration than any that has animated the human race. It insists upon the subordination of national impulses and interests to the imperative claims of a unified world. It repudiates excessive centralization on one hand, and disclaims all attempts at uniformity on the other. Its watchword is unity in diversity."[12]

438.42

The achievement of such ends requires several stages in the adjustment of national political attitudes, which now verge on anarchy in the absence of clearly defined laws or universally accepted and enforceable principles regulating the relationships between nations. The League of Nations, the United Nations, and the many organizations and agreements produced by them have unquestionably been helpful in attenuating some of the negative effects of international conflicts, but they have shown themselves incapable of preventing war. Indeed, there have been scores of wars since the end of the Second World War; many are yet raging.

438.43

The predominant aspects of this problem had already emerged in the nineteenth century when Bahá'u'lláh first advanced his proposals for the establishment of world peace. The principle of collective security was propounded by him in statements addressed to the rulers of the world. Shoghi Effendi commented on his meaning: "What else could these weighty words signify," he wrote, "if they did not point to the inevitable curtailment of unfettered national sovereignty as an indispensable preliminary to the formation of the future Commonwealth of all the nations of the world? Some form of a world superstate must needs be evolved, in whose favor all the nations of the world will have willingly ceded every claim to make war, certain rights to impose taxation and all rights to maintain armaments, except for purposes of maintaining internal order within their respective dominions. Such a state will have to include within its orbit an international executive adequate to enforce supreme and unchallengeable authority on every recalcitrant member of the commonwealth; a world parliament whose members shall be elected by the

438.44

438-12. WOB, pp. 41–42.

438.45 "... A world community in which all economic barriers will have been permanently demolished and the interdependence of Capital and Labor definitely recognized; in which the clamor of religious fanaticism and strife will have been forever stilled; in which the flame of racial animosity will have been finally extinguished; in which a single code of international law—the product of the considered judgment of the world's federated representatives—shall have as its sanction the instant and coercive intervention of the combined forces of the federated units; and finally a world community in which the fury of a capricious and militant nationalism will have been transmuted into an abiding consciousness of world citizenship—such indeed, appears, in its broadest outline, the Order anticipated by Bahá'u'lláh, an Order that shall come to be regarded as the fairest fruit of a slowly maturing age."[13]

438.46 The implementation of these far-reaching measures was indicated by Bahá'u'lláh: "The time must come when the imperative necessity for the holding of a vast, an all-embracing assemblage of men will be universally realized. The rulers and kings of the earth must needs attend it, and, participating in its deliberations, must consider such ways and means as will lay the foundations of the world's Great Peace amongst men."[14]

438.47 The courage, the resolution, the pure motive, the selfless love of one people for another—all the spiritual and moral qualities required for effecting this momentous step towards peace are focused on the will to act. And it is towards arousing the necessary volition that earnest consideration must be given to the reality of man, namely, his thought. To understand the relevance of this potent reality is also to appreciate the social necessity of actualizing its unique value through candid, dispassionate and cordial consultation, and of acting upon the results of this process. Bahá'u'lláh insistently drew attention to the virtues and indispensability of consultation for ordering human affairs. He said: "Consultation bestows greater awareness and transmutes conjecture into certitude. It is a shining light which, in a dark world, leads the way and guides. For everything there is and will continue to be a station of perfection and maturity. The maturity of the gift of understanding is made manifest through consultation."[15] The very attempt to achieve peace through the consultative action he proposed can release such a salutary spirit among the peoples of the earth that no power could resist the final, triumphal outcome.

438-13. WOB, p. 41.
438-14. GWB, p. 249.
438-15. See CC 1:93.

Concerning the proceedings for this world gathering, 'Abdu'l-Bahá, the son 438.48
of Bahá'u'lláh and authorized interpreter of his teachings, offered these
insights: "They must make the Cause of Peace the object of general consultation, and seek by every means in their power to establish a Union of the nations of the world. They must conclude a binding treaty and establish a covenant, the provisions of which shall be sound, inviolable and definite. They must proclaim it to all the world and obtain for it the sanction of all the human race. This supreme and noble undertaking—the real source of the peace and well-being of all the world—should be regarded as sacred by all that dwell on earth. All the forces of humanity must be mobilized to ensure the stability and permanence of this Most Great Covenant. In this all-embracing Pact the limits and frontiers of each and every nation should be clearly fixed, the principles underlying the relations of governments towards one another definitely laid down, and all international agreements and obligations ascertained. In like manner, the size of the armaments of every government should be strictly limited, for if the preparations for war and the military forces of any nation should be allowed to increase, they will arouse the suspicion of others. The fundamental principle underlying this solemn Pact should be so fixed that if any government later violate any one of its provisions, all the governments on earth should arise to reduce it to utter submission, nay the human race as a whole should resolve, with every power at its disposal, to destroy that government. Should this greatest of all remedies be applied to the sick body of the world, it will assuredly recover from its ills and will remain eternally safe and secure."[16]

The holding of this mighty convocation is long overdue. 438.49

With all the ardor of our hearts, we appeal to the leaders of all nations to 438.50
seize this opportune moment and take irreversible steps to convoke this world meeting. All the forces of history impel the human race towards this act which will mark for all time the dawn of its long-awaited maturity.

Will not the United Nations, with the full support of its membership, rise 438.51
to the high purposes of such a crowning event?

Let men and women, youth and children everywhere recognize the eternal 438.52
merit of this imperative action for all peoples and lift up their voices in willing assent. Indeed, let it be this generation that inaugurates this glorious stage in the evolution of social life on the planet.

IV

The source of the optimism we feel is a vision transcending the cessation 438.53
of war and the creation of agencies of international cooperation. Permanent peace among nations is an essential stage, but not, Bahá'u'lláh asserts, the ulti-

---

438-16. SDC, pp. 64–65, and WOB, pp. 37–38.

mate goal of the social development of humanity. Beyond the initial armistice forced upon the world by the fear of nuclear holocaust, beyond the political peace reluctantly entered into by suspicious rival nations, beyond pragmatic arrangements for security and coexistence, beyond even the many experiments in cooperation which these steps will make possible lies the crowning goal: the unification of all the peoples of the world in one universal family.

438.54 Disunity is a danger that the nations and peoples of the earth can no longer endure; the consequences are too terrible to contemplate, too obvious to require any demonstration. "The well-being of mankind," Bahá'u'lláh wrote more than a century ago, "its peace and security, are unattainable unless and until its unity is firmly established."[17] In observing that "mankind is groaning, is dying to be led to unity, and to terminate its agelong martyrdom," Shoghi Effendi further commented that: "Unification of the whole of mankind is the hallmark of the stage which human society is now approaching. Unity of family, of tribe, of city-state, and nation have been successively attempted and fully established. World unity is the goal towards which a harassed humanity is striving. Nation-building has come to an end. The anarchy inherent in state sovereignty is moving towards a climax. A world, growing to maturity, must abandon this fetish, recognize the oneness and wholeness of human relationships, and establish once for all the machinery that can best incarnate this fundamental principle of its life."[18]

438.55 All contemporary forces of change validate this view. The proofs can be discerned in the many examples already cited of the favorable signs towards world peace in current international movements and developments. The army of men and women, drawn from virtually every culture, race and nation on earth, who serve the multifarious agencies of the United Nations, represent a planetary "civil service" whose impressive accomplishments are indicative of the degree of cooperation that can be attained even under discouraging conditions. An urge towards unity, like a spiritual springtime, struggles to express itself through countless international congresses that bring together people from a vast array of disciplines. It motivates appeals for international projects involving children and youth. Indeed, it is the real source of the remarkable movement towards ecumenism by which members of historically antagonistic religions and sects seem irresistibly drawn towards one another. Together with the opposing tendency to warfare and self-aggrandizement against which it ceaselessly struggles, the drive towards world unity is one of the dominant, pervasive features of life on the planet during the closing years of the twentieth century.

438-17. GWB, p. 286.
438-18. WOB, pp. 201, 202.

The experience of the Bahá'í community may be seen as an example of this enlarging unity. It is a community of some three to four million people drawn from many nations, cultures, classes and creeds, engaged in a wide range of activities serving the spiritual, social and economic needs of the peoples of many lands. It is a single social organism, representative of the diversity of the human family, conducting its affairs through a system of commonly accepted consultative principles, and cherishing equally all the great outpourings of divine guidance in human history. Its existence is yet another convincing proof of the practicality of its Founder's vision of a united world, another evidence that humanity can live as one global society, equal to whatever challenges its coming of age may entail. If the Bahá'í experience can contribute in whatever measure to reinforcing hope in the unity of the human race, we are happy to offer it as a model for study. 438.56

In contemplating the supreme importance of the task now challenging the entire world, we bow our heads in humility before the awesome majesty of the divine Creator, who out of His infinite love has created all humanity from the same stock; exalted the gemlike reality of man; honored it with intellect and wisdom, nobility and immortality; and conferred upon man the "unique distinction and capacity to know Him and to love Him," a capacity that "must needs be regarded as the generating impulse and the primary purpose underlying the whole of creation."[19] 438.57

We hold firmly the conviction that all human beings have been created "to carry forward an ever-advancing civilization"; that "to act like the beasts of the field is unworthy of man";[20] that the virtues that befit human dignity are trustworthiness, forbearance, mercy, compassion and loving-kindness towards all peoples. We reaffirm the belief that the "potentialities inherent in the station of man, the full measure of his destiny on earth, the innate excellence of his reality, must all be manifested in this promised Day of God."[21] These are the motivations for our unshakable faith that unity and peace are the attainable goal towards which humanity is striving. 438.58

At this writing, the expectant voices of Bahá'ís can be heard despite the persecution they still endure in the land in which their Faith was born. By their example of steadfast hope, they bear witness to the belief that the imminent realization of this age-old dream of peace is now, by virtue of the transforming effects of Bahá'u'lláh's revelation, invested with the force of divine authority. Thus we convey to you not only a vision in words: we summon the power of deeds of faith and sacrifice; we convey the anxious plea of our coreligionists everywhere for peace and unity. We join with all who are the 438.59

438-19. GWB, p. 65.
438-20. GWB, p. 215.
438-21. GWB, p. 340.

victims of aggression, all who yearn for an end to conflict and contention, all whose devotion to principles of peace and world order promotes the ennobling purposes for which humanity was called into being by an all-loving Creator.

438.60 In the earnestness of our desire to impart to you the fervor of our hope and the depth of our confidence, we cite the emphatic promise of Bahá'u'lláh: "These fruitless strifes, these ruinous wars shall pass away, and the 'Most Great Peace' shall come."[22]

THE UNIVERSAL HOUSE OF JUSTICE

439
Appointment of Continental Boards of Counselors
24 OCTOBER 1985

To the Bahá'ís of the World

Dear Bahá'í Friends,

439.1 It gives us great happiness to announce the membership of the Continental Boards of Counselors as from the Day of the Covenant, 26 November 1985. The number of Counselors has been increased from 63 to 72 and adjustments have been made in their geographical distribution in consonance with the development of the Faith around the world.

439.2 The membership of the Continental Boards of Counselors as now appointed is:

439.2a **AFRICA (18 Counselors):** Mr. Hushang Ahdieh (Trustee of the Continental Fund), Mr. Ḥusayn Ardikání, Mrs. Beatrice O. Asare, Mr. Gila Michael Bahta, Mr. Friday Ekpe, Mr. Oloro Epyeru, Mr. Shidan Fat'he-Aazam, Mr. Kassimi Fofana, Mr. Zekrullah Kazemi, Mr. Muḥammad Kebdani, Mrs. Thelma Khelghati, Mr. Roddy Dharma Lutchmaya, Mr. Daniel Ramoroesi, Dr. Mihdí Samandarí, Mrs. Edith Senoga, Mr. Peter Vuyiya, Mrs. Lucretia Mancho Warren, Mr. Mabuku Wingi.

439.2b **THE AMERICAS (17 Counselors):** Dr. Hidáyatu'lláh Aḥmadíyyih, Mr. Eloy Anello, Dr. Farzam Arbáb (Trustee of the Continental Fund), Dr. Wilma Brady, Mrs. Isabel P. de Calderón, Mr. Rolf von Czekus, Mr. Robert Harris, Mrs. Lauretta King, Dr. Peter McLaren, Mr. Shapoor Monadjem, Mrs. Ruth Pringle, Mr. Donald O. Rogers, Mr. Fred Schechter, Dr. Arturo Serrano, Mr. Alan Smith, Dr. David R. Smith, Mr. Rodrigo Tomás.

439.2c **ASIA (19 Counselors):** Dr. Ṣábir Áfáqí, Mr. Burhání'd-Dín Afshín, Dr. Iraj Ayman, Mr. Bijan Fareed, Dr. John Fozdar, Mr. Zabíḥu'lláh

---

438-22. BNE, p. 40.

Gulmuḥammadí, Mr. Bharat Koirala, Mr. Rúḥu'lláh Mumtází, Mr. S. Nagaratnam, Dr. Perin Olyai, Mrs. Rose Ong, Mr. Khudáraḥm Paymán (Trustee of the Continental Fund), Mr. Masíḥ Rawḥání, Mr. Vicente Samaniego, Dr. Ilhan Sezgin, U Soe Tin, Mrs. Zena Sorabjee, Dr. Chellie J. Sundram, Mr. Michitoshi Zenimoto.

AUSTRALASIA (9 Counselors): Mr. Suhayl 'Alá'í, Mr. Ben Ayala, Justice Richard Benson, Dr. Kamran Eshraghian, Mrs. Tinai Hancock, Mr. Lisiate Maka, Mrs. Gayle Morrison, Dr. Sírús Naráqí, Mrs. Joy Stevenson (Trustee of the Continental Fund). 439.2d

EUROPE (9 Counselors): Dr. Agnes Ghaznavi, Mr. Hartmut Grossmann, Mr. Louis Hénuzet (Trustee of the Continental Fund), Mrs. Ursula Mühlschlegel, Dr. Leo Niederreiter, Mrs. Polin Rafat, Mr. Adib Taherzadeh, Mr. Adam Thorne, Mr. Sohrab Youssefian. 439.2e

The following nineteen devoted believers who are now being relieved of the onerous duties of membership on the Boards of Counselors, will, as distinguished servants of the Cause, continue through their outstanding capacities and experience to be sources of stimulation and encouragement to the friends. 439.3

Mr. A. Owen Battrick, Mr. Erik Blumenthal, Mrs. Shirin Boman, Mrs. Carmen de Burafato, Mr. Athos Costas, Mr. Angus Cowan, Mrs. Dorothy Ferraby, Mr. Aydin Güney, Dr. Dipchand Khianra, Mr. Artemus Lamb, Mr. Kolonario Oule, Dr. Sarah Pereira, Mrs. Betty R. Reed, Dr. Manúchihr Salmánpúr, Mrs. Velma Sherrill, Mr. Hideya Suzuki, Mrs. Bahíyyih Winckler, Mr. Donald Witzel, Mr. Yan Kee Leong. 439.4

We express to each and every one of these dear friends our heartfelt gratitude and assure them of our prayers in the Holy Shrines for the confirmation of their highly meritorious and self-sacrificing services to the Cause of Bahá'u'lláh. 439.5

At this time when the Bahá'í world is facing the challenge of the International Year of Peace, on the point of completing the Seven Year Plan and standing on the threshold of a new Six Year Plan, we have felt it important to call upon the Counselors from all the continents to gather at the World Center for a conference to deliberate on the tasks and opportunities of the years immediately ahead. This conference will take place from 27 December 1985 through 2 January 1986 and is yet one more sign of the rapid advance and consolidation of the institutions of the Cause of God. 439.6

We are profoundly grateful to the Blessed Perfection for His bountiful confirmations which are enabling His strenuously laboring servants in every part of the world to witness the growing influence of His glorious Cause, and to take part in the vitalizing unfoldment of His Administrative Order. 439.7

439.8 It is our fervent prayer at the Sacred Threshold that the followers of Bahá'u'lláh in every land will arise with increased determination and self-abnegation to mirror forth the standards upheld by His potent Faith.

With loving Bahá'í greetings,
THE UNIVERSAL HOUSE OF JUSTICE

440
Reinterment of Bahá'u'lláh's Faithful Half-Brother
17 NOVEMBER 1985

To the Bahá'ís of the World

Beloved Friends,

440.1 It is with a feeling of joy and gratitude that we inform the Bahá'í world of the befitting reinterment of the remains of Mírzá Muḥammad-Qulí, the faithful half-brother and companion in exile of Bahá'u'lláh, and of eleven members of his family, in a new Bahá'í cemetery on a hillside looking across Lake Kinneret and the hills of Galilee towards the Qiblih of the Faith.[1] This historic event, coinciding fortuitously with the first formal presentation of *The Promise of World Peace* to a Head of State, is of especial significance in the annals of the Cause of Bahá'u'lláh.[2]

440.2 On 12 November 1952 the beloved Guardian jubilantly cabled the Bahá'í world his announcement of the acquisition of vitally needed property surrounding the Most Holy Shrine and the Mansion of Bahjí in exchange for land donated by the grandchildren of Mírzá Muḥammad-Qulí.

440.3 The land referred to in this cable had been in the possession of Mírzá Muḥammad-Qulí on the eastern shore of the Sea of Galilee, at a place called Nuqayb. He and his family lived there and farmed the land for many years and on his passing, at the instruction of 'Abdu'l-Bahá, his remains were buried there, as were subsequently those of members of his family.

440.4 In 1937 Kibbutz Ein Gev was established just to the north of the farm, and the two groups of settlers lived as amicable neighbors until the war of 1948 forced the family to leave the land which, lying on the troubled frontier of the new State of Israel, was expropriated by the Government. The grandchildren of Mírzá Muḥammad-Qulí gave their rights in the land to the Faith which received in exchange the much needed land in Bahjí. Thus the little cemetery passed out of Bahá'í hands. It remained untouched until 1972 when

---

440-1. The Qiblih for Bahá'ís is the Most Holy Tomb of Bahá'u'lláh at Bahjí, outside of 'Akká.

440-2. Dr. Rudolph Kirchlaeger, the President of Austria, was the first head of state to receive *The Promise of World Peace* (see message no. 438).

the decision was made to approach the authorities with a view to embellishing the site and maintaining it as a place of historic significance for the Faith. However, plans had already been made for the extension of the plantings of the kibbutz and the eventual development of the land in a way that would not permit the permanent reestablishment of the cemetery in that place. Negotiations were then entered into, as a result of which another plot of land in the immediate neighborhood, but slightly farther from the shore of the Lake on the slope of Tel Susita, was officially designated a Bahá'í cemetery and given over to the Bahá'í Community. The work of fencing it and planting suitable shrubs and trees was then put in hand and preparations were made to reinter the precious remains of this family.

On the morning of Friday 18 October 1985, as the final stage in this process, the remains of Mírzá Muḥammad-Qulí himself were ceremoniously conveyed from the old cemetery to the new and were reinterred there in the presence of the Hands of the Cause Amatu'l-Bahá Rúḥíyyih Khánum and 'Alí-Akbar Furútan, members of the Universal House of Justice and of the International Teaching Center, and a large gathering of World Center friends as well as representatives of the Israeli authorities and of Kibbutz Ein Gev. Mrs. Ḥusníyyih Bahá'í, the granddaughter of Mírzá Muḥammad-Qulí, who is now pioneering in St. Lucia in the West Indies, accompanied by members of her family, had been especially invited to attend the ceremony in honor of her illustrious forebear, to whom 'Abdu'l-Bahá paid eloquent tribute in *Memorials of the Faithful*.³

440.5

THE UNIVERSAL HOUSE OF JUSTICE

441
Execution of 'Azízu'lláh Ashjárí by Firing Squad
24 NOVEMBER 1985

To all National Spiritual Assemblies

WITH DEEP SORROW ANNOUNCE EXECUTION BY FIRING SQUAD ON 19 NOVEMBER 1985 OF 'AZÍZU'LLÁH ASHJÁRÍ, A BAHÁ'Í PRISONER IN TABRÍZ. HE WAS FIFTY YEARS OLD AND HAD BEEN IMPRISONED FOR OVER FOUR YEARS. UNLIKE MANY OTHER EXECUTED BAHÁ'ÍS, HIS BODY WAS GIVEN TO THE FAMILY, AND BURIAL TOOK PLACE 22 NOVEMBER. NO OTHER DETAILS AVAILABLE. LATEST INFORMATION RECEIVED INDICATES 767 BAHÁ'ÍS ARE STILL IN PRISON. . . .

441.1

UNIVERSAL HOUSE OF JUSTICE

440-3. MF, pp. 70–71.

442
Presentation of Peace Statement to the Secretary-General of the United Nations
25 November 1985

To all National Spiritual Assemblies

442.1 ON EVE DAY COVENANT WE ANNOUNCE WITH PROFOUND GRATITUDE SUCCESSFUL PRESENTATION TO SECRETARY-GENERAL UNITED NATIONS STATEMENT PEACE ADDRESSED TO PEOPLES WORLD.[1] PRESENCE AMATU'L-BAHÁ RÚḤÍYYIH KHÁNUM AS HEAD DELEGATION BAHÁ'Í INTERNATIONAL COMMUNITY INVESTED THIS EVENT WITH SIGNIFICANCE THAT WILL WARM HEARTS EVOKE ADMIRATION INSPIRE APPRECIATION FUTURE GENERATIONS. PRESENTATION 22 NOVEMBER TO SECRETARY-GENERAL IN CONJUNCTION ONGOING PRESENTATIONS STATEMENT TO HEADS STATE THROUGHOUT WORLD PRESAGES NEW STAGE IN RELATIONSHIP BAHÁ'Í COMMUNITY TO UNITED NATIONS. IT IMMEDIATELY LENDS ADDED IMPETUS TO ACTIVITIES INTERNATIONAL YEAR PEACE BEING UNDERTAKEN WITH EXEMPLARY VIGOR EFFICIENCY BY NATIONAL LOCAL BAHÁ'Í COMMUNITIES.

442.2 WE TAKE THIS OPPORTUNITY INFORM FRIENDS THAT OF MORE THAN FIFTY HEADS STATE WHO HAVE ALREADY RECEIVED PEACE STATEMENT TWO THUS FAR HAVE DIRECTLY ADDRESSED TO US THEIR POSITIVE THOUGHTFUL RESPONSES, NAMELY, PRESIDENTS COLOMBIA MARSHALL ISLANDS. REPORTS REACHING WORLD CENTER INDICATE INCREASINGLY SYMPATHETIC INTEREST BEING AROUSED BY STATEMENT AMONG PEOPLE ALL LEVELS SOCIETY. WARMLY APPRECIATE ADMIRE CREATIVE ENERGY WHICH FRIENDS EVERYWHERE ARE APPLYING TO STUDY AND DISSEMINATION *THE PROMISE OF WORLD PEACE*.

442.3 REMARKABLE RANGE ACTIVITIES THAT HAVE OCCURRED JUST ONE MONTH SINCE RELEASE STATEMENT IMBUES US WITH FEELINGS SATISFACTION JOY. GRATEFULLY RECOGNIZE BLESSINGS BEING SHOWERED COMMUNITY GREATEST NAME AS DERIVING FROM WORLD-AWAKENING SACRIFICES CHARACTERIZING LIVES OUR BELOVED HEROIC BROTHERS SISTERS WHOSE SPIRITUAL RESOURCES INDOMITABLE COURAGE HAVE WITHSTOOD NO LESS THAN SEVEN YEARS UNRELENTING ASSAULTS UNPRECEDENTED UPHEAVALS PERVADING BAHÁ'U'LLÁH'S NATIVE LAND.

UNIVERSAL HOUSE OF JUSTICE

442-1. The presentation of *The Promise of World Peace* took place on 25 November, the eve of the Day of the Covenant.

1979–1986 • THE SEVEN YEAR PLAN

443
Presentation of Peace Statement to the President of the United States
12 DECEMBER 1985

To the National Spiritual Assembly of the Bahá'ís of the United States

PROFOUNDLY GRATEFUL PRESENTATION PEACE STATEMENT PRESIDENT UNITED STATES RONALD W. REAGAN. THAT THE PRESENTATION OCCURRED AS PART PUBLIC EVENT MARKING HUMAN RIGHTS DAY[1] AND THAT CHIEF EXECUTIVE GREAT REPUBLIC WEST OPENLY CHAMPIONED RIGHTS UNEQUIVOCALLY PROCLAIMED INNOCENCE OUR SORELY WRONGED BRETHREN IN IRAN HAVE INVESTED THIS ACHIEVEMENT WITH SIGNIFICANCE THAT CAN BE APPRECIATED ONLY WITHIN CONTEXT GLORIOUS PROMISES YOUR COUNTRY'S DESTINY RECORDED OUR SACRED SCRIPTURES. ON THIS PORTENTOUS OCCASION WE RECALL WITH HOPE AND JOY PROPHETIC WORDS BELOVED MASTER ADDRESSED YOUR ILLUSTRIOUS COMMUNITY: "THE PRESTIGE OF THE FAITH OF GOD," HE ASSERTED, "HAS IMMENSELY INCREASED. ITS GREATNESS IS NOW MANIFEST. THE DAY IS APPROACHING WHEN IT WILL HAVE CAST A TREMENDOUS TUMULT IN MEN'S HEARTS. REJOICE, THEREFORE, O DENIZENS OF AMERICA, REJOICE WITH EXCEEDING GLADNESS!"[2] 443.1

UNIVERSAL HOUSE OF JUSTICE

444
Adoption of Resolution in Support of Iranian Bahá'ís by the United Nations General Assembly—Affirmation of Emergence from Obscurity
17 DECEMBER 1985

To all National Spiritual Assemblies

WITH JOYFUL GRATEFUL HEARTS WE ACCLAIM UNPRECEDENTED RECOGNITION BAHÁ'Í COMMUNITY THROUGH ADOPTION BY UNITED NATIONS GENERAL ASSEMBLY OF RESOLUTION MAKING SPECIFIC REFERENCE PERSECUTED FRIENDS IRAN. FOR FIRST TIME NAME PRECIOUS FAITH BAHÁ'U'LLÁH MENTIONED HIGHEST MOST WIDELY REPRESENTATIVE INTERNATIONAL FORUM YET ESTABLISHED THUS FULFILLING LONG-CHERISHED WISH BELOVED GUARDIAN. SIGNIFICANCE THIS MOMENTOUS DEVELOPMENT ALSO UNDERSCORED BY FACT THAT ONLY IN THREE INSTANCES BEFORE HAS GENERAL ASSEMBLY ITSELF ADOPTED RESOLUTIONS 444.1

443-1. Human Rights Day, 10 December, was established by the United Nations to commemorate the General Assembly's adoption of the Universal Declaration of Human Rights in 1948.

443-2. Quoted in WOB, p. 79.

CENSURING PARTICULAR COUNTRIES FOR BAD HUMAN RIGHTS RECORDS. PROCESS WHICH RESULTED SUCH A REMARKABLE OUTCOME BEGAN TWO YEARS AGO WITH DECISION UNITED NATIONS COMMISSION ON HUMAN RIGHTS TO SEND REPRESENTATIVE IRAN INVESTIGATE VIOLATIONS HUMAN RIGHTS INCLUDING THOSE DIRECTLY AFFECTING BAHÁ'Í COMMUNITY. THE COMMISSION DETERRED IN ITS INTENTION BY IRANIAN AUTHORITIES REFERRED ISSUE TO GENERAL ASSEMBLY WHERE MATTER WAS DISCUSSED FIRST IN THIRD COMMITTEE WHEN INTERESTS FAITH WERE VIGOROUSLY UPHELD BY REPRESENTATIVES VARIOUS COUNTRIES AND RESOLUTION WAS PROPOSED AND THEN IN PLENARY SESSION WHICH RATIFIED RESOLUTION ON 13 DECEMBER. NOTABLE CONSEQUENCE IS RETENTION ISSUE AGENDA GENERAL ASSEMBLY THUS PERMITTING INTENSIFICATION EFFORTS RELIEVE SITUATION SUFFERING BELIEVERS IRAN IN ANTICIPATION DAY COMPLETE EMANCIPATION FAITH GOD THAT LAND. WARMLY COMMEND ACKNOWLEDGE UNTIRING EFFORTS UNITED NATIONS REPRESENTATIVES BAHÁ'Í INTERNATIONAL COMMUNITY RESOLUTELY SUPPORTED BY NATIONAL SPIRITUAL ASSEMBLIES ALL CONTINENTS.

444.2 OCCURRING IN CONJUNCTION WITH PRESENTATION PEACE STATEMENT BY AMATU'L-BAHÁ RÚḤÍYYIH KHÁNUM TO SECRETARY-GENERAL UNITED NATIONS ONLY FEW WEEKS BEFORE,[1] WITH ONGOING DELIVERY SAME STATEMENT TO HEADS STATE THROUGHOUT WORLD AND WITH UNEQUIVOCAL PUBLIC DEFENSE IRANIAN BAHÁ'ÍS BY PRESIDENT UNITED STATES AT HUMAN RIGHTS DAY CEREMONY, THIS INESTIMABLE ACHIEVEMENT DEFINITELY AFFIRMS EMERGENCE FAITH OBSCURITY HERALDS NEW PHASE IRREPRESSIBLE UNFOLDMENT DIVINELY APPOINTED WORLD ORDER BAHÁ'U'LLÁH.

THE UNIVERSAL HOUSE OF JUSTICE

445

Message to Conference of Counselors in the Holy Land
27 DECEMBER 1985

To the Conference of the Continental Boards of Counselors

Beloved Friends,

445.1 With all the warmth of our hearts we welcome you to this historic Conference, to discuss the challenges and opportunities so swiftly devolving upon the struggling Faith of God.[1]

445.2 The main features of this spiritually potent occasion are evident. Your entry upon your new term of office[2] as the Seven Year Plan approaches its conclu-

---

444-1. See message dated 25 November 1985 (no. 442).

445-1. The Conference was held at the Bahá'í World Center in Haifa, Israel, 27 December 1985 through 2 January 1986.

445-2. In its message dated 24 October 1985 the Universal House of Justice announced the appointment of members of the Continental Boards of Counselors to a new five-year term of service commencing on the Day of the Covenant, 26 November 1985. See message of 29 June 1979 (no. 229) regarding the setting of terms for five-year periods.

sion, with the measures to be taken to ensure its success heavy on your shoulders; consideration of the main features of the new Six Year Plan which will terminate on the eve of the Centenary of the Ascension of Bahá'u'lláh;[3] the golden opportunities offered by the Year of Peace, which must be seized and fully exploited; the dramatic emergence of the Faith from obscurity into the limelight of the world's highest councils, with the attendant enhancement of its status; a tremendous upsurge of zealous activity in the Bahá'í world community as it takes to its heart the recently issued Statement on Peace;[4] the deep and universal commitment already made by that community to a vast variety of social and economic development programs; the widespread and growing awareness among the leaders of mankind that a new stage in human history has opened and that the guidance of the past will not carry it through the emergencies of the present; these, together with the invitation extended to the peoples of the world to examine the Bahá'í community as a working model for the reorganization of the world,[5] are some of the pressures forcing themselves upon the attention of those responsible for the direction, propagation and protection of the Cause of God.

445.3 Who can doubt that we are now entering a period of unprecedented and unimaginable developments in the onward march of the Faith? Some intimation of what these may be can already be deduced and preparations made to deal with them. We know that the present victories will lead to active opposition, for which the Bahá'í world community must be prepared. We know the prime needs of the Cause at the moment: a vast expansion of its numbers and financial resources; a greater consolidation of its community life and the authority of its institutions; an observable increase in those characteristics of loving unity, stability of family life, freedom from prejudice and rectitude of conduct which must distinguish the Bahá'ís from the spiritually lost and wayward multitudes around them. Surely the time cannot be long delayed when we must deal universally with that entry by troops foretold by the Master as a prelude to mass conversion.[6] How are we to prepare ourselves for that examination of the Bahá'í experience which will undoubtedly be made by shrewd and doubting questioners?

445.4 These are some of the subjects before you. The role of the Counselors is all-pervading and vital in the life of the Cause. You must be wise and steadfast, encouraging and brotherly in all your associations with the believers and in your consultation with National Spiritual Assemblies; through your Board members and their assistants you must strengthen and uplift the Local Spiritual Assemblies and rouse the spirits and enlarge the vision of the believers

---

445-3. 28 May 1992.
445-4. See message no. 438.
445-5. The invitation was extended in the message dated October 1985 (no. 438).
445-6. TABA 3:681.

everywhere. Your prayers in the Sacred Shrines during these few days will strengthen and guide you; be assured of our own prayers for a bountiful outpouring of grace and inspiration as you embark upon the deliberations which will enable you to discharge your sacred and historic duties.

445.5 We wish you a happy, exciting and fruitful conference.

THE UNIVERSAL HOUSE OF JUSTICE

446
Release of a Compilation on Women
1 JANUARY 1986

To all National Spiritual Assemblies

Dear Bahá'í Friends,

446.1 The Universal House of Justice requested the Research Department to prepare a compilation on the subject of women, taking into consideration the Bahá'í concepts of the equality of the sexes, the role of education in the development of women, family life, fostering the development of women, etc. The compilation has now been completed and a copy is attached.[1]

446.2 It is left to your discretion to determine how best the contents of this compilation can be shared with the believers in your community.

With loving Bahá'í greetings,
DEPARTMENT OF THE SECRETARIAT

447
Report from Conference in the Holy Land; the Inception of the Fourth Epoch of the Formative Age
2 JANUARY 1986

The Bahá'ís of the World

Dearly loved Friends,

447.1 The eager expectation with which we welcomed to the World Center, on 27 December, sixty-four Counselors from the five continents to discuss, with the International Teaching Center, the challenges and opportunities facing the Bahá'í world community, has, at the conclusion of their historic conference, been transmuted into feelings of deepest joy, gratitude and love.

446-1. See CC 2:355–406.

Graced by the presence of the Hands of the Cause Amatu'l-Bahá Rúḥíyyih 447.2
Khánum, Ugo Giachery, 'Alí-Akbar Furútan, 'Alí-Muḥammad Varqá and Collis Featherstone, the Conference was organized and managed with admirable foresight and efficiency by the International Teaching Center, whose individual members watched over and served untiringly the needs of the participants and the progress of the Conference itself.

Convened in the concourse of the Seat of the Universal House of Justice 447.3
as the Counselors of the Bahá'í world entered upon their new five-year term of office, within months of the termination of the Seven Year Plan and the opening of the new Six Year Plan, its aura heightened by the spiritual potencies of the Holy Shrines and the euphoric sense of victory and blessing now pervading the entire Bahá'í world, the Conference attained such heights of consultative exaltation, spirituality and power as only those serving the Blessed Beauty can enjoy.

The organic growth of the Cause of God, indicated by recent significant 447.4
developments in its life, becomes markedly apparent in the light of the main objectives and expectations of the Six Year Plan: a vast expansion of the numerical and financial resources of the Cause; enlargement of its status in the world; a worldwide increase in the production, distribution and use of Bahá'í literature; a firmer and worldwide demonstration of the Bahá'í way of life requiring special consideration of the Bahá'í education of children and youth, the strengthening of Bahá'í family life and attention to universal participation and the spiritual enrichment of individual life; further acceleration in the process of the maturation of local and national Bahá'í communities and a dynamic consolidation of the unity of the two arms of the Administrative Order; an extension of the involvement of the Bahá'í world community in the needs of the world around it; and the pursuit of social and economic development in well-established Bahá'í communities. These are some of the features of the Six Year Plan which will open on 21 April 1986 and terminate on 20 April 1992.

Riḍván 1992 will mark the inception of a Holy Year, during which the 447.5
Centenary of the Ascension of Bahá'u'lláh will be observed by commemorations around the world and the inauguration of His Covenant will be celebrated, in the City of the Covenant, by the holding of the second Bahá'í World Congress.[1]

The beloved Counselors, strengthened and enriched by their experience in 447.6
the Holy Land, will, as early as possible, consult with all National Spiritual Assemblies on measures to conclude triumphantly the current Plan, and on preparations to launch the Six Year Plan. In anticipation of those consulta-

---

447-1. The first Bahá'í World Congress was held in London during the 1963 Riḍván Festival from 28 April through 2 May. See the first statement from the Universal House of Justice, issued on that occasion and dated 30 April 1963 (message no. 1).

tions, National Spiritual Assemblies will receive the full announcement of the aims and characteristics of that Plan, so that together with the Counselors they may formulate the national plans which will, for each community, establish its pursuit of the overall objectives.[2]

447.7 This new process, whereby the national goals of the next Plan are to be largely formulated by National Spiritual Assemblies and Boards of Counselors, signalizes the inauguration of a new stage in the unfoldment of the Administrative Order. Our beloved Guardian anticipated a succession of epochs during the Formative Age of the Faith; have no hesitation in recognizing that this new development in the maturation of Bahá'í institutions marks the inception of the fourth epoch of that Age.[3]

447.8 Shoghi Effendi perceived in the organic life of the Cause a dialectic of victory and crisis. The unprecedented triumphs, generated by the adamantine steadfastness of the Iranian friends, will inevitably provoke opposition to test and increase our strength. Let every Bahá'í in the world be assured that whatever may befall this growing Faith of God is but incontrovertible evidence of the loving care with which the King of Glory and His martyred Herald, through the incomparable Center of His Covenant and our beloved Guardian, are preparing His humble followers for ultimate and magnificent triumph. Our loving prayers are with you all.

THE UNIVERSAL HOUSE OF JUSTICE

447-2. See message no. 453.

447-3. The Formative Age began in 1921 with the passing of 'Abdu'l-Bahá and with Shoghi Effendi's assumption of the office of Guardian. The first epoch of the Formative Age spanned the years 1921–46 and included the first Seven Year Plan (1937–44) pursued by the Bahá'ís of the United States and Canada. The second epoch, 1946–63, embraced the second Seven Year Plan embarked upon by the Bahá'ís of North America, plans of shorter durations pursued by other national communities, and the Ten Year World Crusade, which for the first time involved the entire Bahá'í world in one common program of expansion and consolidation. The third epoch, 1963–86, encompassed the Nine, Five, and Seven Year Plans inaugurated by the Universal House of Justice. The fourth epoch began in 1986. The epochs of the Formative Age are different from those of the Divine Plan, the first epoch of which began in 1937 and ended in 1963. We continue to be in the second epoch of the Divine Plan. For a fuller explanation of the epochs of the Formative Age, see message no. 451.

448

Women—Their Role in Society and the Establishment of Peace; Membership on the Universal House of Justice
5 January 1986

To an individual Bahá'í

Dear Bahá'í Friend,

The Universal House of Justice has directed us to transmit the following in reply to your letter dated 19 December 1985.

Concerning your request for material relating to the role of women in society and in the establishment of peace, a compilation of extracts from the Writings and Utterances of 'Abdu'l-Bahá, and from letters written on behalf of Shoghi Effendi and the House of Justice, is enclosed.[1] These extracts concern the equality of women, their role, and, either directly or indirectly, their importance to the attainment of peace.

Additionally, from one of 'Abdu'l-Bahá's previously untranslated Tablets, the following extract is taken:

> The handmaidens of God and the bondsmaids in His divine Court should reveal such attributes and attitudes amongst the women of the world as would cause them to stand out and achieve renown in the circles of women. That is, they should associate with them with supreme chastity and steadfast decency, with unshakable faith, articulate speech, an eloquent tongue, irrefutable testimony and high resolve. Beseech God that thou mayest attain unto all these bounties.

'Abdu'l-Bahá also stated:

> The woman is indeed of the greater importance to the race. She has the greater burden and the greater work. Look at the vegetable and the animal worlds. The palm which carries the fruit is the tree most prized by the date grower. The Arab knows that for a long journey the mare has the longest wind. For her greater strength and fierceness, the lioness is more feared by the hunter than the lion.
>
> . . . The woman has greater moral courage than the man; she has also special gifts which enable her to govern in moments of danger and crisis. . . .
>
> (*'Abdu'l-Bahá in London*, pp. 102–03)

448-1. Many of the extracts that were enclosed are included in the compilation on the subject of women sent on 1 January 1986 (see message no. 446). See CC 2:355–406.

448.3d　　Consider, for instance, a mother who has tenderly reared a son for twenty years to the age of maturity. Surely she will not consent to having that son torn asunder and killed in the field of battle. Therefore, as woman advances toward the degree of man in power and privilege, with the right of vote and control in human government, most assuredly war will cease; for woman is naturally the most devoted and staunch advocate of international peace.
(*Promulgation of Universal Peace*, p. 375)

A further extract about women and peace is taken from a letter written on behalf of the Guardian on 24 March 1945:

448.4　　What 'Abdu'l-Bahá meant about the women arising for peace is that this is a matter which vitally affects women, and when they form a conscious and overwhelming mass of public opinion against war there can be no war. The Bahá'í women are already organized through being members of the Faith and the Administrative Order. No further organization is needed. But they should, through teaching and through the active moral support they give to every movement directed towards peace, seek to exert a strong influence on other women's minds in regard to this essential matter.

448.5　　With reference to the membership of the House of Justice being possible only for men, a compilation on this subject is also enclosed.[2] The following extracts are contained in this compilation, but have been selected for separate mention. The first is from *Selections from the Writings of 'Abdu'l-Bahá*, section 38:

448.5a　　The House of Justice, however, according to the explicit text of the Law of God, is confined to men; this for a wisdom of the Lord God's which will erelong be made manifest as clearly as the sun at high noon.

448.6　　And from a letter dated 14 April 1975 written on behalf of the House of Justice to an individual believer, are taken these extracts:

448.6a　　The Universal House of Justice has asked us to assure you that it appreciates the deep concern you express in your recent letter about the membership of the Universal House of Justice being confined to men, and it understands your feeling of frustration at not being able to find an answer that would help you to accept that this is not an injustice being imposed on womankind. The House of Justice agrees with you that our Sacred Writings are replete with passages affirming the equality of both sexes; that from the spiritual point of view, there is no dif-

---

448-2. The extracts are too lengthy to include in this volume.

ference between women and men. In fact, many statements made by 'Abdu'l-Bahá extol women. He has said that "in some respects woman is superior to man." . . .³

448.6b The Universal House of Justice points out that when we accept the Manifestation of God for our time, we must accept what He says though at the moment we may not comprehend the meaning of some of His statements. Some things, such as Bahá'u'lláh's Teachings regarding life after death, we have to accept on faith.

448.7 May you be guided and sustained in promoting the Word of God at the meetings you are planning and in the opportunities that occur in your daily life.

With loving Bahá'í greetings,
DEPARTMENT OF THE SECRETARIAT

449

Message to National Youth Conference, Adelaide, Australia
13 JANUARY 1986

To the Bahá'í Youth of Australia

449.1 ENTRANCE FAITH FOURTH EPOCH FORMATIVE AGE HERALDS ADVENT NEW CHALLENGES OPPORTUNITIES FRIENDS EVERYWHERE EXEMPLIFY DIVINE QUALITIES FIRMLY ESTABLISH BAHÁ'Í COMMUNITY AS MODEL WORTHY EMULATION PEOPLES WORLD INEVITABLY MOVING TOWARD TRANSFORMATION HUMAN SOCIETY AND LAYING FOUNDATIONS WORLD PEACE. CONFIDENT ENERGETIC SPIRIT MEMBERS YOUR GENERATION WILL INSPIRE FELLOW BELIEVERS ALL AGES DISCHARGE THEIR SHARE DUTIES OBLIGATIONS DURING YEAR PEACE AS ALL LABOR UNITEDLY FOR VICTORY SEVEN YEAR PLAN AND CONTEMPLATE NEW LEVELS ACHIEVEMENT COURSE SIX YEAR PLAN.

449.2 OFFERING OUR PRAYERS SUPPLICATIONS YOUR BEHALF. MAY PROMISED BLESSINGS CONFIRMATIONS BAHÁ'U'LLÁH SUSTAIN YOU DURING YOUR DELIBERATIONS AND CROWN YOUR EFFORTS WITH OUTSTANDING VICTORIES.

UNIVERSAL HOUSE OF JUSTICE

448-3. PT, p. 161.

450
Inaugural Broadcast of Radio Bahá'í Panama
31 January 1986

To the National Spiritual Assembly of the Bahá'ís of Panama

450.1 DEEPLY GRATIFIED FIRST TRANSMISSION RADIO BAHÁ'Í PANAMA CULMINATION LONG PERIOD TRAINING DEVOTED LABORS BAHÁ'Í COMMUNITY GUAYMIES MANY INDIVIDUALS. SUGGEST NATIONAL ASSEMBLY SEEK EXTENSION GRADUATED PROGRAMMING PERIOD AT LEAST 6 MONTHS ENABLE NON PROFESSIONAL STAFF ORGANIZE PROCURE MATERIALS GAIN NEEDED TECHNICAL PROFICIENCIES DEVELOP CULTURAL CENTER. CONGRATULATIONS GREAT ACHIEVEMENT.

Universal House of Justice

451
Epochs of the Formative Age
5 February 1986

To all National Spiritual Assemblies

Dear Bahá'í Friends,

451.1 In the letter dated 2 January 1986 written by the Universal House of Justice to the Bahá'ís of the world, reference was made to the inception of the fourth epoch of the Formative Age.[1] In response to questions subsequently put to the House of Justice about the periods related to the earlier epochs of that Age, the Research Department was requested to prepare a statement on the subject. This has now been presented, and a copy is enclosed.

451.2 Kindly share this material of topical interest with the friends, as you deem fit, so that it may be studied in their deepening classes, summer schools, conferences and similar gatherings.

With loving Bahá'í greetings,
Department of the Secretariat

451-1. See message no. 447.

The Epochs of the Formative Age
Prepared by the Research Department of the Universal House of Justice

Introduction:

In disclosing the panoramic vision of the unfoldment of the Dispensation of Bahá'u'lláh, Shoghi Effendi refers to three major evolutionary stages through which the Faith must pass—the Apostolic or Heroic Age (1844–1921) associated with the Central Figures of the Faith;[2] the Formative or Transitional Age (1921–),[3] the "hallmark"[4] of which is the rise and establishment of the Administrative Order, based on the execution of the provisions of 'Abdu'l-Bahá's Will and Testament;[5] and, the Golden Age which will represent the "consummation of this glorious Dispensation."[6] Close examination of the details of Bahá'í history reveals that the individual Ages are comprised of a number of periods—inseparable parts of one integrated whole.[7]

In relation to the Heroic Age of our Faith, the Guardian, in a letter dated 5 June 1947 to the American Bahá'ís, specified that this Age consisted of three epochs and described the distinguishing features of each:

> ... the Apostolic and Heroic Age of our Faith, fell into three distinct epochs, of nine, of thirty-nine and of twenty-nine years' duration, associated respectively with the Bábí Dispensation and the ministries of Bahá'u'lláh and of 'Abdu'l-Bahá. This primitive Age of the Bahá'í Era, unapproached in spiritual fecundity by any period associated with the mission of the Founder of any previous Dispensation, was impregnated, from its inception to its termination, with the creative energies generated through the advent of two independent Manifestations and the establishment of a Covenant unique in the spiritual annals of mankind.[8]

The Formative Age, in which we now live and serve,[9] was ushered in with the passing of 'Abdu'l-Bahá.[10] Its major thrust is the shaping, development and consolidation of the local, national and international institutions of the Faith.[11] It is clear from the enumeration of the tasks associated with the Formative Age that their achievement will require increasingly mature levels of functioning of the Bahá'í community:

451-2. CF, pp. 4–5 Letter dated 5 June 1947 to the American Bahá'ís.
451-3. WOB, p. 98. Letter dated 8 February 1934.
451-4. WOB, p. 156. Letter dated 8 February 1934.
451-5. CF, p. 5. Letter dated 5 June 1947 to the American Bahá'ís.
451-6. WOB, p. 156. Letter dated 8 February 1934.
451-7. GPB, p. xv.
451-8. CF, pp. 4–5. Letter dated 5 June 1947 to the American Bahá'ís.
451-9. WOB, p. 98. Letter dated 8 February 1934.
451-10. GPB, p. xiv.
451-11. GPB, p. 324.

451.5a During this Formative Age of the Faith, and in the course of present and succeeding epochs, the last and crowning stage in the erection of the framework of the Administrative Order of the Faith of Bahá'u'lláh—the election of the Universal House of Justice—will have been completed, the Kitáb-i-Aqdas, the Mother Book of His Revelation, will have been codified and its laws promulgated, the Lesser Peace will have been established, the unity of mankind will have been achieved and its maturity attained, the Plan conceived by 'Abdu'l-Bahá will have been executed, the emancipation of the Faith from the fetters of religious orthodoxy will have been effected, and its independent religious status will have universally recognized, . . .[12]

451.6 The epochs of the Formative Age mark progressive stages in the evolution of the organic Bahá'í community and signal the maturation of its institutions, thus enabling the Faith to operate at new levels and to initiate new functions. The timing of each epoch is designated by the Head of the Faith, and given the organic nature of evolutionary development, the transition from one epoch to another may not be abrupt, but may well occur over a period of time. This is the case, for example, in relation to both the inception of the Formative Age and the end of its first epoch. In relation to the former, the passing of 'Abdu'l-Bahá is the transitional event most often identified with the close of the Heroic Age and the beginning of the Formative Age.[13] However, the Guardian also asserts that the Apostolic Age of the Faith concluded "more particularly with the passing [in 1932] of His well-beloved and illustrious sister, the Most Exalted Leaf—the last survivor of a glorious and heroic age."[14] With regard to the termination of the first epoch of the Formative Age, Shoghi Effendi has placed this between the years 1944[15] and 1946.[16]

451.7 Before describing the individual epochs of the Formative Age, it is important to comment on the use of the term "epoch" in the writings of the Guardian. In a letter dated 18 January 1953, written on his behalf to a National Spiritual Assembly, it is explained that the term is used to apply both to the stages in the Formative Age of the Faith, and to the phases in the unfoldment of 'Abdu'l-Bahá's Divine Plan.[17] We are currently in the fourth—epoch of the

---

451-12. CF, p. 6. Letter dated 5 June 1947 to the American Bahá'ís.
451-13. GPB, p. xiv.
451-14. WOB, p. 98. Letter dated 8 February 1934.
451-15. CF, p. 5. Letter dated 5 June 1947 to the American Bahá'ís.
451-16. MBW, p. 89. Cablegram dated 23 August 1955. See also letter dated 18 January 1953 written on behalf of the Guardian to the National Spiritual Assembly of the United States (reference cited in 451–17 below).
451-17. BN, no. 265 (March 1953): 4. Letter dated 18 January 1953 written on behalf of the Guardian to the National Spiritual Assembly of the United States.

Formative Age[18] and the second epoch of 'Abdu'l-Bahá's Divine Plan.[19] (The first epoch of the Divine Plan began in 1937, with the inception of the First Seven Year Plan of the North American Bahá'í community, and concluded with the successful completion of the Ten Year Crusade in 1963.[20] The second epoch of 'Abdu'l-Bahá's Divine Plan commenced in 1964 with the inauguration of the Nine Year Plan of the Universal House of Justice.)

The primary focus of this statement is on the epochs of the Formative Age of the Dispensation of Bahá'u'lláh.   451.8

The First Epoch of the Formative Age: 1921–1944/46

The first epoch of this Age witnessed the "birth and the primary stages in the erection of the framework of the Administrative Order of the Faith."[21] The epoch was characterized by concentration on the formation of local and national institutions in all five continents,[22] thereby initiating the erection of the machinery necessary for future systematic teaching activities. This epoch was further marked by the launching, at the instigation of the Guardian, of the First Seven Year Plan (1937–1944) by the American Bahá'í community. This Plan, drawing its inspiration from the Tablets of the Divine Plan, represented the first systematic teaching campaign of the Bahá'í community and inaugurated the initial stage of the execution of 'Abdu'l-Bahá's Divine Plan in the Western Hemisphere.[23]   451.9

The Second Epoch of the Formative Age: 1946–1963

This epoch extended the developments of the first epoch by calling for the "consummation of a laboriously constructed Administrative Order,"[24] and was to witness the formulation of a succession of teaching plans designed to facilitate the development of the Faith beyond the confines of the Western Hemisphere and the continent of Europe.[25] This epoch was distinguished, in the first instance, by the simultaneous and often spontaneous prosecution of Bahá'í national plans in both the East and the West.[26] For example, in a letter written at Naw-Rúz 105 [1949] to the Bahá'ís in the East, the beloved Guardian listed the specific plans undertaken by the United States, British, Indian, Persian, Australian and New Zealand, and Iraqi National Spiritual Assemblies, and indicated that this concerted action signalized the transition into the   451.10

451-18. See message no. 447.
451-19. See message no. 14.
451-20. BN, no. 265 (March 1953): 4. Letter dated 18 January 1953 written on behalf of the Guardian to the National Spiritual Assembly of the United States.
451-21. CF, p. 5. Letter dated 5 June 1947 to the American Bahá'ís.
451-22. MBW, p. 19. Cablegram dated 24 December 1951.
451-23. CF, p. 5. Letter dated 5 June 1947 to the American Bahá'ís.
451-24. CF, p. 6. Letter dated 5 June 1947 to the American Bahá'ís.
451-25. CF, p. 6. Letter dated 5 June 1947 to the American Bahá'ís.
451-26. MBW, p. 13. Cablegram dated 25 April 1951.

second epoch of the Formative Age.[27] The internal consolidation and the administrative experience gained by the National Assemblies was utilized and mobilized by the Guardian with the launching of the Ten Year World Crusade[28]—a crusade involving the simultaneous prosecution of twelve national plans. The plans derived their direction from 'Abdu'l-Bahá's Divine Plan, and the goals were assigned by Shoghi Effendi from the World Center of the Faith.[29] A second distinguishing feature of this epoch was the "RISE"[30] and "STEADY CONSOLIDATION"[31] of the World Center of the Faith.

451.11 The second epoch thus clearly demonstrated the further maturation of the institutions of the Administrative Order. It witnessed the appointment of the Hands of the Cause,[32] the introduction of Auxiliary Boards,[33] and the establishment of the International Bahá'í Council.[34] The culminating event of the epoch was the election of the Universal House of Justice in 1963. It further demonstrated the more effective and coordinated use of the administrative machinery to prosecute the goals of the first global spiritual crusade, and the emergence in ever sharper relief of the World Center of the Faith.

The Third Epoch of the Formative Age: 1963–1986

451.12 In addressing the British National Spiritual Assembly in 1951, the Guardian foreshadowed "worldwide enterprises destined to be embarked upon, in future epochs of that same [Formative] Age, by the Universal House of Justice."[35] In announcing the Nine Year Plan, "the second of those world-encircling enterprises destined in the course of time to carry the Word of God to every human soul,"[36] the Universal House of Justice embarked upon the process anticipated by the Guardian and proclaimed the commencement of the third epoch of the Formative Age, an epoch which called the Bahá'ís to a yet more mature level of administrative functioning, consistent with the expected vast increase in the size and diversity of the community, its emergence as a model to mankind, and the extension of the influence of the Faith in the world at large. The House of Justice, in a letter dated October 1963, stated:

451-27. *Tawqí'át-i-Mubárakih, 102–109* B.E. (Tehran: Bahá'í Publishing Trust, 125 B.E.), pp. 99–188. Letter dated Naw-Rúz 105 B.E. to the Bahá'ís in the East.
451-28. CF, p. 140. Letter dated 20 August 1955 to the American Bahá'ís.
451-29. MBW, pp. 151–53. Letter dated 4 May 1953.
451-30. MBW, p. 13. Cablegram dated 25 April 1951.
451-31. MBW, p. 13. Cablegram dated 25 April 1951.
451-32. MBW, pp. 18–20. Cablegram dated 24 December 1951.
451-33. MBW, p. 44. Cablegram dated 8 October 1952. And, pp. 127–28. Letter dated October 1957.
451-34. MBW, pp. 7–8. Cablegram dated 9 January 1951.
451-35. UD, p. 261. Guardian's postscript to a letter dated 25 February 1951, written on his behalf to the National Spiritual Assembly of the British Isles.
451-36. See message no. 6.

> Beloved friends, the Cause of God, guarded and nurtured since its inception by God's Messengers, by the Center of His Covenant and by His Sign on earth, now enters a new epoch, the third of the Formative Age.[37] It must now grow rapidly in size, increase its spiritual cohesion and executive ability, develop its institutions and extend its influence into all strata of society. We, its members, must, by constant study of the lifegiving Word, and by dedicated service, deepen in spiritual understanding and show to the world a mature, responsible, fundamentally assured and happy way of life, far removed from the passions, prejudices and distractions of present day society.[38]

451.12a

The period of the third epoch encompassed three world plans, involving all National Spiritual Assemblies, under the direction of the Universal House of Justice, namely, the Nine Year Plan (1964–1973), the Five Year Plan (1974–1979), and the Seven Year Plan (1979–1986). This third epoch witnessed the emergence of the Faith from obscurity[39] and the initiation of activities designed to foster the social and economic development of communities.[40] The institution of the Continental Boards of Counsellors was brought into existence[41] leading to the establishment of the International Teaching Center.[42] Assistants to the Auxiliary Boards were also introduced.[43] At the World Center of the Faith, the historic construction and occupation of the Seat of the Universal House of Justice was a crowning event.[44]

451.12b

The Fourth Epoch of the Formative Age: 1986–

In a letter dated 2 January 1986 written by the Universal House of Justice to the Bahá'ís of the World, the Supreme Body announced the inception of the fourth epoch of the Formative Age. It highlighted the significant developments that had taken place in the "organic growth of the Cause of God"[45] during the course of the recently completed third epoch, by assessing the readiness of the Bahá'í community to begin to address the objectives of the new Six Year Plan scheduled to begin on 21 April 1986, and, outlined the general aims and characteristics of this new Plan. Whereas national plans had previ-

451.13

451-37. The Center of the Covenant is one of 'Abdu'l-Bahá's titles; the Sign of God is one of the titles 'Abdu'l-Bahá used in His Will and Testament to refer to Shoghi Effendi, the Guardian of the Cause of God.
451-38. See message no. 6.
451-39. See message no. 361.
451-40. See message no. 379.
451-41. See message no. 58.
451-42. See message no. 132.
451-43. See messages no. 132 and 137.
451-44. See message no. 354.
451-45. See message no. 447.

ously derived largely from the World Center, in this new epoch the specific goals for each national community will be formulated, within the framework of the overall objectives of the Plan, by means of consultation between the particular National Spiritual Assembly and the Continental Board of Counselors. As the Universal House of Justice states:

451.13a This new process . . . signalizes the inauguration of a new stage in the unfoldment of the Administrative Order. Our beloved Guardian anticipated a succession of epochs during the Formative Age of the Faith; we have no hesitation in recognizing that this new development in the maturation of Bahá'í institutions marks the inception of the fourth epoch of that Age.[46]

Future Epochs

451.14 The tasks that remain to be accomplished during the course of the Formative Age are many and challenging. Additional epochs can be anticipated, each marking significant stages in the evolution of the Administrative Order and culminating in the Golden Age of the Faith.[47] The Golden Age, itself, will involve "successive epochs"[48] leading ultimately to the establishment of the Most Great Peace, to the World Bahá'í Commonwealth and to the "birth and efflorescence of a world civilization."[49]

452

Plans for the Dedication of the Mother Temple of the Indian Subcontinent, New Delhi

24 FEBRUARY 1986

To all National Spiritual Assemblies

452.1 JOYFULLY ANNOUNCE DEDICATION MOTHER TEMPLE INDIAN SUBCONTINENT WILL BE HELD NEW DELHI 23–27 DECEMBER 1986. AMATU'L-BAHÁ RÚHÍYYIH KHÁNUM WILL REPRESENT UNIVERSAL HOUSE OF JUSTICE THIS HISTORIC OCCASION AND DEDICATE TEMPLE. TEMPLE DEDICATION COMMITTEE FORMED UNDER AEGIS NATIONAL ASSEMBLY INDIA WILL CONVEY NECESSARY INFORMATION.

UNIVERSAL HOUSE OF JUSTICE

451-46. See message no. 447.
451-47. CF, p. 6. Letter dated 5 June 1947 to the American Bahá'ís.
451-48. MBW, p. 155. Letter dated 4 May 1953.
451-49. CF, p. 6. Letter dated 5 June 1947 to the American Bahá'ís.

453

The Six Year Plan—Major Objectives and Setting National Goals
25 February 1986

To all National Spiritual Assemblies

Dear Bahá'í Friends,

The Six Year Plan

453.1 On 2 January 1986, on the closing day of the Counselors' Conference, the Universal House of Justice announced certain features of the Six Year Plan and the methods by which they national goals were to be worked out in consultation between the Counselors and National Spiritual Assemblies.[1] Before Riḍván you will receive a message from the Universal House of Justice to the entire Bahá'í world and also one addressed specifically to the Bahá'ís within the jurisdiction of each National Spiritual Assembly.[2]

453.2 In the meantime the House of Justice wishes you to begin your consultations on the goals of the Six Year Plan for your country. The preliminary steps in goal-setting have already been taken, namely the assessment of each country's strengths and weaknesses which the National Spiritual Assemblies recently made at the request of the Universal House of Justice, and which will undoubtedly be of great assistance to each one of you as you enter the next stage of the process.[3]

453.3 The House of Justice has instructed us to send you the following additional guidelines together with the enclosed statement of the major objectives of the plan at the national level, which includes some suggestions for specific goals to provide a basis for your consultations. You should not, however, confine yourselves to those suggestions.

453.4 A special characteristic of the Six Year Plan is that the conceiving of the detailed national goals is itself to be one of the tasks of the Plan, but this fact should not hold up in any way the activities of your communities. With this letter you are being acquainted with the major objectives of the plan, and every believer, every Local Spiritual Assembly, and all the national committees can pursue immediately, with increasing vigor, many projects towards their attainment, both projects already in process and others which will be newly conceived, so that when the specific national goals for each community are announced they will be received by a united company of devoted followers of Bahá'u'lláh already in the full flood of activity.

453-1. See message no. 447.
453-2. See message no. 456.
453-3. See message no. 415, "Invitation to Participate in the Formulation of the Next Plan" (25 November 1984).

453.5 It is the hope of the Universal House of Justice that each National Assembly will be able to meet before Riḍván with a representative of the Continental Board of Counselors so that from this initial consultation a basis will be laid for consultation on the goals at the National Conventions.

453.6 Other consultations will no doubt continue following the Riḍván Festival. Their duration will depend on the condition of each national community, its size and the complexity of its circumstances. As soon as specific goals have been formulated and agreed, they should be immediately sent to the World Center. They will then be considered by the Universal House of Justice and the International Teaching Center, and, as soon as possible, the National Assembly will be informed of the approval or modification of its proposal. Each submission will be considered on its arrival; the earlier they arrive the better, and in no case should a submission reach the World Center later than 1 November 1986.

453.7 Among the elements of the plan which are not covered by the list of major objectives are the goals for international assistance including pioneering, resident teaching projects, traveling teaching, assistance for development projects, and for the acquisition of properties and vehicles. Notes relating to these elements have been provided to the Continental Boards of Counselors, who will share them with National Assemblies during the process of consultation. Since they are international in nature, these goals will have to be consolidated and approved at the World Center before being generally announced.

453.8 Though the institutions of the Faith are responsible for planning the goals and activities of the Cause, for stimulating and encouraging the believers to arise, and for supporting and unifying them in their services, it is, in the final analysis, through the spiritual decisions and actions of the individual believers that the Faith moves forward on its course to ultimate victory. It is the ardent hope of the Universal House of Justice that every faithful follower of Bahá'u'lláh will search his or her heart and turn with full attention and loving self-sacrifice to the consideration of the goals of the Six Year Plan, and determine how to play a part in their achievement.

453.9 The prayers of the Universal House of Justice and the International Teaching Center at the Sacred Threshold will surround the institutions of the Faith in every continent and nation as you assume your weighty task of conceiving the goals which will guide the national communities of the Faith through the next six years.

With loving Bahá'í greetings,
DEPARTMENT OF THE SECRETARIAT

The Six Year Plan
143–149 • 1986–1992

The Major Objectives

The major objectives of the Six Year Plan include: carrying the healing Message of Bahá'u'lláh to the generality of mankind; greater involvement of the Faith in the life of human society; a worldwide increase in the translation, production, distribution and use of Bahá'í literature; further acceleration in the process of the maturation of national and local Bahá'í communities; greater attention to universal participation and the spiritual enrichment of individual believers; a wider extension of Bahá'í education to children and youth and the strengthening of Bahá'í family life; and the pursuit of projects of social and economic development in well-established Bahá'í communities. 453.10

* * *

Set out below are suggestions for possible ways of achieving the above objectives to act as a basis for consultation and a stimulus for thinking. National Assemblies should not confine themselves to these points if they feel that there are other matters which deserve attention. 453.11

1. Carrying the healing Message of Bahá'u'lláh to the generality of mankind
453.12

- Increase the number of believers from all strata of society, identifying as goals of the plan those specific sectors, minority groups, tribal peoples, etc. which are at present underrepresented in the Bahá'í community and which will, therefore, be given special attention during the plan.

- Increase the number of localities where Bahá'ís reside, opening, in the process, virgin states, provinces, islands or other major civil subdivisions of the country.

- Seize teaching opportunities by planning projects in areas where receptivity is found, aiming at large-scale enrollment and entry by troops where possible.

- Be alert to opportunities for international collaboration with other Bahá'í communities in the promotion of the Faith through: border teaching projects; the sending of traveling teachers; and the teaching of special groups such as those temporarily abroad for study or work, particularly those from countries which are difficult of access, such as China or countries in Eastern Europe.

- Raise up homefront pioneers and traveling or resident teachers to assist in the fulfillment of teaching goals and plans.

- Utilize mass media systems for greater proclamation.
- Make use of drama and singing in the teaching and deepening work and in Bahá'í gatherings, where advisable.

453.13 **2. Greater involvement of the Faith in the life of human society**

- Develop the proper understanding and practice of consultation among members of the Bahá'í community and in the work of Bahá'í institutions, and foster the spirit of consultation in the conduct of human affairs and the resolution of conflicts at all levels of society.
- Foster association with organizations, prominent persons and those in authority concerning the promotion of peace, world order and allied objectives, with a view to offering the Bahá'í teachings and insights regarding current problems and thought.
- Train suitable Bahá'ís to undertake public relations activities.
- Foster appreciation of the Faith in scholarly and academic circles by developing Bahá'í scholarship, by endeavoring to have the Faith included in the curricula and textbooks of schools and universities, and by other means.
- Encourage Bahá'í youth to move towards the front ranks of those professions, trades, arts and crafts necessary to human progress.
- Promote the establishment of Bahá'í clubs in universities and other similar educational institutions.
- Foster the practice of the equality of the sexes both in the life of the Bahá'í community and in society as a whole and, for this purpose, hold special conferences and training programs for women and for men.

453.14 **3. A worldwide increase in the translation, production, distribution and use of Bahá'í literature**

- Foster the use of Bahá'í literature, especially in local languages, supplemented as need be by tape recordings and vidual aids.
- Improve the distribution of Bahá'í literature by taking specific steps, such as the establishment of regional depots where necessary, and the education of Local Spiritual Assemblies in the responsibilities to acquaint the friends with Bahá'í literature and ensure its easy availability.
- Produce greater supplies of Bahá'í literature in accordance with well-thought-out plans of translation, production and distribution.
- Produce, where required for translations into vernacular languages, simplified versions of the Sacred Scriptures, the writings of the Guardian and the statements of the Universal House of Justice.
- Establish Bahá'í lending libraries.

4. Further acceleration in the process of the maturation of local and national Bahá'í communities

- Adopt specific programs to assist and encourage the development of isolated centers into groups, and groups into communities with Local Spiritual Assemblies, resulting in a steady increase of such Assemblies.
- Adopt specific goals and programs to consolidate and strengthen Local Spiritual Assemblies, so that they will:
  - Hold regular meetings with harmonious and productive consultation,
  - Properly organize and conduct the work of their secretariat and treasury,
  - Appoint and coordinate the work of local committees for special aspects of their work, such as teaching, child education, youth activities, literature distribution, etc.,
  - Win the respect and confidence of their local communities so that the believers will turn to them for the resolution of problems and advice in their services to the Cause,
  - Where appropriate, acquire and develop the use of local centers,
  - Obtain incorporation or equivalent recognition as a legal entity,
  - Exercise their responsibilities in relation to marriages, births, transfers of membership, marriages and deaths.
  - Maintain registers of declarations, births, transfers of membership, marriages and deaths.
- Adopt specific goals and programs to consolidate communities with Local Spiritual Assemblies so that the believers will be encouraged to:
  - Attend regularly Nineteen Day Feasts and the observances of Bahá'í Holy Days, and enhance the spiritual quality of such gatherings,
  - Pursue local teaching and deepening activities,
  - Foster the realization of the equality of men and women,
  - Develop local activities for children and youth,
  - Support the fund,
  - Carry out extension teaching projects.
- Develop the functioning of National Spiritual Assemblies, adopting specific plans and programs to:
  - Improve their standard of united, productive, loving consultation,
  - Develop efficiently functioning national secretariats,
  - Enhance the standard of the functioning of national treasuries and promote the goal of financial independence of the national Bahá'í community,

- Appoint strong national committees to carry out, under the general supervision of the National Spiritual Assembly, the many specialized aspects of the work of the Cause, including the detailed planning and prompt execution of the work necessary to achieve all the goals of the Six Year Plan.

• Acquire, where needed and feasible, national and local properties, such as Ḥaẓíratu'l-Quds, teaching institutes, summer schools, Bahá'í cemeteries, etc. and ensure their proper care and maintenance.

• Obtain, where legally possible, official recognition for Bahá'í marriage and holy days and exemption from the payment of taxes on Bahá'í institutions and their activities.

• Ensure the rapid and regular dissemination of news to all believers.

• Hold regular, well-planned and well-run summer and winter schools and conferences at costs and in localities which will permit the largest attendance.

• Encourage collaboration between or amongst Local Spiritual Assemblies in mutually agreed projects.

• Develop and administer correspondence courses for teaching and deepening.

5. Greater attention to universal participation and the spiritual enrichment of individual believers

• Promote universal participation in the life of the Faith and an increased sense of their Bahá'í identity among children, youth and adults.

• Encourage, where feasible, the practice of dawn prayer.

• Encourage individual believers to adopt teaching goals for themselves.

• Carry out activities designed to deepen the believers in both a spiritual and intellectual understanding of the Cause.

• Encourage believers to make greater use of Bahá'í literature.

• Encourage the believers to enhance their command of language to assist them to understand the Bahá'í writings ever more clearly.

• Develop and foster Bahá'í scholarship and lend support to the Associations for Bahá'í Studies.

• Foster obedience to the Bahá'í laws of personal behavior such as abstention from the drinking of alcoholic beverages and from the taking of habit-forming drugs, and inspire the believers to follow the Bahá'í way of life.

6. A wider extension of Bahá'í education to children and youth, and the strengthening of Bahá'í family life

• Encourage the holding of regular classes for the Bahá'í education of children.

- Develop systematic lesson plans and other materials for the Bahá'í education of children.
- Train believers to teach Bahá'í children's classes.
- Establish a program for the guidance of parents, especially mothers, in the care and training of Bahá'í children.
- Sponsor institutes on Bahá'í marriage and family life.
- Encourage community activities involving Bahá'í families.

7. **The pursuit of projects of social and economic development in well-established Bahá'í communities** 453.18
    - Encourage Local Spiritual Assemblies and the rank and file of the believers to consider ways in which they can advance the social and economic development of their communities.
    - Establish tutorial schools[4] and preschools where needed and feasible.
    - Encourage and sponsor adult literacy programs where needed, especially for women.
    - Foster collaboration with other agencies involved in social and economic development in areas where the Bahá'í communities can contribute to the work.

454
Passing of Angus Cowan, Continental Counselor
12 March 1986

The National Spiritual Assembly of the Bahá'ís of Canada

HEARTS GRIEF-STRICKEN PASSING OUTSTANDING PROMOTER CAUSE ANGUS COWAN.[1] 454.1
HIS SERVICES AS PREEMINENT BAHÁ'Í TEACHER OF INDIAN PEOPLES, HIS UNCEASING LABORS THROUGHOUT LONG YEARS DEVOTION RANGED FROM LOCAL AND NATIONAL SPIRITUAL ASSEMBLIES TO AUXILIARY BOARD AND BOARD COUNSELORS. HIS COMPASSION, COURTESY, HUMILITY, MAGNANIMITY UNFORGETTABLE. BAHÁ'Í COMMUNITY CANADA ROBBED OF A DEDICATED NOBLE WORKER WHO BORE HIS SUFFERINGS TO THE VERY END WITH EXEMPLARY FORTITUDE. CONVEY LOVING CONDOLENCES HIS BELOVED WIFE AND FAMILY AND ASSURANCES ARDENT PRAYERS HIS RADIANT SOUL'S PROGRESS THROUGHOUT WORLDS GOD.

UNIVERSAL HOUSE OF JUSTICE

453-4. A Bahá'í tutorial school is a simple school, usually in a rural area where no formal education is available. It often consists of one teacher and a group of students who learn Bahá'í moral and ethical teachings and basic reading, writing, and computation skills.

454-1. For an account of the life and services of Angus Cowan, see BW 19:703–06.

455
Women—Reasons for Exemption from Acts of Worship
17 March 1986

To an individual Bahá'í

Dear Bahá'í Friend,

455.1 The Universal House of Justice has received your letter of 4 March 1986 seeking clarification of a statement concerning certain obligations of women found in the Kitáb-i-Aqdas. We are directed to convey the following.

455.2 There is nothing in the teachings of the Faith to state that women in their courses are exempt from fasting and from offering the obligatory prayers because they are unclean.

455.3 The concept of uncleanness as understood and practiced in the religious communities of the past has been abolished by Bahá'u'lláh. As you are aware, He says that through His revelation "all created things were immersed in the sea of purification" (*God Passes By*, page 154). This should, of course, be understood in the context of His clear instructions about the necessity for all to exemplify immaculate cleanliness, especially when engaged in acts of worship.

455.4 We are to confirm that you are quite right in pointing out that the term used in this regard in the *Synopsis and Codification of the Laws and Ordinances of the Kitáb-i-Aqdas* [p. 36, (4) (c)] is "exemption," not prohibition.

With loving Bahá'í greetings,
DEPARTMENT OF THE SECRETARIAT

456
Riḍván Message 1986
Riḍván 1986

To the Bahá'ís of the World

Dearly loved Friends,

456.1 The Divine Springtime is fast advancing and all the atoms of the earth are responding to the vibrating influence of Bahá'u'lláh's Revelation. The evidences of this new life are clearly apparent in the progress of the Cause of God. As we contemplate, however momentarily, the unfolding pattern of its growth, we can but recognize, with wonder and gratitude, the irresistible power of that Almighty Hand which guides its destinies.

Review of Notable Achievements during the Seven Year Plan

456.2 This progress has accelerated notably during the Seven Year Plan, witnessed by the achievement of many important enterprises throughout the Bahá'í world and vital developments at the heart of the Cause itself. The restoration and opening to pilgrimage of the southern wing of the House of 'Abdu'lláh Páshá; the completion and occupation of the Seat of the Universal House of Justice; the approval of detailed plans for the remaining edifices around the Arc; the expansion of the membership and responsibilities of the International Teaching Center and the Continental Boards of Counselors; the establishment of the offices of Social and Economic Development, and of Public Information; the dedication of the Mother Temple of the Pacific, and dramatic progress with the building of the Temple in India; the expansion of the teaching work throughout the world, resulting in the formation of twenty-three new National Spiritual Assemblies, nearly 8,000 new Local Spiritual Assemblies, the opening of more than 16,000 new localities and representation within the Bahá'í community of 300 new tribes; the issuing of 2,196 new publications, 898 of which are editions of the Holy Text and the enrichment of Bahá'í literature by productions in 114 new languages; the initiation of 737 new social and economic development projects; the addition of three radio stations, with three more soon to be inaugurated—these stand out as conspicuous achievements in a Plan which will be remembered as having set the seal on the third epoch of the Formative Age.[1]

Emergence from Obscurity

456.3 The opening of that Plan coincided with the recrudescence of savage persecution of the Bahá'í community in Iran, a deliberate effort to eliminate the Cause of God from the land of its birth. The heroic steadfastness of the Persian friends has been the mainspring of tremendous international attention focused on the Cause, eventually bringing it to the agenda of the General Assembly of the United Nations, and, together with worldwide publicity in all the media, accomplishing its emergence from the obscurity which characterized and sheltered the first period of its life. This dramatic process impelled the Universal House of Justice to address a Statement on Peace to the Peoples of the World and arrange for its delivery to Heads of State and the generality of the rulers.[2]

Maturation of the Institutions of the Cause

456.4 Paralleling these outstanding events has been a remarkable unfoldment of organic growth in the maturity of the institutions of the Cause. The development of capacity and responsibility on their part and the devolution upon them of continually greater autonomy have been fostered by the encouragement of ever closer cooperation between the twin arms of the Administrative Order.[3] This process now takes a large stride forward as the National Spiri-

---

456-1. The third epoch included the years 1963–86.
456-2. See messages no. 429 and 438.
456-3. The twin arms of the Administrative Order are sometimes designated the "rulers" and the "learned." In this context the "rulers" comprise the members of the Universal House of

tual Assemblies and Counselors consult together to formulate, for the first time, the national goals of an international teaching plan.⁴ Together they must carry them out; together they must implement the world objectives of the Six Year Plan as they apply in each country. This significant development is a befitting opening to the fourth epoch of the Formative Age and initiates a process which will undoubtedly characterize that epoch as national communities grow in strength and influence and are able to diffuse within their own countries the spirit of love and social unity which is the hallmark of the Cause of God.

World Center Goals

456.5   The goals to be achieved at the World Center include publication of a copiously annotated English translation of the Kitáb-i-Aqdas and related texts, education of the Bahá'í world in the law of the Ḥuqúqu'lláh, pursuit of plans for the erection of the remaining buildings on the Arc, and the broadening of the basis of the international relations of the Faith.⁵

456.6   The major world objectives of the Plan have already been sent to National Spiritual Assemblies and Continental Boards of Counselors for their mutual consultation and implementation.

456.7   Dear friends, as the world passes through its darkest hour before the dawn, the Cause of God, shining ever more brightly, presses forward to that glorious break of day when the Divine Standard will be unfurled and the Nightingale of Paradise warble its melody.

With loving Bahá'í greetings,
THE UNIVERSAL HOUSE OF JUSTICE

Justice, National Spiritual Assemblies, and Local Spiritual Assemblies; the "learned" comprise the Hands of the Cause of God, the International Counselors, the Continental Boards of Counselors, and the Auxiliary Board members and their assistants. Both terms also have a wider usage. For example, the term "learned" also embraces Bahá'ís who have attained a prominent position in the teaching work in addition to those who are members of the institutions so designated. For an explanation of the roles of and distinctions between these institutions, see message no. 111.

456-4. See messages no. 447 and 453.

456-5. The laws and ordinances of the Kitáb-i-Aqdas are applied gradually as the Bahá'í community matures. For an explanation of this principle, see message no. 27 on the translation and publication of the Kitáb-i-Aqdas. For a fuller discussion, see the introduction to *A Synopsis and Codification of the Kitáb-i-Aqdas: The Most Holy Book of Bahá'u'lláh*. Over the years the Universal House of Justice has taken steps to prepare the Bahá'í community for further application of such laws and ordinances by publishing the *Synopsis and Codification* (see message no. 125) and by releasing a compilation on the Law of Ḥuqúqu'lláh (see messages no. 404 and 430). The publication of annotated translations in English and Persian of the Kitáb-i-Aqdas represent further such steps.

The buildings to be erected on the Arc are the seats of the International Teaching Center and the Center for the Study of the Texts, and the extension of the International Archives Building, which Shoghi Effendi completed in 1957. Terraces leading to the Shrine of the Báb are being constructed from the foot to the crest of Mount Carmel. The International Bahá'í Library remains to be built at a later date.

Glossary

Glossary

Note: **Boldface** *terms within entries are cross-references to other entries that define or amplify essential terms. A number of entries are based on explanations found in "Definitions of Some of the Oriental Terms Used in Bahá'í Literature" in* The Bahá'í World: An International Record, *Volume XVIII, 1979–1983, pp. 897–904. Other entries are based on explanations found in the notes and glossary in the Kitáb-i-Aqdas. References to Tablets of the Divine Plan cite Tablet and paragraph numbers, e.g., "TDP 6.8." References to messages in the text give the message number, e.g., "message no. 222."*

A

'ABDU'L-'AZÍZ Sulṭán of Turkey, 1861–76. He was responsible for Bahá'u'lláh's banishments to Constantinople, to Adrianople, and to the prison-fortress of 'Akká, Palestine. Willful and headstrong, 'Abdu'l-'Azíz was known for his lavish expenditures. Bahá'u'lláh stigmatized him in the Kitáb-i-Aqdas as occupying the "throne of tyranny." His fall was prophesied in a Tablet (circa 1869) addressed to Fu'ád Pás͟há, Ottoman Foreign Minister while Bahá'u'lláh was imprisoned in 'Akká. As a result of public discontent, heightened by a crop failure in 1873 and a mounting public debt, he was deposed by his ministers on 30 May 1876.

'ABDU'L-BAHÁ *Servant of Bahá:* the title assumed by 'Abbás Effendi (23 May 1844–28 November 1921), eldest son and appointed successor of Bahá'u'lláh and the Center of His Covenant. Upon Bahá'u'lláh's ascension in 1892, 'Abdu'l-Bahá became Head of the Bahá'í Faith in accordance with provisions revealed by Bahá'u'lláh in the Kitáb-i-Aqdas and the Book of the Covenant. Among the titles by which He is known are the Center of the Covenant, the Mystery of God, the Master, and the Perfect Exemplar of Bahá'u'lláh's teachings. See **Tablets of the Divine Plan** and **Will and Testament of 'Abdu'l-Bahá**.

'ABDU'L-ḤAMÍD II Sulṭán of the Ottoman Empire, 1876–1909. He and his uncle, Sulṭan 'Abdu'l-'Azíz, who preceded him, were responsible for forty-six of 'Abdu'l-Bahá's fifty-six years of imprisonment and exile and for Bahá'u'lláh's banishments to Constantinople, Adrianople, and 'Akká. The Young Turks Rebellion in 1908 forced 'Abdu'l-Ḥamíd to reinstate the constitution he had suspended and to free all political and religious prisoners. As a result, 'Abdu'l-Bahá was released from house arrest in September 1908. 'Abdu'l-Ḥamíd was deposed the following year.

ABHÁ *Most Glorious.* See **Alláh-u-Abhá**; **Yá Bahá'u'l-Abhá**.

ABHÁ BEAUTY A translation of *Jamál-i-Abhá*, a title of **Bahá'u'lláh**.

GLOSSARY

ABHÁ KINGDOM *The Most Glorious Kingdom:* the spiritual world beyond this world.

ABHÁ PARADISE See **Abhá Kingdom**.

ADAMIC CYCLE See **Cycle**.

ADMINISTRATIVE ORDER The international system for the administration of the affairs of the Bahá'í community. Ordained by Bahá'u'lláh, it is the agency through which the spirit of His revelation is to exercise its transforming effects on humanity and through which the **Bahá'í World Commonwealth** will be ushered in. Its twin, crowning institutions are the **Guardianship** and the **Universal House of Justice**. The institutions that make it up and the principles by which it operates are set forth in the writings of Bahá'u'lláh and 'Abdu'l-Bahá. Its structure was further clarified and raised up by Shoghi Effendi during his ministry as Guardian of the Faith (1921–57). This process of elucidation continues through guidance from the Universal House of Justice, the supreme governing and legislative body of the Bahá'í Faith, which is supported by National and Local **Spiritual Assemblies** elected by members of the Bahá'í community. These local and national bodies are invested with the authority to direct the Bahá'í community's affairs and to uphold Bahá'í laws and standards. They are also responsible for the education, guidance, and protection of the community. The Administrative Order also comprises the institutions of the **Hands of the Cause of God**, the **International Teaching Center**, and the **Continental Boards of Counselors** and their **Auxiliary Boards** and assistants, who bear particular responsibility for the protection and propagation of the Faith and share with the Spiritual Assemblies the functions of educating, counseling, and advising members of the Bahá'í community. Other institutions of the Administrative Order include **Ḥuqúqu'lláh**, the Bahá'í **Fund**, the **Mas͟hriqu'l-Ad͟hkár**, and the **Nineteen Day Feast**. The present Bahá'í Administrative Order is the precursor of the **World Order of Bahá'u'lláh** and is described by Shoghi Effendi as its "nucleus" and "pattern."

AFNÁN *Twigs:* the Báb's kindred; specifically, descendants of His three maternal uncles and His wife's two brothers.

AGES The Bahá'í Dispensation is divided into three Ages: the Heroic, Formative, and Golden Ages. The Heroic Age, also called the Apostolic or Primitive Age, began in 1844 with the Declaration of the Báb and spanned the ministries of the Báb (1844–53), Bahá'u'lláh (1852–92), and 'Abdu'l-Bahá (1892–1921). The transitional event most often identified with the end of the Heroic Age and the beginning of the Formative Age is the passing of 'Abdu'l-Bahá in 1921. The Formative Age, also known as the Age of Transition or the Iron Age, began in 1921 when Shoghi Effendi, according to instructions in 'Abdu'l-Bahá's Will and Testament, became the Guardian of the Cause of God and began to build Bahá'u'lláh's Administrative Order. The Formative Age is the second and current Age; it is to be followed by the third and final Age, the Golden Age destined to witness the proclamation of the Most Great Peace and the establishment of the Bahá'í World Commonwealth. "The emergence of a world community, the consciousness of world citizenship, the founding of a world civilization and culture," Shoghi Effendi wrote, "—all of which must synchronize with the initial stages in the unfoldment of the Golden Age of the Bahá'í Era—should, by their very nature, be regarded, as far as this planetary life is concerned, as the furthermost limits in the organization of human society, though man, as an individual, will, nay must indeed as a result of such a consummation, continue indefinitely to progress and develop." For a discussion of the significance of the Formative Age, see message no. 95; for an explanation of the epochs of the Formative Age, see message no. 451. See also **Dispensation; Epochs; Bahá'í World Commonwealth.**

AGHṢÁN *Branches:* the sons and male descendants of Bahá'u'lláh.

'AKKÁ A four-thousand-year-old seaport and prison city in northern Israel surrounded by fortress-like walls facing the sea. In the mid-1800s 'Akká became a penal colony to which the worst criminals of the Ottoman Empire were sent. In 1868 Bahá'u'lláh and His family and companions were banished to 'Akká by Sulṭán 'Abdu'l-'Azíz. Bahá'u'lláh was incarcerated within its barracks for two years, two months, and five days. Restrictions were gradually relaxed, and He lived in a series of houses within 'Akká until June 1877, when He moved outside the city walls to the Mansion of Mazra'ih. Bahá'u'lláh named 'Akká "the Most Great Prison."

ALLÁH-U-ABHÁ *God is Most Glorious:* the **Greatest Name,** adopted as a greeting among Bahá'ís during the period of Bahá'u'lláh's exile in Adrianople (1863–68).

AMATU'L-BAHÁ RÚḤÍYYIH KHÁNUM Née Mary Maxwell (b. 1910), also called Rúḥíyyih Rabbání; daughter of May Bolles Maxwell and Sutherland Maxwell of Montreal, and wife of Shoghi Effendi, the Guardian of the Bahá'í Faith. On 26 March 1952, succeeding her illustrious father, she was appointed a Hand of the Cause of God residing in the Holy Land. *Rúḥíyyih* (meaning "spiritual") is a name given to her by Shoghi Effendi on their marriage. *Khánum* is a Persian title meaning "lady," "Madame," or "Mrs." The title *Amatu'l-Bahá* (meaning "Maidservant of Bahá") was used by the Guardian in a cable to a conference in Chicago in 1953. *Rabbání* is a surname given to Shoghi Effendi by 'Abdu'l-Bahá.

ANCIENT BEAUTY A translation of *Jamál-i-Qadím,* a name of God that is also used as a title of Bahá'u'lláh, Who is the latest Manifestation of God to humankind. One cannot always say categorically in any passage whether the reference is to God, to Bahá'u'lláh, or to both.

ANCIENT OF DAYS See **Ancient Beauty.**

APOSTOLIC AGE See **Ages.**

AQDAS *Most Holy.* See **Kitáb-i-Aqdas, The.**

ARC The line of a curved path laid out by Shoghi Effendi on Mount Carmel, stretching across the Bahá'í properties near the Shrine of the Báb and centered on the Monument Gardens. On this Arc the seats of the "world-shaking, world-embracing, world-directing administrative institutions" of the World Order of Bahá'u'lláh are to be located (MA, pp. 32–33). Within the Arc are the resting-places of the Greatest Holy Leaf; her brother, the Purest Branch; and her mother, the Most Exalted Leaf. Edifices already constructed on the Arc include the International Bahá'í

Archives building (completed in 1957), which is to be extended, and the seat of the Universal House of Justice (completed in 1982 and occupied in 1983). Buildings yet to be completed include the International Bahá'í Library and the seats for the International Teaching Center and the Center for the Study of the Texts. See also **Administrative Order; World Order of Bahá'u'lláh.**

ARK The word "ark" means, literally, a boat or ship, something that affords protection and safety, or a chest or box. It is used in two senses in the Bible. In the first sense it refers to the Ark of Noah, which He was bidden to build of gopher wood to preserve life during the Flood. In the second sense it refers to the Ark of the Covenant, the sacred chest representing to the Hebrews God's presence among them. It was constructed to hold the Tablets of the Law in Moses' time and was later placed in the Holy of Holies in the Temple of Jerusalem. The Ark, as a symbol of God's Law and the Divine Covenant that is the salvation of the people in every age and Dispensation, appears in various ways in the Bahá'í writings. Bahá'u'lláh refers to His faithful followers as "the denizens of the Crimson Ark"; He refers to the Ark of the Cause and also to the Ark of His Laws. A well-known passage in which this term is used appears in the Tablet of Carmel: "Ere long will God sail His Ark upon thee, and will manifest the people of Bahá who have been mentioned in the Book of Names." Shoghi Effendi explains that the Ark in this passage refers to the Bahá'í Administrative Center on Mount Carmel and that the dwellers of the Ark are the members of the Universal House of Justice.

ARMY OF LIGHT Generally, the Bahá'í community, but more particularly the "heavenly armies"—"those souls," according to 'Abdu'l-Bahá, "who are entirely freed from the human world, transformed into celestial spirits and have become divine angels. Such souls are the rays of the Sun of Reality who will illumine all the continents" (TDP 8.2).

ÁSÍYIH KHÁNUM *Navváb* (an honorific implying "Grace" or "Highness"); *the Most Exalted Leaf*: wife of Bahá'u'lláh and mother of 'Abdu'l-Bahá, Bahíyyih Khánum, and Mírzá Mihdí. She was married to Bahá'u'lláh in 1835, accompanied Him in His exiles, and died in 1886. Bahá'u'lláh named her His "perpetual consort in all the worlds of God." Her resting-place is in the Monument Gardens on Mount Carmel, next to the tomb of Mírzá Mihdí and near that of the Greatest Holy Leaf.

AUTHOR OF THE BAHÁ'Í REVELATION Bahá'u'lláh.

AUXILIARY BOARDS An institution established by Shoghi Effendi in 1954 to act as "deputies, assistants and advisers" to the **Hands of the Cause of God** as they carry out their twin duties of protection and propagation. With the formation of the **Continental Boards of Counselors** in 1968, the Hands of the Cause of God were freed of responsibility for appointing, supervising, and coordinating the work of the Auxiliary Boards, and these functions were transferred by the Universal House of Justice to the Continental Boards of Counselors. There are two Auxiliary Boards, one for protection and one for propagation; members serve on one of the two boards. In a letter dated 7 October 1973 (see message no. 137) the Universal House of Justice authorized the appointment of assistants to Auxiliary Board members.

B

BÁB, THE *The Gate:* title assumed by Siyyid 'Alí Muḥammad (20 October 1819–9 July 1850) after declaring His mission in Shiraz in 1844. The Báb's station is twofold: He is a Manifestation of God and the Founder of the Bábí Faith, and He is the Herald of Bahá'u'lláh. A detailed, moving, and authoritative work titled *The Dawn-Breakers* (written by Nabíl-i-Zarandí and translated by Shoghi Effendi) recounts the Báb's life and His followers' exploits. See Balyuzi, *The Báb*.

BÁBÍYYIH, THE *The Bábí place* or *the center of the Bábís:* a house in Mashhad, Iran, that served as a residence for Mullá Ḥusayn (the first of the **Letters of the Living**) and Quddús (also a Letter of the Living, whose rank was second only to that of the Báb) and as a place to which inquirers came to learn about the Bábí Faith. The Bahá'í historian Nabíl writes that "A steady stream of visitors, whom the energy and zeal of Mullá Ḥusayn had prepared for the acceptance of the Faith, poured into the presence of Quddús, acknowledged the claim of the Cause, and willingly enlisted under its banner. The all-observing vigilance with which Mullá Ḥusayn labored to diffuse the knowledge of the new Revelation, and the masterly manner in which Quddús edified its ever-increasing number of adherents, gave rise to a wave of enthusiasm which swept over the entire city of Mashhad, and the effects of which spread rapidly beyond the confines of Khurásán. The house of Bábíyyih was soon converted into a rallying center for a multitude of devotees who were fired with an inflexible resolve to demonstrate, by every means in their power, the great inherent energies of their Faith." (DB, p. 267)

BAHÁ'Í ELECTIONS See **Elections, Bahá'í.**

BAHÁ'Í FUND See **Fund.**

BAHÁ'Í INTERNATIONAL COMMUNITY An international body made up of Bahá'í institutions, local and national, continental and international, all closely interrelated, and comprising the worldwide membership of the Bahá'í Faith. Since 1948 the Bahá'í International Community has been affiliated with the United Nations' Office of Public Information. In 1967 the Universal House of Justice assumed the function (shouldered for many years by the National Spiritual Assembly of the Bahá'ís of the United States) of representing the Bahá'í International Community in its capacity as a nongovernmental organization at the United Nations. In 1970 the Bahá'í International Community was granted consultative status with the United Nations Economic and Social Council (ECOSOC), and in 1976 it became affiliated with the United Nations Children's Fund (UNICEF, formerly named the United Nations Children's Emergency Fund). It is also affiliated with the United Nations Environment Program (UNEP). In its work with the United Nations, the Bahá'í International Community participates in meetings of United Nations bodies concerned with such issues as human rights, social development, the status of women, the environment, human settlement, food, science and technology, population, the law of the sea, crime prevention, substance abuse, youth, children, the family, disarmament, and the United Nations University.

BAHÁ'Í WORLD CENTER The world spiritual and administrative centers of the Bahá'í Faith located in the twin cities of 'Akká and Haifa in Israel. See also **Arc; Administrative Order.**

BAHÁ'Í WORLD COMMONWEALTH The future Bahá'í community of nations, Shoghi Effendi explains, that will operate "solely in direct conformity with the laws and principles of Bahá'u'lláh" and will be animated wholly by His spirit. Its "supreme organ" will be the Universal House of Justice functioning in "the plenitude of its power." Its advent will "signalize the long-awaited advent of the Christ-promised Kingdom of God on earth." It will serve as both "the instrument and the guardian of the Most Great Peace." Within the Bahá'í World Commonwealth "all nations, races, creeds and classes" will be "closely and permanently united," and "the autonomy of its state members and the personal freedom and initiative of the individuals that compose them" will be "definitely and completely safeguarded. This commonwealth must, as far as we can visualize it, consist of a world legislature, whose members will, as the trustees of the whole of mankind, ultimately control the entire resources of all the component nations, and will enact such laws as shall be required to regulate the life, satisfy the needs and adjust the relationships of all races and peoples. A world executive, backed by an international Force, will carry out the decisions arrived at, and apply the laws enacted by, this world legislature, and will safeguard the organic unity of the whole commonwealth. A world tribunal will adjudicate and deliver its compulsory and final verdict in all and any disputes that may arise between the various elements constituting this universal system. . . . A world metropolis will act as the nerve center . . . , the focus towards which the unifying forces of life will converge and from which its energizing influences will radiate." The world commonwealth will include a system of international communication; an international auxiliary language; a world script and literature; a uniform and universal system of currency, weights, and measures; and an integrated economic system with coordinated markets and regulated channels of distribution. See also **World Order of Bahá'u'lláh.**

BAHÁ'U'LLÁH *The Glory of God:* title of Mírzá Ḥusayn-'Alí Núrí (12 November 1817–29 May 1892), Founder of the Bahá'í Faith. For accounts of His life, see Shoghi Effendi, *God Passes By;* Nabíl, *Dawn-Breakers;* and Balyuzi, *Bahá'u'lláh: The King of Glory.* Bahá'u'lláh is referred to by a variety of titles, including the Promised One of All Ages, the Blessed Beauty, the Blessed Perfection, the Morn of Truth, the Abhá Luminary, the Dayspring of the Most Divine Essence, the Ancient Beauty, the Ancient Root, the Ancient of Days, the Author of the Bahá'í Revelation, the Mystic Dove, the Sovereign Revealer, the Judge, the Redeemer, the Divine Physician, the Prince of Peace, the Pen of Glory, the Pen of the Most High, the Supreme Pen, the Lord of Hosts, and the Lord of the Age. See also **Book of the Covenant; Hidden Words, The; Kitáb-i-Aqdas, The.**

BAHÍYYIH KHÁNUM (1846–1932) *The Greatest Holy Leaf; the Most Exalted Leaf:* saintly daughter of Bahá'u'lláh and outstanding heroine of the Bahá'í Dispensation. Her death in 1932 marked the final end of the Heroic Age of the Bahá'í Faith, which had drawn to a close with the passing of 'Abdu'l-Bahá in 1921. A monument erected in her memory symbolizes the Bahá'í World Order; its location is Mount Carmel, within the Arc and in close proximity to the resting-places of her brother, Mírzá

Mihdí; her mother, Ásíyih K͟hanúm; and the wife of 'Abdu'l-Bahá, Munírih K͟hánum. Her station as "foremost woman of the Bahá'í Dispensation" and her rank among women are paralleled only by such heroines of previous Dispensations as Sarah, Ásíyih, the Virgin Mary, Fáṭimih, and Ṭáhirih. For a compilation of Bahá'í writings about Bahíyyih K͟hánum and for some of her own letters, see *Bahíyyih K͟hánum: The Greatest Holy Leaf* (1982).

BAHJÍ *Delight, gladness, joy:* the name of the property north of 'Akká where the Shrine of Bahá'u'lláh is situated and where Bahá'u'lláh lived from 1880 until His ascension in 1892. Its extensive gardens were created by Shoghi Effendi and expanded by the Universal House of Justice. The Shrine of Bahá'u'lláh at Bahjí is the **Qiblih** of the Bahá'í world.

BAYÁN *Exposition, explanation, lucidity, eloquence, utterance:* the title given by the Báb to two of His major works, one in Persian, the other in Arabic. It is also used sometimes to denote the entire body of His writings.

B.E. *Bahá'í Era:* denotes the nineteen-month Badí' calendar, which is reckoned from 21 March 1844, the year of the Báb's declaration of His mission.

BEST BELOVED See **Ancient Beauty.**

BLESSED BEAUTY A translation of *Jamál-i-Mubárak*, a title of Bahá'u'lláh. See also **Ancient Beauty.**

BLESSED PERFECTION A translation of *Jamál-i-Mubárak*, a title of Bahá'u'lláh. See also **Ancient Beauty.**

BOOK OF THE COVENANT A translation of *Kitáb-i-'Ahd* (sometimes referred to as *Kitáb-i-'Ahdí*, meaning "the Book of My Covenant"): Bahá'u'lláh's last will and testament, designated by Him as His "Most Great Tablet" and alluded to by Him as the "Crimson Book." The last Tablet revealed before His ascension, it was written in His own hand and entrusted, shortly before His passing, to His eldest son, 'Abdu'l-Bahá. In it Bahá'u'lláh clearly designates 'Abdu'l-Bahá as His successor and as the Center of His **Covenant**, providing for the continuation of divine authority over the affairs of the Faith in the future.

C

CARMEL See **Mount Carmel.**

CENTER OF THE COVENANT A title of **'Abdu'l-Bahá** referring to His appointment by Bahá'u'lláh as the successor to whom all must turn after Bahá'u'lláh's passing. See also **Covenant; Book of the Covenant.**

CENTRAL FIGURES A collective reference to **Bahá'u'lláh**, the Founder of the Bahá'í Faith; the **Báb**, Forerunner of Bahá'u'lláh and Founder of the Bábí Faith; and **'Abdu'l-Bahá**, authorized Interpreter of the Bahá'í writings.

CHIEF STEWARDS See **Hands of the Cause of God.**

COMMUNITY OF THE MOST GREAT NAME The Bahá'í community. See also **Greatest Name.**

CONCLAVE The consultative meetings of the entire body of the **Hands of the Cause of God.** Following Shoghi Effendi's death in 1957, the meetings were held over a period of several days each autumn from 1957 through 1962, when the Hands were responsible for maintaining the unity of the Bahá'í community and for completing the goals of the Ten Year World Crusade. Conclaves continued to be held for a number of years after the election of the Universal House of Justice and were the occasion of consultations between the House of Justice and the body of the Hands on many matters of great significance for the Bahá'í Faith.

CONCOURSE ON HIGH The company of holy souls of the spiritual world.

CONSTITUTION OF THE UNIVERSAL HOUSE OF JUSTICE A document adopted by the Universal House of Justice on 26 November 1972. It consists of two parts: the Declaration of Trust, which sets forth the origins and duties of the Universal House of Justice, and the By-Laws, which specify the terms under which the Universal House of Justice operates and define its relationship to other institutions of the Bahá'í **Administrative Order.**

CONSULTATION In Bahá'í usage, a technical term referring to the process of collective decision-making. The aim of Bahá'í consultation is to arrive at the best solution or to uncover the truth of a matter. Among the requisites for consultation that are set out in the Bahá'í writings are love, harmony, purity of motive, humility, lowliness, patience, and long-suffering. Individuals not only have the right to express their views, but they are expected to express them fully and with the utmost devotion, courtesy, dignity, care, and moderation. If unanimity is not achieved, decisions are arrived at by majority vote. Once a decision is reached, all parties, having had the opportunity to express their views fully, are to work together wholeheartedly to implement it. If the decision is wrong, 'Abdu'l-Bahá says, through unity the truth will become evident and "the wrong made right."

CONTINENTAL FUND See **Fund.**

CONTINENTAL BOARDS OF COUNSELORS An institution of the Bahá'í **Administrative Order** established by the Universal House of Justice in 1968 to extend into the future the functions of protection and propagation of the Faith assigned to the Hands of the Cause of God by 'Abdu'l-Bahá in His Will and Testament. Its members are appointed to five-year terms by the Universal House of Justice and serve in five zones—Africa, the Americas, Asia, Australasia, and Europe. The **International Teaching Center** coordinates the work of the Continental Boards of Counselors, who are assisted in their work by Auxiliary Board members, whom they appoint and supervise. See also **Auxiliary Boards; Hands of the Cause of God; International Teaching Center.**

CONTINENTAL PIONEER COMMITTEES Responsible for gathering and supplying information for and about pioneers and international traveling teachers. Their work complements the functions of the Continental Boards of Counselors and

National Spiritual Assemblies. There are five such committees: one each for Africa, the Americas, Asia, Australasia, and Europe. Members are appointed by the Universal House of Justice; their work is directed by the International Teaching Center.

COVENANT Generally, an agreement or contract between two or more people, usually formal, solemn, and binding. The Universal House of Justice explains, in a letter dated 23 March 1975, that a religious covenant is "a binding agreement between God and man, whereby God requires of man certain behavior in return for which He guarantees certain blessings, or whereby He gives man certain bounties in return for which He takes from those who accept them an undertaking to behave in a certain way." The Universal House of Justice also explains that there are two types of religious covenant: "There is . . . the Greater Covenant which every Manifestation of God makes with His followers, promising that in the fullness of time a new Manifestation will be sent, and taking from them the undertaking to accept Him when this occurs. There is also the Lesser Covenant that a Manifestation of God makes with His followers that they will accept His appointed successor after Him. If they do so, the Faith can remain united and pure. If not, the Faith becomes divided and its force spent." In the Bahá'í Dispensation the Greater Covenant refers to the renewal of God's ancient Covenant through the appearance of the twin Manifestations of God, the **Báb** and **Bahá'u'lláh**, and the promise of another Manifestation to come in the future after the passage of at least one thousand years. The Lesser Covenant, in this case, refers to Bahá'u'lláh's Covenant with His followers, which establishes **'Abdu'l-Bahá** as the Center of the Covenant. It confers upon 'Abdu'l-Bahá the authority to interpret Bahá'u'lláh's writings in order "to perpetuate the influence" of the Faith and to "insure its integrity, safeguard it from schism, and stimulate its world-wide expansion." The Lesser Covenant also establishes the Guardianship and the Universal House of Justice as the twin successors of Bahá'u'lláh and 'Abdu'l-Bahá.

COVENANT-BREAKER A Bahá'í who attempts to disrupt the unity of the Faith by defying and opposing the authority of Bahá'u'lláh as the Manifestation of God for this Age, or His appointed successor, 'Abdu'l-Bahá, or after Him, the Guardian and the Universal House of Justice. Bahá'ís who continue, despite remonstrances, to violate the Covenant are expelled from the Faith by the Universal House of Justice. This provision preserves the unity of the Faith, which is essential to achieving its cardinal purpose of unifying humankind. It also preserves the purity of Bahá'u'lláh's teachings from the disruptive influence of egoistic individuals who, in past Dispensations, have been responsible for dividing every religion into sects, disrupting its mission, and frustrating to a large degree the intention of its Founder. See also **Covenant.**

CRADLE OF THE FAITH Iran, the homeland of the Bábí and Bahá'í Faiths and of Bahá'u'lláh, the Báb, and 'Abdu'l-Bahá.

CRUSADE, TEN YEAR WORLD The international teaching plan inaugurated by Shoghi Effendi in 1953 and completed in 1963, some six years after his death. It was the first global plan in which all national Bahá'í communities pursued their respective goals in one coordinated effort. It culminated with the first election of the Universal House of Justice at Riḍván 1963.

GLOSSARY

CYCLE A unit of time comprising the Dispensations of numerous consecutive Manifestations of God. For example, the Adamic, or Prophetic, Cycle began with Adam and ended with the Dispensation of Muḥammad. The Bahá'í Cycle began with the Báb and is to last at least five hundred thousand years.

D

DANIEL'S PROPHECY The prophecy contained in Daniel 12:12: "Blessed is he that waiteth, and cometh to the thousand three hundred and five and thirty days." 'Abdu'l-Bahá comments in a Tablet to a Kurdish Bahá'í, "Now concerning the verse in Daniel, the interpretation whereof thou didst ask. . . . These days must be reckoned as solar and not lunar years. For according to this calculation a century will have elapsed from the dawn of the Sun of Truth, then will the teachings of God be firmly established upon the earth, and the Divine Light shall flood the world from the East even unto the West. Then, on this day, will the faithful rejoice!" 'Abdu'l-Bahá further explains in the same Tablet that the 1,335 years must be reckoned from 622 A.D., the year of Muḥammad's flight from Mecca to Medina.

Shoghi Effendi associates Daniel's reference to the 1,335 days and 'Abdu'l-Bahá's statements about the prophecy with the centenary of Bahá'u'lláh's declaration of His mission in 1863 and with the worldwide triumph of the Faith. He stressed that the prophecy refers to occurrences within the Bahá'í community, rather than to events in the outside world.

While Shoghi Effendi clearly allied the Faith's triumph with the successful completion of the third teaching plan to be undertaken by the Bahá'ís, in his letters and in those written on his behalf, four specific dates are mentioned as marking the fulfillment of Daniel's prophecy: 1953, 1957, 1960, and 1963. Regarding the year 1960 (derived by a lunar reckoning), Shoghi Effendi anticipated, in *God Passes By*, p. 151, and in a number of his letters, the successful completion of a third Seven Year Plan that was to be inaugurated. Had there been a third Seven Year Plan, it would have concluded in 1960, one hundred lunar years after Bahá'u'lláh's declaration. When the Ten Year Crusade (1953–63) was announced in 1952, Shoghi Effendi linked its completion with the fulfillment of Daniel's prophecy. There are also several references in letters written on Shoghi Effendi's behalf that give 1957 as the date of the prophecy's fulfillment. In still other letters Shoghi Effendi allies the "hundred lunar years" after Bahá'u'lláh's declaration with the year 1953, although the significance of this hundred years is unclear.

Thus it seems the prophecy is not fulfilled by a single date but, rather, by a process that extended over a period of time. A letter dated 7 March 1955 written on Shoghi Effendi's behalf says, "In the Ten Year Crusade, we are actually fulfilling the prophecy of Daniel, because with the completion of the Ten Year Crusade in 1963 we will have established the Faith in every part of the globe." Thus the fulfillment of the prophecy coincided with the period of the Ten Year Crusade, a span of time that included 1953, 1957, 1960, and 1963.

DAWN-BREAKERS The Bábís and early Bahá'ís, many of whom gave their lives as martyrs.

DAWNING PLACE OF REVELATION A title of **Bahá'u'lláh**, or of any **Manifestation of God**.

DAY OF THE COVENANT 26 November, the day 'Abdu'l-Bahá selected for commemorating the inauguration of Bahá'u'lláh's Covenant. The Bahá'ís wished to celebrate 'Abdu'l-Bahá's birthday, but He did not want this because it coincides with the anniversary of the Declaration of the Báb (23 May), when all attention should be given to that sacred event. He gave them instead the Day of the Covenant to celebrate, choosing a date that is six Gregorian months away from the commemoration of Bahá'u'lláh's Ascension. See also **Covenant**.

DAY OF GOD An expression used variously, according to context, to refer to the appearance of a Manifestation of God, to the duration of His life on earth, or to the duration of His Dispensation. It is also used to refer specifically to the advent of Bahá'u'lláh.

DAY OF JUDGMENT The time of the appearance of the Manifestation of God, when the true character of souls is judged according to their response to His Revelation. Also known as the Day of Resurrection.

DAYSPRING OF DIVINE GUIDANCE A title of **Bahá'u'lláh**.

DISPENSATION The period of time during which the laws and teachings of a Prophet of God have spiritual authority. For example, the Dispensation of Jesus Christ lasted until the beginning of the Muḥammadan Dispensation, usually fixed at the year 622 A.D., the year Muḥammad emigrated from Mecca to Medina. The Islamic Dispensation lasted until the advent of the Báb in 1844. The Dispensation of the Báb ended when Bahá'u'lláh experienced the intimation of His mission in the Síyáh-Chál, the subterranean dungeon in Tehran in which He was imprisoned between August and December 1852. The Dispensation of Bahá'u'lláh will last until the advent of the next Manifestation of God, which Bahá'u'lláh asserts will occur in no less than one thousand years.

DIVINE ESSENCE God.

DIVINE PEN A title of **Bahá'u'lláh**.

DIVINE PLAN The Plan for the dissemination of the Faith of Bahá'u'lláh throughout the world, conceived by 'Abdu'l-Bahá and entrusted to the Bahá'ís of North America in fourteen letters called the Tablets of the Divine Plan. The Divine Plan was implemented by Shoghi Effendi and is pursued today under the guidance of the Universal House of Justice. Teaching Plans undertaken within the framework of the Divine Plan include the first Seven Year Plan (1937–44); the second Seven Year Plan (1946–53) pursued at first by the Bahá'ís of the United States and Canada and extended by supplementary plans adopted with the approval or at the behest of Shoghi Effendi by the British Isles, Egypt and the Sudan, Germany, India, Iran, and Iraq; and the Ten Year World Crusade (1953–63), all of which were inaugurated by Shoghi Effendi, and the Nine, Five, Seven, Six, Three, and Four Year Plans launched by the Universal House of Justice. The Divine Plan is divided into epochs. The first epoch included the years 1937–63. We continue to be in the second epoch. The epochs of the Divine Plan are different from those of the Formative Age. See also **Crusade, Ten Year World; Epochs; Plans; Tablets of the Divine Plan.**

DIVINE THRESHOLD See **Sacred Threshold.**

E

EASTERN PILGRIM HOUSE See **Pilgrim House.**

ELECTIONS, BAHÁ'Í Elections conducted according to Bahá'í principles to select individuals to serve as members of Local and National Spiritual Assemblies and the Universal House of Justice. Elections for Local Spiritual Assemblies are generally held on 21 April, the first day of the Riḍván Festival (21 April–2 May), but in certain circumstances can be held on any day during Riḍván. Elections for National Spiritual Assemblies are held annually during Riḍván. Elections for the Universal House of Justice are held every five years. All adult members in good standing in a Bahá'í community may vote for the members of their Local Spiritual Assembly; Bahá'ís in an electoral unit elect one or more delegates who, in turn, elect the members of the National Spiritual Assembly at the national convention. The members of the National Spiritual Assemblies elect the members of the Universal House of Justice at an international convention. Shoghi Effendi advises electors "to consider without the least trace of passion and prejudice, and irrespective of any material consideration, the names of only those who can best combine the necessary qualities of unquestioned loyalty, of selfless devotion, of a well-trained mind, of recognized ability and mature experience." There are no nominations. Campaigning and electioneering are forbidden. Ballots are cast in a prayerful atmosphere, and the nine persons receiving the most votes are considered chosen by God. Members of a minority race or group are given preference when tied for the ninth position; otherwise, ballots are cast to break the tie. A unique and significant aspect of all Bahá'í elections is the fact that voters elect with the understanding that they are free to choose whomever their consciences prompt them to select, and they freely accept the authority of the outcome.

EPISTLE A formal or elegant letter or treatise; a composition in the form of a letter. In Bahá'í usage it refers to certain writings of Bahá'u'lláh—for example, *Epistle to the Son of the Wolf.*

EPOCHS Major units of time used to mark the unfoldment of the **Divine Plan** and the Formative Age. The first epoch of the Divine Plan included the years 1937–63, for whose execution the machinery of the Administrative Order was erected (1921–37) and during which the Faith was expanded throughout the Western Hemisphere (1937–1944/46) and then extended beyond the Western Hemisphere and Europe to the rest of the world. We are now in the second epoch of the Divine Plan.

The first epoch of the Formative Age (1921–46) began with the passing of 'Abdu'l-Bahá in 1921 and ended with the conclusion of the first Seven Year Plan pursued by the Bahá'ís of North America under Shoghi Effendi's direction. The second epoch of the Formative Age (1946–63) began with the launching of the second Seven Year Plan and the adoption of similar plans by other national communities throughout the Bahá'í world and ended with the conclusion of the Ten Year Crusade and the election of the Universal House of Justice. The third epoch of the Formative Age (1963–86) included the Nine, Five, and Seven Year Plans formulated by the Universal House of Justice. The fourth and current epoch of the Formative Age began in 1986. Its inception was marked by the participation of National Spiritual Assemblies in formulating their own goals for the Six Year Plan (1986–92). For more information on the epochs of the Formative Age, see message no. 451. See also **Ages**.

F

FAST, THE A nineteen-day period (2–21 March, the Bahá'í month of 'Alá', or Loftiness) of spiritual renewal and development during which Bahá'ís abstain from food and drink from sunrise to sunset. A symbol of self-restraint, the Fast is a time of meditation, prayer, and spiritual recuperation and readjustment.

FEAST See **Nineteen Day Feast**.

FIRESIDE An informal Bahá'í gathering held for the purpose of discussing the Bahá'í Faith and sharing its teachings.

FOLLOWERS OF THE GREATEST NAME Bahá'ís; followers of **Bahá'u'lláh**. See also **Greatest Name**.

FORMATIVE AGE See **Ages**.

FRIENDS Bahá'ís.

FUND The institution of the Bahá'í Fund, of which there are four main funds, operates on the international, continental, national, and local levels.

The Bahá'í International Fund is administered by the Universal House of Justice and is used to support the work of the Faith at the Bahá'í World Center and to sustain national communities unable to meet their own expenses. The International Deputization Fund, a subsidiary of the Bahá'í International Fund, supports the work of pioneers and traveling teachers and is administered by the International Teaching Center. The Persian Relief Fund, originally established by the National Spiritual Assembly of Iran to assist victims of persecution by the Islamic Republic, is also a subsidiary of the Bahá'í International Fund and is administered by the Universal House of Justice.

The Continental Bahá'í Fund supports the work of the Continental Boards of Counselors and the work of their Auxiliary Boards.

Each National Spiritual Assembly and Local Spiritual Assembly administers its own National and Local Fund, respectively.

The funds of the Bahá'í Faith are managed according to principles laid down by Bahá'u'lláh, 'Abdu'l-Bahá, and Shoghi Effendi. Foremost among the principles are: (1) Except for the portion of the Bahá'í Funds devoted exclusively to charitable, philanthropic, or humanitarian purposes, contributions are accepted only from those who have identified themselves with the Bahá'í Faith and are regarded as its avowed and unreserved supporters. (2) Contributing to the Funds is both a spiritual privilege and a responsibility. (3) All contributions to the Bahá'í Funds are voluntary. (4) The degree of sacrifice and love of the contributor is more important than the amount given. (5) Appeals for donations must be dignified and general in character. (6) Confidentiality of contributions is to be strictly preserved. (7) Receipts are to be issued. Shoghi Effendi referred to the Funds as "the life-blood" of the Bahá'í institutions.

G

GOD'S HOLY MOUNTAIN See **Mount Carmel.**

GOD'S MAJOR PLAN See **Major Plan of God.**

GOD'S MINOR PLAN See **Minor Plan of God.**

GOLDEN AGE See **Ages.**

GREATEST HOLY LEAF See **Bahíyyih Khánum.**

GREATEST NAME The name **Bahá'u'lláh** ("the Glory of God") and its derivatives, such as **Alláh-u-Abhá** ("God is Most Glorious"), **Bahá** ("glory," "splendor," or "light"), and **Yá Bahá'u'l-Abhá** ("O Thou the Glory of the Most Glorious!"). Also referred to as the Most Great Name.

GUARDIANSHIP The institution, anticipated by Bahá'u'lláh in the Kitáb-i-Aqdas and established by 'Abdu'l-Bahá in His Will and Testament, to which Shoghi Effendi was appointed. Shoghi Effendi explains that the Guardianship and the Universal House of Justice constitute the twin pillars of the **World Order of Bahá'u'lláh** and are the twin successors of Bahá'u'lláh and 'Abdu'l-Bahá. The Guardian's chief functions are to interpret the writings of Bahá'u'lláh, the Báb, and 'Abdu'l-Bahá and to be the permanent head of the Universal House of Justice. See also **Shoghi Effendi.**

H

HANDS OF THE CAUSE OF GOD Eminent Bahá'ís appointed by Bahá'u'lláh to stimulate the propagation and ensure the protection of the Faith. 'Abdu'l-Bahá in His Will and Testament conferred authority on the Guardian to appoint Hands of the Cause and specified their duties. Shoghi Effendi, in a message dated October 1957 to the Bahá'í world, called the Hands of the Cause of God "the Chief Stewards of Bahá'u'lláh's embryonic World Commonwealth." After his death on 4 November 1957 the Hands of the Cause of God assumed responsibility for preserving the unity of the Bahá'í Faith and for guiding the Bahá'í world community to the victorious completion of the Ten Year World Crusade planned by Shoghi Effendi. They also called for the election of the Universal House of Justice in 1963 and requested the friends not to elect them, leaving them free to discharge their own specific responsibilities. Following the formation of the Universal House of Justice, five Hands of the Cause of God were selected by fellow Hands of the Cause to serve at the Bahá'í World Center, while the rest continued their continental responsibilities, which included overseeing the work of the Auxiliary Board members. Finding itself unable to appoint or legislate in order to appoint additional Hands of the Cause of God, the Universal House of Justice, in a cable dated 21 June 1968 and a letter dated 24 June 1968, announced the establishment of the institution of the Continental Boards of Counselors to extend the functions of the Hands of the Cause of God into the future. The Hands of the Cause of God were then freed of responsibility for directing the work of Auxiliary Board members and were all given worldwide responsibilities. See also **Auxiliary Boards; Conclave; Continental Boards of Counselors; International Teaching Center.**

ḤARAM-I-AQDAS *The Most Holy Court:* a designation Shoghi Effendi gave to the northwestern quadrant of the gardens surrounding the Shrine of Bahá'u'lláh in Bahjí. It lies immediately outside the entrance to Bahá'u'lláh's tomb.

ḤAẒÍRATU'L-QUDS *The Sacred Fold:* official title designating the headquarters of Bahá'í administrative activity in a particular country or region.

HIDDEN WORDS, THE Bahá'u'lláh's most important ethical work. Revealed circa 1858. Described by Shoghi Effendi as a "marvelous collection of gem-like utterances . . . with which Bahá'u'lláh was inspired, as He paced, wrapped in His meditations, the banks of the Tigris." Originally designated "The Hidden Book of Fáṭimih," the title of this work is an allusion to the Muslim tradition that the Angel Gabriel revealed a Book to Fáṭimih to console her following the death of the Prophet Muḥammad, her Father, and that this Book remained hidden in the spiritual worlds thereafter.

HEROIC AGE See **Ages.**

HOLY DAY A day commemorating a significant Bahá'í anniversary or feast. The nine Bahá'í holy days on which work should be suspended include:

The Feast of **Naw-Rúz** (New Year), 21 March

The first day of **Riḍván,** 21 April

The ninth day of Riḍván, 29 April

The twelfth day of Riḍván, 2 May

The anniversary of the Declaration of the Báb, 23 May

The anniversary of the Ascension of Bahá'u'lláh, 29 May

The anniversary of the Martyrdom of the Báb, 9 July

The anniversary of the Birth of the Báb, 20 October

The anniversary of the Birth of Bahá'u'lláh, 12 November

The **Day of the Covenant**, 26 November, and the Ascension of 'Abdu'l-Bahá, 28 November, are commemorated annually but are not days on which work is to be suspended.

HOLY PLACES Sites in Iran, Iraq, Turkey, and Israel that are associated with significant events in the lives of Bahá'u'lláh, the Báb, and 'Abdu'l-Bahá.

HOLY SHRINES See **Shrine.**

HOLY TEXTS See **Sacred Scriptures.**

HOLY THRESHOLD See **Sacred Threshold.**

HOSTS 'Abdu'l-Bahá explains that "The blessed Person of the Promised One [Bahá'u'lláh] is interpreted in the Holy Book as the Lord of Hosts—the heavenly armies. By heavenly armies those souls are intended who are entirely freed from the human world, transformed into celestial spirits and have become divine angels. Such souls are the rays of the Sun of Reality who will illumine all the continents" (TDP 8.2).

HOUSE OF 'ABDU'LLÁH PÁSHÁ The house in 'Akká that 'Abdu'l-Bahá rented in 1896 and that served as His residence until He moved to Haifa in 1910. For a discussion of the historical significance of the House, see message no. 157.

HOUSES OF WORSHIP See **Mashriqu'l-Adhkár.**

ḤUQÚQU'LLÁH *The Right of God:* one of the fundamental Bahá'í ordinances of the Bahá'í Faith, it is a great law and a sacred institution laid down by Bahá'u'lláh in the Kitáb-i-Aqdas. It is one of the key instruments for constructing the foundation and supporting the structure of the World Order of Bahá'u'lláh. Its far-reaching ramifications extend from enabling individuals to express their devotion to God in a private act of conscience that attracts divine blessings and bounties for the individual, promotes the common good, and directly connects individuals with the Central Institution of the Faith, to buttressing the authority and extending the activity of the Head of the Faith. The law prescribes that each Bahá'í shall pay a certain portion of his accumulated savings after the deduction of all expenses and of certain exempt properties such as one's residence. These payments provide a fund at the disposition of the Head of the Faith for carrying out beneficent activities. Ḥuqúqu'lláh is administered by the Universal House of Justice, and payments are made to trustees appointed by the Universal House of Justice in every country or region. In providing a regular and systematic source of revenue for the Central Institution of the Cause, Bahá'u'lláh has assured the means for the independence and decisive functioning of the World Center of His Faith. The fundamentals of the law of Ḥuqúqu'lláh are promulgated in the Kitáb-i-Aqdas. Further elaborations of its features are found in other writings of Bahá'u'lláh and in those of 'Abdu'l-Bahá, Shoghi Effendi, and the Universal House of Justice. The law was codified in 1987 and made universally applicable as of Riḍván 1992 to all who profess belief in Bahá'u'lláh. For further information, see *Ḥuqúqu'lláh: Extracts from the Writings of Bahá'u'lláh, 'Abdu'l-Bahá, Shoghi Effendi and The Universal House of Justice* (1986).

I

INTERNATIONAL BAHÁ'Í ARCHIVES An institution at the Bahá'í World Center that preserves the writings and sacred relics of the Central Figures of the Faith and Shoghi Effendi as well as other historical documents and items. The International Archives Building, completed in 1957, was the first of five buildings on the **Arc** on Mount Carmel to be constructed.

INTERNATIONAL BAHÁ'Í CONVENTION An event held every five years in Haifa, Israel, at which members of National Spiritual Assemblies from around the world gather to elect the members of the Universal House of Justice.

INTERNATIONAL BAHÁ'Í COUNCIL A nine-member body that served as the precursor of the Universal House of Justice and was first appointed by Shoghi Effendi in the closing months of 1950 and announced to the Bahá'í world in 1951. After the Guardian's passing, it was elected in 1961 by members of National Spiritual Assemblies. Its threefold function, assigned by Shoghi Effendi, was to forge links with the state of Israel, to assist the Guardian in erecting the superstructure of the Shrine of the Báb, and to negotiate with civil authorities about the application of the Bahá'í laws of personal status for Bahá'ís residing in the Holy Land. After Shoghi Effendi's passing, the International Bahá'í Council, at the request of the Hands of the Cause of God, took responsibility for helping to maintain the World Center properties and helping to establish the Universal House of Justice. Those whom Shoghi Effendi appointed to the International Bahá'í Council were Amatu'l-Bahá Rúḥíyyih Khánum, Charles Mason Remey, Amelia Collins, Ugo Giachery, and Leroy Ioas, who were all subsequently appointed Hands of the Cause of God, as well as Jessie Revell, Ethel Revell, Luṭfu'lláh Ḥakím, and Sylvia Ioas. Members elected in 1961 by the National Spiritual Assemblies were Jessie Revell, 'Alí Nakhjavání, Luṭfu'lláh Ḥakím, Ethel Revell, Charles Wolcott, Sylvia Ioas, Mildred Mottahedeh, Ian Semple, and Borrah Kavelin. The International Bahá'í Council ceased to exist upon the election of the Universal House of Justice in 1963.

INTERNATIONAL FUND See **Fund.**

INTERNATIONAL TEACHING CENTER An institution established by the Universal House of Justice in 1973, the members of which are the Hands of the Cause of God and Counselors appointed by the Universal House of Justice to serve at the Bahá'í World Center. Among the institution's many responsibilities are making reports and recommendations to the Universal House of Justice, coordinating and directing the work of the Continental Boards of Counselors, being fully informed of the Faith's condition throughout the world, watching over the security and ensuring the protection of the Faith, and being alert to possibilities for extending the teaching work and for developing social and economic life both inside and outside the Bahá'í community.

K

KING OF GLORY A title of **Bahá'u'lláh.**

KITÁB-I-'AHD, THE See **Book of the Covenant.**

KITÁB-I-AQDAS, THE *The Most Holy Book* (*Kitáb* means "book"; *Aqdas* means "Most Holy"): the chief repository of Bahá'u'lláh's laws and the Mother Book of His revelation, revealed in 'Akká in 1873 and termed by Shoghi Effendi "the Charter of the future world civilization." For a summary of its contents, see GPB, pp. 214–15.

KNIGHTS OF BAHÁ'U'LLÁH The title given by Shoghi Effendi to Bahá'ís who settled in the goal countries enumerated at the outset of the **Ten Year World Crusade** as having no Bahá'ís living in them. All those who settled in such territories during the Holy Year October 1952–October 1953 and, thereafter, the first to settle in the remaining territories were designated Knights of Bahá'u'lláh. The names of the Knights of Bahá'u'lláh are inscribed on a scroll that was laid beneath the floor inside the entrance door of the Shrine of Bahá'u'lláh in May 1992 during the Holy Year commemorating the centenary of Bahá'u'lláh's ascension.

L

LESSER PEACE The first of two major stages in which Bahá'ís believe peace will be established. The Lesser Peace will come about through a binding treaty among the nations for the political unification of the world. It will involve fixing every nation's boundaries, strictly limiting the size of armaments, laying down the principles underlying the relations among governments, and ascertaining all international agreements and obligations. Its inception will synchronize with two processes operating within the Bahá'í Faith—the maturation of local and national Bahá'í institutions and the completion of specified buildings around the **Arc** on Mount Carmel—and will portend the coming of the **Most Great Peace**.

LETTERS OF THE LIVING A translation of *Ḥurúf-i-Ḥáyy*. The first eighteen people who independently recognized and believed in the Báb. Together with Him, they form the first *Váḥid* ("Unit") of the Bábí Dispensation. The word *Ḥáyy*, which is the Name of God "The Living," has the numerical value of eighteen in the *abjad* system of notation in which each letter of the Arabic alphabet is assigned a specific numerical value. The word "Váḥid" has the numerical value of nineteen.

LIFEBLOOD OF THE CAUSE The Bahá'í **Fund**.

LORD OF HOSTS A title of **Bahá'u'lláh**. 'Abdu'l-Bahá explains that "what is meant in the prophecies by the 'Lord of Hosts' and the 'Promised Christ' is the Blessed Perfection [Bahá'u'lláh] and His holiness the Exalted One [the Báb]." See also **Hosts**.

LORD OF THE AGE A designation of the Manifestation of God in each Dispensation. In Islam it was a title given to the promised Qá'im and, therefore, is applied in Bahá'í terminology particularly to the Báb.

M

MAJOR PLAN OF GOD God's plan for humanity that is tumultuous and mysterious in its progress. Its purpose in this cycle is to unify the human race and to establish the Kingdom of God on earth. See also **Minor Plan of God**.

MANIFESTATION OF GOD Designation of a Prophet "endowed with constancy" Who is the Founder of a religious Dispensation, inasmuch as in His words, His person, and His actions He manifests the nature and purpose of God in accordance with the capacity and needs of the people to whom He comes.

MAS͟HRIQU'L-AD͟HKÁR *The Dawning Place of the Praise of God:* a title designating a Bahá'í House of Worship or Temple. Houses of Worship have been constructed in Wilmette, near Chicago, Illinois; Kampala, Uganda; Ingleside, near Sydney, Australia; Langenhain, near Frankfurt am Main, Germany; Panama City, Panama; Apia, Western Samoa; and New Delhi, India. The first Bahá'í House of Worship, built in 1902 in 'Is͟hqábád, Turkmenistan, was damaged by an earthquake in 1948 and, following heavy rains, had to be razed in 1963. For a full description of the institution of the Mas͟hriqu'l-Ad͟hkár, see BW 18:568–88.

MASTER A title of **'Abdu'l-Bahá** referring to the virtues He manifested and to His role as an enduring model for humanity to emulate.

MAZRA'IH A country mansion near the village of Mazra'a, several miles north of the prison city of 'Akká and about a half-mile from the Mediterranean Sea. Bahá'u'lláh lived at Mazra'ih for about two years after leaving 'Akká in 1877. The mansion looks eastward to the hills of Galilee and has a pool and gardens.

MECCA The holy city of Islam, the birthplace of Muḥammad (570 A.D.). In Mecca, the principal place of pilgrimage of the Muslim world, stands the Great Mosque surrounding the Ka'bih (Kaaba), the ancient cubical temple believed to have been built by Abraham and Ishmael, which is the Muslim **Qiblih.**

MIHDÍ, MÍRZÁ *The Purest Branch:* a son of Bahá'u'lláh and brother of 'Abdu'l-Bahá. He died at the age of twenty-two in 1870 when he fell through a skylight while rapt in prayer on the roof of the prison barracks in 'Akká. He asked Bahá'u'lláh to accept his life as a ransom so that pilgrims prevented from attaining Bahá'u'lláh's presence would be enabled to do so. Bahá'u'lláh, in a prayer, made this astounding proclamation: "Glorified art Thou, O Lord my God! Thou seest Me in the hands of Mine enemies, and My son blood-stained before Thy face, O Thou in Whose hands is the kingdom of all names. I have, O My Lord, offered up that which Thou hast given Me, that Thy servants may be quickened and all that dwell on earth be united."

MINOR PLAN OF GOD The part of God's Plan that is revealed by Bahá'u'lláh to His followers and is laid out for them in detailed instructions and successive plans by 'Abdu'l-Bahá, Shoghi Effendi, and the Universal House of Justice. In contrast to the **Major Plan of God,** it proceeds in a methodical, ordered way, disseminating His teachings and raising up the structure of a united world society.

MOST EXALTED LEAF See **Ásíyih Khánum** and **Bahíyyih Khánum.**

MOST GREAT FESTIVAL See **Riḍván.**

MOST GREAT NAME See **Greatest Name.**

MOST GREAT JUBILEE The centenary of the declaration of Bahá'u'lláh's prophetic mission in the Garden of Riḍván in Baghdad, 22 April–3 May 1863. It was commemorated by the first Bahá'í World Congress, held in Royal Albert Hall, London, during the Riḍván Festival (28 April–2 May) 1963. The Most Great Jubilee coincided with the victorious completion of the Ten Year World Crusade Shoghi Effendi launched in April 1953 (fulfilling the prophecy of Daniel 12:12 regarding the spread of the Bahá'í Faith throughout the world) and the establishment of the Universal House of Justice elected a few days earlier in Haifa, Israel. See also **Daniel's Prophecy.**

MOST GREAT OCEAN See **Ancient Beauty.**

MOST GREAT PEACE The second of two major stages in which Bahá'ís believe peace will be established. The Most Great Peace will be the practical consequence of the spiritualization of the world and the fusion of all its races, creeds, classes, and nations. It will rest on the foundation of, and be preserved by, the ordinances of God. See also **Lesser Peace.**

MOST GREAT PRISON The prison city of 'Akká in which Bahá'u'lláh, His family, and companions were confined from 31 August 1868 until June 1877.

MOST HOLY BOOK See **Kitáb-i-Aqdas, The**.

MOST HOLY SHRINE The Shrine of Bahá'u'lláh in **Bahjí**.

MOTHER TEMPLE Refers to the first Bahá'í House of Worship to be built in a hemisphere or continent. For example, the Bahá'í House of Worship outside of Frankfurt am Main, Germany, is known as the Mother Temple of Europe; the House of Worship in Wilmette, Illinois, the Mother Temple of the West.

MOUNT CARMEL A mountain in Israel on which the Shrine of the Báb and the Bahá'í World Center are located. The home of the prophet Elijah, it is referred to by Bahá'u'lláh as "the Hill of God and His Vineyard" and was extolled by Isaiah as the "mountain of the Lord" to which "all nations shall flow." On it Bahá'u'lláh pitched His tent and revealed the **Tablet of Carmel,** the charter of the world spiritual and administrative centers of the Bahá'í Faith. See also **Arc; Bahá'í World Center.**

MYSTERY OF GOD A translation of *Sirru'lláh,* a title Bahá'u'lláh gave to **'Abdu'l-Bahá** referring to His unique spiritual station in which the incompatible characteristics of human nature and superhuman knowledge and perfection are blended and completely harmonized.

N

NATIONAL SPIRITUAL ASSEMBLIES See **Spiritual Assemblies.**

NATIONAL CONVENTION The institution that elects the members of the National Spiritual Assembly during the annual Riḍván Festival (April 21–May 2). At unit or "district" conventions, adult Bahá'ís elect delegates who, in turn, attend the National Convention. There the delegates vote to elect the members of the National Spiritual Assembly, consult about the affairs of the Faith, and offer recommendations to the National Spiritual Assembly. See also **Elections, Bahá'í.**

NAWNAHÁLÁN *Children:* the name of a Bahá'í-owned commercial investment company founded in 1917 in Iran, through 'Abdu'l-Bahá's encouragement, to help Bahá'í children learn to plan for the future and live thriftily by depositing a portion of their allowances each week. It later became the financial arm of the National Spiritual Assembly of Iran. In 1979 the Nawnahálán Company was seized by Iranian authorities, its assets were confiscated, and its staff were denied their salaries and prevented from working.

NAW-RÚZ *New Day:* Bahá'í New Year's Day, the date of the vernal equinox. A Bahá'í holy day on which work is suspended, it is celebrated in the West on 21 March, until such time as the Universal House of Justice fixes the standard for the date throughout the world in accordance with astronomical data.

NIGHTINGALE OF PARADISE See **Ancient Beauty.**

NINE YEAR PLAN The first teaching plan launched by the **Universal House of Justice**. It encompassed the years 1964–73. See also **Plans**.

NINETEEN DAY FEAST A Bahá'í institution inaugurated by the Báb and confirmed by Bahá'u'lláh in the Kitáb-i-Aqdas. It is held on the first day of every Bahá'í month, each consisting of nineteen days and bearing the name of one of the attributes of God. The Feast is the heart of Bahá'í community life at the local level and consists of devotional, consultative, and social elements.

O

ORDER OF BAHÁ'U'LLÁH See **World Order of Bahá'u'lláh**.

P

PEERLESS BELOVED See **Ancient Beauty**.

PEN OF GLORY, PEN OF THE MOST HIGH See **Ancient Beauty**.

PEOPLE OF BAHÁ Generally, the members of the Bahá'í community. Shoghi Effendi explains that in the Tablet of Carmel "the people of Bahá" refers to the members of the Universal House of Justice.

PILGRIMAGE A journey made with the intention of visiting a shrine or holy place. For Bahá'ís it is both a privilege and an obligation, although it is only obligatory for men who are able to make the journey. In the Kitáb-i-Aqdas Bahá'u'lláh specifically ordains pilgrimage to the House of Bahá'u'lláh in Baghdad and to the House of the Báb in Shiraz. On the day of Bahá'u'lláh's ascension, the room where His Holy Dust was laid became a third center of pilgrimage—the most holy spot and the **Qiblih** of the Bahá'í world—for at least the next thousand years. Under current conditions, Bahá'ís assume that the obligation of pilgrimage is satisfied by a visit to the Shrine of Bahá'u'lláh and the Shrine of the Báb in the Holy Land.

The first group of Western pilgrims arrived in 'Akká on 10 December 1898 and included Edward and Lua Getsinger; Phoebe Hearst; Mrs. Hearst's butler, Robert Turner, who was the first African-American in the West to become a Bahá'í; and Mrs. Thornburgh.

PILGRIM HOUSE A house for visiting pilgrims that Mírzá Ja'far Raḥmání built, with 'Abdu'l-Bahá's permission, near the Shrine of the Báb. 'Abdu'l-Bahá composed a dedicatory inscription that appears above its entrance: "This is a spiritual Hostel for Pilgrims, and its founder is Mírzá Ja'far Raḥmání 1327 A.H. [1909]." It was completed in 1909 and was known as the Eastern or Oriental Pilgrim House. In 1969 the increasing number of pilgrims led the Universal House of Justice to decide that pilgrims should be accommodated in hotels, thereby enabling it to convert the pilgrim house into a reception center.

A Western Pilgrim House was built across the street from the House of 'Abdu'l-Bahá in Haifa, shortly after His passing, with funds American Bahá'ís had contributed and in accordance with a design 'Abdu'l-Bahá had selected and modified. In 1963 the Universal House of Justice established its offices in the Western Pilgrim House.

In 1983, after the completion of the Seat of the Universal House of Justice, the Western Pilgrim House became the seat of the International Teaching Center.

Another pilgrim house is located at Bahjí, near the Shrine of Bahá'u'lláh.

PILLARS OF THE UNIVERSAL HOUSE OF JUSTICE National Spiritual Assemblies. See also **Spiritual Assemblies.**

PIONEERS Bahá'ís who leave their hometown or country to reside elsewhere for the purpose of teaching the Bahá'í Faith.

PLANS Refers to the courses of action devised by Shoghi Effendi and, later, by the Universal House of Justice for expanding and consolidating the Bahá'í Faith within the framework of 'Abdu'l-Bahá's **Divine Plan.** Teaching Plans launched by Shoghi Effendi include the first Seven Year Plan (1937–44) and the second Seven Year Plan (1946–53) pursued by the Bahá'ís of the United States; a Six Year Plan pursued by the Bahá'ís of the British Isles (1944–50); plans of varying durations separately pursued between 1947–53 by the National Spiritual Assemblies of Canada, of Central America, of South America, of Australia and New Zealand, of India, Pakistan, and Burma, of Germany and Austria, of Iran, of Iraq, and of Egypt and the Sudan; the Two Year Plan for the development of the Faith in Africa; and the Ten Year World Crusade (1953–63) pursued by the worldwide Bahá'í community. The Universal House of Justice has launched the Nine Year Plan (1964–73), the Five Year Plan (1974–79), the Seven Year Plan (1979–86), the Six Year Plan (1986–92), the Three Year Plan (1993–1996), and the Four Year Plan (1996–2000). See also **Major Plan of God** and **Minor Plan of God.**

PROMISED ONE A title of **Bahá'u'lláh.**

PUREST BRANCH, THE See **Mihdí, Mírzá.**

Q

QIBLIH *"That which one faces; prayer-direction; point of adoration":* the focus to which the faithful turn in prayer. The Qiblih for Muslims is the Ka'bih in **Mecca**; for Bahá'ís it is the Most Holy Tomb of Bahá'u'lláh at **Bahjí,** "the Heart and Qiblih of the Bahá'í world."

R

REMOTE PRISON The city of Adrianople (now Edirne, Turkey), to which Bahá'u'lláh was banished from 12 December 1863 through 12 August 1868. Adrianople is in western Turkey, on its border with Greece and Bulgaria.

REVELATION The conveying of truth from God to humanity. The word is used to refer to the process of divine communication from God to His Manifestation and from the Manifestation to His people; to the words and acts of such communication themselves; and to the entire body of teachings given by a Prophet of God.

RIḊVÁN The Islamic name of the gardener and custodian of Paradise. In Bahá'í terminology the word denotes both "garden" and "paradise"; however, it has also been used to denote God's good-pleasure and His divine acceptance. The Riḍván Festival, the holiest and most significant of all Bahá'í festivals, commemorates Bahá'u'lláh's declaration of His mission to His companions in the Garden of Riḍván in Baghdad in 1863. It is a twelve-day period celebrated annually, 21 April–2 May. It is also called the Most Great Festival. During each Riḍván Festival Local and National Spiritual Assemblies are elected, and, once every five years, the Universal House of Justice is elected.

RUḤÍYYIH KHÁNUM, AMATU'L-BAHÁ See **Amatu'l-Bahá Rúḥíyyih Khánum.**

S

SACRED SCRIPTURES, SACRED TEXTS The Holy Books of the world's religions. Also refers to the writings of Bahá'u'lláh, the Báb, and 'Abdu'l-Bahá.

SACRED THRESHOLD A term used metaphorically and respectfully to denote approach to the Presence of God and, hence, to the precincts of a holy place such as a shrine. It is also sometimes used literally to denote the actual outer or inner threshold of a holy shrine.

SEAL OF THE PROPHETS A title of Muḥammad.

SHOGHI EFFENDI The title by which Shoghi Rabbání (1 March 1897–4 November 1957), great-grandson of Bahá'u'lláh, is generally known to Bahá'ís. (*Shoghi* is an Arabic name meaning "the one who longs"; *Effendi* is a Turkish honorific signifying "sir" or "master.") He was appointed Guardian of the Bahá'í Faith by 'Abdu'l-Bahá in His Will and assumed the office upon 'Abdu'l-Bahá's passing in 1921.

SHRINE The original meaning of the word is a casket or case for books, but it later acquired the special meaning of a casket containing sacred relics, and thence a tomb of a saint, a chapel with special associations, or a place hallowed by some memory. It is used to denote the latter in Bahá'í terminology. The term "Holy Shrines," for example, refers to the burial places of Bahá'u'lláh, the Báb, and 'Abdu'l-Bahá. The House associated with the visit of 'Abdu'l-Bahá to Montreal was designated by Shoghi Effendi as a Bahá'í shrine. Also, when referring to the All-American Convention held in 1944 to commemorate the one hundredth anniversary of the inception of the Bahá'í Faith, the Guardian wrote of the representatives of the American Bahá'í community's being "Gathered within the walls of its national Shrine—the most sacred Temple ever to be reared to the glory of Bahá'u'lláh" (GPB p. 400).

SÍYÁH-CHÁL *Black Pit:* the subterranean dungeon in Tehran in which Bahá'u'lláh was imprisoned August–December 1852. Here, chained in darkness three flights of stairs underground, in the company of his fellow-Bábís and some 150 thieves and assassins, He received the first intimations of His world mission.

SPIRITUAL ASSEMBLIES Administrative institutions of Bahá'u'lláh's World Order that operate at the local and national levels and are elected according to Bahá'í principles. They are responsible for coordinating and directing the affairs of the Bahá'í community in their areas of jurisdiction. The institution of the Local Spiritual Assembly is ordained by Bahá'u'lláh in the Kitáb-i-Aqdas (referred to there as the "House of Justice"); the institution of the National Spiritual Assembly is established by 'Abdu'l-Bahá in His Will and Testament. The term "Spiritual Assembly" was introduced by 'Abdu'l-Bahá so that, while the Faith is still generally unknown, people will not make the erroneous deduction from the term "House of Justice" that it is a political institution. Regional Spiritual Assemblies have been elected in many areas and are gradually reduced in size and eventually replaced by National Spiritual Assemblies as the Faith expands and consolidates itself. See also **Elections, Bahá'í**.

STEWARDS, CHIEF See **Hands of the Cause of God**.

SUPREME HOUSE OF JUSTICE See **Universal House of Justice**.

SUPREME PEN A title of **Bahá'u'lláh**.

SÚRIY-I-MULÚK *Tablet to the Kings:* revealed by Bahá'u'lláh in Adrianople, referred to by Shoghi Effendi as "the most momentous Tablet revealed by Bahá'u'lláh." In it Bahá'u'lláh addresses collectively the monarchs of East and West, the Sultán of Turkey, the kings of Christendom, the French and Persian ambassadors to the Ottoman Empire, the Muslim clergy in Constantinople, the people of Persia, and the philosophers of the world. In the Súriy-i-Mulúk Bahá'u'lláh unequivocally and forcefully proclaims His station. See *Proclamation of Bahá'u'lláh* (pp. 7–12, 47–54, 102–03) for passages of the Súriy-i-Mulúk that have been translated into English.

T

TABLET OF CARMEL The charter of the world spiritual and administrative centers of the Bahá'í Faith, revealed by Bahá'u'lláh in 1890 during one of His visits to Mount Carmel. See GWB, pp. 14–17, or TB, pp. 3–5.

TABLETS Refers to letters revealed by Bahá'u'lláh, the Báb, and 'Abdu'l-Bahá.

TABLETS OF THE DIVINE PLAN Fourteen Tablets revealed by 'Abdu'l-Bahá in 1916 and 1917 and referred to as the charter for propagating the Bahá'í Faith. Addressed to the Bahá'ís of North America, the Tablets convey His mandate for the transmission of the Bahá'í Faith throughout the world. See also **Divine Plan**.

TEN-PART PROCESS A ten-part process of divine revelation described by Shoghi Effendi. It began with Adam and is to end with the erection of the entire machinery of Bahá'u'lláh's Administrative Order and the suffusion of the light of His Revelation, throughout future epochs of the Formative and Golden Ages of the Faith, over the entire planet.

The ten-part process, Shoghi Effendi writes, began "with the planting in the soil of the divine will, of the tree of divine revelation, and which has already passed through certain stages and must needs pass through still others ere it attains its final consummation. The first part of this process was the slow and steady growth of this tree of divine revelation, successively putting forth its branches, shoots and offshoots, and revealing its leaves, buds and blossoms, as a direct consequence of the light and warmth imparted to it by a series of progressive Dispensations associated with Moses, Zoroaster, Buddha, Jesus, Muḥammad and other Prophets, and of the vernal showers of blood shed by countless martyrs in their path. The second part of this process was the fruition of this tree, 'that belongeth neither to the East nor to the West,' when the Báb appeared as the perfect fruit and declared His mission in the Year Sixty [1844] in the city of Shiraz. The third part was the grinding of this sacred seed, of infinite preciousness and potency, in the mill of adversity, causing it to yield its oil, six years later, in the city of Tabríz [1850]. The fourth part was the ignition of this oil by the hand of Providence in the depths and amidst the darkness of the Síyáh-Chál of Tehran a hundred years ago [1852]. The fifth, was the clothing of that flickering light, which had scarcely penetrated the adjoining territory of Iraq, in the lamp of revelation, after an eclipse lasting no less than ten years, in the city of Baghdad [1863]. The sixth, was the spread of the radiance of that light, shining with added brilliancy in its crystal globe in Adrianople [1863–1868], and later on in the fortress town of 'Akká [1868–1877], to thirteen countries in the Asiatic and African continents. The seventh was its projection, from the Most Great Prison, in the course of the ministry of the Center of the Covenant [1892–1921], across the seas and the shedding of its illumination upon twenty sovereign states and dependencies in the American, the European, and Australian continents. The eighth part of that process was the diffusion of that same light in the course of the first, and the opening years of the second, epoch of the Formative Age of the Faith [1921–1953], over ninety-four sovereign states, dependencies and islands of the planet, as a result of the prosecution of a series of national plans, initiated by eleven national spiritual assemblies throughout the Bahá'í world, utilizing the agencies of a newly emerged, divinely appointed Administrative Order, and which has now culminated in the one hundredth anniversary of the birth of Bahá'u'lláh's Mission. The ninth part of this process—the stage we are now entering [1953]—is the further diffusion of that same light over one hundred and thirty-one additional territories and islands in both the Eastern and Western Hemispheres, through the operation of a decade-long world spiritual crusade whose termination will, God willing, coincide with the Most Great Jubilee commemorating the centenary of the declaration of Bahá'u'lláh in Baghdad. And finally the tenth part of this mighty process [1963–] must be the penetration of that light, in the course of numerous crusades and of successive epochs of both the Formative and Golden Ages of the Faith, into all the remaining territories of the globe through the erection of the entire machinery of Bahá'u'lláh's Administrative Order in all territories, both East and West, the stage at which the light of God's triumphant Faith shining in all its power and glory will have suffused and enveloped the entire planet" (MBW, pp. 154–55).

TEN YEAR WORLD CRUSADE See **Crusade, Ten Year World.**

TONGUE OF GRANDEUR See **Ancient Beauty.**

TRANSITION, AGE OF or **PERIOD OF** See **Ages.**

TWIN HOLY CITIES Haifa and 'Akká, Israel.

TWIN HOLY SHRINES The Shrines of Bahá'u'lláh and the Báb. See also **Shrines.**

U

UMÁNÁ COMPANY The Umáná (Trustees') Company was formed in Iran to be the registered owner of all Bahá'í properties, both local and national. Within a few months after the Islamic Republic was established in March 1979, agents of the revolutionary government, acting on the authority of an order from the Office of the Public Prosecutor of the Islamic Republic, seized its assets, properties, and furnishings and dismissed its employees. Among the properties the Umáná Company held were holy places, including the House of the Báb in Shiraz, houses of Bahá'u'lláh in Tehran and Tákur, a temple site, the national Bahá'í headquarters, Bahá'í cemeteries, and a Bahá'í-operated hospital in Tehran.

UNIVERSAL HOUSE OF JUSTICE The supreme governing and legislative body of the Bahá'í Faith. The Guardianship and the Universal House of Justice are the twin, crowning institutions of the Bahá'í Administrative Order. Elected every five years at an international Bahá'í convention, the Universal House of Justice gives spiritual guidance to and directs the administrative activities of the worldwide Bahá'í community. It is the institution Bahá'u'lláh ordained as the agency invested with authority to legislate on matters not covered in His writings. In His Will and Testament 'Abdu'l-Bahá elaborates on its functions and affirms that it is infallibly guided.

V

VAKÍLU'D-DAWLIH A state title, meaning deputy or representative of the government, that refers to Muḥammad Taqí Afnán, cousin of the Báb and chief builder of the 'Ishqábád Temple.

VICAR OF THE PROPHET OF ISLAM The Sulṭán of the Ottoman Empire, alluding to his claim to be the Caliph, or successor of Muḥammad within Sunní Islam.

W

WESTERN PILGRIM HOUSE See **Pilgrim House.**

WILL AND TESTAMENT OF 'ABDU'L-BAHÁ A document that Shoghi Effendi says is "unique in the annals of the world's religious systems," the Will and Testament is the charter of the Administrative Order of the Bahá'í Faith. Written, signed, and sealed by 'Abdu'l-Bahá, the Will and Testament consists of three sections written at three different times between 1901 and the year of His passing. The Will and Testament affirms "the two-fold character of the Mission of the Báb," which was to bring an independent revelation from God and to herald the coming of another, greater revelation through Bahá'u'lláh. It also "discloses the full station of" Bahá'u'lláh as the "Supreme Manifestation of God," declares the fundamental beliefs of the Bahá'í Faith, establishes the institution of the Guardianship, and appoints Shoghi Effendi as Guardian. It provides for the election of the Universal House of Justice and defines its scope. It also creates the institution of the National Spiritual Assembly, provides for the appointment of the Hands of the Cause of God and prescribes their obligations, and exposes the conduct of the Covenant-breakers.

WORLD CENTER See **Bahá'í World Center.**

WORLD CONGRESS See **Most Great Jubilee.**

WORLD CRUSADE See **Crusade, Ten Year World.**

WORLD ORDER OF BAHÁ'U'LLÁH Bahá'u'lláh's "scheme for world-wide solidarity" that is "destined to embrace in the fullness of time the whole of mankind." The current Bahá'í Administrative Order is its nucleus and pattern, providing the "rudiments of the future all-enfolding Bahá'í Commonwealth."

Y

YÁ 'ALÍYYU'L-A'LÁ *O Thou the Exalted, the Most Exalted!* A form of the **Báb**'s name that is used as an invocation.

YÁ BAHÁ'U'L-ABHÁ *O Thou the Glory of the Most Glorious!* A form of **Bahá'u'lláh**'s name (the **Greatest Name**) that is used as an invocation.

Bibliography

Bibliography

Writings of Bahá'u'lláh

A Synopsis and Codification of the Kitáb-i-Aqdas: The Most Holy Book of Bahá'u'lláh. [Compiled by the Universal House of Justice.] Haifa: Bahá'í World Centre, 1973.

Epistle to the Son of the Wolf. 1st ps ed. Translated by Shoghi Effendi. Wilmette, Ill.: Bahá'í Publishing Trust, 1988.

Gleanings from the Writings of Bahá'u'lláh. 1st ps ed. Translated by Shoghi Effendi. Wilmette, Ill.: Bahá'í Publishing Trust, 1983.

The Hidden Words. Translated by Shoghi Effendi. Rev. ed. Wilmette, Ill.: Bahá'í Publishing Trust, 1939.

The Kitáb-i-Aqdas: The Most Holy Book. ps ed. Wilmette, Ill.: Bahá'í Publishing Trust, 1993.

The Kitáb-i-Íqán: The Book of Certitude. 1st ps ed. Translated by Shoghi Effendi. Wilmette, Ill.: Bahá'í Publishing Trust, 1983.

Prayers and Meditations. Translated by Shoghi Effendi. 1st ps ed. Wilmette, Ill.: Bahá'í Publishing Trust, 1987.

The Proclamation of Bahá'u'lláh to the Kings and Leaders of the World. Haifa: Bahá'í World Centre, 1967.

Tablets of Bahá'u'lláh revealed after the Kitáb-i-Aqdas. Compiled by the Research Department of the Universal House of Justice. Translated by Habib Taherzadeh et al. 1st ps ed. Wilmette, Ill.: Bahá'í Publishing Trust, 1988.

Writings of the Báb

Selections from the Writings of the Báb. Compiled by the Research Department of the Universal House of Justice. Translated by Habib Taherzadeh et al. Haifa: Bahá'í World Centre, 1976.

Writings of 'Abdu'l-Bahá

'Abdu'l-Bahá in London: Addresses and Notes of Conversations. Commemorative ed. London: Bahá'í Publishing Trust, 1982.

Foundations of World Unity: Compiled from Addresses and Tablets of 'Abdu'l-Bahá. Wilmette, Ill.: Bahá'í Publishing Trust, 1972.

Memorials of the Faithful. New ed. Translated by Marzieh Gail. Wilmette, Ill.: Bahá'í Publishing Trust, 1996.

Paris Talks: Addresses Given by 'Abdu'l-Bahá in Paris in 1911–1912. 11th ed. London: Bahá'í Publishing Trust, 1969.

The Promulgation of Universal Peace: Talks Delivered by 'Abdu'l-Bahá during His Visit to the United States and Canada in 1912. 2d ed. Compiled by Howard MacNutt. Wilmette, Ill.: Bahá'í Publishing Trust, 1982.

The Secret of Divine Civilization. 1st ps ed. Translated by Marzieh Gail and Ali-Kuli Khan. Wilmette, Ill.: Bahá'í Publishing Trust, 1990.

Selections from the Writings of 'Abdu'l-Bahá. Compiled by the Research Department of the Universal House of Justice. Translated by a Committee at the Bahá'í World Centre and Marzieh Gail. Haifa: Bahá'í World Centre, 1978.

Some Answered Questions. 1st ps ed. Compiled and translated by Laura Clifford Barney. Wilmette, Ill.: Bahá'í Publishing Trust, 1984.

Tablets of Abdul-Baha Abbas. 3 vols. New York: Bahai Publishing Society, 1909–16.

Tablets of the Divine Plan: Revealed by 'Abdu'l-Bahá to the North American Bahá'ís. 1st ps ed. Wilmette, Ill.: Bahá'í Publishing Trust, 1993.

A Traveler's Narrative Written to Illustrate the Episode of the Báb. Translated by Edward G. Browne. New and corrected ed. Wilmette, Ill.: Bahá'í Publishing Trust, 1980.

Will and Testament of 'Abdu'l-Bahá. Wilmette, Ill.: Bahá'í Publishing Trust, 1944.

Writings of Shoghi Effendi

The Advent of Divine Justice. 1st ps ed. Wilmette, Ill.: Bahá'í Publishing Trust, 1990.

Bahá'í Administration: Selected Messages, 1922–1932. 7th ed. Wilmette, Ill.: Bahá'í Publishing Trust, 1974.

Citadel of Faith: Messages to America, 1947–1957. Wilmette, Ill.: Bahá'í Publishing Trust, 1965.

Dawn of a New Day. New Delhi: Bahá'í Publishing Trust, [1970].

Directives from the Guardian. New Delhi: Bahá'í Publishing Trust, [n.d.].

God Passes By. Rev. ed. Wilmette, Ill.: Bahá'í Publishing Trust, 1974.

Guidance For Today and Tomorrow: A Selection from the Writings of Shoghi Effendi, the Late Guardian of the Bahá'í Faith. London: Bahá'í Publishing Trust, 1953.

High Endeavours: Messages to Alaska. Compiled by the National Spiritual Assembly of the Bahá'ís of Alaska. [N.p.]: National Spiritual Assembly of the Bahá'ís of Alaska, 1976.

Letters from the Guardian to Australia and New Zealand, 1923–1957. [Australia]: National Spiritual Assembly of the Bahá'ís of Australia, 1970.

The Light of Divine Guidance: The Messages from the Guardian of the Bahá'í Faith to the Bahá'ís of Germany and Austria. Langenhain, West Germany: National Spiritual Assembly of the Bahá'ís of Germany, 1982.

The Light of Divine Guidance: Letters from the Guardian of the Bahá'í Faith to Individual Believers, Groups and Bahá'í Communities in Germany and Austria. Lan-

genhain, West Germany: National Spiritual Assembly of the Bahá'ís of Germany, 1985.
Messages to America: Selected Letters and Cablegrams Addressed to the Bahá'ís of North America, 1932–1946. Wilmette, Ill.: Bahá'í Publishing Committee, 1947.
Messages to the Bahá'í World, 1950–1957. Rev. ed. Wilmette, Ill.: Bahá'í Publishing Trust, 1971.
Messages to Canada. N.p.: National Spiritual Assembly of the Bahá'ís of Canada, 1965.
The Promised Day Is Come. 1st ps ed. Wilmette, Ill.: Bahá'í Publishing Trust, 1996.
Tawqí'át-i-Mubárakih ["Blessed Letters"]. Vol. 1. Tehran: Bahá'í Publishing Trust, 1972.
The Unfolding Destiny of the British Bahá'í Community: The Messages from the Guardian of the Bahá'í Faith to the Bahá'ís of the British Isles. London: Bahá'í Publishing Trust, 1981.
The World Order of Bahá'u'lláh: Selected Letters. New ed. Wilmette, Ill.: Bahá'í Publishing Trust, 1991.

Writings of the Universal House of Justice

The Constitution of the Universal House of Justice. Haifa: Bahá'í World Centre, 1972.
Messages from the Universal House of Justice, 1968–1973. Wilmette, Ill.: Bahá'í Publishing Trust, 1976.
The Promise of World Peace: To the Peoples of the World. Wilmette, Ill.: Bahá'í Publishing Trust, 1985.
Wellspring of Guidance: Messages, 1963–1968. 2d ed. Wilmette, Ill.: Bahá'í Publishing Trust, 1976.

Compilations of Bahá'í Writings

Bahá'í Holy Places at the World Centre. Haifa: Bahá'í World Centre, 1968.
Bahai Scriptures. 2d ed. Compiled by Horace Holley. New York: Baha'i Publishing Committee, 1928.
Bahá'u'lláh and 'Abdu'l-Bahá. *Bahá'í Prayers and Tablets for the Young.* Wilmette, Ill.: Bahá'í Publishing Trust, 1978.
———. *Let Thy breeze refresh them . . .* [Compiled by the Bahá'í Publishing Trust of the United Kingdom.] Oakham, England: Bahá'í Publishing Trust, 1976.
———. *Tablets Revealed in Honor of the Greatest Holy Leaf.* New York: National Spiritual Assembly of the Bahá'ís of the United States and Canada, 1933.
Bahá'u'lláh, 'Abdu'l-Bahá, and Shoghi Effendi. *Bahá'í Meetings/The Nineteen Day Feast.* Compiled by the Universal House of Justice. Wilmette, Ill.: Bahá'í Publishing Trust, 1976.
———. *Excellence in All Things.* Compiled by the Research Department of the Universal House of Justice. Oakham, England: Bahá'í Publishing Trust, 1981.

———. *The Importance of Deepening our Knowledge and Understanding of the Faith.* Compiled by The Universal House of Justice. Wilmette, Ill.: Bahá'í Publishing Trust, 1983.

———. *The Pattern of Bahá'í Life: A Compilation from Bahá'í Scripture with some passages from the writings of the Guardian of the Bahá'í Faith.* Compiled by the National Spiritual Assembly of the Bahá'ís of the United Kingdom. 3d ed. London: Bahá'í Publishing Trust, 1963.

———. *Spiritual Foundations: Prayer, Meditation, and the Devotional Attitude: Extracts from the Writings of Bahá'u'lláh, 'Abdu'l-Bahá, and Shoghi Effendi.* Compiled by the Research Department of the Universal House of Justice. Wilmette, Ill.: Bahá'í Publishing Trust, 1980.

Bahá'u'lláh, 'Abdu'l-Bahá, Shoghi Effendi, and the Universal House of Justice. *Consultation: A Compilation: Extracts from the Writings and Utterances of Bahá'u'lláh, 'Abdu'l-Bahá, Shoghi Effendi, and The Universal House of Justice.* Compiled by the Research Department of the Universal House of Justice. Thornhill, Ont.: Bahá'í Community of Canada, 1980.

Bahá'u'lláh, the Báb, and 'Abdu'l-Bahá. *Bahá'í Prayers: A Selection of Prayers Revealed by Bahá'u'lláh, the Báb, and 'Abdu'l-Bahá.* New ed. Wilmette, Ill.: Bahá'í Publishing Trust, 1991.

———. *Inspiring the Heart: Selections from the Writings of The Báb, Bahá'u'lláh and 'Abdu'l-Bahá.* [Compiled by the Universal House of Justice.] London: Bahá'í Publishing Trust, n.d.

[Bahá'u'lláh, the Báb, 'Abdu'l-Bahá, Shoghi Effendi, and the Universal House of Justice.] *Bahá'í Marriage and Family Life: Selections from the Writings of the Bahá'í Faith.* [Compiled by the National Spiritual Assembly of the Bahá'ís of Canada.] N.p.: National Spiritual Assembly of the Bahá'ís of Canada, 1983.

Bahá'u'lláh, 'Abdu'l-Bahá, Shoghi Effendi, and Bahíyyih <u>Kh</u>ánum. *Bahíyyih <u>Kh</u>ánum: The Greatest Holy Leaf.* Compiled by the Research Department at the Bahá'í World Centre. Haifa: Bahá'í World Centre, 1982.

The Compilation of Compilations: Prepared by the Universal House of Justice, 1963–1990. 2 vols. Australia: Bahá'í Publications Australia, 1991.

Lights of Guidance: A Bahá'í Reference File. New ed. Compiled by Helen Hornby. New Delhi: Bahá'í Publishing Trust, 1994.

Living the Life: A Compilation. [Compiled by the National Spiritual Assembly of the Bahá'ís of the United Kingdom.] London: Bahá'í Publishing Trust, 1974.

The Mystery of God. Rev. ed. Compiled by Írán Fúrútan Muhájir. London: Bahá'í Publishing Trust, 1979.

Principles of Bahá'í Administration: A Compilation. 4th ed. [Compiled by the National Spiritual Assembly of the Bahá'ís of the United Kingdom.] London: Bahá'í Publishing Trust, 1976.

Other Works

The Bahá'í World: A Biennial International Record, Volume IX, 1940–1944. Compiled by the National Spiritual Assembly of the Bahá'ís of the United States and Canada. Wilmette, Ill.: Bahá'í Publishing Committee, 1945.

The Bahá'í World: A Biennial International Record, Volume X, 1944–1946. Compiled by the National Spiritual Assembly of the Bahá'ís of the United States and Canada. Wilmette, Ill.: Bahá'í Publishing Committee, 1949.

The Bahá'í World: A Biennial International Record, Volume XI, 1946–1950. Compiled by the National Spiritual Assembly of the Bahá'ís of the United States and Canada. Wilmette, Ill.: Bahá'í Publishing Committee, 1952.

The Bahá'í World: A Biennial International Record, Volume XII, 1950–1954. Compiled by the National Spiritual Assembly of the Bahá'ís of the United States. Wilmette, Ill.: Bahá'í Publishing Committee, 1956.

The Bahá'í World: An International Record, Volume XIII, 1954–1963. Compiled by the Universal House of Justice. Haifa: The Universal House of Justice, 1970.

The Bahá'í World: An International Record, Volume XIV, 1963–1968. Compiled by the Universal House of Justice. Haifa: The Universal House of Justice, 1974.

The Bahá'í World: An International Record, Volume XV, 1968–1973. Compiled by the Universal House of Justice. Haifa: Bahá'í World Centre, 1975.

The Bahá'í World: An International Record, Volume XVI, 1973–1976. Compiled by the Universal House of Justice. Haifa: Bahá'í World Centre, 1978.

The Bahá'í World: An International Record, Volume XVII, 1976–1979. Compiled by the Universal House of Justice. Haifa: Bahá'í World Centre, 1981.

The Bahá'í World: An International Record, Volume XVIII, 1979–1983. Compiled by the Universal House of Justice. Haifa: Bahá'í World Centre, 1986.

The Bahá'í World: An International Record, Volume XIX, 1983–1986. Compiled by the Universal House of Justice. Haifa: Bahá'í World Centre, 1994.

Balyuzi, H. M. *'Abdu'l-Bahá: The Centre of the Covenant of Bahá'u'lláh.* London: George Ronald, 1971.

———. *Bahá'u'lláh: The King of Glory.* Oxford: George Ronald, 1980.

———. *Edward Granville Browne and the Bahá'í Faith.* Oxford: George Ronald, 1970.

Blomfield, Lady (Sitárih Khánum). *The Chosen Highway.* Wilmette, Ill.: Bahá'í Publishing Trust, n.d.; repr. 1975.

Chase, Thornton. *In Galilee.* Los Angeles: Kalimát Press, 1985.

Esslemont, J. E. *Bahá'u'lláh and the New Era: An Introduction to the Bahá'í Faith.* 5th rev. ed. Wilmette, Ill.: Bahá'í Publishing Trust, 1980.

Gail, Marzieh. *Khánum: The Greatest Holy Leaf.* Oxford: George Ronald, 1981.

Garis, M. R. *Martha Root: Lioness at the Threshold.* Wilmette, Ill.: Bahá'í Publishing Trust, 1983.

Maxwell, May. *An Early Pilgrimage.* 2d rev. ed. London: George Ronald, 1969.

Moffett, Ruth J. *Du'á: On Wings of Prayer.* Revised and edited by Keven Brown. Happy Camp, Cal.: Naturegraph Publishers, 1984.

Nabíl-i-A'zam (Muhammad-i-Zarandí). *The Dawn-Breakers: Nabíl's Narrative of the Early Days of the Bahá'í Revelation.* Translated and edited by Shoghi Effendi. Wilmette, Ill.: Bahá'í Publishing Trust, 1932.

Nakhjavání, Bahíyyih. *When We Grow Up.* Oxford: George Ronald, 1979.

National Spiritual Assembly of the Bahá'ís of the United States. *Articles of Incorporation, Constitution, and By-Laws of the National Spiritual Assembly of the Bahá'ís of the United States/By-Laws of a Local Spiritual Assembly.* Wilmette, Ill.: Bahá'í Publishing Trust, 1996.

———, comp. *The Bahá'í Community: A Summary of Its Organization and Laws.* Rev. ed. Wilmette, Ill.: Bahá'í Publishing Trust, 1963.

———. *Declaration of Trust and By-Laws of the National Spiritual Assembly of the Bahá'ís of the United States/By-Laws of a Local Spiritual Assembly.* Rev. ed. Wilmette, Ill.: Bahá'í Publishing Trust, 1975.

Rabbaní, Rúhíyyih. *The Priceless Pearl.* London: Bahá'í Publishing Trust, 1969.

Ruhe, David S. *Door of Hope: A Century of the Bahá'í Faith in the Holy Land.* Oxford: George Ronald, 1983.

Sears, William. *A Cry from the Heart: The Bahá'ís in Iran.* Oxford: George Ronald, 1982.

Whitehead, O. Z. *Some Bahá'ís to Remember.* Oxford: George Ronald, 1983.

Index of Names

Index of Names

'Abbásí, Muḥammad, 331.1
'Abbásíyán, Yúsif, 307.3
'Abdu'l-Bahá. *See* entry in General Index.
'Abdu'lláh Páshá, 154
Áfáqí, Ṣábir, 325.1, 439.2c
Afnán, Bahrám, 363.1
Afshín, Burhání'd-Dín, 169.1, 267.3c, 439.2c
Ahdieh, Hushang, 132.8–9, 267.3a, 439.2a
Aḥmadíyyih, Hidáyatu'lláh, 180.1, 267.3b, 439.2b
Aḥrárí, Ḍíyá'u'lláh, 345.1
Akhtar-Khávarí, Núru'lláh, 261.2
'Alá'í, Shu'á'u'lláh, (Hand of the Cause of God) 104.1, 414.1
'Alá'í, Suhayl, 60.1, 130.3, 132.25, 267.3d, 439.2d
'Alavíyán, Buzurg, 286.1
Alexander, Agnes, (Hand of the Cause of God) 90.1, 96.7, 98.3, 403.2
'Alízádih, Maqṣúd, 396.14
Amánat, Ḥusayn, 136.1, 140.1, 210.1
Amín, Mihdí Amín, 307.1
Amíní, Naṣru'lláh, 328.1
Anello, Eloy, 439.2b
Anvarí, Midhí, 277.1
Appa, Seewoosumbur-Jeehoba, 60.1, 132.25, 267.4a
Áqá, Ismá'il, 157.5b
Arbáb, Farzam, 267.3b, 439.2b
Ardebili, Ayatollah Musavi, 309.1, 310.1
Ardikání, Ḥusáyn, 60.1, 132.25, 153.3, 267.3a, 439.25
Armstrong, Alfred, 132.25
Armstrong, Leonora Holsapple, 132.8, 132.25, 185.3
Armstrong, Leonora Stirling, 265.1
Árnadóttir, Hólmfríður, 101.1, 171.2
Asadu'lláhzádih, Ḥusayn, 292.1
Asadu'lláhzádih, Shívá Maḥmúdí, 311.1

Asadyárí, 'Abdu'l-'Alí, 292.1
Asare, Beatrice O., 439.2a
Aṣdaqí, Farhád, 417.1
Ashjárí, 'Azízu'lláh, 441.1
Ashraf, Jadídu'lláh, 331.1
Ashraf, Riḍvánu'lláh, 198.1
Ástání, Yadu'lláh, 254.1
Atharí, Fírúz, 416.1
'Aṭifí, Bahman, 296.1
'Aṭifí, 'Izzat, 296.1
Awjí, Ḥabíbu'lláh, 344.2, 345.1
Ayadi, Dr., 239.2–3
Ayala, Ben, 267.3d, 439.2d
Aydilkhání, 'Abbás, 437.1
Ayman, Iraj, 81.1, 132.25, 267.4a, 439.2c
Ázádí, 'Abdu'l-Ḥusayn, 363.1
A'ẓámí, Ghulám-Ḥusayn, 254.1
Azharra, General Gholam Reza, 215.1
'Azízí, Ḥabíbu'lláh, 296.1
'Azízí, Iskandar, 311.1
'Azízí, Jalál, 307.1

Báb, the. *See* entry in General Index.
Bábázádih, Sa'du'lláh, 328.1
Bahá'í, Ḥusníyyih, 440.5
Bahá'u'lláh. *See* entry in General Index.
Báhirí, Mihdí, 292.1
Bahíyyih Khánum. *See* entry in General Index.
Bahrámsháhí, Rúḥu'lláh, 424.1
Bahta, Gila Michael, 366.1, 439.2a
Baker, Dorothy, (Hand of the Cause of God) 63.7
Bakhtávar, Kamáli'd-Dín, 290
Ballerio, Giovanni, 283.1
Balyuzi, Hasan M., (Hand of the Cause of God) 247.1, 251.3
Banání, Músá, (Hand of the Cause of God) 88.2, 102.1
Bani-Sadr, Abol Hasan (pres., Iran), 262.5

767

INDEX OF NAMES

Baqá, S͟hídruk͟h Amír-Kíyá, 311.1
Barney, Laura. *See* Dreyfus-Barney, Laura.
Bas͟hírí, Aḥmad, 368.2, 413.1–2
Battrick, A. Owen, 169.1, 267.3d, 439.4
Benson, Richard, 132.8, 132.25, 267.3d, 439.2d
Blumenthal, Erik, 60.1, 132.25, 267.3e, 439.4
Boman, Shirin, 70.2, 132.25, 267.3c, 439.4
Bopp, Anneliese, 81.1, 132.25, 230.1, 361.3
Brady, Wilma, 439.2b

Carney, Magdalene, 361.3
Chance, Hugh, p. v, 56.1, 129.1, 209.1, 359.1
Chase, Thornton, 157.10
Chinniah, Inparaju, 200.1, 245.1
Collins, Amelia, (Hand of the Cause of God) 101.1
Costas, Athos, 60.1, 132.25, 267.3b, 439.4
Cowan, Angus, 180.1, 267.3b, 439.4, 454.1

Dávúdí, 'Alímurád, 307.3
Dak͟hílí, Masrúr, 292.1
Dálvand, S͟hahín (S͟hírín), 364.1
De Burafato, Carmen, 60.1, 132.25, 267.3b, 439.4
De Calderón, Isabel P., 382.1, 439.2b
D͟habíḥíyán, 'Azízu'lláh, 261.2
Dihqání, Bahman, 387.1
Dihqání, Hidáyatu'lláh, 277.1
Ḍíyá'í, Nuṣratu'lláh, 392.1
Ḍíyá'u'lláh, Mírzá, 26.1
Dreyfus-Barney, Laura, 150.1, 174.1
Dunbar, Hooper, 60.1, 131.1, 132.5, 361.3
Dunn, Clara, (Hand of the Cause of God) 98.3, 184.2
Dunn, Hyde, (Hand of the Cause of God) 98.3, 184.2

Ekpe, Friday, 132.8, 132.25, 153.3, 267.3a, 439.2a
Elder, Earl E., 251.3
Epyeru, Oloro, 60.1, 132.25, 267.3a, 439.2a
Eshraghian, Kamran, 439.2d

Esslemont, John E., (Hand of the Cause of God) 64.1
Estall, Rowland, 132.8, 132.25, 267.4a

Faizi, Abu'l-Qasim, (Hand of the Cause of God) 24.17, 42.17, 163.1, 269.1
Fareed, Bijan, 439.2c
Farhand, G͟hulám Ḥusayn, 416.1
Farhangí, Masíḥ, 60.1, 132.25, 267.3c, 287.1, 325.1
Farid, Badí'u'lláh, 287.1
Farídání, Firaydún, 261.2
Farnús͟h, Hás͟him, 286.1
Farúhar, Maḥmúd, 323.3
Farúhar, Is͟hráqíyyih, 323.3
Farzánih-Mu'ayyad, Manúc͟hihr, 331.1
Fat'he-Aazam, Shidan, 60.1, 267.3a, 439.2a
Fatheazam, Hushmand, p. v, 56.1, 129.1, 209.1, 359.1
Featherstone, Collis, (Hand of the Cause of God) 91.1, 104.1, 163.1, 288.1, 361.2, 447.2
Ferraby, John, (Hand of the Cause of God) 135.1
Ferraby, Dorothy, 60.1, 132.25, 267.3e, 439.4
Ficicchia, Francesco, 251.1–2
Firdawsí, Fatḥu'lláh, 311.1
Fírúzí, Parvíz, 292.1
Fírúzí, Riḍá, 268.1
Fofana, Kassimi, 366.1, 439.2a
Ford, Bahíyyih, 60.1. *See also* Winckler, Bahíyyih.
Ford, Mary Hanford, 157.11
Fozdar, John, 267.3c, 439.2c
Furúg͟hí, Parvíz, 121.2
Furúḥí, 'Izzatu'lláh, 307.1
Furútan, 'Alí-Akbar, (Hand of the Cause of God) 24.17, 42.17, 104.1, 163.1, 354.2, 361.3, 440.5, 447.2

Gardner, Lloyd, 60.1, 132.25, 267.3b, 423.1
Ghaznavi, Agnes, 267.3e, 439.2e
Ghodoussi, Ayatollah, 262.5
Giachery, Ugo, (Hand of the Cause of God) 24.17, 42.17, 104.1, 110.2, 130.2, 163.1, 177.1, 288.1, 319.1, 447.2

768

INDEX OF NAMES

Gibson, Amoz, p. v, 56.1, 129.1, 209.1, 326.1
Greatest Holy Leaf. *See* entry in General Index.
Grossmann, Hartmut, 267.3e, 439.2e
Grossmann, Hermann, (Hand of the Cause of God) 62.1
Gulmuḥammadí, Zabíḥu'lláh, 267.3c, 439.2c
Gulsẖaní, 'Azízu'lláh, 322.1, 323.3
Güney, Aydin, 267.3c, 439.4

Habíbí, Muḥammad (Suhráb), 285.2
Habíbí, Muḥammad-Báqir (Suhayl), 285.2
Ḥakím, Luṭfu'lláh, p. v, 48; passing of, 64
Ḥakím, Manúcẖihr, 273–274
Ḥakímán, Jalál, 360.1
Ḥakímán, Raḥmatu'lláh, 387.1
Hancock, Tinai, 267.3d, 439.2d
Haney, Paul, (Hand of the Cause of God) 24.17, 42.17, 163.1, 288.1, 346, 361.2
Ḥaqíqí, Akbar, 381.2
Ḥaqíqí, 'Ináyatu'lláh, 416.1
Ḥaqq-Paykar, Badí'u'lláh, 323.3
Harris, Robert, 439.2b
Harwood, Howard, 60.1, 132.25, 267.4a
Ḥasanzádih, Maḥmúd, 261.2, 262.1
Ḥasanzádih-Sẖákirí, Gẖulám-Ḥusayn, 392.1, 393.6
Hassan II (King, Morocco), 12.4
Ḥaṣúrí, Rúḥu'lláh, 421.1
Hénuzet, Louis, 60.1, 132.25, 267.3e, 439.2e
Heohnke, Violet, 132.8, 132.25, 267.4a
Hidáyatí, Jahángír, 368.2, 396.14
Hofman, David H., p. v, 56.1, 129.1, 209.1, 359.1
Hoqbín, Kúrusẖ, 363.1
Hoveida, Abbas Amir, 220.8, 239.2
Húsẖmand, Suhayl, 367–367.1

Ioas, Leroy, (Hand of the Cause of God) 24.17, 25.1
Ioas, Sylvia, 374
Isẖráq-Kẖávarí, 119.1
Isẖráqí, 'Ináyatu'lláh, 363.1, 364.1

Isẖráqí, Ru'yá, 364.1
Isẖráqí, 'Izzat Jánamí, 364.1
Isẖráqí, Muḥammad, 381.3

Kámil-Muqaddam, Asadu'lláh, 396.16
Kásẖáni, Ḥusayn Báqir, 264.2
Kásẖání, Jamál, 416.1
Katẖírí, Iḥsánu'lláh, 400.1
Kátibpúr-Sẖahídí, Ni'matu'lláh, 290.1
Kavelin, H. Borrah, 56.1, 129.1, 209.1, 223.1, 236.1, 359.1
Kazemi, Zekrullah, 132.8, 132.25, 153.3, 267.3a, 439.2a
Kázimí-Mansẖádí, 'Abdu'l-Vahháb, 261.2, 262.1
Kebdaní, Muḥammad, 60.1, 132.25, 153.3, 267.3a, 439.2a
Kermani, Salisa, 267.4a
Kẖádí'í, Manúcẖihr, 292.1
Khadem, Zikrullah, (Hand of the Cause of God) 91.1, 110.2, 177.1, 404.1
Kẖamsí, Mas'úd, 60.1, 70.1, 132.9, 132.25, 267.3b, 361.3
Khan, Peter, 180.1, 267.3d, 361.3
Kẖándil, Ḥusayn, 285.2
Kẖayrkẖáh, Ibráhím, 317.2
Kẖayyámí, Iḥsánu'lláh, 322.1, 323.3
Kẖázeh, Jalál, (Hand of the Cause of God) 104.1
Khelghati, Thelma, 169.1, 267.3a, 439.2a
Khianra, Dipchand, 132.8, 132.25, 267.3c, 439.4
Khomeini, Ayatollah Ruhollah Musavi, 310.1, 378.8n
Kẖusẖkẖú, Sattár, 281.1
Kẖuzayn, Ṭarázu'lláh, 285.2
King, Coretta Scott, 434.3
King, Martin Luther, Jr., 434.2–434.4
King, Lauretta, 267.3b, 439.2b
Koirala, Bharat, 439.2c
Kruka, Josephine, 171.3

Lamb, Artemus, 60.1, 132.25. 267.3b, 439.4
Leong, Yan Kee, 60.1, 132.25, 267.3c, 439.4
Lucas, Paul, 132.8, 132.25, 267.4a
Lutchmaya, Roddy Dharma, 439.2a
Luṭfí, Kámrán, 393.2

INDEX OF NAMES

McHenry, John III, 70.1, 132.25
McLaren, Peter, 132.8, 132.25, 267.3b, 439.2b
McLaughlin, Robert, 115.1
Maḥmúdí, Húshang, 307.3
Maḥmúdí, Zhínús, 307.1
Maḥmúdnizhád, Muná, 364.1
Maḥmúdnizhád, Yadu'lláh, 355.1
Majdhúb, Maḥmúd, 307.1
Maka, Lisiate, 267.3d, 439.2d
Manaan, Mohammad 7.1
Maní'i-Uskú'í, Ḍíyá'u'lláh, 417.1
Manṣúrí, Muḥammad, 331.1
Maqámí, Manúchihr Qa'im, 307.3
Markazí, Shápúr, 411.1, 413.2
Marsella, Elena, 132.8, 132.25, 267.4a
Masehla, William, 169.1, 267.3a
Ma'ṣúmí, Muḥammad-Ḥúsayn, 270.1
Mavaddat, Farhang, 286.1
Maxwell, William Sutherland, (Hand of the Cause of God) 60.1, 132.8
Maxwell, May, 82.2, 150.1, 174.1, 185.3
Mayberry, Florence, 60.1, 131.1, 132.5, 361.3
Mihdí, Mírzá *(The Purest Branch)*. See entry in General Index.
Mihdízádih, Iḥsánu'lláh, 281.1
Miller, William McE., 251.3
Mitchell, Glenford, 332.1, 359.1
Míthaqí, Alláh-Virdí, 292.1
Míthaqíyán, Firaydún, 132.8, 132.25, 200.1
Moffett, Ruth, 214.5
Monadjem, Shapoor, 366.1, 439.2b
Morrison, Gayle, 439.2d
Muhájir, Raḥmátu'lláh, (Hand of the Cause of God) 104.1, 243.1, 337.2
Muḥammad (the Prophet), 220.13, 262.10
Muḥámmad-'Alí, Mírzá, 26.1, 157.7
Muḥammadí, 'Askar, 323.3
Muḥammad-Qulí, Mírzá. *See* entry in General Index.
Muḥammad-Taqí, Ḥájí Mírzá, 4.4
Muhandisí, Khusraw, 311.1
Mühlschlegel, Adelbert, (Hand of the Cause of God) 91.1, 104.1, 256.1
Mühlschlegel, Ursula, 267.3e, 439.2e
Mukhtárí, Mír-Asadú'lláh, 254.1

Mumtází, Rúḥu'lláh, 60.1, 132.25, 267.3c, 439.2c
Muqarrabí, 'Aṭá'u'lláh, 307.3
Muqímí-Abyánih, Zarrín, 364.1
Musavi, Mir Husayn (prime minister, Iran), 310.1
Mushtá'il-Uskú'í, Jalálíyyih, 324.1
Musṭafá, Muḥammad, 153.3, 267.3a
Mustaqím, Jalál, 261.2, 262.1
Muṭahhar 'Abdu'l-Majíd, 387.1
Muṭahharí, 'Alí, 261.2, 262.1
Mutlaq, Ḥusayn, 285.2
Muvaḥḥid, Muḥammad, 307.3

Nádirí, Bahíyyih, 307.3
Nagaratnam, S., 267.3c, 439.2c
Na'ímí, Fírúz, 285.2
Na'ímíyán, 'Alí, 339.1
Nají, Ḥusayn, 307.3
Nakhjavání, 'Alí, 56.1, 129.1, 209.1, 359.1
Nakhjavání, Bahíyyih, 272.1, 272.4
Naráqí, Sírús, 439.2d
Navváb *(Most Exalted Leaf)*. See entry in General Index.
Nawrúzí-Íránzád, Yúnis, 413.1–.2
Nayyirí-Iṣfahání, Ḥusayn, 381.3
Niederreiter, Leo, 267.3e, 439.2e
Nírúmand, Mahshíd, 364.1
Níyákán, 'Alí-Riḍá, 417.1

Olinga, Enoch, (Hand of the Cause of God) 91.1, 104.1, 163.1; assassination of, 237.1, 276.1
Olyai, Perin, 439.2c
Ong, Rose, 439.2c
Osborne, Alfred, 60.1, 132.25, 267.4a
Oule, Kolonario, 60.1, 132.25, 267.3a, 439.4

Pavón, Raúl, 132.8, 132.25, 267.3b, 380.1
Paymán, Khudáraḥm, 60.1, 132.25, 267.3c, 439.2c
Payraví, Jalál, 396.14
Pereira, Sarah, 132.8, 132.25, 267.3b, 439.4
Perks, Thelma, 60.1, 132.25, 267.4a
Pringle, Ruth, 267.3b, 439.2b
Purdil, Fírúz, 413.2
Púr-Ustádkár, Jamshíd, 416.1

INDEX OF NAMES

Pústchí, Yadu'lláh, 287.1

Qadímí, Yúsif, 307.3
Quddús, 176.2
Qurbánpúr, Amínu'lláh, 411.2

Rabbani, Shoghi. *See* Shoghi Effendi.
Radaví, Muhsin, 392.1
Rafat, Polin, 439.2e
Rahímíyán, Rahím, 393.3
Rahímpúr (Khurmá'í), Írán, 381.4
Rahmání, Hádí, 60.1, 104.1, 132.25, 267.4a
Rahmání, Ibráhím, 307.3
Ramoroesi, Daniel, 439.2a
Rawhání, 'Atá'u'lláh, 296.1
Rawhání, Hishmatu'lláh, 307.3
Rawhání, Masíh, 439.2c
Rawhání, Qudratu'lláh, 307.1
Rawshaní, Rúhí, 307.3
Rawshaní, Sírús, 307.1
Reagan, Ronald W. (pres., U.S.), 362.1, 443.1
Reed, Betty R., 81.1, 132.25, 267.3e, 439.4
Remey, Charles Mason, 23.7, 144.1
Revell, Ethel, 388
Revell, Jessie, 33.1
Ridvání, Ahmad, 296.1
Rissanen, Väinö, 171.3
Robarts, John, (Hand of the Cause of God) 320
Rogers, Donald O., 439.2b
Root, Martha, (Hand of the Cause of God) 98.3, 185.3
Rouhani, Mansour, 220.9, 239.2
Rouleau, Eric, 262.13
Ruhe, David, 56.1, 129.1, 209.1, 359.1
Rúhí, Manúchihr, 406
Rúhíyyih Khánum, Amatu'l-Bahá, (Hand of the Cause of God). *See* entry in General Index.
Sabeti, Parviz, 239.2
Sábirán, 'Imád, 153.3, 267.4a
Sábirí, Símín, 364.1
Sábiríyán, Yadu'lláh, 393.4
Sabri, Isobel, 60.1, 132.25, 267.3a, 361.3
Sádiqí, Parvíz, 121.2
Sádiqípúr, 'Abbás-'Alí, 335.1
Sadiqzádih, Kámbíz, 307.3

Safá'í, Suhayl, 360.1
Sahbá, Faríbúrz, 198.1
Salmánpúr, Manúchihr, 60.1, 132.25, 267.3c, 439.4
Samandarí, Mihdí, 60.1, 132.25, 153.3, 267.3a, 439.2a
Samandarí, Farámarz, 254.1
Samandarí, Tarázu'lláh, (Hand of the Cause of God) 42.17, 65.1
Samaniego, Vicente, 60.1, 121.3, 132.25, 267.3c, 439.2c
Samímí, Kámrán, 307.1
Schechter, Fred, 267.3b, 439.2b
Sears, William, (Hand of the Cause of God) 85.1, 104.1, 163.1, 178.1, 179.1, 276.1, 288.1; *A Cry from the Heart*, 312.1
Semple, Ian, p. v, 56.1, 129.1, 209.1, 359.1
Senoga, Edith, 439.2a
Serrano, Arturo, 439.2b
Sezgin, Ilhan, 439.2c
Sherrill, Velma, 132.8, 132.25, 267.3b, 439.4
Shoghi Effendi. *See* entry in General Index.
Sipihr-Arfa', Yadu'lláh, 344.2
Síyávushí, Hidáyat, 351
Síyávushí, Jamshíd, 363.1
Síyávushí, Táhirih, 364.1
Smith, Alan, 439.2b
Smith, David R., 439.2b
Sohrab, Ahmad, 402.8a
Sorabjee, Zena, 132.8–9, 132.25, 267.3c, 439.2c
Stevenson, Joy, 366.1, 439.2d
Stokes, Michael, 304.1
Subhání, Nusratu'lláh, 424.1
Sundram, Chellie J., 60.1, 132.25, 267.3c, 439.2c
Suzuki, Hideya, 169.1, 267.3c, 439.4

Ta'lím, Rúhú'lláh, 416.1
Taherzadeh, Adib, 169.1, 267.3e, 439.2e
Tahqíqí, Habíbu'lláh, 292.1
Talá'í, Kúrush, 311.1
Tanumafili II, Susuga Malietoa, 130, 177, 184.3, 242.1, 341.4, 403.2a
Taslímí, 'Abdu'l-Husayn, 307.3
Taymúrí, Parvín, 193.2
Taymúrí, Rúhu'lláh, 193

INDEX OF NAMES

Thábit, Akhtar, 364.1
Thábit-Rásikh, Gushtásb, 296.1
Thábit-Sarvistání, Ahmad-'Alí, 381.3
Thorne, Adam, 439.2e
Tibyáníyán, Varqá, 287.1
Tillotson, Peter, 110.2
Tin, U Soe, 439.2c
Tízfahm, Ágáhu'lláh, 324.1
Tomás, Rodrigo, 439.2b
True, Edna, 60.1, 132.25, 267.4a
Tumansky, A. G. 251.2

Vafá'í, Násir, 285.2
Vafá'í, Manúchihr, 344.2
Vafá'í, Rahmatu'lláh, 355.1
Vahdat-i-Haqq, Husayn, 317.2
Vahdat, Nusratu'lláh, 399.1
Vahdat, Yadu'lláh, 281.1
Varjávandí, Rustam, 421.1
Varqá, 'Alí-Muhammad, (Hand of the Cause of God) 91.1, 177.1, 361.2, 447.2
Vasudevan, Sankaran-Nair, 60.1
Victoria (Queen, England), 19.4
Von Czekus, Rolf, 439.2b
Vujdání, Bahár, 254.1

Vujdání, Farámarz, 121.2
Vujdání, Rahmatu'lláh, 437.1–.2
Vuyiya, Peter, 132.8, 132.25, 267.3a, 439.2a

Warren, Lucretia Mancho, 439.2a
Winckler, Bahíyyih, 132.25, 267.3a, 439.4.
  See also Ford, Bahíyyih.
Wingi, Mabuku, 382.1, 439.2a
Witzel, Donald, 60.1, 132.25, 267.3b, 439.4
Wolcott, Charles, p. v, 56.1, 129.1, 209.1, 359.1

Yaldá'í, Bahrám, 363.1
Yaldá'í, Nusrat, 364.1
Yávarí, 'Atá'u'lláh, 311.1
Yazdí, 'Azíz, 60.1, 131.1, 132.5, 361.3
Yazdí, Hájí 'Alí, 74.2a
Youssefian, Sohrab, 439.2e

Zá'írpúr, Túbá, 356
Zamání, 'Alí-Muhammad, 396.14
Zenimoto, Michitoshi, 439.2c
Zihtáb, Ismá'íl, 292.1
Zimmer, Hermann, 152.2

General Index

General Index

Ábádih (Iran), 227.12, 277.1, 293.3
'Abbásí, Muḥammad, 331.1
'Abbásíyán, Yúsif, 307.3
'Abdu'l-Bahá: admonition of, to lead one soul, 73.5; appointed Twin Institutions as Successors, 6.2; that which attracts the good pleasure of, 246.6a; authoritative interpretation by, 308.14; 308.17, 402.3, 438.48; Center of Covenant, 13.2, 95.5; Commander of hosts of the Lord, 34.17; comments on bigamy, 251.4b; comments on women, 162.32, 272.6a–6b, 402.9, 448.4; in the days of Shoghi Effendi's childhood, 157.8b; Defender of Spiritual Assemblies, 141.5; divine nature of, 157.5; in England, 48.3; equality of women and men explained by, 145.5, 166.2, 166.6a, 272.2; essence of Bahá'í spirit explained by, 173.9; Exemplar, 68.13; explained what to do in case of attack by robbers, 69.3; fiftieth anniversary of passing of, 95, 96, 99; indicated future importance of Panama Temple site, 42.11; initiated construction of Bahá'í center in 'Iṣhqábád, 4.3, 341.4; injunction to travel like, 34.12; passing of, 95; petitioned for permission to build first Maṣhriqu'l-Aḏhkár of West, 4.5; principle of noninterference in political affairs explained by, 77.4–4e; promotion of Cause by, 14.3; received, concealed, and reinterred remains of the Báb, 157.3; residences of, 157; reference to, by Bahá'u'lláh, 75.15; statements by, about law of inheritance, 166.3; statements by, linking unity of nations to twentieth century, 149.2; statements by, about membership of Universal House of Justice, 166.6–6b; statements by, about the Supreme Tribunal, 422; station of, 46.2, 330; Successor of Bahá'u'lláh and Interpreter of His teachings, 6.3, 251.5, 308.14–15, 308.17, 330.2, 438.48; summons of, to open Arctic and sub-Arctic, 172.1–5; surveillance of, by Governor of 'Akká, 157.2; Tablets of, at World Center, 128.4; use of recordings of voice of, 93; visits of, to France, 174; warning by, of widespread opposition to Faith, 248.6; Will and Testament of, 35.2, 59.1, 75.4, 75.6, 95.4, 95.6, 95.19, 157.12, 188.1, 251.5, 412.3–3a, 451.3. *See also* Divine Plan of 'Abdu'l-Bahá.
'Abdu'lláh Páṣhá, House of, 154, 157, 198, 358.3, 456.1
Abhá Kingdom, 2.3, 269.1, 287.1, 346.2, 388.1, 414.1
Ablutions, 147.3
Adamic cycle, 14.1
Adelaide (Australia), 449

GENERAL INDEX

Ádhirbáyján (Iran), 218.3, 300.1
Administrative Order, 9.4, 14.2, 128.14, 280.2, 397.4; North America as Cradle of, 101.2; unfoldment of, 447.7, 451.3–14. *See also* Auxiliary Boards; Continental Boards of Counselors; Hands of the Cause of God; Local Spiritual Assemblies; National Spiritual Assemblies; Universal House of Justice.
Adrianople, 14.5, 41.1, 46.1, 128.8; and revelation of Súriy-i-Mulúk, 41.1
Adultery, 147.23
Advent of Divine Justice, The (Shoghi Effendi), 224.2, 308.4, 383.3, 383.6
Áfáqí, Ṣábir, 325.1, 439.2c
Afghanistan, 221.5; formation of National Spiritual Assembly of, 104.1
Afnán, Bahrám, 363.1
Africa, 1.2, 77, 88, 162.6, 176, 276, 338.2, 382.1; Continental Boards of Counselors of, 58.1, 60.1, 153.6, 170.3, 267.2–3, 366.1, 439.2a; fortunes of Seven Year Plan in, 338.5; National Spiritual Assemblies of, 71.1, 77, 190.1, 221.13b, 371.1, 394.8; Pioneer Committee for, 31.6; Southern, 169.1; translations of Baháʼí literature in languages of, 128.15; Western, 153.1, 169.1; zones of Continental Boards of Counselors in, 153.6, 267.2
African Campaign, Two Year, 176.3
African Teaching Plan, first, 338.2
Afro-Americans, 97.3
Afshín, Burhániʼd-Dín, 169.1, 267.3c, 439.2c
Afús, Village of, 378.5
Age of maturity. *See* Maturity, age of.

Ages of Baháʼí Faith. *See* Formative Age; Heroic Age.
Aghṣán, 75.13–14, 251.5
Ahdieh, Hushang, 132.8–9, 267.3a, 439.2a
Aḥmadíyyih, Hidáyatu'lláh, 180.1, 267.3b, 439.2b
Aḥrárí, Ḍíyá'u'lláh, 345.1
Akhtar-Khávarí, Núru'lláh, 261.2
ʻAkka, 80.2, 127.1, 128.18, 157.2, 190.4, 192.3, 221.11c; centenary of termination of Baháʼu'lláh's confinement in, 192; House of ʻAbduʼlláh Páshá in, 154, 264.4, 456.2
ʻAláʼí, Shuʻáʼuʼlláh, (Hand of the Cause of God) 104.1; passing of, 414
ʻAláʼí, Suhayl, 60.1, 130.3, 132.25, 267.3d, 439.2d
Alaska, 100.4, 172.2
ʻAlavíyán, Buzurg, 286.1
Aleutians, 100.4
Alexander, Agnes, (Hand of the Cause of God) 96.7, 98.3, 403.2; passing of, 90.1
Algeria, formation of National Spiritual Assembly including, 153.2
ʻAlízádih, Maqṣúd, 396.14
Altiplano, 391.1
Amánat, Ḥusayn, 136.1, 140.1, 210.1
Amazon Basin, 162.6
America, 1.2, 340.2, 346.2, 383
American Baháʼí community: comments and appeals of the Guardian to, 383.2–3; response of, to persecution of Iranian Baháʼís, 383
Americas, 221.13b, 382.1; Continental Boards of Counselors for, 58.1, 60.1, 70.1, 267.3b, 366.1, 439.2b; National Spiritual Assemblies of, 371.1, 394.8;

776

GENERAL INDEX

spiritual destiny of, 340.2, 340.4
Amerindians, 97.3
Amín, Mihdí Amín, 307.1
Amíní, Naṣru'lláh, 328.1
'Ammán, 153.1
Anchorage (Alaska), 141.6, 172, 163.1
Andaman Islands, 371.1
Andes, 308.7
Anello, Eloy, 439.2b
Angola, 221.13b
Antigua, 248.14
Antipodes, 341.6
Anvarí, Mihdí, 277.1
Apia: dedication of Mother Temple of Pacific Islands in, 403; Mashriqu'l-Adhkár of, 341.4
Appa, Seewoosumbur-Jeehoba, 60.1, 132.25, 267.4a
Áqá, Ismá'íl, 157.5b
Aqdas. *See* Kitáb-i-Aqdas, The.
Arbáb, Farzam, 267.3b, 439.2b
Arc, the. *See* Mt. Carmel.
Architects, 136, 143, 198, 210.1
Archives, 409.6; Archives Office of Bahá'í World Center, 409.3; International, 115.1, 164.3; National, 409.7
Arctic Conferences, 171.1
Ardebili, Ayatollah Musavi, 309.1, 310.1
Ardikání, Husáyn, 60.1, 132.25, 153.3, 267.3a, 439.25
Argentina, 185.3, 295.3
Armstrong, Alfred, 132.25
Armstrong, Leonora Holsapple, 132.8, 132.25, 185.3. *See also* Armstrong, Leonora Stirling.
Armstrong, Leonora Stirling, passing of, 265
Army of Light, 6.6, 141.3, 152.1, 338.2, 358.2
Árnadóttir, Hólmfríður, 101.1, 171.2
Arson, 147.23

Art, 252.2
Asadu'lláhzádih, Husayn, 292.1
Asadu'lláhzádih, Shíva Mahmúdí, 311.1
Asadyárí, 'Abdu'l-'Alí, 292.1
Asare, Beatrice O., 439.2a
Aṣdaqí, Farhád, 417.1
Ashjárí, 'Azízu'lláh, 441.1
Ashraf, Jadídu'lláh, 331.1
Ashraf, Riḍvánu'lláh, 198.1
Asia, 169.1, 181.3, 287.1, 341.5, 425.9d; Continental Boards of Counselors for, 58.1, 60.1, 70.1–2, 170.3, 200.1, 267.3c, 325, 439.2c; Mashriqu'l-Adhkár of, 14.6; National Spiritual Assemblies of, 371.1, 394.8; Regional Teaching Conference in, 100.2; Southeast, 89.1–3, 200, 243.1, 245.1
Asian-Australasian Bahá'í Conference, 341
Asiatic U.S.S.R., 341.5
Assemblies. *See* Local Spiritual Assemblies; National Spiritual Assemblies; Spiritual Assemblies.
Ástání, Yadu'lláh, 254.1
Atharí, Fírúz, 416.1
Athens, 169.1
'Aṭifí, Bahman, 296.1
'Aṭifí, 'Izzat, 296.1
Auckland (New Zealand), 141.6, 163.1; International Teaching Conference in, 184
Audio-Visual Center, International. *See* International Bahá'í Audio-Visual Center.
Australasia, 92.2, 98.4, 169.1, 184.2, 248.14, 341.5, 394.8; Continental Board of Counselors for, 58.1, 60.1, 130.3, 170.3, 267.3d, 366.1, 439.2d
Australia, 146.1, 275.9d, 319
Auxiliary Boards, 2.4, 6.4, 13.6, 170.4; appointments to, 170;

Auxiliary Boards *(continued)*
arrest of members of, in Iran,
259, 261; assistants to, 137.3;
clarification of role of, 72.2–10;
consultation with, 182.7;
direction of, by Continental
Boards of Counselors, 58.1, 59.5,
72; collaboration between
National or Regional Teaching
Committees and members of,
194; decisions of, 11.2a–2e;
duties of, 137.4; expansion of,
137; members of, 137.6, 170.3;
members of, serving on National
Spiritual Assemblies, 11; role of,
in relation to Local Spiritual
Assemblies, 137.4, 189.2–3;
terms of service of, 137.4, 394.3
Awjí, Ḥabíbu'lláh, 344.2, 345.1
Axis, spiritual, 341.6
Ayadi, Dr., 239.2
Ayala, Ben, 267.3d, 439.2d
Aydilkhání, 'Abbás, 437.1
Ayman, Iraj, 81.1, 132.25, 267.4a, 439.2c
Ázádí, 'Abdu'l-Ḥusayn, 363.1
A'ẓámí, Ghulám-Ḥusayn, 254.1
Azharra, General Gholam Reza, 215.1
'Azízí, Ḥabíbu'lláh, 296.1
'Azízí, Iskandar, 311.1
'Azízí, Jalál, 307.1

Báb, the, 14.3, 82.3, 116.3, 262.10;
anniversary of birth of, 176.1;
Ḥájí Mírzá Muḥammad-Taqí,
cousin of, 4.4; Holy Shrine of,
1.4, 105.1; House of, 225, 235,
241, 282, 314.1, 358.2; martyrdom of, 157.3, 226.3, 365.1;
pilgrimage to Mecca by, 176.2;
publication of selection of
writings of, 190.1; servant of,
176.2
Bábázádih, Sa'du'lláh, 328.1

Bábulsar (Iran), 317.2
Báft (Iran), 392.1
Baghdad, release of Bahá'í
prisoners in, 233
Bahá'í Administration, 27.3
Bahá'í Administration, 118.2a,
221.5, 378.17. *See also* Administrative Order.
Bahá'í center, first on European
continent, 174.1
Bahá'í children. *See* Children.
Bahá'í clubs, establishment of,
453.13
Bahá'í Community, The (1963), 27.3
Bahá'í community, 195; consolidation of, 2.7, 394.7; consultation
in, 379.3; diverse representation
in, 456.2; example of a unified
community, 77.10–13, 383.4,
427.4, 445.2; efforts of, to enlist
public support for Iranian
Bahá'ís, 358.5; offered as
working model for reorganization
of world, 445.2; registration of
youth as members of, 426.4; selfsufficiency of, 13.5, 379.3;
spiritual center of, 379.2
Bahá'í community life, 195.3a,
275.9k, 289; application of
Bahá'í social teachings in,
308.6–10; tests of, 289
Bahá'í Dispensation, 338.3, 403.2
Bahá'í Educational Center(s),
280.30–32
Bahá'í elections. *See* Elections,
Bahá'í.
Bahá'í Faith, 1.2, 117.4; Administrative Order of, 385.5; attacks on,
218, 227, 234, 251.2, 314;
Central Figures of, 18.4, 46.2,
164.2b, 240.3, 240.5, 330;
emergence of, from obscurity,
284.2, 341.2, 444, 456.3;
epochs of Formative Age of, 451;
expansion of, 13.2, 379.3;

Founders of, 6.2; Heroic Age of, 4.2, 451.3–4a; language used by Central Figures of, 330; moral standards of, 224; opportunities for progress of, 383.6; oppression of, 314.5–6; teachings of, 378.12–17; worldwide publicity about, 314.7–8. *See also* Cause of God.
Bahá'í family life, 280.10, 402; compilation on, 316; relationships between family members in, 272, 402.3–9; roles of parents in, 407
Bahá'í festivals, 147.16
Bahá'í Funds. *See* Funds, Bahá'í.
Bahá'í Holy Days, 280.6, 280.11; children's behavior during observances of, 343
Bahá'í Holy Places at the World Center, 80.4
Bahá'í, Ḥusníyyih, 440.5
Bahá'í institutes. *See* Institutes, Bahá'í.
Bahá'í International Community, 128.9, 321.2, 394.3; appeal to Iranian leaders by, 310; cooperation of, with UNICEF, 376.1; defense of Iranian Bahá'ís, 314.12; European Branch Office of, 283, 321.2; preparations by, for International Youth Year (1985), 376.1–2; Public Information Office of, 432.1; and the United Nations, 49, 68.5, 128.9, 141.5, 284.5, 308.8, 309.1, 376.1–2, 394.3
Bahá'í International Fund, 13.5, 284; demands on, 87.3–5; grave crisis in, 87; needs of, 40; recent victories and immediate challenges, 284; universal participation in, 87.6–8. *See also* Funds, Bahá'í.
Bahá'í life, 14.7, 117.1–3, 126, 195.3a, 214; characteristics of a, 141.12, 224.2; demonstrating concepts of peace through, 427.5; family relationships in, 272; living the, 195.3a; release of compilation on, 121; standards of behavior in, 133, 224
Bahá'í News, 47.1, 54.3, 123, 193, 313.8
Bahá'í Prayers, 402.7
Bahá'í Publishing. *See* Publishing.
Bahá'í Publishing Trust(s), 14.6, 146, 162.20, 275.9g, 295.3; capitalization of, 8, 146.8; establishing and operating, 146; establishment of new, 14.6; financial programs of, 146.12–16; management of, 146.17; name of, 146.3; proceeds of, 146.7; production in, 146.9; publishing program of, 146.10–11; in relation to other committees, 146.5–6
Bahá'í Radio-Television Conference. *See* Hemispheric Bahá'í Radio-Television Conference.
Bahá'í Revelation, 385.7
Bahá'í scholars, qualities of, 217.11–14
Bahá'í scholarship, 217; need for new methodology in, 217.4–14
Bahá'í schools. *See* Schools.
Bahá'í social teachings, 308.6–10
Bahá'ís of Iran. *See* Iranian Bahá'ís.
Bahá'í Studies Seminar (Cambridge, 1978), 217
Bahá'í Unity Conference (Ganado, Arizona, 1972), 112
Bahá'í Women's Conference(s): in New Delhi (1977), 197; first, in South America (1978), 203; first, in United Kingdom (1981), 291; first Danish (1982), 342; in Nairobi, Kenya (1984), 389 in Cotonou, Benin (1984), 398, 401

Bahá'í World, The, 94.3k; Greatest Holy Leaf mentioned in, 313.7
Bahá'í world. *See* Bahá'í world community.
Bahá'í World Center. *See* World Center.
Bahá'í world community, 195, 221.1, 244.1, 358.5, 358.8; conditions of, 24.5–6; diversity in, 383.4; mission of, 221.1; Qiblih of, 26; relationship of, with non-Bahá'í world, 394.8; social and economic development of, 358.6; and the United Nations, 2.10
Bahá'í World Congress, first (1963), 1, 2.4; second (1992), 447.5
Bahá'í Writings, 2.10, 14.5, 141.5, 221.11d, 330; interpretation of, 403.3. *See also* Compilations of Bahá'í Writings, releases of.
Bahá'í youth. *See* Youth, Bahá'í.
Bahá'í Youth Conferences. *See* Youth Conferences.
Bahamas, formation of National Spiritual Assembly of, 141.7, 208.1
Bahá'u'lláh, 1.2, 2.3–4, 2.11, 6.1, 6.3, 13.2, 14.3; acquisition of first residence of, in 'Akká, 12; African peoples referred to by, 77.1; ancestral Houses of, 226.2, 305, 314.1, 358.2; appointment of Successor to, 75.15; arrival of, in Holy Land, 65.1, 67.2, 68.3; Ascension of, 26, 281.3; Bahá'ís should be touched by love of, 18.4; centenary of His summons to the kings, 42.14, 128.8; centenary of termination of confinement of, 192; centenary of voyage from Gallipolli to Most Great Prison, 63.1; central purpose of Dispensation of, 397.3; Dispensation of, 308.14, 451.3–15; exhortation of, to eliminate prejudice, 117.2; fix your love and prayers upon, 289.5d; gaps in legislative ordinances of, 145.4; Holy Tomb of, 26, 74.2a; law of, 77.9; laws not expressly revealed by, 35.3; love for God and, 211; Manifestation of God, 435.2; mankind's response to message of, 195.2; principle mission of, 422.2; Proclamation of, to kings and rulers, 6.8, 14.5, 24.14–16, 42.16, 46.1; purpose of Revelation of, 427.6; rank and station of, 206.3a–3b; references of, to pioneers, 86.2; reinterment of faithful half-brother of, 440; Revelation of, 217.7, 308.14; sacred obligation given by, 35.8; sesquicentennial of birth of, 44, 53; Shoghi Effendi's obedience to, 35.3; Shrine of, 26; sojourn of, on European continent, 17.3; successors of, appointed, 6.2; summons to rulers of America by, 340.1; Shrine of, 26; system ordained by, 75.18; task to build the Order of, 63.4–5; winning good-pleasure of, 248.8; traveling teaching referred to by, 34.10–10a; understanding World Order of, 35.3; voice of, is voice of Great Spirit, 112.3; Writings of, 55.6, 375.5
Bahia (Brazil), Conference in (1977), 141.6, 163.1; message to, 185
Báhirí, Mihdí, 292.1
Bahíyyih Khánum *(The Greatest Holy Leaf)*, 95.6, 157.7, 164.2, 275.8, 313, 334, 337.1, 338.3–4, 338.7, 340.7; fiftieth anniversary of passing of, 313, 334, 358.3; references to, in published

works, 313.4–9
Bahjí, 40.5, 128.7, 192.3, 440.4; acquisition of property adjacent to, 74, 264.2; gardens at, 105, 128.7, 139; Mansion at, 105.1, 157.7, 440.2; Master's teahouse at, 74.2–2a
Bahrámsháhí, Rúḥu'lláh, 424
Bahta, Gila Michael, 366.1, 439.2a
Baker, Dorothy, (Hand of the Cause of God) 63.7
Ba<u>kh</u>távar, Kamáli'd-Dín, 290
Ballerio, Giovanni, 283.1
Balyuzi, Hasan M., (Hand of the Cause of God) 251.3; passing of, 247, 248.1
Bamako (Mali), 169.1
Banání, Músá, (Hand of the Cause of God) 88.2; passing of, 102
Bandar-'Abbás (Iran), 437.2
Bangui (Central African Republic), 71.1
Bani-Sadr, Abol Hasan (pres., Iran), 262.5
Banjul (Gambia), 153.1
Baqá, <u>Sh</u>ídru<u>kh</u> Amír-Kíyá, 311.1
Barney, Laura. See Dreyfus-Barney, Laura.
Ba<u>sh</u>írí, Aḥmad, 368.2, 413.1–2
Battrick, A. Owen, 169.1, 267.3d, 439.4
Behavior, Bahá'í standards of, 133
Belgium, Bahá'í Publishing Trust in, 295.3
Belize, formation of National Spiritual Assembly of, 38.1
Benin, Women's Conference in. See Bahá'í Women's Conferences.
Benson, Richard, 132.8, 132.25, 267.3d, 439.2d
Bermuda, formation of National Spiritual Assembly of, 221.13b, 248.14
Betrothal, 147.7. See also Marriage.
Bigamy, 251.4b–4c

Bírjand (Iran), 254.1, 270.1, 300.1, 378.5
Births, 280.20
Blessed Beauty. See Bahá'u'lláh.
Blumenthal, Erik, 60.1, 132.25, 267.3e, 439.4
Bolivia: Bahá'í radio station in, 391, 394.3; messages to Continental Conference in (1970), 82, 85.1; National Spiritual Assembly of, 304.1
Boman, Shirin, 70.2, 132.25, 267.3c, 439.4
Book of Íqán. See Kitáb-i-Íqán, The.
Book of Laws. See Kitáb-i-Aqdas, The.
Books and materials. See Tutorial schools.
Bophuthatswana, formation of National Spiritual Assembly of, 221.13b, 248.14
Bopp, Anneliese, 81.1, 132.25, 230.1, 361.3
Brady, Wilma, 439.2b
Brazil, 265.1
Brazzaville. See Congo.
British Isles, 135.1a. See also United Kingdom.
British Volunteer Program (BVP), 384.3
Bujnúrd (Iran), 406.1
Búkán (Iran), 323.3
Burial, 147.18, 280.22–23
Burundi, formation of National Spiritual Assembly of, 141.7, 208.1
Bustánábád (Iran), 322.2
BVP. See British Volunteer Program.

Cables, printing of, 94.31
Cairo (Egypt), 338.2
Calamity, the, 149, 246.2–4, 252. See also Catastrophe.
Cameroon, formation of National Spiritual Assembly including, 153.2

Canada, 100.4, 223.1, 419.3; National Spiritual Assembly of, 423.1; unanimous vote by parliament of, against Bahá'í persecution, 255.1
Canadian University Services Overseas (CUSO), 384.3
Canary Islands, formation of National Spiritual Assembly of, 371.1
Canberra, International Conference in (1980), 319, 340.7, 341
Cape Verde Islands: formation of National Spiritual Assembly including, 153.1–2; formation of National Spiritual Assembly of, 221.13b, 371.1
Caracollo (Bolivia), 391.1, 394.3
Caribbean, 187.3, 337.3; message to conference in (1971), 97
Caribbean Conference. *See* Oceanic Conferences.
Carney, Magdalene, 361.3
Caroline Islands: formation of National Spiritual Assembly including, 92.1; National Spiritual Assembly of, 221.13b, 394.8, 403.2b–2c
Casablanca, 390.1
Caspian Sea, 381.2
Catastrophe, 73, 252. *See also* Calamity, the.
Cause of God, 1.3, 1.5, 2.1, 2.11–12, 6.10, 35.10–12, 162.24, 252.4, 394.1, 397.5, 445.2; Ark of, 1.2; development of, 14.7; emergence of, from obscurity, 394.1, 394.5, 394.9, 456.3; financial self-sufficiency in, 141.8; and the individual Bahá'í, 34.15–17; progress of, 6.1; raising God's kingdom on earth, 128.21; relationship of institutions of the Faith to, 453.8; teachers of, 11.26; universal participation in, 19. *See also* Bahá'í Faith.
Celibacy, 126.7a–9
Cemeteries, Bahá'í, 25.1, 227.2
Center of the Covenant. *See* 'Abdu'l-Bahá.
Central African Republic, formation of National Spiritual Assembly including, 71.1
Central America, 187.4, 337.3
Central Figures of the Bahá'í Faith. *See* Bahá'í Faith, Central Figures of.
Chad, 153.2; formation of National Spiritual Assembly including, 71.1; formation of National Spiritual Assembly of, 91.1
Challenging Requirements of the Present Hour, The, 27.3
Chance, Hugh, p. v, 56.1, 129.1, 209.1, 359.1
Chase, Thornton, 157.10
Chastity, 126.7–9, 133.2, 224
Chígán, Village of, 378.5
Children, 162.35–36, 333.5–10, 426.2; behavior of, at Feast, 343; education of, 147.25, 175.25, 222.10, 272.6, 275.9k, 280.6, 394.7; excusing from school on holy days, 29.3–3a; guidelines for registration of, 333; release of compilation of prayers and Tablets for, 167
China, 248.6, 341.5
Chinese race, 89.3, 100.4, 181.8
Chinese-speaking believers, 89.3
Chinniah, Inparaju, 200.1; passing of, 245
Christ, 95.18, 174.1, 217.7, 251.3, 330.7a, 403.2
Christendom, 39.1, 174.1
Christianity, 18.1, 162.6, 217.7, 251.3, 308.14
Christians, 262.7
Ciskei, formation of National

Spiritual Assembly of, 394.8
City of the Covenant. *See* New York City.
Collins, Amelia, (Hand of the Cause of God) 101.1
Colombia, presentation of peace statement to president of, 442.2
Columbus (Ohio), International Youth Conference in (1985), 431
Companionate marriage, 126.7a
Compilations of Bahá'í Writings, releases of: on the assistance of God, 294; on Bahá'í consultation, 204; on Bahá'í education, 175; on Bahá'í Funds, 76; on Bahá'í life, 122; on Bahá'í marriage and family life, 402.2; on Bahá'í meetings and the Nineteen Day Feast, 168; on Bahá'í schools and institutes, 109; on deepening, 353; on divorce, 244; on excellence in all things, 303; on family life, 316; on Ḥuqúqu'lláh, 430, 404.2; on the importance of prayer, meditation, and the devotional attitude, 249; on the individual and teaching, 188; on inspiring the heart, 240; of letters of the Guardian, 162.21; on the Local Spiritual Assembly, 84; on music and singing, 107; on the National Spiritual Assembly, 114; on newsletters, 120; on opposition, 152; on peace, 436; on prayer, 249; of prayers and passages from the Holy Writings, 295; of prayers and Tablets for children and youth, 167; on the Sacred Texts, 280.9; on use of radio and television in teaching, 162.39–51; on women, 446
Conference of Counselors in Holy Land (1985), 445; report from, 447
Conferences. *See* Arctic Conferences; Asian-Australasian Bahá'í Conference; Bahá'í Women's Conferences; Conference of Counselors; Hemispheric Bahá'í Radio-Television Conference; Intercontinental Conferences; International Conferences; International Teaching Conferences; Oceanic Conferences; Youth Conferences.
Confidentiality, elucidation of principle of, 336
Congo (Brazzaville), formation of National Spiritual Assembly including, 71.1, 91.1
Congo Republic, 221.5
Constitution of the Universal House of Justice, 123. *See also* Universal House of Justice.
Consolidation, 2.7, 34.13, 40.2, 42.21, 182, 189.5, 280.4–7, 394.7; of Local Spiritual Assemblies, 189; pioneering and, 36. *See also* Expansion and consolidation.
Consultation, 30.6, 37.9, 79, 453.13; Auxiliary Boards as aid to, 182.7; Bahá'í youth and, 37.9; compilation on, 204; decision-making process of, 79.3–6; practice of art of, 379.3; release of compilation on, 204; spirit of, 79
Continental Boards of Counselors: appointments of, 59, 60, 70, 111, 132, 170, 180, 200, 366, 382, 439; Auxiliary Boards and, 72, 137; establishment of, 58, 59; expansion of roles of, 132, 456.2; first appointments of, 60; further development of, 267; Hands of the Cause of God and, 59.6; International Teaching Center and, 132.6, 394.3;

Continental Boards of Counselors
(*continued*)
membership of, 60, 70, 132.5, 132.25, 170, 170.3, 180, 366; 380.1, 382, 439; message to conference of, 445; nature of, 111; sharing of information between National Spiritual Assemblies and Auxiliary Boards with, 72.5–6; and social and economic development, 379.5; spiritualization campaign launched by, 375.2; terms of service for, 59.4, 229; tribute to, 321.4; work of, 72, 158.2; zones of, 59.3, 132.7, 132.13–24, 153, 267.2

Continental Conferences: in La Paz, Bolivia (1970), 82; in Monrovia, Liberia (1971), 88. *See also* Oceanic Conferences.

Continental Pioneer Committees, 22, 31.2, 280.28–29, 361.6; Continental Counselors and, 158.2, 280.28; functions and responsibilities of, 22.10–15, 148, 158.5

Convention, First International, 2.2. *See also* International Bahá'í Convention(s); National Convention(s); Unit Conventions.

Cook Islands, formation of National Spiritual Assembly of, 221.13b, 394.8, 403.2c

Correspondence courses, 162.14–15, 222.18

Costa Rica, message to Youth Conference in (1983), 357.1

Costas, Athos, 60.1, 132.25, 267.3b, 439.4

Cotonou (Benin), 153.1, 398, 401

Counselors. *See* Continental Boards of Counselors.

Covenant, the, 75.12, 221.14–15, 308, 397.5; of Bahá'u'lláh, 6.3, 308.14, 308.18, 447.5; Center of, 2.1, 6.10, 13.2, 435.4a; City of, 447.5; Day of, 99.2; features of, 6.3; proselytizing, development, and, 308; uniqueness of, 308.14–15

Covenant-breakers, 6.4, 26.1, 144.1, 152.2, 157.7; expulsion and reinstatement of, 6.4, 132.6

Cowan, Angus, 180.1, 267.3b, 439.4, 454.1

Crisis and Victory, 447.8; vision of Bahá'ís amidst, 246.5. *See also* Disaster and triumph, dialectic of.

Crusade. *See* Ten Year World Crusade.

Cry from the Heart, A, 312

CUSO. *See* Canadian University Services Overseas.

Cyprus, formation of National Spiritual Assembly of, 141.7, 208.1

Dahomey (Benin), formation of National Spiritual Assembly of, 153.1–2

Dakar (Cape Verde Islands), formation of National Spiritual Assembly of, 153.1

Dakhílí, Masrúr, 292.1

Dálvand, Shahín (Shírín), 364.1

Danish Women's Seminar, first (1982), 342.1

Dáryún (Iran), 296.1

Dávúdí, 'Alímurád, 307.3

Dawn-Breakers, the, 221.1, 275.3, 314.1

Dawn prayers. *See* Prayers, dawn.

Day of God, 289.3, 358.9, 438.58

Day of Judgment, 378.4

Day of Prayer: for Bahá'ís of Iran, 315, 350; sympathetic organizations invited to, 315

De Burafato, Carmen, 60.1, 132.25,

267.3b, 439.4
De Calderón, Isabel P., 382.1, 439.2b
Decision-making. *See* Consultation.
Declaration of the Báb, 190.2, 403.2
Deepening, 1.5, 35.13, 42.22–26, 137.4, 280.13, 353.1; classes, 280.6; compilation on, 353; by correspondence courses, 162.14–15; as developing a spiritual attitude, 289.4; on Formative Age, 95; on National Spiritual Assemblies, 30.4
Denmark, Bahá'í women's conference in (1982), 342
Development: activities, 379.8; of the Faith, 2.11, 308
Dhabíhíyán, 'Azízu'lláh, 261.2
Dihqání, Bahman, 387.1
Dihqání, Hidáyatu'lláh, 277.1
Disaster and triumph, dialectic of, 248.2–8; 358.2–5. *See also* Crisis and victory.
Dispensation of Bahá'u'lláh, 14.3, 95.19, 99.1, 308.14, 397.3, 451.3–15
Dispensation of Bahá'u'lláh, The, (Shoghi Effendi) 75.1–2, 75.9
Dispensations, 23.16a, 35.11, 95.23, 166.3, 308.14, 308.17, 330.3
Diversity, unity in, 77.9
Divine Plan of 'Abdu'l-Bahá, 1.2, 2.6, 6.5, 14.3, 82.2–3, 95.24, 96.1–2, 97.2, 111.5, 116.1, 141.3, 221.1, 451.5a, 451.7, 451.9–10; first epoch of, 1.2; second epoch of, 2.6, 2.12, 6.5, 14.8; third epoch of, 6.10, 451.12b fourth epoch of, 449.1, 451.13, 456.4; objectives of, 14.8. *See also* Tablets of the Divine Plan; Formative Age.
Divorce, 147.11, 251.4a, 308.13, 378.24; mention of, in *Synopsis and Codification of the Kitáb-i-Aqdas*, 166.5; release of compilation on, 244
Díyá'í, Nuṣratu'lláh, 392.1
Díyá'u'lláh, Mírzá, 26.1
Dizfúl (Iran), 381.4
Djakarta (Indonesia), 68.6
Dominica, formation of National Spiritual Assembly of, 221.13b, 394.3
Dowry, 147.8. *See also* Marriage.
Dreyfus-Barney, Laura, 150, 174.1
Dublin (Ireland), International Conference in (1982), 275.9d, 288.1, 375.2; message to, 329
Dunbar, Hooper, 60.1, 131.1, 132.5, 361.3
Dunn, Clara, (Hand of the Cause of God) 98.3, 184.2
Dunn, John Henry Hyde, (Hand of the Cause of God) 98.3, 184.2

Earthquakes, 4.1, 4.7
Eastern Pilgrim House, 3.3
East Pakistan, formation of National Spiritual Assembly of, 92.1, 104.1
Economic and Social Council (United Nations), 78, 308.8. *See also* United Nations.
ECOSOC. *See* Economic and Social Council.
Ecuador: Bahá'í radio station in, 166.22, 201, 205, 213, 308.7; National Spiritual Assembly of, 278.1, 304.1
Editing, 94.3m
Education, 182.3–5, 272.4, 275.9k; of children, 141.11, 280.6, 280.24–25, 321.2, 333.3, 407.2; release of compilation on, 175; role of parents in, 407; Seven Year Plan achievements in, 338.5; Six Year Plan goals for, 453.17; tutorial schools, 280.30–

Education *(continued)*
41; universal, need for, 438.4
Edward Granville Browne and the Bahá'í Faith (Balyuzi), 251.3
Effendi, Shoghi. *See* Shoghi Effendi; Guardian; Guardianship.
Egypt, 153.2; National Spiritual Assembly of, 338.2
Ekpe, Friday, 132.8, 132.25, 153.3, 267.3a, 439.2a
Elder, Earl E., 251.3
Elections, Bahá'í, 1.4, 103; compilation on, 103; of delegates to National Conventions, 433; electoral units for, 433.1–3; the Guardian's writings on, 103.1; principles governing, 23.8–14; of the Universal House of Justice, p. v, 1.4, 2.1, 6.9, 23, 56, 96.3, 129, 209, 327, 332, 359
Endowments, 142.13, 187.5; national, 14.6, 16.7, 87.1, 275.9f; seizure of, in Iran, 314.1
England, 48.3
Entry by troops, 169.1, 190.3
Epochs. *See* Formative Age.
Epyeru, Oloro, 60.1, 132.25, 267.3a, 439.2a
Equality: racial, 434.2; of men and women, 145.5–6, 166, 272.2, 394.7, 402.6, 407.4, 438.33, 446.1, 453.13; suggested further reading on, 402.9
Equatorial Guinea, 153.2; formation of National Spiritual Assembly of, 134; dissolution of National Assembly of, 169.1; re-formation of National Assembly of, 371.1
Eshraghian, Kamran, 439.2d
Eskimos, 172.3, 340.4
Esslemont, John E., (Hand of the Cause of God) 64.1
Estall, Rowland, 132.8, 132.25, 267.4a

Etelaat, 253.2, 262.9
Ether, 330.4
Europe, 169.1; appointment of Continental Boards of Counselors for, 58.1, 60.1, 170.3, 267.3e, 439.2e; as described by Shoghi Effendi, 174.1; developing dependencies of Mashriqu'l-Adhkár of, 275.9e; Mother Temple of, 2.9; National Spiritual Assemblies in, 371.1, 394.8
European Youth Conference (Innsbruck, 1983), 369
Evin Prison, 396.11
Excellence, 428; release of compilation on, 303
Excellence in All Things, 303
Execution of Bahá'ís in Iran. *See* Martyrdoms; Iranian Bahá'ís.
Expansion and consolidation: of Bahá'í community, 18.6–7, 34.13, 40.2, 42.21, 89.3, 162.3–5, 190.5, 280.4, 394.7, 456.2; Nine Year Plan goals for, 16.5, 19.1; pioneers aiding, 128.3. *See also* Consolidation; Teaching; Traveling Teaching.

Fádílábád (Iran), 193.2
Faizi, Abu'l-Qasim, (Hand of the Cause of God) 24.17, 42.17, 163.1; passing of, 269.1
Family life. *See* Bahá'í family life.
Fareed, Bijan, 439.2c
Farhand, Ghulám Husayn, 416.1
Farhangí, Masíh, 60.1, 132.25, 267.3c, 287.1, 325.1
Faríd, Badí'u'lláh, 287.1
Farídání, Firaydún, 261.2
Farnúsh, Háshim, 286.1
Farúhar, Mahmúd, 323.3
Farúhar, Ishráqíyyih, 323.3
Farzánih-Mu'ayyad, Manúchihr, 331.1

Fast, the (Nineteen Day), prayers for Iranian Bahá'ís during, 218.7, 350.2; youth and, 426.3
Fasting, 147.5–6, 405.2, 426.3
Fat'he-Aazam, Shidan, 60.1, 267.3a, 439.2a
Fatheazam, Hushmand, p. v, 56.1, 129.1, 209.1, 359.1
Feast, Nineteen Day, 137.4, 280.6, 575.7; children's behavior at, 343; of Jamál, 350.1, 352.1; of Mashíyyat, 24.16, 41.1, 42.17–18; of Núr, 190.4; of Qawl, 7.2, 8.8, 12.2; release of compilation on Bahá'í meetings and, 168; of Sulṭán, 73.6
Featherstone, Collis, (Hand of the Cause of God) 91.1, 104.1, 163.1, 288.1, 361.2, 447.2
Ferraby, John, (Hand of the Cause of God) 135
Ferraby, Dorothy, 60.1, 132.25, 267.3e, 439.4
Ficicchia, Francesco, 251.1–2
Fiji Islands, 110.9, 146.1, 403.2b
Firdawsí, Fathu'lláh, 311.1
Fire Tablet, translation of, 258
First Oceanic Conference. *See* Oceanic Conferences.
Fírúzí, Parvíz, 292.1
Fírúzí, Riḍá, 268.1
Five Year Plan (1974–79), 138, 141, 142, 158.1, 159, 162, 169, 179, 190, 207, 221.1, 358.2; activities for women during, 162.32; contact with authorities during, 162.24–25; Continental Boards of Counselors, 207.11; correspondence courses established during, 162.14–15; developing Bahá'í life during, 207.8–10, 142.14, 162.37; elucidation of goals of, 142; external affairs goals for, 142.17; gathering for dawn prayers during, 142.14, 162.38;
goals for children in, 141.11, 162.35–36; goals for Local Spiritual Assemblies during, 162.29–31, 189.1, 199.1; goals and objectives of, 141.4–20, 142, 159, 172.4, 179, 199.1, 207.6–10; holding regular National Teaching Conferences during, 142.15; launching of, 124, 141; literature available during, 162.20; major objectives of, 141.4; major role of youth during, 142.16, 162.33–34; national incorporation goals during, 142.6; National Spiritual Assemblies elected during, 153; opportunities during, 141.20, 207.12; pioneering goals for, 142.18, 162.8–12, 179; proclamation of Faith during, 141.9; progress of, 162, 207.2–5; radio and television, 162.21–23; reaching remote areas during, 162.6; teaching conferences during, 162.16–18; teaching goals for, 162.3–5; teaching tribal peoples and minorities during, 162.7; teaching victories of, 221.3–5; three major objectives of, 141.4; traveling teaching during, 162.12; World Center goals for, 141.5
Florence (Italy), 144.1
Fofana, Kassimi, 366.1, 439.2a
Ford, Bahíyyih, 60.1. *See also* Winckler, Bahíyyih.
Ford, Mary Hanford, 157.11
Formative Age, 95, 99, 451; call for deepening on significance of epochs of, p. xli, 451; extracts from Writings on, 95.4–24; fiftieth anniversary of the opening of, 100.1; first epoch of, 1.2, 451.9; second epoch of, 14.8, 15.1, 451.10–11; signifi-

Formative Age *(continued)* cance of, 95; third epoch of, 6.10, 451.12–12b; fourth epoch of, 449.1, 451.13, 456.4. *See also* Heroic Age.
Fourth heaven, 330.7–7a
Fozdar, John, 267.3c, 439.2c
Frankfurt (West Germany), Intercontinental Conference in (1967), 24.17, 46
Free love, 126.7a
Freetown (Sierra Leone), 153.1
French Antilles, National Spiritual Assembly of, 141.7, 169.1
French Guiana, National Spiritual Assembly of, 141.7, 169.1, 221.13b, 371.1
Frivolous conduct, 224.2
Funafuti (Tuvalu), 248.14
Funds, Bahá'í, 2.11, 13, 76, 321.2, 435; Bahá'í International Fund, 40, 87, 96.6; call for support of international, 419; Continental, 13.6; contributing to, 13.2; crisis of international, 87.3–5, 223.1; development of, 9, 435; increase in resources of, 128.14; International Deputization, 24.11, 34.14, 128.14, 158.5; local, 435; Local Spiritual Assemblies and, 394.7, 435; National, 9, 435; needs of, 13.1, 34.14, 40, 87, 275.10, 385; pioneering financed by, 162.10–12; release of compilation on, 76; requests for financial assistance, 9; role of individual in, 13.3, 14.7, 435.3–4; and teaching, 2.11, 13.2, 34.14; universal participation in, 2.11, 19.3, 87.6–8; use of, 162.26–28, 221.13g; writings of the Guardian on, 13.2, 76.1. *See also* Ḥuqúqu'lláh.
Furúghí, Parvíz, 121.2
Furúhí, 'Izzatu'lláh, 307.1

Furútan, 'Alí-Akbar, (Hand of the Cause of God) 24.17, 42.17, 104.1, 163.1, 354.2, 361.3, 440.5, 447.2
Future, fear of, 252

Gabon, National Spiritual Assembly of, 71.1, 91.1, 153.2, 221.13b, 371.1
Galilee, 440.1, 440.3
Gambia, National Spiritual Assembly of, 153.1–2
Ganado (Arizona), Bahá'í Unity Conference in (1972), 112
Gangtok (Sikkim, India), 38.1
Gardens: at Bahjí, 105, 128.7, 139, 192.2; Monument, 394.8; on Mt. Carmel, 3.3, 14.5, 40.3, 128.7; Riḍván, 192.3; surrounding holy places, 141.5; in Tákur, confiscation and sale of, by Iranian authorities, 305.1
Gardner, Lloyd, 60.1, 132.25, 267.3b; passing of, 423
Gawhar-Dasht Prison, 396.11
Geneva (Switzerland), 263.2, 283.1, 284.5, 321.2, 394.3
Germany, 247, 384.3; House of Worship in, 2.9, 17; National Spiritual Assembly of, 251, 256.1
Ghana, National Spiritual Assembly of, 153.2
Ghaznavi, Agnes, 267.3e, 439.2e
Ghodoussi, Ayatollah, 262.5
Giachery, Ugo, (Hand of the Cause of God) 24.17, 42.17, 104.1, 110.2, 130.2, 163.1, 177.1, 288.1, 319.1, 447.2
Gibson, Amoz, p. v, 56.1, 129.1, 209.1; passing of, 326
Gleanings from the Writings of Bahá'u'lláh, 27.3, 217.12, 308.13, 425.6–7, 425.9c, 425.10a; Bahá'í principles expounded in, 55.10
Global Plan: launching of second, 6.6

God Passes By, 24.15, 27.2, 80.4, 157.3, 192.4, 251.5, 455.3
God's Holy Mountain, 2.1
God's purpose for man, 55.2–4
Golden Age, the, 86.2b, 95.6–7, 95.19, 95.24, 164.5, 451.3, 451.14
Golden Age of Islam, 63.6
Greatest Holy Leaf, 313.1, 337.1, 337.6; commemoration of passing of, 321.3, 334; references to, 313.4–9. *See also* Bahíyyih Khánum.
Greatest Name, 147.20
Greece, National Spiritual Assembly of, 141.7, 169.1
Greenland, 171.2
Grenada, National Spiritual Assembly of, 221.13b, 371.1
Grossmann, Hartmut, 267.3e, 439.2e
Grossmann, Hermann, (Hand of the Cause of God) 62
Guam, 92.1, 100.4
Guardian, 2.1, 340.4, 378.13; comments of, on Universal Court of Arbitration, 412.3b–3c; delineation of Continental Counselors' role by, 10.1–2; divinely guided interpretations of, 6.3, 23.20, 141.3, 308.14; duties and functions of, 23.20, 75.5–8; duties and functions of Local Spiritual Assemblies set out by, 280.3; explanation by, of Lesser Peace, 422.3; explanation by, of role of individual, 14.7; expounding on National Fund, 9.2a; first African Teaching Plan of, 338.2; first Seven Year Plan of, 185.3; last message to Bahá'ís of Australia, 341.6; leadership of, 2.6; letters of, 54, 161, 162.21, 341, 409, 409.5–6, 427.2, letters of, to individual believers, 55, 75, 214, 308, 397, 402, 407, 412, 422, 448; letters of, to National Spiritual Assemblies, 4, 79, 147, 151, 220, 226, 239, 255, 322, 336, 375, 383, 396, 402, 425, 434; letters of, to youth, 428; no way to legislate for second, 5.1, 6.2; passing of, 23.3, 59.1, 75.14, 111.5, 321.3; in relation to the Universal House of Justice, 23.20; resting-place of, 1.4, 2.3; role of, in raising Administrative Order, 14.3, 46.4, 340.3; safeguarding letters of, 54, 161, 409; as "sign of God," 141.3; statement of, on principle of oneness of mankind, 427.2; spiritual authority of, 308.14; summons of, to Bahá'ís of the Americas, 340.4; teaching plans formulated by, 187.3; Ten Year World Crusade of, 2.3, 6.1, 82.2, 97.2, 134.1; writings of, 14.5, 86.2b, 152.3, 206.3c, 313.5, 448.2. *See also* Shoghi Effendi; Guardianship.
Guardianship, 5, 6.2, 35, 75, 164.2a, 251.5; anticipation of, in the Kitáb-i-Aqdas, 251.5; duties and functions of, 75.16; impossibility of appointing a successor to Shoghi Effendi, 5, 6.2; false claim of Charles Mason Remey to, 144.1; institution of, 251.5; and the Universal House of Justice, 35, 75, 412. *See also* Guardian; Shoghi Effendi.
Guaymies, 450.1
Guinea (Equatorial), National Spiritual Assembly of, 134, 153.2
Guinea-Bissau, National Spiritual Assembly of, 153.1–2
Gulmuḥammadí, Zabíḥu'lláh, 267.3c, 439.2c

Gulshaní, 'Azízu'lláh, 322.1, 323.3
Güney, Aydin, 267.3c, 439.4
Gwalior, 308.7

Habíbí, Muḥammad (Suhráb), 285.2
Habíbí, Muḥammad-Báqir (Suhayl), 285.2
Hadíth, 308.14
Haifa, 1.4, 2.5, 53.1, 74.2, 262.11
Ḥakím, Luṭfu'lláh, p. v, 48; passing of, 64
Ḥakím, Manuchihr, 273–274
Ḥakímán, Jalál, 360.1
Ḥakímán, Raḥmatu'lláh, 387.1
Hamadan, 285.1
Hamilton, 248.14
Hancock, Tinai, 267.3d, 439.2d
Hands of the Cause of God, 1.2–3, 34.6, 59.7, 321.4, 394.3; Auxiliary Boards and, 11.1, 20.7, 21.5; impossibility of further appointments of, 111.6–10; and International Teaching Center, 141.18; as representatives of Universal House of Justice at International Conferences, 163; consultation with, 2.6, 5.1, 10.3, 59.6; continental, 10.1; continental zones of, 20.6; development of institution of, 20; functions of, 20.9; in Holy Land, 361.2; London Conclave of, 11.1; as members of International Teaching Center, 132.4, 132.6, 141.18, 163; memorial services for, 25, 90, 102, 104, 135, 237, 243, 247, 256, 269; participation of, in first National Conventions, 91; pilgrims, 3.5; relationship of, to National Spiritual Assemblies, 10, 21; as Standard-Bearers, 6.6, 14.10, 91.1; tribute to, 2.4, 128.10–11, 321.4. *See also* 'Alá'í, Shu'á'u'lláh; Alexander, Agnes;

Baker, Dorothy; Balyuzi, Hasan M.; Banání, Músá; Collins, Amelia; Dunn, Clara; Dunn, Hyde; Esslemont, John E.; Faizi, Abu'l-Qasim; Featherstone, Collis; Ferraby, John; Furútan, 'Alí-Akbar; Giachery, Ugo; Grossmann, Hermann; Haney, Paul; Ioas, Leroy; Khadem, Zikrullah; Kházeh, Jalál; Maxwell, William Sutherland; Muhájir, Raḥmatu'lláh; Mühlschlegel, Adelbert; Olinga, Enoch; Robarts, John; Root, Martha; Rúḥíyyih Khánum, Amatu'l-Bahá; Samandarí, Ṭarázu'lláh; Sears, William; Varqá, 'Alí-Muḥammad.
Haney, Paul, (Hand of the Cause of God) 24.17, 42.17, 163.1, 288.1; passing of, 346, 361.2
Hannoversche Allgemeine Zeitung, 251.1
Happiness, 19.7, 42.23, 126.4, 211.2, 224.2, 246.1, 246.6c, 246.9–10, 289.4
Ḥaqíqí, Akbar, 381.2
Ḥaqíqí, 'Ináyatu'lláh, 416.1
Ḥaqq-Paykar, Badí'u'lláh, 323.3
Ḥaram-i-Aqdas *(Most Holy Court)*, 2.2, 26.1, 139.1, 192.3
Harris, Robert, 439.2b
Harwood, Howard, 60.1, 132.25, 267.4a
Ḥasanzádih, Maḥmúd, 261.2
Ḥasanzádih-Shákirí, Ghulám-Ḥusayn, 392.1, 393.6
Hassan II (King, Morocco), 12.4
Haṣúrí, Rúḥu'lláh, 421.1
Hawaiian Islands, 90, 98.3, 403.2, 403.2b
Ḥaẓíratu'l-Quds: acquisition of, 2.10, 16.6, 87.1, 142.7, 142.10–12, 187.5, 222.19–22; damage to Iranian, 218.2, 226.2, 227.12;

goals for, 14.6, 275.9e
Helsinki (Finland), 141.6, 163.1, 171.3
Hemispheric Bahá'í Radio-Television Conference, 202
Hénuzet, Louis, 60.1, 132.25, 267.3e, 439.2e
Heohnke, Violet, 132.8, 132.25, 267.4a
Heroic Age: epochs of, p. xxxix, 1.2, 4.2, 95.22, 150.1, 192.3
Hidáyatí, Jahángír, 368.2, 396.14
Hidden Words, The, 308.4
Himmatábád (Iran), 293.3
Ḥiṣár (Iran), 322.3
Hofman, David H., p. v, 56.1, 129.1, 209.1, 359.1
Hokkaido (Japan), 100.4
Holy Days: observance of, 29, 137.4, 163.30, 394.7; recognition of, by authorities, 14.6, 110.4, 142.5
Holy Land: attack on Bahá'ís due to location of, 220.15; Centenaries held in, 67.2, 192; Conference of Counselors in, 445, 447; Hands of the Cause of God in, 361.2; International Bahá'í Council in, 388.1; International Teaching Center in, 131.1; Mazra'ih Mansion in, 127.1; pilgrimage to, 3.5; property acquisition in, 264; Seat of the Universal House of Justice in, 115
Holy Places: concern for safety of, 225; extension and beautification of, 40.5, 141.5; in Iran, 221.1; Muslim, 378.8; repairs to, 40.3; seizure and destruction of, by Iranian government, 215, 262.8, 282.1, 305.1, 314.1. See also Ḥaram-i-Aqdas; Holy Shrine(s).
Holy Shrine(s): of Bahá'u'lláh, 1.4, 26, 105.1; beautifying of, 13.5, 40.5, 105.1; location of, 220.15. See also Ḥaram-i-Aqdas.
Holy Year (1992), 447.5
Home for the Aged, 275.9e
Homosexuality, 126.9
Honduras, message to Youth Conference in (1983), 357
Hong Kong, International Teaching Conference in (1976), 141.6, 163.1; message to, 181. See also International Teaching Conferences.
Hoqbín, Kúrush, 363.1
House of 'Abbúd, 192.3
House of 'Abdu'lláh Páshá, 154, 157, 190.1, 198.1, 221.11c, 248.9, 358.3, 456.2
House of the Báb, 225, 235, 241, 282, 314.1, 358.2
House of Bábíyyih, 317.3
House of Bahá'u'lláh in Tákur, 305, 358.2
House of the Master, 361.7
Houses of Worship. See Mashriqu'l-Adhkár(s).
Houston, 223.1
Hoveida, Abbas Amir, 220.8, 239.2
Human Rights Day, 443.1
Human Rights Year (1968), 68.5
Human spirit, 438.12
Human suffering, 425. See also Suffering; Tests.
Hunting, 147.21
Ḥuqúqu'lláh, 147.14, 251.5, 456.5; application of, 430; introduction of, to West, 404; release of compilation on, 430; translation into English of material on, 419.5
Husbands and wives, 147.9–10, 251.4a–4c, 272, 402. See also Bahá'í family life; Marriage.
Húshmand, Suhayl, 367

IBAVC. *See* International Bahá'í Audio-Visual Center.
Iceland, 92.2, 101.1, 104.1, 147, 171.2
Importance of Prayer, Meditation, and the Devotional Attitude, 249
India, Mashriqu'l-Adhkár of. *See* Mashriqu'l-Adhkár(s).
India: teaching success in, 419.3; unit conventions in, 433.8
Indians. *See* Indigenous peoples.
Indigenous peoples, 100.4, 108.2, 172.3, 187.2, 187.6, 326.1, 340.4, 380.1, 454.1
Individual: spiritual enrichment of, 195.3a, 453.16; and teaching, 188
Infallibility, 330.5. *See also* Universal House of Justice, infallibility of.
Inheritance, law of, 147.12, 166.3
In memoriam: Shu'á'u'lláh 'Alá'í, 414; Agnes Alexander, 90; Leonora Stirling Armstrong, 265; Hasan M. Balyuzi, 247; Músá Banání, 102; Inparaju Chinniah, 245; Angus Cowan, 454; Laura Dreyfus-Barney, 150; Abu'l-Qasim Faizi, 269; John Ferraby, 135; Lloyd Gardner, 423; Amoz Gibson, 326; Hermann Grossmann, 62; Luṭfu'lláh Ḥakím, 64; Paul Haney, 346; Leroy Ioas, 25; Sylvia Ioas, 374; Ishráq-Khávarí, 119; Raḥmátu'lláh Muhájir, 243; Adelbert Mühlschlegel, 256; Enoch Olinga, 237; Raúl Pavón, 380; Ethel Revell, 388; Jessie Revell, 33; Ṭarázu'lláh Samandarí, 65
Innsbruck, message to European Youth Conference in (1983), 369
Institute, Louis Gregory Bahá'í, 394.3
Institutes, Bahá'í, 2.10, 117.3

Institutes, teaching, 14.6, 16.8, 87.1, 95.2, 109.1
Institution(s), 1.5, 117.3; aim of, 162.4; of the Auxiliary Board, 11, 137, 162.31; of the Continental Boards of Counselors, 162.31, 229, 267; development of, 3.2, 142.3-4; functions of the, 35.8; of the Guardianship, 5, 35, 75.2; of the Hands of the Cause of God, 6.1, 111; inseparability of, 35.9; of the International Teaching Center, 131.1, 132, 361; of the Local Spiritual Assembly, 118, 141.13-17, 162.29-30, 189; maturation of, 456.4; of the National Spiritual Assembly, 2.4, 14.6; relationship between, of National Spiritual Assembly and Hands of the Cause of God, 6.4-5, 10; responsibility of, 6.4; twin, 6.2, 164.2a, 456.4; of the Universal House of Justice, 2.5, 6.3, 35, 164.2b, 166, 412, 448
Intercontinental Conferences, 14.5, 24.17, 34.7, 42.17-19, 128.8; attendance by Hands of the Cause of God at, 42.17; message to, 46; locations of, 24.17
International Bahá'í Audio-Visual Center (IBAVC), 280.15
International Bahá'í Convention(s), first (1963), 2.2; second (1968), 68.2; third (1973), 116.2; fourth (1978), 207; fifth (1983), 321.3, 415.1
International Bahá'í Council, 33, 64, 361.7, 374, 388.1
International Conference(s) (1982), 341.3; Dublin, Ireland (June), 328; Quito, Ecuador (Aug.), 337; Lagos, Nigeria (Aug.), 338; Canberra, Australia (Sept.), 319, 341; Montreal, Canada (Sept.),

340; attendance of Hands of the Cause at, 288, 320; attendance of rural Bahá'ís at, 278; relocation of Manila conference to Canberra, 318, 319; results of, 358.4; teaching opportunities arising from, 279.9–10. *See also* International Teaching Conferences.
International Federation for Human Rights, 390.1
International Human Development Center, 182.1–3, 195.9
International Teaching Center, 131, 155.1, 169.1, 221.4, 221.11b, 394.3; call for pioneers by, 155.1, 191; Conference of Counselors at, 447.1; development of, 221.11b; duties of, 132, 361.5–6; establishment of, 131; expansion of responsibilities of, 456.2; first gathering in Seat of Universal House of Justice with members of, 354.2; further evolution of, 361; members of, 131.1, 230, 361.3; pioneering goals set by, 302.3, 410.4; plans of, for international teaching, 158.4; Seven Year Plan goals for, 275.9c; social and economic development contributions of, 379.5; term of service in, 141.6, 178.1, 361.4
International Teaching Conferences (1976–77), 141.6, 190.1; Helsinki, Finland (July 1976), 171; Anchorage, Alaska (July 1976), 172; Paris, France (Aug. 1976), 174; Nairobi, Kenya (Sept. 1976), 176, 178.1; Hong Kong (Nov. 1976), 181; Auckland, New Zealand (Jan. 1977), 184; Bahia, Brazil (Jan. 1977), 185; Mérida, Mexico (Feb. 1977), 187; attendance of Hands of the Cause at, 163; results of, 207.3. *See also* International Conferences.
International Travel Teaching. *See* Traveling Teaching.
International Women's Year (1975), 162.32
International Year of Peace (1986), 420, 422.5, 428.3, 434.1, 434.3
International Youth Conference (Columbus, Ohio, 1985), 431
International Youth Year (1985), 376, 386, 428.1; preparations for, 376
Ioas, Leroy, (Hand of the Cause of God) 24.17, 25
Ioas, Sylvia, 374
Iran: alleged Bahá'í involvement in attempted coup, 260.1; arrest of National Spiritual Assembly of, 259, 306; attorney-general of, 377.1, 381.2, 421.1; Bahá'ís considered enemies of Islam in, 297.1; banning of Bahá'í Administration by government in, 377; Central Revolutionary Committee in, 299.1; civil rights in, 231.1; constitution of, 238.1, 239.7, 262.7; conversion of National Bahá'í Headquarters into Islamic University, 231; demolition of House of Bahá'u'lláh in, 305; denial of execution of National Spiritual Assembly members of, 309; destruction of House of the Báb in, 241, 282; disbanding of Bahá'í Administration in, 381.2; dissolution of National Spiritual Assembly of, 377.2; events in Marvda<u>sh</u>t, 218.2; false accusations against Bahá'ís in, 260.1, 377.1; Islamic Revolution in, 378.3; Ministry of Education in, 227.12; National Spiritual Assembly of, 218.6, 234.5,

Iran *(continued)*
377.2; official religions of, 238, 262.9; omission of Bahá'ís from constitution of, 238; persecution of Bahá'ís by government of, 215, 218, 220, 221.13g, 225, 227, 231, 238, 239, 241, 253, 254, 255, 259, 260, 263, 268, 270, 273, 274, 277, 281, 285, 286, 287, 290, 292, 293.1–3, 296, 297, 298, 299, 300, 306, 307, 314, 322, 344, 370, 381, 383, 392, 395, 418, 437; Public Prosecutor of, 378.3, 378.5–25; Rastakhiz Party in, 239.6; release of Bahá'í prisoners in, 328.2; religious minorities of, 220.13, 231.1, 238.1, 239.8, 262.7; revolutionary courts in, 297.1; revolutionary guards in, 395.7; seizure of Bahá'í endowments in, 314.1; situation in, effects of Bahá'í actions on, 234

Iranian Bahá'ís: abductions of, 292.2; arrests of, 259.1, 260.1, 372–3, 322.4, 370.2, 381.2, 387.2, 392.2; assets of, frozen, 293.1, 314.1; calls for day of prayer for, 226, 315, 350; campaign of arrests of, 299; confiscation of homes of, 317.3; defense of, 255, 262, 314, 317.3, 383.8; denial of executions of, 309; disappearances of, 307.3; disbanded Assemblies and organizations of, 378.23, 394.2; dissolution of National Spiritual Assembly of, 377.2; education of youth by, 364.2; false accusations against, 260, 262.2, 281.1, 392.3; forced confessions of, 395, 392.3; harassment of, by fanatical Shí'ah fundamentalists, 227.7; heightened steadfastness of, 317; held in captivity, 368.1; immolation of, in Núk Village, 270; *Le Monde* article on, 262; message to, throughout world, 246; National Spiritual Assembly of, 218.6, 268, 306–7, 378; number of imprisoned, 299.1, 392.2, 421.1, 437.3; omission of, as recognized religious minority in Iranian constitution, 238; open letter to Iranian authorities by, 378; prisoners released, 328.2, 392.2; radio summons of, 297.1; refutation of accusations against, 220, 239; statement by president of United States about, 362; strengths of, 246.11; torture of, 285.1, 395–96; whereabouts of National Spiritual Assembly uncertain, 268; worldwide response to persecutions of, 263, 362, 383. *See also* Iran; Martyrdoms; Persecution, of Bahá'ís in Iran.

Iraq: dissolution of National Spiritual Assembly of, 110.3

Ireland, 92.2, 104.1

Isfahan, 290.1, 296.1, 360.1, 378.5, 381.3, 387.1, 396.13; Episcopalian case in, 378.7

'Ishqábád (Turkmenistan): Bahá'í center in, 4.3; Bahá'í House of Worship in, 4, 157.4; conversion of House of Worship into art gallery, 4.6; earthquakes in, 4.1, 4.7

Ishráq-Khávarí, 119.1

Ishráqí, 'Ináyatu'lláh, 363.1, 364.1

Ishráqí, 'Izzat Jánamí, 364.1

Ishráqí, Muḥammad, 381.3

Ishráqí, Ru'yá, 364.1

Islam, 8.3, 18.1, 82.3, 162.6, 220, 227.7, 262.9, 281.1, 308.14; Golden Age of, 63.6; misrepresentation of Bahá'ís by fanatical sect of, 227.7; reverence of our

Faith for, 314.9; Shí'ih sect of, 330.7a
Islamic Republic of Iran, 225.1
Islamic Universities, 231, 262.9
Israel, 262.11; Bahá'í cemetery in, 440.1, 440.4; negotiations with government of, 74.2
Italy, 39, 151, 169.1
Ívál, Village of, 368.1
Ivory Coast, 91.1, 153.2; establishment of Publishing Trust in, 275.9g

Japan, 100, 146.1, 275.9d
Jerusalem, 74.2, 166.1, 378.8
Jesus Christ. *See* Christ.
Jews, 262.7
Jordan, 141.7, 153.1
Jubilee. *See* Most Great Jubilee; Riḍván.
Judaism, 308.14

Kámil-Muqaddam, Asadu'lláh, 396.16
Kampala, 24.17, 237.1. *See also* Uganda.
Karaj (Iran), 323.3, 413.1
Karbilá (Iraq), 378.8, 378.21
Káshání, Ḥusayn Báqir, 264.2
Káshání, Jamál, 416.1
Káshmar (Iran), 290
Kathírí, Iḥsánu'lláh, 400
Kátibpúr-Shahídí, Ni'matu'lláh, 290.1
Katrina cedar wood, 264.4
Kavelin, H. Borrah, p. v, 56.1, 129.1, 209.1, 223, 236, 359.1
Kayhán Daily, 298.1, 323.3
Kazemi, Zekrullah, 132.8, 132.25, 153.3, 267.3a, 439.2a
Káẓimí-Manshádí, 'Abdu'l-Vahháb, 261.2
Kebdani, Muḥammad, 60.1, 132.25, 153.3, 267.3a, 439.2a
Kenya: message to National

Women's Conference in, 389; International Teaching Conference in, 176
Kermani, Salisa, 267.4a
Khádi'í, Manúchihr, 292.1
Khadem, Zikrullah, (Hand of the Cause of God) 91.1, 110.2, 177.1, 404.1
Khamsí, Mas'úd, 60.1, 70.1, 132.9, 132.25, 267.3b, 361.3
Khan, Peter, 180.1, 267.3d, 361.3
Khándil, Ḥusayn, 285.2
Kháníábád Village (Iran), 328
Khayrkháh, Ibráhím, 317.2
Khayyámí, Iḥsánu'lláh, 322.1, 323.3
Kházeh, Jalál, (Hand of the Cause of God) 104.1
Khelghati, Thelma, 169.1, 267.3a, 439.2a
Khianra, Dipchand, 132.8, 132.25, 267.3c, 439.4
Khomeini, Ayatollah, 310.1, 378.7n
Khurásán (Iran): achievements in, 178.1; persecution of Bahá'ís in, 270.1, 290.1, 300.1, 322.1, 322.3, 406.1
Khushkhú, Sattár, 281.1
Khuy (Iran), 381.2
Khuzayn, Ṭarázu'lláh, 285.2
Kibbutz Ein Gev, 440.4–5
King, Coretta Scott, 434.3
King, Martin Luther, Jr., 434
King, Lauretta, 267.3b, 439.2b
King of Glory. *See* Bahá'u'lláh.
Kingdom of God on earth, 6.5, 6.10, 9.4, 88.4
Kings. *See* Súriy-i-Mulúk.
Kingston (Jamaica), Oceanic Conference held in, 68.6
Kingstown (St. Vincent), seat of National Spiritual Assembly of St. Vincent, 248.14
Kiribati, National Spiritual Assembly of, 403.2b
Kirmán (Iran), 387.1, 392.1

Kirmánsháh (Iran), 416.1
Kitáb-i-'Ahdí, The, 111.3
Kitáb-i-Aqdas, The, 24.20, 27, 75.4, 86.1, 96.11, 128.5, 141.5, 171.1, 251, 330.2, 375.9; centenary of revelation of, 116, 132.1; codification of, 14.5, 125; comments on, 251; Ḥuqúqu'lláh ordained in, 251.5; inheritance, 166.3; laws of, about observance of holy days, 29.1; laws not binding in the West, 147; membership of Universal House of Justice mentioned in, 166; laws of, concerning men and women, 145; obligations of women mentioned in, 455.1; publication of, 125, 128.5; punishments prescribed in, 147.23; reply by Universal House of Justice to questions regarding, 251.3–5; sexual problems mentioned in, 126.9; supplementary material to, 27.4b; *Synopsis and Codification of,* 125, 128.5. *See also* Laws.
Kitáb-i-Íqán, The, 330.7–7a
Knights of Bahá'u'lláh, 2.3–4, 63.7, 82.3, 97.2, 98.3, 237.1, 243.1
Koirala, Bharat, 439.2c
Korea, 100.3, 146.1, 200.1
Kruka, Josephine, 171.3
Kurdistán, 323.3

Lagos (Nigeria), International Conference in, (1982) 275.9d, 288.1; message to, 338
Lake Kinneret, 440.1
Lamb, Artemus, 60.1, 132.25, 267.3b, 439.4
Laos, formation of National Spiritual Assembly of, 38.1
La Paz (Bolivia), conference in, (1970) 68.6; message to, 82.2
Latin America: 337.2–3, 380.1; first Local Spiritual Assembly in, 187.3; Mashriqu'l-Adhkár of, 14.6, 110.2, 337.2; second Bahá'í radio station in, 321.2
Latin America, Mother Temple of. *See* Mashriqu'l-Adhkár(s).
Law(s), 128.5, 308.14, 394.5, 405, 425.7; attitude toward, 126.10; of burial, 280.22–23; concerning men and women, 145; of God, 289.3, 308.14; of Ḥuqúqu'lláh, 404.1, 419.6; of intestacy, 402.3; of the Kitáb-i-Aqdas, 125, 126.9, 145, 147; of marriage, 280.21; of Morocco, 8.6; not yet binding, 147; obedience to, 126.4–7; of personal living, 126, 394.5; understanding of, 405
Leadership, 118.6–6a
Learned, the, 111.3–14
Leeward Islands, formation of National Spiritual Assembly of, 221.13b, 248.14
Le Monde, 253; defense of Iranian Bahá'ís published in, 262.3
Leong, Yan Kee, 60.1, 132.25, 267.3c, 439.4
Lesotho, first national convention of, 91.1
Lesser Peace, 149, 420.1, 422
Letter of the Living, ninth, 82.3
Liberia, 153.1–2
Libya, 153.2
Literature, Bahá'í, 2.10, 94.1, 162.20, 275.9h, 280.8, 453.14; publishing of, 94.1; sale and distribution of, 94.3r; translation of, 14.6
Literacy, 222.15, 280.31–32, 280.37, 280.39, 308.7, 379.6
Local Spiritual Assemblies, 2.7, 14.6, 118, 141.13–17, 162.29–31, 181.6; attendance at meetings of, 266, 280.6, 394.7; consolidation of, 189; consultation by, 79;

development of, 142.3–4, 275.9j; duties and functions of, 18.5, 280.3; election of, 219.1; embryonic nature of, 118; formation of, 113, 189, 199, 219; goals of, 162.30; leadership by, 118.6; objectives of, 118.4–5; organizing of social gatherings by, 280.6; primary duties of, 394.7; release of compilation on, 84; Seven Year Plan goals for, 222.2–4, 280.2–3; Six Year Plan goals for, 453.15
Lomé (Togo), 153.1
London Conclave. *See* Hands of the Cause of God.
London (England), 1.4, 2.1, 2.5, 2.12; Guardian's resting-place in, 156, 177.1; celebration of the Most Great Jubilee in, 2.1, 2.3
London (Ontario), 419.3; Bahá'í Youth Conference in, 408
Long Healing Prayer, translation of, 258
Los Angeles, 223.1
Louis Gregory Bahá'í Institute, 394.3
Loyalty Islands, formation of National Spiritual Assembly including, 403.2b
Lucas, Paul, 132.8, 132.25, 267.4a
Lutchmaya, Roddy Dharma, 439.2a
Luṭfí, Kámrán, 393.2

McHenry, John III, 70.1, 132.25
McLaren, Peter, 132.8, 132.25, 267.3b, 439.2b
McLaughlin, Robert, 115.1
Madhya Pradesh, 308.7
Mahábád (Iran), 254.1
Maḥmúdí, Húshang, 307.3
Maḥmúdí, Zhínús, 307.1
Maḥmúdnizhád, Muná, 364.1
Maḥmúdnizhád, Yadu'lláh, 355.1
Majdhúb, Maḥmúd, 307.1

Majuro (Marshall Islands), 169.1
Maka, Lisiate, 267.3d, 439.2d
Malagasy Republic, 92.1, 104.1
Malaysia, 89, 146.1, 200.1, 245.1
Mali, 141.7, 153.2, 169.1, 221.13b, 394.8
Malietoa, Tanumafili II. *See* Tanumafili II, Susuga Malietoa.
Manaan, Mohammad 7.1
Manifestation(s) of God, 211.2, 214.2, 289.2, 308.11, 330.5–6, 378.11, 425.9d
Maní'i-Uskú'í, Ḍíyá'u'lláh, 417.1
Manila, international conference to be held in, 275.9d, 288.1, 318
Mankind: station and nature of, 425.5b–7a; suffering of, 154, 425. *See also* Human spirit.
Manshád Village (Iran), 255.1, 293.2
Manṣúrí, Muḥammad, 331.1
Maqámí, Manúchihr Qá'im, 307.3
Mariana Islands, formation of National Spiritual Assembly of, 92.1, 208.1, 403.2b
Markazí, Shápúr, 411.1, 413.2
Marriage, Bahá'í, 110.4, 126.7d, 142.5, 280.20–21, 308.13, 378.24; betrothal, 147.7; certificates of, 14.6, 280.21; dowry, 147.8; laws of, 30.5; sex and, 126.7–7d; Tablets on, 402.7
Marsella, Elena, 132.8, 132.25, 267.4a
Marshall Islands, 403.2b; formation of National Spiritual Assembly including, 92.2; formation of National Spiritual Assembly of, 169.1, 184.5; presentation of peace statement in, 442.2
Martinique, formation of National Spiritual Assembly of, 221.13b, 371.1
Martyrdoms: of Iranian Bahá'ís, 193, 218.2–3, 227.9, 227.15, 254,

Martyrdoms *(continued)*
261, 263.1, 268, 277, 281, 285, 286, 287, 290, 292, 296, 297.1, 299, 307, 311, 312, 314, 323, 324, 328, 331, 335, 339, 345, 351, 355, 356, 360, 363, 364, 367, 387, 393, 394.2, 396.3, 399, 400, 406, 411, 413, 416, 417, 421, 424, 437, 441; in Philippine Islands, 121; denial of, 309. *See also* Iranian Bahá'ís; Persecution, of Bahá'ís in Iran.

Martyrs, prayer of 'Abdu'l-Bahá for, 352

Marvdasht (Iran), 218.2

Masehla, William, 169.1, 267.3a

Mashhad (Iran), 262.9, 317.3, 322.1, 322.3, 323.3, 396.11, 399.1, 413.2

Mashriqu'l-Adhkár(s), 128.14, 142.8, 147.15, 265.1, 275.9f, 313.2, 315.4, 326.1, 346.3, 375.7, 379.2; of the Antipodes (Sydney), 184.3, 341.4; architects of, 143, 198.1, 210.1; call for day of prayer at, 315.4; dependencies of, 275.9e; of Europe (Frankfurt am Main), 17, 275.9e; first, of the Bahá'í world, 157.3b, 341.4; first, of the West (Mother Temple of the West, Wilmette), 4.5, 265.1, 340.1; future, on Mt. Carmel, 40.5, 105, 128.19, 164.2a; future, of Tehran, 128.16; in India (Mother Temple of Indian Subcontinent, New Delhi), 4.3, 14.6, 141.6, 143, 196, 198, 221.13a, 242, 250, 257, 275.9b, 284.3, 284.5, 321.2, 385.4, 419.4, 452; of 'Ishqábád, 4, 157.3b, 341.4; Mother Temple of Latin America (Panama City, Panama), 14.6, 24.3, 50, 82.2, 87.1, 97.1, 108; in Samoa, 141.6, 143, 184.5, 210, 221.13a, 284.3, 284.5, 321.2, 385.4, 403; symbolism of, 403.3

Masjid Sulaymán, 293.3

Mass media, proclamation of the Faith through, 141.9, 263.4, 280.13–17, 321.2, 358.6, 453.12

Mass teaching. *See* Teaching the masses.

Master, the, 6.3, 97.1, 340.1, 340.4, 379.2; recordings of the voice of, 93. *See also* 'Abdu'l-Bahá; Divine Plan of 'Abdu'l-Bahá.

Ma'ṣúmí, Muhammad-Ḥúsayn, 270.1

Materialism, 375.3, 438.21–23

Maturity, age of, 147.17, 333.11–14, 426

Mauritania, 153.1–2; formation of National Spiritual Assembly of, 141.7, 208.1

Mauritius, Oceanic Conference in (Rose-Hill,1971), 68.6, 85.1; message to, 82.3–4

Mavaddat, Farhang, 286.1

Maxwell, William Sutherland, (Hand of the Cause of God) 60.1, 132.8

Maxwell, May, 82.2, 150.1, 174.1, 185.3

Mayberry, Florence, 60.1, 131.1, 132.5, 361.3

Mázindarán (Iran), 368.1

Mazra'ih, 190.4, 192.2, 264.3; Mansion of, 127, 128.18, 264.3

Mecca, 82.3, 176.2

Meditation, 375.5, 375.7–9

Mediterranean, 63.6–8

Melanesia, 98.4–5

Melanesians, 403.2b

Memorials of the Faithful ('Abdu'l-Bahá), 440.5

Men and women, laws of the Kitáb-i-Aqdas concerning, 145

Mérida (Mexico), International

Teaching Conference in (1977), 141.6, 163.1; message to, 187
Messages from the Universal House of Justice: 1968–73, 412.2
Micronesia, 98.4–5, 100.4
Micronesians, 403.2b
Mihdí, Mírzá *(the Purest Branch)*, 80, 164.2–2b
Mihdízádih, Iḥsánu'lláh, 281.1
Miller, William McE., 251.3
Mindanao State University, 121.2
Minorities, 77.11, 326.1, 453.12
Mitchell, Glenford, 332.1, 359.1
Mítháqí, Alláh-Virdí, 292.1
Mítháqíyán, Firaydún, 132.8, 132.25, 200.1
Mitháqíyyih Hospital, 227.2
Míyán-Duáb (Iran), 218.3
Mmabatho (Bophuthatswana), 248.14
Moffett, Ruth, 214.5
Monadjem, Shapoor, 366.1, 439.2b
Monogamy, 251.4b–4c
Monrovia (Liberia), 153.1; Continental Conference in (1971), 68.6; message to conference in, 88
Montreal (Quebec), International Conference in (1982), 275.9d, 288; message to, 340
Morocco, 153.2; appeal of Baháʼí prisoners in, 7.1, 8.4, 8.8; death sentence of prisoners in, 8.3, 12.1; embassy in, 12.4; law of, 8.6; persecution of Baháʼís in, 8, 12, 390; release of Baháʼí prisoners in, 12; report on Baháʼí prisoners in, 8; sending of International Federation for Human Rights representative to, 390; Supreme Court of, 12.4
Morrison, Gayle, 439.2d
Mosques, 215.1
Most Exalted Leaf, the, 451.6. *See also* Bahíyyih Khánum.
Most Great Jubilee, 1.2–3, 2.1, 2.3, 2.8, 14.4, 24.1
Most Great Peace, the, 422.2–3, 451.14
Most Holy Book, 4.3. *See also* Kitáb-i-Aqdas, The.
Mother Temple: of the Antipodes, 341.4; of Europe, 2.9, 15.1, 17; of Indian Subcontinent, 196, 456.2; of the Pacific, 456.2; of Latin America, 50, 108, 110.2; of the West, 61, 212. *See also* Mashriqu'l-Adhkár(s).
Mt. Carmel, 3.3, 14.5, 40.5, 115.1, 128.19, 132.2, 164.3, 361.7; Arc on, 128.19, 132.2, 165.1, 275.9a, 394.8; Báb's sepulcher on, 157.3b, 164.2a; excavation of, completed, 169.1; grave of Mírzá Mihdí on, 80.3; obelisk on, 105, 128.19; site of future Mashriqu'l-Adhkár on, 40.5, 128.19; transfer of remains of Holy Family members to, 164.2. *See also* World Center.
Mozambique, formation of National Spiritual Assembly of, 221.13b, 394.8
Muhájir, Raḥmátu'lláh, (Hand of the Cause of God) 104.1, 337.2; passing of, 243, 248.1
Muhájir Teaching Institute, 304.1
Muḥammad (the Prophet), 220.13, 262.10
Muḥámmad-'Alí, Mírzá, 26.1; archbreaker of the Covenant, 157.7
Muḥammadí, 'Askar, 323.3
Muḥammadíyyih, Village of, 387.1
Muḥammad-Qulí, Mírzá, 440
Muḥammad-Taqí, Ḥájí Mírzá, 4.4
Muhandisí, Khusraw, 311.1
Mühlschlegel, Adelbert, (Hand of the Cause of God) 91.1, 104.1; passing of, 256
Mühlschlegel, Ursula, 267.3e, 439.2e

Mukhtárí, Mír-Asadú'lláh, 254.1
Mumtází, Rúḥu'lláh, 60.1, 132.25, 267.3c, 439.2c
Muqarrabí, 'Aṭá'u'lláh, 307.3
Muqímí-Abyánih, Zarrín, 364.1
Murder, 147.23. *See also* Martyrdoms.
Musavi, Mir Husayn (prime minister of Iran), 310.1
Mushtá'il-Uskú'í, Jalálíyyih, 324.1
Music and singing, release of compilation on, 107
Muslim religion. *See* Islam.
Muslims, 8.6, 298.1, 378.8
Muṣṭafá, Muḥammad, 153.3, 267.3a
Mustaqím, Jalál, 261.2
Muṭahhar 'Abdu'l-Majíd, 387.1
Muṭahharí, 'Alí, 261.2
Mutlaq, Husayn, 285.2
Muvaḥḥid, Muḥammad, 307.3
Mystery of God, 96.1, 330.3. *See also* 'Abdu'l-Bahá.

Nádirí, Bahíyyih, 307.3
Nador (Morocco), 7.1
Nagaratnam, S., 267.3c, 439.2c
Na'ímí, Fírúz, 285.2
Na'ímíyán, 'Alí, 339.1
Nairobi (Kenya), 141.6, 163.1, 176, 178, 179.1
Najaf (Iran), 378.8
Najafábád (Iran), 387.1
Nají, Husayn, 307.3
Nakhjavání, 'Alí, p. v, 56.1, 129.1, 209.1, 359.1
Nakhjavání, Bahíyyih, 272.1, 272.4
Namibia, 221.13b, 248.14
Naráqí, Sírús, 439.2d
National Bahá'í communities, financial independence of, 221.13g
National Bahá'í Funds. *See* Funds, Bahá'í.
National Bahá'í Constitution, 308.17

National Bahá'í Women's Conference, Kenya (1984), 389
National Convention(s), 11.2a, 11.2e, 190.2; election of delegates to, 433; first, 91; messages from, 3.1; messages to, (1963) 2, (1964) 15, (1968) 57, (1978) 208
National Fund, the. *See* Funds, Bahá'í.
National Register of Historic Places, 212
National Spiritual Assemblies, 2.4, 14.6; administrative matters of, 336.5–7; attendance of members of, at International Conferences, 278.1; budgets of, 2.9, 9.3, 50.3; call for formation of, 190.2; communications from, 3.1, 222.17; and consolidation activities, 2.7; Constitution of, 333.16; contributing to International Fund, 40.7; Declaration of Trust and By-Laws of, 308.17; election of delegates to National Conventions, 433; electing the Universal House of Justice, 2.1; formation of, 2.10, 6.7, 34.5, 38, 68.8, 71, 85, 92, 104, 134, 141.7, 153.1–2, 190.1, 221.13b, 248.14, 371, 394.3, 394.8; goals of, 280.11, 453; holding orientation courses for pioneers, 384.6; incorporation of, 39; and International Teaching Center, 361.6; members of, as assistants to Auxiliary Board members, 336.5; members of, as Auxiliary Board members, 11; membership records of, 426.4; and National Teaching Committees, 194, 280.5; and Nine Year Plan, 6.6, 14.8; organizing children's activities, 162.36; properties of, 222.19; providing pioneers, 232.3; and relations with

government officials and prominent people, 162.24–25; relationship of, and Hands of the Cause of God, 6.4, 10; relationship of, to Local Spiritual Asemblies, 280.2; release of compilation on, 114; secondary Houses of Justice, 358.4; and Seven Year Plan, 221.12–13; and Six Year Plan, 453; and spiritualization of Bahá'í community, 375.2; statistical reports of, 280.11; and traveling teachers, 51. *See also* National Fund, the.
National Teaching Committees, 194, 280.5
National Youth Conference(s). *See* Youth Conferences.
Nationalism, 77.1, 438.31, 438.35
Navváb *(Ásíyih Khánum, the Most Exalted Leaf)*, 164.2–2a
Nawnahálán Company, 227.21, 239.11
Naw-Rúz messages, (1976) 169, (1980) 248
Nawrúzí-Íránzád, Yúnis, 413.1–2
Nayshábúr (Iran), 293.3
Nayyirí-Isfahání, Husayn, 381.3
Nazareth, 264.2
Nepal, formation of National Spiritual Assembly of, 92.1, 104.1, 358.4; dissolution of National Spiritual Assembly of, 169.1
Netherlands, 384.3
New Caledonia, 403.2b
New Delhi, 24.17; Mashriqu'l-Adhkár (Mother Temple of the Indian Subcontinent) of, 358.3, 452; women's conference in, 197
New Era School (Panchgani), 308.7
New Guinea, 162.6
New Hebrides, formation of National Spiritual Assembly of, 141.7, 169.1, 184.5

Newsletters, Bahá'í, 120, 162.19, 280.18–19
New World Order, 181.2
New York City, 128.9, 432.1; as City of the Covenant, 447.5
New Zealand, 184.3, 402.2
Niamey, formation of National Spiritual Assembly of, 153.1
Nicobar Islands, formation of National Spiritual Assembly of, 371.1
Niederreiter, Leo, 267.3e, 439.2e
Niger, 221.5; formation of National Spiritual Assembly of, 141.7, 153.1–2
Nigeria, 153.2, 278.1; establishment of Publishing Trust in, 275.9g
Nikko (Japan), 100.2
Nine Year Plan (1964–73), 2.10–11, 13.5, 14, 57.1, 86.1, 97.2; acquisition of properties during, 16.6; acquisition of teaching institutes during, 14.6, 16.8, 87.1, 95.2, 109.1; announcement of, 6; commitment to victory during, 68.13; goals and objectives of, 6.8, 10.4, 14.6–9, 16.5, 16.7–10, 66.3–4, 106, 110.9, 125.1; guidelines for, 16; international cooperation during, 110.6, 128.15–16; launching of, 13.1, 14; Local Spiritual Assemblies and, 113; message to conference in Southeast Asia during, 89; national endowments and, 16.7; needs during, 68.11–12; phases of, 34.8, 47; and preparing for celebrations of centenary of proclamation of Bahá'u'lláh, 24.14–16; role of Hands of the Cause as Standard-Bearers during, 6.6, 14.10; tasks of, 14.5, 24.7–13, 34.8, 89.1, 89.3; territories opened during,

Nine Year Plan *(continued)*
16.5, 24.4; themes of, 19.1;
victories of, 128.2–3; youth
fields of service during, 37.4–8
Nineteen Day Feast. *See* Feast,
Nineteen Day.
Nírúmand, Mahshíd, 364.1
Níyákán, 'Alí-Ridá, 417.1
North America: as Cradle of
Administrative Order, 101.2; first
Bahá'í radio station in, 348
North Atlantic Conference. *See*
Oceanic Conferences.
North Pacific Oceanic Conference.
See Oceanic Conferences.
Northwest Pacific, 104.1
Núk (Iran), Village of, 270.1, 378.5
Nuqayb, 440.3
"Nuqtatu'l-Káf," 251.3

Obelisk, 128.19
Oceanic Conferences, 13.5, 14.5,
68.6, 96.5; First (Palermo,
1967), 53, 63, 66, 68.3; in Rose-
Hill, Mauritius (1970), 68.6, 82;
North Pacific (Sapporo, 1971),
68.6, 100; South China Seas
(Singapore [orig. Djakarta],
1971), 68.6, 89; South Pacific
(Suva, Fiji, 1971), 68.6, 98;
Carribean Conference in
Kingston, Jamaica (1971), 97;
North Atlantic (Reykjavik,
1971), 68.6, 100.7, 101, 171.2.
See also Continental Conferences.
Office of Social and Economic
Development, 379.4
Olinga, Enoch, (Hand of the Cause
of God) 91.1, 104.1, 163.1;
assassination of, 237, 248.1,
276.1
Olyai, Perin, 439.2c
Oman, formation of National
Spiritual Assembly of, 208.1

Oneness of mankind, 42.23, 97.3,
379.2, 422.2, 427.2, 438.29,
438.39–41
Ong, Rose, 439.2c
Opposition to the Faith: compilation on, 152; as an opportunity
for progress of Faith, 383.6
Oral Law, 308.14
Osborne, Alfred, 60.1, 132.25,
267.4a
Otavalo (Ecuador), 201.1
Ouagadougou (Upper Volta), 169.1
Oule, Kolonario, 60.1, 132.25,
267.3a, 439.4

Pacific Ocean, 91.1, 221.13b, 403.1–2b
Pacific Islands, 358.3
Palermo (Sicily), 53, 63, 66, 68.3
Panama: Mashriqu'l-Adhkár of,
24.3, 46.5, 50, 82.2, 87.1, 96.4,
108, 110.2; radio station in, 450
Panama City (Panama), 24.17
Panama Temple. *See* Mashriqu'l-Adhkár(s).
Panchgani (India), 308.7
Papacy, 308.14
Papua New Guinea, 403.2b
Paramaribo (French Guiana),
formation of National Spiritual
Assembly of, 169.1
Paris, 141.6, 150.1, 163.1, 174.2
Parliament, European, 263.3
Pars News Agency, 268.1
Pattern of Bahá'í Life, The, 122
Pavón, Raúl, 132.8, 132.25, 267.3b;
passing of, 380
Paymán, Khudárahm, 60.1, 132.25,
267.3c, 439.2c
Payraví, Jalál, 396.14
Peace: barriers to, 438.30–37;
conferences, holding of, 434.3;
International Year of (1986),
427.5, 434.3, 445.2; Most Great,
438.60; people's longing for,

438.7–8; release of statement on, 429; statement on, 429, 438, 442.2; women's role in establishment of, 448; world gathering for, 438.48–52
Peace statement. *See Promise of World Peace, The.*
People of prominence, 248.16
Pereira, Sarah, 132.8, 132.25, 267.3b, 439.4
Perks, Thelma, 60.1, 132.25, 267.4a
Persecution: of Bahá'ís in Iran, 215, 218, 220, 226, 227, 231, 233, 234, 235, 238, 239, 241, 246, 248.1–6, 253, 255, 259, 260, 262, 263, 268, 270, 274, 282, 292, 293, 297, 298, 299, 300, 301, 305, 306, 309, 310, 312, 314, 315, 317, 321.1, 322, 323, 331, 344, 347, 349, 350, 368, 370, 372, 373, 377, 378, 381, 383, 387, 392, 394.2, 395, 396, 418, 437, 447, 456.3; of Bahá'ís of Morocco, 390. *See also* Iranian Bahá'ís; Iran; Martyrdoms.
Persia, 221.1, 227.1, 251.3. *See also* Iran.
Personal transformation, 126, 214.3. *See also* Spiritual development.
Peru, 304, 321.2
Philippine Islands, 146.1, 278.1; martyrdom of Iranian Bahá'í students in, 121
Pilgrimage, 3, 68.9–10, 147.13, 226.2, 456.2; permissions for, 3.5
Pilgrim house, 3, 4.6
Pilgrims, 3.3, 154.1
Pioneer Committees, 24.8–10. *See also* Continental Pioneer Committees.
Pioneering, 162.8–12; goals, 110.7, 142.18, 179, 410.2–7; and consolidation, 36
Pioneers, 2.4, 2.7, 86, 88.2, 121.2, 128.14, 159.1, 162.8, 169.1, 280.26–27; calls for, 28, 31, 83, 155, 191, 232, 302, 394.10, 410; cautions to, against prejudice, 18.3; deployment of, 279; Five Year Plan goals for, 142.18, 179; homefront, 280.26, 453.12; message to, 86; prejudice, 18.3; Seven Year Plan objectives for, 221.13f, 222.5, 275.9l, 279.3; use of, 228. *See also* Continental Pioneer Committees.
Point-à-Pitre (French Antilles), 169.1
Politics, 88.3, 173; principle of noninterference in, 77, 173.8–10; relationship of Bahá'ís to, 55, 173.11–14
Polynesia, 98.4–5
Polynesians, 403.2a
Port Vila (New Hebrides), 169.1
Prayer and Meditation, 375.7–9; release of compilation on, 24
Prayer(s), 1.1, 280.6, 405.2; for Bahá'ís in Iran, 315, 350; for children and youth, 167.1; compilation of, 280.9; daily obligatory, 147.4, 375.7; dawn prayers, 142.14, 162.37, 350.2; for the Dead, 378.24; described by 'Abdu'l-Bahá, 375.7; Fire Tablet, 258; and Local Spiritual Assemblies, 394.7; Long Healing Prayer, 258; Medium Obligatory Prayer, 147.3; for Moroccan prisoners, 7, 12.2; release of compilation on prayer, meditation, and the devotional attitude, 249; revealed by 'Abdu'l-Bahá, 352; steps for guidance in, 214.5–6
Prayers and Meditations (Bahá'u'lláh), 147.3
Prejudice, 77.10–11, 88.3, 117, 438.10

Presbyterian missionaries, 251.2–3
Press, the, 2.3
Priceless Pearl, The, (Rúḥíyyih Rabbani) 157.8
Prince of Peace, 394.11. *See also* Bahá'u'lláh.
Príncipe, 153.2
Principles of Bahá'í Administration, (Shoghi Effendi) 55.9
Principles of the Bahá'í Faith, 77.12; confidentiality, 336; elimination of extremes of wealth and poverty, 438.30; equality of men and women, 438.33; harmony of science and religion, 217.5–6; noninterference in political affairs, 77.4–5, 378.6; oneness of mankind, 427.2, 438.40–47; trustworthiness, 435.7, 438.58; unity of religion, 438.16–17; universal education, 438.34
Pringle, Ruth, 267.3b, 439.2b
Proclamation, 42.21, 45, 128.8–9, 141.9, 280.13; activities, 45.3, 45.7, 68.4; of Bahá'u'lláh to kings and rulers, 6.8, 24.14–16, 42.16; campaign of, 420.3; to Heads of States, 358.2; opportunities for, 383.2; period of, 24.20–23; and publicity, 45.5; and teaching programs, 45.4, 141.9; worldwide, 128.8
Proclamation of Bahá'u'lláh, The, 42.16, 68.4, 130.2
Promise of World Peace, The, 429, 438, 440.1, 442, 443, 444.2, 445.2, 456.3
Promised Day Is Come, The, (Shoghi Effendi) 27.3, 149.2
Promulgation of Universal Peace, The, ('Abdu'l-Bahá) 175.2, 272.6b, 402.9, 448.3d
Propagation: Auxiliary Board members for, 11.2a, 137.2, 153.4; Auxiliary Boards for, 137, 153.4; as function of Continental Boards of Counselors, 58.1; as function of Hands of the Cause, 6.4, 14.5, 58.1
Properties, 24.13, 226.2, 227.2, 231.1, 239.10, 264, 275.9f; attacks on, 241.1; seizure of, 235.2; upkeep of, by Bahá'í community, 222.19–22
Property, lost, 147.22
Prophecy, Daniel's, 1.2
Prophets, 378.11, 425.2
Proselytizing, 308
Protection: Auxiliary Board members for, 11.2a, 137.2, 153.4; Auxiliary Boards for, 137, 153.4; as function of Continental Boards of Counselors, 58.1; as function of Hands of the Cause, 6.4, 14.5, 58.1
Publications: 222.14, 280.9; Sacred Texts, 280.8
Public Information Office: New York, 432.1; World Center, 432, 456.2
Publicity, 2.3, 45.5; about Moroccan situation, 12.3; and proclamation, 45.5
Publishing, Bahá'í: Bahá'í authors, 94.3m, 94.3o–3q; Bahá'í publishers, 94.3i–3n; cables, 94.3l; editing, 94.3m; principles of, 94; requirement of approval before publication, 94.3a–3j, 94.3n–3p; sale and distribution of Bahá'í publications, 94.3r; translations, 94.3d–3e, 128.15; works reviewed elsewhere, 94.3g
Puerto Rico, formation of National Spiritual Assembly of, 104.1
Punishments, 251.4d
Purdil, Fírúz, 413.2
Purest Branch, the. *See* Mihdí, Mírzá.
Púr-Ustádkár, Jam<u>sh</u>íd, 416.1
Pús<u>tch</u>í, Yadu'lláh, 287.1

Qadímí, Yúsif, 307.3
Qad-Ihtaraqa'l-Mukhlisún, 258.1
Qal'ih Malik, Village of, (Iran) 378.5
Qatar, formation of National Spiritual Assembly of, 208.1
Qazvín (Iran), 322.4, 331.1
Qiblih of Bahá'í world, 26, 128.17, 341.4, 440.1
Quito (Ecuador), 275.9d, 288.1, 337.2
Qum (Iran), 322.2
Qur'án, 378.11, 425.9a
Quddús, 176.2
Qurbánpúr, Amínu'lláh, 411.2

Rabbani School (Gwalior, India), 308.7
Rabbani, Shoghi. *See* Shoghi Effendi; Guardian; Guardianship.
Rabbinical Judaism, 308.14
Racial prejudice. *See* Racism.
Racism, 126.5, 308.8, 420.2a, 438.29
Radaví, Muhsin, 392.1
Radio Bahá'í Ecuador. *See* Radio stations, Bahá'í.
Radio and television: programs for, 308.7; publicity for Faith via, 2.3, 162.21–23, 201.1, 202.1, 280.13–17, 391.1; use of, in deepening, 162.21–23, 280.13, 308.7, 391.1; use of, in social and economic development, 358.6; use of, in teaching and proclamation, 162.21–23, 201.1, 202.1, 205.1, 280.13–17, 308.7, 337.3, 391.1
Radio stations, Bahá'í, 275.9i; first, in Otavalo (Ecuador), 201, 205, 213, 308.7, 380.1; del Lago Titicaca (Peru), 304, 321.1; Caracollo (Bolivia), 391, 394.3; at Louis Gregory Bahá'í Institute (WLGI, South Carolina), 348, 394.3; in Panama, 450
Rafat, Polin, 439.2e
Rahímíyán, Rahím, 393.3
Rahímkhán Village (Iran), 323.3
Rahímpúr (Khurmá'í), Írán, 381.4
Rahmání, Hádí, 60.1, 104.1, 132.25, 267.4a
Rahmání, Ibráhím, 307.3
Ramoroesi, Daniel, 439.2a
Ranks and stations, 206
Rawhání, 'Atá'u'lláh, 296.1
Rawhání, Hishmatu'lláh, 307.3
Rawhání, Masíh, 439.2c
Rawhání, Qudratu'lláh, 307.1
Rawshaní, Rúhí, 307.3
Rawshaní, Sírús, 307.1
Reagan, Ronald W. (pres., U.S.), 362.1; presentation of peace statement to, 443.1
Reason and faith, 217.4–14
Recognition of the Faith, 280.11–12
Reconstruction of society, 195, 425
Recordings (audio), 222.15–16, 453.14; of 'Abdu'l-Bahá's voice, 93.1; of Bahá'í talks, 93.1
Reed, Betty R., 81.1, 132.25, 267.3e, 439.4
Regional communities, elected representatives of, 2.1
Regional National Assemblies: dissolution of, 6.7; formation of, 6.7
Regional Spiritual Assemblies, 2.4, 187.3; division of, 2.10
Regional Teaching Committees, 194
Registration: as a Bahá'í youth, 426.4
Religion: disillusionment in, 438.17–20; as greatest means for establishment of order in world, 438.14; harmony of science and, 217.5–6; purpose of, 397.2, 438.13–14; strife in, 438.32

Remey, Charles Mason, 23.7, 144
Responsibilities of Bahá'í life, 246.6–10
Réunion, formation of National Spiritual Assembly of, 92.2, 104.1
Reuters News Agency, 231.1
Revelation, 24.1, 217.7, 308.14, 394.5, 425.9; and development of community life, 403.3; interpretation of, 308.14, 308.17
Revell, Ethel, 388
Revell, Jessie, 33
Reviewing Committees, 94.3f
Revolutionary Courts, 281.1, 297.1, 298.1
Reykjavik (Iceland), 68.6, 101, 101.6, 171.2
Riḍván, 2.1; celebrations of Festival of, 1.2–3, 2.1–3, 2.8; electing Local Spiritual Assemblies during, 189.4
Riḍván messages, (1965) 24, (1966) 34, (1967) 42, (1969) 68, (1970) 81, (1971) 96, (1972) 110, (1973) 128, (1975) 159, (1977) 190, (1982) 321, (1983) 358, (1984) 394, (1985) 427, (1986) 456
Riḍvání, Aḥmad, 296.1
Right of God, the. *See* Ḥuqúqu'lláh.
Rissanen, Väinö, 171.3
Robarts, John, (Hand of the Cause of God) 320
Robbery, 69.3
Rogers, Donald O., 439.2b
Root, Martha, (Hand of the Cause of God) 98.3, 185.3
Rose-Hill (Mauritius), 68.6
Rouhani, Mansour, 220.9, 239.2
Rouleau, Eric, 262.13
Royal Asiatic Society, 251.2
Ruhe, David, 56.1, 129.1, 209.1, 359.1
Rúḥí, Manúchihr, 406
Rúḥíyyih Khánum, Amatu'l-Bahá, (Hand of the Cause of God): attending first gathering of members Universal House of Justice in council chamber of Seat, 354.2; at Bahá'í Women's Conference, 197.1; at ceremony for laying foundation stone of Mother Temple of Indian Subcontinent, 196; at Conference of Counselors, 447.2; at conferences in Bolivia and Mauritius, 85.1; at dedication of Mother Temple of Indian Subcontinent, 452.1; at dedication of Mother Temple of Latin America, 110.2; at dedication of Samoan House of Worship, 394.3; at first national conventions, 91.1, 104.1; at Intercontinental Conferences, 24.17, 42.17, 163.1, 288.1; joined by new counselors moving to Holy Land, 361.3; laying foundation stone of House of Worship in Panama, 24.17, 46.5, 50.1; presenting peace statement to secretary-general of the United Nations, 442.1, 444.2; representing Universal House of Justice, 24.17, 42.17, 91.1, 104.1, 110.2, 163.1, 196, 288.1, 394.3, 452.1; at reinterment of remains of Mírzá Muḥammad-Qulí, 440.5
Rwanda, 104.1
Ryukyus, 100.4

Sabeti, Parviz, 239.2
Sábirán, 'Imád, 153.3
Ṣábirí, Símín, 364.1
Ṣábiríyán, Yadu'lláh, 393.4
Sabri, Isobel, 60.1, 132.25, 267.3a, 361.3
Sacred scriptures of mankind, promise of world peace in, 438.1
Sacred scriptures, Bahá'í: collation

and classification of, 14.5, 54.1, 128.4–5; regular reading of, 375.5
Ṣádiqí, Parvíz, 121.2
Ṣádiqípúr, 'Abbás-'Alí, 335.1
Ṣadiqzádih, Kámbíz, 307.3
Ṣafá'í, Suhayl, 360.1
Ṣahbá, Farìbúrz, 198.1
St. John's, 248.14
St. Lucia, 221.13b
St. Vincent, 221.13b, 248.14
Salmánpúr, Manúchihr, 60.1, 132.25, 267.3c, 439.4
Samandarí, Mihdí, 60.1, 132.25, 153.3, 267.3a, 439.2a
Samandarí, Farámarz, 254.1
Samandarí, Ṭarázu'lláh, (Hand of the Cause of God) 42.17; passing of, 65
Samaniego, Vicente, 60.1, 121.3, 132.25, 267.3c, 439.2c
Ṣamímí, Kámrán, 307.1
Samoa: Mashriqu'l-Adhkár in, 141.6, 143, 210, 275.6, 419.3; National Spiritual Assembly of, 403.2b. *See also* Western Samoa.
San Diego, 223.1
San Francisco, 223.1
Sangsar (Iran), 254.1
Saō Tomé, 153.2
Sapporo (Japan), 68.6, 101.6
Sárí (Iran), 368.1
Sassen (Spitzbergen),171.2
SAVAK (Iranian Secret Police), 220.5, 239.2, 239.5, 262.8, 262.12, 299.1
Saysán, Village of, (Iran) 322.2
Scandinavian volunteer service organizations, 384.3
Schechter, Fred, 267.3b, 439.2b
Scholars, Bahá'í: Hasan M. Balyuzi, 247; Ishráq-Khávarí, 119
Scholarship, 217, 453.13
School(s), 2.10, 4.6, 453.13; New Era Bahá'í, 308.7; Rabbani,

308.7; summer and winter, 95.2, 109.1, 117.3, 222.9, 375.1; tutorial, 222.11–13, 280.25. *See also* Education; Institutes.
Science and religion, 217.5–6
Scotland, 48.3
Scriptures. *See* Bahá'í Writings.
Sears, William, (Hand of the Cause of God) 276; *A Cry from the Heart,* 312; in Ireland, 104.1; in Kenya, 163.1; in Mauritius, 85.1; in Nigeria, 288.1; representing Universal House of Justice, 85.1, 104.1, 163.1, 288.1; return to Africa by, 276
Seat of the Universal House of Justice. *See* Universal House of Justice, seat of.
Selections from the Writings of 'Abdu'l-Bahá, 330.2, 402.7, 448.5
Self-control, 126.7c, 126.9
Self-defense, 69
Semple, Ian, p. v, 56.1, 129.1, 209.1, 359.1
Sénégal, formation of National Spiritual Assembly of, 141.7, 153.2
Senoga, Edith, 439.2a
Serrano, Arturo, 439.2b
Seven Year Plan (1979–86), 216, 221, 275, 280; accomplishments of, 248.9–13, 456.2; Bahá'í literature commissioned during, 280.8; call for pioneers during, 302; comments on aspects of, 280; condition of Bahá'ís in Iran during, 275.3; decision to launch, 216; final year of, 427.1; goals for Bahá'í International Fund during, 284; goals and objectives of, 221, 222, 275, 280, 394.10; goals of National Spiritual Assemblies during, 221.12–13; launching of, 221,

Seven Year Plan *(continued)*
358.2; obstacles and opportunities during, 221.7–10; phases of, 275.5–7, 275.9, 427.1; plans for International Youth Year during, 386.4–5; termination of, 447.3; use of pioneers and traveling teachers during, 228

Sex, 126.7–9, 224.4–5. *See also* Chastity; Marriage.

Sexual immorality, 126.7a–9

Seychelles, formation of National Spiritual Assembly of, 92.1, 104.1

Sezgin, Ilhan, 439.2c

Sherrill, Velma, 132.8, 132.25, 267.3b, 439.4

Shiraz (Iran): arrests of Bahá'ís in, 255.1, 322.4, 364.2–6; Bahá'í cemetery in, 356.1; destruction of Bahá'í homes in, 215.1; House of the Báb in, 157.3b, 225, 226.2, 235, 241, 282, 301.1, 314.1; imprisonment of Bahá'ís in, 347; martyrdoms in, 277, 281, 335, 344.2, 345, 351, 355, 356, 363, 364, 367, 381.3; martyred youth in, 364, 365; persecution of Bahá'ís in, 215, 218.2, 281; Revolutionary Court of, 281.1; torture of Bahá'ís in, 349. *See also* Iran; Iranian Bahá'ís; Martyrdoms; Persecution, of Bahá'ís in Iran.

Shirkat-i-Nawnahálán, 227.5

Shoghi Effendi, 330.3; as author of *God Passes By*, 24.15; birthplace of, 154.1, 157.6, 159.1; childhood and upbringing of, 157.8–8b, 157.12; institution of Hands of the Cause of God fostered by, 13.6; Most Great Jubilee crowning victory of life of, 1.2–3; passing of, 23.3, 59.1, 75.14, 111.6, 321.3; resting-place of, 1.4, 2.3, 177.1; safeguarding letters of, 54, 161, 409; twenty-fifth anniversary of passing of, 275.8, 321.3; writings of, about Greatest Holy Leaf, 313.5. *See also* Guardian; Guardianship.

Sierra Leone, formation of National Spiritual Assembly of, 141.7, 153.1–2

"Sign of God." *See* Guardian, as "sign of God."

Sikkim (India), formation of National Spiritual Assembly of, 38.1

Singapore, formation of National Spiritual Assembly of, 89.4, 92.1, 104.1

Sipihr-Arfa', Yadu'lláh, 344.2

Six Year Plan (1986–92), 427.6, 439.5, 445.2, 447.3, 447.6; emphasis on fostering equality of the sexes during, 453.13; formulation of, 415; major objectives of, 453.10–18; setting national goals for, 447.7, 453; social and economic development goals during, 453.18

Síyáh-Chál, 226.2

Síyávushí, Hidáyat, 351

Síyávushí, Jamshíd, 363

Síyávushí, Ṭáhirih, 364.1

Smith, Alan, 439.2b

Smith, David R., 439.2b

Social and economic development, 358.6, 379.2, 394.6; as field of Bahá'í service in, 379; offices of, 379.4, 456.2; programs of, 385.6, 453.18

Society: involvement by Bahá'ís in life of, 427.4–5; reconstruction of, 397, 425

Soe Tin, U, 439.2c

Sohrab, Ahmad, 402.8a

Solomon Islands: first National Convention of, 91.1; National

Spiritual Assembly of, 403.2b
Somalia, formation of National Spiritual Assembly of, 141.7
Some Answered Questions ('Abdu'l-Bahá): 174.1, 330.4; house where talks were given, 157.4; passing of Laura Dreyfus-Barney, compiler of, 150
Sorabjee, Zena, 132.8–9, 132.25, 267.3c, 439.2c
South America: Bahá'í radio stations in, 201, 205, 213, 275.9i, 304, 308.7, 321.2, 380.1, 391, 394.3; carrying torch of the Faith to, 162.6; continental goals for, 185.7, first International Women's Conference in (1978), 203; as mentioned in message to International Conference in Quito, Ecuador, 337.3; National Spiritual Assemblies of, 190.1; passing of spiritual mother of, 265; indigenous believers of, 82.2
South Carolina, first Bahá'í radio station in, 348.1
South Pacific: first regional National Spiritual Assembly in, 403.2
South Pacific Oceanic Conference. *See* Oceanic Conferences.
Spanish Sahara, 153.2
Spiritual Assemblies: appropriate enforcement of Bahá'í laws by, 405; appropriate intervention by, 289.6, 405.2; differing roles of, and individuals, 289.3; should not pry into people's lives, 405.2. *See also* Local Spiritual Assemblies; National Spiritual Assemblies.
Spiritual authority, 308.14
Spiritual development, 195.3a, 211, 221.6, 375, 397, 425
Spiritual transformation, 425.8
Spitzbergen, 171.2

Standard-Bearers, 6.6, 14.10, 91.1
Star of the West, 27.3, 157.11, 175.2, 330.2; references to Greatest Holy Leaf in, 313.9
Statement on peace. *See Promise of World Peace, The.*
Stevenson, Joy, 366.1, 439.2d
Stokes, Michael, 304.1
Strasbourg (France), 263.3
Study classes, 117.3. *See also* Deepening.
Subḥání, Nuṣratu'lláh, 424.1
Successors, Chosen, 6.2
Sudan, 91.1
Suffering, 55.5, 55.7, 122.3a, 126.10, 151, 173.2, 173.11a, 195.2, 252.3–4, 308.5, 425, 438.6, 438.11, 438.30. *See also* Tests.
Sundram, Chellie J., 60.1, 132.25, 267.3c, 439.2c
Supreme Tribunal, 422
Surinam, 141.7, 169.1
Súriy-i-Mulúk (Tablet to the Kings), 14.5, 20.10, 24.15–16, 128.8; commemoration of revelation of, 41; Intercontinental Conferences to celebrate centenary of revelation of, 45.1
Suva (Fiji), 68.6, 403.2
Suzuki, Hideya, 169.1, 267.3c, 439.4
Sydney (Australia), 24.17, 184.3
Synopsis and Codification of the Kitáb-i-Aqdas, 125, 147, 166, 251, 375.9, 455.4

Tablet of Carmel, 164.2a, 354.1, 358.3
Tablet of the Branch, 330.2
Tablet of the World, 272.6c
Tablet of Visitation, 192.4
Tablets of Abdu'l-Baha Abbas, 23.18b
Tablets of Bahá'u'lláh Revealed after the Kitáb-i-Aqdas, 272.6d

Tablets of the Divine Plan ('Abdu'l-Bahá), 82.1, 98.5, 171.1, 187.2, 340.1; fiftieth anniversary of revelation of, 32, 34.1
Tablets of Visitation, 334.2
Tablet to the Kings, 128.8
Tablighat-i-Islamí, 262.8
Tabríz (Iran), 254.1, 255.1, 268.1, 292.1, 322.2, 396.14, 417.1, 441.1
Taherzadeh, Adib, 169.1, 267.3e, 439.2e
Taḥqíqí, Ḥabíbu'lláh, 292.1
Taiwan, National Spiritual Assembly of, 100.3
Tákur (Iran), House of Bahá'u'lláh in, 226.2; demolition of, 305, 314.1, 358.2
Ṭalá'í, Kúrush, 311.1
Ta'lím, Rúḥu'lláh, 416.1
Tanumafili II, Susuga Malietoa (King, Western Samoa), 130, 242.1; first reigning monarch to accept Faith, 130, 184.3, 341.4, 403.2a; presence of, at future site Samoan House of Worship, 242.1; visit of, to Guardian's resting-place, 177
Taslímí, 'Abdu'l-Ḥusayn, 307.3
Taymúrí, Parvín, 193.2
Taymúrí, Rúḥu'lláh, 193.1–3
Teaching: administrative duties and, not mutually exclusive, 11.2a; appeal to increase efforts in, 14.3–4, 24.12, 34.9–13, 63.9–10, 68.11–12, 73, 100.2, 321.6, 379.9; border, 222.7; and consolidation, 34.13, 221.13e; expanding the work of, 456.2; expansion and consolidation in, 18.6–7; extracts from Bahá"í Writings on, 18.10–18, 52.5–20; Five Year Plan goals for, 88.1, 110.5, 141.4; grassroots, 379.6–7; importance of, 73, 308.11–12; importance of conferences on, 142.15, 162.16–18, 222.8; individual initiative in, 14.7, 34.15–17, 73.5, 221.13d; international goals in, 221.12; Local Spiritual Assembly plans for, 394.7; the masses, 2.8, 18, 20, 30, 43, 51, 52; Nine Year Plan goals for, 14.6–9; and proclamation through use of mass media, 24.20–22a, 42.21, 45, 141.9, 280.13; publishing materials for, 94.1; release of compilation on the individual and, 43; relationship between Bahá'í Fund and, 2.11, 13.2, 34.14; in rural areas, 280.16; Seven Year Plan goals for, 221.12, 221.13c–13f; versus proselytizing, 308.3–5; victories in, 221.3–5; youth and, 37.6, 128.12–13, 141.10, 162.33–34, 222.6, 321.5. *See also* Consolidation; Expansion and Consolidation; International Teaching Center; International Teaching Conferences; Pioneering; Teaching the masses; Traveling Teaching.
Teaching Institutes. *See* Institutes, teaching.
Teaching the masses, 18, 30, 52
Tehran (Iran): Bahá'í cemetery in, 305.1, 381.5; desecration of Bahá'í properties in, 262.9, 305.1, 381.5; martyrdoms in, 286, 311.1, 344.2, 360.1, 381.3, 392.1, 393, 396.14, 400, 416, 421, 424; Mit͟háqíyyih Hospital in, 227.2; persecution of Bahá'ís in, 215.1, 273.1, 297.1, 300.1, 301.1, 317.2, 322.4, 323.3, 368.2, 370.1–2, 392.1; plans for Mas͟hriqu'l-Ad͟hkár in, 128.16; publication of false charges in newspapers of,

268.1. *See also* Iran; Iranian Bahá'ís; Martyrdoms; Persecution, of Bahá'ís in Iran.
Tel Susita, 440.4
Television. *See* Radio and television.
Temples, Bahá'í. *See* Mashriqu'l-Adhkár(s).
Ten Year World Crusade, 1.2, 1.5, 2.8, 2.9, 14.2, 14.6, 14.9, 97.2, 98.3, 134.1, 162.6, 176.3, 403.2
Tests, 289, 425.5–5a; of Bahá'í community life, 289. *See also* Suffering.
Thábit, Akhtar, 364.1
Thábit-Rásikh, Gushtásb, 296.1
Thábit-Sarvistání, Ahmad-'Alí, 381.3
Theft, 147.23
Thorne, Adam, 439.2e
Thule (Greenland), 171.2
Tiapapata (Western Samoa), 403.2h
Tibyáníyán, Varqá, 287.1
Tillotson, Peter, 110.2
Tithes, law of, 147.19
Titicaca, del Lago, (Peru) 304.1
Tízfahm, Ágáhu'lláh, 324.1
Tobago, first National Convention of, 91.1
Togo, formation of National Spiritual Assembly of, 141.7, 153.1–2
Tokyo, 100.2
Tomás, Rodrigo, 439.2b
Tonga, National Spiritual Assembly of, 403.2b
Toronto (Ontario), 280.15
Tradition of the Church, 308.14
Transformation, personal, 182.6, 397.2c, 425.8. *See also* Spiritual development.
Transkei, formation of National Spiritual Assembly of, 221.13b, 248.14
Translation(s): approval of, by Universal House of Justice, 94.3d–3e; of Bahá'í literature, 16.10, 162.20, 295.3, 453.14; of Fire Tablet, 258; by the Guardian, 94.3d–3e; of Kitáb-i-Aqdas, 27, 251.2, 456.5; of *Le Monde* article, 262.5–13; of Long Healing Prayer, 258; of message to Iranian Bahá'ís, 246; of open letter about banning of Bahá'í Administration in Iran, 378.3–25; of prayers and Tablets for children and youth, 167.2; of texts on Ḥuqúqu'lláh, 404.2
Transsexuality, 126.9
Traveling teachers, 30.2, 169.1, 228; call for, 31; deployment of, 279; plan for international collaboration in, 158; selection of, 51; Seven Year Plan goals for, 222.5, 279.4–8; Six Year Plan goals for, 453.12; use of, 228. *See also* Pioneering; Teaching; Traveling Teaching.
Traveling teaching, 158.5, 162.8, 162.13, 162.33–34, 183; launch of program for, 159.1. *See also* Pioneering; Teaching; Traveling Teachers.
Treasury Committees, 435.12–14
Tree of Holiness, 1.4
Tribalism, 77.1, 77.7–11, 88.3, 308.8
Tribal peoples, 162.8, 453.12, 456.2
Trinidad, first National Convention of, 91.1
True, Edna, 60.1, 132.25, 267.4a
Trust Territories, enrollment of people in, 100.4
Tumansky, A. G., 251.2
Tunisia, National Spiritual Assembly of, 153.2
Turkey, 41, 264.4
Turkistan, 4.5–6
Tutorial Schools, 280.30–41; books and materials for, 280.35; definition of, 280.31; program

Tutorial Schools *(continued)*
for, 280.37–39; registration and fees for, 280.36; as a sign of social and economic development, 358.6; teachers for, 280.33–34
Tuvalu, National Spiritual Assembly of, 221.13b, 248.14, 403.2b
Twin institutions (Guardianship and Universal House of Justice), 6.2, 23, 35, 164.2a, 456.4. *See also* Guardianship; Universal House of Justice.

Uganda: dissolution of National Spiritual Assembly of, 221.5; National Spiritual Assembly of, 71.1, 358.4; passing of Hand of the Cause of God Músá Banání in, 102; re-formation of National Spiritual Assembly of, 248.14
Umaná Company, 239.10
Umtata (Transkei), National Spiritual Assembly of, 248.14
UNICEF (United Nations International Children's Emergency Fund), 308.8, 376.1
Unit Conventions, 433.6–8.
United Kingdom, 122.1, 291, 384.3 *See also* British Isles; England; Scotland.
United Nations: Bahá'í consultative status at, 78; Bahá'í International Community and, 49, 68.5, 128.9, 141.5, 284.5, 308.8, 309.1, 376.1–2, 394.3; Decade of Women, 427.5; development of relationship between Bahá'í community and, 2.10, 14.5, 394.9; ECOSOC (Economic and Social Council), 78; Human Rights Commission at, 255.1; International Year of Peace (1986), 420.1, 422.5; International Women's Year (1975), 162.32; International Youth Year (1985), 376, 386, 427.5; permanent Bahá'í representative at, 128.9; presentation of peace statement to secretary-general of, 442; protection of minorities discussed with, 263.2; resolution in support of Iranian Bahá'ís, 314.7, 444; secretary-general of, 255.1. 263.2, 442; support of, for Iranian Bahá'ís, 263.2, 444
United Nations Volunteers, 384.3
United States, 160, 362; destiny of, 443.1; fiftieth anniversary of National Spiritual Assembly of, 160; Louis Gregory Bahá'í Institute in, 394.3; Peace Corps of, 384.3; role of, in support of oppressed peoples, 362; first national youth conference in, 61
Unity: of Assemblies, 336.2–8, 358.7; Bahá'ís' influence toward, 427.4; of Bahá'í world community, 128.15–16, 221.1, 358.7; in diversity, 77.9; of faith and reason, 217.8–10; as fundamental purpose of Faith, 117.4; of nations, 149.2, 422.2; of peoples of the world, 340.4, 438.53–56; political, 422.2; in spirit and action, 397.4–6
Unity Conference (Ganado, Arizona, 1972), 112
Universal Court of Arbitration, 412.3b–3c
Universal House of Justice: authority of, 23.24, 35.8–9, 75.15–17, 308.14; and the Bahá'í International Community, 49; as channel of divine guidance, 6.3; communications of, 2.5; constitution of, 14.5, 23.24, 123, 128.6, 229.1; decisions of, 3.2–3, 5, 111.6–10, 115.1; duties and functions of, 6.2, 75.5–8,

412; election of, 1.4, 2.1, 23, 35, 56, 129, 209, 327, 332, 359; emergence of, 2.8; Guardian's authority to define legislative action of, 75.17; Guardianship and, as twin Successors of 'Abdu'l-Bahá, 6.2, 164.2a; infallibility of, 23, 35.3–7d, 75.2–17, 308.14; legislative process of, 75.10; membership of, 48, 166.6–6b; National Spiritual Assemblies as pillars of, 6.7, 14.6, 190.1; occupation of seat of, 354; powers of elucidation of, 412; powers granted to, by Bahá'u'lláh in Kitáb-i-Aqdas, 75.6; in relation to Guardianship, 5, 6.2, 35, 75, 412; seal of, 2.5; seat of, 3.2, 115, 136.1, 140, 141.5, 159.1, 164, 165, 169.1, 186, 190.1, 334, 354, 447.3; significant meetings with member of, 223, 236; as Trustees of Bahá'í Faith, 438.5
Universal participation, 2.11, 9.2–2a, 19, 87.6–8
Upper Volta, 153.2; formation of National Spiritual Assembly of, 81, 91.1, 141.7, 169.1
Urúmíyyih (Iran), 300.1, 322.1, 324.1, 339.1, 396.13

Vafá'í, Náṣir, 285.2
Vafá'í, Manúchihr, 344.2
Vafá'í, Raḥmatu'lláh, 355.1
Vaḥdat-i-Ḥaqq, Ḥusayn, 317.2
Vaḥdat, Nuṣratu'lláh, 399.1
Vaḥdat, Yadu'lláh, 281.1
Vanuatu, National Spiritual Assembly of, 403.2b
Varjávandí, Rustam, 421.1
Varqá, 'Alí-Muḥammad, (Hand of the Cause of God) 91.1, 177.1, 361.2, 447.2
Vasudevan, Sankaran-Nair, 60.1

Victoria (Queen, England), message of Bahá'u'lláh to, 19.4
Vientiane (Laos), 38.1
Vietnam, disbanding of National Spiritual Assembly of, 221.5
Voice of America, 299
Voluntary Nonsectarian Organizations, service in, 384
Von Czekus, Rolf, 439.2b
Voting, 79.3–6; eligibility for, 333.16; procedures for election of delegates to National Conventions, 433
Vujdání, Bahár, 254.1
Vujdání, Farámarz, 121.2
Vujdání, Raḥmatu'lláh, 437.1–2
Vuyiya, Peter, 132.8, 132.25, 267.3a, 439.2a

War, 438.10, 438.25
Warren, Lucretia Mancho, 439.2a
Wellspring of Guidance, 412.2, 412.4
Western Africa: National Spiritual Assembly of, 153.1; zone of, 153.2. *See also* Africa.
Western Europe, 221.8. *See also* Europe.
Western Hemisphere, 170.3–4, 172.3
Western Pilgrim House, 3.2, 361.7
Western Samoa, 130.2, 184.3, 210.1, 242, 321.1, 403. *See also* Samoa; Tanumafili II, Susuga Malietoa.
West Indies, Bahá'í community of, 187.4
When We Grow Up (Bahíyyih Nakhjavání), 272.1
Will and Testament of 'Abdu'l-Bahá, 23.22c, 35.2, 75.3, 75.18, 412.3–3a, 451.3
Wilmette, 223.1, 265.1, 423.1
Winckler, Bahíyyih, 132.25, 267.3a, 439.4. *See also* Bahíyyih Ford.
Windhoek (Namibia), 248.14

Windward Islands, 104.1, 248.14
Wingi, Mabuku, 382.1, 439.2a
Witzel, Donald, 60.1, 132.25, 267.3b, 439.4
WLGI (radio station), 394.3. *See also* Institute, Louis Gregory Bahá'í; Radio Stations, Bahá'í.
Wolcott, Charles, p. v, 56.1, 129.1, 209.1, 359.1
Women, 147.2, 162.32, 166, 272, 338.5, 448; concept of uncleanness of, abolished, 455.3; emancipation of, 438.33; exemption of, from acts of worship, 455; membership of Universal House of Justice confined to men, 166, 448; mother's role as first educator of child, 407.3; release of compilation on, 446; role in society of, 448; United Nations decade of, 427.5
Women's Conferences. *See* Bahá'í Women's Conferences.
Word of God: Guardian as interpreter of, 75.5–8, 308.14–15, 308.17, 412.2; interpretation of, 308.14–17; most urgent summons of, 14.3; Qu'rán as, 378.11; Revelation of Bahá'u'lláh accepted as, 308.14
World Center, 1.2, 2.2, 3.2, 42.2; commemoration of fiftieth anniversary of passing Greatest Holy Leaf at, 334; communications with, 271.3; developments at, 26, 74, 105, 115, 123, 128.4–7, 128.17–20, 159, 164.3, 165, 169, 262.11, 264, 321.3, 432; embellishment of endowments at, 2.10, 40.5; establishment of International Teaching Center at, 131; extension of gardens at Bahjí, 139; Five Year Plan goals for, 141.5; goals of, 24.2–3, 34.14, 456.5; inauguration of Public Information Office at, 432; International Year of Peace projects initiated at, 420.2; lifeblood of, 68.9; needs of, 40, 385.5, 385.8, 419.4; Nine Year Plan goals for, 14.5; occupation of Seat of Universal House of Justice at, 354; purification of Most Holy Tomb, 26; reinterment of Bahá'u'lláh's faithful half-brother, 440; Seven Year Plan goals for, 221.11–11f; Shoghi Effendi's statement regarding, 385.5; submission of reports to, 280.11; third International Convention at, 116. *See also* Universal House of Justice.
World citizenship, 438.31
World Congress. *See* Bahá'í World Congress.
World Crusade. *See* Ten Year World Crusade.
World order, 438.3–4, 438.26–27, 438.39
World Order of Bahá'u'lláh, the: understanding, 35.3; building, 63.4–5
World Order of Bahá'u'lláh, The, (Shoghi Effendi) 27.3; references to Guardianship in, 251.5
World Religion Day, 434.3
Writings, Bahá'í. *See* Bahá'í Writings.

Yaldá'í, Bahrám, 363.1
Yaldá'í, Nuṣrat, 364.1
Yávarí, 'Aṭá'u'lláh, 311.1
Yazd (Iran), 255.1, 293.1, 396.11, 396.13; Bahá'í situation worsening in, 297.1; executions of Bahá'ís in, 261, 378.7, 421, 424
Yazdí, 'Azíz, 60.1, 131.1, 132.5, 361.3
Yazdí, Ḥájí 'Alí, 74.2a

Year of Peace. *See* Peace.
Yemen (San'a), 371.1
Youssefian, Sohrab, 439.2e
Youth, Bahá'í: activities of, for International Year of Youth (1985), 376, 386, 422.5; declarations of, 333.15; developing skills of consultation, 37.9; and education, 37.7, 67, 453.13; greater participation of, in Bahá'í activities, 321.5, 338.5; martyrdoms of Iranian, 364; messages to, 37, 61, 67, 364, 365, 369, 386, 408, 428, 431, 449; moral standards of, 37.5, 133; noncombatant status of, in military, 333.2a; periods and fields of service for, 37, 142.16, 384; pioneering and, 67, 158.6, 384; registration guidelines for, 333; release of compilation of prayers and Tablets for, 426; responsibilities of, at age of maturity, 426; spiritual maturity of, 333.11–16; and teaching, 37.6, 128.13, 141.10, 162.33–34, 222.6, 321.5–5a; unique opportunities for, 37.3; vital role of, 37.1. *See also* Youth Conferences.

Youth Conferences: first national, in U.S. (1968), 61; in Costa Rica (1983), 357; in Honduras (1983), 357; in Innsbruck, Switzerland (1983), 369; in Ontario, Canada (1984), 408; in Columbus, Ohio (1985), 431; in Adelaide, Australia (1986), 449. *See also* Youth.

Zá'írpúr, Ṭúbá, 356
Zamání, 'Alí-Muḥammad, 396.14
Zanján (Iran), 322.4
Zenimoto, Michitoshi, 439.2c
Zihtáb, Ismá'íl, 292.1
Zimmer, Hermann, 152.2
Zoroastrianism, 24.15, 220.13, 262.7